J. SAPIRSTEIN

PRECISION TESTS OF THE STANDARD ELECTROWEAK MODEL

ADVANCED SERIES ON DIRECTIONS IN HIGH ENERGY PHYSICS

Published

Vol. 1 – High Energy Electron–Positron Physics
(*eds. A. Ali and P. Soding*)

Vol. 2 – Hadronic Multiparticle Production
(*ed. P. Carruthers*)

Vol. 3 – CP Violation
(*ed. C. Jarlskog*)

Vol. 4 – Proton–Antiproton Collider Physics
(*eds. G. Altarelli and L. Di Lella*)

Vol. 5 – Perturbative QCD
(*ed. A. Mueller*)

Vol. 6 – Quark–Gluon Plasma
(*ed. R. C. Hwa*)

Vol. 7 – Quantum Electrodynamics
(*ed. T. Kinoshita*)

Vol. 9 – Instrumentation in High Energy Physics
(*ed. F. Sauli*)

Vol. 10 – Heavy Flavours
(*eds. A. J. Buras and M. Lindner*)

Vol. 11 – Quantum Fields on the Computer
(*ed. M. Creutz*)

Vol. 12 – Advances of Accelerator Physics and Technologies
(*ed. H. Schopper*)

Vol. 13 – Perspectives on Higgs Physics
(*ed. G. L. Kane*)

Vol. 14 – Precision Tests of the Standard Electroweak Model
(*ed. P. Langacker*)

Forthcoming

Vol. 8 – Standard Model, Hadron Phenomenology and Weak Decays on the Lattice
(*ed. G. Martinelli*)

Cover artwork by courtesy of Los Alamos National Laboratory.
"This work was performed by the University of California,
Los Alamos National Laboratory, under the auspices of the
United States Department of Energy."

Advanced Series on
Directions in High Energy Physics — Vol. 14

PRECISION TESTS OF THE STANDARD ELECTROWEAK MODEL

Editor
Paul Langacker
University of Pennsylvania

Published by

World Scientific Publishing Co. Pte. Ltd.
P O Box 128, Farrer Road, Singapore 9128
USA office: Suite 1B, 1060 Main Street, River Edge, NJ 07661
UK office: 57 Shelton Street, Covent Garden, London WC2H 9HE

PRECISION TESTS OF THE STANDARD ELECTROWEAK MODEL

ISBN 981-02-1284-4

Printed in Singapore.

CONTENTS

FOREWORD

The standard model is a mathematically consistent renormalizable field theory which predicts or is consistent with all known aspects of the elementary particles and their interactions over an enormous range of probes and scales. In particular, it is now clear that, with the possible exception of the Higgs sector, the standard electroweak model is the correct theory of nature to an excellent approximation down to a distance scale of 10^{-16} cm. Precision neutral current, charged current, and Z and W pole experiments have established or supported the framework of renormalizable field theory and of gauge theories; have established the $SU_2 \times U_1$ gauge group; have confirmed the fermion representations; have successfully confirmed the predictions for the existence, masses, and properties of the electroweak gauge bosons; have confirmed or led to successful predictions for higher order effects such as the running of the electromagnetic fine structure constant, the top quark mass, and the strong coupling constant; have searched for and excluded large regions of parameter space associated with alternative symmetry breaking mechanisms and other possible new physics beyond the standard model; and have shown that the strong and electroweak coupling constants are consistent with supersymmetric grand unification. In this book, all aspects of the program are considered in detail, including the structure of the standard model; radiative corrections; high precision experiments; and their analysis, results, implications, and prospects.

Paul Langacker
Philadelphia
November 3, 1994

I. INTRODUCTION: PRECISION TESTS AND PARTICLE PHYSICS

INTRODUCTION: PRECISION TESTS AND STANDARD MODEL PHYSICS

PAUL LANGACKER
Department of Physics, University of Pennsylvania,
Philadelphia, Pennsylvania, USA 19104-6396

It is now clear that the standard electroweak model is a correct description of nature to an excellent first approximation. It is also apparent, because of the many questions left unanswered, that there must be new physics underlying the standard model, and most of the activity in the field is devoted to searching for it. Precision experiments have been crucial both in establishing the correctness of the standard model and in searching for new physics. In the future, they are expected to continue to be important, and will be a useful complement to such other methods as high energy colliders, QED tests, searches for CP violation, rare decays, neutrino mass, cosmology, and theory, each of which has its own advantages and range of sensitivity.

The most important tests for establishing electroweak unification have been in the weak neutral current sector, combined with the discovery and properties of the W and Z bosons. It was the discovery of the weak neutral current in 1973 that gave the first experimental verification of what became the standard model [1]. Subsequent detailed studies have established the framework of gauge theories and the correctness of the standard model gauge group and fermion representations. This sector has also been powerful in probing for the presence of new physics that would perturb the standard model predictions. So far, no deviations have been observed, and large ranges of possible new physics have been excluded.

The weak charged current interaction was incorporated into the standard model, not predicted by it. However, the standard model unification greatly improved the theory by making it renormalizable, so that one could calculate higher-order corrections, and also by predicting the W mass. Many of the precise charged current experiments result in the measurement of the elements of the Cabibbo-Kobayashi-Maskawa (CKM) mixing matrix [2]. Tests of the unitarity of the CKM matrix will ultimately lead to significant constraints on many types of new physics which could confuse the determinations of the quantities, such as additional fermions, new gauge bosons, and new sources of CP violation. Some such tests will have to wait until CP violation in B decays has been studied. At present, however, the most sensitive tests for new physics involve muon and beta decay, and it is these which are discussed in this volume. In particular, the weak universality observed in these processes is sensitive to many types of underlying physics.

Quantum electrodynamics [3] is the most precise test. Again, however, QED was incorporated into the standard model, not predicted by it. A future measurement of the muon magnetic moment ($g_\mu - 2$) will constitute a wonderful test of the structure of renormalizable field theory. That should be viewed as the major motivation, although there will be some sensitivity to such new physics as supersymmetry and compositeness.

The various articles in this volume survey most aspects of the history of the precision experiments, the experiments themselves, the theory necessary for their interpretation, and the results and their implications. The section on the standard electroweak model begins with an overview of the standard model, the gauge interactions, and the theoretical problems (Langacker). The apparatus necessary for calculating the higher order corrections is surveyed in detail by Hollik, including a general discussion of renormalization theory and of the different schemes available, including a clarification of the various confusing definitions of the weak angle $\sin^2\theta_W$. Other chapters include the application of this to e^+e^- annihilation at the Z-pole (Hollik), and to low energy processes (Marciano).

The section on precision tests at e^+e^- colliders surveys the history of electroweak precision experiments at these facilities. These are treated first (out of historical order) because they are the most direct tests of the properties of the Z boson and because the LEP experiments are the most precise.

There were many experiments performed below the Z-pole at PEP, PETRA, and TRISTAN. These demonstrated weak-electromagnetic interference, which implies that neutral current processes are not purely S, P, and T, and also that they have a significant axial-vector component. Some limited tests of universality were also made. These experiments measured a number of different reactions, including forward-backward asymmetries and total cross sections in $e^+e^- \to e^+e^-$, $\mu^+\mu^-$, $\tau^+\tau^-$, $c\bar{c}$, $b\bar{b}$, and hadrons. (The first reaction is mainly a QED test.) In particular, the forward-backward asymmetries, $A_{\rm FB}$, at these energies depend only on the axial part of the neutral current interaction, interfering with the vector electromagnetic interaction. In the standard model the axial couplings are an absolute prediction independent of $\sin^2\theta_W$. Therefore, the $A_{\rm FB}$ are absolute predictions, and their measurements are sensitive to deviations from the standard model. The measured asymmetry in b production constrained the weak interactions of the b, implying that the t quark must exist [4, 5]. This evidence was supplemented by arguments from b decays, but was stronger because the b decays are strongly suppressed in the standard model and in principle could be due to new physics. Similarly, theoretical arguments based on the absence of anomalies were inconclusive, due to the existence of such alternate anomaly-killing mechanisms as mirror fermions for the third family. Measurements of the rate and asymmetry for b production at the Z pole have confirmed this conclusion [4, 5].

The total hadronic cross section R (relative to the predicted lowest order $\mu^+\mu^-$ cross section) was long considered a "gold-plated" test of QCD. It turned out, however, that this test was more difficult than expected due to systematic uncertainties and because of the complication of the Z boson annihilation diagram. In fact, the PETRA and TRISTAN results suggested anomalous behavior, especially when confronted with the precise value of the Z mass from LEP. In his article,

Haidt has chosen to emphasize this aspect of the experiments below the Z-pole. He has done a systematic reanalysis of all of the measurements of the hadronic cross section, applying common radiative corrections and treating systematic effects and acceptances as uniformly as possible. The results are in excellent agreement with QCD, and there is no evidence for an anomaly.

The most precise electroweak measurements are those of the ALEPH, DELPHI, L3, and OPAL experiments at LEP, as summarized in the article by Schaile. The experiments have determined the mass of the Z to very high precision and have measured the width and many of the partial widths more accurately than had been anticipated before LEP started running. In addition, several important asymmetries have been measured, including the first observation of the polarization asymmetry at the SLC. The results are generally in spectacular agreement with the predictions of the standard model and have put significant limits on new physics.

In addition to increasing the precision of these existing measurements there is the possibility of measuring polarization asymmetries at LEP and SLC, as described by Blondel. The polarization asymmetry is very clean and is quite sensitive to many types of new physics.

The future high energy phase, LEP II (Treille), will allow measurement of M_W to about 100 MeV, and will also allow for searches for standard model and supersymmetric Higgs bosons and for anomalous trilinear gauge couplings.

The weak neutral current [1], discovered in 1973 by the Gargamelle collaboration at CERN and soon confirmed at Fermilab by the HPW group, was the first confirmation of the predictions of the newly popular $SU_2 \times U_1$ model [6]. The discovery had a difficult birth — it was delayed and complicated by an incorrect earlier experiment that apparently excluded neutral currents. Also the situation was confused by the search (under the wrong lamppost) for flavor changing neutral currents (FCNC). It was historically confusing that flavor-conserving effects could be present if flavor-changing ones were not. Of course, this was explained by the GIM mechanism [7], and now many years later the search for FCNC is extremely important as a probe of many types of new physics, especially compositeness.

After the difficulties of the discovery phase there were new generations of ever more precise neutral current experiments, extending to the present. There were many purely-weak experiments involving νN and νe scattering. In addition, weak-electromagnetic interference was observed in the polarized eD asymmetry at SLAC, in atomic parity violation, and in e^+e^- annihilation experiments. Much theoretical effort has gone into the interpretation of these experiments, including careful analyses of deep inelastic scattering, of radiative corrections, and of how to parametrize new physics. By the end of the 1970's it was possible to begin model-independent analyses of the 4-Fermi interactions which describe low energy processes [8]. In this context, model-independent means assuming a general V and A structure, as could be generated in an arbitrary gauge theory. These studies involved the determination of a number of parameters, and required global analyses [9] of all of the data.

Even now there is not direct experimental proof that the low energy purely-weak neutral current interactions are V and A. For example, the νN interactions could all be described by some combination of S, P, and T [10]. However, there is little theoretical motivation for such models, and the combination of the success of the standard model, the direct discovery of the W and Z, and the fact that so many different types of experiments are successfully described in a simple gauge framework makes it implausible that these processes are mainly S, P, and T. (It is still possible that S, P, and T are present as small perturbations due to new physics.) On the other hand, the weak-electromagnetic interference processes must be dominantly of the V, A type. Similarly, family universality is generally assumed in the analyses, even though it has only been tested in a limited way in the neutral current sector. However, wherever it has been tested it has worked, and the lack of FCNC makes it unlikely that there are any large violations. Within these reasonable assumptions, the combination of the various neutral current experiments and the properties of the W and Z have uniquely established the $SU_2 \times U_1$ gauge group. Furthermore, neutral current and charged current experiments together have, under very general assumptions, uniquely established the standard model assignments of all of the observed fermions, *i.e.*, that the left (right)-handed fermions are SU_2 doublets (singlets) [5].

Deep inelastic scattering $\nu N \to \nu X$ was the major quantitative test of the electroweak model prior to the LEP era, and may continue to be important in the future. The strong interactions are intimately involved, but it was realized early on using isospin arguments [11] that most of the structure function dependence cancels in the ratio of neutral to charged current cross sections, as do many of the systematic uncertainties and much of the sensitivity to the neutrino spectrum. The major residual theoretical uncertainty is the mass m_c of the charm quark, which mainly affects the charged current denominator. These theoretical issues could be resolved at approximately the 1% level, allowing for precise measurements of $\sin^2 \theta_W$ from deep inelastic scattering on (approximately) isoscalar targets. Until the LEP era this was the most accurate determination, and even now deep-inelastic scattering is important for predicting m_t and for searching for deviations from the standard model. Additional experiments on p and n targets were also useful for determining the isospin structure of the neutral current interaction. There may be future higher-energy experiments at Fermilab, which would be sensitive to new physics and for which the high energies would reduce the theoretical uncertainties. The entire subject is described by Perrier, who also carries out a new model-independent analysis of the data.

Elastic $\nu_\mu p \to \nu_\mu p$ scattering has been measured at Brookhaven (Mann). The experiments are difficult but are useful for determining the isospin structure of the weak neutral current. They also gave the first experimental hint of a large s-quark content of the nucleon. There is a possibility of a future higher precision experiment at LAMPF, mainly directed towards probing the s-content.

Other νN experiments, including exclusive and inclusive pion production and deuteron dissociation were historically important [12]. However, these are hard to interpret due to strong interaction effects (one exception is coherent pion production) and therefore no longer play a quantitative role in the analysis.

Elastic $\nu_\mu e \to \nu_\mu e$ scattering (Panman) was the first WNC process observed. It is clean theoretically but difficult experimentally, due to the low cross section. The high precision CHARM II experiment was still running until recently. Most of the experiments have used ν_μ and $\bar{\nu}_\mu$ beams. However, there have also been measurements of $\nu_e e \to \nu_e e$ at LAMPF and of $\bar{\nu}_e e$ at the Savannah River reactor. The ν_μ experiments have yielded precise values for the 4-Fermi interaction, in agreement with the standard model. The $\nu_e e$ experiment has observed the interference between charged current and neutral current contributions to the amplitude (assuming family universality), which implies a V and A structure for the νe interaction. It also implies (and is the only test of) flavor conservation at the neutrino vertex.

The parity violation associated with Z exchange yields a small mixing between atomic S and P wave states. This has led to a beautiful series of measurements of atomic parity violation in atoms, as described by Masterson and Wieman. The field had a difficult beginning during the 1970's, and the original experiments did not see the expected parity violation. In the subsequent years, however, the techniques have been refined, and now there are precise measurements in the cesium atom, originally from Paris and more recently from Boulder. The cesium atom is clean theoretically, since it consists of a single electron outside a tightly bound core. The necessary theoretical calculations of the matrix elements are now good to the 1% level, possibly improvable to a few tenths of a percent, as described by Blundell, Johnson, and Sapirstein. At present the Boulder experiment is accurate to a few percent but will soon go down to much better than 1%. It has a number of quantities which can be reversed to eliminate systematic errors. In the future there may also be measurements on different isotopes of cesium and other atoms, and their ratios will greatly reduce the residual theoretical uncertainties. The atomic parity experiments are especially good for looking for deviations from the standard model. They are the most powerful probes of, *e.g.*, (parity-violating) 4-Fermi operators associated with compositeness or other types of new physics. They are also sensitive to leptoquarks, additional heavy Z' bosons, and mixing with exotic fermions. The existing measurements in cesium are not sensitive to m_t.

There have also been a number of measurements of the parity violating NN interaction involving polarized scattering asymmetries or nuclear transitions [13]. These results, which involve both the charged and neutral current, are not described in this volume because large hadronic uncertainties make them hard to interpret for precision tests.

Charged lepton-hadron asymmetries are described by Souder. The famous SLAC polarized eD experiment established parity violation in the weak neutral current, clearing up confusion from the early (incorrect) atomic parity experiments. The SLAC experiment also established interference between weak and electromagnetic amplitudes, thereby establishing that the neutral current could not be purely S, P, and T. There have been subsequent asymmetry experiments in electron scattering at Mainz and Bates, and in high energy muon scattering at CERN, all of which are in good agreement with standard model predictions. There are proposals for future experiments at CEBAF, BATES, and SLAC which will probably function mainly as a probe of the nucleon, *e.g.*, to measure the strangeness content, but could conceivably yield additional precision tests of the standard model.

HERA will be especially important as a test of QCD and for determining structure functions. However, as described by Spiesberger the HERA measurements also probe some aspects of electroweak physics and of the W mass, and are sensitive to certain types of new physics.

Other important implications of the neutral current are too far removed to be covered here. These include astrophysical and cosmological implications [14], such as their role in supernova explosions (both the neutrino burst and the blow-off), and possible connections with the preferred chiralities of large molecules [15].

The charged current played the dominant historical role in unraveling the nature of the weak interactions. In the section on the weak charged current, selected modern aspects which are connected well with the rest of the volume are considered. Fetscher and Gerber describe the present and ongoing status of high precision muon and leptonic tau decays, which are purely leptonic and thus simple theoretically. The experiments have established the $V - A$ nature of the couplings starting from a completely arbitrary 4-Fermi interaction. Stringent limits on S, P, T, and $V + A$ admixtures have been obtained. In terms of specific types of new physics, strong limits have been obtained on extended gauge structures involving new W_R bosons coupling to $V + A$ currents and on exotic fermions with right-handed interactions.

Deutsch and Quin describe those classes of beta decay experiments that are sensitive to the underlying electroweak physics (rather than to nuclear physics). Here there is also a complete set of experiments which exclude interactions other than the $V - A$. Experiments and theoretical issues are summarized, as are the implications for testing weak universality, searches for W_R, *etc.* There is currently some confusion on the ratio g_A/g_V determined from neutron asymmetries, with one experiment indicating a possible anomaly. There is also concern about weak universality, which is one aspect of CKM unitarity, although there are theoretical ambiguities in the determinations. The beta decay searches are sensitive to extra fermion families, new W_R and Z' gauge bosons, mixing with exotic fermions, and leptoquarks.

Sirlin describes weak universality in more detail. The universality tests depend sensitively on the radiative corrections to beta and muon decay. These calculations are only finite and sensible in the full electroweak theory, not in the Fermi or intermediate vector boson theories. The article describes the history of the development of the radiative corrections and the current status of universality tests, including the implications for new physics and uncertainties from the nuclear mismatch corrections. Herczeg summarizes the implications of muon and beta decay for limiting many types of new physics.

Even after the first round of precise neutral current experiments in the late 1970's it was possible to construct models which duplicated the 4-Fermi interactions of the standard model, but which had very different predictions for the gauge boson masses. These contrived variations were eliminated in 1983 when the W and Z

were observed at CERN and their masses determined. Measurements of the decay distributions tested the $V-A$ couplings of the W, and universality was further tested. The W and Z widths were important in limiting numbers of neutrinos and setting limits on the top quark mass, and QCD was tested in the production cross sections. The current status of precision tests is summarized by Einsweiler. In particular, the current and projected measurements of M_W are described. One anticipates measurements to within 100 MeV at Fermilab, comparable to LEP II but with entirely different systematics. A more precise measurement of M_W is important for comparison with the Z mass and other observables from LEP and elsewhere, both for predicting m_t and for precision tests of the standard model and searches for new physics. Collider searches for heavy W' and Z' bosons are also described. At present, indirect tests and direct searches at colliders are of comparable significance. Other topics include searches for the t and for anomalous triple-gauge vertices.

The various implications of the experiments are discussed throughout the volume. However, the section *Implications of Precision Experiments* summarizes the results of global analyses, in which all of the experimental results are considered simultaneously. A global analysis has the advantage that there is more information than any one experiment. For example, a single measurement can usually be explained simply by shifting the value of $\sin^2\theta_W$ (or of m_t and/or the strong coupling $\alpha_s(M_Z)$), while with several experiments one can compare their results and test the standard model. Another advantage is that one can apply the best possible theory expressions uniformly for similar experiments with properly correlated uncertainties. The disadvantage is that great care is needed with systematic uncertainties and correlations.

Many of the tests of the standard model and searches for new physics are described in the article by Langacker. The current status of the model independent, *i.e.*, general V and A interactions as allowed by an arbitrary gauge theory, are described. The need for a model independent analysis was emphasized in the early 1970's, especially by Bjorken and by Hung and Sakurai [8]. There were many competing gauge models, each of which was described by a different set of parameters. It was useful to parametrize the data in a model-independent framework, allowing one to exclude the models that did not correspond to the allowed regions. The current status of the global analyses in νq, eq, and νe scattering and of the resulting effective 4-Fermi parameters is described.

Other results include the direct standard model tests, the predictions for m_t and $\alpha_s(M_Z)$, and the implications for ordinary and supersymmetric grand unification. In particular, the observed couplings are in dramatic agreement with the possibility of supersymmetric unification.

The precision experiments are sensitive to some but not all possible types of new physics. The possibilities of extended Higgs sectors involving larger representations are described. Such searches are especially interesting in that superstring theories generally do not allow such sectors, whereas theories involving compositeness do. There has been much activity on finding general ways to parametrize arbitrary new physics. In particular, the S, T, and U parameters are described and their current status outlined. These parameters allow for a description of those classes of new physics which only affect gauge boson self-energies, such as nondegenerate heavy fermions or scalars, or degenerate chiral fermions. With the probable direct observation of the t quark by the CDF collaboration [16] it is possible to isolate the new physics contributions to these parameters. Variations on S, T, and U allow arbitrary types of new physics, but only a subset of the observables. These formalisms can be supplemented with a parameter describing anomalous contributions to the $Zb\bar{b}$ vertex, which could be significant in some extensions of the Standard Model. Another possibility are deviation vectors, in which one starts with the Z mass and predicts every other experiment. This has the advantage that it parametrizes all types of new physics and utilizes all experiments.

The precision experiments are sensitive to the presence of additional heavy Z' bosons. There is a competition between the indirect limits and direct searches at colliders. Other possibilities limited by these experiments are new 4-Fermi interactions induced by compositeness, leptoquarks, *etc.* In general the precision experiments are not a good place to search for supersymmetry – the radiative corrections are not sensitive to the superpartner masses except for very limited regions of parameter space. The only exception is that the simple supersymmetric models usually imply a light scalar that acts like a light standard model Higgs. Of course, tests of the coupling constant predictions of ordinary and supersymmetric grand unified theories indirectly search for supersymmetry

Some extensions of the standard model such as E_6 models predict the existence of exotic fermions with unusual weak interactions. By mixing with the ordinary fermions they should lead to observable consequences. London describes the current status of the searches, which involve the simultaneous analyses of neutral and charged current interactions, and the properties of the gauge bosons. It is also possible to consider the simultaneous presence of exotic fermions and Z' bosons.

There are still considerable prospects for future high precision experiments. There have been proposals for νe scattering, deep inelastic scattering, atomic parity violation, and the polarization asymmetry at the Z-pole. The section on the future by Luo updates a comprehensive study [17] of the relative sensitivities of the different proposals to the various possible types of new physics. The article emphasizes the need to have as many precise experiments as possible to maximize sensitivity, confirm deviations, and diagnose their origin. The upshot is that although some types of new physics are better probed by one or another type of experiment, there is no single experiment that is clearly the best for all types of new physics – one should try to carry out as many as are feasible. The future high precision experiments should continue to probe many types of new physics into the TeV range, and should be a useful complement to high energy colliders.

1. The history is described in *Discovery of Weak Neutral Currents: The Weak Interaction Before and After*, ed. D. Cline and A. Mann, AIP Conf. Proc. 300 (AIP, New York, 1994).

2. N. Cabibbo, *Phys. Rev. Lett.* **10**, 531 (1963); M. Kobayashi and M. Maskawa, *Prog. Theor. Phys.* **49**, 652 (1973).
3. For a recent review, see *Quantum Electrodynamics*, ed. T. Kinoshita (World Scientific, Singapore, 1990).
4. For recent discussions, see D. Schaile and P. M. Zerwas, *Phys. Rev.* **D45**, 3262 (1992); G. L. Kane, in *Mexico City High Energy Phenomenology* (QCD161:W568:1991), p241.
5. P. Langacker, *Comm. Nucl. Part. Sci.* **19**, 1 (1989). See also the article *Tests of the Standard Model and Searches for New Physics* in this volume.
6. S. L. Glashow, *Nucl. Phys.* **22**, 579 (1961); S. Weinberg, *Phys. Rev. Lett.* **19**, 1264 (1967); A. Salam in *Elementary Particle Theory*, ed. N. Svartholm (Almqvist and Wiksells, Stockholm, 1969) p 367.
7. S. L. Glashow, J. Iliopoulos, and L. Maiani, *Phys. Rev.* **D2**, 1285 (1970).
8. J. D. Bjorken, *Proc. of the SLAC Summer Inst. on Particle Physics*, ed. M. Zipf, SLAC-195; P. Q. Hung and J. J. Sakurai, *Phys. Lett.* **B63**, 295 (1976); **B69**, 323 (1972); **B72**, 208 (1977); L. M. Sehgal, *Phys. Lett.* **B71**, 91 (1977).
9. L. F. Abbott and R. M. Barnett, *Phys. Rev.* **D18**, 3214 (1978); **D19**, 3230 (1979); I. Liede and M. Roos, *Phys. Lett.* **82B**, 89 (1979); *Nucl. Phys.* **B167**, 397 (1980); J. E. Kim, P. Langacker, M. Levine, and H. H. Williams, *Rev. Mod. Phys.* **53**, 211 (1981); P. Q. Hung and J. J. Sakurai, *ARNPS* **31**, 375 (1981); L. M. Sehgal, *Prog. Nucl. Part. Phys.* **14**, 1 (1985); U. Amaldi, A. Böhm, L. S. Durkin, P. Langacker, A. K. Mann, W. J. Marciano, A. Sirlin, and H. H. Williams, *Phys. Rev.* **D36**, 1385 (1987); G. Costa, J. Ellis, G. L. Fogli, D. V. Nanopoulos, and F. Zwirner, *Nucl. Phys.* **B297**, 244 (1988); G. L. Fogli and D. Haidt, *Z. Phys.* **C40**, 379 (1988); P. Langacker and M. Luo, *Phys. Rev.* **D44**, 817 (1991).
10. B. Kayser *et al.,* *Phys. Lett.* **B52**, 385 (1974); *Phys. Rev.* **D11**, 2547 (1975); E. Fischbach *et al.,* *Phys. Rev.* **D15**, 97 (1977).
11. C. H. Llewellyn-Smith, *Nucl. Phys.* **B228**, 205 (1983).
12. For references, see J. E. Kim *et al.,* [9] and U. Amaldi *et al.,* [9].
13. For a review, see E. G. Adelberger and W. C. Haxton, *ARNPS* **35**, 501 (1985). Other treatments include J. F. Donoghue and B. R. Holstein, *Phys. Rev.* **D31**, 70 (1985); B. Desplanques, J. F. Donoghue, and B. R. Holstein, *Ann Phys.* **124**, 449 (1980).
14. See D. Schramm, in [1], p 469.
15. See D. Kondepudi, in [1], p 491; S. L. Miller and L. E. Orgel, *The Origins of Life on the Earth* (Prentice Hall, New Jersey, 1974); S. W. Fox and K. Dose, *Molecular Evolution and the Origin of Life* (M. Dekker, New York, 1977).
16. CDF: F. Abe *et al., Phys. Rev. Lett.* **73**, 225 (1994).
17. P. Langacker, M. Luo, and A. Mann, *Rev. Mod. Phys.* **64**, 87 (1992).

II. THE STANDARD ELECTROWEAK MODEL

STRUCTURE OF THE STANDARD MODEL

PAUL LANGACKER
Department of Physics, University of Pennsylvania,
Philadelphia, Pennsylvania, USA 19104-6396

Contents

1 The Standard Model Lagrangian

The standard model [1] is a gauge theory [2] of the microscopic interactions. The strong interaction part (QCD [3]) is described by the Lagrangian

$$\mathcal{L}_{SU_3} = -\frac{1}{4}F^i_{\mu\nu}F^{i\mu\nu} + \sum_r \bar{q}_{r\alpha} i \not{D}^\alpha_\beta \, q^\beta_r, \tag{1}$$

where g_s is the QCD gauge coupling constant,

$$F^i_{\mu\nu} = \partial_\mu G^i_\nu - \partial_\nu G^i_\mu - g_s f_{ijk} \, G^j_\mu \, G^k_\nu \tag{2}$$

is the field strength tensor for the gluon fields G^i_μ, $i = 1, \cdots, 8$, and the structure constants f_{ijk} $(i, j, k = 1, \cdots, 8)$ are defined by

$$[\lambda^i, \lambda^j] = 2i f_{ijk} \lambda^k, \tag{3}$$

where the SU_3 λ matrices are defined in Table 1. The F^2 term leads to three and four-point gluon self-interactions. The second term in $\mathcal{L}_{SU_3}$ is the gauge covariant derivative for the quarks: q_r is the rth quark flavor, $\alpha, \beta = 1, 2, 3$ are color indices, and

$$D^\alpha_{\mu\beta} = (D_\mu)_{\alpha\beta} = \partial_\mu \delta_{\alpha\beta} + i g_s \, G^i_\mu \, L^i_{\alpha\beta}, \tag{4}$$

$\lambda^i = \begin{pmatrix} \tau^i & 0 \\ 0 & 0 \end{pmatrix}, \quad i = 1,2,3$
$\lambda^4 = \begin{pmatrix} 0 & 0 & 1 \\ 0 & 0 & 0 \\ 1 & 0 & 0 \end{pmatrix} \qquad \lambda^5 = \begin{pmatrix} 0 & 0 & -i \\ 0 & 0 & 0 \\ i & 0 & 0 \end{pmatrix}$
$\lambda^6 = \begin{pmatrix} 0 & 0 & 0 \\ 0 & 0 & 1 \\ 0 & 1 & 0 \end{pmatrix} \qquad \lambda^7 = \begin{pmatrix} 0 & 0 & 0 \\ 0 & 0 & -i \\ 0 & i & 0 \end{pmatrix}$
$\lambda^8 = \frac{1}{\sqrt{3}} \begin{pmatrix} 1 & 0 & 0 \\ 0 & 1 & 0 \\ 0 & 0 & -2 \end{pmatrix}$

Table 1: The SU_3 matrices.

where the quarks transform according to the triplet representation matrices $L^i = \lambda^i/2$. The color interactions are diagonal in the flavor indices, but in general change the quark colors. They are purely vector (parity conserving). There are no bare mass terms for the quarks in (1). These would be allowed by QCD alone, but are forbidden by the chiral symmetry of the electroweak part of the theory. The quark masses will be generated later by spontaneous symmetry breaking. There are in addition effective ghost and gauge-fixing terms which enter into the quantization of both the SU_3 and electroweak Lagrangians, and there is the possibility of adding an (unwanted) term which violates CP invariance.

The electroweak theory is based on the $SU_2 \times U_1$ Lagrangian [4]

$$\mathcal{L}_{SU_2 \times U_1} = \mathcal{L}_{\text{gauge}} + \mathcal{L}_{\varphi} + \mathcal{L}_f + \mathcal{L}_{\text{Yukawa}}. \tag{5}$$

The gauge part is

$$\mathcal{L}_{\text{gauge}} = -\frac{1}{4} F^i_{\mu\nu} F^{\mu\nu i} - \frac{1}{4} B_{\mu\nu} B^{\mu\nu}, \tag{6}$$

where W^i_μ, $i = 1,\ 2,\ 3$ and B_μ are respectively the SU_2 and U_1 gauge fields, with field strength tensors

$$\begin{aligned} B_{\mu\nu} &= \partial_\mu B_\nu - \partial_\nu B_\mu \\ F_{\mu\nu} &= \partial_\mu W^i_\nu - \partial_\nu W^i_\mu - g\epsilon_{ijk} W^j_\mu W^k_\nu, \end{aligned} \tag{7}$$

where $g(g')$ is the SU_2 (U_1) gauge coupling and ϵ_{ijk} is the totally antisymmetric symbol. The SU_2 fields have three and four-point self-interactions. B is a U_1 field associated with the weak hypercharge $Y = Q - T_3$, where Q and T_3 are respectively the electric charge operator and the third component of weak SU_2. It has no self-interactions. The B and W_3 fields will eventually mix to form the photon and Z boson.

The scalar part of the Lagrangian is

$$\mathcal{L}_\varphi = (D^\mu \varphi)^\dagger D_\mu \varphi - V(\varphi), \tag{8}$$

where $\varphi = \begin{pmatrix} \varphi^+ \\ \varphi^0 \end{pmatrix}$ is a complex Higgs scalar, which is a doublet under SU_2 with U_1 charge $Y_\varphi = +\frac{1}{2}$. The gauge covariant derivative is

$$D_\mu \varphi = \left(\partial_\mu + ig\frac{\tau^i}{2} W_\mu^i + \frac{ig'}{2} B_\mu \right) \varphi, \tag{9}$$

where the τ^i are the Pauli matrices. The square of the covariant derivative leads to three and four-point interactions between the gauge and scalar fields [1].

$V(\varphi)$ is the Higgs potential. The combination of $SU_2 \times U_1$ invariance and renormalizability restricts V to the form

$$V(\varphi) = +\mu^2 \varphi^\dagger \varphi + \lambda(\varphi^\dagger \varphi)^2. \tag{10}$$

For $\mu^2 < 0$ there will be spontaneous symmetry breaking. The λ term describes a quartic self-interaction between the scalar fields. Vacuum stability requires $\lambda > 0$.

The fermion term is

$$\mathcal{L}_F = \sum_{m=1}^{F} \left(\bar{q}_{mL}^0 i \not{D} q_{mL}^0 + \bar{l}_{mL}^0 i \not{D} l_{mL}^0 + \bar{u}_{mR}^0 i \not{D} u_{mR}^0 + \bar{d}_{mR}^0 i \not{D} d_{mR}^0 + \bar{e}_{mR}^0 i \not{D} e_{mR}^0 \right). \tag{11}$$

In (11) m is the family index, $F \geq 3$ is the number of families, and $L(R)$ refer to the left (right) chiral projections $\psi_{L(R)} \equiv (1 \mp \gamma_5)\psi/2$. The left-handed quarks and leptons

$$q_{mL}^0 = \begin{pmatrix} u_m^0 \\ d_m^0 \end{pmatrix}_L \qquad l_{mL}^0 = \begin{pmatrix} \nu_m^0 \\ e_m^{-0} \end{pmatrix}_L \tag{12}$$

transform as SU_2 doublets, while the right-handed fields u_{mR}^0, d_{mR}^0, and e_{mR}^{-0} are singlets. Their U_1 charges are $Y_{q_L} = \frac{1}{6}$, $Y_{l_L} = -\frac{1}{2}$, $Y_{\psi_R} = q_\psi$. The superscript 0 refers to the weak eigenstates, *i.e.*, fields transforming according to definite SU_2 representations. They may be mixtures of mass eigenstates (flavors). The quark color indices $\alpha = r,\ g,\ b$ have been suppressed. The gauge covariant derivatives are

$$\begin{aligned} D_\mu q_{mL}^0 &= \left(\partial_\mu + \tfrac{ig}{2}\tau^i W_\mu^i + i\tfrac{g'}{6} B_\mu\right) q_{mL}^0 & \qquad D_\mu u_{mR}^0 &= \left(\partial_\mu + i\tfrac{2}{3} g' B_\mu\right) u_{mR}^0 \\ D_\mu l_{mL}^0 &= \left(\partial_\mu + \tfrac{ig}{2}\tau^i W_\mu^i - i\tfrac{g'}{2} B_\mu\right) l_{mL}^0 & D_\mu d_{mR}^0 &= \left(\partial_\mu - i\tfrac{g'}{3} B_\mu\right) d_{mR}^0 \\ & & D_\mu e_{mR}^0 &= \left(\partial_\mu - ig' B_\mu\right) e_{mR}^0, \end{aligned} \tag{13}$$

from which one can read off the gauge interactions between the W and B and the fermion fields. The different transformations of the L and R fields (*i.e.*, the symmetry is chiral) is the origin of parity violation in the electroweak sector. The chiral symmetry also forbids any bare mass terms for the fermions.

The last term in (5) is

$$-L_{\text{Yukawa}} = \sum_{m,n=1}^{F} \left[\Gamma_{mn}^u \bar{q}_{mL}^0 \tilde{\varphi} u_{mR}^0 + \Gamma_{mn}^d \bar{q}_{mL}^0 \varphi d_{nR}^0 + \Gamma_{mn}^e \bar{l}_{mn}^0 \varphi e_{nR}^0 \right] + \text{H.C.}, \tag{14}$$

where the matrices Γ_{mn} describe the Yukawa couplings between the single Higgs doublet, φ, and the various flavors m and n of quarks and leptons. One needs representations of Higgs fields with $Y = \frac{1}{2}$ and $-\frac{1}{2}$ to give masses to the down quarks, the electrons, and the up quarks. The representation $\varphi^\dagger$ has $Y = -\frac{1}{2}$, but transforms as the 2^* rather than the 2. However, in SU_2 the 2^* representation is related to the 2 by a similarity transformation, and $\tilde{\varphi} \equiv i\tau^2\varphi^\dagger = \begin{pmatrix} \varphi^{0\dagger} \\ -\varphi^- \end{pmatrix}$ transforms as a 2 with $Y_{\tilde{\varphi}} = -\frac{1}{2}$. All of the masses can therefore be generated with a single Higgs doublet if one makes use of both φ and $\tilde{\varphi}$. The fact that the fundamental and its conjugate are equivalent does not generalize to higher unitary groups. Furthermore, in supersymmetric extensions of the standard model the supersymmetry forbids the use of a single Higgs doublet in both ways in the Lagrangian, and one must add a second Higgs doublet. Similar statements apply to most theories with an additional U_1 gauge factor, *i.e.*, a heavy Z' boson.

2 Spontaneous Symmetry Breaking

Gauge invariance (and therefore renormalizability) does not allow mass terms in the Lagrangian for the gauge bosons or for chiral fermions. Massless gauge bosons are not acceptable for the weak interactions, which are known to be short-ranged. Hence, the gauge invariance must be broken spontaneously [5], which preserves the renormalizability [6]. The idea is simply that the lowest energy (vacuum) state does not respect the gauge symmetry and induces effective masses for particles propagating through it.

Let us introduce the complex vector

$$v = \langle 0|\varphi|0\rangle = \text{constant}, \tag{15}$$

which has components that are the vacuum expectation values of the various complex scalar fields. v is determined by rewriting the Higgs potential as a function of v, $V(\varphi) \to V(v)$, and choosing v such that V is minimized. That is, we interpret v as the lowest energy solution of the classical equation of motion[1]. The quantum theory is obtained by considering fluctuations around this classical minimum, $\varphi = v + \varphi'$.

The single complex Higgs doublet in the standard model can be rewritten in a Hermitian basis as

$$\varphi = \begin{pmatrix} \varphi^+ \\ \varphi^0 \end{pmatrix} = \begin{pmatrix} \frac{1}{\sqrt{2}}(\varphi_1 - i\varphi_2) \\ \frac{1}{\sqrt{2}}(\varphi_3 - i\varphi_4) \end{pmatrix}, \tag{16}$$

[1] It suffices to consider constant v because any space or time dependence $\partial_\mu v$ would increase the energy of the solution. Also, one can take $\langle 0|\psi|0\rangle = \langle 0|A_\mu|0\rangle = 0$, because any non-zero vacuum values would violate Lorentz invariance. These extensions are involved in (higher energy) topological defects, such as monopoles, strings, domain walls, and textures.

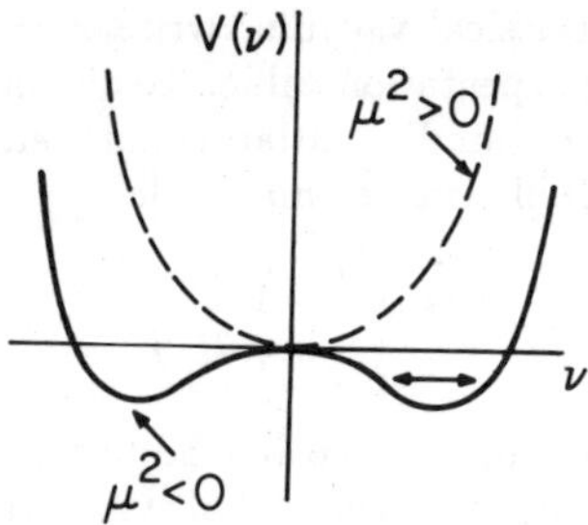

Figure 1: The Higgs potential $V(\nu)$ for $\mu^2 > 0$ (dashed line) and $\mu^2 < 0$ (solid line).

where $\varphi_i = \varphi_i^\dagger$ represent four Hermitian fields. In this new basis the Higgs potential becomes

$$V(\varphi) = \frac{1}{2}\mu^2 \left(\sum_{i=1}^{4} \varphi_i^2\right) + \frac{1}{4}\lambda \left(\sum_{i=1}^{4} \varphi_i^2\right)^2, \tag{17}$$

which is clearly O_4 invariant. Without loss of generality we can choose the axis in this four-dimensional space so that $\langle 0|\varphi_i|0\rangle = 0, \quad i = 1, 2, 4$ and $\langle 0|\varphi_3|0\rangle = \nu$. Thus,

$$V(\varphi) \to V(v) = \frac{1}{2}\mu^2\nu^2 + \frac{1}{4}\lambda\nu^4, \tag{18}$$

which must be minimized with respect to ν. Two important cases are illustrated in Figure 1. For $\mu^2 > 0$ the minimum occurs at $\nu = 0$. That is, the vacuum is empty space and $SU_2 \times U_1$ is unbroken at the minimum. On the other hand, for $\mu^2 < 0$ the $\nu = 0$ symmetric point is unstable, and the minimum occurs at some nonzero value of ν which breaks the $SU_2 \times U_1$ symmetry. The point is found by requiring

$$V'(\nu) = \nu(\mu^2 + \lambda\nu^2) = 0, \tag{19}$$

which has the solution $\nu = (-\mu^2/\lambda)^{1/2}$ at the minimum. (The solution for $-\nu$ can also be transformed into this standard form by an appropriate O_4 transformation.) The dividing point $\mu^2 = 0$ cannot be treated classically. It is necessary to consider the one loop corrections to the potential, in which case it is found that the symmetry is again spontaneously broken [7].

We are interested in the case $\mu^2 < 0$, for which the Higgs doublet is replaced, in first approximation, by its classical value $\varphi \to \frac{1}{\sqrt{2}}\begin{pmatrix} 0 \\ \nu \end{pmatrix} \equiv v$. The generators L^1, L^2, and $L^3 - Y$ are spontaneously broken (*e.g.*, $L^1 v \neq 0$). On the other hand, the vacuum carries no electric charge ($Qv = (L^3 + Y)v = 0$), so the U_{1Q} of electromagnetism is not broken. Thus, the electroweak $SU_2 \times U_1$ group is spontaneously broken down, $SU_2 \times U_{1Y} \to U_{1Q}$.

To quantize around the classical vacuum, write $\varphi = v + \varphi'$, where φ' are quantum fields with zero vacuum expectation value. To display the physical particle content it is useful to rewrite the four Hermitian components of φ' in terms of a new set of variables using the Kibble transformation [8]:

$$\varphi = \frac{1}{\sqrt{2}} e^{i \sum \xi^i L^i} \begin{pmatrix} 0 \\ \nu + H \end{pmatrix}. \tag{20}$$

H is a Hermitian field which will turn out to be the physical Higgs scalar. If we had been dealing with a spontaneously broken global symmetry the three Hermitian fields ξ^i would be the massless pseudoscalar Goldstone bosons [9] that are necessarily associated with broken symmetry generators. However, in a gauge theory they disappear from the physical spectrum. To see this it is useful to go to the unitary gauge

$$\varphi \to \varphi' = e^{-i \sum \xi^i L^i} \varphi = \frac{1}{\sqrt{2}} \begin{pmatrix} 0 \\ \nu + H \end{pmatrix}, \tag{21}$$

in which the Goldstone bosons disappear. In this gauge, the scalar covariant kinetic energy term takes the simple form

$$\begin{aligned} (D_\mu \varphi)^\dagger D^\mu \varphi &= \frac{1}{2}(0\ \nu) \left[\frac{g}{2} \tau^i W_\mu^i + \frac{g'}{2} B_\mu \right]^2 \begin{pmatrix} 0 \\ \nu \end{pmatrix} + H \text{ terms} \\ &\to M_W^2 W^{+\mu} W_\mu^- + \frac{M_Z^2}{2} Z^\mu Z_\mu + H \text{ terms}, \end{aligned} \tag{22}$$

where the kinetic energy and gauge interaction terms of the physical H particle have been omitted. Thus, spontaneous symmetry breaking generates mass terms for the W and Z gauge bosons

$$\begin{aligned} W^\pm &= \frac{1}{\sqrt{2}}(W^1 \mp i W^2) \\ Z &= -\sin\theta_W B + \cos\theta_W W^3. \end{aligned} \tag{23}$$

The photon field

$$A = \cos\theta_W B + \sin\theta_W W^3 \tag{24}$$

remains massless. The masses are

$$M_W = \frac{g\nu}{2} \tag{25}$$

and

$$M_Z = \sqrt{g^2 + g'^2} \frac{\nu}{2} = \frac{M_W}{\cos\theta_W}, \tag{26}$$

where the weak angle is defined by $\tan\theta_W \equiv g'/g$. One can think of the generation of masses as due to the fact that the W and Z interact constantly with the condensate of scalar fields and therefore acquire masses, in analogy with a photon propagating through a plasma. The Goldstone boson has disappeared from the theory but has reemerged as the longitudinal degree of freedom of a massive vector particle.

It will be seen below that $G_F/\sqrt{2} \sim g^2/8M_W^2$, where $G_F = 1.16639(2) \times 10^{-5}\ GeV^{-2}$ is the Fermi constant determined by the muon lifetime. The weak scale ν is therefore

$$\nu = 2M_W/g \simeq (\sqrt{2}G_F)^{-1/2} \simeq 246\ GeV. \tag{27}$$

Similarly, $g = e/\sin\theta_W$, where e is the electric charge of the positron. Hence, to lowest order

$$M_W = M_Z \cos\theta_W \sim \frac{(\pi\alpha/\sqrt{2}G_F)^{1/2}}{\sin\theta_W}, \tag{28}$$

where $\alpha \sim 1/137.036$ is the fine structure constant. Using $\sin^2\theta_W \sim 0.23$ from neutral current scattering, one expects $M_W \sim 78\ GeV$, and $M_Z \sim 89\ GeV$. (These predictions are increased by $\sim (2-3)\ GeV$ by loop corrections.) The W and Z were discovered at CERN by the UA1 [10] and UA2 [11] groups in 1983. Subsequent measurements of their masses and other properties have been in perfect agreement with the standard model expectations (including the higher-order corrections), as is described in the articles of by Schaile and Einsweiler.

After symmetry breaking the Higgs potential becomes

$$V(\varphi) = -\frac{\mu^4}{4\lambda} - \mu^2 H^2 + \lambda\nu H^3 + \frac{\lambda}{4}H^4. \tag{29}$$

The third and fourth terms represent the cubic and quartic interactions of the Higgs scalar. The second term represents a (tree-level) mass

$$M_H = \sqrt{-2\mu^2} = \sqrt{2\lambda}\nu. \tag{30}$$

The weak scale is given in (27), but the quartic Higgs coupling λ is unknown, so M_H is not predicted. A priori, λ could be anywhere in the range $0 \leq \lambda < \infty$. There is now an experimental lower limit $M_H \gtrsim 60$ GeV from LEP [12]. Otherwise, the decay $Z \to Z^* H$ would have been observed. (There are also theoretical lower limits on M_H in the $0 - 10$ GeV range, depending on m_t, when higher-order corrections are included [13].)

There are also plausible theoretical upper limits. If $\lambda > O(1)$ the theory becomes strongly coupled. ($M_H > O(1$ TeV$)$). There is not really anything wrong with strong coupling a priori. However, there are fairly convincing triviality limits, which basically say that the running quartic coupling would become infinite within the domain of validity of the theory if λ and therefore M_H is too large. If one requires the theory to make sense to infinite energy, one may run into problems[2] with the increasing quartic coupling for any λ. However, one only needs for the

[2]This is true for a pure λH^4 theory. The presence of other interactions may eliminate the problems for small λ.

theory to hold up to the next mass scale Λ, at which point the standard model breaks down. In that case [13],

$$M_H < \begin{cases} O(200)\ GeV, \ \Lambda \sim M_P \\ O(600)\ GeV, \ \Lambda \sim 2M_H \end{cases} \tag{31}$$

The more stringent limit of $O(200)$ GeV obtains for Λ of order of the Planck scale $M_P = G_N^{-1/2} \sim 10^{19}$ GeV. If one makes the less restrictive assumption that the scale Λ of new physics can be small, one obtains a weaker limit. Nevertheless, for the concept of an elementary Higgs field to make sense one should require that the theory be valid up to something of order of $2M_H$, which implies that $M_H < O(600)$ GeV. These limits may be relaxed if there are other heavy particles in the theory.

The first term in (29) is the vacuum expectation value

$$\langle 0|V|0\rangle = -\mu^4/4\lambda \tag{32}$$

of the Higgs potential when evaluated at the minimum. This is a c-number which has no significance for the microscopic interactions. However, it assumes great importance when the theory is coupled to gravity, because a constant energy density plays the role of a cosmological constant [14]. The cosmological constant becomes

$$\Lambda_{\text{cosm}} = \Lambda_{\text{bare}} + \Lambda_{\text{SSB}}, \tag{33}$$

where Λ_{bare} is the primordial cosmological constant, which can be thought of as the value of the energy of the vacuum in the absence of spontaneous symmetry breaking. (Eqn. (10) implicitly assumed $\Lambda_{\text{bare}} = 0$.) Λ_{SSB} is the part generated by the Higgs mechanism:

$$|\Lambda_{\text{SSB}}| = 8\pi G_N |\langle 0|V|0\rangle| \sim 10^{50}|\Lambda_{\text{obs}}|. \tag{34}$$

It is some 10^{50} times larger than the observational upper limit Λ_{obs}. This is clearly unacceptable. Technically, one can solve the problem by adding a constant $+\mu^4/4\lambda$ to V, so that V is equal to zero at the minimum (*i.e.*, $\Lambda_{\text{bare}} = 2\pi G_N \mu^4/\lambda$). However, with our current understanding there is no reason for Λ_{bare} and Λ_{SSB} to be related; to have to invoke such an incredibly fine-tuned cancellation to 50 decimal places is a major unsatisfactory feature of the standard model.

The Yukawa interaction in the unitary gauge becomes

$$\begin{aligned} -L_{\text{Yukawa}} &\rightarrow \sum_{m,n=1}^{F} \bar{u}^0_{mL} \Gamma^u_{mn} \left(\frac{\nu + H}{\sqrt{2}}\right) u^0_{mR} + (d,e) \text{ terms } + \text{ H.C.} \\ &= \bar{u}^0_L (M^u + h^u H) u^0_R + (d,e) \text{ terms } + \text{H.C.}, \end{aligned} \tag{35}$$

where in the second form $u^0_L = (u^0_{1L} u^0_{2L} \cdots u^0_{FL})^T$ is an F-component column vector, with a similar definition for u^0_R. M^u is an $F \times F$ fermion mass matrix $M^u_{mn} =$

$\Gamma^u_{mn}\nu/\sqrt{2}$ induced by spontaneous symmetry breaking, and $h^u = M^u/\nu = gM^u/2M_W$ is the Yukawa coupling matrix.

In general M is not diagonal, Hermitian, or symmetric. To identify the physical particle content it is necessary to diagonalize M by separate unitary transformations A_L and A_R on the left- and right-handed fermion fields. (In the special case that M^u is Hermitian one can take $A_L = A_R$). Then,

$$A_L^{u\dagger} M^u A_R^u = M_D^u = \begin{pmatrix} m_u & 0 & 0 \\ 0 & m_c & 0 \\ 0 & 0 & m_t \end{pmatrix} \tag{36}$$

is a diagonal matrix with eigenvalues equal to the physical masses of the charge $\frac{2}{3}$ quarks. Similarly, one diagonalizes the down quark and charged lepton mass matrices by

$$\begin{aligned} A_L^{d\dagger} M^d A_R^d &= M_D^d \\ A_L^{e\dagger} M^e A_R^e &= M_D^e. \end{aligned} \tag{37}$$

In terms of these unitary matrices we can define mass eigenstate fields $u_L = A_L^{u\dagger} u_L^0 = (u_L\ c_L\ t_L)^T$, with analogous definitions for $u_R = A_R^{u\dagger} u_R^0$, $d_{L,R} = A_{L,R}^{d\dagger} d_{L,R}^0$, and $e_{L,R} = A_{L,R}^{e\dagger} e_{L,R}^0$. Assuming the neutrinos are massless, their mass eigenstates are arbitrary. It is convenient to define them in terms of the charged lepton unitary transformation, $\nu_L = A_L^{e\dagger} \nu_L^0$. That is, we define ν_e, ν_μ, ν_τ as the weak interaction partners of the e, μ, and τ. Typical estimates of the quark masses are [15] $m_u = 5.6 \pm 1.1\ MeV$, $m_d = 9.9 \pm 1.1\ MeV$, $m_s = 199 \pm 33\ MeV$, $m_c = 1.35 \pm 0.05\ GeV$, $m_b \sim 4.7\ GeV$, and $m_t > 131\ GeV$ [16] or $m_t = 174 \pm 16\ GeV$ [17]. These are the current masses: for QCD their effects are identical to bare masses in the QCD Lagrangian. They should not be confused with the constituent masses of order 300 MeV generated by the spontaneous breaking of chiral symmetry in the strong interactions. Including QCD renormalizations, the u, d, s and c masses are running masses evaluated at 1 GeV^2, while m_b and m_t are pole masses.

Thus,

$$L_{\text{Yukawa}} = \sum_i \bar{\psi}_i \left(-m_i - \frac{gm_i}{2M_W} H \right) \psi_i. \tag{38}$$

The coupling of the physical Higgs boson to the ith fermion is $gm_i/2M_W$, which is very small except for the top quark. The coupling is flavor-diagonal in the minimal model: there is just one Yukawa matrix for each type of fermion, so the mass and Yukawa matrices are diagonalized by the same transformations. In generalizations in which more than one Higgs doublet couples to each type of fermion there will in general be flavor-changing Yukawa interactions involving the physical neutral Higgs fields [18]. There are stringent limits on such couplings [19]; for example, the $K_L - K_S$ mass difference implies $h/M_H < 10^{-6} GeV^{-1}$, where h is the $\bar{d}s$ Yukawa coupling.

3 The Gauge Interactions

The major quantitative tests of the electroweak standard model involve the gauge interactions of fermions and the properties of the gauge bosons. The charged current weak interactions of the Fermi theory and its extension to the intermediate vector boson theory are incorporated into the standard model, as is quantum electrodynamics. The theory successfully predicted the existence and properties of the weak neutral current. Here I will summarize the structure of the interactions. Later chapters will discuss the phenomenology and tests in more detail.

3.1 *The Charged Current*

The interaction of the W bosons to fermions is given by

$$L = -\frac{g}{2\sqrt{2}}\left(J_W^\mu W_\mu^- + J_W^{\mu\dagger} W_\mu^+\right), \tag{39}$$

where the weak charge-raising current is

$$\begin{aligned} J_W^{\mu\dagger} &= \sum_{m=1}^{F}\left[\bar{\nu}_m^0\gamma^\mu(1-\gamma^5)e_m^0 + \bar{u}_m^0\gamma^\mu(1-\gamma^5)d_m^0\right] \\ &= (\bar{\nu}_e\bar{\nu}_\mu\bar{\nu}_\tau)\gamma^\mu(1-\gamma^5)\begin{pmatrix} e^- \\ \mu^- \\ \tau^- \end{pmatrix} + (\bar{u}\,\bar{c}\,\bar{t})\gamma^\mu(1-\gamma^5)V\begin{pmatrix} d \\ s \\ b \end{pmatrix}. \end{aligned} \tag{40}$$

$J_W^{\mu\dagger}$ has a $V - A$ form, *i.e.*, it violates parity and charge conjugation maximally. The mismatch between the unitary transformations relating the weak and mass eigenstates for the up and down-type quarks leads to the presence of the $F \times F$ unitary matrix $V = A_L^{u\dagger} A_L^d$ in the current. This is the Cabibbo-Kobayashi-Maskawa (CKM) matrix [20], which is ultimately due to the mismatch between the weak and Yukawa interactions. For $F = 2$ families V takes the familiar form

$$V = \begin{pmatrix} \cos\theta_c & \sin\theta_c \\ -\sin\theta_c & \cos\theta_c \end{pmatrix}, \tag{41}$$

where $\sin\theta_c \simeq 0.22$ is the Cabibbo angle. This form gives a good zeroth-order approximation to the weak interactions of the u, d, s and c quarks; their coupling to the third family, though non-zero, is very small. Including these couplings, the 3-family CKM matrix is

$$V = \begin{pmatrix} V_{ud} & V_{us} & V_{ub} \\ V_{cd} & V_{cs} & V_{cb} \\ V_{td} & V_{td} & V_{td} \end{pmatrix}, \tag{42}$$

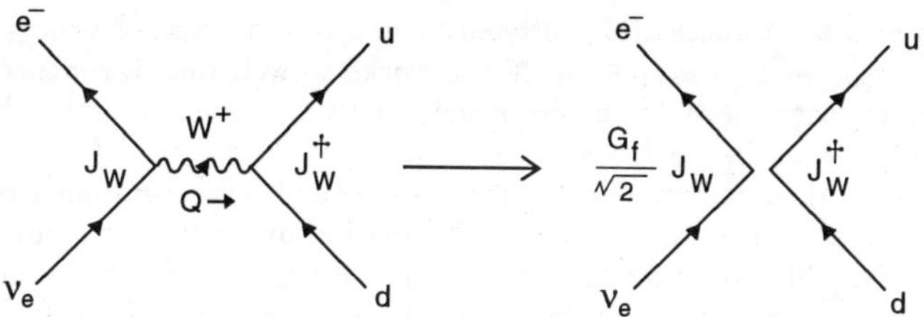

Figure 2: A weak interaction mediated by the exchange of a W and the effective four-fermi interaction that it generates if the four-momentum transfer Q is sufficiently small.

where the V_{ij} may involve a CP-violating phase.

There is nothing to distinguish massless neutrinos except their weak interactions, so one simply defines the ν_e as the weak partner of the electron, and similarly for ν_μ and ν_τ. If there were non-zero neutrino mass then one would have to introduce a leptonic mixing matrix in the current, but its effects would not be important in any process that is not actually sensitive to the masses.

The interaction between fermions mediated by the exchange of a W is illustrated in Figure 2. In the limit $|Q^2| \ll M_W^2$ the momentum term in the W propagator can be neglected, leading to an effective zero-range (four-fermi) interaction

$$-L_{\text{eff}}^{cc} = \frac{G_F}{\sqrt{2}} J_W^\mu J_{W\mu}^\dagger, \tag{43}$$

where the Fermi constant is identified as

$$\frac{G_F}{\sqrt{2}} \simeq \frac{g^2}{8M_W^2} = \frac{1}{2\nu^2}. \tag{44}$$

Thus, the Fermi theory is an approximation to the standard model valid in the limit of small momentum transfer. From the muon lifetime, $G_F = 1.16639(2) \times 10^{-5}$ GeV^{-2}, which implies that the weak interaction scale defined by the VEV of the Higgs field is $\nu = \sqrt{2}\langle 0|\varphi^0|0\rangle \simeq 246$ GeV.

The charged current weak interaction as described by (43) has been successfully tested in a large variety of weak decays [21], including β, K, hyperon, heavy quark, μ, and τ decays. In particular, high precision measurements of β, μ, and τ decays are a sensitive probe of extended gauge groups involving right-handed currents and other types of new physics, as is described in the chapters by Deutsch and Quin; Fetscher and Gerber; and Herczeg. Tests of the unitarity of the CKM matrix are important in searching for the presence of fourth family or exotic fermions and for new interactions, as described by Sirlin and by London. The standard theory has also been successfully probed in neutrino scattering processes such as $\nu_\mu e \to \mu^- \nu_e, \nu_\mu n \to \mu^- p, \nu_\mu N \to \mu^- X$. It works so well that the neutrino-hadron interactions are used more as a probe of the structure of the hadrons and QCD than as a test of the weak interactions.

Weak charged current effects have also been observed in higher orders, such as in the mass difference $M_{K_S} - M_{K_L}$, CP violation in the kaon system [22], and in $B \leftrightarrow \bar{B}$ mixing [23]. For these higher order processes the full theory must be used because large momenta occur within the loop integrals.

3.2 *QED*

The standard model incorporates all of the (spectacular) successes of quantum electrodynamics (QED) [24], which is based on the U_{1Q} subgroup that remains unbroken after spontaneous symmetry breaking. The relevant part of the Lagrangian is

$$L = -\frac{gg'}{\sqrt{g^2 + g'^2}} J_Q^\mu (\cos\theta_W B_\mu + \sin\theta_W W_\mu^3), \tag{45}$$

where the linear combination of neutral gauge fields is just the photon field A_μ. This reproduces the QED interaction provided one identifies the combination of couplings

$$e = g \sin\theta_W \tag{46}$$

as the electric charge of the positron, where $\tan\theta_W \equiv g'/g$. The electromagnetic current is given by

$$\begin{aligned} J_Q^\mu &= \sum_{m=1}^{F} \left[\frac{2}{3}\bar{u}_m^0 \gamma^\mu u_m^0 - \frac{1}{3}\bar{d}_m^0 \gamma^\mu d_m^0 - \bar{e}_m^0 \gamma^\mu e_m^0\right] \\ &= \sum_{m=1}^{F} \left[\frac{2}{3}\bar{u}_m \gamma^\mu u_m - \frac{1}{3}\bar{d}_m \gamma^\mu d_m - \bar{e}_m \gamma^\mu e_m\right]. \end{aligned} \tag{47}$$

It takes the same form when written in terms of either weak or mass eigenstates because all fermions which mix with each other have the same electric charge. Thus, the electromagnetic current is automatically flavor-diagonal.

3.3 *The Neutral Current*

The third class of gauge interactions is the weak neutral current [25], which was predicted by the $SU_2 \times U_1$ model. The relevant interaction is

$$L = -\frac{\sqrt{g^2 + g'^2}}{2} J_Z^\mu \left(-\sin\theta_W B_\mu + \cos\theta_W W_\mu^3\right), \tag{48}$$

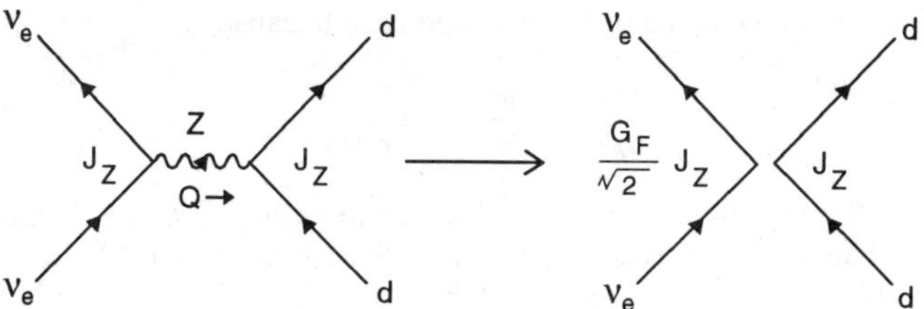

Figure 3: Typical neutral current interaction mediated by the exchange of the Z, which reduces to an effective four-fermi interaction in the limit that the momentum transfer Q can be neglected.

where the combination of neutral fields is the massive Z boson field. The strength is conveniently rewritten as $g/(2\cos\theta_W)$, which follows from $\cos\theta_W = g/\sqrt{g^2+g'^2}$.

The weak neutral current is given by

$$\begin{aligned} J_Z^\mu &= \sum_m \left[\bar{u}^0_{mL}\gamma^\mu u^0_{mL} - \bar{d}^0_{mL}\gamma^\mu d^0_{mL} + \bar{\nu}^0_{mL}\gamma^\mu \nu^0_{mL} - \bar{e}^0_{mL}\gamma^\mu e^0_{mL}\right] - 2\sin^2\theta_W J_Q^\mu \\ &= \sum_m \left[\bar{u}_{mL}\gamma^\mu u_{mL} - \bar{d}_{mL}\gamma^\mu d_{mL} + \bar{\nu}_{mL}\gamma^\mu \nu_{mL} - \bar{e}_{mL}\gamma^\mu e_{mL}\right] - 2\sin^2\theta_W J_Q^\mu \end{aligned} \tag{49}$$

Like the electromagnetic current J_Z^μ is flavor-diagonal in the standard model; all fermions which have the same electric charge and chirality and therefore can mix with each other have the same $SU_2 \times U_1$ assignments, so the form is not affected by the unitary transformations that relate the mass and weak bases. It was for this reason that the GIM mechanism [26] was introduced into the model, along with its prediction of the charm quark. Without it the d and s quarks would not have had the same $SU_2 \times U_1$ assignments, and flavor-changing neutral currents would have resulted. The absence of such effects is a major restriction on many extensions of the standard model involving exotic fermions, as described in the article by London. The neutral current has two contributions. The first only involves the left-chiral fields and is purely $V-A$. The second is proportional to the electromagnetic current with coefficient $\sin^2\theta_W$ and is purely vector. Parity is therefore violated in the neutral current interaction, though not maximally.

In an interaction between fermions in the limit that the momentum transfer is small compared to M_Z one can neglect the Q^2 term in the propagator, and the interaction reduces to an effective four-fermi interaction

$$-L_{\text{eff}}^{NC} = \frac{G_F}{\sqrt{2}} J_Z^\mu J_{Z\mu}. \tag{50}$$

The coefficient is the same as in the charged case because

$$\frac{G_F}{\sqrt{2}} = \frac{g^2}{8M_W^2} = \frac{g^2 + g'^2}{8M_Z^2}. \tag{51}$$

That is, the difference in Z couplings compensates the difference in masses in the propagator. The weak neutral current was discovered at CERN in 1973 by the Gargamelle [27] collaboration and by HPW at Fermilab [28] shortly thereafter, and since that time it has been extensively studied in many interactions, including $\nu e \to \nu e$, $\nu N \to \nu N$, $\nu N \to \nu X$; $e^{\uparrow\downarrow} D \to eX$; atomic parity violation; e^+e^- and Z-pole reactions. These have been the primary quantitative test of the unification part of the standard electroweak model, and all aspects will be discussed extensively in later chapters.

3.4 *Gauge Self-interactions*

The self-interactions of the gauge bosons in the standard model are displayed in Figure 4. Their form is predicted by the underlying gauge invariance, but they have not yet been tested. A sensitive test will have to wait for a study of $e^+e^- \to W^+W^-$ in the second phase, LEP II, of the e^+e^- collider at CERN, as described in the article by Treille [29], and future possible e^+e^- and hadron colliders at higher energy. To lowest order there are three diagrams, as shown in Figure 5. Two of them involve the three-point interaction between a photon or Z boson and W^+W^-. The cross section from any one of these diagrams rises with center of mass energy, but gauge invariance relates these three-point vertices to the couplings of the fermions in such a way that at high energies there is a cancellation. It is another manifestation of the same cancellation which brings higher-order loop integrals under control, leading to a renormalizable theory (otherwise, vector theories would have severe divergences). At LEP II one will be able to observe the cancellation; it would be even more dramatic at possible future colliders at higher energies. Detailed studies of $e^+e^- \to W^+W^-$ should be sensitive to deviations from the standard model, especially those associated with such non-gauge physics as compositeness. In practice, however, many of the types of new physics which could lead to observable effects are already excluded by other observables at LEP I and elsewhere [30].

The processes $\overset{(-)}{q}\, q \to VV'$ would also be sensitive to gauge self-interactions. Finally, one can study the gauge-gauge three and four point vertices in the processes $e^+e^- \to e^+e^-VV'$ and $\overset{(-)}{q}\, q \to \overset{(-)}{q}\, qVV'$. These tests involve the same reactions that would be used to search for a very heavy Higgs boson at a high energy hadron collider and will be important not only for their own sake but as necessary background for the Higgs search.

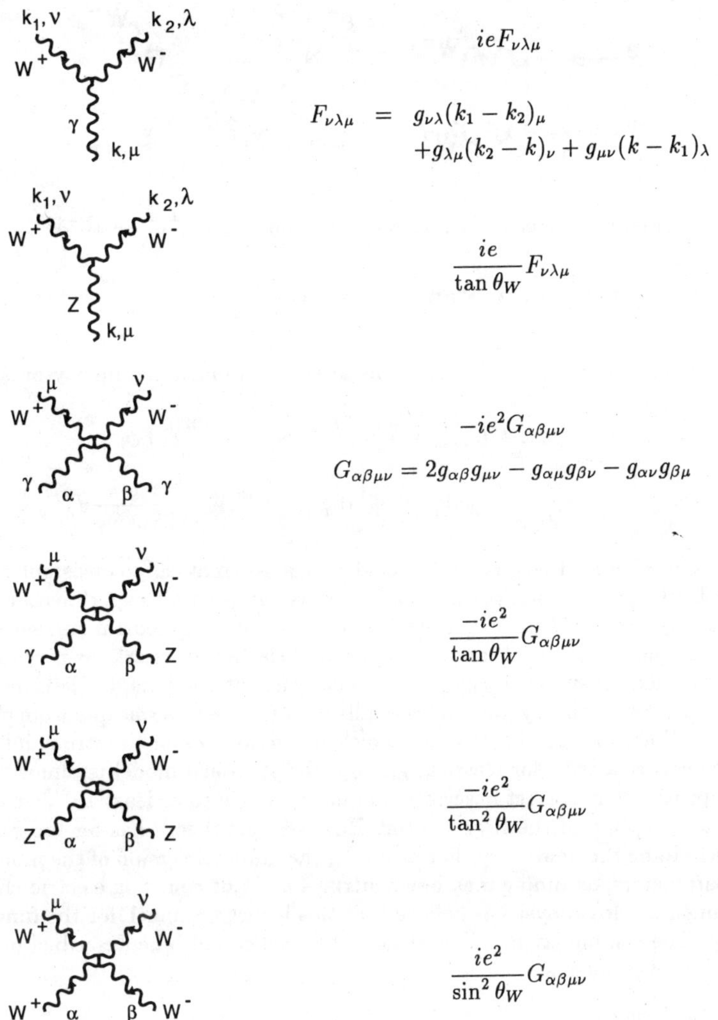

Figure 4: The three and four point-self-interactions of gauge bosons in the standard electroweak model.

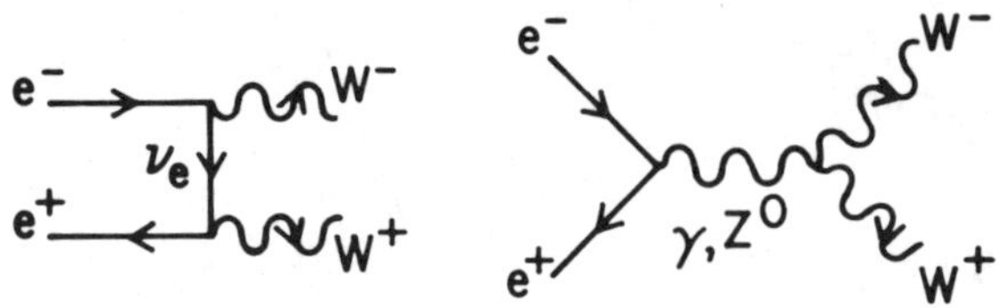

Figure 5: Tree-level diagrams contributing to $e^+e^- \to W^+W^-$.

4 Problems With the Standard Model

The Lagrangian for the standard model after spontaneous symmetry breaking is

$$\begin{aligned} L &= L_{\text{gauge}} + L_{\text{Higgs}} + \sum_i \bar{\psi}_i \left(i \not\partial - m_i - \frac{m_i H}{\nu} \right) \psi_i \\ &\quad - \frac{g}{2\sqrt{2}} \left(J_W^\mu W_\mu^- + J_W^{\mu\dagger} W_\mu^+ \right) - e J_Q^\mu A_\mu - \frac{g}{2\cos\theta_W} J_Z^\mu Z_\mu. \end{aligned} \tag{52}$$

The standard electroweak model is a mathematically-consistent renormalizable field theory which predicts or is consistent with all experimental facts. It successfully predicted the existence and form of the weak neutral current, the existence and masses of the W and Z bosons, and the charm quark, as necessitated by the GIM mechanism. The charged current weak interactions, as described by the generalized Fermi theory, were successfully incorporated, as was quantum electrodynamics. When combined with quantum chromodynamics for the strong interactions and general relativity for classical gravity, the standard model is almost certainly the approximately correct description of nature down to at least 10^{-16}cm, with the possible exception of the Higgs sector. However, the theory has far too much arbitrariness to be the final story. For example, the minimal version of the model has 21 free parameters, assuming massless neutrinos and not counting electric charge (Y) assignments. Most physicists believe that this is just too much for the fundamental theory. The complications of the standard model can also be described in terms of a number of problems.

1. *Gauge Problem*

The standard model is a complicated direct product of three sub-groups, $SU_3 \times SU_2 \times U_1$, with separate gauge couplings. There is no explanation for why only the electroweak part is chiral (parity-violating). Similarly, the standard model incorporates but does not explain another fundamental fact of nature: charge quantization, *i.e.*, why all particles have charges which are multiples of $e/3$. This is important because it allows the electrical neutrality of atoms ($|q_p| = |q_e|$). Possible explanations include: grand unified theories [31], the existence of magnetic monopoles [32], and constraints from the absence or cancellation[3] of anomalies [33].

2. *Fermion Problem*

All matter under ordinary terrestrial conditions can be constructed out of the fermions (ν_e, e^-, u, d) of the first family. Yet we know from laboratory studies that there are ≥ 3 families: (ν_μ, μ^-, c, s) and (ν_τ, τ^-, t, b) are heavier copies of the first family with no obvious role in nature. (The t and ν_τ have not yet been directly observed, although there are candidate t events from CDF [17]. They are assumed to exist because the weak interactions of the b and τ have been well measured and are in agreement with the assumptions that they have SU_2-doublet partners [34].) The standard model gives no explanation for the existence of these heavier families and no prediction for their numbers. Furthermore, there is no explanation or prediction of the fermion masses, which vary over at least five orders of magnitude, or of the CKM mixings. There are many possible suggestions of new physics that might shed light on this, including composite fermions; family symmetries; radiative hierarchies, in which the fermion masses are generated at the loop-level [35], with the lighter families requiring more loops; and the topology of extra space-time dimensions, such as in superstring models [36]. Despite all of these ideas there is no compelling model and none of these yields detailed predictions. The problem is just too complicated. Simple grand unified theories don't help very much with this, except for the prediction of m_b in terms of m_τ in the simplest versions [37].

3. *Higgs/hierarchy Problem*

In the standard model one introduces an elementary Higgs field into the theory to generate masses for the W, Z, and fermions. For the model to be consistent the Higgs mass should not be too different from the W mass, *i.e.*, $M_H^2 = O(M_W^2)$. If M_H were to be larger than M_W by many orders of magnitude there would be a hierarchy problem, and the Higgs self-interactions would be excessively strong. Combining theoretical arguments with laboratory limits one obtains $M_H \lesssim 1$ TeV. (See (31)).

However, there is a complication. The tree-level (bare) Higgs mass receives quadratically-divergent corrections from the loop diagrams in Figure 6. One finds

$$M_H^2 = (M_H^2)_{\text{bare}} + O(\lambda, g^2, h^2)\Lambda^2, \tag{53}$$

where Λ is the next higher scale in the theory. If there were no higher scale one would simply interpret Λ as an ultraviolet cutoff and take the view that M_H is a measured parameter and that $(M_H)_{\text{bare}}$ is not an observable. However, the theory is presumably embedded in some larger theory that cuts off the integral at the finite scale of the new physics[4]. For example, if the next scale is gravity Λ is the Planck

---

[3]The absence of anomalies is not sufficient to determine all of the Y assignments without additional assumptions, such as family universality.

[4]There is no analogous fine-tuning associated with logarithmic divergences, such as those encountered in QED, because $\alpha \ln(\Lambda/m_e) < O(1)$ even for $\Lambda = M_P$.

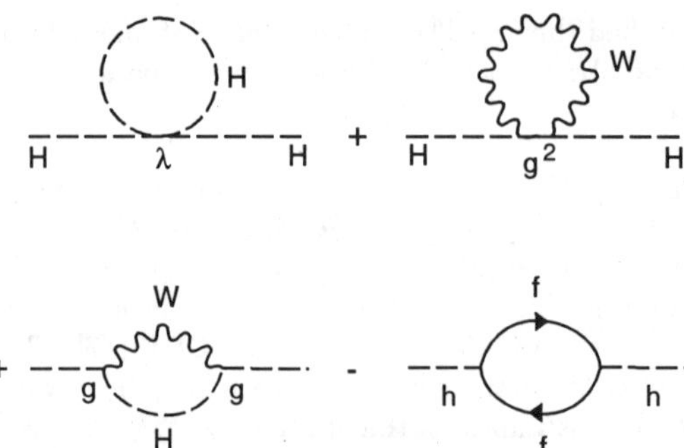

Figure 6: Radiative corrections to the Higgs mass, including self-interactions, interactions with gauge bosons, and interactions with fermions.

scale $M_P = G_N^{-1/2} \sim 10^{19}$ GeV. If there is a simple grand unified theory [31], one would expect Λ to be of order the unification scale $M_X \sim 10^{14}$ GeV. Hence, the natural scale for M_H is $O(\Lambda)$, which is much larger than the expected value. There must be a fine-tuned and apparently highly contrived cancellation between the bare value and the correction, to more than 30 decimal places in the case of gravity. If the cutoff is provided by a grand unified theory there is a separate hierarchy problem at the tree-level. The tree-level couplings between the Higgs field and the superheavy fields lead to the expectation that M_H is equal to the unification scale unless unnatural fine-tunings are done.

One solution to this Higgs/Hierarchy problem is the possibility that the W and Z bosons are composite. However, in this case one would apparently be throwing out the successes of the $SU_2 \times U_1$ gauge theory. Another approach is to eliminate elementary Higgs fields in favor of a dynamical mechanism in which they are replaced by bound states of fermions. Technicolor and composite Higgs models are in this category [38]. The third possibility is supersymmetry [39], which prevents large renormalizations by enforcing cancellations between the various diagrams in Figure 6. However, most grand unified versions do not explain why $(M_W/M_X)^2$ is so small in the first place.

4. *Strong CP Problem*

Another fine-tuning problem is the strong CP problem [40]. One can add an additional term $\frac{\theta}{32\pi^2} g_s^2 F\tilde{F}$ to the QCD Lagrangian which breaks P, T and CP symmetry. $\tilde{F}_{\mu\nu} = \epsilon_{\mu\nu\alpha\beta} F^{\alpha\beta}/2$ is the dual field. This term, if present, would induce an electric dipole moment d_N for the neutron. The rather stringent limits on the dipole moment [41] lead to the upper bound $\theta < 10^{-10}$. The question is, therefore, why is θ so small? It is not sufficient to just say that it is zero because CP violation in the weak interactions leads to a radiative correction or renormalization of θ by $O(10^{-3})$.

Therefore, an apparently contrived fine-tuning is needed to cancel this correction against the bare value. Solutions include the possibility that CP violation is not induced directly by phases in the Yukawa couplings, as is usually assumed in the standard model, but is somehow violated spontaneously [40]. θ then would be a calculable parameter induced at loop level, and it is possible to make θ sufficiently small. However, such models lead to difficult phenomenological and cosmological problems[5]. Alternately, θ becomes unobservable if there is a massless u quark [43]. However, most phenomenological estimates are not consistent with $m_u = 0$ [15, 44]. Another possibility is the Peccei-Quinn mechanism [45], in which an extra global U_1 symmetry is imposed on the theory in such a way that θ becomes a dynamical variable which is zero at the minimum of the potential. Such models imply the existence of very light pseudoscalar particles called axions. Laboratory, astrophysical, and cosmological constaints allow only the range $10^8 - 10^{12}$ GeV for the scale at which the U_1 symmetry is broken.

5. *Graviton Problem*

Gravity is not fundamentally unified with the other interactions in the standard model, although it is possible to graft on classical general relativity by hand. However, this is not a quantum theory, and there is no obvious way to generate one within the standard model context. In addition to the fact that gravity is not unified and not quantized there is another difficulty, namely the cosmological constant. The cosmological constant can be thought of as energy of the vacuum. The energy density induced by spontaneous symmetry breaking is some 50 orders of magnitude larger than the observational upper limit (see Eqns. (33) and (34)). This implies the necessity of severe fine-tuning between the generated and bare pieces, which do not have any a priori reason to be related. Possible solutions include Kaluza-Klein [46] and supergravity theories [39]. These unify gravity but do not solve the problem of quantum gravity or yield renormalizable theories of quantum gravity, nor do they provide any obvious solution to the cosmological constant problem. Superstring theories [36] unify gravity and may yield finite theories of quantum gravity and all the other interactions. It is not clear whether or not they solve the cosmological constant problem.

1. For reviews, see E. S. Abers and B. W. Lee, *Phys. Rev.* **9**, 1 (1973); M. A. B. Beg and A. Sirlin, *ARNPS* **24**, 379 (1974), *Phys. Rep.* **88**, 1 (1982); P. Langacker, *Phys. Rep.* **72**, 185 (1981) and in *TeV Physics*, ed. T. Huang *et al.*, (Gordon and Breach, Philadelphia, 1991), p 53; *Testing the Standard Model*, ed. M. Cvetic and P. Langacker (World, Singapore, 1991); P. Langacker and J. Erler, in *Reviews of Particle Properties*, *Phys. Rev.* **D50**, 1304 (1994).
2. H. Weyl, *Z. Phys.* **56**, 330 (1929); C. N. Yang and R. Mills, *Phys. Rev.* **96**, 191 (1954).

[5]Models in which the CP breaking occurs near the Planck scale may be viable [42].

3. R. D. Field, *Perturbative QCD* (Addison-Wesley, Redwood City, 1989); Yu. L. Dokshitzer *et al., Basics of Perturbative QCD* (Ed. Frontieres, Gif-sur-Yvette, 1991); F. J. Yndurain, *The Theory of Quark and Gluon Interactions* (Springer-Verlag, Berlin, 1993).
4. S. L. Glashow, *Nucl. Phys.* **22**, 579 (1961); S. Weinberg, *Phys. Rev. Lett.* **19**, 1264 (1967); A. Salam in *Elementary Particle Theory*, ed. N. Svartholm, (Almqvist and Wiksells, Stockholm, 1969) p 367.
5. B. W. Lee, *Chiral Dynamics* (Gordon and Breach, NY, 1972); S. Coleman, *Aspects of Symmetry* (Cambridge Univ. Press, Cambridge, 1985).
6. G. 't Hooft and M. Veltman, *Nucl. Phys.* **B50**, 318 (1972), and references therein.
7. S. Coleman and E. Weinberg, *Phys. Rev.* **D7**, 1888 (1973).
8. P. W. Anderson, *Phys. Rev.* **130**, 439 (1963); P. W. Higgs, *Phys. Rev. Lett.* **12**, 132 (1964), **13**, 321 (1964), *Phys. Rev.* **145**, 1156 (1966); F. Englert and R. Brout, *Phys. Rev. Lett.* **13**, 321 (1964); G. S. Guralnik, C. R. Hagen, and T. W. B. Kibble, *Phys. Rev. Lett.* **13**, 585 (1965), *Phys. Rev.* **155**, 1554 (1967).
9. J. Goldstone, *Nuo. Cim.* **19**, 15 (1961); Y. Nambu and G. Jona-Lasinio, *Phys. Rev.* **122**, 345 (1961), **124**, 246 (1961); Y. Nambu, *Phys. Rev. Lett.* **4**, 380 (1960); J. Goldstone, A. Salam, and S. Weinberg, *Phys. Rev.* **127**, 965 (1962).
10. UA1: G. Arnison *et al., Phys. Lett.* **B166**, 484 (1986); C. Albajar *et al., Zeit. Phys.* **C44**, 15 (1989); and references theirin.
11. UA2: R. Ansari *et al., Phys. Lett.* **B186**, 440 (1987); J. Alitti *et al., Phys. Lett.* **B276**, 354 (1992); and references theirin.
12. P. Janot, invited talk at *Neutrino 94*, Eilat, Israel, May 1994, Orsay LAL-94-59.
13. J. F. Gunion *et al., The Higgs Hunters Guide* (Addison-Wesley, Redwood City, 1990); M. Sher, *Phys. Rep.* **179**, 273 (1989); M.S. Chanowitz, *ARNPS* **38**, 323 (1988) and in *TeV Physics*, p 1; H. Haber in *Testing the Standard Model*, p 340; *Perspectives on Higgs Physics*, ed. G. L. Kane (World, Singapore, 1993).
14. For reviews, see S. Weinberg, *Rev. Mod. Phys.* **61**, 1 (1989); M. J. Duncan, in *Testing the Standard Model*, p 743.
15. C. A. Dominguez and E. de Rafael, *Ann. Phys.* **174**, 372 (1987); J. Gasser and H. Leutwyler, *Phys. Rep.* **87**, 777 (1982); S. Narison, *Phys. Lett.* **B216**, 191 (1989); J. F. Donoghue, *ARNPS* **39**, 1 (1989).
16. D∅: S. Abachi *et al., Phys. Rev. Lett.* **72**, 2138 (1994).
17. CDF: F. Abe *et al., Phys. Rev. Lett.* **73**, 225 (1994), *Phys. Rev.* **D50**, 2966 (1994).
18. S.L. Glashow and S. Weinberg, *Phys. Rev.* **D15**, 1958 (1977).
19. See, for example, L. Littenberg and G. Valencia, *ARNPS* **43**, 729 (1993).

20. F. J. Gilman, K. Kleinknecht, and B. Renk, in *Reviews of Particle Properties*, *Phys. Rev.* **D50**, 1315 (1994).
21. For reviews, see G. Barbiellini and G. Santoni, *Riv. Nuo. Cim.* **9** (2), 1 (1986); E. D. Commins and P. H. Buchsbaum, *Weak Interactions of Leptons and Quarks* (Cambridge Univ. Press, Cambridge, 1983).
22. *CP Violation*, ed. C. Jarlskog (World, Singapore, 1989).
23. *B Decays*, ed. S. Stone (World, Singapore, 1992).
24. *Quantum Electrodynamics*, ed. T. Kinoshita (World, Singapore, 1990).
25. *Discovery of Weak Neutral Currents: The Weak Interaction Before and After*, ed. D. Cline and A. Mann, AIP Conf. Proc. 300 (AIP, New York, 1994); J. E. Kim *et al., Rev. Mod. Phys.* **53**, 211 (1981); U. Amaldi *et al., Phys. Rev.* **D36**, 1385 (1987).
26. S. L. Glashow, J. Iliopoulos, and L. Maiani, *Phys. Rev.* **D2**, 1285 (1970).
27. Gargamelle: F. J. Hasert *et al., Phys. Lett.* **B46**, 121, 138 (1973).
28. HPW: A. Benvenuti *et al., Phys. Rev. Lett.* **32**, 800, 1454, 1457 (1974).
29. See also *Physics at LEP*, ed. J. Ellis and R. Peccei, Vol. 2, CERN 86-02.
30. A. De Rújula *et al., Nucl. Phys.* **B384**, 3 (1992); C. P. Burgess and D. London, *Phys. Rev.* **D48**, 4337 (1993); C. P. Burgess *et al., Phys. Rev.* **D49**, 6115 (1994).
31. For reviews, see P. Langacker, *Phys. Rep.* **C72**, 185 (1981); *Ninth Workshop on Grand Unification*, ed. R. Barloutaud (World, Singapore, 1988); G. G. Ross, *Grand Unified Theories* (Benjamin, 1985).
32. See, for example, J. Preskill, *ARNPS* **34**, 461 (1984).
33. J. A. Minahan *et al., Phys. Rev.* **D41**, 715 (1990); C. Q. Geng and R. E. Marshak, *Phys. Rev.* **D41**, 717 (1990); C. Q. Geng, *Phys. Rev.* **D41**, 1292 (1990); K.S. Babu and R. Mohapatra, *Phys. Rev.* **D41**, 271 (1990); S. Rudaz, *Phys. Rev.* **D41**, 2619 (1990).
34. P. Langacker, *Comm. Nucl. Part. Sci.* **19**, 1 (1989); D. Schaile and P. M. Zerwas, *Phys. Rev.* **D45**, 3262 (1992).
35. See, for example, K. S. Babu and R. Mohapatra, *Phys. Rev. Lett.* **66**, 556 (1991); B. S. Balakrishna *et al., Phys. Lett.* **205B**, 345 (1988).
36. M. B. Green, J. H. Schwarz, and E. Witten, *Superstring Theory* (Cambridge Univ. Press, Cambridge, 1987).
37. M. S. Chanowitz, J. Ellis and M. K. Gaillard, *Nucl. Phys.* **B128**, 506 (1977); A. J. Buras, J. Ellis, M. K. Gaillard and D. V. Nanopoulos, *ibid.* **135**, 66 (1978).
38. For a review, see T. Appelquist, in *Mexican School of Particles and Fields*, 1990 (QCD161:M45:1990), p 1.
39. See, for example, H. P. Nilles in *Testing the Standard Model*, p 633 and Phys.

Rep. **C110**, 1 (1984); H. E. Haber and G. Kane, *Phys. Rep.* **C117**, 75 (1985).

40. R. Peccei, in *CP Violation*, p 503.
41. *Reviews of Particle Properties*, L. Montanet *et al.*, *Phys. Rev.* **D50**, 1180 (1994).
42. A. Nelson, *Phys. Lett.* **136B**, 387 (1983), **143B**, 165 (1984); S. Barr, *Phys. Rev.* **D30**, 1805 (1984), **D34**, 1567 (1986).
43. D. B. Kaplan and A. V. Manohar, *Phys. Rev. Lett.* **56**, 2004 (1986).
44. H. Leutwyler, *Nucl. Phys.* **B337**, 108 (1990), Bern BUTP-94/8.
45. R. D. Peccei and H. R. Quinn, *Phys. Rev. Lett.* **38**, 1440 (1977), *Phys. Rev.* **D16**, 1791 (1977); S. Weinberg, *Phys. Rev. Lett.* **40**, 223 (1978); F. Wilczek, *Phys. Rev. Lett.* **40**, 271 (1978).
46. *Modern Kaluza-Klein theories*, ed. T. Appelquist, A. Chodos, and P. G. O. Freund (Addison-Wesley, Menlo Park, 1987).

RENORMALIZATION OF THE STANDARD MODEL

WOLFGANG HOLLIK[1]
Fakultät für Physik
Universität Bielefeld
4800 Bielefeld, Germany

Contents

---

[1]On leave from Max-Planck-Institut für Physik, Föhringer Ring 6, 8000 München 40, Germany

1 Introduction

The e^+e^- colliders LEP and SLC have opened a new era of precision experiments. The determination of the Z resonance parameters with high accuracy by the four LEP experiments [1], together with the measurement of the W mass at $p\bar{p}$ colliders [2] and with the data from neutrino scattering experiments [3] allow precision tests of the electroweak theory never reached before [4]. The present theory of the electroweak interaction, known as the "Standard Model", is the Glashow–Salam-Weinberg model [5] of leptons, extended via the GIM mechanism [6] to the hadronic sector thus incorporating the idea of Cabibbo mixing [7], and made anomaly free through the introduction of the concept of color [8]. As such it is the most comprehensive description of the electroweak phenomena, theoretically consistent and in extraordinary agreement with the experimental data [4].

The possibility of performing precision tests of the electroweak Standard Model is essentially based on its formulation in terms of a quantum field theory with spontaneous symmetry breaking, which is renormalizable [9] and thus allows perturbative calculations for measurable quantities order by order in terms of a few input parameters. The input parameters themselves cannnot be predicted but have to be taken from appropriate experiments. Comparison of the theoretical predictions with the results from precision experiments has confirmed the validity of the Standard Model as a fully fledged quantum field theory, in complete analogy to what has been done for QED. Signals giving need for some significant modifications have not yet been observed.

The still inherent uncertainties in the predictions are related to the as yet unknown mass parameters of the top quark and the Higgs boson. The experimental lower bound for the top mass has improved from 91 GeV [10] to 103 GeV (D0) and 108 GeV (CDF) at 95% C.L. [11], and the existence of a standard Higgs boson can be excluded in the mass range below 60 GeV [4]. According to the principles of quantum field theory, the virtual presence of all physical states in the spectrum shows up in higher order calculations. As a consequence, the experimentally unobserved particles top and Higgs affect the theoretical predictions for the various physical quantities in a calculable way. The same ideas apply, in principle, to all kinds of objects connected with structures beyond the "minimal" standard model (like more Higgs fields, supersymmetric partners, new vector bosons,...) which are too heavy to be observed directly in the experimental search. The higher order contributions hence simultaneously represent an important window to "new physics", in particular to those situations where the decoupling theorem [12] does not hold automatically. One may also consider modifications of the tree level Lagrangian with the interesting possibility of small zeroth order deviations which mimic or cancel the effect of radiative corrections. Since deviations from the Standard Model predictions have not been observed, possible theoretically motivated extensions of the minimal model are subject to significant constraints.

The higher order terms in the perturbative expansion of physical quantities, or radiative corrections, arise from the quantum structure of the underlying theory. Higher order effects related to the presence of the Higgs particle, the top quark, as well as to the self interaction of the gauge fields represent the "genuine" radiative corrections of the electroweak theory. Their typical size in observable quantities is expected to be $\delta_W \leq O(10^{-2})$. Their observation requires therefore that the theoretical uncertainties are not bigger than 0.1%. This requires one to go beyond the one-loop approximation and to include at least the leading contributions from the next order. Also the QED corrections, common to any theory containing the electromagnetic U(1) subgroup, have to be treated carefully and have to be clearly disentangled from the genuine weak corrections in the theoretical predictions and in the analysis of the experiments.

The electroweak Standard Model contains, besides fermion masses, quark mixing angles, and the mass of the Higgs scalar, three free parameters in the gauge sector. In order to make predictions for processes mediated by the exchange of gauge bosons, three independent experimental input data are required for fixing the SU(2) and U(1) gauge coupling constants g_2, g_1, and the vacuum expectation value v of the Higgs field. It is, however, more practical to deal with parameters for which each has a direct relation to a specific experiment and is a well measured quantity. The most accurate set of data points consists of the electromagnetic fine structure constant [13] $\alpha = 137.0359895(61)^{-1}$, the Fermi constant [13] $G_\mu = 1.16639(2) \cdot 10^{-5}$ GeV^{-2}, and the mass of the Z boson $M_Z = 91.187 \pm 0.007$ GeV [1]. The masses of the Higgs boson and the top quark enter the higher order calculations as additional free parameters unless a direct experimental determination of their values is available. Since also the strong interaction is present in the higher order contributions to electroweak quantities, the strong coupling constant α_s is a further independent input parameter.

The calculation of radiative corrections is a lenghty and involved task. Before predictions can be made, a careful discussion of regularization and renormalization is required, together with the extensive use of techniques for the evaluation of loop diagrams. It is the purpose of this presentation to give a comprehensive description of the basic entries and concepts for the calculation of electroweak observables beyond the lowest order and to provide the required expressions for practical calculations, with special emphasis on the vector boson masses and e^+e^- processes. To begin with, we briefly recall the basic Lagrangian of the electroweak Standard Model and its parametrizations, as the starting point for perturbative calculations (section 2).

Section 3 is concerned with the concept of renormalization and with a detailed discussion of the renormalization in the on-shell scheme, which treats the particle masses together with α as the basic parameters in the perturbation expansion. The results can be considered as the building blocks to be used for the computation of amplitudes for electroweak processes at the 1-loop level. Technical details for the evaluation of electroweak 1-loop diagrams are collected in section 4.

In section 5 we provide the explicit 1-loop results for the Standard Model self energies of fermions and gauge bosons. This completes the 1-loop renormalization and allows us to discuss the impact of parameter renormalization on the correlation between the various electroweak quantities. The detailed interdependence of the electroweak parameters in terms of the Fermi constant is presented in section 6. It enables us to predict the M_W-M_Z correlation and to perform comparisons with existing data, thereby setting bounds to the range of the unknown top and Higgs masses. The necessity of including higher than 1-loop order terms is discussed, together with the description of how the 1-loop results have to be modified in order to incorporate all the next order terms available from existing calculations. This is accompanied by a discussion of the remaining uncertainties in theoretical predictions.

Section 7 contains a review of other renormalization schemes. In particular we describe the $\overline{MS}$ renormalization scheme in some more detail and give the relation between the $\overline{MS}$ and the on-shell parameters, together with numerical results. The last section 8 is devoted to renormalizable extensions of the minimal model.

Applications to e^+e^- processes are considered separately in the subsequent extra chapter.

2 The tree level Lagrangian

2.1 *The classical Lagrangian*

The phenomenological basis for the formulation of the Standard Model is given by the following empirical facts:

- The SU(2)×U(1) family structure of the fermions:
  The fermions appear as families with left-handed doublets and right-handed singlets:

$$\begin{pmatrix} \nu_e \\ e \end{pmatrix}_L, \quad \begin{pmatrix} \nu_\mu \\ \mu \end{pmatrix}_L, \quad \begin{pmatrix} \nu_\tau \\ \tau \end{pmatrix}_L, \quad e_R, \quad \mu_R, \quad \tau_R$$

$$\begin{pmatrix} u \\ d \end{pmatrix}_L, \quad \begin{pmatrix} c \\ s \end{pmatrix}_L, \quad \begin{pmatrix} t \\ b \end{pmatrix}_L, \quad u_R, \quad d_R, \quad c_R, \cdots$$

  They can be characterized by the quantum numbers of the weak isospin I, I_3, and the weak hypercharge Y.

- The Gell-Mann-Nishijima relation:
  Between the quantum numbers classifying the fermions with respect to the group SU(2)×U(1) and their electric charges Q the relation

$$Q = I_3 + \frac{Y}{2} \tag{1}$$

  is valid.

- The existence of vector bosons:
  There are 4 vector bosons as carriers of the electroweak force

$$\gamma, \quad W^+, \quad W^-, \quad Z$$

  where the photon is massless and the $W^\pm$, Z have masses $M_W \neq 0$, $M_Z \neq 0$.

This empirical structure can be embedded in a gauge invariant field theory of the unified electromagnetic and weak interactions by interpreting SU(2)×U(1) as the group of gauge transformations under which the Lagrangian is invariant. This full symmetry has to be broken by the Higgs mechanism down to the electromagnetic gauge symmetry; otherwise the $W^\pm$, Z bosons would also be massless. The minimal formulation, the Standard Model, requires a single scalar field (Higgs field) which is a doublet under SU(2).

According to the general principles of constructing a gauge invariant field theory with spontaneous symmetry breaking, the gauge, Higgs, and fermion parts of the electroweak Lagrangian

$$\mathcal{L}_{cl} = \mathcal{L}_G + \mathcal{L}_H + \mathcal{L}_F \tag{2}$$

are specified in the following way:

Gauge fields
SU(2)×U(1) is a non-Abelian group which is generated by the isospin operators I_1, I_2, I_3 and the hypercharge Y (the elements of the corresponding Lie algebra). Each of these generalized charges is associated with a vector field: a triplet of vector fields $W_\mu^{1,2,3}$ with $I_{1,2,3}$ and a singlet field B_μ with Y. The isotriplet W_μ^a, $a = 1, 2, 3$, and the isosinglet B_μ lead to the field strength tensors

$$\begin{aligned} W_{\mu\nu}^a &= \partial_\mu W_\nu^a - \partial_\nu W_\mu^a + g_2\, \epsilon_{abc}\, W_\mu^b W_\nu^c, \\ B_{\mu\nu} &= \partial_\mu B_\nu - \partial_\nu B_\mu. \end{aligned} \tag{3}$$

g_2 denotes the non-Abelian SU(2) gauge coupling constant and g_1 the Abelian U(1) coupling. From the field tensors (3) the pure gauge field Lagrangian

$$\mathcal{L}_G = -\frac{1}{4}\, W_{\mu\nu}^a W^{\mu\nu,a} - \frac{1}{4}\, B_{\mu\nu} B^{\mu\nu} \tag{4}$$

is formed according to the rules for the non-Abelian case.

Fermion fields and fermion-gauge interaction
The left-handed fermion fields of each lepton and quark family (colour index is suppressed)

$$\psi_j^L = \begin{pmatrix} \psi_{j+}^L \\ \psi_{j-}^L \end{pmatrix}$$

with family index j are grouped into SU(2) doublets with component index $\sigma = \pm$, and the right-handed fields into singlets

$$\psi_j^R = \psi_{j\sigma}^R .$$

Each left- and right-handed multiplet is an eigenstate of the weak hypercharge Y such that the relation (1) is fulfilled. The covariant derivative

$$D_\mu = \partial_\mu - i\, g_2\, I_a W_\mu^a + i\, g_1 \frac{Y}{2} B_\mu \tag{5}$$

induces the fermion-gauge field interaction via the minimal substitution rule:

$$\mathcal{L}_F = \sum_j \bar{\psi}_j^L i\gamma^\mu D_\mu \psi_j^L + \sum_{j,\sigma} \bar{\psi}_{j\sigma}^R i\gamma^\mu D_\mu \psi_{j\sigma}^R \tag{6}$$

Higgs field, Higgs - gauge field and Yukawa interaction
For spontaneous breaking of the SU(2)×U(1) symmetry leaving the electromagnetic gauge subgroup $U(1)_{em}$ unbroken, a single complex scalar doublet field with hypercharge $Y = 1$

$$\Phi(x) = \begin{pmatrix} \phi^+(x) \\ \phi^0(x) \end{pmatrix} \tag{7}$$

is coupled to the gauge fields

$$\mathcal{L}_H = (D_\mu \Phi)^+ (D^\mu \Phi) - V(\Phi) \tag{8}$$

with the covariant derivative

$$D_\mu = \partial_\mu - i\, g_2\, I_a W_\mu^a + i \frac{g_1}{2} B_\mu .$$

The Higgs field self-interaction

$$V(\Phi) = -\mu^2\, \Phi^+\Phi + \frac{\lambda}{4} (\Phi^+\Phi)^2 \tag{9}$$

is constructed in such a way that it has a non-vanishing vacuum expectation value v, related to the coefficients of the potential V by

$$v = \frac{2\mu}{\sqrt{\lambda}} . \tag{10}$$

The field (7) can be written in the following way:

$$\Phi(x) = \begin{pmatrix} \phi^+(x) \\ (v + H(x) + i\chi(x))/\sqrt{2} \end{pmatrix} \tag{11}$$

where the components ϕ^+, H, χ now have vacuum expectation values zero. Exploiting the invariance of the Lagrangian one notices that the components ϕ^+, χ can be gauged away which means that they are unphysical (Higgs ghosts or would-be Goldstone bosons). In this particular gauge, the unitary gauge, the Higgs field has the simple form

$$\Phi(x) = \frac{1}{\sqrt{2}} \begin{pmatrix} 0 \\ v + H(x) \end{pmatrix}.$$

The real part of ϕ^0, $H(x)$, describes physical neutral scalar particles with mass

$$M_H = \mu\sqrt{2}. \tag{12}$$

The Higgs field components have triple and quartic self couplings following from V, and couplings to the gauge fields via the kinetic term of Eq. (8).

In addition, Yukawa couplings to fermions are introduced in order to make the charged fermions massive. The Yukawa term is conveniently expressed in the doublet field components (7). We write it down for one family of leptons and quarks:

$$\begin{aligned} \mathcal{L}_{Yukawa} \;=\; & -g_l\,(\overline{\nu}_L\,\phi^+\,l_R + \bar{l}_R\,\phi^-\,\nu_L + \bar{l}_L\,\phi^0\,l_R + \bar{l}_R\,\phi^{0*}\,l_L) \\ & - g_d\,(\overline{u}_L\,\phi^+\,d_R + \overline{d}_R\,\phi^-\,u_L + \overline{d}_L\,\phi^0\,d_R + \overline{d}_R\,\phi^{0*}\,d_L) \\ & - g_u\,(-\overline{u}_R\,\phi^+\,d_L - \overline{d}_L\,\phi^-\,u_R + \overline{u}_R\,\phi^0\,u_L + \overline{u}_L\,\phi^{0*}\,u_R)\,. \end{aligned} \tag{13}$$

ϕ^- denotes the adjoint of ϕ^+.
By $v \neq 0$ fermion mass terms are induced. The Yukawa coupling constants $g_{l,d,u}$ are related to the masses of the charged fermions by Eq. (23). In the unitary gauge the Yukawa Lagrangian is particularly simple:

$$\mathcal{L}_{Yukawa} = -\sum_f m_f\,\bar{\psi}_f\psi_f - \sum_f \frac{m_f}{v}\,\bar{\psi}_f\psi_f\,H\,. \tag{14}$$

As a remnant of this mechanism for generating fermion masses in a gauge invariant way, Yukawa interactions between the massive fermions and the physical Higgs field occur with coupling constants proportional to the fermion masses.

Physical fields and parameters

The gauge invariant Higgs-gauge field interaction in the kinetic part of Eq. (8) gives rise to mass terms for the vector bosons in the non-diagonal form

$$\frac{1}{2}\left(\frac{g_2}{2}v\right)^2 (W_1^2 + W_2^2) + \frac{v^2}{4}\left(W_\mu^3, B_\mu\right) \begin{pmatrix} g_2^2 & g_1 g_2 \\ g_1 g_2 & g_1^2 \end{pmatrix} \begin{pmatrix} W_\mu^3 \\ B_\mu \end{pmatrix}. \tag{15}$$

The physical content becomes transparent by performing a transformation from the fields W_μ^a, B_μ (in terms of which the symmetry is manifest) to the "physical" fields

$$W_\mu^\pm = \frac{1}{\sqrt{2}}(W_\mu^1 \pm iW_\mu^2) \tag{16}$$

and

$$\begin{aligned} Z_\mu &= \quad \cos\theta_W\, W_\mu^3 + \sin\theta_W\, B_\mu \\ A_\mu &= -\sin\theta_W\, W_\mu^3 + \cos\theta_W\, B_\mu \end{aligned} \tag{17}$$

In these fields the mass term (15) is diagonal and has the form

$$M_W^2\, W_\mu^+ W^{-\mu} + \frac{1}{2}(A_\mu, Z_\mu)\begin{pmatrix} 0 & 0 \\ 0 & M_Z^2 \end{pmatrix}\begin{pmatrix} A^\mu \\ Z^\mu \end{pmatrix} \tag{18}$$

with

$$\begin{aligned} M_W &= \frac{1}{2}g_2 v \\ M_Z &= \frac{1}{2}\sqrt{g_1^2 + g_2^2}\, v \end{aligned} \tag{19}$$

The mixing angle in the rotation (17) is given by

$$\cos\theta_W = \frac{g_2}{\sqrt{g_1^2 + g_2^2}} = \frac{M_W}{M_Z}. \tag{20}$$

Identifying A_μ with the photon field which couples via the electric charge $e = \sqrt{4\pi\alpha}$ to the electron, e can be expressed in terms of the gauge couplings in the following way

$$e = \frac{g_1 g_2}{\sqrt{g_1^2 + g_2^2}} \tag{21}$$

or

$$g_2 = \frac{e}{\sin\theta_W}, \quad g_1 = \frac{e}{\cos\theta_W}. \tag{22}$$

Finally, from the Yukawa coupling terms in Eq. (13) the fermion masses are obtained:

$$m_f = g_f \frac{v}{\sqrt{2}} = \sqrt{2}\frac{g_f}{g_2} M_W\,. \tag{23}$$

The relations above allow one to replace the original set of parameters

$$g_2,\ g_1,\ \lambda,\ \mu^2,\ g_f \tag{24}$$

by the equivalent set of more physical parameters

$$e,\ M_W,\ M_Z,\ M_H,\ m_f \tag{25}$$

where each of them can (in principle) be measured directly in a suitable experiment.

An additional very precisely measured parameter is the Fermi constant G_μ which is the effective 4-fermion coupling constant in the the Fermi model, measured by the muon lifetime:

$$G_\mu = 1.16639(2) \cdot 10^{-5}\,\mathrm{GeV}^{-2}$$

Consistency of the Standard Model at $q^2 \ll M_W^2$ with the Fermi model requires the identification (see section 6)

$$\frac{G_\mu}{\sqrt{2}} = \frac{e^2}{8\sin^2\theta_W M_W^2}, \tag{26}$$

which allows us to relate the vector boson masses to the parameters α, G_μ, and $\sin^2\theta_W$ as follows:

$$\begin{aligned} M_W^2 &= \frac{\pi\alpha}{\sqrt{2}G_\mu} \cdot \frac{1}{\sin^2\theta_W} \\ M_Z^2 &= \frac{\pi\alpha}{\sqrt{2}G_\mu} \cdot \frac{1}{\sin^2\theta_W \cos^2\theta_W} \end{aligned} \tag{27}$$

and thus to establish also the $M_W - M_Z$ interdependence:

$$M_W^2\left(1 - \frac{M_W^2}{M_Z^2}\right) = \frac{\pi\alpha}{\sqrt{2}G_\mu}. \tag{28}$$

2.2 *Gauge fixing and ghost fields*

Since the S matrix element for any physical process is a gauge invariant quantity it is possible to work in the unitary gauge with no unphysical particles in internal lines. For a systematic treatment of the quantization of $\mathcal{L}_{cl}$ and for higher order calculations, however, it is better to refer to a renormalizable gauge. This can be done by adding to $\mathcal{L}_{cl}$ a gauge fixing Lagrangian, for example

$$\mathcal{L}_{fix} = -\frac{1}{2}\left(F_\gamma^2 + F_Z^2 + 2F_+F_-\right) \tag{29}$$

with linear gauge fixings of the 't Hooft type:

$$\begin{aligned} F_\pm &= \frac{1}{\sqrt{\xi^W}}\left(\partial^\mu W_\mu^\pm \mp iM_W\xi^W\phi^\pm\right) \\ F_Z &= \frac{1}{\sqrt{\xi^Z}}\left(\partial^\mu Z_\mu - M_Z\xi^Z\chi\right) \\ F_\gamma &= \frac{1}{\sqrt{\xi^\gamma}}\partial^\mu A_\mu \end{aligned} \tag{30}$$

with arbitrary parameters $\xi^{W,Z,\gamma}$. In this class of 't Hooft gauges, the vector boson propagators have the form

$$\frac{i}{k^2 - M_V^2}\left(-g^{\mu\nu} + \frac{(1-\xi^V)k^\mu k^\nu}{k^2 - \xi^V M_V^2}\right)$$

$$= \frac{i}{k^2 - M_V^2}\left(-g^{\mu\nu} + \frac{k^\mu k^\nu}{k^2}\right) + \frac{i\xi^V}{k^2 - \xi^V M_V^2}\,\frac{k^\mu k^\nu}{k^2}, \tag{31}$$

the propagators for the unphysical Higgs fields are given by

$$\frac{i}{k^2 - \xi^W M_W^2} \quad \text{for} \quad \phi^\pm \tag{32}$$

$$\frac{i}{k^2 - \xi^Z M_Z^2} \quad \text{for} \quad \chi^0, \tag{33}$$

and Higgs-vector boson transitions do not occur.

For completion of the renormalizable Lagrangian the Faddeev-Popov ghost term $\mathcal{L}_{gh}$ has to be added [14] in order to balance the undesired effects in the unphysical components introduced by $\mathcal{L}_{fix}$:

$$\mathcal{L} = \mathcal{L}_{cl} + \mathcal{L}_{fix} + \mathcal{L}_{gh} \tag{34}$$

where

$$\mathcal{L}_{gh} = \bar{u}^\alpha(x)\,\frac{\delta F^\alpha}{\delta\theta^\beta(x)}\,u^\beta(x) \tag{35}$$

with ghost fields u^γ, u^Z, $u^\pm$, and $\frac{\delta F^\alpha}{\delta\theta^\beta}$ being the change of the gauge fixing operators (35) under infinitesimal gauge transformations characterized by $\theta^\alpha(x) = \{\theta^a(x), \theta^Y(x)\}$.

In the 't Hooft-Feynman gauge ($\xi = 1$) the vector boson propagators (31) become particularly simple: the transverse and longitudinal components, as well as the propagators for the unphysical Higgs fields $\phi^\pm$, χ and the ghost fields $u^\pm$, u^Z have poles which coincide with the masses of the corresponding physical particles $W^\pm$ and Z.

2.3 *Feynman rules*

Expressed in terms of the physical parameters we can write down the Lagrangian

$$\mathcal{L}(A_\mu, W_\mu^\pm, Z_\mu, H, \phi^\pm, \chi, u^\pm, u^Z, u^\gamma;\, M_W, M_Z, e, \ldots)$$

in a way which allows us to read off the propagators and the vertices most directly. We specify them in the $R_{\xi=1}$ gauge where the vector boson propagators have the simple algebraic form $\sim g_{\mu\nu}$.

$$\mathcal{L}_G + \mathcal{L}_H =$$

$$\frac{1}{2} A_\mu \Box A^\mu \qquad \frac{-i\, g_{\mu\nu}}{q^2}$$

$$+ W^-_\mu (\Box + M_W^2)\, W^{+\mu} \qquad \frac{-i\, g_{\mu\nu}}{q^2 - M_W^2}$$

$$+ \frac{1}{2} Z_\mu (\Box + M_Z^2)\, Z^\mu \qquad \frac{-i\, g_{\mu\nu}}{q^2 - M_Z^2}$$

$$+ \frac{1}{2} H (\Box + M_H^2)\, H \qquad \frac{i}{q^2 - M_H^2}$$

+ interaction terms
VV, VH, HH

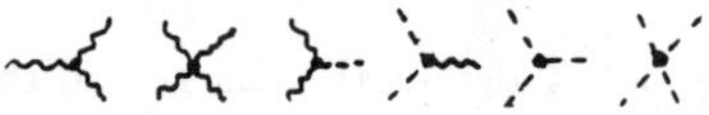

+ (unphysical degrees of freedom)

$$\mathcal{L}_F + \mathcal{L}_{Yukawa} =$$

$$\sum_f \bar{f}\,(i\partial\!\!\!/ - m_f)\, f \qquad \frac{i}{q\!\!\!/ - m_f}$$

$$+ J^\mu_{em} A_\mu \qquad \gamma \qquad -i\, e\, Q_f \gamma_\mu$$

$$+ J^\mu_{NC} Z_\mu \qquad \mathrm{Z} \qquad i\, \frac{e}{2 \sin\theta_W \cos\theta_W}\, \gamma_\mu (v_f - a_f \gamma_5)$$

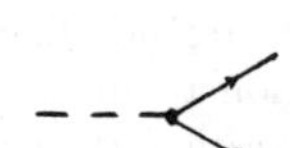

$$+ J^\mu_{CC} W_\mu \qquad \mathrm{W} \qquad i\, \frac{e}{2\sqrt{2} \sin\theta_W}\, \gamma_\mu (1 - \gamma_5)\, V_{jk}$$

$$- \frac{g_f}{\sqrt{2}} \bar{f} f H \qquad -i\, \frac{g_f}{\sqrt{2}} = i\, \frac{e}{2 \sin\theta_W} \frac{m_f}{M_W}$$

+ (unphysical degrees of freedom) (36)

These Feynman rules provide the ingredients to calculate the lowest order amplitudes for fermionic processes. For the complete list of all interaction vertices we refer to the literature [15].

In order to describe scattering processes between light fermions in lowest order we can, in most cases, neglect the exchange of Higgs bosons because of their small Yukawa couplings to the known fermions. The standard processes accessible by the experimental facilities are basically 4-fermion processes. These are mediated by the gauge bosons and, sufficient in lowest order, defined by the vertices for the fermions interacting with the vector bosons. They are given in the Lagrangian above for the electromagnetic, neutral and charged current interactions. The neutral current coupling constants in (36) read

$$\begin{aligned} v_f &= I_3^f - 2Q_f \sin^2\theta_W \\ a_f &= I_3^f \,. \end{aligned} \tag{37}$$

Q_f and I_3^f denote the charge and the third isospin component of f_L.

The quantities V_{jk} in the charged current vertex are the elements of the unitary 3×3 matrix

$$U_{KM} = \begin{pmatrix} V_{ud} & V_{us} & V_{ub} \\ V_{cd} & V_{cs} & V_{cb} \\ V_{td} & V_{ts} & V_{tb} \end{pmatrix} \tag{38}$$

which describes family mixing in the quark sector [7]. Its origin is the diagonalization of the quark mass matrices from the Yukawa coupling which appears since quarks of the same charge have different masses. For massless neutrinos no mixing in the leptonic sector is present. Due to the unitarity of U_{KM} the mixing is absent in the neutral current.

For a proper treatment of the charged current vertex at the one-loop level, the matrix U_{KM} has to be renormalized as well. As was shown in [16], where the renormalization procedure was extended to U_{KM}, the resulting effects are completely negligible for the known light fermions. We therefore skip the renormalization of U_{KM} in our discussion of radiative corrections.

3 Renormalization

3.1 *General remarks*

The tree level Lagrangian (2) of the minimal SU(2)×U(1) model involves a certain number of free parameters which are not fixed by the theory. The definition of these parameters and their relation to measurable quantities is the content of a renormalization scheme. The parameters (or appropriate combinations) can be determined from specific experiments with help of the theoretical results for cross sections and lifetimes. After this procedure of defining the physical input, other observables can be predicted allowing verification or falsification of the theory by comparison with the corresponding experimental results.

In higher order perturbation theory the relations between the formal parameters and measurable quantities are different from the tree level relations in general. Moreover, the procedure is obscured by the appearence of divergences from the loop integrations. For a mathematically consistent treatment one has to regularize the theory, e.g. by dimensional regularization (performing the calculations in D dimensions). But then the relations between physical quantities and the parameters become cutoff dependent. Hence, the parameters of the basic Lagrangian, the "bare" parameters, have no physical meaning. On the other hand, relations between measurable physical quantities, where the parameters drop out, are finite and independent of the cutoff. It is therefore in principle possible to perform tests of the theory in terms of such relations by eliminating the bare parameters [17, 73].

Alternatively, one may replace the bare parameters by renormalized ones by multiplicative renormalization for each bare parameter g_0

$$g_0 = Z_g\, g = g + \delta g \tag{39}$$

with renormalization constants Z_g different from 1 by a 1-loop term. The renormalized parameters g are finite and fixed by a set of renormalization conditions. The decomposition (39) is to a large extent arbitrary. Only the divergent parts are determined directly by the structure of the divergences of the one-loop amplitudes. The finite parts depend on the choice of the explicit renormalization conditions.

This procedure of parameter renormalization is sufficient to obtain finite S-matrix elements when wave function renormalization for external on-shell particles is included. Off-shell Green functions, however, are not finite by themselves. In order to obtain finite propagators and vertices, the bare fields in $\mathcal{L}$ also have to be redefined in terms of renormalized fields by multiplicative renormalization

$$\phi_0 = Z_\phi^{1/2}\, \phi\,. \tag{40}$$

Expanding the renormalization constants according to

$$Z_i = 1 + \delta Z_i$$

the Lagrangian is split into a "renormalized" Lagrangian and a counter term Lagrangian

$$\mathcal{L}(\phi_0, g_0) = \mathcal{L}(Z_\phi^{1/2}\phi, Z_g g) = \mathcal{L}(\phi, g) + \delta\mathcal{L}(\phi, g, \delta Z_\phi, \delta g) \tag{41}$$

which renders the results for all Green functions in a given order finite.

The simplest way to obtain a set of finite Green functions is the "minimal subtraction scheme" [18] where (in dimensional regularization) the singular part of each divergent diagram is subtracted and the parameters are defined at an arbitrary mass scale μ. This scheme, with slight modifications, has been applied in QCD where due to the confinement of quarks and gluons there is no distinguished mass scale in the renormalization procedure.

The situation is different in QED and in the electroweak theory. There the classical Thomson scattering and the particle masses set natural scales where the parameters can be defined. In QED the favoured renormalization scheme is the on-shell scheme where $e = \sqrt{4\pi\alpha}$ and the electron, muon, ... masses are used as input parameters. The finite parts of the counter terms are fixed by the renormalization conditions that the fermion propagators have poles at their physical masses, and e becomes the $ee\gamma$ coupling constant in the Thomson limit of Compton scattering. The extraordinary meaning of the Thomson limit for the definition of the renormalized coupling constant is elucidated by the theorem that the exact Compton cross section at low energies becomes equal to the classical Thomson cross section. In particular this means that e resp. α is free of infrared corrections, and that its numerical value is independent of the order of perturbation theory, only determined by the accuracy of the experiment.

This feature of e is preserved in the electroweak theory. In the electroweak Standard Model a distinguished set for parameter renormalization is given in terms of e, M_Z, M_W, M_H, m_f with the masses of the corresponding particles. This electroweak on-shell scheme is the straight-forward extension of the familiar QED renormalization, first proposed by Ross and Taylor [19] and used in many practical applications [15, 20, 21, 22, 23, 24, 25, 26, 27, 28, 29]. For stable particles, the masses are well defined quantities and can be measured with high accuracy. The masses of the W and Z bosons are related to the resonance peaks in cross sections where they are produced and hence can also be accurately determined. The masses of the Higgs boson and the top quark, as long as they are experimentally unknown, are treated as free input parameters. The light quark masses can only be considered as effective parameters. In the cases of practical interest they can be replaced in terms of directly measured quantities like the cross section for $e^+e^- \to$ hadrons.

The electroweak mixing angle is related to the vector boson masses in general by

$$\sin^2\theta_W = 1 - \frac{M_W^2}{\rho_0 M_Z^2} \tag{42}$$

where $\rho_0 \neq 1$ at the tree level in case of a Higgs system more complicated than with doublets only. We want to restrict our discussion of radiative corrections primarily to the minimal model with $\rho_0 = 1$. For $\rho_0 \neq 1$ see section 8.2.

Instead of the set e, M_W, M_Z as basic free parameters one may alternatively use as basic parameters α, G_μ, M_Z [30] or α, G_μ, $\sin^2\theta_W$ with the mixing angle deduced from neutrino-electron scattering [31] or perform the loop calculations in the $\overline{MS}$ scheme [32, 33, 34, 35]. The so-called $*$-scheme [36, 37] is a different way of book-keeping in terms of effective running couplings. We will discuss the various schemes after the detailed discussion of the on-shell renormalization.

A full treatment of one-loop renormalization has to comprise also the unphysical sector. Since we are interested only in the calculation of physical amplitudes for light fermions at the one-loop level we drop the discussion of the unphysical sector.

3.2 *The renormalization transformation and counter terms*

Following the general principles discussed above we attach multiplicative renormalization constants to each free parameter and each symmetry multiplet of fields in the symmetric Lagrangian:

$$
\begin{aligned}
W_\mu^a &\to \left(Z_2^W\right)^{1/2} W_\mu^a \\
B_\mu &\to \left(Z_2^B\right)^{1/2} B_\mu \\
\psi_j^L &\to \left(Z_L^j\right)^{1/2} \psi_j^L \\
\psi_{j\sigma}^R &\to \left(Z_R^{j\sigma}\right)^{1/2} \psi_{j\sigma}^R \\
\Phi &\to \left(Z^\Phi\right)^{1/2} \Phi \\
g_2 &\to Z_1^W \left(Z_2^W\right)^{-3/2} g_2 \\
g_1 &\to Z_1^B \left(Z_2^B\right)^{-3/2} g_1 \\
v &\to \left(Z^\Phi\right)^{1/2} (v - \delta v) \\
g_f &\to \left(Z^\Phi\right)^{-1/2} Z_1^f g_f \\
\mu^2 &\to \left(Z^\Phi\right)^{-1} (\mu^2 - \delta\mu^2) \\
\lambda &\to \left(Z^\Phi\right)^{-2} Z_\lambda \lambda
\end{aligned}
\tag{43}
$$

The r.h.s. represent the bare fields and parameters, the quantities without the Z-factors are the corresponding renormalized fields and parameters.

Field renormalization ensures that we end up with finite Green functions. The field renormalization in (43) is performed in a way that it respects the gauge symmetry by introducing the minimal number of field renormalization constants. Therefore also the counter term Lagrangian and the renormalized Green functions reflect the gauge symmetry. The price for this, however, is that not all residues of the propagators can be normalized to unity. As a consequence, any calculation with the renormalized Lagrangian will have to include finite multiplicative wave function renormalization factors for some of the external lines in S matrix elements.

It is of course possible to perform the renormalization in such a way that these finite corrections do not appear [23, 24, 25, 27]. But then the Lagrangian will contain many constants which have to be calculated in terms of the few fundamental parameters.

The independent renormalization of the Higgs vacuum expectation value v absorbs the linear term in the Higgs potential, which is induced by the appearance of tadpole diagrams in one-loop order, in such a way that the relation

$$v = \frac{2\mu}{\sqrt{\lambda}}$$

remains valid for the renormalized parameters with v being the minimum of the Higgs potential at the one-loop level. As a practical consequence of this tadpole renormalization, all tadpole graphs can be omitted in the renormalized amplitudes and Green functions. They are, however, necessary to make the mass counter terms gauge independent.

The systematic way for obtaining results for physical amplitudes in one-loop order is scheduled as follows: The expansion (41) yields the renormalized Lagrangian $\mathcal{L}$ which can now be re-parametrized in terms of the physical parameters (25) and the physical fields A_μ, Z_μ, $W_\mu^\pm$, H (also the unphysical Higgs field components $\phi^\pm$, χ, and the ghost fields u are present in the R_ξ gauge), and the counter term Lagrangian $\delta\mathcal{L}$. From $\delta\mathcal{L}$ the counter term Feynman rules are derived. After rewriting them in terms of (25) the counter term graphs have to be added to the 1-loop vertex functions calculated from $\mathcal{L}$. The renormalization constants in (43) are fixed afterwards by imposing the appropriate renormalization conditions. The results are finite Green functions in terms of the parameter set (25) from which the S matrix elements for all processes of interest can be obtained.

In order to perform mass and field renormalization we have to dress the propagators of the vector bosons and fermions by the self energies, i.e. the amputated 1-particle irreducible 2-point functions.

Vector boson self energies:
The self energies $\Sigma^{j\ell}$ enter the transverse components of the vector boson propagators $D_{\mu\nu}$ as follows ($V = \gamma,\ Z,\ W$):

$$\begin{aligned} D_{\mu\nu}^V(k) &= -ig_{\mu\nu}\left(\frac{1}{k^2-M_V^2} - \frac{1}{k^2-M_V^2}\,\Sigma^{VV}(k^2)\,\frac{1}{k^2-M_V^2}\right) \\ D_{\mu\nu}^{\gamma Z}(k) &= +ig_{\mu\nu}\,\frac{1}{k^2-M_Z^2}\,\Sigma^{\gamma Z}(k^2)\,\frac{1}{k^2}. \end{aligned} \qquad (44)$$

We can drop the longitudinal components $\sim k_\mu k_\nu$ since they only yield terms which are suppressed by m_f^2/M_V^2 in physical amplitudes.

The Σ's represent the sum of all contributing one-loop diagrams. The corresponding renormalized self energies are obtained by adding the counter terms derived from $\delta\mathcal{L}$. It is convenient to introduce the following linear combinations of the SU(2) and the U(1) field renormalization constants $\delta Z_2^{W,B}$ and the coupling renormalization constants $\delta Z_1^{W,B}$ $(i = 1, 2)$:

$$\begin{pmatrix} \delta Z_i^\gamma \\ \delta Z_i^Z \end{pmatrix} = \begin{pmatrix} s_W^2 & c_W^2 \\ c_W^2 & s_W^2 \end{pmatrix} \begin{pmatrix} \delta Z_i^W \\ \delta Z_i^B \end{pmatrix}, \tag{45}$$

with the abbreviations

$$s_W^2 = \sin^2\theta_W, \quad c_W^2 = \cos^2\theta_W. \tag{46}$$

Using the notation (45) we can write down the renormalized self energies as follows (all renormalized quantities are denoted by the same symbols as the corresponding unrenormalized ones in connection with a superscript $\hat{}$):

$$\begin{aligned}
\hat{\Sigma}^{\gamma\gamma}(k^2) &= \Sigma^{\gamma\gamma}(k^2) + \delta Z_2^\gamma\, k^2 \\
\hat{\Sigma}^{ZZ}(k^2) &= \Sigma^{ZZ}(k^2) - \delta M_Z^2 + \delta Z_2^Z\,(k^2 - M_Z^2) \\
\hat{\Sigma}^{WW}(k^2) &= \Sigma^{WW}(k^2) - \delta M_W^2 + \delta Z_2^W\,(k^2 - M_W^2) \\
\hat{\Sigma}^{\gamma Z}(k^2) &= \Sigma^{\gamma Z}(k^2) - \delta Z_2^{\gamma Z}\, k^2 + (\delta Z_1^{\gamma Z} - \delta Z_2^{\gamma Z})\, M_Z^2 .
\end{aligned} \tag{47}$$

In the last line the combinations $(i = 1, 2)$

$$\delta Z_i^{\gamma Z} = \frac{c_W s_W}{c_W^2 - s_W^2}\,(\delta Z_i^Z - \delta Z_i^\gamma) \tag{48}$$

have been introduced.

The mass counter terms $\delta M_{W,Z}^2$ following from (19) and (43) are related to the fundamental renormalization constants by

$$\frac{\delta M_Z^2}{M_Z^2} - \frac{\delta M_W^2}{M_W^2} = \frac{s_W}{c_W}\,(3\,\delta Z_2^{\gamma Z} - 2\,\delta Z_1^{\gamma Z}) . \tag{49}$$

This relation allows us to express finally the $\delta Z_i^{Z,W}$ in terms of the unrenormalized on-shell vector boson self energies.

Fermion self energies:
The fermion self energy Σ^f is related to the fermion propagator in the following way:

$$S_F^f(k) = \frac{i}{\not{k} - m_f} - \frac{i}{\not{k} - m_f}\,\Sigma^f(k)\,\frac{1}{\not{k} - m_f}. \tag{50}$$

Σ^f can be decomposed according to

$$\begin{aligned}
\Sigma^f(k) &= \not{k}\,\Sigma_V^f(k^2) + \not{k}\,\gamma_5\,\Sigma_A^f(k^2) + m_f\,\Sigma_S^f(k^2) \\
&= \not{k}\,\frac{1-\gamma_5}{2}\,\Sigma_L^f(k^2) + \not{k}\,\frac{1+\gamma_5}{2}\,\Sigma_R^f(k^2) + m_f\,\Sigma_S^f(k^2)
\end{aligned} \tag{51}$$

with scalar functions $\Sigma^f_{V,A,S}$ resp. $\Sigma^f_{L,R,S}$ related via

$$\Sigma_L = \Sigma_V - \Sigma_A, \quad \Sigma_R = \Sigma_V + \Sigma_A .$$

By adding the counter terms as derived from our renormalization transformation (43) we obtain the renormalized fermion self energies:

$$\begin{aligned}\hat{\Sigma}^f(k) &= \not{k}\left(\Sigma^f_V(k^2) + \delta Z^f_V\right) + \not{k}\gamma_5\left(\Sigma^f_A(k^2) - \delta Z^f_A\right) \\ &\quad + m_f\left(\Sigma^f_S(k^2) - \delta Z^f_V - \frac{\delta m_f}{m_f}\right) \end{aligned} \tag{52}$$

with

$$\delta Z^f_V = \frac{\delta Z_L + \delta Z^f_R}{2}, \quad \delta Z^f_A = \frac{\delta Z_L - \delta Z^f_R}{2} . \tag{53}$$

δZ_L is the left-handed renormalization constant for the whole doublet; therefore not all of the $\delta Z^f_{V,A}$ are independent for the members of a family. We have dropped the family index in the formulae.

The mass renormalization

$$\frac{\delta m_f}{m_f} = \delta Z^f_1 - \frac{\delta v}{v} \tag{54}$$

contains, besides δv, the Yukawa coupling renormalization constant δZ^f_1. Consequently, fermion mass renormalization fixes the renormalization of the Yukawa couplings which is of interest e.g. for the discussion of the fermionic Higgs boson decays at the one-loop level [38].

Vertex corrections

For coupling constant renormalization and for dressing the fermion gauge boson vertices we have to add the counter terms following from (43) in order to obtain the renormalized vertices. With the coupling constants v_f, a_f of the fermion f to the Z specified in (36) and (37) we get the renormalized electromagnetic vertex as

$$\begin{aligned}\hat{\Gamma}^{\gamma ff}_\mu &= \Gamma^{\gamma ff}_\mu - i\,e\,Q_f\,\gamma_\mu\,(\delta Z^\gamma_1 - \delta Z^\gamma_2 + \delta Z^f_V - \delta Z^f_A\,\gamma_5) \\ &\quad - i\,\frac{e}{2s_W c_W}\,\gamma_\mu(v_f - a_f\gamma_5)\,(\delta Z^{\gamma Z}_1 - \delta Z^{\gamma Z}_2) \end{aligned} \tag{55}$$

with the unrenormalized vertex

$$\Gamma^{\gamma ff}_\mu = -\,i\,e\,Q_f\,\gamma_\mu + i\,e\,\Lambda^{\gamma f}_\mu . \tag{56}$$

$\Lambda^{\gamma f}_\mu$ denote the one-loop vertex corrections. For light on-shell fermions ($f \neq b, t$)) with momenta p, p' and $k^2 = (p - p')^2 \gg m_f^2$ they essentially consist of vector and axial vector form factors only:

$$\Lambda^{\gamma f}_\mu = i\,e\,\gamma_\mu\,[\Lambda^{\gamma f}_V(k^2) - \gamma_5\,\Lambda^{\gamma f}_A(k^2)] . \tag{57}$$

The renormalized weak neutral current vertex has the form

$$\begin{aligned}\hat{\Gamma}_\mu^{Zff} &= \Gamma_\mu^{Zff} + i\,\frac{e}{2s_W c_W}\,\gamma_\mu(v_f - a_f\gamma_5)\,(\delta Z_1^Z - \delta Z_2^Z) \\ &\quad + i\,e\,Q_f\,\gamma_\mu\,(\delta Z_1^{\gamma Z} - \delta Z_2^{\gamma Z}) \\ &\quad + i\,\frac{e}{2s_W c_W}\,\gamma_\mu\,(v_f\delta Z_V^f + a_f\delta Z_A^f) \\ &\quad - i\,\frac{e}{2s_W c_W}\,\gamma_\mu\gamma_5\,(v_f\delta Z_A^f + a_f\delta Z_V^f)\,. \end{aligned} \tag{58}$$

The renormalized charged current vertex reads:

$$\hat{\Gamma}_\mu^{CC} = \Gamma_\mu^{CC} + i\,\frac{e}{2\sqrt{2}s_W}\,\gamma_\mu(1-\gamma_5)\,(1 + \delta Z_1^W - \delta Z_2^W + \delta Z_L) \tag{59}$$

with δZ_L for the corresponding lepton or quark doublet. Γ_μ always denote the unrenormalized vertices:

$$\begin{aligned}\Gamma_\mu^{Zff} &= i\,\frac{e}{2s_W c_W}\,\gamma_\mu(v_f - a_f\gamma_5) + i\,e\,\Lambda_\mu^{Zf} \\ \Gamma_\mu^{CC} &= i\,\frac{e}{2\sqrt{2}s_W}\gamma_\mu(1-\gamma_5) + i\,e\,\Lambda_\mu^{CC}\,. \end{aligned} \tag{60}$$

The one-loop vertex corrections Λ_μ have a similar decomposition into form factors as given above for the photon case.

3.3 *On-shell renormalization conditions and renormalization constants*

The renormalization conditions can be separated into two classes: the on-shell subtraction of the self energies which makes the particle content of the theory evident, and the generalization of the QED charge renormalization. Since we have introduced more renormalization constants than physical parameters we are free to fix the supernumerary ones by the requirement of residue $= 1$ for a corresponding number of propagators. In order to be as close as possible to the common QED renormalization these residue conditions are imposed on the photon and the charged lepton propagators.

The on-shell subtraction conditions can be written in the following way:

$$\mathrm{Re}\,\hat{\Sigma}^{WW}(M_W^2) = \mathrm{Re}\,\hat{\Sigma}^{ZZ}(M_Z^2) = \mathrm{Re}\,\hat{\Sigma}^f(p = m_f) = 0\,. \tag{61}$$

The "QED-like" conditions read explicitly:

$$\hat{\Gamma}_\mu^{\gamma ee}(k^2 = 0, \not{p} = \not{q} = m_e) = ie\gamma_\mu$$

$$
\begin{aligned}
\hat{\Sigma}^{\gamma Z}(0) &= 0 \\
\frac{\partial \hat{\Sigma}^{\gamma\gamma}}{\partial k^2}(0) &= 0 \\
\lim_{\not k \to m_-} \frac{1}{\not k - m_-} \hat{\Sigma}^f(k)\, u_-(k) &= 0
\end{aligned} \tag{62}
$$

if u_- is the wave function for the $I_3 = -1/2$ particle.

The last condition is formulated for charged leptons and quarks with $I_3 = -1/2$. It means a condition for the left and right handed fermion field renormalization constants Z_L, Z_R^-. In the case of leptons Z_L also determines the neutrino field renormalization. For the $I_3 = +1/2$ quarks an additional Z_R^+ for the right handed fields is at our disposal. This constant can be adjusted in a way that the renormalized left and right handed parts of the up-type propagators have equal residues at $k^2 = m_+^2$ (but $\neq 1$).

The solution of the system (61) and (62) yields all those renormalization constants which we need for the vector boson propagators, the fermion - gauge boson vertex corrections, and the fermion wave function renormalization. We write them down in terms of the unrenormalized expressions.

The mass counter terms for the W and Z self energies follow immediately from the unrenormalized on-shell values by means of Eq. (47) and (61):

$$
\begin{aligned}
\delta M_W^2 &= \mathrm{Re}\, \Sigma^{WW}(M_W^2) \\
\delta M_Z^2 &= \mathrm{Re}\, \Sigma^{ZZ}(M_Z^2)\,.
\end{aligned} \tag{63}
$$

Their dependence on the $Z_i^{\gamma,Z}$ $(i = 1, 2)$ in (49) together with the set of equations (62) yields:

$$
\begin{aligned}
\delta Z_2^\gamma &= -\Pi^\gamma(0) \equiv -\frac{\partial \Sigma^\gamma}{\partial k^2}(0) \\
\delta Z_1^\gamma &= -\Pi^\gamma(0) - \frac{s_W}{c_W}\frac{\Sigma^{\gamma Z}(0)}{M_Z^2} \\
\delta Z_2^Z &= -\Pi^\gamma(0) - 2\,\frac{c_W^2 - s_W^2}{s_W c_W}\frac{\Sigma^{\gamma Z}(0)}{M_Z^2} + \frac{c_W^2 - s_W^2}{s_W^2}\left(\frac{\delta M_Z^2}{M_Z^2} - \frac{\delta M_W^2}{M_W^2}\right) \\
\delta Z_1^Z &= -\Pi^\gamma(0) - \frac{3c_W^2 - 2s_W^2}{s_W c_W}\frac{\Sigma^{\gamma Z}(0)}{M_Z^2} + \frac{c_W^2 - s_W^2}{s_W^2}\left(\frac{\delta M_Z^2}{M_Z^2} - \frac{\delta M_W^2}{M_W^2}\right) \\
\delta Z_2^W &= -\Pi^\gamma(0) - 2\,\frac{c_W}{s_W}\frac{\Sigma^{\gamma Z}(0)}{M_Z^2} + \frac{c_W^2}{s_W^2}\left(\frac{\delta M_Z^2}{M_Z^2} - \frac{\delta M_W^2}{M_W^2}\right) \\
\delta Z_1^W &= -\Pi^\gamma(0) - \frac{3 - 2s_W^2}{s_W c_W}\frac{\Sigma^{\gamma Z}(0)}{M_Z^2} + \frac{c_W^2}{s_W^2}\left(\frac{\delta M_Z^2}{M_Z^2} - \frac{\delta M_W^2}{M_W^2}\right).
\end{aligned} \tag{64}
$$

The last two constants δZ_i^W are not independent but are linear combinations of δZ_i^γ and δZ_i^Z. They are given here for completeness.

The mass renormalization term for a massive fermion f is determined by

$$\frac{\delta m_f}{m_f} = \Sigma_V^f(m_f^2) + \Sigma_S^f(m_f^2). \tag{65}$$

The doublet field renormalization constants δZ_L and the singlet renormalization constants δZ_R^- in the $I_3 = -1/2$ states follow from (62) to be:

$$\begin{aligned} \delta Z_L &= -\Sigma_L(m_-^2) - m_-^2 \left[\Sigma_L'(m_-^2) + \Sigma_R'(m_-^2) + 2\Sigma_S'(m_-^2)\right] \\ \delta Z_R^- &= -\Sigma_R(m_-^2) - m_-^2 \left[\Sigma_L'(m_-^2) + \Sigma_R'(m_-^2) + 2\Sigma_S'(m_-^2)\right] \end{aligned} \tag{66}$$

The $\Sigma_{L,..}$ are the invariant functions in Eq. (51), and $\Sigma_{L,..}'$ denotes the derivative

$$\Sigma_{L,..}'(k^2) = \frac{\partial \Sigma_{L,..}}{\partial k^2}.$$

By means of (53) one can rewrite (66) as follows:

$$\begin{aligned} \delta Z_V^- &= -\Sigma_V(m_-^2) - 2m_-^2 \left[\Sigma_V'(m_-^2) + \Sigma_S'(m_-^2)\right] \\ \delta Z_A^- &= +\Sigma_A(m_-^2). \end{aligned} \tag{67}$$

In the case of leptons δZ_L renormalizes simultaneously the neutrino propagator with the consequence that its residue is different from 1 by the finite amount

$$\hat{\Pi}^\nu = -\Sigma_L^\nu(0) - \delta Z_L. \tag{68}$$

Therefore, in external ν lines a finite wave function renormalization factor

$$1 - \frac{1}{2}\left[\Sigma_L^\nu(0) + \delta Z_L\right] \tag{69}$$

has to be inserted.

For the right handed u-type quarks an additional condition has to be imposed in order to fix δZ_R^+. We will treat the $I_3 = +1/2$ quarks in a way that the residues for their left and right handed propagators become equal. This looks somewhat arbitrary. In S matrix elements, however, in the sum

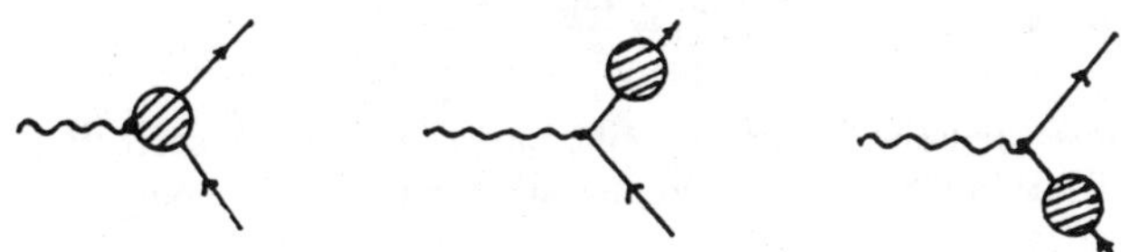

this unsymmetric treatment is compensated by the corresponding renormalized 3-point vertices.

We have skipped the Higgs mass on-shell renormalization since it is not needed for our purpose. The on-shell subtraction for the Higgs self energy together with the tadpole conditions fix also the scalar sector at the one-loop level.

Charge renormalization:

It is instructive to have a look at the direct counter term δe for the renormalization of the electric charge, which can easily be derived from the basic renormalization constants given above in Eq. (64). Making use of the definition (21) and the renormalization transformation (RT) in Eq. (43) we obtain

$$e^2 = \frac{g_1^2 g_2^2}{g_1^2 + g_2^2} \quad \xrightarrow{RT} \quad \frac{g_1^2 g_2^2}{g_1^2 + g_2^2}\,(1 + 2\delta Z_1^\gamma - 3\delta Z_2^\gamma) \equiv e^2\left(1 + 2\frac{\delta e}{e}\right) \tag{70}$$

where

$$\frac{\delta e}{e} = \delta Z_1^\gamma - \frac{3}{2}\delta Z_2^\gamma = \frac{1}{2}\Pi^\gamma(0) - \frac{s_W}{c_W}\frac{\Sigma^{\gamma Z}(0)}{M_Z^2}\,. \tag{71}$$

This demonstrates that δe, although fixed in terms of the electron specific condition (62), is fermion independent. The first of the two universal terms in Eq. (71) is the photon vacuum polarization, as in QED (but including also bosonic loops), the second term contains the mixing between the photon and the Z boson. As we will see in the next section, the fermion loops vanish for $k^2 = 0$. Only the non-abelian bosonic loops yield $\Sigma^{\gamma Z}(0) \neq 0$.

The reason for the universality of δe is the $U(1)_Y$ Ward identity, which is formally identical to the QED Ward identity [39], yielding the following relation between the field and coupling renormalization constants for the U(1) part:

$$\delta Z_1^B = \delta Z_2^B\,.$$

Exploiting this relation together with the second condition of (62) allows us to write down the following identity:

$$\delta Z_1^\gamma - \delta Z_2^\gamma = -\frac{s_W}{c_W}\frac{\Sigma^{\gamma Z}(0)}{M_Z^2} \tag{72}$$

which immediately leads to (71).

The charge renormalization condition in (62) is an explicit condition for the vector part of the electron vertex:

$$\Lambda_V^{\gamma e}(0) + \delta Z_V^e + \delta Z_1^\gamma - \delta Z_2^\gamma + \frac{4s_W^2 - 1}{4s_W c_W}\frac{\Sigma^{\gamma Z}(0)}{M_Z^2} = 0\,. \tag{73}$$

The identity (72) shows that the fermion specific part of the vertex corrections and of the field renormalization cancel in the combination entering the condition (73):

$$\Lambda_V^{\gamma e}(0) + \delta Z_V^e = \frac{1}{4s_W c_W}\frac{\Sigma^{\gamma Z}(0)}{M_Z^2}\,. \tag{74}$$

Another consequence of the Ward identity is the absence of an electromagnetic axial vector coupling for real photons. The renormalized electromagnetic axial vector vertex in Eq. (55) is finite and vanishes for $k^2 \to 0$. Simultaneously, this also holds for the electromagnetic form factor of the neutrino.

Renormalization of $\sin^2\theta_W$:

The counter term for the electroweak mixing angle can be derived in a similar way as done for δe on the basis of the definition (22), the transformation (43), and the relation (49):

$$s_W^2 = \frac{g_1^2}{g_1^2 + g_2^2} \xrightarrow{RT} s_W^2 + c_W^2 \left(\frac{\delta M_Z^2}{M_Z^2} - \frac{\delta M_W^2}{M_W^2} \right) \equiv s_W^2 + \delta s_W^2 \qquad (75)$$

with

$$s_W^2 = 1 - \frac{M_W^2}{M_Z^2} \,.$$

This becomes obvious also in a more direct way from expanding the exact relation between the bare quantities up to the 1-loop order:

$$\begin{aligned} s_0^2 = 1 - \frac{M_W^{0\,2}}{M_Z^{0\,2}} &= 1 - \frac{M_W^2 + \delta M_W^2}{M_Z^2 + \delta M_Z^2} \\ &= 1 - \frac{M_W^2}{M_Z^2} + \frac{M_W^2}{M_Z^2} \left(\frac{\delta M_Z^2}{M_Z^2} - \frac{\delta M_W^2}{M_W^2} \right) . \end{aligned} \qquad (76)$$

Concluding this discussion we summarize the principal structure of electroweak calculations:

- The treel level Lagrangian $\mathcal{L}(e, M_W, M_Z, \ldots)$ is sufficent for lowest order calculations and the parameters can be identified with the physical parameters.
- For higher order calculations, $\mathcal{L}$ has to be considered as the bare Lagrangian of the theory $\mathcal{L}(e_0, M_W^0, M_Z^0, \ldots)$ with bare parameters which are related to the physical ones by

  $$e_0 = e + \delta e, \quad M_W^{0\,2} = M_W^2 + \delta M_W^2, \quad M_Z^{0\,2} = M_Z^2 + \delta M_Z^2 \,.$$

  The counterterms are fixed in terms of a specific set of 1-loop self energies via Eq.s (63), (71), (76). For any 4-fermion process the S-matrix element with the corresponding loop diagrams and the counter terms is finite after external wave function renormalization.
- When field renormalization is performed, also the individual self energies and vertex corrections are finite.

4 Calculation of one-loop integrals

In this section we provide the technical details for the calculation of radiative corrections for electroweak precision observables. The methods used are essentially based on the work of [20] and [40].

4.1 *Dimensional regularization*

The diagrams with closed loops occuring in higher order perturbation theory involve integrals over the loop momentum. These integrals are in general divergent for large integration momenta (UV divergence). For this reason we need a regularization, which is a procedure to redefine the integrals in such a way that they become finite and mathematically well-defined objects. The widely used regularization procedure for gauge theories is that of dimensional regularization [41], which is Lorentz and gauge invariant: replace the dimension 4 by a lower dimension D where the integrals are convergent:

$$\int \frac{d^4k}{(2\pi)^4} \to \mu^{4-D} \int \frac{d^Dk}{(2\pi)^D} \tag{77}$$

An (arbitrary) mass parameter μ has been introduced in order to keep the dimensions of the coupling constants in front of the integrals independent of D. After renormalization the results for physical quantities are finite in the limit $D \to 4$.
The metric tensor in D dimensions has the property

$$g^{\mu}_{\mu} = g_{\mu\nu}g^{\nu\mu} = \mathrm{Tr}(1) = D\,. \tag{78}$$

The Dirac algebra in D dimensions

$$\{\gamma_\mu, \gamma_\nu\} = 2\, g_{\mu\nu}\, \mathbf{1} \tag{79}$$

has the consequences

$$\begin{aligned}
\gamma_\mu\gamma^\mu &= D\,\mathbf{1} \\
\gamma_\rho\gamma_\mu\gamma^\rho &= (2-D)\,\gamma_\mu \\
\gamma_\rho\gamma_\mu\gamma_\nu\gamma^\rho &= 4g_{\mu\nu}\,\mathbf{1} - (4-D)\,\gamma_\mu\gamma_\nu \\
\gamma_\rho\gamma_\mu\gamma_\nu\gamma_\sigma\gamma^\rho &= -2\,\gamma_\sigma\gamma_\nu\gamma_\mu + (4-D)\,\gamma_\mu\gamma_\nu\gamma_\sigma
\end{aligned} \tag{80}$$

A consistent treatment of γ_5 in D dimensions is more subtle [42]. In theories which are anomaly free like the Standard Model we can use γ_5 as anticommuting with γ_μ:

$$\{\gamma_\mu, \gamma_5\} = 0\,. \tag{81}$$

4.2 *One- and two-point integrals*

In the calculation of self energy diagrams the following types of one-loop integrals appear:

1-point integral:

$$\mu^{4-D}\int\frac{d^Dk}{(2\pi)^D}\frac{1}{k^2-m^2} =: \frac{i}{16\pi^2}A(m) \tag{82}$$

2-point integrals:

$$\mu^{4-D}\int\frac{d^Dk}{(2\pi)^D}\frac{1}{[k^2-m_1^2][(k+q)^2-m_2^2]} =: \frac{i}{16\pi^2}B_0(q^2,m_1,m_2) \tag{83}$$

$$\mu^{4-D}\int\frac{d^Dk}{(2\pi)^D}\frac{k_\mu;\,k_\mu k_\nu}{[k^2-m_1^2][(k+q)^2-m_2^2]} =: \frac{i}{16\pi^2}B_{\mu;\,\mu\nu}(q^2,m_1,m_2)\,. \tag{84}$$

The vector and tensor integrals B_μ, $B_{\mu\nu}$ can be expanded into Lorentz covariants and scalar coefficients:

$$\begin{aligned}B_\mu &= q_\mu\,B_1(q^2,m_1,m_2)\\ B_{\mu\nu} &= g_{\mu\nu}B_{22}(q^2,m_1,m_2)\,+\,q_\mu q_\nu B_{21}(q^2,m_1,m_2)\,.\end{aligned} \tag{85}$$

The coefficient functions can be obtained algebraically from the scalar 1- and 2-point integrals A and B_0. Contracting (84) with q^μ, $g^{\mu\nu}$ and $q^\mu q^\nu$ yields:

$$\begin{aligned}\int\frac{kq}{[k^2-m_1^2][(k+q)^2-m_2^2]} &= \frac{i}{16\pi^2}q^2B_1\\ \int\frac{k^2}{[k^2-m_1^2][(k+q)^2-m_2^2]} &= \frac{i}{16\pi^2}\left(DB_{22}+q^2B_{21}\right)\\ \int\frac{(kq)^2}{[k^2-m_1^2][(k+q)^2-m_2^2]} &= \frac{i}{16\pi^2}\left(q^2B_{22}+q^4B_{21}\right)\,.\end{aligned} \tag{86}$$

Solving these equations and making use of the decompositions

$$\int\frac{k^2}{[k^2-m_1^2][(k+q)^2-m_2^2]} = \int\frac{1}{k^2-m_2^2}+m_1^2\int\frac{1}{[k^2-m_1^2][(k+q)^2-m_2^2]}$$

$$\int \frac{kq}{[k^2-m_1^2][(k+q)^2-m_2^2]} = \frac{1}{2}\int \frac{1}{k^2-m_1^2} - \frac{1}{2}\int \frac{1}{k^2-m_2^2} + \frac{m_2^2-m_1^2-q^2}{2}\int \frac{1}{[k^2-m_1^2][(k+q)^2-m_2^2]}$$

$$\int \frac{(kq)^2}{[k^2-m_1^2][(k+q)^2-m_2^2]} = \frac{1}{2}\int \frac{kq}{k^2-m_1^2} - \frac{1}{2}\int \frac{kq}{(k+q)^2-m_2^2} + \frac{m_2^2-m_1^2-q^2}{2}\int \frac{kq}{[k^2-m_1^2][(k+q)^2-m_2^2]}$$

and of the definition (82,83) we obtain:

$$\begin{aligned}
B_1(q^2,m_1,m_2) &= \frac{1}{2q^2}\left[A(m_1)-A(m_2)+(m_1^2-m_2^2-q^2)B_0(q^2,m_1,m_2)\right] \\
B_{22}(q^2,m_1,m_2) &= \frac{1}{6}\Big[A(m_2)+2m_1^2B_0(q^2,m_1,m_2) \\
&\quad +(q^2+m_1^2-m_2^2)B_1(q^2,m_1,m_2)+m_1^2+m_2^2-\frac{q^2}{3}\Big] \\
B_{21}(q^2,m_1,m_2) &= \frac{1}{3q^2}\Big[A(m_2)-m_1^2B_0(q^2,m_1,m_2) \\
&\quad -2(q^2+m_1^2-m_2^2)B_1(q^2,m_1,m_2)-\frac{m_1^2+m_2^2}{2}+\frac{q^2}{6}\Big] \, . \qquad (87)
\end{aligned}$$

Finally we have to calculate the scalar integrals A and B_0. With help of the Feynman parametrization

$$\frac{1}{ab} = \int_0^1 dx \, \frac{1}{[ax+b(1-x)]^2}$$

and after a shift in the k-variable, B_0 can be written in the form

$$\frac{i}{16\pi^2}\, B_0(q^2,m_1,m_2) = \int_0^1 dx \, \frac{\mu^{4-D}}{(2\pi)^D}\int \frac{d^Dk}{[k^2-x^2q^2+x(q^2+m_1^2-m_2^2)-m_1^2]^2} \, . \qquad (88)$$

The advantage of this parametrization is a simpler k-integration where the integrand is only a function of $k^2=(k^0)^2-\vec{k}^2$. In order to transform it into a Euclidean integral we perform the substitution [2]

$$k^0 = i\,k_E^0, \;\; \vec{k} = \vec{k_E}, \;\; d^Dk = i\,d^Dk_E$$

[2]The $i\epsilon$-prescription in the masses ensures that this is compatible with the pole structure of the integrand.

where the new integration momentum k_E has a definite metric:

$$k^2 = -k_E^2, \quad k_E^2 = (k_E^0)^2 + \cdots + (k_E^{D-1})^2 .$$

This leads us to a Euclidean integral over k_E:

$$\frac{i}{16\pi^2} B_0 = i \int_0^1 dx \frac{\mu^{4-D}}{(2\pi)^D} \int \frac{d^D k_E}{(k_E^2 + Q)^2} \tag{89}$$

where

$$Q = x^2 q^2 - x(q^2 + m_1^2 - m_2^2) + m_1^2 - i\varepsilon \tag{90}$$

is a constant with respect to the k_E-integration.

Also the 1-point integral A of (82) can be transformed into a Euclidean integral:

$$\frac{i}{16\pi^2} A(m) = -i \frac{\mu^{4-D}}{(2\pi)^D} \int \frac{d^D k_E}{k_E^2 + m^2} . \tag{91}$$

Both k_E- integrals are of the general type

$$\int \frac{d^D k_E}{(k_E^2 + L)^n}$$

of rotational invariant integrals in a D-dimensional Euclidean space. They can be evaluated in D-dimensional polar coordinates ($k_E^2 = R$)

$$\int \frac{d^D k_E}{(k_E^2 + L)^n} = \frac{1}{2} \int d\Omega_D \int_0^\infty dR\, R^{\frac{D}{2}-1} \frac{1}{(R+L)^n} ,$$

yielding

$$\frac{\mu^{4-D}}{(2\pi)^D} \int \frac{d^D k_E}{(k_E^2 + L)^n} = \frac{\mu^{4-D}}{(4\pi)^{D/2}} \cdot \frac{\Gamma(n - \frac{D}{2})}{\Gamma(n)} \cdot L^{-n+\frac{D}{2}} . \tag{92}$$

The singularities of our initially 4-dimensional integrals are now recovered as poles of the Γ-function for $D = 4$ and values $n \le 2$.

Although the l.h.s. of Eq. (92) as a D-dimensional integral is sensible only for integer values of D, the r.h.s. has an analytic continuation in the variable D: it is well defined for all complex values D with $n - \frac{D}{2} \neq 0, -1, -2, \ldots$, in particular for

$$D = 4 - \epsilon \quad \text{with } \epsilon > 0 .$$

For physical reasons we are interested in the vicinity of $D = 4$. Hence we consider the limiting case $\epsilon \to 0$ and perform an expansion around $D = 4$ in powers of ϵ. For this task we need the following properties of the Γ-function at $x \to 0$:

$$\Gamma(x) = \frac{1}{x} - \gamma + O(x) ,$$

$$\Gamma(-1 + x) = -\frac{1}{x} + \gamma - 1 + O(x) \tag{93}$$

with

$$\gamma = -\Gamma'(1) = 0.577\ldots$$

known as Euler's constant.

$n = 1$:
Combining (91) and (92) we obtain the scalar 1-point integral for $D = 4 - \epsilon$:

$$\begin{aligned} A(m) &= -\frac{\mu^\epsilon}{(4\pi)^{-\epsilon/2}} \cdot \frac{\Gamma(-1+\frac{\epsilon}{2})}{\Gamma(1)} \cdot \left(m^2\right)^{1-\epsilon/2} \\ &= m^2\left(\frac{2}{\epsilon} - \gamma + \log 4\pi - \log\frac{m^2}{\mu^2} + 1\right) + O(\epsilon) \\ &\equiv m^2\left(\Delta - \log\frac{m^2}{\mu^2} + 1\right) + O(\epsilon) \end{aligned} \tag{94}$$

Here we have introduced the abbreviation for the singular part

$$\Delta = \frac{2}{\epsilon} - \gamma + \log 4\pi\,. \tag{95}$$

$n = 2$:
For the scalar 2-point integral B_0 we evaluate the integrand of the x-integration in Eq. (89) with help of Eq. (92) as follows:

$$\begin{aligned} \frac{\mu^\epsilon}{(4\pi)^{2-\epsilon/2}} \cdot \frac{\Gamma(\frac{\epsilon}{2})}{\Gamma(2)} \cdot Q^{-\epsilon/2} &= \frac{1}{16\pi^2}\left(\frac{2}{\epsilon} - \gamma + \log 4\pi - \log\frac{Q}{\mu^2}\right) + O(\epsilon) \\ &= \frac{1}{16\pi^2}\left(\Delta - \log\frac{Q}{\mu^2}\right) + O(\epsilon)\,. \end{aligned} \tag{96}$$

Since the $O(\epsilon)$ terms vanish in the limit $\epsilon \to 0$ we skip them in the following formulae. Insertion into Eq. (89) with Q from Eq. (90) yields:

$$B_0(q^2, m_1, m_2) = \Delta - \int_0^1 dx\, \log\frac{x^2q^2 - x(q^2 + m_1^2 - m_2^2) + m_1^2 - i\varepsilon}{\mu^2} \tag{97}$$

The explicit analytic result can be found in [15].

4.3 *Three-point integrals*

In the calculation of vertex corrections the following scalar, vector, and tensor 3-point integrals occur:

$$\mu^{4-D}\int\frac{d^Dk}{(2\pi)^D}\frac{1;\ k_\mu;\ k_\mu k_\nu}{[k^2 - m_1^2][(k+p_1)^2 - m_2^2][(k+p_1+p_2)^2 - m_3^2]} = \frac{i}{16\pi^2}C_{0;\,\mu;\,\mu\nu}\,. \tag{98}$$

Expanding into Lorentz covariants

$$
\begin{aligned}
C^{\mu} &= p_1^{\mu} C_{11} + p_2^{\mu} C_{12} \\
C^{\mu\nu} &= g^{\mu\nu} C_{20} + p_1^{\mu} p_1^{\nu} C_{21} + p_2^{\mu} p_2^{\nu} C_{22} \\
&\quad + (p_1^{\mu} p_2^{\nu} + p_1^{\nu} p_2^{\mu}) C_{23}
\end{aligned} \tag{99}
$$

and performing all possible tensor contractions yields the coefficient functions in terms of the scalar 3-point integral C_0 and the 2-point integrals. The vector coefficients read:

$$
\begin{aligned}
C_{11} &= [p_2^2 R_1 - (p_1 p_2) R_2]/\kappa \\
C_{12} &= [-(p_1 p_2) R_1 + p_1^2 R_2]/\kappa
\end{aligned} \tag{100}
$$

where

$$\kappa = p_1^2 p_2^2 - (p_1 p_2)^2$$

and

$$
\begin{aligned}
R_1 &= [B_0(p_3^2, m_1, m_3) - B_0(p_2^2, m_2, m_3) - (p_1^2 + m_1^2 - m_2^2) C_0]/2 \\
R_2 &= [B_0(p_1^2, m_1, m_2) - B_0(p_3^2, m_1, m_3) + (p_1^2 - p_3^2 - m_2^2 + m_3^2) C_0]/2
\end{aligned} \tag{101}
$$

with the notation

$$p_3^2 = (p_1 + p_2)^2 \,.$$

The tensor coefficients are given by

$$
\begin{aligned}
C_{20} &= [B_0(p_2^2, m_2, m_3) + r_1 C_{11} + r_2 C_{12} + 2 m_1^2 C_0 + 1]/4 \\
C_{21} &= [p_2^2 R_3 - (p_1 p_2) R_5]/\kappa \\
C_{23} &= [-(p_1 p_2) R_3 + p_1^2 R_5]/\kappa \\
C_{22} &= [-(p_1 p_2) R_4 + p_1^2 R_6]/\kappa
\end{aligned} \tag{102}
$$

with

$$r_1 = p_1^2 + m_1^2 - m_2^2, \quad r_2 = p_3^2 - p_1^2 + m_2^2 - m_3^2$$

and

$$
\begin{aligned}
R_3 &= -C_{20} - [r_1 C_{11} - B_1(p_3^2, m_1, m_3) - B_0(p_2^2, m_2, m_3)]/2 \\
R_5 &= -[r_2 C_{11} - B_1(p_1^2, m_1, m_2) + B_1(p_3^2, m_1, m_3)]/2 \\
R_4 &= -[r_1 C_{12} - B_1(p_3^2, m_1, m_3) + B_1(p_2^2, m_2, m_3)]/2 \\
R_6 &= -C_{20} - [r_2 C_{12} + B_1(p_3^2, m_1, m_3)]/2
\end{aligned} \tag{103}
$$

The genuine new element in the expressions above is the scalar 3-point integral

$$\frac{i}{16\pi^2} C_0(p_1, p_2; m_1, m_2, m_3) = \int \frac{1}{[k^2 - m_1^2][(k+p_1)^2 - m_2^2][(k+p_1+p_2)^2 - m_3^2]} \,.$$

After Feynman parametrization

$$\frac{1}{D_1 D_2 D_3} = \int_0^1 dx \int_0^1 dy \frac{1}{\{(1-x)D_1 + x[yD_2 + (1-y)D_3]\}^3}$$

and Wick rotation the momentum integration can be performed applying Eq. (92) for $n = 3$ and $D = 4$. The result is a 2-parameter integral for C_0:

$$C_0(p_1, p_2; m_1, m_2, m_3) = -\int_0^1 dx \int_0^x dy \frac{1}{ax^2 + by^2 + cxy + dx + ey + f} \qquad (104)$$

with

$$\begin{aligned} &a = p_3^2, \quad b = p_2^2, \quad c = p_1^2 - p_2^2 - p_3^2, \\ &d = m_3^2 - m_1^2 - p_3^2, \\ &e = m_2^2 - m_1^2 + p_3^2 - p_1^2, \\ &f = m_1^2 - i\varepsilon\,. \end{aligned} \qquad (105)$$

For real solutions α of the quadratic equation

$$b\alpha^2 + c\alpha + a = 0$$

the integral can be expressed as a sum over dilogarithms:

$$C_0 = \frac{1}{c + 2\alpha b} \sum_{l=1}^{3} \sum_{j=1}^{2} (-1)^l \left\{ \mathrm{Li}_2\left(\frac{x_l}{x_l - y_{lj}}\right) - \mathrm{Li}_2\left(\frac{x_l - 1}{x_l - y_{lj}}\right) \right\} \qquad (106)$$

together with

$$\begin{aligned} x_1 &= -\frac{d + 2a + c\alpha}{c + 2\alpha b}, \\ x_2 &= -\frac{d}{(1-\alpha)(c + 2\alpha b)}, \\ x_3 &= \frac{d}{\alpha(c + 2\alpha b)}, \\ y_{1j} &= \frac{-c \pm \sqrt{c^2 - 4b(a + d + f)}}{2b}, \\ y_{2j} = y_{3j} &= \frac{-d \pm \sqrt{d^2 - 4f(a + b + c)}}{2a}. \end{aligned} \qquad (107)$$

The condition for α being real is always fulfilled for vertices with two particles on-shell, in particular for 2-particle decay and scattering processes.

For the special situation of vertex corrections for the fermion-gauge boson vertices with light fermions the expressions become considerably simpler. The analytic results for the Zff vertices will be given in the next chapter on e^+e^- annihilation. For low energy processes, like muon decay or neutrino scattering, where the external momenta can be neglected in view of the internal gauge boson masses, the 3-point integrals (98) can immediately be reduced to 2-point integrals.

A similar comment applies to the 4-point functions. The general expressions are very lengthy and involved. We do not want to list them here but refer to the literature [20, 43]. For the situation of $e^+e^- \to f\bar{f}$ with light fermions they again become much simpler and will also be given in the corresponding places of the next chapter. In low energy processes the 4-point integrals shrink essentially to 2-point functions.

5 Standard Model one-loop expressions

As an application of the previous section we give here the explicit results for the fermion and vector boson self energies in the Standard Model and discuss their impact on the electroweak parameters. The self energies are of particular importance since they determine all the renormalization constants and hence provide the complete 1-loop renormalization of the Standard Model. The results are given in the $R_{\xi=1}$ gauge.

5.1 *Fermion self energies*

The diagrams contributing $\Sigma^f(k^2)$ for a given fermion species are displayed in Figure 1. f' denotes the isospin partner of the fermion f.

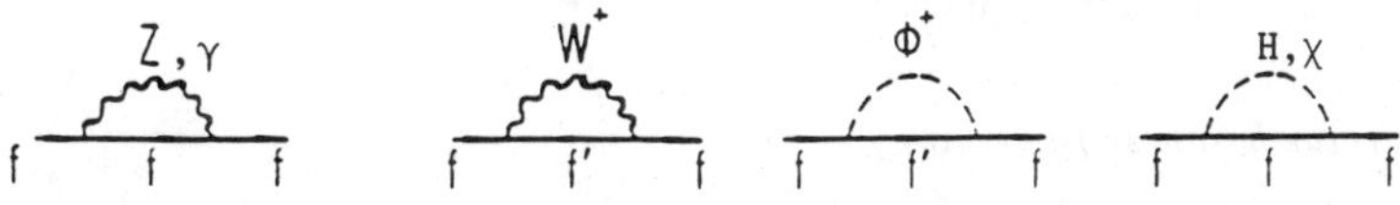

Fig. 1: *Fermion self energies*

We list the scalar coefficients $\Sigma^f_{V,A,S}$ in the decomposition of Σ^f in Eq. (51):

$$\begin{aligned}\Sigma^f_V(k^2) &= -\frac{\alpha}{4\pi}\,[\,Q_f^2(2B_1(k^2,m_f,\lambda)+1) \\ &\quad +\frac{v_f^2+a_f^2}{4s_W^2c_W^2}(2B_1(k^2,m_f,M_Z)+1) \\ &\quad +\frac{1}{4s^2M_W^2}(m_f^2+m_{f'}^2)B_1(k^2,m_{f'},M_W)\end{aligned}$$

$$
\begin{aligned}
& +\frac{1}{4s^2}(2B_1(k^2,m_{f'},M_W)+1) \\
& +\frac{m_f^2}{4s^2M_W^2}\,(B_1(k^2,m_f,M_H)+B_1(k^2,m_f,M_Z))] \\
\Sigma_A^f(k^2) \;=\;& -\frac{\alpha}{4\pi}\,[-\frac{2v_fa_f}{4s_W^2c_W^2}(2B_1(k^2,m_f,M_Z)+1) \\
& +\frac{1}{4s^2M_W^2}(m_{f'}^2-m_f^2)B_1(k^2,m_{f'},M_W) \\
& -\frac{1}{4s^2}(2B_1(k^2,m_{f'},M_W)-1)] \\
\Sigma_S^f(k^2) \;=\;& -\frac{\alpha}{4\pi}\,[\,Q_f^2(4B_0(k^2,m_f,\lambda)-2) \\
& +\frac{v_f^2-a_f^2}{4s_W^2c_W^2}(4B_0(k^2,m_f,M_Z)-2)+\frac{m_{f'}^2}{2s^2M_W^2}B_0(k^2,m_{f'},M_W) \\
& +\frac{m_f^2}{4s^2M_W^2}(B_0(k^2,m_f,M_Z)-B_0(k^2,m_f,M_H))]
\end{aligned}
\tag{108}
$$

Inserting these expressions into Eq.s (65-67) fixes the mass and field renormalization counter terms of the fermions.

In Eq. (108) the quantity λ has been introduced as a fictitious small photon mass in order to regularize the IR singularity from the diagram with photon exchange. All other diagrams are IR finite. For light fermions with m_f, $m_{f'} \ll M_{W,Z}$ one can neglect the Higgs contributions. For b quarks, only the diagrams with H and χ^0 can be neglected, but not the one where the charged Higgs component ϕ^+ goes together with the top quark.

5.2 *Vector boson self energies*

The diagrams contributing to the self energies of the photon, W, Z and the photon-Z transition are shown in Figure 2. We first consider the fermion loops.

Fermionic contributions:

Photon self energy:

We give the expression for a single fermion with charge Q_f and mass m. The total contribution is obtained by summing over all fermions. Evaluating the fermion loop diagram we obtain in the notation of section 4.2:

$$
\Sigma^{\gamma\gamma}(k^2) \;=\; \frac{\alpha}{\pi}\,Q_f^2\,\{-A(m)+\frac{k^2}{2}B_0(k^2,m,m)+2\,B_{22}(k^2,m,m)\}
$$

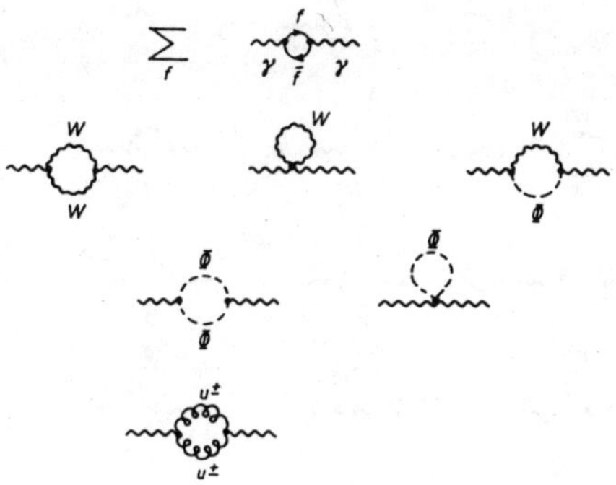

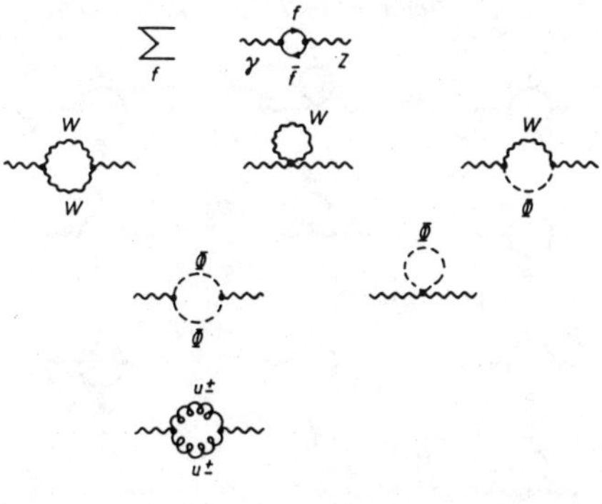

Fig. 2a: *Photon self energy and photon-Z transition*

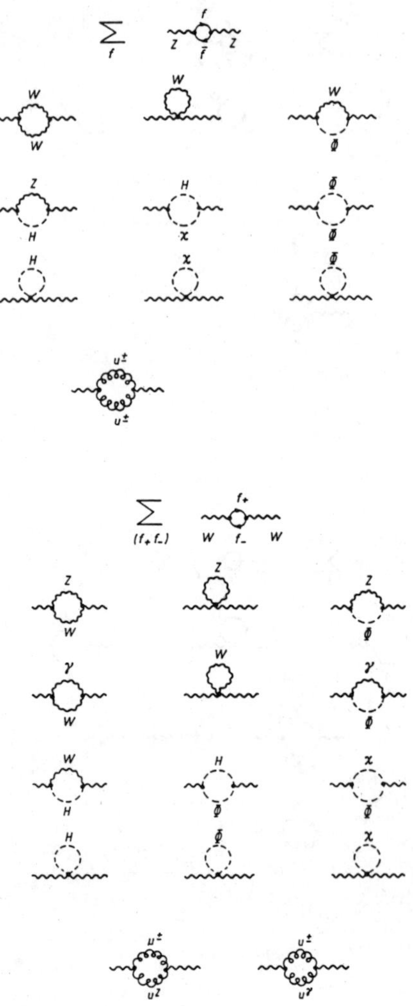

Fig. 2b: *W and Z self energies*

$$= \frac{\alpha}{3\pi} Q_f^2 \left\{ k^2 \left(\Delta - \log \frac{m^2}{\mu^2} \right) + (k^2 + 2m^2)\, \bar{B}_0(k^2, m, m) - \frac{q^2}{3} \right\} . \quad (109)$$

$\bar{B}_0$ denotes the finite function

$$\bar{B}_0(k^2, m, m) = -\int_0^1 dx \, \log \left(\frac{x^2 k^2 - x k^2 + m^2}{m^2} - i\varepsilon \right) \quad (110)$$

in the decomposition

$$B_0(k^2, m, m) = \Delta - \log \frac{m^2}{\mu^2} + \bar{B}_0(k^2, m, m) \,. \quad (111)$$

The dimensionless quantity

$$\Pi^\gamma(k^2) = \frac{\Sigma^{\gamma\gamma}(k^2)}{k^2} \quad (112)$$

is usually denoted as the photon "vacuum polarization". We list two simple expressions arising from Eq. (109) for special situations of practical interest:

- light fermions ($| k^2 | \gg m^2$):

$$\Pi^\gamma(k^2) = \frac{\alpha}{3\pi} Q_f^2 \left(\Delta - \log \frac{m^2}{\mu^2} + \frac{5}{3} - \log \frac{| k^2 |}{m^2} + i\pi\, \theta(k^2) \right) \quad (113)$$

- heavy fermions ($| k^2 | \ll m^2$):

$$\Pi^\gamma(k^2) = \frac{\alpha}{3\pi} Q_f^2 \left(\Delta - \log \frac{m^2}{\mu^2} + \frac{k^2}{5m^2} \right) \quad (114)$$

Photon - Z mixing:

Each charged fermion yields a contribution

$$\Sigma^{\gamma Z}(k^2) = -\frac{\alpha}{3\pi} \frac{v_f Q_f}{2 s_W c_W} \left\{ k^2 \left(\Delta - \log \frac{m^2}{\mu^2} \right) + (k^2 + 2m^2)\, \bar{B}_0(k^2, m, m) - \frac{q^2}{3} \right\} . \quad (115)$$

As in the photon case, the fermion loop contribution vanishes for $k^2 = 0$.

Z and W self energies:

We give the formulae for a single doublet, leptons or quarks, with $m_\pm$, $Q_\pm$, $v_\pm$, $a_\pm$ denoting mass, charge, vector and axial vector coupling of the up(+) and the down(-) member. At the end, we have to perform the sum over the various doublets, including color summation.

$$\Sigma^{ZZ}(k^2) = \frac{\alpha}{\pi} \sum_{f=+,-} \{ \frac{v_f^2 + a_f^2}{4s_W^2 c_W^2} \left[2B_{22}(k^2, m_f, m_f) + \frac{k^2}{2} B_0(k^2, m_f, m_f) - A(m_f) \right] - \frac{m_f^2}{8s_W^2 c_W^2} B_0(k^2, m_f^2, m_f^2) \}$$

$$\Sigma^{WW}(k^2) = \frac{\alpha}{\pi} \cdot \frac{1}{4s_W^2} \{ 2B_{22}(k^2, m_+, m_-) - \frac{A(m_+) + A(m_-)}{2} + \frac{k^2 - m_+^2 - m_-^2}{2} B_0(k^2, m_+, m_-) \} \quad (116)$$

Again, the following two cases are of particular practical interest:

- Light fermions:

In the light fermion limit $k^2 \gg m_\pm^2$ the$Z$ and W self-energies simplify considerably:

$$\Sigma^{ZZ}(k^2) = \frac{\alpha}{3\pi} \cdot \frac{v_+^2 + a_+^2 + v_-^2 + a_-^2}{4s_W^2 c_W^2} k^2 \left(\Delta - \log \frac{k^2}{\mu^2} + i\pi \right),$$

$$\Sigma^{WW}(k^2) = \frac{\alpha}{3\pi} \cdot \frac{k^2}{4s_W^2} \left(\Delta - \log \frac{k^2}{\mu^2} + i\pi \right). \quad (117)$$

- Heavy fermions:

Of special interest is the case of a heavy top quark which yields a large correction $\sim m_t^2$. In order to extract this part we keep for simplicity only those terms which are either singular or quadratic in the top mass $m_t \equiv m_+$ ($N_C = 3$):

$$\Sigma^{ZZ}(k^2) = N_C \frac{\alpha}{3\pi} \left\{ \frac{v_+^2 + a_+^2 + v_-^2 + a_-^2}{4s_W^2 c_W^2} k^2 - \frac{3m_t^2}{8s_W^2 c_W^2} \right\} \left(\Delta - \log \frac{m_t^2}{\mu^2} \right) + \cdots$$

$$\Sigma^{WW}(k^2) = N_C \frac{\alpha}{3\pi} \left\{ \frac{k^2}{4s_W^2} \left(\Delta - \log \frac{m_t^2}{\mu^2} \right) - \frac{3m_t^2}{8s_W^2} \left(\Delta - \log \frac{m_t^2}{\mu^2} + \frac{1}{2} \right) \right\} + \cdots \quad (118)$$

The quantity

$$\Delta\rho = \frac{\Sigma^{ZZ}(0)}{M_Z^2} - \frac{\Sigma^{WW}(0)}{M_W^2} \quad (119)$$

is finite as far as the heavy fermion contribution is considered which yields for the top quark:

$$\Delta\rho = N_C \frac{\alpha}{16\pi s_W^2 c_W^2} \frac{m_t^2}{M_Z^2}. \quad (120)$$

	$\Sigma^{\gamma\gamma}(k^2) = \frac{\alpha}{4\pi s_W^2}\cdot$	$\Sigma^{ZZ}(k^2) = \frac{\alpha}{4\pi s_W^2}\cdot$	$\Sigma^{\gamma Z}(k^2) = \frac{\alpha}{4\pi s_W^2}\cdot$	$\Sigma^{WW}(k^2) = \frac{\alpha}{4\pi s_W^2}\cdot$
	$A_1(k^2, M_W, M_W)\cdot s_W^2$	$c_W^2 A_1(k^2, M_W, M_W)$	$s_W c_W A_1(k^2, M_W, M_W)$	$c_W^2 A_1(k^2, M_W, M_W)$ $+s_W^2 A_1(k^2, 0, M_W)$
	$2s_W^2 A_2(M_W)$	$2c_W^2 A_2(M_W)$	$2c_W s_W A_2(M_W)$	$A_2(M_W) + c_W^2 A_2(M_Z)$
	$2s_W^2 M_W^2 B_0(k^2, M_W, M_W)$	$2s_W^4 M_Z^2 B_0(k^2, M_W, M_W)$	$-2\frac{s_W^3}{c_W} M_W^2 B_0(k^2, M_W, M_W)$	$s_W^4 M_Z^2 B_0(k^2, M_Z, M_W)$ $+s_W^2 M_W^2 B_0(k^2, 0, M_W)$
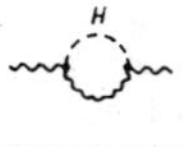 H	0	$\frac{M_Z^2}{c_W^2} B_0(k^2, M_H, M_Z)$	0	$M_W^2 B_0(k^2, M_H, M_W)$
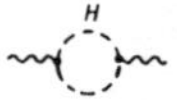 H	0	$-\frac{1}{c_W^2} B_{22}(k^2, M_H, M_Z)$	0	$-B_{22}(k^2, M_H, M_W)$
	$-4s_W^2 B_{22}(k^2, M_W, M_W)$	$-\frac{(c_W^2 - s_W^2)^2}{c_W^2} B_{22}(k^2, M_W, M_W)$	$-2\frac{s_W}{c_W}(c_W^2 - s_W^2) B_{22}(k^2, M_W, M_W)$	$-B_{22}(k^2, M_Z, M_W)$
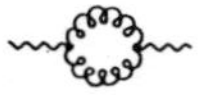	$2s_W^2 B_{22}(k^2, M_W, M_W)$	$2c_W^2 B_{22}(k^2, M_W, M_W)$	$2s_W c_W B_{22}(k^2, M_W, M_W)$	$2c_W^2 B_{22}(k^2, M_Z, M_W)$ $+2s_W^2 B_{22}(k^2, 0, M_W)$
	$-2s_W^2 A(M_W)$	$-\frac{1}{4c_W^2}[A(M_H) + A(M_Z)]$ $-\frac{(c_W^2 - s_W^2)^2}{2c_W^2} A(M_W)$	$-\frac{s_W}{c_W}(c_W^2 - s_W^2) A(M_W)$	$-\frac{1}{4}[A(M_H) + A(M_Z)]$ $-\frac{1}{2} A(M_W)$

Table 1: Bosonic contributions to the vector boson self energies

Bosonic contributions:

The bosonic contributions to the vector boson self energies consist of the loop diagrams involving the gauge boson self interactions, the Higgs boson together with its unphysical components, and the Faddeev-Popov ghost fields. They are listed synoptically in Table 1 for the γ, Z, W self energies and the $\gamma - Z$ transition. The result for each part is the sum of the entries in the corresponding column, times the factor indicated in the first line. The functions A_1, A_2 appearing in table 1, are abbreviations for the following combinations of 1- and 2-point integrals:

$$\begin{aligned} A_1(k^2, m_1, m_2) &= A(m_1) + A(m_2) - (m_1^2 + m_2^2 + 4\,k^2)\,B_0(k^2, m_1, m_2) \\ &\quad -10\,B_{22}(k^2, m_1, m_2) \,+\, 2\,(m_1^2 + m_2^2 - \frac{k^2}{3}), \\ A_2(m) &= -3\,A(m) - 2\,m^2\,. \end{aligned}$$

5.3 *Electroweak parameter shifts*

We can now apply the results of 5.2 to discuss the contributions to the electroweak parameter shifts at the 1-loop level via the renormalization procedure. Such shifts are essentially the counter terms for the electric charge in Eq. (71) and for the electroweak mixing angle in Eq. (76). Since these counter terms are universal, they appear everywhere where in the lowest order expressions e resp. $\sin^2\theta_W$ is present. The shifts by the counter terms are not finite. However, their finite parts contain large terms which turn out to be the dominating contributions in the 1-loop corrections to the relations between the physical parameters.

$\Delta\alpha$ and effective charge:

The charge counter term in Eq. (71) contains the photon vacuum polarization at $k^2 = 0$. We split off the subtracted part evaluated at M_Z^2:

$$\begin{aligned} \Pi^\gamma(0) &= -\mathrm{Re}\Pi^\gamma(M_Z^2) + \Pi^\gamma(0) + \mathrm{Re}\,\Pi^\gamma(M_Z^2) \\ &= -\mathrm{Re}\,\hat{\Pi}^\gamma(M_Z^2) + \mathrm{Re}\,\Pi^\gamma(M_Z^2)\,. \end{aligned} \tag{121}$$

The subtracted quantity $\hat{\Pi}^\gamma(M_Z^2)$, which is the renormalized vacuum polarization according to Eq.s (112,47,64), is UV finite. Its fermionic content can be split into a leptonic and a hadronic part:

$$\mathrm{Re}\,\hat{\Pi}^\gamma_{ferm}(M_Z^2) = \mathrm{Re}\,\hat{\Pi}^\gamma_{lept}(M_Z^2) + \mathrm{Re}\,\hat{\Pi}^\gamma_{had}(M_Z^2)\,.$$

Heavy top quarks decouple from the subtracted vacuum polarization, as can be seen immediately from Eq. (114):

$$\hat{\Pi}^\gamma_{top}(M_Z^2) = \frac{\alpha}{\pi}\,Q_t^2\,\frac{M_Z^2}{5\,m_t^2}\,. \tag{122}$$

Whereas the leptonic content can easily be obtained from

$$\mathrm{Re}\,\hat{\Pi}^{\gamma}_{lept}(M_Z^2) = \sum_{l=e,\mu,\tau} \frac{\alpha}{3\pi}\left(\frac{5}{3} - \log\frac{M_Z^2}{m_l^2}\right), \tag{123}$$

no light quark masses are available as reasonable input parameters for the hadronic content. Instead, the 5 flavor contribution to $\hat{\Pi}^{\gamma}_{had}$ can be derived from experimental data with the help of a dispersion relation

$$\hat{\Pi}^{\gamma}_{had}(M_Z^2) = \frac{\alpha}{3\pi} M_Z^2 \int_{4m_\pi^2}^{\infty} ds' \, \frac{R^{\gamma}(s')}{s'(s' - M_Z^2 - i\varepsilon)} \tag{124}$$

with

$$R^{\gamma}(s) = \frac{\sigma(e^+e^- \to \gamma^* \to hadrons)}{\sigma(e^+e^- \to \gamma^* \to \mu^+\mu^-)}$$

as an experimental quantity up to a scale s_1 and applying perturbative QCD for the tail region above s_1. Using e^+e^- data for the energy range below 40 GeV the integral (124) yields [44]

$$\mathrm{Re}\,\hat{\Pi}^{\gamma}_{had}(s_0) = -0.0282 \pm 0.0009 \tag{125}$$

for $s_0 = (92\,\mathrm{GeV})^2$. The error is almost completely due to the experimental data. Combining this result with the leptonic part one obtains

$$\mathrm{Re}\,\hat{\Pi}^{\gamma}_{ferm}(M_Z^2) = -0.0595 \pm 0.0009 \;\text{ for }\; M_Z = 91.187\,\mathrm{GeV}. \tag{126}$$

This finite quantity arising from the light fermion loops is independent of the structure of the electroweak model. It corresponds to a QED induced shift

$$\Delta\alpha = -\,\mathrm{Re}\,\hat{\Pi}^{\gamma}_{ferm}(M_Z^2) \tag{127}$$

in the electromagnetic fine structure constant:

$$\alpha \to \alpha(1 + \Delta\alpha)$$

which can be resummed according to the renormalization group accommodating all the leading logarithms of the type $\alpha^n \log^n(M_Z/m_f)$. The result is an effective fine structure constant at the Z mass scale:

$$\alpha(M_Z^2) = \frac{\alpha}{1 - \Delta\alpha} = \frac{1}{128.9 \pm 0.1}. \tag{128}$$

The ρ-parameter:

The ρ-parameter, which in the Standard Model is unity at the tree level, gets a deviation $\Delta\rho$ from 1 by radiative corrections. ρ has been defined as the ratio of the neutral to the charged current strength in neutrino scattering. It is modified by the quantity $\Delta\rho$ in Eq. (119) yielding the expression (120) for the contribution of the top quark. A general doublet of fermions with masses m_1, m_2 causes a shift of ρ by [45]

$$\Delta\rho_{ferm} = N_C \frac{\alpha}{16\pi s_W^2 c_W^2 M_Z^2}\left(m_1^2 + m_2^2 - \frac{2m_1^2 m_2^2}{m_1^2 - m_2^2}\log\frac{m_1^2}{m_2^2}\right). \tag{129}$$

For the (t,b)-doublet, neglecting m_b, Eq. (120) is recovered.

It is important to note that this potentially large fermionic contribution to $\Delta\rho$ simultaneously constitutes the leading shift for the electroweak mixing angle according to Eq. (76) since there are no other terms $\sim m_f^2$ in the mass counter terms $\delta M_Z^2, \delta M_W^2$ besides those which are k^2-independent, hence leading to:

$$\frac{\delta M_Z^2}{M_Z^2} - \frac{\delta M_W^2}{M_W^2} \simeq \frac{\Sigma^{ZZ}(0)}{M_Z^2} - \frac{\Sigma^{WW}(0)}{M_W^2} = \Delta\rho\,. \tag{130}$$

There is also a Higgs contribution to $\Delta\rho$ which, however, is not UV finite by itself when derived from the diagrams involving the physical Higgs boson only. From table 1 one obtains:

$$\begin{aligned}\Delta\rho_H \;=\; & \frac{g_2^2}{16\pi^2}\cdot\frac{3}{4M_W^2}\left[M_W^2(\Delta - \log\frac{M_W^2}{\mu^2}) - M_Z^2(\Delta - \log\frac{M_Z^2}{\mu^2}) + \frac{5}{6}(M_Z^2 - M_W^2)\right. \\ & \left. + \frac{M_Z^2 M_H^2}{M_Z^2 - M_H^2}\log\frac{M_H^2}{M_Z^2} - \frac{M_W^2 M_H^2}{M_W^2 - M_H^2}\log\frac{M_H^2}{M_W^2}\right].\end{aligned} \tag{131}$$

From this expression the dependence on M_H for large Higgs masses $M_H \gg M_{W,Z}$ can be derived which, in contrast to heavy fermions, is only logarithmic [46]:

$$\Delta\rho_H \simeq \frac{g_2^2}{16\pi^2}\cdot\frac{3s_W^2}{4c_W^2}\log\frac{M_H^2}{M_W^2} + \cdots \tag{132}$$

In the limit $s_W^2 \to 0, M_Z \to M_W$, where the $U(1)_Y$ is switched off, one finds $\Delta\rho_H = 0$. This is the consequence of the global $SU(2)_R$ symmetry of the Higgs Lagrangian ('custodial symmetry'), which is broken by the $U(1)_Y$ group. $\Delta\rho_H$ is thus a measure of the $SU(2)_R$ breaking by the weak hypercharge.

In contrast to the top term $\sim m_t^2$, the Higgs boson enters the shift of $\sin^2\theta_W$ not exclusively through $\Delta\rho$. There are additional M_H-dependent terms in $\delta M_Z^2, \delta M_W^2$ besides the ones in $\Sigma^{WW}(0)$ and $\Sigma^{ZZ}(0)$. The same remark holds for logarithmic top terms $\sim \log(m_t/M_W)$ which are not present in $\Delta\rho$.

The $M_W - M_Z$ interdependence:

Incorporating the parameter shifts

$$\alpha \to \alpha(1+\Delta\alpha), \quad s_W^2 \to s_W^2 + c_W^2\Delta\rho$$

with $\Delta\alpha, \Delta\rho$ from Eq.s (120), (127) into the relation (28), we obtain the approximate correlation at 1-loop

$$M_W^2\left(1-\frac{M_W^2}{M_Z^2}\right)=\frac{\pi\alpha}{\sqrt{2}G_\mu}\left(1+\Delta\alpha-\frac{c_W^2}{s_W^2}\Delta\rho+\cdots\right) \tag{133}$$

between the vector boson masses and the other electroweak parameters α and G_μ, taking into account the large contributions from light and heavy fermions. The $\cdots$ indicate the residual terms belonging to the full calculation discussed in the next section.

<u>The NC couplings:</u>

In a similar way as done above, we obtain a universal shift in the overall normalization of the NC coupling constants in Eq.s (36,37)

$$\begin{aligned}\frac{e}{2s_Wc_W} &\rightarrow \frac{e}{2s_Wc_W}\left[1+\frac{1}{2}\left(\Delta\alpha-\frac{c_W^2-s_W^2}{s_W^2}\Delta\rho\right)\right]\\ &= (\sqrt{2}G_\mu M_Z^2)^{1/2}\left[1+\frac{\Delta\rho}{2}\right]\\ &\rightarrow (\sqrt{2}G_\mu M_Z^2\rho_f)^{1/2}\end{aligned} \tag{134}$$

The complete expressions for the normalization factor

$$\rho_f = 1+\Delta\rho+\cdots$$

and the effective mixing angle

$$s_f^2 = s_W^2 + c_W^2\Delta\rho+\cdots$$

in the Zff vertex between on-shell Z bosons and fermions will be be presented and discussed in the subsequent chapter on the Z resonance.

6 The muon lifetime and the gauge boson masses

6.1 *One-loop corrections to the muon lifetime*

The interdependence between the gauge boson masses is established through the accurately measured muon lifetime or the Fermi coupling constant G_μ, respectively. The μ-lifetime τ_μ is related to G_μ by the defining equation, including QED corrections to the 4-point Fermi interaction [47]:

$$\frac{1}{\tau_\mu}=\frac{G_\mu^2 m_\mu^5}{192\pi^3}\left(1-\frac{8m_e^2}{m_\mu^2}\right)\left(1+\frac{3m_\mu^2}{5M_W^2}\right)\left[1+\frac{\alpha}{2\pi}(\frac{25}{4}-\pi^2)\right]. \tag{135}$$

The leading $2nd$ order correction is obtained by replacing

$$\alpha \to \alpha \left(1 + \frac{2\alpha}{3\pi} \log \frac{m_\mu}{m_e}\right).$$

This formula is used as the defining equation for G_μ in terms of the experimental μ-lifetime. In lowest order, the Fermi constant is given by the Standard Model expression (26) for the decay amplitude. In 1-loop order, $G_\mu/\sqrt{2}$ is identified with the expression

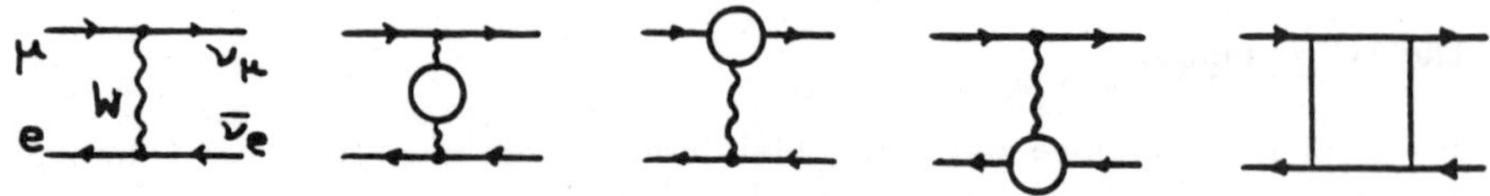

$$\begin{aligned} \frac{G_\mu}{\sqrt{2}} &= \frac{e^2}{8s_W^2 M_W^2}\left[1 + \frac{\hat{\Sigma}^{WW}(0)}{M_W^2} + \delta_{VB}\right] \\ &\equiv \frac{e^2}{8s_W^2 M_W^2}\left[1 + \Delta r\right]. \end{aligned} \tag{136}$$

The quantity $\Delta r(e, M_W, M_Z, M_H, m_f)$ is the UV and IR finite electroweak 1-loop correction to the muon decay amplitude in the Standard Model. Eq. (136) is the correlation between the vector boson masses and the other electroweak precision parameters α and G_μ. Due to the presence of m_t, M_H in Δr, this correlation becomes dependent on experimentally unknown quantities at the 1-loop level.

$\hat{\Sigma}^{WW}(0)$ is the renormalized W self energy from (47), evaluated at $k^2 = 0$, with the counter terms specified in (64). The term

$$\delta_{VB} = \frac{\alpha}{4\pi s_W^2}\left(6 + \frac{7 - 4s_W^2}{2s_W^2} \log c_W^2\right) \tag{137}$$

summarizes the vertex corrections and box diagrams in the decay amplitude, more explicitly shown in Figure 3. A set of infra-red divergent "QED correction" graphs has been removed from this class of diagrams. These left-out diagrams, together with the real bremsstrahlung contributions, reproduce the QED correction factor of the Fermi model result in Eq. (135) and therefore have no influence on the relation between G_μ and the Standard Model parameters.

δ_{VB} has the structure

$$\delta_{VB} = \hat{F}^{We\nu} + \hat{F}^{W\mu\nu} - \frac{1}{2}\hat{\Pi}^{\nu_e} - \frac{1}{2}\hat{\Pi}^{\nu_\mu} + \delta_{box} \tag{138}$$

with the following ingredients:

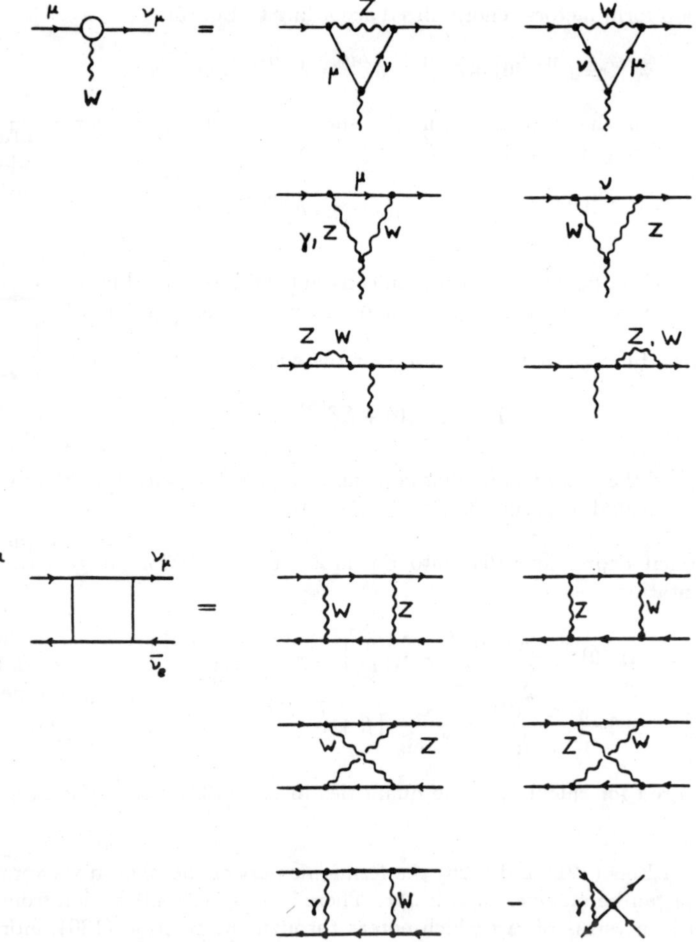

Fig. 3: *Vertex corrections with external self energies and box diagrams contributing to the 1-loop amplitude for $\mu \to \nu_\mu e \bar{\nu}_e$. For the $We\nu$-vertex the analogous sample of vertex corrections is present as well. Omitted are the "QED" diagrams with a photon in the external charged lepton lines, and the photonic vertex correction to the Fermi amplitude is subtracted from the box diagram with photon exchange*

- the CC 1-loop form factors renormalized according to Eq. (59)

$$\hat{F}^{W\ell\nu} = F^{W\ell\nu}(0) + \delta Z_1^W - \delta Z_2^W + \delta Z_L^{(\ell)}, \quad \ell = e, \mu$$

  where $F^{W\ell\nu}(0)$ is the form factor in the sum of the CC vertex correction diagrams evaluated at $k^2 = 0$:

$$\Lambda_\mu^{W\ell\nu} = i\,\frac{e}{2\sqrt{2}s_W}\,\gamma_\mu(1-\gamma_5)\,F^{W\ell\nu}(0)\,.$$

  $\delta Z_L^{(\ell)}$ is the doublet field renormalization constant (66), evaluated from the e- and μ self energies without the virtual photon contribution (section 5.1).

- the finite wave function renormalization for the e and μ neutrino with

$$\hat{\Pi}^{\nu_\ell} = \Sigma_L^{\nu_\ell}(0) + \delta Z_L^{(\ell)}\,,$$

- the sum δ_{box} of the massive box diagrams and the γW-box with the IR subtraction as indicated in Figure 3.

Inserting the explicit expressions (64) into Eq. (136), the result for Δr gets the following representation:

$$\begin{aligned}\Delta r \;=\; & \Pi^\gamma(0) - \frac{c_W^2}{s_W^2}\left(\frac{\delta M_Z^2}{M_Z^2} - \frac{\delta M_W^2}{M_W^2}\right) + \frac{\Sigma^{WW}(0) - \delta M_W^2}{M_W^2} \\ & +2\,\frac{c_W}{s_W}\,\frac{\Sigma^{\gamma Z}(0)}{M_Z^2} + \frac{\alpha}{4\pi s_W^2}\left(6 + \frac{7-4s_W^2}{2s_W^2}\log c_W^2\right). \end{aligned} \tag{139}$$

All formulae required for calculating the quantities in Eq. (139) can be found in section 5.2.

By means of Eq.s (121) and (126) the fermionic part of the photon vacuum polarization $\Pi^\gamma(0)$ can be made explicit in Δr. There is also the contribution from a heavy top quark in terms of $\Delta\rho$ which enters through the relation (130) with $\Delta\rho$ given in Eq. (120). As can be seen from Eq. (139), there are no other large or potentially large terms in Δr besides those associated with $\Delta\alpha$ and $\Delta\rho$. When restricting to these two entries only, we recover the result already found in (133) by the discussion in section 5.3. Now we are able to write the following complete expression for Δr giving special emphasis on the dominant fermionic contributions:

$$\Delta r = \Delta\alpha - \frac{c_W^2}{s_W^2}\,\Delta\rho + (\Delta r)_{remainder}\,. \tag{140}$$

$\Delta\alpha$, Eq. (127), contains the large logarithmic corrections from the light fermions and $\Delta\rho$ the leading quadratic correction from a large top mass. All other terms are collected in the $(\Delta r)_{remainder}$. It should be noted that the remainder also contains a term logarithmic in the top mass with a large coefficient:

$$(\Delta r)^{top}_{remainder} = -\frac{\alpha}{4\pi s_W^2}\left(\frac{c_W^2}{s_W^2} - \frac{1}{3}\right)\log\frac{m_t}{M_Z} + \cdots \qquad (141)$$

Also the Higgs boson contribution is part of the remainder. For large M_H, it increases only logarithmically as it was already observed in the discussion of the ρ-parameter:

$$(\Delta r)^{Higgs}_{remainder} \simeq \frac{\alpha}{16\pi s_W^2}\cdot\frac{11}{3}\left(\log\frac{M_H^2}{M_W^2} - \frac{5}{6}\right). \qquad (142)$$

The typical size of $(\Delta r)_{remainder}$ is of the order ~ 0.01.

6.2 *Higher order contributions*

For a top mass of 90 GeV the 1-loop quantity Δr is of the size 0.06 - 0.07 and we expect a 2-loop contribution typically of the order $(\Delta r)^2 \simeq 0.005$. This corresponds to a shift in the W mass of about 90 MeV which is the precision of the M_W measurement at LEP 200 and signals the need of going beyond the first order corrections in Eq. (136).

(i) Summation of large $\Delta\alpha$ terms:

The replacement of the $\Delta\alpha$-part

$$1 + \Delta\alpha \to \frac{1}{1 - \Delta\alpha}$$

of the 1-loop result in Eq. (140) was already discussed in the context of the effective electromagnetic charge (128). It correctly takes into account all orders in the leading logarithmic corrections $(\Delta\alpha)^n$, as can be shown by renormalization group arguments [48]. The evolution of the electromagnetic coupling with the scale μ is described by the renormalization group equation

$$\mu\frac{d\alpha}{d\mu} = -\frac{\beta_0}{2\pi}\alpha^2 \qquad (143)$$

with the coefficient of the 1-loop β-function in QED

$$\beta_0 = -\frac{4}{3}\sum_{f\neq t} Q_f^2. \qquad (144)$$

The solution of the RGE contains the leading logarithms in the resummed form as given in Eq. (128). It corresponds to a resummation of the iterated 1-loop vacuum polarization to all orders. The non-leading QED-terms of next order are numerically not significant. Thus, in a situation where large corrections are only due to the evolution of the electromagnetic charge between two very different scales set by m_f and M_Z, the resummed form

$$G_\mu = \frac{\pi\alpha}{\sqrt{2}M_W^2 s_W^2}\frac{1}{1-\Delta r} = \frac{\pi\alpha}{\sqrt{2}M_Z^2 c_W^2 s_W^2}\frac{1}{1-\Delta r} \tag{145}$$

with Δr in Eq. (140) represents a good approximation to the full result.

(ii) Summation of large $\Delta\rho$ terms:

In case of a heavy top, where also $\Delta\rho$ is large, the powers $(\Delta\rho)^n$ are not correctly resummed in Eq. (145). A result correct in the leading terms up to $O(\alpha^2)$ is instead given by [49] by the independent resummation

$$\frac{1}{1-\Delta r} \to \frac{1}{1-\Delta\alpha}\cdot\frac{1}{1+\frac{c_W^2}{s_W^2}\Delta\overline{\rho}} + (\Delta r)_{remainder} \tag{146}$$

where

$$\Delta\overline{\rho} = N_C\,\frac{G_\mu m_t^2}{8\pi^2\sqrt{2}}\cdot\left[1+\frac{G_\mu m_t^2}{8\pi^2\sqrt{2}}\,\rho^{(2)}\right] \tag{147}$$

incorporates the result from 2-loop 1-particle irreducible diagrams. For light Higgs bosons $M_H \ll m_t$, where M_H can be neglected, the coefficient

$$\rho^{(2)} = 19 - 2\pi^2 \tag{148}$$

was first calculated by Hoogeveen and van der Bij [50] and was recently confirmed by Barbieri et al. [51]. The general function $\rho^{(2)}$, valid for all Higgs masses, has been derived in [51]. For large Higgs masses $M_H > 2m_t$, a good approximation is given by the asymptotic expression with $r = (m_t/M_H)^2$ [51]

$$\begin{aligned}\rho^{(2)} &= \frac{49}{4} + \pi^2 + \frac{27}{2}\log r + \frac{3}{2}\log^2 r \\ &+\frac{r}{3}\left(2 - 12\pi^2 + 12\log r - 27\log^2 r\right) \\ &+\frac{r^2}{48}\left(1613 - 240\pi^2 - 1500\log r - 720\log^2 r\right).\end{aligned} \tag{149}$$

Figure 4 shows the function $\rho^{(2)}$ together with the asymptotic formula. Except for very small Higgs masses, the results deviate significantly from the approximation for $M_H \to 0$.

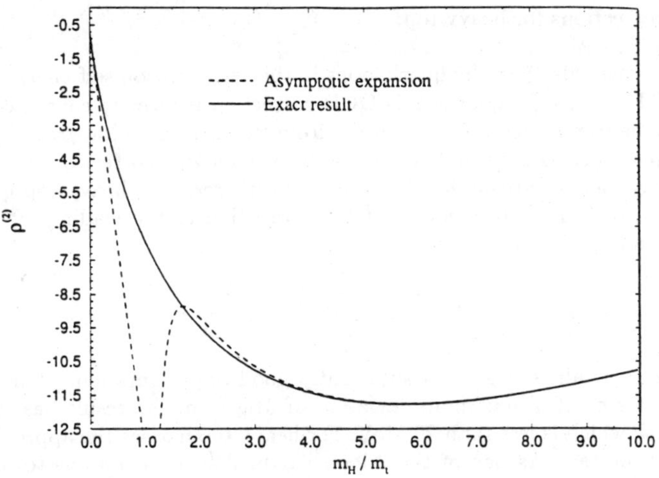

Fig. 4: *The function $\rho^{(2)}$ in Eq. (147), from [51].*

With the resummed ρ-parameter

$$\rho = \frac{1}{1 - \Delta\bar{\rho}} \tag{150}$$

Eq. (146) is compatible with the following form of the $M_W - M_Z$ interdependence

$$G_\mu = \frac{\pi}{\sqrt{2}} \frac{\alpha(M_Z^2)}{M_W^2 \left(1 - \frac{M_W^2}{\rho M_Z^2}\right)} \cdot \left[1 + (\Delta r)_{remainder}\right] . \tag{151}$$

It is interesting to compare this result with the corresponding lowest order $M_W - M_Z$ correlation in a more general model with a tree level ρ-parameter $\rho_0 \neq 1$: the tree-level ρ_0 enters in the same way as the ρ from a heavy top in the minimal model. The same applies for the quadratic mass terms from other particles like scalars or additional heavy fermions in isodoublets with large mass splittings. Hence, up to the small quantity $(\Delta r)_{remainder}$, they are indistinguishable from an experimental point of view ($\Delta\alpha$ is universal). In the minimal model, however, ρ is calculable in terms of m_t, M_H whereas ρ_0 is an *additional* free parameter.

(iii) QCD corrections for heavy top:

Virtual gluons contribute to the quark loops in the vector boson self-energies at the 2-loop level. For the light quarks this QCD correction is already contained in the result for the hadronic vacuum polarization from the dispersion integral, Eq. (124). Fermion loops involving the top quark get additional $O(\alpha\alpha_s)$ corrections which have been calculated perturbatively [52]. The dominating term for heavy top quarks is of the form $\alpha_s \alpha m_t^2$ and represents the QCD correction to the leading m_t^2 term of the ρ-parameter:

$$\Delta\rho \rightarrow \Delta\rho^{\alpha\alpha_s} = -\,\Delta\rho \cdot \frac{\alpha_s(m_t^2)}{\pi} \cdot \frac{2}{3}\left(\frac{\pi^2}{3} + 1\right). \tag{152}$$

This leading term already gives a sufficiently good approximation. This can be quantified in terms of a maximum deviation of M_W from the result based on the exact formulae which is less than 20 MeV. For heavy top masses the approximation becomes even better. As one of the 2-loop irreducible contributions to ρ, $\Delta\rho^{\alpha\alpha_s}$ has to be incorporated into $\Delta\overline{\rho}$ and resummed together with the electroweak 2-loop irreducible term as indicated in Eq. (151). Non-perturbative QCD effects in the gauge boson self energies associated with the $t\bar{t}$ threshold can be estimated with help of dispersive methods [53]. Expressed in terms of M_W, they shift the perturbative result by about +40 MeV for $m_t = 250$ GeV; for $m_t < 200$ GeV the influence on M_W is smaller than 25 GeV.

(iv) Non-leading higher order terms:

The modification of Eq. (146) by placing $(\Delta r)_{remainder}$ into the denominator

$$\frac{1}{1-\Delta r} \rightarrow \frac{1}{(1-\Delta\alpha)\cdot(1+\frac{c_W^2}{s_W^2}\Delta\overline{\rho}) \;-\; (\Delta r)_{remainder}} \tag{153}$$

correctly incorporates the non-leading higher order terms containing mass singularities of the type $\alpha^2 \log(M_Z/m_f)$ [54]

The treatment of the higher order reducible terms in Eq. (153) can be further refined by performing in $(\Delta r)_{remainder}$ the following substitution

$$\frac{\alpha}{s_W^2} \rightarrow \frac{\sqrt{2}}{\pi} G_\mu M_W^2 (1-\Delta\alpha) \tag{154}$$

in the expansion parameter of the combination

$$\left(\frac{\delta M_Z^2}{M_Z^2} - \frac{\delta M_W^2}{M_W^2}\right) - \Delta\rho$$

after cancellation of the UV singularity in the combination (139) or in the $\overline{MS}$ scheme with $\mu = M_Z$. This is discussed in [55] and is equivalent to the method described in [28] as well as to the recipe given at the end of ref. [49]. Numerically this modification is of some importance in the M_W-M_Z correlation for very heavy top quarks above 250 GeV. As an example, for $m_t = 300$ GeV one obtains a change in M_W by about 40 MeV.

A general comment, however, is in order: The refined treatment of the non-leading reducible higher order terms can be considered as an improvement only in case that the 2-loop irreducible non-leading terms are essentially smaller in size. Irreducible contributions of the type $\alpha G_\mu m_t^2 \log(m_t/M_Z)$ are unknown, and one has to rely on the assumption that the suppression by $1/N_C$ relative to the 2-loop reducible term is not compensated by a large coefficient. For bosonic 2-loop terms reducible and irreducible contributions are a priori of the same size and one does not gain from resumming 1-loop terms. In order to be on the safe side, the differences caused by the summation of non-leading reducible terms should be considered as a theoretical uncertainty at the level of 1-loop calculations improved by higher order leading terms.

6.3 *Numerical results and experimental data*

The correlation of the electroweak parameters, complete at the one-loop level and with the proper incorporation of the leading higher order effects, is given by the following equation:

$$\begin{aligned} M_W^2\left(1-\frac{M_W^2}{M_Z^2}\right) &= \frac{\pi\alpha}{\sqrt{2}G_\mu}\cdot\frac{1}{(1-\Delta\alpha)\cdot(1+\frac{c_W^2}{s_W^2}\Delta\overline{\rho})-(\Delta r)_{remainder}} \\ &\equiv \frac{\pi\alpha}{\sqrt{2}G_\mu}\cdot\frac{1}{1-\Delta r}\,. \end{aligned} \tag{155}$$

The Δr in Eq. (155) is an effective quantity beyond the 1-loop order, introduced to obtain the formal analogy to the naively resummed first order result in Eq. (145). $\Delta\overline{\rho}$ includes the 2-loop irreducible electroweak and QCD corrections to the ρ-parameter:

$$\begin{aligned} \Delta\overline{\rho} &= \Delta\rho^{(1)}+\Delta\rho^{(2)} \\ &= N_C\frac{G_\mu m_t^2}{8\pi^2\sqrt{2}}\cdot\left[1+\frac{G_\mu m_t^2}{8\pi^2\sqrt{2}}\rho^{(2)}-\frac{\alpha_s(m_t^2)}{\pi}\cdot\frac{2}{3}\left(\frac{\pi^2}{3}+1\right)\right]. \end{aligned} \tag{156}$$

The correlation (155) allows us to predict a value for the W mass after the other parameters have been specified. These predicted values for M_W are put together in table 2 for various Higgs and top masses. The present experimental value for the W mass from the combined UA2 and CDF results [2] is

$$M_W^{exp} = 80.14 \pm 0.26\,\text{GeV}\,. \tag{157}$$

m_t	$M_H = 60$	100	300	1000
90	79.952	79.925	79.854	79.760
120	80.109	80.082	80.010	79.915
150	80.275	80.248	80.173	80.078
180	80.462	80.433	80.355	80.257
210	80.674	80.643	80.557	80.454
240	80.912	80.879	80.783	80.671

Table 2: The W mass M_W as predicted by the Standard Model for $M_Z = 91.187$ GeV and various top and Higgs masses, based on Eq. (155). The refinement described in Eq. (154) was taken into account. Nonperturbative QCD effects associated with the $t\bar{t}$ threshold have been neglected. All masses are in GeV

We can define the quantity Δr also as a physical observable by

$$\Delta r = 1 - \frac{\pi\alpha}{\sqrt{2}G_\mu} \frac{1}{M_W^2\left(1 - \frac{M_W^2}{M_Z^2}\right)}. \tag{158}$$

Experimentally, it is determined by M_Z and the ratio M_W/M_Z. Theoretically, it can be computed from M_Z, G_μ, α after specifying the masses M_H, m_t by solving Eq. (155). In Figure 5 we display the prediction for Δr as a function of m_t in various steps: the first order calculation based on Eq. (145) with the lowest order Δr, then including the electroweak higher order terms on the basis of Eq. (146), and finally including also the QCD corrections related to m_t. Both electroweak and QCD higher order effects yield a positive shift to Δr and thus diminish the slope of the first order dependence on m_t for large top masses. The effect on Δr coming from the modified $\rho^{(2)}$ in Eq. (147) for large M_H is shown in Figure 6. It causes an additional weakening of the sensitivity to m_t for large Higgs masses.

The theoretical prediction for Δr for various Higgs and top masses is displayed in Figure 7. For comparison with data, the experimental 1σ limits from the direct measurements of M_Z at LEP and M_W/M_Z in $p\bar{p}$ are indicated. The experimental input from LEP [1, 4] and from the combined UA2 and CDF results [2] is

$$M_Z = 91.187 \pm 0.007\text{GeV}, \quad s_W^2 = 0.2275 \pm 0.0052\,.$$

For $M_H < 1$ TeV these results constrain the top mass to the range $m_t < 203$ GeV at the 1σ level. The present experimental error does not allow a sensitivity to the Higgs mass. Precision measurements of M_W at LEP 200 will pin down the error to $\delta\Delta r = 0.006$ (0.004 with high luminosity). This would determine m_t with an accuracy of about $\delta m_t = 10$ GeV. A still inherent uncertainty from the unknown Higgs mass (with $M_H > 60$ GeV), however, would give an an additional theoretical error of ± 17 GeV. The expected precision in the determination of Δr matches the

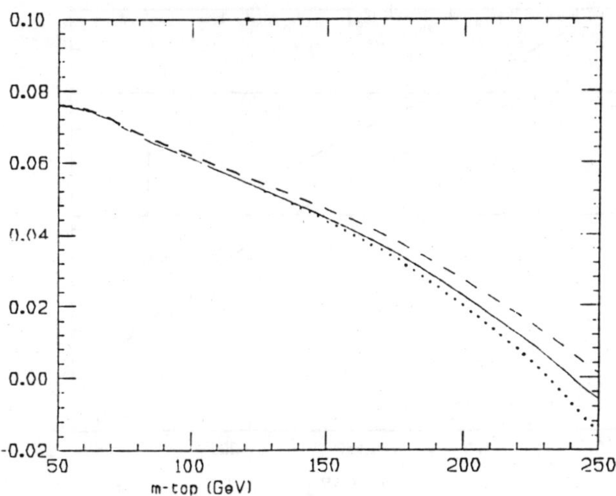

Fig. 5: Δr *in* $O(\alpha)$ *(dotted), in* $O(\alpha^2)$ *(full), and in* $O(\alpha^2 + \alpha\alpha_s)$ *(dashed).* $M_Z = 91.187$ *GeV,* $M_H = 300$ *GeV.*

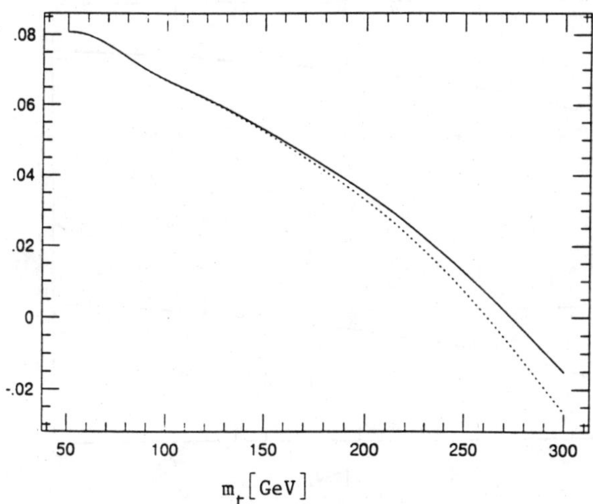

Fig. 6: Δr *in* $O(\alpha^2 + \alpha\alpha_s)$ *for* $M_H = 1$ *TeV with the Higgs dependent* ρ*-parameter (full) and the approximation (148) (dashed).* $M_Z = 91.187$ *GeV*

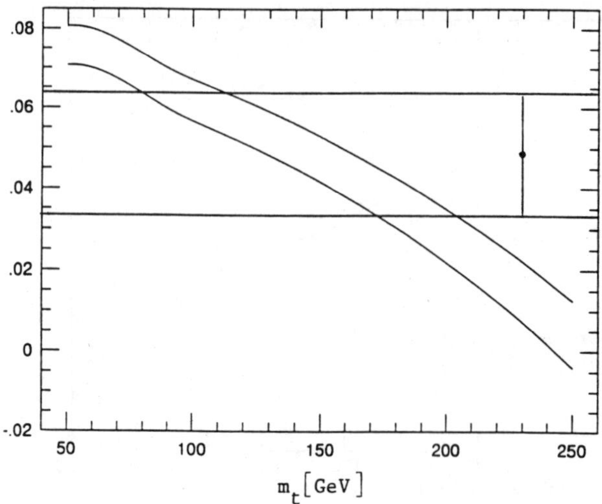

Fig. 7: *Δr as a function of the top mass for $M_H = 60, 1000$ GeV (lower, upper line). $M_Z = 91.187 \pm .007$ GeV. 1σ bounds with $s_W^2 = 0.2275 \pm 0.0052$ from combined UA2 and CDF results [2].*

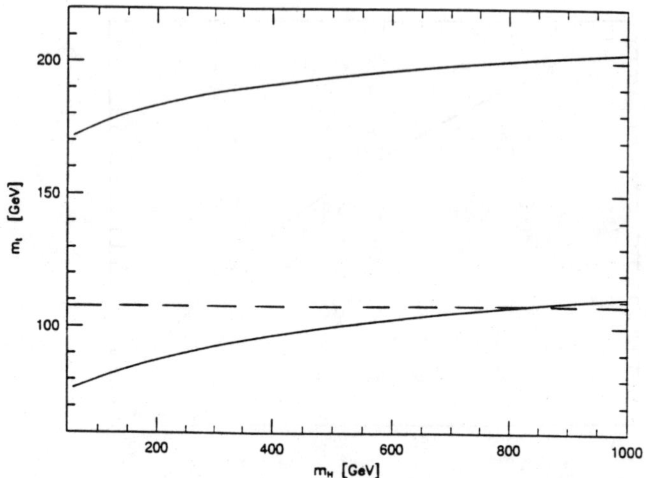

Fig. 8: *Sensitivity of the top mass bounds from Δr on the Higgs mass. The allowed m_t range is between the curves. The bound on m_t from the direct search is also indicated.*

size of $(\Delta r)_{remainder}$ and thus will provide some sensitivity also to the Higgs mass in case that the top quark would be discovered experimentally. For virtual Higgs effects, however, the observables from the Z resonance are more suitable.

The bounds on m_t, following from the experimental constraint

$$(\Delta r)_{exp} = 0.0489 \pm 0.0153$$

depend on the Higgs mass. This dependence is illustrated in Figure 8 . The allowed m_t range is further restricted by the bound [10, 11] from the direct search.

7 Renormalization schemes

In a Quantum Field Theory like the electroweak standard model the starting point for perturbative calculations is the Lagrangian with a set of free mass and coupling parameters. The general discussion of renormalization in Quantum Field Theory has shown that the freedom in parametrizing the theory can be used to introduce convenient renormalization constants, or counter terms, equivalently, and to absorb the divergences in the calculation of S-matrix elements or Green functions. It is also possible to deal with the bare parameters of the theory for relating measurable quantities to each other, but the bare parameters are cutoff dependent and hence have no simple physical interpretation.

A renormalization scheme is a choice of definite procedures for dealing with the parameters of the theory, together with the infinities from the loop amplitudes, in terms of measurable physical quantities. In a more general sense, it comprises the choice of the regularization procedure, the way of treating field renormalization, the specification of the gauge fixing terms and the respective FP ghost part, and a set of prescriptions how the formal parameters can be measured.

Before one can make predictions from the theory, a set of independent parameters has to be determined from experiment. This can either be done for the bare quantities or for renormalized parameters which have a simple physical interpretation. In a more restrictive sense, a renormalization scheme characterizes a specific choice of experimental data points to be used as input defining the basic parameters of the Lagrangian in terms of which the perturbative calculation of physical amplitudes is performed.

Predictions for the relations between physical quantities do not depend on the choice of a specific renormalization scheme if we perform the calculation to all orders in the perturbative expansion. Practical calculations, however, are obtained from truncated perturbation series, making the predictions depend on the chosen set of basic parameters and thus leading to a scheme dependence.

Differences between various schemes are formally of higher order than the one under consideration. To make this obvious, we consider a simplified model with only a single coupling paramater α. Calculation of a 1-loop amplitude for a process with the lowest order amplitude $M^{(0)} = \alpha^2 A_0$ yields

$$M^{(1)} = \alpha^2 A_0 \, [1 + b\alpha] \, .$$

In another scheme with α' different from α by a 1-loop term

$$\alpha' = \alpha \, [1 + a\alpha]$$

the result is

$$M'^{(1)} = \alpha'^2 A_0 \, [1 + b'\alpha'] \, .$$

After insertion of α', with $b' = b - 2a$, one gets

$$M'^{(1)} = M^{(1)} + \alpha^2 A_0 \, [3b' a \, \alpha^2 + O(\alpha^3)].$$

Without an explicit calculation of the $O(\alpha^2)$ correction the difference $M'^{(1)} - M^{(1)}$ has to be considered as an uncertainty. The study of the scheme dependence of the perturbative results, after improvement by resumming the leading terms, allows us to estimate the missing higher order contributions.

Parametrizations or 'renormalization schemes' frequently used in electroweak calculations are:

1. the on-shell (OS) scheme with

$$\alpha, \; M_W, \; M_Z, \; m_f, \; M_H$$

2. the G_μ scheme with the basic parameters

$$\alpha, \; G_\mu, \; M_Z, \; m_f, \; M_H$$

3. the low energy scheme with the mixing angle as a basic parameter defined in neutrino-electron scattering:

$$\alpha, \; G_\mu, \; \sin^2\theta_{\nu e}, \; m_f, \; M_H$$

4. the $*$ scheme where the bare parameters e_0, G_μ^0, s_0^2 are eliminated and replaced in terms of dressed running (k^2-dependent) parameters

$$e_*^2(k^2), \; G_{\mu *}(k^2), \; s_*^2(k^2); \; m_f, \; M_H$$

5. the $\overline{MS}$-scheme.

Some details on the $\overline{MS}$ scheme will be given in the next subsection, followed by a brief discussion of the other renormalization schemes.

7.1 *The $\overline{MS}$-scheme*

The modified minimal subtraction scheme ($\overline{MS}$-scheme) [32, 33, 34, 35] is one of the simplest ways to obtain finite 1-loop expressions by performing the substitution

$$\frac{2}{\epsilon} - \gamma + \log 4\pi + \log \mu^2 \;\rightarrow\; \log \mu^2_{\overline{MS}}$$

in the divergent parts of the loop integrals, Eq. (95). Formally, the $\overline{MS}$ self energies and vertex corrections are obtained by splitting the bare masses and couplings into $\overline{MS}$ parameters and counter terms

$$M_0^2 = \hat{M}^2 + \delta\hat{M}^2, \quad e_0 = \hat{e} + \delta\hat{e}, \tag{159}$$

where the counter terms together with field renormalization constants

$$1 + \delta\hat{Z}_i$$

are defined in such a way that they absorb the singular parts proportional to

$$\Delta = \frac{2}{\epsilon} - \gamma + \log 4\pi\,.$$

As a consequence, self energies and vertex corrections in the $\overline{MS}$-scheme depend on the arbitrary scale μ.

Perturbative calculations start from the Lagrangian with the formal $\overline{MS}$ parameters

$$\mathcal{L}(\hat{e}, \hat{M}_W, \hat{M}_Z, \ldots).$$

The $\overline{MS}$ parameters fulfill the same relations as the corresponding bare parameters. In particular, the mixing angle in the $\overline{MS}$-scheme, denoted by $\hat{s}^2$, can be expressed in terms of the $\overline{MS}$ masses of W and Z in the following way:

$$\hat{s}^2 = 1 - \frac{\hat{M}_W^2}{\hat{M}_Z^2}\,. \tag{160}$$

The relation of the $\overline{MS}$ parameters to the conventional OS-parameters is obtained by calculating the dressed vector boson propagators and the dressed electron-photon vertex in the Thomson limit in the $\overline{MS}$-scheme and identifying the poles with the OS masses and the electromagnetic coupling with the classical charge.

- The $\overline{MS}$ charge:

The $\overline{MS}$ analog of the OS charge renormalization condition Eq. (71) reads:

$$\hat{e}\left[1-\frac{1}{2}\,\Pi^{\gamma}_{\overline{MS}}(0)+\frac{\hat{s}}{\hat{c}}\,\frac{\Sigma^{\gamma Z}_{\overline{MS}}(0)}{\hat{M}_Z^2}\right]=e. \tag{161}$$

The l.h.s. is the coupling constant of the dressed electromagnetic vertex in the Thomson limit which has to be identified with the classical charge.
The $\overline{MS}$ self energies in Eq. (161) read explicitly:

$$\begin{aligned}
\Pi^{\gamma}_{\overline{MS}}(0) &= \frac{\hat{e}^2}{16\pi^2}\,A^{\gamma}(0),\\
A^{\gamma}(0) &= \frac{4}{3}\sum_f Q_f^2\,\log\frac{\mu^2}{m_f^2}+3\,\log\frac{M_W^2}{\mu^2}-\frac{2}{3},\\
\frac{\hat{s}}{\hat{c}}\,\frac{\Sigma^{\gamma Z}_{\overline{MS}}(0)}{\hat{M}_Z^2} &= -\frac{\hat{e}^2}{8\pi^2}\,\log\frac{M_W^2}{\mu^2}.
\end{aligned} \tag{162}$$

A natural scale for electroweak physics is given by $\mu = M_Z$. Hence, the correlation between e and $\hat{e}$ involves large logarithms from the light fermions which can be resummed according to the RGE (143). The bosonic terms are small. Resummation leads to the relation

$$e^2=\frac{\hat{e}^2}{1+\frac{\hat{e}^2}{16\pi^2}\left[A^{\gamma}(0)+4\,\log\frac{M_W^2}{\mu^2}\right]}. \tag{163}$$

Inverting this equation yields the $\overline{MS}$ charge expressed in terms of the OS charge

$$\hat{e}^2=\frac{e^2}{1-\frac{e^2}{16\pi^2}\left[A^{\gamma}(0)+4\,\log\frac{M_W^2}{\mu^2}\right]}. \tag{164}$$

Choosing $\mu = M_Z$ we can evaluate the expression in (165) to obtain the $\overline{MS}$ fine structure constant at the Z mass scale

$$\hat{\alpha}=\frac{\alpha}{1-\Delta\hat{\alpha}} \tag{165}$$

with the value

$$\Delta\hat{\alpha}=0.0684\pm0.0009-\frac{8\alpha}{9\pi}\log\frac{m_t}{M_Z}+\frac{\alpha}{2\pi}\left(\frac{7}{2}\log c_W^2-\frac{1}{3}\right). \tag{166}$$

The first term is due to the light fermions. It can be obtained from the quantity in Eq. (127) by adding the constant term

$$\frac{\alpha}{\pi}\left(\frac{5}{3}+\frac{55}{27}\,(1+\frac{\alpha_s}{\pi})\right).$$

The uncertainty in Eq. (166) is the hadronic uncertainty of $\Delta\alpha$ in Eq. (127).

$\hat{\alpha}$ has to be distinguished from the effective charge at the Z scale introduced in Eq. (128) which contains only the light fermion contributions. A heavy top quark decouples in $\Delta\alpha$ according to Eq. (122), but does not decouple in $\Delta\hat{\alpha}$. Numerically one finds

$$\begin{aligned} \alpha(M_Z^2)^{-1} &= 128.8 \pm 0.1 \\ (\hat{\alpha})^{-1} &= 127.8 \; - \; 128.0 \pm 0.1 \end{aligned} \tag{167}$$

The variation in $\hat{\alpha}$ in Eq. (167) corresponds to a top mass range from $m_t = 90$ GeV to 250 GeV.

- The $\overline{MS}$ masses:

The $\overline{MS}$ mass parameters $\hat{M}_W^2, \hat{M}_Z^2$ enter the corresponding transverse propagators together with the self energies as follows ($V = W, Z$):

$$D_V = \frac{1}{k^2 - \hat{M}_V^2 + \Sigma_{\overline{MS}}^{VV}(k^2)} \tag{168}$$

The OS-masses fulfill the pole conditions

$$M_V^2 - \hat{M}_V^2 + \mathrm{Re}\,\Sigma_{\overline{MS}}^{VV}(M_V^2) = 0 \tag{169}$$

yielding $\hat{M}_V^2$ expressed in terms of the OS-masses:

$$\hat{M}_V^2 = M_V^2 + \mathrm{Re}\,\Sigma_{\overline{MS}}^{VV}(M_V^2). \tag{170}$$

The mass parameters $\hat{M}_V^2$ are μ-dependent. We can choose $\mu = M_Z$ as the natural scale for electroweak calaculations, as done also for $\hat{\alpha}$.

The self energies $\Sigma_{\overline{MS}}$ are obtained from the expressions given in section 5.2 by dropping everywhere the singular term Δ and substituting

$$e \to \hat{e}, \quad s_W \to \hat{s}, \quad c_W \to \hat{c}$$

in the couplings, with $\hat{c}^2 = 1 - \hat{s}^2$. It is convenient to remove the overall normalization factors and to write for the real parts:

$$\begin{aligned} \mathrm{Re}\,\Sigma_{\overline{MS}}^{WW} &= \frac{\hat{e}^2}{\hat{s}^2} A_W(k^2), \\ \mathrm{Re}\,\Sigma_{\overline{MS}}^{ZZ} &= \frac{\hat{e}^2}{\hat{s}^2\hat{c}^2} A_Z(k^2). \end{aligned} \tag{171}$$

- The $\overline{MS}$ mixing angle:

The mixing angle $\hat{s}^2$ in the $\overline{MS}$-scheme, defined in Eq. (160), can be related to the OS mixing angle $s_W^2 = 1 - M_W^2/M_Z^2$ by substituting $\hat{M}_{W,Z}^2$ from Eq. (170), yielding

$$\hat{s}^2 = s_W^2 + c_W^2\, X_{\overline{MS}}, \quad \hat{c}^2 = c_W^2\,(1 - X_{\overline{MS}}) \tag{172}$$

with

$$X_{\overline{MS}} = \frac{\hat{e}^2}{\hat{s}^2}\left(\frac{A_W(M_W^2)}{M_W^2} - \frac{A_Z(M_Z^2)}{\hat{c}^2 M_Z^2}\right)\left(1 - \frac{\hat{e}^2}{\hat{s}^2}\,\frac{A_Z(M_Z^2)}{\hat{c}^2 M_Z^2}\right)^{-1}. \tag{173}$$

Making use of the property

$$X_{\overline{MS}} = \frac{\hat{e}^2}{\hat{s}^2}\,\frac{A_W(M_W^2)}{M_W^2} - (1 - X_{\overline{MS}})\,\frac{\hat{e}^2}{\hat{s}^2}\,\frac{A_Z(M_Z^2)}{\hat{c}^2 M_Z^2}$$

the relation (172) can simplified:

$$\hat{s}^2 = s_W^2 + \frac{\hat{e}^2}{\hat{s}^2}\,\frac{A_Z(M_Z^2) - A_W(M_W^2)}{M_Z^2}. \tag{174}$$

The leading 2-loop irreducible contributions are incorporated by adding in (174) the extra term $c_W^2\,\Delta\rho^{(2)}$ with $\Delta\rho^{(2)}$ from Eq. (156).

Eq. (174) determines $\hat{s}^2$ in terms of the OS parameters. $\hat{e}^2$ has to be taken from Eq. (164) or (165), respectively, for $\mu = M_Z$. Numerical values for $\hat{s}^2$ (with $\mu = M_Z$) are listed in table 3 together with the corresponding values for the OS counter part s_W^2. Note that there are slightly different definitions of $\hat{s}^2$ in the literature, depending on the treatment of the logarithmic m_t terms (e.g. [34] versus [66]).

One can obtain $\hat{s}^2$ also in a more direct way from the experimental data points α, G_μ, M_Z, without passing first through the OS-calculation, by deriving the effective Fermi constant in the $\overline{MS}$-scheme

$$\frac{G_\mu}{\sqrt{2}} = \frac{\hat{e}^2}{8\,\hat{s}^2\hat{c}^2\hat{\rho}\,M_Z^2}\cdot\frac{1}{1 - \Delta\hat{r}} \tag{175}$$

where

$$\begin{aligned} \Delta\hat{r} &= \frac{\hat{e}^2}{\hat{s}^2}\,\frac{A_W(0) - A_W(M_W^2)}{M_W^2} + \hat{\delta}_{VB}, \\ \hat{\delta}_{VB} &= \frac{\hat{\alpha}}{4\pi\hat{s}^2}\left[6 + \frac{7 - 5s_W^2 + \hat{s}^2(3c_W^2/\hat{c}^2 - 10)}{2s_W^2}\,\log c_W^2\right], \end{aligned} \tag{176}$$

m_t (GeV)	M_H (GeV)	s_W^2	$\hat{s}^2$
90	60	0.2312	0.2335
90	300	0.2331	0.2343
90	1000	0.2349	0.2350
120	60	0.2282	0.2329
120	300	0.2301	0.2338
120	1000	0.2319	0.2345
150	60	0.2250	0.2322
150	300	0.2270	0.2330
150	1000	0.2288	0.2337
180	60	0.2214	0.2312
180	300	0.2235	0.2321
180	1000	0.2254	0.2328
210	60	0.2173	0.2301
210	300	0.2196	0.2311
210	1000	0.2215	0.2318
240	60	0.2127	0.2289
240	300	0.2152	0.2299
240	1000	0.2173	0.2307

Table 3: The mixing angles s_W^2 and $\hat{s}^2$ in the on-shell and in the $\overline{MS}$-scheme for $M_Z = 91.187$ GeV and various top and Higgs masses.

together with

$$\begin{aligned} M_W^2 &= \hat{c}^2 \hat{\rho} M_Z^2, \\ \hat{\rho} &= \frac{1}{1 - X_{\overline{MS}}}. \end{aligned} \tag{177}$$

For given parameters $\alpha, G_\mu, M_Z, m_t, M_H$ the solution of this set of equations yields the quantities $\hat{s}^2, \hat{\rho}$ together with M_W. $\Delta\hat{r}$ is a small correction and has only a mild dependence on the top and Higgs masses. For the m_t, M_H range allowed in Figure 8 one has

$$\Delta\hat{r} = 0.0050 \pm 0.0034 \tag{178}$$

where the variation is due to the unknown mass parameters.

The term $\hat{\delta}_{VB}$ in $\Delta\hat{r}$ is the vertex and box correction to the muon decay amplitude in the $\overline{MS}$-scheme [34]. The given expression refers to a mixed $\overline{MS}$ - on-shell calculation of the loop diagrams where $\overline{MS}$-couplings are used but on-shell masses in the propagators. Numerically the differences to the corresponding expression exclusively with $\overline{MS}$ parameters is insignificant ($< 3 \cdot 10^{-4}$). The main difference to the on-shell quantity δ_{VB} in Eq. (137) (besides the parametrization) is the extra additive term

$$-\frac{\hat{\alpha}}{\pi}\log c_W^2 \equiv -\frac{\hat{\alpha}}{\pi}\log\frac{M_W^2}{\mu^2} \quad \text{for } \mu = M_Z$$

arising from the UV singularity in the sum of the diagrams.

The Standard Model prediction for $\hat{s}^2$ following from the mass range $m_t > 90$ GeV and $60\,\text{GeV} < M_H < 1000\,\text{GeV}$ together with the constraint from the experimental values for M_Z, M_W is given by

$$\hat{s}^2 = 0.2330 \pm 0.0016\,. \tag{179}$$

This includes the uncertainty induced by the hadronic vacuum polarization in Eq. (166) resp. (126).

The $\overline{MS}$ quantities $\hat{\alpha}$, $\hat{s}^2$ are formal parameters which have no simple relation to physical quantities. The interest in these parameters is based on two important features:

- They are universal, i.e. process independent, and take into account the universal large effects from fermion loops. Expressing the NC coupling constants for the Zff vertices in terms of $\hat{\alpha}$, $\hat{s}^2$ yields a good approximation to the complete results (134):

$$\begin{aligned}\sqrt{2}G_\mu M_Z^2\,\rho_f &= \frac{\hat{e}^2}{4\,\hat{s}^2\hat{c}^2}(1+\delta\hat{\rho}_f),\\ s_f^2 &= \hat{s}^2 + \delta\hat{s}_f^2\,.\end{aligned} \tag{180}$$

  The flavor dependent residual corrections $\delta\hat{\rho}_f$ and $\delta\hat{s}_f^2$ are small and practically independent of m_t and M_H. An exception is the Zbb vertex, where also non-universal large top terms are present [56].

- The knowledge of the values for $\hat{\alpha}$ and $\hat{s}^2$ at the Z scale allows the extrapolation of the SU(2) and U(1) couplings

$$\hat{\alpha}_1(\mu^2) = \frac{\hat{\alpha}(\mu^2)}{\hat{c}^2(\mu^2)}, \quad \hat{\alpha}_2(\mu^2) = \frac{\hat{\alpha}(\mu^2)}{\hat{s}^2(\mu^2)} \tag{181}$$

  to large mass scales and, together with the strong coupling constant $\alpha_s(\mu^2)$ in the $\overline{MS}$-scheme, to test scenarios of Grand Unification. In particular the minimal SU(5) model of Grand Unfication predicts with α and α_s as input [57]:

$$\hat{s}^2_{SU(5)}(M_Z^2) = 0.2102^{+0.0037}_{-0.0031}$$

  which is in disagreement with the result (179). Supersymmetric models of Grand Unification, however, are in favor [57, 58].

7.2 *Other renormalization schemes*

We briefly address the other renormalization schemes mentioned in the beginning of section 7. We restrict this discussion to parameter renormalization only.

- The G_μ-scheme:

The G_μ-scheme [30] with the parameters

$$\alpha,\ G_\mu,\ M_Z,\ m_f,\ M_H$$

treats G_μ as a basic parameter to be renormalized instead of the W mass. The counter terms, which appear in the bare quantities

$$e_0 = e + \delta e,\ \ M_Z^{0\,2} = M_Z^2 + \delta M_Z^2,\ \ G_\mu^0 = G_\mu + \delta G_\mu \tag{182}$$

after separating off the renormalized parameters, are determined by the on-shell conditions for e and M_Z as in the OS-scheme. The renormalization condition for G_μ, which replaces the on-shell condition for M_W, defines G_μ as the experimental Fermi constant, thus fixing the counter term δG_μ by the requirement of absorbing the 1-loop contribution to the μ-decay amplitude:

$$\frac{\delta G_\mu}{G_\mu} = -\frac{\Sigma^{WW}(0)}{M_W^2} - \frac{G_\mu M_W^2}{2\pi^2\sqrt{2}}\left[4\,(\Delta - \log\frac{M_W^2}{\mu^2}) + 6 + \frac{7-4s^2}{2s^2}\log c^2\right]. \tag{183}$$

The mixing angle is a derived quantity following from the exact relation between the bare quantities

$$s_0^2 c_0^2 = \frac{e_0^2}{4\sqrt{2}G_\mu^0 M_Z^{0\,2}} \tag{184}$$

by the one-loop expansion according to (182)

$$s^2 = \frac{1}{2}\left(1 - \sqrt{1 - \frac{4\pi\alpha}{\sqrt{2}M_Z^2 G_\mu}}\right) \tag{185}$$

with the counter term in the decomposition $s_0^2 = s^2 + \delta s^2$

$$\delta s^2 = \frac{c^2 s^2}{c^2 - s^2}\left(2\frac{\delta e}{e} - \frac{\delta M_Z^2}{M_Z^2} - \frac{\delta G_\mu}{G_\mu}\right). \tag{186}$$

The physical Wmass is obtained from the pole condition for the W-propagator as the solution of the equation

$$M_W^2 - m_W^2 - \delta M_W^2 + \mathrm{Re}\,\Sigma^{WW}(M_W^2) = 0 \tag{187}$$

with

$$\begin{aligned} m_W^2 &= \frac{M_Z^2}{2}\left(1+\sqrt{1-\frac{4\pi\alpha}{\sqrt{2}M_Z^2G_\mu}}\right), \\ \frac{\delta M_W^2}{M_W^2} &= \frac{s^2}{c^2-s^2}\left(\frac{\delta G_\mu}{G_\mu}-2\frac{\delta e}{e}+\frac{c^2}{s^2}\frac{\delta M_Z^2}{M_Z^2}\right). \end{aligned} \tag{188}$$

Because of the large effects associated with the renormalization of e, s^2 is a bad approximation for the mixing angle in the perturbative expansion. The improved mixing angle

$$\tilde{s}^2 = \frac{1}{2}\left(1-\sqrt{1-\frac{4\pi\alpha(M_Z^2)}{\sqrt{2}M_Z^2G_\mu}}\right) \tag{189}$$

with $\alpha(M_Z^2)$ from Eq. (128) includes the resummed large contribution from the light fermions and is hence a better starting point for perturbative calculations. Making use of $\tilde{s}^2$, one has simultaneously to subtract $\Delta\alpha$ from the charge renormalization counter term and replace in Eq. (187)

$$2\frac{\delta e}{e} \to 2\frac{\delta e}{e} - \Delta\alpha\,. \tag{190}$$

- The low energy scheme:

The scheme with $e, G_\mu, \sin^2\theta_{\nu e}$ [31, 32] exclusively deals with parameters related to low energy experiments. The mixing angle $\sin^2\theta_{\nu e} = s_{\nu e}^2$ is treated as a fundamental parameter determined from ν-e scattering in terms of the ratio

$$R = \frac{\sigma(\nu_\mu e \to \nu_\mu e)}{\sigma(\bar{\nu_\mu} e \to \bar{\nu_\mu} e)} = \frac{(1-4s_{\nu e}^2)^2+(1-4s_{\nu e}^2)+1}{(1-4s_{\nu e}^2)^2-(1-4s_{\nu e}^2)+1}. \tag{191}$$

The renormalization of e and G_μ is the same as in the G_μ-scheme. For renormalizing $s_{\nu e}^2$, the counter term in $s_0^2 = s_{\nu e}^2 + \delta s_{\nu e}^2$ is fixed by the condition that $\delta s_{\nu e}^2$ absorbs the 1-loop contribution $\delta R^{(1)}$ to the ratio R:

$$R(s_{\nu e}^2 + \delta s_{\nu e}^2) + \delta R^{(1)} = R(s_{\nu e}^2)\,. \tag{192}$$

Taking the experimental result R_{exp} yields a numerical value for $s_{\nu e}^2$.

Both vector boson masses are derived quantities following from the pole conditions for W and Z:

$$\begin{aligned} M_W^2 - m_W^2 - \delta M_W^2 + \mathrm{Re}\,\Sigma^{WW}(M_W^2) &= 0, \\ M_Z^2 - m_Z^2 - \delta M_Z^2 + \mathrm{Re}\,\Sigma^{ZZ}(M_Z^2) &= 0 \end{aligned} \tag{193}$$

with

$$m_W^2 = \frac{\pi\alpha}{\sqrt{2}G_\mu s_{\nu e}^2}, \quad m_Z^2 = \frac{\pi\alpha}{\sqrt{2}G_\mu s_{\nu e}^2 c_{\nu e}^2} \tag{194}$$

and

$$\begin{aligned}\frac{\delta M_W^2}{M_W^2} &= 2\frac{\delta e}{e} - \frac{\delta G_\mu}{G_\mu} - \frac{\delta s_{\nu e}^2}{s_{\nu e}^2}, \\ \frac{\delta M_Z^2}{M_Z^2} &= 2\frac{\delta e}{e} - \frac{\delta G_\mu}{G_\mu} - \frac{c_{\nu e}^2 - s_{\nu e}^2}{c_{\nu e}^2}\frac{\delta s_{\nu e}^2}{s_{\nu e}^2}.\end{aligned} \tag{195}$$

A slightly modified version of this scheme was used in [32]. There the condition (192) was imposed in the $\overline{MS}$ renormalization prescription

$$R(\hat{s}_{\nu e}^2) + \delta\hat{R}^{(1)} = R_{exp}$$

yielding the $\overline{MS}$- version $\hat{s}_{\nu e}^2$ of the low energy mixing angle.

- The $*$ scheme:

The bare parameters of the Standard Model can be eliminated in a formally different way [36, 37] by introducing a set of 4 effective parameters

$$e_*(s),\ s_*^2(s),\ G_{\mu*}(s),\ \rho_*(s), \tag{196}$$

where in the minimal model only three are independent. These running parameters ($s = k^2$) contain the real parts of the self energies. They are arranged in such a way that the amplitude for a 4-fermion process with self energy corrections is obtained from the Born amplitude by the formal replacement

$$(e,\ s_W^2,\ G_\mu,\ \rho) \to (e_*,\ s_*^2,\ G_{\mu*},\ \rho_*), \tag{197}$$

supplemented by the corresponding imaginary parts. When the physical input is taken from the experimental data points α, G_μ, M_Z, the result for the 4-fermion scattering amplitudes is identical to that of the conventional on-shell scheme with self energy corrections after the 2-loop 1-particle irreducible leading contributions are built in.

In the following we give the relation between the conventional expressions of section 3.3 and the corresponding ones in terms of the $*$ -parameters. For a more detailed dicussion of the propagator corrections in the on-shell scheme we refer to the section on e^+e^- annihilation.

$$\begin{array}{rcl} \text{on-shell} & \leftrightarrow & * \\[4pt] \dfrac{e^2}{1+\mathrm{Re}\,\hat{\Pi}^{\gamma}(s)} & \leftrightarrow & e_*^2(s) \\[4pt] s_W^2 - s_W c_W \,\mathrm{Re}\,\dfrac{\hat{\Pi}^{\gamma Z}(s)}{1+\hat{\Pi}^{\gamma}(s)} & \leftrightarrow & s_*^2(s) \\[4pt] \dfrac{e^2}{s_W^2}\cdot\dfrac{1}{s-M_W^2+\hat{\Sigma}^{WW}(s)} & \leftrightarrow & \dfrac{e_*^2}{s_*^2}\cdot\dfrac{1}{s-\frac{e_*^2}{s_*^2}\frac{1}{4\sqrt{2}G_{\mu*}}+i\sqrt{s}\,\Gamma_{*W}(s)} \\[4pt] \dfrac{e^2}{s_W^2 c_W^2}\cdot\dfrac{1}{s-M_Z^2+\hat{\Sigma}^{Z}(s)} & \leftrightarrow & \dfrac{e_*^2}{s_*^2 c_*^2}\cdot\dfrac{1}{s-\frac{e_*^2}{s_*^2 c_*^2}\frac{1}{4\sqrt{2}G_{\mu*}\rho_*}+i\sqrt{s}\,\Gamma_{*Z}(s)} \end{array} \tag{198}$$

The quantities $\Gamma_{*Z}(s)$, $\Gamma_{*W}(s)$ correspond to the imaginary parts of the Z and W self energies. The relation to the physical Z width (and similar for W) is given by

$$\Gamma_Z = \frac{\Gamma_{*Z}(M_Z^2)+\Delta\Gamma_Z}{1+\kappa_*}$$

where $\Delta\Gamma_Z$ denotes the corrections to the Z width in $O(\alpha^2)$ not of the self energy type (vertex, QED and QCD corrections), discussed in the next chapter, and κ_* is determined by the residue of the Z propagator in (198):

$$s - \frac{e_*^2}{s_*^2 c_*^2}\,\frac{1}{4\sqrt{2}G_{\mu*}\rho_*} = (s-M_Z^2)\cdot(1+\kappa_*)\,.$$

The zero of the l.h.s. corresponds to the physical Z mass.

The * star arrangement as well as the on-shell one with resummation of the self energies contain higher order terms which are in general not gauge invariant. The leading terms, however, arise from light and heavy fermions which belong to the gauge invariant subclass of fermion loops, and the resummation yields the reducible higher order terms to all orders. The bosonic loop contributions on the other hand give gauge invariant results only when they are combined with vertex and box diagrams of the same order in a physical matrix element. They have always to be understood as expanded to one-loop order when appearing in formally higher order expressions. In the 't Hooft-Feynman gauge the numerical differences are irrelevant; in the unitary gauge, however, the individual contributions become divergent.

7.3 *Uncertainties of theoretical predictions*

In order to establish in a significant manner possibly small effects from unknown physics we have to know the uncertainties of our theoretical predictions which have to be confronted with the experiments.

The sources of uncertainties in theoretical predictions are the following:

- the experimental errors of the parameters used as an input. With the choice α, G_μ, and M_Z from LEP we can keep these errors as small as possible. The errors from this source are then determined by δM_Z since the errors of α and G_μ are negligibly small. For any of the mixing angles with $s_W^2, \hat{s}^2, s_f^2$

$$\frac{\delta s^2}{s^2} = \frac{2\,c^2}{c^2 - s^2}\,\frac{\delta M_Z}{M_Z} \tag{199}$$

  one finds

$$\delta s^2 \simeq 5 \cdot 10^{-5}\,.$$

- the uncertainties from quark loop contributions to the radiative corrections . Here, we have to distinguish two cases: the uncertainties from the light quark contributions to $\Delta\alpha$ and the uncertainties from the heavy quark contributions to $\Delta\rho$. In both cases the uncertainties are due to strong interaction effects, which are not sufficiently under control theoretically. The problems are due to:
  (i) the QCD parameters. The scale of α_s and the definition and scale of quark masses to be used in the calculation of a particular quantity are quite ambiguous in many cases.
  (ii) the bad convergence and/or breakdown of perturbative QCD. In particular at low q^2 and in the resonance regions theoretically poorly known nonperturbative effects are non-negligible.

  The theoretical problems with the hadronic contributions of the 5 known light quarks to $\Delta\alpha$ can be circumvented by using the experimental e^+e^--annihilation cross-section $\sigma_{tot}(e^+e^- \to \gamma^* \to hadrons)$. The error [44]

$$\delta(\Delta\alpha) = \pm 0.0009$$

  is dominated by the large experimental errors in the continuum contributions to $\sigma_{tot}(e^+e^- \to \gamma^* \to hadrons)$ below the Υ threshold, and can be improved only by more precise measurements of hadron production in e^+e^--annihilation in the corresponding low energy region. This uncertainty leads to an error in the W-mass prediction

$$\frac{\delta M_W}{M_W} = \frac{s_W^2}{c_W^2 - s_W^2}\,\frac{\delta(\Delta r)}{2(1-\Delta r)}$$

of $\delta M_W = 17$ MeV and $\delta \sin^2\theta = 0.0003$ in the prediction of the various weak mixing parameters $s_W^2, \hat{s}^2, s_f^2$.

The contribution to $\Delta\rho$ from quark doublets with large mass splitting exhibits large QCD corrections of the weak current quark loops. For a heavy top one finds

$$\Delta\rho = \frac{\sqrt{2}G_\mu}{16\pi^2} 3m_t^2 K_{QCD} + \cdots$$

with

$$K_{QCD} = 1 - \frac{2\pi^2+6}{9}\frac{\alpha_s}{\pi}$$

for asymptotically large m_t [52]. The corrections obtained are not well determined numerically because it remains unclear which scale should be chosen for α_s. Also the ambiguity in the definition of m_t has not been taken into account.
Again, the problem can be controlled better by using dispersion relations. In this approach, the remaining uncertainties in Δr

$$\delta(\Delta r)_{QCD} \simeq \begin{cases} 0.0005 & m_t < 150GeV \\ 0.0015 \cdot (m_t/250GeV)^2 & m_t > 150GeV \end{cases} \tag{200}$$

have been estimated in [59]. In the heavy top region, where the errors of Δr and $\Delta\rho$ are correlated by $\delta(\Delta r) \simeq c_W^2/s_W^2 \delta(\Delta\rho)$, the uncertainties in the NC couplings in Eq. (134) can be estimated in terms of $\delta\Delta\rho$. The error of the normalization turns out to be smaller than $5 \cdot 10^{-4}$, and for the mixing angle one finds

$$\delta s^2 < 0.00015\,.$$

- the uncertainties from omission of higher order effects. After resummation of the leading terms, how large are the omitted higher order effects? Since a complete two-loop calculation has not been done, we only can guess how large such effects could be. In the calculation of Δr the difference is given, in the approximation we consider, by using different parameters in the evaluation of $\Delta r_{remainder}$ defined in Eq. (140). A supposedly conservative estimate of the error made by omitting the higher order effects has been given in [60].

$$\delta(\Delta r)_{higher-order} = \pm 0.001 \tag{201}$$

which can be added quadratically to the hadronic errors. Explicit comparisons between OS and $\overline{MS}$ calculations [35] as well as between different versions of the OS scheme [61] for the Z resonance observables have shown to be well below the experimental uncertainties. The typical size of the theoretical uncertainty of improved one-loop calculations is thus around 0.001.

8 Extension to larger theories

We want to conclude this chapter with an outlook on renormalizable generalizations of the minimal model and their effect on electroweak observables. Extended models can be classified in terms of the following categories:

(i) extensions within the minimal gauge group SU(2)×U(1) with $\rho_{tree} = 1$

(ii) extensions within SU(2)×U(1) with $\rho_{tree} \neq 1$

(iii) extensions with larger gauge groups SU(2)×U(1)×G and respective extra gauge bosons.

Extensions of the class (i) are, for example, models with additional (sequential) fermion doublets, more Higgs doublets, and the minimal supersymmetric version of the Standard Model.

8.1 *Parametrization of self energy corrections*

If "new physics" would be present in the form of new particles which couple to the gauge bosons but not directly to the external fermions in a 4-fermion process, the formulae for the self energies in section 3.3 are general enough that those effects can be built in by calculating the additonal loop diagrams.

In order to have a description which is as far as possible independent of the special type of extra heavy particles, it is convenient to introduce a parametrization of the radiative corrections from the vector boson self-energies in terms of the static ρ-parameter

$$\Delta\rho(0) = \frac{\Sigma^{ZZ}(0)}{M_Z^2} - \frac{\Sigma^{WW}(0)}{M_W^2} - 2\frac{s_W}{c_W}\frac{\Sigma^{\gamma Z}(0)}{M_Z^2} \tag{202}$$

and the combinations

$$\begin{aligned} \Delta_1 &= \frac{1}{s_W}\Pi^{3\gamma}(M_Z^2) - \Pi^{33}(M_Z^2) \\ \Delta_2 &= \Pi^{33}(M_Z^2) - \Pi^{WW}(M_W^2) \\ \Delta\alpha &= \Pi^{\gamma\gamma}(0) - \Pi^{\gamma\gamma}(M_Z^2)\,. \end{aligned} \tag{203}$$

The quantities in Eq. (203) are the isospin components of the self-energies

$$\begin{aligned} \Sigma^{\gamma Z} &= -\frac{1}{c_W}\left(\Sigma^{3\gamma} - s_W^2\Sigma^{\gamma\gamma}\right) \\ \Sigma^{ZZ} &= \frac{1}{c_W^2}\left(\Sigma^{33} - 2s_W\Sigma^{3\gamma} + s_W^2\Sigma^{\gamma\gamma}\right) \end{aligned} \tag{204}$$

in the expansions

$$\mathrm{Re}\,\Sigma^{ij}(k^2) = \Sigma^{ij}(0) + k^2\Pi^{ij}(k^2)\,. \tag{205}$$

The Δ-notation above has been introduced in [62]. Several other conventions are used in the literature:

- The S,T,U parameters of [63] are related to (203) by

$$S = \frac{4s_W^2}{\alpha}\Delta_1,\;\; T = \frac{1}{\alpha}\Delta\rho(0),\;\; U = \frac{4s_W^2}{\alpha}\Delta_2, \tag{206}$$

- the ϵ-parameters of [64] by

$$\epsilon_1 = \Delta\rho,\;\; \epsilon_2 = -\Delta_2,\;\; \epsilon_3 = \Delta_1, \tag{207}$$

- the h-parameters of [65] by

$$h_V = \frac{1}{\alpha}\Delta\rho(0),\;\; h_{AZ} = \frac{4\pi}{\sqrt{2}G_\mu M_W^2}\Delta_1,\;\; h_{AW} = h_{AZ} + \frac{4\pi}{\sqrt{2}G_\mu M_W^2}\Delta_2, \tag{208}$$

- and the parameters of [67] by

$$\Delta_\rho(0) = \frac{1}{4\sqrt{2}G_\mu}\Delta\rho,\;\; \Delta_3 = -\frac{1}{4\sqrt{2}G_\mu c_W^2}\Delta_1,\;\; \Delta_\pm = c_W^2\Delta_3 - \frac{1}{4\sqrt{2}G_\mu}\Delta_2. \tag{209}$$

The combinations (203) of self energies contribute in a universal way to the electroweak parameters (the residual corrections not from self-energies are dropped since they are identical to the Standard Model ones):

1. the $M_W - M_Z$ correlation in terms of Δr:

$$\Delta r = \Delta\alpha - \frac{c_W^2}{s_W^2}\Delta\rho(0) - \frac{c_W^2 - s_W^2}{s_W^2}\Delta_2 + 2\Delta_1 \tag{210}$$

2. the normalization of the NC couplings at M_Z^2

$$\Delta\rho_f = \Delta\rho(0) + \Delta_Z \tag{211}$$

   where the extra quantity

$$\Delta_Z = M_Z^2\,\frac{d\,\Pi^{ZZ}}{dk^2}(M_Z^2)$$

   in (211) is from the residue of the Z propagator at the peak. Heavy particles decouple from Δ_Z.

3. the effective mixing angles

$$s_f^2 = (1+\Delta\kappa')\,\tilde{s}^2, \quad \tilde{s}^2 = \frac{1}{2}\left(1-\sqrt{1-\frac{4\pi\alpha(M_Z^2)}{\sqrt{2}G_\mu M_Z^2}}\right), \tag{212}$$

with

$$\Delta\kappa' = -\frac{c_W^2}{c_W^2-s_W^2}\Delta\rho(0) + \frac{\Delta_1}{c_W^2-s_W^2}\,. \tag{213}$$

The finite combinations of self energies (202) and (203) are of practical interest since they can be extracted from precision data in a fairly model independent way. An experimental observable particular sensitive to Δ_1 is the weak charge Q_W which determines the atomic parity violation in Cesium [66]

$$Q_W = -73.20 \pm 0.13 - 0.82\Delta\rho(0) - 102\Delta_1 \tag{214}$$

being almost independent of $\Delta\rho(0)$.

The theoretical interest in the Δ's is based on their selective sensitivity to different kinds of new physics.

- $\Delta\alpha$ gets contributions only from light charged particles whereas heavy objects decouple.

- $\Delta\rho(0)$ is a measure of the violation of the custodial $SU(2)$ symmetry. It is sensitive to particles with large mass splittings in multiplets. As an example, we have already encountered fermion doublets with different masses, see Eq. (129). Another example are the Higgs bosons of a 2-Higgs doublet model [69, 70, 71, 72] with masses M_{H^+}, M_h, M_H, M_A and mixing angles β, α for the charged $H^\pm$ and the neutral h^0, H^0, A^0 Higgs bosons, yielding

$$\begin{aligned}\Delta\rho(0) &= \frac{G_\mu}{8\pi^2\sqrt{2}}\Big[\sin^2(\alpha-\beta)F(M_{H^+}^2, M_A^2, M_H^2) \\ &\quad + \cos^2(\alpha-\beta)F(M_{H^+}^2, M_A^2, M_h^2)\Big]\end{aligned} \tag{215}$$

with

$$F(x,y,z) = x + \frac{yz}{y-z}\log\frac{y}{z} - \frac{xy}{x-y}\log\frac{x}{y} - \frac{xz}{x-z}\log\frac{x}{z}\,.$$

For either $M_{H^+} \gg M_{neutral}$ or vice versa one finds a positive contribution

$$\Delta\rho(0) \simeq \frac{G_\mu M_{H^+}^2}{8\pi^2\sqrt{2}} \quad \text{or} \quad \frac{G_\mu M_{neutral}^2}{8\pi^2\sqrt{2}} > 0\,. \tag{216}$$

Also a negative contribution

$$\Delta\rho(0) < 0 \quad \text{for} \quad M_{h,H} < M_{H^+} < M_A \quad \text{and} \quad M_A < M_{H^+} < M_{h,H}$$

is possible in the unconstrained 2-doublet model.

- Δ_1 is sensitive to chiral symmetry breaking by masses. In particular, a doublet of mass degenerate heavy fermions yields a contribution

$$\Delta_1 = N_C^f \frac{G_\mu M_W^2}{12\pi^2\sqrt{2}}, \tag{217}$$

whereas the contribution of degenerate heavy fermions to $\Delta\rho(0)$ is zero. Hence, Δ_1 can directly count the number N_{deg} of mass degenerate fermion doublets:

$$\Delta_1^f = 4.5 \cdot 10^{-4} \cdot N_{deg}.$$

Δ_1 also gets sizeable contributions from models with a large number of additional fermions like in technicolor models. For example, $\Delta_1 \simeq 0.017$ for $N_{TC} = 4$ and one family of technifermions [63, 68].

8.2 *Models with* $\rho_{tree} \neq 1$

One of the basic relations of the minimal Standard Model is the tree level correlation between the vector boson masses and the electroweak mixing angle

$$\rho_{tree} = \frac{M_W^2}{M_Z^2 \sin^2\theta_W} = 1\,.$$

Many extensions of the minimal model, like those discussed in the previous section, preserve this feature.

The formulation of the electroweak theory in terms of a local gauge theory requires at least a single scalar doublet for breaking the electroweak symmetry SU(2)×U(1) → U(1)$_{em}$. In contrast to the fermion and vector boson part, very little is known empirically about the scalar sector. Without the assumption of minimality, quite a lot of options are at our disposal, including more complicated multiplets of Higgs fields. In general models the tree level ρ-parameter $\rho_{tree} = \rho_0$ is determined by

$$\rho_0 = \frac{\sum_i v_i^2[I_i(I_i+1) - I_{3i}^2]}{2\sum_i v_i^2 I_{3i}^2}$$

where v_i, I_{3i} are the vacuum expectation values and third isospin component of the neutral component of the i-th Higgs multiplet in the representation with isospin I_i. The presence of at least a triplet of Higgs fields gives rise to $\rho_0 \neq 1$. As a consequence, the tree level relations between the electroweak parameters have to be generalized according to

$$\sin^2\theta_W \to s_\theta^2 = 1 - \frac{M_W^2}{\rho_0 M_Z^2} \tag{218}$$

and

$$\frac{G_\mu}{\sqrt{2}} = \frac{e^2}{8s_\theta^2 M_W^2} = \frac{e^2}{8s_\theta^2 c_\theta^2 \rho_0 M_Z^2} \tag{219}$$

Writing $\rho_0 = (1-\Delta\rho_0)^{-1}$, we obtain for the mixing angle:

$$s_\theta^2 = 1 - \frac{M_W^2}{M_Z^2} + \frac{M_W^2}{M_Z^2}\Delta\rho \equiv s_W^2 + c_W^2 \Delta\rho_0\,, \tag{220}$$

for the overall normalization factor in the NC vertex:

$$\frac{e}{2s_\theta c_\theta} = \left(\sqrt{2} G_\mu M_Z^2 \rho_0\right)^{1/2}, \tag{221}$$

and for the $M_W - M_Z$ interdependence:

$$M_W^2\left(1 - \frac{M_W^2}{\rho_0 M_Z^2}\right) = \frac{e^2}{4\sqrt{2}G_\mu}, \tag{222}$$

in complete analogy to what we have found from the top quark loops.

At the level of radiative corrections, a small $\Delta\rho_0$ may be included by

$$\Delta r \to \Delta r - \frac{c_W^2}{s_W^2}\Delta\rho_0 \tag{223}$$

for the M_W-M_Z correlation, and

$$\rho_f \to \rho_f + \Delta\rho_0, \quad s_f^2 \to s_f^2 + c_W^2 \Delta\rho_0 \tag{224}$$

for the normalization and the effective mixing angles of the Zff couplings.

A complete discussion of radiative corrections requires not only the calculation of the extra loop diagrams from the non-standard Higgs sector but also an extension of the renormalization procedure [73, 74]. Since M_W, M_Z and $\sin^2\theta_W$ (or ρ_0, eqivalently) are now independent parameters, one extra renormalization condition is required. A natural condition would be to define the mixing angle for electrons s_e^2 in terms of the ratio of the dressed coupling constants at the Z peak

$$\frac{g_V^e}{g_A^e} =: 1 - 4s_e^2$$

which is measurable in terms of the left-right or the forward-backward asymmetries. This fixes the counter term for s_e^2 by

$$\frac{\delta s_e^2}{s_e^2} = \frac{c_e}{s_e}\frac{\mathrm{Re}\,\Sigma^{\gamma Z}(M_Z^2)}{M_Z^2} + 2\frac{c_e}{s_e}\frac{\Sigma^{\gamma Z}(0)}{M_Z^2} + \Delta\kappa_e \tag{225}$$

with the finite part $\Delta\kappa_e$ of the electron-Z vertex correction. The counter terms for the other parameters α, M_Z are treated as usual. With this input, we obtain a renormalized ρ-parameter and the corresponding counter term for the bare ρ-parameter $\rho_0^b = \rho + \delta\rho$ as follows:

$$\begin{aligned} \rho &= \frac{M_W^2}{M_Z^2 c_e^2}, \\ \frac{\delta\rho}{\rho} &= \frac{\delta M_W^2}{M_W^2} - \frac{\delta M_Z^2}{M_Z^2} + \frac{\delta s_e^2}{c_e^2}. \end{aligned} \tag{226}$$

Other derived quantities are:

- The relation between M_W and G_μ:

$$M_W^2 = \frac{\pi\alpha}{\sqrt{2}G_\mu s_e^2} \cdot \frac{1}{1-\Delta r} \tag{227}$$

with

$$\Delta r = \frac{\Sigma^{WW}(0) - \delta M_W^2}{M_W^2} + \Pi^\gamma(0) - \frac{\delta s_e^2}{s_e^2} + 2\frac{c_e}{s_e}\frac{\Sigma^{\gamma Z}(0)}{M_Z^2} + \delta_{VB}\,. \tag{228}$$

- The normalization of the Zff couplings at 1-loop:

$$\begin{aligned} &\frac{e^2}{4s_e^2c_e^2}\left[1 + \Pi^\gamma(0) - \frac{c_e^2 - s_e^2}{c_e^2}\frac{\delta s_e^2}{s_e^2} + 2\frac{c_e^2 - s_e^2}{c_e s_e}\frac{\Sigma^{\gamma Z}(0)}{M_Z^2} + \Delta\rho_f\right] \qquad (229)\\ = \;& \sqrt{2}G_\mu M_Z^2\rho\left[1 - \frac{\Sigma^{WW}(0) - \delta M_W^2}{M_W^2} + \frac{\delta s_e^2}{c_e^2} - 2\frac{s_e}{c_e}\frac{\Sigma^{\gamma Z}(0)}{M_Z^2} - \delta_{VB} + \Delta\rho_f\right] \end{aligned}$$

where $\Delta\rho_f$ denotes the finite part of the Zff vertex correction.

- The effective mixing angles of the Zff couplings:

$$s_f^2 = s_e^2(1 - \Delta\kappa_e + \Delta\kappa_f).$$

These relations predict the Z boson couplings, M_W and ρ in terms of the data points $\alpha, G_\mu, M_Z, s_e^2$. By this procedure, the leading m_t^2-dependence of the self energy corrections to theoretical predictions is absorbed into the renormalized ρ-parameter, leaving a $\sim \log m_t/M_Z$ term as an observable effect. For the Zbb-vertex, an additional m_t^2 dependence is found in the non-universal vertex corrections $\Delta\rho_b$ and $\Delta\kappa_b$. This makes observables containing this vertex the most sensitive top indicators in the class of models with $\rho_{tree} \neq 1$.

In the minimal Standard Model, the quantity equivalent to (226) can be calculated in terms of the data points α, G_μ, M_Z and the parameters m_t, M_H. With the experimental constraints from M_W in section 5.3 and and $s_e^2 = 0.2328 \pm 0.0007$ from LEP data [1, 4] we obtain

$$\rho_{SM} = 1.0069 \pm 0.0040 \,. \tag{230}$$

In the extended models we can calculate ρ from

$$\rho = \frac{\pi\alpha}{\sqrt{2}G_\mu M_Z^2 s_e^2 c_e^2} \cdot \frac{1}{1-\Delta r} \tag{231}$$

in terms of the input data $\alpha, G_\mu, M_Z, s_e^2$ together with m_t and the parameters of the Higgs sector. Such a complete calculation, however, does not exist as yet. Instead, we can get a value for ρ from directly using the data on M_W^2/M_Z^2 and $s_e^2 = 0.2324 \pm 0.0011$ from forward-backward asymmetries at LEP [1, 4] yielding

$$\rho = 1.0064 \pm 0.0069 \,. \tag{232}$$

The difference $\rho - \rho_{SM}$ can be interpreted as a measure for a deviating tree level structure. The data imply that it is compatible with zero.

8.3 *Extra Z bosons*

The existence of additional vector bosons is predicted by GUT models based on groups bigger than SU(5), like E_6 and $SO(10)$, by models with symmetry breaking in terms of a strongly interacting sector, and composite scenarios. Typical examples of extended gauge symmetries are the $SU(2) \times U(1) \times U(1)_{\chi,\psi,\eta}$ models following from E_6 unification, or LR-symmetric models. In the following we consider only models with an extra $U(1)$.

The mixing between the mathematical states Z_0 of the minimal gauge group and Z_0' of an extra hypercharge form the physical mass eigenstates Z, Z', where the lighter Z is identified with the resonance at LEP. The mass eigenstates are obtained by a rotation

$$\begin{aligned} Z &= \cos\theta_M \, Z_0 + \sin\theta_M \, Z_0' \\ Z' &= -\sin\theta_M \, Z_0 + \cos\theta_M \, Z_0' \end{aligned} \tag{233}$$

with a mixing angle θ_M related to the mass eigenvalues by

$$\tan^2\theta_M = \frac{M_{Z_0}^2 - M_Z^2}{M_{Z'}^2 - M_{Z_0}^2}, \quad M_{Z_0}^2 = \cos^2\theta_M \, M_Z^2 + \sin^2\theta_M \, M_{Z'}^2 \,. \tag{234}$$

$M_{Z_0}^2$ denotes the nominal mass of Z_0. In constrained models with the Higgs fields in doublets and singlets only, the usual Standard Model relation holds

$$\sin^2\theta_W = 1 - \frac{M_W^2}{M_{Z_0}^2}$$

between the masses and the mixing angle in the Lagrangian

$$\mathcal{L}_{NC} = \frac{g_2}{\cos\theta_W} J^\mu_{Z_0} Z^\mu_0 + g' J^\mu_{Z'_0} Z'^\mu_0 \tag{235}$$

with

$$J^\mu_{Z_0} = J^\mu_L - \sin^2\theta_W J^\mu_{em}.$$

It is convenient to introduce the quantity

$$s^2_W = 1 - \frac{M^2_W}{M^2_Z}, \quad c^2_W = 1 - s^2_W \tag{236}$$

with the physical mass of the lower eigenstate. For small mixing angles θ_M we have the following relation:

$$\sin^2\theta_W = s^2_W + c^2_W \Delta\rho_{Z'} \tag{237}$$

with

$$\Delta\rho_{Z'} = \sin^2\theta_M \left(\frac{M^2_{Z'}}{M^2_Z} - 1\right). \tag{238}$$

The W mass is obtained from

$$M^2_W = \frac{\pi\alpha}{\sqrt{2}G_\mu \sin^2\theta_W(1-\Delta r)}$$

after the substitution (237):

$$M^2_W = \frac{M^2_Z}{2}\left(1 + \sqrt{1 - \frac{\pi\alpha}{\sqrt{2}G_\mu M^2_Z \rho_{Z'}(1-\Delta r)}}\right) \tag{239}$$

with $\rho_{Z'} = (1 - \Delta\rho_{Z'})^{-1}$. Formally, $\rho_{Z'}$ appears as a non-standard tree level ρ-parameter. In all present practical applications the radiative correction Δr was approximated by the standard model correction.

The mass mixing has two implications for the NC couplings of the Z boson:

- $\Delta\rho_{Z'}$ contributes to the overall normalization by a factor

$$\rho^{1/2}_{Z'} \simeq 1 + \frac{1}{2}\Delta\rho_{Z'}$$

  and to the mixing angle by a shift

$$s^2_W \to s^2_W + c^2_W \Delta\rho_{Z'}.$$

  Both effects are universal, parametrized by $M_{Z'}$ and the mixing angle θ_M in a model independent way,

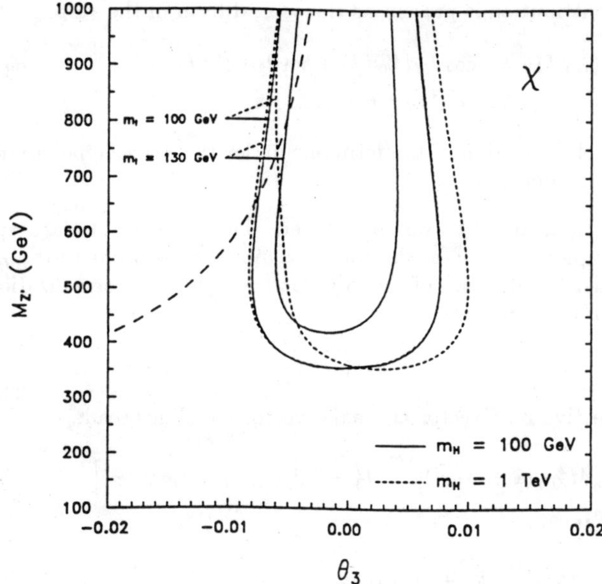

Fig. 9: *90% C.L. contours for mass and mixing angle ($\theta_3 = \theta_M$) of the extra Z' in the $SU(2) \times U(1) \times U(1)_\chi$ model, from [76]*

- A non-universal contribution is present as the second term in the vertex

$$\begin{aligned}(Zff) &= \cos\theta_M(Z_0ff)+\sin\theta_M(Z_0'ff)\\ &\simeq (Z_0ff)+\theta_M(Z_0'ff)\,.\end{aligned}$$

  It depends on the classification of the fermions under the extra hypercharge and is strongly model dependent.

Complete 1-loop calculations are not available as yet. The present standard approach consists in the implementation of the standard model corrections to the Z_0 parts of the coupling constants in terms of the form factors ρ_f for the normalization and κ_f for the effective mixing angles

$$s_W^2 \to s_f^2 = \kappa_f s_W^2.$$

In this approach the effective Zff vector and axial vector couplings read:

$$\begin{aligned}v_Z^f &= \left[\sqrt{2}G_\mu M_Z^2\rho_f(1+\Delta\rho_{Z'})\right]^{1/2}\left[I_3^f-2Q_f(\kappa_f s_W^2+c_W^2\Delta\rho_{Z'})\right]\\ &\quad+\sin\theta_M\, v_{Z_0'}^f\,,\\ a_Z^f &= \left[\sqrt{2}G_\mu M_Z^2\rho_f\right]^{1/2} I_3^f+\sin\theta_M\, a_{Z_0'}^f\,.\end{aligned}\tag{240}$$

The quantities $a_{Z_0'}^f\, v_{Z_0'}^f$ denote the extra $U(1)$ couplings between the fermion f and the Z_0'.

From an analysis of the electroweak precision data the mixing angle is constrained typically to $|\,\theta_M\,| < 0.01$, not very much dependent on the specification of the model [75, 76]. An example is shown in Figure 9.

References

1. R. Tanaka, talk presented at the *XXVI International Conference on High Energy Physics*, Dallas 1992. Saclay preprint X-LPNHE/92-3, to appear in the proceedings
2. H. Plothow-Besch, in: *Proceeedings of the LP-HEP 91 Conference*, Geneva 1991, eds. S. Hegarty, K. Potter, E. Quercigh, World Scientific, Singapore 1992
3. G.L. Fogli and D. Haidt, *Z. Phys.* **C40** (1988) 379;
   CDHS Collaboration, H. Abramowicz et al., *Phys. Rev. Lett.* **B57** (1986) 298;
   A. Blondel et al, *Z. Phys.* **C45** (1990) 361;
   CHARM Collaboration, J.V. Allaby et al., *Phys. Lett.* **B177** (1987) 446;
   Z. Phys. 36 (1987) 611;
   CHARM-II Collaboration, D. Geiregat et al., *Phys. Lett.* **B 247** (1990) 131;
   Phys. Lett. **B259** (1991) 499;
   CCFR Collaboration, talk presented by T. Bolton at the *XXVI International Conference on High Energy Physics*, Dallas 1992, to appear in the proceedings
4. G. Rolandi, plenary talk presented at the *XXVI International Conference on High Energy Physics*, Dallas 1992. CERN-PPE/92-175 (1992), to appear in the proceedings
5. S.L. Glashow, *Nucl. Phys.* **B22** (1961) 579;
   S. Weinberg, *Phys. Rev. Lett.* **19** (1967) 1264;
   A. Salam, in: *Proceedings of the 8th Nobel Symposium*, p. 367, ed. N. Svartholm, Almqvist and Wiksell, Stockholm 1968
6. S.L. Glashow, I. Iliopoulos, L. Maiani, *Phys. Rev.* **D2** (1970) 1285;
7. N. Cabibbo, *Phys. Rev. Lett.* **10** (1963) 531;
   M. Kobayashi, K. Maskawa, *Prog. Theor. Phys.* **49** (1973) 652
8. H.Y. Han, Y. Nambu, *Phys. Rev.* **139** (1965) 1006;
   C. Bouchiat, I. Iliopoulos, Ph. Meyer, *Phys. Lett.* **B138** (1972) 652
9. G. 't Hooft, *Nucl. Phys.* **B33** (1971) 173; *Nucl. Phys.* **B35** (1971) 167
10. CDF Collaboration, M. Contreras, *Proceeedings of the LP-HEP 91 Conference*, Geneva 1991, eds. S. Hegarty, K. Potter, E. Quercigh, World Scientific, Singapore 1992
11. D0 Collaboration, talk presented by S. Protopopescu at the *XXVIIIth Rencontres de Moriond: Electroweak Interactions and Unified Theories*, Les Arcs 1993;
    CDF Collaboration, talk presented by B. Harral at the *XXVIIIth Rencontres de Moriond: Electroweak Interactions and Unified Theories*, Les Arcs 1993
12. T. Appelquist, J. Carazzone, *Phys. Rev.* **D11** (1975) 2856
13. Particle data group, *Phys. Rev.* **D45** (1992) Number 11, Part II
14. L.D. Faddeev, V.N. Popov, *Phys. Lett.* **B25** (1967) 29

15. M. Böhm, W. Hollik, H. Spiesberger, *Fortschr. Phys.* **34** (1986) 687
16. A. Denner, T. Sack, *Nucl. Phys.* **B347** (1990) 203
17. G. Passarino, in: *Proceeedings of the LP-HEP 91 Conference*, Geneva 1991, eds. S. Hegarty, K. Potter, E. Quercigh, World Scientific, Singapore 1992
18. G. 't Hooft, *Nucl. Phys.* **B61** (1973) 455; *Nucl. Phys.* **B62** (1973) 444
19. D.A. Ross, J.C. Taylor, *Nucl. Phys.* **B51** (1973) 25
20. G. Passarino, M. Veltman, *Nucl. Phys.* **B160** (1979) 151
21. M. Consoli, *Nucl. Phys.* **B160** (1979) 208
22. A. Sirlin, *Phys. Rev.* **D22** (1980) 971;
W. J. Marciano, A. Sirlin, *Phys. Rev.* **22** (1980) 2695;
A. Sirlin, W. J. Marciano, *Nucl. Phys.* **189** (1981) 442
23. D.Yu. Bardin, P.Ch. Christova, O.M. Fedorenko, *Nucl. Phys.* **B175** (1980) 435; *Nucl. Phys.* **B197** (1982)1
24. J. Fleischer, F. Jegerlehner, *Phys. Rev.* **D23** (1981) 2001
25. K.I. Aoki, Z. Hioki, R. Kawabe, M. Konuma, T. Muta, *Suppl. Prog. Theor. Phys.* **73** (1982) 1;
Z. Hioki, *Phys. Rev. Lett.* **65** (1990) 683, E:1692; *Z. Phys.* **C49** (1991) 287
26. M. Consoli, S. LoPresti, L. Maiani, *Nucl. Phys.* **B223** (1983) 474
27. D.Yu. Bardin, M.S. Bilenky, G.V. Mithselmakher, T. Riemann, M. Sachwitz, *Z. Phys.* **C44** (1989) 493
28. W. Hollik, *Fortschr. Phys.* **38** (1990) 165
29. M. Consoli, W. Hollik, F. Jegerlehner, in: *Z Physics at LEP 1*, eds. G. Altarelli, R. Kleiss and C. Verzegnassi, CERN 89-08 (1989)
30. G. Passarino, R. Pittau, *Phys. Lett.* **B228** (1989) 89;
V.A. Novikov, L.B. Okun, M.I. Vysotsky, CERN-TH.6538/92 (1992)
31. M. Veltman, *Phys. Lett.* **B91** (1980) 95;
M. Green, M. Veltman, *Nucl. Phys.* **B169** (1980) 137, E: *Nucl. Phys.* **B175** (1980) 547;
F. Antonelli, M. Consoli, G. Corbo, *Phys. Lett.* **B91** (1980) 90;
F. Antonelli, M. Consoli, G. Corbo, O. Pellegrino, *Nucl. Phys.* **B183** (1981) 195
32. G. Passarino, M. Veltman, *Phys. Lett.* **B237** (1990) 537
33. W.J. Marciano, A. Sirlin, *Phys. Rev. Lett.* **46** (1981) 163;
A. Sirlin, *Phys. Lett.* **B232** (1989) 123
34. G. Degrassi, S. Fanchiotti, A. Sirlin, *Nucl. Phys.* **B351** (1991) 49
35. G. Degrassi, A. Sirlin, *Nucl. Phys.* **B352** (1991) 342
36. D.C. Kennedy, B.W. Lynn, *Nucl. Phys.* **B322** (1989) 1
37. M. Kuroda, G. Moultaka, D. Schildknecht, *Nucl. Phys.* **B350** (1991) 25
38. D.Yu. Bardin, B.M. Vilensky, P.Ch. Christova, *Sov. J. Nucl. Phys.* **53** (1991)

152;
A. Dabelstein, W. Hollik, *Z. Phys.* **C53** (1992) 25;
B.A. Kniehl, *Nucl. Phys.* **B376** (1992) 3

39. J.C. Ward, *Phys. Rev.* **78** (1950) 1824
40. G. 't Hooft, M. Veltman, *Nucl. Phys.* **B135** (1979) 365
41. C. Bollini, J. Giambiagi *Nuovo Cim.* **12B** (1972) 20;
J. Ashmore, *Nuovo Cim. Lett.* **4** (1972) 289;
G. 't Hooft, M. Veltman, *Nucl. Phys.* **B44** (1972) 189
42. P. Breitenlohner, D. Maison, *Comm. Math. Phys.* **52** (1977) 11, 39, 55
43. A. Denner, U. Nierste, R. Scharf, *Nucl. Phys.* **B367** (1991) 637
44. H. Burkhardt, F. Jegerlehner, G. Penso, C. Verzegnassi, *Z. Phys.* **C43** (1989) 497;
F. Jegerlehner, preprint Paul-Scherrer-Institut, PSI-PR-91-16 (1991), in: *Progress in Particle and Nuclear Physics*, ed. A. Fässler, Pergamon Press, Oxford, U.K.
45. M. Veltman, *Nucl. Phys.* **B123** (1977) 89;
M.S. Chanowitz, M.A. Furman, I. Hinchliffe, *Phys. Lett.* **B78** (1978) 285
46. M. Veltman, *Acta Phys. Polon.* **B8** (1977) 475.
47. R.E Behrends, R.J. Finkelnstein, A. Sirlin, *Phys. Rev.* **101** (1956) 866;
T. Kinoshita, A. Sirlin, *Phys. Rev.* **113** (1959) 1652
48. W.J. Marciano, *Phys. Rev.* **D20** (1979) 274
49. M. Consoli, W. Hollik, F. Jegerlehner, *Phys. Lett.* **B227** (1989) 167.
50. J.J. van der Bij, F. Hoogeveen, *Nucl. Phys.* **B283** (1987) 477
51. R. Barbieri, M. Beccaria, P. Ciafaloni, G. Curci, A. Vicere, *Phys. Lett.* **B288** (1992) 95; CERN-TH.6713/92 (1992)
52. A. Djouadi, C. Verzegnassi, *Phys. Lett.* **B195** (1987) 265;
A. Djouadi, *Nuovo Cim.* **100A** (1988) 357;
D. Yu. Bardin, A.V. Chizhov, Dubna preprint E2-89-525 (1989);
B.A. Kniehl, J.H. Kühn, R.G. Stuart, *Phys. Lett.* **B214** (1988) 621;
B.A. Kniehl, *Nucl. Phys.* **B347** (1990) 86;
F. Halzen, B.A. Kniehl, *Nucl. Phys.* **B353** (1991) 567
53. B.A. Kniehl, A. Sirlin, DESY 92-102 (1992);
S. Fanchiotti, B.A. Kniehl, A. Sirlin, CERN-TH.6749/92 (1992)
54. A. Sirlin, *Phys. Rev.* **D29** (1984) 89
55. S. Fanchiotti, A. Sirlin, New York University preprint NYU-Th-91/02/04 (1991), in: *Beg Memorial Volume*, eds. A. Ali and P. Hoodbhoy, World Scientific, Singapore 1991
56. A.A. Akhundov, D.Yu. Bardin, T. Riemann, *Nucl. Phys.* **B276** (1986) 1;
W. Beenakker, W. Hollik, *Z. Phys.* **C40** (1988) 141;
J. Bernabeu, A. Pich, A. Santamaria, *Phys. Lett.* **B200** (1988) 569; CERN-

TH.5931/90 (1990)

57. P. Langacker, M. Luo, *Phys. Rev.* **D44** (1991) 817

58. U. Amaldi, W. de Boer, H. Fürstenau, *Phys. Lett.* **B260** (1991) 447;
J. Ellis, S. Kelley, D.V. Nanopoulos, *Phys. Lett.* **B260** (1991) 131; *Phys. Lett.* **B287** (1992) 725;
G.G. Ross, R.G. Roberts, *Nucl. Phys.* **B377** (1992) 571

59. B.A. Kniehl, J.H. Kühn, R.G. Stuart, in: *Polarization at LEP*, eds. G. Alexander et al., CERN 88-06 (1988)

60. W. Hollik, H.J. Timme, *Z. Phys.* **C32** (1986) 125;
F. Jegerlehner, *Z. Phys.* **C32** (1986) 425

61. D.Yu. Bardin, W. Hollik, T. Riemann, *Z. Phys.* **C49** (1991) 485

62. G. Burgers, F. Jegerlehner, in: *Z Physics at LEP 1*, eds. G. Altarelli, R. Kleiss and C. Verzegnassi, CERN 89-08 (1989)

63. M.E. Peskin, T. Takeuchi, *Phys. Rev. Lett.* **65** (1990) 964

64. G. Altarelli, R. Barbieri, *Phys. Lett.* **B253** (1991) 161;
G. Altarelli, R. Barbieri, S. Jadach, *Nucl. Phys.* **B269** (1992) 3; E: *Nucl. Phys.* **B276** (1992) 444

65. D.C. Kennedy, P. Langacker, *Phys. Rev. Lett.* **65** (1990) 2967

66. W.J. Marciano, J.L. Rosner, *Phys. Rev. Lett.* **65** (1990) 2963

67. B.W. Lynn, M.E. Peskin, R.G. Stuart, in: *Physics with LEP*, eds. J. Ellis and R. Peccei, CERN 86-02 (1986)

68. B. Holdom, J. Terning, *Phys. Lett.* **B247** (1990) 88;
M. Golden, L. Randall, *Nucl. Phys.* **B361** (1991) 3;
C. Roiesnel, T.N. Truong, *Phys. Lett.* **B256** (1991) 439

69. D. Toussaint, *Phys. Rev.* **D18** (1978) 1626;
J.M Frere, J. Vermaseren, *Z. Phys.* **C19** (1983) 63

70. S. Bertolini, *Nucl. Phys.* **B272** (1986) 77;
W. Hollik, *Z. Phys.* **C32** (1986) 291; *Z. Phys.* **C37** (1988) 569

71. A. Denner, R. Guth, J.H. Kühn, *Phys. Lett.* **B240** (1990) 438

72. A. Denner, R. Guth, W. Hollik, J.H. Kühn, *Z. Phys.* **C51** (1991) 695

73. G. Passarino, Torino preprint DFTT/G-90-3 (1990);
G. Passarino, in *Proceedings of the XXVIth Rencontre de Moriond: '91 Electroweak Interactions and Unified Theories*, ed. Tran Thanh Van

74. B.W. Lynn, E. Nardi, *Nucl. Phys.* **B381** (1992) 467

75. G. Altarelli et al., *Nucl. Phys.* **B342** (1990) 15; *Phys. Lett.* **B245** (1990) 669;
M.C. Gonzalez-Garcia, J.W.F. Valle, *Phys. Lett.* **B259** (1991) 365;
J. Layssac, F.M. Renard, C. Verzegnassi, *Z. Phys.* **C53** (1992) 97

76. F. del Aguila, W. Hollik, J.M. Moreno, M. Quiros, *Nucl. Phys.* **B372** (1992) 3

PREDICTIONS FOR e^+e^- PROCESSES

WOLFGANG HOLLIK[1]
Fakultät für Physik
Universität Bielefeld
4800 Bielefeld, Germany

Contents

[1]on leave from Max-Planck-Institut für Physik, Föhringer Ring 6, 8000 München 40, Germany

1 Introduction

In the previous chapter [1], in the following always denoted as I, a description has been given of the basic entries and concepts for the calculation of electroweak observables beyond the lowest order. As a special application of particular importance, the relation between the vector boson masses following from the precisely known muon lifetime has been discussed. This chapter is dedicated to the detailed description of the cross section for the basic e^+e^- processes of fermion pair production $e^+e^- \to f\bar{f}$ and the related measurable quantities, together with the Standard Model predictions for the Z pole observables which are investigated at LEP and SLC with high accuracy. Since polarization effects play an essential role around the Z peak [2], we also include polarization for the incoming $e^\pm$ states.

The measurement of the Z mass from the Z line shape provides us with an additional precise input parameter besides α and G_μ. Other observable quantities around the Z peak, like total and partial decay widths, forward-backward asymmetries, ..., allow us to perform precision tests of the theory by comparing the experimental results with the theoretical predictions. In lowest order, the Z observables are completely fixed in terms of α, G_μ, M_Z. The dependence of the loop contributions on the masses of the experimentally unknown particles top and Higgs prevents the Standard Model predictions for the observables from being unique. On the other hand, this gives us a tool for constraining the allowed mass range from the experimental precision data. Since 1-loop terms are of the order α/π and typically enhanced by factors $\sim \log M_Z^2/m_f^2$ or $\sim m_t^2/M_Z^2$, the size $\alpha/\pi = 0.0023$, in view of experimental precisions of a few 10^{-3}, makes the need for dressing the Born amplitudes by next order contributions obvious.

For the purpose of comparison with experimental data, we give numerical predictions for all relevant Z observables (Z width, partial widths, forward-backward asymmetries, left-right asymmetries) including the electroweak higher order corrections as far as they are available at the present stage. Formulae for QED corrections for the various observables as well as QCD corrections for quantities involving hadronic final states are also presented together with a discussion of their effects on the measureable quantities.

Section 2 contains the lowest order formulae for the $e^+e^- \to f\bar{f}$ cross section including beam polarization. In section 3 we give a description of the higher order contributions and present the general formulae for the dressed cross sections including the electroweak radiative corrections. A particular useful concept for physics around the Z pole is that of effective NC coupling constants which will be discussed in Section 4. The effective couplings incorporate the non-QED electroweak corrections in terms of dressed vector and axial vector couplings or, equivalently, overall normalization factors and effective mixing angles. They allow a compact notation for the γ and Z exchange amplitudes and thus a transparent presentation and discussion of the Z observables. The various observables are discussed in detail in Section 5, split into the inclusive quantities: line shape, Z width and partial widths, and the asymmetries: forward-backward and left-right asymmetries, for the various fermion species. Numerical predictions from the Standard Model are given, showing the top and Higgs mass dependence. QED and QCD corrections are separately discussed, and compact formulae are given for calculating the observables with the accuracy required by the experimental precision.

2 The cross section for $e^+e^- \to f\bar{f}$ and lowest order results

The basic processes in e^+e^- collisions below the WW threshold are the annihilation processes into lepton and quark pairs $e^+e^- \to f\bar{f}$, which are described in lowest order by the photon and Z exchange diagrams. Since polarization effects play an essential role around the Z peak we will discuss the e^+e^- cross section including polarization of the incoming $e^\pm$.

The degrees of longitudinal polarization of $e^\pm$ are denoted by $P_L^\pm$ ($P_L = 1$: right handed, $P_L = -1$: left handed, for both electron and positron), the degrees of transverse polarization by $P_T^\pm$. Since the transverse polarization is the "natural polarization" in a storage ring we want to include this situation for completeness. The principle physical insight gained from transverse polarization does not exceed that from unpolarized beams; information is transferred from the polar angle to the azimuthal angle distributions. Measurements of azimuthal observables, however, are normally less sensitive to systematic uncertainties than polar angle distributions or charge asymmetries, which makes the transverse polarization also an interesting and practically important tool.

Without neglecting terms from the final fermion mass m_f the differential cross section can be written in the following way, where the color factor $N_C^f = 1$ (leptons), $N_C^f = 3$ (quarks) distinguishes between the final state fermions ($\theta = \angle(e^-, f)$, $\mu_f = m_f^2/s$, $s = (p_- + p_+)^2$):

$$\frac{d\sigma}{d\Omega} = \frac{\alpha^2}{4s} N_C^f \sqrt{1-4\mu_f}\, \left\{(1 - P_L^+ P_L^-)\, X_U + (P_L^+ - P_L^-)\, X_L + P_T^+ P_T^-\, X_T\right\} \tag{1}$$

with

$$\begin{aligned} X_U &= G_1(s)\,(1+\cos^2\theta) + 4\mu_f\, G_2(s)\,\sin^2\theta + \sqrt{1-4\mu_f}\, G_3(s)\cdot 2\cos\theta \\ X_L &= H_1(s)\,(1+\cos^2\theta) + 4\mu_f\, H_2(s)\,\sin^2\theta + \sqrt{1-4\mu_f}\, H_3(s)\cdot 2\cos\theta \\ X_T &= (1-4\mu_f)\, [F_1(s)\,\cos 2\phi + F_2(s)\,\sin 2\phi]\,\sin^2\theta\,. \end{aligned} \tag{2}$$

In lowest order, the functions in Eq. (2) are determined by the vector and axial vector coupling constants in Eq. I (37) and the Z propagator in the lowest order

Breit-Wigner approximation in the normalization

$$\chi_Z^0(s) = \frac{\sqrt{2}G_\mu M_Z^2}{4\pi\alpha} \cdot \frac{s}{s - M_Z^2 + iM_Z\Gamma_Z^0} \tag{3}$$

as follows:

$$\begin{aligned}
G_1(s) &= Q_e^2Q_f^2 + 2v_ev_fQ_eQ_f\,\mathrm{Re}\,\chi_Z^0(s) + (v_e^2+a_e^2)(v_f^2+a_f^2-4\mu_f a_f^2) \mid \chi_Z^0(s)\mid^2 \\
G_2(s) &= Q_e^2Q_f^2 + 2v_ev_fQ_eQ_f\,\mathrm{Re}\,\chi_Z^0(s) + (v_e^2+a_e^2)v_f^2 \mid \chi_Z^0(s)\mid^2 \\
G_3(s) &= 2a_ea_fQ_eQ_f\,\mathrm{Re}\,\chi_Z^0(s) + 4v_ea_ev_fa_f \mid \chi_Z^0(s)\mid^2 \\
\\
H_1(s) &= 2a_ev_fQ_eQ_f\,\mathrm{Re}\,\chi_Z^0(s) + 2v_ea_e(v_f^2+a_f^2-4\mu_f a_f^2) \mid \chi_Z^0(s)\mid^2 \\
H_2(s) &= 2a_ev_fQ_eQ_f\,\mathrm{Re}\,\chi_Z^0(s) + 2v_ea_ev_f^2 \mid \chi_Z^0(s)\mid^2 \\
H_3(s) &= 2v_ea_fQ_eQ_f\,\mathrm{Re}\,\chi_Z^0(s) + 2(v_e^2+a_e^2)v_fa_f \mid \chi_Z^0(s)\mid^2 \\
\\
F_1(s) &= Q_eQ_f^2 + 2v_ev_fQ_eQ_f\,\mathrm{Re}\,\chi_Z^0(s) + (v_e^2-a_e^2)(v_f^2+a_f^2) \mid \chi_Z^0(s)\mid^2 \\
F_2(s) &= 2v_ea_e(v_f^2+a_f^2)\,\mathrm{Im}\,\chi_Z^0(s)
\end{aligned} \tag{4}$$

The lowest order Z width in (3) is the sum over the partial widths

$$\Gamma_Z^0 = \sum_f \Gamma^0(Z \to f\bar{f}) \tag{5}$$

with

$$\Gamma^0(Z \to f\bar{f}) = N_C^f \frac{\sqrt{2}G_\mu M_Z^3}{12\pi}\sqrt{1-\frac{4m_f^2}{M_Z^2}}\left[v_f^2\left(1+\frac{2m_f^2}{M_Z^2}\right)+a_f^2\left(1-\frac{4m_f^2}{M_Z^2}\right)\right]. \tag{6}$$

3 Radiative corrections

3.1 *General description*

The radiative corrections to the process $e^+e^- \to f\bar{f}$ can be subdivided quite naturally into the following subclasses:

- "QED corrections", which consist in one-loop of those diagrams with an extra photon added to the Born diagrams either as a real bremsstrahlung photon or a virtual photon inside a loop, depicted in Figure 1. The QED corrections are common to any theory containing the electromagnetic $U(1)_{em}$ as a subgroup. Around the Z resonance they are large (typically 40%) and hence require one to go beyond the one-loop approximation (full two-loop QED and summation of multiple soft photon radiation).

- "Weak corrections", which collect all the other diagrams without virtual photons: The subset which involves corrections to the vector boson propagators γ, Z, the set of vertex corrections (where the virtual photon contributions have been removed), and box diagrams with two massive boson exchange (Figure 2).

The separation of the QED corrections is sensible in a twofold respect: (i) they form a gauge invariant subset, (ii) they depend on the details of the experiments via the cuts applied to the final state photon. This requires a precise experimental control of detector properties and phase space cuts, most conveniently with the help of Monte Carlo techniques. The proper treatment of the QED corrections constitutes the link between data taking and physics analysis. The other class of the infrared finite weak corrections is independent of the experimental set-up; it includes the more subtle part of the electroweak theory beyond the tree level and is thus sensitive also to the presence of "new physics".

Due to the smallness of the electron mass the lowest order Higgs exchange diagram can be neglected. For the same reason also diagrams with Higgs - gauge boson mixing and box diagrams where one or both of the internal vector bosons are replaced by Higgs scalars are negligible; they are suppressed by at least a factor $\frac{\alpha}{\pi}\frac{m_e}{M_W}$. The propagator corrections, however, involve all particles of the model, in particular the Higgs boson and the top quark, and thus depend on M_H and m_t. For light final fermions ($f \neq b, t$) the vertex corrections contain only W and Z in virtual states. Vertex corrections to heavy fermions depend also on the Higgs-fermion Yukawa couplings.

The tree level formulae (2) are valid also for heavy particle final states like $t\bar{t}$. The complete treatment of the radiative corrections for such a heavy fermion production process, however, is more involved and will not be discussed here (see [3]). In this article we want to restrict ourselves to the situation where only the known fermions appear as external particles for which the one-loop corrections can be cast into a compact and transparent form. Since the finite m_f-terms give only a very small contribution at the tree level, we can neglect all fermion mass effects in the loop diagrams with two exceptions:

- mass terms from a virtual top quark are always treated without approximation in vertex corrections involving the (t, b) doublet and the WW box contribution for $b\bar{b}$ final states;
- in the QED corrections all mass terms (also from light fermions) have to be kept which would lead to mass singularities for $m_f \to 0$.

In this approximation the vertex corrections can be represented in terms of form factors $F_{V,A}^{Z(\gamma)f}(k^2)$ for the vector and axial vector currents only. Also the box diagrams can be written as a sum over terms like

$$(\text{initial current}) \cdot (\text{final current}) \cdot (\text{formfactor}).$$

where the currents have only vector and axial vector contributions. This simple structure arises from helicity conservation at the vector boson - fermion vertices in

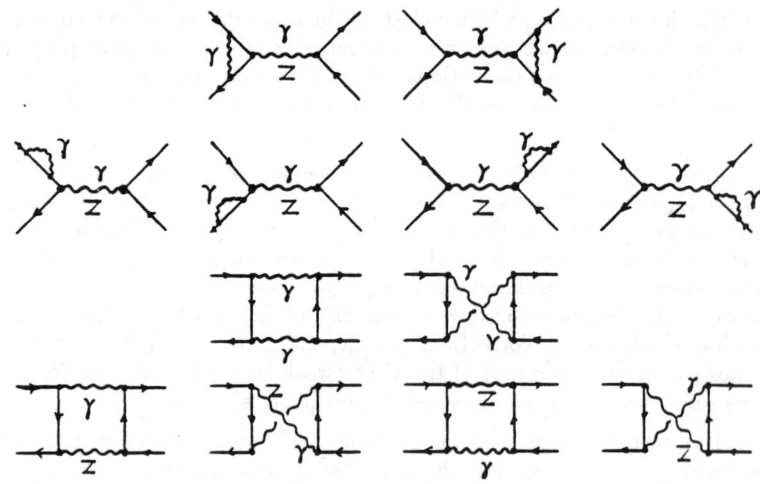

Fig. 1: *Virtual QED corrections for* $e^+e^- \rightarrow f\bar{f}$

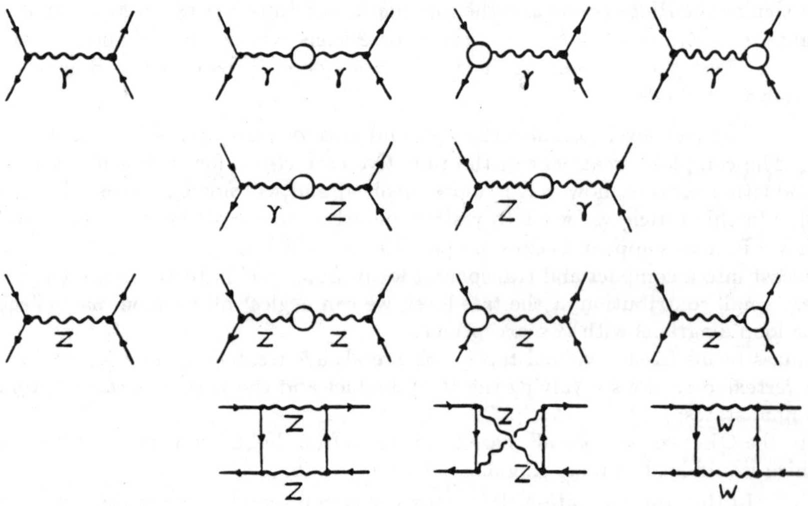

Fig. 2: *Non-QED corrections for* $e^+e^- \rightarrow f\bar{f}$

the small mass limit. It allows one to express the corrected cross section in the same way as given in Eq.s (1) and (2); only the invariant functions have to be substituted by their corrected versions.

3.2 *The photon - Z propagators*

In lowest order, after diagonalization of the neutral boson mass matrix, the propagator matrix is diagonal. But mixing due to quantum corrections prohibits the photon and Z boson from propagating independently of each other in higher orders. Consequently, the propagator of the γZ system has to be considered as a 2×2 matrix. The radiative corrections to the propagator system can be obtained by inversion of the matrix (we consider the transverse parts only)

$$(\mathbf{D}_{\mu\nu})^{-1} = i\, g_{\mu\nu} \begin{pmatrix} k^2 + \hat{\Sigma}^{\gamma\gamma}(k^2) & \hat{\Sigma}^{\gamma Z}(k^2) \\ \hat{\Sigma}^{\gamma Z}(k^2) & k^2 - M_Z^2 + \hat{\Sigma}^{ZZ}(k^2) \end{pmatrix} \tag{7}$$

with the 1-particle irreducible (1PI) renormalized self energies specified in I (47), (64), yielding:

$$\mathbf{D}_{\mu\nu} = -i\, g_{\mu\nu} \begin{pmatrix} D_\gamma & D_{\gamma Z} \\ D_{\gamma Z} & D_Z \end{pmatrix} \tag{8}$$

where ($s = k^2$)

$$D_\gamma(s) = \frac{1}{s + \hat{\Sigma}^{\gamma\gamma}(s) - \frac{\hat{\Sigma}^{\gamma Z}(s)^2}{s - M_Z^2 + \hat{\Sigma}^{ZZ}(s)}} \tag{9}$$

$$D_Z(s) = \frac{1}{s - M_Z^2 + \hat{\Sigma}^{ZZ}(s) - \frac{\hat{\Sigma}^{\gamma Z}(s)^2}{s + \hat{\Sigma}^{\gamma\gamma}(s)}} \tag{10}$$

$$D_{\gamma Z}(s) = -\frac{\hat{\Sigma}^{\gamma Z}(s)}{[s + \hat{\Sigma}^{\gamma\gamma}(s)]\,[s - M_Z^2 + \hat{\Sigma}^{ZZ}(s)] - \hat{\Sigma}^{\gamma Z}(s)^2}\,. \tag{11}$$

The matrix (8) can be diagonalized only for one specific value of k^2. This has been done by fixing the mixing counter term in such a way that (8) is diagonal for $k^2 = 0$ (I, section 3.3).

The presence of the $\hat{\Sigma}^{\gamma Z}(s)^2$-term in the Z propagator requires a modification of the Z mass renormalization condition I (62) which fixes the Z mass counter term δM_Z^2. Instead of I (63) we have the condition

$$\delta M_Z^2 = \mathrm{Re}\left[\Sigma^{ZZ}(M_Z^2) - \frac{\left(\hat{\Sigma}^{\gamma Z}(M_Z^2)\right)^2}{M_Z^2 + \hat{\Sigma}^{\gamma\gamma}(M_Z^2)}\right]\,. \tag{12}$$

The condition for the W mass counter term δM_W^2 is unchanged and so are all the other renormalization conditions formulated in section 3.3 of I. But the set of equations I (63) and (64) defining the renormalization constants and the renormalized 2-point functions are implicit equations since they contain the renormalized mixing on the r.h.s. However, it is straightforward to solve them for the renormalized quantities in terms of the non-renormalized ones.

The photon field renormalization constant δZ_2^γ (and also the charge renormalization constant δZ_1^γ) thus is not changed by the mixing term.

For Z and W, the additional mass counter term in Eq. (12) enters the field and coupling renormalization constants and hence the renormalized self energies I (47). In other places like the boson-fermion vertices this additional term does not show up since the corresponding counter terms contain only differences $\delta Z_1 - \delta Z_2$ which are independent of $\delta M_{Z,W}^2$. Vertex renormalization is therefore not affected.

The modification of the W propagator via δZ_2^W has an interesting consequence: The quantity Δr in 1-loop, given by Eq. I (139), is modified according to

$$\Delta r \quad \rightarrow \quad \Delta r + \frac{c_W^2}{s_W^2} \mathrm{Re} \left(\frac{\left(\hat{\Sigma}^{\gamma Z}(M_Z^2)\right)^2 / M_Z^2}{M_Z^2 + \hat{\Sigma}^{\gamma\gamma}(M_Z^2)} \right) . \tag{13}$$

The geometrical resummation of this modified Δr, which corresponds to the W-propagator

$$D_W(s) = \frac{1}{s - M_W^2 + \hat{\Sigma}^{WW}(s)},$$

leads automatically to the factorization property I (146) when the leading terms with $\Delta\alpha$ and $\Delta\rho$ are considered, thus taking account of the proper summation of all the leading higher order reducible terms. The 1PI leading 2-loop term $\Delta\rho^{(2)}$ from I (156) has to be added explicitly to the irreducible contribution to the ρ-parameter.

The Z propagator (10) is still incomplete in a twofold sense:

(i) The 2-loop irreducible contribution $\Delta\rho^{(2)}$, Eq. I (156), has to be built in by means of substituting

$$\frac{\delta M_Z^2}{M_Z^2} - \frac{\delta M_W^2}{M_W^2} \rightarrow \frac{\delta M_Z^2}{M_Z^2} - \frac{\delta M_W^2}{M_W^2} + \Delta\rho^{(2)} \tag{14}$$

in the counter terms, in accordance with the property I (130).

(ii) The width has to completed by the higher order contributions $\Delta\Gamma_Z$ resulting from the weak vertex corrections, the QED and QCD corrections.

With these improvements, the Z propagator has the form

$$\begin{aligned} D_Z(s) &= \frac{1}{s - M_Z^2 + \hat{\Sigma}^Z(s) + i\,\frac{s}{M_Z}\Delta\Gamma_Z} \\ &\simeq \frac{1}{1+\hat{\Pi}^Z(M_Z^2)} \cdot \frac{1}{s - M_Z^2 + i\,\frac{s}{M_Z}\Gamma_Z} \end{aligned} \tag{15}$$

around the Z peak, with

$$\hat{\Sigma}^Z(s) = \hat{\Sigma}^{ZZ}(s) - \frac{\left(\hat{\Sigma}^{\gamma Z}(s)\right)^2}{s + \hat{\Sigma}^{\gamma\gamma}(s)} \tag{16}$$

and

$$\begin{aligned} \hat{\Pi}^Z(s) &= \mathrm{Re}\,\frac{\hat{\Sigma}^Z(s)}{s - M_Z^2} \\ &= \mathrm{Re}\,\frac{d\hat{\Sigma}^Z}{ds}(M_Z^2) \quad \text{for} \quad s = M_Z^2\,. \end{aligned} \tag{17}$$

The quantity Γ_Z, determined by

$$M_Z\Gamma_Z = \frac{1}{1+\hat{\Pi}^Z(M_Z^2)}\left[\mathrm{Im}\left(\hat{\Sigma}^{ZZ}(M_Z^2) - \frac{\left(\hat{\Sigma}^{\gamma Z}(M_Z^2)\right)^2/M_Z^2}{M_Z^2 + \hat{\Sigma}^{\gamma\gamma}(M_Z^2)}\right) + \Delta\Gamma_Z\right] \tag{18}$$

is the width of the Z resonance. With the help of the effective couplings to be discussed in section 4.2, we will give a simpler and more compact representation for Γ_Z in section 5.2.

3.3 *Vertex corrections*

As mentioned in the general description of section 3.1 the vertex corrections can be summarized in terms of s-dependent vector and axial vector form factors if the masses m_f of the external fermions are small compared to M_W, both for the electromagnetic and the weak NC vertex. In our terminology, "vertex corrections" denote the renormalized $\gamma(Z)ff$ 3-point functions in one-loop order according to I (55) and (58), together with the finite wave function renormalizations for external fermions. The form factors given here can be treated as additive terms to the effective V and A coupling constants in the amplitudes.

In contrast to the propagator corrections the vertex corrections are not universal, depending in general on the fermion species. For this reason we have to list them separately for ν, e, u, d type fermions. In addition, the b quark is exceptional due to the virtual top contributions in the vertex. The vertex correction diagrams are listed in Figure 3. For $f \neq b$ only (a) - (c) have to be taken into account; (d)

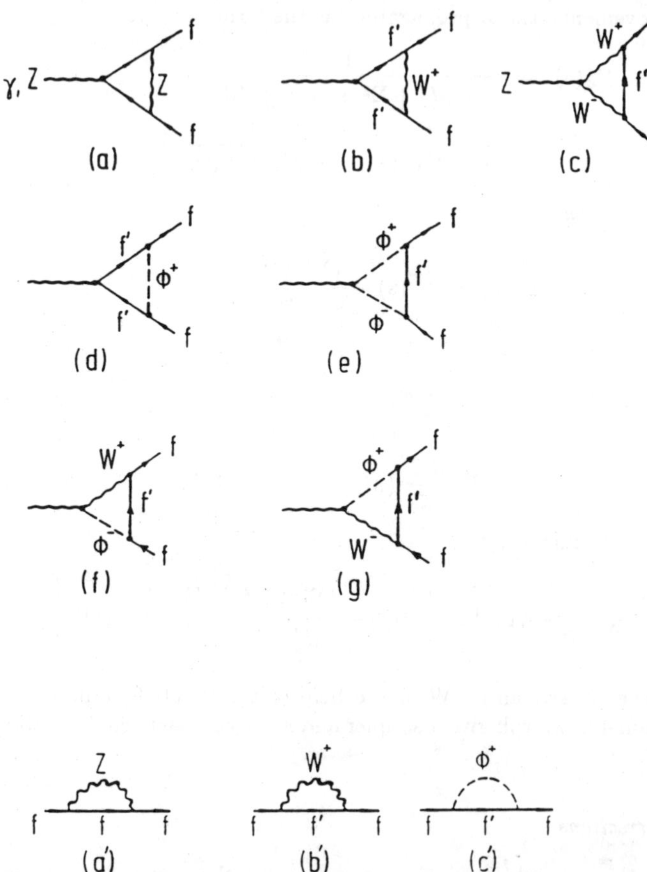

Fig. 3: *Non-QED vertex corrections for the γ, Z-ff vertices and fermion self-energies. The diagrams with the physical Higgs and the neutral Goldstone boson are neglected. The diagrams (d-g) and (c') are negligible for light fermions $\neq b$. The self-energies have to be understood as inserted in the external lines with a factor 1/2. f' denotes the isospin doublet partner of f.*

- (g) are negligible due to small Yukawa couplings (as well as the dropped graphs with internal neutral Higgs states).

Our terminology is as follows:

$F_{V,A}^{Zf}$ and $F_{V,A}^{\gamma f}$ denote the IR finite weak (i.e. without the virtual photon diagrams) form factors for the Zff and γff vertex which, together with the lowest order terms yield the dressed vertices:

$$\begin{aligned}
\hat{\Gamma}_\mu^{Zff} &= i\frac{e}{2s_W c_W}\gamma_\mu\left[v_f + F_V^{Zf}(s) - \gamma_5\left(a_f + F_A^{Zf}(s)\right)\right], \\
\hat{\Gamma}_\mu^{\gamma ff} &= -i\,e\,Q_f\,\gamma_\mu - i\,e\,\gamma_\mu\left[F_V^{\gamma f}(s) - F_A^{\gamma f}(s)\,\gamma_5\right]. \qquad (19)
\end{aligned}$$

The weak form factors in (19) are explicitly given by the following set of formulae:

Neutral current vertex:

neutrinos:

$$\begin{aligned}
F_V^{Z\nu} &= F_A^{Z\nu} \qquad (20) \\
&= \frac{\alpha}{4\pi}\left[\frac{1}{8c_W^2 s_W^2}\Lambda_2(s,M_Z) + \frac{2s_W^2-1}{4s_W^2}\Lambda_2(s,M_W) + \frac{3c_W^2}{2s_W^2}\Lambda_3(s,M_W)\right]
\end{aligned}$$

charged fermions:

$$\begin{aligned}
F_V^{Zf} &= \frac{\alpha}{4\pi}\left[\frac{v_f(v_f^2+3a_f^2)}{4s_W^2c_W^2}\Lambda_2(s,M_Z) + F_L^f\right] \\
F_A^{Zf} &= \frac{\alpha}{4\pi}\left[\frac{a_f(3v_f^2+a_f^2)}{4s_W^2c_W^2}\Lambda_2(s,M_Z) + F_L^f\right] \qquad (21)
\end{aligned}$$

with

$$\begin{aligned}
F_L^\ell &= \frac{1}{4s_W^2}\Lambda_2(s,M_W) - \frac{3c_W^2}{2s_W^2}\Lambda_3(s,M_W) \\
F_L^u &= -\frac{1-\frac{2}{3}s_W^2}{4s_W^2}\Lambda_2(s,M_W) + \frac{3c_W^2}{2s_W^2}\Lambda_3(s,M_W) \\
F_L^d &= \frac{1-\frac{4}{3}s_W^2}{4s_W^2}\Lambda_2(s,M_W) - \frac{3c_W^2}{2s_W^2}\Lambda_3(s,M_W). \qquad (22)
\end{aligned}$$

Electromagnetic vertex:

$$\begin{aligned}
F_V^{\gamma f} &= \frac{\alpha}{4\pi}\left[\frac{Q_f(v_f^2+a_f^2)}{4s_W^2c_W^2}\Lambda_2(s,M_Z) + G_L^f\right] \\
F_A^{\gamma f} &= \frac{\alpha}{4\pi}\left[\frac{Q_f\,2v_f a_f}{4s_W^2c_W^2}\Lambda_2(s,M_Z) + G_L^f\right] \qquad (23)
\end{aligned}$$

with

$$\begin{aligned} G_L^\ell &= -\frac{3}{4s_W^2}\Lambda_3(s,M_W) \\ G_L^u &= -\frac{1}{12s_W^2}\Lambda_2(s,M_W)+\frac{3}{4s_W^2}\Lambda_3(s,M_W) \\ G_L^d &= \frac{1}{6s_W^2}\Lambda_2(s,M_W)-\frac{3}{4s_W^2}\Lambda_3(s,M_W)\,. \end{aligned} \tag{24}$$

The functions Λ_2, Λ_3 have the form

$$\begin{aligned} \Lambda_2(s,M) &= -\frac{7}{2}-2w-(2w+3)\log(-w) \\ &\quad +2(1+w)^2\left[\mathrm{Li}_2(1+\frac{1}{w})-\frac{\pi^2}{6}\right] \\ \Lambda_3(s,M) &= \frac{5}{6}-\frac{2w}{3}+\frac{(2w+1)}{3}\sqrt{1-4w}\log(x)+\frac{2}{3}w(w+2)\log^2(x) \end{aligned} \tag{25}$$

with

$$w=\frac{M^2}{s+i\varepsilon},\quad x=\frac{\sqrt{1-4w}-1}{\sqrt{1-4w}+1}.$$

The functions F_L^d and G_L^d cannot be used for b quarks. We list the more involved expressions for F_L^b, G_L^b in the appendix. They depend explicitly on the top mass. F_L^b contains a term $\sim m_t^2/M_Z^2$ which is s-independent:

$$F_L^b=\frac{\alpha}{4\pi}\frac{1}{4s_W^2c_W^2}\frac{m_t^2}{M_Z^2}+\cdots$$

3.4 *The box contributions*

The last ingredient in the one-loop corrected amplitudes are the box diagrams with two gauge boson exchange. The genuine weak box diagrams are those with ZZ and WW exchange. They are UV and IR finite (UV divergent only in the unitary gauge). They depend in addition also on the scattering angle θ, resp. the Mandelstam variables

$$t=-\frac{s}{2}(1-\cos\theta),\quad u=-\frac{s}{2}(1+\cos\theta)$$

and give the following contributions to the weak amplitude (for $m_f^2\ll M_W^2$):

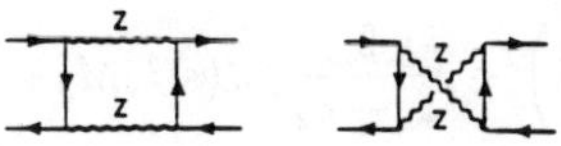

$$= \frac{e^2}{(4s_W^2c_W^2)^2}\frac{J_{NC}^{(e)}\cdot J_{NC}^{(f)}}{s}\cdot\frac{\alpha}{2\pi}\left[I(s,t,M_Z)-I(s,u,M_Z)\right]$$

$$+\frac{e^2}{(4s_W^2c_W^2)^2}\frac{J_{NC}^{(e)5}\cdot J_{NC}^{(f)5}}{s}\cdot\frac{\alpha}{2\pi}\left[I_5(s,t,M_Z)+I_5(s,u,M_Z)\right] \qquad (26)$$

where

$$\begin{aligned} J_{NC}^{(e)} &= \bar v_e\,\gamma_\mu(v_e^2+a_e^2-2v_ea_e\gamma_5)\,u_e \qquad (27)\\ J_{NC}^{(f)} &= \bar u_f\,\gamma_\mu(v_f^2+a_f^2-2v_fa_f\gamma_5)\,v_f\\ J_{NC}^{(e)5} &= \bar v_e\,\gamma_\mu\gamma_5(v_e^2+a_e^2-2v_ea_e\gamma_5)\,u_e\\ J_{NC}^{(f)5} &= \bar u_f\,\gamma_\mu\gamma_5(v_f^2+a_f^2-2v_fa_f\gamma_5)\,v_f \end{aligned}$$

and the functions I, I_5 defined below;

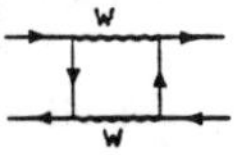

$$= \frac{e^2}{(4s_W^2)^2}\frac{J_{CC}^{(e)}\cdot J_{CC}^{(f)}}{s}\cdot\frac{\alpha}{2\pi}\left[I(s,t,M_W)+I_5(s,t,M_W)\right] \text{ for } I_3^f=-1/2$$

(28)

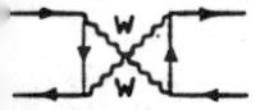

$$= \frac{e^2}{(4s_W^2)^2}\frac{J_{CC}^{(e)}\cdot J_{CC}^{(f)}}{s}\cdot\frac{\alpha}{2\pi}\left[-I(s,u,M_W)+I_5(s,u,M_W)\right] \text{ for } I_3^f=+1/2$$

with

$$\begin{aligned} J_{CC}^{(e)} &= \bar v_e\gamma_\mu(1-\gamma_5)u_e\\ J_{CC}^{(f)} &= \bar u_f\gamma_\mu(1-\gamma_5)v_f\,. \end{aligned}$$

The analytic expressions for I and I_5 read ($f\neq b,t$):

$$\begin{aligned} I_5(s,t,M) &= \frac{s}{s+t}\left\{\frac{s+2t+2M^2}{2(s+t)}\left[\mathrm{Li}_2\left(1+\frac{t}{M^2}\right)-\frac{\pi^2}{6}-\log^2\left(-\frac{y_1}{y_2}\right)\right]\right.\\ &\quad +\frac{1}{2}\log\left(-\frac{t}{M^2}\right)+\frac{y_2-y_1}{2}\log\left(-\frac{y_1}{y_2}\right)\\ &\quad \left.+\frac{s+2t-4M^2t/s+2M^4/t-2M^4/s}{2(s+t)(x_2-x_1)}\cdot J(s,t,M)\right\}, \end{aligned}$$

$$
\begin{aligned}
I(s,t,M) &= I_5(s,t,M) + 2\log^2\left(-\frac{y_1}{y_2}\right) + \frac{2}{x_1-x_2}\cdot J(s,t,M)\,, \\
J(s,t,M) &= \mathrm{Li}_2\frac{x_1}{x_1-y_1} + \mathrm{Li}_2\frac{x_1}{x_1-y_2} - \mathrm{Li}_2\frac{x_2}{x_2-y_1} - \mathrm{Li}_2\frac{x_2}{x_2-y_2}
\end{aligned}
\tag{29}
$$

with

$$
\begin{aligned}
x_{1,2} &= \frac{1}{2}\left(1 \pm \sqrt{1-\frac{4M^2}{s}\left(1+\frac{M^2}{t}\right)}\right)\,, \\
y_{1,2} &= \frac{1}{2}\left(1 \pm \sqrt{1-\frac{4M^2}{s}}\right)\,.
\end{aligned}
$$

For $f = b$ the top mass dependence in the WW box has to be taken into account. In this case we have to replace the expression $I + I_5$ in Eq. (28) by a more complicated term, which can be written in terms of a parameter integral:

$$
I(s,t,M)+I_5(s,t,M) = -\int_0^1 dx \int_0^1 dy \int_0^1 dz\, z(1-z)\left(\frac{s}{L} + \frac{(1-z+z^2y(1-y))st}{L^2}\right)
\tag{30}
$$

with

$$
L = -(1-z)^2x(1-x)s \;-\; z^2y(1-y)t \;+\; (1-z)M^2 \;+\; z(1-y)m_t^2\,.
$$

We do not give the complicated analytical expression here but refer to the systematic treatment of the heavy fermion case to the literature [3].

Since these box diagrams are non-resonant their contribution near the Z peak is of the order of the $\frac{\alpha}{\pi}$ -corrections to the photon exchange amplitude. This is the reason why their numerical effects in physical quantities on the Z (or not far away) are negligibly small. Their contribution to the differential cross section at $s = M_Z^2$ is smaller than 0.02%. Their influence is slightly bigger at energies which are several GeV below or above M_Z, but still negligible.

The smallness of the box contributions to the Z cross section is based on their suppression by the large resonance factor. They are in general not negligible for processes which are off resonance. As an example we refer to the box contributions to the quantity Δr (the box contributions in muon decay)

$$
(\Delta r)_{box} = -\frac{\alpha}{4\pi}\left(1-\frac{5}{s_W^2}+\frac{5}{2s_W^4}\right)\log c_W^2 \;= 0.004
$$

which corresponds to a shift $\delta s_W^2 = 0.001$ in the mixing angle.

3.5 *The dressed cross section*

In order to obtain the full result for the differential cross section (1) we have to specify the invariant functions in (2). We do this by introducing a compact notation in terms of s-dependent coupling constants and propagator functions according to table 1. The coupling constants v_f, a_f are those of Eq. I (37); the form factors $F_V^{\gamma e} \cdots F_A^{Zf}$ are from the vertex corrections (19).

The functions χ are defined as follows:

$$\begin{aligned} \chi_1 &= s\, D_\gamma(s) \\ \chi_2 &= s\, D_Z(s)\, (\frac{1}{4s_W^2 c_W^2})^2 \\ \chi_3 &= -\, s\, D_{\gamma Z}(s)\, \frac{1}{4s_W^2 c_W^2} \end{aligned} \tag{31}$$

with the propagators $D_{\gamma,Z,\gamma Z}$ from section 3.2. $\chi_{9,10}$ belong to the box diagrams with ZZ exchange and χ_{11} to that with WW exchange. In the notation of section 3.4 they read:

$$\begin{aligned} \chi_9 &= \frac{\alpha}{2\pi}(\frac{1}{4s_W^2 c_W^2})^2\, [I(s,t,M_Z) - I(s,u,M_Z)] \\ \chi_{10} &= \frac{\alpha}{2\pi}(\frac{1}{4s_W^2 c_W^2})^2\, [I_5(s,t,M_Z) + I_5(s,u,M_Z)] \\ \chi_{11} &= \frac{\alpha}{2\pi}\left\{ \begin{array}{ll} [I(s,t,M_W) + I_5(s,t,M_W] & \text{for } I_3^f = -\frac{1}{2} \text{ fermions} \\ [-I(s,u,M_W) + I_5(s,u,M_W] & \text{for } I_3^f = +\frac{1}{2} \text{ fermions.} \end{array} \right. \end{aligned} \tag{32}$$

Now we have collected everything for specifying the invariant functions as to be inserted into the cross section formulae (2):

$$\begin{aligned} G_1(s,t) &= \text{Re} \sum_{j,k=1}^{11} (V_j^e V_k^{e*} + A_j^e A_k^{e*})\, (V_j^f V_k^{f*} + A_j^f A_k^{f*})\, \chi_j \chi_k^* \\ G_3(s,t) &= \text{Re} \sum_{j,k=1}^{11} (V_j^e A_k^{e*} + A_j^e V_k^{e*})\, (V_j^f A_k^{f*} + A_j^f V_k^{f*})\, \chi_j \chi_k^* \\ H_1(s,t) &= \text{Re} \sum_{j,k=1}^{11} (V_j^e A_k^{e*} + A_j^e V_k^{e*})\, (V_j^f V_k^{f*} + A_j^f A_k^{f*})\, \chi_j \chi_k^* \\ H_3(s,t) &= \text{Re} \sum_{j,k=1}^{11} (V_j^e V_k^{e*} + A_j^e A_k^{e*})\, (V_j^f A_k^{f*} + A_j^f V_k^{f*})\, \chi_j \chi_k^* \end{aligned} \tag{33}$$

j	V_j^e	A_j^e	V_j^f	A_j^f	χ_j
1	Q_e	0	Q_f	0	χ_1
2	v_e	a_e	v_f	a_f	χ_2
3	Q_e	0	v_f	a_f	χ_3
4	v_e	a_e	Q_f	0	$\chi_4 = \chi_3$
5	$F_V^{\gamma e}(s)$	$F_A^{\gamma e}(s)$	Q_f	0	$\chi_5 = \chi_1$
6	Q_e	0	$F_V^{\gamma f}(s)$	$F_A^{\gamma f}(s)$	$\chi_6 = \chi_1$
7	$F_V^{Ze}(s)$	$F_A^{Ze}(s)$	v_f	a_f	$\chi_7 = \chi_2$
8	v_e	a_e	$F_V^{Zf}(s)$	$F_A^{Zf}(s)$	$\chi_8 = \chi_2$
9	$(v_e^2 + a_e^2)$	$2v_e a_e$	$(v_f^2 + a_f^2)$	$2v_f a_f$	χ_9
10	$2v_e a_e$	$(v_e^2 + a_e^2)$	$2v_f a_f$	$(v_f^2 + a_f^2)$	χ_{10}
11	$\frac{1}{(2s_W)^2}$	$\frac{1}{(2s_W)^2}$	$\frac{1}{(2s_W)^2}$	$\frac{1}{(2s_W)^2}$	χ_{11}

Table 1: Couplings and propagators

$$F_1(s,t) = \text{Re} \sum_{j,k=1}^{11} (V_j^e V_k^{e*} - A_j^e A_k^{e*})(V_j^f V_k^{f*} + A_j^f A_k^{f*}) \chi_j \chi_k^*$$

$$F_2(s,t) = -\text{Im} \sum_{j,k=1}^{11} (V_j^e A_k^{e*} - A_j^e V_k^{e*})(V_j^f V_k^{f*} + A_j^f A_k^{f*}) \chi_j \chi_k^* \,.$$

For the functions G_2 and H_2 the lowest order expressions in (4) are sufficient; the same applies to the small mass term in G_1, H_1.

The cross section for unpolarized beams requires only G_1 and G_3. H_1, H_3 yield the longitudinal and F_1, F_2 the transverse polarization part of the cross section.

4 Effective couplings at the Z peak

Around the Z resonance, the amplitude for $e^+e^- \to f\bar{f}$ of the previous section can be cast into a simpler form closer to the lowest order amplitude:

$$A(e^+e^- \to f\bar{f}) = A_\gamma + A_Z + (box) \tag{34}$$

where A_γ denotes the dressed photon and A_Z the dressed Z exchange amplitude. (box) summarizes the terms from the massive box diagrams which, however, can be neglected around the Z.

4.1 *Electromagnetic couplings*

The dressed photon exchange amplitude can be written in the following way:

$$A_\gamma = \frac{e^2}{1+\hat{\Pi}^\gamma(s)} \cdot \frac{Q_e Q_f}{s} \cdot [(1+F_V^{\gamma e})\gamma_\mu - F_A^{\gamma e}\gamma_\mu\gamma_5] \otimes \left[(1+F_V^{\gamma f})\gamma^\mu - F_A^{\gamma f}\gamma^\mu\gamma_5\right] \,. \tag{35}$$

$\hat{\Pi}^\gamma$ is the subtracted vacuum polarization $\hat{\Pi}^\gamma(s) = \Pi^\gamma(s) - \Pi^\gamma(0)$ with

$$\Pi^\gamma(s) = \frac{\Sigma^{\gamma\gamma}(s)}{s} \,.$$

Writing it in the denominator takes into account the resummation of the leading log's from the light fermions. The bosonic contributions have to be understood as expanded to first order. The form factors $F_{V,A}(s)$ from the vertex corrections vanish for real photons: $F_{V,A}^{\gamma e,f}(0) = 0$. The typical size of the various corrections is (real parts):

$$\begin{aligned} \hat{\Pi}^\gamma(M_Z^2) &= -0.06 \\ F_V^{\gamma e}(M_Z^2) &\simeq F_A^{\gamma e}(M_Z^2) \simeq 10^{-3} \,. \end{aligned}$$

For the region around the Z peak, the photon vertex form factors are negligibly small. With sufficient accuracy, we can approximate (35) by

$$A_\gamma = \frac{e^2}{1+\hat{\Pi}^\gamma_{ferm}(s)} \cdot \frac{Q_e Q_f}{s} \cdot \gamma_\mu \otimes \gamma^\mu \tag{36}$$

with the fermionic vacuum polarization only. For the Z resonance region, we can take the result I (126) keeping a small s-dependence:

$$\hat{\Pi}^\gamma_{ferm}(s) = -0.0595 - \frac{40\alpha}{18\pi}\log\frac{s}{91.187\text{GeV}} \pm 0.0009 + i\frac{\alpha}{3}\sum_{f\neq t} Q_f^2 N_C^f \,. \tag{37}$$

4.2 *Neutral current couplings*

More important is the weak dressing of the Z exchange amplitude. Without the box diagrams the corrections factorize and we obtain a result quite close to the Born amplitude:

$$\begin{aligned} A_Z \;=\; & \sqrt{2}G_\mu M_Z^2(\rho_e\rho_f)^{1/2}\cdot \\ & \cdot\frac{[\gamma_\mu(I_3^e - 2Q_e s_W^2\kappa_e) - I_3^e\gamma_\mu\gamma_5]\otimes[\gamma^\mu(I_3^f - 2Q_f s_W^2\kappa_f) - I_3^f\gamma^\mu\gamma_5]}{s - M_Z^2 + i\frac{s}{M_Z^2}\cdot M_Z\Gamma_Z}\,. \end{aligned} \tag{38}$$

The weak corrections appear in terms of fermion-dependent form factors ρ and κ in the coupling constants and in the width in the denominator.

The s-dependence of the imaginary part is due to the s-dependence of $\mathrm{Im}\hat{\Sigma}^Z$; the linearization is completely sufficient in the resonance region. We postpone the discussion of the Z width for the moment and continue with the form factors.

The form factors ρ and κ in Eq. (38) have *universal* parts (i.e. independent of the fermion species) and *non-universal* parts which explicitly depend on the type of the external fermions. The universal parts arise from the counterterms and the boson self-energies, the non-universal parts from the vertex corrections and the fermion self-energies in the external lines:

$$\begin{aligned} \rho_f &= \rho_{univ}[1 + (\Delta\rho)^f_{non-univ}] \\ \kappa_f &= \kappa_{univ}[1 + (\Delta\kappa)^f_{non-univ}]\,. \end{aligned} \tag{39}$$

with

$$\begin{aligned} \rho_{univ} &= \frac{1-\Delta r}{1+\hat{\Pi}^Z(M_Z^2)} \\ (\Delta\rho)_{non-univ} &= (1 + F_A^{Zf}(M_Z^2)/a_f)^2 - 1 \end{aligned}$$

$$
\begin{aligned}
\kappa_{univ} &= 1 - \frac{c_W}{s_W} \frac{\hat{\Pi}^{\gamma Z}(M_Z^2)}{1+\hat{\Pi}^{\gamma}(M_Z^2)} \\
(\Delta\kappa)_{non-univ} &= \frac{1}{2\bar{s}_W^2 Q_f} \frac{v_f F_A^{Zf}(M_Z^2) - a_f F_V^{Zf}(M_Z^2)}{a_f + F_A^{Zf}(M_Z^2)}
\end{aligned} \tag{40}
$$

with

$$
\bar{s}_W^2 = s_W^2\, \kappa_{univ}\,. \tag{41}
$$

The building blocks are: Δr with the higher order terms as given in Eq. I (155), or equivalently in Eq. (13); $\hat{\Pi}^Z$ in Eq. (17); the vertex correction form factors in Eq. (21); and the quantity

$$
\hat{\Pi}^{\gamma Z}(M_Z^2) = \frac{\Sigma^{\gamma Z}(M_Z^2) - \Sigma^{\gamma Z}(0)}{M_Z^2} - \frac{c_W}{s_W}\left(\frac{\delta M_Z^2}{M_Z^2} - \frac{\delta M_W^2}{M_W^2}\right) + 2\,\frac{\Sigma^{\gamma Z}(0)}{M_Z^2}. \tag{42}
$$

The form factors ρ and κ listed above correspond, up to small terms which are negligible near the Z peak, to those by Bardin et al. [4, 5] used in the code ZFITTER [6]. In eq. (40) we have kept the form factors in the non-expanded version in order to have all the leading 2-loop terms also for the Zbb case correct. The non-leading terms have to be understood as expanded. We prefer the non-expanded version because it is more compact and the numerical differences are below the theoretical uncertainties (see section 7.3 of I).

Evaluating the complicated expressions in the leading terms, we obtain the simple result

$$
\begin{aligned}
\rho_{univ} &= \frac{1}{1-\Delta\bar{\rho}} + \cdots \\
\kappa_{univ} &= 1 + \frac{c_W^2}{s_W^2}\,\Delta\bar{\rho} + \cdots
\end{aligned}
$$

with the irreducible $\Delta\bar{\rho}$ from Eq. I (156). This result contains the summation of all reducible contributions to the ρ-parameter together with the irreducible contribution up to the 2-loop order. Expanding up to 1-loop order we recover the expressions found in I (134) by the discussion of the electroweak parameter shifts.

With ρ_f and κ_f we can define NC vertices at the Z resonance with effective coupling constants $g_{V,A}^f$:

$$
\begin{aligned}
J_\mu^{NC} &= \left(\sqrt{2}G_\mu M_Z^2 \rho_f\right)^{1/2} \left[(I_3^f - 2Q_f s_W^2 \kappa_f)\gamma_\mu - I_3^f \gamma_\mu\gamma_5\right] \\
&= \left(\sqrt{2}G_\mu M_Z^2\right)^{1/2} \left[g_V^f\,\gamma_\mu - g_A^f\,\gamma_\mu\gamma_5\right].
\end{aligned} \tag{43}
$$

Due to the imaginary parts of the self energies and vertices, the form factors and the effective couplings, respectively, are complex quantities. The approximation, where the couplings are taken as real, is called the "improved Born approximation".

The Zbb couplings:

The separation of a universal part in the effective couplings is sensible for two reasons: for the light fermions ($f \neq b, t$) the non-universal contributions are small, and (practically) independent of the unknown parameters m_t, M_H which enter only the universal part. This is, however, not true for the b-quark where also the non-universal parts have a strong dependence on m_t [7, 8, 9] resulting from the virtual top quark in the vertex corrections. For the leading term $\sim m_t^2$, next order corrections have been calculated recently:

(a) electroweak 2-loop corrections $O(G_\mu^2 m_t^4)$ [10, 11]

(b) QCD corrections $O(\alpha_s G_\mu m_t^2)$ [12, 13, 14]

The result for the form factors can be written as follows:

$$\begin{aligned} (\Delta\rho)^b_{non-univ} &= (\Delta\rho)^d_{non-univ} + \Delta\rho^b(m_t), \\ (\Delta\kappa)^b_{non-univ} &= (\Delta\kappa)^d_{non-univ} + \Delta\kappa^b(m_t) \end{aligned} \tag{44}$$

with

$$\begin{aligned} \Delta\rho^b(m_t) &= -4\frac{G_\mu m_t^2}{8\pi^2\sqrt{2}}\left[1 + \frac{G_\mu m_t^2}{8\pi^2\sqrt{2}}(\tau^{(2)} - 1) - \frac{\alpha_s}{\pi}\frac{\pi^2 - 3}{3}\right] \\ & -\frac{\alpha}{4\pi s_W^2}\left(\frac{8}{3} + \frac{1}{6c_W^2}\right)\log\frac{m_t^2}{M_W^2} + \cdots \\ \Delta\kappa^b(m_t) &= -\frac{1}{2}\Delta\rho^b(m_t) + 6\left(\frac{G_\mu m_t^2}{8\pi^2\sqrt{2}}\right)^2 . \end{aligned} \tag{45}$$

The 2-loop coefficient $\tau^{(2)}$, calculated in [10], is shown in Figure 4. For small Higgs masses $M_H \ll m_t$ the result is

$$\tau^{(2)} = 9 - \frac{\pi^2}{3} \tag{46}$$

This result has been confirmed recently by an independent calculation [11]. For $M_H > 2m_t$ a good approximation is given by the asymptotic expression [10] with $r = (m_t/M_H)^2$:

$$\begin{aligned} \tau^{(2)} &= \frac{1}{144}\Big[311 + 24\pi^2 + 282\log r + 90\log^2 r \\ & -4r(40 + 6\pi^2 + 15\log r + 18\log^2 r) \\ & +\frac{3r^2}{100}(24209 - 6000\pi^2 - 45420\log r - 18000\log^2 r)\Big] \end{aligned} \tag{47}$$

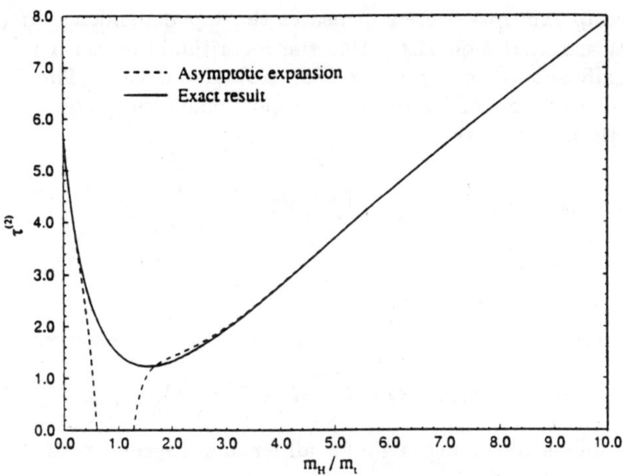

Fig. 4: *The function $\tau^{(2)}$ in Eq. (45), from [10].*

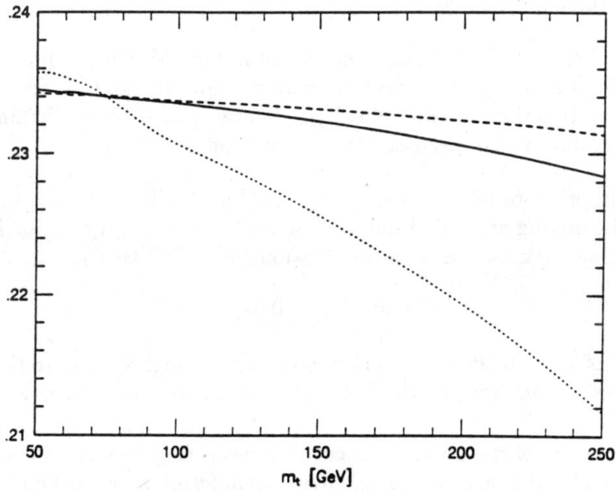

Fig. 5: $s_W^2 = 1 - M_W^2/M_Z^2$ *(dotted) and the on-resonance effective mixing angles for leptons (full) and for b quarks (dashed).* $M_H = 100\,GeV$, $M_Z = 91.187\ GeV$.

The appearance of $\Delta\rho^b(m_t)$ overcompensates the top dependence of $(\Delta\rho)_{univ}$ in amplitudes with external b-quarks. Also the logarithmic m_t-term in Eq. (45) is numerically significant for top quark masses around 150 GeV. The complete expression for the 1-loop part of $\Delta\rho^b(m_t)$ not proportional to m_t^2 is obtained from F_L^b, appendix Eq. (102), as follows:

$$\Delta\rho^b(m_t)_{NL} = -\frac{\alpha}{4\pi}\left(4\,F_L^b(M_Z^2) - \frac{m_t^2}{s_W^2 c_W^2 M_Z^2}\right).$$

Effective mixing angles:

Of particular interest are the effective mixing angles in the vector couplings:

$$\sin^2\theta_{eff}^f := s_W^2\,\mathrm{Re}\,\kappa_f = \mathrm{Re}\left(\overline{s}_W^2 + \overline{s}_W^2\,(\Delta\kappa)_{non-univ}^f\right). \tag{48}$$

$\overline{s}_W^2$ denotes the universal mixing angle for all fermion species; in the leading term it is given by

$$\overline{s}_W^2 = s_W^2 + c_W^2\,\Delta\overline{\rho} + \cdots \tag{49}$$

It is equivalent to s_*^2 of [15] and, up to non-leading terms, to s'^2 of [16] and to the $\overline{MS}$ mixing angle $\hat{s}^2$ of [17], discussed in section 7.1 of I. Various mixing angles are displayed in Figure 5. Compared to s_W^2, the on-resonance couplings are less sensitive to m_t than the W mass.

In table 2 and 3 we put together the Standard Model predictions for the normalization factors ρ_f and the effective mixing angles in Eq. (43) for the various fermion types as functions of m_t and M_H. These quantities determine the on-resonance observables to be discussed later in section 5.

As one can read off from table 3 and from table 3 in I, the difference between the effective leptonic mixing angle and the $\overline{MS}$ mixing angle $\hat{s}^2$ is very small. Neglecting the little variation with m_t, one has the relation ($m_t < 200$ GeV)

$$\hat{s}^2 \simeq \sin^2\theta_{eff}^l + 0.0002. \tag{50}$$

An uncertainty of $\delta s^2 = 0.0003$ is common to all mixing angles is due to the hadronic error in the photon vacuum polarization, Eq. I (125) (see also the discussion in I, section 7.3).

The advantage of the universal mixing angles is that they are flavor independent and incorporate the large universal effects from fermion loops. One has to keep in mind, however, that they are theoretical constructs and do not simply relate to physical quantities, like s_W^2 or $\sin^2\theta_{eff}^l$. Their precise extraction from experimental data requires a clean treatment of the residual corrections which are different for the various definitions.

m_top	M_H = 60	300	1000
	neutrinos		
100.	1.0034-i.0001	1.0027-i.0001	1.0015-i.0000
120.	1.0049-i.0001	1.0042-i.0001	1.0030-i.0001
150.	1.0075-i.0002	1.0068-i.0001	1.0056-i.0001
180.	1.0108-i.0002	1.0100-i.0002	1.0088-i.0002
210.	1.0146-i.0003	1.0138-i.0003	1.0125-i.0002
240.	1.0191-i.0004	1.0181-i.0003	1.0168-i.0003
	leptons		
100.	1.0001-i.0047	0.9994-i.0046	0.9982-i.0046
120.	1.0013-i.0047	1.0006-i.0047	0.9994-i.0046
150.	1.0035-i.0047	1.0027-i.0047	1.0015-i.0047
180.	1.0062-i.0048	1.0053-i.0048	1.0041-i.0047
210.	1.0094-i.0049	1.0084-i.0048	1.0070-i.0048
240.	1.0131-i.0050	1.0119-i.0049	1.0104-i.0049
	up-quarks		
100.	1.0007-i.0035	1.0001-i.0035	0.9988-i.0035
120.	1.0020-i.0036	1.0013-i.0035	1.0000-i.0035
150.	1.0042-i.0036	1.0034-i.0036	1.0022-i.0035
180.	1.0069-i.0037	1.0060-i.0036	1.0047-i.0036
210.	1.0101-i.0037	1.0091-i.0037	1.0077-i.0037
240.	1.0138-i.0038	1.0126-i.0037	1.0111-i.0037
	down-quarks		
100.	1.0017-i.0020	1.0010-i.0020	0.9997-i.0020
120.	1.0029-i.0020	1.0022-i.0020	1.0009-i.0020
150.	1.0051-i.0021	1.0043-i.0020	1.0031-i.0020
180.	1.0078-i.0021	1.0069-i.0021	1.0056-i.0021
210.	1.0110-i.0022	1.0100-i.0021	1.0086-i.0021
240.	1.0148-i.0022	1.0135-i.0022	1.0120-i.0022
	b-quarks		
100.	0.9985+i.0017	0.9977+i.0017	0.9965+i.0017
120.	0.9975+i.0017	0.9967+i.0017	0.9955+i.0017
150.	0.9956+i.0018	0.9948+i.0017	0.9935+i.0017
180.	0.9934+i.0018	0.9925+i.0018	0.9912+i.0018
210.	0.9908+i.0018	0.9899+i.0018	0.9884+i.0018
240.	0.9879+i.0019	0.9869+i.0019	0.9852+i.0018

Table 2: Standard Model values for the overall normalization factors ρ_f in the neutral current couplings of Eq. (43) for the various fermion species. The input values for m_t and M_H are in GeV. Other parameters used for the calculation are: $M_Z = 91.187$ GeV, $\alpha_s = 0.12$.

m_top	M_H = 60	300	1000
		leptons	
100.	0.2333+i.0031	0.2341+i.0031	0.2348+i.0030
120.	0.2328+i.0031	0.2336+i.0031	0.2343+i.0030
150.	0.2320+i.0031	0.2328+i.0031	0.2335+i.0031
180.	0.2310+i.0031	0.2319+i.0031	0.2326+i.0031
210.	0.2299+i.0031	0.2308+i.0031	0.2315+i.0031
240.	0.2286+i.0031	0.2296+i.0031	0.2303+i.0031
		up-quarks	
100.	0.2332+i.0029	0.2340+i.0029	0.2347+i.0029
120.	0.2327+i.0029	0.2335+i.0029	0.2342+i.0029
150.	0.2319+i.0029	0.2327+i.0029	0.2334+i.0029
180.	0.2309+i.0029	0.2318+i.0029	0.2325+i.0029
210.	0.2298+i.0029	0.2307+i.0029	0.2314+i.0029
240.	0.2285+i.0030	0.2295+i.0029	0.2303+i.0029
		down-quarks	
100.	0.2331+i.0027	0.2339+i.0027	0.2346+i.0027
120.	0.2326+i.0027	0.2334+i.0027	0.2341+i.0027
150.	0.2318+i.0027	0.2326+i.0027	0.2333+i.0027
180.	0.2308+i.0027	0.2317+i.0027	0.2323+i.0027
210.	0.2297+i.0028	0.2306+i.0027	0.2313+i.0027
240.	0.2284+i.0028	0.2294+i.0028	0.2301+i.0027
		b-quarks	
100.	0.2334+i.0023	0.2343+i.0023	0.2349+i.0022
120.	0.2332+i.0023	0.2341+i.0023	0.2347+i.0023
150.	0.2329+i.0023	0.2337+i.0023	0.2344+i.0023
180.	0.2325+i.0023	0.2333+i.0023	0.2340+i.0023
210.	0.2320+i.0023	0.2329+i.0023	0.2336+i.0023
240.	0.2315+i.0023	0.2324+i.0023	0.2332+i.0023

Table 3: Standard Model values for the effective mixing angles $\kappa_f s_W^2$ in the neutral current couplings of Eq. (43) for the various fermion species. The input values for m_t and M_H are in GeV. Other parameters used for the calculation are: $M_Z = 91.187$ GeV, $\alpha_s = 0.12$.

4.3 *Cross sections around the Z resonance*

With help of the amplitudes A_γ in the sufficient approximation of Eq. (36) and A_Z in Eq. (38) with the NC couplings (43) we can write down a Born-like expression for the $e^+e^- \to f\bar{f}$ cross section around the Z resonance. The general expressions (1) and (2) for the differential cross section are still valid. We can take over the expressions in Eq. (4) for the invariant functions with the following substitutions:

$$
\begin{aligned}
& v_e^2 \to |\, g_V^e \,|^2, \quad a_e^2 \to |\, g_A^e \,|^2 \\
& v_f^2 \to |\, g_V^f \,|^2, \quad a_f^2 \to |\, g_A^f \,|^2 \\
& \chi_Z^0 \to \chi_Z = \frac{\sqrt{2}G_\mu M_Z^2}{4\pi\alpha} \, \frac{1}{s - M_Z^2 + i\frac{s}{M_Z}\Gamma_Z} \\
& \{v_e v_f, a_e a_f, a_e v_f, v_e a_f\} \,\mathrm{Re}\,\chi_Z^0 \to \mathrm{Re}\,(\,\{g_V^e g_V^f, g_A^e g_A^f, g_A^e g_V^f, g_V^e g_A^f\}\, \chi_\gamma^* \chi_Z\,) \\
& v_e a_e, v_e a_e v_f a_f \to \mathrm{Re}(g_V^e g_A^{e*}),\ \mathrm{Re}\,(g_V^e g_A^{e*})\,\mathrm{Re}\,(g_V^f g_A^{f*}) \\
& Q_e^2 Q_f^2 \to Q_e^2 Q_f^2 \,|\, \chi_\gamma \,|^2
\end{aligned}
\tag{51}
$$

with

$$
\chi_\gamma = \frac{1}{1 + \hat{\Pi}^\gamma_{ferm}(s)}
\tag{52}
$$

from Eq. (37). Note that the imaginary part of the photon vacuum polarization is of some significance for the leptonic forward-backward asymmetry on-resonance where it contributes $\simeq +0.002$.

In the way described above, we obtain the pure electroweak cross section- which still has to be dressed by QED corrections and final state QCD corrections in case of quark pair production before comparisons with experimental data can be performed.

5 Z peak observables

5.1 *The Z line shape*

The integrated cross section $\sigma(s)$ for $e^+e^- \to f\bar{f}$ around the Z resonance with unpolarized beams is obtained from the formulae of the previous section in a straight forward way, expressed in terms of the effective vector and axial vector coupling constants. It is, however, convenient to rewrite $\sigma(s)$ in terms of the Z width and the partial widths Γ_e, Γ_f in order to have a more model independent parametrization. The following form [18, 19] includes final state photon radiation and QCD corrections in case of quark final states: [2]

$$\sigma(s) = \frac{12\pi\Gamma_e\Gamma_f}{|\, s - M_Z^2 + iM_Z\Gamma_Z(s)\,|^2}\left\{\frac{s}{M_Z^2} + R_f\,\frac{s-M_Z^2}{M_Z^2} + I_f\,\frac{\Gamma_Z}{M_Z} + \cdots\right\} + \frac{4\pi\,\alpha(s)^2}{3s}\,Q_f^2 N_C^f(1+\delta_{QCD})(1+\delta_{QED}) \qquad (53)$$

with

$$\Gamma_Z(s) = \Gamma_Z\left\{\frac{s}{M_Z^2} + \epsilon\,\frac{s-M_Z^2}{M_Z^2} + \cdots\right\} \qquad (54)$$

and $N_C^f = 1$ for leptons and $= 3$ for quarks. The QCD correction in case $f = q$ is given in Eq. (71). The terms R_f, I_f, ϵ are small quantities calculable in terms of the basic parameters. R_f and I_f describe the γ-Z interference (improved Born approximation)

$$R_f = \frac{2\,Q_e Q_f g_V^e g_V^f}{[(g_V^e)^2+(g_A^e)^2][(g_V^f)^2+(g_A^f)^2]}\,\frac{4\pi\alpha(s)}{\sqrt{2}G_\mu M_Z^2}$$

$$I_f = \frac{2\,Q_e Q_f g_V^e g_V^f}{[(g_V^e)^2+(g_A^e)^2][(g_V^f)^2+(g_A^f)^2]}\,\frac{4\pi\alpha(s)}{\sqrt{2}G_\mu M_Z^2}\cdot\frac{s}{M_Z^2}\,\mathrm{Im}\,\hat{\Pi}^\gamma\,, \qquad (55)$$

and the last term is the QED background from pure photon exchange with

$$\alpha(s) = \frac{\alpha}{1+\mathrm{Re}\,\hat{\Pi}^\gamma_{ferm}(s)}$$

and $\hat{\Pi}^\gamma_{ferm}$ from Eq. (37). The small correction

$$\epsilon = \sum_f \epsilon_f, \quad \epsilon_f \simeq \frac{6m_f^2}{M_Z^2}\frac{\Gamma_f}{\Gamma_Z}\frac{(g_A^f)^2}{(g_V^f)^2+(g_A^f)^2}$$

is due to finite fermion mass effects in the final states. In the Standard Model. I_f and ϵ have negligible influence on the line shape.

The s-dependent width gives rise to a dislocation of the peak maximum by $\simeq -34$ GeV [20, 21] The first term in the expansion (53) is the pure Z resonance. It differs from a Breit-Wigner shape by the s-dependence of the width:

$$\sigma_{res}(s) = \sigma_0\,\frac{s\Gamma_Z^2}{(s-M_Z^2)^2 + s^2\Gamma_Z^2/M_Z^2}, \quad \sigma_0 = \frac{12\pi}{M_Z^2}\cdot\frac{\Gamma_e\Gamma_f}{\Gamma_Z^2}. \qquad (56)$$

[2] Since initial state photon radiation is treated separately the QED correction factor in Eq. (53) has to be removed in Γ_e.

By means of the substitution [21]

$$s - M_Z^2 + is\Gamma_Z/M_Z = (1+i\gamma)(s - \hat{M}_Z^2 + i\hat{M}_Z\hat{\Gamma}_Z) \tag{57}$$

with

$$\hat{M}_Z = M_Z(1+\gamma^2)^{-1/2},\ \hat{\Gamma}_Z = \Gamma_Z(1+\gamma^2)^{-1/2},\ \gamma = \frac{\Gamma_Z}{M_Z} \tag{58}$$

a Breit-Wigner resonance shape is recovered:

$$\sigma_{res}(s) = \sigma_0 \frac{s\hat{\Gamma}_Z^2}{(s-\hat{M}_Z^2)^2 + \hat{M}_Z^2\hat{\Gamma}_Z^2} \tag{59}$$

Numerically one finds: $\hat{M}_Z - M_Z \simeq -34$ MeV, $\hat{\Gamma}_Z - \Gamma_Z \simeq -1$ MeV. σ_0 is not changed. $\hat{M}_Z$ corresponds to the real part of the S-matrix pole of the Z-resonance [22].

QED corrections:

The observed cross section is the result of convoluting Eq. (53) with the initial state QED corrections consisting of virtual photon and real photon bremsstrahlung contributions:

$$\sigma_{obs}(s) = \int_0^{k_{max}} dk\, H(k)\, \sigma(s(1-k))\,. \tag{60}$$

k_{max} denotes a cut to the radiated energy. Kinematically it is limited by $1 - 4m_f^2/s$ or $1 - 4m_\pi^2/s$ for hadrons, respectively. For the required accuracy, multi-photon radiation has to be included. The radiator function $H(k)$ with soft-photon resummation and the exact $O(\alpha^2)$ result [23] for initial state QED corrections is given by [18]

$$H(k) = \beta\, k^{\beta-1}(1 + \delta_1^{V+S} + \delta_2^{V+S}) + \delta_1^H + \delta_2^H \tag{61}$$

with

$$\begin{aligned}
\beta &= \frac{2\alpha}{\pi}(L-1), \quad L = \log\frac{s}{m_e^2}\,, \\
\delta_1^{V+S} &= \frac{\alpha}{\pi}\left(\frac{3}{2}L + \frac{\pi^2}{3} - 2\right), \\
\delta_2^{V+S} &= \left(\frac{\alpha}{\pi}\right)^2\left[\left(\frac{9}{8} - \frac{\pi^2}{3}\right)L^2 + s_{21}L + s_{20}\right], \\
\delta_1^H &= \frac{\alpha}{\pi}(L-1)(k-2)\,, \\
\delta_2^H &= \left(\frac{\alpha}{\pi}\right)^2\left(h_{22}L^2 + h_{21}L + h_{20}\right), \\
h_{22} &= -\frac{1+(1-k)^2}{k}\log(1-k) \\
&\quad +(2-k)\left[\frac{1}{2}\log(1-k) - 2\log k - \frac{3}{2}\right] - k\,.
\end{aligned} \tag{62}$$

The other coefficients of the non-leading-log terms can be found in the Z Line Shape section of 'Z Physics at LEP 1' [18]. Usually the k-integration is performed numerically (see the codes ZSHAPE [18], ZFITTER [6], MIZA [24]) or by Monte Carlo simulation as in KORALZ [25]. Analytic expressions are given in [18, 19], and simple approximate formulae can be found in [26].

The QED corrections have two major impacts on the line shape:

- a reduction of the peak height of the resonance cross section by

$$\sigma_{obs}^{peak} \simeq \sigma_{res}^{peak} \left(\frac{\Gamma_Z}{M_Z}\right)^{\beta} (1 + \delta_1^{V+S}) \simeq 0.74\, \sigma_{res}^{peak}\,, \tag{63}$$

- a shift in the peak position compared to the non-radiative cross section by

$$\Delta\sqrt{s_{max}} = \frac{\beta\pi}{8}\Gamma_Z \tag{64}$$

  resulting in the relation between the peak position and the nominal Z mass:

$$\begin{aligned} \Delta\sqrt{s_{max}} &\simeq M_Z + \frac{\beta\pi}{8}\Gamma_Z - \frac{\Gamma_Z^2}{4M_Z} \\ &\simeq 89\,\mathrm{MeV}\,. \end{aligned} \tag{65}$$

It is important to note that, to high accuracy, these effects are practically universal, depending only on M_Z and Γ_Z as parameters. This allows a model independent determination of these parameters from the measured line shape. The effect of the initial state QED corrections on the line shape is depicted in Figure 6.

A final remark concerns the QED corrections resulting from the interference between initial and final state radiation. They are not included in the treatment above, but they can be added in $O(\alpha)$ since they are small anyway. For not too tight cuts, as it is the case for practical applications, these interference corrections to the line shape are negligible and we do not list them here.

From line shape measurements one obtains the parameters M_Z, Γ_Z, σ_0 or the partial widths, respectively. Whereas M_Z is used as a precise input parameter, together with α and G_μ, the width and partial widths allow comparisons with the predictions of the Standard Model to be discussed next.

5.2 *Z widths and partial widths*

The total Z width Γ_Z can be calculated as the sum over the partial decay widths

$$\Gamma_Z = \sum_f \Gamma(Z \to f\bar{f}) + \cdots \tag{66}$$

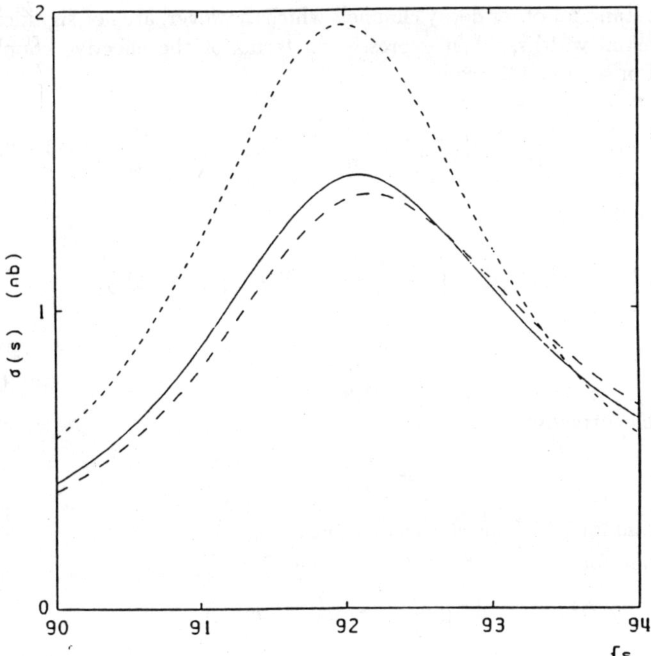

Fig. 6: *The Z line shape for muon pair production (from [18]). The various stages of the calculation are: only with weak corrections, no photonic corrections (- - -), with first order QED corrections (— —), and with second order exponentiated QED corrections (———).*

where the ellipses stand for other decay channels which, however, are not significant. The fermionic partial widths, when expressed in terms of the effective coupling constants defined in section 4.2, read:

$$\begin{aligned}\Gamma(Z\to f\bar{f}) &= \Gamma_0\sqrt{1-\frac{4m_f^2}{M_Z^2}}\left[\mid g_V^f\mid^2\left(1+\frac{2m_f^2}{M_Z^2}\right)+\mid g_A^f\mid^2\left(1-\frac{4m_f^2}{M_Z^2}\right)\right]\cdot(1+\delta_{QED})\\ &\quad +\Delta\Gamma_{QCD}^f \qquad (67)\\ &\simeq \Gamma_0\left[\mid g_V^f\mid^2+\mid g_A^f\mid^2\left(1-\frac{6m_f^2}{M_Z^2}\right)\right]\cdot(1+\delta_{QED})+\Delta\Gamma_{QCD}^f\end{aligned}$$

with

$$\Gamma_0 = N_C^f\,\frac{\sqrt{2}G_\mu M_Z^3}{12\pi}. \qquad (68)$$

The photonic QED correction

$$\delta_{QED} = Q_f^2\,\frac{3\alpha}{4\pi} \qquad (69)$$

is very small, maximum 0.17% for charged leptons.
The QCD correction for hadronic final states is given by

$$\Delta\Gamma_{QCD}^f = \Gamma_0\left[\mid g_V^f\mid^2+\mid g_A^f\mid^2\right]\cdot K_{QCD} \qquad (70)$$

with [27]

$$\begin{aligned}K_{QCD} &= 1+\frac{\alpha_s}{\pi}+1.41\left(\frac{\alpha_s}{\pi}\right)^2-12.8\left(\frac{\alpha_s}{\pi}\right)^3 \qquad (71)\\ &= 1.039\pm0.003 \quad \text{for } \alpha_s(M_Z^2)=0.12\pm0.01\end{aligned}$$

for the light quarks with $m_q \simeq 0$.

For b quarks the QCD corrections are different due to finite b mass terms [28] and to top quark dependent 2-loop diagrams for the axial part [29]:

$$\begin{aligned}\Delta\Gamma_{QCD}^b &= \Gamma_0\left[\mid g_V^b\mid^2+\mid g_A^b\mid^2\right]\cdot K_{QCD}\\ &\quad +\Gamma_0\mid g_V^b\mid^2\cdot\frac{12m_b^2}{M_Z^2}\left[\frac{\alpha_s}{\pi}+\left(\frac{\alpha_s}{\pi}\right)^2(6.07-2\ell)\right.\\ &\qquad\qquad \left.+\left(\frac{\alpha_s}{\pi}\right)^3(2.38-24.29\ell+0.083\ell^2)\right]\\ &\quad +\Gamma_0\mid g_A^b\mid^2\cdot\frac{6m_b^2}{M_Z^2}\left[\frac{\alpha_s}{\pi}(2\ell-1)+\left(\frac{\alpha_s}{\pi}\right)^2(17.96+14.14\ell-0.083\ell^2)\right]\\ &\quad +\Gamma_0\mid g_A^b\mid^2\cdot\frac{1}{3}\left(\frac{\alpha_s}{\pi}\right)^2 I(M_Z^2/4m_t^2) \qquad (72)\end{aligned}$$

with

$$\ell = \log \frac{M_Z^2}{m_b^2}$$

and

$$I(x) \simeq -9.250 + 1.037x + 0.632x^2 + 6\,\log(2\sqrt{x})\,.$$

The finite b-mass terms contribute +2 MeV to the partial Z width into b quarks. Moreover, the top mass dependent correction at the 2-loop level yields an additional, but negative, contribution. For large m_t this top-dependent term tends to cancel the positive and constant correction resulting from $m_b \neq 0$ in $O(\alpha_s)$.

By the error of α_s, as taken in Eq. (71), an uncertainty in the total Z width of $\Delta\Gamma_Z = 7$ MeV is induced.

In table 4 the Standard Model predictions for the various partial widths and the total width of the Z boson are collected. They include all the electroweak, QED and QCD corrections discussed above.

Figure 7a contains the total width prediction of the Standard Model together with the average from the LEP experiments [30]. The hadronic and the leptonic widths are separately displayed in Figures 7b and 7c together with the LEP averages [30]. All data are in agreement with the Standard Model predictions and can be interpreted as constraints to the allowed top mass range.

An instructive way of pointing out the top dependent vertex corrections is the ratio Γ_b/Γ_{had} which is shown in Figure 8. It is practically independent of the Higgs mass. The experimental result [31]

$$\frac{\Gamma_b}{\Gamma_{had}} = 0.2200 \pm 0.0031$$

yields a stringent bound on the top quark mass which is largely independent of the ρ parameter.

Another observable of particular interest is the ratio of the hadronic to the leptonic width of the Z boson [32] (see comment after table 5)

$$R_Z = \frac{\Gamma_{had}}{\Gamma_e} = 20.76 \pm 0.08\,. \qquad (73)$$

R_Z is practically a constant for given M_Z in the minimal model due to (accidental) cancellations between the universal and b-vertex contributions. The error is almost completely the QCD uncertainty. Non-standard terms would spoil this cancellation and exhibit a deviation from the value given in Eq. (73) which makes R_Z a sensitive indicator of possible New Physics. The present LEP average $R_Z^{exp} = 20.83 \pm 0.06$ [30] however, does not show deviations from the Standard Model.

m_t	M_H	Γ_ν	Γ_e	Γ_u	Γ_d	Γ_b	Γ_{had}	Γ_{tot}	R_Z
100	60	166.3	83.47	296.2	382.3	377.9	1734.4	2483.6	20.79
	100	166.3	83.45	296.0	382.2	377.7	1733.7	2482.9	20.78
	500	166.1	83.32	295.2	381.3	376.8	1729.4	2477.6	20.76
	1000	166.0	83.24	294.8	380.8	376.3	1727.1	2474.6	20.75
120	60	166.5	83.59	296.8	383.0	377.5	1736.7	2486.9	20.78
	100	166.5	83.57	296.7	382.9	377.3	1736.0	2486.2	20.78
	500	166.3	83.44	295.8	382.0	376.4	1731.7	2480.9	20.76
	1000	166.2	83.36	295.4	381.5	375.9	1729.4	2477.9	20.75
150	60	166.9	83.81	297.9	384.3	376.8	1740.9	2492.8	20.78
	100	166.9	83.80	297.8	384.2	376.6	1740.1	2492.0	20.77
	500	166.7	83.65	296.9	383.3	375.7	1735.7	2486.6	20.75
	1000	166.6	83.57	296.5	382.8	375.2	1733.4	2483.6	20.75
180	60	167.4	83.09	299.2	385.9	376.0	1745.8	2500.0	20.77
	100	167.3	84.07	299.1	385.7	375.8	1745.1	2499.2	20.76
	500	167.1	83.92	298.2	384.8	374.9	1740.4	2493.4	20.75
	1000	167.0	83.83	297.8	384.3	374.4	1738.0	2490.3	20.74
210	60	167.9	84.42	300.8	387.7	375.1	1751.7	2508.4	20.75
	100	167.9	84.40	300.6	387.5	375.0	1750.9	2507.5	20.75
	500	167.6	84.23	299.7	386.5	373.9	1745.9	2501.3	20.73
	1000	167.5	84.13	299.2	386.0	373.4	1743.3	2498.0	20.73
240	60	168.5	84.81	302.5	389.8	374.1	1758.4	2518.2	20.74
	100	168.5	84.78	302.4	389.6	374.0	1757.5	2517.2	20.74
	500	168.2	84.58	301.3	388.5	372.8	1752.0	2510.2	20.72
	1000	168.1	84.48	300.8	387.9	372.2	1749.2	2506.5	20.71

Table 4: Partial and total Z widths in MeV for various top and Higgs masses (in GeV). $M_Z = 91.187$ GeV, $\alpha_s = 0.12$. Not listed are the values for $\Gamma_\tau = 0.9977\,\Gamma_e$ and $\Gamma_c = 0.9986\,\Gamma_u$.

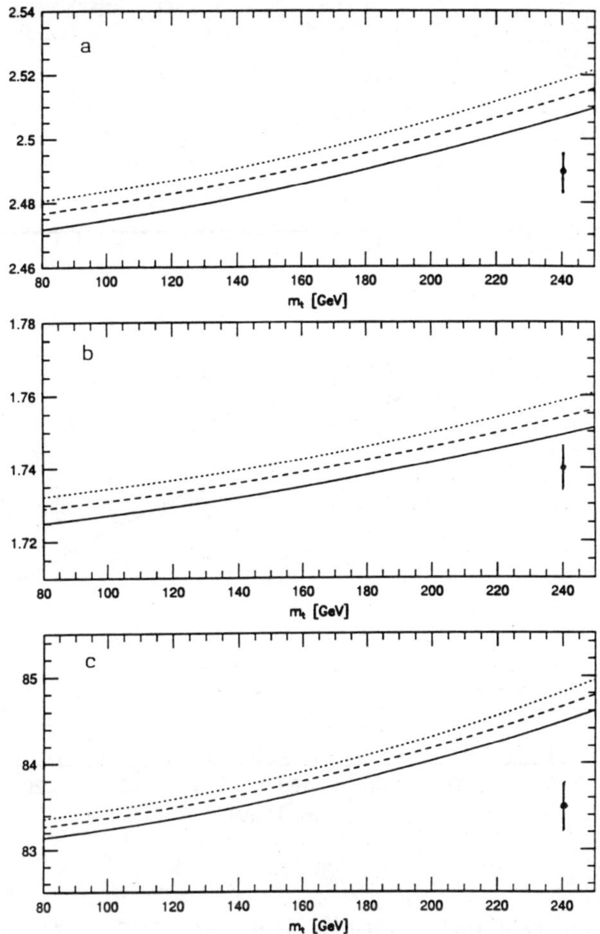

Fig. 7: *Standard Model predictions for* a: *the total Z width in GeV,* b: *the hadronic Z width in GeV,* c: *the leptonic Z width in MeV.* $M_H = 60$ *GeV (dotted), 300 GeV (dashed), 1000 GeV (full).* $M_Z = 91.187$ *GeV.*

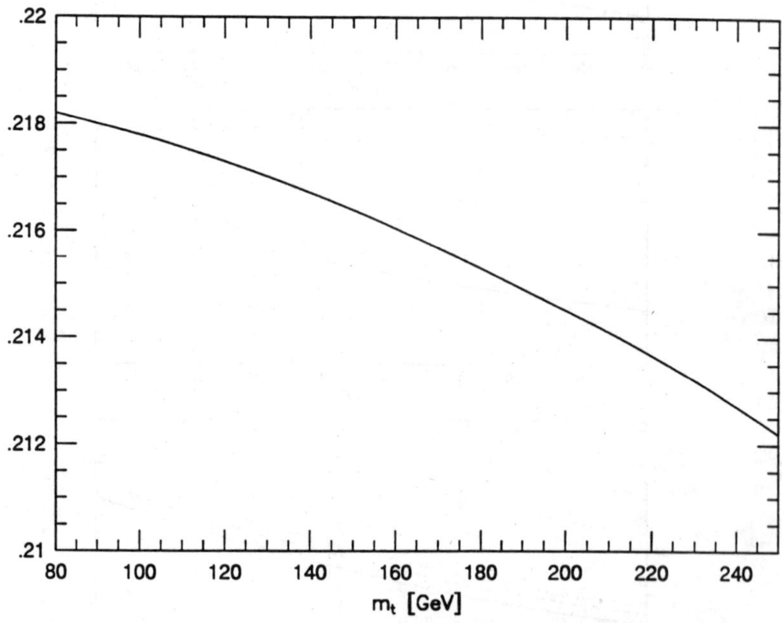

Fig. 8: *The Standard Model prediction for the ratio* Γ_b/Γ_{had}. $M_Z = 91.187$ *GeV. The dependence on* M_H *in the predicted curve is not visible.*

observable	LEP 1993	Standard Model prediction
M_Z (GeV)	91.187 ± 0.007	input
Γ_Z (GeV)	2.488 ± 0.007	$2.486 \pm 0.010 \pm 0.007$
σ_0^{had} (nb)	41.45 ± 0.17	$41.45 \pm 0.03 \pm 0.06$
R_Z	20.83 ± 0.06	$20.76 \pm 0.03 \pm 0.07$
Γ_e (MeV)	83.5 ± 0.3	83.7 ± 0.4
Γ_{inv} (MeV)	496.2 ± 8.8	500.3 ± 1.9
Γ_b/Γ_{had}	0.2200 ± 0.0031	0.2160 ± 0.0016
ρ_l	1.0002 ± 0.0035	1.0026 ± 0.0038
$\sin^2\theta_{eff}^l$	0.2322 ± 0.0008	0.2332 ± 0.0013

Table 5: LEP results and Standard Model predictions for the Z parameters.

In table 5 the Standard Model predictions for Z pole observables are put together. The first error is the variation with m_t, M_H in the range allowed by Fig. 8 in I, the second error is the hadronic uncertainty from α_s. The recent combined LEP results on the Z resonance parameters, under the assumption of lepton universality, are also shown in table 5, according to Schaile [30]. The value for Γ_b/Γ_{had} is from Tenchini [31]. The experimental value for $\sin^2\theta_{eff}^l$ is from the forward-backward asymmetry for leptons and b quarks and the τ polarization.

5.3 *Asymmetries*

5.3.1 *Left-right asymmetry*

The left-right asymmetry is defined as the ratio

$$A_{LR} = \frac{\sigma_L - \sigma_R}{\sigma_L + \sigma_R} \tag{74}$$

where $\sigma_{L(R)}$ denotes the integrated cross section for left (right) handed electrons. A_{LR}, in case of lepton universality, is equal to the final state polarization in τ-pair production:

$$A_{pol}^{\tau} = A_{LR}\,. \tag{75}$$

Expressed in terms of the functions G_i, H_i from (4) with the substitution (51) using the effective coupling constants one finds

$$A_{LR} = \frac{H_1(s) + 2\mu_f H_2(s)}{G_1(s) + 2\mu_f G_2(s)}, \tag{76}$$

including the set of weak corrections. For the on-resonance asymmetry ($s = M_Z^2$)

in the improved Born approximation this simplifies to

$$A_{LR}(M_Z^2) = A_e + \Delta A_{LR}^I + \Delta A_{LR}^Q \tag{77}$$

where the combination

$$A_e = \frac{2g_V^e g_A^e}{(g_V^e)^2 + (g_A^e)^2} = \frac{2(1-4\sin^2\theta_{eff}^l)}{1+(1-4\sin^2\theta_{eff}^l)^2} \tag{78}$$

depends only on the effective mixing angle Eq. (48) for the electron. The small contributions from the interference with the photon exchange

$$\Delta A_{LR}^I = \frac{2Q_e Q_f\, g_A^e g_V^f}{(g_V^{e\,2} + g_A^{e\,2})\,(g_V^{f\,2} + g_A^{f\,2})} \frac{4\pi\alpha(M_Z^2)}{\sqrt{2}G_\mu M_Z^2} \cdot \frac{\Gamma_Z}{M_Z} \cdot \mathrm{Im}\,\hat{\Pi}^\gamma \tag{79}$$

and from the pure photon exchange part

$$\Delta A_{LR}^Q = -\frac{A_e\, Q_e^2 Q_f^2}{(g_V^{e\,2} + g_A^{e\,2})\,(g_V^{f\,2} + g_A^{f\,2})} \left(\frac{4\pi\alpha(M_Z^2)}{\sqrt{2}G_\mu M_Z^2}\right)^2 \left(\frac{\Gamma_Z}{M_Z}\right)^2 . \tag{80}$$

are listed in table 6 for the various final state fermions. Except from lepton final states, they are negligibly small. Mass effects from final fermions practically cancel. The same holds for QCD corrections in the case of quark final states, final state QED corrections, and QED corrections from the interference of initial-final state photon radiation. Initial state QED corrections can be treated in complete analogy to Eq. (60) applied to $\sigma_{L,R}(s)$. Their net effect in the asymmetry (74) is also very small and practically independent of cuts [33, 34]. A_{LR} thus represents a unique laboratory for testing the non-QED part of the electroweak theory. Measurements of A_{LR} are essentially measurements of $\sin^2\theta_{eff}^l$ or of the ratio g_V^e/g_A^e. Table 7 contains the Standard Model predictions for A_{LR} based on the complete expression (76). They depend significantly on m_t and M_H, for the various flavors. Graphically, the left-right asymmetry for leptonic final states (no QED corrections) is depicted in Figure 9.

5.3.2 *Forward-backward asymmetries*

The forward-backward asymmetry is defined by

$$A_{FB} = \frac{\sigma_F - \sigma_B}{\sigma_F + \sigma_B} \tag{81}$$

with

$$\sigma_F = \int_{\theta>\pi/2} d\Omega\, \frac{d\sigma}{d\Omega}, \quad \sigma_B = \int_{\theta<\pi/2} d\Omega\, \frac{d\sigma}{d\Omega} . \tag{82}$$

f	A_e	ΔA_{LR}^I	ΔA_{LR}^Q
μ	0.1511	0.0002	-0.0009
τ	0.1511	0.0002	-0.0009
c	0.1511	0.0005	-0.0003
b	0.1511	0.0004	-0.0001

Table 6: Contributions to the on-resonance left-right asymmetry for various final state fermions. $\sin^2\theta_{eff}^l = 0.231$, $M_Z = 91.187$ GeV.

This gives for the asymmetry without QED and QCD corrections:

$$A_{FB} = \frac{3}{4}\frac{G_3(s)}{G_1(s)+2\mu_f G_2(s)}\sqrt{1-4\mu_f} \tag{83}$$

with the functions G_i from (4) and the substitution (51) incorporating the set of weak corrections in terms of the effective coupling constants.

For the on-resonance asymmetry ($s = M_Z^2$) we get in the improved Born approximation:

$$A_{FB}(M_Z^2) = \frac{3}{4}\cdot A_e\,A_f\left(1-4\mu_f+6\mu_f\frac{(g_A^f)^2}{(g_V^f)^2+(g_A^f)^2}\right) + \Delta A_{FB}^I + \Delta A_{FB}^Q\,. \tag{84}$$

A_f is defined as

$$A_f = \frac{2g_V^f g_A^f}{(g_V^f)^2+(g_A^f)^2} = \frac{2(1-4\mid Q_f\mid s_f^2)}{1+(1-4\mid Q_f\mid s_f^2)^2} \tag{85}$$

with a short-hand notation for the effective mixing angle in Eq. (48):

$$s_f^2 \equiv \sin^2\theta_{eff}^f\,.$$

The small contributions $\Delta A_{FB}^{I,Q}$ result from the the interference with the photon exchange

$$\Delta A_{FB}^I = \frac{3}{4}\frac{2Q_e Q_f\, g_A^e g_A^f}{(g_V^{e\,2}+g_A^{e\,2})\,(g_V^{f\,2}+g_A^{f\,2})}\frac{4\pi\alpha(M_Z^2)}{\sqrt{2}G_\mu M_Z^2}\cdot\frac{\Gamma_Z}{M_Z}\cdot \mathrm{Im}\,\hat{\Pi}^\gamma \tag{86}$$

	m-top	MH=60	100	500	1000
leptons					
	100	0.1307	0.1287	0.1220	0.1189
	130	0.1363	0.1343	0.1276	0.1245
	150	0.1406	0.1386	0.1319	0.1288
	200	0.1537	0.1516	0.1445	0.1413
	250	0.1699	0.1677	0.1595	0.1562
u-quarks					
	100	0.1327	0.1307	0.1240	0.1209
	130	0.1383	0.1363	0.1296	0.1265
	150	0.1426	0.1407	0.1339	0.1308
	200	0.1558	0.1537	0.1465	0.1433
	250	0.1720	0.1698	0.1615	0.1582
d-quarks					
	100	0.1332	0.1313	0.1246	0.1214
	130	0.1389	0.1369	0.1302	0.1271
	150	0.1433	0.1413	0.1345	0.1314
	200	0.1564	0.1544	0.1472	0.1439
	250	0.1728	0.1706	0.1622	0.1589
b-quarks					
	100	0.1332	0.1313	0.1246	0.1214
	130	0.1387	0.1368	0.1301	0.1270
	150	0.1430	0.1410	0.1342	0.1312
	200	0.1554	0.1534	0.1464	0.1432
	250	0.1704	0.1682	0.1602	0.1570

Table 7: Standard Model predictions for the left-right asymmetries A_{LR} for various flavors (without QED corrections). Final state QCD corrections are negligible. Input parameters: m_t, M_H in GeV, $M_Z = 91.187$ GeV, $\alpha_s = 0.12$.

f	$\frac{3}{4}A_eA_f$	mass correction	ΔA_{FB}^I	ΔA_{FB}^Q
μ	0.0171	$< 10^{-6}$	0.0018	-0.0001
τ	0.0171	$1.3 \cdot 10^{-5}$	0.0018	-0.0001
c	0.0758	$2.5 \cdot 10^{-5}$	0.0011	-0.0002
b	0.1061	$1.5 \cdot 10^{-5}$	0.0004	$-5 \cdot 10^{-5}$

Table 8: On-resonance forward-backward asymmetries for $s_{eff}^2 = 0.231$, $M_Z = 91.187$ GeV.

and from the pure photon exchange part:

$$\Delta A_{FB}^Q \;=\; -\frac{3}{4}\,\frac{A_eA_f\,Q_e^2Q_f^2}{(g_V^{e\,2}+g_A^{e\,2})\,(g_V^{f\,2}+g_A^{f\,2})}\left(\frac{4\pi\alpha(M_Z^2)}{\sqrt{2}G_\mu M_Z^2}\right)^2\left(\frac{\Gamma_Z}{M_Z}\right)^2 . \tag{87}$$

The on-resonance asymmetries are essentially determined by the values of the effective mixing angles for e and f entering the product A_eA_f. Through $s_{e,f}^2$ also the dependence of the asymmetries on the basic Standard Model parameters m_t, M_H is fixed. The small corrections from finite mass effects, interference and photon exchange can be considered practically independent of the details of the model. For demonstrational purpose we list in table 8 the various terms in the on-resonance asymmetries according to Eq. (84) for a common value of the effective mixing angle $s_e^2 = s_f^2 = 0.231$.

Table 9 contains the Standard Model predictions for A_{FB} based on the complete expression (83) for the various flavors, depending on m_t and M_H. Graphically, the asymmetries are displayed in Figure 10.

Final state QED and QCD corrections:

According to the representation of A_{FB} as the ratio of the antisymmetric to the symmetric part of the cross section the effects can be summmarized as follows:

If **no cuts** are applied, only the symmetric part

$$\sigma \;=\; \sigma_F + \sigma_B$$

	m-top	MH=60	100	500	1000
muons					
	100	0.0147	0.0143	0.0131	0.0125
	130	0.0158	0.0154	0.0141	0.0135
	150	0.0166	0.0162	0.0149	0.0143
	200	0.0194	0.0189	0.0174	0.0167
	250	0.0232	0.0227	0.0207	0.0199
c-quarks					
	100	0.0668	0.0658	0.0622	0.0606
	130	0.0699	0.0688	0.0652	0.0635
	150	0.0722	0.0711	0.0675	0.0658
	200	0.0794	0.0782	0.0743	0.0725
	250	0.0884	0.0872	0.0825	0.0807
s-quarks					
	100	0.0936	0.0922	0.0875	0.0852
	130	0.0976	0.0962	0.0914	0.0892
	150	0.1007	0.0993	0.0945	0.0923
	200	0.1101	0.1086	0.1035	0.1012
	250	0.1217	0.1201	0.1142	0.1118
b-quarks					
	100	0.0940	0.0926	0.0879	0.0857
	130	0.0979	0.0965	0.0918	0.0896
	150	0.1009	0.0995	0.0947	0.0925
	200	0.1097	0.1083	0.1032	0.1010
	250	0.1202	0.1187	0.1129	0.1106

Table 9: Standard Model predictions for the forward-backward asymmetries A_{FB} for various flavors (without QED and QCD corrections). Input parameters: m_t, M_H in GeV, $M_Z = 91.187$ GeV, $\alpha_s = 0.12$.

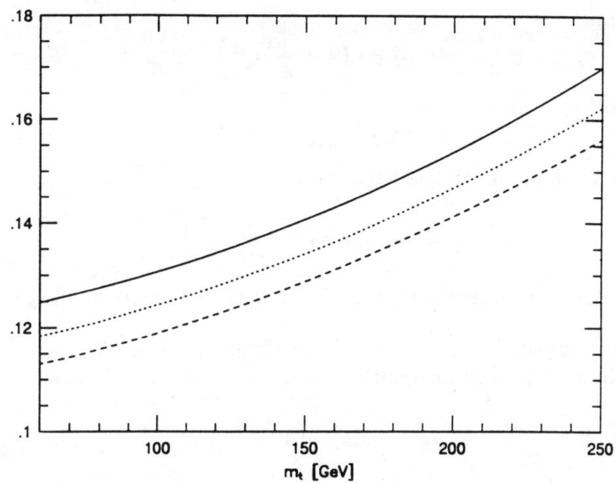

Fig. 9: *The on-resonance ($s = M_Z^2$) left-right asymmetry A_{LR} in muon pair production. M_Z = 91.187 GeV. M_H = 60 GeV (full), 300 GeV (dotted), 1000 GeV (dashed). No QED corrections.*

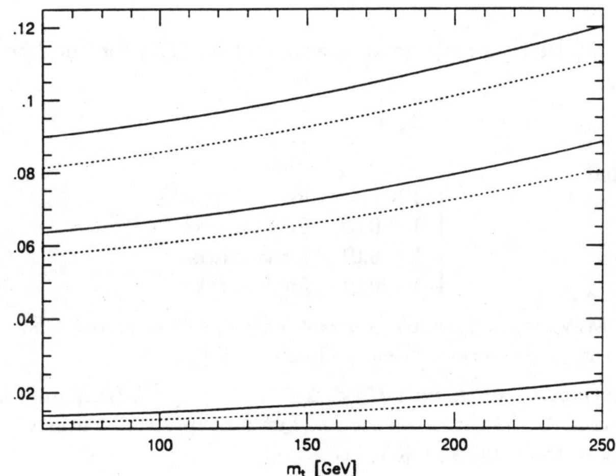

Fig. 10: *The on-resonance forward-backward asymmetries for: leptons (lower), $c\bar{c}$ (middle), $b\bar{b}$ (upper curves). M_Z = 91.187 GeV. M_H = 60 GeV (full curves), M_H = 1000 GeV (dotted curves). No QED and QCD corrections.*

gets a correction:

$$\sigma \rightarrow \sigma \cdot \left(1 + \frac{3\alpha}{4\pi} Q_f^2\right) \tag{88}$$

whereas [35, 36]

$$\delta(\sigma_F - \sigma_B) = 0\,. \tag{89}$$

This results in a correction to the asymmetry

$$A_{FB} \rightarrow A_{FB} \cdot \left(1 - \frac{3\alpha}{4\pi} Q_f^2\right) \tag{90}$$

which is a very small negative contribution (< 0.17 % relative to A_{FB}).

Quite in analogy, for the QCD single gluon emission [37, 38] the following correction to the asymmetry for quark final states with $m_q \rightarrow 0$ arises:

$$A_{FB} \rightarrow A_{FB} \cdot \left(1 - \frac{\alpha_s}{\pi}\right)\,. \tag{91}$$

For massive quarks, the QCD final state corrections can be included by multiplying the purely electroweak asymmetry by a factor

$$1 - \frac{\alpha_s}{\pi}\Delta_q\,.$$

The coefficient Δ_q is, to a very good approximation (1%) for the known quarks given by [38, 39]

$$\Delta_q = 1 - \frac{4}{3}\pi\frac{m_q}{M_Z} \tag{92}$$

which means that

$$\Delta_q = \begin{cases} 1 & \text{for } u, d\text{-quarks} \\ 1 - 0.02 & \text{for } s\text{-quarks} \\ 1 - 0.07 & \text{for } c\text{-quarks} \\ 1 - 0.21 & \text{for } b\text{-quarks} \end{cases} \tag{93}$$

with $m_s = 500$ MeV, $m_c = 1.5$ GeV, $m_b = 4.5$ GeV, $M_Z = 91.187$ GeV. The exact formulae are given in the report "Heavy Quarks" [40].

More **restrictive cuts** with $E_\gamma < \Delta E \ll E = \sqrt{s}/2$ (soft photon domain) yield larger corrections to the cross section but multiply the symmetric and anti-symmetric parts in the same way [33, 41, 42]:

$$(\sigma_F \pm \sigma_B) \cdot \left[1 + \frac{2\alpha}{\pi} Q_f^2 \beta_f \log\frac{\Delta E}{E} + \frac{3\alpha}{2\pi} Q_f^2 \beta_f + \frac{\alpha}{\pi} Q_f^2 \left(\frac{\pi^2}{3} - \frac{1}{2}\right)\right] \tag{94}$$

with

$$\beta_f = \log\frac{s}{m_f^2} - 1\,.$$

These corrections therefore cancel in the asymmetry:

$$\delta(A_{FB}) \;=\; 0\,. \tag{95}$$

This remains valid also after the soft photon contributions have been summed to all orders.

In general, with or without cuts, the asymmetry is affected only very little by the final state QED corrections.

QED corrections from initial - final state interference:

In the contributions resulting from the interference of initial and final state radiation together with the QED box diagrams (Fig. 1) no large logarithms $\log(s/m_f^2)$ are present. Instead, angular dependent terms like

$$\log\frac{t}{u} \;=\; \log\frac{1-\cos\theta}{1+\cos\theta}$$

enter the result, which induce an additional asymmetry. In the Z region the resonant diagrams are the most important ones; the others are associated with the photon exchange and hence are suppressed by an additional factor Γ_Z/M_Z.

In the resonance region the interference contributions show a behaviour different from the continuum and can be easily understood in terms of the resulting correction factor [33, 41, 42] for the angular distribution close to the peak:

$$1+Q_eQ_f\left\{\frac{4\alpha}{\pi}\log\frac{t}{u}\,\log\mid\epsilon\frac{s}{M_R^2-s+s\epsilon}\mid+X^B_{fin}(\theta)+V^{\gamma Z}_{fin}(\theta)\right\}\equiv 1+C_{int}(\theta) \tag{96}$$

with

$$\epsilon=\frac{\Delta E}{E}=\frac{2\Delta E}{\sqrt{s}},\quad M_R^2=M_Z^2-iM_Z\Gamma_Z\,,$$

which is antisymmetric and multiplies the symmetric part of the resonant angular distribution, yielding

$$\delta(\sigma_F-\sigma_B)=\frac{\alpha^2}{4s}\cdot 2\int_0^1 d\cos\theta\,(g_V^{e\,2}+g_A^{e\,2})(g_V^{f\,2}+g_A^{f\,2})\mid\chi_Z\mid^2\cdot(1+\cos^2\theta)\cdot C_{int}(\theta) \tag{97}$$

$$\simeq\frac{\alpha^2}{4s}\cdot(g_V^{e\,2}+g_A^{e\,2})(g_V^{f\,2}+g_A^{f\,2})\mid\chi_Z\mid^2\cdot\frac{2\alpha}{\pi}Q_eQ_f(1+8\log 2)\log\mid\frac{M_R^2-s+s\epsilon}{s\epsilon}\mid.$$

X^B_{fin} denotes the finite (non-IR) part of the real photon and $V^{\gamma Z}_{fin}$ that of the virtual photon contributions. For $s\approx M_Z^2$ the finite virtual term has the property [3]

$$V^{\gamma Z}_{fin}(\theta) \;=\; -X^B_{fin}(\theta)\,+\,O\left(\frac{\alpha}{\pi}\frac{\Gamma_Z}{M_Z}\right)\,; \tag{98}$$

consequently, the log-term in (96) is practically the only part generating an additional antisymmetric contribution. Its size depends crucially on the cuts applied to the radiated photon:

[3] for the complete expression see e.g. [43]

(i) For very loose cuts ($\epsilon \to 1$) this term becomes also of order $O\left(\frac{\alpha}{\pi}\frac{\Gamma_Z}{M_Z}\right)$ which means that the total contribution to the asymmetry is negligibly small. This corresponds exactly to the situation of a very narrow resonance [44] where the interference contributions are practically zero.

(ii) For tight cuts ($\epsilon \approx 0.01$) this log term yields a positive contribution of a few percent to the asymmetry ($f = \mu$). The difference between (97) and the exact result is negligible for $\epsilon < 0.1$.

The proper inclusion of hard photons for the case (i) does not change the result that the interference contribution is very small [35, 36, 45]. For cuts $\epsilon > 0.3$ the interference contributes less than 0.0006 to A_{FB} for muons; for $\epsilon \to 1$ one finds $\simeq 0.0003$. For quark final states, the corresponding δA_{FB} from the interference is even smaller due to the charge factors.

Initial state QED corrections:

As we know from the integrated cross section, the initial state corrections give rise to a significant reduction of the peak height which is due to the rapid variation of $\sigma(s)$ with the energy. Since the asymmetry $A_{FB}(s)$ is a steeply increasing function around the Z the energy loss from initial-state radiation $s \to s' < s$ leads to a reduction in the asymmetry as well:

$$A_{FB}(s') < A_{FB}(s) .$$

Quantitatively, the $O(\alpha)$ correction to A_{FB} for muons close to the peak $\delta A_{FB} \simeq -0.02$ is of the order of the on-resonance asymmetry itself. Therefore it is obvious that also the higher order QED contributions have to be taken into account carefully.

We can express the initial state QED corrections to A_{FB} in a compact form, quite in analogy to the convolution integral for the integrated cross section $\sigma(s)$ in Eq. (60):

$$A_{FB}(s) = \frac{1}{\sigma(s)} \int_0^{k_{max}} dk \, \frac{4(1-k)}{(2-k)^2} \, \tilde{H}(k) \, \sigma_{FB}(s') , \quad k_{max} \leq 1 - \frac{4m_f^2}{s} . \tag{99}$$

The basic ingredients are the expression for the non-radiative antisymmetric cross section with inclusion of the weak corrections

$$\sigma_{FB}(s) \equiv \sigma_F(s) - \sigma_B(s) = \frac{\pi\alpha^2}{s} \cdot G_3(s) \sqrt{1 - 4\mu_f}$$

resulting from Eq. (4) and the substitutions (51), and the radiator function $\tilde{H}(k)$. The quantity

$$s' = (1-k)\, s = (p_f + p_{\bar{f}})^2$$

is the invariant mass of the outgoing fermion pair. The effect of the change in the scattering angle by the boost from the $f\bar{f}-$cms to the laboratory frame is taken into account by the kinematical factor in front of $\tilde{H}$ in the convolution integral.

$\tilde{H}$ is different from the radiator function H for the symmetric cross section in Eq. (60) in the hard photon terms. According to the present status of the calculation, $\tilde{H}$ contains the exact $O(\alpha)$ contribution [35, 36, 46], the $O(\alpha^2)$ contributions in the leading-log approximation [47], and the resummation of soft photons to all orders [39]:

$$\begin{aligned}\tilde{H}(k) &= \beta\, k^{\beta-1}\left[1+\frac{\alpha}{\pi}\left(\frac{3}{2}L+\frac{\pi^2}{3}-2\right)+\left(\frac{\alpha}{\pi}\right)^2 L^2\left(\frac{9}{8}-\frac{\pi^2}{3}\right)\right] \\ &+\frac{\alpha}{\pi}\left[(L-1)(k-2)-\frac{1+(1-k)^2}{k}\log\frac{4(1-k)}{(2-k)^2}\right] \\ &+\left(\frac{\alpha}{\pi}\right)^2 L^2\left[h_{22}+\tilde{h}_{22}\right] \end{aligned} \tag{100}$$

with β, L, h_{22} from Eq. (62), and

$$\begin{aligned}\tilde{h}_{22} &= \frac{1}{4}\left[\frac{k^3}{2(1-k)}-(2-k)\log(1-k)+2k\right. \\ &\left.+\frac{k^2}{\sqrt{1-k}}\left(\arctan\frac{1}{\sqrt{1-k}}-\arctan\sqrt{1-k}\right)\right].\end{aligned} \tag{101}$$

Figure 11 displays the effect of the QED corrections on A_{FB} for muons for the case where no cuts are applied. The behaviour of A_{FB} is qualitatively similar to that of the integrated cross section where the higher order QED contributions bring the prediction closer to the lowest order result compared to the $O(\alpha)$ corrections. The figure includes both initial and final state radiation with soft photon resummation, but with Born like couplings: taking (43) for leptons with $\rho_l = 1$ and an effective leptonic mixing angle $s_l^2 = 0.232$. The influence of final state radiation, however, is practically invisible.

Summarizing the QED corrections, the most important contributions come from initial state radiation. The explicit calculation of the complete $O(\alpha)$ and leading $O(\alpha^2)$ corrections together with the resummation of soft photons to all orders is available. The QED corrections from final state radiation and initial-final interference, are very small ($|\,\delta A_{FB}\,| < 0.001$) if no tight cuts to the photon phase space are applied. More restrictive cuts make the interference contributions to A_{FB} important exceeding the level of 0.01 (for muons) when the photon is restricted to energies below 1 GeV. The complete set of QED corrections is available in (semi-) analytic form, exact in $O(\alpha)$ and with leading higher order terms, also for situations with cuts, covering: energy or invariant mass cuts, accollinearity cuts, acceptance cuts

[46, 48, 49]. Compact analytic and semi-analytic formulae for the QED corrections are implemented in the FORTRAN codes ZFITTER [6] and MIZA [50]. Recently, an independent code TOPAZ0 has been developped [49] showing agreement with ZFITTER within 0.2%. A Monte Carlo version is given by KORALZ [25].

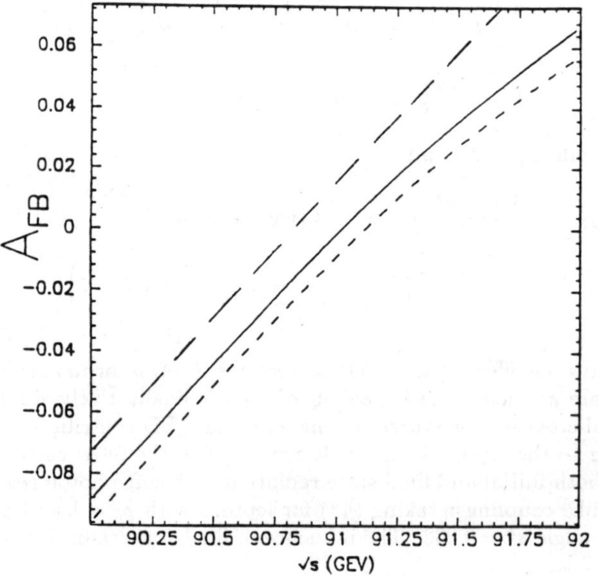

Fig. 11: *A_{FB} for $e^+e^- \to \mu^+\mu^-$ (no cuts). The curves display the cases: no QED corrections (— — —), with $O(\alpha)$ QED corrections (- - -) and $O(\alpha^2)$ QED corrections with soft photon resummation (—). For the non-radiative part Born-like couplings are taken with $s_l^2 = 0.232$, $M_Z = 91$ GeV, $\Gamma_Z = 2.47$ GeV (from [39])*

6 Concluding remarks

The basic processes in e^+e^- collisions at LEP/SLC are the production of fermion pairs $e^+e^- \rightarrow f\bar{f}$. At LEP2, W pair production, which has not been discussed here, will be the process of main interest. For a recent comprehensive review on the calculation techniques, results and predictions for W production and decay in e^+e^- collisions see [51].

A crucial role in the interpretation of electroweak precision experiments is played by the radiative corrections. A careful treatment is required for the QED corrections which form the link between data taking and physics analysis. In $e^+e^- \rightarrow f\bar{f}$ they can be separated in a gauge invariant and ultraviolet finite way from the residual part of genuine weak corrections. For hadronic final states also QCD corrections have to be taken into account. The theoretical understanding together with the experimental control of detector properties and phase space cuts is important for extracting the information contained in the purely electroweak part of the precision observables.

The calculation of radiative corrections is theoretically well established for the observables in $e^+e^- \rightarrow f\bar{f}$ around the Z peak, including contributions beyond the 1-loop order for the QED and QCD parts and also for the purely electroweak part in its leading terms. After taking the measured Z mass, besides α and G_μ, for completion of the input, each other precision observable provides a test of the electroweak theory. The theoretical predictions of the Standard Model depend in general on the masses of the as yet experimentally unknown top quark and Higgs boson through the virtual presence of these particles in the loops. As a consequence, precision data can be used to pin down the allowed range of the mass parameters. The consistency of the Standard Model beyond the tree level is visualized in terms of a common range for m_t, M_H derived from a great variety of experimental data for all kinds of electroweak observables.

Theoretical uncertainties in the Standard Model predictions have their origin in the uncertainties of the hadronic vacuum polarization of the photon, of the value of the strong coupling constant α_s, and from the unknown higher order contributions. Whereas the QCD uncertainty is responsible for the main theoretical error in the hadronic part of the Z width (5-7 MeV), the uncertainty in the electroweak mixing angle is dominated by the error of the vacuum polarization yielding $\delta s_W^2 = 0.0003$. Other uncertainties from the unknown non-leading 2-loop terms can be estimated to be of the order of 0.1% in the measureable quantities.

Since all searches for new particles have been negative, the precise knowledge of the radiative corrections gets another important aspect: the indirect search for new physics in terms of deviations between the measured quantities and the Standard Model predictions. Up to now such deviations could not be established, but they still may show up in the further increase of the accuracy, in particular when the mass of the top quark will be known as an experimental quantity.

A Vertex corrections for b quarks

In this appendix we list the explicit formulae for the 1-loop top dependent parts F_L^b (Zbb vertex) and G_L^b (γbb vertex) of the vertex corrections in Eq. (21,23).

F_L^b is the sum of the top dependent Zbb vertex diagrams (see Figure 3) and the counter term from the external self energies (without the Z contribution to the b quark self energy). The result is

$$F_L^b = \sum_{i=b}^{g} F_i + \frac{1}{2}\left(\frac{2}{3}s_W^2 - 1\right)\delta Z_L^{fin} \tag{102}$$

where

$$\delta Z_L^{fin} = \frac{1}{2s_W^2}(2 + \frac{m_t^2}{M_W^2})\left(\bar{B}_1(m_b^2, m_t, M_W) + m_b^2 \bar{B}_1'(m_b^2, m_t, M_W)\right). \tag{103}$$

For the function $\bar{B}_1$ see equation (....); $\bar{B}_1'$ denotes the derivative

$$\bar{B}_1'(s, m_1, m_2) = \frac{\partial}{\partial s}\bar{B}_1(s, m_1, m_2).$$

The F_i in (102) are the expressions corresponding to the diagrams in Figure 3 after subtracting those (divergent) parts which are cancelled by the vertex counter term after renormalization. The read with $s = (p_b + p_{\bar{b}})^2$:

$$\begin{aligned}
F_b &= \frac{v_t + a_t}{4s_W^2}\left\{-\frac{3}{2} + 2\log\frac{M_W}{m_t} + 4\,C_2^0(s, m_t, m_t, M_W)\right. \\
&\quad -2s\left[C_2^+(s, m_t, m_t, M_W) - C_2^-(s, m_t, m_t, M_W)\right] \\
&\quad \left. +4s\,C_1^+(s, m_t, m_t, M_W) - 2s\,C_0(s, m_t, m_t, M_W)\right\} \\
&\quad -\frac{v_t - a_t}{4s_W^2}\,2m_t^2\,C_0(s, m_t, m_t, M_W), \\
F_c &= -\frac{c_W^2}{2s_W^2}\left\{-\frac{3}{2} + 12\,C_2^0(s, M_W, M_W, m_t)\right. \\
&\quad -2s\left[C_2^+(s, M_W, M_W, m_t) - C_2^-(s, M_W, M_W, m_t)\right] \\
&\quad \left. +4s\,C_1^+(s, M_W, M_W, m_t)\right\}, \\
F_d &= \frac{v_t - a_t}{4s_W^2}\left(\frac{m_t}{M_W}\right)^2\left\{-\frac{3}{4} + \log\frac{M_W}{m_t} + 2\,C_2^0(s, m_t, m_t, M_W)\right.
\end{aligned}$$

$$
\begin{aligned}
& -s \left[C_2^+(s,m_t,m_t,M_W) - C_2^-(s,m_t,m_t,M_W)\right]\Big\} \\
& -\frac{v_t + a_t}{4s_W^2}\left(\frac{m_t}{M_W}\right)^2 m_t^2\, C_0(s,m_t,m_t,M_W), \\
F_e \;&=\; \frac{s_W^2 - c_W^2}{4s_W^2 c_W^2}\left(\frac{m_t}{M_W}\right)^2 \left\{-\frac{1}{4} + 2\,C_2^0(s,M_W,M_W,m_t)\right\}, \\
F_f \;&=\; F_g = -\frac{m_t^2}{2}\,C_0(s,M_W,M_W,m_t). \qquad (104)
\end{aligned}
$$

The functions C_1^+, C_2^+, C_2^-, C_2^0 are specified in terms of $\bar{B}_0$, $\bar{B}_1$ defined below and

$$
C_0(s,M,M,M') = C_0(-p_b, p_b + p_{\bar{b}}, M', M, M),
$$

which is the scalar 3-point integral I Eq. (104) with equal external masses $p_b^2 = p_{\bar{p}}^2 = m_b^2$, as follows:

$$
\begin{aligned}
(4m_b^2 - s)\,C_1^+(s,M,M,M') \;&=\; \log\frac{M'}{M} + \bar{B}_0(s,M,M) - \bar{B}_0(m_b^2,M,M') \\
&\quad +(M'^2 - M^2 + m_b^2)\,C_0(s,M,M,M'), \\
C_2^0(s,M,M,M') \;&=\; \frac{1}{4}\left[\bar{B}_0(s,M,M) + 1\right] \\
&\quad +\frac{1}{2}(M^2 - M'^2 - m_b^2)\,C_1^+(s,M,M,M') \\
&\quad +\frac{1}{2}M'^2\,C_0(s,M,M,M'), \\
(4m_b^2 - s)\,C_2^+(s,M,M,M') \;&=\; \frac{1}{2}\bar{B}_0(s,M,M) + \frac{1}{2}\left[\bar{B}_1(m_b^2,M',M) - \frac{1}{4}\right] \\
&\quad +(M'^2 - M^2 + m_b^2)\,C_1^+(s,M,M,M') \\
&\quad -C_2^0(s,M,M,M'), \qquad (105) \\
s\,C_2^-(s,M,M,M') \;&=\; -\frac{1}{2}\left[\bar{B}_1(m_b^2,M',M) - \frac{1}{4}\right] - C_2^0(s,M,M,M').
\end{aligned}
$$

For the electromagnetic γbb vertex we have to use the following G_L^b, calculated in analogy to F_L:

$$
G_L^b = \sum_{i=b}^{g} G_i - \frac{1}{6}\,\delta Z_L^{fin} \qquad (106)
$$

with δZ_L^{fin} from (103) and

$$
\begin{aligned}
G_b &= \frac{1}{6s_W^2}\Big\{-\frac{3}{2}+2\log\frac{M_W}{m_t}+4\,C_2^0(s,m_t,m_t,M_W) \\
&\quad -2s\left[C_2^+(s,m_t,m_t,M_W)-C_2^-(s,m_t,m_t,M_W)\right] \\
&\quad +4s\,C_1^+(s,m_t,m_t,M_W)-2s\,C_0(s,m_t,m_t,M_W) \\
&\quad -2m_t^2\,C_0(s,m_t,m_t,M_W)\Big\}, \\
G_c &= -\frac{1}{4s_W^2}\Big\{-\frac{3}{2}+12\,C_2^0(s,M_W,M_W,m_t) \\
&\quad -2s\left[C_2^+(s,M_W,M_W,m_t)-C_2^-(s,M_W,M_W,m_t)\right] \\
&\quad +4s\,C_1^+(s,M_W,M_W,m_t)\Big\}, \\
G_d &= \frac{1}{6s_W^2}\left(\frac{m_t}{M_W}\right)^2\Big\{-\frac{3}{4}+\log\frac{M_W}{m_t}+2\,C_2^0(s,m_t,m_t,M_W) \\
&\quad -s\left[C_2^+(s,m_t,m_t,M_W)-C_2^-(s,m_t,m_t,M_W)\right] \\
&\quad -m_t^2\,C_0(s,m_t,m_t,M_W)\Big\}, \\
G_e &= -\frac{1}{4s_W^2}\left(\frac{m_t}{M_W}\right)^2\Big\{-\frac{1}{4}+2\,C_2^0(s,M_W,M_W,m_t)\Big\}, \\
G_f &= G_g = \frac{m_t^2}{4s_W^2}\,C_0(s,M_W,M_W,m_t). \qquad (107)
\end{aligned}
$$

The C functions are the same as in (105).

Finally we have to specify the functions $\bar{B}_0$ and $\bar{B}_1$.

$\bar{B}_0$ is the finite part of the scalar 2-point integral B_0 in I Eq. (97) with the following subtraction:

$$\bar{B}_0(s,M,M') = B_0(s,M,M') - \Delta + \frac{1}{2}\log\frac{MM'}{\mu^2} \qquad (108)$$

with Δ defined in I Eq. (95).

The function $\bar{B}_1$ is given by

$$\bar{B}_1(s,M,M') = -\frac{1}{4}+\frac{M^2}{M^2-M'^2}\log\frac{M}{M'}+\frac{M'^2-M^2-s}{2s}F(s,M,M'). \qquad (109)$$

with the subtracted B_0 function

$$F(s,M,M') = B_0(s,M,M') - B(0,M,M').$$

References:

1. W. Hollik: Renormalization of the Standard Model, this volume
2. *Polarization at LEP*, CERN-88/06 (1988), eds. G. Alexander et al.
3. W. Beenakker, W. Hollik, S. van der Marck, *Nucl. Phys.* **B365** (1991) 24
4. D.Yu. Bardin, M.S. Bilenky, G.V. Mithselmakher, T. Riemann, M. Sachwitz, *Z. Phys.* **C44** (1989) 493
5. D. Yu. Bardin, W. Hollik, T. Riemann, *Z. Phys.* **C49** (1991) 485
6. D. Yu. Bardin et al., CERN-TH.6443/92 (1992)
7. A.A. Akhundov, D.Yu. Bardin, T. Riemann, *Nucl. Phys.* **B276** (1986) 1
8. W. Beenakker, W. Hollik, *Z. Phys.* **C40** (1988) 141
9. J. Bernabeu, A. Pich, A. Santamaria, *Phys. Lett.* **B200** (1988) 569; *Nucl. Phys.* **B363** (1991) 326
10. R. Barbieri, M. Beccaria, P. Ciafaloni, G. Curci, A. Vicere, *Phys. Lett.* **B288** (1992) 95; CERN-TH.6713/92 (1992)
11. A. Denner, W. Hollik, B. Lampe, CERN-TH.6874/93 (1993), to appear in *Z. Phys.* **C**
12. J. Fleischer, O.V. Tarasov, F. Jegerlehner, P. Rączka, *Phys. Lett.* **B293** (1992) 437
13. G. Buchalla, A.J. Buras, Munich preprint MPI-PTh 111/92, TUM-T31-36/92 (1992)
14. G. Degrassi, Padova preprint DFPD 93/TH/03 (1993)
15. D.C. Kennedy, B.W. Lynn, *Nucl. Phys.* **B322** (1989) 1
16. M. Consoli, S. LoPresti, L. Maiani, *Nucl. Phys.* **B223** (1983) 474
17. W.J. Marciano, A. Sirlin, *Phys. Rev. Lett.* **46** (1981) 163;
    A. Sirlin, *Phys. Lett.* **B232** (1989) 123;
    G. Degrassi, S. Fanchiotti, A. Sirlin, *Nucl. Phys.* **B351** (1991) 49;
    G. Degrassi, A. Sirlin, *Nucl. Phys.* **B352** (1991) 342
18. F.A. Berends et al., in: *Z Physics at LEP 1*, CERN 89-08 (1989), eds. G. Altarelli, R. Kleiss, C. Verzegnassi, Vol. I, p. 89;
    W. Beenakker, F.A. Berends, S.C. van der Marck, *Z. Phys.* **C46** (1990) 687
19. A. Borelli, M. Consoli, L. Maiani, R. Sisto, *Nucl. Phys.* **B333** (1990) 357
20. G. Burgers, F.A. Berends, W. Hollik, W.L. van Neerven, *Phys. Lett.* **B203** (1988) 177
21. D. Yu. Bardin, A. Leike, T. Riemann, M. Sachwitz, *Phys. Lett.* **B206** (1988) 539
22. G. Valencia, S. Willenbrock, *Phys. Lett.* **B 259** (1991) 373;
    R.G. Stuart, *Phys. Lett.* **B272** (1991) 353
23. G. Burgers, F.A. Berends, W.L.van Neerven, *Nucl. Phys.* **B297** (1988) 429; E: *Nucl. Phys.* **B304** (1988) 921

24. L. Garrido, M. Martinez, R. Miquel, J. Harton, R. Tanaka, *Z. Phys.* **C49** (1991) 645

25. B.F.L. Ward, S. Jadach, Z. Wąs, *Comp. Phys. Commun.* **66** (1991) 276

26. W. Beenakker, F.A. Berends, S.C. van der Marck, *Z. Phys.* **C46** (1990) 687

27. K.G. Chetyrkin, A.L. Kataev, F.V. Tkachov, *Phys. Lett.* **B85** (1979) 277;
M. Dine, J. Sapirstein, *Phys. Rev. Lett.* **43** (1979) 668;
W. Celmaster, R. Gonsalves, *Phys. Rev. Lett.* **44** (1980) 560;
S.G. Gorishny, A.L. Kataev, S.A. Larin, *Phys. Lett.* **B259** (1991) 144;
L.R. Surguladze, M.A. Samuel, *Phys. Rev. Lett.* **66** (1991) 560

28. T.H. Chang, K.J.F Gaemers, W.L. van Neerven, *Nucl. Phys.* **B202** (1982) 407;
K.G. Chetyrkin, J.H. Kühn, *Phys. Lett.* **B248** (1992) 359; *Phys. Lett.* **B282** (1992) 221;
K.G. Chetyrkin, J.H. Kühn, A. Kwiatkowski, Karlsruhe preprint TTP92-07 (1992);
K.G. Chetyrkin, A. Kwiatkowski, Karlsruhe preprint TTP92-39 (1992)

29. J.H. Kühn, B.A. Kniehl, *Phys. Lett.* **B224** (1990) 229; *Nucl. Phys.* **B329** (1990) 547

30. D. Schaile, this volume and private communication

31. R. Tenchini, talk at the XXIIIth Rencontre de Moriond on *Electroweak Interactions and Unified Theories*, Les Arcs 1993 (to appear in the Proceedings)

32. G. Girardi, W. Hollik, C. Verzegnassi, *Phys. Lett.* **B240** (1990) 492

33. M. Böhm, W. Hollik *Nucl. Phys.* **B204** (1982) 45; *Z. Phys.* **C 23** (1984) 31

34. S. Jadach, J.H. Kühn, R.G. Stuart, Z. Wąs, *Z. Phys.* **C38** (1988) 609;
J.H. Kühn, R.G. Stuart, *Phys. Lett.* B200 (1988) 360

35. D. Bardin, M.S. Bilenky, O.M. Fedorenko, T. Riemann, Dubna preprint JINR-E2-88-324 (1988)

36. D. Bardin, M.S. Bilenky, A. Chizhov, A. Sazonov, Yu. Sedykh, T. Riemann, M. Sachwitz, *Phys. Lett.* **B229** (1989) 405

37. J. Jersak, E. Laerman, P.M. Zerwas, *Phys. Rev.* **D25** (1980) 1218

38. A. Djouadi, *Z. Phys.* **C39** (1988) 561

39. M. Böhm, W. Hollik et al., in: *Z Physics at LEP 1*, CERN 89-08 (1989), eds. G. Altarelli, R. Kleiss, C. Verzegnassi, Vol. I, p. 203;
W. Beenakker, F.A. Berends, S.C. van der Marck, *Phys. Lett.* **B252** (1990) 299

40. J.H. Kühn, P.Zerwas et al., in: *Z Physics at LEP 1*, CERN 89-08 (1989), eds. G. Altarelli, R. Kleiss, C. Verzegnassi, Vol. I, p. 267

41. M. Greco, G. Pancheri, Y. Srivastava, *Nucl. Phys.* **B171** (1980) 118; E: *Nucl. Phys.* **B197** (1982) 543

42. F.A. Berends, R. Kleiss, S. Jadach, *Nucl. Phys.* **B202** (1982) 63

43. W. Hollik, *Fortschr. Phys.* **38** (1990) 165

44. M. Greco, G. Pancheri, Y. Srivastava, *Nucl. Phys.* **B101** (1975) 234; F.A. Berends, G.J. Komen, *Nucl. Phys.* **B115** (1976) 114

45. S. Jadach, Z. Wąs, *Phys. Lett.* **B219** (1989) 103

46. D. Bardin, M.S. Bilenky, A. Chizhov, A. Sazonov, O. Fedorenko, T. Riemann, M. Sachwitz, *Nucl. Phys.* **B351** (1991) 1

47. W. Beenakker, F.A. Berends, W.L. van Neerven, in: Proceedings of the 1989 Ringberg Workshop *Radiative Corrections for* e^+e^- *Collisions*, p. 3, ed. J.H. Kühn, Springer, Berlin - Heidelberg - New York 1989

48. D. Bardin, L. Vertogradov, Yu. Sedykh, T. Riemann, CERN-TH.5434/89 (1989)

49. G. Montagna, F. Piccinini, O. Nicrosini, G. Passarino, R. Pittau, Pavia-Torino preprint FNT/T-92/02, DFTT/G-93-1 (1993)

50. M. Martinez, in: *Proceedings of the XVIII International Meeting on Fundamental Physics on " Precision Tests of the Standard Model at High Energy Colliders "*, Santander 1990. World Scientific, Singapore 1991, p. 385, eds. F. del Aguila,
Appendix of M. Martinez with the ALEPH Collaboration, *Z. Phys.* **C53** (1992) 1

51. A. Denner, *Fortschr. Phys.* **41** (1993) 307

RADIATIVE CORRECTIONS TO NEUTRAL CURRENT PROCESSES

William J. Marciano
Brookhaven National Laboratory
Upton, New York 11973, USA

Contents

1. Introduction

The standard $SU(3)_c \times SU(2)_L \times U(1)_Y$ model of strong and electroweak interactions has been extremely successful. It correctly predicted the existence and observed properties of weak neutral currents as well as the $W^\pm$ and Z bosons.[1] Furthermore, because that gauge theory is renormalizable, its predictions can be scrutinized at the quantum loop level by high precision measurements. A deviation from expectations would signal the presence of "new physics". Remarkably, at this time there are no solid experimental results which cannot be accommodated by the standard model at the 1 or 2 sigma level. That accomplishment is very impressive when one considers the wealth of experimental data which must be confronted. Nevertheless, we do anticipate the emergence of "new physics" as higher energy and precision are probed. That conviction stems from a general dissatisfaction with electroweak symmetry breaking and mass generation via the simple Higgs mechanism. Although that mechanism can accommodate all known particle masses and mixing (even CP violation), it does not explain their origin. One hopes that a truly fundamental theory would have no free parameters and would elucidate the origin of mass either through additional symmetry, new dynamics, or by some as yet "unknown". Uncovering those missing ingredients is the goal of high energy physics.

This chapter presents a general overview of electroweak radiative corrections and their implications for the standard model. As illustrative examples, the focus will be on low energy weak neutral current processes, but the discussion will be somewhat general.

Radiative corrections depend to some extent on the as yet unknown top quark and Higgs scalar masses. Global comparisons between theory and all high precision experiments have thereby been used to infer (for the top quark pole mass)[2]

$$m_t(pole) = 143 \pm 26 \text{ GeV} \tag{1.1}$$

for an assumed central value of $m_H \simeq 200$ GeV and the range 57 GeV $< m_H <$ 800 GeV. Fortunately (or unfortunately), the dependence on the Higgs mass, m_H, is small. When top is discovered and m_t measured, the comparison with Eq. (1.1) will be a beautiful test of the standard model at the loop level. The present experimental bound[3]

$$m_t > 108 \text{ GeV} \qquad \text{(CDF)} \tag{1.2}$$

suggests that Eq. (1.1) is likely to be on the mark.

After m_t is directly measured, one can go back and use precision measurements to differentiate light and heavy Higgs scenarios or search for hints of "new physics". Those pursuits will compare experiments at the few tenths of a percent level and will be limited to some extent by theoretical uncertainties. For that reason, our discussion will often emphasize underlying theory uncertainties from hadronic loops and higher order effects.

Precision tests of the standard model not only probe for signals of "new physics", but also provide important model building constraints. Already, they suggest significant constraints on Z' bosons,[4] heavy fermions, technicolor,[5] etc. Should a "new physics" signal be directly uncovered at the next generation of high energy colliders, precision measurements carried out at low energies will be invaluable for deciphering it.

The contents of this overview are organized as follows. In section 2, a particular electroweak renormalization scheme is described. That prescription is based on the physical observables α, G_μ, m_Z, m_W, fermion masses, and m_H (as well as quark mixing angles which will not be discussed). Those quantities are not completely independent. They are connected by the lowest order natural relation

$$\frac{e_0^2}{g_{2_0}^2} = 1 - (m_W^0/m_Z^0)^2 \tag{1.3}$$

which is maintained for renormalized physical observables, up to finite calculable corrections.[6] Those radiative corrections exhibit a sensitive dependence on m_t and thus play an important role in constraining its value.[7] It is also useful to introduce the bare weak mixing angle, θ_W^0, which depends on the other parameters via

$$\sin^2\theta_W^0 = \frac{e_0^2}{g_{2_0}^2} = 1 - (m_W^0/m_Z^0)^2. \tag{1.4}$$

A specific renormalized quantity

$$\sin^2\theta_W(m_Z)_{\overline{MS}} \equiv \frac{e^2(m_Z)_{\overline{MS}}}{g_2^2(m_Z)_{\overline{MS}}} \tag{1.5}$$

defined by $\overline{MS}$ (modified minimal subtraction) is advocated here.[8,9] That definition is particularly convenient for discussing weak neutral current processes and testing grand unified theories.

It can be perturbatively related to experimental observables or more physical definitions of $\sin^2\theta_W$ such as[8,10,11]

$$\sin^2\theta_W \equiv 1 - m_W^2/m_Z^2 \tag{1.6}$$

$$\sin^2\theta_W(0) \equiv \pi\alpha/\sqrt{2}m_W^2 G_\mu \tag{1.7}$$

$$\sin^2\bar{\theta}_W \text{ LEP Asymmetry Definition} \tag{1.8}$$

using the natural relation in Eq. (1.4) and calculated loop corrections. Section 3 illustrates the form of one loop radiative corrections to several weak neutral current processes. Those include atomic parity violation, polarized electron scattering asymmetries, and neutrino-electron scattering. Modulo some QED effects such as bremsstrahlung and Coulombic interactions, it turns out that radiative corrections to each of those reactions can be parametrized by two (process dependent) quantities ρ and κ each of which is $1 + \mathcal{O}(\alpha)$ terms. In section 4, two examples of "new physics" are considered. Z' bosons motivated by E_6 grand unification and the Peskin-Takeuchi[5] S, T, and U parametrization of gauge boson loop corrections are discussed. Their effects on various observables are illustrated. Finally, in section 5, conclusions and an outlook on the future is given.

2. Renormalization Prescription

The basic renormalized parameters of the standard model can be categorized as masses, couplings, and mixing angles. Masses are traditionally defined[12] as the real part of the propagator pole $m^2 = \mathrm{Re}s_0$. That definition is manifestly gauge independent and directly related to physical observables. However, in the case of the Z resonance, a slightly different definition, m_Z, has been employed in Breit-Wigner fits[12]

$$m_Z^2 = m_Z^2(pole) + \Gamma_Z^2. \tag{2.1}$$

Those definitions differ by about 34 MeV which is much larger than the experimental uncertainty

$$m_Z = 91.187 \pm 0.007 \text{ GeV}. \tag{2.2}$$

Similarly, in the case of the $W^{\pm}$ bosons, a renormalized W mass, m_W, is more appropriate for low energy studies than the pole mass; so here

$$m_W^2 \equiv m_W^2(pole) + \Gamma_W^2 \tag{2.3}$$

will be employed. That W mass definition differs from the pole by about 25 MeV, while ± 40 MeV is the goal of future experiments. At hadron colliders, m_W will be normalized using $m_Z = 91.187$ GeV; so, m_W as defined in Eq. (2.3) is quite natural. At $e^+e^- \to W^+W^-$ facilities, such as LEPII, the W mass extraction will be set by definition and inclusion of the relevant radiative corrections.

In the case of quark masses, running $\overline{MS}$ defined masses are sometimes more convenient than physical masses, e.g. renormalization group studies. The distinction for light quarks will not concern us. However, the top quark mass definition should be properly specified in any discussion of precision electroweak studies, particularly if two loop QCD corrections are included. The top quark pole and $\overline{MS}$ masses are related by[13]

$$m_t(pole) = m_t(m_t)_{\overline{MS}}\left(1 + \frac{4}{3}\frac{\alpha_s(m_t)}{\pi} + 10.9\left(\frac{\alpha_s(m_t)}{\pi}\right)^2 + \cdots\right) \tag{2.4}$$

In addition, the leading $\mathcal{O}(G_\mu m_t^2)$ corrections are (for $m_H \simeq 200$ GeV) approximately $-\ 5/4\ \frac{G_\mu m_t^2}{8\sqrt{2}\pi^2} \simeq -0.3\%$ and will be neglected. For $\alpha_s(m_t) \simeq 0.10$–$0.11$, the QCD corrections give rise to a 5–6% difference, which can be sizable ~ 8 GeV for $m_t \simeq 150$ GeV. In our discussion, we employ $m_t \equiv m_t(m_t)_{\overline{MS}}$ in all one loop radiative corrections. As we shall illustrate, that prescription has the advantage of not inducing large $\mathcal{O}(\alpha\alpha_s m_t^2/m_W^2)$ two loop effects. Hence, one can largely ignore two loops, modulo leading log summations which are controlled by the renormalization group. Eq. (2.4) can be used to translate to the top pole mass definition.

2.1 α, G_μ, m_Z, and m_W

In addition to fermion masses and mixing angles, m_H and three other independent parameters are needed to specify the standard electroweak model. The three parameters can be any combination of α, G_μ, m_Z, and m_W. Since the first three are already precisely measured, they are generally used as input to predict m_W. We shall follow that approach; however, for some discussions it can be cumbersome or misleading. The introduction of an auxiliary mixing angle via $\sin^2\theta_W(m_Z)_{\overline{MS}}$ often facilitates a simple closed form presentation of radiative corrections.

The usual fine structure constant, α, is defined in terms of the physical electric charge

$$\alpha \equiv \frac{e^2}{4\pi} = Z_3 \frac{e_0^2}{4\pi}\mu^{n-4} \tag{2.5}$$

where μ is the 't Hooft mass unit[14] introduced to keep α dimensionless in n dimensions. The Z_3 photonic wavefunction renormalization is conventionally defined using the vacuum polarization function $\Pi_{\gamma\gamma}(q^2)$

$$Z_3^{-1} = 1 + \Pi_{\gamma\gamma}(0) \tag{2.6}$$

$$\begin{aligned}\Pi_{\gamma\gamma}(q^2) = -\frac{e_0^2}{6\pi^2}\Bigg\{&\sum_f Q_f^2\left(\frac{1}{n-4} + \frac{\gamma}{2} - \ell n\sqrt{4\pi} + 3\int_0^1 dx(x-x^2)\ell n(m_f^2 + q^2(x^2-x))\right) \\ &-\frac{21}{4}\left(\frac{1}{n-4} + \frac{\gamma}{2} - \ell n\sqrt{4\pi} - \frac{1}{21} + 3\int_0^1 dx(x-x^2)\ell n(m_W^2 + q^2(x^2-x))\right)\Bigg\}\end{aligned} \tag{2.7}$$

with the sum taken over all charged fermions. Of course, at $q^2 = 0$, one must actually employ non-perturbative techniques or real data to obtain the hadronic contribution.

Experimentally, one knows α very precisely[15]

$$\alpha^{-1} = 137.0359895\,(61) \tag{2.8}$$

That atomic physics parameter is, however, not directly applicable for weak interaction studies. Instead, one can employ $\overline{MS}$ renormalization of $\Pi_{\gamma\gamma}(q^2)$ by subtracting only $(\frac{1}{n-4} + \frac{\gamma}{2} - \ell n\sqrt{4\pi})$ terms, to define $\alpha(\mu)_{\overline{MS}}$. Using a dispersion relation to evaluate the hadronic contributions, one finds

$$\alpha^{-1}(m_Z)_{\overline{MS}} = 127.9 \pm 0.1 \tag{2.9}$$

where the error stems from uncertainties in $e^+e^- \to$ hadrons data.[16] Even though α has essentially no uncertainty, the error in Eq. (2.9) in extrapolating to short-distances represents a limitation for precision tests of the standard model. It can be reduced by better measurements of low energy $e^+e^- \to$ hadrons or eliminated by employing some short-distance (precision) parameter in place of α. For perturbative expansions with $g_2^2/4\pi$ we will use $\alpha(m_Z)_{\overline{MS}}/\sin^2\theta_W(m_Z)_{\overline{MS}}$.

The Fermi constant, G_μ, is obtained from the muon lifetime via the defining equation

$$\tau_\mu^{-1} = \Gamma(\mu \to all) = \frac{G_\mu^2 m_\mu^5}{192\pi^3} f\left(\frac{m_e^2}{m_\mu^2}\right)\left(1 + \frac{3}{5}\frac{m_\mu^2}{m_W^2}\right)\left[1 + \frac{\alpha(m_\mu)}{2\pi}\left(\frac{25}{4} - \pi^2\right)\right]$$

$$f(x) = 1 - 8x + 8x^3 - x^4 - 12x^2 \ell n x$$

$$\alpha^{-1}(m_\mu) \simeq \alpha^{-1} - \frac{2}{3\pi}\ell n\left(\frac{m_\mu}{m_e}\right) + \frac{1}{6\pi} = 136 \tag{2.10}$$

Employing $\tau_\mu = 2.197035 \pm 0.000040 \times 10^{-6} s$ then gives

$$G_\mu = 1.16639(2) \times 10^{-5}\ \mathrm{GeV}^{-2} \tag{2.11}$$

with essentially no theoretical uncertainty. Most of the radiative corrections to muon decay, including m_t dependent loops, have been absorbed into G_μ. Those effects will reappear when we examine the natural relations in Eqs. (1.3) & (1.4) or use G_μ to normalize weak neutral current amplitudes.

The Z boson mass in Eq. (2.2) has been precisely measured at LEP. The experimental uncertainty is insignificant, but the specific definition of m_Z is important for comparison with m_W and $\sin^2\theta_W(m_Z)_{\overline{MS}}$ for the purpose of extracting m_t.

Given the value of α, G_μ, and m_Z, the $W^\pm$ mass is determined by using Eq. (1.3) in conjunction with the one loop corrections to $\Pi_{\gamma\gamma}(0)$, muon decay, and the Z and $W^\pm$ masses. Those loop effects can be collected into a quantity traditionally called Δr and used in the relation[10,17]

$$1 - m_W^2/m_Z^2 = \frac{\pi\alpha}{\sqrt{2}G_\mu m_W^2(1-\Delta r)} \tag{2.12}$$

where

$$\Delta r = \frac{2\delta e}{e} - \frac{2\delta g_2}{g_2} - \cot^2\theta_W\left(\frac{\delta m_W^2}{m_W^2} - \frac{\delta m_Z^2}{m_Z^2}\right). \tag{2.13}$$

Normalizing at $m_t = 140$ GeV and $m_H = 200$ GeV, one finds[18,19]

$$\Delta r = 0.0469 \pm 0.0011 \tag{2.14}$$

which corresponds to the central prediction $m_W = 80.18$ GeV. The dependence of Δr on m_t is illustrated in table 1 and the implication for m_W is shown in table 2. In the asymptotic limit of large m_t and m_H, the contribution to Δr is given by δ

$$\delta = \frac{\alpha(m_Z)_{\overline{MS}}}{\pi \sin^2\theta_W(m_Z)_{\overline{MS}}}\left[\frac{-3}{16}\frac{m_t^2}{m_W^2}\cot^2\theta_W(m_Z)_{\overline{MS}} + \frac{11}{48}\ell n\left(\frac{m_H^2}{m_Z^2}\right)\right] \tag{2.15}$$

Because they contribute with opposite sign, large m_t effects can be offset by very large m_H. So, if m_t turns out to be much larger than the 143 GeV central value in

Table 1 Standard model predictions for Δr, $\Delta r(m_Z)_{\overline{MS}}$, and $\Delta\hat{r}$ as functions of $m_t(m_t)_{\overline{MS}}$ for $m_H \simeq 200$ GeV. The uncertainty comes from hadronic vacuum polarization and 2 loop effects. These values incorporate the updates of QCD and top threshold effects given by Fanchiotti, Kniehl and Sirlin.[20]

$m_t(m_t)_{\overline{MS}}$ (GeV)	Δr ± 0.0011	$\Delta r(m_Z)_{\overline{MS}}$ ± 0.0011	$\Delta\hat{r}$ ± 0.0011
100	0.0596	0.0693	0.0664
110	0.0565	0.0694	0.0656
120	0.0534	0.0698	0.0649
130	0.0503	0.0700	0.0642
140	0.0469	0.0702	0.0633
150	0.0435	0.0704	0.0625
160	0.0398	0.0705	0.0616
170	0.0360	0.0706	0.0605
180	0.0319	0.0707	0.0594
190	0.0277	0.0708	0.0585
200	0.0232	0.0709	0.0574
210	0.0186	0.0710	0.0562
220	0.0137	0.0711	0.0551
230	0.0086	0.0712	0.0539
240	0.0033	0.0712	0.0525
250	−0.0021	0.0712	0.0510

Eq. (1.1), it could signal a very large m_H. The exact dependence of Δr on m_t and m_H can be found in ref. 17.

2.2 $\sin^2\theta_W(m_Z)_{\overline{MS}}$

A variety of renormalized weak mixing angles have been advocated in the literature. Some physical definitions are given in Eqs. (1.6)–(1.8). Two of those quantities are related through Δr

$$\sin^2\theta_W \equiv 1 - m_W^2/m_Z^2 = \sin^2\theta_W(0)/(1-\Delta r) \tag{2.16}$$

Each of those definitions has its own drawback. The quantity $\sin^2\theta_W(0) \equiv \pi\alpha/\sqrt{2}m_W^2 G_\mu$ is the ratio of a long-distance parameter, α, defined at zero momentum transfer and short-distance quantities G_μ and m_W^2. It is, therefore, not a good expansion parameter for weak neutral current amplitudes. When it is employed, spurious vacuum polarization corrections of $\mathcal{O}(7\%)$ are induced[8] (approximately the difference between α and $\alpha(m_Z)_{\overline{MS}}$). Since one strives to employ Born level parameters which do not introduce significant higher order effects, $\sin^2\theta_W(0)$ is rarely used except in discussions of Δr.

Table 2 Standard model predictions for m_W and $\sin^2\theta_W(m_Z)_{\overline{MS}}$ as functions of $m_t(m_t)_{\overline{MS}}$. Those predictions are based on $m_Z = 91.187$ GeV, $\alpha = 1/137.036$, $G_\mu = 1.16639 \times 10^{-5}$ GeV^{-2}, and the assumption $m_H \simeq 200$ GeV. The quoted uncertainties correspond to hadronic vacuum polarization and two loop effects.

$m_t(m_t)_{\overline{MS}}$ (GeV)	$m_t(pole)$ (GeV)	m_W(GeV) ± 0.02	$\sin^2\theta_W(m_Z)_{\overline{MS}}$ ± 0.0004
100	106.2	79.96	0.2336
110	116.7	80.01	0.2333
120	127.2	80.07	0.2331
130	137.7	80.12	0.2328
140	148.2	80.18	0.2325
150	158.7	80.24	0.2322
160	169.2	80.30	0.2319
170	179.7	80.37	0.2315
180	190.2	80.44	0.2311
190	200.6	80.51	0.2308
200	211.1	80.58	0.2304
210	221.6	80.65	0.2300
220	232.0	80.73	0.2296
230	242.5	80.81	0.2292
240	252.9	80.89	0.2287
250	263.5	80.97	0.2282

The mass definition $\sin^2\theta_W \equiv 1 - m_W^2/m_Z^2$ in Eq. (1.6) does not induce large distance vacuum polarization effects. It does, however, introduce weak isospin breaking loop effects. An important example is the top-bottom mass difference. When neutral current amplitudes are expressed in terms of $\sin^2\theta_W$, radiative corrections of $\mathcal{O}(\alpha m_t^2/m_W^2)$ are induced. Such an effect can be misleading since it gives the appearance of m_t sensitivity, e.g. in Z decay asymmetries, when none intrinsically exists. For that reason, $\sin^2\theta_W$ has fallen out of favor as m_t has been growing. An exception is the use of $\sin^2\theta_W \equiv 1 - m_W^2/m_Z^2$ in the ratio of deep-inelastic neutrino neutral to charged current scattering $R_\nu \equiv \sigma_{NC}(\nu_\mu N \to \nu_\mu X)/\sigma_{CC}(\nu_\mu N \to \mu X)$. That ratio has an intrinsic dependence on $\alpha m_t^2/m_W^2$ which cancels with the $\alpha m_t^2/m_W^2$ dependence induced by the use of $\sin^2\theta_W$ in the neutral current amplitude. Therefore, to a very good approximation, R_ν measures $\sin^2\theta_W$ independent of m_t and when used in conjunction with m_Z can be viewed as an indirect determination of m_W.[20]

An alternative weak mixing angle has been employed by LEP workers.[11] They define an effective $\sin^2\bar{\theta}_W$ at the Z pole via the ratio of vector and axial vector couplings. In principle, that definition should allow a direct determination of $\sin^2\bar{\theta}_W$ via forward-backward lepton or tau polarization asymmetry measurements at the Z. Unfortunately, there are still radiative corrections that must be applied to the asymmetries; so, the definition of $\sin^2\bar{\theta}_W$ is somewhat buried in computer codes. Fortunately, $\sin^2\bar{\theta}_W$ is numerically very close to the $\overline{MS}$ definition (to be discussed below)

$$\sin^2\theta_W(m_Z)_{\overline{MS}} \simeq \sin^2\bar{\theta}_W, \tag{2.17}$$

and the two can be used interchangeably for now. (The correction to Eq. (2.17) is typically of order 0.0002.) When the errors on $\sin^2\bar{\theta}_W$ are reduced to the level of ± 0.0005, the approximate relationship in Eq. (2.17) should be scrutinized and refined.

Rather than employing any of the physical weak mixing angles described above, it is more convenient to define an unphysical angle via modified minimal subtraction ($\overline{MS}$)

$$\sin^2\theta_W(\mu)_{\overline{MS}} \equiv e^2(\mu)_{\overline{MS}}/g_2^2(\mu)_{\overline{MS}} \tag{2.18}$$

Each of the couplings $e^2(\mu)_{\overline{MS}}$ and $g_2^2(\mu)_{\overline{MS}}$ is individually defined by $\overline{MS}$ with μ the mass scale of dimensional regularization. When μ is greater than m_t, the $\overline{MS}$ prescription is very simple. It corresponds to subtracting all $(\frac{1}{n-4} + \frac{\gamma}{2} - \ell n\sqrt{4\pi})$ loop corrections. For the standard model with 3 generations and one Higgs doublet, that translates to[8]

$$\sin^2\theta_W(\mu)_{\overline{MS}} = \sin^2\theta_W^0\left[1 + \frac{e_0^2\mu^{n-4}}{8\pi^2}\left(\frac{11}{3} + \frac{19}{6\sin^2\theta_W^0}\right)\left(\frac{1}{n-4} + \frac{\gamma}{2} - \ell n\sqrt{4\pi}\right) + \cdots\right] \tag{2.19}$$

where $\cdots$ indicates higher order terms. Because of its simple definition, $\sin^2\theta_W(\mu)_{\overline{MS}}$ is very convenient for grand unified theories (GUTS). For example, in leading log approximation, the minimal SU(5) model predicts[8] ($\sin^2\theta_W^0 = 3/8$)

$$\sin^2\theta_W(\mu)_{\overline{MS}} = \frac{3}{8}\left[1 - \frac{109}{18}\frac{\alpha(\mu)_{\overline{MS}}}{\pi}\ell n\left(\frac{m_X}{\mu}\right)\right], \mu > m_t \tag{2.20}$$

where m_X is the heavy X boson mass, $\mathcal{O}(10^{15}$ GeV).

For most low-energy processes as well as measurements at the Z pole, it is convenient to keep μ fixed at $\mu = m_Z$. That requirement causes a slight complication because $m_t > m_Z$ (from CDF bounds). To deal with that issue is a bit subtle since $g_2(\mu)_{\overline{MS}}$ can be defined by charged or neutral current couplings and top affects them differently. It is convenient to subtract an extra[21,22]

$$\frac{\alpha}{\pi}\left(\frac{1}{3} - \frac{8}{9}\sin^2\theta_W\right)\ell n\left(\frac{m_t}{\mu}\right) \tag{2.21}$$

from Eq. (2.19) for $\mu < m_t$. That additional finite subtraction keeps $\gamma - Z$ mixing independent of $\ell n m_t$ in neutral current amplitudes and allows $\sin^2\theta_W(\mu)_{\overline{MS}}$ to be continuous at $\mu = m_t$. Numerically, the additional subtraction is very small, ~ 0.0001 at $\mu = m_Z$, but its inclusion is important for strict adherence to the $\overline{MS}$ formalism.

Predictions for $\sin^2\theta_W(m_Z)_{\overline{MS}}$ can be obtained from α, G_μ, and m_Z using the relation

$$\sin^2 2\theta_W(m_Z)_{\overline{MS}} = \frac{4\pi\alpha}{\sqrt{2}G_\mu m_Z^2(1-\Delta\hat{r})} \tag{2.22}$$

where the radiative correction $\Delta\hat{r}$ is given in table 1 as a function of m_t. Note that $\Delta\hat{r}$ is much less sensitive to m_t than Δr. From the values of $\Delta\hat{r}$, the predictions for $\sin^2\theta_W(m_Z)_{\overline{MS}}$ in table 2 follow. For GUT extrapolations

$$\sin^2\theta_W(m_t)_{\overline{MS}} = \sin^2\theta_W(m_Z)_{\overline{MS}} + \frac{\alpha(m_Z)_{\overline{MS}}}{2\pi}\left[\frac{23}{6}+\frac{17}{9}\sin^2\theta_W\right]\ell n\left(\frac{m_t}{m_Z}\right) \tag{2.23}$$

In section 3, we will employ $\sin^2\theta_W(m_Z)_{\overline{MS}}$ to parametrize weak neutral current amplitudes. In terms of that quantity, the electroweak radiative corrections will be simplified.

2.3 Top Quark Mass

Electroweak observables are sensitive to the top quark mass via loop corrections to the W and Z self-energies. As illustrated in table 2, when α, G_μ, and m_Z are given, m_W and $\sin^2\theta_W(m_Z)_{\overline{MS}}$ become sensitive determinators of m_t. Those two latter quantities are related by

$$m_W^2 = \frac{\pi\alpha}{\sqrt{2}G_\mu \sin^2\theta_W(m_Z)_{\overline{MS}}(1-\Delta r(m_Z)_{\overline{MS}})} \tag{2.24}$$

where the loop corrections $\Delta r(m_Z)_{\overline{MS}}$ are given in table 1. Note that they are very insensitive to the value of m_t. From (2.24), one can see that the combination $m_W^2\sin^2\theta_W(m_Z)_{\overline{MS}}$ is rather insensitive to m_t. Therefore, a $\pm 0.05\%$ determination of m_W (i.e. ± 40 MeV) is equal to a ± 0.0002 measurement of $\sin^2\theta_W(m_Z)_{\overline{MS}}$ for the purpose of determining m_t.

At present, direct measurements of m_W give[15]

$$m_W = 80.22 \pm 0.26 \text{ GeV} \tag{2.25}$$

which suggests (from table 2) $m_t(pole) \simeq 152$ GeV. Indirect determinations of m_W via R_ν imply[2,23]

$$m_W = 80.00 \pm 0.13 \pm 0.27 \text{ GeV} \qquad (R_\nu \;\&\; m_Z) \tag{2.26}$$

which corresponds to a lower m_t.

Measurements of $\sin^2\theta_W(m_Z)_{\overline{MS}}$ at LEP and SLAC via decay asymmetries give[2]

$$\sin^2\theta_W(m_Z)_{\overline{MS}} = 0.2326 \pm 0.0011 \qquad \text{(LEP Average)} \tag{2.27}$$
$$\sin^2\theta_W(m_Z)_{\overline{MS}} = 0.2372 \pm 0.0056 \qquad \text{(SLD)} \tag{2.28}$$

From table 2, we see that those results suggest $m_t(pole) \simeq 141$ GeV which is quite consistent with the m_W implications. A more powerful LEP constraint on m_t follows from measurements of the Z partial and total decay widths. Taken together with the above results, one finds globally

$$\begin{aligned} m_t(m_t)_{\overline{MS}} &= 135 \pm 26 \text{ GeV} \\ m_t(pole) &= 143 \pm 26 \text{ GeV} \end{aligned} \qquad \text{(Global Fit)} \tag{2.29}$$

where the central values correspond to $m_H \simeq 200$ GeV and the errors allow for 57 GeV $\leq m_H \leq 800$ GeV.

Given the good agreement between theory and experiment, we will assume $m_t(m_t)_{\overline{MS}} \simeq 140$ GeV, $m_H \simeq 200$ GeV, and $\sin^2\theta_W(m_Z)_{\overline{MS}} = 0.2325$ as central values in discussions of radiative corrections which follow. (Preliminary updates from LEP suggest that increases in Eq. (2.29) are imminent.) For the examples we have chosen as illustrations, the sensitivity to shifts in m_t and m_H is very small; so, they serve primarily as probes of "new physics" rather than refinements of the above constraints.

3. Radiative Corrections and Neutral Currents

Electroweak radiative corrections have been calculated at the one loop level for a variety of weak neutral current processes. Here, I will discuss three relevant examples which are particularly easy to illustrate, atomic parity violation, polarized electron scattering and neutrino-electron scattering. In all cases, the renormalization prescription entails choosing G_μ and $\sin^2\theta_W(m_Z)_{\overline{MS}}$ as the lowest order expansion parameters and explicitly computing the one loop quantum corrections. As we shall see, the effect of those corrections can be (almost entirely) parameterized by the replacements $G_\mu \to \rho G_\mu$ and $\sin^2\theta_W(m_Z)_{\overline{MS}} \to \kappa \sin^2\theta_W(m_Z)_{\overline{MS}}$ where ρ and κ are both $1 + \mathcal{O}(\alpha)$ process dependent quantities. In addition to ρ and κ, there are generally bremsstrahlung effects which require a careful study of detector characteristics and experimental cuts. We will not discuss bremsstrahlung in any detail, but when necessary present radiative inclusive cross-sections.

3.1 Atomic Parity Violation

The interference of weak neutral current and electromagnetic amplitudes gives rise to parity violation in all physical systems. At low energies, such effects are generally much too small to observe. However, under special circumstances they may be significantly enhanced. For example, Bouchiat and Bouchiat[24] pointed out that parity violating effects in heavy atoms are enhanced by roughly $Z^3 \sim 10^5 - 10^6$ relative to hydrogen. To date, atomic parity violation has been observed in bismuth, thallium, cesium, and lead.[25] Those experiments confirm standard model expectations and constrain "new physics" appendages. In the case of cesium, progress has been particularly impressive. An experiment in Boulder, Colorado has measured atomic parity violation with $\pm 2\%$ precision.[26] Concurrently, the atomic theory[27] uncertainty has been lowered to about $\pm 1\%$. Hopefully, both errors will be further reduced in the near future. Anticipating a $\pm 0.5\%$ total error, implies a sensitive test of the standard model at the level of its radiative corrections and probe of "new physics" such as Z' bosons[28] and technicolor loop effects.[21]

In addition, the possibility of scrutinizing atomic theory and even eliminating most of its uncertainty by comparing parity violation in chains of isotopes has been advocated.[29] Given the impressive history of atomic physics experiments, one might envision a time when those presently difficult measurements become routine laboratory procedures and reach $\pm 0.1\%$ accuracy. Given such possibilities, a thorough understanding of electroweak radiative corrections is required. Here, we review the status of such calculations.[30,31,32,33]

The low energy electron-quark parity violating Hamiltonian (due to Z exchange) is conventionally parametrized by[30]

$$\begin{aligned} H_{PV} = \frac{G_\mu}{\sqrt{2}} \Big(& C_{1u} \bar{e}\gamma_\mu\gamma_5 e \bar{u}\gamma^\mu u + C_{2u} \bar{e}\gamma_\mu e \bar{u}\gamma^\mu\gamma_5 u \\ & + C_{1d} \bar{e}\gamma_\mu\gamma_5 e \bar{d}\gamma^\mu d + C_{2d} \bar{e}\gamma_\mu e \bar{d}\gamma^\mu\gamma_5 d \\ & + C_{1s} \bar{e}\gamma_\mu\gamma_5 e \bar{s}\gamma^\mu s + C_{2s} \bar{e}\gamma_\mu e \bar{s}\gamma^\mu\gamma_5 s \\ & + \cdots \Big) \end{aligned} \tag{3.1}$$

where $\cdots$ represent heavy quark terms $q = c, b, t$. Following our renormalization prescription, G_μ has been used to normalize the effective interaction and the C's are given by (in lowest order)

$$\begin{aligned} C_{1u} &= \frac{1}{2}\left(1 - \frac{8}{3}\sin^2\theta_W(m_Z)_{\overline{MS}}\right) \\ C_{1d} = C_{1s} &= -\frac{1}{2}\left(1 - \frac{4}{3}\sin^2\theta_W(m_Z)_{\overline{MS}}\right) \\ C_{2u} &= \frac{1}{2}\left(1 - 4\sin^2\theta_W(m_Z)_{\overline{MS}}\right) \\ C_{2d} = C_{2s} &= -\frac{1}{2}\left(1 - 4\sin^2\theta_W(m_Z)_{\overline{MS}}\right). \end{aligned} \tag{3.2}$$

The C_{2q} terms are suppressed by $1 - 4\sin^2\theta_W(m_Z)_{\overline{MS}} \simeq 0.07$, due to the electron's neutral current vector coupling, and thus naturally small. That suppression is fortunate, because those amplitudes undergo strong interaction renormalizations at low energies and are therefore less clean theoretically than the C_{1q} which are not renormalized by strong interactions[31] (due to CVC) at $q^2 = 0$. In fact, for heavy atoms, the C_{1q} are of prime importance because they add coherently over all quarks in the nucleus and parity violating effects are proportional to the so-called weak charge Q_W

$$Q_W(Z,A) \equiv 2[(A+Z)C_{1u} + (2A-Z)C_{1d}] \tag{3.3}$$

Compared to Q_W, C_{2q}, C_{1s} and electron-electron parity violating effects in heavy atoms are quite negligible[27] (e.g. $< 0.1\%$ for cesium); hence, we ignore them in this discussion. More important are nuclear anapole moment effects which can be several percent relative to Q_W and are not reliably computable.[25] Fortunately, those contributions to atomic parity violating transitions can be eliminated by averaging transitions involving pairs of hyperfine lines.[26]

To obtain the one loop radiative corrections to Q_W, we proceed in steps. At the quark level, one combines the radiative corrections to muon decay and the electron-quark neutral current amplitudes (at $q^2 = 0$) to give (perturbatively)[31]

$$\begin{aligned} C_{1u} &= \frac{1}{2}\rho'_{PV}\left(1 - \frac{8}{3}\kappa'_{PV}(0)\sin^2\theta_W(m_Z)_{\overline{MS}}\right) \\ C_{1d} &= -\frac{1}{2}\rho'_{PV}\left(1 - \frac{4}{3}\kappa'_{PV}(0)\sin^2\theta_W(m_Z)_{\overline{MS}}\right) \end{aligned} \tag{3.4}$$

where

$$\begin{aligned} \rho'_{PV} = 1 - \frac{\alpha}{2\pi} + \frac{\alpha(m_Z)_{\overline{MS}}}{2\pi}\Bigg\{ & \frac{3}{8s^2}\frac{m_t^2}{m_W^2} + \frac{3}{8s^4}\ell n c^2 - \frac{7}{8s^2} + \frac{3\xi}{8s^2}\left(\frac{\ell n(c^2/\xi)}{c^2-\xi} + \frac{1}{c^2}\frac{\ell n\xi}{1-\xi}\right) \\ & - \frac{1}{s^2} - 4(1-4s^2)\left[\ell n\left(\frac{m_Z^2}{M^2}\right) + \frac{3}{2}\right] - \frac{9}{16s^2c^2}\left(1 - \frac{16}{9}s^2\right)\left(1 + (1-4s^2)^2\right)\Bigg\} \end{aligned} \tag{3.5a}$$

$$\begin{aligned} \kappa'_{PV}(0) = 1 - \frac{\alpha}{2\pi s^2}\Bigg\{ & \frac{1}{3}\sum_i (T_{3i}Q_i - 2s^2Q_i^2)\ell n\left(\frac{m_i^2}{m_Z^2}\right) - \left(\frac{7}{2}c^2 + \frac{1}{12}\right)\ell n c^2 + \frac{7}{9} - \frac{s^2}{3} \\ & - \frac{1}{6}(1-4s^2)\left(\ell n\left(\frac{m_Z^2}{m_e^2}\right) + \frac{1}{6}\right)\Bigg\} - \frac{\alpha(m_Z)_{\overline{MS}}}{2\pi s^2}\Bigg\{\frac{9-8s^2}{8s^2} \\ & + \left(\frac{9}{4} - 4s^2\right)(1-4s^2)\left(\ell n\left(\frac{m_Z^2}{M^2}\right) + \frac{3}{2}\right) + \frac{9}{16s^2c^2}\left(\frac{1}{2} - 2s^2 + \frac{16}{9}s^4\right)\left(1 + (1-4s^2)^2\right)\Bigg\} \end{aligned} \tag{3.5b}$$

In those expressions $s^2 \equiv \sin^2\theta_W(m_Z)_{\overline{MS}}$, $c^2 \equiv \cos^2\theta_W(m_Z)_{\overline{MS}}$, $\xi = m_H^2/m_Z^2$, and M is a hadronic cutoff $\simeq 0.7 \sim 1$ GeV used in γZ box diagrams. The M dependence signals hadronic uncertainties which we subsequently address at the nucleon level.[32] Fortunately, that uncertainty is suppressed by a $1-4s^2$ factor and rendered negligible.

In the perturbative expressions of Eq. (3.5), we have judiciously employed $\alpha = 1/137$ and $\alpha(m_Z)_{\overline{MS}} = 1/127.9$ for different contributions. (The difference is of two loop order.) The first coupling is employed for QED loops and low momentum photon exchange (e.g. γZ mixing) while the latter is more appropriate for short-distance electroweak loops such as box diagrams. Our choice reduces two loop uncertainties since it incorporates most leading logs.

The corrections in Eq. (3.5a) have the following origins.[31] The first term is due to a QED renormalization of the electron's axial-vector current at $q^2 = 0$. That term as well as some other low energy photonic effects should actually be recalculated for heavy atoms using bound state electron propagators.[34] However, those modifications are expected to be small; so, we neglect them. The next four terms in ρ'_{PV} result from self-energy and vertex corrections to the neutral current amplitudes[17] and muon decay corrections (via G_μ normalization). The dependence on m_t and m_H reflects a difference in Z and W self-energies. It is similar to the dependence exhibited in Eq. (2.15) from the comparison of m_W and m_Z. The $\mathcal{O}(\alpha m_t^2/m_W^2)$ term is potentially important for large m_t; so, it is of interest to also examine two loop corrections of that form. Employing $m_t \equiv m_t(m_t)_{\overline{MS}}$ and assuming $m_H \simeq 1.4 m_t$, one finds (roughly)

$$m_t^2 \to m_t^2\left(1 - \frac{2\alpha_s(m_t)}{3\pi}\left(\frac{\pi^2}{3} - 3\right) - 0.4\frac{\alpha(m_Z)_{\overline{MS}}}{\pi s^2}\frac{m_t^2}{m_W^2} + \cdots\right) \tag{3.6}$$

where $\cdots$ represent non-enhanced two loop effects. As previously mentioned, the QCD correction is very small[19] because an $\overline{MS}$ definition of m_t is employed. Effects that grow as m_t^4 are also fairly insignificant if m_t is not too large. Since other uncalculated two loop corrections are potentially just as large as those in Eq. (3.6), we ignore all such effects and instead assign an uncertainty to ρ'_{PV} which reflects their neglect. The last three terms in Eq. (3.5a) come from WW, γZ and ZZ box diagrams.

The corrections to $\kappa'_{PV}(0)$ in Eq. (3.5b) also exhibit hadronic sensitivities. The sum in that expression is over all fermions with mass $< m_Z$ and Q_i = electric charge, $T_{3i} = \pm 1/2$ (a color factor of 3 must be included for quarks). That term and its neighbor (from $W^\pm$ loops) are induced by γZ vacuum polarization mixing. Such mixing involves quarks at $q^2 = 0$ and hence cannot be reliably obtained from the naive perturbative result in Eq. (3.5b). Instead, a dispersion relation must be used in conjunction with $e^+e^- \to$ hadrons data to properly determine the quark contribution.[32] We subsequently give the results of such an analysis. The remaining corrections to κ'_{PV} are from the electron charge radius and box diagrams.

We now focus on hadronic effects and uncertainties in Eq. (3.5) when applied to heavy atoms. The following discussion summarizes and updates the original analysis of ref. 32 where details can be found.

The hadronic vacuum polarization corrections to γZ mixing have been extracted from $e^+e^- \rightarrow$ hadrons data most recently by Jegerlehner.[16] Employing his results and the analysis of ref. 32, one finds (for $s^2 = 0.2325$)

$$\frac{1}{3}\sum_{\text{quarks}}(T_{3i}Q_i - 2s^2Q_i^2)\ell n\left(\frac{m_i^2}{m_Z^2}\right) \rightarrow -6.85 \pm 0.50 \tag{3.7}$$

The uncertainty in that contribution will turn out to be the dominant theoretical uncertainty (modulo atomic theory). In principle, it could be reduced somewhat by better $e^+e^- \rightarrow$ hadrons data.

The non-perturbative result in Eq. (3.7) can be mimicked by employing effective quark masses[16] $m_u \simeq 62$ MeV, $m_d \simeq 83$ MeV, $m_s \simeq 215$ MeV, $m_c \simeq 1.5$ GeV, $m_b \simeq 4.5$ GeV and multiplying by the QCD correction $(1 + \alpha_s/\pi) \simeq 1.042$. We subsequently use those masses to estimate changes in $\kappa'_{PV}(q^2)$ for $q^2 \neq 0$.

A second hadronic uncertainty enters ρ'_{PV} and κ'_{PV} via low frequency contributions to γZ box diagrams. Those effects were studied in ref. 32 for nucleons using a form factor analysis. There, it was shown that the electron-nucleon parity violating Hamiltonian was given by

$$H_{PV} = \frac{G_\mu}{\sqrt{2}}\bar{e}\gamma_\mu\gamma_5 e(C_{1p}\bar{p}\gamma^\mu p + C_{1n}\bar{n}\gamma^\mu n) + \cdots \tag{3.8}$$

where

$$\begin{aligned} C_{1p} &= \frac{1}{2}\rho'_{PV}(1 - 4\kappa'_{PV}(0)\sin^2\theta_W(m_Z)_{\overline{MS}}) \\ C_{1n} &= -\frac{1}{2}\rho'_{PV} - 0.3\frac{\alpha(m_Z)_{\overline{MS}}}{\pi}(1 - 4s^2) \end{aligned} \tag{3.9}$$

with ρ'_{PV} and $\kappa'_{PV}(0)$ as given previously, but the γZ box contribution replaced by

$$\ell n\left(\frac{m_Z^2}{M^2}\right) + \frac{3}{2} \rightarrow 11.0 \pm 1.0 \tag{3.10}$$

(That value is slightly updated compared with ref. 32.) Here, we again have some hadronic uncertainty. Fortunately, it is suppressed by $1-4s^2$ in Eq. (3.5). That suppression factor will be maintained in going from nucleons to heavy nuclei and thus render additional two nucleon correlation effects (of the type discussed in ref. 35) small and presumably unimportant.

In terms of C_{1p} and C_{1n}, the weak charge is given by

$$Q_W(A, Z) = 2(A - Z)C_{1n} + 2ZC_{1p} \tag{3.11}$$

From the above formulas, one finds (for $m_t = 140$ GeV, $m_H = 200$ GeV, $\sin^2\theta_W(m_Z)_{\overline{MS}} = 0.2325$)

$$\rho'_{PV} = 0.9857 \pm 0.0004 \pm 0.0004 \quad (3.12a)$$
$$\kappa'_{PV}(0) = 1.0029 \pm 0.0025 \pm 0.0005 \quad (3.12b)$$

where the first error comes from 1 loop hadronic uncertainties and the second corresponds to (uncalculated) higher order effects. From those results, one finds

$$C_{1p} = 0.03317 \pm 0.00013 \quad (3.13a)$$
$$C_{1n} = -0.49290 \pm 0.00028 \quad (3.13b)$$
$$Q_W(Z,A) = (0.9858 \pm 0.0006)(Z - A) + (0.06634 \pm 0.0024)Z \quad (3.13c)$$

In the case of ${}^{133}_{55}\mathrm{Cs}$ where the most precise measurements have been carried out

$$Q_W({}^{133}_{55}\mathrm{Cs})^{\mathrm{theory}} = -73.24 \pm 0.14 \quad (3.14)$$

That prediction is extremely insensitive to the values of m_t and m_H employed.[21,36] Changing those parameters modifies ρ'_{PV} via Eq. (3.5a) but that effect is almost completely canceled by the change in $\sin^2\theta_W(m_Z)_{\overline{MS}}$ (see table 2). Therefore, Eq. (3.14) represents a testable prediction with very little uncertainty. It is to be compared with the experimental result[26]

$$Q_W({}^{133}_{55}\mathrm{Cs})^{\mathrm{exp.}} = -71.04 \pm 1.58 \pm 0.88 \quad (3.15)$$

where the errors are experimental and atomic theory.[27] The agreement is quite good and must be viewed as a success for the standard model. As the errors in Eq. (3.15) are further reduced, we should see an interesting confrontation with theory at the level of electroweak radiative corrections. One can even probe for "new physics". Two examples, Z' bosons and technicolor models will be subsequently discussed.

Rather than comparing Q_W with a $\sin^2\theta_W(m_Z)_{\overline{MS}}$ derived from α, G_μ, and m_Z, one could employ alternative strategies. For example, a $\sin^2\theta_W(m_Z)_{\overline{MS}}$ directly measured in Z decay asymmetries or derived from α, G_μ, and m_W (see Eq. (2.24)) could be used in the Q_W predictions. In that case, $Q_W({}^{133}_{55}\mathrm{Cs})$ will exhibit the dependence on m_t and m_H given by the ρ'_{PV} in Eq. (3.5a). To be competitive with other loop constraints on m_t and m_H, one must aim for a $\pm 0.2\%$ measurement of $Q_W({}^{133}_{55}\mathrm{Cs})$. Such an advance represents a significant challenge for experiment and atomic theory.

As mentioned before, one might envision measuring $Q_W(Z,A)$ for a chain of isotopes and comparing those values to eliminate most atomic theory uncertainties.[29] (Neutron distributions may cause variations between isotopes.[37]) The ρ'_{PV} dependence also cancels in such a comparison; so, it can be considered merely as an alternate way to measure $\sin^2\theta_W(m_Z)_{\overline{MS}}$. To be competitive with Z asymmetry measurements, ratios of weak charges for widely separated isotopes would need to be measured to $\pm 0.1\%$ precision.

3.2 Polarized Electron Scattering

Low energy polarized electron scattering experiments also probe the weak neutral current and extensions of the standard model. By measuring the parity violating asymmetry

$$A_{pol} = \frac{d\sigma_R - d\sigma_L}{d\sigma_R + d\sigma_L} \tag{3.16}$$

for right and left-handed electron scattering, one detects the interference between γ and Z exchange amplitudes. Various asymmetry experiments have already been carried out or are planned. They range from the classic SLAC deep-inelastic e-D experiment[38] to lower energy measurements at BATES[39] and CEBAF.

One possibility suggested by G. Feinberg[40] is particularly simple. It involves the elastic scattering of polarized electrons on a spinless isoscalar target such as ^{12}C or ^{4}He. For those cases, axial-vector hadronic matrix elements are negligible and the vector current matrix elements are parametrized by a single form factor which is identical for the electromagnetic and weak neutral current amplitudes. (For now, we neglect strange quark radii[41] and nuclear isospin impurity effects[42] which can modify the asymmetry. Such subtleties will be subsequently addressed.) Therefore, hadronic matrix element uncertainties cancel in the A_{pol} ratio and the predicted lowest order asymmetry depends on $C_{1u} + C_{1d}$ (or $C_{1p} + C_{1n}$). One finds (in lowest order)

$$A_{pol} = \frac{1}{2}\frac{G_\mu q^2}{\sqrt{2}\pi\alpha}(C_{1p} + C_{1n}) = -\frac{G_\mu q^2}{\sqrt{2}\pi\alpha}\sin^2\theta_W(m_Z)_{\overline{MS}} \quad (0^+ \to 0^+) \tag{3.17}$$

where q is the momentum transfer. An experiment[39] at BATES on ^{12}C gave $\sin^2\theta_W(m_Z) = 0.200 \pm 0.051$ where the error is mainly statistical. One could envision a reduction in the experimental uncertainty to $\pm 1\%$ by longer runs. At that level, eC is competitive with atomic parity violation experiments; so, it becomes important to assess all theoretical uncertainties and include electroweak radiative corrections.

Discussion of the radiative corrections is facilitated by considering the quantity

$$\frac{A_{pol}}{\left(-\frac{G_\mu}{\sqrt{2}\pi}\frac{q^2}{\alpha}\right)} = \sin^2\theta_W(m_Z)_{\overline{MS}} \tag{3.18}$$

Modulo nuclear isospin impurity effects (which are $< 1\%$ for ^{4}He and ^{12}C and should be corrected separately[42]), in the limit $q^2 \to 0$, radiative corrections to that ratio have the same form as in the atomic parity violation discussion. They modify the r.h.s. of Eq. (3.18) to[32]

$$-\frac{1}{2}(C_{1p} + C_{1n}) = \rho'_{PV}\kappa'_{PV}(0)\sin^2\theta_W(m_Z)_{\overline{MS}} + 0.15\frac{\alpha(m_Z)_{\overline{MS}}}{\pi}(1 - 4s^2) \tag{3.19}$$

where ρ'_{PV} and $\kappa'_{PV}(0)$ were previously given in Eqs. (3.5) & (3.12). Numerically, one finds the prediction

$$\frac{A_{pol}}{\left(-\frac{G_\mu}{\sqrt{2}\pi}\, q^2/\alpha\right)} \xrightarrow[q^2\to 0]{} 0.2299 \pm 0.0003 \qquad (0^+ \to 0^+) \tag{3.20}$$

which is applicable to ^{4}He and ^{12}C (after correcting for nuclear isospin impurities).

Of course, real scattering asymmetry measurements must be made at $q^2 \neq 0$. Varying q^2 will modify the asymmetry prediction in Eq. (3.20) due to changes in the radiative corrections and the presence of nuclear strange quark charge radii effects. (Nuclear isospin impurity effects will also exhibit some q^2 dependence. We continue to assume that they are treated separately or are measured experimentally under the phenomenological prescription we subsequently give.)

Collecting all q^2 dependent modifications of Eq. (3.20) into a correction factor $1 + R(q^2)$, one expects

$$\frac{A_{pol}}{\left(\frac{-G_\mu}{\sqrt{2}\pi}\,\frac{q^2}{\alpha}\right)} = 0.2299(1 + R(q^2)) \tag{3.21}$$

$$R(0) = 0$$

where $R(q^2)$ will exhibit some process dependence. For low q^2, one expects contributions to $R(q^2)$ to be small. QED corrections from bremsstrahlung and low momentum photon exchange largely factorize and cancel in the asymmetry ratio. Residual effects can be computed and should not introduce an error larger than $\pm 0.5\%$ in $R(q^2)$. In the case of ρ'_{PV} and κ'_{PV}, they are slightly modified for $q^2 \neq 0$ primarily from changes in γZ mixing and γZ box diagrams. For low q^2, those effects are small and tend to cancel.[32] In the case of the BATES kinematics, $|q^2| \simeq 3 \times 10^4$ MeV^2, we expect $R(q^2)$ to be modified by about $0.001 \sim 0.002$ due to changes in ρ'_{PV} and κ'_{PV}.

More controversial is the effect of strange quarks on $R(q^2)$. For $q^2 \neq 0$, the strange radius of the nucleus can give rise to a shift in the asymmetry. For example, Musolf and Donnelly[43] found that strange quarks can give a q^2/m_N^2 contribution to $R(q^2)$ for small $|q^2|$ when Jaffe's estimate[41] of the strange radius is employed. That corresponds to about a 3% reduction in the asymmetry for BATES' kinematics. Given the uncertainty in such a prediction (the strange radius has not been experimentally observed), it represents an obstacle for interpreting precision measurements of the asymmetry at the $\pm 1\%$ level. One could overcome that uncertainty somewhat by first measuring the strange radius of a nucleon. Such experiments are planned.[43] Alternatively, we propose to overcome this problem by a direct experimental approach to all $R(q^2)$ effects. If one can measure the polarization asymmetry with high experimental precision for several values of q^2, then the $R(q^2)$ dependence should be ascertained and one can extrapolate to $q^2 = 0$ for comparison with the standard model prediction in Eq. (3.20). With such a procedure, it seems feasible that a confrontation between experiment and theory can be made at the $\pm 1\%$ level. We shall assume such a determination is possible in our later discussion of "new physics" sensitivity.

3.3 Neutrino-Electron Scattering

High statistics neutrino scattering experiments have been carried out for deep-inelastic $\overset{(-)}{\nu}_\mu N$, elastic $\overset{(-)}{\nu}_\mu p$, and $\overset{(-)}{\nu} e$. Here, we discuss $\overset{(-)}{\nu}_\mu e$ scattering, including effects of electroweak radiative corrections. The results are taken from Sarantakos, Sirlin, and Marciano.[44] For $\overset{(-)}{\nu}_\mu$-hadron scattering, see [17,20].

Consider the differential cross-sections for $\nu_\mu e \to \nu_\mu e$ and $\bar{\nu}_\mu e \to \bar{\nu}_\mu e$. Neglecting Z propagator effects, the lowest order cross-sections are given by[44]

$$\begin{aligned}
\frac{d\sigma(\nu_\mu e \to \nu_\mu e)}{dz} &= \frac{2G_\mu^2 p_1 \cdot p_2}{\pi}\left\{\varepsilon_-^2 + \varepsilon_+^2(1-z)^2 - \varepsilon_-\varepsilon_+ \frac{m_e^2}{p_1 \cdot p_2} z\right\} \\
\frac{d\sigma(\bar{\nu}_\mu e \to \bar{\nu}_\mu e)}{dz} &= \frac{2G_\mu^2 p_1 \cdot p_2}{\pi}\left\{\varepsilon_+^2 + \varepsilon_-^2(1-z)^2 - \varepsilon_-\varepsilon_+ \frac{m_e^2}{p_1 \cdot p_2} z\right\} \\
\varepsilon_- &= \frac{1}{2}(1 - 2\sin^2\theta_W(m_Z)_{\overline{MS}}) \\
\varepsilon_+ &= -\sin^2\theta_W(m_Z)_{\overline{MS}}
\end{aligned} \tag{3.22}$$

where p_1 and p_2 are the initial moments of the neutrino and electron, and

$$z = \frac{-q^2}{2p_1 \cdot p_2} = \frac{E_e' - m_e}{E_\nu}, \quad 0 \le z < \left(1 + \frac{m_e}{2E_\nu}\right)^{-1} \tag{3.23}$$

where E_e' and E_ν are the energies of the final state electron and initial state neutrino in the lab (electron rest) frame.

Consistent with our renormalization scheme, we have normalized the lowest order cross sections in terms of G_μ and $\sin^2\theta_W(m_Z)_{\overline{MS}}$. The $\overset{(-)}{\nu}_e e \to \overset{(-)}{\nu}_e e$ cross-sections are obtained from Eq. (3.22) via the replacement

$$\varepsilon_- \to -\frac{1}{2}(1 + 2\sin^2\theta_W(m_Z)_{\overline{MS}}) \tag{3.24}$$

where charged current amplitudes reduce ε_- by -1 due to destructive interference with the weak neutral current.

Incorporating electroweak radiative corrections is straightforward. Modulo QED correction, the electroweak loop corrections lead to the replacements

$$\begin{aligned}
G_\mu &\to \rho_{NC}^{(\nu_\mu e)} G_\mu \\
\sin^2\theta_W(m_Z)_{\overline{MS}} &\to \kappa^{(\nu_\mu e)}(q^2)\sin^2\theta_W(m_Z)_{\overline{MS}}
\end{aligned} \tag{3.25}$$

where[44]

$$\rho_{NC}^{(\nu_\mu e)} = 1 + \frac{\alpha(m_Z)_{\overline{MS}}}{2\pi}\left\{\frac{3}{8s^2}\frac{m_t^2}{m_W^2} + \frac{3}{8s^4}\ell n c^2 - \frac{7}{8s^2} + \frac{3\xi}{8s^2}\left(\frac{\ell n(c^2/\xi)}{c^2-\xi} + \frac{1}{c^2}\frac{\ell n\xi}{1-\xi}\right)\right.$$
$$\left. + \frac{1}{c^2s^2}\left(\frac{19}{8} - \frac{7}{2}s^2 + 3s^4\right)\right\} \tag{3.26a}$$

$$\kappa^{(\nu_\mu e)}(q^2) = 1 - \frac{\alpha}{2\pi s^2}\left\{2\sum_i(T_{3i}Q_i - 2s^2Q_i^2)J_i(q^2) - \left(\frac{7}{2}c^2 + \frac{1}{12}\right)\ell n c^2 - 2J_\mu(q^2)\right.$$
$$\left. + \frac{1}{3}\ell n c^2 + \frac{1}{3}c^2 + \frac{1}{2}\right\} - \frac{\alpha(m_Z)_{\overline{MS}}}{2\pi s^2}\left\{\frac{1}{c^2}\left(\frac{19}{8} - \frac{17}{4}s^2 + 3s^4\right)\right\} \tag{3.26b}$$

$$J_i(q^2) = \int_0^1 dx\, x(1-x)\ell n\left(\frac{m_i^2 - q^2x(1-x)}{m_Z^2}\right)$$

Notice the similarity with Eq. (3.5). Many of the corrections are universal, having been introduced via G_μ normalization or from γZ mixing. The sum in Eq. (3.26b) is over all charged fermions (with a color factor of 3 for quarks). For low q^2, quark loops must be evaluated using a dispersion relation in conjunction with $e^+e^- \to$ hadrons data. Alternatively, one can employ the effective quark masses given in section 3.1. Using Eq. (3.7) and the formulas given above with $m_t = 140$ GeV, $m_H = 200$ GeV, and $s^2 = 0.2325$, one finds at $q^2 = 0$

$$\rho_{NC}^{(\nu_\mu e)} = 1.0104$$
$$\kappa^{(\nu_\mu e)}(0) = 0.9966 \pm 0.0025 \tag{3.27}$$

where the uncertainty comes from hadronic vacuum polarization effects in γZ mixing.

For $q^2 \neq 0$, $\kappa^{(\nu_\mu e)}(q^2)$ is somewhat modified due to the q^2 dependence of the $J_i(q^2)$ terms in Eq. (3.26b). Those effects are, however, very small for two reasons. The electron vacuum polarization contribution to γZ mixing, which is the most q^2 dependent (at low q^2) is suppressed by $(1-4s^2)$. The light quark mass contributions (u, d, s) to γZ mixing are to a large extent canceled by the neutrino charge radius (the J_μ term) and that cancellation carries over to the q^2 dependence.[44] Therefore, the error in neglecting the q^2 dependence of $\kappa^{(\nu_\mu e)}(q^2)$ for moderate E_ν, 1–100 GeV, is insignificant compared to the error in Eq. (3.27). For very high energy neutrinos, the shift in κ can be obtained from the formulas in ref. 44. For most applications, however, employing $\kappa^{(\nu_\mu e)}(0)$ is perfectly adequate.

Turning to the QED corrections, we consider the extreme relativistic electron limit such that terms proportional to $m_e^2/p_1 \cdot p_2$ and m_e^2/q^2 can be dropped. Integrating over all bremsstrahlung, one finds that the scattering cross sections are shifted by[44]

$$\frac{\Delta\sigma(\nu_\mu e \to \nu_\mu e(\gamma))}{\sigma(\nu_\mu e \to \nu_\mu e)} = \frac{\alpha}{\pi}\left\{\varepsilon_-^2 F_- + \frac{1}{3}\varepsilon_+^2 F_+\right\}$$
$$\frac{\Delta\sigma(\bar{\nu}_\mu e \to \bar{\nu}_\mu e(\gamma))}{\sigma(\bar{\nu}_\mu e \to \bar{\nu}_\mu e)} = \frac{\alpha}{\pi}\left\{\varepsilon_+^2 F_- + \frac{1}{3}\varepsilon_-^2 F_+\right\} \qquad (3.28)$$

where

$$F_+ = F_- + 1 = -\frac{2}{3}\ell n\left(\frac{2p_1 \cdot p_2}{m_e^2}\right) - \frac{1}{6}\left(\pi^2 - \frac{43}{4}\right) \qquad (3.29)$$

Corrections to the spectrum are given in ref. 44.

In total, we find that $\overset{(-)}{\nu}e$ cross-sections are predicted up to about the $\pm 0.3\%$ level with the dominant error coming from hadronic vacuum polarization in γZ mixing. As in the case of atomic parity violation, those uncertainties could be reduced by better measurements of $e^+e^- \to$ hadrons. At present, however, that uncertainty is negligible compared to experimental errors.

To reduce systematic errors, high precision experiments actually measure ratios such as $R \equiv \sigma(\nu_\mu e)/\sigma(\bar{\nu}_\mu e)$ which is predicted (modulo very small q^2 and $p_1 \cdot p_2$ dependent $\mathcal{O}(\alpha)$ effects) to be

$$R = \frac{3\left\{1 - 4\kappa^{(\nu_\mu e)}(0)\sin^2\theta_W(m_Z)_{\overline{MS}} + \frac{16}{3}(\kappa^{(\nu_\mu e)}(0)\sin^2\theta_W(m_Z)_{\overline{MS}})^2\right\}}{\{1 - 4\kappa^{(\nu_\mu e)}(0)\sin^2\theta_W(m_Z)_{\overline{MS}} + 16(\kappa^{(\nu_\mu e)}(0)\sin^2\theta_W(m_Z)_{\overline{MS}})^2\}} \qquad (3.30)$$

The quantity $R' \equiv \sigma(\nu_\mu e)/[\sigma(\nu_e e) + \sigma(\bar{\nu}_\mu e)]$ has also been discussed[45] as a vehicle for precision studies.

At present, the best measurement of R comes from the CHARM II collaboration.[46] Using their results, one finds

$$\sin^2\theta_W(m_Z)_{\overline{MS}} = 0.2328 \pm 0.0092 \pm 0.0006 \qquad \text{(CHARM II)} \qquad (3.31)$$

where the first error is experimental and the second comes from the uncertainty in $\kappa^{(\nu_\mu e)}(0)$. The central value is very close to the world average of 0.2325 ± 0.0006; but, the experimental error is not competitive with more precise measurements of Z decay asymmetries which presently give $\sin^2\theta_W(m_Z)_{\overline{MS}}$ to about ± 0.0011 and ultimately may reduce the error to ± 0.0004. Nevertheless, precision measurements of $\overset{(-)}{\nu}_\mu e$ and $\overset{(-)}{\nu}_e e$ scattering can be used to search for "new physics". In the next section, we discuss the sensitivities of atomic parity violation, polarized electron-Carbon, and $\overset{(-)}{\nu}e$ scattering to Z' bosons and technicolor effects and show how improvements in those low energy experiments could probe "new physics" up to about the TeV scale.

4. New Physics Signatures

Precision electroweak measurements test the standard model at the level of its radiative corrections and probe for "new physics". The "new physics" may arise at the tree level, as in the case of extra Z' bosons or be induced via quantum loop effects, e.g. technicolor scenarios. Whatever, the source, we already know that existing experiments allow at most from a few tenths of a percent to perhaps a few percent deviations from the standard model. Therefore, clear discovery of some "new physics" is unlikely to come simply from improved precision in a single experimental measurement. Instead, one will check various scenarios by global fits to all data. Also, should some new phenomenon be directly observed at high energies, low energy constraints will allow us to sort out its properties.

To illustrate the utility of low energy electroweak measurements, we consider here two possible scenarios: 1) The effect of Z' bosons on low energy neutral current phenomena and 2) Modifications of low energy observables due to heavy loop effects, e.g. technicolor models. Our discussion for both cases will be somewhat general; however, we will focus primarily on atomic parity violation, polarized e-C scattering, and $\overset{(-)}{\nu}_\mu e$ scattering, since they were extensively reviewed in section 3. For concreteness, we will assume a future set of experiments in those areas will lead to combined theoretical and experimental errors

$$\begin{aligned} \frac{\Delta Q_W}{Q_W}({}^{133}_{55}\mathrm{Cs}) &= \pm 0.5\% \\ \Delta A(eC)_{pol}/A_{pol} &= \pm 1\% \\ \Delta R/R &= \pm 1\% \end{aligned} \tag{4.1}$$

Those hopes are to be compared with the existing best experiments which give $\Delta Q_W(Cs)/Q_W(Cs) = \pm 2.5\%$, $\Delta A/A = \pm 25\%$, and $\Delta R/R = \pm 7\%$. Achievement of those goals is most likely for atomic parity violation but the other two improvements also seem possible if dedicated efforts are made. The values in Eq. (4.1) were chosen because, as we shall see, they probe new physics at roughly the same level.

4.1 Z' Bosons

Extra neutral gauge bosons (generically called Z' bosons) can be easily appended to the standard model via additional $U'(1)$ gauge symmetries. They arise quite naturally in grand unified theories (GUTS) and some superstring models. For example, the SO(10) model has one such additional boson which is denoted by Z_χ while E_6 has Z_χ as well as a second flavor diagonal neutral boson, Z_ψ.[47]

The discovery of additional Z' bosons would have profound implications. Besides signaling new forces in nature, their specific properties could point to a particular GUT, Left-Right model, compositeness, etc.

If such bosons exist, they may eventually be produced and studied at high energy colliders. For example, the Fermilab tevatron $p\bar{p}$ collider CDF group has given the bounds (at 95% CL)[15]

$$\begin{aligned} m_{Z_\chi} &> 340 \text{ GeV} \\ m_{Z_\psi} &> 320 \text{ GeV} \end{aligned} \tag{4.2}$$

We anticipate those searches to continue up to about 500 GeV as the Fermilab luminosity and energy (1.8 TeV $\rightarrow$ 2.0 TeV) increase. At the SSC, the potential for finding Z' bosons should extend to the 5–10 TeV region depending on their specific properties.

To complement Z' production searches, it is useful to look for indirect hints of a Z''s existence via low energy phenomenology. Here, we will give a general discussion and then focus on atomic parity violation, polarized e-C scattering and R as specific probes of Z' bosons.

We consider the appendage of N weak neutral gauge bosons $Z_i, i = 1, 2\ldots$ to the standard model. The fermionic sector of the electroweak neutral current interaction Lagrangian is given by[48]

$$\mathcal{L}_{int} = -eA^\mu J_\mu^{em} - \frac{g_2}{\cos\theta_W} Z^\mu J_\mu^{NC} - \sqrt{\frac{3}{8}} g_2 \tan\theta_W \sum_i \sqrt{\lambda_i} Z_i^\mu J_\mu^i \tag{4.3}$$

where

$$J_\mu^{em} = \sum_f Q_f (\bar{f}_R \gamma_\mu f_R + \bar{f}_L \gamma_\mu f_L) \tag{4.4a}$$

$$J_\mu^{NC} = \sum_f (-\sin^2\theta_W Q_f \bar{f}_R \gamma_\mu f_R + (T_{3f} - Q_f \sin^2\theta_W) \bar{f}_L \gamma_\mu f_L \tag{4.4b}$$

$$J_\mu^i = \sum_f (Q_{f_R}^i \bar{f}_R \gamma_\mu f_R + Q_{f_L}^i \bar{f}_L \gamma_\mu f_L) \tag{4.4c}$$

with Q_f the fermion electric charge ($Q_e = -1$), T_{3f} = weak isospin. The λ_i in Eq. (4.3) are arbitrary but the normalization is chosen such that $\lambda_i \simeq 1$ in many GUTS. In fact, for E_6 models

$$\begin{aligned} 3Q_{e_R}^\chi &= 3Q_{u_R}^\chi = -Q_{d_R}^\chi = +1 \\ Q_{e_L}^\chi &= Q_{\nu_{e_L}}^\chi = -3Q_{u_L}^\chi = -3Q_{d_L}^\chi = +1 \end{aligned} \tag{4.5a}$$

$$\begin{aligned} Q_{e_R}^\psi &= Q_{u_R}^\psi = Q_{d_R}^\psi = -\sqrt{5/27} \\ Q_{e_L}^\psi &= Q_{\nu_L}^\psi = Q_{u_L}^\psi = Q_{d_L}^\psi = \sqrt{5/27} \end{aligned} \tag{4.5b}$$

In general one would expect all of the Z_i to mix with one another as well as the ordinary Z. However, we know from LEP physics that the Z couplings exhibit little if any mixing. Therefore, we do not consider ordinary Z mixing in this discussion. For E_6-inspired models, the combinations

$$\begin{aligned} Z_\beta &= Z_\chi \cos\beta + Z_\psi \sin\beta \\ Z'_\beta &= -Z_\chi \sin\beta + Z_\psi \cos\beta \\ &- \pi/2 \le \beta \le \pi/2 \end{aligned} \tag{4.6}$$

will be taken as mass eigenstates with $m_{Z_\beta} < m_{Z'_\beta}$.

Given the above interactions, one can easily work out the modifications of weak neutral current amplitudes due to Z' bosons. Here we give the results for $Q_W(Cs)$, $A_{pol}(eC)$, and $\sigma(\overset{(-)}{\nu}_\mu e)$ in the case of E_6 inspired models with $\lambda_\chi = \lambda_\psi = 1$ and the couplings in Eq. (4.5). One finds[49]

$$\begin{aligned} Q_W(^{133}_{55}\mathrm{Cs}) = -73.24 \pm 0.14 + 65.41\Bigg\{ &\frac{m_Z^2}{m_{Z_\beta}^2}\left(\cos^2\beta + \sqrt{\frac{5}{3}}\sin\beta\cos\beta\right) \\ &+ \frac{m_Z^2}{m_{Z'_\beta}^2}\left(\sin^2\beta - \sqrt{\frac{5}{3}}\sin\beta\cos\beta\right)\Bigg\} \end{aligned} \tag{4.7}$$

$$\begin{aligned} \frac{A_{pol}(eC)}{\left(\frac{-G_\mu}{\sqrt{2}\pi}\frac{q^2}{\alpha}\right)} = (0.2299 \pm 0.0003)\Bigg(1 + R(q^2) - &\frac{m_Z^2}{m_{Z_\beta}^2}\left(\cos^2\beta + \sqrt{\frac{5}{3}}\sin\beta\cos\beta\right) \\ &- \frac{m_Z^2}{m_{Z'_\beta}^2}\left(\sin^2\beta - \sqrt{\frac{5}{3}}\sin\beta\cos\beta\right)\Bigg) \end{aligned} \tag{4.8}$$

while for $\overset{(-)}{\nu}_\mu e$ scattering one finds the ε_- and ε_+ in Eq. (3.22) are changed to[50]

$$\begin{aligned} \varepsilon_- &= \frac{1}{2}(1-2s^2) - \frac{3}{4}s^2\left[\frac{m_Z^2}{m_{Z_\beta}^2}\left(\cos\beta + \sqrt{\frac{5}{27}}\sin\beta\right)^2 + \frac{m_Z^2}{m_{Z'_\beta}^2}\left(-\sin\beta + \sqrt{\frac{5}{27}}\cos\beta\right)^2\right] \\ \varepsilon_+ &= -s^2 - \frac{3}{4}s^2\Bigg[\frac{m_Z^2}{m_{Z_\beta}^2}\left(\cos\beta + \sqrt{\frac{5}{27}}\sin\beta\right)\left(\frac{1}{3}\cos\beta - \sqrt{\frac{5}{27}}\sin\beta\right) \\ &\quad + \frac{m_Z^2}{m_{Z'_\beta}^2}\left(\sin\beta - \sqrt{\frac{5}{27}}\cos\beta\right)\left(\frac{1}{3}\sin\beta + \sqrt{\frac{5}{27}}\sin\beta\right)\Bigg] \end{aligned} \tag{4.9}$$

The first two quantities are most sensitive for $\beta = 0$, the SO(10) model where $Z_\beta = Z_\chi$; so, we focus on that scenario. In that case, the above expressions become (ignoring Z_ψ) (including radiative corrections)

$$Q_W({}^{133}_{55}\mathrm{Cs}) = -73.24 \pm 0.14 + 65.4\frac{m_Z^2}{m_{Z_\chi}^2}$$

$$\frac{A_{pol}}{\left(\frac{-G_\mu}{\sqrt{2}\pi}\frac{q^2}{\alpha}\right)} = (0.2299 \pm 0.0003)\left(1 + R(q^2) - \frac{m_Z^2}{m_{Z_\chi}^2}\right)$$

$$\varepsilon_- = 0.2683\left(1 - 0.65\frac{m_Z^2}{m_{Z_\chi}^2}\right)$$

$$\varepsilon_+ = -0.2317\left(1 + 0.25\frac{m_Z^2}{m_{Z_\chi}^2}\right) \tag{4.10}$$

One sees that a $\pm 0.5\%$ measurement of $Q_W(Cs)$ probes M_{Z_χ} at the 1.2 TeV level. Indeed, if we were to attribute the central values spread in Eqs. (3.14) and (3.15), i.e. $\Delta Q_W = 2.2$ to a Z_χ, it would correspond to $m_\chi \simeq 500$ GeV. Amazingly, such a massive Z_χ would not have been detected in any other precision measurement or collider experiment (see Eq. (4.2)). A new measurement of $Q_W(Cs)$ to ± 0.35 would see a 500 GeV Z_χ at the 6σ level. Also, for other cesium isotopes, one finds

$$Q_W({}^{133+X}_{55}\mathrm{Cs}) = -73.24 - 0.986X + (65.4 + 0.62X)\frac{m_Z^2}{m_{Z_\chi}^2} \tag{4.11}$$

Measuring Q_W for chains of Cs isotopes and other atoms could reveal the couplings of an extra Z'.

To probe the $m_{Z_\chi} \simeq 1$ TeV level in polarized eC scattering requires a measurement of A_{pol} to $\pm 0.8\%$ which is hard but not impossible. At issue is the control of $R(q^2)$ in Eq. (4.10) at that level. If a deviation from the standard model is found in atomic parity violation, low energy polarized electron scattering experiments may provide our best means of confirmation.

In the case of $\overset{(-)}{\nu}_\mu e$ scattering, one finds from Eq. (4.10) that a 1 TeV Z_χ can be observed (at the 1σ) level) by a $\pm 0.8\%$ measurement of $\sigma(\nu_\mu e)$ or a similar measurement of R (since $\sigma(\bar{\nu}_\mu e)$ is rather insensitive to a Z_χ). Alternatively, $R' \equiv \sigma(\nu_\mu e)/[\sigma(\nu_e e) + \sigma(\bar{\nu}_\mu e)]$ is somewhat more sensitive to Z_χ

$$R' = R'_{\text{standard model}}\left(1 - 1.3\frac{m_Z^2}{m_{Z_\chi}^2}\right) \tag{4.12}$$

A 1% measurement of R' would probe the TeV scale.

4.2 S, T, and U

A nice formalism for studying heavy physics effects on gauge boson self-energies was introduced by Peskin and Takeuchi,[5] the S, T and U parametrization. Here, we describe the implementation of that formalism and current experimental constraints on those parameters.

Heavy new fermions, such as technifermions, a heavy fourth generation, etc. enter low-energy phenomenology through gauge boson self-energies. In the electroweak sector those include the vacuum polarization functions $\Pi_{\gamma\gamma}(q^2)$, $\Pi_{\gamma Z}(q^2)$, $\Pi_{ZZ}(q^2)$, and $\Pi_{WW}(q^2)$. Using $\overline{MS}$ subtraction removes heavy particle contributions to the first two by absorbing such effects into the definitions $\alpha(m_Z)_{\overline{MS}}$ and $\sin^2\theta_W(m_Z)_{\overline{MS}}$. There are, however, residual effects in Π_{ZZ} and Π_{WW} that remain after renormalization and are observable as corrections to the natural relationships in Eq. (1.4).

The heavy new fermion loop effects can be parametrized by three observables S_W, S_Z and T defined by[21]

$$\left.\frac{\Pi_{WW}^{new}(m_W^2)-\Pi_{WW}^{new}(0)}{m_W^2}\right|_{\overline{MS}} = \frac{\alpha(m_Z)_{\overline{MS}}}{4\sin^2\theta_W(m_Z)_{\overline{MS}}} S_W = Z_W^{new}-1 \tag{4.13}$$

$$\left.\frac{\Pi_{ZZ}^{new}(m_Z^2)-\Pi_{ZZ}^{new}(0)}{m_Z^2}\right|_{\overline{MS}} = \frac{\alpha(m_Z)_{\overline{MS}}}{\sin^2 2\theta_W(m_Z)_{\overline{MS}}} S_Z = Z_Z^{new}-1 \tag{4.14}$$

$$\frac{\Pi_{WW}^{new}(0)}{m_W^2}-\frac{\Pi_{ZZ}^{new}(0)}{m_Z^2} = \alpha(m_Z)_{\overline{MS}}T = \rho(0)^{new}-1 \tag{4.15}$$

where "new" means only heavy new particle loops are included, $|_{\overline{MS}}$ means a modified minimal subtraction is applied, and $\alpha(m_Z)_{\overline{MS}}$ has been factored out. In terms of the Peskin-Takeuchi parameters

$$\begin{aligned} S &= S_Z \\ T &= T \\ U &= S_W - S_Z \end{aligned} \tag{4.16}$$

The T and U correspond to isospin violating effects, while S is isospin conserving.

Even without new physics, S_W, S_Z, and T provide a convenient means for approximating deviations from our assumed $m_t = 140$ GeV and $m_H = 200$ GeV values. Deviations give rise to[22,51]

$$\begin{aligned} S_W &\simeq \frac{1}{6\pi}\ell n\left(\frac{m_H}{200\text{ GeV}}\right)+\frac{2}{3\pi}\ell n\left(\frac{m_t}{140\text{ GeV}}\right) \\ S_Z &\simeq \frac{1}{6\pi}\ell n\left(\frac{m_H}{200\text{ GeV}}\right)-\frac{1}{3\pi}\ell n\left(\frac{m_t}{140\text{ GeV}}\right) \\ T &\simeq \frac{3}{16\pi s^2}\left(\frac{m_t^2-(140\text{ GeV})^2}{m_W^2}\right)-\frac{3}{8\pi c^2}\ell n\left(\frac{m_H}{200\text{ GeV}}\right) \end{aligned} \tag{4.17}$$

Note that $U = \frac{1}{\pi}\ell n(m_t/140 \text{ GeV})$ is much smaller than T and can be generally ignored.

S_W, S_Z, and T are measured in the following way. Employing α, G_μ, m_Z, $m_t = 140$ GeV, $m_H = 200$ GeV (along with light fermion masses and mixing) completely specifies the standard model, modulo "new physics" or deviations in m_t or m_H. If such effects are present, they lead to shifts in the radiative corrections previously discussed (see section 2)[21]

$$\begin{aligned} \Delta r &= 0.0469 \pm 0.0011 - 0.0195 S_W + 0.0361 S_Z - 0.0258 T \\ \Delta r(m_Z)_{\overline{MS}} &= 0.0702 \pm 0.0011 + 0.0084 S_W \\ \Delta \hat{r} &= 0.0633 \pm 0.0011 + 0.011 S_Z - 0.00782 T \end{aligned} \tag{4.18}$$

as well as a change in the ρ parameters of section 3

$$\rho^{new}(0) = 1 + 0.00782T \tag{4.19}$$

used in all weak neutral current amplitudes.

From the changes given above, one finds the new predictions

$$\sin^2\theta_W(m_Z)_{\overline{MS}} = 0.2325 \pm 0.0004 + 0.00365 S_Z - 0.00261 T \tag{4.20a}$$

$$m_W = 80.18 \pm 0.02 + 0.45T - 0.63 S_Z + 0.34 S_W \text{ GeV} \tag{4.20b}$$

$$Q_W({}^{133}_{55}\text{Cs}) = -73.24 \pm 0.14 - 0.8 S_Z - 0.005 T \tag{4.20c}$$

$$\frac{A_{pol}(eC)}{\left(\frac{-G_\mu}{\sqrt{2}\pi} q^2/\alpha\right)} = 0.2299(1 + R(q^2) + 0.016 S_Z - 0.003 T) \tag{4.20d}$$

$$R = \frac{\sigma(\nu_\mu e)}{\sigma(\bar{\nu}_\mu e)} = 1 - 0.029 S_Z + 0.021 T \tag{4.20e}$$

$$\Gamma_Z/\Gamma_Z^{\text{Standard Model}} = 1 - 0.0038 S_Z + 0.0105 T \tag{4.20f}$$

$$\Gamma(Z \to e^+e^-)/\Gamma^{S.M.}_{e^+e^-} = 1 - 0.0021 S_Z + 0.0093 T \tag{4.20g}$$

From those expressions, we see that most precision experiments are sensitive to a linear combination of S_Z and T. The exception is m_W which depends on both S_W and S_Z. In fact, one can isolate S_W by comparing m_W and $\sin^2\theta_W(m_Z)_{\overline{MS}}$ (measured via a Z decay asymmetry)[21]

$$S_W = 118\left\{2\left(\frac{m_W - 80.18\text{ GeV}}{80.18\text{ GeV}}\right) + \frac{\sin^2\theta_W(m_Z)_{\overline{MS}} - 0.2325}{0.2325}\right\} \tag{4.21}$$

In addition, one notes that $Q_W(Cs)$ and $A_{pol}(eC)$ are rather insensitive to T and thus provide direct measurements of S_Z.[21]

By comparing experiments with the theoretical predictions, including its S & T dependence, one obtains constraints on those parameters. Various existing constraints along with anticipated future advances are illustrated in table 3. Taking $S_W = S_Z = S$, one finds[2]

$$S = -0.05 \pm 0.26 + 1.15T \tag{4.22}$$

The data is very consistent with S & $T \simeq 0$, but could accommodate an $S \simeq 1$ if T is also $\mathcal{O}(1)$.

Table 3 Present experimental constraints on S_W, S_Z, and T along with anticipated future sensitivities. Central values of $m_t = 140$ GeV and $m_H = 200$ GeV are assumed.

Experiment	Current Constraint	Future
$m_W = 80.22 \pm 0.26$ GeV	$T - 1.4S_Z + 0.76S_W = 0.09 \pm 0.60$	± 0.09
LEP Asymmetries	$S_Z - 0.69T = 0.03 \pm 0.32$	± 0.15
$(\sin^2\theta_W(m_Z)_{\overline{MS}} = 0.2326 \pm 0.0011)$		
A_{LR} (SLD)	$S_Z - 0.69T = 1.33 \pm 1.5$	± 0.10
$Q_W({}^{133}_{55}\text{Cs}) = -71.04 \pm 1.81$	$S_Z - 0.005T = -2.7 \pm 2.0 \pm 1.1$	± 0.5
$\Gamma_Z = 2492 \pm 7$ MeV	$T - 0.36S_Z = 0.27 \pm 0.27$	± 0.27
$\Gamma_{\ell^+\ell^-} = 83.33 \pm 0.29$ MeV	$T - 0.23S_Z = -0.35 \pm 0.37$	± 0.37
$R = \frac{\sigma(\nu_\mu e)}{\sigma(\bar{\nu}_\mu e)}$	$S_Z - 0.69T = 0.08 \pm 2.5$	?
$R_\nu \equiv \sigma(\nu_\mu N)_{NC}/\sigma(\nu_\mu N)_{CC}$	$T - 0.4S_Z = -0.22 \pm 0.67$	± 0.24
$R_{\bar{\nu}} = \sigma(\bar{\nu}_\mu N)_{NC}/\sigma(\bar{\nu}_\mu N)_{CC}$	$T - 0.028S_Z = 1.4 \pm 1.3$	± 0.65

Why is S interesting? Each new heavy chiral $\text{SU}(2)_L$ doublet contributes $1/6\pi$ to S;[52] so, a signal for heavy fermions could be $S > 0$. Technicolor models have many heavy doublets. Indeed, one expects (including enhancements)

$$S \simeq (0.05 - 0.10)N_T N_D + 0.12 \quad \text{(technicolor)} \tag{4.23}$$

where N_T is the number of technicolors and N_D is the number of flavor doublets. (The +0.12 comes from a heavy effective Higgs in such a scenario.) For a generic one generation model with $N_D = 4$ and $N_T = 4$, one expects[5] $S \simeq 0.9 \sim 1.7$. So far, there is no evidence for $S > 0$, but $S \sim 1$ is not completely ruled out.

In the future, individual experiments are expected to reach a sensitivity of $0.1 \sim 0.5$ for S and T. At that level, technicolor should unveil itself if it is real. We note that supersymmetric theories do not introduce additional chiral fermions and thus naturally give very small S and T. So, as long as their values remain close to zero, it bodes well for supersymmetry enthusiasts.

5. Outlook

The discovery of weak neutral currents marked a tremendous success for the $SU(2)_L \times U(1)$ model. Now, that theory has been tested at the level of its quantum corrections via 1% (or better) measurements. So far, no hints of "new physics" have emerged, but m_t has been constrained to 143 ± 26 GeV. Confronting that prediction with a direct determination of m_t, after its discovery, will be very exciting. An intriguing outcome would be a very heavy m_t, well beyond the loop constraint. Indeed, the heavier top turns out to be, the more interesting it becomes.

After m_t is known, precision measurements may provide some constraint on m_H. At least the light, $m_H \simeq 100$ GeV, and heavy $m_H \simeq 800$ GeV, cases may be differentiated. Confirming the Higgs' existence, will require direct discovery at high energies and exploration of its properties. That occurrence must wait for the next generation of colliders.

To the dismay of technicolor advocates, there is no evidence for $S > 0$; however, $S \sim \mathcal{O}(1)$ is not ruled out. The next generation of precision measurements will probe S down to about ± 0.2. At that level, technicolor or some other heavy fermion effect might well emerge. In addition, we saw that Cs atomic parity violation is already sensitive to $m_{Z_\chi} \simeq 500$ GeV and is capable of reaching to ~ 1 TeV. Similar sensitivities are possible with polarized eC and $\nu_\mu e$ scattering experiments; but reaching that level will not be easy. Nevertheless, such pursuits are worth the effort. Those low energy measurements complement precision determinations of m_W, $\sin^2\theta_W(m_Z)_{\overline{MS}}$ etc., searches for rare or forbidden processes, and high energy collider explorations. Collectively, they test the standard model and provide guidance for our scientific imaginations, as we strive to decipher and appreciate the laws of Nature.

References

1. S. Weinberg, Phys. Rev. Lett. 19 (1967) 1264; A. Salam, In "Elementary Particle Theory", ed. N. Svartholm (Almquist & Wiksells, 1968) p. 367.
2. See W. Marciano in proceedings of DPF 92, Fermilab, Batavia, Illinois (World Scientific) p. 185.
3. CDF, Abe *et al.*, unpublished.

4. P. Langacker and M. X. Luo, Phys. Rev. D45 (1992) 278. P. Langacker, M. Luo, and A. Mann, Rev. Mod. Phys. 64 (1992) 87.
5. M. Peskin and T. Takeuchi, Phys. Rev. Lett. 65 (1990) 964; Phys. Rev. D46 (1992) 381.
6. C. Bollini, J. Giambiagi, and A. Sirlin, Nuovo Cimento 16A (1973) 423; W. Marciano, Nucl. Phys. B84 (1975) 132.
7. M. Veltman, Nucl. Phys. B123 (1977) 89.
8. W. Marciano, Phys. Rev. D20 (1979) 274.
9. W. Marciano and A. Sirlin, Phys. Rev. Lett. 46 (1981) 163; see also proceedings "The Second Workshop on Grand Unification", eds. Leveille, Sulak and Unger (Berkhäuser, Boston 1981) p. 151.
10. A. Sirlin, Phys. Rev. D22 (1980) 971.
11. See M. Böhm and W. Hollik in "Z Physics at LEP I", CERN89-08, p. 203.
12. S. Willenbrock and G. Valencia, Phys. Lett. B259 (1991) 373; A. Sirlin, Phys. Rev. Lett. 67 (1991) 2127; R. Stuart, Phys. Lett. B262 (1991) 113.
13. W. Marciano, Phys. Rev. Lett. 62 (1989) 2793; H. Arason *et al.*, Phys. Rev. D46 (1992) 3945.
14. G. 't Hooft, Nucl. Phys. B61 (1973) 455.
15. Review of Particle Properties, Phys. Rev. D45 (1992) part II.
16. F. Jegerlehner, in "Testing the Standard Model", eds. M. Cvetic and P. Langacker, (World Scientific, Singapore 1991) p. 476.
17. W. Marciano and A. Sirlin, Phys. Rev. D22 (1980) 2695.
18. G. Degrassi, S. Fanchiotti, and A. Sirlin, Nucl. Phys. B351 (1991) 49.
19. S. Fanchiotti, B. Kniehl, and A. Sirlin, CERN preprint Th. 6749/92 (1992).
20. A. Sirlin and W. Marciano, Nucl. Phys. B189 (1981) 442; C. H. Llewellyn Smith and J. F. Wheater, Phys. Lett. B105 (1981) 486.
21. W. Marciano and J. Rosner, Phys. Rev. Lett. 65 (1990) 2963.
22. W. Marciano, Annu. Rev. Nucl. Part. Sci. 41 (1991) 469.
23. U. Amaldi, *et al.*, Phys. Rev. D36 (1987) 1385.
24. M. A. Bouchiat and C. Bouchiat, J. Phys. 35 (1974) 899.
25. I. B. Khriplovich, "Parity Nonconservation in Atomic Phenomena", (Gordon and Breach, 1991).
26. M. C. Noecker, B. P. Masterson, and C. E. Wieman, Phys. Rev. Lett. 61 (1988) 310.
27. S. A. Blundell, J. Sapirstein, and W. R. Johnson, Phys. Rev. D45 (1992) 1602.
28. D. London and J. Rosner, Phys. Rev. D34 (1986) 1530.
29. V. A. Dzuba, V. Flambaum, and I. B. Khriplovich, Z. fur. Phys. D1 (1986) 243; C. Monroe *et al.*, Phys. Rev. Lett. 65 (1990) 1571.
30. W. Marciano and A. Sanda, Phys. Rev. D17 (1978) 3055.
31. W. Marciano and A. Sirlin, Phys. Rev. D27 (1983) 552.
32. W. Marciano and A. Sirlin, Phys. Rev. D29 (1984) 75.
33. Bryan W. Lynn, Columbia University report, 1982 (unpublished); 1982 Les Houches Summer School Lectures.

34. J. Sapirstein and D. Yennie, in "Quantum Electrodynamics", ed. T. Kinoshita, (World, Singapore 1990) p. 560.
35. W. Jaus and G. Rasche, Phys. Rev. D41 (1990) 166.
36. P. Sandars, J. Phys. B23 (1990) L655.
37. E. Forston, Y. Pang, and L. Wilets, Phys. Rev. Lett. 65 (1990) 2857; S. Pollock, E. Forston, and L. Wilets, Univ. of Washington preprint 1992.
38. C. Y. Prescott *et al.*, Phys. Lett. B77 (1978) 347; 84B (1979) 524.
39. P. A. Souder *et al.*, Phys. Rev. Lett. 65 (1990) 694.
40. G. Feinberg, Phys. Rev. D12 (1975) 3575.
41. R. L. Jaffe, Phys. Lett. B229 (1989) 275.
42. T. Donnelly, J. Dubach, and I. Sick, Nucl. Phys. A503 (1989) 589.
43. M. Musolf and T. Donnelly, Z. Phys. C57 (1993) 559.
44. S. Sarantakos, A. Sirlin, and W. Marciano, Nucl. Phys. B217 (1983) 84.
45. LAMPF Proposal LA-11300-P, D. H. White (spokesman).
46. G. Rädel in proceedings of DPF92, Fermilab, Batavia, Ill. (1992) p. 415.
47. J. Rosner, Comm. Nucl. Part. Phys. 14 (1985) 229.
48. W. Marciano, Nucl. Phys. B11 (1989) 5.
49. W. Marciano in "1987 DPF Salt Lake City Meeting", eds. C. DeTar and J. Ball, p. 319.
50. S. Godfrey and W. Marciano, "1987 BNL Neutrino Workshop Proceedings", ed. M. Murtagh, p. 195.
51. D. Kennedy and P. Langacker, Phys. Rev. Lett. 65 (1990) 2967.
52. S. Bertolini and A. Sirlin, Nucl. Phys. B248 (1984) 589.

III. PRECISION TESTS AT e^+e^- COLLIDERS

e^+e^- Annihilation from $\sqrt{s}$ = 20 to 65 GeV

D. HAIDT
DESY
2000 Hamburg 52, Germany

Contents

1 Introduction

The process $e^+e^- \to hadrons$ is the long range and observable manifestation of the fundamental process $e^+e^- \to q\bar{q}$ (for q = u,d,s,c,b) and includes all higher order processes $e^+e^- \to q\bar{q}\gamma$ and $e^+e^- \to q\bar{q}g$ with one or more photons or gluons, because these higher order contributions cannot be distinguished experimentally from the tree diagram. In lowest order QFD (Quantum Flavor Dynamics) the interaction is due to the exchange of a neutral vector boson (γ, Z), as shown in Figure 1.

The experimental data obtained at PETRA, PEP, TRISTAN, SLC and LEP (Figure 2) cover the energy range governed by the dominance of photon exchange at the lowest energies right up to the dominance of Z exchange, and this allows a full study of the weak neutral current of the electroweak theory. Furthermore, the process $e^+e^- \to hadrons$ receives significant contributions from QCD and provides information on the strong coupling constant over a large energy range. The observable total cross sections can be reliably predicted in terms of both their electroweak and strong properties.

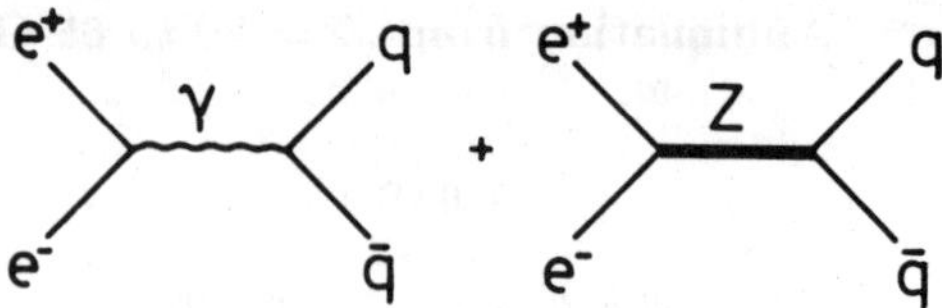

Figure 1: Tree diagrams for $e^+e^- \rightarrow$ *hadrons*

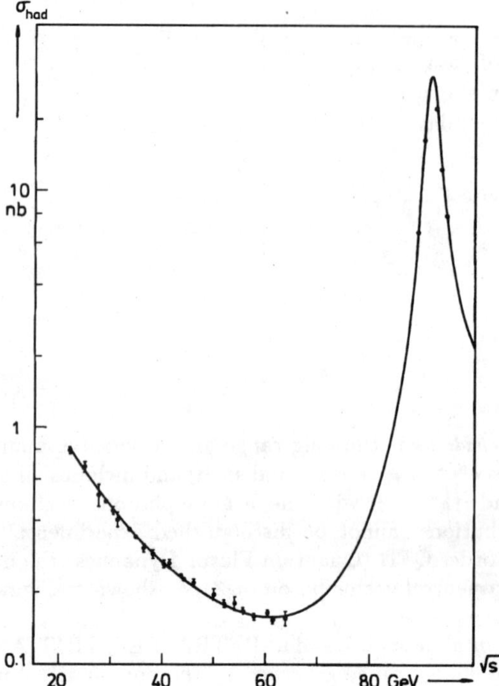

Figure 2: Measured total hadron cross sections versus center of mass energy in comparison (line) with the theoretical prediction for the reference parameter choice (see text)

In this report a coherent analysis of all the total hadron cross sections below the Z resonance is presented and this brings out all the salient theoretical and experimental details and the relations between them. The data from the e^+e^-colliders PETRA,PEP,TRISTAN have been collected in many different experiments for more than one decade and have extended the understanding and validity of the Standard Model deeper and deeper into the timelike region. The data published over the years reflect the considerable progress in experimentation and phenomenology of the e^+e^- annihilation process. The simultaneous progress in the theoretical understanding of higher order processes in QFD [12, 13] and QCD offer an incentive to analyse the data all together in a consistent fashion. Such a treatment of all data within the common theoretical framework provided by the program package ZFITTER ensures the extraction of Standard Model parameters with the smallest possible systematic errors and also allows, as a natural next step, the inclusion of the total hadron cross sections measured by the LEP experiments.

Since only the study of total hadron cross sections is addressed, some hints are added concerning other topics. In the Landoldt-Börnstein handbook article on electroweak interactions [4] the whole subject is covered including data compilations up to the end of 1986. The PETRA groups have demonstrated 1981 that the angular distribution in $e^+e^- \to \mu\mu$ is not symmetric. This observation stands out as a milestone in establishing the electroweak theory in a new energy regime. For the first time, a deviation from QED had been put in evidence. The deviation pattern from QED fitted well in the embracing theory QFD. It was then possible to specify quantitatively the range of validity of QED [5]. The various processes $e^+e^- \to l^+l^-$ in the PETRA and PEP energy range have provided valuable tests of QED up to $O(\alpha^4)$ [6]. Such tests were important in order to extract reliably the small electroweak effects and to determine the relevant electroweak parameters. The precisely measured axial vector constant a_b by the JADE group supported the assignment of the top quark as the weak isospin partner of the b quark [7]. The TRISTAN e^+e^- collider came into operation at the end of 1986, shortly after the final shut down of PETRA, and extended the energy range up to 65 GeV. The TRISTAN data provide unique tests of the Standard Model in the (γ, Z) interference region, where the lepton [25, 26, 27] and quark [28, 29, 30] asymmetries are large. Summaries can be found in the proceedings of the recent international high energy conferences.

2 Data

2.1 Collection of Data

The published data from the e^+e^- colliders PETRA, PEP, TRISTAN are summarized in Table 1. Previous summaries on the low energy data can be found in [1, 2, 3, 4] and [15]. The latter was the first attempt to put all PETRA data on the same footing and to fit them simultaneously. Here a more refined analysis and com-

Experiment	Year of Publication	Remarks
PETRA		
CELLO	1987 [15]	JS
JADE	1987 [16]	JS QED
MARK J	1986 [17]	JS
PLUTO	1983 [19]	not included in fit
TASSO	1990 [18]	JS QED
PEP		
HRS		not included in fit
MAC	1985 [20]	
MARK II	1990 [21]	JS
TRISTAN		
AMY	1990 [22]	JS and FS
TOPAZ	1990 [23]	JS and FS
VENUS	1990 [24]	JS and FS

Table 1: Collection of experimental data.

parison with theory is presented. Some entries are not included in the subsequent fits because the experimental uncertainties are large. Since their first publication the TRISTAN groups have collected much more data, particularly at 58 GeV. Their evaluation is still in progress.

2.2 Comparison of Experiments

The substance of the published data analyses is not touched. However, some ingredients entering the interpretation of the data must be revised in the light of progress achieved up to date, otherwise a meaningful comparison among the experiments is not possible.

The basic formula used to evaluate the quantity R_{had} is :

$$R_{had} = \frac{N - N_B}{L\epsilon(1+\delta)} \frac{1}{\sigma_\mu} \tag{1}$$

The hadron cross section is normalized to the lowest order QED $e^+e^- \to \mu^+\mu^-$ cross section, i.e. $\sigma_\mu = 4\pi\alpha^2/3s$. In this equation N is the observed number of multihadron events satisfying the selection criteria, and contains radiative events (i.e. events of genuine electroweak origin, real bremsstrahlung and QCD type events). N_B is the number of background events, L is the luminosity determined from Bhabha events, while $\epsilon(1+\delta)$ are the acceptance and radiative correction factors taking account of the actual experimental conditions. When comparing different experiments three aspects are important :

- Radiative corrections :
  The meaning is not unique. In fact, some experiments considered only QED corrections, while others included also partial or full electroweak contributions reflecting the state of art at the time of publication. During the preparation phase of LEP the anticipated high statistics led to a careful review of the radiative effects associated with the processes $e^+e^- \rightarrow hadrons$ and $e^+e^- \rightarrow e^+e^-$. This activity resulted in new analytic, semianalytic and Monte Carlo programs aimed at the interpretation of the Z-resonance data [13, 12]. One of these program packages is ZFITTER [14]. In a special effort [31] the range of validity of this program originally aimed at the Z resonance region has been extended in order to cover also the PETRA energy range with the same intrinsic precision.

- Input parameters :
  The evaluation of corrections depended upon the numerical knowledge of quantities such as the mass of the Z-boson or the mass of the, as yet unobserved, top quark. Publications at the beginning of the 80ies usually assumed values for the top quark of around 20 GeV and for the Z mass around 92 GeV. Almost every experiment had its own choice of parameters.

- Hadronization :
  The measurement of the total cross section is not performed in the full phase space requiring an acceptance calculation which was done by a Monte Carlo simulation of the detector response (denoted by A) to the multihadron events (see below). The final publications are based on a rather similar treatment of the hadronic final state using the LUND JETSET (denoted by JS) program package [8]. As a matter of fact, all experimental groups have explicitly demonstrated the excellent agreement of various predicted final state distributions with the measurements.

In the comparison such differences are unified as far as possible a posteriori resulting in slightly reduced systematic uncertainties.

The total cross section of $e^+e^- \rightarrow hadrons$ at $\sqrt{s}$ is defined in equation 2 :

$$\sigma_{tot} = \int_0^{k_{max}} H(z)\sigma_1((1-z)s)dz \qquad (2)$$

The corresponding R value is defined by $R = \sigma_{tot}/\sigma_\mu$. In this equation appears σ_1, the full electroweak 1-loop cross section, convoluted with the radiator function H. The Standard Model parameters enter in σ_1. The integral depends explicitly on k_{max}, the maximum relative photon energy considered, which was chosen by most groups to be 0.99.

All experimental groups have evaluated the correction factor relating the observed number of multihadron events (after background subtraction) with the Born cross section, thus excluding radiative events. The relevant quantity is :

$$\epsilon(1+\delta) = \frac{\int_0^{k_{max}} A(z)H(z)\sigma_1((1-z)s)dz}{\sigma_{born}} \tag{3}$$

In practice, however, this quantity cannot be determined analytically because of the complicated detector geometry, the detector performance and the event selection criteria. Instead of calculating

$$\epsilon = \frac{\int_0^{k_{max}} A(z)H(z)\sigma_1((1-z)s)dz}{\int_0^{k_{max}} H(z)\sigma_1((1-z)s)dz} \tag{4}$$

the average acceptance ϵ is obtained by Monte Carlo methods. Both geometry and performance of the whole experiment are simulated. Then complete events are generated. This means that the individual configurations reflect both the electroweak properties as well as the full hadron final state. The actual procedure of the event generation in the JETSET package is based upon the prescription set up at the beginning of the 80-ies by Berends, Kleiss and others [10, 11]. A generated event (N_{gen}) is accepted, if it satisfies the event selection criteria (N_{acc}), leading to the definition :

$$\epsilon = \frac{N_{acc}}{N_{gen}} \tag{5}$$

The required quantity $\epsilon(1+\delta)$ is the product of the two equations 5 and 6 with

$$(1+\delta) = \frac{\int_0^{k_{max}} H(z)\sigma_1((1-z)s)dz}{\sigma_{born}} \tag{6}$$

Obviously, the equation 3 can always be factorised into the equations 4 and 6 as long as the same theoretical cross sections are applied. This is the case for the PETRA experiments, but not for the TRISTAN experiments in their presently published form. The three TRISTAN groups were the first in applying full 1-loop calculations as provided by Fujimoto, Kato and and Shimizu [9]. The quantity $\epsilon(1+\delta)$ is understood as the product of equation 5 obtained from JETSET with 1+δ obtained from Fujimoto, Kato and Shimizu. Since the TRISTAN groups have also included in their publication the actually used values for $1+\delta$, this difference can be, and is, taken into account.

Some qualitative features of the apparatus response function A appearing in the convolution integral 3 are easli recognised. The event selection criteria entail a characteristic shape of $A(z)$ looking, roughly speaking, like a smoothed step function in z. The low z region consists mainly of $q\bar{q}$ events. Their acceptance is limited by the geometry of the detector. The high z region, on the contrary, consists of $q\bar{q}\gamma$ events, where the photon carries an appreciable energy fraction ($E_\gamma = z\frac{1}{2}\sqrt{s}$). As z is large the effective energy $\sqrt{1-z}\sqrt{s}$ of the $q\bar{q}$ system in the $q\bar{q}\gamma$ events is small implying a decreasing detection efficiency (low multiplicity) for z approaching its (flavor dependent) kinematic limit. In addition, the angular distribution of $q\bar{q}\gamma$ events is steeper compared to $q\bar{q}$ events.

The convolution integral 2, as well as equations 4 and 6, weights hard $q\bar{q}\gamma$ events with the cross section taken at the effective mass $\sqrt{1-z}\sqrt{s}$ of the $q\bar{q}$ pair, rather than at $\sqrt{s}$. This boils down to an enhancement factor $\frac{1}{1-z}$ with three consequences :

- The integral of equation 3 is nearly independent of k_{max} as long as it is larger than about 0.8.
- The integral of equation 6 increases with increasing integration limit k_{max}.
- The integral of equation 6 is sensitive to the $q\bar{q}$ resonance region when the effective energy in the argument of σ_1 is below about 12 GeV. None of the programs used aimed at more than a rough average description of this region. The values for the quark masses used in ZFITTER result from a fit satisfying certain dispersion relations. Their numerical values differ from the choice made in the JETSET program with the net effect of overestimating the average acceptance (equation 5), since usually k_{max} was chosen to be 0.99.

It is probably due to the remarkable constancy of equation 3 for large enough k_{max} that all experimental groups preferred to publish Born cross sections. This avoids the problems mentioned above, but has the drawback of eliminating that part of the physics which is in radiative events.

2.3 Comparison of Data with Theory

It is the aim of this article to compare data directly to the Standard Model as it is programmed in ZFITTER [14]. Then the data are only to be corrected for the average acceptance. Equation 4 shows that this quantity, being a ratio, is only little dependent on electroweak physics. For instance, the s-dependence of the cross sections used in JETSET and ZFITTER are rather similar, although different in absolute value. There is, however, a caveat related to event configurations with hard photons. Any difference in describing these events in the two programs will modify the average value. One such difference is surely the threshold behaviour at low effective energies.

Making use of the above observation that the integral of equation 3 is hardly sensitive to k_{max} it is safer to exclude the resonance region by requiring $k_{max} = 1-(12\,GeV)^2/s$ instead of the notorious 0.99. For the published data at energies above 20 GeV k_{max} is larger than 0.8 and the constancy of the integral in equation 3 applies. Energy points below 20 GeV are ignored. The events corresponding to

k-values above the limit are not relevant for determining QFD parameters.

As discussed above the procedure to quantify the acceptance is basically the same for all experiments. The event generator in JETSET treats separately (not so in ZFITTER) soft and hard photon processes showing strong peaking in various parts of the phase space. Due to the approximations inherent to the calculation of ϵ it is impossible to reduce the intrinsic uncertainty without reanalysing each experiment including all its details and using a more refined event generator. Therefore, no attempt is made to modify ϵ (eq. 5) other than taking account of the excluded resonance region, and the sizeable uncertainty on ϵ estimated by the experimental groups is left unaltered. For similar reasons, also the luminosities and their uncertainties are taken over from the publications.

3 Results

3.1 Consistency Test

Figure 2 summarizes impressively what has been achieved since the first PETRA runs up to the early LEP runs. The consistency of the corrected data with each other and with theory is most easily investigated by referring each experiment to the theoretical prediction for a suitably fixed set of input parameters :

$$\vec{p}_{ref} : M_z = 91.187\,GeV\ , m_{top} = 130\,GeV\ , M_H = 100\,GeV\ and\ \Lambda_{QCD} = 0.200\,GeV$$

The value for $\alpha_s(M_z^2)$ corresponding to this value for Λ (second order QCD and 5 flavours) is 0.117.

The data from PETRA and TRISTAN cover each a whole energy range (as opposed to the PEP data) and can be representated in the form :

$$\frac{\sigma_{exp}(E_i)}{\sigma_{th}(E_i|\vec{p}_{ref})} = u_0 + u_1(E_i - < E >) \tag{7}$$

This is an adequate representation as seen in Figure 3. The parameters u_0 and u_1 are expected to be compatible with 1 and 0 respectively. They are per constructionem uncorrelated, but depend upon the a priori choice $\vec{p}_{ref}$ (see Figure 3).

A two-parameter fit to the four PETRA and the three TRISTAN experiments according to equation 7 gives the results summarised in Table 2. It is instructive to look at the distribution of

$$\chi_i = \left(\frac{\sigma_{exp}(E_i)}{\sigma_{th}(E_i|\vec{p}_{ref})} - u_0\right) / \frac{\delta\,\sigma_{exp}(E_i)}{\sigma_{th}(E_i|\vec{p}_{ref})}$$

for all points in a given experiment. The dispersion D_χ, defined by $\sqrt{<\chi^2> - <\chi>^2}$, is expected to scatter around 1, provided the correction procedure is unbiased. The numbers in Table 2 (columns 3,5) show that the s-dependence of the data agree well

Experiment	$(u_0-1)/\delta u_0$	$u_1/\delta u_1$	$< E >$ (GeV)	$< \chi > \pm D_\chi$	points
PETRA					
CELLO	−1.60	−0.25	39.1	$+0.07 \pm 0.41$	8
JADE	+2.35	−1.07	35.4	-0.09 ± 0.82	18
MARK J	−0.83	+1.39	36.2	$+0.05 \pm 0.97$	16
TASSO	+2.49	+0.43	35.1	-0.11 ± 0.93	3
TRISTAN					
AMY	+0.43	−0.10	57.6	$+0.24 \pm 1.00$	16
TOPAZ	+1.60	−0.03	57.3	$+0.16 \pm 0.66$	13
VENUS	−0.03	−0.87	57.9	$+0.03 \pm 1.04$	13

Table 2: Two-parameter fits

with each other and with the predicted behaviour. This test is quite encouraging as it does not reveal any significant bias in the estimate of the systematics related to the s-dependence. On the other hand the dispersion of the fit values of the 7 $u_0/\delta u_0$ is significantly larger than 1, as expected, since the overall normalisations uncertainties were not yet included. After inclusion, however, the dispersion is significantly less than 1 indicating an overestimate of the normalisation by about 25 %.

The combined fit including the overall uncertainties gives :

$$< u_0 >= 1.003 \pm 0.009 \; and \; < u_1 >= 0.0001 \pm 0.0006 \; at \; < E >= 42.4 \; GeV$$

The theoretical prediction $\sigma_{th}(E|\vec{p}_{ref})$ agrees remarkably well with the low energy data, and also with the PEP data, which yield $\sigma_{exp}(29\,GeV)/\sigma_{theor}(29\,GeV|\vec{p}_{ref}) = 1.018 \pm 0.017$.

The theoretical sensitivity of the total cross section to small changes in the input parameter values around p_{ref} is visualised in Figure 3. The curves result from the ratio :

$$\frac{\sigma_{th}(p_{ref})}{\sigma_{th}(p_{ref}+dp)}$$

where p_{ref} stands for the above reference values, while dp denotes a small variation around p_{ref}. The curves show the effect of varying just one parameter, i.e. curve (a) variation of Λ by 0.2 GeV, curve (b) variation of the top mass by 40 GeV and curve (c) the variation of the Z mass by 0.010 GeV. These are typical uncertainties as of today. The variation of the Higgs mass is not relevant presently and has been left fixed at 100 GeV.

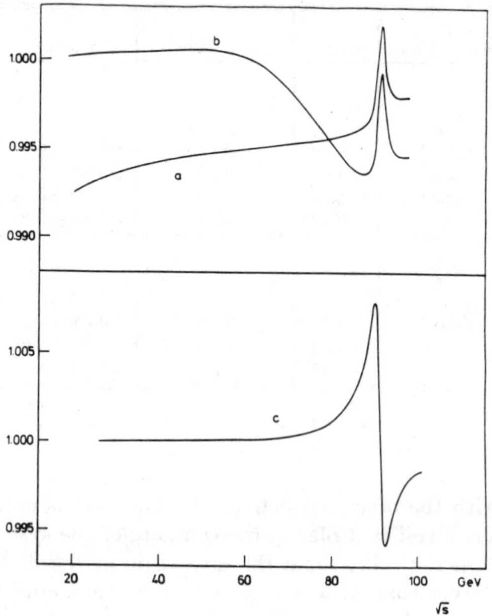

Figure 3: Theoretical sensitivity of the total cross section to small parameter changes (see text)

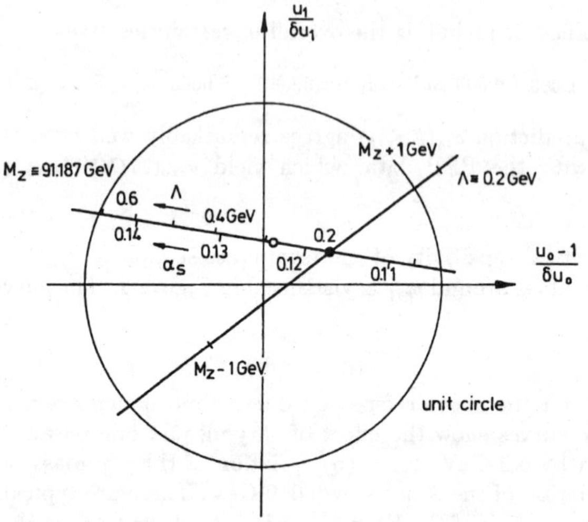

Figure 4: 2-parameter fit to $e^+e^- \to$ *hadrons*

3.2 Fit Results

The total hadron cross section in the energy range of 20 - 65 GeV is characterised by not depending upon the values of the top quark (see Figure 3) and the Higgs mass, thus leaving the dependences upon the Z propagator mass and the strong interaction parameter Λ. The two-parameter representation of equation 7 is convenient to study the dependence upon the Z-mass and Λ, as illustrated in Figure 4. The line for constant Λ, fixed at 0.200 GeV, shows that the best value for the Z propagator mass, i.e. the closest distance of approach to (0,0), agrees well within the large uncertainty with the precise measurement of the Z resonance value obtained by the LEP experiments, namely 91.187 ± 0.007 GeV. Fixing the Z propagator mass at that value makes the confrontation with theory a 1-parameter problem with Λ_{QCD} being the only parameter and yields the result :

$$\alpha_s(M_z^2) = 0.124 \pm 0.021$$

using the 2nd order formula in ZFITTER for 5 flavours. The fit spanning the energy range from 20 to 65 GeV is quoted at the Z resonance. The result from the data below the Z agrees well with the value obtained at the Z by the LEP experiments from the hadronic width [32] : $\alpha_s(M_z^2) = 0.130 \pm 0.012$.

The forthcoming analysis of the high statistics runs of TRISTAN with the upgraded detectors and analysis methods are expected to contribute a more precise test of the Z propagator and an improved value for α_s.

4 Acknowledgements

It is a pleasure to thank Prof. R.Marshall for his cooperation in part of the work. The confrontation with theory profited from the help of Dr. D.Bardin, who rendered ZFITTER applicable down to the PETRA energies. I enjoyed discussions with Drs. D.Bardin, T.Sjöstrand and H.Spiesberger on the treatment of the acceptance and radiative corrections.

5 References

1. A.Ali and P.Söding, *High Energy e^+e^- Physics - Advanced Series on directions in High Energy Physics* Vol. 1, World Scientific 1988
2. R.Marshall : see article in [1], p.142
3. D.Haidt : *Status of the electroweak Standard Model, Weak Interactions and Neutrinos 1989 at Ginosar, Israel*
Nucl. Phys. (Proc. Suppl.) **13** (1990) 3 and DESY Preprint 89-73

4. D.Haidt and H.Pietschmann : *Landolt-Börnstein New Series* **I/10**, Springer (1988)
5. See ref. [4] p. 93.
6. See ref. [4] p. 93-109.
7. See ref. [4] p. 130-146 and 162-169.
8. T.Sjöstrand : *CPC* **27** (1982) 243 and *CPC* **28** (1983) 229
9. J.Fujimoto, K.Kato and Y.Shimizu : *Prog. Theor. Phys.* **79** (1988) 701
10. F.A.Berends and A.Böhm : see article [1], p.28.
11. F.A.Berends, R.Kleiss and S.Jadach : *Nucl. Phys.* **B202** (1982) 63 *CPC* **43** (1983) 185
12. F.A.Berends : *The Z line shape, Proc. of the International Workshop on radiative corrections for* e^+e^- *collisions at Schloß Ringberg*, Springer Verlag 1989, p.55 ;
See also ref 28 therein
13. *Z Physics at LEP 1*, CERN Yellow report 89-08 (1989), edited by G.Altarelli, R.Kleiss and C.Verzegnassi.
14. D.Bardin et al.: CERN-TH 6443/92 (1992) and references therein
15. H.J.Behrend et al.: *Phys. Lett.* **183B** (1987) 400
16. B.Naroska : *Phys. Rep.* **148** (1987) 67
17. B.Adeva et al.: *Phys. Rev.* **D34** (1986) 681 ; PhD Thesis by D.Linnhöfer (kindly provided by Prof. R.Rau)
18. W.Braunschweig et al.: *Z. Phys.* **C47** (1990) 187
19. L.Crigee and G.Knies : *Phys. Rep.* **83** (1983) 153
20. E.Fernandez et al.: *Phys. Rev.* **D31** (1985) 1537
21. C.von Zanthier et al.: *Phys. Rev.* **D43** (1990) 34
22. T.Kumita : *Phys. Rev.* **D42** (1990) 1339
23. I.Adachi et al.: *Phys. Lett.* **234B** (1990) 525
24. K.Abe et al.: *Phys. Lett.* **234B** (1990) 382 and *Phys. Lett.* **246B** (1990) 297
25. K.Abe et al.: *Z. Phys.* **C48** (1990) 13
26. A.Bacala et al.: *Phys. Lett.* **218B** (1989) 112
27. I.Adachi et al.: *Phys. Lett.* **208B** (1988) 319
28. A.Okamoto et al.: *Phys. Lett.* **278B** (1992) 393
29. M.Shirakata et al.: *Phys. Lett.* **278B** (1992) 499
30. K.Nagai et al.: *Phys. Lett.* **278B** (1992) 506
31. D.Bardin : private communication; the version 4.5 is used.
32. S.Bethke : *Talk at XXVI Int. Conf. on High Energy Physics, Dallas, Texas, 6-12 August 1992*; preprint HD-PY 92/13 (1992)

Z^0-pole EXPERIMENTS

D. SCHAILE
Albrecht-Ludwigs-Universität Freiburg
W-7800 Freiburg, Germany

Contents

1 Introduction

Electroweak precision tests in e^+e^--collisions at the Z^0 resonance can be compared with tests of QED like the g-2 experiments: they have the accuracy necessary for probing the quantum structure of the theory. In contrast to pure QED, radiative corrections in the electroweak theory are also sensitive to particles with masses far beyond the energy available for direct production. In particular observables depend on the masses of the up to now elusive top quark, M_t, and the Higgs Boson, M_H. A consistent description of all our data with a unique value of M_t and M_H constitutes a stringent test of the theory and a sensitive probe of new physics.

In section 2 of this report we first give a short overview of the e^+e^- colliders operating at the Z^0-pole and the major experiments installed at these machines. In section 3 we present the precision measurements at the Z^0-pole. The aim of this section is to provide the reader with the experimental and theoretical background for a basic set of results. This basic set of results will then be used in section 4 to look at selected aspects, by applying the relevant parameter transformations. Finally section 5 summarizes the results presented and gives an outlook to the future of electroweak precision tests.

2 Colliders and Experiments

In summer 1989 two e^+e^--machines started operation at the Z^0-pole: the Stanford Linear Collider (**SLC**) at SLAC shortly followed by the Large Electron-Positron collider (**LEP**) at CERN. These machines differ substantially in design.

The construction of the SLC, a linear accelerator 3 km in length, made use of the existing 20 GeV linear electron accelerator, which was upgraded to achieve a maximum energy of about 50 GeV. Electron and positron bunches are accelerated within the same machine cycle and then diverted by a large dipole magnet into separate arcs, which guide electrons and positrons to a straight section with a head-on collision point. The final focus in the straight section causes the beams to be squeezed to sizes of a few microns at the interaction point. The bunches are then ejected by means of kicker-magnets into beam-dumps. The SLC has been constructed not only for Z^0-pole physics operation but also to develop the accelerator technology necessary for the design of future linear e^+e^--colliders.

The SLC has succeeded in making the first measurement of the mass of the Z^0, M_Z, in e^+e^--collisions at the Z^0-pole [1]. The luminosity delivered by the machine in successive running periods up to the end of 1991 stayed substantially below the expectation, however. In 1992 SLC has started a physics program with longitudinally polarized e^+e^- beams. First results are based on 10 000 Z^0 decays with an average beam polarization of 22%.

LEP is an e^+e^- storage ring with a circumference of about 27 km. In normal operation the luminosity is about 10^{31} cm^{-2} s^{-1}. The luminosity is likely to improve further due to a better understanding of the machine and due to a doubling of the number of e^+ and e^- bunches from initially four of each. A first test of eight on eight bunches at the end of 1992 has shown that both machine and experiments can be operated equally well with this increased number of bunches. By the end of 1992 the four LEP collaborations had recorded $4.7 \cdot 10^6$ hadronic Z^0 decays. One of the major goals of LEP is a precise determination of M_Z and the total width of the Z^0, Γ_Z. For this purpose several energy scans with centre-of-mass energies within ± 3 GeV of the Z^0 mass have been performed.

The present centre-of-mass energy of LEP is limited to 110 GeV. An energy upgrade is in progress in order to allow LEP to operate above the W^+W^- pair production threshold. The goal is to observe the first W^+W^- pairs by the end of 1994. In circular e^+e^--colliders under favorable conditions a natural transverse polarization may build up due to the interaction of the e^+ and e^- with the magnetic guide field, a phenomenon referred to as the Sokolov-Ternov effect [2]. Since the first observation of transverse polarizarion at LEP in August 1990 a lot of progress has been made in exploiting polarization as a tool to calibrate the LEP energy scale. Turning transverse into longitudinal polarization by means of spin-rotators is an option for future LEP running and is discussed in detail in this volume [3].

SLC, with only a single interaction region, has started operation with the Mark II detector [4], which is now replaced by the SLD detector [5]. In LEP 4 out of the 8 interaction zones are capable of accomodating large experiments. They are occupied by the ALEPH [6], DELPHI [7], L3 [8] and OPAL [9] detectors. All the above are general purpose detectors with almost 4π solid angle coverage. A typical detector is shown in Figure 1.

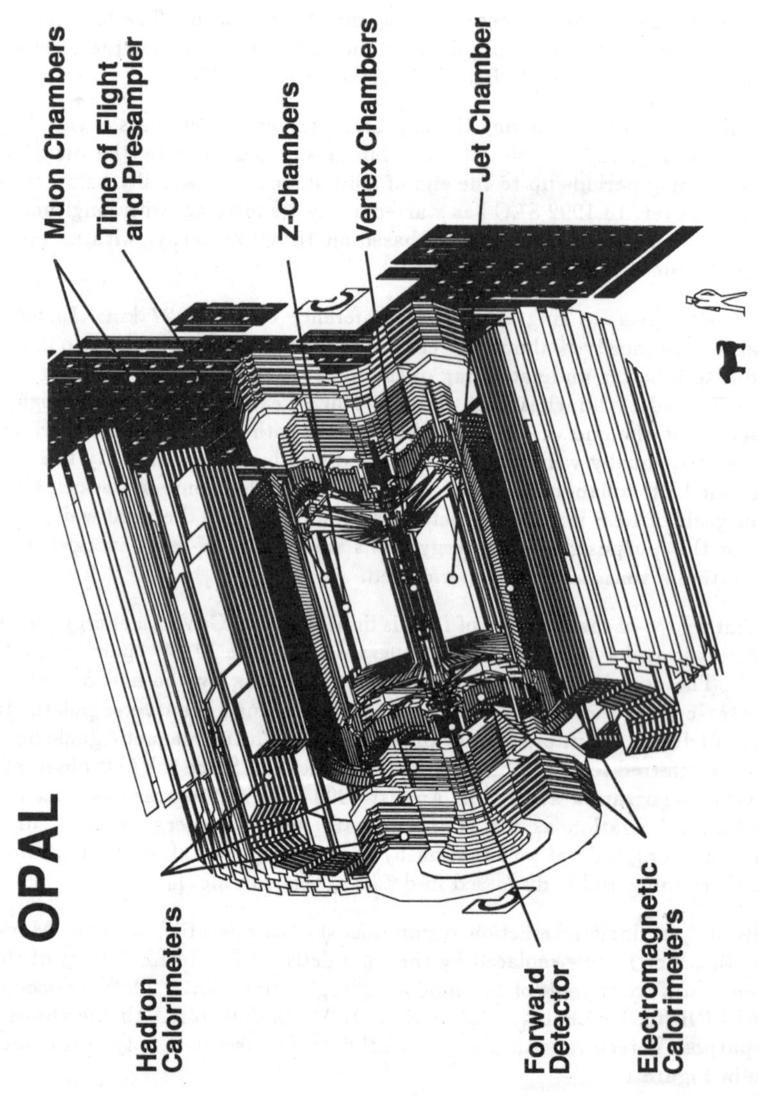

Figure 1: An example of a detector for Z^0-pole physics.

The interaction region is in the centre of a high resolution vertex chamber. Most of the experiments started with dedicated drift chambers, which have now been complemented by silicon microvertex chambers, capable of reaching an impact parameter resolution of typically 10 μm. The vertex chamber is surrounded by a large volume tracking chamber within a magnetic field of 0.4-1.5 Tesla oriented along the beam axis. The reconstruction of tracks within these chambers allows the determination of the momentum of charged particles with an accuracy of typically $\Delta p/p^2 \approx 10^{-3}\text{GeV}^{-1}$. The specific energy loss of charged particles can be determined by multiple sampling with an accuracy of up to 3%, allowing particle identification also in the region of the relativistic rise of the energy loss-momentum relation. The design of the inner tracking chamber differs among the experiments: ALEPH and DELPHI choose a TPC, OPAL a Jet Chamber with a z-coordinate determination by charge division, L3 has only a small inner tracking chamber which operates in the time expansion mode. The Mark II and SLD drift chambers have a cell structure with both wires parallel to the beam axis and wires with a small inclination to provide stereo information.

Between central tracking detectors and the calorimetry some experiments have introduced a layer of time of flight hodoscopes. DELPHI and SLD also have a ring imaging Čerenkov system providing additional particle identification capabilities.

All experiments emphasize the detection of electromagnetic energy. The calorimeters surrounding the inner detectors are therefore subdivided into a high resolution electromagnetic calorimeter and a hadronic calorimeter of several interaction lengths. There are various approaches to the detection of electromagnetic energy: Liquid Argon calorimeters (Mark II and SLD), sandwiched layers of lead and proportional wire planes (ALEPH), a high density projection chamber, where ionization produced between lead layers is drifted to the ends of 90 cm long modules (DELPHI), BGO crystals (L3) or a lead glass calorimeter (OPAL).

In most experiments the iron of the return yoke is interlaced with streamer tubes and serves as hadron calorimeter. The outer layer of each detector is again a system of tracking chambers for muon identification. L3 differs in design from the other experiments as the coil surrounds the entire detector to allow a precise momentum measurement in the muon chambers.

The luminosity is determined by measuring small angle Bhabha scattering in a system of tracking chambers and calorimeters located in the inner part of the detector endcaps. The LEP experiments are currently supplementing their luminosity detectors with silicon devices to improve the acceptance determination.

3 Precision Measurements at the Z^0-pole

In this section we will discuss the analysis of electroweak precision tests at the Z^0-pole and provide a basic set of results based on the statistics collected at LEP up to the end of 1991 and during the first run with longitudinal polarization at SLC in 1992.

3.1 Z^0 *Lineshape and Leptonic Forward-Backward Asymmetries*

3.1.1 Experimental Aspects

The design of the LEP detectors allows a trigger on hadrons and leptons with high redundancy resulting in an acceptance of 100% with an uncertainty of less than 0.1% over the solid angle considered in the analysis.

The selection of hadrons and lepton pairs at LEP is conceptually easy, as they can be discriminated by a few simple cuts like cluster or track multiplicities, deposited energy and energy balance against backgrounds which are $\mathcal{O}(0.1\%)$ for hadrons, e^+e^- and $\mu^+\mu^-$ and $\mathcal{O}(1\%)$ for $\tau^+\tau^-$ pairs. The challenge of the analysis is motivated by the aim to match the systematic error of efficiency and acceptance corrections to the statistics available. The tools used to reach this goal involve both elaborate detector simulations and cross-checks with the data themselves. Tables 1 and 2 summarize statistics and systematics of the event selection (see [10, 11, 12, 13] for the 1989/90 data and [14, 15, 16] for the 1991 data).

	'89/'90	'91	'92
ALEPH	165 k	286 k [1)]	739 k [1)]
DELPHI	125 k	243 k [1)]	751 k [1)]
L3	115 k	310 k [1)]	688 k [1)]
OPAL	166 k	314 k	767 k [1)]
LEP	575 k	1153 k	2945 k

Table 1: Number of selected hadronic events from the 4 LEP collaborations. Preliminary numbers are marked by [1)]. The LEP results of this report are based on the statistics collected up to the end of 1991 only, as the analysis of the '92 data is not completed yet.

The absolute luminosity is obtained from the ratio of the number of events measured in small angle Bhabha scattering and the theoretical prediction for the cross section of this process within the acceptance of the luminosity monitor. The polar

	hadrons '90 [%]	'91 [%]	e^+e^- '90 [%]	'91 [%]	$\mu^+\mu^-$ '90 [%]	'91 [%]	$\tau^+\tau^-$ '90 [%]	'91 [%]
ALEPH	0.2	0.2 [1)]	0.5	0.4 [1)]	0.5	0.5 [1)]	0.9	0.8 [1)]
DELPHI	0.4	0.4 [1)]	0.7	0.8 [1)]	0.8	0.5 [1)]	1.2	1.2 [1)]
L3	0.4	0.2 [1)]	1.0	0.5 [1)]	0.8	0.5 [1)]	2.1	0.7 [1)]
OPAL	0.4	0.20	0.7	0.45	0.5	0.25	1.3	0.76

Table 2: Systematic errors of the event selection. Preliminary numbers are marked by [1)].

acceptance of the forward detectors typically ranges from 40–120 mrad. At small polar angles the cross section for Bhabha scattering is given by :

$$\frac{d\sigma}{d\theta} \approx \frac{32\pi\alpha^2}{s}\frac{1}{\theta^3}$$

resulting in an integrated cross section:

$$\sigma_{Bhabha} \approx \frac{16\pi\alpha^2}{s}\left(\frac{1}{\theta^2_{min}} - \frac{1}{\theta^2_{max}}\right)$$

where θ_{min} and θ_{max} define the limits of the acceptance in polar angle. The most important task of the luminosity determination is therefore the precise monitoring of the edge of the acceptance at the inner radius*. The theoretical uncertainties are related to missing higher order calculations. The current theoretical uncertainty of the absolute luminosity determination amounts to about 0.3% [17], depending on the specific experimental cuts involved. Table 3 summarizes the status of the experimental systematic errors for the luminosity determination. As a consequence of new detector upgrades based on silicon technology the experimental systematic error may decrease to 0.1% in the near future†.

Very important for the measurement of the Z^0 mass is the accuracy of the LEP centre-of-mass energy determination. The techniques applied for the published results for the 1990 data [10, 11, 12, 13] were based on flip coil measurements in a reference magnet in sequence with the magnets in the LEP ring, the calibration of the magnetic field by flux loops mounted on the pole faces of all dipoles and the measurement of the revolution frequency of protons circulating on the same orbit as the positrons. With the observation of transverse polarization at LEP [18] at the end of the 1990 running period a new method based on the resonant depolarization by a weak oscillating B-field started to be applied in 1991. Resonant depolarization

*As a rule of thumb, the precision of the inner edge affects the accuracy of the luminosity $\mathcal{L}$ like: $\Delta\mathcal{L}/\mathcal{L} \approx 2\Delta\theta_{min}/\theta_{min}$, i.e. $\Delta\mathcal{L}/\mathcal{L} \approx 2\% \cdot \Delta r_{min}$ [mm] for an inner radius of the luminosity monitor $r_{min} = 10$ cm.

†Before LEP start-up an experimental accuracy of 2% for the luminosity determination was considered as a sonic barrier.

	'90 [%]	'91 [%]
ALEPH	0.6	0.45 [1)]
DELPHI	0.8	0.55 [1)]
L3	0.7	0.52 [1)]
OPAL	0.7	0.60

Table 3: The experimental systematic error of the luminosity determination. Preliminary numbers are marked by [1)].

occurs, if the frequency of the B-field applied matches the spin precession frequency of the electrons. The feasibility of the method has been proven by several successful measurements [19, 20] and is described in detail in this volume [3].

Besides the total cross section, experiments also measure the distribution of the production angle of the final state fermions, which is defined as the angle between the incoming e^+-direction and the outgoing antifermion $\bar{f}$ (Figure 2). Theory predicts the angular distribution for the process $e^+e^- \to f\bar{f}$, which is:

$$\frac{d\sigma}{d(\cos\theta)} \propto 1 + \cos^2\theta + B\cos\theta \tag{1}$$

for $f \neq e$. Having verified that the data follow the theoretical prediction the information content of the angular distribution can be summarized in a single number, the 'forward-backward asymmetry' (A_{FB}) given by:

$$A_{FB} = \frac{N_F - N_B}{N_F + N_B} = \frac{3}{8}B \ . \tag{2}$$

Here the number of 'forward' events (N_F) is defined to be the number of events for which $\theta < \frac{\pi}{2}$. Similarly, N_B is the number for which $\frac{\pi}{2} < \theta < \pi$.

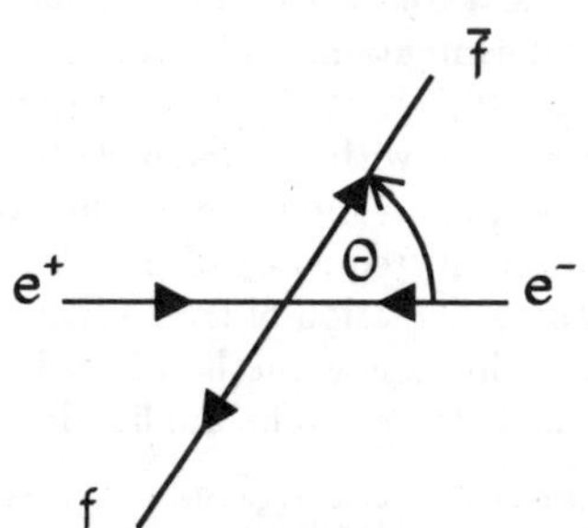

Figure 2: Definition of the scattering angle θ in e^+e^- annihilation.

To determine the lepton forward-backward asymmetries $A_{FB}^{\ell\ell}$, the same event selection as for the cross sections is applied. The measurement requires charge identification. It is, however, not sensitive to overall normalization factors, nor to a symmetric θ dependence of the acceptance, provided A_{FB} is determined from a maximum likelihood fit of eqn. (1) to the angular distribution. The systematic errors for $A_{FB}^{\ell\ell}$ are due to charge misidentification, angular resolution and the displacement of the event vertex. Up to now these errors could be kept small compared to lepton statistics.

3.1.2 Parametrization

In this subsection we will first give a few simple expressions for the lowest order cross sections for the process $e^+e^- \to f\bar{f}$. For $f \neq e$ we only have to consider the annihilation of the initial e^+e^- pair via a photon or a Z^0 into the final $f\bar{f}$ state (Figure 3 a)), whereas the treatment of the process $e^+e^- \to e^+e^-$ is complicated by the presence of t-channel diagrams (Figure 3 b)). We will then discuss how these expressions are modified by radiative corrections. In the Minimal Standard Model,

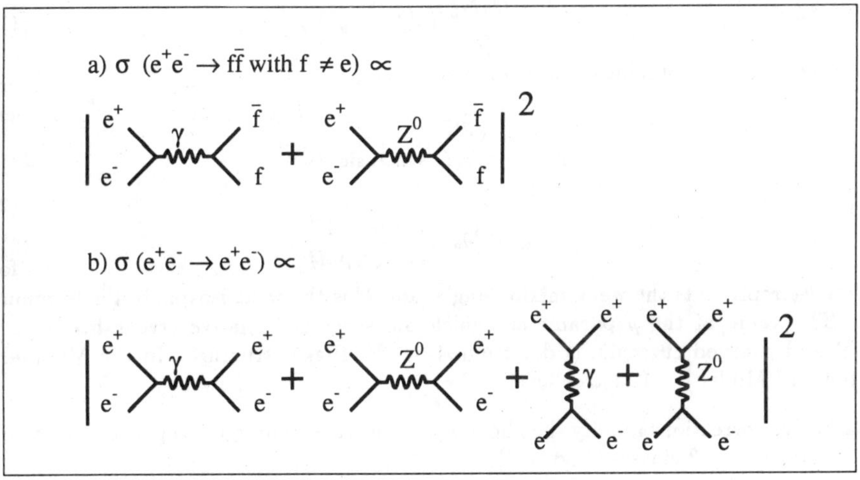

Figure 3: Lowest order cross sections for the process $e^+e^- \to f\bar{f}$.

neglecting fermion masses, the tree level predictions for the process $e^+e^- \to f\bar{f}$ can be described with only 3 free parameters, which have to be determined from measurements. These 3 free parameters are commonly expressed in terms of the most accurate measurements available. At the Z^0-pole the most natural choice for these measurements is:

$$\alpha,\ G_F \text{ and } M_Z\ ,$$

where α is the electromagnetic coupling constant and G_F the Fermi constant.

For the process $e^+e^- \rightarrow f\bar{f}$ with $f \neq e$ the total cross section in the vicinity of $\sqrt{s} = M_Z$ is dominated by Z^0 exchange. At Born level the cross section can therefore be written as:

$$\sigma(s) = \sigma_{f\bar{f}}^{\text{pole}} \frac{s\Gamma_Z^2}{(s - M_Z^2)^2 + M_Z^2\Gamma_Z^2} + '\gamma Z^0' + '\gamma' \tag{3}$$

where $\sigma_{f\bar{f}}^{\text{pole}}$ represents the cross section for the process $e^+e^- \rightarrow f\bar{f}$ at $\sqrt{s} = M_Z$, 'γ' and 'γZ^0' represent small $\mathcal{O}(1\%)$ contributions from pure photon exchange and the γZ^0-interference. The pole cross section $\sigma_{f\bar{f}}^{\text{pole}}$ can be written in terms of the Z^0 partial decay widths into initial and final state, Γ_{ee} and $\Gamma_{f\bar{f}}$:

$$\sigma_{f\bar{f}}^{\text{pole}} = \frac{12\pi}{M_Z^2}\frac{\Gamma_{ee}\Gamma_{f\bar{f}}}{\Gamma_Z^2}\ . \tag{4}$$

In the Standard Model the partial widths of the Z^0 are not free parameters but can be written in terms of vector and axial-vector coupling constants g_v^f and g_a^f of the Z^0:

$$\Gamma_{f\bar{f}} = \frac{G_F M_Z^3}{6\pi\sqrt{2}}((g_a^f)^2 + (g_v^f)^2) \tag{5}$$

At tree level the couplings can be expressed as

$$g_a^f = \sqrt{\rho} I_f^3 \tag{6}$$
$$g_v^f = \sqrt{\rho}(I_f^3 + 2Q_f \sin^2\theta_W) \tag{7}$$

with

$$\sin^2\theta_W \cos^2\theta_W = \frac{\pi\alpha}{G_F\sqrt{2}}\frac{1}{\rho M_Z^2}\ . \tag{8}$$

Here θ_W represents the weak mixing angle, and I_f^3 is the weak isospin of the fermion f. The value of the ρ parameter, which measures the relative strength of neutral and charged currents, is determined by the Higgs structure. In the Minimal Standard Model $\rho = 1$ at tree level.

Radiative corrections modify the above relations. It is common to separate radiative corrections into 3 classes (Figure 4):

i. Photonic corrections:
Photonic corrections refer to all diagrams with real or virtual photons added to the Born diagram. These corrections are large ($\mathcal{O}(30\%)$) and depend on experimental cuts. The dominant contribution arises from diagrams where a photon is radiated off the initial state, thus modifying the effective centre-of-mass energy, which has a substantial effect on cross-sections close to a resonance. Photonic corrections are taken into account by convoluting the cross section for the hard scattering process by

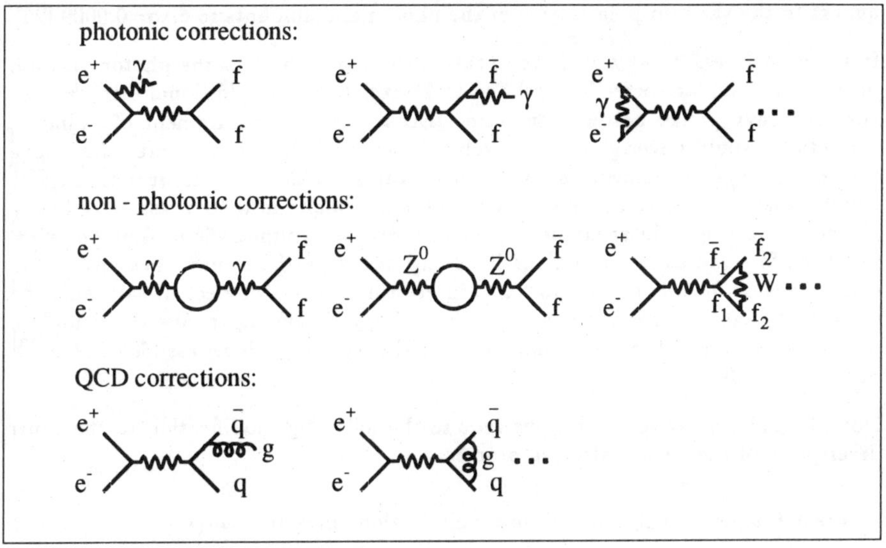

photonic corrections:

non - photonic corrections:

QCD corrections:

Figure 4: Radiative corrections to the process $e^+e^- \rightarrow f\bar{f}$.

a radiator function, which can be calculated within the framework of QED. Thanks to a substantial theoretical effort (see [22] and references therein) photonic corrections to the s-channel are well understood. The theoretical accuracy is estimated to be 0.1% or better, thus well matching the experimental systematics. For the process $e^+e^- \rightarrow e^+e^-$ the situation before the start of the Z^0-pole experiments was considered as unsatisfactory [23]. A significant improvement for large angle Bhabha scattering was achieved by the work of [24]. The authors quote an uncertainty of 0.5% for these calculations. A lot of work has also been invested into higher order corrections to small angle Bhabha scattering [25]. More work, however, is needed to exploit fully the anticipated experimental accuracy of forthcoming LEP data.

ii. Non-photonic corrections:
Non-photonic corrections denote the electroweak complement[‡] to photonic corrections. A familiar example of non-photonic diagrams is the vacuum-polarization of the photon, which leads to an s-dependent correction of the electromagnetic coupling constant:

$$\alpha \rightarrow \alpha(s) = \frac{\alpha}{1 - \Delta\alpha(s)}$$

At present the dominant uncertainty of $\alpha(M_Z^2)$ is due to the contribution of light

[‡]A separation of electroweak radiative corrections into photonic and non-photonic diagrams is rigorous only in $\mathcal{O}(\alpha)$; it can be justified, however, by the smallness of higher order corrections.

quarks to the vacuum-polarization of the photon and amounts to $\delta\Delta\alpha$=0.0009 [26].

In the electroweak theory we have to take into account besides the photon vacuum polarization similar corrections related to Z^0-exchange and additional diagrams involving heavy gauge bosons. In pure QED a precise measurement of radiative corrections would never give us any hint of particles which have a mass far above the energy scale of the process under consideration. This is a consequence of exact charge conservation, as the associated symmetry results in a suppression of heavy physics appearing in internal loops. The electroweak symmetry is broken, however, and therefore radiative corrections involving heavy particles may have observable consequences. This is one of the most interesting aspects of electroweak radiative corrections and electroweak precision tests: They potentially probe the complete particle spectrum of the theory and not only the part which is accessible at energies presently available.

Non-photonic radiative corrections require the following modification to the Born description of the hard scattering process:

- An s-dependent photon vacuum polarization correction $\Delta\alpha(s)$
- An s-dependence of the Z^0 width in the propagator which can be approximated well by:
$$\Gamma_Z(s) = s/M_Z^2 \cdot \Gamma_Z(s = M_Z^2)\,.$$
In the following we will use $\Gamma_Z \equiv \Gamma_Z(s = M_Z^2)$.
- Effective vector and axial-vector couplings denoted by $\hat{g}_v$ and $\hat{g}_a$. The s-dependence of $\hat{g}_v$ and $\hat{g}_a$, however, is small in the vicinity of the peak.

These modifications to the Born description of the hard scattering process essentially retain the Born structure and are therefore referred to as the 'improved Born approximation'.

iii. QCD corrections:
QCD corrections to the process $e^+e^- \rightarrow f\bar{f}$ account for gluon radiation off real and virtual quarks. The dominant effect of QCD corrections can be expressed as a modification of the partial widths for Z^0 decays into $q\bar{q}$-pairs. Here QCD corrections are applied as a factor which can be written as an expansion of the strong coupling constant:

$$\Gamma_{q\bar{q}} \rightarrow \Gamma_{q\bar{q}}(1 + \delta_{QCD})$$

with:

$$\delta_{QCD} = A\left(\frac{\alpha_s(M_Z^2))}{\pi}\right) + B\left(\frac{\alpha_s(M_Z^2)}{\pi}\right)^2 + C\left(\frac{\alpha_s(M_Z^2)}{\pi}\right)^3 + \ldots\,.$$

QCD corrections to the Z^0 hadronic partial width have recently been calculated to $\mathcal{O}(\alpha_s^3)$ [27], superseding an earlier erroneous calculation. The effects of quark masses on these corrections are non-negligible [28]. Besides corrections to the final $q\bar{q}$ state, gluon radiation also induces corrections to internal loops, which have been calculated to $\mathcal{O}(\alpha\alpha_s)$.

The determination of the Z^0 resonance parameters at LEP is based on calculations which have been described in detail elsewhere [29, 30, 31, 32]. Photonic corrections to the hard scattering process are calculated complete in $\mathcal{O}(\alpha)$, including leading $\mathcal{O}(\alpha^2)$ contributions and soft photon exponentiation. The t-channel and s-t-interference contributions in large angle $e^+e^- \to e^+e^-$ scattering are treated as a correction to the s-channel based on [24]. For the s-channel hard scattering process without photonic corrections two approaches are followed frequently:

One approach is based on the precise calculation of radiative corrections within the Minimal Standard Model. The programs mentioned above include a full one-loop calculation with leading $\mathcal{O}(\alpha^2 M_t^4)$ and $\mathcal{O}(\alpha\alpha_s)$ terms, using M_Z, M_t, M_H and α_s as input parameters. QCD corrections to the final $q\bar{q}$ states are calculated in $\mathcal{O}(\alpha_s^3)$, taking into account $b\bar{b}$ mass effects.

Substituting appropriate values for the effective couplings $\hat{g}_v^f$ and $\hat{g}_a^f$ the deviations between the improved Born approximation and the full Standard Model calculation are much smaller than the present experimental accuracy. This motivates the second approach, which uses the improved Born approximation as a starting point for a model independent test of radiative corrections. To facilitate the discussion of results we give below the differential cross section for the process $e^+e^- \to f\bar{f}$ in the improved Born approximation (without photonic and QCD corrections and neglecting fermion masses):

$$\begin{aligned}\frac{2s}{\pi\alpha^2}\frac{d\sigma}{d\cos\theta}(e^+e^- \to f\bar{f}) &= \left|\frac{1}{1-\Delta\alpha}\right|^2 (1+\cos^2\theta) \qquad (9)\\ &+ 4\mathrm{Re}\left\{\frac{2}{1-\Delta\alpha}\chi(s)\left[\hat{g}_v^e\,\hat{g}_v^f\,(1+\cos^2\theta) + 2\,\hat{g}_a^e\,\hat{g}_a^f\cos\theta\right]\right\}\\ &+ 16|\chi(s)|^2\left[(\hat{g}_a^{e\,2}+\hat{g}_v^{e\,2})(\hat{g}_a^{f\,2}+\hat{g}_v^{f\,2})(1+\cos^2\theta) + 8\,\hat{g}_a^e\,\hat{g}_a^f\,\hat{g}_v^e\,\hat{g}_v^f\cos\theta\right],\end{aligned}$$

with

$$\chi(s) = \frac{G_F M_Z^2}{8\pi\alpha\sqrt{2}}\frac{s}{s - M_Z^2 + is\Gamma_Z/M_Z}. \qquad (10)$$

Here G_F is the Fermi constant, $\hat{g}_v$ and $\hat{g}_a$ denote effective vector and axial-vector coupling constants of the Z^0 to fermions, the superscripts 'e' and 'f' refer to initial state electron and the final state fermion.

The introduction of effective couplings $\hat{g}_v^f$ and $\hat{g}_a^f$ also motivates the definition of an effective ρ-parameter and an effective weak mixing angle, the two sets of effective parameters being related via eqn. (6) and (7). The definition of these effective quantities in the literature differs slightly by the treatment of higher order corrections and the inclusion of vertex and box diagrams (see e.g. [32]). Frequently used for the effective weak mixing angle are the following notations:

$$\sin^2\theta_{\rm W}^{eff,\rm f} \equiv \kappa_{\rm f} \sin^2\theta_{\rm W} \tag{11}$$

with $\sin^2\theta_{\rm W} \equiv 1 - \frac{M_{\rm W}^2}{M_{\rm Z}^2}$ and $\kappa_{\rm f}$ representing a flavour dependent form factor containing the non-photonic corrections. Secondly a universal quantity $\sin^2\overline{\theta}_{\rm W}$ can be defined which is related to $\sin^2\theta_{\rm W}^{eff,\rm f}$ via:

$$\sin^2\theta_{\rm W}^{eff,\rm f} = (1 + \Delta\kappa_{\rm f,vertex}) \sin^2\overline{\theta}_{\rm W} \tag{12}$$

Here $\Delta\kappa_{\rm f,vertex}$ is the non-universal part of the form factor $\kappa_{\rm f}$ originating from vertex corrections. Within the Standard Model $\Delta\kappa_{\rm f,vertex}$ can be calculated to a good approximation as a function of α, $G_{\rm F}$, $M_{\rm Z}$ and known fermion masses, with the exception of the b quark, where vertex corrections are top mass dependent. For leptons we have $\sin^2\theta_{\rm W}^{eff,\ell} \approx \sin^2\overline{\theta}_{\rm W} + 0.0007$.

The effective weak mixing angle $\sin^2\theta_{\rm W}^{eff,\ell}$ can be determined directly from measured leptonic forward-backward asymmetries at the $\rm Z^0$-pole without any knowledge of the effective ρ parameter[§]. There are also ways to express the effective ρ-parameter in terms of the effective weak mixing angle (loosely speaking by using eqn. (8) with the substitution $\alpha \to \alpha(s)$). To a certain level of accuracy the effect of electroweak radiative corrections can be condensed into one single parameter, designating the effective weak mixing angle as a convenient quantity to compare the sensitivity of various measurements to non-photonic corrections. The combined set of electroweak data, however, determines the effective weak mixing angle with an accuracy where different definitions matter (c.f. section 4.5).

In this report we tend to deemphasize the presentation of results in terms of an effective weak mixing angle for two reasons: i) The combination of the full harvest of electroweak precision tests into one consistent definition of the effective weak mixing angle requires small corrections to the improved Born expressions which have to be calculated within the Standard Model framework. The model independence of such a combined effective weak mixing angle is therefore questionable. ii) The effective weak mixing angle in the full Standard Model calculation is not a primary input but a derived quantity and therefore looses its distinction among other observables.

3.1.3 Results

Figures 5–7 give examples of measured cross sections and leptonic forward-backward asymmetries as a function of centre-of-mass energy. The LEP experiments deter-

[§]Neglecting small contributions from photon exchange and the $\gamma\rm Z^0$-interference.

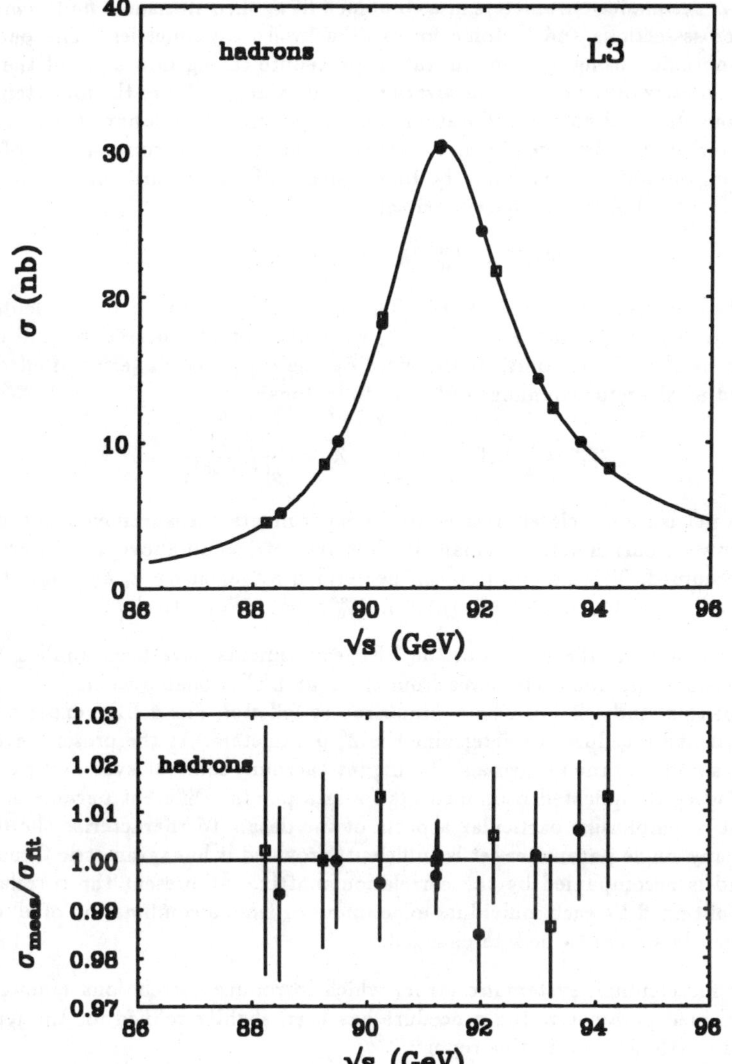

Figure 5: The measured hadronic cross section (top) and the distribution of residuals from a fit to the Standard Model prediction (bottom). Circular and square symbols refer to the '91 and '90 data, respectively.

mine the Z^0 resonance parameters in a combined fit to their measured hadronic and leptonic cross-sections and leptonic forward-backward asymmetries. The parameters are obtained using a χ^2 minimization procedure taking into account the full covariance matrix of the data. The covariance matrix also includes the uncertainties arising from the LEP energy calibration. Several parameter sets have been chosen to characterize the measurements. In order to facilitate the combination of the results from the four experiments, the LEP collaborations also present, in addition to their favorite choices, one parameter set:

$$M_Z,\ \Gamma_Z,\ \sigma^{\text{pole}}_{\text{had}},\ R_\ell \text{ and } A^{\text{pole}}_{\text{FB}}(\ell^+\ell^-).$$

In the above parameter set $R_\ell \equiv \Gamma_{\text{had}}/\Gamma_{\ell\ell}$ and $A^{\text{pole}}_{\text{FB}}(\ell^+\ell^-)$ refers to the leptonic forward-backward asymmetry at $\sqrt{s} = M_Z$ without photonic corrections and originating from Z^0 exchange only. It therefore can be expressed in terms of effective vector and axial-vector couplings of the Z^0 to leptons:

$$A^{\text{pole}}_{\text{FB}} = \frac{3}{4}\mathcal{A}_e\mathcal{A}_\ell \qquad \text{with} \qquad \mathcal{A}_f = \frac{2\,\hat{g}^f_a\,\hat{g}^f_v}{(\hat{g}^{f\,2}_a + \hat{g}^{f\,2}_v)}\,.$$

This parametrization is closely related to the experimental measurements and correlations between parameters are small. In the parametrization above, lepton universality is assumed. This assumption can be verified by replacing R_ℓ and $A^{\text{pole}}_{\text{FB}}(\ell^+\ell^-)$ by R_e, R_μ, R_τ and $A^{\text{pole}}_{\text{FB}}(e^+e^-)$, $A^{\text{pole}}_{\text{FB}}(\mu^+\mu^-)$, $A^{\text{pole}}_{\text{FB}}(\tau^+\tau^-)$, respectively.

Methods to combine the results of the LEP experiments have been studied by a working group with representatives from the four LEP Collaborations. The conclusions of this study [33] can be summarized as follows: The 4 LEP experiments use equivalent procedures to determine the Z^0 parameters. At the present level of accuracy, simple parameter averages taking into account common systematic errors are as good as complicated combined fits to data points. Different parameter sets are useful to emphasize particular aspects of the data. To characterize the data, however, any single parameter set is sufficient, provided it has symmetric Gaussian errors and is accompanied by the correlation matrix. At present the correlation matrices obtained by each individual experiment or from a combination of all data are sufficiently similar to be interchanged.

To allow for common systematic errors which introduce correlations among the LEP combined parameters the procedure has been slightly refined for the results presented in [15, 34] and in this report:

- Each experiment provides the sets of five and nine parameters described above, accompanied by its correlation matrix.
- From the covariance matrices for the 5 or 9 parameters of the individual experiments a 20×20 or a 36×36 covariance matrix $\mathcal{V}$ is constructed, which also

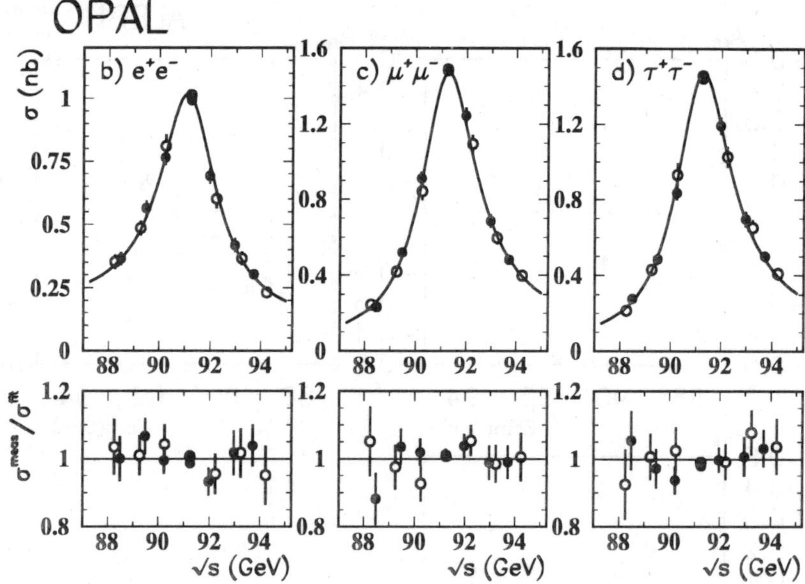

Figure 6: Measured leptonic cross sections (top) and the distribution of residuals from a fit to the Standard Model prediction (bottom). Full and open symbols refer to the '91 and '90 data, respectively.

includes the effect of a common 0.3% theoretical uncertainty of the theoretical calculation of the Bhabha cross-section for the luminosity measurement on $\sigma_{\rm had}^{\rm pole}$ and the correlated common uncertainty due to the LEP energy calibration on $M_{\rm Z}$ and $\Gamma_{\rm Z}$ (see Table 19 in the Appendix).

- Then a combined parameter set is obtained by minimizing $\chi^2 = \Delta^T \mathcal{V}^{-1} \Delta$, where Δ denotes the vector of residuals of the combined parameter set to the results of the individual experiments.

The results of the individual LEP experiments and the LEP combined result for the parameter sets without and with lepton universality are given in Tables 4 and 5. The bar charts in Figures 8–10 give a graphical overview for the results obtained assuming lepton universality. In these bar charts the contribution from common systematic errors to each parameter has been removed from the error of each individual experiment. It has been derived by performing the fit which combines the LEP data with and without common systematic errors. With the measurement of $M_{\rm Z}$ the tree level prediction of the Standard Model is fixed. We therefore

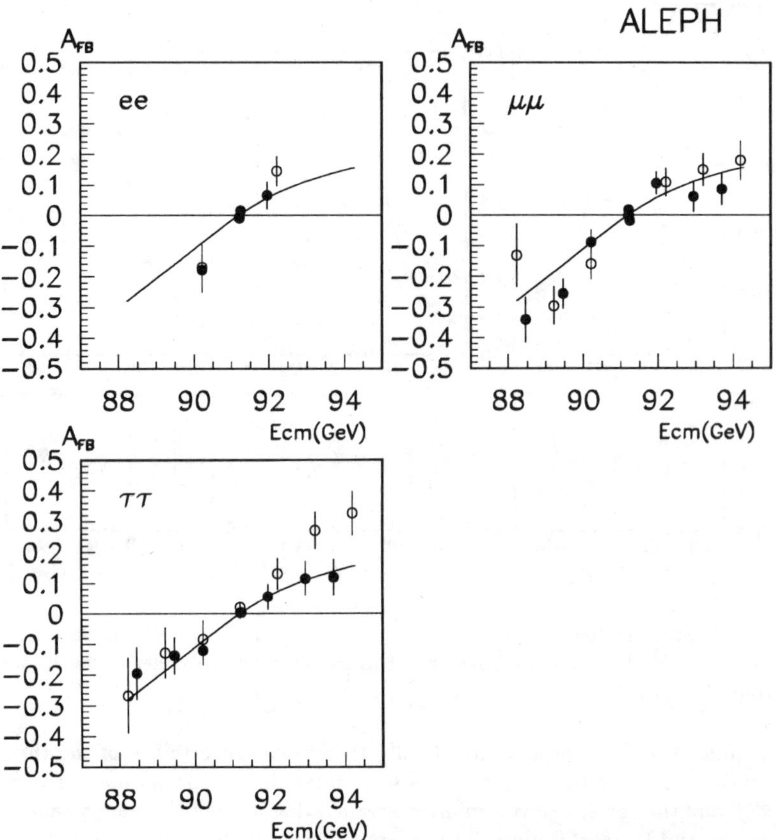

Figure 7: Measured leptonic forward-backward asymmetries. For the process $e^+e^- \to e^+e^-$ the contributions from the t-channel and the s-t-interference have been subtracted, and only the data within ± 1 GeV of the Z^0 peak, which have been included in the fitting pocedure by the ALEPH experiment, are shown. The solid line displays the result of a fit in the framework of the improved Born approximation imposing lepton universality. Full and open symbols refer to the '91 and '90 data, respectively.

can compare all further measurements with the Standard Model prediction, taking into account the variations introduced by radiative corrections, which require the knowledge of M_t, M_H and α_s.

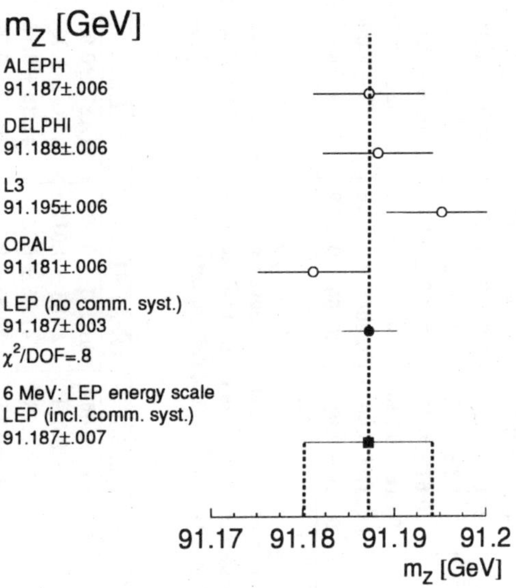

Figure 8: LEP results for M_Z.

3.2 *Polarization Asymmetries*

The Standard Model predicts parity violation not only for charged currents but also for the neutral current. Parity violation in neutral currents was first observed in the scattering of polarized electrons on deuterium at SLAC [35]. Parity violation in neutral currents also allows the verification of tiny effects of the γZ^0-interference in atomic transitions. Today parity violation in neutral currents is well established. For the process $e^+e^- \to f\bar{f}$ it manifests itself by

- A final state fermion polarization
- An asymmetry of the production cross section with respect to left-handed and right-handed polarization of the incoming electron beam

The measurement of polarization asymmetries in e^+e^- collisions at the Z^0-pole serves as a precision test of lepton couplings.

	ALEPH [10, 15][1)]	DELPHI [11, 15][1)]	L3 [12, 16][1)]	OPAL [13, 14]	LEP
M_Z	91.187 ± 0.009	91.187 ± 0.009	91.195 ± 0.009	91.181 ± 0.009	91.187 ± 0.007
Γ_Z	2.501 ± 0.012	2.488 ± 0.012	2.491 ± 0.011	2.483 ± 0.012	2.491 ± 0.007
σ_{had}^{pole}	41.60 ± 0.27	40.86 ± 0.28	41.35 ± 0.28	41.45 ± 0.31	41.33 ± 0.18
R_e	20.69 ± 0.13	20.79 ± 0.28	21.09 ± 0.25	20.99 ± 0.25	20.88 ± 0.12
R_μ	20.88 ± 0.13	20.92 ± 0.22	21.13 ± 0.22	20.65 ± 0.17	20.86 ± 0.10
R_τ	20.77 ± 0.13	20.69 ± 0.30	20.67 ± 0.27	21.22 ± 0.25	20.85 ± 0.13
$A_{FB}^{pole}(e^+e^-)$	0.0140 ± 0.0093	0.013 ± 0.013	0.016 ± 0.013	-0.002 ± 0.012	0.0102 ± 0.0057
$A_{FB}^{pole}(\mu^+\mu^-)$	0.0074 ± 0.0072	0.0148 ± 0.0083	0.0237 ± 0.0095	0.0047 ± 0.0076	0.0113 ± 0.0040
$A_{FB}^{pole}(\tau^+\tau^-)$	0.0269 ± 0.0082	0.033 ± 0.010	0.028 ± 0.013	0.0165 ± 0.0082	0.0248 ± 0.0047

Table 4: Results from a parametrization of hadronic and leptonic cross sections and leptonic forward-backward asymmetries without the assumption of lepton universality. The LEP average parameter set has been obtained with a $\chi^2/NDOF = 24.1/(36-9)$. Preliminary results are marked by [1)]. The correlation matrix for the LEP combined parameter set is given in Table 20 of the Appendix.

	ALEPH [10, 15][1)]	DELPHI [11, 15][1)]	L3 [12, 16][1)]	OPAL [13, 14]	LEP
M_Z	91.187 ± 0.009	91.187 ± 0.009	91.195 ± 0.009	91.181 ± 0.009	91.187 ± 0.007
Γ_Z	2.501 ± 0.012	2.488 ± 0.012	2.491 ± 0.011	2.483 ± 0.012	2.491 ± 0.007
σ_{had}^{pole}	41.60 ± 0.27	40.86 ± 0.28	41.36 ± 0.28	41.45 ± 0.31	41.33 ± 0.18
R_ℓ	20.78 ± 0.13	20.82 ± 0.16	21.00 ± 0.15	20.88 ± 0.13	20.87 ± 0.07
$A_{FB}^{pole}(\ell^+\ell^-)$	0.0154 ± 0.0048	0.0202 ± 0.0059	0.0225 ± 0.0066	0.0076 ± 0.0050	0.0152 ± 0.0027

Table 5: Results from a parametrization of hadronic and leptonic cross sections and leptonic forward-backward asymmetries with the assumption of lepton universality. The LEP average parameter set has been obtained with a $\chi^2/NDOF = 15.3/(20-5)$. Preliminary results are marked by [1)]. The correlation matrix for the LEP combined parameter set is given in Table 21 of the Appendix.

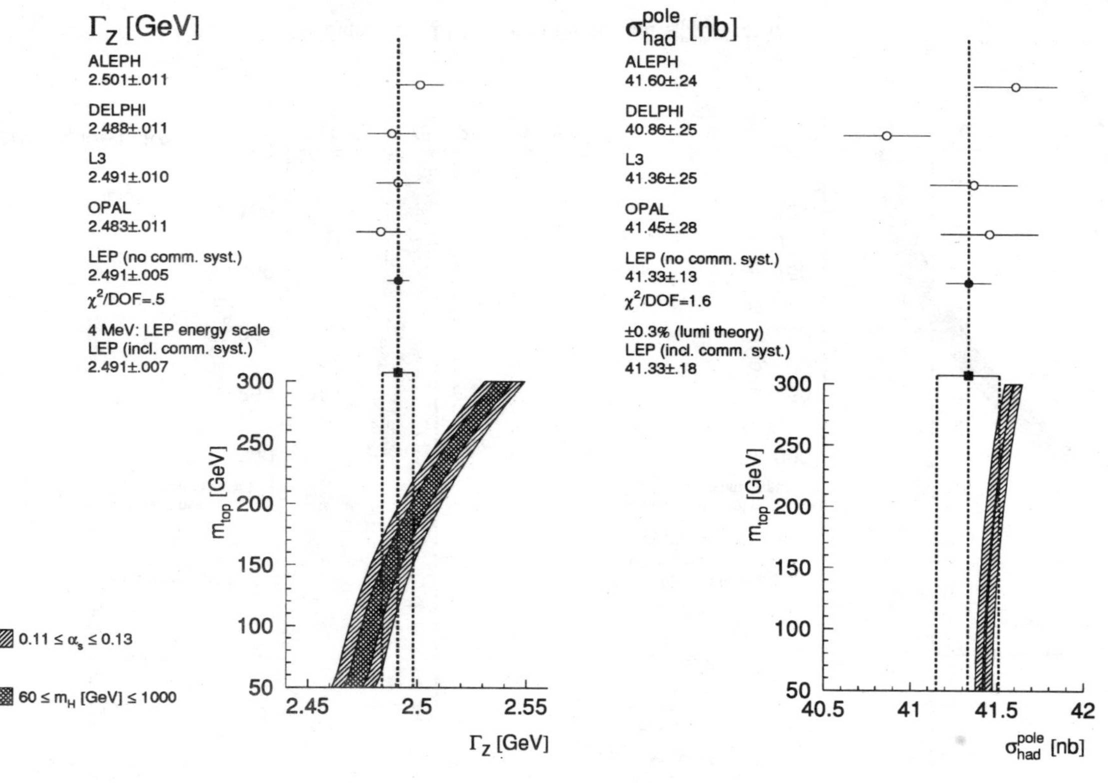

Figure 9: LEP results for Γ_Z and σ^{pole}_{had}.

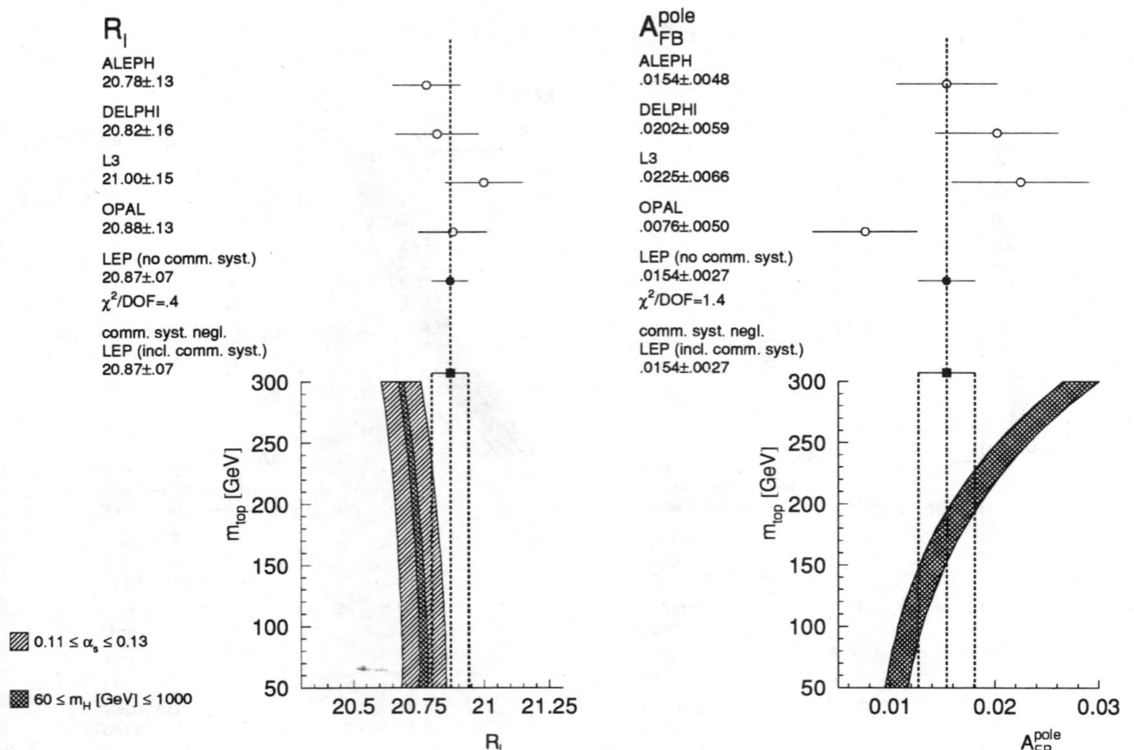

Figure 10: LEP results for R_ℓ and $A_{FB}^{pole}(\ell^+\ell^-)$.

For unpolarized e^+e^- beams the polarization $\mathcal{P}_f$ of the final state fermions is defined as:

$$\mathcal{P}_f = \frac{1}{\sigma_f^{tot}} \left(\sigma_f(h=+1) - \sigma_f(h=-1)\right)$$

where $\sigma_f(h=+1)$ and $\sigma_f(h=-1)$ refer to the production cross section of the process $e^+e^- \to f\bar{f}$ for positive $(h=+1)$ and negative $(h=-1)$ helicity fermions, respectively, and σ_f^{tot} denotes the total fermion production cross section. A negative value for $\mathcal{P}_f$ means that fermions produced in neutral current reactions are preferentially left-handed, as is observed in charged current reactions. From an analysis of the angular distribution of the final state fermion polarization a forward-backward asymmetry $A_{FB}^{\mathcal{P}_f}$ can be defined:

$$A_{FB}^{\mathcal{P}_f} = \langle \mathcal{P}_f \rangle_{\cos\theta>0} - \langle \mathcal{P}_f \rangle_{\cos\theta<0} \,.$$

Up to now the polarization of the final state fermions has only been measured for τ leptons. For polarized e^+e^- beams the left-right asymmetry A_{LR} is defined as:

$$A_{LR} = \frac{1}{\sigma_f^{tot}} \left(\sigma_L - \sigma_R\right)$$

where σ_L (σ_R) denotes the total production cross-section $e^+e^- \to f\bar{f}$ for a left-handed (right-handed) polarization of the incoming electrons. From an analysis of the angular distribution of the final state fermions a polarized forward-backward asymmetry can be derived as:

$$A_{FB}^{pol} = \langle A_{LR} \rangle_{\cos\theta>0} - \langle A_{LR} \rangle_{\cos\theta<0} \quad .$$

The general formalism of how to include fermion helicities into the description of the differential cross section for the process $e^+e^- \to f\bar{f}$ can be found in [36]. The formulae simplify considerably if only Z-exchange is considered. At the Z^0-pole the asymmetries defined above then have a simple relation to the vector and axial-vector couplings of the Z^0:

$$\mathcal{P}_f(s = M_Z^2) = -\mathcal{A}_f \tag{13}$$

$$A_{FB}^{\mathcal{P}_f}(s = M_Z^2) = \frac{3}{4}\mathcal{A}_e \tag{14}$$

$$A_{LR}(s = M_Z^2) = \mathcal{A}_e \tag{15}$$

$$A_{FB}^{pol}(s = M_Z^2) = \frac{3}{4}\mathcal{A}_f \,. \tag{16}$$

The effects of the inclusion of the full $e^+e^- \to f\bar{f}$ amplitude and of photonic corrections are small, each effect increases the absolute value of the predicted asymmetries at the Z^0-pole by approximately 0.002.

3.2.1 The τ Polarization

The τ lepton plays an exceptional role in the investigation of final state fermion polarizations in e^+e^--collisions because fermion and antifermion can easily be discriminated, the τ has a short lifetime and parity is violated in its weak decays.

Assuming the $V-A$ structure of the weak charged current the decay products can therefore be used as spin analyzers.

The classical example for the determination of the τ helicity from its decay products is the $\tau \to \pi(\mathrm{K})\nu$ decay (Figure 11). The π is spinless and therefore the ν_τ has to

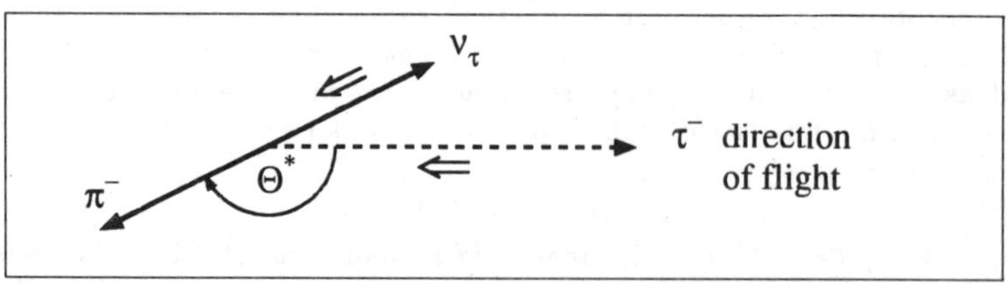

Figure 11: Helicity considerations for the $\tau \to \pi(\mathrm{K})\nu$ decay.

carry the spin of the τ. The $V-A$ structure of the weak charged current interactions requires the ν_τ to be left-handed and therefore the angular distribution in the τ-restframe depends on the polarization state of the τ lepton and is given by:

$$W(\cos\theta^*) = \frac{1}{N}\frac{dN}{d\cos\theta^*} = \frac{1}{2}\left(1+\mathcal{P}_\tau\cos\theta^*\right) ,$$

where $\cos\theta^*$ refers to the decay angle of the π in the τ rest frame. The angular distribution of the decay products in the τ rest frame has a one-to-one correspondence to the energy spectra in the e^+e^- centre-of-mass frame. Neglecting mass terms this relation is for the $\tau \to \pi(\mathrm{K})\nu$ decay:

$$\cos\theta^* \approx 2\frac{E_\pi}{E_\tau} - 1 .$$

Figure 12 shows the π energy spectrum measured in τ decays. Most of the experiments determine the τ polarization by fitting a linear combination of the $h = +1$ and $h = -1$ simulated distributions to their data. The simulation includes an event generation taking into account radiative corrections, the detector response and the effect of event selection cuts. A complementary approach has been pursued by the OPAL collaboration. They first deconvolute the data for the effects of event selection, detector resolution and photonic corrections and then fit the expected distribution of kinematic variables directly to the data to obtain $\mathcal{P}_\tau$.

The branching fraction of the classical $\tau \to \pi(\mathrm{K})\nu$ channel being only 12%, the τ polarization measurement benefits significantly by also considering other decay channels. The purely leptonic decay modes $\tau \to e\nu\nu$ and $\tau \to \mu\nu\nu$ have a large branching fraction, but the polarization signal is diluted as the decay angles cannot be reconstructed because of the two neutrinos in these decays. Only the lepton energy can

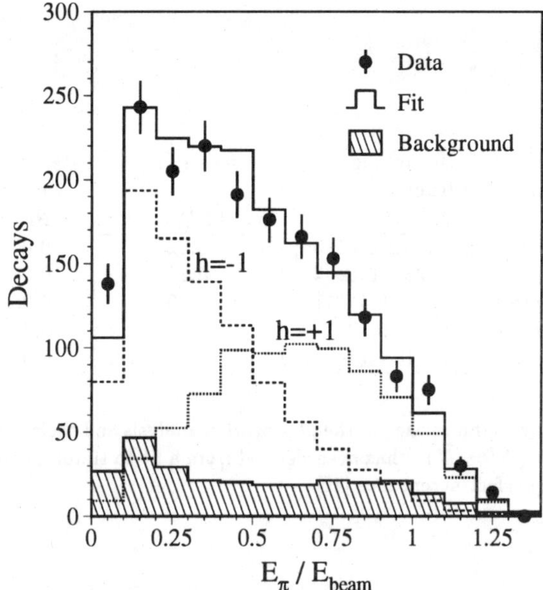

Figure 12: The spectrum of $\tau \rightarrow \pi(\mathrm{K})\nu$ decays as a function of the normalized energy E_π/E_{beam} (L3 collaboration). Also shown is the fitted contribution from each helicity including backgrounds for that helicity. The hatched histogram shows the total background.

be measured. Assuming the V–A-structure for the charged current and neglecting terms $\mathcal{O}(M_\tau/M_Z)$ and $\mathcal{O}(M_\ell/M_\tau)$ the expected distribution of the lepton energies is

$$W(x) = \frac{1}{N}\frac{dN}{dx} = \frac{1}{3}\left[4x^3 - 9x^2 + 5 + \mathcal{P}_\tau(8x^3 - 9x^2 + 1)\right]$$

with $x = E_\ell/E_\tau$. Among the semileptonic decay modes the $\tau \rightarrow \rho\nu$ and the $\tau \rightarrow \mathrm{a}_1\nu$ decays have been analyzed. The ρ and the a_1 meson, however, have spin 1 allowing for two possible spin configurations in τ decays (Figure 13). To fully exploit the information available, experiments therefore also take into account the angular distribution of the ρ and a_1 decay products in the ρ and a_1 restframe. Table 6 compares the statistical weights for the various τ decay channels. The τ-polarization measurements of the four LEP collaborations are summarized in Table 7.

Figure 14 displays the average values of $\mathcal{A}_\tau$ from each of the four LEP experiments and compares the combined average with the Standard Model prediction. For this combination we use the average result for $\mathcal{P}_\tau$ as quoted by the experiments to allow for correlations between the individual decay channels. We apply a small correction ($\mathcal{P}_\tau \rightarrow \mathcal{P}_\tau - 0.003$ for ALEPH, DELPHI and L3, $\mathcal{P}_\tau \rightarrow \mathcal{P}_\tau - 0.002$ for OPAL) to

τ decay mode	branching fraction [%]	sensitivity $S = (\Delta\mathcal{P}_\tau\sqrt{N})^{-1}$	statistical weight $\propto Br \cdot S^2$
$\tau \to e\nu\nu$	17.71 ± 0.12 [34]	0.22	0.07
$\tau \to \mu\nu\nu$	17.23 ± 0.12 [34]	0.22	0.07
$\tau \to \pi(K)\nu$	12.3 ± 0.4 [37]	0.59	0.36
$\tau \to \rho\nu$	23.7 ± 0.6 [37]	0.48	0.46
$\tau \to a_1\nu$	≈ 7 [38, 39]	0.22	0.03

Table 6: The τ decay channels used in the polarization analysis and their statistical weights. The sensitivity $S = (\Delta\mathcal{P}_\tau\sqrt{N})^{-1}$ has been derived from a fit to simulated data in the limit $N \to \infty$, assuming perfect acceptance [40].

experiment	$\tau \to e\nu\nu$	$\tau \to \pi(K)\nu$
ALEPH 1)	$-.161 \pm .089 \pm .045$	$-.149 \pm .032 \pm .019$
DELPHI	$-.120 \pm .220 \pm .080$	$-.350 \pm .110 \pm .070$
L3	$-.127 \pm .097 \pm .062$	$-.148 \pm .046 \pm .033$
OPAL 1)	$.053 \pm .073 \pm .061$	$-.093 \pm .057 \pm .052$
LEP	$-.073 \pm .057$	$-.151 \pm .028$
experiment	$\tau \to \mu\nu\nu$	$\tau \to \rho\nu$
ALEPH 1)	$-.153 \pm .068 \pm .029$	$-.127 \pm .032 \pm .030$
DELPHI	$-.050 \pm .180 \pm .070$	$-.240 \pm .090 \pm .070$
L3	$-.020 \pm .101 \pm .055$	$-.152 \pm .035 \pm .029$
OPAL 1)	$-.178 \pm .078 \pm .061$	$-.180 \pm .060 \pm .050$
LEP	$-.126 \pm .051$	$-.151 \pm .028$
experiment		$\tau \to a_1\nu$
ALEPH 1)		$-.150 \pm .150 \pm .070$
DELPHI		
L3		$.105 \pm .164 \pm .093$
OPAL		
LEP		$-.039 \pm .124$

Table 7: Summary of τ-polarization measurements [41, 42, 43]. The DELPHI results are based on the 1990 data only. Preliminary numbers are marked by 1). The combined LEP result has been obtained from a simple weighted average neglecting possible correlations.

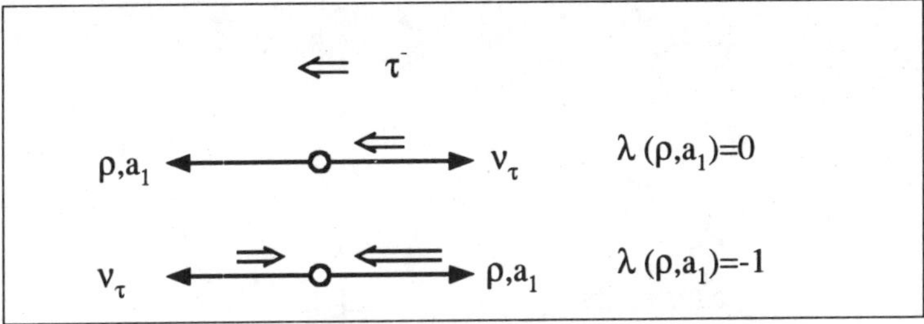

Figure 13: The two possible spin configurations for the $\tau \to \rho\nu$ and $\tau \to a_1\nu$ decays.

obtain $\mathcal{P}_\tau(s = M_Z^2)$ deconvoluted for photonic corrections and taking into account Z^0 exchange only; i.e. to allow the averages to be parametrized as $\mathcal{P}_\tau = -\mathcal{A}_\tau$. The value of $\mathcal{A}_\tau$ can then be compared directly with other measurements of polarization asymmetries.

An analysis of the angular distribution ot $\mathcal{P}_\tau$, which is shown in Figure 15, allows the determination of $\mathcal{A}_e$. Two experiments have performed such an analysis, the results are also displayed in Figure 14.

3.2.2 The Left-Right Asymmetry

The left-right asymmetry has been measured at $E_{cm} = 91.55$ GeV with the SLD detector at SLC using a longitudinally polarized electron beam [44]. Polarized e^- beams are produced by injecting the linac with electrons of $\approx$28% polarization produced by the illumination of a GaAs photocathode with circularly polarized laser light. The electron helicities are reversed by changing the laser helicity on a cycle-by-cylcle basis. During the process of acceleration and transport the electrons undergo partial depolarization arriving at the interaction point with an average polarization of $\approx$22%. The polarization is monitored by measuring the Compton scattering of the electrons on a circularly polarized laser beam.

As the left-right asymmetry close to the Z^0-pole does not depend on the final state, the event selection inclusively aims at Z^0 decays and is primarily based on the energy deposit in the liquid Argon calorimeter. Only wide-angle Bhabhas, which exhibit a different left-right asymmetry than the other final states due to the interference with t-channel amplitudes, are removed from the event sample by limiting the sum of the two largest energy deposits within a localized region.

The resulting left-right asymmetry is

$$A_{LR} = 0.100 \pm 0.044(\text{stat.}) \pm 0.003(\text{syst.})$$

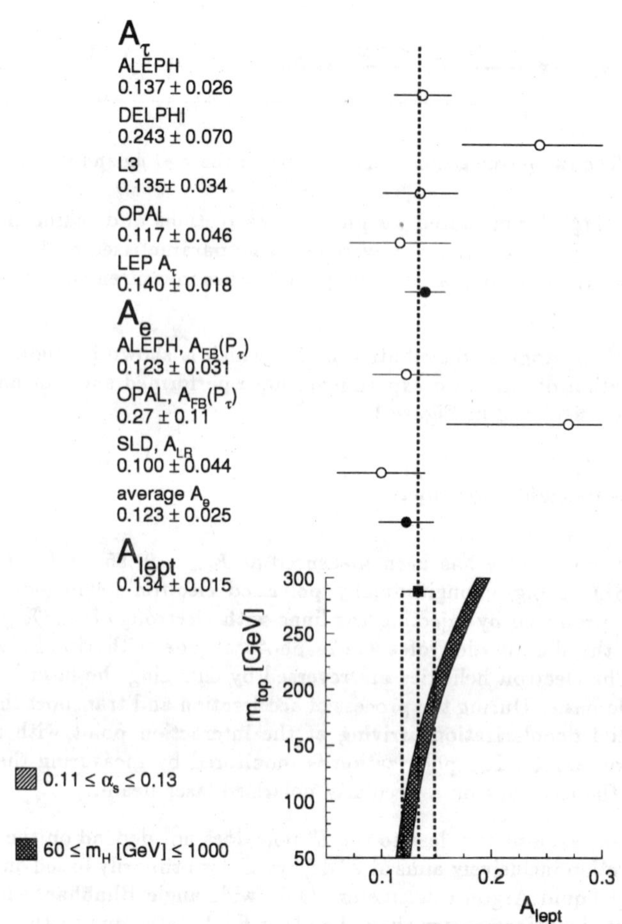

Figure 14: Results from polarization asymmetries.

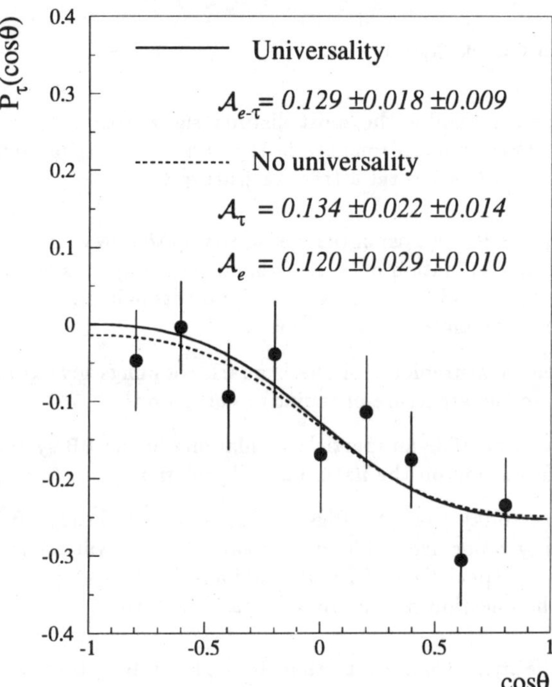

Figure 15: The angular distribution of the τ polarization.

based on a sample of ≈10 000 Z^0 candidates. The systematic uncertainty is dominated by the 3% systematic error on the mean polarization.

For the upcoming 1993 run SLC is expected to provide the luminosity for a data sample of at least 50 000 Z^0s with a mean electron polarization of 40%. This would allow a statistical error $\Delta A_{LR} = 0.011$. There is some hope to make use of the recent development of strained lattice cathodes to reach source polarizations in excess of 80% [45].

3.3 *Determination of Quark Couplings*

The determination of quark couplings requires the separation of primary quark flavours in a sample of hadronic events. In the following we will present several

methods to determine quark partial widths and forward-backward asymmetries.

3.3.1 The b Quark Sector

Primary b quarks exhibit the most distinct signatures as they have a hard fragmentation, b flavoured hadrons are heavy and have long lifetimes. The ability to separate b quarks has several attractive features:

- The b quark is the isospin partner of the up to now elusive top quark. Already a rough determination of its neutral current couplings is sufficient to establish its Standard Model weak isospin assignments [46], and thereby experimentally prove the existence of the top quark.
- Precision measurements of the b quark couplings give complementary information on the structure of radiative corrections.
- A study of particle-antiparticle oscillations in the $B\overline{B}$-system can give important information on the nature of CP violation.
- We know today that the Higgs - if it exists - is heavy. If its mass is within the energy range accessible in the near future, it will predominantly couple to b quarks. Especially at LEP200 the identification of b quarks will be vital to assess the question of the origin of particle masses.

Up to now the most common method to separate b quarks uses the semileptonic decays $b \to \ell X$. Leptons from primary b-flavoured hadron decays are characterized by a high momentum p because of the hard b quark fragmentation. They also have a relatively high transverse momentum p_T with respect to the nearest jet because of the large b-flavoured hadron mass.

In addition to the prompt semi-leptonic decay $b \to \ell X$ a number of other sources contribute to the high p, high p_T lepton sample, which are shown in Figure 16. The inclusive leptons from the various sources differ in their p and p_T spectra, as shown in Figure 17. Therefore the observed p versus p_T distribution of inclusive leptons is fitted to an expression of the form:

$$N(p, p_T) = 2N_{had} \sum_i C_i f_i(p, p_T) \ .$$

In the expression above, N_{had} denotes the total number of hadronic events in the data, C_i the total fraction of events of class i in the lepton sample and the $f_i(p, p_T)$ are the probability densities for each contribution as a function of p and p_T. For the contribution originating from primary b quarks:

$$C_i \equiv C_{b\bar{b}} = \frac{\Gamma_{b\bar{b}}}{\Gamma_{had}} Br(b \to \ell X)\eta_{b\bar{b}} \ .$$

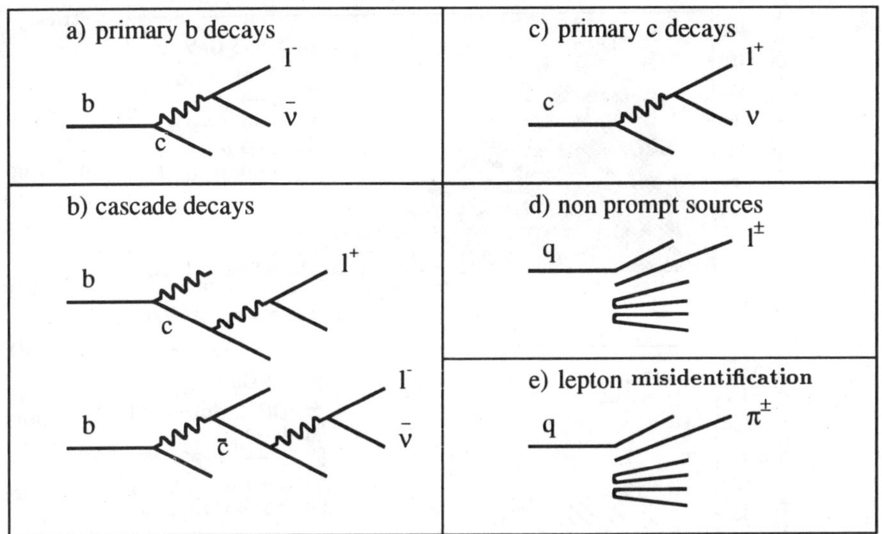

Figure 16: Sources of inclusive leptons in hadronic events.

Here $Br(\mathrm{b} \to \ell X)$ refers to the leptonic branching ratio for b-decays and $\eta_{\mathrm{b\bar{b}}}$ to the efficiency for detecting the associated lepton.

The four LEP collaborations choose to fit the lepton spectra with different combinations of assumptions and free parameters. Most crucial for the determination of $\Gamma_{\mathrm{b\bar{b}}}$ is the leptonic branching ratio $Br(\mathrm{b} \to \ell X)$. Some experiments base their determination of $\Gamma_{\mathrm{b\bar{b}}}$ on measurements of $Br(\mathrm{b} \to \ell X)$ at the $\Upsilon(4S)$ resonance or in the e^+e^- continuum at PEP/PETRA. The determination of $Br(\mathrm{b} \to \ell X)$ requires the use of a model of the semileptonic decay spectrum $f_{\mathrm{b\bar{b}}}(p, p_T)$ to extrapolate to low momenta where the presence of secondary decays $\mathrm{b} \to \mathrm{c} \to \ell\nu_\ell X$ and experimental limitations in lepton identification prevent a direct measurement of the rate of $\mathrm{b} \to \ell\nu_\ell X$ decays. $Br(\mathrm{b} \to \ell X)$ has also been measured directly at LEP [47, 48] using the ratio of hadronic events with one (n_ℓ) and two $(n_{\ell\ell})$ inclusive leptons:

$$Br(\mathrm{b} \to \ell X) \approx \frac{2n_{\ell\ell}/n_\ell}{1 + 2n_{\ell\ell}/n_\ell} \ .$$

Note that the above formula neglects lepton sources other than primary b quark decays. The dominant systematic error of this method again arises from the modelling of B decays.

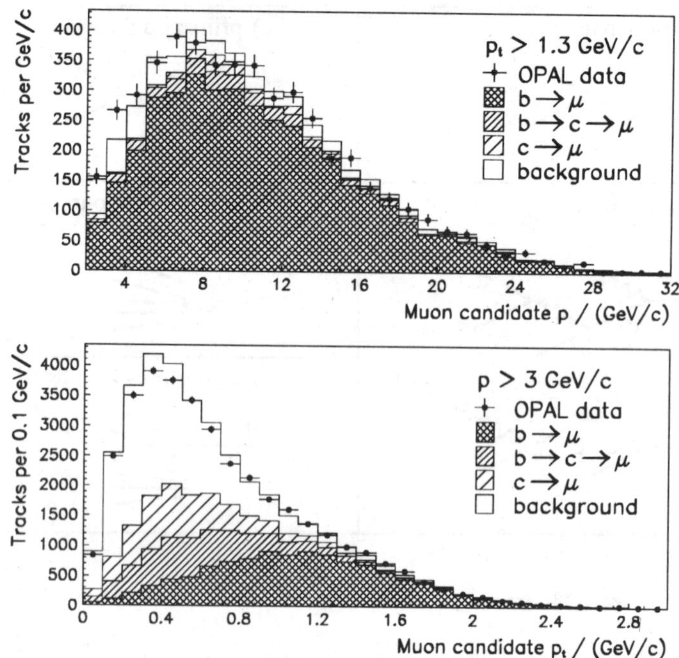

Figure 17: The distribution of momentum p (top) and transverse momentum p_T (bottom) for muon candidates passing all the selection cuts except that on p or p_T respectively.

Other approaches to separate b quarks are based on global event shape variables such as the boosted sphericity product [49]. As there are many variables which allow a partial separation of event classes, also neural networks have been used to discriminate $b\bar{b}$ and $c\bar{c}$ events from the hadronic sample. With the advent of high resolution microvertex chambers a new field of b quark discrimination starts to open. Decays of b-flavoured hadrons are expected to have on average several charged tracks with a large positive impact parameter. OPAL has presented a mixed tag method, which uses the lepton tag as well as a tag which requires a certain number of tracks in one hemisphere with a significantly large impact parameter [50]. The tags which are applied separately to each hemisphere define a number of event classes. The simultaneous analysis of these event classes allows the reduction of the decay model dependence of the determination of $\Gamma_{b\bar{b}}$. At present the dominant systematic error arises from the variation of cuts for the vertex tag. A summary of measurements of $\Gamma_{b\bar{b}}$ is shown in Figure 18.

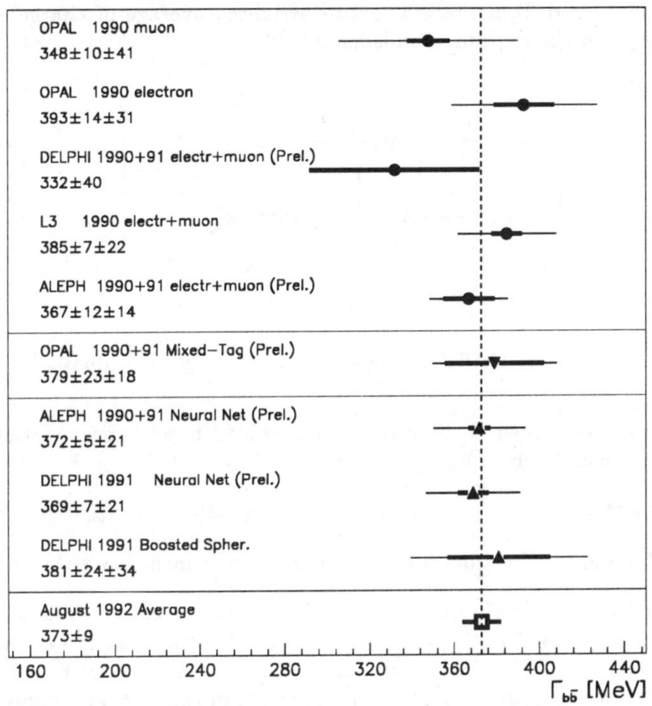

Figure 18: Status of $\Gamma_{b\bar{b}}$ from LEP [50]. Results are based on $\Gamma_{had} = 1740 \pm 12$ MeV [33] and $Br(b \to \ell X) = 0.113 \pm 0.010 \pm 0.006$ [47].

For the measurement of quark forward-backward asymmetries the jet originating from the quark must be discriminated against the jet originating from the antiquark. The inclusive lepton analysis can be naturally extended to determine the $b\bar{b}$ forward-backward asymmetry $A_{FB}^{b\bar{b}}$. For the total asymmetry the contributions in Figure 16 have to be considered separately. Furthermore it is very important to consider the mixing of neutral B mesons which contributes to wrong sign leptons (see Figure 19). The amount of mixing can be expressed by a parameter χ_B defined as:

$$\chi_B = \frac{Br(b \to \overline{B}^0 \to B^0 \to \ell^+ X)}{Br(b \to \ell^{\pm} X)} .$$

$B\overline{B}$ mixing at LEP has been determined using the ratio of like-sign and unlike-sign inclusive dilepton events [51]. Alternatively the charge of one b quark has been tagged by a high p, high p_T electron or muon and the charge of the other b quark has been measured from the momentum weighted average of charges of particles within the jet in the opposite hemisphere [52].

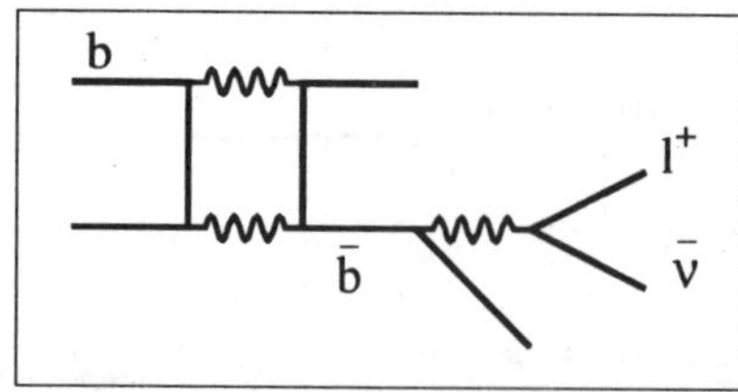

Figure 19: $B\overline{B}$-mixing as source of wrong-sign leptons.

The contributions shown in Figures 16 and 19 lead to an experimentally observed asymmetry which is given by:

$$A_{\rm FB}^{raw} = (f_{\rm b\to\ell} - f_{\rm b\to c\to\ell} + f_{\rm b\to\bar{c}\to\ell})(1-2\chi_{\rm B})A_{\rm FB}^{\rm b\bar{b}} - f_{\rm c\to\ell}A_{\rm FB}^{\rm c\bar{c}} + f_{bg}A_{\rm FB}^{bg}$$

where f_i denotes the fraction of contribution i in the inclusive lepton sample.

The upper left part of Figure 20 shows recent LEP measurements [53, 54] of the observed $\rm b\bar{b}$ asymmetry $A_{\rm FB}^{\rm b\bar{b}}(obs.) = (1-2\chi_{\rm B})A_{\rm FB}^{\rm b\bar{b}}$. Correlated systematic errors arise from the imprecise understanding of the production and decay of hadrons containing b- and c quarks. A detailed evaluation of common systematic errors has been performed in [55]. The study estimates the present correlated systematics to $\Delta A_{\rm FB}^{\rm b\bar{b}}/A_{\rm FB}^{\rm b\bar{b}} \approx 5\%$. To compare the LEP average with the Standard Model prediction a mixing correction with $\chi_{\rm B} = 0.126 \pm 0.012$ [56] has been applied in Figure 20.

3.3.2 The c Quark Sector

Several methods to tag events originating from primary $\rm c\bar{c}$ quarks have been used. One method is to extend the inclusive lepton analysis in the region of low p and $p_{\rm T}$ [53, 57, 58] where a sizable fraction of the events can be traced to primary c quarks. For the determination of $\Gamma_{\rm c\bar{c}}/\Gamma_{\rm had}$ the branching ratio $Br({\rm c}\to \ell X)$ is needed, which has been determined at energies below the $\rm Z^0$-pole. ALEPH has determined $\Gamma_{\rm b\bar{b}}$, $\Gamma_{\rm c\bar{c}}$, $Br({\rm b}\to\ell X)$, $Br({\rm c}\to\ell X)$, the fragmentation parameters $\langle x_E({\rm b})\rangle$ and $\langle x_E({\rm c})\rangle$, and the mixing parameter $\chi_{\rm B}$ in a a combined fit to inclusive single and dilepton distributions [48].

A complementary method is based on the reconstruction of $\rm D^{*\pm}$ decays. The mass difference between the $\rm D^{*\pm}$ and the $\rm D^0$ is only a few MeV larger than the mass of the $\pi^\pm$ so that the decays $\rm D^{*+}\to D^0\pi^+$ and $\rm D^{*-}\to \overline{D^0}\pi^-$ result in an excess of tracks with

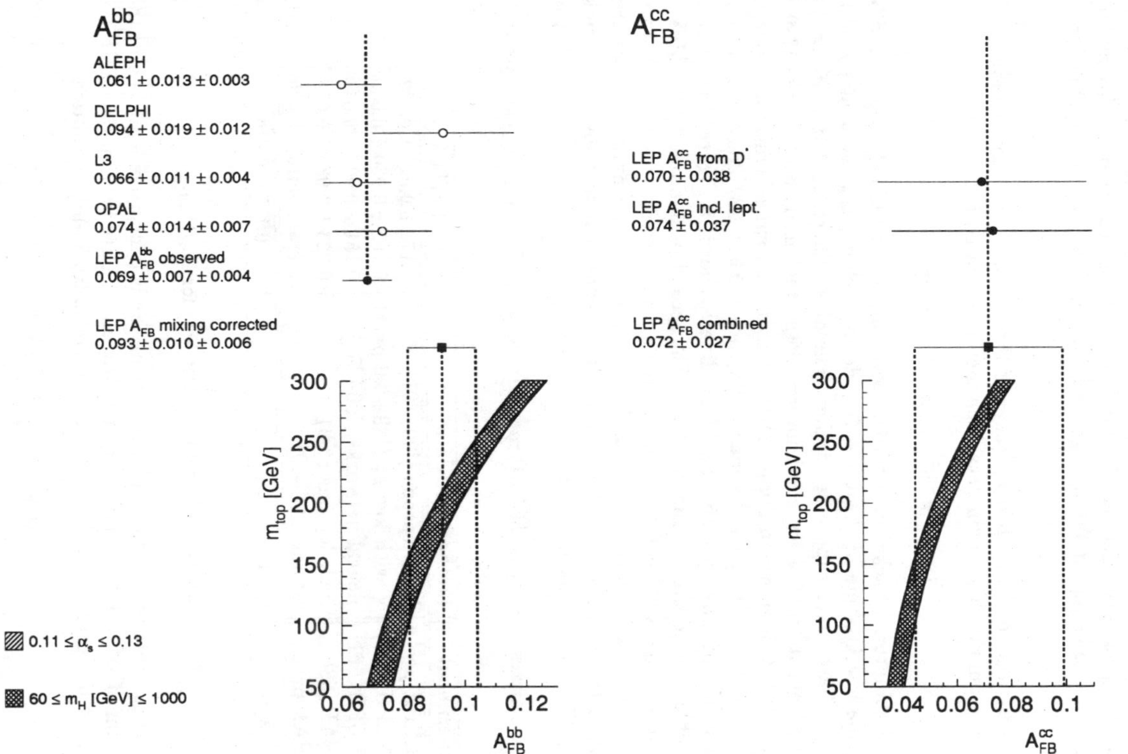

Figure 20: Status of $A_{FB}^{b\bar{b}}$ and $A_{FB}^{c\bar{c}}$ from LEP [55]. For the results of the individual experiments the first error quoted is statistical the second error is the systematic error quoted by the experiment. The LEP average has been obtained by first removing from each experiment the quoted uncertainties which were considered as correlated.

very low transverse momentum p_T with respect to the jet axis. DELPHI has analysed the p_T^2 distribution of charged tracks with momentum $1.5 < p$ [GeV] < 2.5 [59]. Together with the probability for a c quark to fragment into a $D^{*\pm}$ from measurements at $\sqrt{s} = 10.55$ GeV and assuming this probability to be energy independent, the ratio $\Gamma_{c\bar{c}}/\Gamma_{had}$ has been derived. The dominant systematic error of this method arises from the modelling of the background from non $c\bar{c}$ reactions to the low p_T^2 region.

Also the reconstruction of exclusive decays of the $D^{*\pm}$ has been used to study the properties of c quarks. The main decay channel investigated is:

$$D^{*+} \to \pi^+ D^0 \quad \text{with} \quad D^0 \to K^- \pi^+$$

The high combinatorial background for the $D^0\pi$ mode can be largely suppressed by exploiting the low Q-value of the $D^* \to D^0\pi$ decay, yielding a prominent signal in an otherwise phase space suppressed region. The mass difference $\Delta M = M(K\pi\pi) - M(K\pi)$ for all events with a $K\pi$ combination in the region of the D^0 mass $1790 < M(K\pi)$ [MeV] < 1940 is shown in Figure 21b). A significant enhancement is apparent at $\Delta M = 146$ MeV. Cutting on the mass difference $142.5 < \Delta M$ [MeV] < 148.5 one obtains the invariant mass spectrum for D^0 candidates shown in Figure 21a). A clear signal is visible around the nominal D^0 mass. The second peak around 1600 MeV is attributed to the decay $D^0 \to K^-\pi^+\pi^0$ where the π^0 is not reconstructed. Using the product branching ratio that a D^* meson is produced from a primary c quark and then decays in the channel $D^* \to D^0\pi \to \pi(\pi K)$ from experiments below the Z^0-pole the ratio $\Gamma_{c\bar{c}}/\Gamma_{had}$ can be determined.

Table 8 summarizes the status of LEP measurements for $\Gamma_{c\bar{c}}/\Gamma_{had}$. As the two most

experiment	method and data	$\Gamma_{c\bar{c}}/\Gamma_{had}$
ALEPH prel.	incl. leptons ('90+'91)	$0.170 \pm 0.010 \pm 0.022$
DELPHI [59]	π with low p_T^2 ('89+'90 part)	$0.162 \pm 0.030 \pm 0.050$
DELPHI [61]	neural networks ('90+'91)	$0.151 \pm 0.008 \pm 0.041$
OPAL [58]	incl. leptons ('90)	$0.223 \pm 0.032 \pm 0.059$
OPAL prel.	D^* tag ('90+'91)	$0.188 \pm 0.015 \pm 0.026$
average		0.175 ± 0.016

Table 8: Status of LEP measurements for $\Gamma_{c\bar{c}}/\Gamma_{had}$.

precise measurements result from independent methods (inclusive leptons and the reconstruction of $D^* \to D^0\pi$) common systematic errors have been neglected in the average. Using the present LEP average $\Gamma_{had} = 1741.2 \pm 6.6$ MeV we obtain

$$\Gamma_{c\bar{c}} = 305 \pm 28 \text{ MeV}$$

in good agreement with the Standard Model prediction [60] $\Gamma_{c\bar{c}} = 297^{+4}_{-3}$ MeV.

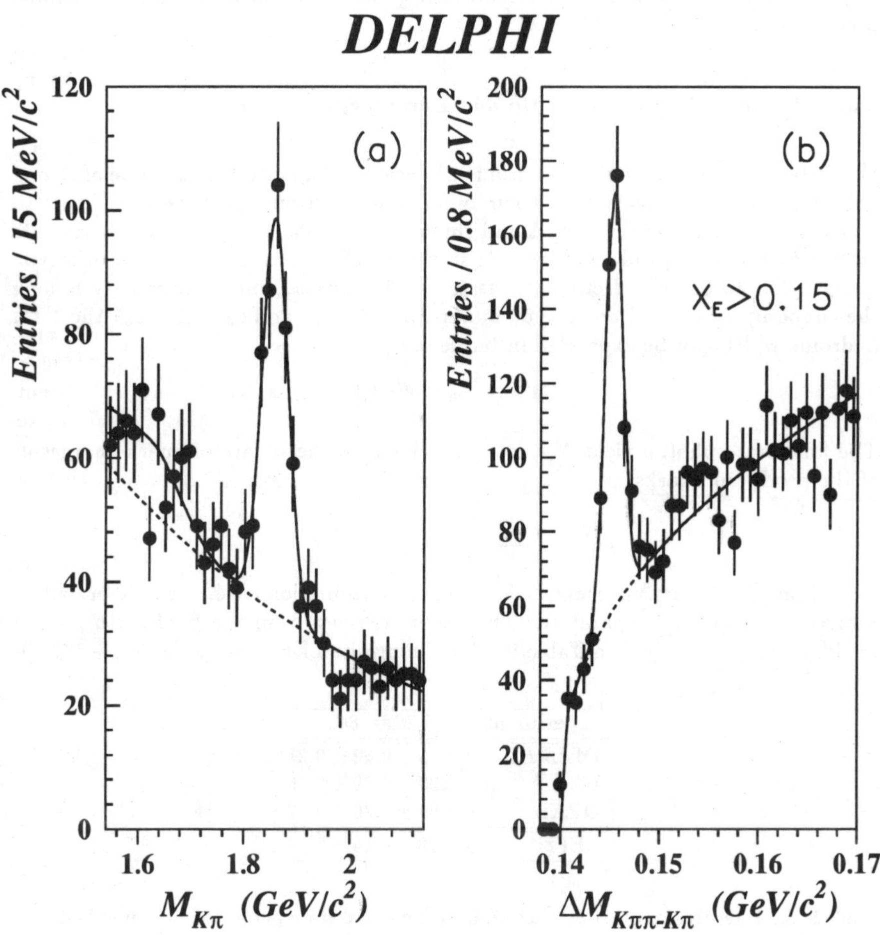

Figure 21: Reconstruction of exclusive D* decays. a) The Kπ invariant mass distribution for all events with $142.5 < \Delta M = M(\mathrm{K}\pi\pi) - M(\mathrm{K}\pi)$ [MeV] < 148.5. b) The distribution of ΔM for all events with a Kπ combination with an invariant mass $1790 < M(\mathrm{K}\pi)$ [MeV] $<$ 1940. The solid line represents a fit to the signal over a smooth background. The background contribution is indicated as dashed line.

The inclusive lepton tag as well as the reconstruction of exclusive $D^{*\pm}$ decays have also been used to determine the forward-backward asymmetry of c quarks $A_{FB}^{c\bar{c}}$. Figure 20 compares LEP measurements of $A_{FB}^{c\bar{c}}$ [53, 57, 62] with the Standard Model prediction.

3.3.3 Inclusive Separation of Up and Down Type Quarks

Final state photons emitted in hadronic Z^0 decays are an interesting probe of strong and electroweak interactions. Their differential distributions have been used to study the interplay between quark and gluon radiation [63]–[71]. Three experiments have used the total production rate to study the Z^0 couplings of up and down type quarks [63, 66, 68, 71]. Defining $C_q \equiv \hat{g}_a^{q\,2} + \hat{g}_v^{q\,2}$ and assuming universality among the couplings of up and down type quarks; i.e. $C_u = C_c$ and $C_d = C_s = C_b$¶ the total hadronic width can be expressed in terms of C_q:

$$\Gamma_{had} \propto \sum_q C_q = 3C_d + 2C_u\,.$$

The final state photon yield $N_{q\bar{q}\gamma}$ is proportional to the square of quark charges e_q^2 of the produced quarks:

$$N_{q\bar{q}\gamma} \propto \sum_q e_q^2 C_q = \frac{1}{9}(3C_d + 8C_u)\,.$$

Combining both measurements allows the determination of C_u and C_d or, alternatively, $\Gamma_{u\bar{u}}$ and $\Gamma_{d\bar{d}}$ separately. The results derived from the final state photon analysis are summarized in Table 9. Combining the LEP average for $3C_d + 8C_u$ obtained from final state photons with the LEP average $\Gamma_{had} = 1741.2 \pm 6.6$ MeV we obtain

experiment	$3C_d + 8C_u$
DELPHI	$4.15 \pm 0.62 \pm 0.21$
L3	$2.97 \pm 0.29 \pm 0.16$
OPAL	$3.09 \pm 0.20 \pm 0.17$
LEP	$3.10 \pm 0.16 \pm 0.17$

Table 9: LEP results for $3C_d + 8C_u$ determined from final state photon radiation in hadronic events [66, 68, 71]. The first error quoted includes statistics and uncorrelated experimental systematics. The second error accounts for theoretical uncertainties in the determination of the final state photon yield and is treated as fully correlated in the averaging procedure.

$$C_d = 0.405 \pm 0.026 \qquad \Gamma_{d\bar{d}} = 419 \pm 27 \text{ MeV}$$
$$C_u = 0.235 \pm 0.038 \qquad \Gamma_{u\bar{u}} = 244 \pm 39 \text{ MeV}$$

¶In the Standard Model small deviations of 2-3% are expected for the couplings of the b quark.

in good agreement with the Standard Model prediction [60] for $\frac{1}{3}(\Gamma_{d\bar{d}} + \Gamma_{s\bar{s}} + \Gamma_{b\bar{b}}) = 381 \pm 3$ MeV and $\frac{1}{2}(\Gamma_{u\bar{u}} + \Gamma_{c\bar{c}}) = 297^{+4}_{-3}$ MeV. Note that the results for $\Gamma_{u\bar{u}}$ and $\Gamma_{d\bar{d}}$ are fully anticorrelated.

3.3.4 The q$\bar{\text{q}}$ Charge Asymmetry

For any fermion f the average charge produced in the forward and backward hemispheres is given by $< Q^f_F >= q_f A^{f\bar{f}}_{FB}$ and $< Q^f_B >= -q_f A^{f\bar{f}}_{FB}$ leading to a charge flow $< Q^f_{FB} >= 2q_f A^{f\bar{f}}_{FB}$. Summed over 5 quarks the charge flow is

$$< Q_{FB} >= \sum_{f=(u,d,s,c,b)} 2q_f A^{f\bar{f}}_{FB} \frac{\Gamma_{f\bar{f}}}{\Gamma_{had}}$$

To measure a charge flow in hadronic events, the hadron charges within jets have to be related to the charge of the primary q$\bar{\text{q}}$ pair produced in the hard scattering process. The algorithms to recover these primary quark charges are based on the idea that the primary parton charge manifests itself in the leading hadrons. The interpretation of the measured distributions requires a detailed modelling of the fragmentation process, which introduces significant systematics into this measurement. For this observable, not all experiments provide a value of the average charge asymmetry deconvoluted for detector and fragmentation effects, but rather quote an effective weak mixing angle, which has been derived from fitting the observed asymmetry to the Monte Carlo prediction. The LEP results are summarized in Table 10 in terms of $\sin^2\theta_W^{eff,\ell}$. Using the Standard Model calculation with

experiment	$\sin^2\theta_W^{eff,\ell}$
ALEPH[1)]	$0.2295 \pm 0.0029 \pm 0.0040$
DELPHI	$0.2345 \pm 0.0030 \pm 0.0027$
OPAL	$0.2321 \pm 0.0028 \pm 0.0020$
LEP	$0.2329 \pm 0.0017 \pm 0.0026$

Table 10: Results for the effective weak mixing angle from the measurement of the q$\bar{\text{q}}$ charge asymmetry [55, 72, 73]. The first error quoted includes statistics and uncorrelated experimental systematics. The second error accounts for uncertainties arising from fragmentation and B$\bar{\text{B}}$-mixing and is treated as fully correlated in the averaging procedure. Preliminary results are marked by [1)].

$M_Z = 91.187 \pm 0.007$ GeV and $\alpha_s(M_Z^2) = 0.120 \pm 0.006$ this value corresponds to an average charge asymmetry

$$< Q_{FB} >= 0.0387 \pm 0.0035$$

The average value $\sin^2\theta_W^{eff,\ell} = 0.2329 \pm 0.0017 \pm 0.0026$ determined from the q$\bar{\text{q}}$ charge asymmetry is in good agreement with $\sin^2\theta_W^{eff,\ell} = 0.2320 \pm 0.0016$ resulting from

$A_{FB}^{pole}(\ell^+\ell^-)$. The comparison, however, also shows that the common systematic error due to fragmentation uncertainties makes it uninteresting to pursue the measurement of the $q\bar{q}$ charge asymmetry with higher statistics. Nonetheless the method itself remains an interesting tool for studies which are more severely limited by statistics, if a charge discrimination is performed with less inclusive channels.

4 Selected Aspects of the Results

4.1 *The Invisible Width of the* Z^0

One of the major goals of e^+e^- physics at the Z^0-pole is the determination of the partial width of the Z^0 into invisible channels, Γ_{inv}. In the Standard Model the only channels contributing to Γ_{inv} are massless neutrinos. The number of generations and therefore the number of light neutrino species, N_ν, are not predicted by the Standard Model. Upper limits from experiments before the start of Z^0-pole physics were ranging between 4–5 [74]. The first results of N_ν from LEP [75] were based on an integrated luminosity of $\approx$ 190 nb^{-1}, collected during the first 15 days of physics running. At that time N_ν was determined from the hadronic line shape only assuming the Standard Model prediction for all partial widths resulting in:

$$\sigma_{had}^{pole} = \frac{12\pi}{M_Z^2}\frac{\Gamma_{\ell\ell}^{SM}\Gamma_{had}^{SM}}{\Gamma_Z^2} \qquad \text{with} \qquad \Gamma_Z = \Gamma_{had}^{SM} + 3\Gamma_{\ell\ell}^{SM} + N_\nu\Gamma_{\nu\nu}^{SM}\,.$$

With this method, the dominant sensitivity to N_ν was given by the absolute normalization, reflected in σ_{had}^{pole}, rather than the measurement of the total width. With increased statistics the partial widths for hadrons and leptons could be determined separately, allowing a model independent determination of Γ_{inv}:

$$\Gamma_{inv} = \Gamma_Z^{meas} - \Gamma_{had}^{meas} - 3\Gamma_{\ell\ell}^{meas}\,.$$

Table 11 summarizes the current status of the determination of partial widths at LEP. Any partial width shows a non-negligible top mass dependence due to radiative corrections. As most of these corrections are universal, the bulk of this dependence cancels in the ratio of partial widths. Within the Standard Model framework and extensions which only introduce universal corrections to partial widths, it is therefore more advantageous to derive N_ν from the ratio $\Gamma_{inv}/\Gamma_{\ell\ell}$. Comparing the present LEP average:

$$\Gamma_{inv}/\Gamma_{\ell\ell} = 5.987 \pm 0.070$$

to the Standard Model prediction [60] $\Gamma_{\nu\nu}/\Gamma_{\ell\ell} = 1.993^{+0.002}_{-0.003}$ results in

$$N_\nu = 3.004 \pm 0.035\,.$$

This result is supported by direct measurements of the Z^0 invisible width at LEP by counting the number of events with only a single photon but otherwise no further

Without Lepton Universality:	
Γ_{ee}	83.41±0.35
$\Gamma_{\mu\mu}$	83.49±0.51
$\Gamma_{\tau\tau}$	83.53±0.61
With Lepton Universality:	
$\Gamma_{\ell\ell}$	83.43±0.29
Γ_{had}	1741.2±6.6
Γ_{inv}	499.5 ±5.6

Table 11: Z^0 partial decay widths [MeV].

activity in the detector [76, 77]. We do not necessarily expect an integer number for N_ν. A heavy neutrino with Standard Model couplings contributes to N_ν with:

$$\delta N_\nu = \left(1 - 4\frac{M_\nu^2}{s}\right)^{1/2} \left(1 - \frac{M_\nu^2}{s}\right).$$

With the present accuracy such a heavy neutrino can be excluded for $M_\nu < 45.4$ GeV at the 95% confidence level. The experimental determination of Γ_{inv} at LEP is a powerful constraint on various extensions of the Standard Model predicting new particles that couple to the Z^0. The resulting limits on light and heavy neutrinos have an important impact on our understanding of cosmological and astrophysical models.

4.2 *Lepton Universality*

Lepton universality predicts that the gauge couplings of the three known lepton doublets are equal. Deviations from lepton universality may arise from the mixing of the conventional leptons with new lepton species.

Precision tests at the Z^0-pole are a unique tool to test the hypothesis of lepton universality for the neutral current couplings of charged leptons. For their determination we combine the Z^0-pole averages for hadronic and leptonic cross sections and leptonic forward-backward asymmetries (Table 4) with $\mathcal{A}_\tau$ and $\mathcal{A}_e$ from polarization asymmetries (Figure 14). The results of a fit to the effective vector and axial-vector couplings of leptons with and without the assumption of universality are summarized in Table 12. Figure 22 shows the one standard deviation contours in the $\hat{g}_a^\ell$-$\hat{g}_v^\ell$ plane. The axial-vector couplings are essentially determined by the leptonic partial widths, whereas the asymmetry data measure $\hat{g}_v^\ell$ / $\hat{g}_a^\ell$. The error on the electron axial-vector coupling is roughly $\sqrt{2}$ times smaller than the others as the electron couplings enter in both initial and final state for the process $e^+e^- \to e^+e^-$. The accuracy of the vector coupling for the τ lepton significantly profits from the

	$\hat{g}_a^\ell$	$\hat{g}_v^\ell$
e^+e^-	-0.4998 ± 0.0011	-0.0344 ± 0.0042
$\mu^+\mu^-$	-0.5005 ± 0.0016	-0.027 ± 0.010
$\tau^+\tau^-$	-0.4999 ± 0.0019	-0.0386 ± 0.0042
$\ell^+\ell^-$	-0.49988 ± 0.00090	-0.0351 ± 0.0025

Table 12: The effective vector and axial-vector couplings of charged leptons determined in a fit with the following input data: M_Z, Γ_Z, σ_{had}^{pole}, R_e, R_μ, R_τ, $A_{FB}^{pole}(e^+e^-)$, $A_{FB}^{pole}(\mu^+\mu^-)$ and $A_{FB}^{pole}(\tau^+\tau^-)$ with their correlation matrix, together with $\mathcal{P}_\tau$ and $\mathcal{A}_e$ from $A_{FB}^{\mathcal{P}_\tau}$ and A_{LR}. Free parameters of the fit are also M_Z, Γ_Z and σ_{had}^{pole} constrained by their input values. Without the hypothesis of lepton universality we obtain $\chi^2/NDOF = 5.0/(11-9)$, and $\chi^2/NDOF = 6.5/(11-5)$ with this hypothesis.

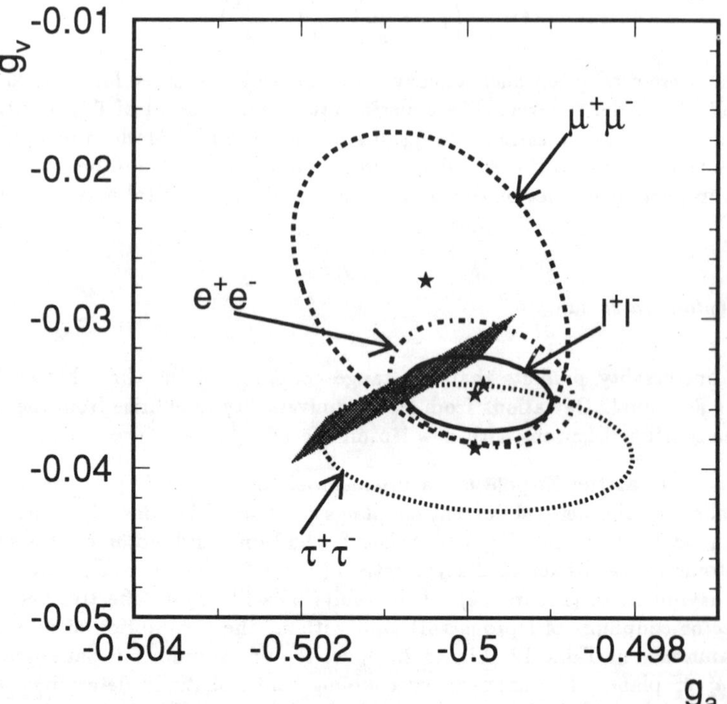

Figure 22: One standard deviation contours in the $\hat{g}_v^\ell$ - $\hat{g}_a^\ell$ plane. The shaded band represents the Standard Model prediction [60].

measurement of $\mathcal{P}_\tau$. The results are in very good agreement with the hypothesis of lepton universality in the neutral current sector.

Universality in charged current reactions between electron and muon has been established with high accuracy in low energy experiments. High statistics and modern detector technology enable LEP to make significant contributions in exploring this question for the τ lepton. Lepton universality for the τ lepton can be tested by comparing the decay widths for muons and tau leptons into an electron and two neutrinos:

$$\begin{aligned} \Gamma(\mu \to e\nu\nu) = \frac{1}{\tau_\mu} &= \frac{G_\mu^2}{192}\frac{m_\mu^5}{\pi^3}(1+\delta) \\ \Gamma(\tau \to e\nu\nu) = \frac{Br(\tau \to e\nu\nu)}{\tau_\tau} &= \frac{G_\tau^2}{192}\frac{m_\tau^5}{\pi^3}(1+\epsilon)\,. \end{aligned} \tag{17}$$

Here G_μ represents the Fermi constant and G_τ the equivalent for τ decays; τ_μ (τ_τ) and m_μ (m_τ) the lifetime and the mass of the muon (tau); δ and ϵ are small quantities arising from phase space and radiative corrections. Neglecting phase space and radiative corrections eqns. (17) result in:

$$\left(\frac{G_\tau}{G_\mu}\right)^2 = \left(\frac{\tau_\mu}{\tau_\tau}\right)\left(\frac{m_\mu}{m_\tau}\right)^5 Br(\tau \to e\nu\nu)\,. \tag{18}$$

Going from the Fermi theory to the Glashow-Salam-Weinberg model, $G_\mu \propto g_\mu g_e$ and $G_\tau \propto g_\tau g_e$, resulting in $(G_\tau/G_\mu) = (g_\tau/g_\mu)$. Earlier results [78] report a ratio of g_τ/g_μ which is consistent with unity at the two standard deviation level. Since then the accuracy of m_τ has been improved substantially by a measurement of BES [79] and new results on $Br(\tau \to e\nu\nu)$ [80, 81] became available from CLEO and LEP. Furthermore LEP contributed with a very accurate determination of the τ lifetime. The new measurements of the τ parameters result in [34]:

$$(g_\tau/g_\mu) = 0.987 \pm 0.006.$$

The accuracy on the ratio of coupling constants has improved by more than a factor of two and the result is again consistent with unity at the 2.2 standard deviation level.

4.3 *Combined Analysis of Quark and Lepton Couplings*

An exclusive separation of primary quark flavours in e^+e^- reactions at the Z^0-pole has been achieved so far only for b and c quarks. The determination of the neutral current couplings of quarks relies on the precise knowledge of the couplings of the electron in the initial state, as up to now there are no measurements of the final state polarization of quarks. Lepton universality being established to a high level of accuracy, we will impose this assumption in the following analysis. In a first step, we will combine the measurements used previously to determine the leptonic couplings with the partial widths and the forward-backward asymmetries of b and c quarks, respectively.

In the determination of effective vector and axial-vector couplings we encounter a technical problem: Partial widths as well as the forward-backward asymmetry at the peak are symmetric with respect to an interchange of $\hat{g}_v^f$ and $\hat{g}_a^f$. For the charged lepton couplings this problem no longer manifests itself as the precision of the leptonic data is high enough to allow the fit to find a stable local minimum. The choice between the local minima can be based on information from the Z^0-pole experiments alone, by considering the effect of the γZ^0-interference in the vicinity of the Z^0-pole (see subsection 4.4). The situation is different for quarks, especially for the b-quark, as $\hat{g}_v^b$ and $\hat{g}_a^b$ are close in value to each other as compared to the present experimental accuracy. We therefore perform a transformation to effective chiral couplings:

$$\hat{g}_L^f \equiv (\hat{g}_v^f + \hat{g}_a^f)/2 \quad (19)$$

$$\hat{g}_R^f \equiv (\hat{g}_v^f - \hat{g}_a^f)/2 \quad (20)$$

Expressing partial widths and forward-backward asymmetries at the Z^0-pole in terms of effective chiral couplings:

$$\Gamma_{f\bar{f}} \propto \hat{g}_L^{f\,2} + \hat{g}_R^{f\,2} \quad (21)$$

$$A_{FB}^{pole}(f\bar{f}) \propto \frac{\hat{g}_L^{f\,2} - \hat{g}_R^{f\,2}}{\hat{g}_L^{f\,2} + \hat{g}_R^{f\,2}}. \quad (22)$$

The measurement of the forward-backward asymmetry of quarks therefore allows a discrimination between $\hat{g}_L^{f\,2}$ and $\hat{g}_R^{f\,2}$ as soon as the data are accurate enough to determine the sign of the forward-backward asymmetries.

Table 13 gives the results of a fit to the effective chiral couplings of leptons and quarks. The experimental results are in good agreement with the Standard Model prediction, also given in Table 13. The comparison of the measurement errors for quark couplings with the uncertainty of the Standard Model prediction shows that the sensitivity of the measurements is not yet sufficient to probe radiative corrections to the final state quark couplings. This is because the quark forward-backward asymmetries depend more strongly on $\mathcal{A}_e$ than on $\mathcal{A}_q$.

Figure 23 displays the one standard deviation contours for the effective chiral couplings of charged leptons, c and b quarks. The input data do not constrain the sign of $\hat{g}_L^f$ and $\hat{g}_R^f$. For convenience we have chosen the solution corresponding to the Standard Model prediction, with the exception of the b quark, where the two mirror solutions $\pm\,\hat{g}_R^b$ do not form disjunct contours.

The fit also determines $\hat{g}_L^{\nu\,2} + \hat{g}_R^{\nu\,2}$. Within the Standard Model context $\hat{g}_R^{\nu\,2} = 0$ and the result can be interpreted as a precision determination of $\hat{g}_L^{\nu\,2}$. In the $\hat{g}_L^f$-$\hat{g}_R^f$

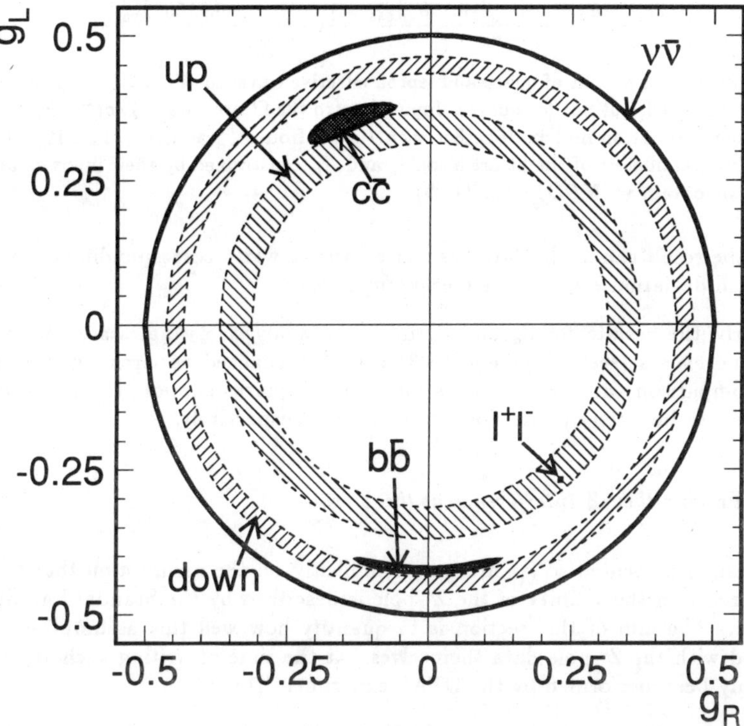

Figure 23: Effective chiral couplings from a combined analysis of hadronic and leptonic data. For leptons, b and c quarks the resulting one standard deviation contours are shown. These are compared to the measurements of the partial Z^0 widths $\Gamma_{\rm inv}$, $\Gamma_{d\bar{d}}$ and $\Gamma_{u\bar{u}}$, which are represented by circles in the $\hat{g}_L^f$ - $\hat{g}_R^f$ plane.

	Z^0-pole data	Standard Model [60]
$\hat{g}_L^\ell$	-0.2673 ± 0.0012	$-0.2675^{+0.0030}_{-0.0034}$
$\hat{g}_R^\ell$	0.2325 ± 0.0014	$0.2333^{+0.0016}_{-0.0020}$
$\hat{g}_L^{\nu\,2} + \hat{g}_R^{\nu\,2}$	0.2510 ± 0.0028	$0.2514^{+0.0015}_{-0.0013}$
$\hat{g}_L^{\mathrm{b}}$	$-0.420^{+0.016}_{-0.009}$	$-0.4209^{+0.0010}_{-0.0009}$
$\hat{g}_R^{\mathrm{b}}$	$0.061^{+0.065}_{-0.19}$	$0.0777^{+0.0006}_{-0.0006}$
$\hat{g}_L^{\mathrm{c}}$	0.349 ± 0.034	$0.3454^{+0.0028}_{-0.0024}$
$\hat{g}_R^{\mathrm{c}}$	$-0.158^{+0.096}_{-0.055}$	$-0.1555^{+0.0013}_{-0.0011}$

Table 13: Results of a fit of quark and lepton effective chiral couplings. For leptons universality is assumed, but not for quarks. Input data to the fit are: M_Z, Γ_Z, $\sigma^{\text{pole}}_{\text{had}}$, R_ℓ and $A^{\text{pole}}_{\text{FB}}$ with their correlation matrix, together with $\mathcal{P}_\tau$, $\mathcal{A}_e$ from $A^{\mathcal{P}_\tau}_{\text{FB}}$ and A_{LR}, $\Gamma_{\text{b}\bar{\text{b}}}$, $\Gamma_{\text{c}\bar{\text{c}}}$, $A^{\text{b}\bar{\text{b}}}_{\text{FB}}$ and $A^{\text{c}\bar{\text{c}}}_{\text{FB}}$. Free parameters of the fit are also M_Z and R_ℓ, constrained by their input values. For this fit we obtain $\chi^2/NDOF$=0.5/(11-9).

plane the result can be displayed as a circle with a width corresponding to the ±one standard deviation error of the measurement.

Similarly the results for C_d and C_u from the analysis of final state radiation in hadronic events (see subsection 3.3.3) can be visualized as circles in this plane. The comparison with the contours for b and c quarks support the hypothesis of universality between up and down type quarks, respectively.

4.4 *Probing the γZ-Interference at the Z^0-pole*

The results presented so far are based implicitly on the assumption that the γZ^0-interference in the vicinity of the Z^0-pole is described by the Standard Model prediction. The aim of this section is to quantify how well this assumption can be verified with the Z^0-pole data themselves. At the time of writing such an analysis has only been performed by the OPAL experiment [13, 14].

Figure 24 a) displays the Standard Model prediction for the fractional contribution of the γZ^0-interference to the total hadron and muon cross section. The contribution vanishes close to the Z^0 peak ‖ and varies by $\approx 0.5\%$ over the energy range which has been scanned by LEP so far.

The OPAL experiment has performed a fit to their hadronic cross sections, where the Standard Model contribution of the γZ^0-interference term has been multiplied by a scale factor which has been treated as an additional free parameter. They obtain for this scale factor -1.1 ± 3.6, which is in good agreement with unity. Note,

‖The γZ^0 interference contribution is nonzero for $s = M_Z^2$ due to both photonic and non-photonic radiative corrections.

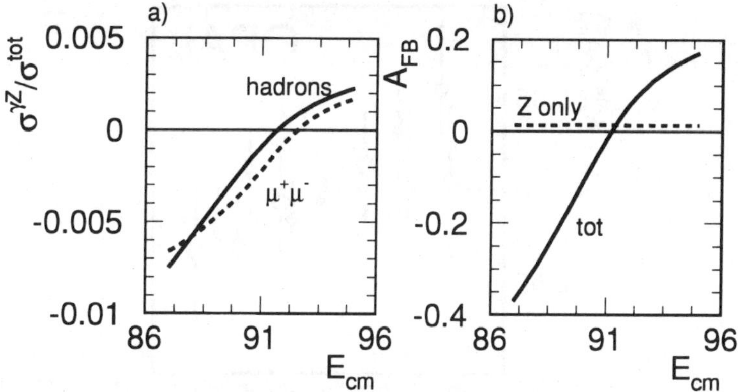

Figure 24: Standard Model prediction for the effect of the γZ^0-interference term.
a) The fractional contribution of the interference term to total cross sections for the processes $e^+e^- \to$ hadrons and $e^+e^- \to \mu^+\mu^-$.
b) Full Standard Model-prediction for the forward-backward asymmetry of the process $e^+e^- \to \mu^+\mu^-$ and the contribution arising from the Z^0 exchange term only.

however, that the detailed shape of the interference term is highly correlated with the value of M_Z. The above uncertainty on the scale factor, which is dominated by statistics, translates into an additional error of ≈ 10 MeV on M_Z.

Figure 24 b) compares the full Standard Model prediction of the forward-backward asymmetry for the process $e^+e^- \to \mu^+\mu^-$ with the contribution arising from Z^0-exchange only. Figure 24 demonstrates that the energy dependence of the leptonic forward-backward asymmetries probes the contribution to the asymmetry from the interference term and not from the pure Z^0 exchange term. The energy dependence of the forward-backward asymmetry, however, can be tested with high precision by the LEP experiments. Figure 25 shows the difference averaged over all 3 lepton species between measured forward-backward asymmetries and the Standard Model fit result.

The OPAL collaboration has also performed a fit to their measured hadronic and leptonic cross sections and leptonic forward-backward asymmetries parametrizing the differential cross section as:

$$\frac{2s}{\pi\alpha^2}\frac{d\sigma}{d\cos\theta}(e^+e^- \to \ell^+\ell^-) = \left|\frac{1}{1-\Delta\alpha}\right|^2 (1+\cos^2\theta) \tag{23}$$

$$+ 4\mathrm{Re}\left\{\frac{2}{1-\Delta\alpha}\chi(s)\left[C^s_{\gamma Z}(1+\cos^2\theta) + 2C^a_{\gamma Z}\cos\theta\right]\right\}$$

$$+ 16|\chi(s)|^2\left[C^s_{ZZ}(1+\cos^2\theta) + 8C^a_{ZZ}\cos\theta\right] .$$

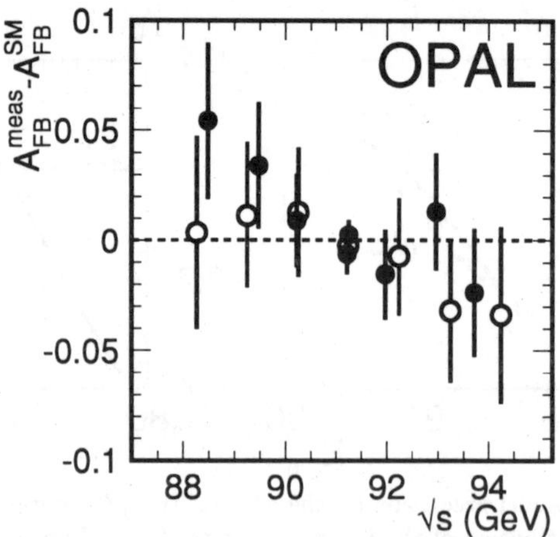

Figure 25: The difference averaged over all 3 leptonic species between the measured forward-backward asymmetries and the Standard Model fit result. Full and open symbols refer to the '91 and '90 data, respectively.

The propagator function $\chi(s)$ is defined as in eqn. (10). For the hadronic total cross sections the Z^0 exchange term has been parametrized by a Breit-Wigner with an s-dependent width, the γ exchange and the interference term have been calculated within the Standard Model framework. Besides M_Z, Γ_Z and σ_{had}^{pole} the four coefficients $C_{\gamma Z}^s$, $C_{\gamma Z}^a$, C_{ZZ}^a and C_{ZZ}^s are treated as free parameters, where the superscripts 's' and 'a' refer to terms symmetric and antisymmetric in $\cos\theta$. The coefficients C_{ZZ}^s and C_{ZZ}^a are essentially fixed by the high statistics data at the Z^0 peak. The coefficient $C_{\gamma Z}^s$ is sensitive to the energy dependence of cross sections and $C_{\gamma Z}^a$ to the energy dependence of the forward-backward asymmetries.

Comparing eqn. (9) and eqn. (23) the C-parameters can be expressed in terms of effective vector and axial-vector couplings. Any pair of C-parameters can be used to determine $\hat{g}_v^{\ell\,2}$ and $\hat{g}_a^{\ell\,2}$. Table 14 compares the values determined from the γZ^0-interference and from Z^0 exchange to the Standard Model prediction. All values of $\hat{g}_v^{\ell\,2}$ and $\hat{g}_a^{\ell\,2}$ in Table 14 are consistent with each other. The largest deviation is in the parameter $\hat{g}_a^{\ell\,2}$ determined from the γZ^0-interference, which differs by two standard deviations from the value resulting from Z^0 exchange terms only and the Standard Model prediction.

The values of $\hat{g}_v^{\ell\,2}$ and $\hat{g}_a^{\ell\,2}$ derived from the γZ^0-interference provide an experimental verification of the assignment $\hat{g}_a^{\ell\,2} \gg \hat{g}_v^{\ell\,2}$, which remains ambiguous if only data at $\sqrt{s} = M_Z$ are taken into account.

Determination Related to	Parameters Used	$\hat{g}_a^{\ell\,2}$	$\hat{g}_v^{\ell\,2}$
Z-exchange only	$C_{ZZ}^s \equiv (\hat{g}_a^{\ell\,2} + \hat{g}_v^{\ell\,2})^2$ $C_{ZZ}^a \equiv \hat{g}_a^{\ell\,2}\,\hat{g}_v^{\ell\,2}$	0.2500 ± 0.0017	0.00064 ± 0.00044
γZ-interference only	$C_{\gamma Z}^a \equiv \hat{g}_a^{\ell\,2}$ $C_{\gamma Z}^s \equiv \hat{g}_v^{\ell\,2}$	0.215 ± 0.018	0.013 ± 0.015
Standard Model prediction [60]		0.2508 ± 0.0014	$0.0011^{+0.0004}_{-0.0003}$

Table 14: Effective leptonic couplings, $\hat{g}_a^{\ell\,2}$ and $\hat{g}_v^{\ell\,2}$, derived from the C parameters of eqn. (23) determined by a fit to measured hadronic and leptonic cross sections and leptonic forward-backward asymmetries of the OPAL experiment as compared to the Standard Model prediction.

4.5 *Global Analysis of Z⁰-pole Data within the Standard Model Framework*

The data presented so far show no deviation from the prediction of the Standard Model. In this section we first want to show that all available data are consistent with a unique set of Standard Model parameters and then combine the data to constrain the model's unknown input parameters.

Table 15 summarizes the measurements included in the following analysis. Among the Z^0-pole data we have chosen only those measurements where the errors are comparable to the expected top mass variation and which are expected to improve in the future. The result from the Z^0-pole data will then be compared to further electroweak precision tests, which are also listed in Table 15.

In the Standard Model framework any observable X can be parametrized as:

$$X = f(\alpha, G_F, M_Z, M_t, M_H, \alpha_s) \tag{24}$$

Relation (24) can be used to map all observables onto a common scale, e.g. each observable X predicts a value of M_Z:

$$M_Z = g(\alpha, G_F, X, M_t, M_H, \alpha_s) \tag{25}$$

Figure 26 confronts the direct measurement of M_Z with the prediction for M_Z from three groups of observables: i) hadronic and leptonic cross sections and leptonic forward backward asymmetries; ii) polarization and quark forward-backward asymmetries; iii) the non-Z^0-pole precision measurements listed in Table 15. The resulting ± 1 standard deviation bands are shown as a function of M_t for M_H fixed to 300 GeV. The value of the strong coupling constant has been constrained to $\alpha_s(M_Z^2) = 0.120 \pm 0.006$, determined from an analysis [82] in $\mathcal{O}(\alpha_s^2)$ of event shapes,

measurement	result
a) Z^0-pole experiments	
line-shape and lepton asymmetries:	
M_Z	91.187 ± 0.007 GeV
Γ_Z	2.491 ± 0.007 GeV
σ_{had}^{pole}	41.33 ± 0.18 nb
R_ℓ	20.87 ± 0.07
A_{FB}^{pole}	0.0152 ± 0.0027
+ correlation matrix (Table 21)	
polarization asymmetries:	
$\mathcal{P}_\tau$, A_{LR}:	
$\mathcal{A}_\tau$	0.140 ± 0.018
$\mathcal{A}_e$	0.123 ± 0.025
quark asymmetries:	
$A_{FB}^{b\bar{b}}$	0.093 ± 0.012
$A_{FB}^{c\bar{c}}$	0.072 ± 0.027
topology of hadronic events:	
α_s [82]	0.120 ± 0.006
b) p$\bar{\text{p}}$, νN and PV in atomic transitions	
M_W [83, 84]	80.15 ± 0.27
$\sin^2\theta_W(\nu N)$ [85]	0.2283 ± 0.0052
Q_w [38]	71.04 ± 1.81

Table 15: Summary of measurements included in the combined analysis of Standard Model parameters. Section a) summarizes Z^0-pole averages, section b) electroweak precision tests from hadron colliders, νN-scattering and parity violation in atomic transitions.

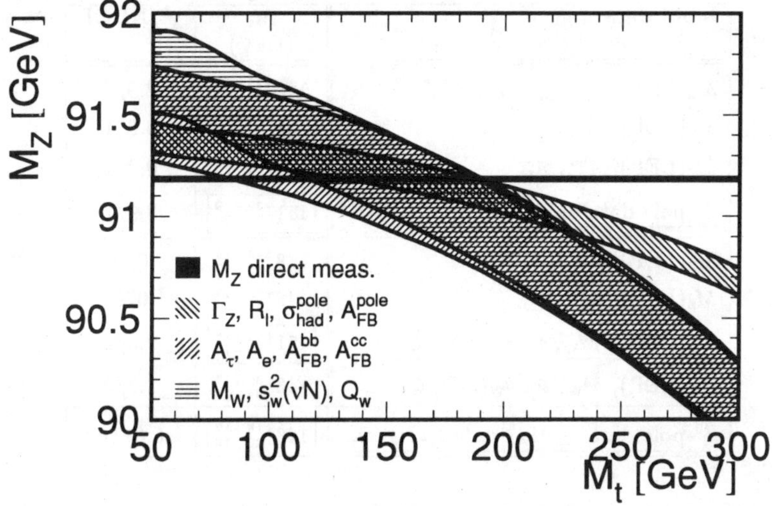

Figure 26: Comparison of the direct measurement of M_Z at LEP with indirect determinations from various observables assuming the Standard Model prediction. The bands display the ± 1 standard deviation variation when fixing the value of M_H to 300 GeV and allowing the strong coupling constant to vary within $\alpha_s(M_Z^2) = 0.120 \pm 0.006$.

jet rates and energy correlations at LEP and SLC. All bands overlap for a top mass ranging from 130–190 GeV.

The purpose of Figure 26 is to illustrate the consistency of measurements, but does not allow the derivation of confidence intervals for M_t. The one standard deviation confidence intervals for M_t resulting from fits to various subsamples of data are given in Table 16. As illustrated in Figure 26 a constraint on M_t needs a measurement of M_Z. Therefore Table 16 contains only two truly independent subsamples: the combination of Z^0-pole data and the combination of non-Z^0-pole experiments, if combined with $M_Z = 90.90 \pm 0.3 \pm 0.2$ as determined by the CDF experiment [86] at the Tevatron.

Having shown that all data are consistent with a unique set of Standard Model parameters we will present our final result for two sets of input data: Z^0-pole data only and the combination of all electroweak precision tests in Table 15. For these two input data sets we have performed two different fits: one with and one without the constraint on the strong coupling constant $\alpha_s(M_Z^2) = 0.120 \pm 0.006$. The results of

Input data	M_t (GeV)	χ^2/NDOF
M_Z(LEP), Γ_Z, σ_{had}^{pole}, R_ℓ, A_{FB}^{pole}	$146^{+26}_{-32}{}^{+18}_{-19}$	2.9/3
M_Z(LEP), $\mathcal{A}_\tau$, $\mathcal{A}_e$	$139^{+54}_{-83}{}^{+21}_{-30}$	0.3/1
M_Z(LEP), $A_{FB}^{b\bar{b}}$, $A_{FB}^{c\bar{c}}$	$189^{+48}_{-65}{}^{+15}_{-22}$	0.2/1
Z^0-pole data only	$151^{+22}_{-25}{}^{+19}_{-20}$	3.9/7
M_Z(LEP), $\sin^2\theta_W(\nu N)$	$138^{+43}_{-49}{}^{+15}_{-17}$	0/0
M_Z(LEP), M_W	$146^{+42}_{-49}{}^{+15}_{-16}$	0/0
M_Z(CDF), M_W, $\sin^2\theta_W(\nu N)$, Q_w	$151^{+44}_{-52}{}^{+14}_{-16}$	2.3/2
M_Z(LEP), M_W, $\sin^2\theta_W(\nu N)$, Q_w	$143^{+30}_{-34}{}^{+15}_{-17}$	1.7/2
Z^0-pole data + M_W, $\sin^2\theta_W(\nu N)$, Q_w	$148^{+18}_{-20}{}^{+17}_{-19}$	5.6/10

Table 16: Consistency checks for top mass fits from combined data sets. The central value and the 1st error of M_t refer to M_H = 300 GeV, fixed. The 2nd error gives the variation of the central value for Higgs mass values spanning the interval $60 < M_H(\text{GeV}) < 1000$. For all fits the value of α_s has been constrained to $\alpha_s(M_Z^2) = 0.120 \pm 0.006$.

these four fits are summarized in Table 17. Relaxing the constraint on α_s does not significantly alter the value of M_t. The fitted value of $\alpha_s(M_Z^2)$, which is determined essentially from the Z^0-pole cross-section data, is slightly higher but still consistent with the results derived from the analysis of event topologies and τ-decays [82]. The absolute χ^2 value for all fits is very good as anticipated by the consistency of the various data sets. The shape of the χ^2 curve as a function of M_t is shown in Figure 27 for the fit to all data with and without α_s constraint. The individual measurements presented in this report have been compared to a wide range of Standard Model parameters. The combined set of measurements, however, imposes a significant restriction on the Standard Model prediction as illustrated in Table 18 for selected observables. Note that the indirect determination of M_W from the Z^0-pole data is already more precise than the direct determination at hadron colliders.

In the preceding discussion we have assumed that we have no sensitivity to M_H by fixing the value of M_H to 300 GeV and specifying an additional error for all results from the variation of M_H in the interval $60 \leq M_H$ (GeV) ≤ 1000. This situation is close to reality as can be seen from Figure 28 displaying χ^2-curves as a function of M_H. With and without constraint on $\alpha_s(M_Z^2)$ the present data favour a Higgs mass which is at the boundary of exclusion limits from direct searches. The lower χ^2 curve originates from a simulation based on the current experimental accuracies and the Standard Model predictions for the data with M_H=50 GeV. The current

Input data	M_t [GeV]	$\alpha_s(M_Z^2)$	χ^2/NDOF
a) with α_s constraint			
Z^0-pole data	$151^{+22}_{-25}{}^{+19}_{-20}$		3.9/7
all data	$148^{+18}_{-20}{}^{+17}_{-19}$		5.6/10
b) without α_s constraint			
Z^0-pole data	$143^{+23}_{-28}{}^{+18}_{-20}$	$0.134 \pm 0.008 \pm 0.002$	2.0/6
all data	$144^{+19}_{-21}{}^{+17}_{-19}$	$0.134 \pm 0.008 \pm 0.002$	3.7/9

Table 17: Results of top mass fits to the combined data sets. The central value and the 1st error of M_t and $\alpha_s(M_Z^2)$ refer to $M_H = 300$ GeV, fixed. The 2nd error gives the variation of the central value for Higgs mass values spanning the interval $60 < M_H(\text{GeV}) < 1000$. For the fits in section a) the value of α_s has been constrained to $\alpha_s(M_Z^2) = 0.120 \pm 0.006$.

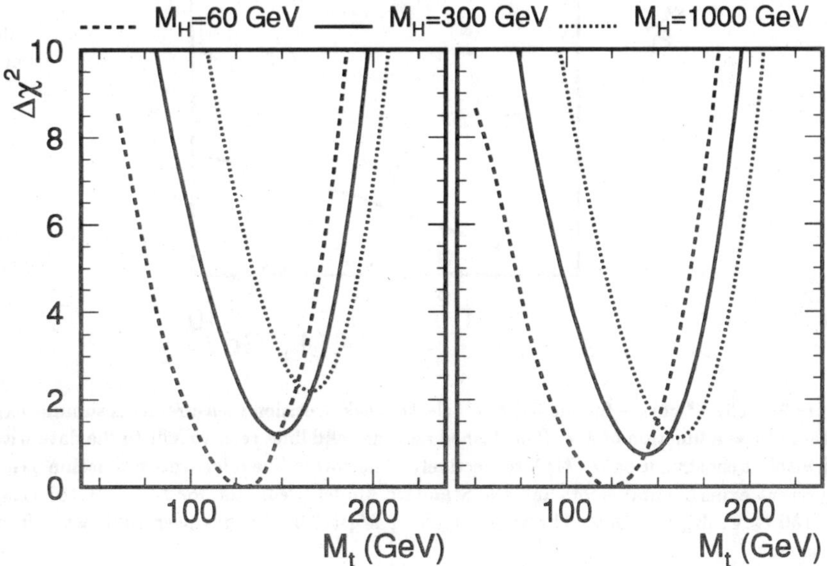

Figure 27: The χ^2 curves for the fits to all electroweak precision measurements summarized in Table 15 as a function of M_t for three different Higgs mass values spanning the interval $60 < M_H\,(\text{GeV}) < 1000$. The left Figure refers to a fit where the strong coupling constant has been constrained to $\alpha_s(M_Z^2) = 0.120 \pm 0.006$, the right Figure refers to a fit with $\alpha_s(M_Z^2)$ unconstrained. The minimum value of χ^2 from the $M_H = 60$ GeV curve has been subtracted in both cases.

	Z^0-pole data	all data
M_W [GeV]	$80.12 \pm 0.15 \pm 0.01$	$80.13 \pm 0.12 \pm 0.01$
$\sin^2\theta_W$	$0.2279^{+0.0031}_{-0.0027} \pm 0.0003$	$0.2278^{+0.0024}_{-0.0022} \pm 0.0003$
$\sin^2\theta_W{}^{lept}_{eff}$	$0.2332 \pm 0.0008 \pm 0.0002$	$0.2332 \pm 0.0006 \pm 0.0002$

Table 18: Derived quantities from the fits without α_s constraint above. The central values and the 1st errors refer to $M_H = 300$ GeV, fixed. The 2nd error gives the variation of the central value for Higgs mass values spanning the interval $60 < M_H(\text{GeV}) < 1000$.

steepness of the experimental χ^2 curves has to be attributed to the fact that the statistical fluctuations of the data around the Standard Model prediction are still large as compared to the expected effects due to a variation of M_H.

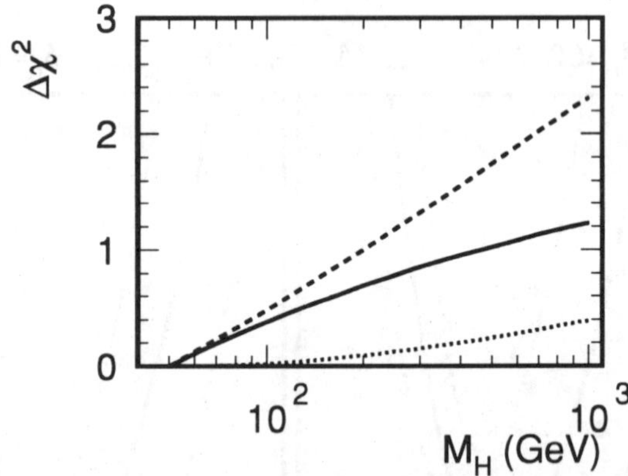

Figure 28: The χ^2 curves for the fits to all electroweak precision measurements summarized in Table 15 as a function of M_H. The dashed and the solid lines refer to a fit to the data with and without constraint on $\alpha_s(M_Z^2)$, respectively. The dotted line refers to a simulation based on present experimental errors and the Standard Model prediction for $M_Z = 91.187$ GeV, M_t=150 GeV, M_H=50 GeV. The value of χ^2 for $M_H = 50$ GeV has been subtracted from all curves.

5 Summary and Outlook

A wealth of electroweak precision tests has become available from the Z^0-pole experiments, some already reaching today an accuracy at the permille level. All measurements presented show good agreement among the experiments and with the Standard Model prediction. The data allow tests of lepton universality in neutral current reactions with an unprecedented precision. In the first running phase of LEP the measurement of the invisible width of the Z^0 has determined unambiguously the number of light neutrino species. Today this measurement constitutes a precision determination of the Z^0 coupling to neutrinos, allowing tight limits on exotic channels to be set. The measurement of production rates and forward-backward asymmetries of b and c quarks has reached a sensitivity which allows the separation of quark and lepton couplings. The comparison with measurements of final state radiation in hadronic events provides an experimental verification of universality between up and down type quarks. Contrary to lepton couplings, the accuracy of quark couplings is still low compared to the effects expected from radiative corrections in the Standard Model framework.

The combination of available measurements provides a significant constraint on the mass of the top quark. The best current estimate is:

$$M_t = 144^{+19}_{-21}{}^{+17}_{-19} \text{ GeV},$$

where the central value and the first error refer to a mass of the Higgs boson of $M_H = 300$ GeV, fixed, and the second error refers to Higgs masses spanning the interval $60 \leq M_H\,[\text{GeV}] \leq 1000$. The same set of electroweak precision tests provides simultaneously a determination of $\alpha_s(M_Z^2)$ which is complementary in all aspects of experimental and theoretical uncertainties to the methods used in traditional QCD analyses and of similar accuracy:

$$\alpha_s(M_Z^2) = 0.134 \pm 0.008 \pm 0.002,$$

the second error reflecting again the ignorance of M_H.

Up to the end of 1994 LEP is expected to increase its present statistics of $4.7 \cdot 10^6$ hadronic Z^0 events by a factor of two to three. This corresponds to an increase of statistics by an order of magnitude compared to this report, which is restricted to the analysis of data recorded up to the end of 1991. To match the statistics accumulated off peak to the progress in the LEP energy calibration, LEP will continue scanning, hoping to reach a precision of 2–3 MeV on the accuracy of both M_Z and Γ_Z. Note that the attempted relative precision on M_Z compares to that of the Fermi constant G_F. The initial success of the polarization program at SLC gives excellent prospects for complementary measurements testing the helicity structure of the neutral current. After 1994 LEP plans to run above the W-pair production threshold.

This will also be an interesting domain for direct particle searches, especially the search for the Higgs boson.

Up to now electroweak precision tests do not allow any significant indirect constraint to be placed on the mass of the Higgs boson. At the end of this report we wish to give a glimpse into the future (a detailed discussion has been presented in [87]). In Figure 29 we display the direct measurement of M_Z and the indirect determination of M_Z from all other electroweak precision tests available today, for fixed $M_H = 300$ GeV. This band as a function of M_t has been derived as discussed

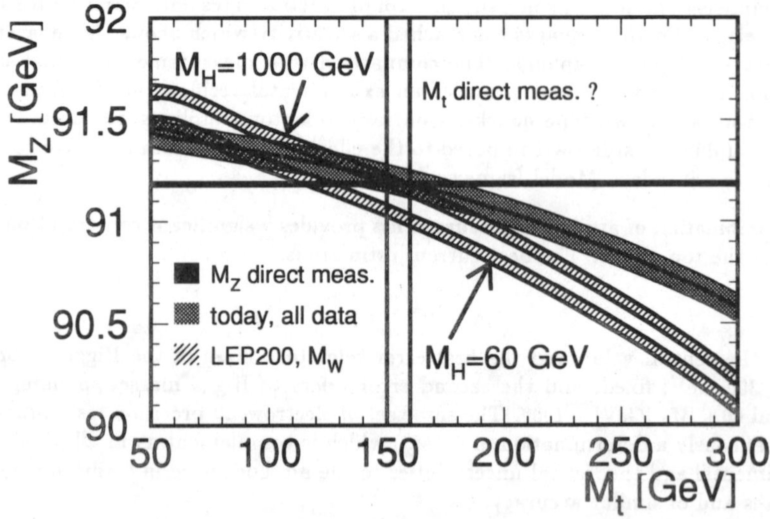

Figure 29: Comparison of the direct measurement of M_Z at LEP with an indirect determination from the sum of all other electroweak precision tests summarized in table 15, assuming the Standard Model. The band reflects the ± 1 standard deviation errors of the measurements and a variation of the strong coupling constant within $\alpha_s(M_Z^2) = 0.120 \pm 0.006$. For this band the value of M_H has been fixed to 300 GeV. Also shown are the bands which would result from a measurement of M_W at LEP 200 with an accuracy of 37 MeV for two different Higgs masses. The vertical band corresponds to a direct measurement of M_t with an accuracy of 5 GeV.

in subsection 4.5. A decrease of the width of this band by a factor of two within the next two years is probably a conservative estimate. It has been shown that a direct measurement of M_W at LEP200 could be performed with an accuracy of 37 MeV [88]. The indirect determination of M_Z resulting from such a measurement is also indicated in Figure 29 for $M_H = 60$ GeV and $M_H = 1000$ GeV. Finally, we expect a direct determination of the mass of the top quark with a precision of 10–5 GeV at the Tevatron or future hadron colliders, corresponding to the vertical band in Figure 29. At future e^+e^- colliders this precision might even be improved to 0.5 GeV.

The message of Figure 29 is twofold: Firstly, we will become sensitive to M_H and might produce interesting indirect constraints on M_H, provided a direct measurement of M_t becomes available. And last but not least: **There are chances to observe the break down of the Standard Model in electroweak precision tests!**

Acknowledgements
I would like to thank my colleagues from the LEP and SLC experiments for providing me with preliminary data and for interesting discussions. I am grateful to C. Hawkes for a critical reading of the manuscript and useful comments. I would also like to thank P. Langacker for encouraging me to write this article and many suggestions.

6 References

1. G.S. Abrams et al., Mark II Coll., Phys. Rev. Lett. **63** (1989) 724.
2. A. Sokolov and I.M. Ternov, Sov. Phys. Doklady **8** (1964) 1203.
3. Polarization in e^+e^- Annihilation, A. Blondel, in this volume.
4. G.S. Abrams et al., Mark II Coll., Nucl. Instr. and Meth. **A281** (1989) 55.
5. The SLD Design Report, SLAC Report 273, 1984.
6. D. Decamp et al., ALEPH Coll., Nucl. Instr. and Meth. **A294** (1990) 121.
7. P. Aarnio et al., DELPHI Coll., Nucl. Instr. and Meth. **A303** (1991) 233.
8. B.Adeva et al., L3 Coll., Nucl. Instr. and Meth. **A289** (1990) 35.
9. K. Ahmet et al., OPAL Col., Nucl. Instr. and Meth. **A305** (1991) 275.
10. D. Decamp et al., ALEPH Coll., Z. Phys. **C53** (1992) 1.
11. P. Abreu et al., DELPHI Coll., Nucl. Phys. **B367** (1991) 511.
12. B. Adeva et al., L3 Coll., Z. Phys. **C51** (1991) 179.
13. G. Alexander et al., OPAL Coll., Z. Phys. **C52** (1991) 175.
14. Precision Measurements of the Neutral Current from Hadron and Lepton Production at LEP, P.D. Acton et al., OPAL Coll., CERN-PPE-93-03, 13 January 1993.
15. Electroweak Results from LEP, R. Tanaka, presented at the XXVI International Conference on High Energy Physics, Dallas, August 6-12, 1992; X-LPNHE/92-3.
16. L3 preliminary results for the Z^0 line shape and leptonic forward-backward

asymmetry, Dec. 1992; LEP electroweak results, T. Kawamoto, presented at the Aspen Winter Conference in Particle Physics, 11 January 1993.

17. W. Beenakker and B. Pietrzyk, Phys. Lett. **B296** (1992) 244.
18. L. Knudsen et al., Phys. Lett. **B270** (1991) 97.
19. Measurement of LEP Beam Energy by Resonant Spin Depolarization, L. Arnaudon et al., CERN-PPE/92-49.
20. The Energy Calibration of LEP in 1991, L. Arnaudon et al., CERN-PPE/92-125 and CERN-SL/92-37(DI).
21. Proceedings of the Workshop on Z Physics at LEP1, CERN 89-08, Sept. 1989, ed. G.Altarelli et al..
22. Z line shape, F. Berends et al. in [21], Vol. 1, p. 89.
23. Monte Carlo for electroweak physics, R. Kleiss et al. in [21], Vol. 3, p. 1.
24. W. Beenakker, F.A. Berends and S.C. Van der Marck, Nucl. Phys. **B349**, 323(1991).
25. S. Jadach, E. Richter-Was, B.F.L. Ward, Z. Was, Phys. Lett. **B253** (1991) 469;
    S. Jadach, E. Richter-Was, B.F.L. Ward, Z. Was, Phys. Lett. **B260** (1991) 438;
    W. Beenakker, F.A. Berends, S.C. van der Marck, Nucl. Phys. **B355** (1991) 281.
26. H. Burkhardt et al., Z. Phys. **C43** (1989) 497.
27. S.G. Gorishny, A.L. Kataev and S.A. Larin, Phys. Lett. **B259** (1991) 144.;
    L.R. Surguladze and M.A. Samuel, Phys. Rev. Lett. **66** (1991) 560.
28. K.G. Chetyrkin and J.H. Kühn, Phys. Lett. **B248** (1990) 359;
    K.G. Chetyrkin, J.H. Kühn and A.Kwiatkowski, Phys. Lett. **B282** (1992) 221.
29. D. Bardin et al., Z. Phys. **C44** (1989) 493; Nucl. Phys. **B351** (1991) 1; Phys. Lett. **B229** (1989) 405; Phys. Lett. **B255** (1991) 290;
    ZFITTER, an Analytical Program for Fermion Pair Production in e^+e^- Annihilation, D. Bardin et al., CERN-TH 6443/92 (May 1992).
30. M. Martinez et al., Z. Phys. **C49** (1991) 645.
31. Electroweak radiative corrections for Z Physics, M. Consoli, W. Hollik and F. Jegerlehner in [21], Vol. 1, p. 7.
32. Δr, or the relation between electroweak couplings and the weak vector boson masses, G. Burgers et al., in [21], Vol. 1, p. 55.
33. The LEP Collaborations: ALEPH, DELPHI, L3 and OPAL, Phys. Lett. **B276** (1992) 247.
34. Precision Tests of the Electroweak Interaction, L. Rolandi, presented at the XXVI International Conference on High Energy Physics, Dallas, August 6-12, 1992.
35. C.Y. Prescot et al., Phys. Lett. **B77** (1978) 347 and **B84** (1979) 524.

36. D. Bardin et al., Nucl. Phys. **B351** (1991) 1; Z. Phys. **C44** (1989) 493.
37. M. Roney, presented at the XXVI International Conference on High Energy Physics, Dallas, August 6-12, 1992.
38. Particle Data Group, Phys. Rev. **D45** (1992) 1.
39. D. Decamp et al., ALEPH Coll., Phys. Lett. **276B** (1992) 247; H. Albrecht et al., ARGUS Coll., DESY 92-125, Sept. 1992.
40. A. Rouge, Z. Phys. **C48** (1990) 75.
41. A Study of the Decays of τ Leptons Produced on the Z^0 Resonance at LEP, P. Abreu et al., DELPHI Coll., CERN/PPE 92-060 (15 April 92).
42. A Measurement of tau Polarisation in Z^0 Decays, O. Adriani et al., L3 Coll., CERN-PPE/92-132 (08 August 1992).
43. Measurement of τ Polarization in Z^0 Decays, K.S. Kumar, presented at the XXVI International Conference on High Energy Physics, Dallas, August 6-12, 1992.
44. The SLD Coll., presented by B.A. Schumm at the DPF, 7th Meeting of the American Physical Society, 10-14 November 92, SLAC-PUB-5995.
45. T. Maruyama et al., Phys. Rev. **B46** (1992) 4261.
46. Measuring the weak Isospin of b Quarks. D. Schaile and P.M. Zerwas, Phys. Rev. **D45** (1992) 3262.
47. B. Adeva et al., L3 Coll., Phys. Lett. **B261** (1991) 177.
48. ALEPH Coll., preliminary result, presented by W. Venus at the Conference on Results and Perspectives in Particle Physics, La Tuile, March 92.
49. P. Abreu et al., DELPHI Coll., Phys. Lett. **B281** (1992) 383.
50. A Novel Method to Measure $\Gamma_{b\bar{b}}/\Gamma_{had}$, C. Moisan and S. de Jong, presented at the XXVI International Conference on High Energy Physics, Dallas, August 6-12, 1992.
51. D. Decamp et al., ALEPH Coll., Phys. Lett. **B258** (1991) 237; B. Adeva et al., L3 Coll., Phys. Lett. **B288** (1992) 395; B. Adeva et al., L3 Coll., Phys. Lett. **B252** (1990) 703; P.D. Acton et al., OPAL Coll., Phys. Lett. **B276** (1992) 379.
52. D. Buskulic et al., ALEPH Coll., Phys. Lett. **B284** (1992) 177.
53. Measurement of the $e^+e^- \to b\bar{b}$ and $e^+e^- \to c\bar{c}$ Forward-Backward Asymmetries at the Z^0 Resonance, O. Adriani et al., L3 Coll., CERN-PPE/92-121 (21 July 1992).
54. ALEPH, DELPHI and OPAL Colls., preliminary results contributed to the XXVI International Conference on High Energy Physics, Dallas, August 6-12, 1992.
55. Measurements at LEP of the Forward-Backward Asymmetries of Quarks, T. Wyatt, presented at the XXVI International Conference on High Energy Physics, Dallas, August 6-12, 1992.

56. Presented at the XII International Conference on Physics in Collision, Boulder, June 1992.
57. D. Decamp et al., ALEPH Coll., Phys. Lett. **B263** (1991) 325.
58. M.Z. Akrawy et al., OPAL Coll., Phys. Lett. **B263** (1991) 311.
59. P. Abreu et al., DELPHI Coll., Phys. Lett. **B252** (1990) 140.
60. In this report we quote the Standard Model prediction based on a calculation of [29]. If not stated otherwise the central value refers to $M_Z = 91.187$ GeV, $M_t = 150$ GeV, $M_H = 300$ GeV and $\alpha_s(M_Z^2) = 0.12$ and the quoted uncertainty allows for variations $90 \leq M_t$ [GeV] ≤ 200, $60 \leq M_H$ [GeV] ≤ 1000 and $0.114 \leq \alpha_s(M_Z^2) \leq 0.126$.
61. Classification of the Hadronic Decays of the Z^0 into b and c Quark Pairs using a Neural Network, P. Abreu et al., DELPHI Coll, CERN/PPE 92-151 (15 September 1992).
62. D. Decamp et al., ALEPH Coll., Physics Letters, **B266** (1991) 218; preliminary results ALEPH, DELPHI and OPAL Colls., contributed to the XXVI International Conference on High Energy Physics, Dallas, August 6-12, 1992.
63. M.Z. Akrawy et al., OPAL Coll., Physics Lett. **B246** (1990) 285.
64. G. Alexander et al., OPAL Coll., Phys. Lett. **B264** (1991) 219.
65. D. Decamp et al., ALEPH Coll., Phys. Lett. **B264** (1991) 476.
66. P. Abreu et al., DELPHI Coll., Z. Phys. **C53** (1992) 555.
67. Isolated Hard Photon Emission in Hadronic Z^0 Decays, O. Adriani et al., L3 Coll., CERN-PPE/92-131 (30 July 1992).
68. Determination of Quark Electroweak Couplings from Direct Photon Production in Hadronic Z Decays, O. Adriani et al., L3 Coll., CERN-PPE/92-203 (4 December 1992).
69. P.D. Acton et al., OPAL Coll., Z. Phys. **C54** (1992) 193.
70. Measurement of Prompt Photon Production in Hadronic Z Decays, D. Buskulic et al., ALEPH Coll., CERN-PPE/92-143 (31 August 1992)
71. Studies of Strong and Elektroweak Interactions Using Final State Photon Emission in Hadronic Z^0 Decays, P.D. Acton et al., OPAL Coll., CERN-PPE/92-215 (18 December 1992).
72. P. Abreu et al., DELPHI Coll., Phys. Lett. **B277** (1992) 371.
73. P.D. Acton et al., OPAL Coll., Phys. Lett. **B294** (1992) 436.
74. D. Denegri, B. Sadoulet and M. Spiro, Rev. Mod. Phys. **62** (1990) 1.
75. D. Decamp et al., ALEPH Coll., Phys. Lett. **B231** (1989) 519. P. Aarnio et al., DELPHI Coll., Phys. Lett. **B231** (1989) 539. B.Adeva et al., L3 Coll., Phys. Lett. **B231** (1989) 509. M.Z. Akrawy et al., OPAL Coll., Phys. Lett. **B231** (1989) 530.
76. B. Adeva et al., L3 Coll., Phys. Lett. **B275** (1992) 209.

77. M.Z. Akrawy et al., OPAL Coll., Z. Phys. **C50** (1991) 373.
78. Heavy Flavour Physics, M. Danilov, Proceedings of the Joint International Lepton-Photon Symposium and Europhysics Conference on High Energy Physics, Geneva, 25th July - 1st August 1991, ed. S. Hegarty, K. Potter and E. Quercigh.
79. Measurement of the mass of the τ lepton, J.Z. Bai et al., SLAC-PUB 5870 and BEBC-EP-92-01.
80. τ Physics at CLEO, K. Gan, presented at the XXVI International Conference on High Energy Physics, Dallas, August 6-12, 1992.
81. Lifetime and Branching Ratios of the τ Lepton, M. McCubbin, presented at the XXVI International Conference on High Energy Physics, Dallas, August 6-12, 1992.
82. Tests of QCD, S. Bethke, presented at the XXVI International Conference on High Energy Physics, Dallas, August 6-12, 1992; HD-PY 92/13, Oct. 1992.
83. J. Alitti et al., UA2 Coll., Phys. Lett. **B276** (1992) 354.
84. F. Abe et al., CDF Coll., Phys. Rev. Lett. **65** (1990) 2243; Phys. Rev. **D43** (1991) 2070.
85. A. Blondel et al., CDHS Coll., Z. Phys. **C45** (1990) 361;
J.V. Allaby et al., CHARM Coll., Z. Phys. **C36** (1987) 611;
Results from the CCFR Collaboration, T. Bolton, presented at the XXVI International Conference on High Energy Physics, Dallas, August 6-12, 1992.
86. F. Abe et al., CDF Coll., Phys. Rev. Lett. **63** (1989) 720.
87. D. Schaile, Z. Phys. **C54** (1992) 387.
88. The Measurement of M_W at LEP200, L. Camilleri, report of the W-mass working group to the LEPC, 2-4 November 1992.

Appendix: Correlation Matrices

Parameter	1	2
1 M_Z	6^2	-3^2
2 Γ_Z	-3^2	4^2

Table 19: Covariance matrix elements originating from uncertainties in the centre-of-mass energy, in MeV^2.

Parameter	1	2	3	4	5	6	7	8	9
1 M_Z	1.000	.019	.012	−.002	−.002	−.151	.015	.036	.024
2 σ_{had}^{pole}	.019	1.000	.063	.094	.074	−.171	.014	−.004	−.002
3 R_e	.012	.063	1.000	.076	.051	−.005	−.027	.004	.003
4 R_μ	−.002	.094	.076	1.000	.053	.013	.000	.010	−.001
5 R_τ	−.002	.074	.051	.053	1.000	.013	.000	−.001	.012
6 Γ_Z	−.151	−.171	−.005	.013	.013	1.000	.002	.014	.001
7 $A_{FB}^{pole}(e^+e^-)$	.015	.014	−.027	.000	.000	.002	1.000	.008	.014
8 $A_{FB}^{pole}(\mu^+\mu^-)$	.036	−.004	.004	.010	−.001	.014	.008	1.000	.008
9 $A_{FB}^{pole}(\tau^+\tau^-)$	.024	−.002	.003	−.001	.012	.001	.014	.008	1.000

Table 20: The parameter correlation matrix for the standard LEP parametrization of hadronic and leptonic cross sections and leptonic forward-backward asymmetries without assuming lepton universality. The results of this fit are summarized in Table 4.

Parameter	1	2	3	4	5
1 M_Z	1.000	.019	.002	−.157	.041
2 σ_{had}^{pole}	.019	1.000	.129	−.169	.004
3 R_ℓ	.002	.129	1.000	.004	.003
4 Γ_Z	−.157	−.169	.004	1.000	.000
5 A_{FB}^{pole}	.041	.004	.003	.000	1.000

Table 21: The parameter correlation matrix of the LEP average for the standard LEP parametrization of hadronic and leptonic cross sections and leptonic forward-backward asymmetries assuming lepton universality. The results of this fit are summarized in Table 5.

BEAM POLARIZATION IN e^+e^- ANNIHILATION

Alain Blondel
L.P.N.H.E., Ecole Polytechnique, Palaiseau, 91128 France

Contents

1 Introduction

This article will address two types of beam polarization experiments: firstly, transverse beam polarization, which is the natural polarization that occurs in storage rings; secondly, longitudinal polarization, that can be obtained directly from a source of longitudinally polarized electrons in a linear accelerator, or by spin rotation of the transverse polarization in a storage ring.

The relevance of beam polarization for precision measurements comes from the possibility of changing one – important – property of the initial state, the spin, without modifying any other. The resulting improvement in systematic errors is often decisive when precision experiments are performed.

In the particular case of e^+e^- annihilation at the Z pole, polarization plays a principal role for several important observables:

- Transverse polarization is precious for the beam energy calibration by resonant depolarization, which contributes sizeably to the reduction of systematic errors on the Z mass M_Z and width Γ_Z. In the future, it will also be useful in the determination of the W mass.

- Longitudinal polarization allows a very precise determination of the effective weak mixing angle $\sin^2\theta_w^{\rm eff}$ via the measurement of the Left-Right Polarization asymmetry, and a precise determination of the weak couplings of quarks by the Forward-Backward polarized asymmetries.

The combination of these observables, in particular Γ_Z and $\sin^2\theta_w^{\rm eff}$, allows bounds to be placed on the top quark mass and on the Higgs boson mass – or on whatever is playing its role. It can be argued that the improvement in precision brought about by polarized beams is the best way to make this game meaningful.

I will first discuss how one measures beam polarization, as it is an essential ingredient in both types of experiments. Then I will describe the observation of transverse beam polarization in the Large Electron Positron collider (LEP) at CERN and other storage rings, and its application for beam energy calibration, with present (M_Z) and future (Γ_Z, M_W) results. The longitudinal polarization experiments will be described next, with the first important results coming from the Stanford Linear Collider (SLC). The possibility of longitudinal polarization experiments at LEP will also be addressed. I will conclude with remarks on the potential physics output from a successful programme.

2 Beam Polarization Measurement Techniques

Two main techniques are used to measure the polarization of electron beams. Møller scattering of the beam on atomic electrons, and Compton scattering on polarized photons, usually produced by a laser. In both cases the spin dependent cross-sections are isolated by varying the relevant polarization of the target.

2.1 *Møller scattering*

The Møller polarimeters make use of the polarization asymmetries of the electron-electron scattering cross-section. At tree level, and in the limit $m_e = 0$, the differential cross-section for this process is:

$$\frac{d\sigma}{d\Omega} = \frac{\alpha^2}{s}\frac{(3+\cos^2\theta)^2}{\sin^4\theta}\left[1 - P_z^1 P_z^2 A_z(\theta) - P_t^1 P_t^2 A_t(\theta)\cos(2\phi - \phi_1 - \phi_2)\right] \quad (1)$$

where s is the center-of-mass energy squared; θ and ϕ are the center-of-mass angles of the scattered electron ($\phi = 0$ is arbitrary); P_z^1, P_z^2 (resp. P_t^1, P_t^2) are the longitudinal (resp. transverse) polarizations of the incident beam (1) and of the target (2); ϕ_1, ϕ_2 are the azimuths of the transverse polarization vectors; the longitudinal and transverse asymmetry functions $A_z(\theta), A_t(\theta)$ determine the analyzing power of the polarimeter, and are given by:

$$A_z(\theta) = \frac{(7+\cos^2\theta)\sin^2\theta}{(3+\cos^2\theta)^2} \quad (2)$$

$$A_t(\theta) = \frac{\sin^4\theta}{(3+\cos^2\theta)^2} \quad (3)$$

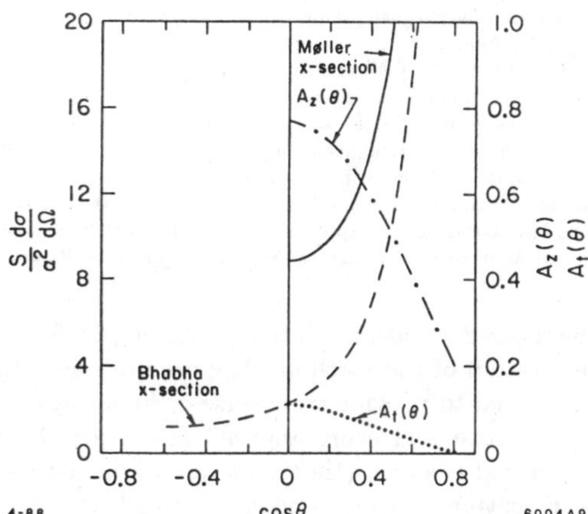

Figure 1:
Unpolarized differential cross-sections for Møller and Bhabha scattering as a function of the cm scattering angle. The longitudinal and transverse asymmetries for both processes are also shown. From [2].

The values of the cross-section and analyzing power as a function of θ are shown in figure 1. Both asymmetry functions are maximal for 90° scattering, and decrease rapidly for small angles.

In a practical design such as that of the SLC [1, 2], the target is a thin iron foil, placed in the beam line. The interesting scattering occurs on the two external electrons (out of 26) of the iron atoms, that can be fully polarized by submitting the foil to an external magnetic field (around 100 gauss) parallel to the foils axis. The maximum polarization of the iron electrons is thus 8%. The target polarization can

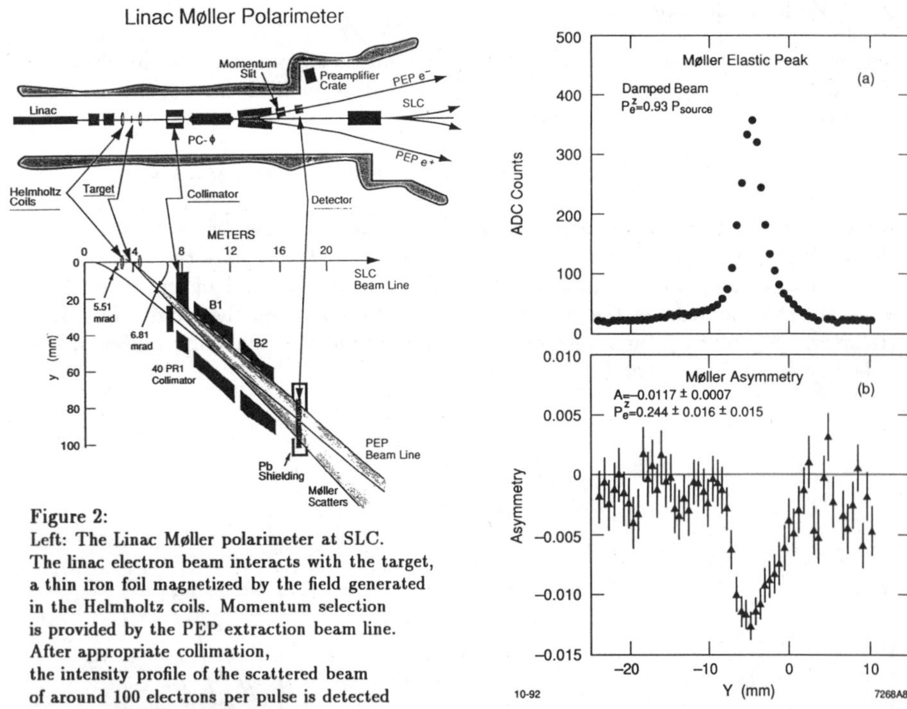

Figure 2:
Left: The Linac Møller polarimeter at SLC. The linac electron beam interacts with the target, a thin iron foil magnetized by the field generated in the Helmholtz coils. Momentum selection is provided by the PEP extraction beam line. After appropriate collimation, the intensity profile of the scattered beam of around 100 electrons per pulse is detected in an array of silicon strip detectors.
Right: Measurement of the beam polarization at the end of the SLC linac with the Møller polarimeter. a) the Møller elastic peak appears at the angle given by two-body kinematics. b) the asymmetry upon beam polarization reversal measures the polarization to be $\mathbf{P}_z = 0.244 \pm 0.016 \pm 0.015$. From [3].

be measured with a relative precision of a few percent. This constitutes the present limitation of the method. The orientation of the foil with respect to the beam can be varied to provide transverse or longitudinal polarization.

The scattering angles θ and ϕ are defined by momentum selection on the scattered electron (there is a one-to-one correspondence $P' = P/2(1+\cos\theta)$ between the scattered electron momentum P' and θ) followed by azimuth selection. The Møller scattering is identified from the background of elastic and quasi-elastic nuclear interactions by the two-body kinematics, that fixes the laboratory scattering angle once the momentum is fixed: $\theta_{lab} = 2m_e(1/P' - 1/P)$. The extraction line Møller polarimeter of the SLC is shown in figure 2. The extraction line of SLC to PEP is used as momentum spectrometer at 15 GeV, corresponding to a center-of-mass angle of 110 degrees, and the scattered electrons are detected in an array of silicon strip detectors. The Møller scattering signal and asymmetry upon reversal of the beam polarization are shown in figure 2. A few hundred scatters occur at each beam passage, allowing a determination of the beam polarization to a few % per minute.

The main advantage of Møller polarimeter is that, in principle, it allows full determination of the polarization vector with a unique set-up. Its limitation, however, lies in understanding the target polarization. Also, the statistics of scatters are limited by the maximum thickness of the target.

2.2 *Compton scattering*

The use of a polarized laser beam as polarized target offers several advantages, as powerful beams with light polarization of nearly 100% are readily available, with practical spin reversal.

The differential polarized compton scattering cross-section is given by [4, 5]:

$$\frac{d\sigma}{d\Omega} = \frac{1}{2}(r_e \frac{q}{q_0})^2 \left(\Phi_0 + \Phi_1 + \Phi_2\right) \tag{4}$$

where q_0 is the incident photon energy in the *electron* center-of-mass system, q the outgoing photon energy and r_e the classical electron radius. One has

$$\Phi_0 = (1 + \cos^2\theta) + (q_0 - q)(1 - \cos\theta) \tag{5}$$

$$\Phi_1 = (\xi_1 \cos 2\phi + \xi_2 \sin 2\phi) \sin^2\theta \tag{6}$$

$$\Phi_2 = \xi_3(1 - \cos\theta)\vec{\zeta}[\vec{q_0} \cos\theta + \vec{q}] \tag{7}$$

where q, θ, ϕ are the photon momentum and scattering angles (in the electron center-of-mass) ; $\vec{\zeta} = (\zeta_1, \zeta_2, \zeta_3)$ is the electron polarization vector and $\vec{\xi} = (\xi_1, \xi_2, \xi_3)$ is the photon polarization vector [6] represented on the Pointcarré sphere, namely: $\xi_3 = \pm 1$ corresponds to right- and left-circular polarization; $\xi_1 = \pm 1$ corresponds to linear polarization along the x and y axes respectively; $\xi_2 = \pm 1$ corresponds to linear polarization at $\pm 45°$ to the x axis in the x-y plane. The collision is assumed here to take place head-on, with the incident electron moving in the +z direction. The term Φ_0 corresponds to the total unpolarized cross-section. Φ_1 corresponds to effects dependent on the photon linear polarization and can be useful for calibration purposes – or be a nuisance for systematic errors. Finally Φ_2 is the term of interest for polarimetry, as it depends on the product of the electron and photon polarizations. One can see that only ξ_3 enters in Φ_2, so that *circular* photon polarization is used in electron beam polarimetry.

The kinematics of Compton scattering of 45 GeV electrons off laser photons is rather unusual. Both scattering products are swept forward at very small angles to the incident electron beam direction. The relation between the center-of-mass scattering angle θ and the laboratory angle and energy of the recoil photon is shown in figure 3, for the particular parameters of the LEP polarimeter, with a green laser ($\lambda = 530$ nm). High energy photons from 0 to 30 GeV are produced at very small angle, $\theta_{lab} = 1/\gamma \simeq 10^{-5}$ radians for 90° scattering.

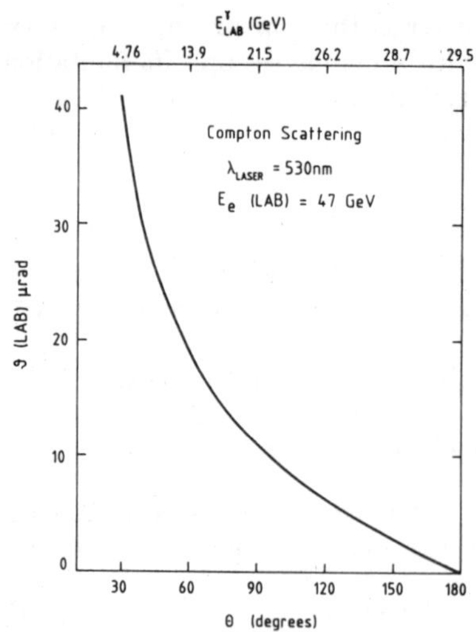

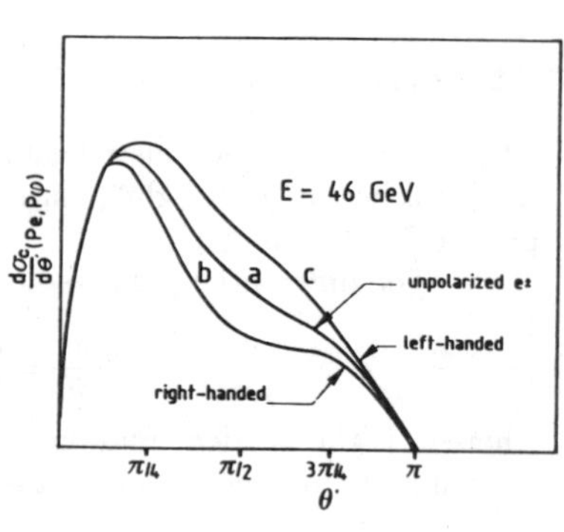

Figure 3: Compton kinematics:
a) the relation between the electron center-of-mass scattering angle and the photon energy and angle in the lab. frame; (from [7])
b) Spin dependent Compton cross-section vs scattering angle for transversely polarized electron beam. Line a shows the unpolarized case, while lines b and c show the right- and left-handed circular light on fully polarized electrons. (From [17]).

2.2.1 *Transverse polarization measurement*

Originally suggested by Baier and Khoze [8], the laser polarimeter has become a standard equipment in e^+e^- storage rings. The first compton polarimeter worked at SPEAR [9] followed by similar devices at VEPP2 and VEPP4 in Novosibirsk (U.S.S.R.) [10, 11, 12], DORIS in Hamburg (Germany) [13], CESR in Cornell (U.S.A) [14], HERA [15] and TRISTAN [16]. The LEP polarimeter [17] is of similar design.

Transverse electron polarization corresponds to $\vec{\zeta} = (0, \mathbf{P}_e, 0)$ illuminated with circular polarized light $\vec{\xi} = (0, 0, \pm\xi)$. The asymmetry upon reversal of the light circular polarization from ξ to $-\xi$ is given by:

$$A(\mathbf{P}_e, \xi) = \frac{d\sigma/d\Omega_R - d\sigma/d\Omega_L}{d\sigma/d\Omega_R + d\sigma/d\Omega_L} \tag{8}$$

$$= \frac{\Phi_2}{\Phi_0} = \mathbf{P}_e \xi \sin\phi F(\theta, q_0) \tag{9}$$

It materializes as a ϕ asymmetry, resulting in an up-down asymmetry. The function F is shown in figure 3b; it represents the analyzing power, maximum for $\cos\theta = 0$.

In a practical design, a powerful laser is shined at the electron beam at low angle. The wave length chosen is a compromise between practicality and the analyzing power which increases with the photon energy. A wave length of 532 nm is used at LEP, and obtained from a frequency doubled Yag-Nd pulsed laser.

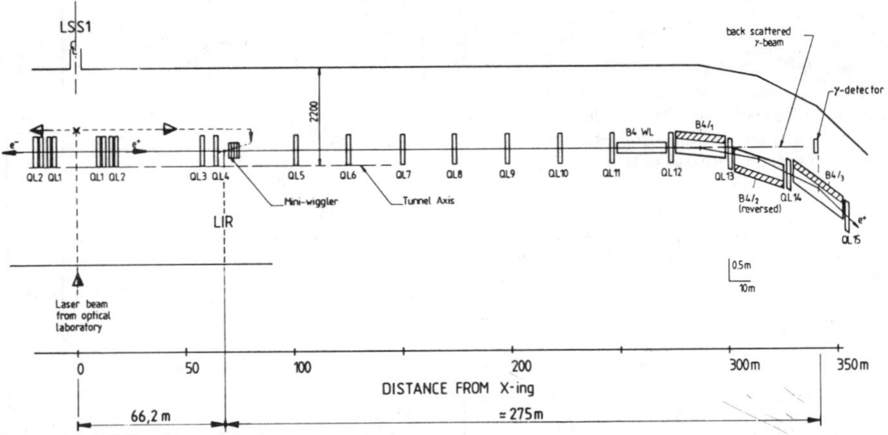

Figure 4: Set-up of the LEP polarimeter in LEP straight section 1. The laser beam is guided to the Laser Interaction Region (LIR) where it interacts with an angle of 3 mrad. The backscattered photons are separated from the electron beam 254 meters downstream at the beginning of the arc, and detected in a silicon strip calorimeter. (From [17]).

Because of the very small diffusion angle, a very long lever arm is needed to make the effect measurable. The LEP polarimeter set-up is shown in figure 4.

Two modes of operation can be envisaged for a Compton polarimeter.

– The ***single photon*** mode is used in HERA, where a low power laser continuously shines the electron beam. The backscattered photons are detected one by one, and a full analysis of the photon angular distribution versus its energy can be performed [18]. In particular the up-down asymmetry can be measured as a function of the photon energy, figure 5. The difficulty of the single photon method is that the background from beam gas bremsstrahlung is far from negligible. It can be tolerated in HERA with a short (29 m) specially evacuated straight section. It

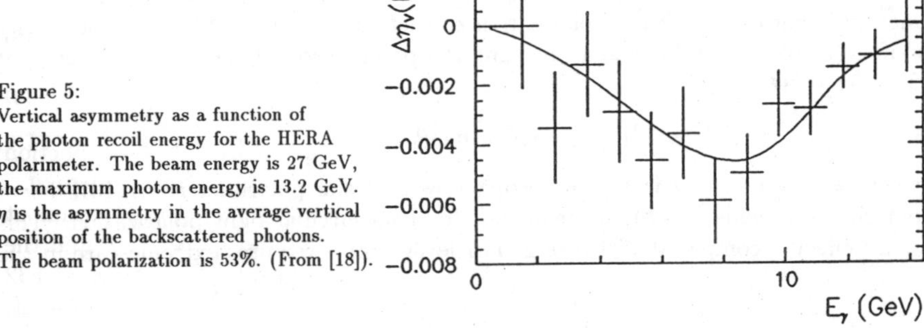

Figure 5:
Vertical asymmetry as a function of the photon recoil energy for the HERA polarimeter. The beam energy is 27 GeV, the maximum photon energy is 13.2 GeV. η is the asymmetry in the average vertical position of the backscattered photons. The beam polarization is 53%. (From [18]).

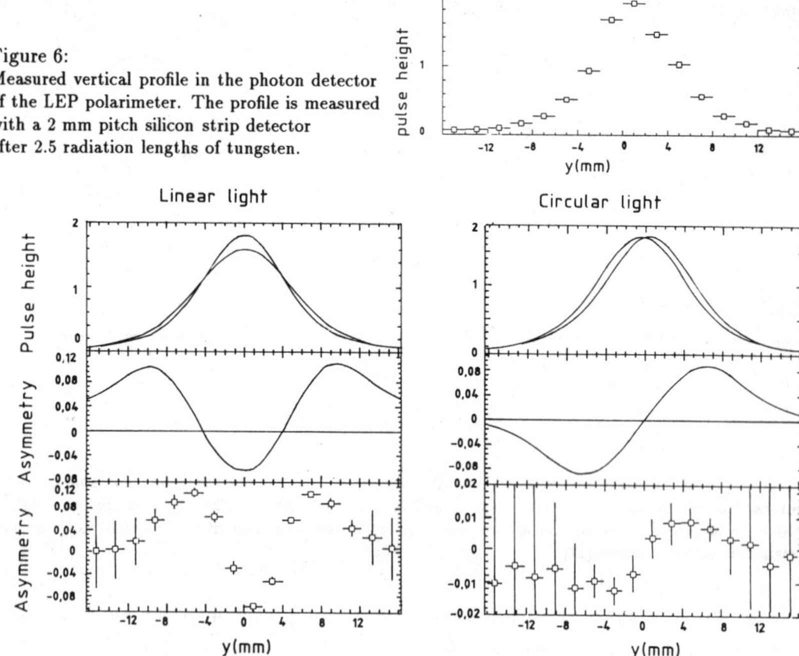

Figure 6:
Measured vertical profile in the photon detector of the LEP polarimeter. The profile is measured with a 2 mm pitch silicon strip detector after 2.5 radiation lengths of tungsten.

Figure 7: Simulated and measured asymmetries in the vertical profiles of backscattered photons.
Top left: simulated profile and asymmetry for linear light;
Bottom left: measured asymmetry for linear light;
Top right: simulated profile and asymmetry for circular light on 100% polarized electrons;
Bottom right: measured asymmetry for circular light. The beam polarization is 10%. From [19].

would become completely unpractical for the 250 m path at LEP.

– The *multi-photon* mode uses a high peak power laser beam working at a low repetition rate (30 to 100 Hz for the LEP polarimeter, while the beam passage rate is 10 kHz). In this case the background is essentially negligible with respect to the several 10^3 backscattered photons per laser pulse. The integrated vertical profile of the backscattered photon pulse is recorded in the photon detector, figure 6. The measurement of transverse polarization consists in detecting the center-of-gravity shift $\Delta\langle Y\rangle$ of this profile when reversing the polarization of the circular laser light from $-\xi$ to $+\xi$:

$$\Delta\langle Y\rangle = \kappa\ \xi\ P_e \tag{10}$$

where $\kappa = 500 \pm 30\,\mu m$ is the analyzing power of the polarimeter. A 10% beam polarization yields a $\sim 50\,\mu m$ mean-shift for 100% circular light polarization. This small effect – compared with the 250 m lever arm – requires extreme care in the light polarization reversal mechanism.

In LEP, polarization reversal is provided by a half-wavelength plate rotating synchronously with the laser pulsing [19]. By changing the time delay between the plate axis and the laser pulse one can switch to linearly polarized light, for which no vertical shift is expected but a change of shape in the distribution, due to the term Φ_1 in equation 7. This provides useful calibration and cross-check against systematic errors arising from possible geometrical differences between the two helicities. The expected and observed effects are shown in figure 7.

Transverse polarization at LEP can be measured to a precision of 1% every minute, with systematic errors estimated to be $\Delta P_e = (0.5 + 12P_e)\%$.

2.2.2 *Longitudinal polarization measurement*

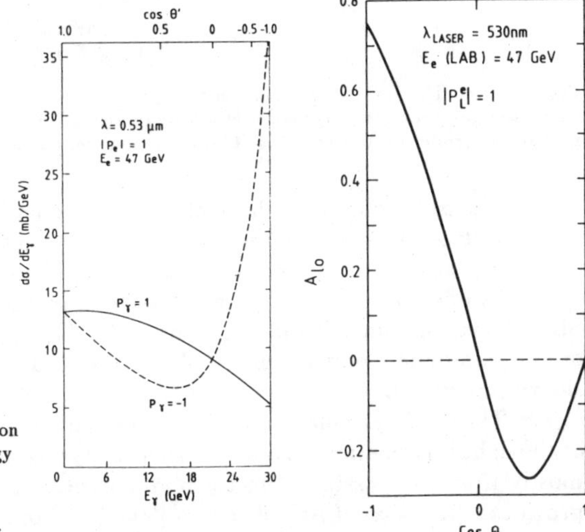

Figure 8:
a) The Compton differential cross-section as a function of the recoil photon energy in the laboratory system for longitudinally polarized electrons. The photon polarization is denoted P_γ.
b) The corresponding asymmetry.

For longitudinally polarized electrons illuminated by circularly polarized light, the equations become:

$$\Phi_1 = 0 \tag{11}$$

$$\Phi_2 = \xi P_e(1 - \cos\theta)[q_0 + q]\cos\theta \tag{12}$$

so that the polarization effects show up only in the energy distribution of the outgoing photon – or electron. The resulting distributions for fully polarized electrons are shown in figure 8a and the corresponding asymmetry in figure 8b. The asymmetry depends strongly on the momentum of the recoil photon, or equivalently of the recoil electron.

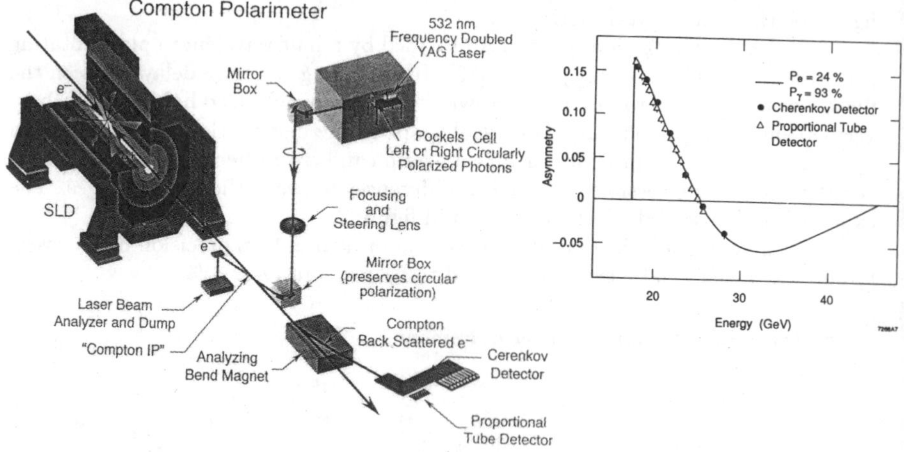

Figure 9: left: The SLC compton polarimeter layout.
right: the measured asymmetry upon light helicity reversal as a function of the scattered electron momentum. Each point corresponds to a cell of the Čerenkov counter or a channel of the proportional tube array.

The SLC compton polarimeter, shown in figure 9, analyses magnetically the recoil electrons to measure their spectrum in an integrated mode, using Čerenkov detectors, backed up by proportional tubes. The laser polarization is reversed at will by a pockell cell, which is well matched to the 120 Hz repetition rate of the SLC. The maximum asymmetry is for $\cos\theta = -1$, e.g. the maximum photon energy, or the minimum electron energy of 17 GeV. An example of the measurement is shown in figure 9.

The main systematic error in the measurement comes from the understanding of the light polarization and of the exact analyzing power of the detector, stemming mostly from the possible detector non-linearity. At present, the SLC polarimeter provides a precision of $\Delta\mathbf{P}_e/\mathbf{P}_e = 2.7\%$ [20]. It is hoped to reduce this uncertainty to $\pm1\%$ eventually.

Various setups have been proposed [21, 7] to measure longitudinal polarization in LEP. The recoil photon spectrum has to be analyzed. This can be done either by using the single photon mode, or, better, by using a converter/sweeping magnet combination to measure the spectrum of electrons from pair conversions in a way similar to the SLC polarimeter.

3 Physics with Transverse Beam Polarization

3.1 *Polarized beams in electron storage rings*

The complex topic of how to generate, store and manipulate beam polarization in storage rings is masterfully described in [22]. Transverse polarization builds up in a e^+e^- storage ring by the Sokolov-Ternov effect[23]: synchrotron radiation emission has a small spin-flip probability, with a large asymmetry in favor of orienting the particles' magnetic moment along the guiding magnetic field. In a perfect accelerator a large asymptotic transverse polarization ($-8/5\sqrt{3} \simeq 92.4\%$) would build up with a rise time τ_p:

$$\tau_p = \left(\frac{5\sqrt{3}}{8} \frac{\hbar r_e {E_{\text{beam}}}^5}{m_e^6 \rho^3} \right)^{-1} \tag{13}$$

The rise time depends very strongly on the beam energy E_{beam} and on the radius ρ of the accelerator. In LEP at 46 GeV per beam, the polarization time is 310 minutes.

In a real accelerator, depolarizing resonances occur, reducing the asymptotic degree of polarization P_∞ and its effective rise time τ_p^{eff} in the same ratio. More specifically, a depolarizing time τ_d competes with the polarization time τ_p so that

$$P_\infty = \quad 0.924 \quad \times \frac{1}{1+\frac{\tau_p}{\tau_d}} \tag{14}$$

$$\tau_p^{\text{eff}} = \quad \tau_p \quad \times \frac{1}{1+\frac{\tau_p}{\tau_d}} \,. \tag{15}$$

The calculation of depolarizing effects, and of the attainable polarization degree, is extremely difficult. The problem lies in the combination of several unfavorable factors:

- The extreme sensitivity of the spin vector to transverse magnetic fields: the precession of the polarization vector around a transverse field is amplified with respect to the rotation of the particle by a factor ν called spin-tune, directly related to the beam energy by the anomalous magnetic moment $a_e = \frac{g_e - 2}{2} = 1.1596521884(43)10^{-3}$ [24] and the mass $m_e c^2 = 0.51099906(15)$ MeV of the electron:

$$\nu = a_e \gamma = \frac{g_e - 2}{2} \frac{E_{\text{Beam}}}{m_e c^2} = \frac{E_{\text{Beam}}}{0.4406486(1)} \,. \tag{16}$$

  The spin tune is also equal to the number of precessions over one turn of the machine. A spin tune $\nu = 103.5$ corresponds to the Z pole beam energy of 45.5 GeV. Any imperfection in the planarity of the storage ring is amplified accordingly.

- Of course, most of these imperfections are by essence unknown. This renders the predictions statistical – and the expected results fluctuating.

- The very long "polarization damping time" is equal to the polarization time – hours. As a consequence, the effect of imperfections is memorized by the polarization vector over typically 10^8 turns of the machine.
- Spin resonances occur each time the spin tune is in phase with the basic motions of the beam particles: turn around the machine (integer resonances); betatron oscillations and synchrotron oscillations (side bands of the integer resonances). These resonances correspond to specific particles energies. The spacing between resonances is energy independent, whereas the beam energy spread is a rapidly increasing function of energy. For instance, the distance in energy between integer resonances is 440 MeV, not comfortably large with respect to the beam energy spread of 40 MeV in LEP at the Z.

The motion of the polarization vector in a storage ring is given by the Thomas-BMT equation [25].

$$\frac{d\vec{P}}{ds} = \frac{e}{m_e c\gamma}\vec{P} \times \left[(1+a_e\gamma)\vec{B}_\perp + (1+a_e)\vec{B}_\parallel\right] \tag{17}$$

Where s is the coordinate along the ring, and $B_\perp$ and $B_\parallel$ are the components of the magnetic field transverse and longitudinal to the orbit. Unless the spin tune is an integer, there exists a stable periodic solution to this equation for an electron in periodic motion in a storage ring, denoted $\hat{n}(s)$. In a perfect, flat, machine, with $\vec{B}$ uniform, $\hat{n}$ is parallel to $\vec{B}$. Because of stochastic synchrotron radiation, in particular, not all particles have the same trajectory, but in a perfect ring they all have the same $\hat{n}$ vector, and polarization can quietly build up.

In a realistic situation, however, vertical kicks do occur, due in particular to quadrupole misalignments, and the $\hat{n}$ vector is no longer vertical. Furthermore, its orientation depends on the trajectory (*spin-orbit coupling*). The spin vector of any given particle will precess around the $\hat{n}$ vector corresponding to its present trajectory and spin diffusion will occur, resulting in depolarization. This phenomenon is described by the Derbenev-Kondratenko formula [26] which we give here in a simplified form:

$$P_\infty = -\frac{8}{5\sqrt{3}} \frac{\sum_j |B_j|^3 L_j}{\sum_j |B_j|^3 L_j \left(1 + \frac{11}{18}|\mathbf{\Gamma}_j|^2\right)} \tag{18}$$

The sum runs over the magnets in the ring. Their strength is B_j and their length L_j. This equation can be identified with equation 15 with:

$$\frac{1}{\tau_p} \propto \sum_j |B_j|^3 L_j \tag{19}$$

$$\frac{1}{\tau_d} \propto \sum_j |B_j|^3 L_j \frac{11}{18} |\mathbf{\Gamma}_j|^2 \tag{20}$$

Note that the synchrotron radiation driving term $|B_j|^3 L_j$ is responsible for both polarization and depolarization.

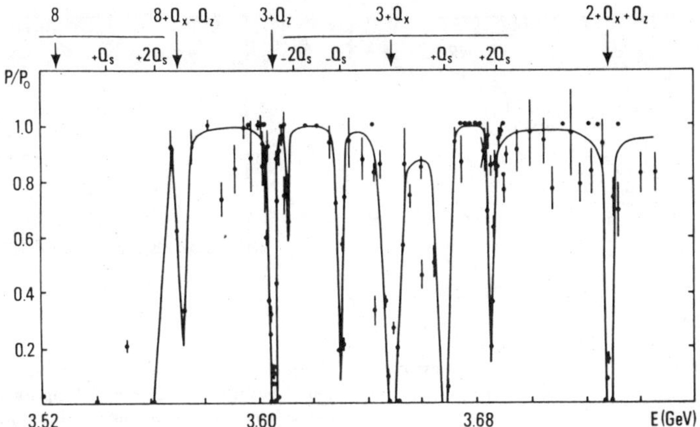

Figure 10: Polarization measurements from SPEAR. The possible location of several higher order resonances is indicated.

The vector Γ_j is the spin orbit coupling vector at the magnet j:

$$\frac{1}{\gamma}\Gamma = \frac{\partial \hat{n}}{\partial \gamma} \tag{21}$$

It can be decomposed along the normal modes of motion, horizontal (x, x') and vertical (y, y') betatron motion and synchrotron oscillations(s):

$$\frac{1}{\gamma}\Gamma = \frac{\partial \hat{n}}{\partial s} + \frac{\partial \hat{n}}{\partial x}\eta_x + \frac{\partial \hat{n}}{\partial x'}\eta'_x + \frac{\partial \hat{n}}{\partial y}\eta_y + \frac{\partial \hat{n}}{\partial y'}\eta'_y \tag{22}$$

$$= \Gamma_s + \Gamma_x + \Gamma_y \tag{23}$$

The sum over the ring of these components are called spin-orbit coupling integrals [27, 28]. They have a resonant structure, so that strong depolarization takes place for:

$$\nu = m_0 + m_s Q_s + m_x Q_x + m_y Q_y \tag{24}$$

where $m_{0,s,x,y,}$ are integers and $Q_{s,x,y}$ are the synchrotron and horizontal and vertical betatron tunes. Several of these resonances were observed at SPEAR [29], figure 10.

The calculation of the vector Γ can be done rigorously to first order using the formalism introduced by Chao [30], and implemented in the computer programme SLIM and several adaptations of it for more realistic simulations of defects [31]. These only calculate resonances with $|m_s| + |m_x| + |m_y| = 1$. The calculation of higher order resonances calls for analytical methods [32, 33], or complete spin-tracking [34], which requires enormous computing power, however. It is generally accepted that the driving terms of the first order resonances are likely to determine the strength of the higher order ones. For this reason, correction algorithms are based on first order calculations.

The most salient feature of spin simulation results for LEP is the dominance of the synchrotron resonances over betatron resonances. The driving terms of these

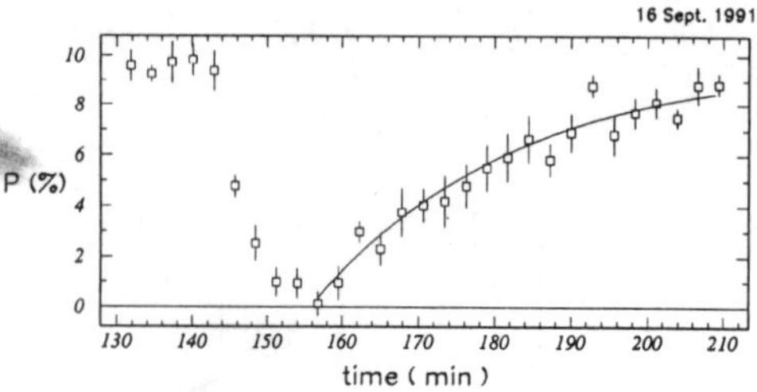

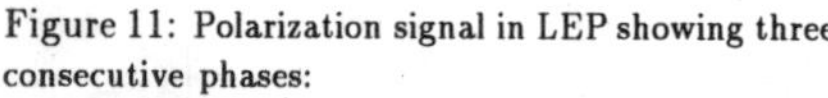

Figure 11: Polarization signal in LEP showing three consecutive phases:
1) stable polarization $\mathbf{P} \simeq 9.6\%$;
2) depolarization by a static harmonic bump $\nu = 106$;
3) polarization rise.
The solid curve shows a fit to the polarization rise time of 35±10 minutes corresponding to $\mathbf{P}_\infty = (10.7 \pm 3.1)\%$.

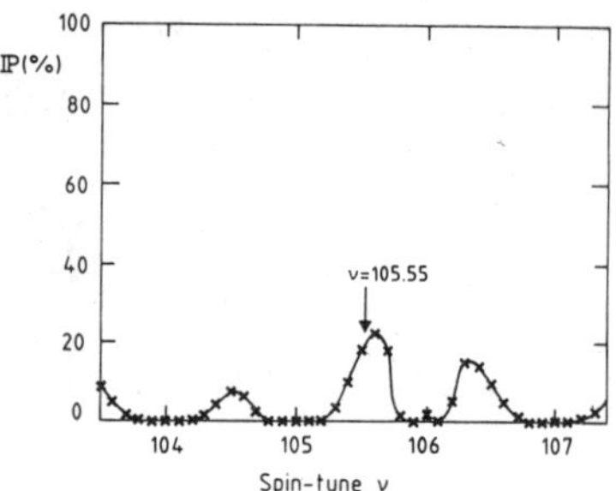

Figure 12: Expected polarization degree as a function of beam energy in the LEP optics used in the first polarization measurements. From [35].

synchrotron resonances are proportional to $|\delta\hat{n}|^2$ where $\delta\hat{n}$ is the tilt of the equilibrium spin vector with respect to the normal to the plane of the accelerator. The immediate consequence is that depolarization increases like the square of the beam energy, and like the square of the size of the defects. A general strategy to improve polarization is therefore to align the orbit to the best possible precision.

Spin simulations have been used for i) choice of the appropriate working point; ii) compensation of known defects; iii) improvement of polarization.

The first order predictions of the polarization behavior show the dominance of integer resonances, and among them, of 'systematic' integer resonances [36]:

$$\nu = 8k \pm \text{int}[Q_{x/y}] \,. \tag{25}$$

Consequently, a beam energy $E = 46.5\,\text{GeV}$ ($\nu = 105.55$) was chosen for polarization studies, and the closest betatron resonances were moved away by choosing smaller values of the fractional parts of the betatron tunes. This choice, and careful vertical orbit corrections down to 0.5 mm RMS, allowed observation of 10% polarization at LEP in 1990 [19] and 1991, as shown in figure 11. The first order prediction was 20%, as shown in figure 12, but higher order effects were expected to reduce it somewhat.

An example of correction of known defects is the compensation of the spin rotation induced by the experimental solenoids. This is of practical importance since the first polarization experiments in LEP required turning off the experimental solenoids, a lengthy and dangerous procedure. The scheme adopted in LEP [39] was first suggested at DESY [37, 38]. A rotation opposed to that created by the solenoid is generated in the arcs on each side of the experimental straight section, by vertical closed bumps, figure 13. As a result the spin vector is vertical everywhere else in the machine. Spin simulations show that, indeed, the integer resonances

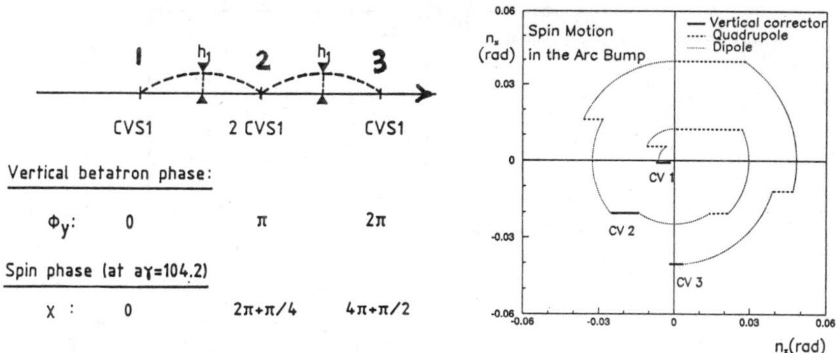

Figure 13: Left: the arc bumps used to spin-compensate the experimental solenoids in LEP. Right: motion of the tip of the $\hat{n}$ vector in the bump. The solenoid then creates a rotation of n_x by +66mrad. A second bump, opposite in to the first one, situated on the other side of the straight section, brings the spin back to vertical ($n_x = n_z = 0$).

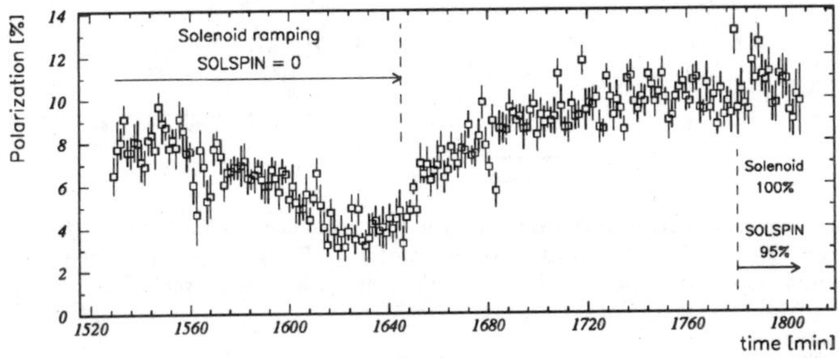

Figure 14: Experimental verification of solenoid spin compensation at LEP. The polarization level is shown as function of time, while the solenoid of the ALEPH experiment was ramping from 0 to full field. At first no compensation is excited and the polarization degree drops. At time 1655 the solenoid compensation is turned on and polarization rises back up to the initial level.

are compensated when the tilt of the $\hat{n}$ vector is compensated. This was verified experimentally in 1992 [40], figure 14, and used since.

A similar *spin-matching* exercise is necessary if one wants to implement the spin 90° spin rotation required for obtaining longitudinally polarized beam, (see section 4.5). This was also performed with first order spin theory.

Improvement of polarization is based on the so-called *harmonic spin-matching* procedure. Here again, the fact that integer resonances and their Q_s side bands dominate the spectrum leads to simplification. The improvement of polarization can be obtained by reducing the tilt of $\hat{n}$, by appropriate spin rotations. It can be shown that, if the defects of the machine are dominated by the misalignment of the quadrupoles and the corresponding orbit correctors, the tilt of the $\hat{n}$ vector can be decomposed in the Fourier components (a_k, b_k) of the orbit, calculated in the spin

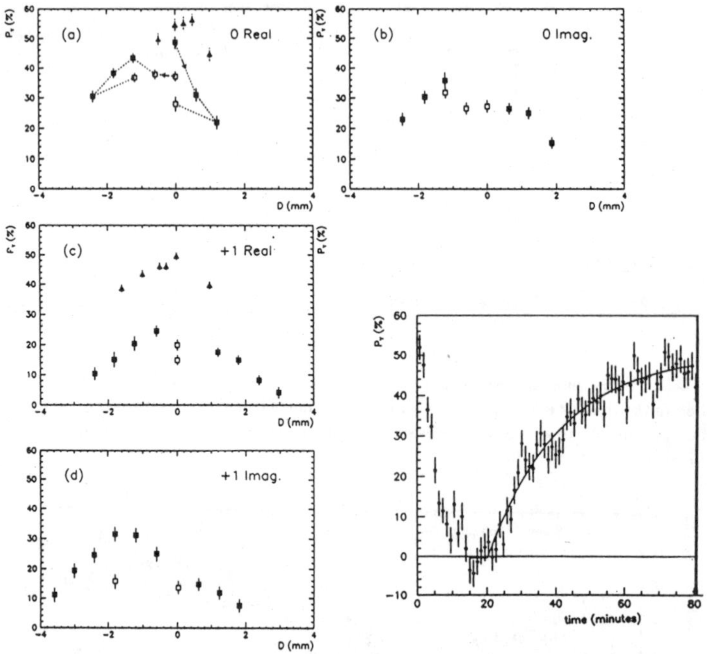

Figure 15: Improvement of polarization with harmonic spin matching. Left: over 50% (from about 15%) polarization was achieved at HERA by empirical spin-matching. Figures a) b) c) d) show the scan of the harmonic correctors. The bottom right picture shows the polarization rising over 50%.

precession frame [41]:

$$|d\hat{n}|^2 \propto \sum_k \frac{1}{(\nu - k)^2}(a_k^2 + b_k^2) \tag{26}$$

It is possible to construct combinations of bumps – similar to those shown above for the solenoid compensation – which generate exactly the desired harmonics. One can see from equation 26 that the most enhanced harmonics are those close to the spin tune, Int(ν)=$k, k+1$. Correcting for the real and imaginary components of each of these close harmonics requires four bumps. One beauty of the technique is that the harmonics are orthogonal to each other: the formalism provides a set of independent spin correctors.

This technique was first applied at PETRA [42], with improvements in the polarization degree from 20% to over 70%. The correctors were varied in turn to find the optimum polarization. The same procedure was successfully performed on HERA [18] in 1992, with polarization reaching 56%, figure 15a.

For LEP, the polarization rise time is much too long for this procedure to be successful. A more deterministic approach is necessary. If the beam orbit monitors (BOM) were infinitely precise, one could measure directly the harmonic components of the orbit. The precision available was not sufficient in 1990-1992. The beam monitors were recently upgraded, and the machine realigned, so that the deterministic harmonic spin-matching has become possible. A polarization degree of 30 to 40% (from 10%) has been reproducibly obtained in 1993 by this method.

Further improvement for LEP could come from the use of dedicated polarization wigglers [43] designed to reduce the polarization time from $\simeq$ 300 minutes to 30 minutes and simplify the optimization procedures. It is feared, however, that the unavoidable increase in the beam energy spread will at least partially cancel the benefits [33].

So far, the only source of depolarization considered were the energy jumps due to synchrotron radiation in the arcs. When the beams enter in collision, beam-beam collisions provide another source of stochastic orbit jumps, in the form of vertical kicks at the IP's. It is fair to say that this aspect has not been as carefully studied as the previous one. A rudimentary model based on random orbit kicks was implemented by Chao [30] in SLIM. Spin-matching conditions were derived by Buon [44], for both transverse and longitudinal polarization. They establish relations between the betatron tunes and the energy, so that one expects perfect spin matching only at discrete energies. Polarization in collisions was observed at SPEAR, CESR, VEPP and PETRA, but the phenomenon was sometimes irreproducible. Observations at PETRA indicate that depolarization only sets in when one approaches the beam-beam limit where vertical emittance is substantially increased. Clearly, this issue requires substantially more work. Polarization has not yet been observed in collisions at LEP.

3.2 *Transverse polarization asymmetry*

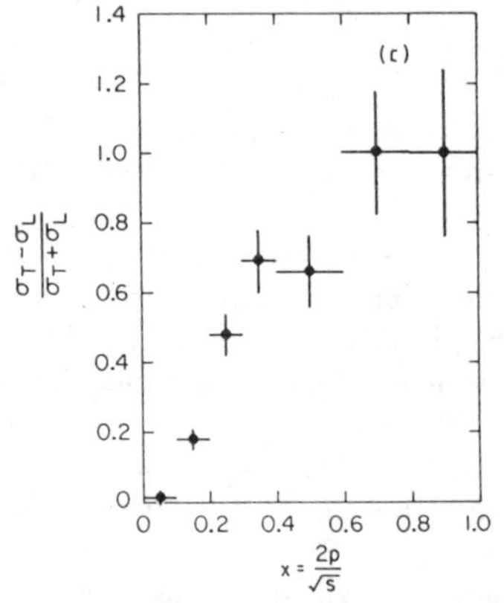

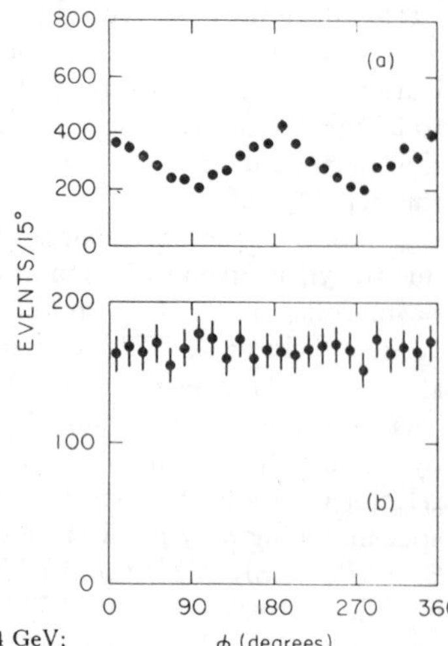

Figure 16: Transverse polarization asymmetry at SPEAR.
a) Observed hadron yield for all particles with $x > 0.3$ at $\sqrt{s}$=7.4 GeV;
b) Same as a) at the spin-depolarization resonance $\sqrt{s}$=6.2 GeV;
c) Transverse polarization asymmetry as a function of particle momentum, showing leading particle effect.

Transverse polarization has been observed in every electron storage ring where it was searched for and a few beautiful experiments were done. At SPEAR at SLAC (U.S.A.), observation of the transverse polarization asymmetry [45] –an azimuthal modulation of particle production dependent on the spin of the particle– was a clear confirmation of the conjecture that jets of hadrons originate from the production of spin $\frac{1}{2}$ partons (Figure 16). This experiment was performed with over 70% beam polarization in collision mode.

3.3 *Beam energy calibration and hadronic resonances masses*

Extremely precise calibration of the beam energy, to one part in 10^5, can be done by exciting an artificial depolarizing resonance, using the electron spin in nearly the same way as the nuclear spins in a Nuclear Magnetic Resonance probe.

Accurate measurements of the masses of the ω, ϕ, J/ψ, ψ', Υ, Υ', and Υ'' resonances have been performed using this technique at VEPP2 and VEPP4 in Novosibirsk (U.S.S.R.) [10, 11, 12], at DORIS in Hamburg (Germany) [13], and CESR in Cornell (U.S.A) [14]. Figure 17 shows the observation of depolarization in VEPP4 and DORIS. For all these experiments the measurement of energy was

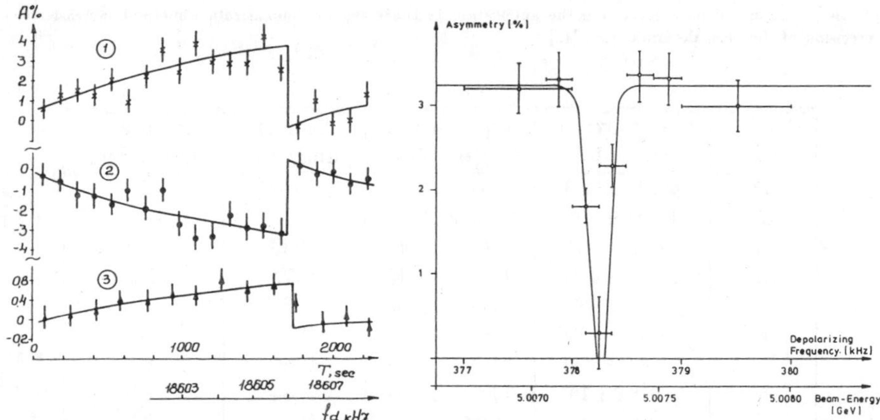

Figure 17: Examples of depolarization curves: The beam polarization is measured as an up-down asymmetry of back scattered light on the polarized beam. A small tunable RF magnetic field is applied on the beam, and polarization measured as a function of its frequency. a) VEPP4 in Novosibirsk [11]. 1 and 2: back scattering of synchrotron radiation light from the e^+ beam on the e^- beam and vice versa. 3: back scattering of polarized laser light on the e^- beam. τ_p = 50 minutes, P_∞=80%. b) at DORIS in Hambourg [13] with back scattered laser light. τ_p= 4 minutes, P_∞=80%, luminosity up to $1.5 \cdot 10^{31}$ cm^2/s.

performed while the accelerator was in normal colliding conditions with detectors taking data. Polarization degrees of up to 80% and luminosities approaching half the peak luminosity were obtained. The increase in precision obtained by experiments at VEPP2 and VEPP4 are listed in table 1.

3.4 *Beam energy calibration for LEP*

This section will describe how this powerful method has been applied to improve the determination of the Z boson mass, and is hoped to improve also the Z width measurement in the future.

3.4.1 *Effect of beam energy uncertainties on the Z line shape parameters*

The Z mass and width are extracted from a fit to the measured cross-section in $e^+e^- \to hadrons$ as a function of center-of-mass energy, as shown in figure 18. The determination of the Z mass is sensitive to the knowledge of the absolute energy scale, while the Z width is affected by possible errors in the differences between the energies of the scan points. Also sensitive to the relative energies is the forward-backward asymmetry for $e^+e^- \to \ell^+\ell^-$, as shown in figure 19, from which one of the most precise [1] measurement of the effective weak mixing angle $\sin^2\theta_w^{\rm eff}$ is obtained. The Z line shape was scanned in 1990 and 1991, but not in 1992.

Energy errors are classified in four categories.

[1] In absence of longitudinally polarized beams.

Table 1: A list of particles where the application of the resonant depolarization method increased the precision of the mass determination [12].

Particle	World average value (MeV)	Experimental results (MeV)	Year of publication	Accuracy improvement
$K^{\pm}$	493.84 ± 0.13	493.670 ± 0.029	1979	5
K^0	497.67 ± 0.13	497.661 ± 0.033	1987	4
ω	782.40 ± 0.20	781.780 ± 0.10	1983	2
Φ	1019.7 ± 0.24	1019.52 ± 0.13	1975	2.5
J/Ψ	3097.1 ± 0.90	3096.93 ± 0.09	1981	10
Ψ'	3685.3 ± 1.20	3686.00 ± 0.10	1981	10
Υ	9456.2 ± 9.50	9460.59 ± 0.12	1986	80
Υ'	$10016.0 \pm 10.$	10023.6 ± 0.50	1984	20
Υ"	$10347.0 \pm 10.$	10355.3 ± 0.50	1984	20

- The absolute energy scale error, $(\frac{\Delta E}{E})_{abs}$.
- The non-linearity $\alpha \pm \Delta\alpha$ of the response of the magnets to the exciting current.
- Point-to-point errors that account for possible higher order effects in the relation between dipole current and beam energy, $(\frac{\Delta E}{E})^{setting}_{p-t-p}$.
- Non-reproducibility errors, $(\frac{\Delta E}{E})_{rep}$, coming from several possible sources of variability, such as temperature, tidal effects, corrector settings, unknown variations in the RF status. The error from these sources average out with N_{scan}, the number of times the Z line shape is scanned. (7 times in 1990, 5 times in 1991).

The systematic errors on $\mathrm{M_Z}, \Gamma_\mathrm{Z}, \sin^2\theta_w^{\mathrm{eff}}$ resulting from energy uncertainties are as follows.

$$\frac{\Delta \mathrm{M_Z}}{\mathrm{M_Z}} = \left(\frac{\Delta E}{E}\right)_{abs} \oplus 0.5 \left[\left(\frac{\Delta E}{E}\right)^{setting}_{p-t-p} \oplus \frac{1}{\sqrt{N_{scan}}}\left(\frac{\Delta E}{E}\right)_{rep}\right] \tag{27}$$

$$\frac{\Delta \Gamma_\mathrm{Z}}{\Gamma_\mathrm{Z}} = \left(\frac{\Delta E}{E}\right)_{abs} \oplus 0.5\frac{\mathrm{M_Z}}{\Gamma_\mathrm{Z}} \left[\left(\frac{\Delta E}{E}\right)^{setting}_{p-t-p} \oplus \frac{1}{\sqrt{N_{scan}}}\left(\frac{\Delta E}{E}\right)_{rep}\right] \oplus \Delta\alpha \tag{28}$$

$$\Delta \sin^2\theta_w^{\mathrm{eff}} = 0.0004 \cdot \frac{\mathrm{M_Z}}{10\mathrm{MeV}} \left[\left(\frac{\Delta E}{E}\right)^{setting}_{p-t-p} \oplus \frac{1}{\sqrt{N_{scan}}}\left(\frac{\Delta E}{E}\right)_{rep}\right] \tag{29}$$

where $\oplus$ represents a quadratic sum: $a \oplus b = \sqrt{a^2+b^2}$. The coefficient 0.5 in front of the non-reproducibility and point-to-point errors is a numerical factor which can vary – slightly – depending on the number of points in the scan and on their energies.

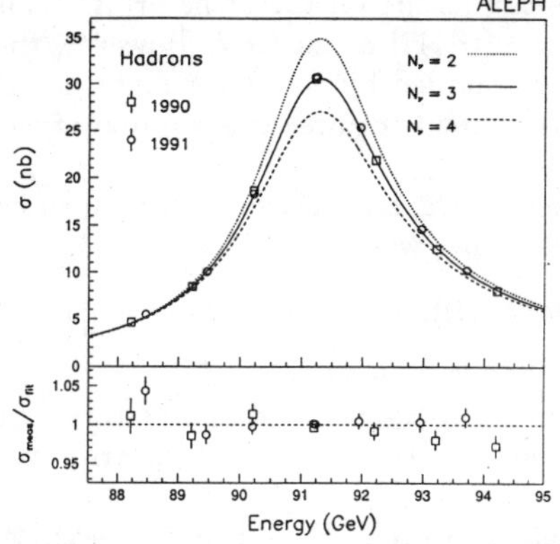

Figure 18: The Z line shape. A fit to a theoretical formula gives the parameters M_Z and Γ_Z. The standard model predictions for $N_\nu = 2, 3, 4$ are shown. The lower plot shows the agreement of the theoretical shape to the measured one. The experiments (here ALEPH) are responsible for the vertical error bars, LEP and the LEP energy working group for the horizontal ones.

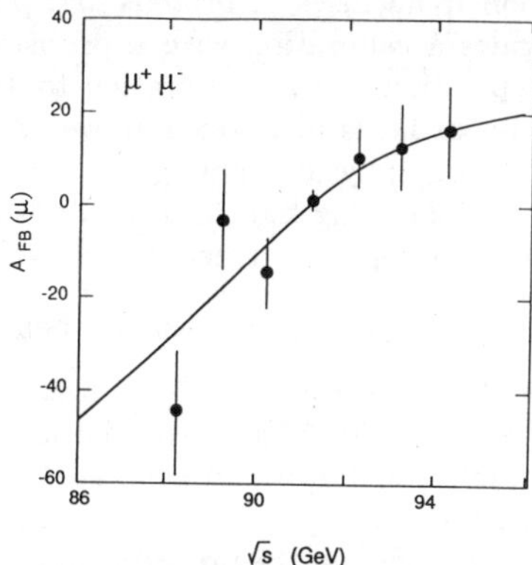

Figure 19: The forward backward asymmetry for leptons (from L3). The energy dependence is rather steep. The most important physics information is contained in the offset of the curve at $E_{c.m.} = M_Z$ and is affected by uncertainties in the beam energy.

3.4.2 *Magnetic measurements*

The determination of energies before polarization was available is described in [46], and is based on magnetic measurements. The beam energy, which is proportional to the field integral around the LEP ring, is derived from the measurement of the magnetic field in a LEP reference dipole magnet. A flipping coil is used to measure the field in a pure iron magnet powered in series with the ring's dipoles. This measurement is continuously available and is labeled E_{FD} (for Field Display). However, the reference magnet is of different nature than the LEP concrete-iron dipoles and situated in different conditions of temperature and humidity. An absolute calibration by other methods is necessary.

A direct measurement of the field generated by the dipoles is provided by cycling the magnets and measuring the induced current in a closed electrical loop, imbedded in the LEP dipoles. This "flux-loop" method is in principle rather accurate ($\pm 10^{-4}$), but cannot be applied while beams circulate. Furthermore it only measures the field produced by the dipoles, and not the additional fields which influence the beam energy, such as the earth magnetic field, the possible permanent magnetic properties of the beam pipe, as well as the dipole components of the orbit correctors and quadrupoles. An in-situ calibration with circulating beams is thus required.

An elegant solution is provided by comparing, at injection energy, the revolution frequencies of protons and positrons circulating on the same orbit. This provides a calibration with a precision of $\pm 3 \cdot 10^{-5}$ [47] at 20 GeV. However, the extrapolation of this calibration to 45 GeV is entailed by the uncertainty in the non-linear fields mentioned above. Also, this method requires interruption of the high energy physics running.

Combining the three methods described above, an average correction to be applied to E_{FD} is obtained, which, for the 1990 data, was:

$$E_{Beam} = E_{FD} \times \left[1 - (6.4 \pm 2.4)10^{-4}\right] . \tag{30}$$

This allows an estimate of the absolute energy scale with an error of $(\frac{\Delta E}{E})_{abs} = \pm 2.2 \cdot 10^{-4}$. However, the non-reproducibility and point-to-point errors were essentially unknown and estimated to be about 10 MeV each. Non-linearity was not considered.

Altogether, the precision on M_Z was limited to ± 21 MeV, and on Γ_Z to $\pm 5 MeV$, while the statistical errors for the 1990 data set were [48] ± 5 MeV and ± 9 MeV, respectively. The use of resonant depolarization was necessary to improve this result.

3.4.3 *Resonant depolarization*

The spin tune ν, or number of spin precessions in one turn around the ring, is proportional to the beam energy, equation (16). It is determined by resonant depolarization [49].

Resonant depolarization is produced by exciting the beam with an oscillating magnetic field generated by a vertical kicker magnet. This exciting field is thus perpendicular to the beam axis and situated in the plane (e.g. horizontal) of the ring. If the frequency of the resulting spin kick is in phase with the spin precession, a resonance condition occurs. The electron spins are coherently swept away from the vertical direction, and polarization disappears.

Because the beam encounters the transverse exciting field only once per turn, the frequency at resonance is determined by the fractional part of the spin tune, δ_s. The standard magnetic measurements are precise enough to provide the integer part without ambiguity. Furthermore, because the exciting field is a transverse field perpendicular to the beam axis, and not a rotating field around the vertical axis, a resonant condition occurs also for the mirror frequency $1-\delta_s$. The mirror ambiguity is resolved by a second measurement after inducing a small change of beam energy with a change of RF frequency.

The actual spin precession frequency is the product of the spin tune by the revolution frequency f_{rev} of the particles around the LEP ring. The revolution frequency is obtained by dividing the RF frequency f_{RF}, nominally set to f_{RF}=352,254,170 Hz and measured to better than 1 Hz, by the RF harmonic number, 31324. This yields, in nominal conditions, f_{rev}=11,245.5041(1).

To find the resonance, the horizontal field of maximum integrated strength $b\ell$ is excited at a slowly varying frequency, in such a way as to cross the resonance

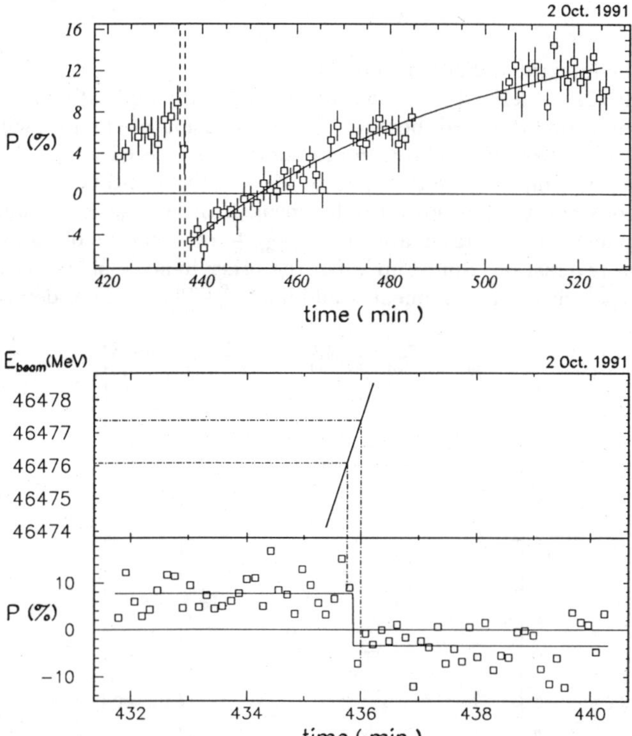

Figure 20: Polarization signal on 2 October 1991, showing the localization of the depolarizing frequency within the sweep.
Top: display of data points, with the frequency sweep indicated with vertical dashed lines. The full line represents the result of a fit with starting polarization $(-4.9 \pm 1.)\%$, polarization rise-time (60 ± 13) minutes, asymptotic polarization $(18.4 \pm 4.1)\%$.
Bottom: expanded view of the sweep period, with the individual data sets displayed (there are 10 sets per point). The frequency sweep lasted 7 data sets. The corresponding beam energy is shown in the upper box. Spin flip occurred between the two vertical dash-dotted lines.

condition. The expected signal for resonance crossing is a steep variation of the polarization degree, in a way that depends on the rapidity with which the equivalent spin tune is swept, $\frac{\Delta\nu}{\Delta t}$ [50]:

$$\frac{P(final)}{P(initial)} = 2e^{-\chi} - 1 \tag{31}$$

$$\chi = \frac{(\pi\nu.\frac{b\ell}{BL})^2 \times f_{rev}}{\frac{\Delta\nu}{\Delta t}}, \tag{32}$$

where BL is the integrated guide field of LEP.

Formula (31) is derived for proton accelerators. Its application for e^+e^- storage rings is not straightforward, because of the excitation of energy oscillations of particles due to emission of synchrotron radiation. This effect leads to decoherence of the polarization component in the plane of the ring (horizontal component), with a decay time constant which is not yet well known. If this decay time is much longer than the resonance crossing time, a value of $\frac{P(final)}{P(initial)}$ of -1 can be obtained, leading to polarization reversal or "Spin-Flip". On the other hand, a short decay time of the horizontal polarization component would limit $\frac{P(final)}{P(initial)}$ to 0, i.e. depolarization.

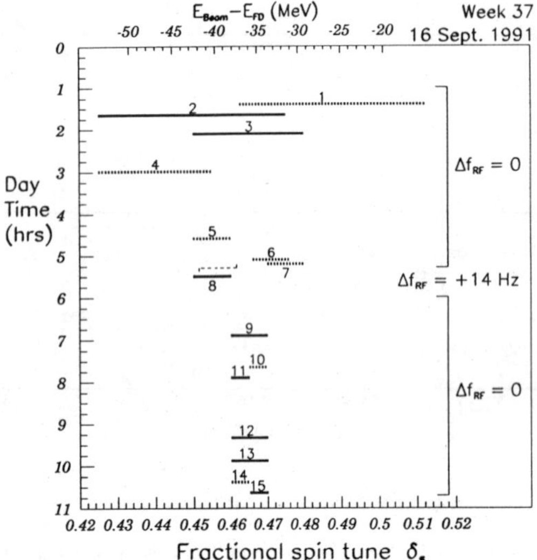

Figure 21: Graphical representation of the depolarizations on 16 September 1991. Full line: depolarization or spin flip; dashed line: No depolarization.

An example of resonant depolarization is shown in figure 20. The resulting energy measurement is E_{beam} = 46,466.6 ± 0.6 MeV, e.g. precise to ± 1.5 10^{-5}. Partial spin-flip took place, indicating that the decay time of the horizontal polarization component is of the same order of magnitude as the resonance crossing time, a few seconds.

In practice, the resonance is searched by successive frequency sweeps of varying range, as shown in figure 21. In 1991 the resonance could be located within a sweep range of $\Delta\delta_s = 0.005$, leading to an energy error of ± 0.6 MeV. This was improved in 1992 to $\Delta\delta_s = 0.002$, or a precision of ± 0.25 MeV.

Several tests were made to ascertain that the observed resonance was not a spurious one. In similarity to the static resonances of equation 24, some depolarization was observed at frequencies corresponding to the synchrotron side-bands of the main resonance. This confusion could cause an error of $\Delta E_{beam} = Q_s \times E_{beam}/\nu = 27$ MeV. However these side-bands were readily discarded by changing the synchrotron tune.

Other systematic errors were considered, in particular the possible interference of the resonance with the static depolarizing resonances present in the machine [51]. This effect was estimated to be less than 0.5 MeV.

In conclusion, the resonant depolarization of the electron beam could be located unambiguously and the energy measured at a given instant with a precision of ± 0.6 MeV. We will see, however, that other uncertainties limit the usefulness of the result.

3.5 *Energy variations and the tidal effect*

The beam energy measured during the polarization runs has to be traced over several weeks to the period where the LEP experiments are taking data. This is done with the field display, which has a more limited resolution of ± 1 MeV.

Furthermore, as can already be seen by comparing measurements 11 and 15 in figure 21, the position of the resonance varies, on a rather short time scale. When compiling the results of the 1991 calibrations, a scatter of ± 3 MeV in beam energy was noticed.

Variation of the magnetic field in the LEP dipoles with respect to the field display can originate from temperature variations, and more generally from other environmental parameters. The temperature coefficient of the LEP magnets was measured and compared with flux-loop measurements [52]. Correcting the depolarization data for the measured temperatures increased the scatter to ± 3.7 MeV!

It was suggested by G. Fischer and A. Hofmann [53] that earth tides could be the cause of these variations. Earth tides, generated by the combined effect of the gravitational fields of the moon and the sun have been known since the middle of the 19th century and precisely measured for more than 30 years [54]. Contrary to oceanic tides, earth tides are not a resonant phenomenon, and can be calculated successfully as a global deformation of our planet. They result in a linear dilatation of dimensions, with a maximum swing of $\pm\ 3.10^{-8}$.

How do earth tides affect the beam energy? The "central" orbit of LEP – the orbit that passes through the center of the quadrupoles and sextupoles – is also expanded (or shrunk) in this proportion, corresponding to a total change of the 27 km circumference by 600 μm. However, the true orbit of superelativistic electrons is constrained by the fixed RF frequency to a fixed length. As a result,

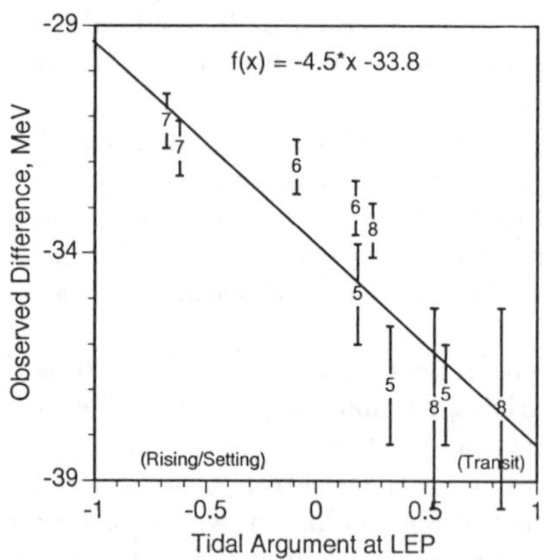

Figure 22: Correlation between the energy measured with resonant depolarization in 1991 and the earth tide amplitude. The numbering on the measurements represents their sequence in time, and the error bars give the range in beam energy within which resonant depolarization was observed.

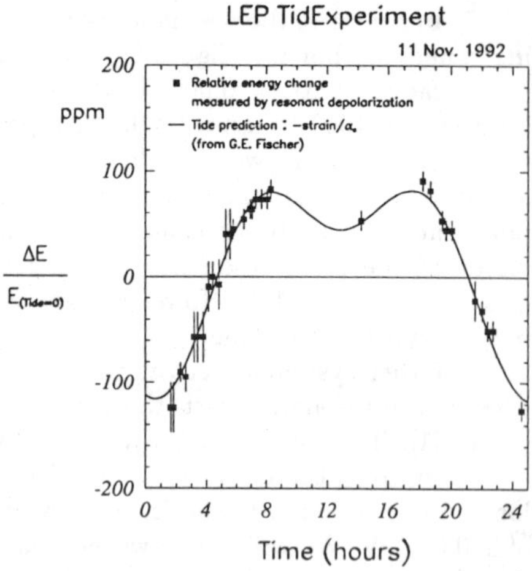

Figure 23: Beam energy variations measured over 24 hours compared to the expectation from the tidal LEP deformation.

when the machine expands (shrinks), the beam passes through the quadrupoles off center towards the inside (outside) of the ring. Since the quadrupoles are focusing on the average, the beam sees less (more) integrated magnetic field. This results in an energy change of

$$\frac{\Delta E}{E} = -\frac{1}{\alpha}\frac{\Delta R_{LEP}}{R_{LEP}} \tag{33}$$

where α is the *momentum compaction factor*; α depends on the horizontal betatron tune, so that the energy swing is larger, ± 7 MeV, for the 90° lattice used since 1992, than for the 60° lattice used in 1991, where the energy swing was ± 3.5 MeV.

The beam energy measurements of 1991 are plotted in figure 22 against the tide amplitude. A correlation of $\simeq 2\ \sigma$ significance is visible. The calibrations sampled the tide effect sufficiently well, nevertheless, for their average to be applicable to the physics data.

A dedicated experiment was performed in 1992 to ascertain the effect [55]. The results, shown in figure 23, agree with the tide prediction with a precision, and stability, of better than 0.5 MeV over 24 hours.

3.6 *Determination of the Z mass and width*

Resonant depolarization measures the sum of spin precessions along the trajectory, which is proportional to the average energy in the arcs (the energy is not constant due the energy loss by synchrotron radiation). One need to relate this to the energy at the interaction points during the physics runs. This implies corrections to the average physics conditions, in particular for temperature and tidal effect. In addition a correction has to be made for the (known) asymmetry in the accelerating cavities, which shifts the center of mass energies in OPAL and L3 by 12.8 and 12.7 MeV respectively.

A full account of the steps leading to the energy calibration of LEP in 1991 is given in [52], and the determination of the Z mass in [57]. The following values are found for the error components entering equation 29.

– Absolute energy error: $(\frac{\Delta E}{E})_{abs} = \pm\ 5.7\ 10^{-5}$

– Non-linearity : $\Delta\alpha = \pm\ 1.5$ MeV/GeV

– Further point-to-point error: $(\frac{\Delta E}{E})^{setting}_{p-t-p} = \pm\ 310^{-5}$

– Non-reproducibility: $(\frac{\Delta E}{E})_{rep} = \pm\ 10\ 10^{-5}$.

The Z mass and width are extracted by a fit of the experimental cross-sections [56] to the theoretical line-shape [58]. The average of the four LEP experiments is:

$$M_Z = \ 91.187 \ \ \pm 0.0035_{stat} \pm 0.0063_{LEP} \text{ GeV} \tag{34}$$

$$\Gamma_Z = \ 2.488 \ \ \pm 0.0054_{stat} \pm 0.0045_{LEP} \text{ GeV} \tag{35}$$

The systematic error of 6.3 MeV on the Z mass is dominated by the absolute energy scale error, while the error on the Z width is dominated by the poorly known non-linearity.

3.7 *Prospects for improvement on the Z line shape*

Even though it is measured to a relative precision of 6 10^{-5}, the Z mass is a fundamental parameter to be measured with the best possible precision. The Z width is even more interesting since it provides, by comparison with M_Z, a powerful test of the Standard Model, sensitive to electroweak radiative effects.

The error on the width is still statistics limited. Furthermore the systematic error is substantial solely because the energy calibration was available at one energy only.

The 1993 running of LEP is dedicated to a new scan of the Z resonance. The LEP chamonix workshop [59] discussed the strategy in detail. Several improvements to the machine setup were successful, in particular the solenoid spin-compensation, and the observation of polarization on the same optics used for physics data taking. Polarization was observed – and energy calibration performed – at the two most favorable points on the line shape, E_{beam} = 89.42 ("peak - 2"), and E_{beam} = 93.0 ("peak + 2"). Energy calibration can be performed at the end of physics fills, and lasts only 3-4 hours – instead of typically 24 hours previously. Two calibrations are performed every week.

For a total integrated luminosity of 40-50 pb^{-1}, a total error of ± 3 MeV or less on both M_Z and Γ_Z seems achievable.

3.8 *The W mass*

The energy of LEP will be upgraded to 90 GeV per beam towards the end of 1994. One of the important goals is the determination of the W boson mass. Several methods have been suggested to measure it [60], all require knowledge of the beam energy at LEPII, with a precision better than the expected statistical precision of the measurement, 25 MeV. If shown practical and feasible with the advertised accuracy, the proposal for measuring the beam energy by using the kinematical end-point in Møller scattering [61] could be interesting. In view of the present improvement in the polarization with realignment of LEP and harmonic spin-matching it is not impossible that energy calibration by resonant depolarization will be achievable at LEPII energies, or at least at a close enough energy. In any case, the LEPI energy measurement system, with the present polarimeter and the flux loop, will need to be kept operational for the LEPII era.

4 Physics with longitudinally polarized beams

In the energy range of LEP weak interactions dominate. The Standard Electroweak Model $SU(2)_L \times U(1)$ being left-right asymmetric, helicity effects are expected to play an important role. This is in contrast with lower energy e^+e^- accelerators where the QED-dominated physics leaves limited interest for longitudinally polarized beams.

The SLC was designed from the start to be a polarized machine, building on the success of the deep-inelastic experiments with polarized electrons [62]. The early LEP studies were quite aware of the importance of beam polarization [63]. The exact benefits were not realized, however, and the difficulties in obtaining polarized beams seemed overwhelming. The subject of longitudinally polarized beams for LEP has been extensively studied in 1988 [64].

4.1 *Helicity effects in* e^+e^- *annihilation*

Longitudinal polarization effects manifest themselves in many reactions. Most of all, longitudinally polarized beams allow measurements of the weak couplings of fermions to the Z in a clean and precise way. Helicity effects in $e^+e^- \to Z \to f\bar{f}$ are sketched in figure 24.

The fact that the Neutral Current Couplings are different for left-handed and right-handed fermions leads to polarization and forward-backward asymmetries which can be expressed in terms of the coupling asymmetries:

$$\mathcal{A}_f \equiv \frac{g_{Lf}^2 - g_{Rf}^2}{g_{Lf}^2 + g_{Rf}^2} = \frac{2g_{Vf}g_{Af}}{g_{Vf}^2 + g_{Af}^2} \tag{36}$$

The couplings are related to the weak mixing angle $\sin^2\theta_w^{\text{eff}}$ by the well known relations:

$$g_{Lf} = I_{3f} - Q_f \cdot \sin^2\theta_w^{\text{eff}} \tag{37}$$

Initial state helicity	cross-section	Final state helicity
Left e^-, right e^+, $\mathcal{P} = -1$ $\overset{\Leftarrow}{\underset{\longrightarrow}{e^-}}$ $\underset{\cdot}{\mathrm{Z}}$ $\overset{\Leftarrow}{\underset{\longleftarrow}{e^+}}$	$\propto g_{Le}^2$	Forward: $\overset{\Leftarrow}{\underset{\longleftarrow}{\bar{f}}}$ $\overset{\Leftarrow}{\underset{\longrightarrow}{f}}$ $\propto g_{Lf}^2$ Backward: $\overset{\Leftarrow}{\underset{\longleftarrow}{f}}$ $\overset{\Leftarrow}{\underset{\longrightarrow}{\bar{f}}}$ $\propto g_{Rf}^2$
Right e^-, left e^+, $\mathcal{P} = +1$ $\overset{\Rightarrow}{\underset{\longrightarrow}{e^-}}$ $\underset{\cdot}{\mathrm{Z}}$ $\overset{\Rightarrow}{\underset{\longleftarrow}{e^+}}$	$\propto g_{Re}^2$	Forward: $\overset{\Rightarrow}{\underset{\longleftarrow}{\bar{f}}}$ $\overset{\Rightarrow}{\underset{\longrightarrow}{f}}$ $\propto g_{Rf}^2$ Backward: $\overset{\Rightarrow}{\underset{\longleftarrow}{f}}$ $\overset{\Rightarrow}{\underset{\longrightarrow}{\bar{f}}}$ $\propto g_{Lf}^2$
$\overset{\Leftarrow}{\underset{\longrightarrow}{e^-}}$ $\underset{\cdot}{\mathrm{Z}}$ $\overset{\Rightarrow}{\underset{\longleftarrow}{e^+}}$	0	
$\overset{\Rightarrow}{\underset{\longrightarrow}{e^-}}$ $\underset{\cdot}{\mathrm{Z}}$ $\overset{\Leftarrow}{\underset{\longleftarrow}{e^+}}$	0	

Figure 24: Helicity effects in $e^+e^- \to Z$ production. The arrows $\Leftarrow$ and $\Rightarrow$ indicate the helicity of the particles.

$$g_{Rf} = -Q_f \cdot \sin^2\theta_w^{\mathrm{eff}} \tag{38}$$

$$g_{Af} = 2(g_{Lf} - g_{Rf}) \tag{39}$$

$$g_{Vf} = 2(g_{Lf} + g_{Rf}) \tag{40}$$

These relations are nowadays used as *definition* for $\sin^2\theta_w^{\mathrm{eff}}$. The values of Neutral Current Couplings and their sensitivity to $\sin^2\theta_w^{\mathrm{eff}}$ are given in table 2.

The observables obtainable with longitudinal polarization are:

- the left-right asymmetry of Z production [65, 66, 67, 68, 1]:

$$A_{LR} = -\frac{1}{\mathcal{P}}\frac{\sigma_{\mathcal{P}} - \sigma_{-\mathcal{P}}}{\sigma_{\mathcal{P}} + \sigma_{-\mathcal{P}}} \simeq \mathcal{A}_e; \tag{41}$$

- the forward-backward polarized asymmetries [69]:

$$A_{FB}^{pol(f)} = -\frac{1}{\mathcal{P}}\frac{(\sigma_{\mathcal{P},F} - \sigma_{-\mathcal{P},F}) - (\sigma_{\mathcal{P},B} - \sigma_{-\mathcal{P},B})}{(\sigma_{\mathcal{P},F} + \sigma_{-\mathcal{P},F}) + (\sigma_{\mathcal{P},B} + \sigma_{-\mathcal{P},B})} \simeq \frac{3}{4}\mathcal{A}_f; \tag{42}$$

Table 2: Numerical values of quantum numbers, Neutral Current Couplings, chiral coupling asymmetry $\mathcal{A}_f$, and sensitivity of $\mathcal{A}_f$ for the four types of fermions. The value of $\sin^2\theta_w^{\rm eff}$ is 0.23.

Fermion type	I_{3f}	Q_f	g_{Af}	g_{Vf}	$\mathcal{A}_f$	$\frac{\partial \mathcal{A}_f}{\partial \sin^2\theta_w^{\rm eff}}$
ν	1/2	0	1	1	1	0
e^-	-1/2	-1	-1	-0.08	0.16	-7.9
u	1/2	2/3	1	0.39	0.69	-3.5
d	-1/2	-1/3	-1	-0.69	0.94	-0.6

where

$$\mathcal{P} = \frac{\mathbf{P}_{e^-} - \mathbf{P}_{e^+}}{1 - \mathbf{P}_{e^-}\mathbf{P}_{e^+}} \tag{43}$$

is the polarization of the e^+e^- system. In the case where only the e^- are polarized $\mathcal{P} = \mathbf{P}_e$ The cross-sections for beam polarization $\pm\mathcal{P}$ are denoted $\sigma_{\pm\mathcal{P}}$. The indices F, B indicate integration over the forward or backward hemisphere.

Without polarization the forward-backward asymmetry can be measured:

$$A_{FB}^{(f)} \simeq \frac{3}{4}\mathcal{A}_e\mathcal{A}_f. \tag{44}$$

In either case, for the tau lepton, the polarization of the final state fermion is measurable. The tau polarization is a function of polar angle,

$$\mathcal{P}_\tau(cos\theta) \simeq -\frac{\mathcal{A}_\tau - \frac{2cos\theta}{1+cos^2\theta}\mathcal{P}_Z}{1 - \frac{2cos\theta}{1+cos^2\theta}\mathcal{P}_Z\mathcal{A}_\tau}, \tag{45}$$

from which one can derive the average, integrated over polar angle,

$$\langle\mathcal{P}_\tau\rangle \simeq -\mathcal{A}_\tau, \tag{46}$$

or another version of the forward-backward polarization asymmetry [70],

$$A_{FB}^{pol(\tau)} = \frac{(\sigma_{\mathcal{P}_\tau=1,F} - \sigma_{\mathcal{P}_\tau=-1,F}) - (\sigma_{\mathcal{P}_\tau=1,B} - \sigma_{\mathcal{P}_\tau=-1,B})}{(\sigma_{\mathcal{P}_\tau=1,F} + \sigma_{\mathcal{P}_\tau=-1,F}) + (\sigma_{\mathcal{P}_\tau=1,B} + \sigma_{\mathcal{P}_\tau=-1,B})} \simeq \frac{3}{4}\mathcal{P}_Z, \tag{47}$$

where $\mathcal{P}_\tau$ is now the polarization of the tau lepton, $\mathcal{P}_Z$ is the longitudinal polarization of the produced Z,

$$\mathcal{P}_Z = -\frac{\mathcal{A}_e - \mathcal{P}}{1 - \mathcal{A}_e\mathcal{P}} \tag{48}$$

and $\mathcal{P}$ is the polarization of the e^+e^- system, equation 43. If the beams are unpolarized, $\mathcal{P}_Z = -\mathcal{A}_e{}^2$.

If longitudinal polarization is available, measurements of A_{LR} and $A_{FB}^{pol(f)}$ for the various types of fermions allow a complete determination of the fermion couplings [69, 72, 73].

[2] Many unnatural minus signs come in these equations because $\mathcal{A}_f$ are defined as Left-Right asymmetries – this is the natural convention for the $SU(2)_L \times U(1)$ standard model –, while polarizations are traditionally defined as +1 for right-handed particles.

The Left-Right polarization asymmetry A_{LR} is a rare example of an observable that combines nearly all advantages. A_{LR} is simply obtained by comparing the total cross-section for producing a Z from a left-handed (σ_L) or right-handed (σ_R) e^+e^- system. It is obvious (Figure 24) that A_{LR} at the Z pole will not depend on the way the Z decays, but only on the electron Neutral Current Couplings [66, 68]. Detailed calculations show that this is affected only by minor and well understood corrections [68, 71].

The statistical precision on A_{LR} and $\sin^2\theta_w^{\rm eff}$ from an exposure with alternating polarization $\mathbf{P}_e$ collecting N Z decays is:

$$\Delta A_{LR} = \frac{1}{\mathbf{P}_e\sqrt{N}} \tag{49}$$

$$\Delta\sin^2\theta_w^{\rm eff} = \frac{\Delta A_{LR}}{7.9}. \tag{50}$$

Given the large cross-section at the top of the resonance a remarkable statistical precision can be envisaged: $\Delta A_{LR} = 0.002$ (statistical) for 10^6 events and 50% polarization. This corresponds to $\Delta\sin^2\theta_w^{\rm eff} = 0.0003$! More realistically for SLC, 10^5 events and 60% polarization are sufficient to match *with one single measurement* the precision of $\Delta\sin^2\theta_w^{\rm eff} = 0.0007$ obtained by combining all the unpolarized forward-backward asymmetries and τ polarization data from the LEP data at the time of the 1993 winter conferences [74].

4.2 *Polarized beams at SLC*

4.2.1 *General layout*

The overall SLC layout with polarized beams is shown in figure 25. A more complete description can be found in [1, 2, 75]. The orientation of the electron spin is shown along its transport from the electron gun, where polarization is generated, to the interaction point.

Pulses of longitudinally polarized electrons are produced by photoemission from a polarized cathode [76], at a rate of 120 Hz. The positrons are created by interaction of a second ("scavenger") electron bunch on a radiator situated 2/3 down the linac, and cannot be polarized. In order to reduce the emittance, the beam is transported to damping rings. The polarization is preserved by combination of spin precession in the transfer line and spin rotation in a superconducting solenoid, so that it is aligned along the damping ring magnetic field. Until 1992, two more superconducting solenoids were used to tune the spin orientation in the transfer line from the damping ring to the linac, so that , taking into account the precession in the arc, the polarization is longitudinal at the interaction point. In 1993, a new optical configuration with flat – instead of round previously – beam profile has been commissioned. The two solenoids situated after the damping ring not only act on the spin, but would generate betatron coupling which is detrimental to flat beam operation. They were replaced by a spin-rotation system based on bumps in the

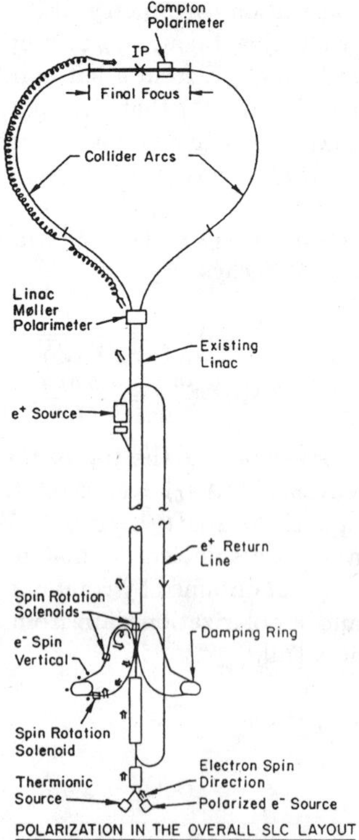

Figure 25

SLC arcs [78], in a way similar to the bumps described for solenoid compensation in LEP (section 3.1, figure 13).

The direction of the polarization vector is monitored at the end of the linac by a Møller polarimeter, and after the IP by a compton polarimeter, as already described in section 2. The performance of SLC has been improving regularly in the last two years, both in luminosity and polarization: the 1992 analysis [20] is based on 10224 hadronic Z decays, with 22% typical polarization. As of July 1993, more than 40000 hadronic Z decays have been recorded with up to 64% longitudinal polarization.

4.2.2 *The SLC polarized source*

The first successful acceleration of polarized electrons to high energy was demonstrated at SLAC in 1974 using an atomic beam source [77]. In 1976, Pierce and Meier [79] observed polarized photoemission from negative affinity Gallium Arsenide. Most polarized electron sources that have been used in accelerators are based on this technique [80]. Such a photocathode polarized electron source was successfully employed in the atomic parity violation experiment [62]. The SLC polarized source is an improved version of it.

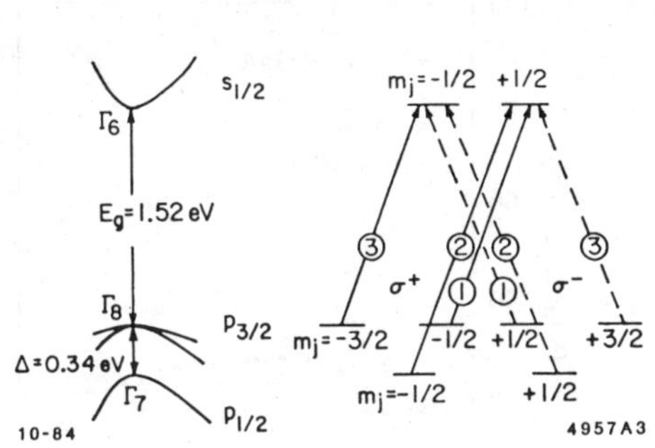

Figure 26: Left: the band structure of GaAs near gap minimum.
Right: the corresponding energy levels of the $S_{\frac{1}{2}}$, $P_{\frac{1}{2}}$ and $P_{\frac{3}{2}}$ states. The allowed transitions for right-handed (left-handed) photons are marked by full (dashed) lines. The angular momentum state m_j indicates the projection on the incident photon direction. The relative transition rates are indicated by the circled numbers.

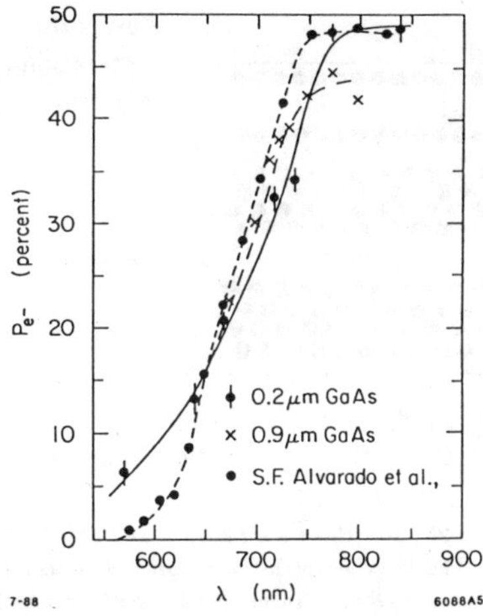

Figure 27: Measured Dependence of the polarization at the output of CsF coated GaAs photocathodes on the incident laser wavelength.

Gallium Arsenide is a semi-conductor with the band structure sketched on figure 26. The absorption of a photon with pumping of an electron from the valence band into the conduction band obeys the selection rule $\Delta m_j = \pm 1$ for right-/left handed photons. If a right-handed photon has an energy that allows the transition ($P_{\frac{3}{2}}$ to $S_{\frac{1}{2}}$), but not ($P_{\frac{1}{2}}$ to $S_{\frac{1}{2}}$), which has an energy gap larger by 0.34 eV, the electron will end up in the $m_j = -\frac{1}{2}$ state with a probability of 3 and in the $m_j = \frac{1}{2}$ state with a probability of 1. m_j is the projection of the angular momentum on the incident photon direction. In the S state, the electron spin carries the total angular momentum, so that $m_j = -\frac{1}{2}$ corresponds to a spin opposite to the photon direction. In a practical set-up where the laser hits a GaAs photocathode at normal incidence, the electron will be accelerated opposite to the incident photon direction. It will therefore be right-handed, with a polarization:

$$\mathbf{P}_e = \frac{3-1}{3+1} = 0.5. \tag{51}$$

In order to make this a source one must allow the electron to come out of the material. This is not normally possible with pure GaAs, and a coating is necessary to reduce the free electron energy to below the conduction band ("negative affinity"). The best coatings are Cs_2O, and CsF. Keeping good quantum efficiency (1-5%) requires regular cesiation [76].

The actual polarization of the electron beam depends strongly on the energy of the incident photon, as shown in figure 27. It is difficult to obtain a sufficiently powerful laser at the low wavelength required by high polarization. In 1992 the laser was a Dye laser at 715 nm. The actual polarization of generated electrons was not 50% but rather lower, slightly below 30%.

It is possible to improve the polarization of the electron beam from GaAs if

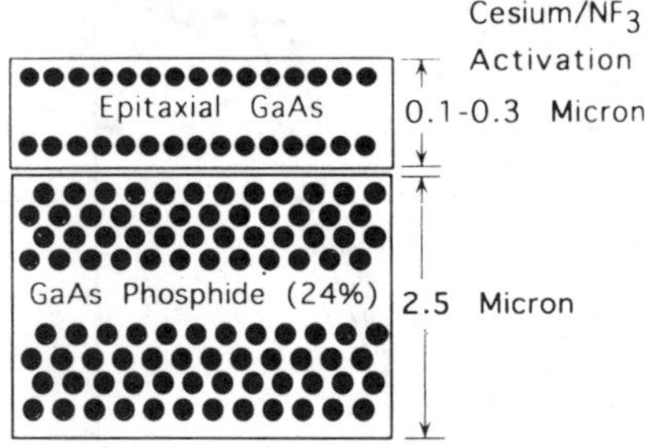

Figure 28: Sketch of a strained GaAs photocathode. The laser beam hit the upper part, coated with Cesium/NF_3. An epitaxial layer of GaAs is grown on a support of $GaAs_{1-x}P_x$ (x=0.24). The slight mismatch in the lattice structure results in a mechanical strain.

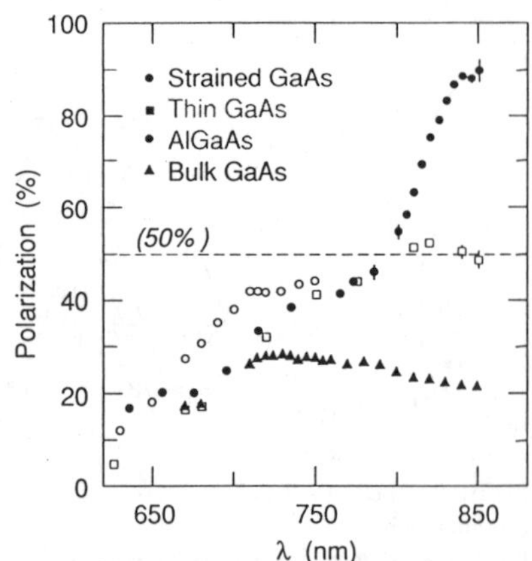

Figure 29: Measured Dependence of the polarization at the output of strained GaAs photocathodes on the incident laser wavelength.

one is able to split the energies of the $P_{\frac{3}{2},\frac{1}{2}}$ and $P_{\frac{3}{2},\frac{3}{2}}$ states so that the $P_{\frac{3}{2},\frac{3}{2}}$ state have higher energy –e.g. the gap is smaller. Then, provided the laser is tuned to a low enough energy, the m= $\frac{3}{2}$ state can be selected and the polarization can reach a theoretical maximum of 100%. This breakthrough was achieved [81] with strained GaAs photocathodes.

It is a well known fact[3] that the band structure of solids is altered by mechanical strain. The band splitting can actually go one way or the other depending whether the material is under compression or tension. Strain photocathodes with high electron polarization were first obtained in a thin epitaxial layer of $In_{1-y}Ga_yAs$ grown on a GaAs substrate, and later from epitaxial GaAs grown on a thick GaAsP buffer layer. The strain is due to a small lattice mismatch of the epitaxial layer with respect to the substrate. The strain can be varied by changing the concentration y or the thickness of the layer. Thinner layers lead to optimal polarization at a lower wavelength. High degrees of electron polarization have been obtained with an epitaxial layer of 0.15 microns of GaAs on a 2.5 microns layer of $GaAs_{1-x}P_x$ (x=0.24) shown in figure 28. The wavelength dependence of the polarization is shown in figure 29. Polarization of up to 90% were obtained.... for a laser wavelength of 850 microns.

This technique is extremely demanding on the laser. In absence of commercial lasers meeting the specifications, a new Ti:Sapphire, optically pumped by two Nd:YAG lasers, was specially developed [82]. It is presently able to run at high power at 850 nm wavelength. Both the strained photocathodes and the laser came

[3] In solid state physics!

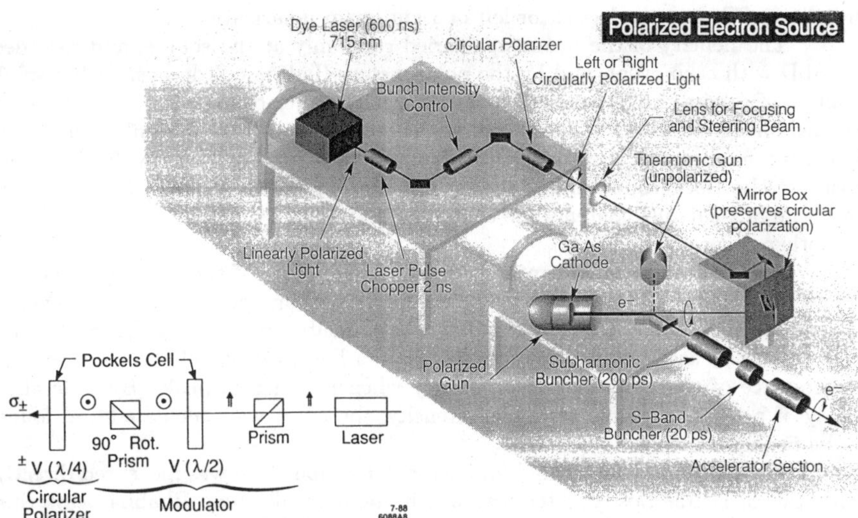

Figure 30: Layout of the SLC polarized electron source. Detail of the circular polarizer for fast helicity reversal is shown on the lower left corner.

into operation in 1993, yielding useable polarization of up to 70%.

The sign of the polarization of the beam is controlled by the helicity of the laser. This is controlled by pockell cells and can be reversed on a pulse basis. The overall layout of the polarized electron source and the detail of the laser helicity reversal system are shown in figure 30. The SLC polarized source was able, in 1992, to deliver 4.10^{10} electrons per bunch at a repetition rate of 120 Hz with a polarization of 29%. In 1993 these numbers have been considerably improved to 7.10^{10} and 70%. The beam transport of the polarized beam to the IP results in some further loss, to 22% and 64% respectively.

4.3 *Measurement of A_{LR} with SLD*

The first measurement of A_{LR} [20] was performed at SLC with the SLD detector [83]. The measurement is based on 10244 Z decays recorded in 1992 with an average beam polarization of (22.4±0.6)%. The center of mass energy was 91.55±0.04 GeV, measured with magnetic beam spectrometers [84].

The left-right asymmetry is measured as:

$$A_{LR} = \frac{1}{|P_e|}\frac{N_L - N_R}{N_L + N_R} \tag{52}$$

where P_e is the average beam polarization of the e^- beam, and N_L, N_R are the numbers of Z decays selected for left or right beam polarization.

The helicity of the beam was varied randomly at the source, and recorded by SLD with each event, and by the polarimeter. Hadronic Z decays, with a small content of τ pairs (1.5%) were selected with a calorimeter-based selection. The beam related backgrounds, as well as the physics backgrounds, "two-photon" interactions and e^+e^- final states, were efficiently reduced to a level of (<0.7%, 0.1%, 0.7%) respectively. Backgrounds dilute the asymmetry and must be corrected for. Because of its large t-channel content, the e^+e^- final state has a different asymmetry than the other Z decays.

The helicity reversal can potentially change some of the beam parameters, namely the energy and, more critically, the luminosity and polarization. The energy difference might come from slightly different intensities resulting in different beam loading in the accelerating cavities. It matters here because of the steep energy dependence of the cross-section on energy when running off-peak. By a detailed study of possible asymmetry in the intensity, the effect was calculated and found to be negligible.

The luminosity integrated for each polarization state was measured directly with the SLD luminosity detectors, which count small angle Bhabha scattering events. Low angle Bhabha scattering has a very small left-right asymmetry, 3 10^{-4} for the angular range of the SLD luminosity detectors, due to the small interference of $Z \to e^+e^-$ decays. The measured luminosity asymmetry was $(1.9 \pm 6.2)10^{-3}$. This uncertainty would add significantly to the experimental error on A_{LR}. A better estimate of the luminosity asymmetry could, however, be inferred from a detailed study of possible asymmetries in the relevant beam parameters, intensity, beam position and beam size, to be $(1.8\pm4.2)10^{-4}$.

The value of A_{LR} is inferred from the measured asymmetry $A_m = (N_L - N_R)/(N_L + N_R) = (2.23 \pm 0.99)\%$ by the formula

$$A_{LR} = \frac{A_m}{\mathbf{P}_e} + \frac{1}{\mathbf{P}_e}\left[A_m f_b + A_m^2 A_{\mathbf{P}_e} - \frac{E}{\sigma}\frac{d\sigma}{dE}A_E - A_\epsilon - A_\mathcal{L}\right] \tag{53}$$

where f_b is the background fraction, $A_{\mathbf{P}_e}$ is the asymmetry in the beam polarization, and $A_E, A_\epsilon, A_\mathcal{L}$ are the asymmetries in the beam energy, detector acceptance and luminosity. The result is:

$$A_{LR} = 0.100 \pm 0.044_{\mathrm{stat.}} \pm 0.004_{\mathrm{syst.}}. \tag{54}$$

The systematic error is a quadratic sum of uncertainties stemming from the beam polarization measurement (2.7%, see section 2.2.2) the luminosity asymmetry (1.9%) and the background contamination (1.4%). These errors are given in relative fraction of the measured A_{LR}.

From A_{LR}, after a very small (0.0003) correction for QED effects for the beam energy being shifted from the Z pole, one can derive the value of $\sin^2\theta_w^{\mathrm{eff}}$:

$$\sin^2\theta_w^{\mathrm{eff}} = 0.2378 \pm 0.0056_{\mathrm{stat.}} \pm 0.0005_{\mathrm{syst.}}. \tag{55}$$

4.4 *Prospects for improvements in the measurement of A_{LR} with SLD*

From the smallness of i) the corrections, and ii) the systematic error on the result, one can deduce that the measurement of $\sin^2\theta_w^{\text{eff}}$ from A_{LR} should soon be considerably improved. As mentioned in earlier sections, major improvements in SLC have taken place in all aspects of the project: integrated luminosity, beam polarization degree and polarimetry. The precision on A_{LR} is given by examination of formula 53 as a quadratic sum ($\oplus$) of the following terms:

$$\Delta A_{LR} = \frac{1}{\mathbf{P}_e}\frac{1}{\sqrt{N}} \oplus A_{LR}\frac{\Delta \mathbf{P}_e}{\mathbf{P}_e} \oplus \frac{1}{\mathbf{P}_e}\Delta A_{\mathcal{L}} \oplus A_{LR}\Delta f_b. \tag{56}$$

It is clear that increasing the degree of polarization brings direct improvement of the measurement. The SLD detector [83] has recorded so far in 1993 more than 40000 Z decays, with an average polarization of 64%. The measurement should still be statistics dominated and reach a precision of $\Delta\sin^2\theta_w^{\text{eff}} = \pm\ 0.0010$. This is already better than the measurement at LEP from the best channel, the b forward-backward asymmetry, where the average from the four LEP experiments (roughly 4 10^6 Z decays) yields an error of $\Delta\sin^2\theta_w^{\text{eff}} = \pm\ 0.0011$, with a larger component of experimental and theoretical systematic errors, moreover. Eventually, it is hoped that SLD will record 750000 Z decays, with a polarization of 75%, yielding

$$\Delta\sin^2\theta_w^{\text{eff}} = \pm 0.00025_{\text{stat.}} \pm 0.00012_{\text{syst.}}. \tag{57}$$

4.5 *Longitudinal polarization at LEP*

In view of the great statistical power and systematic quality of the measurement of A_{LR}, and of the fundamental importance of a high precision measurement of $\sin^2\theta_w^{\text{eff}}$, the feasibility of a longitudinal polarization program at LEP has been studied. The physics potential was reviewed in [64], and the feasibility of spin rotators in [85]. It is not expected that longitudinal polarization experiments will take place before the high energy of LEP, which pushes them to after 1997. By then, high luminosity should be available in LEP by multibunching [86], so that exposures of 5 10^6 Z decays per experiment per year can be envisaged.

Running longitudinal polarization experiments at LEP requires:

- 1. the ability to build-up a large degree of polarization with colliding beams;

- 2. spin rotation;

- 3. control of the polarization of each bunch and polarization measurement;

- 4. measurement of A_{LR} in the experiments.

The polarization build-up has been abundantly discussed in section 3.1. At present 30-40% polarization is obtained by harmonic spin-matching, in orbits that are fully compatible with high luminosity running. To go further would require correction for the only parameter that remains unknown in the orbit measurement, the offset between the position of the beam pick-ups and the center of the quadrupoles. This offset can be measured by survey, which would be very lengthy and not infinitely valid in time, or with a very elegant method that was developed at CESR [87], and recently tested at LEP [88].

The absolute center of a quadrupole is determined by exciting it with a slow frequency (50 Hz) AC current while scanning the beam position with an orbit bump. The response of the beam to the excitation, monitored with a pick-up, shows a steep minimum when the beam goes through the center of the quadrupole, where the magnetic field vanishes. This method allows absolute calibration of the offset with a precision of better than 50 microns. The practical implementation is still to be worked out. Given sufficient stability and reliability of the pick-ups, this should allow transverse polarization in excess of 50% to be easily obtained.

In order to run physics experiments, the high polarization has to be kept with high luminosity. The full compatibility of Polarization with the Pretzel optics necessary for multi-bunch operation is expected and was demonstrated experimentally [40, 89]. The depolarization due to beam-beam interaction, however, has not yet been studied.

Until a high degree of polarization with colliding beams has been observed, it is not possible to assert that longitudinal polarization experiments will be feasible at LEP.

The natural polarization mechanism in the arcs and wigglers is transverse. Spin rotators are needed to make it longitudinal locally in the interaction regions, while keeping it transverse in the arcs. This last point is crucial, fault of which very strong depolarization would take place. The principle sketch of spin rotators is shown in figure 31. After consideration of various schemes [90, 91], the scheme retained for LEP is the Richter-Schwitters spin rotator [92]. The spin-matching conditions [93, 94] for this rotator, shown in figure 31, are rather simple and a practical implementation is feasible [95, 96, 85] without substantial loss of polarization. The complete spin-compensation in presence of solenoids and beam-beam kicks remains to be fully worked out, however, as well as a fully satisfactory scheme to shield the experiments from the strong synchrotron radiation in the last bend of the rotator. A test spin-rotator is proposed [85] to study these issues.

Once the spin is rotated, one finds oneself in a situation where the electrons and positrons have the same helicity, and the net e^+e^- polarization, equation 43, is zero. Furthermore the cross-section is reduced. However, as suggested by [91], the helicity of the e^+e^- system can be made either positive or negative by depolarizing one beam or the other, with, e.g. the kicker used for energy calibration by resonant depolarization. In fact, any one of the bunches in the machine can be depolarized at will to a negligible level [97] offering the following succession of e^+e^- helicity states [98]:

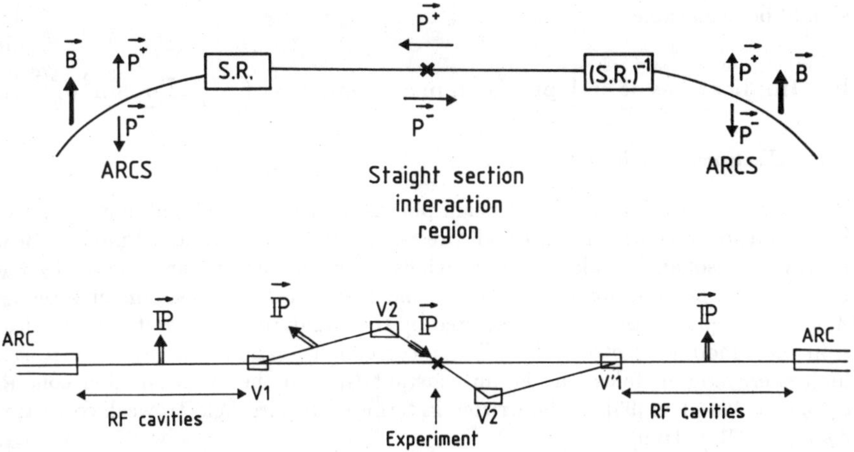

Figure 31: Top: principle of spin rotators for LEP. The spin orientation of electrons (P^-) and positrons (P^+) is indicated.
Bottom: the Richter-Schwitters spin-rotator, showing the direction of the positron spin and the location of the radio-frequency cavities.

$$
\begin{array}{cccccccc}
e^- & & \Rightarrow & & \Rightarrow & & & \\
\longrightarrow & \ominus & \ominus & \ominus & \ominus & \cdots & & \\
 & \Leftarrow & & & \Leftarrow & & & e^+ \\
 & \oplus & \oplus & \oplus & \oplus & \cdots & & \longleftarrow \\
 & 1 & 2 & 3 & 4 & & &
\end{array}
$$

The comparison of the four respective total cross-sections:

$$\sigma_1 = \sigma_u(1 + P_{e^+}A_{LR}) \tag{58}$$

$$\sigma_2 = \sigma_u(1 - P_{e^-}A_{LR}) \tag{59}$$

$$\sigma_3 = \sigma_u \tag{60}$$

$$\sigma_4 = \sigma_u(1 - P_{e^+}P_{e^-} + (P_{e^+} - P_{e^-})A_{LR}) \tag{61}$$

allows a measurement of A_{LR} but also of P_{e^+} and P_{e^-} from the data. The role of the polarimeter is to monitor the evolution of the polarization with time and the possible differences between one bunch and an other, but its absolute calibration is obtained from the data.

This scheme can certainly help reduce the experimental errors on the measurement of A_{LR}. These were studied in great detail in [64]. From this study, and from the knowledge acquired on cross-section measurements at LEP, one can safely assert that, *given sufficient polarization*, a measurement of $\sin^2\theta_w^{\text{eff}}$ to a precision of

$$\Delta\sin^2\theta_w^{\text{eff}} = \pm 0.0001 \tag{62}$$

should be achievable.

5 Implications of high precision measurements of $M_Z, \Gamma_Z, \sin^2\theta_w^{eff}, M_W$

5.1 *Electroweak radiative corrections*

We have seen in this article that beam polarization plays, and will play even more in the future, a central role in high precision tests of the standard model. Beam energy by resonant depolarization provides accurate measurements of M_Z, Γ_Z, M_W. Longitudinal polarization gives access to the most sensitive measurement of $\sin^2\theta_w^{eff}$, A_{LR}. These quantities are not predicted by the standard model, but their relationships are, modulo electroweak radiative corrections. The relations between these high energy observables, the Fermi constant G_F and the QED running constant $\alpha(M_Z^2)^{-1} = 128.8 \pm 0.1$ [99] can be written in terms of universal electroweak corrections $\Delta\rho, \Delta_{3Q}, \Delta r^{ew}$ as [100]:

$$\Gamma_\ell = \frac{G_F M_Z^3}{24\sqrt{2}\pi}[1+\Delta\rho]\left[1+(\frac{g_{V\ell}}{g_{A\ell}})^2\right](1+\frac{3}{4}\frac{\alpha}{\pi}); \tag{63}$$

$$\Gamma_\ell = \frac{\alpha(M_Z{}^2)\ M_Z}{48\sin^2\theta_w^{eff}\cos^2\theta_w^{eff}(1+\Delta_{3Q})}\left[1+(\frac{g_{V\ell}}{g_{A\ell}})^2\right](1+\frac{3}{4}\frac{\alpha}{\pi}); \tag{64}$$

$$M_Z^2 = \frac{\pi\alpha(M_Z^2)}{\sqrt{2}G_F(1+\Delta\rho)(1+\Delta_{3Q})\sin^2\theta_w^{eff}\cos^2\theta_w^{eff}}; \tag{65}$$

$$M_W^2 = \frac{\pi\alpha(M_Z^2)}{\sqrt{2}G_F(1-\Delta r^{ew})(1-\frac{M_W^2}{M_Z^2})}; \tag{66}$$

$$\frac{\sin^2\theta_w^{eff}\cos^2\theta_w^{eff}}{(1-\frac{M_W^2}{M_Z^2})(\frac{M_W^2}{M_Z^2})} = \frac{1-\Delta r^{ew}}{1+\Delta\rho+\Delta_{3Q}}. \tag{67}$$

5.2 *On the importance of* $\alpha(M_Z^2)$

It can be seen from these relations that the error on $\alpha(M_Z^2)$ does not affect *all* of them. In particular, the relations

$$M_Z \leftrightarrow \Gamma_Z \quad \text{and} \quad \frac{M_W}{M_Z} \leftrightarrow \sin^2\theta_w^{eff}$$

are unaffected by $\alpha(M_Z^2)$. Therefore,

the present limitation on the error in $\alpha(M_Z^2)$ ***should not be considered a intrinsic limit to the measurement of electroweak radiative corrections.***

This being said, it is clear that a better knowledge of $\alpha(M_Z^2)$ would bring substantial value. The error on $\alpha(M_Z^2)$ is presently given by the experimental uncertainties in the measurement of the e^+e^- hadronic cross-sections between 1 and

10 GeV [99]. This error could certainly be reduced in two steps: i) a new compilation of e^+e^- cross-sections using the additional data now available; ii) further cross-section measurements using more modern detectors should easily improve on the available data, which often date back more than 15 years. The measurement of these cross-sections will soon become as important for precision physics than the measurement of any of the aforementioned quantities.

5.3 *Higgsometry*

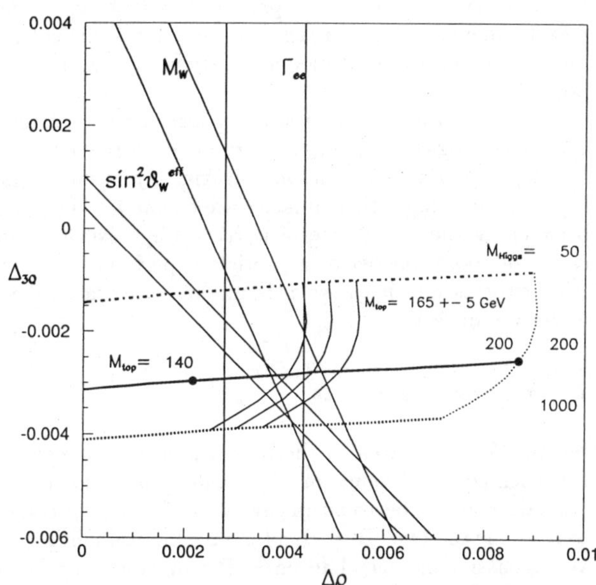

Figure 32: A possible scenario for the ultimate precision of Electroweak measurements at LEP. Minimal Standard Model predictions in the $(\Delta\rho, \Delta_{3Q})$ plane.
MSM predictions: dash-dotted line: M_t free, M_H= 50 GeV; full line: M_t free, M_H= 200 GeV; dotted line: M_t free, M_H= 1000 GeV;
Experimental constraints: vertical band: $\Delta\Gamma_\ell = \pm 0.07$ MeV. 25° band: $\Delta M_W = \pm$ 60 MeV. 45° band: $\Delta \sin^2\theta_w^{\rm eff} = \pm$ 0.0001. The MSM prediction for $M_t = 165 \pm 5$ GeV as expected from direct measurement of the top mass is also indicated.

As an example of the impact of the precision measurements concerned by beam polarization, the constraints in the $\Delta\rho$ vs. Δ_{3Q} plane were derived in [101] for the following set of measurements: Γ_Z to $\pm$ 2 MeV; M_W to $\pm$ 30 MeV; $\sin^2\theta_w^{\rm eff}$ to $\pm$ 0.0001. The standard model relation between Δr^{ew} and $\Delta\rho, \Delta_{3Q}$ was assumed. As can be seen in figure 32, combined with a measurement of the top quark mass, these measurements would certainly place significant limits on the mass of the Higgs boson, or on whatever plays its role.

6 Conclusions

Beam polarization greatly enhances the precision experiments that probe the standard electroweak model and its radiative corrections. So far, it has contributed an extremely precise measurement of the Z boson mass, and is soon to provide substantial improvements to the Z width and $\sin^2\theta_w^{\rm eff}$ measurements. The W mass measurement will also benefit from it. One certainly hopes that, before the end of this century, and to a great part thanks to beam polarization experiments, these measurements will have reached a precision sufficient to significantly constrain the most nebulous sector of modern particle theory, the spontaneous symmetry breaking.

Meanwhile, polarization provides the experimentalist with exquisitely delicate technological challenges, from orbit corrections to 50 microns over 27 km, to the deposition of 0.15 micron epitaxial layers on GaAs substrates. Subtle effects by earth tides and other lunacies are encountered on the way to extreme precision, broadening the experience of particle physicists to sometimes exotic fields. This close collaboration between particle experimentalists, accelerator physicists, solid state experts and laser wizards might be a practical version of the unification of forces we all seek.

7 Acknowledgements

Martin Veltman revealed the beauties of electroweak radiative corrections for me. Friedrich Dydak gave me my first taste of precision experiments. Bryan Lynn, Paul Langacker and Alberto Sirlin gave me different, but convincing, explanations on how to combine the two. Burt Richter gave me brilliant ideas and my old friend Claudio Verzegnassi some very bad ones. Bryan Montague introduced me to spin rotators and Jean Buon to their spin matching equations. Jean-Pierre Koutchouk went with me through some tough politics. Massimo Placidi cooked pasta at 3:00 am while polarization rose. Ken Moffeit taught me clever mirrors to preserve light polarization in the midst of a deep white water adventure. Bernd Dehning insisted, after 47 hours of tide-watching in the control room, to take the last, decisive, point of figure 23. To them and John Thresher, Jean Badier, Herb Steiner, Walter Blum, John Jowett, Eberhard Keil, Charlie Prescott, Marty Breidenbach, Morris Swartz, the LEP polarization and energy calibration teams and all my ALEPH colleagues, many thanks for enjoyable collaboration.

8 References

1. D. Blockus et al., "Proposal for polarization at SLC", SLAC-Prop-1, Stanford (1986).
2. M. L. Swartz, in "Polarization at LEP", G. Alexander et al., eds., CERN Yellow report 88-06 (1988) Vol. II, p. 163.
3. K. Moffeit, Private communication.
4. U. Fano, J. Op. Soc. Am 39 (1949) p. 859.
5. F. W. Lipps and H. A. Tolhoek, Physica, XX (1954) pp. 85 and 395.
6. The photon polarization vector is defined in [5] in terms of the potential vector components. This definition is closely related to the well known Stockes parameters, see e.g. the following reference books:
   W. A. Shurcliff, "Polarized Light", Harvard University Press (1962);
   J. M. Bennett and H. A. Bennett, in "Handbook of Physics", chapt. 10, W. G. Driscoll and W. Vaughan ed., McGraw-Hill Book Company (1978).
7. G. Alexander et al., in "Polarization at LEP", G. Alexander et al., eds., CERN Yellow report 88-06 (1988) Vol. II, p. 3.
8. V. N. Baier and V. A. Khoze, Sov. J. Nucl. Phys. 9 (1969) p. 238.
9. D. B. Gustavson et al., Nucl. Instr. Meth. 165 (1979) p. 177.
10. A. A. Zholentz et al., Phys. Lett. 96B (1980) p. 214.
11. A. S. Artamonov et al., Phys. Lett. 118B (1982) p. 225.
12. A. N. Skrinskii and Yu. M. Shatunov, Sov. Phys. Usp. 32(6) (1989) p. 548.
13. D. P. Barber et al., Phys. Lett. 135B (1984) p. 498.
14. W. W. MacKay et al., Phys. Rev. D29 (1984) p. 2483.
15. D. P. Barber et al., DESY 92-136 (1992) to appear in Nucl. Instr. Meth.
16. K. Nakajima (The TRISTAN Polarization Study Group) "Polarization Measurements at TRISTAN", proc. of the 3d European Particle Accelerator Conf. Berlin, March 1992, H. Henke et al. eds., Editions frontières (1992).
17. M. Placidi and R. Rossmanith, Nucl. Instr. Meth. A274 (1989) p. 79.
18. D. P. Barber et al., "High Spin Polarization at the HERA Storage Ring", DESY 93-038.
19. L. Knudsen et al. (The LEP polarization collab.), Phys. Lett. B270 (1991) p. 97.
20. K. Abe et al. (SLD Collab.) Phys. Rev. Lett. 70 (1993) p. 2515.
21. J. Badier et al., ALEPH Note 87-17 (1987).
22. B. W. Montague, "Polarized beams in high energy storage rings", Phys. Rep. 113 (1984) p. 1.

23. A. A. Sokolov and I. M. Ternov, Sov. Phys. Doklady 8 (1964) p. 1203.
24. R.S. Van Dyck Jr. et al., Phys. Rev. Lett. 59 (1987) p. 26.
25. L. H. Thomas, Philos. Mag. 3 (1927) p. 1;
V. Bargmann, L. Michel and V. L. Telegdi, Phys. Rev. Lett. 2 (1959) p. 435.
26. Ya. S. Derbenev and A. M. Kondratenko, Sov. Phys. JETP 37 (1973) p. 968.
27. A. W. Chao and K. Yokoya, DESY M-82/09.
28. J. Buon and K. Steffen, Nucl. Instr. Meth. A245 (1986) p. 248.
29. J. R. Johnson et al., Nucl. Instr. Meth. 204 (1983) p. 261.
30. A. W. Chao, Nucl. Instr. Meth. 180 (1981) p. 29.
31. A. Ackermann, J. Kewisch, T. Limberg, to be published.
J.-P. Koutchouk and T. Limberg in "Polarization at LEP", G. Alexander et al., eds., CERN Yellow report 88-06 (1988) Vol. II, p. 204.
32. K. Yukoya, KEK report 92-6 (1992);
S. Mane, Phys. Rev. A36 (1987) pp. 105 and 120;
ibid, Nucl. Instr. Meth. A292 (1990) p. 52;
ibid, in "Polarization at LEP", G. Alexander et al., eds., CERN Yellow report 88-06 (1988) Vol. II, p. 233.
33. J. Buon, in "Polarization at LEP", G. Alexander et al., eds., CERN Yellow report 88-06 (1988) Vol. II, p. 243.
34. J. Kewisch et al., Phys. Rev. Lett. 62 (1989) p. 419.
M. Böge, DESY HERA 92-07 (1992) p. 211.
35. J.-P. Koutchouk, in proc. 9 Int. Conf. on High Energy Spin Physics, Bonn, Germany, W. Meyers et al. eds., Springer-Verlag (1990) Vol. II, p. 90.
36. J.-P. Koutchouk, 8th Int. Symp. on H. E. Spin Physics, Minneapolis 1988.
37. D. P. Barber et al., DESY 82-076 (1982);
K. Steffen, Internal report DESY PET-82 (1982), unpublished.
38. R. Rossmanith, LEP-note 525 (1985).
39. A. Blondel, LEP-Note 629 (1990).
40. M. Placidi, "polarization results and future perspectives", proc. third Chamonix workshop on LEP performance, J. Poole editor, CERN-SL/93-19(DI) (1993) p. 281.
41. R. Rossmanith and R. Schmidt, Nucl. Instr. Meth. A236 (1985) p. 231.
42. H. D. Bremer et al., DESY-M 82-026 (1982)
43. A. Blondel and J. M. Jowett, LEP-note 606 (1988).
44. J. Buon, proc. 12th Int Conf. on High energy accelerators, Batavia (1984), and LAL-RT/83-12 (1983); J. Buon, Journal de Physique 46 (1985) C2-637.
45. R. F. Schwitters et al., Phys. Rev. Lett. 35 (1975) p. 1320.
G. Hanson et al., Phys. Rev. Lett. 35 (1975) p. 1611.

46. V. Hatton et al., CERN LEP performance note 12 (1990).

47. A. Hofmann et al., 2nd European Particle Acc. Conf., Nice 1990.

48. D. Decamp et al. (ALEPH Coll.) Z. Phys. C53 (1992) p. 1;
P. Aarnio et al. (DELPHI Coll.), Nucl. Phys. B367 (1991) p. 511;
B. Adeva et al. (L3 Coll.), Z. Phys. C51 (1991) p. 179;
G. Alexander et al. (OPAL Coll.), Z. Phys. C52 (1991) p. 175;
(The LEP collaborations) Phys. Lett. B276 (1992) p. 247.

49. L. Arnaudon et al. (The LEP polarization collaboration), Phys. Letters B284 (1992) p. 431.

50. M. Froissard and R. Stora, Nucl. Inst. Meth. 7 (1960) p. 297.

51. J. Buon, "Interference effect between depolarization resonances...", LAL-RT 87-09 (1987).

52. L. Arnaudon et al. (The working group on LEP energy) "The energy calibration of LEP in 1991" CERN-PPE/92-125, CERN-SL/92-37(DI) (1992).

53. G. Fischer and A. Hofmann, in second workshop on LEP performance, J. Poole ed., CERN-SL/92-29 (DI) (1992), p. 337.

54. A very complete survey of the subject of earth tides, covering both theory and experiment can be found in:
P. Melchior, "The tides of planet earth", 2nd edition, Pergammon press (1983).
see also G. Berger and R. H. Lovberg, Science 170 (oct. 1970) p. 296.

55. L. Arnaudon et al., "effect of tidal forces on the beam energy of LEP", presented by M. Placidi at the 1993 PAC conference, Washington, to appear in the proceedings.

56. D. Buskulic et al. (ALEPH coll.), "Update of Electroweak Parameters from Z Decays", CERN-PPE/93-40 (1993), to be published in Phys. Lett. B. (1993);
P. Abreu et al. (DELPHI coll.), " Measurements of the lineshape of the Z and determination of electroweak parameters from its hadronic and leptonic decays", to be published in Nucl. Phys. B (1993);
O. Adriani et al. (L3 coll.), " Results from the L3 Experiment at LEP", CERN-PPE/93-031, subm. to Physics Reports (1993);
P.D. Acton et al. (OPAL Coll.), "Precision Measurements of the Neutral Current from Hadron and Lepton Production at LEP", CERN-PPE/93-003 (1993), subm. to Z. Phys. C.

57. The Working group on Lep Energy and the LEP collaborations ALEPH, DELPHI, L3 and OPAL, Phys. Lett. B307 (1993) p. 187.

58. G. Burgers, in "Polarization at LEP", G. Alexander et al., eds., CERN Yellow report 88-06 (1988) Vol. I, p. 121;
G. Burgers, F. Jegerlehner, B. Kniehl and J. Kühn, Proceedings of the Workshop on Z physics at LEP, CERN Report 89-08 Vol.I, 55;
M. Consoli, W. Hollik and F. Jegerlehner, ibid, p. 7;
M. Martinez et al., Z. Phys. C49 (1991) 645;

D. Bardin et al., Z. Phys. C44 (1989) 493; Phys. Lett. B229 (1989) 405; Nucl. Phys. B351 (1991) 1; Phys. Lett. B255 (1991) 290 and CERN-TH.6443/92 (1992).

59. There was a full session on Energy calibration and polarization at the third Workshop on LEP performance, J. Poole editor, CERN-SL/93-19(DI) (1993) pp. 277-374.

60. D. Treille, this volume.

61. P. Galumian et al., "A method for precise energy calibration of electron beam energy", Université de Lausanne preprint, IPNL 92-2 (Aout 1992, add. Nov, 1992).

62. C. Y. Prescott et al., Phys. Lett. B77 (1978) p. 347.

63. P. Darriulat, in 'proceedings of the LEP summer study', CERN 79-01 (1979), p. 219.

64. "Polarization at LEP", G. Alexander et al., eds., CERN Yellow report 88-06 (1988).

65. C.Y. Prescott, proc. Int. Symp. on High-Energy Physics with Polarized Beams and Polarized Targets, Lausanne, 1980, C. Joseph and J. Soffer (ed.), Birkhäuser Verlag, Basel (1981) p. 34.
M. Böhm and W. Hollik, Nucl. Phys. B204 (1982) p. 45.

66. B.W. Lynn and R.G. Stuart, Nucl. Phys. B253 (1985) p. 84;
B.W. Lynn, in [64], Vol.I, p. 24;
D.C. Kennedy et al., Nucl. Phys. B321 (1989) p. 83;
see also W. Hollik, this volume.

67. B.W. Lynn, M.E. Peskin and R.G. Stuart, SLAC-PUB 3725 (1985), 'Physics at LEP' CERN 86-02, Geneva (1986) p. 90.

68. B.W. Lynn and C. Verzegnassi, Phys. Rev. D35 (1987) p. 3326.

69. A. Blondel, B.W. Lynn, F.M. Renard and C. Verzegnassi, Nucl. Phys. B304 (1988) p. 438.

70. S. Jadach and Z. Wąs, *in* 'Z Physics at LEP I', G. Altarelli, R. Kleiss and C. Verzegnassi eds., CERN 89-08 (1989) Vol I, p. 235.

71. B.A. Kniehl, J.H. Kühn and R.G. Stuart, in [64], Vol.I, p. 158.

72. D. Treille, in [64], vol.I, p. 265.

73. J. Drees, K. Mönig, H. Staeck and S. Überschär, in [64], vol.I, p. 317;
P.J. Dornan, in [64], vol.I, p. 344.

74. D. Schaile, this volume;
V. Innocente, presentation at the XXVII th Rencontres de Moriond, Electroweak Interactions and Unified Theories, Les Arcs, March 1993, to appear in the proceedings.
See also: M. Pepe-Altarelli, invited talk at "Les rencontres de physique de la vallee d'aoste", Results and Perspectives in Particle Physics (1993), preprint

LNF-93/019 (P) (1993).

75. N. Phinney, "Review of SLC performance", SLAC-Pub-5864 (1992).
76. D. Schultz et al.,"Polarized source performance for SLC-SLD", SLAC-Pub 5768 (1992); ibid., SLAC-Pub 6060 (1993).
77. P. S. Cooper et al., Phys. Rev. Lett. 34 (1975) p. 1589.
78. T. Limberg, P. Emma, and R. Rossmanith, "The north arc of SLC as a spin rotator", Proc. of the Particle Accelerator Conference, Washington, USA, (1993).
79. D. T. Pierce and F. Meier, Physical Review B13 (1976) p. 5484.
80. A revue of low energy polarized electron sources can be found in: C. K. Sinclair, SLAC-Pub-3505 (1984); also published in proc. VIth international conference on high energy spin physics, J. Soffer ed., Les Editions de Physique, Les Ulis, France (1985).
81. T. Maruyama et al., Phys. Rev. Lett. 66 (1991) p. 2376; T. Nakanishi et al., Phys. Lett. A158 (1991) p. 345; T. Maruyama et al., Phys. Rev. B46 (1992) p. 4261; A. Aoyagi et al., Phys. Lett. A167 (1992) p. 415.
82. J. Frisch, M. Woods, Z. Zolotorev, "A New Ti:Sapphire laser for the new polarized electron source", SLAC-Pub-5950.
83. "The SLD design report", SLAC-Report 273, 1984.
84. J. Kent et al., SLAC-Pub 4922 (1989).
85. C. Bovet et al., "A study of longitudinal polarization at LEP", E. Keil and J.-P. Koutchouk eds., CERN-SL/92-10(AP) (1992).
86. "Report of the working group on high luminosities at LEP", E. Blucher et al., CERN 91-02 (1991).
87. D. Rice et al., IEEE trans. Nucl. Sci., Vol. NS-20, 4 (1983).
88. R. Schmidt, "Misalignments from K-modulation", proc. 3d Chamonix workshop on LEP performance, J. Poole editor, CERN-SL/93-19(DI) (1993) p. 139.
89. J. M. Jowett, proc. 3d Chamonix workshop on LEP performance, J. Poole editor, CERN-SL/93-19(DI) (1993) p. 293.
90. B. W. Montague, CERN-ISR-Th/80-39 (1980); J. Buon, Journal de Physique 46 (1985) C2-631.
91. M. Placidi and R. Rossmanith, LEP-note 545 (1985).
92. R. Schwitters and B. Richter, PEP Note 87 (1974);
93. A. Blondel, LEP-note 603 (1988)
94. J. Buon, LAL-RT 88-02 (1988)
95. A. Blondel and E. Keil, in [64], vol.II, p. 250.
96. E. Söderström, CERN/LEP-TH/89-58, (1989).
97. J. Buon and J.M. Jowett, LEP Note 584 (1987)

98. A. Blondel, Phys. Lett. 202B (1988), p. 145.

99. H. Burkhardt, F. Jegerlehner, G. Penso and C. Verzegnassi, Z. Phys. C43 (1989) p. 497.

100. This exercise has been performed by many different authors. see, e.g. P. Langacker, this volume. The notations here are those of ref [101], and A. Blondel and C. Verzegnassi, CERN-PPE/93-81 (1993) to appear in Phys. Lett. B.

101. A. Blondel, F. M. Renard and C. Verzegnassi, Phys. Lett. B269 (1991) p. 419.

THE LEP 200 PROGRAMME

D. Treille
CERN, CH-1211 Geneva 23, Switzerland

Introduction

One of the objectives of the LEP programme is the study of e^+e^- collisions at center of mass energies around 200 GeV, an entirely unexplored physics domain. This goal, foreseen since the earliest studies, was one of the reasons for choosing such a large radius for LEP. Once a few millions Z^0 per experiment will have been registered, raising the energy is the option unanimously preferred by the LEP community. However the ultimate performance of LEP 200 was never precisely defined. The very first studies (1) were quoting energies up to 130 GeV per beam for a larger machine. The slightly reduced radius retained should still make it possible, if the adequate means are provided, to approach in a later stage 120 GeV per beam, which corresponds to the limit of the magnets. Presently the firm orders concern a given number (192) of radiofrequency cavities ; depending on the accelerating voltage they will achieve, this should provide a CM energy between 165 and 175 GeV. The hope of the community is to reach at least 190 GeV CM and accumulate a luminosity of 500 pb^{-1} per experiment.

1. The present situation of LEP

The LEP machine had a quite successful start. The 4 experiments have already registered altogether about 4 million Z^0s. Machine development has been actively pursued as well. It led to the observation of transverse polarization of the beams and to its use for a very accurate determination of their energy through the resonant depolarization method (2). From now on LEP will exploit a 8 bunch scheme (instead of 4) made possible by the addition of separators and which in principle should double the luminosity.

The Standard Model (SM) of electroweak interactions has already been tested at LEP with an unprecedented accuracy. The mass of the Z^0, now known to about one part in

10^4, has reached the prestigious status of becoming the third fundamental constant of the SM, which, with the fine structure constant α and the weak coupling constant G_μ, constitute the basic input to the calculation of other SM observables.

Such predictions are also dependent on the still unknown top mass and, to a lesser extent, Higgs mass. This happens because these particles, even if they are not directly produced yet, intervene as virtual states in the electroweak processes. Z^o lineshape and asymmetry observables, measured with accuracy and confronted with the SM expectations, have led, with the help of former results from neutrino physics and hadron colliders, to an indirect determination of the top mass, as we discuss later.

Direct searches have excluded the existence of the Higgs boson in its standard version up to a mass of about 55 GeV.

For all these studies early results of SLC and ingenious methods pioneered around this machine have been of great help.

What about the next steps ? Eagerly waited for is the discovery of the top and a direct and semi accurate measurement of its mass (± 5 GeV) : this should happen in the near future at the Fermilab collider if the limit quoted below makes sense and is not an artifact of a more complicated physical world. At LEP 1 the accumulated statistics and a better control of the systematical errors will lead to still better measurements of the Z^o observables. As for LEP 200, it will bring at least three major breakthroughs :

- a very accurate (better than 10^{-3}) measurement of the W mass, a key observable of the SM, measurement to which the Fermilab collider should contribute as well.
- the first direct measurement of the triple boson couplings.
- the exploration of the Higgs sector up to a mass about equal to

$$M_H = \sqrt{s} - 100 \text{ GeV}$$

as we will explain later.

2. The LEP 200 machine

It is well known that electron circular machines are limited in energy by the rapid growth of the synchrotron radiation (SR) emission. The energy loss per turn is, in MeV

$$U_s = 0.0885\ \frac{E_{GeV}^4}{\rho_m}$$

One notices the dependence in the 4th power of the energy at a given radius of curvature ρ. While at LEP 1 the loss per turn is a modest 130 MeV, it will become 2.3 GeV at E_{beam} = 95 GeV, i.e. 2.5 % of the particle energy.

The critical energy

$$E_c = \frac{3}{2}\hbar\, c\, \gamma^3/\rho\ ,$$

where $\gamma = E/m_e$, gives an idea of the typical energy of the emitted photons. At LEP 1 it is about 100 keV, at LEP 200 it will reach the MeV level.

The lost energy has to be restored by using radiofrequency cavities. The warm copper cavities presently used at LEP 1 will be totally insufficient at high energies and the LEP energy upgrade depends primarily on the realization of a large set of high-performance superconducting cavities. This is the main challenge of LEP 200 and we will see in the next section where we stand in this respect.

This is not however the only problem to be solved. One has also to implement in the LEP machine a new beam optics, more tightly focused. To operate LEP at high energies in the neighbourhood of 100 GeV per beam, a lattice with 90^0 phase advance per cell is needed to maximize luminosity by keeping the colliding beams near the beam beam limit.

Such an optics was successfully commissioned up to 46.5 GeV during the machine development of last year (3). A new 90^0 optics, which required substantial changes to all the insertions to make it compatible with all known requirements, in particular with a scheme having more bunches per beam, is now the basic option for all LEP operation. Enormous progress was made in 91. By increasing the bunch length it was possible to accumulate a current of more than 600 μA per bunch. For small beam currents the specific luminosity (the luminosity divided by the product of stored currents) was about a factor 2 higher that the maximum ever achieved with the 60 degrees optics.

Hardware-wise a number of modifications of the LEP lattice affecting about 10% of its circumference must be performed for LEP 200. For instance some of the quadrupole magnets in the straight sections would presently run out of focusing strength above 65 GeV. The lattice modifications and the installations of new components will happen in the long winter shut downs.

3. The RF cavities

But the main issue is to provide the required RF accelerating voltage. What matters is not only its total amount, but also its distribution around the ring. At 95 GeV the energy loss per turn due to synchroton radiation is, as we saw, 2.5 % of the beam energy : this loss occurs continuously in the arcs while the energy is restored at discrete locations, and the mean energy of particles actually varies in a sawtooth manner around the ring. The reflection symmetry present in the LEP layout is destroyed by the closed orbit distortions remaining after correction and all beam parameters are slightly different for e^+ and e^- bunches. In particular the beams may miss each other at the interaction point if their positions are not corrected by electrostatic separators. Such effects are proportional to

$$\frac{E^3}{n_{RF}}$$

(where n_{RF} is the number of RF stations) and the desire to minimize these effects led to the choice of installing RF stations around all even interaction points (IP) as shown in fig. 1.

In order to operate LEP 200 in an economical way, superconducting cavities (SCC) have been developed since the earliest design (4). The scheme presently retained for LEP 200 is the one presented in fig. 1. It consists of 192 SCC cavities which, together with the existing copper cavities, should upgrade the beam energy above the W pair threshold by 1994 (5).

Fig. 2 shows the number of cavities (SCC alone, no contribution from copper ones) and the power needed to reach a given energy. Two assumptions about the accelerating field are shown : 6 and 8 MV/m. A slightly higher number of cavities is actually needed to cope with eventual equipment failures. The possibility to add more cavities in LEP exists : the decision to do so will depend on their achieved performances and on the general policy of the laboratory.

4. The physics at LEP 200

The physics capabilities of LEP 200 have been thoroughly studied in the past (6). However ideas and focuses have changed, new motivations have appeared. The competition of the Fermilab hadron collider is more precisely defined. The potential of the future large machines (hadron super colliders or e^+e^- linear collider) has been sharpened, so that it is possible now to provide a broader perspective. I will concentrate on three main topics : the W mass, coupling measurements and the Higgs search. I will use abundantly the results of a very recent Workshop organized by the LEPC on LEP 200 physics (5).

4.1. The measurement of M_W (7)

In the Standard Model, two bosons, the $W^\pm$, carry the electroweak charged current. They were discovered at CERN in 1983 together with their neutral partner, the Z^0. But, unlike in the Z^0 case, their properties have not yet been measured with great precision.

Presently the hadron colliders at CERN and Fermilab have provided the W mass with an uncertainty of about $\pm$ 300 MeV. The first aim of LEP 200 will be to determine this quantity with the highest possible accuracy.

4.1.1. The exposure of LEP 200

With 500 pb^{-1} per experiment at $\sqrt{s}$ = 190 GeV the number of expected W pairs is ~ 8000. This corresponds to ~ 4000 purely hadronic decays and ~ 1300 decays containing a lepton of a given family. These numbers will be still reduced by acceptance problems. The results of this section are obtained from simulations where such a population of W pairs is generated and processed with realistic detector acceptance and properties.

4.1.2. Why should one measure M_W precisely?

As described in this book, the top quark and the Higgs boson of the SM, even if they are too heavy to be directly observable at LEP200 affect all other parameters through radiative corrections. The measurement of each quantity will thus yield a relation between $\sin^2\theta_w^{eff}$ and M_t. M_H will also intervene, but weakly. Simultaneous fits to several observables can then provide M_t and, potentially, M_H.

Fixing M_H (300 GeV), one gets for M_t the allowed range $M_t = 137 \pm {}^{+22}_{-25}{}^{+18}_{-22}$. The present electroweak data have been used. The second error is for M_H = 1000 GeV/c^2 (upper error) and 50 GeV/c^2 (lower error). A constraint $\bar{\alpha}s = 0.114 \pm 0.007$ has been imposed. If one fixes the central values to M_t = 140 GeV, M_H = 300 GeV and $\alpha_s = 0.114$, the sensitivity with current data ($\leq$ 1991) is

$$\Delta M_t = {}^{+47}_{-53}\ \mathrm{GeV}$$

but M_H can still be in a region as large as (50 $\rightarrow$ 1000 GeV).

LEP 1 errors will decrease with more statistics but the improvement will not be sufficient to lead to interesting bounds per se on the Higgs mass. The two major steps to be expected are the measurements of M_W and M_t.

Fig. 3 illustrates the fact that the M_W measurement, even if it is accurate ($\pm$ 50 MeV), will not improve much the sensitivity on M_H until M_{top} is obtained as well. If M_{top} is measured to $\pm$ 5 GeV, the uncertainty on M_H in the SM will be, for a central value of 300 GeV:

$$M_H = 300\ {}^{+220\ \mathrm{GeV}}_{-140\ \mathrm{GeV}} \qquad \text{(fig. 4)}$$

an already very interesting result.

4.1.3. Present situation and future competition to LEP 200

The present knowledge of M_W is obtained through a direct measurement of $W \rightarrow l\upsilon$ at hadron colliders. Table 1 gives the UA2 and CDF results and their combined value.

Future improvements are expected from the CDF and DØ detectors at the Fermilab collider, which in principle will have collected 100 pb^{-1} by the end of 1993. For CDF, with an increased acceptance, this should multiply the number of existing $W \rightarrow e\upsilon$ and $W \rightarrow \mu\upsilon$ by ~ 25 and ~ 40, respectively. By that time CDF hopes to reach an error

$$\Delta M_W = \pm\ 100\ \mathrm{MeV/c^2}$$

about 40% statistical and 60% systematic. DØ should get a similar accuracy.

4.1.4. How to measure M_W at LEP 200

Three methods have been considered.

A) the direct reconstruction of W's using either hadronic events, where both W decay into jets, or hadronic-leptonic events, where one W decays into jets and the other one into $l\upsilon$. In both cases the beam energy and the equality of both masses are used as constraints.

B) the shape of the excitation curve, for which a scan is performed near and above the threshold: from the measured shape of the cross section one deduces the value of M_W.

C) the lepton end point energy: the measurement of M_W is obtained from the shape of the lepton spectrum of leptonic decays. The lepton energy is bounded by

$$\frac{1}{2}\left(E_{beam} - \sqrt{E_{beam}^2 - M_W^2}\right) < E_\ell < \frac{1}{2}\left(E_{beam} + \sqrt{E_{beam}^2 - M_W^2}\right)$$

A first remark is that all three methods require the knowledge of E_{beam}: one should find a way to measure it accurately. A second conclusion is that since photons emitted during the interaction take away some energy and reduce the effective E_{beam}, one should know very accurately the spectrum of these radiated photons.

A) The direct reconstruction method

We can distinguish three categories of W pair events:

- all hadronic decay (a)
- hadronic-leptonic decay (b)
- all leptonic decay (c)

Invariant masses are reconstructed from the decay products using the energy-momentum conservation and the equality of the two masses as constraints. Neutrinos are undetected and the energy of taus is not measured. One should therefore ignore $W \to \tau \upsilon_\tau$ decays, as well as purely leptonic events. On the other hand, jet directions and e, μ energy and directions are, in principle, correctly measured.

(a) For hadronic events the procedure is the following:

- one rejects events with too small total energy or too much missing momentum.
- one reconstructs events as 4 jets: one way is to maximize a generalization of thrust to 4 jets axes, the 4-thrust T4. By rejecting events with T4 < 0.94, one eliminates broad events associated to hard gluon radiation.
- associating the 4 jets 2 by 2, one performs a constrained fit for each of the 3 combinations: the jet energies and momenta are corrected using the available constraints.
- if E_{min} is the energy of the lowest energy jet and θ_{min} is the smallest interjet angle, one rejects events with

$$E_{min} \times \theta_{min} < 10 \text{ GeV} \cdot \text{rad}$$

which are likely to correspond to QCD background.
- one selects the favoured combination of jets on the basis of χ^2.

Fig. 5 (DELPHI) shows the result for the fitted W mass. The flat background comes from bad combinations, not from the apparatus performance. The reconstruction efficiency is ~ 34%, the loss being due to the elimination of hard gluon radiation. The statistical error, for 500 pb^{-1}, is $\pm$ 70 MeV/c^2.

The fitted mass is 422 MeV/c^2 larger than the input mass. This fact, due mostly to the effect of initial state radiation (ISR), exhibits the main difficulty of the reconstruction method.

To check biases in this method, one can run the Monte Carlo without initial state radiation; then one finds that:

i) due to the constraints applied, energy measurement errors do not contribute to the offset.
ii) on the other hand, since the two jets are nearly back to back, the mismeasurement of a particle tends on the average to decrease the angle beween jets and therefore to decrease M_W.

The net result is that, without ISR, M_W is reconstructed ~ 130 MeV/c^2 lower than the generated value.

As for the effects of ISR, they can be decomposed in two components:

i) the photon, if detected, is assigned to a jet: this leads to a mismeasurement of the jet direction and decreases M_W.
ii) if the photon is not detected, the fit assumes a nominal energy, while the effective energy is in fact lowered by the radiation of E_γ; this leads to a positive offset

$$\Delta M = E_\gamma \frac{M_W}{\sqrt{s}}$$

At E_{beam} = 88 GeV, the mean radiated energy is 1.4 GeV and ΔM= +600 MeV/c^2. Combining i) and ii) one finds that the shift due to ISR is $\Delta M \cong$ +500 MeV/c^2.

(b) Hadronic-leptonic events have been studied similarly (ALEPH), using Pythia and Galeph. The neutrino momentum is defined as:

$$P_{\upsilon} = -P_{miss}$$

Angles are kept as they are measured. Jet energies come from a fit with the constraint

$$(E_{J1} + E_{J2} + E_{l} - \sqrt{s})^2 - E_{\upsilon}^2 = 0$$

The masses M_{q1q2} and $M_{e\upsilon}$ are calculated separately and their average is taken. The result is shown in fig. 6. Only e and μ are used in this exercice.

The statistical error on the mass, from that method, is 89 ± 11 MeV/c^2 for 500 pb^{-1}.

Combining hadronic events and hadronic-leptonic ones, one finds for $E_b = 88$ GeV and 500 pb^{-1} a statistical error of $\pm$ 55 MeV/c^2. OPAL has performed a similar analysis, with the same results.

The dependance of the statistical error on E_{beam} is not strong. Fig. 7 (ALEPH) shows that it may get ~ 20% worse in going from $\sqrt{s}$ = 176 to $\sqrt{s}$ = 190 GeV. This is not a decisive fact and the choice of energy has thus to be made by considering other physics topics in addition.

B) The excitation curve method

The sensitivity of the WW cross section as a function of M_w is given by fig. 8. One sees that, since the expected statistical accuracy is $\pm$ 1% at best, the excitation curve can only be a good estimator of M_w near threshold, around 164 GeV.

But the low value of the cross section near threshold will limit the potential of other W physics. Furthermore the width and radiative correction effects are important in that region.

Given a total integrated luminosity of 500 pb^{-1}, one can as an example consider two energies, 180 and 164 GeV, and vary the sharing of luminosities between them. The result is given by fig. 9. One sees that the best result is obtained by adopting an equal sharing and corresponds to ± 100 MeV/c^2 of statistical error. Combined with the direct reconstruction method, this excitation curve method would only give a ~ 10% overall improvement.

C) The electron end point method

The spectra corresponding to different values of M_W start being different only above $E_l \simeq 65$ GeV. The statistics is low (fig. 10). The energy calibration of the detector must be excellent, since 1% miscalibration corresponds to a shift of 0.7 GeV in mass.

One finds at best $\Delta M_W \gtrsim 150$ MeV/$c^2 \sqrt{\frac{500 \text{ pb}^{-1}}{\varepsilon L}}$ where ε is the lepton detection efficiency and L the integrated luminosity.

In summary the reconstruction method is statistically more promising than the other ones. One should now consider in detail the systematic errors associated to that method.

4.1.5. Systematic errors of the reconstruction method

A) The beam energy

We saw that an accurate knowledge of the beam energy is necessary, since a procedure of constraint is used. At LEP 1 this is possible through the resonant depolarization method (2). Unfortunately it is likely that polarization will not be obtained at LEP 200: the expected energy spread of the beam is $\Delta E \simeq 150$ MeV, a value dangerously comparable to the interval of 440 MeV between polarization resonances. From magnetic measurements only, one can expect an uncertainty of ~ ± 26 MeV.

With a calibration at 47 GeV through resonant depolarization and an extrapolation to 90 GeV using the flux loop method (2), one can obtain ~ ± 17 MeV.

The best way would be to install a H_2 gas jet target in the machine and detect Moeller scattering of the beam electrons by the gas jet electrons. From the measurement of the minimum opening angle between the two outgoing electrons, one can expect an accuracy of ~ ± 2 MeV

B) The radiative correction (9)

Much effort has been devoted to the study of radiative corrections and their implementation.

The Dubna-Zeuthen group has now written a program to calculate initial state radiative corrections, taking into account the finite width of the W and higher order terms (table 2). The biggest uncertainty in the calculation comes from the non inclusion of the $0(\alpha^2)$ non universal t-channel contribution. This is evaluated to be:

$$\Delta M_W \sim \pm 5 \text{ MeV/c}^2$$

However an ideal Monte Carlo, needed to estimate the shift, should include a correct description of initial and final state radiation, in particular:

- the correct angular and energy distribution of the radiated photon.
- the correlation between the angle and energy of the photon.
- the correlation between the radiated photon and the W directions.
- higher order effects.

No Monte Carlo has all these effects implemented yet: to take this into account, the systematic error resulting from ISR and FSR has been increased to:

$$10 \text{ MeV/c}^2$$

However, such event generators providing the correlations needed are on their way (Dubna-Zeuthen group, KEK-LAPP Collaboration).

C) Control of systematics (10)

Once the mass is reconstructed from real data, the result will have to be shifted by the offset found in the Monte Carlo. The problem is to check the Monte Carlo.

A method using mixed Lorentz boosted Z^0 events, from real data, to simulate W pairs, has been proposed by DELPHI (8): the mass shift in the reconstructed Z^0 mass is: -140 MeV/c^2. The similarity of this result with the shift found in the MC generated W^+W^- events means that one can trust the shift prediction to ~ 10 MeV/c^2.

The exact knowledge of detector performances is not critical. Results on hadronic events are essentially independent of the energy and momentum resolution of the detectors, because of the beam energy constraint. Systematic shifts in energy will not lead to more than a 10 MeV/c^2 systematic error (fig. 11). Background effects are not expected to contribute appreciably in any category of events. Table 3 summarizes the systematic errors.

4.2. Measurement of the triple boson vertex (11)

LEP 200 is the machine where 3-boson couplings will be directly observed for the first time. The W pair production involves graphs like the ones of fig. 12a,b.

4.2.1. The general picture (12,13)

The SM predicts a uniquely defined Yang Mills form with a given coupling constant for γWW and ZWW at tree level. New forms, associated with q^2 dependent form factors, are generated by higher order terms. One must check whether these effects agree with standard model computations. Indeed new phenomena, like a non standard mass generation mechanism or an extension of the SM through an additional group factor, would lead to modifications of the multi boson couplings.

Table 4 lists the independent VWW couplings and shows their SM values. The Yang Mills couplings $g_{\gamma}^{SM} = 1$ and $g_{ZWW}^{SM} = \cot\theta_w$ can be affected by an additional term δ_V (q^2). The other forms do not exist at tree level in the SM. One can define another set of coefficients, such that they intervene linearly in the extra piece of the Lagrangian, as shown in table 5. The physical meaning of the couplings is suggested by table 6, which gives their contributions to the various helicity amplitudes of the W pair production.

4.2.2. Restrictions of the number of parameters

By using symmetry relations or adopting special model assumptions one can reduce the number of free parameters and form factors.

Symmetry restrictions can be obtained from space time or internal symmetries as shown in table 4. With for instance C,P and CP invariances there are 3 free parameters left for the ZWW case.

The impact of theoretical restrictions varies with the severity of the assumptions.

The SM predicts all 5 parameters.

$$\kappa_\gamma = \kappa_Z = 1 \qquad \lambda_\gamma = \lambda_Z = 0 \qquad g_{ZWW} = \cot\theta_W$$

The requirement that local SU(2) x U(1) is preserved leads to the equality:

$$\lambda_\gamma = \lambda_Z$$

If one refuses the possibility of an instrinsic SU(2) violation, one has the further relation:

$$(\kappa_Z - 1)\, g_{ZWW} = -\,(\kappa_\gamma - 1)\, \mathrm{tg}\theta_W$$

Finally one can impose the requirement that the s dependence of the WW process is well behaved: for instance the absence of s^2 terms in WW scattering implies that:

$$\delta_Z = g_{ZWW} - \cot\theta_W = (\kappa_\gamma - 1)/s_W c_W$$

Another attitude is to construct various effective lagrangians involving standard fields and satisfying the SU(2) x U(1) gauge invariance. After spontaneous symmetry breaking, these lagrangians produce particular anomalous 3 boson couplings. Such dimension 6 operators, as listed in (14) correspond each to a 1 parameter model contributing to several anomalous couplings. An arbitrary superposition of them leads to 4 free parameters.

Couplings of a higher dimensionality would give further possibilities.

4.2.3. Indirect limits on the size of anomalous couplings

If a model gives relations between VWW and $Vf\bar{f}$ couplings, a precise measurement of the later, at the Z peak, will give restrictions on the former as well.

An extreme case is the SM, totally constrained at tree level: $g_Z = \cot\theta_W$.

Among the effective lagrangians, there are cases, like for O_{WB}, (table 7) where precise relations exist between the departures from the SM in $Zf\bar{f}$ on one hand and in ZWW and γWW on the other hand: one can then expect that the impact of LEP 200 will be minor, as shown by fig. 13.

On the contrary, other operators, not directly related to $Zf\bar{f}$, are not constrained at tree level; they correspond to what one calls "blind" directions in the parameter space. LEP 200 will bring a decisive information on their size (fig. 14). (14,15)

If one goes beyond the SM, for instance by adding one extra SU(2), the constraints from the measurements at LEP 1 get much looser, and room is left for observations at LEP 200 (16).

Another type of constraints on 3 boson couplings exploits their virtual contribution through radiative corrections to low energy processes and processes accurately measured at LEP I.

The experimental constraint on an anomalous coupling appears then as a sum rule (13):

$$\int_0^\infty L_{VWW}(q^2) f(q^2) dq^2 < \varepsilon_{exp}$$

In this formula, the effective lagrangian may be taken as q^2 independent or as a function of q^2. In the former case, the left hand side may be a divergent integral: the only way to cure it is to add to L_{VWW} terms which cancel the divergence. Each anomalous 3-boson coupling becomes then part of a gauge invariant operator involving other couplings. New diagrams are generated and cancel the bad high energy behaviour of the 3-boson ones. However the procedure is not unique and provides no well defined constraint on anomalous 3 boson couplings.

In the more natural case where a q^2 dependence is considered, the Lagrangian changes, as q^2 increases and approaches the scale Λ of new physics; the integral computed with the low q^2 expression has no reason to be a good approximation of the total contribution. Such an integral cannot therefore constrain the local value of the integrand and a direct measurement is necessary.

4.2.4. *Present limits*

One set of limits comes from the hadron colliders. From $\bar{p}p \to e\upsilon\gamma + X$ at $\sqrt{s}$ = 630 GeV (UA2) and 1 TeV (CDF) one gets (12):

$$-3.1 < \kappa < 4.2 \qquad \text{at 95\% CL}$$
$$-3.6 < \lambda < 3.5$$

LEP I and υ-scattering, through loop effects, give also:

$$-1.7 < \lambda < 1.6 \qquad \text{at 95\% CL}$$
$$0.9 < \kappa < 2.3 \qquad \text{at 68\% CL}$$

4.2.5. *Expectations from LEP 200*

At LEP 200, the s channel in ee → WW is the only one contributing to the Triple Boson Vertex. As for the mass measurements, one can consider the mixed decay (~ 30%) and the hadronic one (~ 50%).

The background channels are those of fig. 15. Appropriate cuts allow to eliminate most of the background for the mixed decay mode of W pairs.

Experimentally, one has to reconstruct the W production angle $\cos\theta$ and the decay angles θ_l and ϕ_l of the charged lepton. The angular range is divided in several bins; each one is corrected for acceptance and radiative corrections are applied. A maximum likelihood or χ^2 fit is finally performed.

The generators used presently are: (17)

i) WWGENE, a modified Kleiss generator.

ii) Phythia 5.6/JETSET 7-3.

iii) WWIBA.

All three include initial state radiation. Fig. 16 shows that they do not fully agree and more effort is needed to ensure that they will match the statistical precision of the data. However this does not affect the conclusions drawn below.

As emphasized previously, observables like the production angular distribution of the W pair or the angular distribution of its decay products can be fitted with one or more free parameters. The following plots are given for the W pair mixed decay mode and 500 pb^{-1} of integrated luminosity.

Single parameter fit

Fig. 17 shows the result of a single parameter fit to the angular distribution. Fig. 17a gives the uncertainty on the free parameter κ_γ under two assumptions: κ_Z fixed to 1, or κ_Z given as a function of κ_γ as required by the assumption of no intrinsic SU(2) violation. Fig. 17b gives the uncertainty on δ_Z, the departure of g_{ZWW} from the SM value, and on λ_γ under the assumption $\lambda_Z = \lambda_\gamma$,which corresponds to the preservation of local SU(2)xU(1).

2 parameter fits

In fig. 18 the relation $\lambda_\gamma = \lambda_Z$ is assumed. 95% confidence limit contour lines for the W production angular distribution are shown for pairs of parameters, the others parameters being kept at their standard value. Three energies are shown in each case. The results are obtained with generated data but would be quite similar with reconstructed ones.

The addition of the information on the W decay angles leads to a substantial improvement of the accuracy, as shown in fig. 19 for the couples (δ_{gWW}, $\lambda_Z = \lambda_\gamma$) and (κ_γ-1, $\lambda_Z = \lambda_\gamma$).

3 parameter fits (17)

We choose as free parameters:

$$g_{ZWW} \qquad \lambda_\gamma = \lambda_Z \qquad \kappa_\gamma \quad ,$$

κ_Z being related to κ_γ and g_{ZWW} by the assumption of the absence of intrinsic SU(2) violation. One has also set to zero the contribution ε_{WB} from O_{WB}: the restrictions due to current data, as emphazised by (14), are thus taken into account, and one checks the possible contributions of the other sources ($O_{W\phi}$, $O_{B\phi}$,and O_W) listed in table 7.

The gain in accuracy when $\sqrt{s}$ increases is substantial, as shown by fig. 20.

4 parameter fits (18)

If one assumes that local SU(2) x U(1) is not preserved, λ_γ and λ_Z are now two free parameters. However κ_Z is still a function of κ_γ and g_{ZWW} as previously. The result shows again the rapid gain of precision with $\sqrt{s}$ for any couple of parameters (fig. 21).

The exploitation of the 4 jet channel, in which the charge of the W is deduced on a statistical basis from the jet charges, is considered as well (8).

4.3. Higgs search at LEP 200 (19)

This has been studied at the Aachen workshop (6). However, at that time, the SM Higgs was the only one to be studied. Furthermore the impact of b tagging in Higgs searches was not yet considered. The conclusion of the Workshop was that the visibility limit for the SM Higgs at LEP 200 would be ~ 80 GeV/c^2.

In the following we will explain why b tagging is important, quote what can be expected at LEP in this respect and describe the results of Monte Carlo studies performed recently. The result is that, with the use of b tagging, the most difficult domain corresponding to $M_H \sim M_Z$ can be explored as well with the luminosities expected at LEP 200. In fact the reach for a Higgs boson can be summarized by the approximate formula

$$M_H \sim \sqrt{s} - 100 \text{ GeV}$$

While the 100 GeV mass region has nothing special for a SM Higgs boson, although theorists tend to favour a light one, we will show that this is different for the lightest boson of SUSY models. In particular, for the minimal version (MSUSY) (20) and provided the top is not too heavy, we will show that an upgraded LEP 200 (~ 210 GeV) could bring a decisive answer.

4.3.1. Signal and background

At LEP 200 the SM Higgs is produced by the bremsstrahlung process

$$ee \rightarrow HZ \qquad \text{(fig. 22a)}$$

(Ioffe-Khoze-Bjorken)

The fusion processes (fig. 22b) are not yet strong enough, although an interference term is already present in the $\upsilon_e\upsilon_e H$ channel (21).

In the Bremsstrahlung process, the cross section for a given Higgs mass rises rapidly at threshold and approaches its maximum for $\sqrt{s} \sim M_H + 100$ GeV(fig. 23). The value of this maximum, for increasing M_H, is a slowly decreasing function.

The Higgs boson decays ~ 95% hadronically; the associated Z^o can go to $\upsilon\bar{\upsilon}$, l^+l^-, $q\bar{q}$. One has therefore the three search channels as at LEP I:

I	ZH	$\rightarrow$	4 jets
II	ZH	$\rightarrow$	2 jets $\upsilon\bar{\upsilon}$
III	ZH	$\rightarrow$	2 jets l^+l^-

4.3.2. The interest of b tag

If one considers the signal and background channels shown in fig. 15, one sees that, for the case $M_H \sim M_Z$, the channel ee $\rightarrow$ ZZ appears as an irreducible background from the kinematical point of view. However since

$$\text{BR}\,(H^o \rightarrow b\bar{b}) \sim 85\%$$
$$\text{BR}\,(Z^o \rightarrow b\bar{b}) \sim 15\%$$

an efficient and pure b tagging can increase the signal/background ratio. From Table 8 one can see that it provides indeed a decisive step in the significance of the signal.

The WW background can always be distinguished from the HZ channel by kinematical features. But its cross section is so high that b tagging turns out to be very useful as well to complement the kinematical rejection. It is particularly efficient since

$$\text{BR}\,(W \rightarrow b...) \sim 0$$

If, instead of the SM HZ channel, we consider the SUSY channel of associated production

$$ee \rightarrow hA$$

with $M_h = M_A = M_W$, the kinematical rejection against W pairs disappears. Due to the small cross section of the signal (fig. 24), an extremely powerful rejection of W's through b tagging is the only chance to observe such a channel.

4.3.3. The Monte Carlo results

The four LEP experiments have performed the relevant simulation work to demonstrate the visibility of the SM Higgs boson. A pionneering study from ALEPH is described in (22).

The kinematical rejection against background channels is obtained by rejecting events which fit the WW and ZZ hypotheses. The analysis has much in common with the one leading to the determination of the W mass. As an order of magnitude, the rejection against W pairs, when the signal under consideration is ZH (90), amounts to ~ 50 in a realistic simulation.

The main experimental input is the performance of the b tagging procedure. It can be summarized by quoting its efficiency ε_b and its purity P, defined for the hadronic population of the Z final state. Fig. 25 (DELPHI) presents the achieved and expected values, quite characteristic of the LEP experiments.

With such kind of input, fig. 26 (L3) shows that b tag is indeed the key to the visibility of the Higgs signal in all three search channels.

4.3.4. The discovery limit

From the combined study of the three channels as described above, one can deduce the discovery limit for the SM Higgs at LEP 200. An effect of 5σ,with at least 5 events observed, is assumed.

Fig. 27 shows such limits (22), plotted as the integrated luminosity required versus M_H, at various CM energies. The sharp rise corresponds to the drop of the cross section when the available energy becomes insufficient to produce a given mass. The agreement between various studies is satisfactory. One sees that the approximate rule

$$M_H \simeq \sqrt{s} - 100\,\text{GeV}$$

is verified.

Since LEP I will allow the exploration of masses up to ~ 65 GeV(23) (fig. 28), it is clear that a CM energy of 175 GeV (corresponding to the present "official" version of LEP 200) will not bring a dramatic improvement of the coverage of Higgs search. A CM energy of at least 190 GeV, unanimously desired by the LEP community, would guarantee an extension of the explored range by ~ 1.5 as compared to LEP I. It would allow to explore the 90 GeV mass region, which is the most difficult one and for which other methods at future machines ($H \to 2\gamma$ at LHC/SSC), may have problems of their own.

The curve in fig. 29a is given for$\sqrt{s}$ = 240 GeV. One can consider this either as a preview for the NLC or as an extremely optimistic point of view on the ultimate phase of LEP. A more realistic attitude is to assume that LEP could a some stage reach 210 GeV (see fig. 1): fig. 29b gives the mass reach under such an assumption.

While the 80–110 GeV mass region has no special meaning for the SM Higgs boson, although some theories tends to favour a light one, we show now that it is a crucial domain for the SUSY Higgs sector.

4.3.5. The SUSY Higgs sector

In the Minimal version of SUSY, the Higgs sector is strongly constrained. At tree level the lightest scalar boson h^o has to be lighter than the Z^o. Loop corrections involving the top and its spartners raise the value of that upper bound by a quantity which increases rapidly with M_t (fig. 30) and slowly with M_{sq}. All statements about M_h depend therefore crucially on the still unknown M_t. We assume in the following M_t = 140 GeV, unless stated otherwise, since this corresponds to the central value of the indirect LEP determination (admittedly within the SM). We also choose M_{sq} = 1 TeV.
Fig. 31 (24) gives the iso-mass curves for the lightest scalar, h^o, in the MSUSY plane: (tgβ, M_A), where tgβ is the ratio of the two vacuum expectation values and M_A is the pseudoscalar Higgs A^o mass. One should notice the quasi-stationarity of M_{ho} in a vast region on the upper right of the plane: since the discovery limit corresponds to a given mass, one can understand that a small step in the discovery limit, i.e a small step in the available CM energy, can induce a large and possibly decisive step in the coverage of the plane toward the upper right.

Fig. 32 (24) gives the production ratio of h^o and $H^o{}_{SM}$ through the bremsstrahlung process. It shows that, beyond $M_A \cong$ 100 GeV, the h^o is produced practically as the Standard Higgs. Since it also decays in a standard way both searches are identical.

The heaviest scalar H^o whose mass is given in fig. 33, is produced proportionally to $\cos^2(\alpha-\beta)$. The hA reaction goes also like $\cos^2(\alpha-\beta)$ and has but is moreover the penalty of a p-wave β^3 factor: these facts limit the usefulness of the hA channel to a vertical band below $M_A \sim$ 90 GeV, even if $\sqrt{s}$ is raised to 210 GeV.

4.3.6. Implications of LEP 200 searches for SUSY

The study of the visibility of SUSY Higgses is very similar to the one performed for the SM Higgs boson, since in the mass range considered, they decay mostly (~ 85%) to $b\bar{b}$. In some cases the $\tau^+\tau^-$ mode can bring a useful complement as well.

With $\sqrt{s} = 190$ GeV, the discovery limit is ~ 90 GeV, one can expect, and it is shown in fig. 34 that LEP 200 cannot, by far, explore completely the MSUSY plane.

At $\sqrt{s} = 240$ GeV the exercise leading to fig. 29a can be interpreted in terms of MSUSY Higgses: they would be excluded up to ~ 125 GeV with 500 pb^{-1}. Going back to fig. 30 one can deduce that the MSUSY model could then be rejected if the unknown parameters (M_t, M_{sq}) lie on the left of the curve $M_{ho} = 120$ GeV.

For $\sqrt{s} = 210$ GeV, fig. 35 shows the impacts of the three channels on the exclusion of the SUSY plane ($M_t = 140$ GeV, $M_{sq} = 1$ Tev). The plane can be covered completely: but a vertical region, shaped as a tie, along $M_A \sim 90$ GeV is only marginally excluded. However the hole which will be left by the future hadron colliders in that plane (24) would be filled.

Such conclusions are valid in the frame of MSUSY. Many other versions of SUSY do exist. They all require one light boson, at least, but do not provide such sharp bounds on its mass. For instance a theoretical variant, allowing the inclusion of singlet Higgses in arbitrary number, would correspond to bounds on M_{ho} as shown in fig. 36 (25): one sees that, in the limit of large tgβ one recovers the bound of the MSUSY model.

4.4. Search for other SUSY particles (26)

This subject has been studied extensively in the past and we will refer to (6) for details.

For other SUSY particles there are no such constraints as for the Higgs sector. The naturalness arguments suggest that some sparticles, in particular gauginos, could be light as well.

We consider briefly the chargino search

$$e^+e^- \to \chi^+\chi^-$$

proceeding through the diagrams of fig. 37. Usually the cross section is confortably large (several pb), but, for specific parameter choices, it may becomes extremely low through destructive interference between the two processes .

The signatures, suggested by fig. 37, are

$(\chi l \upsilon)$	$(\chi l \upsilon)$	$\rightarrow$	acoplanar lepton pairs	10%
$(\chi$hadr)	$(\chi$hadr)	$\rightarrow$	hadron + $\not{E}$	50%
$(\chi l \upsilon)$	$(\chi$hadr)	$\rightarrow$	isol. lepton + $\not{E}$	40%

and the background comes mostly from W pairs.

Fig. 38 gives, in the (M versus μ) plane, the various exclusion contours to be expected (27).

For some values of the SUSY parameters, neutralino searches could bring a valuable information as well (27).

4.5. Searches through classical measurements (26)

At LEP 200, one will perform the usual measurements on fermion pairs:

$$\sigma_{f\bar{f}},\ A^{FB}_{f\bar{f}},\ R$$

Non Standard Models (E6, LR models) could lead to deviations from the SM predictions.

At $\sqrt{s} = 190$ GeV and for 500 pb^{-1}, one expects

		σ(pb)		Nb of events
$l\bar{l}$	→	3 x 3.36	←	5040
hadrons		25.39		12700

This evaluation was done with full O(α) radiative corrections. A cut is included to suppress γγ interactions and the γZ final state.

The systematic errors adopted are:

$$\frac{\Delta L}{L} = 0.5\%, \quad \frac{\Delta\varepsilon_\ell}{\varepsilon_\ell} = 0.5\%, \quad \frac{\Delta\varepsilon_h}{\varepsilon_h} = 1\%$$

where L, ε_l, ε_h are the luminosity and the efficiencies to leptonic and hadronic channels, respectively.

Fig. 39 shows, under such conditions, the sensivity to an hypothetical Z' as a function of the key parameters of the E_6 (cosβ) and LR model (α_{LR}). The combination of the three measurements gives bounds on such objects ranging between 0.4 and 1.0 TeV. For a given value of the parameter, the scaling law with luminosity is

$$M_{Z'} \sim \left[\int L dt\right]^{1/4}$$

The dependence on systematic errors is not critical. The mass reach improves slowly with energy (~ $\sqrt{s}$).

Conclusion

At LEP 200 the W mass will be measured with a precision of ~ 60 MeV/c^2 by each of the four experiments. The 3–gauge boson couplings will be studied with a sensitivity of ~ 10%. The study of the Higgs channel will be extended to the 100 GeV/c^2 range and, if the top quark mass is low, a small increase in the CM energy could have a dramatic impact on the SUSY Higgs sector.

Acknowledgements

I would like to thank P. Darriulat, C. Wyss, J. Jowett, F. Renard for illuminating discussions and the contributors to the LEPC LEP 200 Workshop who produced most of the results quoted here.

REFERENCES

(1) The Pink Book, CERN/ISR-LEP/79-33, August 79

(2) The energy calibration of LEP in 1991, CERN PPE/92-125

(3) Proceedings of the Second Workshop on LEP performances, Chamonix, Jan. 92, CERN-SL/92-29

(4) Proceedings of the ECFA Workshop on LEP 200, CERN 87-08, P. Bernard et al., p 29.

(5) LEPC Workshop on LEP 200, CERN, Nov. 4th 1992, talk by C. Wyss

(6) Ref. 4, p 49-452

(7) Ref. 5, talk by L. Camilleri

(8) Report from the DELPHI working group on LEP 200 Physics, DELPHI, note in preparation.

(9) D. Bardin, M. Bilenky, A. Olchevski, T. Riemann, in preparation.

(10) N.J. Kjaer, R. Moeller, DELPHI 91-17 Phys 88

(11) Ref. 5, talk by R. Eichler
See also ref. (4) M. Davier et al., p 120
D. Treille et al., p 414

(12) F.M. Renard, PM/92-28 and ref. theirin

(13) G.J. Gounaris and F.M Renard, THES/TP 92-11 - PM/92-31

(14) A. de Rujula et al., CERN TH-6272/91 - FTUAM 91-31

(15) K. Hagiwara et al., MAD/PH 690 (Feb. 92), PLB283 (1992) 353
R. Szalapski et al., MAD/PH/719
P. Hernandez et al., CERN-TH6670

(16) M. Bilenky et al., BI-TP 92/25 - PM/92-29

(17) See in particular ref. 8

(18) M. Bilenky et al., BI-TP 92/44 (oct. 92)

(19) Ref. 5, talk by D. Treille
See also ref., 4 Sau Lan Wu et al., p 312

(20) See for instance H.E. Haber and G.L. Kane, Phys. Rep. 117 (85) p 75
R. Barbieri, Riv. Nuovo Cimento 11 (88) p1

(21) M. Dubinin, private communication

(22) P. Janot, LAL 92-27, May 1992

(23) E. Gross, CERN-PPE, 92-91 (1992)

(24) Z. Kunszt, and F. Zwirner, CERN TH 6150/91

(25) M. Quiros, J.P. Espinosa, Madrid Theory Preprint, 1992

(26) Ref. 5, talk by G. Coignet
See also ref. 4, C. Dionisi et al., p 380

(27) J.F. Grivaz, LAL 92-64

Table 1 = Present data on M_W

		(STAT)	(SYST)
UA2 ($e\upsilon$)	$M_W / M_Z = 0.8813$	$\pm$ 0.0036	$\pm$ 0.0019
	$M_W = 80.35 \pm 0.37$ GeV (M_Z from LEP)		
CDF ($e\upsilon$, $\mu\upsilon$)	$M_W = 79.91 \pm 0.39$ GeV		
Combined	$M_W = 80.13 \pm 0.27$ GeV		

Table 2 = Radiative Energy loss at E_b = 88 GeV

		<Erad>GeV
Energy loss at $O(\alpha)$		1.520
+ soft photon expon. on s channel and universal terms on t channel	–354 MeV	1.166
+ $O(\alpha)$ of non universal t channel terms	+3 MeV	1.169
+ Coulomb singularity (FSR)	+11 MeV	1.180
+ L^2+L+C of $O(\alpha)$ + fermion pairs	+32 MeV	1.212

SUMMARY OF SYSTEMATIC ERRORS

Table 3

● BEAM ENERGY	± 17 MeV/C^2
● RADIATIVE CORRECTIONS UNCERTAINTY AND IMPLEMENTATION IN MONTE-CARLO	± 10 MeV/C^2
● RECONSTRUCTION METHOD	± 10 MeV/C^2
● DETECTOR PERFORMANCE	± 10 MeV/C^2
	± 24 MeV/C^2

Table 4 = The possible anomalous couplings and their value in the SM

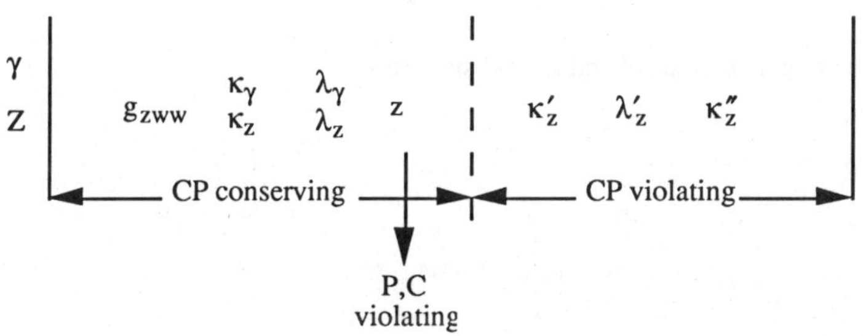

In the SM:

$g_{ZWW} = \cot\theta_W$

$\kappa_\gamma = \kappa_Z = 1$

$\lambda_\gamma = \lambda_Z = z = \kappa'_Z = \lambda'_Z = \kappa''_Z = 0$

Table 5 = Alternative set of anomalous couplings and the C, P and CP conserving Lagrangian

$\delta_Z \equiv g_{ZWW} - \cot\theta_W$

$x_\gamma \equiv \kappa_\gamma - 1$

$x_Z \equiv (\kappa_Z - 1)(\cot\theta_W + \delta_Z) \simeq (\kappa_Z - 1)\cot\theta_W$

$y_\gamma \equiv \lambda_\gamma$

$y_Z \equiv \lambda_Z \cot\theta_W$

$$
\begin{aligned}
L_1 &= L_1^{SM} + \\
&ie\,\delta_Z \left[Z_\mu \left(W^{-\mu\upsilon} W^{+}_{\upsilon} - W^{+\mu\upsilon} W^{-}_{\upsilon} \right) + Z_{\mu\upsilon} W^{+\mu} W^{-\upsilon} \right] \\
&+ ie\, x_\gamma\, F_{\mu\upsilon} W^{+\mu} W^{-\upsilon} \\
&+ ie\, x_Z\, Z_{\mu\upsilon} W^{+\mu} W^{-\upsilon} \\
&+ ie \frac{y_\gamma}{M_W^2} F^{\upsilon\lambda} W^{-}_{\lambda\mu} W^{+\mu}{}_{\upsilon} + ie \frac{y_Z}{M_Z^2} Z^{\upsilon\lambda} W^{-}_{\lambda\mu} W^{+\mu}{}_{\upsilon}
\end{aligned}
$$

$W_{\mu\upsilon}$ and $Z_{\mu\upsilon}$ are the non-abelian W and Z field strengths.

Table 6 = Contributions of the anomalous couplings to the various helicity states

	$\tau = \tau' = \pm 1$	$-\tau = \tau' = \pm 1$	$\tau = \tau' = 0$	$\tau = 0, \tau' = \pm$
$-\frac{2}{s} + 2\frac{\cot\theta_W + \delta_Z}{s - M_Z^2} A$	$-\beta$	0	$-\beta\left(1 + \frac{s}{2M_W^2}\right)$	$-\beta\frac{\sqrt{s}}{M_W}$
$-\frac{x_\gamma}{s} + \frac{y_Z}{s - M_Z^2} A$	0	0	$-\beta\frac{s}{M_W^2}$	$-\beta\frac{\sqrt{s}}{M_W}$
$-\frac{x_\gamma}{s} + \frac{y_Z}{s - M_Z^2} A$	$-\beta\frac{s}{M_W^2}$	0	0	$-\beta\frac{\sqrt{s}}{M_W}$

$$A = \frac{-1 + 4s_W^2 + 2\lambda}{4s_W c_W} \qquad \beta = \sqrt{1 - \frac{4M_W^2}{s}}$$

λ is the electron helicity. τ and τ' are the W^- and W^+ helicities

Table 7 = Elements of the d = 6 basis {Oi} and blind directions

Basic bosonic grafts	Basic fermion grafts	Blind directions $\not\subset \{O_i\}$
$O_{WB} \equiv \Phi\vec{\sigma}\Phi\vec{W}_{\mu\nu}B^{\mu\nu}$	$O_{e\mu} \equiv \vec{J}_\rho(L_e)\vec{J}^\rho(L_\mu)$	$O_W \equiv (1/3!)\, \vec{W}^\nu_\mu \times \vec{W}^\lambda_\nu\, \vec{W}^\mu_\lambda$
$O_\Phi \equiv J_\rho(\Phi)\, J^\rho(\Phi)$	$O_e \equiv \vec{J}_\rho(\Phi)\vec{J}^\rho(L_e)$	$O_{B\Phi} \equiv i\, B^{\mu\nu}\left(D_\mu\Phi\right) D_\nu\Phi$
$O_{DW} \equiv \left[D^\rho\vec{W}_{\mu\nu}\right]\left[D_\rho W^{\mu\nu}\right]$	$O_\mu \equiv \vec{J}_\rho(\Phi)\vec{J}^\rho(L_\mu)$	$O_{W\Phi} \equiv i\, \vec{W}^{\mu\nu}\left(D_\mu\Phi\right) \vec{\sigma}\, D_\nu\Phi$

A graft is an addition to the SM Lagrangian. Φ is the Higgs field; $W_{\nu\mu}$ and $B_{\nu\mu}$ are non-abelian field strengths. D is the covariant derivative.

Table 8 = Rates for the channel indicated (190 GeV, 500 pb^{-1})

Events	WW	ZZ	HZ
I. 4-JETS:			
produced	4500	170	115
after topol. and kin. cuts	90	45	33
after b-tagging	–4	11	–16
II. JET-JET -ν - $\bar{\nu}$			
produced	0	98	34
after b-tagging	0	–12	17
III. JET-JET–ℓ^+–ℓ^- ($\ell \neq \tau$)			
produced	0	33	12
after b-tagging	0	–4	6

FIGURES

Fig. 1 The layout of RF cavities around LEP. What is shown is the officially approved set up. By filling the room available at the even-points one could ultimately reach ~ 210 GeV CM energy, if the cavities work at 6 MV/m. Each box can contain 8 cavities. Dots are for SCC, bars for warm ones.

Fig. 2 The number of cavities needed to reach a given beam energy, under two assumptions for the accelerating field. On the right is shown the RF power needed. On top the cross section for W pair production is indicated, with the same horizontal scale.

Fig. 3 M_W versus M_t in the SM, for three Higgs masses. The present data and the impact of an accurate W mass measurement with no M_t measurement are shown (8).

Fig. 4 The same as fig. 3 once the top mass is known (8).

Fig. 5 The distribution of fitted minus generated mass and fitted mass, for all hadronic WW events (DELPHI, ref. 8).

Fig. 6 The average of hadronic and leptonic masses for mixed decay events (ALEPH) (7).

Fig. 7 The statistical error on M_w versus $\sqrt{s}$ (ALEPH) (7).

Fig. 8 Cross section dependence on M_w for $M_H = 200$ GeV/c^2.
Exact one loop result and Born approximation (8).

Fig. 9 M_w resolution from line shape fit and rescaling (reconstruction) method (8).

Fig. 10 The distribution of the generated lepton energy for $M_w = 80.2$ GeV/c^2 and $E_b =$ 88 GeV (8).

Fig. 11 The W mass shift due to various energy shifts (upper curve for neutral hadrons, middle one for photons, lower one for charged particles) (ALEPH) (7).

Fig. 12 The graphs of W pair production.

Fig. 13 χ^2 test of the significance of $\varepsilon_{WB} \neq O$ on $d\sigma/d\cos\theta$

Fig. 14 Individual 2 σ constraints and allowed combined contours in the (δ_W, m_t) plane. The dashed band is for LEP 200, the grey area for lower energy measurements.

Fig. 15 The relevant processes versus $\sqrt{s}$, in LEP 200 and NLC domains.

Fig. 16 The W pair cross section from various generators versus $\sqrt{s}$ (11).

Fig. 17 Result of single parameter fits to the W pair angular cross section. The uncertainty on the varied parameter is given versus $\sqrt{s}$. The choices made for the other parameters, kept fixed, are indicated. (11)

Fig. 18 95% confidence limit contour lines for the W production angular distribution (8).

Fig. 19 95% confidence limit contour lines obtained from the distributions of $\cos\theta_W$, ($\cos\theta_W$, $\cos\theta_\ell$) and ($\cos\theta_W$, $\cos_\ell$, ϕ_ℓ). (8)

Fig. 20 95% confidence limit contour lines obtained from the production angular distribution when contributions from anomalous operators $O_{W\phi}$, $O_{B\phi}$ and O_W are considered. (8)

Fig. 21 Contours plots from fits to 4 free parameters (18).

Fig. 22 The bremsstrahlung process (a) and fusion ones (b) for Higgs production.

Fig. 23 Cross section versus $\sqrt{s}$ of Higgs production, for various Higgs masses (from 60 GeV upward, by steps of 20 GeV).

Fig. 24 Details on Higgs production cross sections at LEP 200.

Fig. 25 Efficiency versus purity of b tagging, defined for Z^o final states. (8)
Lower curve: presently achieved performances.
Intermediate curve: as expected in 93.
Upper curve: as expected in 95.

Fig. 26 The mass distribution of the Higgs candidates for the various backgrounds and for the sum of signal and backgrounds; the three search channels are shown. In each case the left figure is before b tagging, the right one after b tagging. (L3) (19)

Fig. 27 Discovery limits (integrated luminosity needed versus M_H) for two CM energies (22).

Fig. 28 "History" of Higgs search at LEP I and its extrapolation. (23)

Fig. 29 Discovery limits at higher CM energies (22,19).

Fig. 30 The upper limit of the h^o mass as a function of M_t and M_{sq} (for large $tg\beta$). (24)

Fig. 31 The isomass curves for h^o in the $tg\beta - M_A$ plot of MSUSY. $M_t = 140$ GeV, $M_{sq} = 1$ TeV. (24)

Fig. 32 The ratio of $h^o Z^o$ and $H^o_{SM} Z^o$ production in the MSUSY plane. (24)

Fig. 33 The isomass curves for the heaviest scalar of MSUSY, H^o. (24)

Fig. 34 Exclusion domain of the MSUSY plane for $\sqrt{s}$ = 190 GeV. (22)

Fig. 35 The coverage of LEP 210 in the MSUSY plane, for Mt = 140 GeV, M_{sq} = 1 TeV. The tie shaped region is only covered at the 3 σ level while the remaining of the plane is covered at the 5 σ level. (22,19)

Fig. 36 The upper limit of the h^o mass in non minimal models, where scalar Higgses are added. In the limit of large tgβ one recovers the results of MSUSY. (25)

Fig. 37 Diagrams for chargino production and decay.

Fig. 38 The (M vs μ) plane, in GeV/c^2 unit, for tan β = 2. Domain excluded at LEP 1 from chargino searches (1) and from neutralino searches, also using the Z width measurements (2).
Top: for m_o = 1 TeV/c^2, domain which can be explored by chargino searches at LEP 200 with 100 pb^{-1} at $\sqrt{s}$ = 175 GeV (3), and its extension if $\sqrt{s}$ = 190 GeV (4); in both cases, the kinematic limit is reached.
Bottom: for m_o = 100 GeV/c^2, domain which can be explored with 100 pb^{-1} at $\sqrt{s}$ = 190 GeV (a), and its extension with 500 pb^{-1} (b).
M is the $SU(2)_L$ gaugino mass, μ is the supersymmetric Higgs mixing mass. m_o is an additional parameter needed to calculate the masses of the scalar leptons and quarks (from ref. 27).

Fig. 39 Sensitivity limit in TeV of an hypothetical Z' as a function of the key parameters of the E_6 and LR models. (26)

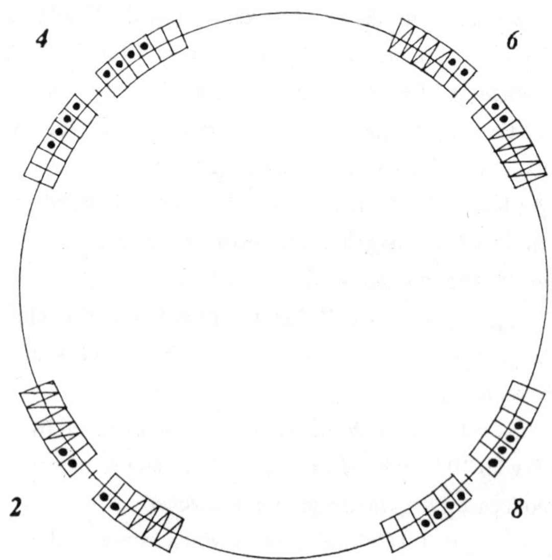
4
6
2
8

Fig. 1

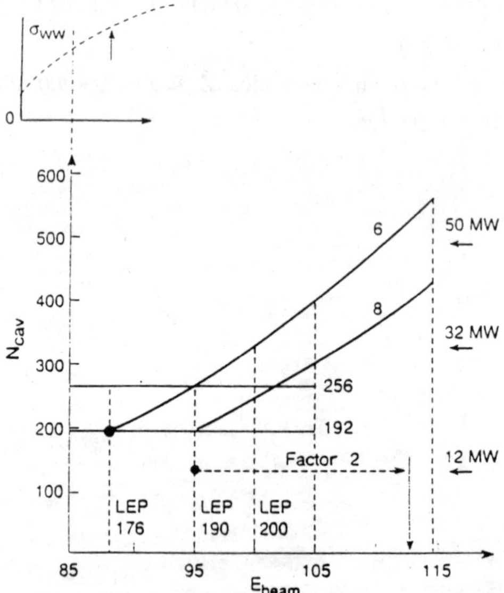
σWW
0
600
500
400
300
200
100
Ncav
6
8
50 MW
32 MW
12 MW
256
192
Factor 2
LEP
176
LEP
190
LEP
200
85
95
105
115
Ebeam

Fig. 2

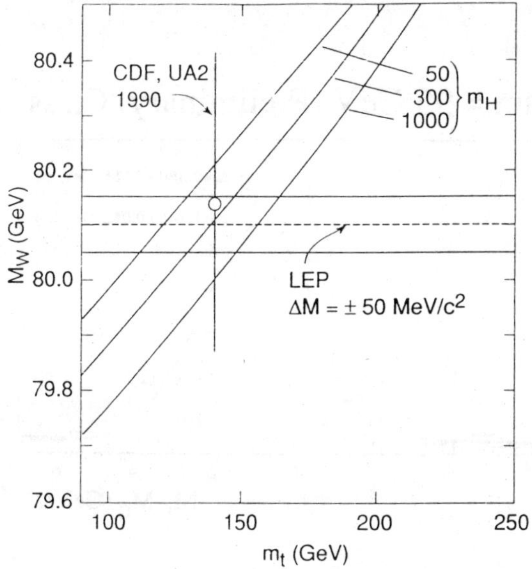

80.4
80.2
80.0
79.8
79.6
100
150
200
250
M_W (GeV)
m_t (GeV)
CDF, UA2
1990
50
300
1000
m_H
LEP
$\Delta M = \pm 50$ MeV/c^2

Fig. 3

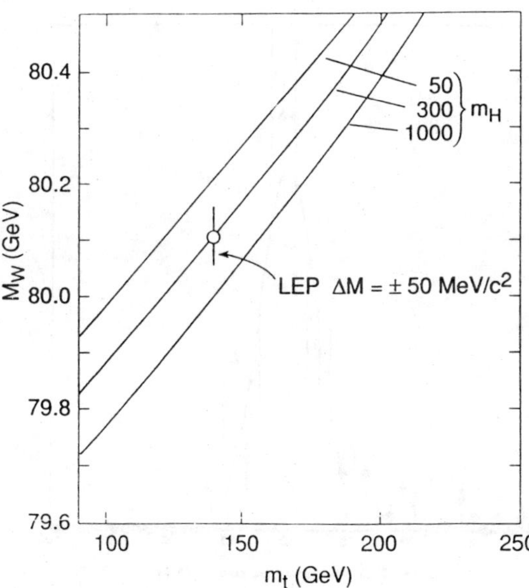

80.4
80.2
80.0
79.8
79.6
100
150
200
250
M_W (GeV)
m_t (GeV)
50
300
1000
m_H
LEP $\Delta M = \pm 50$ MeV/c^2

Fig. 4

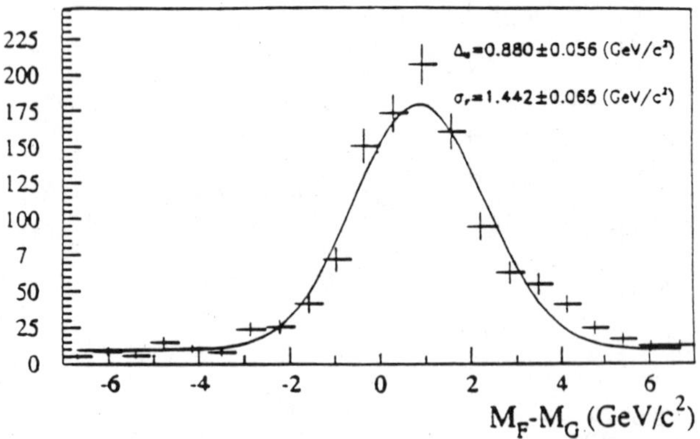
Ebeam = 95 GeV (Preliminary) Class II
Δ=0.880±0.056 (GeV/c²)
σ=1.442±0.065 (GeV/c²)
225
200
175
150
125
100
7
50
25
0
-6
-4
-2
0
2
4
6
M_F-M_G (GeV/c²)

Fig. 5

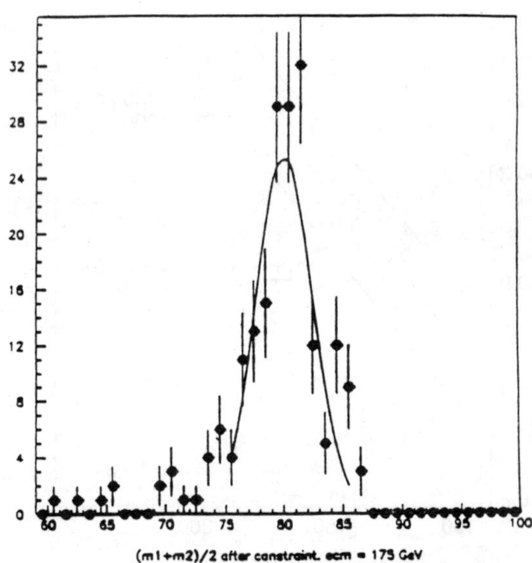
32
28
24
20
16
12
8
4
0
60
65
70
75
80
85
90
95
100
(m1+m2)/2 after constraint. ecm = 175 GeV

Fig. 6

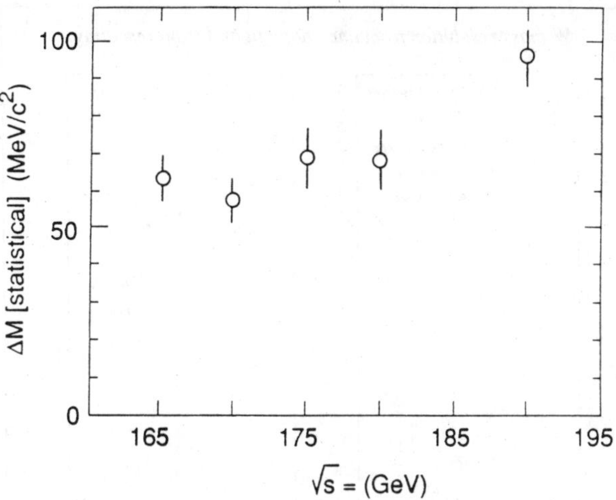
ΔM [statistical] (MeV/c²)
100
50
0
165
175
185
195
√s = (GeV)

Fig. 7

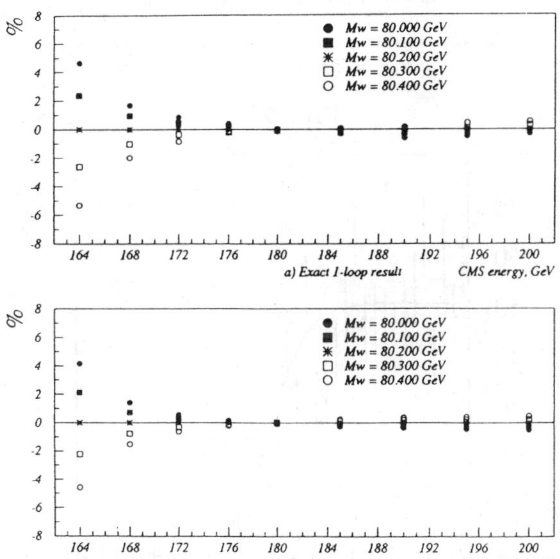
%
Mw = 80.000 GeV
Mw = 80.100 GeV
Mw = 80.200 GeV
Mw = 80.300 GeV
Mw = 80.400 GeV
a) Exact 1-loop result
CMS energy, GeV
b) Born approximation
CMS energy, GeV

Fig. 8

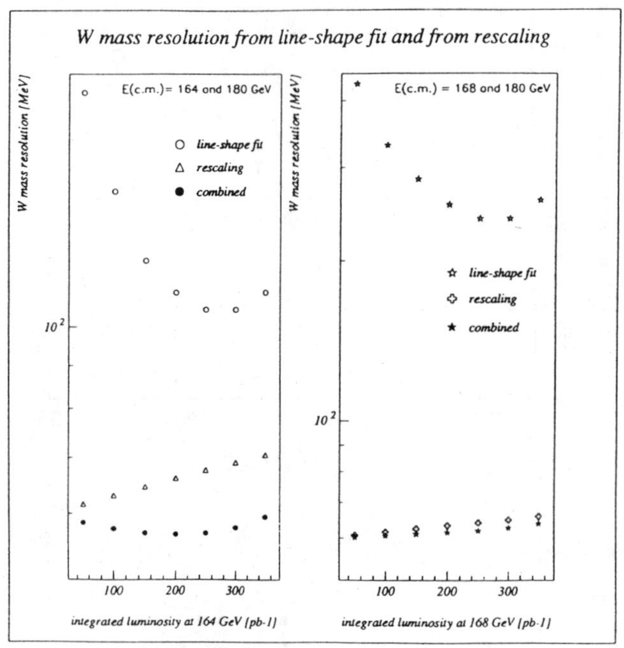
W mass resolution from line-shape fit and from rescaling
E(c.m.)= 164 and 180 GeV
line-shape fit
rescaling
combined
W mass resolution [MeV]
10^2
100
200
300
integrated luminosity at 164 GeV [pb-1]
E(c.m.) = 168 and 180 GeV
line-shape fit
rescaling
combined
W mass resolution [MeV]
10^2
100
200
300
integrated luminosity at 168 GeV [pb-1]

Fig. 9

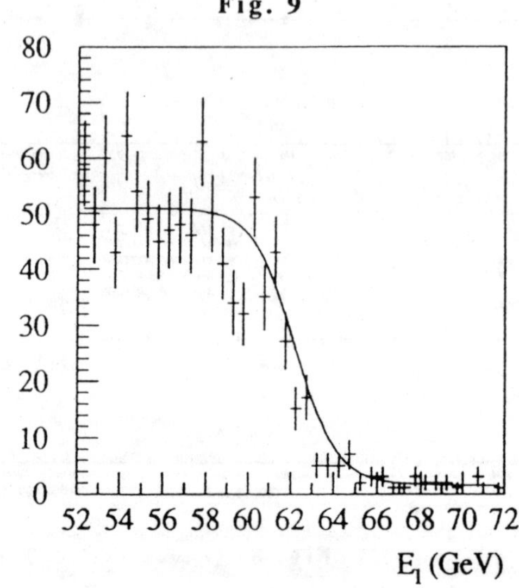
80
70
60
50
40
30
20
10
0
52 54 56 58 60 62 64 66 68 70 72
E_l (GeV)

Fig. 10

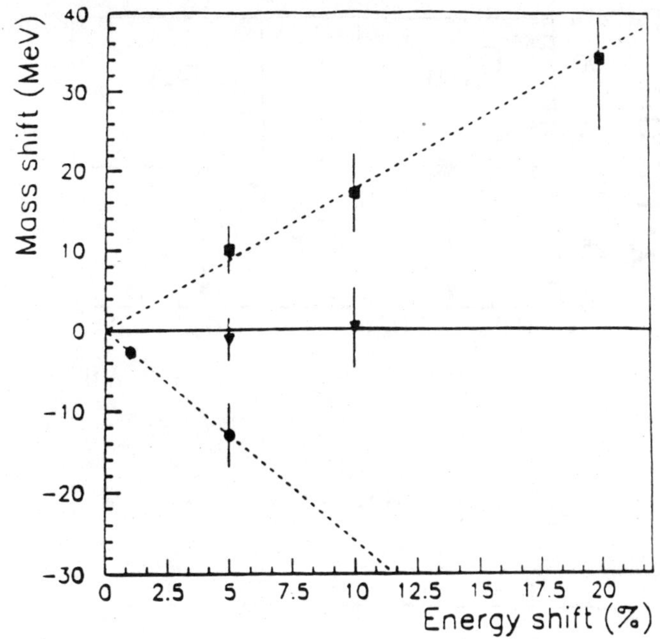

Mass shift (MeV)
40
30
20
10
0
-10
-20
-30
0
2.5
5
7.5
10
12.5
15
17.5
20
Energy shift (%)

Fig. 11

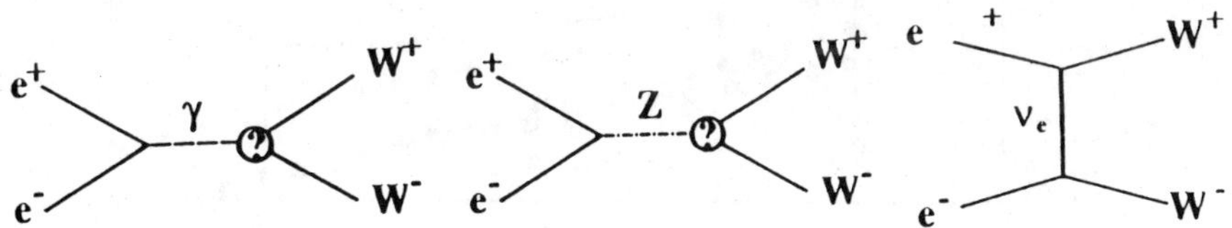

e+
e-
γ
W+
W-
e+
e-
Z
W+
W-
e +
e-
νe
W+
W-

Fig. 12

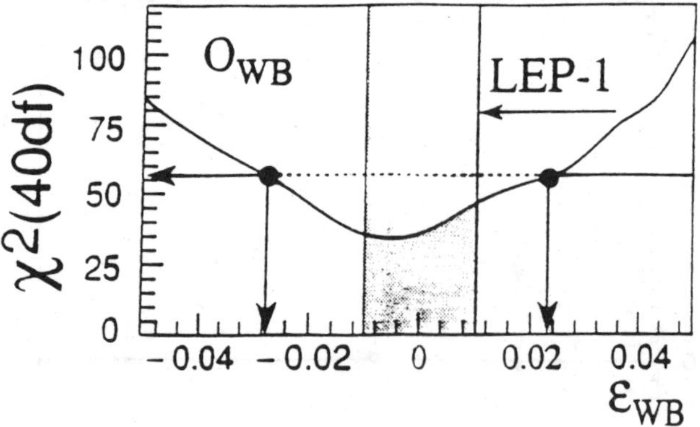

O_{WB}
LEP-1
$\chi^2(40df)$
100
75
50
25
0
- 0.04
- 0.02
0
0.02
0.04
ε_{WB}

Fig. 13

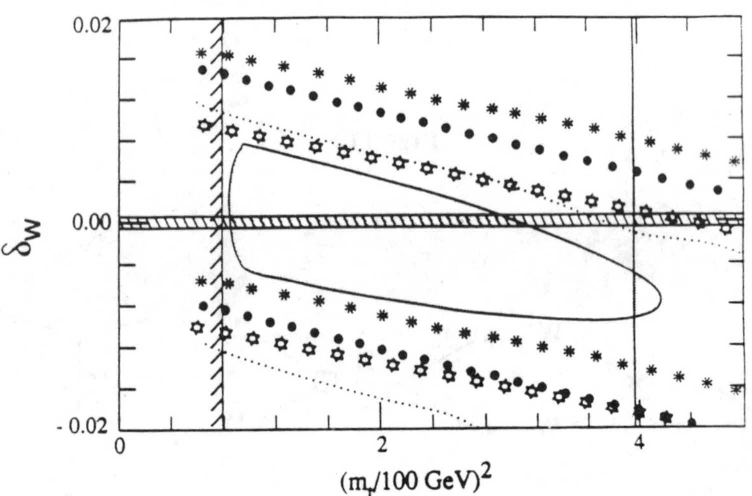

0.02
0.00
- 0.02
δ_W
0
2
4
$(m_t/100\ GeV)^2$

Fig. 14

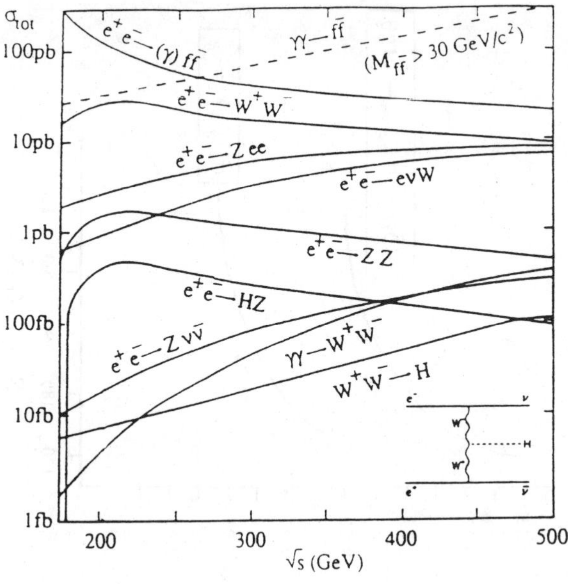
σ_{tot}
100pb
10pb
1pb
100fb
10fb
1fb
$e^+e^- \to (\gamma) f\bar{f}$
$\gamma\gamma \to f\bar{f}$
($M_{f\bar{f}} > 30$ GeV/c^2)
$e^+e^- \to W^+W^-$
$e^+e^- \to Zee$
$e^+e^- \to e\nu W$
$e^+e^- \to ZZ$
$e^+e^- \to HZ$
$e^+e^- \to Z\nu\bar{\nu}$
$\gamma\gamma \to W^+W^-$
$W^+W^- \to H$
200
300
400
500
$\sqrt{s}$ (GeV)

Fig. 15

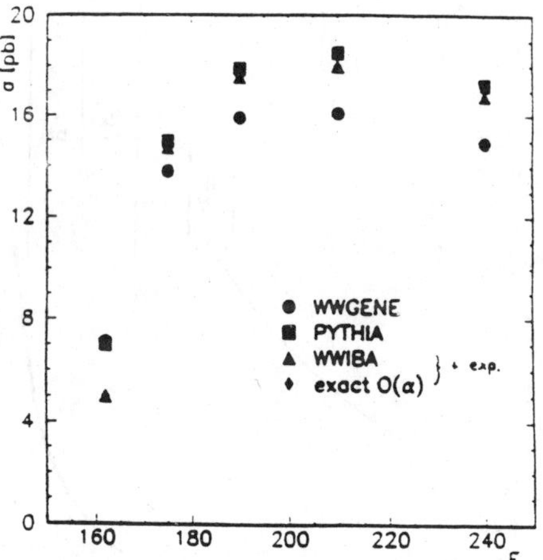
σ [pb]
20
16
12
8
4
0
160
180
200
220
240
WWGENE
PYTHIA
WWIBA
exact O(α)
+ exp.

Fig. 16

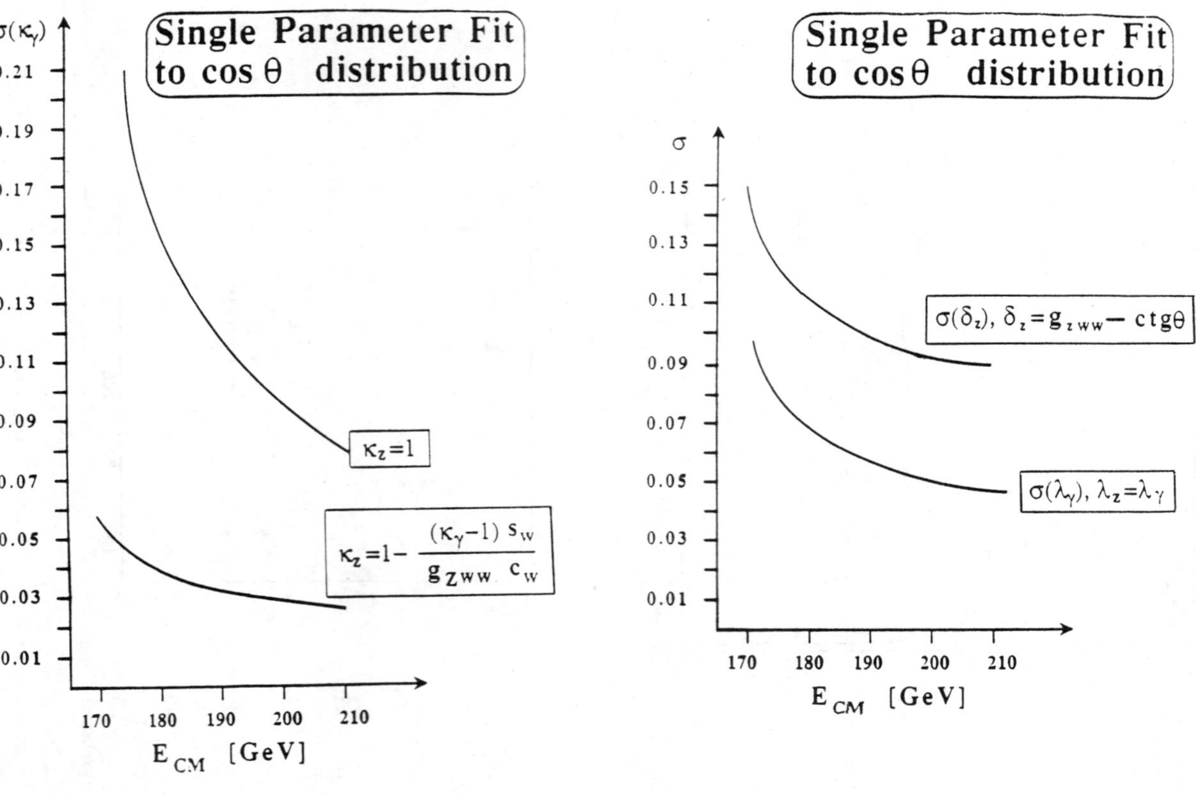

Single Parameter Fit to cos θ distribution
σ(κ_γ)
0.21
0.19
0.17
0.15
0.13
0.11
0.09
0.07
0.05
0.03
0.01
κ_z=1
κ_z=1− (κ_γ−1) s_W / g_ZWW c_W
170
180
190
200
210
E_CM [GeV]
Single Parameter Fit to cos θ distribution
σ
0.15
0.13
0.11
0.09
0.07
0.05
0.03
0.01
σ(δ_z), δ_z=g_zww − ctgθ
σ(λ_γ), λ_z=λ_γ
170
180
190
200
210
E_CM [GeV]

Fig. 17

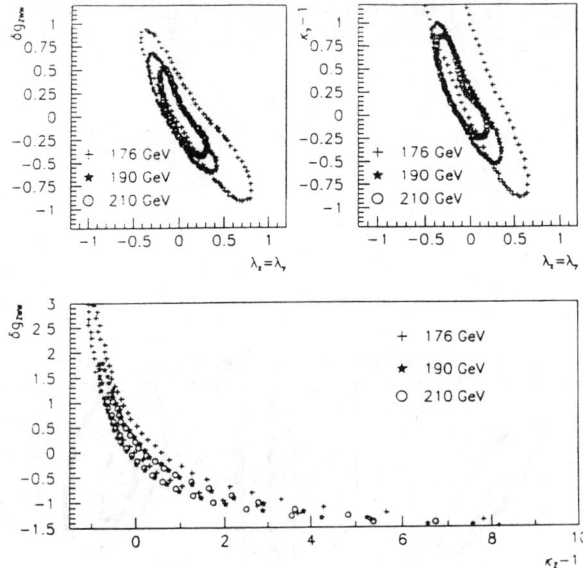

δg_ZWW
κ_γ − 1
+ 176 GeV
★ 190 GeV
○ 210 GeV
λ_Z = λ_γ
κ_Z − 1

Fig. 18

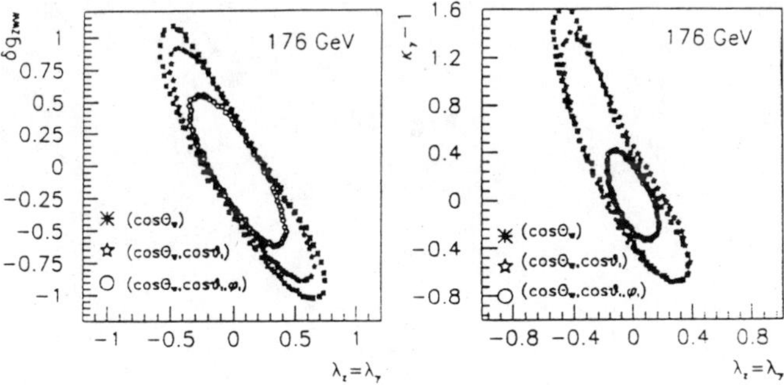

δg_ZWW
κ_γ − 1
176 GeV
(cosΘ_W)
(cosΘ_W, cosϑ_l)
(cosΘ_W, cosϑ_l, φ_l)
λ_Z = λ_γ

Fig. 19

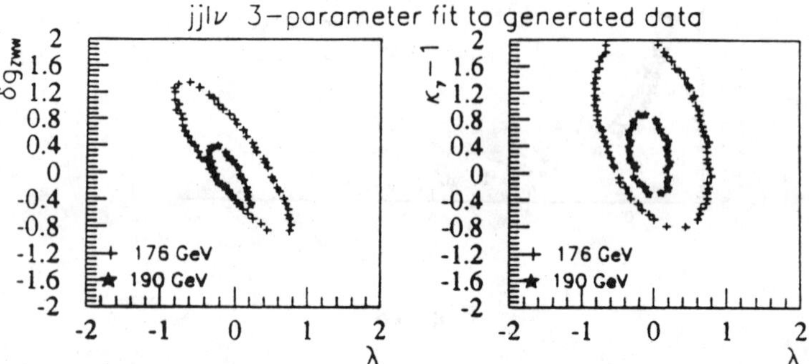
jjlν 3−parameter fit to generated data
δg_ZWW
κ_γ − 1
2
1.6
1.2
0.8
0.4
0
-0.4
-0.8
-1.2
-1.6
-2
-2 -1 0 1 2
λ
+ 176 GeV
★ 190 GeV

Fig. 20

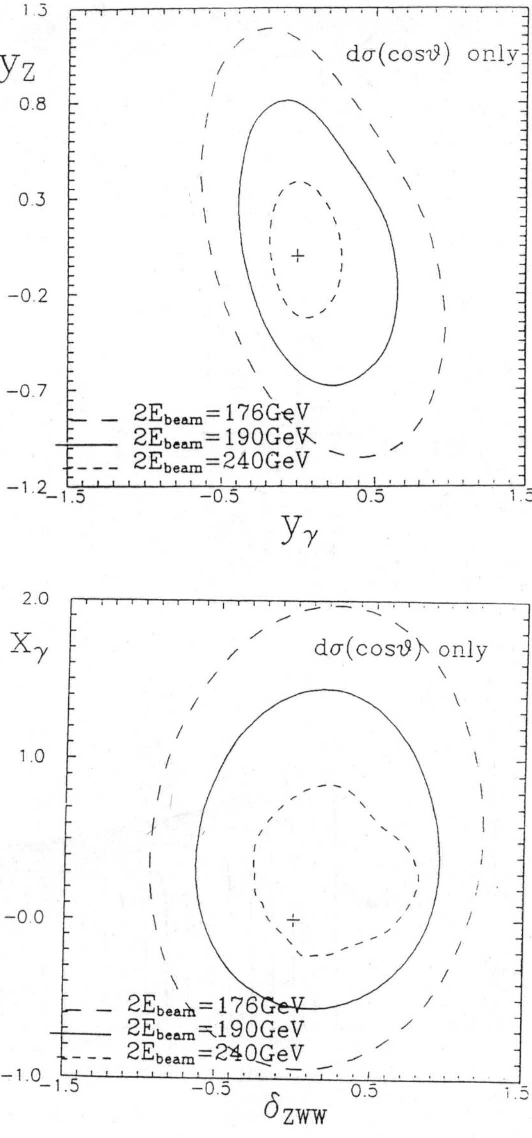

dσ(cosϑ) only
yZ
1.3
0.8
0.3
−0.2
−0.7
−1.2
−1.5
−0.5
0.5
1.5
yγ
2Ebeam=176GeV
2Ebeam=190GeV
2Ebeam=240GeV
dσ(cosϑ) only
Xγ
2.0
1.0
−0.0
−1.0
δZWW
2Ebeam=176GeV
2Ebeam=190GeV
2Ebeam=240GeV

Fig. 21

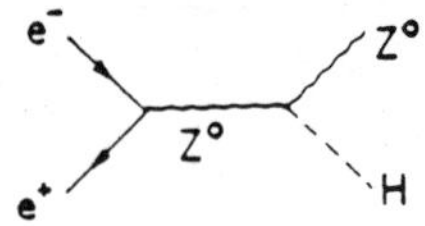

e⁻
e⁺
Z°
Z°
H

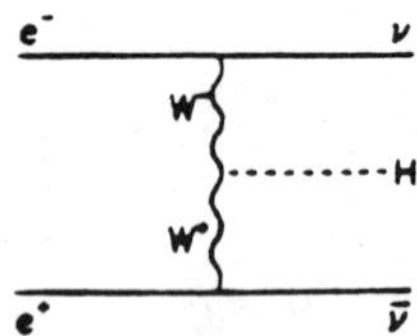

e⁻
ν
W
H
W*
e⁺
ν̄

Fig. 22

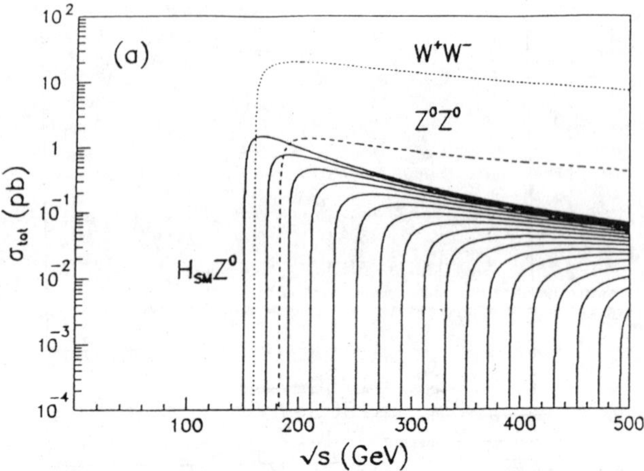

(a)
W⁺W⁻
Z⁰Z⁰
H_SM Z⁰
σ_tot (pb)
√s (GeV)
10²
10
1
10⁻¹
10⁻²
10⁻³
10⁻⁴
100
200
300
400
500

Fig. 23

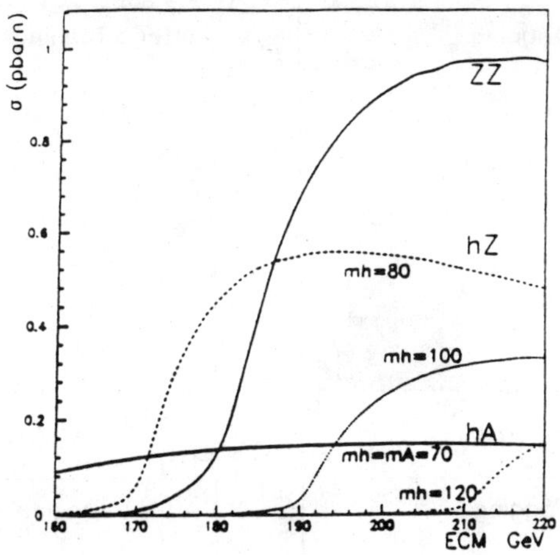
σ (pbarn)
ZZ
hZ
mh=80
mh=100
hA
mh=mA=70
mh=120
ECM GeV

Fig. 24

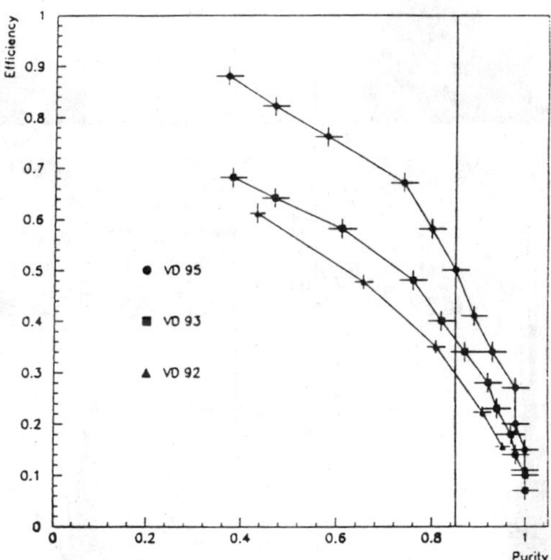
Efficiency
VD 95
VD 93
VD 92
Purity

Fig. 25

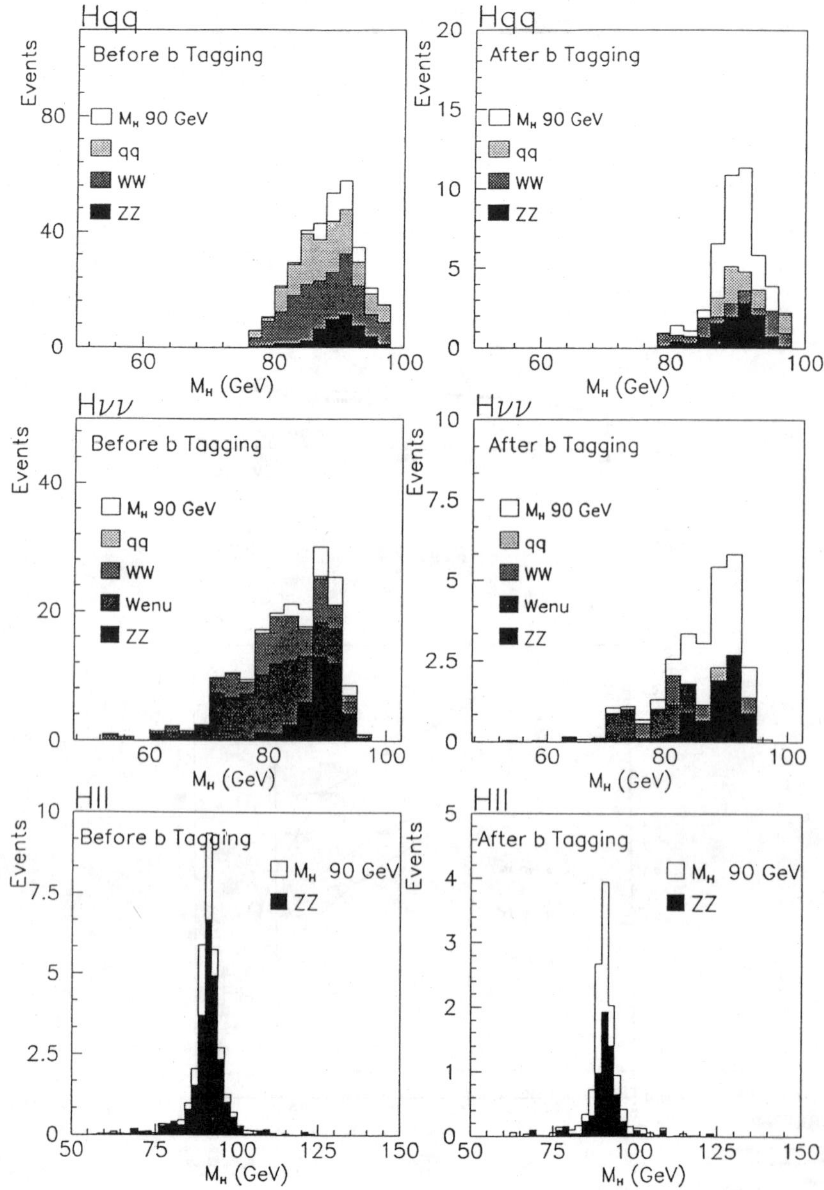

Hqq
Before b Tagging
After b Tagging
M_H 90 GeV
qq
WW
ZZ
Events
M_H (GeV)
Hνν
Wenu
Hll

Fig. 26

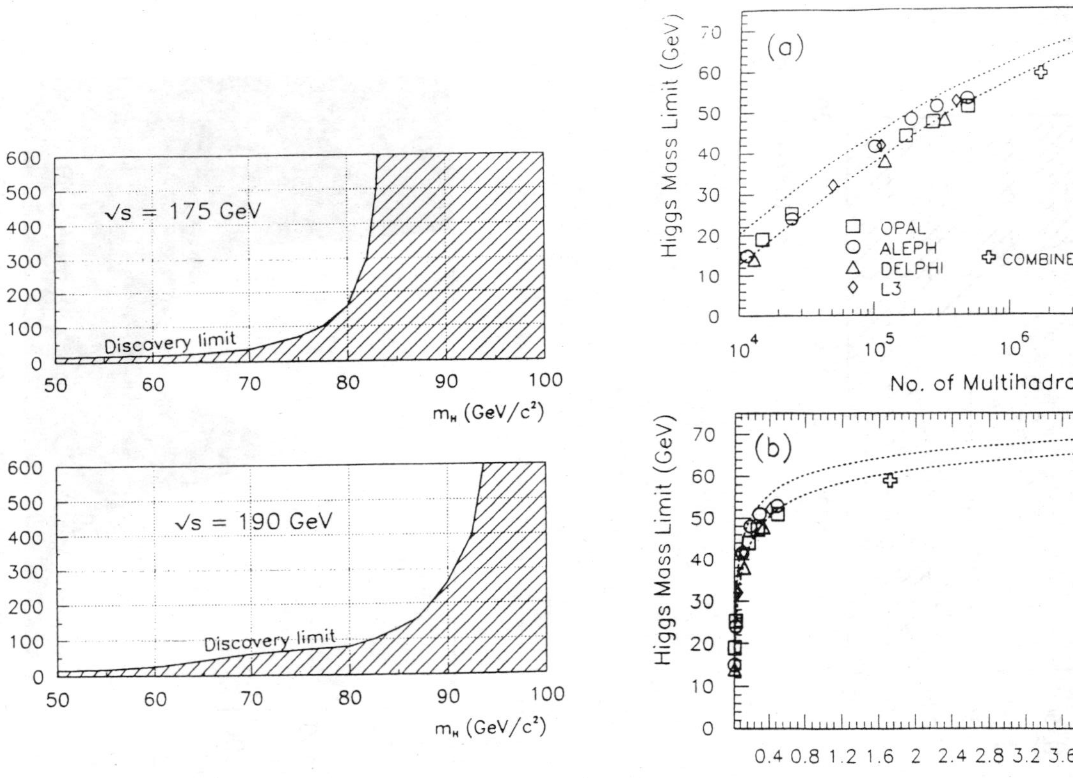

√s = 175 GeV
Discovery limit
L_{min} (pb^{-1})
m_H (GeV/c^2)
√s = 190 GeV
Discovery limit
L_{min} (pb^{-1})
m_H (GeV/c^2)
(a)
OPAL
ALEPH
DELPHI
L3
COMBINED
Higgs Mass Limit (GeV)
No. of Multihadrons
(b)
Higgs Mass Limit (GeV)
No. of Multihadrons (Millions)

Fig. 27

Fig. 28

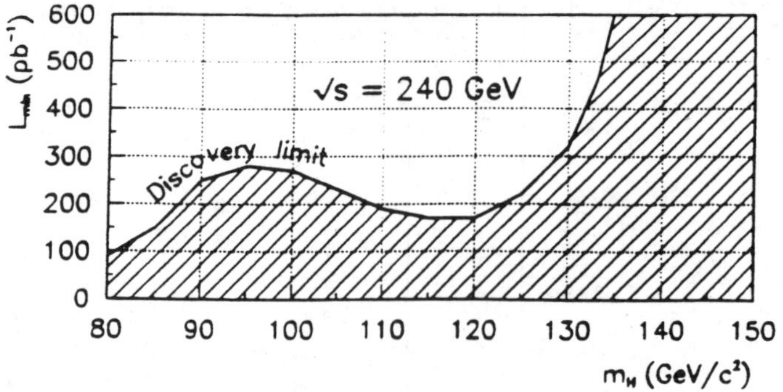
√s = 240 GeV
Discovery limit
L_{min} (pb^{-1})
m$_H$ (GeV/c^2)

Fig. 29

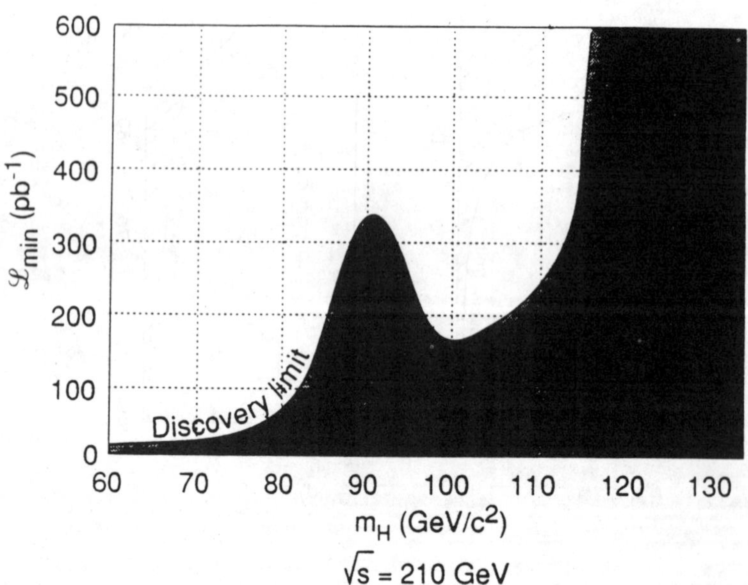
Discovery limit
$\mathcal{L}_{min}$ (pb^{-1})
m$_H$ (GeV/c^2)
√s = 210 GeV

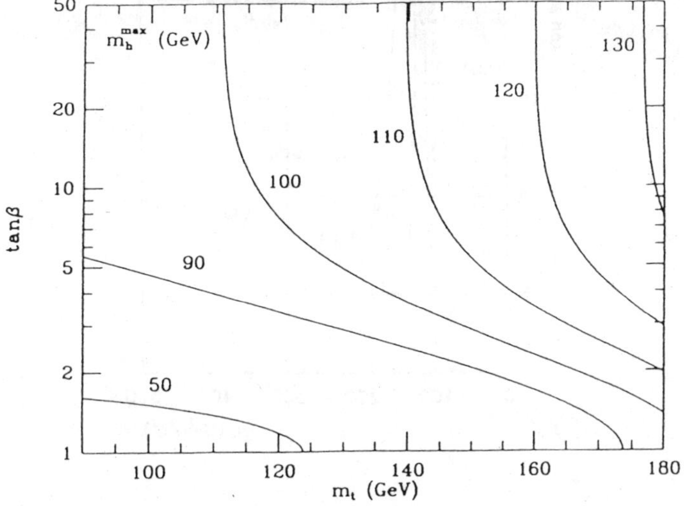

m_h^{max} (GeV)
130
120
110
100
90
50
tanβ
m_t (GeV)

Fig. 30

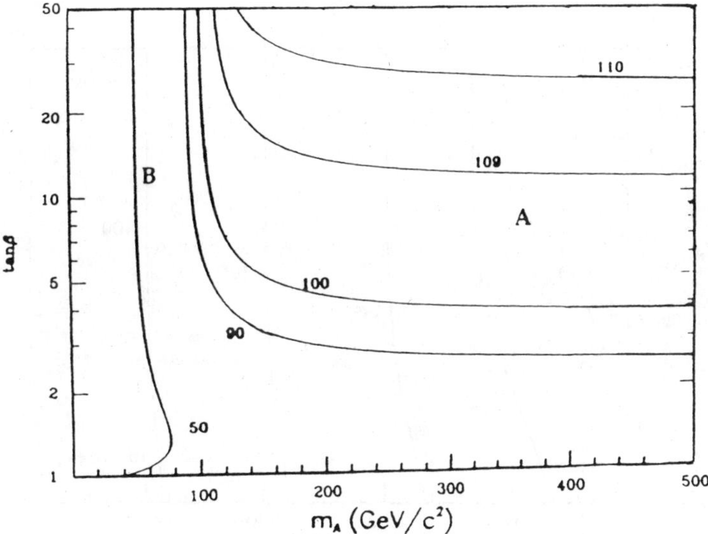

110
109
B
A
100
90
50
tanβ
m_A (GeV/c^2)

Fig. 31

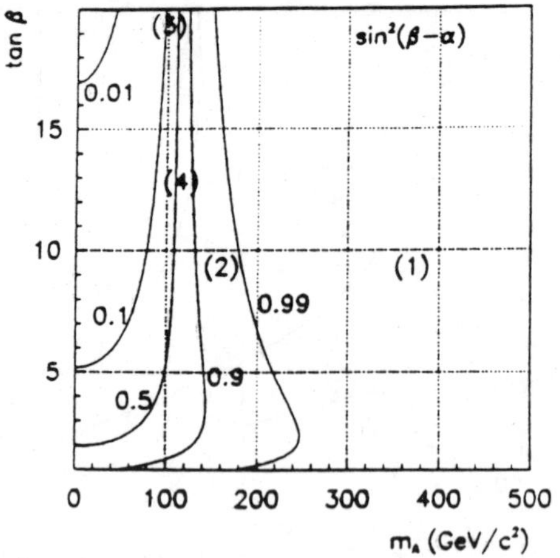
tan β
sin²(β−α)
(3)
0.01
15
(4)
10
(2)
(1)
0.1
0.99
5
0.9
0.5
0
100
200
300
400
500
m_A (GeV/c²)

Fig. 32

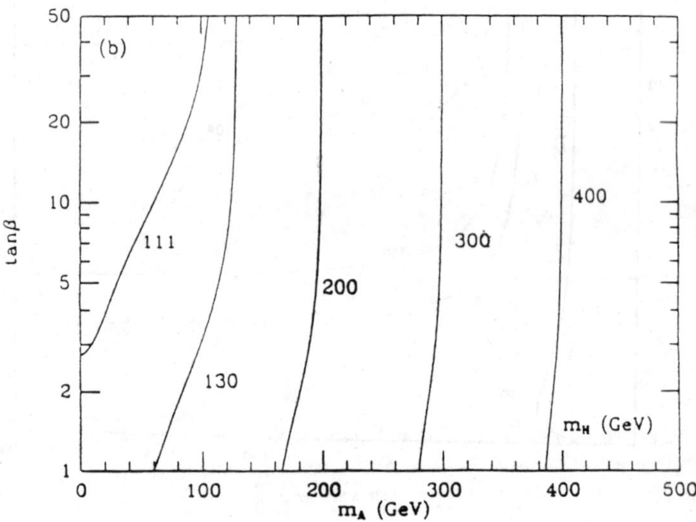
50
(b)
20
10
400
tanβ
111
300
5
200
2
130
m_H (GeV)
1
0
100
200
300
400
500
m_A (GeV)

Fig. 33

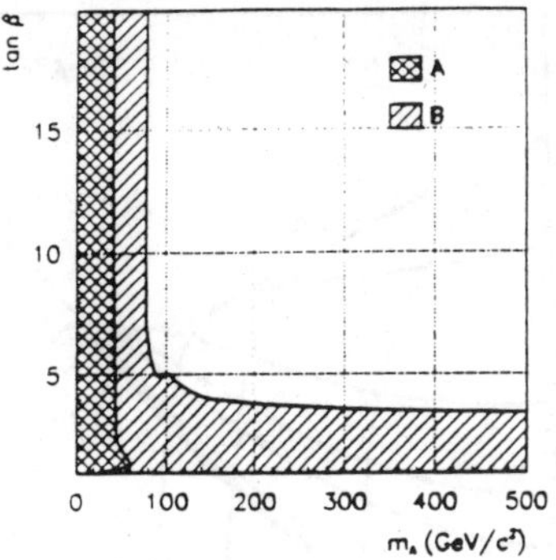
tan β
15
10
5
A
B
0
100
200
300
400
500
m_A (GeV/c²)

Fig. 34

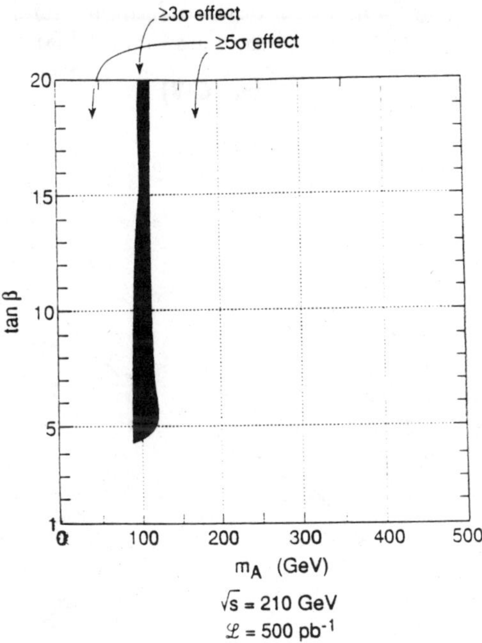
≥3σ effect
≥5σ effect
20
15
10
5
1
tan β
0
100
200
300
400
500
m_A (GeV)
√s = 210 GeV
ℒ = 500 pb⁻¹

Fig. 35

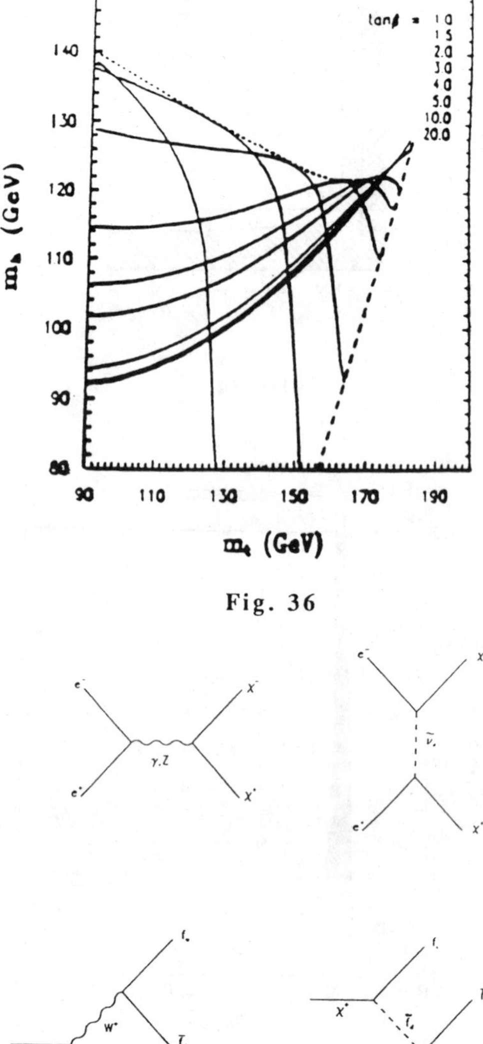
tanβ = 1.0
1.5
2.0
3.0
4.0
5.0
10.0
20.0
m_h (GeV)
m_t (GeV)
150
140
130
120
110
100
90
80
90
110
130
150
170
190
γ,Z
W*

Fig. 36

Fig. 37

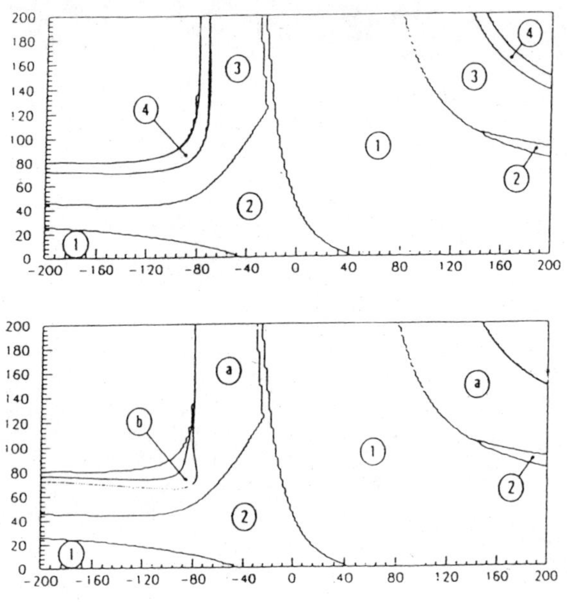
200
180
160
140
120
100
80
60
40
20
0
-200
-160
-120
-80
-40
0
40
80
120
160
200
1
2
3
4
a
b

Fig. 38

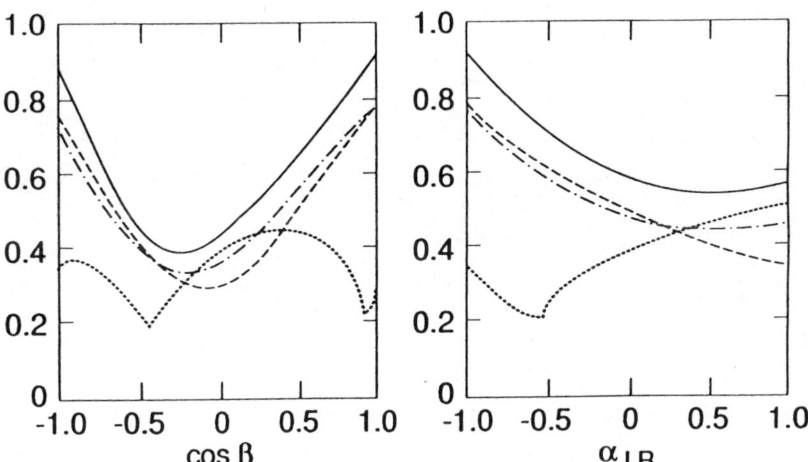
1.0
0.8
0.6
0.4
0.2
0
-1.0
-0.5
0
0.5
1.0
cos β
α LR

Fig. 39

IV. THE WEAK NEUTRAL CURRENT

THE MEASUREMENT OF ELECTROWEAK PARAMETERS FROM DEEP INELASTIC NEUTRINO SCATTERING

Frédéric Perrier

Département d'Astrophysique, de physique des Particules, de physique Nucléaire et de l'Instrumentation Associée
CE-Saclay, 91191 Gif-sur-Yvette, France

Contents

1 Introduction

The experimental study of neutrino interactions played an important role in the unification of the electromagnetic and the weak interactions. In 1973, the discovery of neutral-current interactions of muon-neutrinos in the Gargamelle bubble chamber at CERN was the first experimental success of the Electroweak Theory of Glashow, Weinberg and Salam [1]. Neutral-current interactions were later observed in other processes but, with the availability of high-energy neutrino beams and large neutrino detectors at CERN and Fermilab, the scattering of muon-neutrinos off nucleons remained the most powerful tool to measure the main free parameter of the Standard Electroweak Theory[1], the electroweak mixing parameter $\sin^2\theta_W$.

Indeed, in 1983, the determination of $\sin^2\theta_W$ from the comparison of the rate of neutral-current interactions with the rate of charged-current interactions had reached a precision $\Delta\sin^2\theta_W$ around ± 0.012. At this level of accuracy, it became relevant to

[1]Throughout this paper, "Standard Electroweak Theory" refers to the Electroweak Theory with the assumption $\rho = 1$ at tree level.

address the theoretical uncertainties associated to the Quark-Parton Model (QPM) used to interpret the experimental results. Already in 1979, Quantum Chromodynamics (QCD) corrections had been introduced into the framework of the QPM [2] but further strong interaction effects could potentially affect the determination of $\sin^2\theta_W$. However, Llewellyn-Smith demonstrated that using isospin symmetry, the theoretical uncertainties in the determination of $\sin^2\theta_W$ from deep inelastic neutrino scattering could in fact be maintained at ± 0.005, a precision matching the precision expected from the measurement of W and Z boson masses at the $p\bar{p}$ colliders.

Therefore, a new experimental effort was undertaken both at CERN and Fermilab to bring the experimental accuracy of the deep inelastic scattering neutrino experiments to the level of the theoretical uncertainties ± 0.005.

In 1986, the results from these experiments allowed a detailed test of the Standard Electroweak Theory by comparing them with the results from $p\bar{p}$ colliders. A first investigation of electroweak radiative corrections could be made and a first indication on the top mass $m_{top} < 200$ GeV was then obtained. Also, the results were precise enough to start discriminating between the predictions of supersymmetric and non supersymmetric theories, favoring the supersymmetric theories.

This paper is devoted to the measurements of electroweak parameters achieved during these epical years of deep inelastic neutrino experiments. After a historical overview of the basic experimental methods and the experimental achievements before 1983, the phenomenology of deep inelastic neutrino scattering will be presented. The precise measurements of the neutral-current to charged-current cross-section ratios performed after 1983 will then be described in detail. The determination of electroweak parameters will be the subject of the following section. Finally, possible experimental challenges for the future will be addressed in the conclusion[2].

2 Historical Overview of the Experimental Methods

2.1 *Basic Features of Deep Inelastic Neutral-Current and Charged-Current Neutrino Scattering*

It is useful to recall some simple relations to keep in mind while discussing the deep inelastic neutrino scattering experiments.

A neutrino interacts with matter through the emission of a W boson (charged-current interaction) or the emission of a Z boson (neutral-current interaction). The boson is absorbed by nuclei, producing hadrons in the final state. The charged-current interactions and neutral-current interactions of muon-neutrinos are distinguished by the presence of a muon in the final state in the charged-current interactions (figure 1).

[2]The experimental results obtained with free proton and neutron targets will not be described in this paper. A review of this topic can be found in [27].

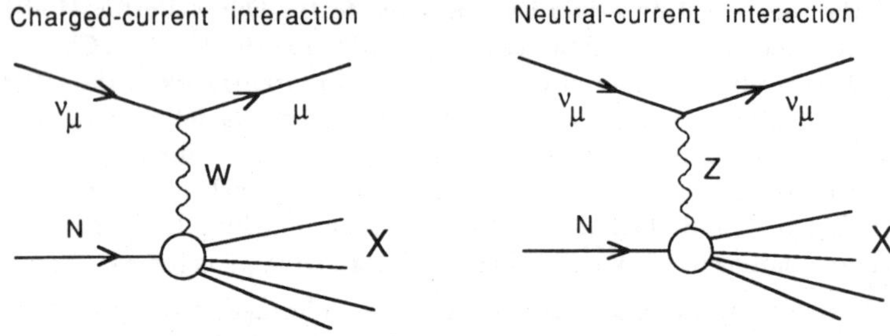

Figure 1: Interactions of muon-neutrinos with nuclei.

Muons are indeed rarely produced by hadrons. In addition, they are penetrating particles. For example, in iron, the penetration length of a muon of momentum p_μ (in GeV) is approximately $0.64\, p_\mu$ meters while hadrons are absorbed exponentially with an absorption length of 17 cm. Therefore, muons are easy to separate from hadrons provided they have more than 1 GeV energy or so. The separation between charged-current and neutral-current interactions will therefore be easier at high energies.

Another reason to work at high energies is the fact that the charged-current total cross-section of a neutrino off nucleons is proportional to the incoming neutrino energy E_ν :

$$\sigma_{tot}^{CC}(\nu_\mu N) = \sigma_0\, E_\nu \tag{1}$$

$$\text{where :} \quad \sigma_0 \approx 0.68\, 10^{-38}\,\text{cm}^2/\text{Nucleon/GeV} \approx 4.1\, 10^{-13}\,\text{m}^2/\text{ton/GeV} \tag{2}$$

The total number of charged-current interactions observed in a detector of mass M and section S inserted in a neutrino beam with a certain energy spectrum $\Phi(E_\nu)$ is therefore :

$$N^{CC}(\nu_\mu N) = \frac{M}{S}\sigma_0 \int \Phi(E_\nu) E_\nu \mathrm{d}E_\nu = \frac{M}{S}\sigma_0 N_\nu \langle E_\nu \rangle \tag{3}$$

where N_ν is the total number of incident neutrinos and $\langle E_\nu \rangle$ is the average neutrino energy.

The charged-current total cross-section for antineutrinos is measured to be approximately :

$$\sigma_{tot}^{CC}(\overline{\nu}_\mu N) \approx \frac{1}{2}\, \sigma_{tot}^{CC}(\nu_\mu N) \tag{4}$$

In principle, neutrinos also interact with the electrons in matter. However, as the cross-section of the neutrino scattering is proportional to the target mass, the scattering off electrons is reduced by three to four orders of magnitude compared with the scattering off nuclei and is negligible.

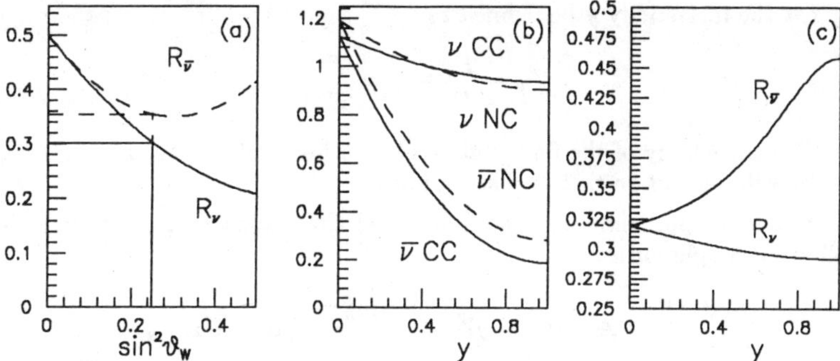

Figure 2: (a) The cross-section ratios R_ν and $R_{\bar{\nu}}$ as a function of $\sin^2\theta_W$. (b) The y-distributions of charged-current and neutral-current neutrino and antineutrino interactions off nuclei. The neutral-current cross-sections are normalized to the corresponding charged-current cross-sections. (c) The ratios R_ν and $R_{\bar{\nu}}$ as a function of the inelasticity y.

In the Standard Electroweak Theory, the neutral-current cross-section is related to the charged-current cross-section by the following formulae :

$$R_\nu = \frac{\sigma^{NC}(\nu_\mu N)}{\sigma^{CC}(\nu_\mu N)} = \frac{1}{2} - \sin^2\theta_W + \frac{5}{9}\sin^4\theta_W\,(1+r) = \frac{1}{2} - \sin^2\theta_W + \frac{5}{6}\sin^4\theta_W$$
$$R_{\bar{\nu}} = \frac{\sigma^{NC}(\bar{\nu}_\mu N)}{\sigma^{CC}(\bar{\nu}_\mu N)} = \frac{1}{2} - \sin^2\theta_W + \frac{5}{9}\sin^4\theta_W\left(1+\frac{1}{r}\right) = \frac{1}{2} - \sin^2\theta_W + \frac{5}{3}\sin^4\theta_W \quad (5)$$

where $r = \sigma^{CC}(\bar{\nu}N)/\sigma^{CC}(\nu N) = \frac{1}{2}$. The ratios R_ν and $R_{\bar{\nu}}$ are shown in figure 2a as a function of $\sin^2\theta_W$. The neutral-current cross-sections are of the same order of magnitude and smaller than the charged-current cross-sections[3]. For a toy value of $\sin^2\theta_W = \frac{1}{4}$, which is actually close to the actual value, one has $R_\nu = \frac{29}{96} = 0.302$ and $R_{\bar{\nu}} = \frac{17}{48} = 0.354$.

The sensitivity of the ratios R_ν and $R_{\bar{\nu}}$ to $\sin^2\theta_W$ is given by :

$$\frac{\Delta R_\nu}{\Delta \sin^2\theta_W} \approx -\frac{7}{12} \qquad \frac{\Delta R_{\bar{\nu}}}{\Delta \sin^2\theta_W} \approx -\frac{1}{6} \quad (6)$$

hence R_ν is about three times more sensitive to $\sin^2\theta_W$ than $R_{\bar{\nu}}$.

In practice, the values of the ratios are modified by the presence of experimental cuts. In particular, in order to observe the neutral-current interaction, the energy E_h of the final hadronic system must be larger than a certain threshold E_h^{cut}.

[3]Before the neutral-current was actually observed, it was known that $\sin^2\theta_W$ could not be large because otherwise the process $\bar{\nu}_e e \to \bar{\nu}_e e$ would have been seen.

Let the inelasticity y be defined as :

$$y = \frac{E_h}{E_\nu} = 1 - \frac{E'}{E_\nu} \tag{7}$$

where E' is the energy of the final state lepton. The effect on the cross-sections of a cut on E_h will depend on their y-distributions.

In a first approximation, the differential charged-current cross-sections have the following simple form :

$$\frac{\mathrm{d}\sigma_\nu^{CC}}{\mathrm{d}y} = A + B(1-y)^2 \qquad \frac{\mathrm{d}\sigma_{\bar{\nu}}^{CC}}{dy} = B + A(1-y)^2 \tag{8}$$

The value of the ratio B/A is fixed by the experimental measurement of the ratio $\sigma_{\bar{\nu}}^{CC}/\sigma_\nu^{CC} = 0.5$ and is $1/5$. Hence :

$$\frac{1}{\sigma_\nu^{CC}}\frac{\mathrm{d}\sigma_\nu^{CC}}{\mathrm{d}y} = \frac{3}{16}\left[5 + (1-y)^2\right] \qquad \frac{1}{\sigma_\nu^{CC}}\frac{\mathrm{d}\sigma_{\bar{\nu}}^{CC}}{\mathrm{d}y} = \frac{3}{16}\left[1 + 5(1-y)^2\right] \tag{9}$$

One can see in figure 2b that the neutrino charged-current cross-section is almost independent of y and that the antineutrino cross-section is a steeply falling function of y.

A cut on the hadronic energy E_h will correspond to a cut in the small regions of y. The sensitivity of the charged-current cross-sections to y_{min}, the minimum y value, is given by :

$$\frac{\Delta\sigma_\nu^{CC}}{\sigma_\nu^{CC}} \approx -\frac{9}{8}\Delta y_{min} \qquad \frac{\Delta\sigma_{\bar{\nu}}^{CC}}{\sigma_{\bar{\nu}}^{CC}} \approx -\frac{9}{4}\Delta y_{min} \tag{10}$$

The antineutrino cross-section is therefore decreasing two times faster than the neutrino cross-section if the threshold E_h^{cut} is increasing or, equivalently, if the neutrino energy is decreasing.

The high-y region corresponds to small outgoing muon momenta. In that region, the outgoing muon will be hidden in the hadronic final state and the charged-current interaction will be confused experimentally with a neutral-current interaction. One can see that the fraction of charged-current events mistaken as neutral-current events is smaller for higher neutrino energies and that this fraction is also much smaller for antineutrino scattering than for neutrino scattering.

The y-distributions of the neutral-current cross-sections are given by the following approximate relations :

$$\begin{aligned}
\frac{\mathrm{d}\sigma_\nu^{NC}}{\mathrm{d}y} &= \left(\frac{1}{2} - \sin^2\theta_W + \frac{5}{9}\sin^4\theta_W\right)\frac{\mathrm{d}\sigma_\nu^{CC}}{\mathrm{d}y} + \frac{5}{9}\sin^4\theta_W\frac{\mathrm{d}\sigma_{\bar{\nu}}^{CC}}{\mathrm{d}y} \\
\frac{\mathrm{d}\sigma_{\bar{\nu}}^{NC}}{\mathrm{d}y} &= \left(\frac{1}{2} - \sin^2\theta_W + \frac{5}{9}\sin^4\theta_W\right)\frac{\mathrm{d}\sigma_{\bar{\nu}}^{CC}}{\mathrm{d}y} + \frac{5}{9}\sin^4\theta_W\frac{\mathrm{d}\sigma_\nu^{CC}}{\mathrm{d}y}
\end{aligned} \tag{11}$$

which will be discussed in detail in section 3.

Using Eq. 9 and $\sin^2\theta_W = \frac{1}{4}$, one obtains :

$$\frac{1}{\sigma_\nu^{CC}}\frac{d\sigma_\nu^{NC}}{dy} = \frac{21}{384}\left[5 + \frac{11}{7}(1-y)^2\right] \qquad \frac{1}{\sigma_\nu^{CC}}\frac{d\sigma_{\bar{\nu}}^{NC}}{dy} = \frac{21}{384}\left[\frac{11}{7} + 5(1-y)^2\right] \quad (12)$$

The neutral-current y-distributions are similar to the corresponding charged-current y-distributions (see figure 2b) but the neutrino neutral-current y-distribution is slightly steeper than the neutrino charged-current y-distribution whereas the antineutrino neutral-current y-distribution is slightly flatter than the antineutrino charged-current y-distribution. In other words, R_ν is decreasing slowly with y whereas $R_{\bar{\nu}}$ is increasing slowly with y (see figure 2c). Another consequence of this fact is that, if the energy of the incident neutrino is increased or, equivalently, if the threshold E_h^{cut} is decreased, the value of the ratio R_ν will increase whereas the value of $R_{\bar{\nu}}$ will decrease.

The sensitivity of R_ν and $R_{\bar{\nu}}$ to a change in y_{min} is given by :

$$\frac{\Delta R_\nu}{\Delta y_{min}} \approx -\frac{5}{256} \approx -0.02 \qquad \frac{\Delta R_{\bar{\nu}}}{\Delta y_{min}} \approx \frac{5}{64} \approx 0.08 \quad (13)$$

The actual value of the ratio R_ν is rather insensitive to y_{min}, hence to the hadron energy threshold or the neutrino energy, whereas larger effects are expected for antineutrinos.

To summarize, high energy neutrinos are favored to study neutral-current interactions because the cross-sections are increasing linearly with energy, the fraction of events with $E_h > E_h^{cut}$ is larger and the separation between neutral-current and charged-current is easier. Therefore, the progress in the study of neutral-current interactions in history were directly related to the progress in the making of neutrino beams.

2.2 *Principles of Neutrino Beams*

Neutrinos are copiously produced in the decay of charged pions and kaons, the dominating decay modes being the two body decays :

$$\pi^+ \to \mu^+\nu_\mu(100\%) \qquad K^+ \to \mu^+\nu_\mu(62.5\%)$$

with a small contribution from the so called Kμ3 and Ke3 decays :

$$K^+ \to \pi^0\mu^+\nu_\mu(3.2\%)\ (\text{K}\mu3) \qquad K^+ \to \pi^0 e^+\nu_e(4.8\%)\ (\text{Ke3})$$

The principle of a neutrino beam is depicted in figure 3.

The charged hadrons emerging from a target hit by a proton beam are focused to form a beam which is transported by a magnetic system to an evacuated region

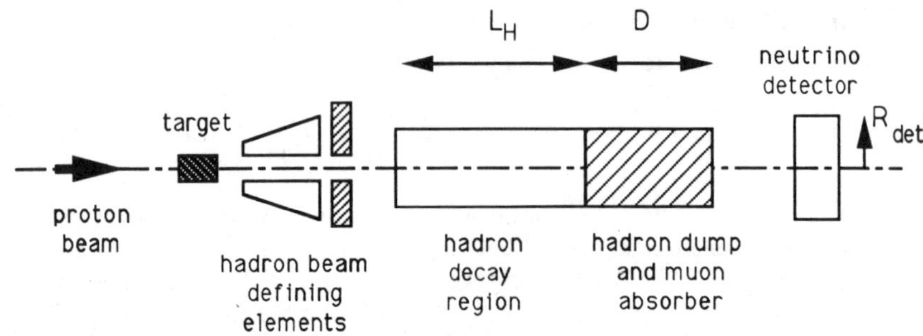

Figure 3: The principle of a neutrino beam.

pointing to the neutrino detector. In this region, the hadrons decay freely. The remaining hadrons are absorbed at the end of the decay region. The neutrino detector is installed after a large amount of shielding material which absorbs the muons coming from hadron decays. From Eq. 2, one can see that in order to see one charged-current interaction of a 10 GeV energy neutrino in a detector of 10 tons and 1 m^2 section, about 10^{11} hadron decays are needed.

The properties of the neutrino beam crossing the detector are defined by the properties of the hadron beam. Since the dominating decays of pions and kaons are two-body decays, this relationship is simple.

Let us consider the decay $H \rightarrow \mu\nu_\mu$ of a hadron H of mass m_H and energy $\mathcal{E}_H$. The energy E_ν^H of the outgoing neutrino has to be less than the kinematical limit $E_{max}^H = \mathcal{E}_H(1 - \frac{m_\mu^2}{m_H^2})$. For a pion, $E_{max}^\pi/\mathcal{E}_\pi \approx 0.427$. For a kaon, $E_{max}^K/\mathcal{E}_K \approx 0.954$.

Because of the two-body kinematics, the energy is related to the decay angle θ by:

$$E_\nu^H = \mathcal{E}_H \frac{m_H^2 - m_\mu^2}{m_H^2 + \mathcal{E}_H^2\theta^2} \tag{14}$$

Let us assume that the hadron H decays at a distance x from the beginning of the decay region. The smallest neutrino energy is reached when the neutrino is hitting the outer edge of the detector whose radius is R_{det}. This happens for a decay angle $\theta(x)$ such that:

$$\theta(x) = \frac{R_{det}}{L_H + D - x} \tag{15}$$

where L_H is the length of the decay region and D the distance between the end of the decay region and the neutrino detector (see figure 3). In addition, the two-body kinematics ensures that the distribution of E_ν^H is flat between $E_\nu^H(x)$ and E_{max}^H.

The distance of flight of a pion of energy $\mathcal{E}_H$ (in GeV) is $55.92\,\mathcal{E}_H$ meters. For

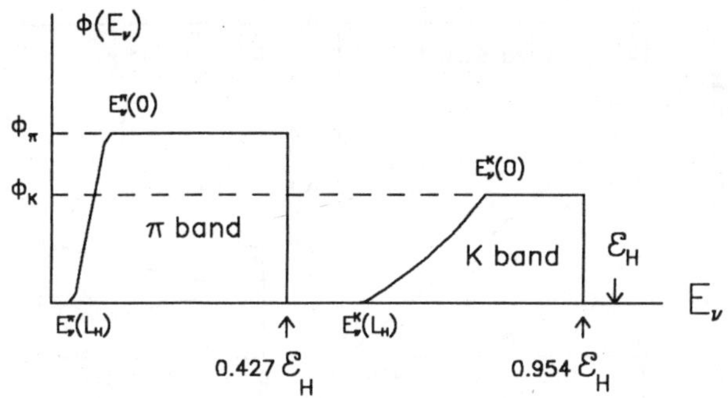

Figure 4: The ideal energy distribution of neutrinos through the detector from the two-body decays of pions and kaons of energy $\mathcal{E}_H$.

a kaon, it is 7.513 $\mathcal{E}_H$ meters. The decay probability is therefore much smaller for a pion than for a kaon. For example, for a 300 m long decay region and $\mathcal{E}_H = 200$ GeV, the decay length is 11 km for a pion, 1.5 km for a kaon and the decay probability P_H 3% for a pion, 18% for a kaon.

Assuming a flat distribution of the decay point x, an assumption valid for both pions and kaons for the energies that will be considered here, the energy spectrum of the neutrinos hitting the detector is depicted in figure 4. This spectrum has a flat component Φ_0^H between $E_\nu^H(0)$ and E_{max}^H and a rising component between $E_\nu^H(L_H)$ and $E_\nu^H(0)$ given by :

$$\Phi^H(E_\nu) = \Phi_0^H \left[\frac{R_{det}}{L_H} \frac{\mathcal{E}_H}{\sqrt{\frac{\mathcal{E}_H}{E_\nu}(m_H^2 - m_\mu^2) - m_H^2}} - D \right] \tag{16}$$

The neutrino spectrum has a dichromatic structure : neutrinos from pion decay (referred to later as the "pion band" of the spectrum) are concentrated at low energies whereas neutrinos from kaon decay (the "kaon band" of the spectrum) are concentrated at high energies. In figure 4, the pion and kaon bands are well separated. The rising part of each band comes from the finite size of the neutrino source as seen from the neutrino detector. This effect, sometimes called "parallax" of the neutrino beam, is more important for the kaon band. It can be reduced if for example the distance D is increased or if the length of the decay region is reduced but this also decreases the number of neutrinos crossing the detector.

The relative height of the pion and kaon bands in the energy spectrum is given by :

$$\frac{\Phi_K}{\Phi_\pi} = \frac{E_{max}^\pi}{E_{max}^K} \frac{N_K}{N_\pi} \frac{P_K}{P_\pi} \frac{B_K}{B_\pi} \tag{17}$$

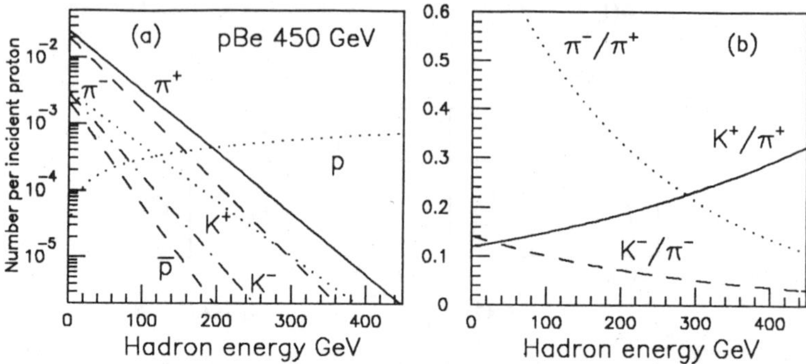

Figure 5: (a) The energy distribution of the hadrons emerging from a Beryllium target hit by a 450 GeV proton beam [3]. (b) Ratios π^-/π^+, K^+/π^+ and K^-/π^- versus the energy of the hadron.

N_H is the number of hadrons H and B_H the two-body decay branching ratio of H. The actual shape of the neutrino spectrum is then given by the composition and energy spectrum of the hadron beam.

The energy distribution of the hadrons emerging from a target hit by a proton beam of energy $\mathcal{E}_p$ can be empirically parametrized in the following way [3]:

$$\frac{dN}{d\mathcal{E}_H} = a\left[\frac{b}{\mathcal{E}_p} e^{-b\frac{\mathcal{E}_H}{\mathcal{E}_p}}\right] \tag{18}$$

where the constants a and b are measured experimentally. These distributions for a 450 GeV proton beam are shown in figure 5a. They are peaked at low energy.

The nature of the beam (neutrino or antineutrino) is defined by a magnetic device which defocuses the particles with the wrong sign. The number of negative hadrons emerging from the target is of the order of one third to one tenth of the number of positive hadrons (figure 5b). It is therefore more difficult to make intense antineutrino beams. One can also notice that the ratio of the number of kaons to the number of pions is of the order of 0.1 to 0.2 (see figure 5b). The high energy part of the neutrino spectrum (the kaon band) is therefore reduced compared with the low energy part. This effect, which is more pronounced for antineutrino beams than for neutrino beams, is partly compensated by the fact that the decay probability is about six times larger for a kaon than for a pion. In addition, the interaction cross-section is proportional to the neutrino energy, and finally the number of interactions from the kaon and the pion bands are rather similar.

If no selection on the energy of the outgoing hadrons is made, the energy distribution of neutrinos is exponentially falling with energy. In such a beam, called a "Wide Band Beam" (WBB), the only beam defining element is a sign focusing

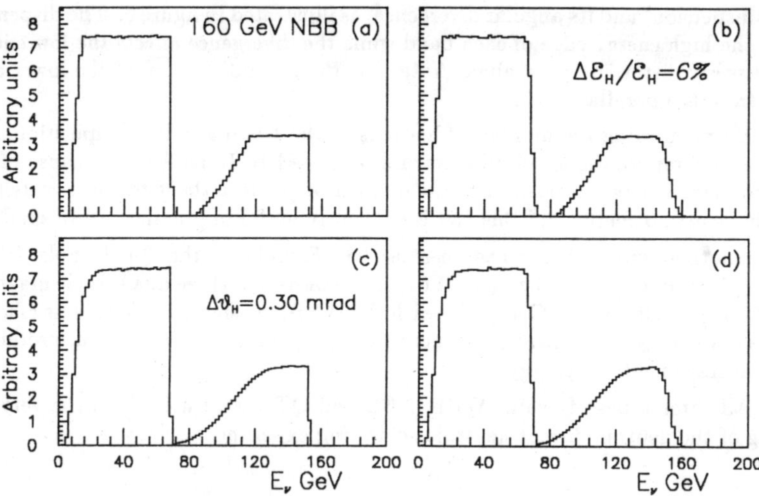

Figure 6: The effect of the angular divergence and the energy dispersion of the hadron beam on the shape of the ideal NBB neutrino energy spectrum : (a) no divergence and no dispersion (b) with energy dispersion only (c) with angular divergence only (d) with both dispersion and divergence.

system. This system, which consists of a pulsed toroidal magnet with the shape of a horn, is very efficient but a significant contamination of positive hadrons usually remains in negative hadron beams, because of the large number of wrong sign hadrons in that case. In a WBB, the average energy of the neutrinos is low but the intensity is large. When the mass of the detector is small, as it was the case for the Gargamelle bubble chamber, it is necessary to use such a beam to observe neutrino interactions.

For many measurements, however, the WBB is a difficult beam to handle. For example, for the measurement of the total neutrino interaction cross-section, the absolute spectrum of the neutrinos crossing the detector must be known and this is very difficult to estimate in a WBB. For the measurement of the electroweak parameters, a detailed knowledge of the shape of the spectrum is mandatory and the contamination of the wrong type of neutrinos in the beam must be negligible.

For such measurements, a system of magnetic quadrupoles and dipoles is used to select hadrons around a given energy. In such a beam, called a "Dichromatic Beam" or a "Narrow Band Beam" (NBB), the neutrino spectrum is similar to the ideal case discussed before but is modified by the finite energy spread of the hadron beam (dispersion) and its angular divergence, as illustrated in figure 6. The dispersion affects the high energy edge of each band while the divergence affects the low energy edge, already distorted by parallax. Note that the main distortion of the lower edge comes from the parallax.

By monitoring the number of hadrons in the beam and its composition, it is possible to compute the absolute spectrum of the NBB. In addition, because of the two-body kinematics, the position of the interaction point in the detector is correlated with the neutrino energy and this fact is extremely useful in practice, as we shall see.

Another type of beam has been used in Fermilab : the Quadrupole Triplet Beam (QTB). In this case, the magnetic system consists of three quadrupole magnets. It is a high intensity beam. Compared with the WBB, the sign selection is less efficient but the low energy component of the neutrino energy spectrum is reduced, producing a high average neutrino energy.

All three types of beams WBB, NBB and QTB were used from the very beginning of the neutral-current neutrino scattering experiments.

2.3 *The Early Experiments : from Gargamelle to the Fermilab Dichromatic Beam*

The Gargamelle bubble chamber (GGM), filled with freon, had a mass of 10 tons, 4.5 tons of which were useful for detecting clearly neutral-current interactions (fiducial mass). In 1973, it was exposed to a WBB, obtained from the 26 GeV proton beam of the PS at CERN, with an average neutrino energy of 3.5 GeV. Nearly one hundred interactions without muon and with a hadronic final state with more than 1 GeV were observed. These events could not be attributed to a neutron background and they were interpreted as neutrino neutral-current interactions. In addition, one candidate event corresponding to the unambiguous process $\overline{\nu}_\mu e \to \overline{\nu}_\mu e$ was observed in the same data sample, bringing substantial support to the claim that neutral currents were actually present [4].

This discovery was soon confirmed by counter experiments at Fermilab using higher energy beams. The neutrino beam line at Fermilab is depicted in figure 7. The energy of the proton beam was 300 GeV, 380 GeV and 400 GeV.

Both QTB and WBB beams were used to study neutral-current interactions in the Harvard-Pennsylvania-Wisconsin-Fermilab detector (HPWF), which was located in LabC, 1035 m from the hadron beam dump. This detector (figure 8) consisted of 60 tons of liquid scintillator with a fiducial mass of 8 tons. The energy deposited in the modules of liquid scintillator was read-out by phototubes in order to measure the hadronic energy E_h. Muons were detected by optical spark chambers (WGSC) inserted every four modules and by a scintillation counter and a spark chamber placed around a toroidal iron magnet with which the momentum could be determined. Interactions were separated into "no-muon" and "with muon" categories by visual scanning

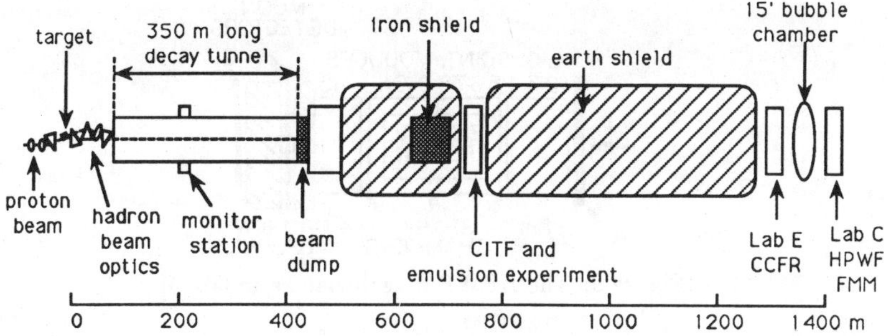

Figure 7: Layout of the neutrino beam at Fermilab.

of the recorded events. A few hundred interactions were observed in neutrino beams.

The HPWF detector had an active target material and was able to detect slow recoils ($E_h^{cut} = 4\,\mathrm{GeV}$). However, the interaction length (84 cm) and the radiation length (53 cm) being very large, hadrons could be confused with muons and a large fraction of the shower energy was not measured in the calorimeter.

In the California Institute of Technology-Fermilab detector (CITF), a dense material, iron, was used instead.

This detector, depicted in figure 9, was a 140 ton sandwich of iron plates and scintillation counters with a fiducial mass of 85 tons. The interaction length in iron being 17 cm, the hadronic shower is contained within 1.5 m. The hadronic energy is measured by the energy deposition in liquid scintillator slabs inserted between each iron slab. Muons are detected in spark chambers inserted every second iron slab and the momentum is determined by the deflection in a toroidal iron magnet located behind the target calorimeter.

The CITF detector represented a significant increase in the target mass and it could be exposed to a NBB with $\mathcal{E}_H = 160\,\mathrm{GeV}$.

Neutrino interactions in the CITF detector are shown in figure 9. Neutral-current and charged-current interactions are separated on the basis of the penetration, defined as the length travelled in iron by the most penetrating particle in the final state. The observed distributions of the penetration are shown in figure 10 for $E_h^{cut} > 6\,\mathrm{GeV}$. Most of the charged-current interactions are observed with large penetrations. An excess of events at small penetration reveals the presence of the neutral-current interaction. The number of charged-current interactions mistaken as neutral-current interactions can be estimated by extrapolating the measured penetration distribution below the neutral-current peak. It was estimated that about half of the charged-current interactions below the neutral-current peak were actually due to high momentum muons emitted at large angle and leaving the target calorime-

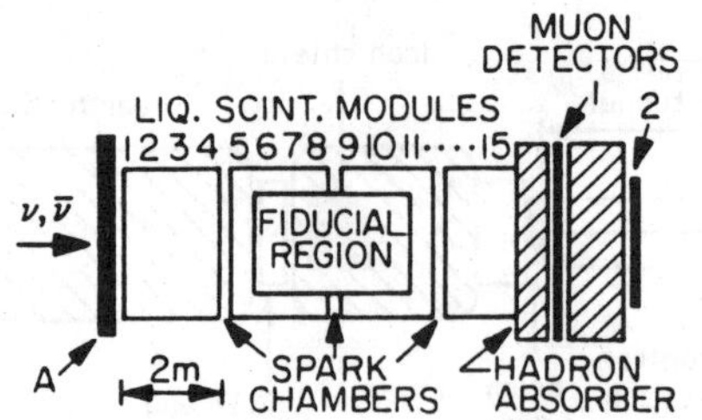

Figure 8: The HPWF detector at Fermilab [5].

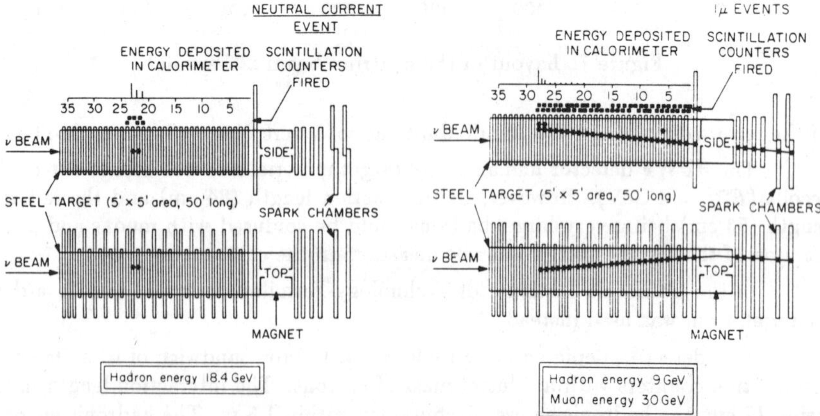

Figure 9: Neutrino interactions in the CITF detector at Fermilab [6].

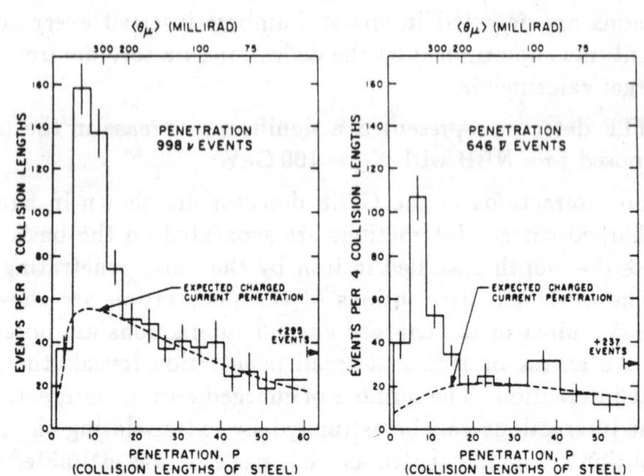

Figure 10: Event-length distribution in the CITF detector at Fermilab [6].

	E_h^{cut} GeV	$\langle E_\nu \rangle$ GeV	ν beam Candidates NC	CC	$\langle E_{\overline{\nu}} \rangle$ GeV	$\overline{\nu}$ beam Candidates NC	CC	$\sin^2\theta_W$
GGM [9]	1	3.5	175	223	3.5	151	273	0.31±0.06
HPWF [10]	4	85	724	2498	41	69	198	0.23±0.06
CITF [11]	12	85	1033		45	239		0.33±0.07

Table 1: Results from the early neutrino deep inelastic scattering experiments.

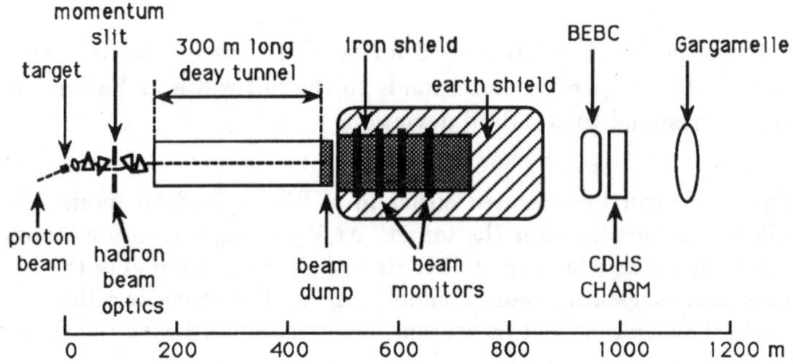

Figure 11: Layout of the neutrino beam at CERN.

ter by the side. This method was going to be extensively used in the subsequent experiments.

From these early experiments, the existence of neutral-current interactions was firmly established. Neutral currents were also observed in other processes like elastic scattering $^{(}\overline{\nu}^{)}\mathrm{p} \rightarrow ^{(}\overline{\nu}^{)}\mathrm{p}$ or single pion production $\nu N \rightarrow \nu N'\pi^0$ and $\overline{\nu}_e e \rightarrow \overline{\nu}_e e$ [7, 8]. However, in deep inelastic scattering experiments, the value of the ratios R_ν and $R_{\overline{\nu}}$ could be determined. The results were found in agreement with the Standard Electroweak Theory and first values of the electroweak mixing parameter could be given (see table 1).

However, the measurements were limited in statistics and new detectors were built at CERN and Fermilab to investigate in detail the new interaction.

2.4 *The Second Generation of Deep Inelastic Neutrino Experiments*

In 1976, a neutrino line was set up at CERN using 400 GeV protons from the SPS accelerator. The CERN neutrino beam line is depicted in figure 11.

	Fermilab	CERN before 1984	CERN 1984
Length of decay tunnel (m)	350	292	292
Distance between the decay tunnel and the detector (m)	350 (CITF) to 1035 (HPWF)	456	456
Proton beam energy (GeV)	400	400	450
Hadron beam energy (GeV)	100 to 250	200	160
Hadron beam dispersion (%)	10	6	9
Horizontal divergence (mrad)	0.15	0.16	0.28
Vertical divergence (mrad)	0.15	0.21	0.34

Table 2: Parameters of the NBBs used for neutral-current measurements at CERN and Fermilab. The dispersion corresponds to the full width at half maximum. The divergences correspond to standard deviations.

The beam can be operated either in WBB or in NBB mode. Both beams are installed permanently after the target. At Fermilab, the various types of beams are arranged on trains which are rolled after the target. Changing the beam type is not an easy and necessarily reproducible process. The characteristics of the CERN beams used for neutral-current measurements are compared with the Fermilab beams in table 2.

The distance between the detector and the end of the decay tunnel is larger at Fermilab. Therefore, the fraction of the kaon band sampled by the detector is smaller, and the parallax is smaller, at the cost of event rate as discussed before. The energy spread at CERN is smaller than at Fermilab but the angular divergence is larger. The shapes of the NBB at CERN and at Fermilab are therefore different, as illustrated in figure 12.

Four detectors were installed in the CERN neutrino beam : Gargamelle, moved from the PS site, a large European bubble chamber BEBC (Big European Bubble Chamber) and the two large electronic detectors of the CDHS (CERN-Dortmund-Heidelberg-Saclay) and CHARM (CERN-Hamburg-Rome-Moscow) collaborations.

The characteristics and results of these experiments are collected in table 3.

The BEBC bubble chamber, filled with a mass of 13 tons of a mixture of Hydrogen and Neon, offered a substantial progress compared with Gargamelle. The chamber was installed in a vertical 3 Tesla magnetic field allowing the reconstruction of charged particle momenta with a resolution of 7 %. Muons were also detected in an External Muon Identifier (EMI) made of 150 m^2 of proportional wire chambers. Recorded interactions were classified as charged-current if a muon with a momentum larger than 5 GeV was seen in the EMI and as a neutral-current otherwise. In addition, the total transverse momentum of the hadrons with respect to the neutrino axis was

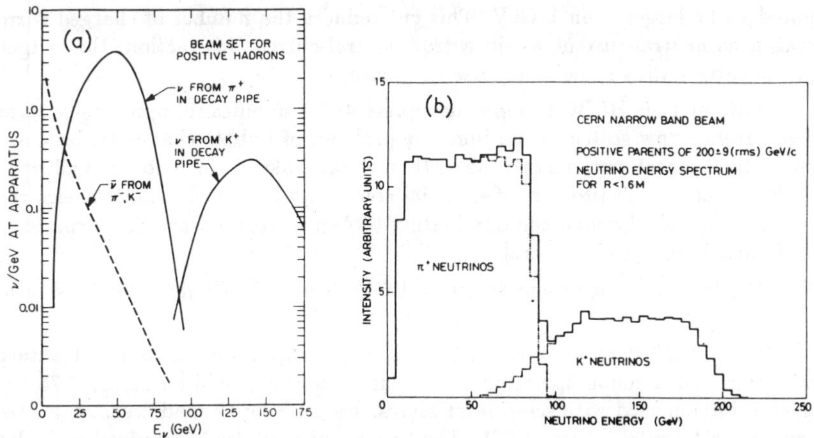

Figure 12: (a) The spectrum of the 160 GeV NBB through the CITF detector at Fermilab [8]. (b) The spectrum of the 200 GeV NBB through the CDHS detector at CERN [16].

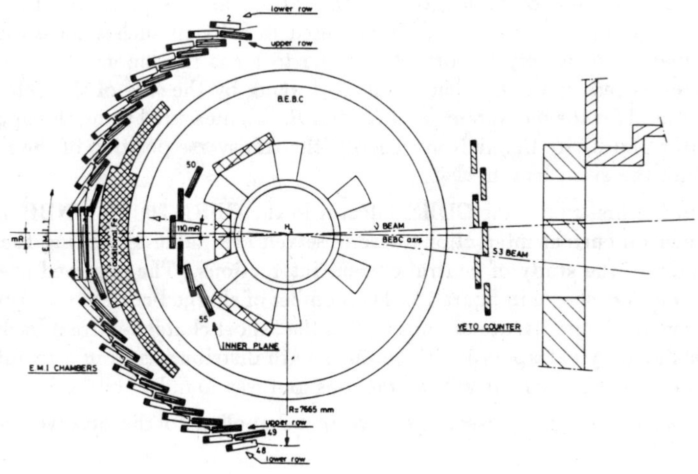

Figure 13: The BEBC bubble chamber at CERN [12].

required to be larger than 1 GeV. This cut reduces the number of charged-current mistaken as neutral-current as, in a true neutral-current interaction, the outgoing neutrino carries away some transverse momentum.

Although the BEBC experiment represented a significant improvement, it still suffered from a low statistics, an inherent problem of bubble chambers, because of both the light target material and the method of data taking which does not allow high interaction rates. This problem of statistics could only be solved with large electronic detectors and they became the dominating tool of neutrino scattering experiments and of particle physics in general.

The first large electronic detector at CERN was the CDHS detector shown in figure 14.

The CDHS detector combines in a single unit both functions of a target calorimeter and a muon spectrometer. It is a sandwich of iron slabs, 3.75 m in diameter, interspersed with scintillator plates. Each of the 19 modules is magnetized to produce a toroidal field of 1.6 T. The first seven modules have scintillator plates inserted every 5 cm and the other modules every 15 cm. Each iron module is followed by a large drift chamber measuring coordinates with a precision of 2 mm.

The CDHS detector had a total mass of 1250 tons and a fine longitudinal segmentation. These two features are important for the measurement of the ratios R_ν and $R_{\bar{\nu}}$, for which the neutral-current and charged-current interactions are separated on the basis of their event length or penetration, as was pioneered by the CITF detector. The muon spectrometer is not used directly in such a measurement but the magnetic field is very important in order to focus the muons on the beam axis and therefore reduce the number of muons leaving by the side of the detector. The variables used for the measurement of R_ν and $R_{\bar{\nu}}$, all measured using the signals in the scintillator plates, are the hadronic energy, the transverse position of the interaction vertex and the event length.

In the first run of the CDHS detector in the CERN 200 GeV NBB, more than 10,000 neutral-current interactions were observed (see table 3), setting the stage for a new area in the study of neutral-current interactions. The observed event length distributions are shown in figure 15. The number of charged-current interactions mistaken as neutral-current interactions (called the "short charged-current background") was calculated by extrapolating the event length distribution from a monitor region containing only charged-current interactions and was found to be 14%.

Two other large corrections have to be applied to the observed number of events.

The first correction comes from the contribution of the so called "WBB neutrinos" accompanying the NBB neutrino beam. These neutrinos are produced by off-momentum hadrons produced before the entrance of the decay tunnel. The number of such neutrinos of unknown spectrum is very difficult to estimate. This is a serious problem because these neutrinos have a low average energy and the muons in charged-current interactions will be hidden in the shower, faking a neutral-current

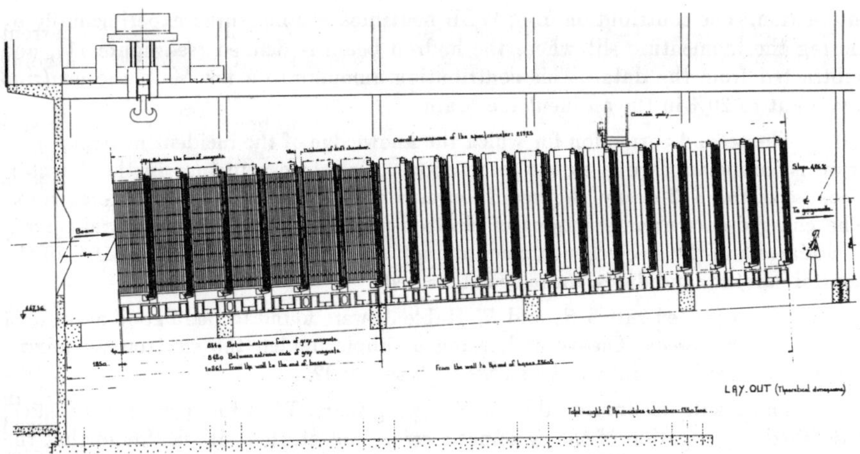

Figure 14: The CDHS detector at CERN [13].

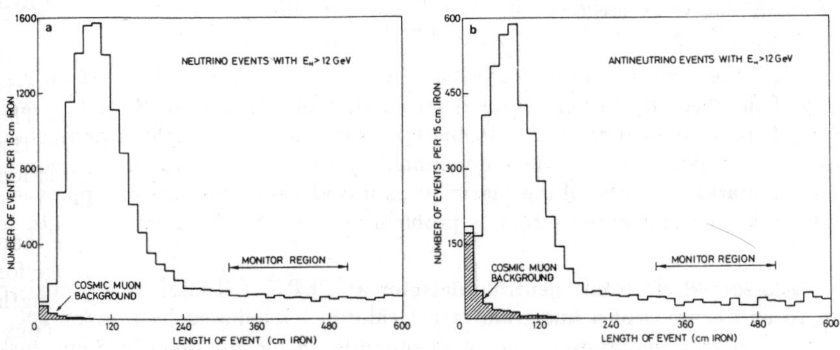

Figure 15: Event length distributions observed in the first run of the CDHS detector in the CERN 200 GeV Narrow Band Beam [16] with the neutrino beam (a) and with the antineutrino beam (b).

interaction. The contribution from WBB neutrinos was measured experimentally by closing the momentum slit where the hadron beam is defined (see figure 11), and subtracted from the data. This contribution amounts to a few % in the neutrino beam but to 20% in the antineutrino beam.

The second correction for which the knowledge of the incident neutrino spectrum is important is due to the interactions of electron-neutrinos in the detector. Because of the Ke3 decay, the beam contains a small fraction of electron-neutrinos. The charged-current interactions of electron-neutrinos are seen as neutral-current candidates and the electron-neutrino interactions, estimated to be about 10% of the neutral-current candidates, must be subtracted.

The corrected ratios R_ν and $R_{\bar{\nu}}$ (table 3) were found to be in good agreement with the Electroweak Theory and, using a simple model, the electroweak mixing parameter was determined to be $\sin^2\theta_W = 0.24 \pm 0.02$.

This result was improved in the following years. With the increased statistics, a detailed investigation of the systematic errors was started. It was found that the dominating systematic error was coming from the incomplete knowledge of the WBB neutrino background in the NBB. Between the momentum slit and the decay tunnel, the hadron beam is passing through a small amount of matter where secondary hadrons can be produced. The contribution of the neutrinos produced by these secondary hadrons could not be measured and a systematic error of 1.2 % on R_ν and 2.5 % on $R_{\bar{\nu}}$ had to be assigned. In total, the statistical error had reached the level of the systematic error of 1.7 % on R_ν and 3 % on $R_{\bar{\nu}}$.

With the experimental errors being reduced to the level of ± 0.010 on $\sin^2\theta_W$, a study of the theoretical errors in the determination of $\sin^2\theta_W$ from R_ν and $R_{\bar{\nu}}$ was started. This theoretical error, due to the QPM used to interpret the experimental ratios, was estimated to be ± 0.006, dominated by the uncertainties on the mass of the charm quark. The mass of the charm quark introduces a kinematical suppression in the charged-current cross-section, a problem that will be discussed at length in section 5.

The second electronic neutrino detector at CERN, CHARM, is depicted in figure 16 and consists of a fine-grain target calorimeter followed by a muon spectrometer. The calorimeter was made of 78 subunits, each composed of a 8 cm thick marble plate equiped with scintillators and proportional drift tubes, and surrounded by a magnetized iron frame. The available fiducial mass of the CHARM detector, 65 tons for a radial distance R_{det} to the detector axis less than 1.2 m, was much smaller than the fiducial mass in the CDHS detector but this is partly compensated by the fact that the hadronic energy can be measured down to 2 GeV.

The CHARM detector, actually designed for the study of neutrino scattering off electrons, had a high potential of pattern recognition as illustrated in figure 17. The high spatial resolution of the detector allows neutral-current and charged-current interactions to be separated on the basis of their topology, looking in the event for the pattern of hits typical of a muon. However, the neutral-current candidate sample

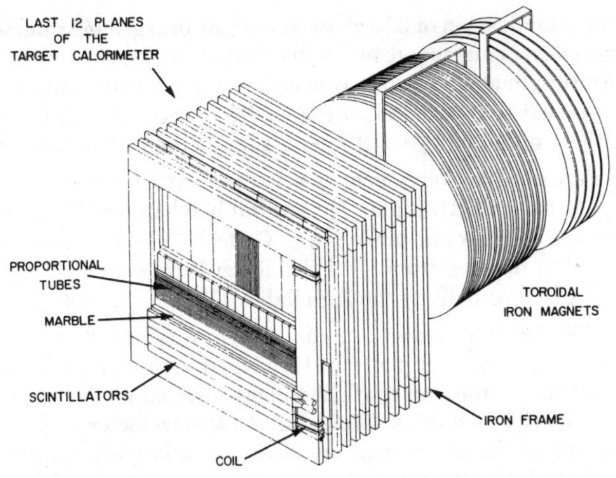

Figure 16: The CHARM detector at CERN [14].

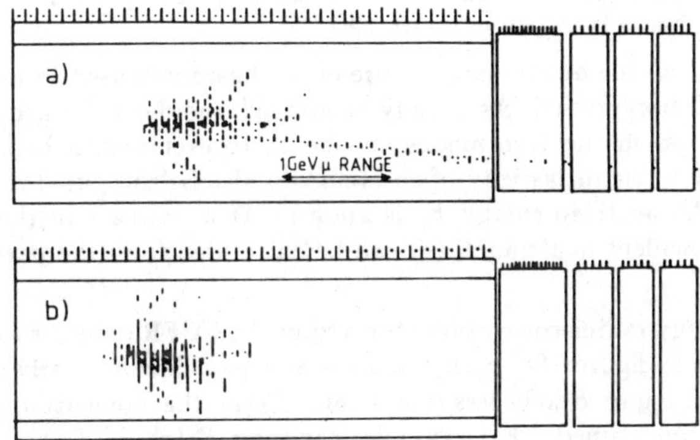

Figure 17: A charged-current interaction (a) and a neutral-current interaction (b) in the CHARM detector at CERN [43].

still contains a contamination of 6 % charged-current background interactions. These charged-current events correspond to muons leaving by the side of the detector and very low momentum muons. The WBB neutrino and electron-neutrino backgrounds have to be subtracted as in the case of the CDHS experiment. The results (table 3) are similar to those obtained by CDHS, with slightly larger errors due to statistics.

The pattern of high energy neutrino detectors at Fermilab in the eighties was similar to the pattern at CERN (figure 7) : one bubble chamber (the 15" bubble chamber), one massive iron detector, built by the Caltech-Columbia-Fermilab-Rochester-Rockfeller (CCFR[4]), inspired from the old Caltech-Fermilab detector, and one light finely segmented detector built by the Fermilab-MIT-Michigan collaboration (FMM).

The CCFR detector, shown in figure 18, consists of a 690 ton non-magnetic target calorimeter made of steel plates instrumented with liquid scintillator counters inserted every 10 cm of iron and spark chambers inserted every 20 cm of iron. The target calorimeter is followed by a toroidal muon spectrometer which is not used in the neutral-current to charged-current cross-section ratio measurements. The transverse position of the interaction vertex is measured using the spark chambers with a resolution of 4.3 cm.

This detector was exposed to various NBBs with energies ranging from 100 GeV to 250 GeV, in two sets of exposures, E616 and E701. In order to ensure a 100% efficient trigger, unbiased for neutral-current and charged-current interactions, the hadron energy was required to be larger than 20 GeV. The event length distribution in the neutrino beam is shown in figure 19.

In addition, the dichromatic nature of the beam was used to reduce the short charged-current background. As already mentioned, the short charged-current events correspond to low momentum muons and hence to events with high inelasticity y (Eq. 7). However, the inelasticity of an event can also be computed from the hadron energy E_h if the neutrino energy E_ν is known. At a radius r in the detector, the energy of the incident neutrino $E_\nu(r)$ can be calculated from the parameters of the hadron beam.

The energy-radius correlation observed in the CCFR detector for the 200 GeV beam is shown in figure 19c, with the kaon and pion bands clearly visible. If the hadron energy is required to be less than a value $E_h(r)$, the momentum of the outgoing muon will be constrained. For example, choosing $E_h(r) = E_\nu(r) - 3.3\,\text{GeV}$, the muon momentum $p_\mu = E_\nu - E_h$ will be larger than 3.3 GeV on average and have a penetration length of more than 2 m, larger than the length of the hadron shower. The distribution of the event length for events with an optimum $E_h(r)$ cut is shown in figure 19c. The contribution of the short charged-current is reduced from 22% to 5%. The remaining contribution comes from finite resolution effects and events with muons leaving by the side of the detector. The price that has to be paid is that the radius-dependent hadron energy cut requires a detailed knowledge of the beam

[4]Although the contributing institutions have changed over the years, the names of the collaborations will be kept as CDHS, CHARM and CCFR in this paper.

	Target Mass η (Eq. 89)	Cuts Fiducial Mass	Recorded Events	Results
BEBC [15]	Ne/H_2 14 tons 0.034	$E_h^{cut} = 9\,\text{GeV}$ $p_\perp^{min} > 1\,\text{GeV}$ 9.3 tons	956 NCν 2222 CCν 288 NC$\bar{\nu}$ 653 CC$\bar{\nu}$	$R_\nu = 0.345 \pm 0.015 \pm 0.009$ $R_{\bar{\nu}} = 0.364 \pm 0.029 \pm 0.009$ $x = 0.182 \pm 0.023$
CDHS 1977 [16]	Fe 1250 tons 0.07	$E_h^{cut} = 12\,\text{GeV}$ $R_{det} < 1.6\,\text{m}$ 500 tons	10770 NCν 26509 CCν 3314 NC$\bar{\nu}$ 6483 CC$\bar{\nu}$	$R_\nu = 0.293 \pm 0.006 \pm 0.008$ $R_{\bar{\nu}} = 0.35 \pm 0.02 \pm 0.02$ $x = 0.24 \pm 0.02$
CDHS 1979/81 [17]	Fe 1250 tons 0.07	$E_h^{cut} = 10\,\text{GeV}$ $R_{det} < 1.4\,\text{m}$ $R_{beam} < 1.2\,\text{m}$ 260 tons	18176 NCν 44274 CCν 6852 NC$\bar{\nu}$ 14434 CC$\bar{\nu}$	$R_\nu = 0.301 \pm 0.005 \pm 0.005$ $R_{\bar{\nu}} = 0.363 \pm 0.010 \pm 0.011$ $x = 0.227 \pm 0.012 \pm 0.006$ $R.C.\ m_c(m_c) = 1.27\,\text{GeV}$
CHARM 1981 [18]	Marble 180 tons -0.005	$E_h^{cut} = 2\,\text{GeV}$ $R_{det} < 1.2\,\text{m}$ 65 tons	2361 NCν 6503 CCν 1126 NC$\bar{\nu}$ 2751 CC$\bar{\nu}$	$R_\nu = 0.320 \pm 0.010$ $R_{\bar{\nu}} = 0.377 \pm 0.020$ $x = 0.220 \pm 0.014$
CCFR E616 [19]	Fe 690 tons 0.07	$E_h^{cut} = 20\,\text{GeV}$ $R_{det} < 1\,\text{m}$	22744 CCν	$R^+_{corr} = 0.317 \pm 0.007$ $R^-_{corr} = 0.249 \pm 0.015$ $x = 0.242 \pm 0.011 \pm 0.005$ $R.C.\ m_c = 1.5\,\text{GeV}$
CCFR E616 +E701 [20]	Fe 690 tons 0.07	$E_h^{cut} = 20\,\text{GeV}$	55024 CCν	$R_\nu = 0.291 \pm 0.006$ $R_{\bar{\nu}} = 0.384 \pm 0.018$ $x = 0.239 \pm 0.010 \pm 0.006$ $R.C.\ m_c = 1.5\,\text{GeV}$
FMM [21]	Sand/Steel 340 tons 0.01	$E_h^{cut} = 10\,\text{GeV}$ $R_{det} < 1\,\text{m}$ 55 tons	2244 NCν 7512 CCν 723 NC$\bar{\nu}$ 1945 CC$\bar{\nu}$	$R_\nu = 0.307 \pm 0.007$ $R_{\bar{\nu}} = 0.384 \pm 0.017$ $x = 0.247 \pm 0.017 \pm 0.005$ $R.C.\ m_c = 1.5\,\text{GeV}$

Table 3: Summary of the second generation of neutrino neutral-current experiments with nearly isoscalar targets, all done with NBBs. The first error on the ratios R_ν and $R_{\bar{\nu}}$ is statistical and the second error, when given, systematical. The first error on $x = \sin^2\theta_W$ is the total experimental error and the second error, when given, the theoretical error. The radiative corrections, when applied, are defined in the Sirlin scheme.

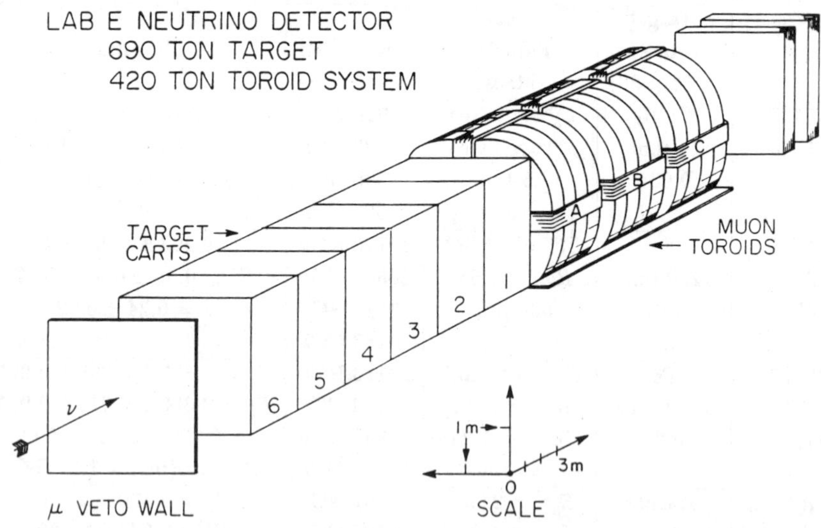

Figure 18: The CCFR detector at Fermilab [22].

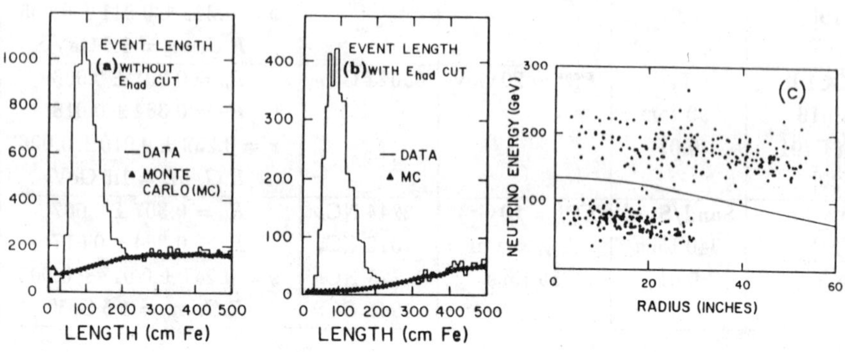

Figure 19: The event length distributions observed in the CCFR detector without (a) and with (b) a radius dependent hadron energy cut. (c) The energy-radius correlation of the NBB observed in the CCFR detector [20].

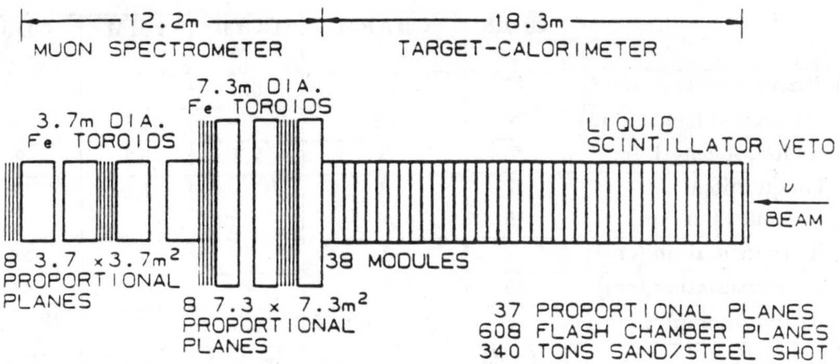

Figure 20: The FMM detector at Fermilab [24].

parameters.

The results from the first data taking E616, given in table 3, were presented in the form of the Paschos-Wolfenstein ratios [23] :

$$R_{\pm} = \frac{\mathrm{d}\sigma_{\nu}^{NC} \pm \mathrm{d}\sigma_{\bar{\nu}}^{NC}}{\mathrm{d}\sigma_{\nu}^{CC} \pm \mathrm{d}\sigma_{\bar{\nu}}^{CC}} \tag{19}$$

Using Eq. 11, these ratios take the following simple forms :

$$R_{+} = \frac{1}{2} - \sin^2\theta_W + \frac{10}{9}\sin^4\theta_W \quad R_{-} = \frac{1}{2} - \sin^2\theta_W, \tag{20}$$

independent of y.

The measured values (table 3) are in good agreement with the Standard Electroweak Theory with $\sin^2\theta_W = 0.242 \pm 0.011$. The experiment was repeated using the same experimental method and the values of R_ν and $R_{\bar{\nu}}$ were used to get $\sin^2\theta_W$ from Eq. 11. A similar result was found : $\sin^2\theta_W = 0.239 \pm 0.010$, slightly higher, although compatible with the results from the CERN experiments.

In 1981, a second large neutrino detector was installed in Lab C, in the location of the HPWF detector, by the FMM collaboration. This detector, shown in figure 20, is a high granularity calorimeter followed by a muon spectrometer and is similar to the CHARM detector, both in principle and performances. It is made of 608 alternating planes of plastic extrusions filled with sand and steel shot interleaved with planes of flash chambers. The transverse granularity is 5 mm. The fiducial mass is 55 tons. As for the CHARM detector, the neutral-current and charged-current candidates are separated on the basis of their topology, but as more statistics was available than in the case of CHARM, the hadron energy cut could be raised to 10

	CDHS	CHARM	CCFR	FMM	CDHS upgraded
Target density (g/cm^3)	6.7	1.3	6.5	1.4	6.7
Interaction length (cm)	20	93	20	85	20
Radiation length (cm)	2	20	2.2	12	2
Longitudinal segmentation (cm)	5 15	8	10	1.6	15
H. segmentation (cm)	–	1	–	0.5	15
V. segmentation (cm)	45	1	–	0.5	15
Radial resolution (cm)	22	$\frac{15}{\sqrt{E}} + 0.006E$	4.3	2	$\frac{12}{\sqrt{E}} + 3.7$
σ_h/E_h (%)	$\frac{70}{\sqrt{E}}; \frac{100}{\sqrt{E}}$	$\frac{49}{\sqrt{E}} + 1.3$	$\frac{81}{\sqrt{E}} + \frac{72}{E}$	$\frac{50}{\sqrt{E}} + 5$	$\frac{60}{\sqrt{E}} + 2$
σ_e/E_e (%)	$\frac{25}{\sqrt{E}}$	$\frac{20}{\sqrt{E}}$	$\frac{25}{\sqrt{E}}$	$\frac{25}{\sqrt{E}}$	$\frac{25}{\sqrt{E}}$
σ_θ (mrad)	–	$\frac{520}{E} + \frac{141}{\sqrt{E}}$	–	$\frac{900}{E} + 14$	–

Table 4: Main characteristics of the high energy electronic neutrino detectors at CERN and Fermilab. In the resolution formulae, the energy is expressed in GeV.

GeV. Another difference compared with CHARM is that, like in CCFR, a hadron energy cut is imposed to correspond to a cut $y < 0.7$. The results obtained using the FMM detector are similar to the results obtained by the other experiments (table 3).

The characteristics of the high energy detectors are compared in table 4.

The performances of the two massive detectors CDHS and CCFR are similar. The longitudinal segmentation is smaller for a large part of the CDHS detector but it should be noted that, if the longitudinal segmentation guarantees that the event length is measured better on an event by event basis, a precise determination of the cross-section ratios depends on the accuracy on average only. This relies more on the knowledge of the detection efficiency for a minimum ionizing particle and on independent cross-checks than on the longitudinal segmentation per se.

The hadron energy resolution is similar in CCFR and in CDHS but is significantly better in the fine-grain calorimeters CHARM and FMM. The hadron energy resolution is not important as far as the measurement of the cross-section ratio is concerned. The y-distributions of neutral-current and charged-current are similar so the effect of energy resolution on these ratios is negligible. This is also true for the resolution on the radial position of the vertex. It is of the order of 2 cm in fine-grain calorimeters and of the order of 22 cm in CDHS but this has no impact on the measurement of R_ν and $R_{\bar{\nu}}$. However, resolution is more important for the measurement of the y-distributions and in any case, if a radius-dependent energy cut is made, the knowledge of the resolution becomes important.

In the fine-grain calorimeters, the direction of the hadron shower with respect to the beam axis can be measured with a precision of about 30 mrad for a 50 GeV hadron shower. With such a precision, the measurement of neutral-current structure functions becomes possible, a measurement that can not be considered in the massive iron detectors.

2.5 *The Motivations for Precise Deep Inelastic Neutrino Experiments*

In 1983 the experimental study of neutral-current neutrino interactions was ready to take a new turn. The ratios R_ν and $R_{\bar{\nu}}$ were measured by the experiments to be compatible with the Standard Electroweak Theory for $\sin^2\theta_W \approx 0.228$ and the electroweak mixing parameter was measured with an accuracy around ± 0.012, including theoretical errors.

This experimental accuracy was such that radiative corrections had to be taken into account.

In principle, radiative corrections have to be considered as soon as an observable is computed in a renormalisable quantum field theory such as the Standard Electroweak Theory.

Any observable quantity A like R_ν can be written [54] :

$$A = A_B(\alpha, G_\mu, \sin^2\theta^r_W)\left[1 + \delta^r_A(\alpha, G_\mu, \sin^2\theta^r_W, m_t, m_H, ...)\right] \tag{21}$$

where A_B is a function computed at the Born approximation like Eq. 11. The constants α and G_μ are the fine structure constant and the muon decay constant respectively. They are fixed by very precise experiments. The parameter $\sin^2\theta^r$, which depends on the renormalisation scheme, is the universal mixing parameter of the theory. The radiative correction term δ^r_A depends both on the observable and the renormalisation scheme. In addition, the radiative corrections, through virtual loops, depend on unknown parameters of the Electroweak Theory like the top quark mass m_t and the Higgs boson mass m_H. The radiative corrections could also be sensitive to some New Physics, as discussed in detail by Langacker in [25] and in other contributions to the present book [61].

The renormalisation scheme is arbitrary and so is the definition of the electroweak mixing parameter. However, if the renormalisation scheme is changed, the radiative correction term δ^r_A will change to compensate for the change in the electroweak parameter and the value of the observable will remain the same[5].

In general, one can write :

$$\sin^2\theta^r = \left(1 - \frac{m_W^2}{m_Z^2}\right)(1 + \delta^r_s) \tag{22}$$

[5]Up to corrections of order higher than the order considered.

A convenient choice is the scheme introduced by Sirlin [26] where :

$$\sin^2\hat{\theta} = 1 - \frac{m_W^2}{m_Z^2} \qquad \left(\hat{\delta}_s^r = 0\right) \tag{23}$$

In the case of R_ν , one can then write :

$$R_\nu = R_\nu^B(\sin^2\hat{\theta}_W)\left[1 + \hat{\delta}_\nu\right] \tag{24}$$

where R_ν^B is a function computed in a given model like Eq. 11. The radiative correction term $\hat{\delta}_\nu$ is small and largely independent of m_t and M_H. This term is in fact mostly due to the trivial electromagnetic photon radiation from the muon in charged-current interactions.

In 1983, the neutral-current interaction had also been observed in other processes. In particular, a value of the electroweak mixing parameter had been derived from polarized electron scattering on Deuterium at SLAC with a comparable accuracy [27] :

$$\sin^2\hat{\theta}_W(eD) = 0.221 \pm 0.015(\text{exp.}) \pm 0.013(\text{theo.})$$

This value is in agreement with the results of the deep inelastic scattering neutrino experiments (table 3). This fact already was a great success for the Standard Electroweak Theory, as was the discovery of the W and Z bosons with the expected masses shortly afterwards. But the Standard Electroweak Theory could be put to an even more stringent test.

The W and Z boson masses are related to the electroweak mixing parameter through the following relations, similar to Eq. 21 :

$$m_W = \frac{A}{\sin\theta_W}\frac{1}{\sqrt{1-\Delta r}} \qquad m_Z = \frac{2A}{\sin 2\theta_W}\frac{1}{\sqrt{1-\Delta r}} \tag{25}$$

$$\text{where :} \quad A = \sqrt{\frac{\pi\alpha}{\sqrt{2}G_\mu}} = 37.28\ \text{GeV} \tag{26}$$

Δr is a radiative correction term depending quadratically on the top mass and logarithmically on the Higgs mass. In the Standard Electroweak Theory , for m_t =45 GeV and M_H =100 GeV, one has $\Delta r \approx 0.07$. The effect of the radiative correction term is to increase the prediction of the W and Z masses by approximately 3 GeV, an effect sometimes referred to as the "radiative shift" of the intermediate vector boson masses.

From Eq. 25, one has :

$$\delta\left[\Delta r(m_W)\right] \approx 1.9\,\frac{\Delta m_W}{m_W} \qquad \delta\left[\Delta r(m_Z)\right] \approx 1.9\,\frac{\Delta m_Z}{m_Z}$$

$$\delta\left[\Delta r(m_W)\right] \approx 4.2\,\Delta\sin^2\theta_W \qquad \delta\left[\Delta r(m_Z)\right] \approx 2.9\,\Delta\sin^2\theta_W \tag{27}$$

Assuming an error on the W and Z mass of order 1%, a measurement of $\sin^2\theta_W$ with an accuracy of order 0.006 would therefore give an error $\delta[\Delta r]$ on Δr of order 0.025. Such a precision would allow a first investigation of the radiative correction term Δr and of the top mass.

Hence the question was raised whether the error on the determination of $\sin^2\theta_W$ from deep inelastic scattering neutrino scattering could be reduced to the desired level of 0.006. From the experimental point of view, the systematic errors, dominated by the understanding of the neutrino beam, could be reduced to the level of 0.005. A statistical error of 0.005 could also be attained by increasing the intensity of the beam.

From the theoretical point of view, it was stressed that the deep inelastic scattering process involves the understanding of the hadronic part of the interaction and it was not guaranteed that the model describing the structure of hadrons at high energies, the QPM, was adequate for the determination of a fundamental parameter like $\sin^2\theta_W$. However it was shown by Llewellyn-Smith that Eq. 11 did not actually depend on the QPM nor on any detailed understanding of the strong interaction as they could be derived assuming isospin symmetry [28]. Corrections like the presence of the charm mass need to be applied to Eq. 11 to describe the real world but these corrections were not expected to introduce an uncertainty on $\sin^2\theta_W$ larger than 0.005.

It was therefore decided to launch a new program of deep inelastic neutrino scattering experiments both at CERN and Fermilab.

Before these experiments are described in some detail, the phenomenology of deep inelastic scattering will now be presented.

3 The Phenomenology of Deep Inelastic Neutrino Scattering and the Quark-Parton Model

3.1 *The Effective Lagrangian and the Basic Neutrino-Fermion Processes*

Let us assume that the interaction between neutrinos and fermions involves only vector and axial couplings at tree level, an assumption valid in gauge theories. The effective lagrangian density describing the neutral-current interaction between a left handed neutrino[6] and a fermion f has the following general form :

$$\mathcal{L}_{NC}(\nu f) = -\frac{G_\mu}{\sqrt{2}}\bar{\nu}\gamma_\alpha(1-\gamma_5)\nu\,\bar{f}\gamma^\alpha\left[\epsilon_L(f)\frac{1-\gamma_5}{2}+\epsilon_R(f)\frac{1+\gamma_5}{2}\right]f \qquad (28)$$

where G_μ is the muon decay constant and $\epsilon_L(f)$ and $\epsilon_R(f)$ are the effective chiral coupling constants of fermion f.

If, like in the Standard Electroweak Theory, the neutral-current interaction between neutrinos and fermions $\nu(k)\,f(p) \to \nu(k')\,f(p')$ is mediated by a vector boson

[6]Because of their production mechanism, neutrinos in accelerator beams have always a left-handed chirality and antineutrinos a right-handed chirality.

B of mass m_B, the effective chiral couplings can be expressed in terms of the vector and axial couplings $g^f_{L,R}$ of the fermion f to the boson B :

$$\epsilon_L(f) = \rho_0 \frac{g^f_V + g^f_A}{2} (g^\nu_V + g^\nu_A) \frac{m_B^2}{Q^2 + m_B^2}$$
$$\epsilon_R(f) = \rho_0 \frac{g^f_V - g^f_A}{2} (g^\nu_V + g^\nu_A) \frac{m_B^2}{Q^2 + m_B^2} \tag{29}$$

$$\text{where } \rho_0 = \frac{m_W^2}{m_B^2 \cos^2\theta_W} \quad \text{and} \quad Q^2 = -(k-k')^2 \tag{30}$$

In the Standard Electroweak Theory, the vector and axial couplings are given by :

$$g^f_A = T^f_3 \qquad g^f_V = T^f_3 - 2Q_f \sin^2\theta_W \tag{31}$$

$$\text{giving :} \quad \epsilon_L(f) = T^f_3 - Q_f \sin^2\theta_W \qquad \epsilon_R(f) = -Q_f \sin^2\theta_W \tag{32}$$

where T^f_3 and Q_f are the weak isospin and the electric charge of fermion f.

The differential cross-sections are given by :

$$\frac{\mathrm{d}\sigma^{NC}}{\mathrm{d}\cos\theta^\star}(\nu f) = \frac{2G_\mu^2}{\pi s}\frac{\kappa'}{\kappa}\left[\epsilon_L(f)^2 p.k\, p'.k' + \epsilon_R(f)^2 p'.k\, p.k'\right] = \frac{\mathrm{d}\sigma^{NC}}{\mathrm{d}\cos\theta^\star}(\bar\nu\bar f)$$
$$\frac{\mathrm{d}\sigma^{NC}}{\mathrm{d}\cos\theta^\star}(\bar\nu f) = \frac{2G_\mu^2}{\pi s}\frac{\kappa'}{\kappa}\left[\epsilon_R(f)^2 p.k\, p'.k' + \epsilon_L(f)^2 p'.k\, p.k'\right] = \frac{\mathrm{d}\sigma^{NC}}{\mathrm{d}\cos\theta^\star}(\nu\bar f) \tag{33}$$

where $\theta^\star$ is the scattering angle of the neutrino in the center-of-mass frame, $s = (k+p)^2$ and $\kappa(\kappa')$ is the initial (final) momentum in the center-of-mass frame. The initial mass of the fermion f is neglected but a finite mass m_f is allowed in the final state, for example in order to describe the production of a heavy charmed meson from the sea. In that case : $\kappa'/\kappa = (s - m_f^2)/s$.

The charged-current process $\overset{(-)}{\nu}(k)\, f(p) \to \ell^\mp(k')\, f'(p')$ where $\ell^\mp$ is a charged lepton and (f,f') is a weak isospin doublet, can also be described by an effective lagrangian density of the type of Eq. 28, with the effective coupling constants :

$$\epsilon_L^{CC}(f) = \frac{m_W^2}{Q^2 + m_W^2} \qquad \epsilon_R^{CC}(f) = 0 \tag{34}$$

The propagator terms will be omitted in the following and the cross-sections are :

$$\frac{\mathrm{d}\sigma^{CC}}{\mathrm{d}\cos\theta^\star}(\nu f) = \frac{2G_\mu^2}{\pi s}\frac{\kappa'}{\kappa} p.k\, p'.k' = \frac{\mathrm{d}\sigma^{CC}}{\mathrm{d}\cos\theta^\star}(\bar\nu\bar f)$$
$$\frac{\mathrm{d}\sigma^{CC}}{\mathrm{d}\cos\theta^\star}(\bar\nu f) = \frac{2G_\mu^2}{\pi s}\frac{\kappa'}{\kappa} p'.k\, p.k' = \frac{\mathrm{d}\sigma^{CC}}{\mathrm{d}\cos\theta^\star}(\nu\bar f) \tag{35}$$

If the fermions f,f' are quarks, the cross-section has to be multiplied by the element $|U_{ff'}|^2$ of the Cabibbo-Kobayashi-Maskawa (CKM) mixing matrix. The scale of the interaction is given by $G_\mu^2/\pi \approx 1.6810^{-38}\mathrm{cm}^2\mathrm{GeV}^{-2}$.

3.2 *The General Form of the Semi-Inclusive Scattering Cross-Section*

Let us consider the inclusive scattering of a lepton ℓ off a nucleon $\ell(k)\,N(P) \rightarrow \ell'(k')\,X(P')$. To describe this process, the following variables are useful :

- E : energy of the incoming lepton in the lab frame.
- E' : energy of the outgoing lepton in the lab frame.
- θ : angle of the outgoing lepton in the lab frame with respect to the incoming lepton direction.
- $q = k - k'$: 4-vector of the momentum transfer.
- $Q^2 = -q^2$: virtuality of the momentum transfer.
- $\nu = P.q/M$ is the energy transferred to the hadronic system. M is the mass of the nucleon. The hadron energy E_h measured in a calorimeter is by definition ν and not the energy of the 4-vector P'[7].
- $W^2 = P'^2$: invariant mass of the hadronic system
- The two Bjorken variables x and y (the inelasticity) :

$$x = \frac{Q^2}{2P.q} = \frac{Q^2}{2MyE} \qquad y = \frac{P.q}{P.k} = \frac{\nu}{E} \tag{36}$$

If polarisation effects are not considered, only three of these variables are independent.

The invariant mass W^2 has to be larger than a nucleon mass :

$$W^2 = P'^2 = (P + k - k')^2 = M^2 - Q^2 + 2P.q \geq M^2$$

$$\text{Hence :} \qquad x \leq 1 \tag{37}$$

In addition, a kinematical bound for the inelasticity has to be satisfied. In the case of the charged-current neutrino scattering, one has :

$$y \leq 1 - \frac{m^2}{m^2 + Q^2} - \frac{Q^2 + m^2}{4E^2} \tag{38}$$

where m is the muon mass. For neutral-current scattering, one has :

$$y \leq \frac{1}{1 + \frac{Mx}{2E}} \tag{39}$$

[7]Because of baryon number conservation, a nucleon mass is not converted into pions and therefore not accounted for in the energy sum in a calorimeter calibrated with pions.

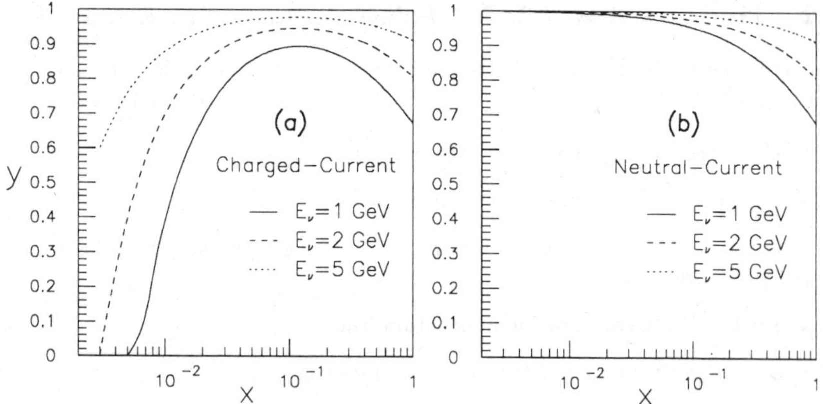

Figure 21: Kinematical boundaries in the $x-y$ plane for neutrino charged-current (a) and neutral-current scattering (b).

The kinematical boundaries in the $x-y$ plane described by Eq. 38 and 39 are depicted in figure 21. The upper right corner of the $x-y$ plane is excluded by the presence of the target mass M. In addition, for charged-current scattering, the finite muon mass forbids y to be larger than $1-\frac{m}{E}$ and x smaller than $\frac{m^2}{2ME}$.

The cross-section of the inclusive neutrino scattering is obtained by summing the contributions from all final states X comprising n hadrons with momenta p'_k with $1 \leq k \leq n$:

$$\frac{\mathrm{d}^2\sigma}{\mathrm{d}E'\mathrm{d}\Omega} = \frac{1}{4ME}\sum_X \sum_{spins} |T_{fi}|^2 (2\pi)^4 \delta^4(P'-q-P)\frac{E'}{(2\pi)^3 2}\prod_{k=1}^{n}\frac{\mathrm{d}^3 p'_k}{(2\pi)^3 2E'_k} \tag{40}$$

The transition matrix element is :

$$T_{fi} = \frac{G_\mu}{\sqrt{2}}\,\overline{u}(\vec{k}')\gamma^\alpha(1-\gamma_5)u(\vec{k})\,\langle X|J_\alpha(0)|N\rangle \tag{41}$$

Averaging over the initial spin states, the cross-section can be expressed as the contraction of two tensors :

$$\begin{aligned}\frac{\mathrm{d}^2\sigma}{\mathrm{d}x\mathrm{d}y}(\nu N) &= \frac{G_\mu^2 M y}{16\pi}\,L^{\alpha\beta}W_{\alpha\beta}\\ \frac{\mathrm{d}^2\sigma}{\mathrm{d}x\mathrm{d}y}(\overline{\nu} N) &= \frac{G_\mu^2 M y}{16\pi}\,\overline{L}^{\alpha\beta}\overline{W}_{\alpha\beta}\end{aligned} \tag{42}$$

with :

$$\begin{aligned}
L^{\alpha\beta} &= k'_\sigma k_\rho Tr\left[\gamma^\sigma\gamma^\alpha\gamma^\rho\gamma^\beta(1-\gamma_5)\right] = 8\left[k'^\alpha k^\beta - k.k' g^{\alpha\beta} + k^\alpha k'^\beta - i\epsilon^{\sigma\alpha\rho\beta}k'_\sigma k_\rho\right] \\
\overline{L}^{\alpha\beta} &= k_\sigma k'_\rho Tr\left[\gamma^\sigma\gamma^\alpha\gamma^\rho\gamma^\beta(1-\gamma_5)\right] = 8\left[k'^\alpha k^\beta - k.k' g^{\alpha\beta} + k^\alpha k'^\beta + i\epsilon^{\sigma\alpha\rho\beta}k'_\sigma k_\rho\right] \\
W_{\alpha\beta} &= \frac{1}{4\pi M}\sum_X\sum_{spins}(2\pi)^4\delta^4(P-q-P)\langle X|J_\alpha(0)|N\rangle\langle N|J^\dagger_\beta(0)|X\rangle \\
\overline{W}_{\alpha\beta} &= \frac{1}{4\pi M}\sum_X\sum_{spins}(2\pi)^4\delta^4(P-q-P)\langle X|J^\dagger_\alpha(0)|N\rangle\langle N|J_\beta(0)|X\rangle
\end{aligned} \quad (43)$$

where $g_{\alpha\beta}$ is the metric tensor and $\epsilon_{\alpha\beta\sigma\rho}$ is the totally antisymmetric tensor. Applying a translation on one of the current, the final state X disappears and the $W_{\alpha\beta}$ tensor can be written [29] :

$$W_{\alpha\beta} = \frac{1}{4\pi M}\int dx e^{iqx}\langle N|J^\dagger_\beta(x)J_\alpha(0)|N\rangle = \frac{1}{4\pi M}\int dx e^{-iqx}\langle N|J^\dagger_\beta(0)J_\alpha(x)|N\rangle \quad (44)$$

where the bracket includes a summation over the nucleon spins. Unfortunately, the $W_{\alpha\beta}$ tensor, which contains all the information about the nucleon structure, can not be computed from first principles. Many properties of the cross-section, however, can be derived without the knowledge of the functional form of $W_{\alpha\beta}$.

From Eq. 44, one can see that the tensor $W_{\alpha\beta}$ is hermitian.

$$W^\star_{\alpha\beta} = W_{\beta\alpha} \quad (45)$$

The most general form of a hermitian tensor, function of two 4-vectors q and P, is :

$$\begin{aligned}
W_{\alpha\beta} &= -g_{\alpha\beta}W_1 + \frac{P_\alpha P_\beta}{M^2}W_2 + i\epsilon_{\alpha\beta\sigma\rho}\frac{P^\sigma P^\rho}{2M^2}W_3 + \frac{q_\alpha q_\beta}{M^2}W_4 \\
&+ \frac{P_\alpha q_\beta + P_\beta q_\alpha}{2M^2}W_5 + i\frac{P_\alpha q_\beta - P_\beta q_\alpha}{2M^2}W_6
\end{aligned} \quad (46)$$

where the six functions W_i are real functions of ν and Q^2 of dimension $[\text{Energy}]^{-1}$.

The function W_3 violates parity and the function W_6 violates time reversal and hence CP.

The positivity of the tensor $W_{\alpha\beta}$ gives constraints on the functions W_i [30] :

$$\begin{aligned}
& W_2\left(1+\frac{\nu^2}{Q^2}\right) \geq W_1 \geq \frac{\sqrt{\nu^2+Q^2}}{2M}|W_3| \\
& W_1 + \frac{\nu^2}{Q^2}W_2 + \frac{Q^2}{M^2}W_4 \geq \frac{\nu}{M}W_5 \\
& \frac{W_5^2+W_6^2}{4} \leq W_2W_4 + \frac{M^2}{Q^2+\nu^2}W_1\left(W_2 + \frac{\nu}{M}W_5 - W_1 - \frac{Q^2}{M^2}W_4\right)
\end{aligned} \quad (47)$$

The tensors for the various deep inelastic scattering processes are :

$$
\begin{aligned}
W^{NC}_{\alpha\beta} &= \frac{1}{4\pi M}\int \mathrm{d}x e^{iqx}\langle N|J^{NC\dagger}_{\beta}(x)J^{NC}_{\alpha}(0)|N\rangle \\
\overline{W}^{NC}_{\alpha\beta} &= \frac{1}{4\pi M}\int \mathrm{d}x e^{iqx}\langle N|J^{NC}_{\beta}(x)J^{NC\dagger}_{\alpha}(0)|N\rangle \\
W^{CC}_{\alpha\beta} &= \frac{1}{4\pi M}\int \mathrm{d}x e^{iqx}\langle N|J^{CC\dagger}_{\beta}(x)J^{CC}_{\alpha}(0)|N\rangle \\
\overline{W}^{CC}_{\alpha\beta} &= \frac{1}{4\pi M}\int \mathrm{d}x e^{iqx}\langle N|J^{CC}_{\beta}(x)J^{CC\dagger}_{\alpha}(0)|N\rangle
\end{aligned}
\tag{48}
$$

The tensor for the deep inelastic scattering of a charged lepton can also be considered :

$$
W^{EM}_{\alpha\beta} = \frac{1}{4\pi M}\int \mathrm{d}x e^{iqx}\langle N|J^{EM\dagger}_{\beta}(x)J^{EM}_{\alpha}(0)|N\rangle \tag{49}
$$

In the Standard Electroweak Theory, the form of the currents is known :

$$
\begin{aligned}
J^{CC}_{\alpha} &= J^{1}_{\alpha} + iJ^{2}_{\alpha} & J^{NC}_{\alpha} &= J^{3}_{\alpha} - 2\sin^2\theta_W J^{EM}_{\alpha} \\
\vec{J}_{\alpha} &= \sum_{fermions} \overline{\psi}_i\vec{T}\gamma_\alpha(1-\gamma_5)\psi_i & J^{EM}_{\alpha} &= \sum_{fermions} \overline{\psi}_i\mathcal{Q}\gamma_\alpha\psi_i
\end{aligned}
\tag{50}
$$

where $\vec{T}$ is the weak isospin operator and $\mathcal{Q}$ the electric charge operator. The currents J^{NC}_{α} and J^{EM}_{α} are hermitian operators.

$$
\text{One can see that :} \quad \overline{W}^{NC}_{\alpha\beta} = W^{NC}_{\alpha\beta} \tag{51}
$$

$$
\text{However, in general :} \quad \overline{W}^{CC}_{\alpha\beta} \neq W^{CC}_{\alpha\beta} \tag{52}
$$

In practice, dimensionless structure functions F_i are used instead of the W_i functions :

$$
F_1(x,Q^2) = M W_1(\nu,Q^2) \qquad F_i(x,Q^2) = \nu W_i(\nu,Q^2) \;\; i \geq 2 \tag{53}
$$

Neglecting the T violating effects (F_6), the expression of the cross-section is[8] :

$$
\begin{aligned}
\frac{\mathrm{d}^2\sigma}{\mathrm{d}x\mathrm{d}y}(\overset{(-)}{\nu}N) = & \frac{G_\mu^2 ME}{\pi}\left[xy^2F_1\left(1+\frac{m^2}{Q^2}\right) + F_2\left(1-y-\frac{Mxy}{2E}-\frac{m^2}{4E^2}\right)\right. \\
& \left. \pm F_3\left[xy\left(1-\frac{y}{2}\right) - y\frac{m^2}{4EM}\right] + F_4xy\frac{m^2}{2ME} - F_5\frac{m^2}{2ME}\right]
\end{aligned}
\tag{54}
$$

In the case of the electromagnetic scattering, neglecting the muon mass :

$$
\frac{\mathrm{d}^2\sigma}{\mathrm{d}x\mathrm{d}y}(\ell^{\pm}N) = \frac{8\pi\alpha^2 ME}{Q^4}\left[xy^2F_1^\gamma + F_2^\gamma\left(1-y-\frac{Mxy}{2E}\right)\right] \tag{55}
$$

[8] The minus sign in front of F_3 corresponds to the antineutrino scattering.

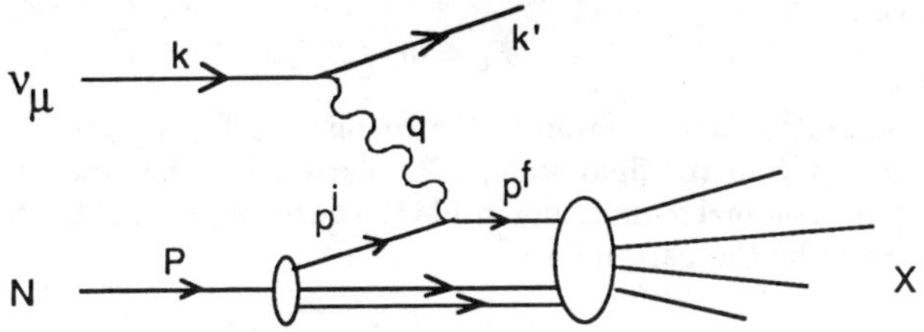

Figure 22: The deep inelastic process in the Quark-Parton Model.

In terms of the structure functions, the inequalities of Eq. 47 become :

$$F_2\left(1+\frac{4M^2x^2}{Q^2}\right) \geq 2xF_1 \geq \sqrt{1+\frac{4M^2x^2}{Q^2}}x|F_3|$$
$$F_1+\frac{1}{2x}F_2+2xF_4 \geq F_5$$
$$\frac{F_5^2+F_6^2}{4} \leq F_2F_4+\frac{F_1}{1+\frac{4M^2x^2}{Q^2}}\left[\frac{2M^2x}{Q^2}F_2+F_5-F_1-2xF_4\right] \tag{56}$$

3.3 *The Scaling Hypothesis and the Quark-Parton Model*

In 1967, Bjorken argued that the structure functions have a non-zero limit at large Q^2 : $F_i(x,Q^2) \to F_i^\infty(x)$. This behaviour is to be contrasted for example with the behaviour of the proton form factor which vanishes like $1/Q^2$. This fact can be interpreted if it is assumed that the nucleon contains point-like objects called "partons". The structure functions are then combinations of the momentum distributions of the partons inside the nucleon.

More precisely, the following picture of the interaction emerges. At low Q^2, the leptonic current of 4-vector q will interact with the nucleon, producing some resonant states in the invariant mass W^2. At large Q^2 (deep inelastic domain) the current will interact with one parton, the recombination of the partons into hadrons in the final state leading to a smooth distribution in the invariant mass squared (figure 22).

Some basic facts can be deduced from these assumptions.

First, it can be shown that the structure function F_4 in neutrino scattering is related to chiral symmetry breaking [29], in other words to strong interaction effects producing terms like m_p^2/Q^2 where m_p is a parton mass. These terms will vanish for high Q^2. Hence :

$$F_4 = 0 \tag{57}$$

Let us consider the scattering of the current q off a parton i in the nucleon, producing a parton f in the final state. The parton i is characterized by $i(\xi)$, its distribution of longitudinal momentum inside the nucleon. ξ is the fraction of nucleon momentum carried by the parton i :

$$\xi = \frac{p_0^i + p_3^i}{P_0 + P_3} = \frac{\sqrt{m_i^2 + p_3^{i2}} + p_3^i}{\sqrt{M^2 + P_3^2} + P_3} \tag{58}$$

where p^i is the 4-vector of the parton i, the third coordinate axis being aligned with q, and m_i is the mass of parton i. Defined in this way, ξ is a Lorentz-invariant quantity.

The variable ξ can be evaluated in the frame, called the Breit frame, where the space-like 4-vector q has the form (0,0,0,-Q). In that frame :

$$x = \frac{Q^2}{2P.q} = \frac{Q^2}{2P_3Q} \quad \text{hence} \quad P_3 = \frac{Q}{2x} \tag{59}$$

The value of p_3^i is constrained by the fact that the final parton f, with mass m_f, is on mass shell :

$$\begin{aligned} m_f^2 = p^{f2} = (p^i + q)^2 &= -Q^2 + 2p^i.q + m_i^2 = -Q^2 + 2Qp_3^i + m_i^2 \\ \text{hence} \quad p_3^i &= \frac{Q^2 + m_f^2 - m_i^2}{2Q} \end{aligned} \tag{60}$$

If the masses are neglected, $p_3^i = Q/2$. The Breit frame is the frame where the struck parton direction is reversed. From this fact, a relation between F_2 and F_1 can already be derived by considering the absorption of a helicity 0 current. This contribution is indeed proportional to :

$$F_2\left(1 + \frac{4M^2x^2}{Q^2}\right) - 2xF_1 \tag{61}$$

However, if the partons are assumed to be fermions, their helicity is conserved by the interaction. Since their direction is reversed, they must absorb a unit of spin and therefore the helicity 0 absorption cross-section vanishes. In the QPM, the assumption that partons are fermions therefore imposes :

$$q_L = F_2\left(1 + \frac{4M^2x^2}{Q^2}\right) - 2xF_1 = 0 \tag{62}$$

The function q_L, constrained to be positive by Eq. 56, is often called the "longitudinal structure function" and Eq. 62 is known as the Callan-Gross relation.

From either the Callan-Gross relation or Eq. 57, it follows from Eq. 56[9] that :

$$F_5 = 2F_1 \quad \text{and} \quad F_6 = 0 \tag{63}$$

In the QPM, one can derive other relations on general grounds [29]. But, if the partons are assumed to be quarks, the expression of the structure function in terms of the momentum distributions can be calculated.

The general expression for ξ can be obtained from Eq. 58, Eq. 59 and Eq. 60 :

$$\xi = \left[1 + \frac{m_f^2 - m_i^2}{Q^2} + \sqrt{1 + 2\frac{m_f^2 + m_i^2}{Q^2} + \frac{(m_f^2 - m_i^2)^2}{Q^4}}\right] \frac{x}{1 + \sqrt{1 + \frac{4M^2x^2}{Q^2}}} \tag{64}$$

At the impulse approximation, the contribution of the subprocess $qi \to f$ to the cross-section is :

$$\int_0^1 \mathrm{d}z\, i(z)\, \frac{\mathrm{d}^2\sigma}{\mathrm{d}x\mathrm{d}y}(qi \to f) \tag{65}$$

Using the kinematical constraint (Eq. 64), the elementary cross-section reads :

$$\left(\frac{\partial \xi}{\partial x}\right)_y \left(\frac{\partial \cos\theta^\star}{\partial y}\right)_\xi \frac{\mathrm{d}\sigma}{\mathrm{d}\cos\theta^\star}(s(\xi), \cos\theta^\star)\delta(z - \xi) \tag{66}$$

$$\text{where : } s(\xi) = m_i^2 + \xi(S - M^2) \qquad \text{and } S = (k + P)^2 = 2ME \tag{67}$$

The contribution of parton i to the cross-section is therefore :

$$i(\xi)\left(\frac{\partial \xi}{\partial x}\right)_y \left(\frac{\partial \cos\theta^\star}{\partial y}\right)_\xi \frac{\mathrm{d}\sigma}{\mathrm{d}\cos\theta^\star}(s(\xi), \cos\theta^\star) \tag{68}$$

The elementary scattering cross-sections have been given in Eq. 33 and Eq. 35.

Eq. 68 will first be applied to the case of the charged-current process where the masses of partons and the target are neglected. In that case (Eq. 58), $\xi = x$, $s = xS = 2MEx$ and :

$$y = 1 - \frac{P.k'}{P.k} = 1 - \frac{p.k'}{p.k} = 1 - \frac{s - m^2}{s}\frac{1 + \cos\theta^\star}{2} \tag{69}$$

and the cross-section is (Eq. 35) :

$$\begin{aligned} \frac{\mathrm{d}\sigma}{\mathrm{d}y}(\nu i \to \ell^- f) &= \frac{G_\mu^2}{\pi}(s - m^2)|U_{fi}|^2 \\ \frac{\mathrm{d}\sigma}{\mathrm{d}y}(\bar{\nu} i \to \ell^+ f) &= \frac{G_\mu^2}{\pi}\left[(1 - y^2)s - (1 - y)m^2\right]|U_{fi}|^2 \end{aligned} \tag{70}$$

[9] At high Q^2, the last inequality in Eq. 56 reads : $F_6^2 + (F_5 - 2F_1)^2 \leq 4(F_2 - 2xF_1)F_4$.

The contribution of parton i to the cross-section is then :

$$\frac{d^2\sigma^{CC}}{dxdy}(\nu i) = \frac{G_\mu^2 ME}{\pi}\left[2x - \frac{m^2}{ME}\right]|U_{fi}|^2 i(x)$$
$$\frac{d^2\sigma^{CC}}{dxdy}(\overline{\nu} i) = \frac{G_\mu^2 ME}{\pi}\left[2(1-y)^2 x - (1-y)\frac{m^2}{ME}\right]|U_{fi}|^2 i(x) \quad (71)$$

which can be written :

$$\frac{d^2\sigma^{CC}}{dxdy}(\nu i) = \frac{G_\mu^2 ME}{\pi}\left[xy^2\left(1+\frac{m^2}{Q^2}\right) + 2x(1-y)\right.$$
$$\left. +2\left(xy(1-\frac{y}{2}) - y\frac{m^2}{4ME}\right) - \frac{m^2}{ME}\right]|U_{fi}|^2 i(x)$$
$$\frac{d^2\sigma^{CC}}{dxdy}(\overline{\nu} i) = \frac{G_\mu^2 ME}{\pi}\left[xy^2\left(1+\frac{m^2}{Q^2}\right) + 2x(1-y)\right.$$
$$\left. -2\left(xy(1-\frac{y}{2}) - y\frac{m^2}{4ME}\right) - \frac{m^2}{ME}\right]|U_{fi}|^2 i(x) \quad (72)$$

Identifying with Eq. 54, one finds that the contribution of parton i to the charged-current structure functions are :

$$F_1^{\nu CC} = F_1^{\overline{\nu} CC} = i(x) \qquad F_4^{\nu CC} = F_4^{\overline{\nu} CC} = 0$$
$$F_2^{\nu CC} = F_2^{\overline{\nu} CC} = 2xi(x) \qquad F_5^{\nu CC} = F_5^{\overline{\nu} CC} = 2i(x)$$
$$F_3^{\nu CC} = F_3^{\overline{\nu} CC} = 2i(x) \quad (73)$$

In this simple case, one can indeed check Eq. 57, Eq. 62 and Eq. 63. The contribution of an antiparton to the structure functions are also given by Eq. 73 but with a minus sign for F_3.

Similarly, the contributions of a parton (or antiparton) i to the electromagnetic structure functions are :

$$F_1^{EM} = Q_i^2 i(x)$$
$$F_2^{EM} = 2xQ_i^2 i(x) \quad (74)$$

The contributions of parton i (of distribution $i(x)$) and antiparton i (of distribution $\overline{i}(x)$) to the neutral-current structure functions can be found using the same method and the result is :

$$F_1^{\nu NC} = F_1^{\overline{\nu} NC} = \left(i(x) + \overline{i}(x)\right)\left(\epsilon_L(i)^2 + \epsilon_R(i)^2\right)$$
$$F_2^{\nu NC} = F_2^{\overline{\nu} NC} = 2x\left(i(x) + \overline{i}(x)\right)\left(\epsilon_L(i)^2 + \epsilon_R(i)^2\right)$$
$$F_3^{\nu NC} = F_3^{\overline{\nu} NC} = 2\left(i(x) - \overline{i}(x)\right)\left(\epsilon_L(i)^2 - \epsilon_R(i)^2\right) \quad (75)$$

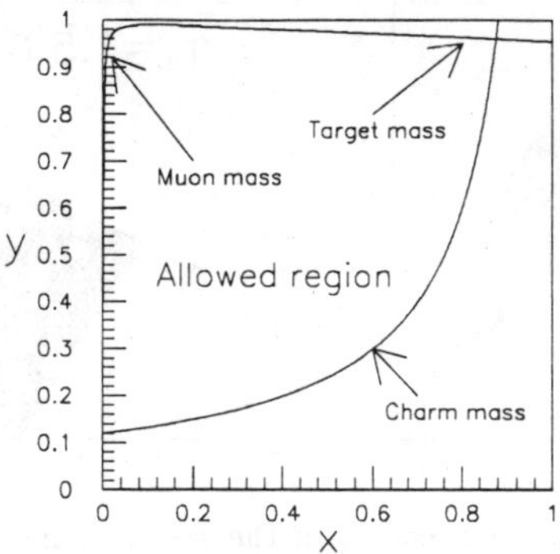

Figure 23: The phase space restriction in the plane $x - y$ if a charm quark with mass 1.5 GeV is produced in the final state by a 10 GeV energy neutrino.

When a charm quark is produced in the final state, its mass m_c of about 1.5 GeV can not be neglected.

The charm quark can be produced in charged-current processes (νd $\rightarrow \mu^- c$, νs $\rightarrow \mu^- c$ and charge conjugates) and in neutral-current processes (νc $\rightarrow \nu^- c$ and charge conjugates). In the initial state of neutral-current processes, the charm mass will be neglected as it would be in contradiction with the target mass. If the muon mass is neglected (we already know how it modifies the cross-section), Eq. 58 gives :

$$\xi = x\left(1 + \frac{m_c^2}{Q^2}\right) \tag{76}$$

Requiring $\xi \leq 1$ (or equivalently $W^2 \geq m_c^2$), the phase space in the $x - y$ plane is restricted, as depicted in figure 23. Since :

$$y = 1 - \frac{s - m_c^2}{s}\frac{1 + \cos\theta^\star}{2}, \tag{77}$$

the contribution of the process $qi \rightarrow$ c to the cross-section is :

$$\frac{\mathrm{d}\sigma}{\mathrm{d}y}(qi \rightarrow \mathrm{c}) = \frac{G_\mu^2 s}{\pi}\left(1 - \frac{m_c^2}{s}\right)|U_{ci}|^2 \tag{78}$$

$$\text{However :} \quad \frac{m_c^2}{s} = \frac{(\xi - x)y}{\xi} \quad \text{and} \quad s = \xi S = \xi 2ME \tag{79}$$

Hence :

$$\frac{\mathrm{d}\sigma}{\mathrm{d}y}(qi \rightarrow \mathrm{c}) = \frac{G_\mu^2 ME}{\pi}\left[1 - y + \frac{xy}{\xi}\right]|U_{ci}|^2 2\xi i(\xi) \tag{80}$$

$$\frac{\mathrm{d}\sigma}{\mathrm{d}y}(qi \rightarrow \mathrm{c}) = \frac{G_\mu^2 ME}{\pi}\left[xy^2 i(\xi) + 2\xi i(\xi)(1 - y) + 2xy(1 - \frac{y}{2})i(\xi)\right]|U_{ci}|^2 \tag{81}$$

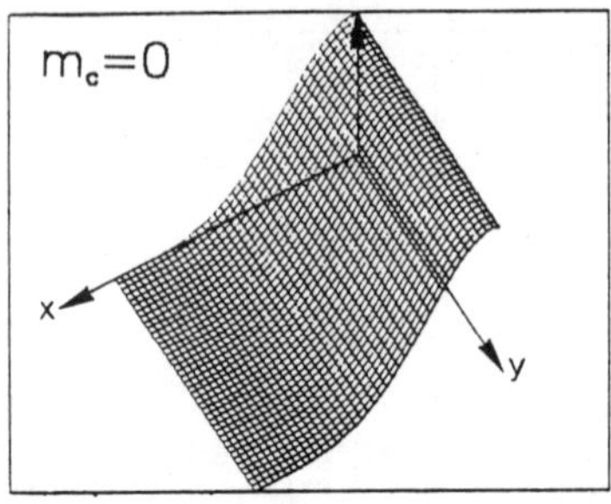

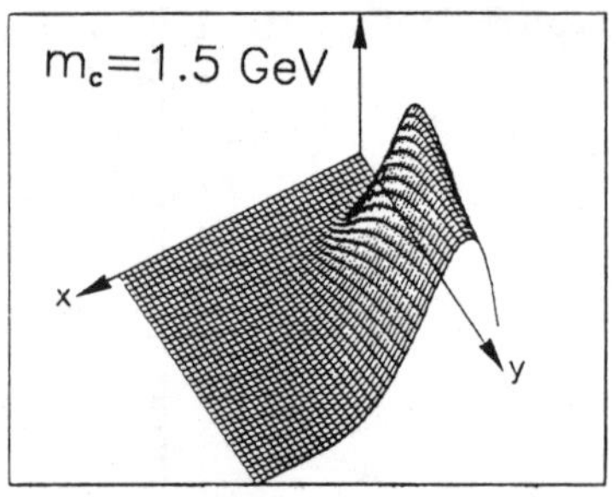

Figure 24: Distortion of the cross-section in the $x - y$ plane if a charm quark with mass 1.5 GeV is produced from the scattering of 10 GeV energy neutrino off a d quark.

Identifying with Eq. 54, the contributions of the subprocess $qi \rightarrow$ c to the charged-current neutrino structure functions are:

$$\begin{aligned} F_1^{\nu CC} &= |U_{ci}|^2 i(\xi) \\ F_2^{\nu CC} &= 2\xi|U_{ci}|^2 i(\xi) \\ F_3^{\nu CC} &= 2|U_{ci}|^2 i(\xi) \end{aligned} \tag{82}$$

The distortion of the cross-section due to a charm mass of 1.5 GeV is illustrated in figure 24. It is a large effect at small y and small x. This description of charm production is known as the "Slow Rescaling Model" [31].

The Callan-Gross relation is not valid any more, as expected. The longitudinal structure function has a contribution :

$$q_L = F_2 - 2xF_1 = 2\frac{m_c^2}{Q^2} i(\xi)|U_{ci}|^2 \tag{83}$$

The contribution of the charm (or anti-charm) to the neutral-current structure functions is obtained with the same method :

$$\begin{aligned} F_1^{\nu NC} &= F_1^{\bar{\nu} NC} = (c(\xi) + \bar{c}(\xi)) \left(\epsilon_L(c)^2 + \epsilon_R(c)^2\right) \\ F_2^{\nu NC} &= F_2^{\bar{\nu} NC} = 2\xi\left(c(\xi) + \bar{c}(\xi)\right) \left(\epsilon_L(c)^2 + \epsilon_R(c)^2\right) \\ F_3^{\nu NC} &= F_3^{\bar{\nu} NC} = 2\left(c(\xi) - \bar{c}(\xi)\right) \left(\epsilon_L(c)^2 - \epsilon_R(c)^2\right) \end{aligned} \tag{84}$$

In addition, the cross-section is constrained by the kinematical requirement of producing at least one D meson in charged-current scattering or two D mesons in neutral-current scattering. The effect of this purely kinematical suppression is negligible compared with the effects of slow rescaling.

With Eq. 73, Eq. 74, Eq. 75, Eq. 82 and Eq. 84, we know the contributions of each subprocess to the various structure functions.

3.4 *Building a Quark-Parton Model*

To build a QPM, the content of a nucleon N has to be given in terms of the various quark momentum distributions $i_N(x)$. At the energies considered here, the u,d,s and c quarks play a role. Hence, sixteen functions are needed to describe the available experiments : $u_p(x)$, $d_p(x)$, $s_p(x)$, $c_p(x)$, $\overline{u}_p(x)$, $\overline{d}_p(x)$, $\overline{s}_p(x)$, $\overline{c}_p(x)$, $u_n(x)$, $d_n(x)$, $s_n(x)$, $c_n(x)$, $\overline{u}_n(x)$, $\overline{d}_n(x)$, $\overline{s}_n(x)$ and $\overline{c}_n(x)$.

These distributions can be determined by studying experimentally the scattering of charged leptons and the charged-current scattering. However, the data are not precise enough to allow the determination of so many functions and some additional symmetries are imposed :

$$\begin{array}{rll}
\text{Isospin symmetry :} & u_n(x) = d_p(x) = d(x) & \overline{u}_n(x) = \overline{d}_p(x) = \overline{d}(x) \\
& d_n(x) = u_p(x) = u(x) & \overline{d}_n(x) = \overline{u}_p(x) = \overline{u}(x) \\
\text{Strange sea universality :} & s_n(x) = s_p(x) = s(x) & \overline{s}_n(x) = \overline{s}_p(x) = \overline{s}(x) \\
\text{Charm sea universality :} & c_n(x) = c_p(x) = c(x) & \overline{c}_n(x) = \overline{c}_p(x) = \overline{c}(x)
\end{array} \tag{85}$$

These distributions are constrained by the proton and neutron quantum numbers :

$$\begin{array}{rl}
\text{Charge :} & 1 = \frac{2}{3}\int \mathrm{d}x\,(u(x) - \overline{u}(x)) - \frac{1}{3}\int \mathrm{d}x\,\left(d(x) - \overline{d}(x)\right) \\
\text{Baryon number :} & 1 = \frac{1}{3}\int \mathrm{d}x\,(u(x) - \overline{u}(x)) + \frac{1}{3}\int \mathrm{d}x\,\left(d(x) - \overline{d}(x)\right) \\
\text{Strangeness :} & 0 = \int \mathrm{d}x\,(s(x) - \overline{s}(x)) \\
\text{Charm :} & 0 = \int \mathrm{d}x\,(c(x) - \overline{c}(x))
\end{array} \tag{86}$$

The distributions $u_V(x) = u(x) - \overline{u}(x)$ and $d_V(x) = d(x) - \overline{d}(x)$ define the quantum numbers of the proton and the neutron and they are referred to as the "valence quarks" distributions. The other distributions reflect the presence of quark-antiquark pairs and are called "sea quark distributions".

Since the targets used experimentally in the precise experiments are close to isoscalarity, it is convenient to write :

$$\frac{\mathrm{d}^2\sigma}{\mathrm{d}x\mathrm{d}y} = \frac{\mathrm{d}^2\sigma}{\mathrm{d}x\mathrm{d}y}(N) + \eta\frac{\mathrm{d}^2\sigma}{\mathrm{d}x\mathrm{d}y}(n-p) \tag{87}$$

$$\text{where}: \quad \frac{d^2\sigma}{dxdy}(N) = \frac{1}{2}\left(\frac{d^2\sigma}{dxdy}(p) + \frac{d^2\sigma}{dxdy}(n)\right)$$

$$\frac{d^2\sigma}{dxdy}(n-p) = \frac{1}{2}\left(\frac{d^2\sigma}{dxdy}(n) - \frac{d^2\sigma}{dxdy}(p)\right) \tag{88}$$

$$\text{and}: \quad \eta = (N_n - Z)/(N_n + Z) \quad (\text{non}-\text{isoscalarity}) \tag{89}$$

where N_n is the number of neutrons and Z the number of protons.

Summing all subprocesses according to the formulae given above, the expressions for the charged-current structure functions are :

$$\begin{aligned}
F_1^{CC}(\nu N) &= \frac{1}{2}\left[(u(x)+d(x))\,|U_{ud}|^2 + (u(\xi)+d(\xi))\,|U_{cd}|^2\right. \\
&\quad +2s(x)|U_{us}|^2 + 2s(\xi)|U_{cs}|^2 \\
&\quad \left.+\left(\overline{u}(x)+\overline{d}(x)\right)\left(|U_{ud}|^2+|U_{us}|^2\right) + 2\overline{c}(x)\left(|U_{cd}|^2+|U_{cs}|^2\right)\right] \\
F_2^{CC}(\nu N) &= x\,(u(x)+d(x))\,|U_{ud}|^2 + \xi\,(u(\xi)+d(\xi))\,|U_{cd}|^2 \\
&\quad +2xs(x)|U_{us}|^2 + 2\xi s(\xi)|U_{cs}|^2 \\
&\quad +x\left(\overline{u}(x)+\overline{d}(x)\right)\left(|U_{ud}|^2+|U_{us}|^2\right) + 2x\overline{c}(x)\left(|U_{cd}|^2+|U_{cs}|^2\right) \\
F_3^{CC}(\nu N) &= (u(x)+d(x))\,|U_{ud}|^2 + (u(\xi)+d(\xi))\,|U_{cd}|^2 \\
&\quad +2s(x)|U_{us}|^2 + 2s(\xi)|U_{cs}|^2 \\
&\quad -\left(\overline{u}(x)+\overline{d}(x)\right)\left(|U_{ud}|^2+|U_{us}|^2\right) - 2\overline{c}(x)\left(|U_{cd}|^2+|U_{cs}|^2\right) \\
F_1^{CC}(\nu n-p) &= \frac{1}{2}\left[(u(x)-d(x))\,|U_{ud}|^2 + (u(\xi)-d(\xi))\,|U_{cd}|^2\right. \\
&\quad \left.+\left(\overline{d}(x)-\overline{u}(x)\right)\left(|U_{ud}|^2+|U_{us}|^2\right)\right] \\
F_2^{CC}(\nu n-p) &= x\,(u(x)-d(x))\,|U_{ud}|^2 + \xi\,(u(\xi)-d(\xi))\,|U_{cd}|^2 \\
&\quad +x\left(\overline{d}(x)-\overline{u}(x)\right)\left(|U_{ud}|^2+|U_{us}|^2\right) \\
F_3^{CC}(\nu n-p) &= (u(x)-d(x))\,|U_{ud}|^2 + (u(\xi)-d(\xi))\,|U_{cd}|^2 \\
&\quad -\left(\overline{d}(x)-\overline{u}(x)\right)\left(|U_{ud}|^2+|U_{us}|^2\right) \\
F_1^{CC}(\overline{\nu} N) &= \frac{1}{2}\left[(u(x)+d(x))\left(|U_{ud}|^2+|U_{us}|^2\right) + 2c(x)\left(|U_{cd}|^2+|U_{cs}|^2\right)\right. \\
&\quad +\left(\overline{u}(x)+\overline{d}(x)\right)|U_{ud}|^2 + \left(\overline{u}(\xi)+\overline{d}(\xi)\right)|U_{cd}|^2 \\
&\quad \left.+2\overline{s}(x)|U_{us}|^2 + 2\overline{s}(\xi)|U_{cs}|^2\right] \\
F_2^{CC}(\overline{\nu} N) &= x\,(u(x)+d(x))\left(|U_{ud}|^2+|U_{us}|^2\right) + 2xc(x)\left(|U_{cd}|^2+|U_{cs}|^2\right) \\
&\quad +x\left(\overline{u}(x)+\overline{d}(x)\right)|U_{ud}|^2 + \xi\left(\overline{u}(\xi)+\overline{d}(\xi)\right)|U_{cd}|^2 \\
&\quad +2x\overline{s}(x)|U_{us}|^2 + 2\xi\overline{s}(\xi)|U_{cs}|^2 \\
F_3^{CC}(\overline{\nu} N) &= (u(x)+d(x))\left(|U_{ud}|^2+|U_{us}|^2\right) + 2c(x)\left(|U_{cd}|^2+|U_{cs}|^2\right)
\end{aligned}$$

$$
\begin{aligned}
& - \left(\overline{u}(x) + \overline{d}(x)\right) |U_{ud}|^2 - \left(\overline{u}(\xi) + \overline{d}(\xi)\right) |U_{cd}|^2 \\
& -2\overline{s}(x)|U_{us}|^2 - 2\overline{s}(\xi)|U_{cs}|^2 \\
F_1^{CC}(\overline{\nu}n - p) &= \frac{1}{2}\left[(d(x) - u(x))\left(|U_{ud}|^2 + |U_{us}|^2\right)\right. \\
& \left. + \left(\overline{u}(x) - \overline{d}(x)\right)|U_{ud}|^2 + \left(\overline{u}(\xi) - \overline{d}(\xi)\right)|U_{cd}|^2\right] \\
F_2^{CC}(\overline{\nu}n - p) &= x\,(d(x) - u(x))\left(|U_{ud}|^2 + |U_{us}|^2\right) \\
& +x\left(\overline{u}(x) - \overline{d}(x)\right)|U_{ud}|^2 + \xi\left(\overline{u}(\xi) - \overline{d}(\xi)\right)|U_{cd}|^2 \\
F_3^{CC}(\overline{\nu}n - p) &= (d(x) - u(x))\left(|U_{ud}|^2 + |U_{us}|^2\right) \\
& - \left(\overline{u}(x) - \overline{d}(x)\right)|U_{ud}|^2 - \left(\overline{u}(\xi) - \overline{d}(\xi)\right)|U_{cd}|^2 \qquad (90)
\end{aligned}
$$

The expressions for the neutral-current structure functions are :

$$
\begin{aligned}
F_1^{NC}(\nu N) = F_1^{NC}(\overline{\nu}N) &= \frac{1}{2}\left[\left(u(x) + d(x) + \overline{u}(x) + \overline{d}(x)\right)\left(\Sigma^u + \Sigma^d\right)\right. \\
& \left. +2\,(s(x) + \overline{s}(x))\,\Sigma^d + 2\,(c(\xi) + \overline{c}(\xi))\,\Sigma^u\right] \\
F_2^{NC}(\nu N) = F_2^{NC}(\overline{\nu}N) &= x\left(u(x) + d(x) + \overline{u}(x) + \overline{d}(x)\right)\left(\Sigma^u + \Sigma^d\right) \\
& +2x\,(s(x) + \overline{s}(x))\,\Sigma^d + 2\xi\,(c(\xi) + \overline{c}(\xi))\,\Sigma^u \\
F_3^{NC}(\nu N) = F_3^{NC}(\overline{\nu}N) &= \left(u(x) + d(x) - \overline{u}(x) - \overline{d}(x)\right)\left(\Delta^u + \Delta^d\right) \\
& +2\,(s(x) - \overline{s}(x))\,\Delta^d + 2\,(c(\xi) - \overline{c}(\xi))\,\Delta^u \\
F_1^{NC}(\nu n - p) = F_1^{NC}(\overline{\nu}n - p) &= \frac{1}{2}\left(d(x) - u(x) + \overline{d}(x) - \overline{u}(x)\right)\left(\Sigma^u - \Sigma^d\right) \\
F_2^{NC}(\nu n - p) = F_2^{NC}(\overline{\nu}n - p) &= x\left(d(x) - u(x) + \overline{d}(x) - \overline{u}(x)\right)\left(\Sigma^u - \Sigma^d\right) \\
F_3^{NC}(\nu n - p) = F_3^{NC}(\overline{\nu}n - p) &= \left(d(x) - u(x) - \overline{d}(x) + \overline{u}(x)\right)\left(\Delta^u - \Delta^d\right) \qquad (91)
\end{aligned}
$$

$$
\begin{aligned}
\text{where :} \quad & \Sigma^f = \epsilon_L(f)^2 + \epsilon_R(f)^2 \\
& \Delta^f = \epsilon_L(f)^2 - \epsilon_R(f)^2 \qquad (92)
\end{aligned}
$$

Finally, the expressions for the electromagnetic structure functions are :

$$
\begin{aligned}
F_1^{\gamma N} &= \frac{5}{36}\left(u(x) + d(x) + \overline{u}(x) + \overline{d}(x)\right) + \frac{1}{18}(s(x) + \overline{s}(x)) + \frac{4}{18}(c(\xi) + \overline{c}(\xi)) \\
F_2^{\gamma N} &= \frac{5}{18}x\left(u(x) + d(x) + \overline{u}(x) + \overline{d}(x)\right) + \frac{1}{9}(xs(x) + x\overline{s}(x)) + \frac{4}{9}(\xi c(\xi) + \xi\overline{c}(\xi)) \\
& \qquad (93)
\end{aligned}
$$

If the mass of the charm quark is neglected and if the mixing matrix is assumed to be two-dimensional (Cabibbo mixing), the formulae for charged-current structure functions are simpler :

$$
\begin{aligned}
F_2^{CC}(\nu N) = 2xF_1^{CC}(\nu N) &= x\left(u(x)+d(x)+\overline{u}(x)+\overline{d}(x)\right) \\
&\quad +2x\left(s(x)+\overline{c}(x)\right) \\
F_3^{CC}(\nu N) &= u(x)+d(x)+2s(x)-\overline{u}(x)-\overline{d}(x)-2\overline{c}(x) \\
F_2^{CC}(\nu n-p) = 2xF_1^{CC}(\nu n-p) &= x\left(u(x)-d(x)+\overline{d}(x)-\overline{u}(x)\right) \\
F_3^{CC}(\nu n-p) &= u(x)-d(x)-\overline{d}(x)+\overline{u}(x) \\
F_2^{CC}(\overline{\nu} N) = 2xF_1^{CC}(\overline{\nu} N) &= x\left(u(x)+d(x)+\overline{u}(x)+\overline{d}(x)\right) \\
&\quad +2x\left(c(x)+\overline{s}(x)\right) \\
F_3^{CC}(\overline{\nu} N) &= u(x)+d(x)+2c(x)-\overline{u}(x)-\overline{d}(x)-2\overline{s}(x) \\
F_2^{CC}(\overline{\nu} n-p) = 2xF_1^{CC}(\overline{\nu} n-p) &= x\left(d(x)-u(x)+\overline{u}(x)-\overline{d}(x)\right) \\
F_3^{CC}(\overline{\nu} n-p) &= d(x)-u(x)-\overline{u}(x)+\overline{d}(x) \qquad (94)
\end{aligned}
$$

The following simple relation between electromagnetic and charged-current structure functions is then satisfied :

$$F_2^{\gamma N} = \frac{5}{18} F_2^{CCN}\left[1-\frac{3}{5}\frac{s(x)+\overline{s}(x)-c(x)-\overline{c}(x)}{q(x)+\overline{q}(x)}\right] \qquad (95)$$

$$\text{where :} \quad F_2^{CCN}(x) = \frac{1}{2}\left(F_2^{CC}(\nu N)+F_2^{CC}(\overline{\nu} N)\right) = x\left(q(x)+\overline{q}(x)\right) \qquad (96)$$

$$\text{and :} \quad q(x) = u(x)+d(x)+s(x)+c(x) \qquad (97)$$

With the same conditions, one has :

$$F_3^{CCN}(x) = \frac{1}{2}\left(F_3^{CC}(\nu N)+F_3^{CC}(\overline{\nu} N)\right) = q(x)-\overline{q}(x) \qquad (98)$$

3.5 *Scaling Violations and QCD Corrections in the Quark-Parton Model*

Up to now, quarks were considered as non-interacting objects and momentum distributions and structure functions were scale invariant, namely independent of Q^2. However, quarks are colored objects and one can expect that the presence of the strong interaction is going to modify this picture. Quarks will interact before and after the interaction with the incoming current. As a result of this interaction, the momentum distributions and the structure functions will depend on the scale Q^2.

Strong interaction corrections to the free QPM can indeed be calculated in Perturbative QCD (PQCD). At first order of the strong coupling constant α_s, the diagrams involved are depicted in figure 25.

It can be shown that the momentum distribution $i(x)$ of parton i is replaced by an effective distribution $i(x, Q^2)$ and the functional form of the Q^2 dependence

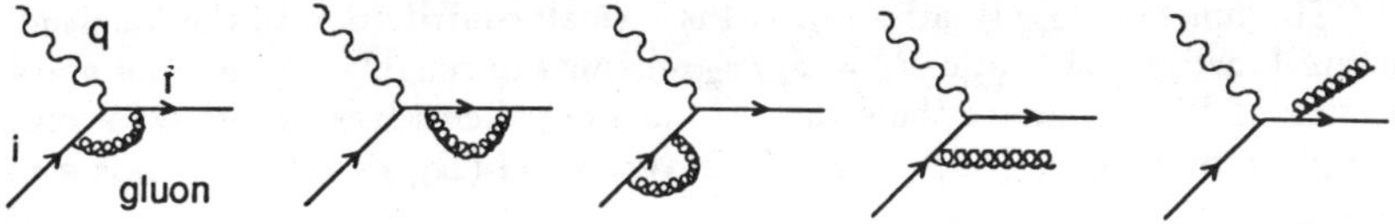

Figure 25: First order QCD corrections to the Quark-Parton Model.

(called "Q^2 evolution") is given by the Altarelli-Parisi equations :

$$\frac{\mathrm{d}}{\mathrm{d}LogQ^2} i(x,Q^2) = \frac{\alpha_s(Q^2)}{2\pi} \int_x^1 \frac{\mathrm{d}y}{y} \left[i(y,Q^2) P_{qq}(\frac{x}{y}) + g(y,Q^2) P_{qg}(\frac{x}{y}) \right] \tag{99}$$

where $g(y, Q^2)$ is the gluon distribution function, $P_{qq}(x)$ and $P_{qg}(x)$ are known functions of x called "splitting functions". The strong coupling constant is given at Leading order by :

$$\alpha_s^{LO}(Q^2) = \frac{12\pi}{33 - 2N_f} \frac{1}{LogQ^2/\Lambda^2} \tag{100}$$

where N_f is the number of quark flavors and Λ is a free parameter of the theory.

Eq. 99 predicts smooth, logarithmic scaling violations for the structure functions. The Q^2 evolution of the quark distributions can be understood if the quarks are viewed as dynamical objects instead of point-like fermions, or, in more involved terms, as fractal objects whose fractal dimension governs the slope of the Q^2 evolution.

The above conclusions are basically the same at Next-to-Leading order, but the expressions of the structure functions in terms of quark distributions (given in Eq. 90 to 93) are slightly modified. In fact, at Next-to-Leading order, the definition of the effective quark distributions is somewhat arbitrary [32]. One possible definition is to identify the content of the F_1 structure function with the quark distributions :

$$F_1(q(x), \bar{q}(x)) \rightarrow F_1 \left[q(x,Q^2), \bar{q}(x,Q^2) \right] \tag{101}$$

The F_2 and F_3 distributions are then modified according to :

$$F_2(q(x), \bar{q}(x)) \rightarrow F_2 \left[q(x,Q^2), \bar{q}(x,Q^2) \right] + F_L(x,Q^2) \tag{102}$$

$$F_3(q(x), \bar{q}(x)) \rightarrow F_3 \left[q(x,Q^2), \bar{q}(x,Q^2) \right] - D_3(x,Q^2) \tag{103}$$

$$\text{where : } \quad F_L(x,Q^2) = \frac{\alpha_s(Q^2)}{2\pi} x^2 \int_x^1 \frac{\mathrm{d}y}{y^3} \left[\frac{8}{3} F_2(y,Q^2) + 4N_f g(y,Q^2)(1 - \frac{x}{y}) \right]$$

$$\text{and : } \quad D_3(x,Q^2) = \frac{\alpha_s(Q^2)}{2\pi} \frac{4}{3} \int_x^1 \frac{\mathrm{d}y}{y} F_3(y,Q^2) \left(1 - \frac{x}{y} \right) \tag{104}$$

The function F_L given by Eq. 104 is a small contribution to the longitudinal structure function : the ratio $R_L = F_L/F_2$ amounts to roughly 0.1. It is an absolute prediction of PQCD, unlike the structure functions themselves, whose Q^2 evolution only is given by PQCD. The function D_3 is very small ($D_3/F_2 \approx 0.01$) and is usually neglected.

Finally, it is interesting to note that, since the F_3 structure function is a difference of quark and antiquark distributions (it is called "non-singlet"), its Q^2 evolution, given by the Altarelli-Parisi equations (Eq. 99), does not depend on the gluon distribution :

$$\frac{\mathrm{d}xF_3}{\mathrm{d}LogQ^2} = \frac{\alpha_s(Q^2)}{2\pi}x\int_x^1 \frac{\mathrm{d}y}{y^2}yF_3(y,Q^2)P_{qq} \tag{105}$$

Parasitic Q^2 evolution can exist because of nuclear effects like Fermi motion. These effects are small at high energies and will not be considered here. But, in addition to the logarithmic Q^2 dependence arising from perturbative corrections, terms in $(1/Q^2)^n$ are also expected in the structure functions from non-perturbative effects.

One type of such effects has already been encountered : the parton and target masses do introduce corrections in m_p^2/Q^2 or M^2x^2/Q^2. Dynamical effects can also generate power terms.

In general, the $W_{\alpha\beta}$ tensor can be written in the form of an expansion of functions of Q^2 and local operators independent of Q^2 (Operator-Product Expansion) [33] :

$$MW_{\alpha\beta} = \sum_k w_k(Q^2)\langle O^k_{\alpha\beta}(0)\rangle \tag{106}$$

The functions w_k exhibit a Q^2 dependence according to the dimension of the operator $O^k_{\alpha\beta}$:

$$w_k(Q^2) = \frac{c_k(Q^2)}{(Q^2)^{\tau-2}} \tag{107}$$

The exponent τ is referred to as the "twist" of the operator k. The dimensionless function $c_k(Q^2)$ shows a slow logarithmic dependence which can be calculated in principle in PQCD.

The contributions encountered up to now in the QPM are due to twist 2 operators. Higher twist contributions correspond to non-perturbative effects and are not calculable, although many models exist for estimating their size. These unknown higher twist operators are a potential worry for the measurement of a fundamental parameter like $\sin^2\theta_W$. However, the effect of higher twist operators can be kept small. First of all, the validity of the QPM can be demonstrated by the measurement of the structure functions in deep inelastic neutrino and charged lepton scattering.

3.6 *The Measurement of Structure Functions and the Parametrization of Parton Densities*

The measurement of structure functions is a very delicate topic and only the main results will be listed here. The reader is referred to specific papers [34] for a detailed discussion.

The structure functions are best measured with neutrino and antineutrino charged-current and charged lepton scattering off isoscalar or heavy targets.

One of the most precise tests of the QPM is the "Gross-Llewellyn-Smith" sum rule [35] which states that[10] :

$$S_{GLS} = \int_0^1 \frac{xF_3^{CCN}}{x} dx = 3\left(1 - \frac{\alpha_s(Q^2)}{\pi}\right) \tag{108}$$

which is $S_{GLS} = 2.62$ at $Q^2 = 3$ GeV^2 for $\Lambda = 250$ MeV.

A precise measurement of the sum rule has been published by the CCFR collaboration [36] : $S_{GLS} = 2.66 \pm 0.029(\text{stat.}) \pm 0.075(\text{syst.})$. This result, which is in good agreement with the prediction of PQCD, provides a test of the QPM at the 3% level.

In figure 26, the most precise measurements of the F_2 structure functions are compared [34]. The structure function $F_2^{\gamma N}$ has been measured precisely at SLAC with electron scattering at low Q^2. Precise measurements at high Q^2 are provided by the BCDMS and NMC muon scattering experiments. The structure function $F_2^{CCN}(x, Q^2)$ has been measured recently by the CCFR collaboration with high statistics. In figure 26, it has been corrected according to Eq. 95 using a previous estimate of the strange sea. The agreement between $F_2^{\gamma N}$ and the corrected F_2^{CCN} is impressive over the full range of x and Q^2. This fact demonstrates the universality of the parton distributions and is a strong evidence for the validity of the QPM.

In figure 26, clear scaling violations are observed. It can be shown that the observed Q^2 dependence agrees with the PQCD prediction of Eq. 99. For example, in figure 26, a value of the Λ parameter of α_s and a gluon distribution function $g(x, Q^2)$ has been fitted to the high statistics data from SLAC and BCDMS. The result of this fit, shown in figure 26, is in very good agreement with the observed Q^2 evolution. This result is in fact one of the most significant tests of PQCD and one of the most precise measurements of the QCD coupling constant. In principle, higher twist terms can also be fitted to the data. The result of a fit including twist 4 terms is that behaviours in $1/Q^2$ are actually not observed at high Q^2. Experimentally, there is no evidence for large higher twist terms.

In the case of the structure function F_2, the scaling violations depend on the gluon distribution function. This is not the case for the structure function xF_3 whose Q^2 evolution is governed by the more simple Eq. 105. The evolution of F_3 is

[10] This is in fact Eq. 86 corrected by PQCD.

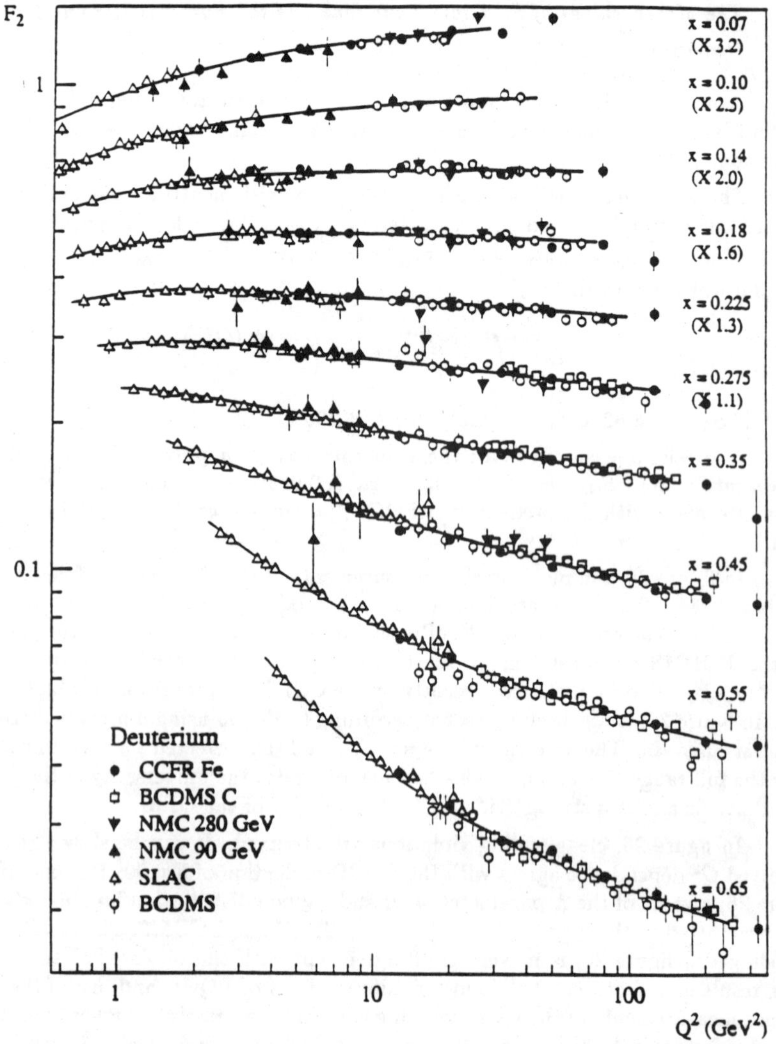

Figure 26: Measurements of $F_2(x, Q^2)$. $F_2^{\gamma N}$ has been measured precisely at SLAC with electron scattering and CERN with muon scattering (BCDMS and NMC). The most recent measurement of F_2^{CCN} from neutrino scattering (CCFR) is also shown after applying the correction of Eq. 95. The line corresponds to a QCD fit [34].

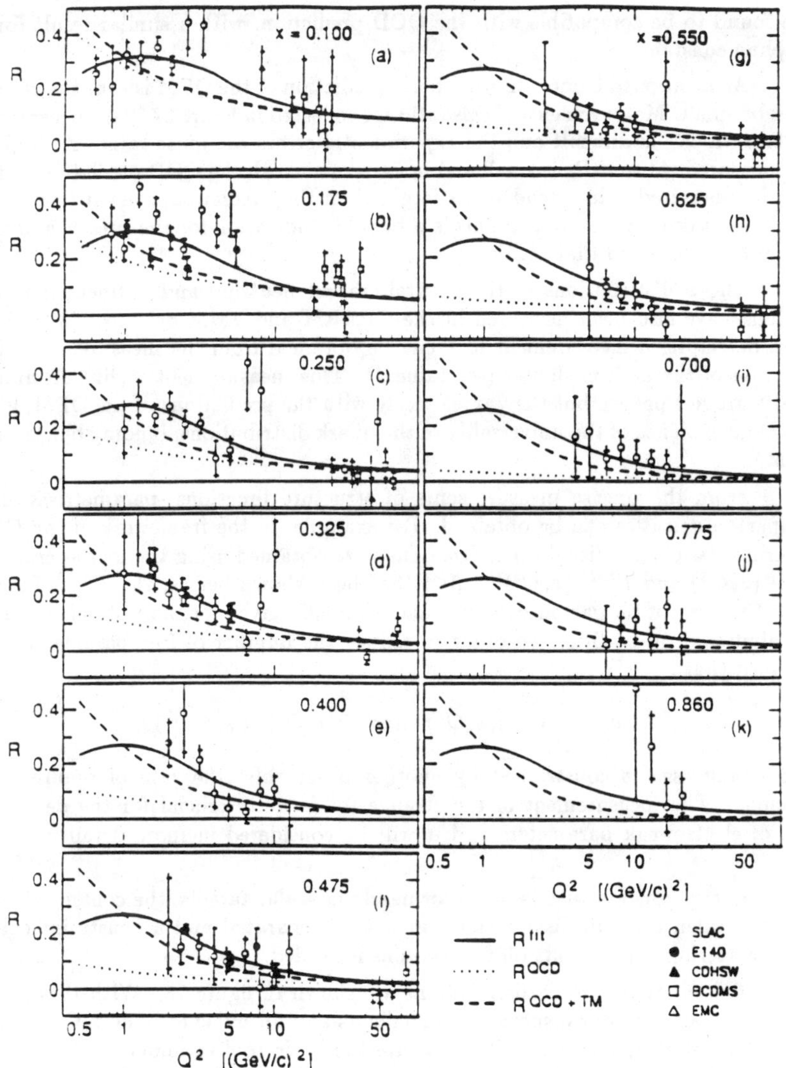

Figure 27: The measurement of the Callan-Gross violation in deep inelastic scattering. Data from electron, muon and neutrino scattering are collected and compared with theoretical predictions [38].

also found to be compatible with the QCD prediction, with a similar result for the coupling constant.

As mentioned before, another basic prediction of the QPM is that the ratio R_L must be small. Measurements of this ratio are collected in figure 27. The measurement of the ratio R_L is difficult and the experimental errors are quite large. At high Q^2, the measured value of R_L is small and in agreement with the QCD prediction. At low Q^2, the measured values tend to be larger than the prediction, even if target mass effects are taken into account. This is a possible indication for the presence of small higher twist terms in that region.

The x distributions of the neutral-current neutrino and antineutrino cross-sections have also been measured in the CHARM and FMM detectors [39, 40]. In these detectors, a measurement of x can be obtained from the measurement of the direction of the hadron shower (see table 4). This measurement is difficult and the results are not precise but they agree nicely with the predictions of the QPM, giving one more evidence of the universality of the quark distributions, buiding blocks of the QPM.

From the precise measurements of structure functions, parametrisations of the parton densities can be obtained. For example, in the framework of the CDHS experiments, parton distribution functions were obtained using the measurements of $F_2^{CCN}(x, Q^2)$ and $F_3^{CCN}(x, Q^2)$[11]. As it has been shown before, the F_2^{CCN} function gives the sum of all parton densities and $F_3(x, Q^2)$ gives the sum of valence quark distributions. In addition, in order to reduce the number of free parameters, it is assumed that :

$$\overline{u}(x, Q^2) = \overline{d}(x, Q^2) \quad and \quad \overline{s}(x, Q^2) = s(x, Q^2) \tag{109}$$

The strange sea is constrained by another observable, the rate of opposite sign dimuons. The measurement of the strange sea is quite crucial for the determination of electroweak parameters and it will be considered in more detail in section 4.

In experiments with isoscalar or nearly isoscalar targets, the content of u and d in the valence quark distribution can not be separated and a constraint coming from scattering experiments off free protons is needed.

The obtained quark distributions are shown in figure 28. With these distributions, we have at our disposal all ingredients and formulae to build a QPM in the framework of PQCD and it will constitute the basic tool in understanding physics effects in the experimental measurements. It is a reliable tool and it has been checked in detail experimentally. However, as far as the measurement of electroweak parameters is concerned, a framework which is independent of dynamical assumptions is needed.

[11] A discussion of recent parametrisations of parton distributions can be found in [37].

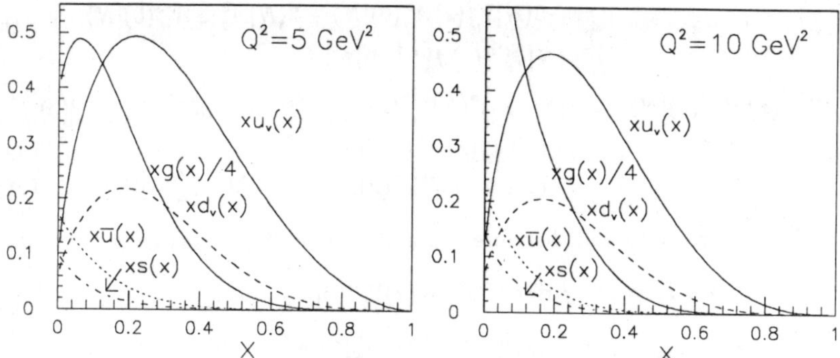

Figure 28: Parametrisations of parton distributions [41].

3.7 *The Llewellyn-Smith Equations*

As noticed by Llewellyn-Smith [28], reviving arguments given by Paschos and Wolfenstein [23], most of the neutral-current and charged-current cross-sections can be related without any dynamical hypothesis, assuming only isospin symmetry. The proof for this statement will be sketched now.

The currents given in Eq. 50 can be rewritten as :

$$\vec{J}_\alpha = \vec{V}_\alpha - \vec{A}_\alpha$$

$$J_\alpha^{NC} = (1-2z)V_\alpha^3 - zV_\alpha^0 - A_\alpha^3 \tag{110}$$

$$\text{where}: \quad \vec{V}_\alpha = \sum_{fermions} \overline{\psi}_i \vec{T} \gamma_\alpha \psi_i \quad \vec{A}_\alpha = \sum_{fermions} \overline{\psi}_i \vec{T} \gamma_\alpha \gamma_5 \psi_i$$

$$V_\alpha^0 = \sum_{fermions} \overline{\psi}_i Y \gamma_\alpha \psi_i \qquad z = \sin^2\theta_W \tag{111}$$

$$\text{and}: \quad \mathcal{Q} = T_3 + \frac{Y}{2} \tag{112}$$

where Y is the weak hypercharge.

However, some simple relations exist for average values of operator products on an isoscalar state N :

$$\langle N|X_+Y_-|N\rangle = 2\langle N|X_3Y_3|N\rangle$$

$$\langle N|X_3B|N\rangle = 0 \tag{113}$$

where $\vec{X}$ and $\vec{Y}$ are vector operators and B is a scalar operator. Eq. 113 can be applied to the charged-current and neutral-current case if the weak isospin operator $\vec{T}$ is identified with the strong isospin operator. Then :

$$\langle N|J_\beta^{CC\dagger}(x)J_\alpha^{CC}(0)|N\rangle = 2\langle N|V_\beta^3(x)V_\alpha^3(0)|N\rangle + 2\langle N|A_\beta^3(x)A_\alpha^3(0)|N\rangle$$

$$
\begin{aligned}
& & -2\langle N|V_\beta^3(x)A_\alpha^3(0)|N\rangle - 2\langle N|A_\beta^3(x)V_\alpha^3(0)|N\rangle \\
& = & \langle N|J_\beta^{CC}(x)J_\alpha^{CC\dagger}(0)|N\rangle \\
\langle N|J_\beta^{NC\dagger}(x)J_\alpha^{NC}(0)|N\rangle & = & (1-2z)^2\langle N|V_\beta^3(x)V_\alpha^3(0)|N\rangle + z^2 2\langle N|V_\beta^0(x)V_\alpha^0(0)|N\rangle \\
& & +\langle N|A_\beta^3(x)A_\alpha^3(0)|N\rangle \\
& & -(1-2z)\langle N|V_\beta^3(x)A_\alpha^3(0) + A_\beta^3(x)V_\alpha^3(0)|N\rangle \qquad (114)
\end{aligned}
$$

The cross-sections will therefore contain the following terms :

$$
\begin{aligned}
V.VI_3^2 &= L^{\alpha\beta}\int \mathrm{d}x e^{iqx}\langle N|V_\beta^3(x)V_\alpha^3(0)|N\rangle \\
A.AI_3^2 &= L^{\alpha\beta}\int \mathrm{d}x e^{iqx}\langle N|A_\beta^3(x)A_\alpha^3(0)|N\rangle \\
V.VY^2 &= L^{\alpha\beta}\int \mathrm{d}x e^{iqx}\langle N|V_\beta^0(x)V_\alpha^0(0)|N\rangle \\
2V.AI_3^2 &= L^{\alpha\beta}\int \mathrm{d}x e^{iqx}\langle N|V_\beta^3(x)A_\alpha^3(0) + A_\beta^3(x)V_\alpha^3(0)|N\rangle \qquad (115)
\end{aligned}
$$

The charged-current and neutral-current cross-sections are then given by (a common factor being omitted) :

$$
\begin{aligned}
\mathrm{d}\sigma^{CC}(\overset{(-)}{\nu}N) &= 2\left[V.VI_3^2 + A.AI_3^2 \pm 2V.AI_3^2\right] \qquad (116)\\
\mathrm{d}\sigma^{NC}(\overset{(-)}{\nu}N) &= \frac{1}{2}\left[1+(1-2z)^2\right]\left[V.VI_3^2 + A.AI_3^2 \pm 2V.AI_3^2\right] \pm (1-2z)2V.AI_3^2 \\
& +\frac{1}{2}\left[(1-2z)^2 - 1\right]\left[V.VI_3^2 - A.AI_3^2\right] + z^2 V.VY^2 \qquad (117)
\end{aligned}
$$

The first two terms in Eq. 117 can be written in terms of charged-current cross-sections. The third term breaks the chiral symmetry of strong interaction and is therefore a contribution due to quark masses. These contributions will be neglected at high energies.

To interpret the fourth term, it will now be assumed that the nucleon contains only u and d quarks. In the QPM, it can be shown that :

$$
V.VY^2 = \frac{2}{9}\left[V.VI_3^2 + A.AI_3^2\right] \qquad (118)
$$

This relation remains valid even if perturbative corrections are included. The only dynamical assumption which will be made here is that Eq. 118 remains also valid at all orders in the strong interaction. Since the $V.VY^2$ term in Eq. 117 is weighted by $\sin^4\theta_W$, potential higher-twist contributions in Eq. 118, whose presence can not be excluded, will have a negligible effect on the neutral-current cross-sections.

The final expressions of the neutral-current cross-sections with these assumptions is then :

$$
\mathrm{d}^{NC}(\overset{(-)}{\nu}N) = \frac{1}{2}\left[1+(1-2z)^2\right]\left[V.VI_3^2 + A.AI_3^2 \pm 2V.AI_3^2\right] \pm (1-2z)2V.AI_3^2
$$

$$+\frac{2}{9}z^2\left[V.VI_3^2 + A.AI_3^2\right] \qquad (119)$$

which gives, using Eq. 116, the so-called Llewellyn-Smith relations :

$$\begin{aligned} d\hat{\sigma}^{NC}(\nu N) &= \left[\frac{1}{2} - \sin^2\theta_W + \frac{5}{9}\sin^4\theta_W\right] d\hat{\sigma}^{CC}(\nu N) + \frac{5}{9}\sin^4\theta_W d\hat{\sigma}^{CC}(\overline{\nu} N) \\ d\hat{\sigma}^{NC}(\overline{\nu} N) &= \left[\frac{1}{2} - \sin^2\theta_W + \frac{5}{9}\sin^4\theta_W\right] d\hat{\sigma}^{CC}(\overline{\nu} N) + \frac{5}{9}\sin^4\theta_W d\hat{\sigma}^{CC}(\nu N) (120) \end{aligned}$$

where the hat refers to the conditions of validity of the equations.

The Paschos-Wolfenstein relations Eq. 19 follow from the Llewellyn-Smith relations. However, it is important to note that the prediction for the ratio R_- rely only on isospin symmetry. In the case of R_-, the hypothesis of Eq. 118 for the hypercharge term $V.VY^2$ is not needed as it disappears in the difference of neutrino and antineutrino neutral-current cross-sections.

In order to get a precise value of $\sin^2\theta_W$, the Llewellyn-Smith relations have to be corrected for the presence of strange quark, the charm mass effects, non-isoscalarity of some heavy nuclei like Fe and higher order radiative corrections. But these corrections are small and the QPM will be a well tested and reliable framework for estimating them. Thus, the determination of $\sin^2\theta_W$ is largely independent of dynamical assumptions and this justifies the precise neutrino scattering experiments that were launched after 1983.

4 The Precise Measurement of the Neutral-Current to Charged-Current Interaction Ratio

In this section, some experimental details of the precise measurements of the neutral-current to charged-current cross-section ratio will be presented. Three experiments will be described : the measurements by the CDHS [42] and CHARM [43] collaborations in the 1984 CERN NBB and the experiment E770 [44] performed by the CCFR collaboration in the 1988 Fermilab QTB.

A precision measurement needs a lot of careful work and also relies on quite a number of tricks to reduce the systematics errors. Some of these experimental subtleties will be pointed out.

4.1 *The Neutrino Beams*

Since the dominating uncertainty in the previous experiment by the CDHS collaboration was coming from the knowledge of the beam and in particular the WBB background, the CERN NBB set-up (figure 11) was carefully modified for the 1984 experiment (see figure 29).

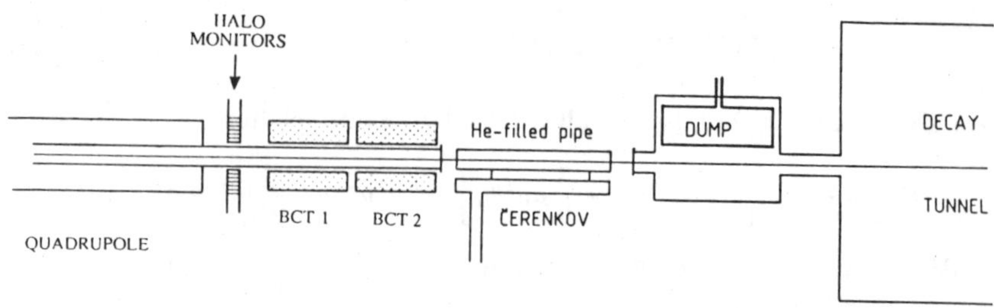

Figure 29: The Narrow Band Beam set-up at CERN in 1984. Details of the section before the decay tunnel showing the monitors in front of the dump [42].

The hadron beam was defined by a set of four quadrupoles whose parameters were adjusted to maximize the event rate in the detector. The optimum corresponded to a nominal beam energy of 160 GeV. Compared with previous beams (table 2), the energy dispersion and the angular divergences were increased by 50%.

The beam line was instrumented with several types of monitoring devices. The absolute intensity of the primary proton beam and of the secondary hadron beam was measured with beam current transformers (BCTs). The transverse position of the proton beam was measured regularly in order to control its impact point on the target. The position and the profile of the hadron beam was measured by a set of segmented ionisation chambers. Eight ionisation chambers were placed at several locations upstream of the BCTs to control the halo in the hadron beam. The position and intensity of the beam was also monitored by a set of solid-state detectors inserted in the shielding iron after the decay tunnel. The composition of the hadron beam was measured by a Cherenkov counter placed after the BCTs.

Various changes were introduced in the beam in order to reduce the WBB component. First, the number of neutrinos produced in the hadron beam at the end of the decay tunnel could be estimated using the results of a previous experiment where the proton beam was transported down to the end of the decay tunnel without interactions (beam dump experiment). The number of neutrinos from this source is found to be 0.4 % of the total neutrino flux and is therefore negligible. Second, the pressure in the decay tunnel was constantly monitored and kept below 0.13 Torr, which ensured that the number of interactions of the hadron beam in the decay tunnel was negligible. The hadrons producing neutrinos therefore can only be produced before the decay tunnel.

The production of hadrons before the decay tunnel could be reduced through various means. The various monitoring devices around the beam ensured that the

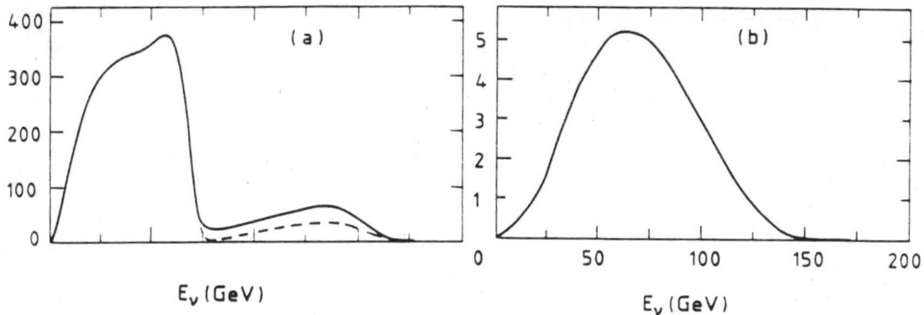

Figure 30: (a) The spectra of muon-neutrino (solid line) and muon-antineutrino (dashed line) beams inside the acceptance of the CDHS detector (CERN 1984 NBB). (b) The electron-neutrino spectrum in the neutrino beam. The vertical scale is in arbitrary units [42].

beam was not mistuned and scraping the walls of the beam pipe. The hadron beam was kept in a single vacuum pipe and met walls only at the place where the Cherenkov counter was inserted. In addition, the Cherenkov counter was replaced most of the time by a helium pipe.

Finally, an experimental measurement of the WBB was done by inserting in the hadron beam a 1.5 m long iron dump. The NBB component of the beam was switched off and the WBB component, which was coming from hadrons produced before the dump, remained. Furthermore, the position of this movable beam dump was chosen in order to produce approximately the same amount of neutrinos as in the beam dump at the end of the decay tunnel under normal conditions. With this set-up, after subtraction of the experimentally measured WBB background, the events in the detector are produced by a beam which is an excellent approximation of an ideal NBB.

The simulated spectrum of this NBB beam through the CDHS detector is shown in figure 30 in the neutrino beam. The distributions of the reconstructed neutrino energy and of the hadron energy of charged-current interactions are shown in figure 31. The average incident energy in the neutrino beam was 55 GeV. The average energy of recorded charged-current interactions in the neutrino beam was 85 GeV.

In the Fermilab QTB, the emphasis was laid on the intensity. A very high statistics of neutrino interactions in the CCFR detector was recorded with this beam. This very high statistics allowed tighter cuts to be applied on the data and the parameters of the beam to be extracted from the data themselves. The neutrino spectrum in the detector was found using reconstructed charged-current interactions

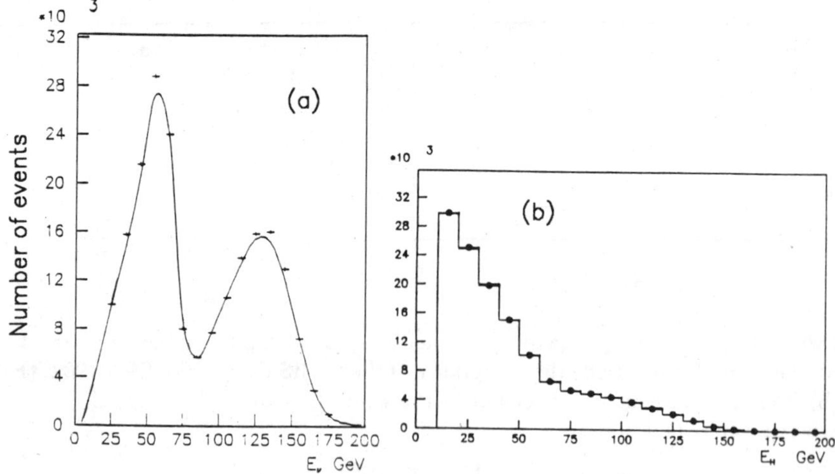

Figure 31: (a) The reconstructed energy distribution of charged-current interactions in the CDHS detector. The line corresponds to the Monte-Carlo simulation of the beam and the detector. (b) The measured hadron energy distribution of charged-current candidates in the CDHS detector.

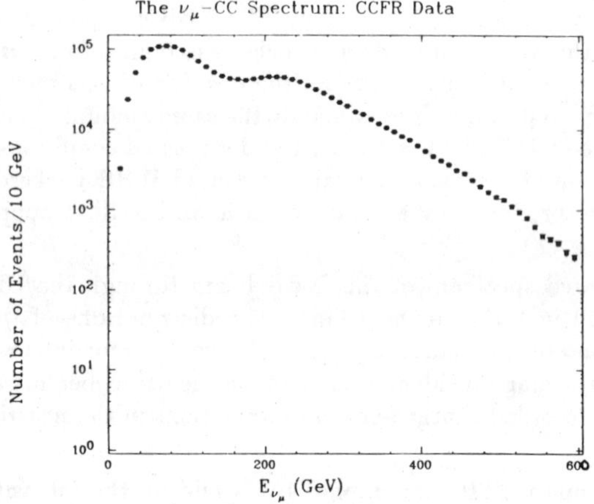

Figure 32: The reconstructed energy distribution of charged-current interactions recorded in the Fermilab 1988 QTB [44].

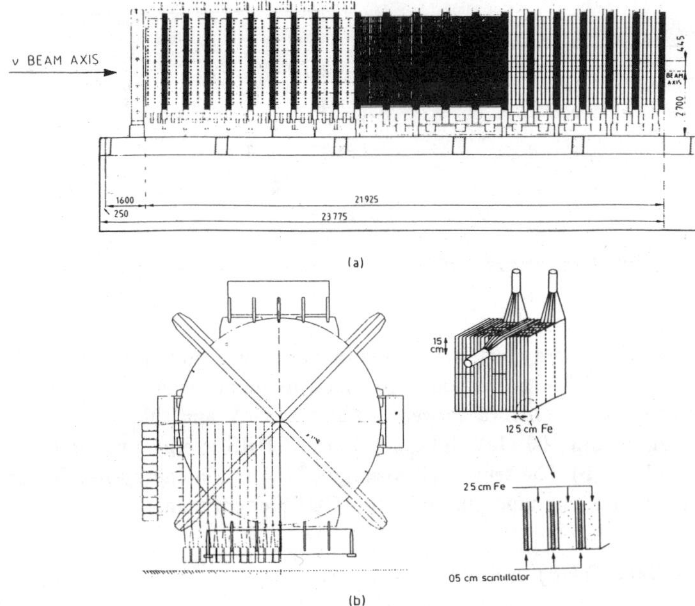

Figure 33: (a) Layout of the upgraded CDHS detector. (b) Structure of the new type of CDHS modules [42].

whose energy distribution is shown in figure 32. The average energy of the incident neutrinos was 80 GeV and the average energy of recorded charged-current interactions 166 GeV, thus significantly larger than in the CERN NBB.

4.2 *The Neutrino Detectors*

The CHARM and CCFR detectors have already been described. Compared with its original configuration (figure 14), the CDHS detector was modified in 1983, as shown in figure 33. Ten new calorimeter modules were installed. In these new modules, in order to obtain a better measurement of the transverse position of the shower, the scintillator plates, inserted between 2.5 cm thick iron plates, were split in 24 strips of 15 cm width (figure 33b). Five consecutive strips were viewed by a single photomultiplier, producing a longitudinal segmentation of 12.5 cm of iron. The strips in consecutive read-out planes were oriented orthogonally. This configuration leads to a radial resolution of about 5 cm on the shower vertex.

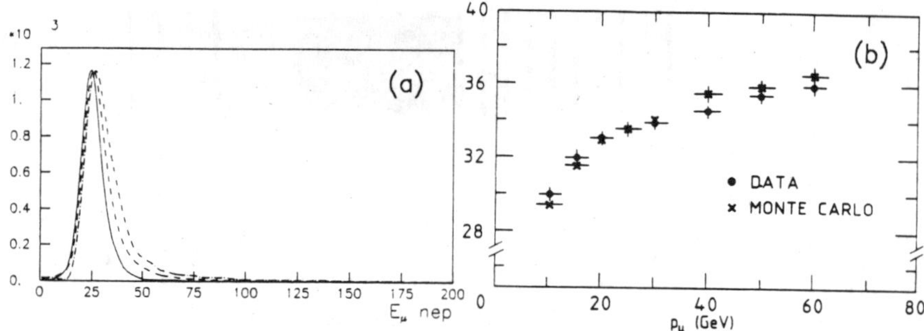

Figure 34: (a) The energy lost by a muon in the shower box measured in the CDHS detector. The solid line corresponds to muon momenta smaller than 20 GeV, the dashed line to muon momenta between 30 and 80 GeV and the dotted line to muon momenta larger than 100 GeV. The nep is the CDHS energy unit and amounts to roughly 0.1 GeV. (b) The truncated average of the muon energy loss in the shower box as a function of the muon momentum (CDHS experiment).

4.3 *The Data Taking and the Data Samples*

The data taking for high statistics experiments requires a great care over a period of time of the order of six months or more. Over this period, the quality of the beam must be permanently monitored and any drift in its properties must be reported to the beam control teams. In addition, the detector must be surveyed in order to maintain its properties constant. The thoroughness of this process is the key of precision experiments. All three detectors CDHS, CHARM and CCFR had a long experience in running and all this experience was put to use to guarantee the quality of the measurement of the electroweak mixing parameter.

The principle of the measurement of the neutral-current to charged-current cross-section ratio is the same as in the old experiments. For each hadron shower recorded in the detector, the hadron energy and the position of the interaction vertex has to be measured. In addition, a variable separating neutral-current and charged-current has to be reconstructed. This variable is the event length in the case of iron detectors and the number of identified muon tracks in the case of the CHARM detector.

In all three detectors, the hadron energy is obtained from the scintillator signals. This raw energy will be referred to as the "shower energy". For example, in the case of the CDHS experiment, the shower energy is defined as the energy sum of the scintillator signals in a box with a 1.5 m length and a 1.2 m width.

For charged-current interactions, the hadron energy E_h is obtained from the

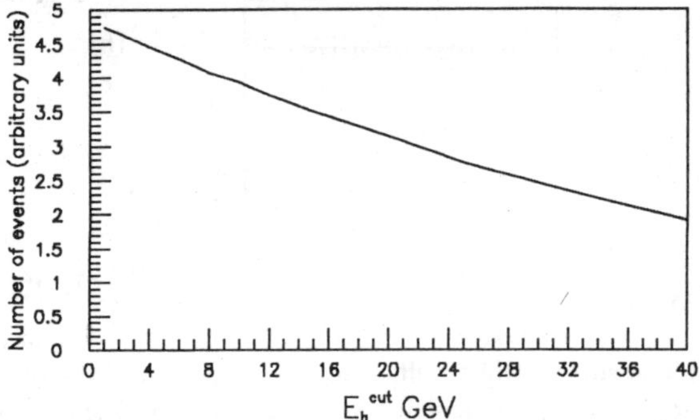

Figure 35: The number of charged-current interactions in a 1.3 m radius ideal detector in the CERN 160 GeV NBB beam as a function of the hadron energy cut-off E_h^{cut}.

shower energy E_{sho} after subtracting the energy lost by the muon in the shower box. This energy lost by the muon in the box E_μ^B can be measured experimentally (figure 34a) and amounts to roughly 3 GeV. In figure 34b, it is shown that the measured average energy loss is in good agreement with the simulation. However, the energy loss of particles is subject to large fluctuations, especially towards high energies (Landau tail), and the average value of the distribution is not well defined and depends on the truncation value of the distribution. The actual value E_μ^{eff} which has to be subtracted from the observed shower energy E_{sho} is constrained by the requirement that the subtraction must be correct on average :

$$N^{CC}(E_h > E_h^{cut}) = N^{CC}(E_{sho} - E_\mu^{eff} > E_h^{cut}) \tag{121}$$

The value E_μ^{eff} is obtained by folding the measured distributions into the charged-current energy distribution and is found to be in between the peak and the truncated average value. The effective subtracted energy loss depends slightly on the hadron energy cut-off E_h^{cut}.

In the CDHS experiment, the value of E_μ^{eff} was determined with an uncertainty of $\pm$150 MeV. This translates (see Eq. 10 for a rough estimate) to an uncertainty of 0.4% on R_ν, and 1.2% on $R_{\bar{\nu}}$. In the CCFR experiment, the error on R_ν is reduced to 0.15%, mostly because of the higher beam energy.

In the fine-grain calorimeter of the CHARM experiment, the cells hit by the muon are separated from the shower (see figure 17) and these cells are removed when the shower energy is computed. Only a small correction remains to be done.

The cut-off E_h^{cut} applied on the hadron energy is different for the three experiments. From the physics point of view, it is safer to use higher values of E_h^{cut} as non-perturbative effects could contaminate the low hadron energy sample and introduce problems in the interpretation. In addition, many systematic errors on the

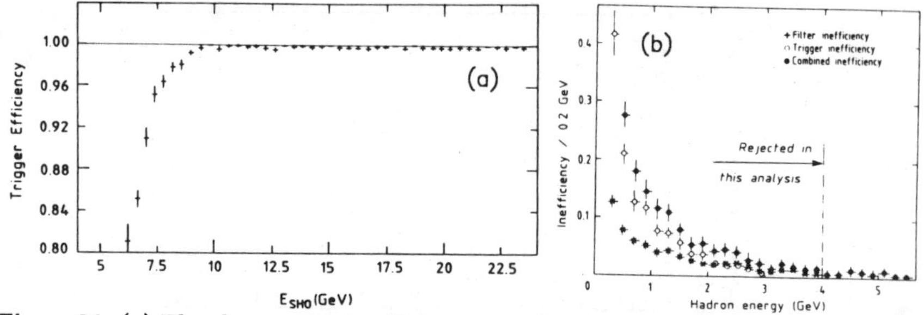

Figure 36: (a) The shower trigger efficiency as a function of the shower energy in the CDHS detector [42]. (b) The inefficiency of the trigger and of the cosmic ray filter as a function of shower energy in the CHARM detector [43].

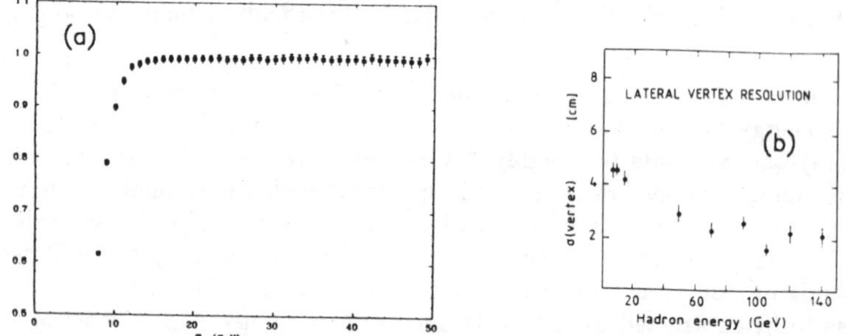

Figure 37: (a)The trigger efficiency as a function of the shower energy in the CCFR detector. (b) The resolution on the transverse position of the interaction vertex in the CHARM detector.

value of $\sin^2\theta_W$ are reduced with higher values of E_h^{cut}. However, the event statistics is reduced rapidly if E_h^{cut} is increased. For example, figure 35 shows the number of recorded charged-current interactions in the CERN beam as a function of E_h^{cut}. In practice, the experiments try to use the lowest possible cut-off allowed by the performances of the calorimeters and in particular by the trigger efficiency.

Two main independent trigger conditions were used in the experiments : one local energy deposition trigger, sensitive to a shower energy deposition and called "shower trigger", and one "muon trigger", sensitive to the propagation in the detector of a minimum ionizing particle. The shower trigger does not introduce any bias between neutral-current and charged-current interactions as both have a shower in the final state. This trigger is used for the measurement of the neutral-current to charged-current cross-section ratio. The muon trigger selects charged-current interactions and is used for measuring the efficiency of the shower trigger as a function of shower energy. The measured shower trigger efficiencies as a function of the hadron energy

	CDHS	CHARM	CCFR
E_h^{cut} (GeV)	10	4	30
Transverse vertex cut (m)	Radius<1.3	2.4×2.4	Radius<0.762
Transverse size of the detector (m)	ϕ3.75	3×3	3×3
Fiducial Mass (ton)	141	87	69
Total Target Mass (ton)	1150	180	690

Table 5: Summary of experimental cuts used for the high statistics neutrino experiments.

are shown in figure 36 and 37a for the three experiments. The trigger is fully efficient if E_{sho} >10 GeV, E_{sho} >4 GeV and E_{sho} >30 GeV for the CDHS, CHARM and CCFR experiments respectively. The lower hadron energy cut-off for the CHARM experiment, due to the fine-grain structure of the calorimeter and a better energy resolution, partly compensates for the loss of statistics due to the smaller detector mass.

In fact, what actually matters for the measurement of the neutral-current to charged-current cross-section ratio is to check that the trigger efficiency is the same for neutral-current and charged-current interactions. However, since the muon of charged-current interactions leaves some energy in the hadron shower, the hadron energy scale of neutral-current and charged-current is slightly different. It is therefore important to make sure that the trigger is indeed fully efficient.

Finally, fiducial cuts are applied to retain events having occurred in regions of the detectors allowing a good understanding of the systematic errors. A cut on the longitudinal position of the vertex is imposed and, more importantly, a cut on the radial position of the vertex. In the CDHS experiment, the radial position of the vertex is given by the radial position of the barycenter of the scintillators hits in the shower box. This measurement leads to a transverse resolution of 5 cm. In the CCFR experiment, using a weighted average of the hits in the spark chambers immediatly downstream of the vertex, a resolution of 4 cm is achieved. In the CHARM experiment, the measurements from all detector components (scintillators, proportional drift tubes and streamer tubes) are used. The resolution, given in figure 37b, reaches a precision better than 3 cm for E_{sho} > 40 GeV. The neutral-current to charged-current cross-section ratio is not sensitive to the radial resolution and the performances of the detectors are adequate.

The radial cut used in the CCFR experiment is significantly smaller than for CERN experiments (table 5). This small radial cut causes a big loss in statistics (figure 38) but is important in reducing the number of charged-current events with a muon leaving by the side and also the contribution of electron-neutrinos.

The events remaining after the fiducial cuts are classified as neutral-current

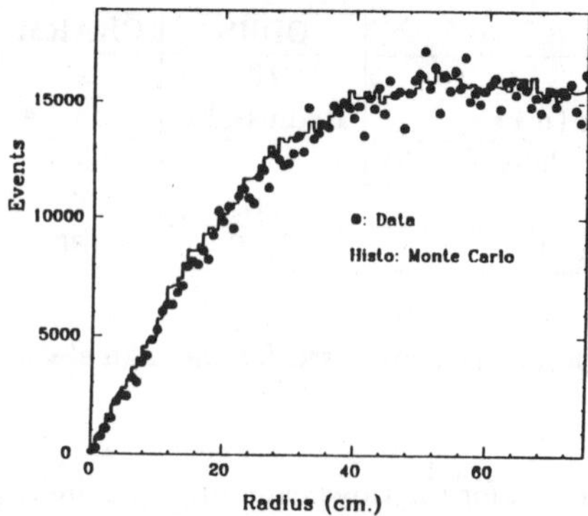

Figure 38: The distribution of the event radius in the CCFR experiment [44].

candidates (NC) or charged-current candidates (CC).

In the iron detectors CDHS and CCFR, the event classification is based on the visible event length L, defined as the thickness of iron between the read-out plane containing the event vertex and the last read-out plane hit by the most penetrating particle of the event.

In the CDHS experiment, a cut-off length L_c is defined such that practically all neutral-current events have $L < L_c$ while minimizing the number of short charged-current events. The cut-off L_c is chosen to follow the longitudinal shower profile and depends logarithmically on E_{sho} :

$$L_c(\mathrm{m}) = 0.75 + 0.38 Log E_{sho}(GeV) \qquad (122)$$

The NC candidate sample is defined by the conditions $E_{sho} > 10$ GeV and $L \leq L_c$ and the CC candidate sample is defined by the conditions $E_{sho} - E_\mu^{eff} > 10$ GeV and $L > L_c$. A NC candidate and a CC candidate in the CDHS detector are displayed in figure 39.

In the CCFR detector, the events are classified as "short events" (NC) if L <3 m and as "toroid events" (CC) if some energy is recorded in the three most downstream counters in the calorimeter.

Interactions in the CHARM detector are classified by an automatic pattern recognition program, a CC interaction being an interaction where a muon is found originating from the event vertex. The muon is defined as a set of related hits isolated from the hadron shower (see figure 17). In order to reject hadronic tracks, the candidate muon track is required to have a total range corresponding to an energy loss larger than 1 GeV and to be visible over a range corresponding to an energy loss

(a) Neutrino Beam	CDHS		CHARM		CCFR	
	NC	CC	NC	CC	NC	CC
Number of candidates (10^3)	61	138	39	108	151	293
Cosmic background (%)	−1.8	−0.0	−0.8	−0.0	0	0
WBB background (%)	−1.5	−1.2	−5.1	−4.0		
CC misidentified as NC (%)	−16.	+6.9	−9.5	+3.4	−22.	
NC misidentified as CC (%)	+0.3	−0.1	+4.8	−1.7		
ν_e correction (%)	−5.0	+1.8	−5.9	−0.1	−8.0	

(b) Antineutrino Beam	CDHS		CHARM	
	NC	CC	NC	CC
Number of candidates (10^3)	3.0	5.6	2.3	5.5
Cosmic background (%)	−10.	−0.1	−4.0	−0.0
WBB background (%)	−4.1	−2.6	−12.	−8.2
CC misidentified as NC (%)	−6.2	+3.0	−1.9	+0.8
NC misidentified as CC (%)	+0.2	−0.1	+4.5	−1.8
ν_e correction (%)	−3.5	+0.1	−3.2	−0.1

Table 6: Neutral-current and charged-current candidate samples and the corrections in the neutrino (a) and antineutrino beams (b). The results from CCFR have not been published yet and the numbers given in (a) are indicative only.

	CDHS		CHARM		CCFR
	$\frac{\Delta R_\nu}{R_\nu}$	$\frac{\Delta R_{\bar\nu}}{R_{\bar\nu}}$	$\frac{\Delta R_\nu}{R_\nu}$	$\frac{\Delta R_{\bar\nu}}{R_{\bar\nu}}$	$\frac{\Delta R_\nu}{R_\nu}$
Data Statistics	0.81	3.9	0.84	3.5	0.40
Hadron energy measurement	0.4	1.2	0.1	0.3	0.15
Cosmic background	0.1	0.3	0.0	0.1	0.00
Beam spectrum	0.1	0.3	0.3	0.6	0.17
CC misidentified as NC	0.45	0.6	0.3	0.2	0.23
NC misidentified as CC	0.2	0.2	0.1	0.2	0.00
ν_e correction	0.2	0.2	0.3	0.2	0.45
Total Systematic error	0.68	1.5	0.54	0.8	0.55
Total Experimental error	1.0	4.2	1.0	3.6	0.68

Table 7: Summary of experimental errors (in %) on the value of the neutral-current to charged-current cross-section ratios. The errors for the CCFR experiment have been estimated from the preliminary errors given for $\sin^2\theta_W$.

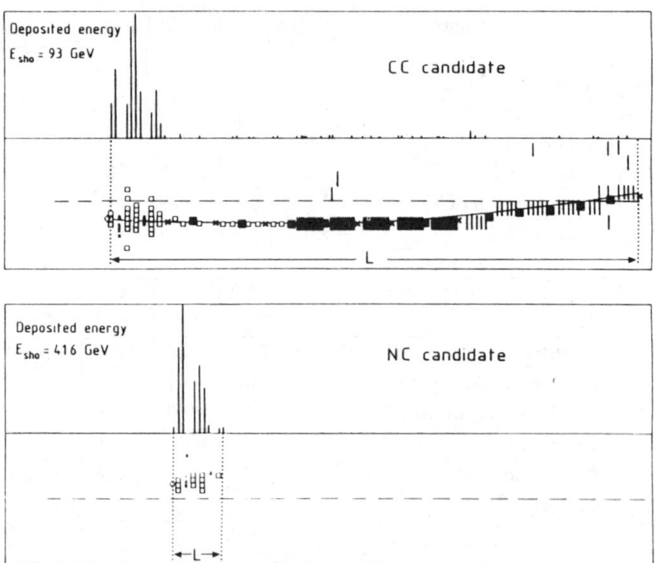

Figure 39: Display of a neutral-current candidate and of a charged-current candidate in the CDHS detector [42].

larger than 0.67 GeV. The extrapolated intercept of the muon with the vertex plane has to be within 30 cm of the vertex.

The number of NC and CC candidates in the various detectors are indicated in table 6. Various corrections have to be applied to obtain the true numbers of neutral-current and charged-current interactions with $E_h > E_h^{cut}$. The uncertainties on R_ν and $R_{\bar{\nu}}$ introduced by the corrections are collected in table 7 and in table 10 in terms of the error on the value of $\sin^2\theta_W$.

First, the interactions due to cosmics or to neutrinos with unknown spectrum have to be subtracted.

Cosmic rays are recorded in the detector and contaminate mostly the NC sample (figure 40). The number of cosmic interactions was measured outside the beam gate and subtracted. The measured rate however has to be corrected for dead-time losses. Showers produced by cosmic rays tend to have small shower energies and the cosmic background is very small for a hadron energy cut-off of 30 GeV. In the case of the CHARM experiment where, because of the 4 GeV hadron energy cut-off, the cosmic background would be the largest, the peculiar topology of most of the cosmic ray interactions (as in figure 40) was used to reduce the cosmic background

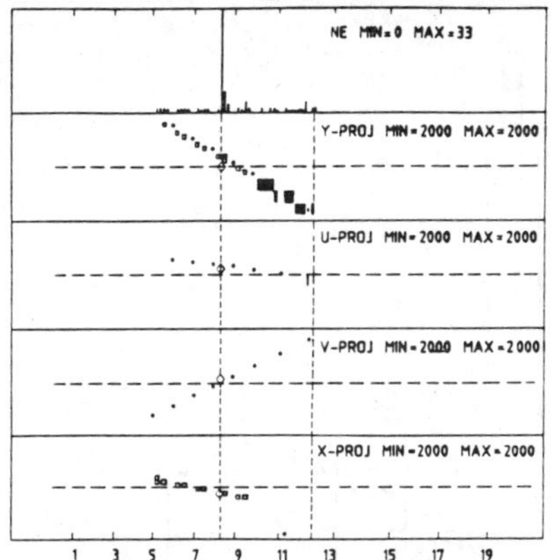

Figure 40: A cosmic ray interaction in the CDHS detector.

to a small contamination of 0.8 %[12]. In all cases, the subtraction of the cosmic ray background introduces a very small error on the cross-section ratio measurement.

In the case of the CERN experiments, the contribution of WBB neutrinos was measured when the dump was inserted into the hadron beam line. This was done 28% of the time in the neutrino beam and 37% of the time in the antineutrino beam. The relative normalisation of the two data conditions was done using the measurement of the BCTs in the hadron beam. From table 6, one can see that the contribution of WBB neutrinos in the neutrino beam is small for $E_h > 10$ GeV, as the value of the ratio is almost unchanged by the subtraction. In the case of the CHARM experiment for which $E_h > 4$ GeV, the correction is larger, as expected for a contribution decreasing logarithmically with energy. The systematic error in this correction, negligible in the case of the CDHS experiment, small in the case of the CHARM experiment (0.3% on the cross-section ratio), is mainly due to possible contributions from hadrons in the decay tunnel or in the dump at the end of the decay tunnel. After this correction is done, the interactions in the detector are due only to NBB neutrinos of known spectrum.

In the case of the CCFR experiment, as already mentioned, the spectrum of the incident neutrinos is extracted from the data themselves. Using the high statistics data sample, the uncertainty on the value of $\sin^2\theta_W$ coming from the knowledge of the beam spectrum could be maintained at the level of 0.001.

But the most important corrections to the number of candidates are due to the misclassification of NC and CC interactions and the subtraction of electron-neutrino interactions and these corrections will be examined in more detail.

[12]This is the cosmic ray filter mentioned in figure 36b.

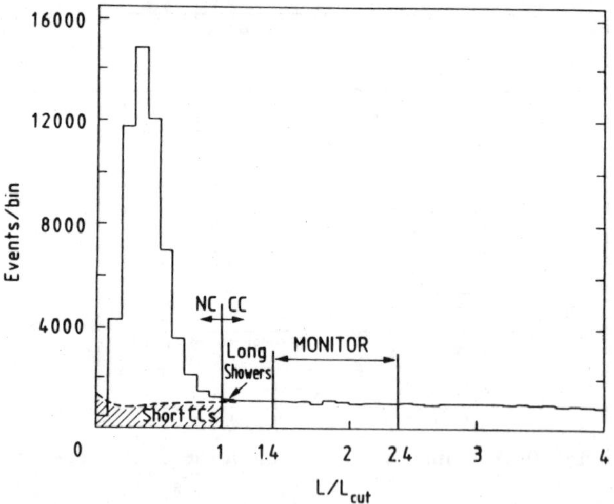

Figure 41: The distribution of the event length L, in units of the cut-off L_c for neutrino interactions with E_h > 10 GeV. The background from cosmic rays and WBB neutrinos has been subtracted [42].

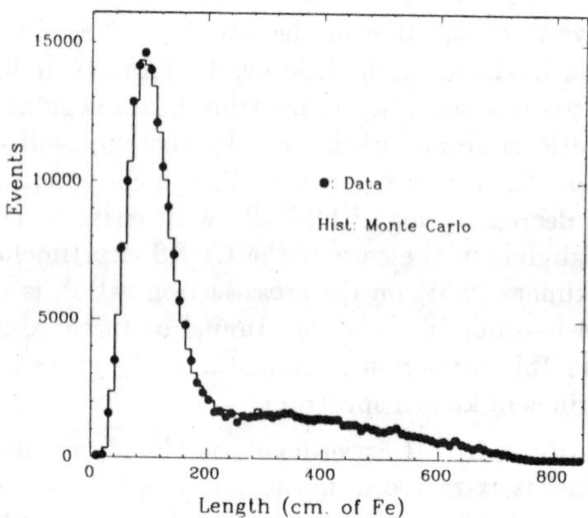

Figure 42: Event length for the data (solid histogram) and the simulation (plotted points) in the CCFR detector [44].

4.4 *The Contamination of the Neutral-Current Sample by Charged-Current Interactions*

The subtraction of short charged-current interactions in the iron detectors is based on the event length distribution (figures 41 and 42). The number of short charged-current events is extrapolated from the observed number of events in a reference region containing only charged-current interactions. In the CDHS experiment, this reference region is called "monitor region" and is defined by $1.4L_c < L < 2.4L_c$. In the CCFR experiment, the reference region consists of the "intermediate" events which have $L > 30$ counters and their last plane before the third last counter of the calorimeter.

The calculation of the short charged-current interactions is done using a simulation in the detector of only the final state muon of charged-current interactions. No attempt is made to simulate the hadron shower in the final state. Ratios of short charged-current to the number of charged-current in the monitor region are obtained from this simulation, one for $E_{sho} > E_h^{cut}$ GeV (to be applied for the subtraction of short charged-current from the NC sample) and one for $E_h > E_h^{cut}$ GeV (to be applied for the correction of the CC sample). These corrections are large (16% of the NC candidates are short charged-current in the CDHS experiment and 22% in the CCFR experiment). However, the uncertainty in these extrapolations is rather small. The various error sources on the extrapolation factors are collected in table 8 in the case of the CDHS detector.

Charged-current interactions can be short for two reasons : because the muon momentum is low (high-y events) and because the muon is emitted with a large angle and leaves the detector by the side. Neglecting the muon mass, the angle of emission of the muon with respect to the incoming neutrino direction θ_μ is given by :

$$\cos\theta_\mu = 1 - \frac{M}{E}\frac{xy}{1-y} \tag{123}$$

Lines of constant emission angle in the $x - y$ plane are shown in figure 43 for incident neutrino energies of 10 and 80 GeV. Events with large emission angles tend to have high y also. However, if the number of low momentum muons depend only on the y distribution, the number of events leaving by the side is sensitive to the x distribution as well.

In addition, the detection efficiency of muons near the edge of the detector is known poorly and the precise measurement of the exit point is difficult. Furthermore, the number of muons leaving by the side is sensitive to beam parameters such as the divergence. It is therefore desirable to reduce as much as possible the number of events leaving by the side of the detector.

In the CCFR detector, there is no magnetic field in the calorimeter and a rough estimate of the number of charged-current interactions leaving by the side can

Error Sources		ν	$\overline{\nu}$
Beam	NBB beam energy ±2 GeV	0.02	
	K/π ratio ± 5 %	0.08	
	Position of the beam center ± 1 cm	0.08	
	Beam divergence ± 0.02 mrad	0.10	
	Total error from the beam	0.15	0.2
Detector	Hadron energy calibration	0.10	0.4
	Hadron energy resolution	0.10	0.3
	Transverse vertex position	0.06	
	Muon energy loss	0.40	0.4
	Event length ± 1.5 cm	0.80	0.8
	Total error from the detector	1.00	1.0
Theory	Longitudinal Structure Function	0.60	4.0
	Fraction of antiquarks	0.00	1.5
	Total cross-section versus energy	0.16	
	Q^2 evolution	0.05	
	Mass of charm quark	0.02	0.8
	Radiative Corrections	0.10	
	Total error from the theory	0.60	4.4
	Monte-Carlo statistics	0.75	2.2
Total Error		1.40	5

Table 8: Summary of the systematic errors (in %) on the extrapolation factors, ratios of the number of short charged-current interactions divided by the number of charged-current interations in the monitor region (CDHS experiment).

be obtained using Eq. 123 as

$$1-\frac{1}{1+\frac{M}{E}\frac{\langle x\rangle}{1-\cos\theta_{\mu}}} \tag{124}$$

Taking an average x of 0.22 and an average exit angle of 10° gives a fraction of about 25% for an incident neutrino energy of 40 GeV. To keep the systematic error small, it is therefore important to restrict the fiducial volume to small radii.

In the case of the CDHS detector, the focusing magnetic field reduces strongly the fraction of muons leaving by the side. Only 7% of the short charged-current interactions are leaving the detector by the side, for a radius cut with respect to the center of the detector of 1.3 m. Under these conditions, the distributions of the event radius with respect to the beam axis (figure 44a) and of the shower energy (figure 44b) in the monitor region are well reproduced by the Monte-Carlo simulation.

Keeping under control the number of events leaving by the side, the extrapolation is dominated by the y-distribution of charged currents. This dependence

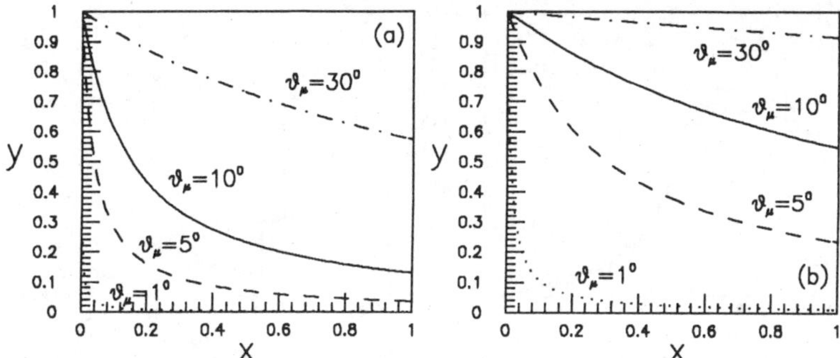

Figure 43: Lines of constant muon angle emission in the $x-y$ plane for a 10 GeV incident neutrino (a) and a 80 GeV incident neutrino (b).

however is weak because the events in the reference region already have high y. For example, in the CDHS experiments, the events in the monitor region have $\langle y \rangle = 0.85$ compared with $\langle y \rangle = 0.94$ in the NC signal region. This extrapolation is therefore largely insensitive to uncertainties about detector response and physics assumptions (table 8).

The main uncertainty comes from the measurement of the event length. The event length can be affected by backscattering at the vertex or noise and inefficiency at the end of the muon track. In the CDHS experiment, the position of the vertex was checked by varying the threshold used to define the beginning of the shower. The position of the last visible plane of the event could be cross-checked using the hits in the drift chambers and a total uncertainty of ±1.5 cm was assigned to the event length measurement. This error translates to a 1% error on the extrapolation factor and a 0.3% error on the cross-section ratio in the neutrino beam.

In the CCFR experiment, the position of the vertex could be checked in dimuon events where the vertex position measured in the calorimeter can be compared with the intersection point of the two reconstructed muon tracks, a very cute cross-check in fact.

The main theoretical error source comes from the longitudinal structure function which introduces a linear term in y in the cross-section (Eq. 54).

The total systematic uncertainty on the short charged-current subtraction amounts to 0.45% on the cross-section ratio in the CDHS experiment. In the CCFR experiment, this uncertainty is reduced to 0.23%.

In the CHARM experiment, charged-current are mistaken as neutral-current candidates if the final state muon has not been identified by the pattern recognition program. This program requires the muon momentum to be larger than 1 GeV, so

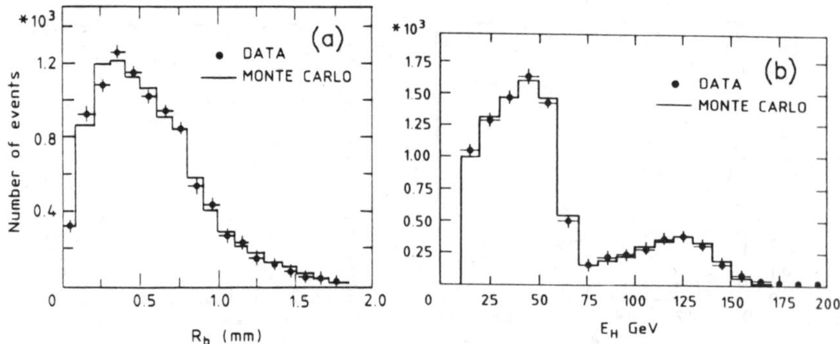

Figure 44: The distribution of the event radius with respect to the beam axis (a) and of the shower energy (b) for events in the monitor region (CDHS experiment) [42].

already muons with less than 1 GeV are lost. For a muon with a momentum larger than 1 GeV, losses can also occur if it leaves by the side of the detector or if it is hidden inside the hadronic shower. These two contributions are of equal importance and two methods were used to estimate these losses. In both methods, no attempt is made to simulate the hadronic shower.

In the first method, real charged-current events were used. The hits along the reconstruced muon track were removed and a new muon, generated according to the distributions of charged-current in the NBB, was added to the event. These semi-simulated events were then passed through the muon finding algorithm and the reconstruction efficiency was measured.

In the second method, the first visible point outside the hadron shower along the reconstructed muon track was measured as a function of the shower energy and the opening angle of the muon and the hadron shower. The inefficiency is then defined as the fraction of the time the distance between the muon end-point and the first visible point is smaller than the threshold used to define a visible muon track.

The results given by the two methods agree and the muon finding efficiency is shown in figure 45a as a function of muon momentum. It varies rapidly below 3 GeV. The total inefficiency for charged-current interactions in the CHARM detector is calculated to be 3.5%, resulting in a contamination of 9.5% of the NC candidate sample.

To compute the contamination of the NC sample, in order to cancel some of the systematic errors, the number of events predicted by the simulation was normalised

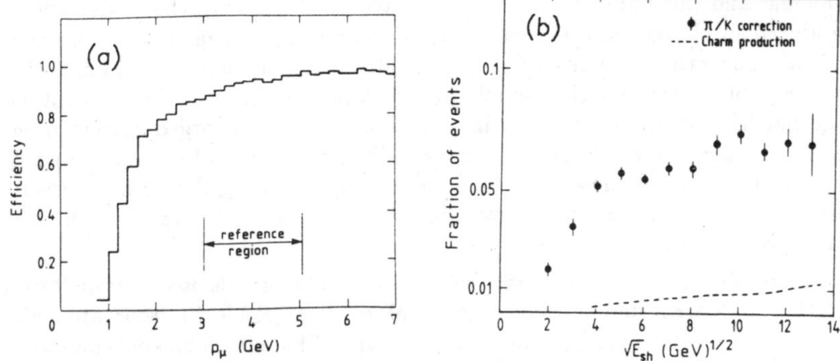

Figure 45: (a) The muon finding efficiency in the CHARM detector as a function of muon momentum. (b) The fraction of neutral-current interactions mistaken as charged-current interactions in the CHARM detector as a function of hadron energy [43].

to the number of events observed in a reference region shown in figure 45a. This correction was estimated to induce a systematic error of 0.32% on the cross-section ratio.

4.5 *The Contamination of the Charged-Current Sample by Neutral-Current Interactions*

In the case of the iron detectors, a NC can be classified as a CC if there is a fluctuation in the shower producing a shower length larger than L_c or if a hadron of the shower decays into a muon. These two contributions are small and only 0.3% of the NC events are classified as CC interactions in the CDHS detector.

In the CHARM detector however, a NC interaction will be identified as a CC candidate if a hadron from the shower is mistaken as a muon track or if a real muon is actually produced in the hadron shower. Because of the low threshold of the muon recognition (1 GeV), these two effects are significantly more important than in the iron detectors.

The fraction of events with a shower containing a reconstructed muon track is calculated using the data. Using a sample of well identified charged-current interactions, the fraction of the time a second muon was identified by the pattern recognition program was measured to be $6.02 \pm 0.08(\text{stat.}) \pm 0.04(\text{syst.})$ %. However, the hadronic showers in neutral-current events and in charged-current events are different and this number has to be corrected. In particular, charm particles can be produced in charged-current interactions, decaying into muons with a semileptonic branching ratio of about 10%. This contribution amounts to 0.54 ± 0.08 and has to be subtracted. Another small correction is the difference in the content of strange particles of hadron showers in neutral-current and charged-current interactions. The final correction is shown in figure 45b as a function of shower energy. The total fraction of neutral-current interactions classified as charged-current interactions amounts to 4.8%. This correction introduces a systematic uncertainty of 0.25% on the cross-section ratio.

From table 6, one can see that the experimental methods used in iron detectors like CDHS and CCFR and in the fine-grain calorimeter CHARM to separate neutral-current and charged-current interactions are quite different. The contamination of the neutral-current sample by charged-current interactions is smaller in the CHARM experiment but the contamination of the charged-current sample by neutral-current interactions is larger. The overall correction of the cross-section ratio is quite similar in the two techniques.

4.6 *The Correction for Electron-Neutrino Interactions*

Finally the NC candidate sample must be corrected for the contamination of neutral-current and charged-current electron-neutrino interactions. For charged-current electron-neutrino interactions, the measured shower energy corresponds to the total incident neutrino energy.

In the CERN NBB, electron-neutrinos are produced by the Ke3 decay of charged kaons. The contribution coming from the decay of muons produced in the beam is negligible. In the Fermilab QTB, the primary proton beam is pointing towards the CCFR detector to maximize the neutrino interaction rate and electron-neutrino can also be produced in the decay of neutral kaons $K_L^0 \to \pi^- e^+ \nu_e$. This source is important : it contributes roughly 17% of all electron-neutrino interactions in the CCFR detector.

In the NBB, the calculation of the electron-neutrino contamination relies on a simulation of the beam and the knowledge of the calorimeter performances with electrons. The calorimeters from CHARM and CDHS were calibrated with pion and electron beams of various energies and the detector response to electron was measured precisely. The calculation of the incident electron-neutrino spectrum in the detectors depends on the beam parameters and in particular on the beam composition. The ratio K/π in the hadron beam is measured by a differential Cherenkov counter with a systematic error of $\pm 3\%$. This uncertainty translates to a 0.2% uncertainty on the cross-section ratio.

However, the beam parameters and especially the beam composition can be constrained with a high precision with the data themselves, as demonstrated by the CDHS experiment. In the sample of muon-neutrino charged-current events where the

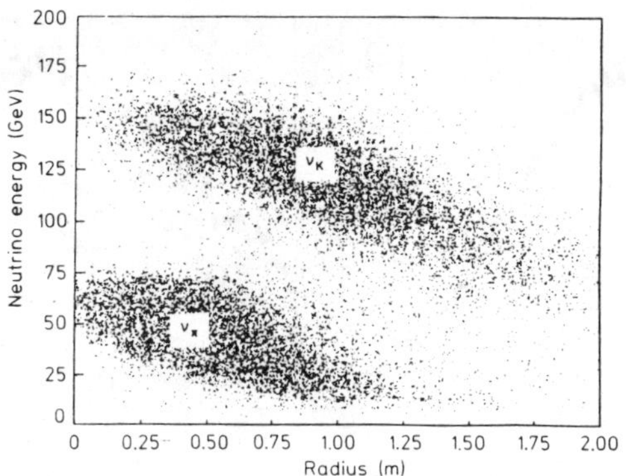

Figure 46: The reconstructed neutrino energy $E_h + p_\mu$ versus the event radius with respect to the beam axis in the CDHS detector [45].

momentum of the final state muon is reconstructed in the spectrometer, the energy of the incident neutrino is measured as $E_h + p_\mu$. In figure 46 where the neutrino energy is shown versus the event radius with respect to the beam axis, the dichromatic nature of the beam is very clearly visible and the pion and kaon bands clearly separated. Using this measurement, the uncertainty coming from the experimental measurement of the K/π ratio is largely reduced. Two other systematic uncertainties in the calculation of the neutrino-electron contamination then need to be considered. First there is an error coming from the 1% uncertainty on the Ke3 branching ratio. Then the simulation of the electron-neutrino charged-current interactions involve processes outside the deep-inelastic domain like quasi-elastic interactions whose cross-section is not well known. This introduces a theoretical error on the correction. The total systematic error on the cross-section ratio coming from the electron-neutrino correction is estimated to be 0.2% in the CDHS experiment.

The correction on the cross-section ratio is similar in the CDHS and CHARM experiments and amounts to about 6%. The contamination of the neutral-current sample by electron-neutrino charged-current interactions is increasing at high shower energies as the shower energy spectrum of the electron-neutrino charged-current is shifted to high energies (reflecting the ν_e spectrum from figure 30b) with an average shower energy of 80 GeV whereas the average shower energy of the NC sample is 49 GeV.

In the case of the CCFR experiment, the beam parameters can be determined accurately with the high-statistics charged-current data sample reconstructed in the spectrometer. However, the K_L^0 component can not be constrained with the data and must be calculated using experimental data on K_L^0 production.

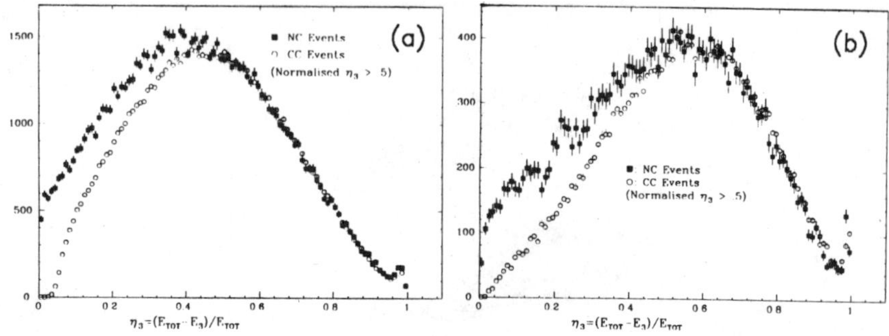

Figure 47: The distribution of the η_3 variable in the CCFR experiment for neutral-current and charged-current candidates with 50 GeV $< E_h <$ 100 GeV (a) and 150 GeV $< E_h <$ 200 GeV (b). For each figure, the neutral-current events are normalized to the number of charged-current observed with $\eta_3 > 0.5$. The excess of events at low η_3 in the neutral-current sample is a measurement of the fraction of electron-neutrino interactions [44].

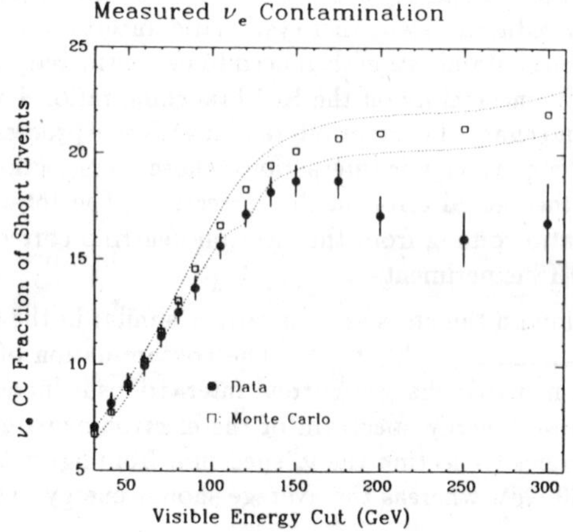

Figure 48: The fraction of short events which are attributed to charged-current electron-neutrino interactions in the CCFR detector. The points are from the η_3 determination, while the band represents the prediction of the beam simulation [44].

An independent check of the electron-neutrino flux can be obtained from the data themselves using the different longitudinal shower energy distribution of charged-current electron-neutrino interactions. The longitudinal shape of showers is characterized by a variable η_3 defined by :

$$\eta_3 = 1 - \frac{E_3}{E_{sho}} \tag{125}$$

where E_3 is the energy measured in the three most upstream counters of the shower. Neutral-current interactions have a broad distribution in η_3, reflecting the large fluctuations in hadronic showers. Electron-neutrino charged-current events, by contrast, have a η_3 distribution peaked towards small values since a substantial fraction of the energy is produced by an electron shower which is a concentrated energy deposition in iron. Measured η_3 distributions in the short event sample are shown in figure 47. The η_3 distribution of neutral-current interactions is obtained from the data using identified muon-neutrino charged-current interactions after removing the final state muon. The expected distribution from electron-neutrino interactions is obtained by folding the hadron and electron shower shapes obtained in test beams.

It is therefore possible to extract from the measured η_3 distribution the fraction of electron-neutrino interactions in the short event sample. This fraction is shown in figure 48 as a function of the hadron energy cut-off, comparing it with the fraction calculated from a simulation of the beam. The two independent calculations are consistent and a systematic uncertainty of 3.8% is assigned to the electron-neutrino flux. This error is dominated by the 20% uncertainty in the K^0 source of electron-neutrinos. This error is the dominating experimental systematic error on the determination of electroweak parameters in the CCFR experiment.

4.7 *Results and Summary of Experimental Errors*

The experimental errors on the measurements of the cross-section ratios are collected in table 7.

The experimental accuracies of the CDHS and CHARM experiments are comparable.

The dominating systematic uncertainties on the R_ν measurement by the CDHS experiment are the subtraction of the muon energy loss in the shower box and the subtraction of the short charged-current contribution in the neutral-current sample. In the CHARM experiment, the subtraction of the WBB background, the short charged-current contamination of the neutral-current sample and the subtraction of electron-neutrino interactions each introduce a 0.3% systematic error on the measurement of R_ν.

Most of the experimental errors of the two experiments are independent. In particular, the subtraction of the electron-neutrino interactions is done using independent methods in the two experiments. However, it should be noted that, although

Experiment	R_ν	$R_{\bar{\nu}}$	r
CDHS	0.3072±0.0033	0.382±0.016	0.393±0.014
CDHS corrected for non-isoscalarity	0.3135	0.376	0.409
CDHS corrected for non-isoscalarity and extrapolated to E_h >4 GeV	0.3167	0.371	0.453
CHARM E_h >4 GeV	0.3093±0.0031	0.390±0.014	0.456±0.011
CHARM E_h >9 GeV	0.3052±0.0033	0.397±0.015	0.429±0.010

Table 9: Comparison of the experimental results from the CDHS and CHARM experiments.

the experimental methods are quite different in the CDHS and the CHARM experiments, the subtraction of short charged-current in the neutral-current sample in both cases relies on some assumptions on the high-y behaviour of the charged-current cross-section and this correction is in principle not done completely independently in the two experiments.

The methods for subtracting the WBB background are the same in the two experiments but this is only a small contribution to the error, in both cases. In conclusion, it can be assumed that the errors of the two experiments are in a first approximation independent and the results of the experiments can be averaged accordingly.

Some systematic errors on the cross-section ratio $R_{\bar{\nu}}$ are slightly larger than the corresponding errors on R_ν. For example, the contribution of cosmic events and WBB background are larger in the antineutrino beam, which is significantly more difficult to produce than the neutrino beam. The contamination of the neutral-current sample by short charged-current is smaller for antineutrinos than for neutrinos because of the y-distribution of the cross-section. However, this subtraction is more sensitive to the uncertainty coming from the longitudinal structure function and the systematic error is increased compared with the neutrino case. The error on the cross-section ratio $R_{\bar{\nu}}$ anyway is largely dominated by the statistical error in both experiments. About 20% of the data only were taken with the antineutrino beam.

The results from the CERN experiments are compared in table 9. The table also shows the result for the antineutrino to neutrino charged-current cross-section ratio r. To obtain this number, the antineutrino data have to be corrected for the different energy spectrum. The error on the ratio r is dominated by the statistical error on the antineutrino sample.

In table 9, the results from the CDHS experiments are corrected for the difference in the target between CDHS (iron) and CHARM (marble), and for the different hadron energy cut-off (a rough estimate can be obtained from Eq. 13). After corrections, the two experiments are in acceptable agreement.

	CDHS	CHARM	CCFR Preliminary
Data Statistics	0.0041	0.0040	0.0024
Monte-Carlo Statistics			0.0016
Hadron energy measurement	0.0020	0.0006	0.0009
Cosmic background	0.0005	0.0000	0.0000
Beam spectrum	0.0005	0.0015	0.0010
CC misidentified as NC	0.0023	0.0016	0.0014
NC misidentified as CC	0.0010	0.0010	
ν_e correction	0.0010	0.0015	0.0027
Total Systematic error	0.0034	0.0030	0.0033
Total Experimental error	0.0053	0.0050	0.0044

Table 10: Summary of experimental errors on the value of $\sin^2\theta_W$.

The experimental errors of the CCFR experiment can better be compared with the errors of the CERN experiments in terms of the corresponding error on the extracted value of $\sin^2\theta_W$. Owing to the high intensity of the QTB, the statistical uncertainty of the CCFR experiment is roughly two times smaller than in the CERN experiments. The higher energy of the beam and the larger hadron energy cut-off allow also a reduction of the error in the subtraction of short charged-current in the neutral-current sample. The dominating uncertainty of the CCFR experiment is the subtraction of the electron-neutrino background, especially because of the poorly known contribution of electron-neutrinos coming from K_L^0 decay. Considering the experimental errors of the CCFR experiment, possibilities for improvements do appear. The real question now is whether it is actually interesting to push the limits of the experimental accuracy further in a future experiment. This depends on the limitations in the interpretation and this is the topic of the next section.

5 The Interpretation of the Deep Inelastic Neutrino Scattering Experiments

The interpretation of the experimental results relies on corrections obtained using the QPM. These corrections will first be studied in the framework of the Standard Electroweak Theory and the value of the electroweak mixing parameter $\sin^2\theta_W$ will be extracted. In the case of the CDHS and CHARM experiments, the determination of $\sin^2\theta_W$ will be done using the Llewellyn-Smith equations (Eq. 120 in section 3.7). Then, some extensions of the Standard Electroweak Theory will be considered.

5.1 *The Quark-Parton Model Corrections*

Under the conditions of validity of the Llewellyn-Smith equations (Eq. 120), one has, for $E_h > E_h^{cut}$:

$$R_\nu^0 = \frac{\int \mathrm{d}E\,\Phi_\nu(E)\hat{\sigma}^{NC}(\nu N)}{\int \mathrm{d}E\,\Phi_\nu(E)\hat{\sigma}^{CC}(\nu N)} = \frac{1}{2} - \sin^2\theta_W + \frac{5}{9}\sin^4\theta_W\left(1 + r^0\right)$$

$$R_{\bar\nu}^0 = \frac{\int \mathrm{d}E\,\Phi_{\bar\nu}(E)\hat{\sigma}^{NC}(\bar\nu N)}{\int \mathrm{d}E\,\Phi_{\bar\nu}(E)\hat{\sigma}^{CC}(\bar\nu N)} = \frac{1}{2} - \sin^2\theta_W + \frac{5}{9}\sin^4\theta_W\left(1 + \frac{1}{\bar r^0}\right) \quad (126)$$

$$\text{where : } \quad r^0 = \frac{\int \mathrm{d}E\,\Phi_\nu(E)\hat{\sigma}^{CC}(\bar\nu N)}{\int \mathrm{d}E\,\Phi_\nu(E)\hat{\sigma}^{CC}(\nu N)} \quad \text{and} \quad \bar r^0 = \frac{\int \mathrm{d}E\,\Phi_{\bar\nu}(E)\hat{\sigma}^{CC}(\bar\nu N)}{\int \mathrm{d}E\,\Phi_{\bar\nu}(E)\hat{\sigma}^{CC}(\nu N)} \quad (127)$$

However, the conditions of validity of Eq. 120 (exact isospin symmetry with only u and d quarks at tree level and no mass effects) do not correspond to the real world and the quantities R_ν^0, $R_{\bar\nu}^0$, r^0 and $\bar r^0$ have to be extracted from the experimental numbers R_ν, $R_{\bar\nu}$ and r using the QPM discussed in section 3.4. More precisely, the following correction factors are calculated in the QPM :

$$f_R = \left(\frac{R_\nu^0}{R_\nu}\right)_{\rm QPM} \qquad \bar f_R = \left(\frac{R_{\bar\nu}^0}{R_{\bar\nu}}\right)_{\rm QPM} \qquad f_r = \left(\frac{r^0}{r}\right)_{\rm QPM} \qquad \bar f_r = \left(\frac{\bar r^0}{r}\right)_{\rm QPM} \quad (128)$$

The corrected ratios are then obtained by multiplying the experimental number by the correction obtained in the QPM. In addition, one can consider predictions of the experimental ratios in the framework of the Llewellyn-Smith equations, namely :

$$R_\nu^{LS}(\sin^2\theta_W, {\rm QPM}) = \frac{1}{f_R}\left[\frac{1}{2} - \sin^2\theta_W + \frac{5}{9}\sin^4\theta_W\,(1 + f_r r)\right]$$

$$R_{\bar\nu}^{LS}(\sin^2\theta_W, {\rm QPM}) = \frac{1}{\bar f_R}\left[\frac{1}{2} - \sin^2\theta_W + \frac{5}{9}\sin^4\theta_W\left(1 + \frac{1}{\bar f_r r}\right)\right] \quad (129)$$

The QPM is built according to the recipes given in section 3.4, using the quark distributions obtained from the measurement of structure functions in the CDHS detector [41]. The parameters of the quark CKM mixing matrix have been taken from [46][13]. A charm mass value of 1.31 GeV, as measured by the CCFR collaboration, has been used [47]. The one-loop electroweak radiative corrections have been calculated using a program from Bardin [49] with m_t =100 GeV and m_H =100 GeV.

The results are given in table 11 for the CDHS and CHARM experiments. For the CDHS experiment, the result $R_{\bar\nu} = 0.363 \pm 0.015$(exp.) obtained in the previous 200 GeV NBB (table 3) was quite precise. In table 11, it has been corrected for the conditions of the 1984 experiment and averaged with the 1984 result to yield a combined value of 0.374 ± 0.011(exp.).

[13] $|U_{ud}|^2 = 0.9512 \pm 0.0012$, $|U_{ud}|^2 = 0.0488 \pm 0.0013$, $|U_{ud}|^2 = 0.0488 \pm 0.0013$ and $|U_{ud}|^2 = 0.9493 \pm 0.0016$.

		Experimental Results	Correction Factors	Corrected Ratios
CDHS	R_ν	0.3072 ± 0.0033	1.027	$0.3155 \pm 0.0034 - 0.008\,(m_c - 1.31)$
	$R_{\overline{\nu}}$	0.374 ± 0.011	0.996	$0.372 \pm 0.011 - 0.017\,(m_c - 1.31)$
	r	0.393 ± 0.014	0.982	$0.386 \pm 0.014 + 0.002\,(m_c - 1.31)$
	$\overline{r}$		0.947	$0.372 \pm 0.013 + 0.002\,(m_c - 1.31)$
CHARM	R_ν	0.3093 ± 0.0031	1.000	$0.3094 \pm 0.0031 - 0.008\,(m_c - 1.31)$
	$R_{\overline{\nu}}$	0.390 ± 0.014	1.000	$0.390 \pm 0.014 - 0.016\,(m_c - 1.31)$
	r	0.456 ± 0.011	0.956	$0.436 \pm 0.011 + 0.001\,(m_c - 1.31)$
	$\overline{r}$		0.939	$0.428 \pm 0.011 + 0.001\,(m_c - 1.31)$

Table 11: Correction factors from the QPM and corrected cross-section ratios for the CDHS and CHARM results. The corrections have been calculated for $\sin^2\theta_W = 0.225$, $m_c = 1.31$ GeV, $m_t = 100$ GeV and $m_H = 100$ GeV.

The various corrections and the related uncertainties are given separately in table 12 for the case of the CDHS experiment. For each correction, this table shows the size of the correction and its uncertainty, separately for $R_\nu^{\rm QPM}$, $R_{\overline{\nu}}^{\rm QPM}$, R_ν^{LS}, $R_{\overline{\nu}}^{LS}$ and $\sin^2\theta_W^{LS}$, which is the value of $\sin^2\theta_W$ obtained by comparing the observed value of R_ν with $R_\nu^{LS}(\sin^2\theta_W)$. The corrections and their uncertainties are estimated when all the other parameters of the QPM are at their default value.

The correlations between R_ν and $R_{\overline{\nu}}$ are displayed in figure 49, when the QPM is used alone (figure 49a) and when it is used together with the Llewellyn-Smith equations (figure 49b).

Each correction will now be considered and commented in turn. The CDHS experiment will be taken as an example unless it is specified otherwise. The corrections are calculated using the full QPM expressions of the structure functions (Eq. 90). However, in order to get a feeling for the corrections, it is useful to have simplified formulae at hand.

Neglecting the longitudinal structure function, the charm mass threshold effect and using a two-dimensional quark mixing matrix, the total cross-sections are given by the following relations where a common factor $G_\mu^2 ME/\pi$ has been omitted :

$$\begin{aligned}
\sigma^{CC}(\nu N) &= U + D + 2S + \overline{\not{U}} + \overline{\not{D}} + 2\,\overline{\not{C}} \\
\sigma^{CC}(\nu n - p) &= U - D + \overline{\not{D}} - \overline{\not{U}} \\
\sigma^{CC}(\overline{\nu} N) &= \overline{U} + \overline{D} + 2\overline{S} + \not{U} + \not{D} + 2\,\not{C} \\
\sigma^{CC}(\overline{\nu} n - p) &= \overline{U} - \overline{D} + \not{D} - \not{U} \\
\sigma^{NC}(\nu N) &= \alpha_L\left(U + D + \overline{\not{U}} + \overline{\not{D}}\right) + \alpha_R\left(\overline{U} + \overline{D} + \not{U} + \not{D}\right) + 2\epsilon_L(u)^2\left(C + \overline{\not{C}}\right) \\
&\quad + 2\epsilon_L(d)^2\left(S + \overline{\not{S}}\right) + 2\epsilon_R(u)^2\left(\overline{C} + \not{C}\right) + 2\epsilon_R(d)^2\left(\overline{S} + \not{S}\right)
\end{aligned}$$

Corrections	$\frac{\Delta R_\nu}{R_\nu}$ (%)		$\frac{\Delta R_{\bar\nu}}{R_{\bar\nu}}$ (%)		$\Delta \sin^2\theta_W^{LS}$
	QPM	LS	QPM	LS	
Muon mass	+0.24	+0.25	+0.43	+0.38	+0.0012
W and Z propagator	+0.14	+0.13	+0.08	+0.08	+0.0006
Unitary CKM matrix $\lvert U_{ud}\rvert^2 = 0.9512 \pm 0.0012$ [46]	+0.47 ±0.00	+0.46 ±0.00	+0.27 ±0.00	+0.31 ±0.00	+0.0022 ±0.0000
Non-isoscalarity $D_V/U_V = 0.39 \pm 0.04$	−2.06 ±0.20	−1.92 ±0.18	+1.57 ±0.19	+0.76 ±0.09	−0.0091 ±0.0009
Longitudinal structure function QCD prediction ±50 %	+0.37 ±0.19	+0.12 ±0.07	−1.36 ±0.57	−0.13 ±0.07	+0.0006 ±0.0003
Quark sea $(\overline{U}+\overline{D})/(U+D) = 0.13 \pm 0.02$	+1.43 ±0.26	+0.05 ±0.09	−8.30 ±0.93	−0.59 ±0.08	+0.0024 ±0.0004
QCD evolution $\Lambda_{QCD} = 200 \pm 100$ MeV	- ±0.14	- ±0.07	- ±0.14	- ±0.03	- ±0.0003
Strange sea $\overline{S}/\overline{D} = 0.38 \pm 0.05$ [47]	+0.65 ±0.11	−0.07 ±0.08	+0.46 ±0.02	−0.96 ±0.16	+0.0023 ±0.0004
Charm sea $C/S = 0.2 \pm 0.2$	+0.08 ±0.10	+0.07 ±0.08	+0.05 ±0.05	+0.10 ±0.12	+0.0003 ±0.0004
Strange sea asymmetry $S/\overline{S} = 1 \pm 0.1$	- ±0.05	- ±0.04	- ±0.15	- ±0.06	- ±0.0002
Charmed sea asymmetry $C/\overline{C} = 1 \pm 0.3$	- ±0.06	- ±0.06	- ±0.12	- ±0.09	- ±0.0003
Isospin breaking in the sea $\overline{U}/\overline{D} = 1 \pm 1$	- ±0.04	- ±0.06	- ±0.83	- ±0.28	- ±0.0003
Error from $r = 0.393 \pm 0.014$	-	±0.13	-	±0.74	±0.0006
Radiative corrections m_t =100 GeV and m_H =100 GeV	−2.22 ±0.44	−2.23 ±0.44	−2.15 ±0.43	−2.13 ±0.43	−0.0106 ±0.0020
Total Theoretical Error (fixed m_c)	±0.64	±0.54	±1.52	±0.95	±0.0025
Charm mass $m_c = 1.31 \pm 0.24$	+1.71 ±0.63	+1.58 ±0.61	+2.02 ±1.00	+2.79 ±1.08	+0.0077 ±0.0029
Total Theoretical Error	±0.9	±0.8	±1.8	±1.4	±0.0038

Table 12: The corrections and uncertainties on the values of R_ν and $R_{\bar\nu}$ predicted by the QPM and the QPM together with the Llewellyn-Smith equations. The last column refers to the value of $\sin^2\theta_W$ extracted from the corrected R_ν and r using the Llewellyn-Smith equations. The first number gives the size of the correction with all other parameters at their default value. The number below corresponds to the uncertainty produced by the indicated change of parameter.

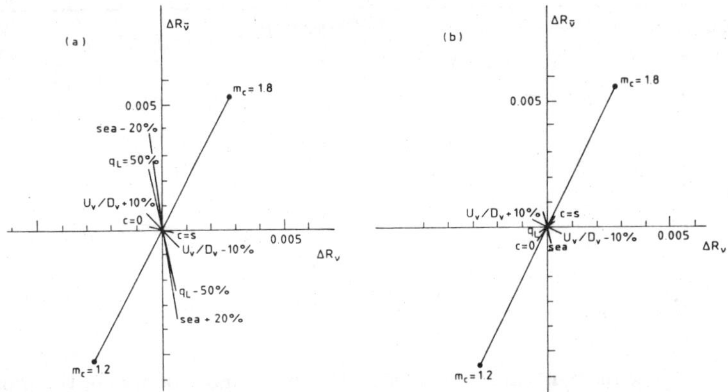

Figure 49: The correlation between the theoretical errors on R_ν and $R_{\bar\nu}$ if the QPM is used alone (a) and if the QPM is used together with the Llewellyn-Smith equations (b) [42].

$$\begin{aligned}
\sigma^{NC}(\nu n - p) &= \beta_L \left(D - U + \overline{\not{D}} - \overline{\not{U}}\right) + \beta_R \left(\overline{D} - \overline{U} + \not{D} - \not{U}\right) \\
\sigma^{NC}(\bar\nu N) &= \alpha_R \left(U + D + \overline{\not{U}} + \overline{\not{D}}\right) + \alpha_L \left(\overline{U} + \overline{D} + \not{U} + \not{D}\right) + 2\epsilon_R(u)^2 \left(C + \overline{\not{C}}\right) \\
&\quad + 2\epsilon_R(d)^2 \left(S + \overline{\not{S}}\right) + 2\epsilon_L(u)^2 \left(\overline{C} + \not{C}\right) + 2\epsilon_L(d)^2 \left(\overline{S} + \not{S}\right) \\
\sigma^{NC}(\bar\nu n - p) &= \beta_R \left(D - U + \overline{\not{D}} - \overline{\not{U}}\right) + \beta_L \left(\overline{D} - \overline{U} + \not{D} - \not{U}\right)
\end{aligned} \tag{130}$$

where A and $\not{A}$ refer to integrals of the quark distribution function $a(x, Q^2)$:

$$A = \int \mathrm{d}x\mathrm{d}y\, xa(x, Q^2) \qquad \not{A} = \int \mathrm{d}x\mathrm{d}y\, xa(x, Q^2)(1-y)^2 \tag{131}$$

$$\begin{aligned}
\text{and} \,:\, \alpha_L &= \epsilon_L(u)^2 + \epsilon_L(d)^2 \qquad \beta_L = \epsilon_L(u)^2 - \epsilon_L(d)^2 \\
\alpha_R &= \epsilon_R(u)^2 + \epsilon_R(d)^2 \qquad \beta_R = \epsilon_R(u)^2 - \epsilon_R(d)^2
\end{aligned} \tag{132}$$

The integrals U, D, S and $\overline{S}$ are shown in figure 50 as a function of the neutrino energy. In simple calculations, one can use : $U = 0.22$, $D = 0.09$, $S = \overline{S} = 0.007$, $\overline{U} = \overline{D} = 0.018$, $C = \overline{C} = 0.2S$ and $\not{A} = A/3$. For a 10 GeV hadron energy cut-off, the effective momentum fractions are reduced by about 20 % and $\not{A} \approx A/5$. One should also keep in mind the following values of the neutral-current couplings :

$$\begin{aligned}
\epsilon_L(u)^2 = 0.12 \quad \epsilon_L(d)^2 = 0.18 \quad \epsilon_R(u)^2 = 0.024 \quad \epsilon_R(d)^2 = 0.006 \\
\alpha_L = 0.3 \quad \beta_L = -0.06 \quad \alpha_R = 0.03 \quad \beta_R = 0.018
\end{aligned} \tag{133}$$

and the following approximate values of the neutrino-nucleon cross-sections (in units of $G_\mu^2 ME/\pi$) :

$$\tilde\sigma^{CC}(\nu N) = 0.34 \quad \tilde\sigma^{CC}(\bar\nu N) = 0.15 \quad \tilde\sigma^{NC}(\nu N) = 0.10 \quad \tilde\sigma^{NC}(\bar\nu N) = 0.055 \tag{134}$$

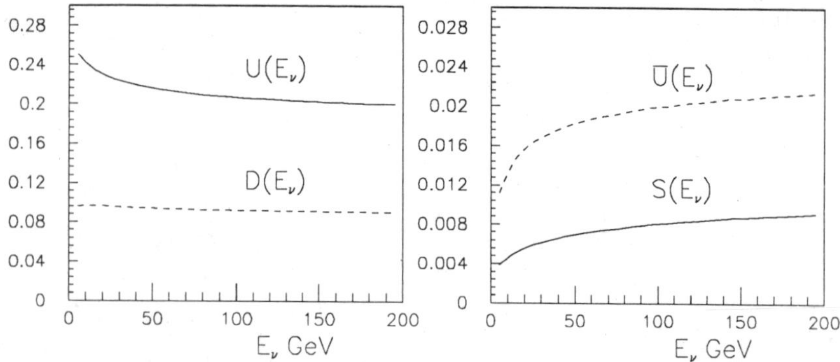

Figure 50: The momentum fraction carried by u,d,s and ū quarks in the proton as a function of the incident neutrino energy [41].

The muon mass introduces a small reduction of the charged-current cross-section : -0.24% for neutrinos and -0.43% for antineutrinos. The correction on $\sin^2\theta_W^{LS}$ is $+0.0012$, of which $+0.0005$ is due to the muon mass terms in the charged-current cross-section (Eq. 54) and $+0.0008$ to the kinematical limit described in section 3.2 and figure 21.

Another trivial correction of the cross-sections is the propagator effect due to the massive intermediate boson masses. The shift of the cross-section ratios is given by :

$$\frac{\Delta R_\nu}{R_\nu} = \frac{\Delta R_{\bar{\nu}}}{R_{\bar{\nu}}} \approx 2\sin^2\theta_W \frac{\langle Q^2 \rangle}{m_W^2} \tag{135}$$

The average Q^2 in neutrino interactions is 21 GeV^2 and 11 GeV^2 in antineutrino interactions. R_ν is therefore increased by 0.14% and $R_{\bar{\nu}}$ by 0.08%, and $\sin^2\theta_W^{LS}$ by 0.0006.

The charged-current cross-sections are largely insensitive to the value of the parameters of the CKM matrix which is assumed to be a 3×3 unitary matrix. This is because the mixing matrix of the first two generations is close to the 2×2 Cabibbo unitary matrix. In that limit, for a zero mass of the charm quark, the charged-current cross-sections are exactly independent of the mixing angles. The cross-sections are sensitive to the mixing angles only because the charm mass is non-zero and the udsc mixing matrix not exactly unitary.

A large correction however, present only in the case of the CDHS and CCFR experiments, is the correction for the non-isoscalarity of the target. An order of magnitude of this correction can be obtained from Eq. 130 :

$$\begin{aligned}
\frac{\Delta\sigma_\nu^{CC}}{\sigma_\nu^{CC}} &\approx \eta\frac{U-D}{\tilde{\sigma}_\nu^{CC}} \approx 0.38\eta \\
\frac{\Delta\sigma_\nu^{NC}}{\sigma_\nu^{NC}} &\approx -\eta\left(\beta_L+\frac{1}{3}\beta_R\right)\frac{U-D}{\tilde{\sigma}_\nu^{NC}} \approx 0.07\eta \\
\frac{\Delta\sigma_{\bar\nu}^{CC}}{\sigma_{\bar\nu}^{CC}} &\approx -\eta\frac{1}{3}\frac{U-D}{\tilde{\sigma}_{\bar\nu}^{CC}} \approx -0.29\eta \\
\frac{\Delta\sigma_{\bar\nu}^{NC}}{\sigma_{\bar\nu}^{NC}} &\approx -\eta\left(\beta_R+\frac{1}{3}\beta_L\right)\frac{U-D}{\tilde{\sigma}_{\bar\nu}^{NC}} \approx 0.005\eta \qquad (136)
\end{aligned}$$

The correction is positive except for the antineutrino charged-current cross-section, and much larger for the charged-current than for the neutral-current cross-sections. For iron, $\eta \approx 0.07$ and the correction is -2.2% for R_ν and $+2.1\%$ for $R_{\bar\nu}$[14]. The correction for $\sin^2\theta_W^{LS}$ amounts to -0.0090, a large shift compared with the experimental error of 0.005. The uncertainty on this correction comes from the rather poor knowledge of $U-D$. An experimental determination of the ratio $(U_V - D_V)/(U_V + D_V)$ is obtained by comparing the structure function $F_2^{\gamma n}$ with $F_2^{\gamma p}$ or $F_2^{\nu n}$ with $F_2^{\nu p}$. Experiments off free protons (liquid hydrogen) suffer from low statistics and the ratio $(U_V - D_V)/(U_V + D_V)$ is known to a precision of $\pm 10\%$ only. This translates into a ± 0.0009 error on $\sin^2\theta_W^{LS}$. However, recently, high statistics data from the New Muon Collaboration (NMC) became available and the experimental precision can be reduced significantly in the future [50].

The cross-section ratios are sensitive to the longitudinal structure function and the amount of sea quarks (table 12), especially $R_{\bar\nu}$. The dependence on the longitudinal structure function is given by :

$$\begin{aligned}
&\frac{\Delta\sigma_\nu^{CC}}{\sigma_\nu^{CC}} \approx \frac{R_L}{2}\frac{Q+\overline{Q}}{\tilde{\sigma}_\nu^{CC}} \approx 0.51R_L \qquad \frac{\Delta\sigma_\nu^{NC}}{\sigma_\nu^{NC}} \approx \frac{R_L}{2}\frac{(Q+\overline{Q})(\alpha_L+\alpha_R)}{\tilde{\sigma}_\nu^{NC}} \approx 0.57R_L \\
&\frac{\Delta\sigma_{\bar\nu}^{CC}}{\sigma_{\bar\nu}^{CC}} \approx \frac{R_L}{2}\frac{Q+\overline{Q}}{\tilde{\sigma}_{\bar\nu}^{CC}} \approx 1.16R_L \qquad \frac{\Delta\sigma_{\bar\nu}^{NC}}{\sigma_{\bar\nu}^{NC}} \approx \frac{R_L}{2}\frac{(Q+\overline{Q})(\alpha_L+\alpha_R)}{\tilde{\sigma}_{\bar\nu}^{NC}} \approx 1.04R_L \\
&\text{making :} \quad \frac{\Delta r}{r} \approx 0.65R_L \qquad \frac{\Delta R_\nu}{R_\nu} \approx 0.06R_L \qquad \frac{\Delta R_{\bar\nu}}{R_{\bar\nu}} \approx -0.12R_L \qquad (137)
\end{aligned}$$

with $R_L \approx 0.10$. This sensitivity is reduced to a negligible level if the Llewellyn-Smith equations are used because the large variation of r compensates the small change in R_ν and $R_{\bar\nu}$.

Large corrections however remain even in the formalism of Llewellyn-Smith equations : the amount of strange sea and the mass of the charm quark. As these two effects are correlated, they will be considered together and, because of their

[14] In the CHARM experiments, the experimental results are already corrected for a small non-isoscalarity corresponding to $\eta = -0.005$.

importance, in some detail. The uncertainty on the value of the charm mass is indeed the dominating theoretical error in the determination of electroweak parameters.

Charm quarks can be produced in charged-current neutrino scattering off d and s quarks ($\bar{c}$ by antineutrinos off $\bar{d}$ and $\bar{s}$ quarks). They are observed in the final state through their semileptonic decay into a muon, producing opposite-sign dimuon events in the detector, spectacular events that were first observed in 1975 in the HPWF detector.

In such events, the leading muon comes from the neutrino charged-current interaction vertex, while the second muon comes from the semileptonic decay of the charmed hadron. The following kinematical variables are defined :

$$\begin{aligned} E_{vis} &= E_{\mu_1} + E_{\mu_2} + E_h \\ x_{vis} &= Q^2_{vis}/2M(E_h + E_{\mu_2}) \quad \text{with :} \quad Q^2_{vis} = 4E_{vis}E_{\mu_1}\sin^2(\theta_{\mu_1}/2) \end{aligned} \tag{138}$$

In the framework of the slow rescaling, the production of the charmed meson and its subsequent semileptonic decay is described by (omitting a factor $G^2_\mu M/\pi$) :

$$\begin{aligned} \frac{d^3\sigma^{-+}}{dxdydz}(\nu N) &= E\left[\xi\left(u(\xi) + \xi d(\xi)\right)|U_{cd}|^2 + 2\xi s(\xi)|U_{cs}|^2\right]\left[1 - \frac{m_c^2}{2ME\xi}\right] D(z)B_c \\ \frac{d^3\sigma^{+-}}{dxdydz}(\bar{\nu} N) &= E\left[\xi\left(\bar{u}(\xi) + \xi\bar{d}(\xi)\right)|U_{cd}|^2 + 2\xi\bar{s}(\xi)|U_{cs}|^2\right]\left[1 - \frac{m_c^2}{2ME\xi}\right] D(z)B_c \end{aligned} \tag{139}$$

where z is the ratio of the meson momentum to the charm quark momentum, $D(z)$ is the fragmentation function of the charm-quark and B_c is the average semileptonic branching ratio of the charm mesons in the final state.

The fragmentation function of the charm quark can be modelled using the Peterson function :

$$D(z) \propto \frac{1}{z\left(1 - \frac{1}{z} - \frac{\epsilon}{1-z}\right)} \tag{140}$$

with the value $\epsilon = 0.20 \pm 0.04$. Taking the electroweak mixing angles from the unitarity constraint, the dimuon production is determined by the mass of the charm quark, the strange sea and the semileptonic branching ratio B_c. By combining the samples obtained in E744 and E770, the CCFR collaboration recently obtained a precise determination of the charm mass and the strange sea [47].

The strange sea is supposed to have the form $xs(x) \propto (1-x)^\beta$ with a normalisation given by $\kappa = 2S/(\overline{U} + \overline{D})$. The parameters m_c, β, κ and B_c are obtained from a χ^2 minimization between the data and the simulated samples binned in five E_{vis} and ten x_{vis} bins. The results are :

$$\begin{aligned} m_c &= 1.31 \pm 0.24 \qquad \kappa = 0.373 \pm 0.05 \\ \beta &= 9.45 \pm 0.70 \qquad B_c = 0.105 \pm 0.009 \end{aligned} \tag{141}$$

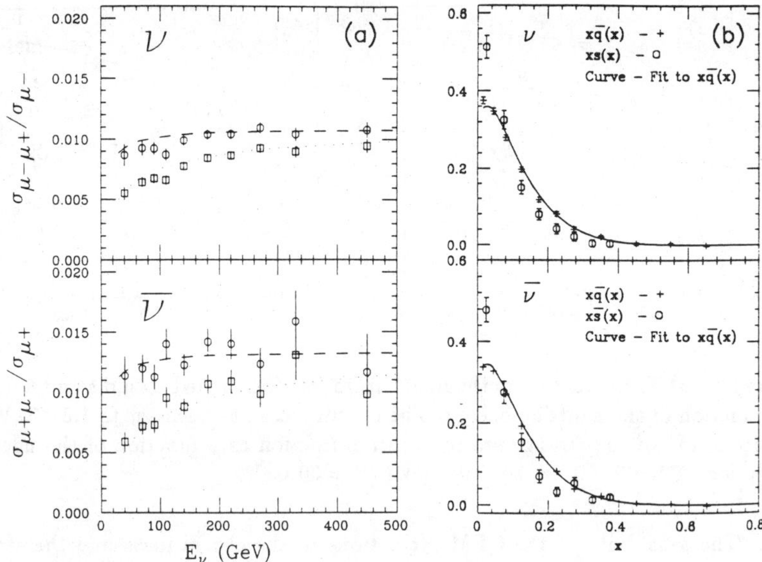

Figure 51: (a) The opposite-sign dimuon rates versus neutrino energy observed in the CCFR experiment. The rates corrected for acceptance, smearing and kinematical cuts are indicated by squares. The rates corrected for slow rescaling with $m_c = 1.31$ GeV are given by circles. The curves indicate the slow rescaling model prediction with $m_c = 1.31$ GeV (dotted) and $m_c = 0$ (dashed) [47]. (b) Comparison of the strange and the antistrange distribution functions with the sea quark distribution function as a function of x [48].

The errors include the systematic uncertainties but are still dominated by statistical errors. The χ^2 of 42.5 for 46 degree of freedom reveals that the data are very well reproduced by the simple slow rescaling model. This can be seen from figure 51 showing the opposite-sign dimuon cross-section as a function of the neutrino energy. In addition of supporting the slow rescaling model, these data give a value for the charm mass 1.31 ± 0.24 GeV which is more precise and slightly lower than the previous estimate (1.5 ± 0.3 GeV) used in [42, 43][15]. The momentum fraction carried by strange quarks is also smaller than the previous determinations ($\kappa \approx 0.45$). The data also show evidence for a softer distribution of the strange quarks in the nucleon compared with sea quarks (figure 51b).

[15]Most of the experiments presented in table 3 actually used this value of 1.5 GeV, except the CDHS collaboration for the 1979/1981 data analysis where a QCD running mass was used with an average of 1.1 GeV.

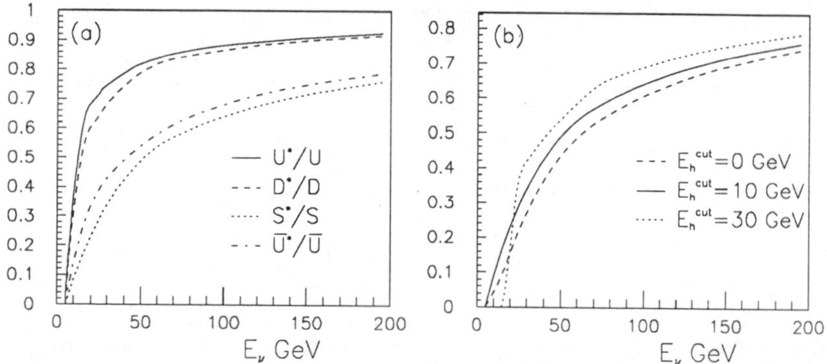

Figure 52: (a) The reduction of the momentum fraction carried by u,d,s and ū quarks as a function of the neutrino energy. The charm mass has been set to 1.31 GeV. (b) The reduction of the strange sea momentum fraction as a function of the neutrino energy for $E_h^{cut} = 0$, $E_h^{cut} = 10$ GeV and $E_h^{cut} = 30$ GeV.

The sensitivity of the QPM predictions to the charm mass and the strange sea is shown in table 12. In evaluating the effect of changing the charm mass, it is important to also include the fact that the strange sea depends on the charm mass, as this is a combination of both which is fixed by the dimuon production. The same is true for the $\overline{u}$ and $\overline{d}$ distributions in antineutrino dimuon production.

Here, the following correlations have been used :

$$\begin{aligned} \kappa(m_c) &= 0.373\,(1 + 0.37(m_c - 1.31)) \\ \overline{U}(m_c) &= \overline{U}(1.31)\,(1 + 0.08(m_c - 1.31)) \end{aligned} \tag{142}$$

where m_c is given in GeV. Therefore, both the neutral-current and the charged-current depend on the charm mass. In the case of the CDHS experiment :

$$\begin{aligned} \frac{\Delta N_\nu^{CC}}{N_\nu^{CC}} &\approx 0.033\Delta m_c \qquad \frac{\Delta N_\nu^{NC}}{N_\nu^{NC}} \approx 0.057\Delta m_c \\ \frac{\Delta N_{\overline{\nu}}^{CC}}{N_{\overline{\nu}}^{CC}} &\approx 0.044\Delta m_c \qquad \frac{\Delta N_{\overline{\nu}}^{NC}}{N_{\overline{\nu}}^{NC}} \approx 0.085\Delta m_c \end{aligned} \tag{143}$$

where Δm_c is given in GeV.

The sensitivity on the charm mass is reduced for larger hadron energy cut-off and larger neutrino energies, as illustrated in figure 52 showing $A^\star/A$ for u,d,s and ū quarks, where :

$$A^\star = \int \mathrm{d}x\mathrm{d}y\, \xi\, a(\xi, Q^2) \left(1 - \frac{m_c^2}{2ME\xi}\right) \tag{144}$$

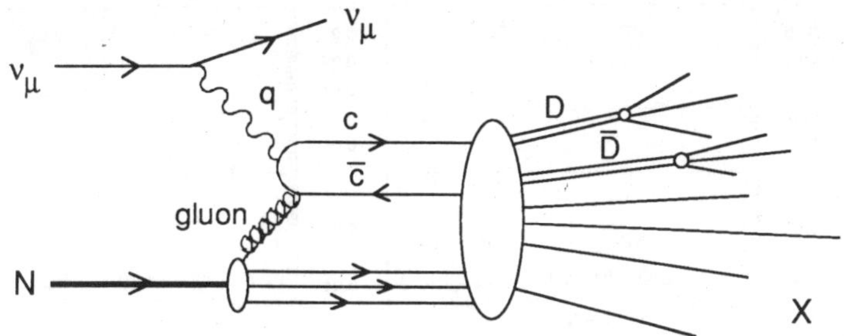

Figure 53: The production of $c\bar{c}$ pairs from gluons in deep inelastic scattering.

as a function of the neutrino energy.

Note that the neutral-current cross-sections are indeed more sensitive to m_c than the charged-current cross-sections. The presence of a non-zero charm mass in the QPM increases the value of $\sin^2\theta_W^{LS}$ from the CDHS experiment by 0.0077 and introduces an uncertainty of ± 0.0029.

The uncertainty in the QPM due to the charm sea content of the nucleon is not easy to quantify. Since the charm quark is heavier than the nucleon, the very concept of charm sea is questionable. The concept of charm sea is rather to take into account in some effective way the production of $c\bar{c}$ pairs from the gluons in the nucleon as depicted in figure 53. This so-called charm sea can be defined as : $xc(x) = 9/8F_2^{\gamma c\bar{c}}$. $F_2^{\gamma c\bar{c}}$ corresponds to the production of $c\bar{c}$ pairs in deep inelastic muon scattering, a process signed by the presence of dimuons and trimuons in the final state [53]. This measurement is difficult and the measurement of the charm sea is not precise. However the cross-section ratios predicted by the QPM are not very sensitive to the charm sea. Assuming $C/S = 0.2 \pm 0.2$, the error on $\sin^2\theta_W^{LS}$ is 0.0005.

In the QPM, it has been assumed that $S = \overline{S}$, $C = \overline{C}$ and $\overline{U} = \overline{D}$.

Although in principle only the integrals are constrained (Eq. 86), a large difference between the strange and the antistrange sea is not expected theoretically (in particular in the type of diagram depicted in figure 53) and is not seen experimentally (see figure 51). However, a difference between strange and antistrange distribution has potentially a big effect on the cross-section ratios as the neutrino charged-current cross-section contains only strange quark contributions and the antineutrino charged-current cross-section contains only antistrange quark contributions (Eq. 130). The change in the cross-section ratios expected by allowing $\chi_s = S/\overline{S}$ to be different from 1 is given approximately by :

$$\frac{\Delta\sigma_\nu^{CC}}{\sigma_\nu^{CC}} \approx \frac{\chi_s - 1}{\chi_s + 1}\frac{2S^\star}{\tilde{\sigma}_\nu^{CC}} \approx 0.054\frac{\chi_s - 1}{\chi_s + 1}$$

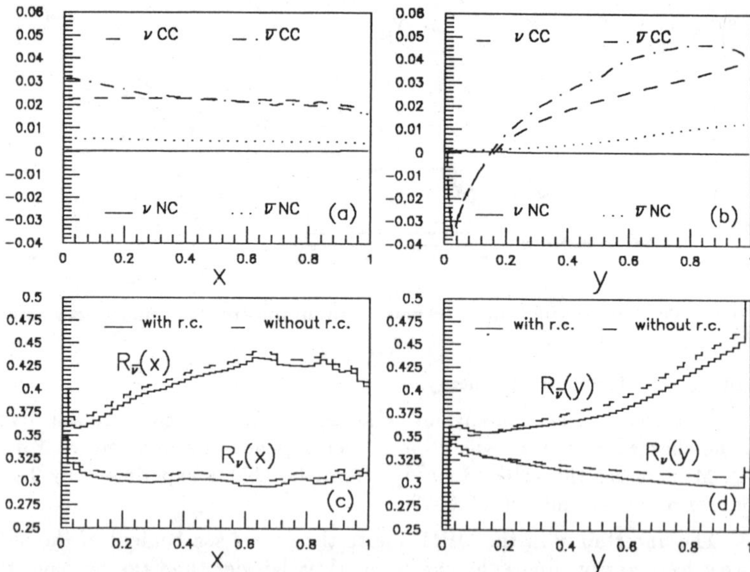

Figure 54: The correction of the cross-sections due to one-loop radiative corrections $(\sigma^{1loop} - \sigma^{tree})/\sigma^{tree}$ as a function of x (a) and y (b). Also shown are the ratios R_ν and $R_{\bar\nu}$ with and without radiative corrections as a function of x (c) and y (d). The corrections have been computed using a program by Bardin [49] assuming $m_t = 100$ GeV and $m_H = 100$ GeV.

$$\frac{\Delta\sigma_\nu^{NC}}{\sigma_\nu^{NC}} \approx \frac{4}{3}\frac{\chi_s - 1}{\chi_s + 1}\left(\epsilon_L(d)^2 - \epsilon_R(d)^2\right)\frac{S}{\tilde\sigma_\nu^{NC}} \approx 0.019\frac{\chi_s - 1}{\chi_s + 1} \tag{145}$$

Taking $\chi_s = 1.1$ decreases R_ν^{QPM} by 0.05 % and $\sin^2\theta_W^{LS}$ by 0.0002. Similarly, allowing $\chi_c = C/\overline{C} = 1.3$, introduces a shift of the cross-sections given by :

$$\begin{aligned} \frac{\Delta\sigma_\nu^{CC}}{\sigma_\nu^{CC}} &\approx \frac{\chi_c - 1}{\chi_c + 1}\frac{2S^\star}{\tilde\sigma_\nu^{CC}} \approx -0.016\frac{\chi_c - 1}{\chi_c + 1} \\ \frac{\Delta\sigma_\nu^{NC}}{\sigma_\nu^{NC}} &\approx \frac{4}{3}\frac{\chi_c - 1}{\chi_c + 1}\left(\epsilon_L(d)^2 - \epsilon_R(d)^2\right)\frac{S}{\tilde\sigma_\nu^{NC}} \approx 0.0033\frac{\chi_c - 1}{\chi_c + 1} \end{aligned} \tag{146}$$

Note that in the case of the charm, the shifts of the neutral-current and charged-current cross-sections add up in the ratios while they cancel partly for the strange quarks. One can check that the errors induced on $\sin^2\theta_W^{LS}$ are rather small.

In contrast to the strange sea and charm sea asymmetries, there is no reason to expect the u and d quark sea distribution to be equal. In contrary, there are a number of theoretical arguments [51] and experimental results [52] indicating that they should indeed be quite different. However, from Eq. 130, it appears that the difference between the u and d quark sea can only affect the predictions of the QPM through the non-isoscalar part of the cross-section, a contribution reduced by $\eta = 0.07$ in iron and $\eta = 0.005$ in the case of the CHARM experiment. Allowing $\overline{\mathrm{U}}/\overline{\mathrm{D}} = 2$ has a negligible effect indeed (table 12).

Finally, the QPM cross-sections have to be corrected for radiative effects. These radiative effects cause relatively large corrections : 1.8% on R_ν and 2.5% on $R_{\bar{\nu}}$. The first order calculation by Bardin [49] in the Sirlin scheme is used with $m_t = 100$ GeV and $m_H = 100$ GeV. The size of the radiative corrections of the number of events seen in the CDHS experiments are shown in figure 54 as a function of x and y.

The radiative corrections are of two types : "purely electromagnetic" and "electroweak" radiative corrections [54].

Purely electromagnetic corrections, which correspond to the emission of real or virtual photons by a fermion, make the bulk of the correction (2.3% on R_ν and 2.6% on $R_{\bar{\nu}}$). The dominant contribution comes from collinear photon emission by the muon in charged-current interactions, an effect which has no counterpart in neutral-current interactions. In first approximation, this effect corresponds to reducing the momentum of the outgoing muon, hence to shifting y to larger values and x to smaller values, as can be seen in figure 54.

The electroweak radiative corrections describe the electroweak propagators, the purely weak box and vertex diagrams. The loop corrections to the propagators introduce a Q^2 evolution of the electroweak coupling constants. Purely weak box and vertex diagrams modify the form of the neutrino-nucleon cross-section. At the energies of the neutrino experiments, these effects are small, as can be seen from the correction of the neutral-current interaction rates in figure 54.

Electroweak radiative corrections have however an important feature : they depend on the unknown parameters of the Standard Electroweak Theory m_t and m_H. This dependence is illustrated in figure 55 for the conditions of the CDHS experiments.

Because of a fortuitous cancellation, the ratio R_ν varies only within 0.5% of its value if the top mass is changed from 30 to 400 GeV. The corresponding change of $\sin^2\theta_W^{LS}$ is small compared with the experimental error[16]. In contrast, the ratio $R_{\bar{\nu}}$ is quite sensitive to the top mass.

The effect of the radiative corrections is to increase $\sin^2\theta_W^{LS}$ by 0.0106 in the CDHS experiment. This result agrees with those obtained by various other authors [55] whithin 0.002, which is the error assigned to the correction. For the recent result of the CCFR collaboration, radiative corrections were checked extensively and an error of 0.010 was assigned.

[16]Note that the CDHS [42], CHARM [43] and CCFR [44] collaborations have published their results assuming $m_t = 60$ GeV, $m_t = 45$ GeV and $m_t = 150$ GeV respectively.

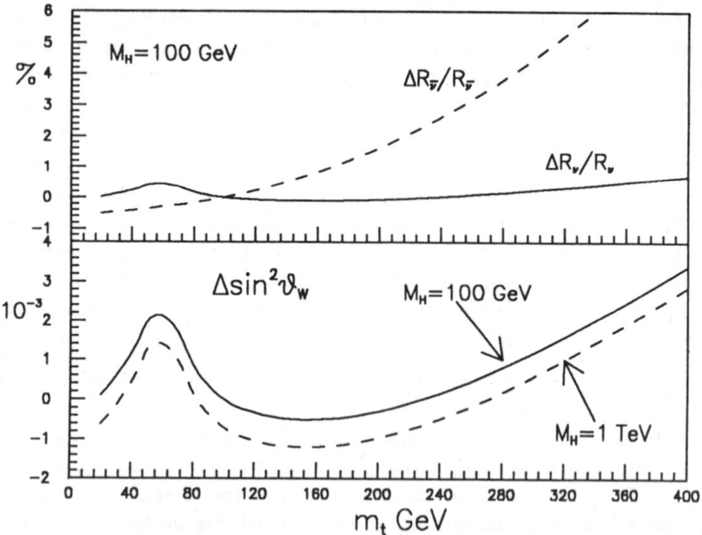

Figure 55: The cross-section ratios R_ν^{LS} and $R_{\bar\nu}^{LS}$ and the electroweak parameter in the Sirlin scheme (Eq. 23) determined from R_ν and r using the Llewellyn-Smith equations as a function of the top quark mass. The full line corresponds to $m_H = 100$ GeV while the dashed line to $m_H = 1$ TeV.

5.2 *The Determination of Electroweak Parameters*

5.2.1 $\sin^2\theta_W$ *from* R_ν *and* r

The values of $\sin^2\theta_W^{LS}$ extracted from the corrected experimental results of the CDHS and CHARM experiments (table 11) are collected in table 13. These results from the QPM are very similar to previously published values of $\sin^2\theta_W^{LS}$ by the experiments themselves [42, 43, 44] or others [27], the change compared with those values coming mostly from the new values chosen for the charm and top quark masses. The preliminary value of $\sin^2\theta_W$ from the E770 experiment is taken as given by the CCFR collaboration [44], correcting it to correspond to a charm quark mass of 1.31 GeV and a top quark mass of 100 GeV.

In principle, a value of $\sin^2\theta_W^{LS}$ can also be extracted from the result of the CCFR experiment. A value of r for the conditions of the E770 experiment could be extrapolated from the measurements obtained at CERN. However, the CCFR collaboration so far has not given the corrected experimental value of R_ν. Once this value becomes available, the same method can be applied to determine the value of

Experiment	$\sin^2\theta_W$	Charm mass dependence
CDHS	$0.2228 \pm 0.0053 \pm 0.0038$	$0.012(m_c - 1.31)$
CHARM	$0.2345 \pm 0.0050 \pm 0.0033$	$0.013(m_c - 1.31)$
CCFR Preliminary	$0.2250 \pm 0.0044 \pm 0.0038$	$0.010(m_c - 1.31)$
Combined Value	$0.2274 \pm 0.0028 \pm 0.0037$	$0.012(m_c - 1.31)$

Table 13: The value of $\sin^2\theta_W$ extracted from the experimental cross-section ratios. The Llewellyn-Smith equation for R_ν is used in the case of CDHS and CHARM. The charm mass is $m_c = 1.31$ GeV and the radiative corrections have been calculated for $m_t = 100$ GeV and $m_H = 100$ GeV. The result of the CCFR experiment [44], given for $m_c = 1.34$ GeV and $m_t = 150$ GeV, was modified accordingly. The first error on $\sin^2\theta_W$ is experimental, the second error is the theoretical error from the QPM.

Errors Sources	CDHS	CHARM	CCFR
Statistical Error	0.0041	0.0040	0.0029
Experimental Systematical Error from R_ν	0.0034	0.0031	0.0033
Experimental Error from r	0.0006	0.0006	
Total Experimental Error	0.0054	0.0051	0.0044
Unitary CKM matrix $\lvert U_{ud}\rvert^2 = 0.9512 \pm 0.0012$	0.0000	0.0000	0.0000
Non-isoscalarity $D_V/U_V = 0.39 \pm 0.04$	0.0009	0.0001	0.0011
q_L QCD prediction ± 50 %	0.0003	0.0004	0.0010
Quark Sea $(\overline{U} + \overline{D})/(U + D) = 0.13 \pm 0.02$	0.0004	0.0002	0.0016
QCD evolution $\Lambda_{QCD} = 200 \pm 100$ MeV	0.0003	0.0003	0.0003
Higher Twist effects	0.0003	0.0003	0.0003
Strange Sea $\overline{S}/\overline{D} = 0.38 \pm 0.05$ [47]	0.0004	0.0004	0.0006
Charm Sea $C/S = 0.2 \pm 0.2$	0.0004	0.0004	0.0016
Strange Sea Asymmetry $S/\overline{S} = 1 \pm 0.1$	0.0002	0.0001	0.0005
Charmed Sea Asymmetry $C/\overline{C} = 1 \pm 0.3$	0.0003	0.0002	0.0004
Isospin Breaking in the sea $\overline{U}/\overline{D} = 1 \pm 1$	0.0003	0.0000	0.0003
Radiative Corrections	0.0020	0.0020	0.0010
Charm Mass $m_c = 1.31 \pm 0.24$	0.0029	0.0030	0.0024
Total Theoretical Error	0.0038	0.0037	0.0038
Total Error	0.0066	0.0063	0.0058

Table 14: Summary of experimental and theoretical errors on the value of $\sin^2\theta_W^{LS}$ (CDHS and CHARM) and $\sin^2\theta_W$ determined from R_ν (CCFR). These errors have been estimated using the QPM described in this paper. Note that the total errors (including theoretical errors) of the experiments are correlated. The theoretical errors for the CCFR experiment (except the error from the radiative corrections) have been estimated by the author of this paper and are directly comparable with the theoretical errors of CDHS and CHARM.

$\sin^2\theta_W^{LS}$ from the experimental results of CDHS, CHARM and CCFR[17].

The errors from the various experiments, collected in table 14, are comparable. In the CCFR experiment, the q_L and quark sea uncertainties are larger than in CDHS and CHARM because the Llewellyn-Smith equations were not used. The sensitivity to the charm mass is reduced in the CCFR experiment because of the larger neutrino energy and the larger hadron energy cut-off.

The results from the CDHS and CHARM experiments are in acceptable agreement, as their difference corresponds to 1.6 standard deviations. The result from the CCFR experiment agrees well with the result from the CDHS experiment.

The average value[18] of the three experiments yields a value of :

$$\sin^2\hat{\theta}_W = 0.2274 \pm 0.0028(\text{exp.}) \pm 0.0037(\text{theo.}) \tag{147}$$

The total error amounts to 0.0046.

If a top quark mass of 150 GeV would be chosen, the value of $\sin^2\theta_W$ would be decreased by 0.0005 (figure 55). If a Higgs mass of 1 TeV would be chosen, keeping the top mass at 100 GeV, the value of $\sin^2\theta_W$ would be decreased by 0.0007.

5.2.2 *The top mass from R_ν, $R_{\bar{\nu}}$ and r*

Since the ratio $R_{\bar{\nu}}$ is sensitive to the top mass and rather insensitive to the value of $\sin^2\theta_W$, one can derive a limit on the top mass by comparing $R_{\bar{\nu}}^{LS}$ from the QPM with the experimental value, using the top mass dependence shown in figure 55.

The result $R_{\bar{\nu}} = 0.374 \pm 0.011(\text{exp.})$ from the CDHS experiment is in good agreement with the prediction from the QPM $R_{\bar{\nu}}^{LS} = 0.380 \pm 0.005(\text{theo.})$ and a rather stringent upper limit on the top mass can be derived :

$$m_t < 280 \text{ GeV } (90\% \; c.l.) \tag{148}$$

However, the significance of this result has to be taken with caution. The result $R_{\bar{\nu}} = 0.390 \pm 0.014(\text{exp.})$ from the CHARM experiment is not in good agreement with the prediction from the QPM $R_{\bar{\nu}}^{LS} = 0.367 \pm 0.005(\text{theo.})$ and no interesting limit on the top mass at 90% confidence level can be derived.

5.2.3 *ρ and $\sin^2\theta_W$ from R_ν, $R_{\bar{\nu}}$ and r*

In extensions of the Electroweak Theory, the electroweak couplings are in general characterised by an overall scale factor ρ at tree level. The definition of the renor-

[17]The value of R_ν from the E770 experiment corrected for all experimental effects being unavailable at the time of publication of this paper, the author apologizes for not being able to extract his own value of $\sin^2\theta_W$ from the CCFR result.

[18]Since the results of the CDHS, CHARM and CCFR collaborations are not extracted with the same method, it is not very justified however to average the results of the three experiment.

malised electroweak mixing parameter in the Sirlin scheme becomes [26] :

$$\sin^2\hat{\theta}_W = 1 - \frac{m_W^2}{\rho m_Z^2} \tag{149}$$

and the Llewellyn-Smith equations are then written :

$$\begin{aligned} R_\nu^0 &= \rho^2\left[\frac{1}{2} - \sin^2\theta_W + \frac{5}{9}\sin^4\theta_W\left(1+r^0\right)\right] \\ R_{\bar{\nu}}^0 &= \rho^2\left[\frac{1}{2} - \sin^2\theta_W + \frac{5}{9}\sin^4\theta_W\left(1+\frac{1}{r^0}\right)\right] \end{aligned} \tag{150}$$

Actually, in a first approximation, some contributions of the radiative corrections can be interpreted as effective parameters ρ^{eff} and $\sin^2\theta_W^{eff}$:

$$\begin{aligned} \rho^{eff} &= 1 + \Delta\rho = 1 + \frac{3G_\mu}{8\pi^2\sqrt{2}}m_t^2 \approx 1 + 0.02\left(\frac{m_t}{250GeV}\right)^2 \\ \sin^2\theta_W^{eff} &= \sin^2\hat{\theta}_W\left(1+\hat{\delta}(m_t)\right) \end{aligned} \tag{151}$$

A measurement of ρ and $\sin^2\theta_W$ can be therefore be interpreted as a measurement of the tree level ρ parameter and of all the radiative effects not taken into account[19].

A common fit of Eq. 150 to the data yields the results given in table 15. The value of ρ is found to be compatible with 1, with an error of 2% dominated by the statistical error on $R_{\bar{\nu}}$.

These results depend strongly on the top mass, as expected (figure 56). One can however conclude that, at the 0.14% level, the results for ρ are compatible with the predictions of the Standard Electroweak Theory ($\rho = 1$).

5.2.4 m_W/m_Z from R_ν, $R_{\bar{\nu}}$ and r

Remaining in the Electroweak Theory, the free parameters can be changed from ρ and $\sin^2\hat{\theta}_W$ to m_W/m_Z and $\sin^2\hat{\theta}_W$:

$$\begin{aligned} R_\nu^0 &= \left(\frac{m_W}{m_Z}\right)^4 \frac{\frac{1}{2} - \sin^2\theta_W + \frac{5}{9}\sin^4\theta_W\left(1+r^0\right)}{\left(1-\sin^2\theta_W\right)^2} \\ R_{\bar{\nu}}^0 &= \left(\frac{m_W}{m_Z}\right)^4 \frac{\frac{1}{2} - \sin^2\theta_W + \frac{5}{9}\sin^4\theta_W\left(1+\frac{1}{r^0}\right)}{\left(1-\sin^2\theta_W\right)^2} \end{aligned} \tag{152}$$

Numerically, it appears that R_ν is a weak function of $\sin^2\theta_W$. Hence, R_ν represents a measurement of the mass ratio m_W/m_Z independent of $\sin^2\theta_W$, and from

[19]The m_t dependences of ρ^{eff} and $\sin^2\theta_W^{eff}$ cancel approximately in R_ν but not in $R_{\bar{\nu}}$ as $R_{\bar{\nu}}$ depends only weakly on $\sin^2\theta_W^{eff}$.

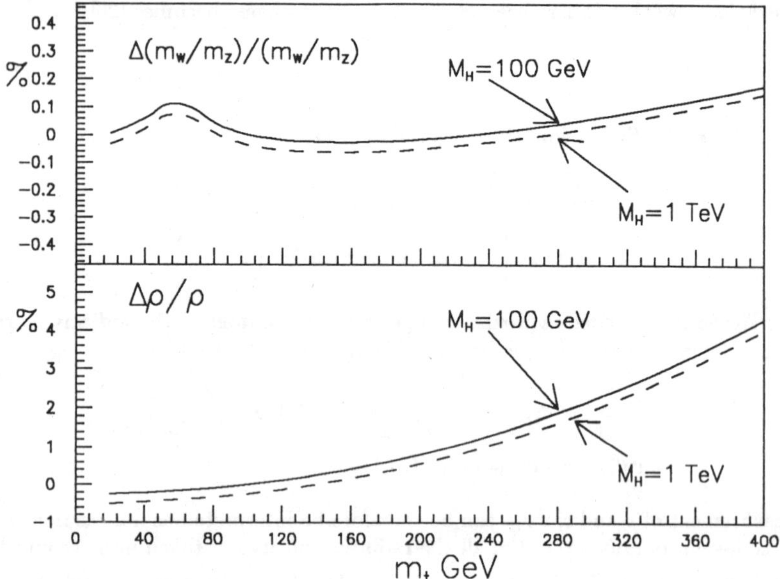

Figure 56: The measured values of ρ and m_W/m_Z as a function of the top mass.

the values of the unknown parameters of the theory ρ, m_t and m_H. The dependence of m_W/m_Z on the top mass is for example shown in figure 56. The results are given in table 15. The result from CDHS and CHARM combined is :

$$m_W/m_Z = 0.8777 \pm 0.0031 - 0.005(m_c - 1.31) \tag{153}$$

The mass ratio m_W/m_Z is thus determined with a precision of 0.3%, a precision which can be improved slightly by adding the recent CCFR result.

This result for the mass ratio m_W/m_Z can be compared with the direct measurements from the observation of W and Z bosons at CERN and Fermilab. The average value is [57, 58] :

$$m_W/m_Z = 0.8790 \pm 0.0030 \tag{154}$$

with a precision similar to the one achieved in neutrino scattering[20]. The two determinations of the mass ratio m_W/m_Z agree remarkably, providing a consistency test of the Electroweak Theory at the 0.5% level, independently of assumptions on ρ or

[20]The error on the measurement of the W mass will be reduced to ±100 MeV in the near future so an error on the measurement of the mass ratio of ±0.12% can be expected from the collider experiments.

	Fit results
CDHS	$\rho = 0.989 \pm 0.019 - 0.023(m_c - 1.31)$ $\sin^2\theta_W = 0.212 \pm 0.020 - 0.010(m_c - 1.31)$
	$m_W/m_Z = 0.8827 \pm 0.0041 - 0.005(m_c - 1.31)$
	$\alpha_L = 0.3060 \pm 0.0053 - 0.007(m_c - 1.31)$ $\alpha_R = 0.0245 \pm 0.0055 - 0.003(m_c - 1.31)$
CHARM	$\rho = 1.034 \pm 0.020 - 0.022(m_c - 1.31)$ $\sin^2\theta_W = 0.267 \pm 0.021 - 0.009(m_c - 1.31)$
	$m_W/m_Z = 0.8702 \pm 0.0050 - 0.004(m_c - 1.31)$
	$\alpha_L = 0.2909 \pm 0.0059 - 0.006(m_c - 1.31)$ $\alpha_R = 0.0424 \pm 0.0082 - 0.005(m_c - 1.31)$
CDHS and CHARM combined	$\rho = 1.004 \pm 0.014 - 0.022(m_c - 1.31)$ $\sin^2\theta_W = 0.233 \pm 0.014 - 0.009(m_c - 1.31)$
	$m_W/m_Z = 0.8777 \pm 0.0031 - 0.005(m_c - 1.31)$
	$\alpha_L = 0.2998 \pm 0.0039 - 0.007(m_c - 1.31)$ $\alpha_R = 0.0304 \pm 0.0046 - 0.004(m_c - 1.31)$

Table 15: Summary of results for ρ-$\sin^2\theta_W$, m_W/m_Z, and α_L-α_R obtained from R_ν and $R_{\bar{\nu}}$ assuming $m_t = 100$ GeV and $m_H = 100$ GeV.

m_t. One can therefore conclude that the observed W and Z bosons are indeed the only mediators of neutrino interactions, up to a precision 0.5% at least.

The combined result for the mass ratio reads :

$$m_W/m_Z = 0.8784 \pm 0.0022 \tag{155}$$

$$\text{corresponding to : } \sin^2\theta_W(m_W/m_Z) = 0.2284 \pm 0.0039 \tag{156}$$

5.2.5 α_L *and* α_R *from* R_ν, $R_{\bar{\nu}}$ *and* r

In a more general framework, one can chose α_L and α_R as free parameters :

$$R^0_\nu = \alpha_L + \alpha_R\, r^0 \qquad R^0_{\bar{\nu}} = \alpha_L + \alpha_R\, \frac{1}{r^0} \tag{157}$$

The results are given in table 15. Note that α_L is determined with a precision of 1.3% and α_R with a precision of 15%. The large error on α_R is dominated by the statistical error on $R_{\bar{\nu}}$.

These results are in agreement with the Standard Electroweak Theory predictions $\alpha_L = 0.3016$ and $\alpha_R = 0.0286$ for $\sin^2\theta_W = 0.227$.

By using deep inelastic neutrino scattering results off protons and neutrons, one can disentangle the up-type quark and down-type quark contributions to α_L and

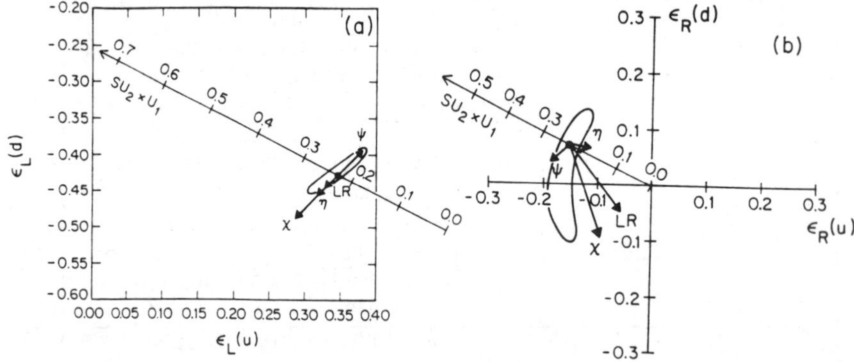

Figure 57: The results of a combined determination of the effective couplings $\epsilon_{L,R}(u)$, $\epsilon_{L,R}(d)$ of the neutrino-quark neutral-current interaction including results from deep inelastic proton and neutron scattering, neutrino-nucleon elastic scattering and neutrino π^0 coherent production [27]. The prediction of the Standard Electroweak Theory is shown, and the changes induced by various additional Z bosons [61].

α_R, and determine values for $\epsilon_L(u)^2$, $\epsilon_R(u)^2$, $\epsilon_L(d)^2$ and $\epsilon_R(d)^2$. In addition, measurements from elastic neutrino-nucleon scattering [56] and coherent π^0 production fix linear combinations of $\epsilon_L(u)$, $\epsilon_R(u)$, $\epsilon_L(d)$ and $\epsilon_R(d)$, resolving sign ambiguities.

The results of a combined determination [27] of the effective neutral-current neutrino-quark couplings are shown in figure 57. A good agreement of the data with the Standard Electroweak Theory is observed.

6 Conclusion and Outlook

In this paper, the measurement of electroweak parameters in deep inelastic neutrino scattering experiments has been presented. In the framework of the Standard Electroweak Theory, the value of the electroweak mixing parameter in the Sirlin renormalisation scheme is determined to be :

$$\sin^2\hat{\theta}_W(\nu N) = 0.227 \pm 0.005 \tag{158}$$

for $m_t = 100$ GeV and $m_H = 100$ GeV.

This value of $\sin^2\theta_W$ can be compared with the predictions of Grand Unified Theories (GUTs). This comparison favors supersymmetric GUTs over non supersymmetric GUTs, a trend which was felt as soon as the results from the CDHS and CHARM experiments were published in 1986 [61].

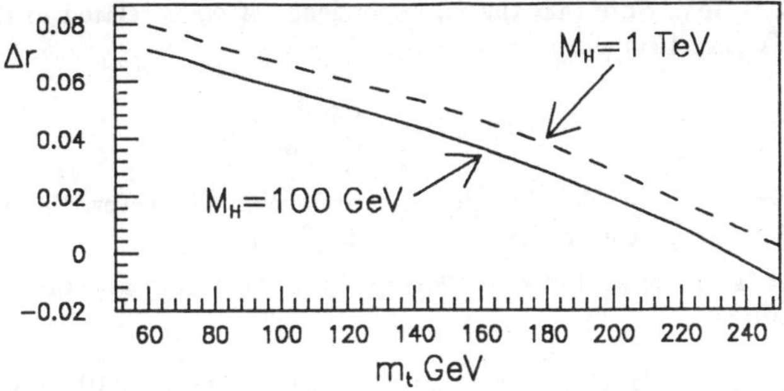

Figure 58: The electroweak correction for the W and Z masses Δr as a function of the top mass [59].

The result for $\sin^2\theta_W$ can be used to obtain a measurement of electroweak radiative correction term Δr in Eq. 25.

Taking the combined result for the Z mass from the LEP collaborations $m_Z = 91.175 \pm 0.021$ GeV [58], yields :

$$\Delta r(m_Z, \sin^2\theta_W(\nu N)) = 0.048 \pm 0.000(m_Z) \pm 0.015(\nu N) \tag{159}$$

A precise result for the W mass can be obtained by multiplying the result for m_W/m_Z obtained by the UA2 collaboration by the very precise value of the Z mass obtained at LEP. This result, combined with the precise result for the W mass obtained by the CDF collaboration yields an average value $m_W = 80.14 \pm 0.27$ GeV [57]. From this value and the value $\sin^2\theta_W$ from deep inelastic neutrino scattering, one obtains :

$$\Delta r(m_W, \sin^2\theta_W(\nu N)) = 0.048 \pm 0.006(m_W) \pm 0.022(\nu N) \tag{160}$$

This impressive measurement of Δr, performed as soon as 1986 with slightly larger errors from the colliders, represented the first investigation of the radiative corrections in the Standard Electroweak Theory, and a turning point in the phenomenology of the Standard Model.

These results are in good agreement with the prediction of the Standard Electroweak Theory [59] :

$$\Delta r = 0.0574 \pm 0.0013(\text{theo.}) \tag{161}$$

for $m_Z = 91.1$ GeV, $m_t = 100$ GeV and $m_H = 100$ GeV.

The variation of Δr with the top mass is shown in figure 58. From the comparison of the neutrino scattering result with the measurement of the vector boson masses, one can conclude that :

$$m_t < 180 \text{ GeV } (90\% \; c.l.) \tag{162}$$

for $m_H = 100$ GeV. Note that the m_t dependence of Δr is related to the m_t dependence of ρ^{eff} (Eq. 151) as :

$$\Delta r \approx \Delta\alpha - \frac{\cos^2\theta_W}{\sin^2\theta_W}\Delta\rho \tag{163}$$

where $\Delta\alpha \approx 0.06$ is an electromagnetic contribution. However, the limit given by Eq. 162 is independent from the previous one (Eq. 148).

If Δr is determined from $\sin^2\theta_W(m_W/m_Z)$ (Eq. 156) and m_Z, the errors are slightly smaller :

$$\Delta r(m_Z, \sin^2\theta_W(m_W/m_Z) = 0.051 \pm 0.000(m_Z) \pm 0.012(\nu N) \tag{164}$$

and one can even obtain a rough estimate of the top mass :

$$m_t = 135 \pm 45 \text{ GeV} \tag{165}$$

This result, although it can be improved by using the precise LEP electroweak measurements [60], underlines the important role played by the precise neutrino deep inelastic scattering experiments.

The errors in the determination of Δr however are now dominated by the error in the determination of $\sin^2\theta_W$ in neutrino scattering. In addition, the neutral-current neutrino scattering appears as a unique sensitive probe of some extensions of the Standard Electroweak Theory [61], as also illustrated in figure 57. One might therefore investigate the ways to improve significantly the measurement of $\sin^2\theta_W$ in neutrino scattering.

Following [62], a new neutrino scattering experiment will be considered, taking as reference the precise measurements achieved by the CDHS, CHARM and CCFR collaborations.

More precisely, let us consider a very massive (3000 tons for example) iron detector exposed to a Fermilab-like QTB. An increase in the intensity of the proton beam as considered for the Tevatron would allow very high intensities and we will assume that 15 million neutrino charged-current interactions and 800,000 antineutrino charged-current interactions have been recorded in the detector. We will investigate the induced error on $\sin^2\theta_W$ obtainable in such a toy experiment, assuming that a hadron energy cut-off of 150 GeV is used.

Observables beyond R_ν and $R_{\bar{\nu}}$ can be considered. The Paschos-Wolfenstein ratios R_+ and R_- (Eq. 19) are very interesting ratios to measure, as we shall see. In addition, one can study the ratio r_{NC} and the double ratio R_2 defined as[21] :

$$r_{NC} = \frac{\sigma_{\bar{\nu}}^{NC}}{\sigma_\nu^{NC}} \qquad R_2 = \frac{r_{NC}}{r} = \frac{R_{\bar{\nu}}}{R_\nu} \tag{166}$$

[21] The antineutrino data are supposed to have been corrected to correspond to the neutrino beam energy spectrum.

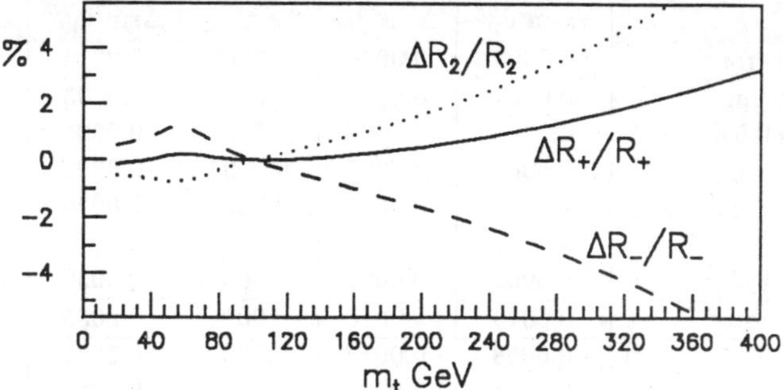

Figure 59: The sensitivity of the Paschos-Wolfenstein ratios R_+ and R_-, and of the double ratio R_2 to the value of the top quark mass, assuming $m_H = 100$ GeV.

A value of $\sin^2\theta_W$ can be extracted from R_ν and r ($\sin^2\theta_W^{LS}$), from R_+ ($\sin^2\theta_W^+$), from R_- ($\sin^2\theta_W^-$), from r_{NC} ($\sin^2\theta_W^{NC}$) and from the double ratio R_2 ($\sin^2\theta_W^{DR}$). $\sin^2\theta_W^+$ and $\sin^2\theta_W^-$ will be extracted by comparing the experimental value of the ratios with their QPM predictions. In the case of $\sin^2\theta_W^{DR}$, we will in addition use the Llewellyn-Smith equations formalism, correcting the experimental ratio R_2 with the QPM to compare it with :

$$R_2^0 = \frac{\frac{1}{2} - \sin^2\theta_W + \frac{5}{9}\sin^4\theta_W\left(1 + \frac{1}{r^0}\right)}{\frac{1}{2} - \sin^2\theta_W + \frac{5}{9}\sin^4\theta_W\left(1 + r^0\right)} \tag{167}$$

From the theoretical point of view, R_- is the most robust observable. The Paschos-Wolfenstein prediction (Eq. 19) for R_- follows from isospin invariance only and is broken only by mass effects, like charm quark mass effects.

The physics contents of these ratios are quite different. In particular, r_{NC} and R_2 are independent of the value of ρ. The sensitivity of the ratios R_+, R_- and R_2 to the top mass is indicated in figure 59. In the following, the values of $\sin^2\theta_W$ extracted from R_ν, R_+ and R_- rely on the assumption $\rho = 1$.

The statistical errors on the ratios are given by :

$$\frac{\Delta R_\nu}{R_\nu} \approx \frac{1.9}{\sqrt{N_\nu^{CC}}} \qquad \frac{\Delta R_+}{R_+} \approx 1.4\sqrt{\frac{1}{N_\nu^{CC}} + \frac{0.18}{N_{\bar\nu}^{CC}}} \qquad \frac{\Delta R_-}{R_-} \approx 3.8\sqrt{\frac{1}{N_\nu^{CC}} + \frac{0.18}{N_{\bar\nu}^{CC}}}$$

$$\frac{\Delta R_{\bar\nu}}{R_{\bar\nu}} \approx \frac{1.9}{\sqrt{N_\nu^{CC}}} \qquad \frac{\Delta r_{NC}}{r_{NC}} \approx 1.8\sqrt{\frac{1}{N_\nu^{CC}} + \frac{0.8}{N_{\bar\nu}^{CC}}} \qquad \frac{\Delta R_2}{R_2} \approx 2\sqrt{\frac{1}{N_\nu^{CC}} + \frac{1}{N_{\bar\nu}^{CC}}} \tag{168}$$

The corresponding errors on $\sin^2\theta_W$, given by :

$$\Delta\sin^2\theta_W^{LS} \approx -0.5\frac{\Delta R_\nu}{R_\nu} \qquad \Delta\sin^2\theta_W^+ \approx -0.7\frac{\Delta R_+}{R_+} \qquad \Delta\sin^2\theta_W^- \approx -0.3\frac{\Delta R_-}{R_-}$$

	$\Delta\sin^2\theta_W^{LS}$	$\Delta\sin^2\theta_W^{+}$	$\Delta\sin^2\theta_W^{-}$	$\Delta\sin^2\theta_W^{NC}$	$\Delta\sin^2\theta_W^{DR}$
Statistical Error	0.0003	0.0005	0.0006	0.0014	0.0011
Beam spectrum	0.0003	0.0003	0.0003	0.0003	0.0003
Short CC $\pm0.5\%$	0.0007	0.0008	0.0006	0.0004	0.0005
ν_e correction $\pm1\%$	0.0004	0.0005	0.0004	0.0002	0.0002
Relat. norm. $\pm0.5\%$		0.0001	0.0004	0.0030	
Error from r	0.0001				0.0005
Total Syst. Error	0.0009	0.0009	0.0009	0.0030	0.0008
Total Exp. Error	0.0010	0.0010	0.0011	0.0033	0.0014
$m_c\pm0.10$ GeV	0.0008	0.0013	0.0002	0.0021	0.0007
Strange Sea $\pm6\%$	0.0001	0.0001	0.0000	0.0021	0.0001
$C/S = 0.2 \pm 0.1$	0.0006	0.0011	0.0000	0.0021	0.0008
$U_V/D_V \pm 5\%$	0.0004	0.0003	0.0006	0.0001	0.0007
$q_L = q_L^{QCD} \pm 20\%$	0.0001	0.0002	0.0001	0.0056	0.0002
$S/\overline{S} = 1 \pm 0.1$	0.0003	0.0004	0.0007	0.0018	0.0013
$C/\overline{C} = 1 \pm 0.1$	0.0001	0.0000	0.0002	0.0001	0.0003
Radiative Corrections	0.0005	0.0005	0.0005	0.0001	0.0001
Total Theoretical Error	0.0012	0.0019	0.0011	0.0066	0.0019
Total Error	0.0016	0.0022	0.0015	0.0074	0.0024

Table 16: The uncertainty in the determination of $\sin^2\theta_W$ assuming a toy experiment using an iron detector exposed to the Fermilab QTB and having recorded 15 million neutrino charged-current interactions and 800,000 antineutrino neutral-current interactions. The theoretical error from the change of the strange sea is given as an indication but is not summed in quadrature with the other error as its contribution is already included in the error coming from the charm quark mass.

$$\Delta \sin^2\theta_W^{NC} \approx -0.6\frac{\Delta r_{NC}}{r_{NC}} \qquad \text{and } \Delta \sin^2\theta_W^{DR} \approx -0.6\frac{\Delta R_2}{R_2}, \tag{169}$$

are collected in table 16. The measurement from R_ν is the most favored statistically but all observables have an adequate statistical error in the toy experiment.

The theoretical errors on $\sin^2\theta_W$ are collected in table 16. These errors rely on assumptions which ought to be commented.

With the large event sample collected in the toy experiment, precise measurement of structure functions would be undertaken. The structure functions could be determined independently in the neutrino and the antineutrino beams. This measurement would allow a very detailed investigation of the QPM as discussed in [62], in particular of the higher twist effects. It will be also assumed that the value of U_V/D_V will be constrained by the recent result of the NMC collaboration to an accuracy better than 5%.

In addition, the very large sample of dimuon events would allow one to reach the limit of systematic uncertainties in the determination of the charm mass and the strange sea. Conservative errors of ± 0.1 GeV on m_c and 3% on the amount of strange sea will be assumed here. This error, together with the fact that a hadron energy cut-off of 150 GeV is used, reduces the theoretical error coming from the charm mass, which is the largest theoretical error in the current results, to very small numbers. In particular, the Paschos-Wolfenstein ratio R_- is insensitive to the charm mass. At such high energies, the amount of strange sea is not correlated with the value of the charm mass and the charm mass error corresponds indeed to the change of the strange sea, as can be seen in table 16 for the case of $\sin^2\theta_W^{NC}$.

With a precise measurement of structure functions, a precise measurement of the longitudinal structure function will be possible with an accuracy of about 20% of the QCD prediction. This is certainly a difficult measurement. Most of the determinations of $\sin^2\theta_W$ are insensitive to the longitudinal structure function. However, the ratio r_{NC} is quite sensitive to it (see Eq. 137) and this feature makes this ratio quite uninteresting for the determination of $\sin^2\theta_W$.

At the level of accuracy of ± 0.0010, one has to worry about many effects which did not need to be considered before. For example, the uncertainty due to a possible asymmetry between the strange and the antistrange sea has to be analyzed with care. If one assumes that the high statistics dimuon sample permits the comparison of the strange and the anti-strange quark distribution to an accuracy of $\pm 10\%$, a quite conservative assumption, the uncertainty induced on $\sin^2\theta_W^-$ is of 0.0007, the largest uncertainty on $\sin^2\theta_W^-$ in table 16.

In the toy experiment, the charm sea becomes an important contribution to the theoretical error. The possibility of constraining the charm sea by a precise comparison of the electromagnetic structure functions with the neutrino structure functions (Eq. 95) should be envisaged. In addition, the assumption that the charm sea is proportional to the strange sea is probably not adequate, as a softening of the strange sea with respect to the u and d sea was observed (figure 51b). The charm production should in fact be treated in the framework of PQCD and not only as a naive effective intrinsic quark distribution. Actually, a more detailed framework than the QPM should be envisaged as effects due to light quark masses and target mass effects could also be significant at the level of precision considered here.

Pending a more detailed theoretical analysis, it appears that the theoretical error on $\sin^2\theta_W$ is of the order of 0.0020 for $\sin^2\theta_W^+$ and $\sin^2\theta_W^{DR}$ and of the order of 0.0010 for $\sin^2\theta_W^{LS}$ and $\sin^2\theta_W^-$. It needs then to be checked whether the experimental systematic errors can reach this level of accuracy.

With the high statistic data sample envisaged here, the beam spectrum can be reconstructed from the data with an adequate accuracy and the uncertainties coming from the shape of the neutrino spectra can be assumed to be small.

The largest contamination of the NC sample comes from the short charged-current interactions, with fractions estimated to 20% and 8% in the neutrino and antineutrino beams respectively[22]. One can assume that the theoretical uncertainty in the extrapolation from the monitor region will become negligible. Assuming a systematic uncertainty on the event length of 1 cm on average produces an uncertainty of 0.5% on the short CC contaminations. The induced errors on $\sin^2\theta_W$ are indicated in table 16 and they are at the level of 0.0008 or below.

The largest uncertainty in the CCFR experiment was coming from the 3.8% uncertainty on the ν_e flux, a large part of this uncertainty coming from the unkown K^0 component of the beam. This component can be eliminated by targeting the proton beam at a small angle with respect to the detector. One can also imagine that the charged particles can be swept away in the beam during special data taking periods, leaving only the K_L^0 component. One can therefore assume that, using the η_3 technique of the CCFR collaboration, the ν_e contamination can be determined with an uncertainty of 1%. Assuming a contamination of 8% in the neutrino beam and 4% in the antineutrino beam, the errors induced on $\sin^2\theta_W$ are small.

For the observables involving both neutrino and antineutrino interactions (R_+, R_- and r_{NC}) the systematic error from the relative normalisation of the two beams has to be estimated. This relative normalisation can be obtained from various methods [62] and an accuracy of 0.5% is deemed feasible, although it is a rather delicate measurement.

The total systematic uncertainties appear to be smaller than the corresponding theoretical errors. In the toy experiment, the value of $\sin^2\theta_W$ can then be determined to an accuracy of ± 0.0016 from R_ν and R_- and to an accuracy of ± 0.0024 from R_+ and R_2. As the relative contributions of statistical, systematical and theoretical uncertainties are quite different in these four cases, a comparison of the four results will provide a very important cross-check of the experiment, as well as a test of the Standard Electroweak Theory.

If ρ is left as a free parameter, its value will be determined from R_ν with an accuracy of $\pm 0.25\%$, using the value of $\sin^2\theta_W$ obtained from R_2,

As a conclusion, an experiment having recorded 15 million neutrino charged-current interactions and 800,000 antineutrino interactions with $E_h > 150$ GeV would determine the value of $\sin^2\theta_W$ with a precision of ± 0.0016, assuming $\rho = 1$, and the value of ρ with a precision of $\pm 0.25\%$.

Such an experiment is actually proposed at Fermilab [63], using an upgraded version of the CCFR detector and making use of the progressive increase of intensity in the Tevatron[23]. A precision of ± 0.003 on $\sin^2\theta_W$ is expected in the first phase of this program, and a precision of ± 0.0015 could be reached after the completion of the Tevatron accelerator upgrade program.

---

[22] These fractions would be reduced by a factor 2 if a 400 GeV NBB would be used. However, it appears quite difficult to get very large data samples in such a NBB.

[23] With the need for very high statistics, it is not guaranteed that high energy accelerators like SSC and LHC can deliver interesting high energy beams for deep inelastic scattering.

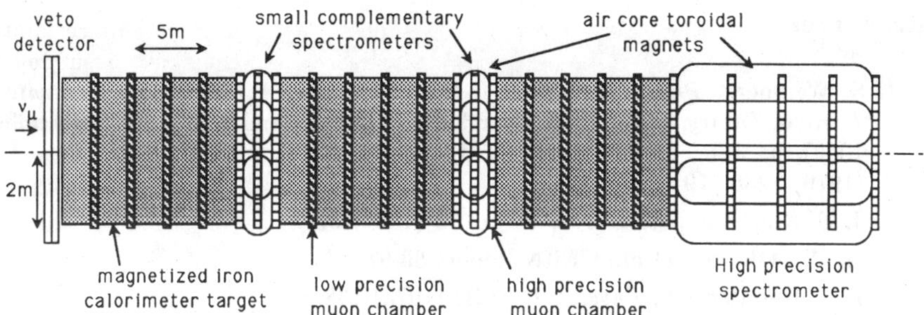

Figure 60: A possible detector for a very high precision deep inelastic neutrino scattering experiment.

Since this experimental program is actually a real challenge, one may design a new detector with significantly improved performances over the performances of CDHS, CHARM and CCFR. This detector shoud have a large mass in order to collect the necessary data sample. A precision on the muon momentum of 1% is also required in order to obtain a very precise measurement of structure functions, to allow for a detailed investigation of the QPM. Such a precision can be achieved in an Air Core toroidal magnet [64]. A new neutrino detector could consist in magnetized iron modules of the type of CDHS as a calorimeter target, followed by a 10m long spectrometer based on air core toroids. Smaller air core toroids would also be inserted in the calorimeter section to obtain a precise measurement of low momentum muons, especially important for the high-y part of the cross-section. The spectrometers would be instrumented with $\pm 100\mu$ tracking chambers. The necessary magnetic field in the air core toroids would be achieved by a set of eight superconducting coils extending from 30 cm to 2 m radius and providing 200,000 Ampere-turns.

Such a futuristic detector would certainly open new possibilities in the measurement of electroweak parameters from deep inelastic neutrino scattering. Other types of detectors are certainly possible and, to take up the new challenge, the best is to let free our imagination!

Acknowledgements

The author is indebted to Paul Langacker and Patrice Perez for comments on the manuscript, as well as to Claude Guyot, Alain Milsztajn and Marc Virchaux for many discussions over the years. The author thanks Sanjib Mishra for providing him with unpublished recent information and figures from the CCFR experiment at Fermilab and for sharing with him his insight and enthusiasm about neutrino physics.

References

1. S. Weinberg, *Phys. Rev. Lett.* **19** (1967) 1264; A. Salam, in *Elementary Particle Theory*, edited by N. Svartholm (Almquist and J. Wiksell, Sotckholm, 1969), p. 367 ; S. L. Glashow, J. Iliopoulos and L. Maiani, *Phys. Rev.* **D2** (1970) 1285. **19** (1967) 1264,
2. J. E. Kim et al., *Rev. Mod. Phys.* **53** (1981) 211.
3. H. W. Atherton et al., CERN Report 80-07.
4. F. J. Hasert et al., *Phys. Lett.* **46B** (1973) 138.
5. A. Benvenuti et al., *Phys. Rev. Lett.* **32** (1974) 1457.
6. B. C. Barish et al., *Phys. Rev. Lett.* **34** (1975) 1679.
7. P. Musset and J.-P. Vialle, *Phys. Rep.* **39C** (1978) 1.
8. B. C. Barish, *Phys. Rep.* **39C** (1978) 279.
9. J. Blietschau et al., *Nucl. Phys.* **B118** (1977) 218.
10. P. Wanderer at al., *Phys. Rev.* **D17** (1978) 1679.
11. B. C. Barish et al., *Phys. Rev.* **D17** (1978) 2199.
12. R. Beuselinck et al., *Nucl. Instr. Meth.* **154** (1978) 445.
13. M. Holder et al., *Nucl. Instr. Meth.* **148** (1978) 235.
14. A. N. Diddens et al., *Nucl. Instr. Meth.* **178** (1980) 27.
15. P. C. Bosetti et al., *Nucl. Phys.* **B217** (1983) 1.
16. M. Holder et al., *Phys. Lett.* **71B** (1977) 1.
17. H. Abramowicz et al., *Z. Phys.* **C28** (1985) 51.
18. M. Jonker et al., *Phys. Lett.* **99B** (1981) 265.
19. P. G. Reutens et al., *Phys. Lett.* **152B** (1985) 404.
20. P. G. Reutens et al., *Z. Phys.* **C45** (1990) 539.
21. D. Bogert et al., *Phys. Rev. Lett.* **55** (1985) 1969.
22. B. C. Barish et al., *IEEE Trans. Nucl. Sci.* **25** (1978) 532.
23. E. A. Paschos and L. Wolfenstein, *Phys. Rev.* **D7** (1973) 91.
24. D. Bogert et al., *IEEE Trans. Nucl. Sci.* **29** (1982) 363.
25. P. Langacker, M. Luo and A. K. Mann, *Rev. Mod. Phys.* **64** (1992) 87.
26. A. Sirlin, *Phys. Rev.* **D22** (1980) 971.
27. U. Amaldi et al., *Phys. Rev.* **D36** (1987) 1385.
28. C. H. Llewellyn-Smith, *Nucl. Phys.* **B228** (1983) 205.
29. C. H. Llewellyn-Smith, *Phys. Rep.* **C3** (1972) 261; and references therein.
30. M. G. Doncel and E. de Rafael, *Nuov. Cim.* **4A** (1971) 363.
31. R. M. Barnett, *Phys. Rev.* **D14** (1976) 70.
   H. Georgi and H. D. Politzer, *Phys. Rev.* **D14** (1976) 1829.

32. G. Altarelli, *Phys. Rep.* **81** (1982) 1.

33. L. F. Abbott and R. M. Barnett, *Ann. Phys.* **125** (1980) 309.

34. M. Virchaux, Preprint DAPNIA/SPP 92-30 (1992), to appear in the proceedings of the Workshop "QCD-20 years later", Aachen, 9-13 june 1992.

35. D. J. Gross and C. Llewellyn-Smith, *Nucl. Phys.* **B14** (1969) 337.

36. W. C. Leung et al., Preprint NEVIS-1460 (1992).

37. W. J. Stirling, Proceedings of the Lepton-Photon Symposium, Geneva, 25 July-1 August 1992.

38. L. W. Whitlow et al., *Phys. Lett.* **B250** (1990) 193.

39. J. V. Allaby et al., *Phys. Lett.* **B213** (1988) 554.

40. T. S. Mattison et al., *Phys. Rev.* **B42** (1990) 1311.

41. H. D. Brummel, Diplomarbeit, Institut für Physik der Universität Dortmund, 1984.

42. H. Abramowicz et al., *Phys. Rev. Lett.* **57** (1986) 298; A. Blondel et al., *Z. Phys.* **C45** (1990) 361.

43. J. V. Allaby at al., *Phys. Lett.* **177B** (1986) 446, J. V. Allaby et al., *Z. Phys.* **C36** (1987) 611.

44. T. Bolton et al., Proceedings of the XXVIIth Rencontre de Moriond, Les Arcs, Savoie, France, March 15-22 1992, Editions Frontières, and S. R. Mishra, private communication.

45. P. Berge et al., *Z. Phys.* **C35** (1987) 443.

46. Particle Data Group, *Phys. Lett.* **239B** (1990) II,1.

47. S. A. Rabinowitz et al., *Phys. Rev. Lett.* **70** (1993) 134.

48. M. H. Shaevitz, *Nucl. Phys. B (Proc. Suppl)* **19** (1991) 270.

49. D. Yu. Bardin et al., *Nucl. Phys.* **B175** (1980) 435, *Nucl. Phys.* **B197** (1982) 435, Dubna Preprint E2-86-260 (1986).

50. D. Allasia et al., *Phys. Lett.* **249B** (1990) 336.

51. G. Preparata et al., *Phys. Rev. Lett.* **66** (1991) 687.

52. P. Amaudruz et al., *Phys. Rev. Lett.* **66** (1991) 2712.

53. J.-J. Aubert et al., *Nucl. Phys.* **B213** (1983) 31.

54. See the contributions of W. Hollik and W. Marciano to this book for a professional description of radiative corrections.

55. J. F. Wheater, C. H. Llewellyn-Smith, *Nucl. Phys.* **B208** (1982) 27; *Nucl. Phys.* **B226** (1983) 547 (erratum); A. Sirlin, W. J. Marciano, *Nucl. Phys.* **B189** (1981) 442; L. Liede et al., Helsinki University Preprint HU-TFP-8345 (1983).

56. See the contribution of A. K. Mann to this book.

57. J. Alitti et al., *Phys. Lett.* **B276** (1992) 354, F. Abe et al., *Phys. Rev.* **D43**

(1991) 2070.

58. The LEP Collaborations, *Phys. Lett.* **B276** (1992) 247.

59. G. Burgers and F. Jegerlehner, "Physics at LEP 1", CERN Report 89-08, Volume 1, 55.

60. See the contributions of D. Schaile and A. Blondel to this book.

61. See the contributions of P. Langacker to this book.

62. S. R. Mishra, "A Second Generation Neutrino Deep Inelastic Scattering Experiment at the FNAL Tevatron", presented at the workshop, "Fermilab in the 1900's", Beckenridge, Colorado; S. R. Mishra, Nevis-Preprint 1437, 1991.

63. "Precision Measurements of Neutrino Neutral Current Interactions Using a Sign-Selected Beam", Fermilab Proposal P815, 1991.

64. C. Guyot at al., DPHPE Saclay Print 88-741, 1988; H. Desportes al., Preprint DAPNIA/STCM 92-04, October 1992.

NEUTRINO-PROTON ELASTIC SCATTERING

ALFRED K. MANN
Department of Physics, University of Pennsylvania, 209 S. 33rd Street
Philadelphia, PA 19104 U.S.A.

Contents

1 Introduction

Apart from purely leptonic reactions, the simplest scattering reactions with which to study weak neutral currents are the semileptonic elastic reactions

$$\nu_\mu p \rightarrow \nu_\mu p \tag{1}$$

and

$$\bar{\nu}_\mu p \rightarrow \bar{\nu}_\mu p \quad . \tag{2}$$

The cross sections for these reactions are in principle well known within small theoretical uncertainty and are largely independent of detailed assumptions concerning the quark constituents of the proton. (However, see later for a particularly interesting exception to this statement.) Radiative corrections to the cross sections are small ($< 1\%$), and, experimentally, reactions (1) and (2) give rise to a clean distinctive signal. Furthermore, reactions (1) and (2) are sufficiently similar so that, when both are measured in the same detector and analyzed simultaneously, uncertainties in several experimental quantities are correlated, leading to reduced systematic errors on final results.

On the other hand, for good physics reasons, relatively few experiments in neutrino-proton elastic scattering have been done. At neutrino energies below about 0.5 GeV—as at the Los Alamos Meson Facility (LAMPF) and the Rutherford Laboratory spallation source—the low final state proton kinetic energies (T_p) make detection difficult in the face of other backgrounds. At neutrino energies in the vicinity of 1 GeV—e.g., at the CERN PS and the Brookhaven National Laboratory (BNL) AGS—the proton kinetic energies are also relatively low, $150 \lesssim T_p \lesssim 500$ MeV, and backgrounds from other reactions, including pion production and neutrino-neutron elastic scattering, must be identified and subtracted from the desired signal. The solution has been to construct massive, active, particle sensitive detectors which serve both as the neutrino target and to identify the proton and measure its energy and angle relative to the incident neutrino while discriminating against backgrounds. At higher neutrino energies—greater than about 10 GeV from the CERN SPS and Fermilab—the elastic (and quasielastic, i.e., $\nu_\mu n \to \mu^- p$) cross sections are small compared with the total cross section and hence those events are difficult to extract from the totality of final states. Reactions (1) and (2) have thus far only been studied with neutrinos in the 1 GeV region.

Nevertheless, experimental studies of neutrino-proton elastic scattering have contributed significantly to our understanding of the nature of the weak neutral current and the now well established Electroweak Theory. Neutrino-proton elastic scattering is one of seven physical phenomena that lead directly to precise evaluation of the weak neutral coupling strength, $\sin^2\theta_W$, the single undetermined parameter in the Electroweak Theory [1, 2]. In establishing the validity of the theory initially, each of the seven measurements played an important part by yielding essentially the same value of $\sin^2\theta_W$ despite the substantive differences among the phenomena. These measurements extended over the remarkably large range in Q^2 from roughly 10^{-11} to 10^4 GeV^2, where Q^2 is the four-momentum transfer within the reaction, and effectively certified $\sin^2\theta_W$ as a universal constant [3]. In the future, as experimental precisions improve, the same neutral scattering phenomena—including neutrino-proton elastic scattering—will be used to test the Electroweak Theory at a higher level of precision and to probe for departures from the physics covered in that theory [4].

In what follows, a brief summary of the earlier experiments on $\nu_\mu p$ is given, which leads to a more detailed description of the method and results of the latest, most precise measurement carried out at BNL by a BNL–Brown-Hiroshima–KEK–Osaka–Pennsylvania–Stony Brook collaborations referred to as E-734 [5]. The next section (3) provides an abbreviated theoretical framework with which to extract $\sin^2\theta_W$ and other parameters of interest from those data. The results are discussed in Section 4. In Section 5 the question of strange quark contributions to neutrino-proton elastic scattering is addressed in the light of the non-zero value of the strange quark form factor G_s found in E-734. G_s is a measure of the strength of the matrix element $< N|\bar{s}\gamma_\mu\gamma_5 s|N >$ and therefore a measure of the strange quark contribution to the nucleon spin. Finally, the present experimental situation in neutrino-proton elastic scattering and prospects for its future development are discussed in Section 6.

2 Measurements of Neutrino-Proton Elastic Scattering

The earliest search for $\nu_\mu p \to \nu_\mu p$, published in 1964 [6], set an upper limit on the ratio $R_\nu = \sigma(\nu_\mu p \to \nu_\mu p)/\sigma(\nu_\mu n \to \mu^- p)$ of 0.03. This was corrected in 1970 when many of the same authors, again using a CERN heavy liquid bubble chamber, gave a new upper limit for R_ν of $\leq 0.12 \pm 0.06$, for $150 \leq T_p < 500$ MeV and $1 < E_\nu < 4$ GeV [7]. The reaction $\nu_\mu p \to \nu_\mu p$ was first positively observed in two independent experiments [8, 9] at BNL six years later, and confirmed at CERN [10] in 1978. The corresponding reaction $\bar{\nu}_\mu p \to \bar{\nu}_\mu p$ was measured at BNL in 1976 [11] and again in 1981 [12]. Apart from an occasional conference report, e.g., from the Argonne National Laboratory bubble chamber in 1974 [13], which yielded $R_\nu = 0.08 \pm 0.20$ for $Q^2 \geq 0.43$ GeV^2, the

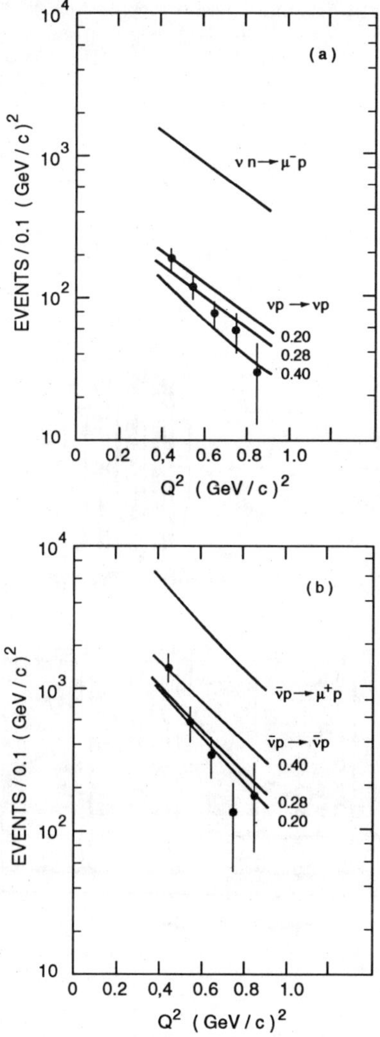

Figure 1: Differential cross section for (a) $\nu_\mu p \to \nu_\mu p$ and (b) $\bar{\nu}_\mu p \to \bar{\nu}_\mu p$. The solid curves are the predictions for the WS-GIM model with the indicated values of $\sin^2 \theta_W$ [14]

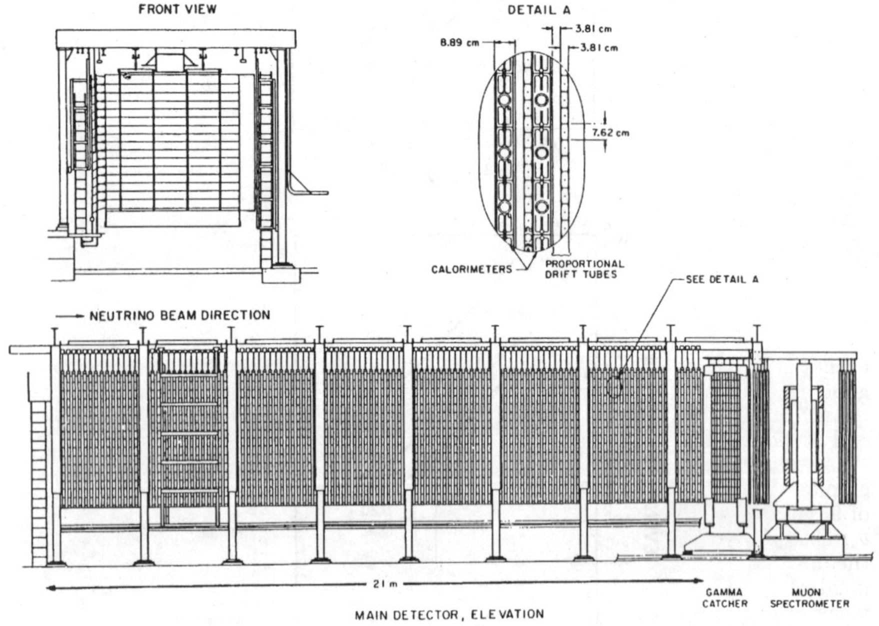

Figure 2: A schematic drawing of the BNL-Brown-KEK-Osaka-Pennsylvania-Stony Brook neutrino detector (E-734) [5]

citations above constitute the totality of experimental publications on neutrino-proton elastic scattering prior to 1982.

Higher statistics samples of $\nu_\mu p \to \nu_\mu p$ and $\bar{\nu}_\mu p \to \bar{\nu}_\mu p$ data were reported and analysed in a paper published in 1982 [14]. In that paper, which included data from earlier publications [9, 11], 240 $\nu_\mu p$ and 119 $\bar{\nu}_\mu p$ events determined the ratios

$$R_\nu = \frac{\sigma(\nu_\mu p \to \nu_\mu p)}{\sigma(\nu_\mu n \to \mu^- p)} = 0.11 \pm 0.015 \tag{3}$$

$$R_{\bar{\nu}} = \frac{\sigma(\bar{\nu}_\mu p \to \bar{\nu}_\mu p)}{\sigma(\bar{\nu}_\mu p \to \mu^+ n)} = 0.19 \pm 0.035 \tag{4}$$

and

$$R_{NC} = \frac{\sigma(\bar{\nu}_\mu p \to \bar{\nu}_\mu p)}{\sigma(\nu_\mu p \to \nu_\mu p)} = 0.41 \pm 0.09 \quad , \tag{5}$$

all for $0.4 < Q^2 < 0.9\ GeV^2$, with $Q^2 = -q^2 = 2M_pT_p$, and using

$$R_{CC} = \frac{\sigma(\bar{\nu}_\mu p \to \mu^+ n)}{\sigma(\nu_\mu n \to \mu^- p)} = 0.239 \quad . \tag{6}$$

The value of R_{CC} in the interval $0.4 < Q^2 < 0.9(GeV/c)^2$ in Eq. (6) was obtained by calculation with normalization provided by a measured sample of low Q^2 quasielastic events extrapolated to cover the entire Q^2 region. The systematic errors assigned to $R_\nu, R_{\bar{\nu}}$, and R_{NC} were approximately 15%, 20%, and 25%, respectively, intended by the authors to be added in quadrature with the statistical errors. In addition, the differential cross sections $d\sigma(\nu_\mu p)/dQ^2$ and $d\sigma(\bar{\nu}_\mu p)/dQ^2$ were measured in the same Q^2 interval as the ratios above. These are shown in Fig. 1.

A contemporary experiment with results based on 94 $\nu_\mu p$ events and 104 $\bar{\nu}_\mu p$ events [12] yielded $R_\nu = 0.11 \pm 0.03$, and $R_{NC} = 0.44 \pm 0.12$, and Q^2-distributions in good agreement with the values in Eqs. (3) and (5). Hence confirmed, approximate quantitative results for neutrino-proton elastic scattering were available before the end of 1982.

A more massive, finer grained detector, shown in Fig. 2, was subsequently employed in a neutrino scattering program (E-734) at BNL to yield data samples of 951 $\nu_\mu p$ and 776 $\bar{\nu}_\mu p$ events [15] This detector was also used to study the $\nu_\mu e$ and $\bar{\nu}_\mu e$ elastic reactions [16]. Most importantly, by means of the muon spectrometer at the downstream end of Fig 2, it was possible to study in detail the properties of the incident ν_μ and $\bar{\nu}_\mu$ beams, including contaminations therein [5], as well as to use the measured rates for the quasielastic reactions for carefully determined normalization of the neutral current rates.

To see the evolution of the experimental technique in the interval between the earlier experiments and the latest one, we give the values and associated errors of the measured ratios $R_\nu, R_{\bar{\nu}}$ and R_{NC} obtained from E-734 [15] for comparison with the corresponding quantities in Eqs. (3), (4), and (5). From [15] one finds

$$R_\nu = 0.153 \pm 0.007(\text{stat}) \pm 0.017(\text{syst}) \tag{7}$$

$$R_{\bar{\nu}} = 0.218 \pm 0.012(\text{stat}) \pm 0.023(\text{syst}) \tag{8}$$

and

$$R_{NC} = 0.302 \pm 0.019(\text{stat}) \pm 0.037(\text{syst}) \tag{9}$$

all for $0.5 < Q^2 < 1.0\ GeV^2$.

One sees agreement within errors among the results of [12, 14] and [15], but the statistical errors are lower in Eqs. (7), (8), and (9) by a factor of about 3. The systematic errors, representing improved understanding of the ν_μ and $\bar{\nu}_\mu$ beams, the absolute normalization, and semi-empirical knowledge of the backgrounds, are appreciably smaller and well quantified in the later experiment. This is plainly shown by the Q^2-distributions in Fig. 3 when compared with those in Fig 1.

The results for $\sin^2\theta_W$ and the axial-vector form factor $G_A(Q^2)$ obtained from these experiments, and the implications therefrom, are discussed below, after a brief summary in the next section of the theoretical framework for understanding neutrino-proton elastic scattering.

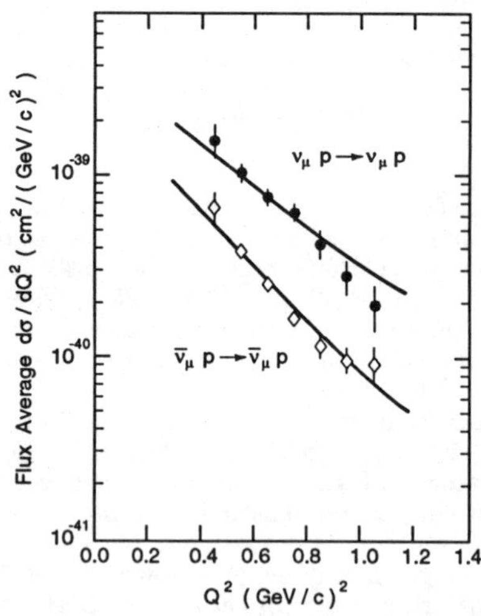

Figure 3: The data points are the measured flux-averaged differential cross sections for $\nu_\mu p \rightarrow \nu_\mu p$ and $\bar{\nu}_\mu p \rightarrow \bar{\nu}_\mu p$ from the E-734 experiment. The solid curves are the best fits to the combined data with the values $M_A = 1.06$ GeV / c^2 and $\sin^2\theta_W = 0.220$. This fitting procedure imposes adjustment of solid curves by scale factors of 1.05 for $\nu_\mu p$ and 1.09 for $\nu_\mu p$ consistent with the absolute scale uncertainty of approximately 11% in each of the individual cross sections, which was included in the fitting procedure. The errors bars represent statistical errors and also include Q^2 -dependent systematic errors.

3 Theory of Neutrino-Proton Elastic Scattering

Knowing the incident neutrino direction and assuming a fixed proton target, a $\nu_\mu p$ elastic scattering event is completely described by two independent kinematic variables. The directly measured quantitites are the range of the recoiling proton, its rate of energy loss, and the angle θ_p with respect to the direction of the incident neutrino. Range and energy loss measurements identify the particle as a proton and determine T_p. Track measurements in wire chambers determine θ_p. The data analysis of recent experiments has been based on the measured variables θ_p and Q^2, which also specify the incident neutrino energy through the relation

$$E_\nu = \frac{M_P}{\cos\theta_p(1+2M_p/T_p)^{1/2}-1} \quad . \tag{10}$$

In experiments [14] and [15], 80% of the $\nu - p$ elastic events were from protons bound in carbon nuclei. In the E_ν and Q^2 regions of the experiments, the impulse approximation is valid so that the protons could be treated as free, apart from effects due to Fermi momentum, Pauli exclusion, and free particle nuclear scattering which were taken into account.

In the Electroweak Theory the effective Lagrangian for neutrino-hadron elastic scattering is [17]

$$L^{\nu p} = \frac{G_F}{\sqrt{2}}\, \bar{\nu}\gamma^\mu(1+\gamma_5)\nu J^h_\mu \tag{11}$$

where the hadronic current J^h also has polar-vector V and axial-vector A components. In Eq. (11) G_F is the Fermi (weak) constant and γ^μ and γ_5 are Dirac matrices. J^h may be written as

$$J^h_\mu = \alpha V^{(3)}_\mu + \beta A^{(3)}_\mu + \gamma V^{(0)}_\mu + \delta A^{(0)}_\mu + ... \tag{12}$$

with $V^{(3)}_\mu (V^{(0)}_\mu)$ the isovector (isoscalar) polar vector currents, and $A^{(3)}_\mu (A^{(0)}_\mu)$ the isovector (isoscalar) axial-vector currents. The ellipses represent heavy quark $(s, c, b \ldots)$ currents which (for s-quarks) will be introduced explicitly below. In the standard model

$$\begin{aligned} \alpha = &\; 1 - 2\sin^2\theta_W \quad , \quad \beta = 1 \\ \gamma = &\; -\tfrac{2}{3}\sin^2\theta_W \quad , \quad \delta = 0 \text{ (usually)} \quad . \end{aligned} \tag{13}$$

The parameter $\sin^2\theta_W$ may also be taken as the strength of coupling of the electromagnetic and weak neutral currents, and hence only polar vector currents contain $\sin^2\theta_W$.

Assuming time reversal invariance, isospin invariance of the hadronic current, and no second class currents [18], the polar-vector part of the matrix element of J^h between nucleon states is

$$V_\mu = \bar{u}_f(p') \left[\gamma_\mu F_1(Q^2) + \frac{i\sigma_{\mu\nu}}{2M_p} q^\nu F_2(Q^2)\right] u_i(p) \quad , \tag{14}$$

where q^ν is the momentum transfer. The form factors $F_1(Q^2)$ and $F_2(Q^2)$ in Eq. 14 are equated with the corresponding electromagnetic form factors obtained from electron-hadron elastic scattering [19], in which both the isovector and isoscalar form factors have the same Q^2-dependence. One can write

$$F_i(Q^2) = \left(\frac{1}{2} - \sin^2\theta_W\right) F_i^{(3)}(Q^2) - \sin^2\theta_W F_i^{(o)}(Q^2) - \frac{1}{2}F_i^S(Q^2) \ , \tag{15}$$

where the form factors F_1^S and F_2^S which represent, respectively, the charge radius and magnetism of the strange quark sea with the $\sin^2\theta_W$ dependence suppressed are now explicitly included. A dipole form is customarily assumed for the Q^2-dependence of the non-strange F_i, so that

$$\begin{aligned} F_1(Q^2) &= \tau F_2(Q^2) + \frac{F_1(0)}{(1+Q^2/M_V^2)^2} \\ F_2(Q^2) &= \frac{F_2(0)}{(1+\tau)(1+Q^2/M_V^2)^2} \end{aligned} \tag{16}$$

where $\tau = Q^2/4M_p^2$.

The neutral axial-vector current is

$$A_\mu = \bar{u}_f(p')\gamma_\mu\gamma_5 G_A(Q^2)u_i(p) \tag{17}$$

with

$$G_A(Q^2) = \frac{\frac{1}{2}g_A}{(1+Q^2/M_A^2)^2} + \frac{1}{2}G_S \tag{18}$$

where G_S represents the explicit contribution to the proton spin $(\bar{s}\gamma_\mu\gamma_5 s)$ of the strange quark sea. In Eqs. (16) and (18), $M_V = 0.84$ GeV, the world average value of $M_A = 1.032 \pm 0.036$ GeV (but see below), and $g_A = 1.26$, as determined from neutron decay.

Assuming $F_1(Q^2)$ and $F_2(Q^2)$ to be well approximated by Eq. (16) with $F_1(0)$ and $F_2(0)$ for $\sin^2\theta_W = 0.233$ given by $F_1(0) = \frac{1}{2} - 2\sin^2\theta_W = 0.034$ ($F_1^S(0)$ is identically zero), and $F_2(0) = [(\frac{1}{2} - \sin^2\theta_W)(\mu_p - \mu_n) - \sin^2\theta_W(\mu_p + \mu_n) - F_2^S(0)/2] = 1.017 - F_2^S(0)/2$, then the quantities to be determined from $\nu_\mu p$ and $\bar{\nu}_\mu p$ elastic scattering measurements are $\sin^2\theta_W, M_A$ (if an independent determination is desired), and, when possible, the strange quark form factors F_1^S, F_2^S, and G_S,

4 Results

In the pre-1982 experiments described above the aims varied with the year in which the experiment was performed and the statistical precision that could be achieved. Early on, for example, rough confirmation of the "Weinberg-Salam" model was the goal [20]. Later, when larger samples of $\nu_\mu p$ and $\bar{\nu}_\mu p$ events became available, it was possible to exploit the specific properties of those samples, viz., that the differential cross section for $\nu_\mu p$ scattering is sensitive to both $\sin^2\theta_W$ and M_A, while the differential cross section for the $\bar{\nu}_\mu p$ reaction is primarily sensitive to M_A. These properties are exhibited in Fig. 4, taken from reference [15]. The simultaneous fit to the $d\sigma(\nu_\mu p)/dQ^2$ and $d\sigma(\bar{\nu}_\mu p)/dQ^2$ data in Fig. 3 yielded

$$\sin^2\theta_W = 0.218^{+0.039}_{-0.047}$$

and

$$M_A = 1.06 \pm 0.05 \text{ GeV} \tag{19}$$

with a χ^2 for the fit of 15.8/14 dof. The best fit curves are shown in Fig. 3.

In those data, for the first time, a search was made for an additional term in the axial-vector current, i.e., for the presence of a form factor, possibly G_S, in

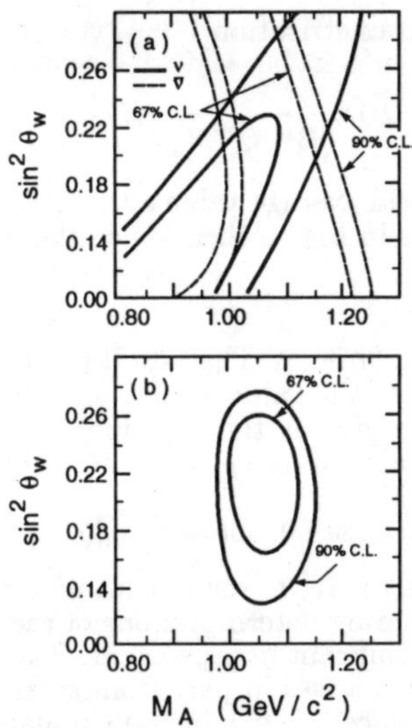

Figure 4: Final-fit confidence-level contours in the $\sin^2\theta_W$, M_A space for (a) $d\sigma / dQ^2$ fit individually for the neutrino and antineutrino elastic scattering samples. Note the relative insensitivity to $\sin^2\theta_W$ of the antineutrino. (b) A simultaneous fit to both neutrino and antineutrino differential cross sections.

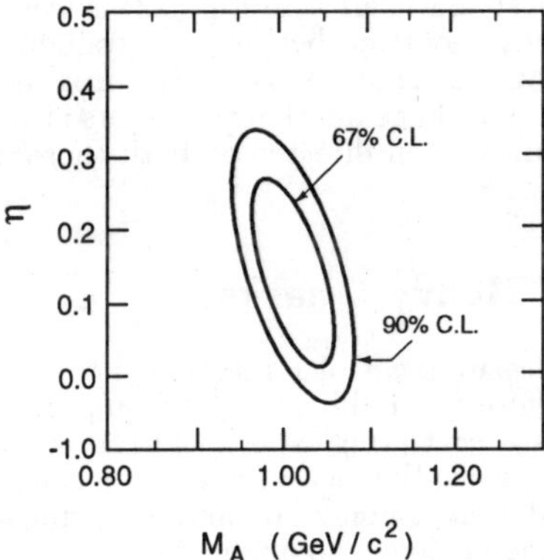

Figure 5: Simultaneous fit of $d\sigma / dQ^2$ for the neutrino and antineutrino elastic scattering samples in the M_A, η space with $\sin^2\theta_W$ fixed at 0.220. M_A has been constrained to the world average value.

addition to G_A by using the parametrization

$$G_A(Q^2) = \frac{\frac{1}{2}g_A}{(1+Q^2/M_A^2)^2}(1+\eta) \tag{20}$$

and constraining M_A to the world average value and $\sin^2\theta_W$ to the value 0.22. A fit to the differential cross sections in Fig. 3 then led to the result

$$\eta = 0.12 \pm 0.07 \tag{21}$$

or, equivalently, $0.00 \leq \eta \leq 0.25$ at 90% c.l. This result is independent of the assumed value of $\sin^2\theta_W$, but correlated to the value of M_A, as indicated in Fig. 5. For this reason the authors preferred to give as the principal results of their analysis the value of η in Eq. (21) and the value of $\sin^2\theta_W$ when M_A is fixed at the world average value, namely

$$\sin^2\theta_W = 0.220 \pm 0.016(\text{stat})^{+0.023}_{-0.031}(\text{syst}) \quad . \tag{22}$$

Statistically, Eq. (22) represents a 7% measurement of $\sin^2\theta_W$, comparable in precision with other neutrino scattering determinations of the same quantity [3]. This result for $\sin^2\theta_W$ is capable of significant improvement. The limitation in Eq. (22) is the systematic uncertainty which arises in part from small Q^2-independent effects, e.g., proton identification and uncertainties in the simulation of nuclear processes and the neutrino beam shape; and in part from small Q^2-dependent uncertainties which might be introducing distortions of the distributions in Fig. 3. In the entire Q^2 region of [15], 16% of the final $\nu_\mu p$ sample was calculated to be due to the reaction $\nu_\mu n \to \nu_\mu n$ followed by an $n-p$ collision, although $\sigma(\nu_\mu n \to \nu_\mu n)$ is expected to be approximately $1.5\sigma(\nu_\mu p \to \nu_\mu p)$. Similarly, in the $\bar{\nu}_\mu p$ final sample the fraction was 11%. These were subtracted statistically using the Monte Carlo simulation of the experiment after detailed comparison of the simulated neutron scattering with observed neutron scattering in the $\bar{\nu}_\mu p \to \mu^+ n$ channel. Reduction or possibly elimination of these contributions to the systematic error in Eq. (22) is feasible with a finer grain detector and the more intense neutrino beams now available. In addition to the value of a more precise ($\sim 1\%$) determination of $\sin^2\theta_W$ from neutrino-proton elastic scattering [4], the study of the strange quark form factors is also a recommendation for further intensive experimentation directed at both those goals, as discussed in the next section.

5 Contributions of Heavy Quarks

The somewhat primitive measurement of η mentioned toward the end of the last section occurred shortly before the publication of results from a polarized muon-polarized proton deep inelastic scattering experiment [21] on quark contributions to the proton spin. As a consequence the measurement of η elicited only mild interest. It was, however, recognized that a non-zero value of η might be due either to a heavy (s) quark current in the proton or to a non-standard axial vector term. Early estimates of η, based on a contribution to the proton spin from the $s\bar{s}$ sea ranged from 0.04 to as much as 0.25 [22] (apart from sign), and so were consistent with the experimental value in Eq. (21). Accordingly, the quantity η has for simplicity been equated with the form factor G_S in much of the literature. However, see the discussion in reference [23].

Note that the presence of G_S in neutrino-proton elastic scattering and the correspondence of that scattering process with the deep inelastic scattering of polarized charged leptons on polarized protons are accounted for by the unit chirality of the neutrino. For example, for backward (180^0) scattering in the center of mass frame, ν_μ with negative helicity are forbidden to scatter from protons with positive helicity values because the scattering is only S-wave in nature. The result is the scattering of polarized leptons by effectively polarized protons in $\nu_\mu p$ scattering as well as in the polarized deep inelastic scattering.

The measurements of the EMC collaboration [21] have been interpreted to suggest that a relative contribution to the proton spin from the strange quarks is

$$\Delta s = -0.23 \pm 0.08 \tag{23}$$

where $\Delta s = \int_0^1 [s^\uparrow(x) + \bar{s}^\uparrow(x) - s^\downarrow(x) - \bar{s}^\downarrow(x)]dx$, with x the fraction of momentum carried by the $s(\bar{s})$-quarks. With similar definitions of Δu and Δd, the values are

$$\Delta u = 0.74 \pm 0.08 \tag{24}$$

and

$$\Delta d = -0.51 \pm 0.08 \quad . \tag{25}$$

The negative signs in Eqs. (23) and (25) indicate that the mean z-component of the s- and d- quarks is oppositely directed to the proton spin. Consequently, Eqs. (23)–(25) suggest that the total quark (plus antiquark) contribution to the proton spin is approximately zero. This is a completely unexpected result in the naive non-relativistic quark model, and indeed the source of the proton spin is not yet accounted for quantitatively in non-perturbative quantum chromodynamics.

The quantity Δs may be equated directly with the form factor G_S, so that the value of η found from $\nu_\mu p$ elastic scattering has been taken as direct, albeit weakly, supporting evidence for the conclusion that $\Delta s \neq 0$ [23].

6 Future Experiments

The question of the parton origin of the nucleon spin has been and continues to be addressed at length in the theoretical literature. It is also the goal of several experiments with polarized electrons and muons that are now in the planning or construction stage [24], and has been discussed as an outcome of more precise parity violation atomic physics measurements [25]. A new measurement of $\nu_\mu p$ elastic scattering is projected [26] at low ν_μ energies which will strictly constrain Q^2 to values below about $0.1\ Gev^2$. In that circumstance the expression for the differential cross section becomes

$$d\sigma(\nu_\mu p)/dQ^2 \cong \frac{G_F^2}{2\pi}\left[\frac{1}{4}(g_A + G_S)^2\left(1 + \frac{Q^2}{4E_\nu^2}\right) + F_1(0)^2\left(1 - \frac{Q^2}{4E_\nu^2}\right)\right] \quad , \tag{26}$$

with $g_A = 1.26$, and $F_1(0) = 0.034 - F_1^S(0)/2$. At $Q^2 = 0$, the strange charge radius $F_1^S(0) = 0$, and the only unknown in Eq. (26) is G_S. This experiment has the potential to measure Δs with an error of about 0.04, an improvement of a factor of approximately 2 relative to the errors in Eqs. (21) and (23), providing that neutrino scattering on bound protons can be disentangled from that on the free protons in the target, and account taken of final state protons that are the secondary products of $\nu_\mu n \to \nu_\mu n$ in the detector [27].

At higher neutrino energies, where higher Q^2 reactions take place, measurement of the strange polar vector form factors, F_1^S and F_2^S, might be made as suggested by a recent re-analysis of the Q^2-distributions in Fig. 3 [28]. Interest in the strange quark effects in the nucleon, particularly as important tests of non-perturbative QCD, is likely to encourage the performance of future precise $\nu_\mu p$ measurements from which improved precision on the value of $\sin^2\theta_W$ would also be forthcoming.

7 Summary

Measurements of neutrino-proton elastic scattering have served as one of the several precision tests of the Standard Electroweak Theory. Elastic semileptonic scattering depends on the form factors of the hadron involved. Hence the equivalence of the polar-vector form factors in ep and $\nu_\mu p$ elastic scattering, as demonstrated in the $\nu_\mu p$ measurements that came many years after the ep experiments, is an important quantitative confirmation of the theory. Perhaps more directly, the value of $\sin^2\theta_W$ extracted from $\nu_\mu p$ measurements, in conjunction with the values obtained from other very different weak neutral current phenomena, serves to confirm the universality of that fundamental parameter specifically and the theory generally.

As the precision of $\nu_\mu p$ measurements has improved, it has been possible to search for and evaluate a heavy quark contribution to the scattering process, in particular, to indicate a non-zero strange quark contribution to the proton spin. The prospect of solidifying this result in a more precise measurement of $\nu_\mu p$ elastic scattering, and the possibility of extending the measurements to include other strange quark form factors are strong motivations for further study of the $\nu_\mu p$ reaction.

Finally, it is worth emphasizing that there are available the results of only one high statistics experiment involving both $\nu_\mu p$ and $\bar{\nu}_\mu p$ with carefully studied systematic errors and backgrounds, from which most of the information described in this article has been obtained. At least one additional experiment, aimed at the highest precision possible with current techniques, and motivated by the need to measure $\sin^2\theta_W$ to the order of 1%, and to study the strange quark influence on the properties of the proton would be well-justified in the future.

I am indebted to many colleagues and collaborators, past and present, who contributed to my appreciation of neutrino scattering. This research was supported in part by the U.S. Department of Energy.

1. S. Glashow, Nucl. Phys. 22, 579 (1961); S. Weinberg, Phys. Rev. Lett. 19, 1264 (1967); A. Salam, in *Elementary Particle Theory*, edited by N. Svartholm (Almquist & Witsells, Stockholm, 1968, p. 367.
2. The explicit form of the parameter, $\sin^2\theta_W$, is conventional in the sense that $\sin\theta_W \to 0$ when $\theta_W \to 0$, and $\sin^2\theta_W$ is always positive in accord with the definition of coupling strength.
3. U. Amaldi et al. Phys. Rev. D36, 1385 (1987).
4. P. Langacker, M. Luo, and A. K. Mann, Rev. Mod. Phys. 64, 87 (1992).
5. L. A. Ahrens et al., Nucl. Instrum. Methods A254, 515 (1987); Phys. Rev. D34, 75 (1986).
6. M. M. Block et al. Phys. Lett. 12, 281 (1964).
7. D.C. Cundy et al., Phys. Lett. B31, 478 (1970).
8. W. Lee et al., Phys. Rev. Lett. 37, 186 (1976).

9. D. Cline et al. Phys. Rev. Lett. 37, 252 (1976).
10. M. Pohl et al. Phys. Lett. B72, 489 (1978); H. Faissner et al., Phys. Rev. D21, 555 (1980).
11. D. Cline et al., Phys. Rev. Lett. 37, 648 (1976).
12. P. Coteus et al., Phys. Rev. D24, 1420 (1981).
13. P. Schreiner, in Proc. XIV Int'l Conf. on High Energy Physics, London, 1974, edited by J.R. Smith, (Rutherford Laboratory), p. IV-126.
14. J. Horstkotte et al. Phys. Rev. D25, 2743 (1982).
15. L. A. Ahrens et al., Phys. Rev. D35, 785 (1987).
16. L. A. Ahrens et al., Phys. Rev. D41, 3297 (1990); see also J. Panman, this volume.
17. E. Fischbach et al., Phys. Rev. Lett. 37, 582 (1976); Phys. Rev. D15, 97 (1977); J. E. Kim et al., ibid. 18, 123 (1978); Rev. Mod. Phys. 53, 211 (1981).
18. See, for example, the papers in ref. 17.
19. M. Olsson et al., Phys. Rev. D17, 2938 (1978); B. Bartoli et al., Riv. Nuovo Cimento 2, 2411 (1972).
20. See, for example, A. Entenberg et al., Phys. Rev. Lett. 42, 1198 (1979).
21. J. Ashman et al. (European Muon Collaboration), Phys. Lett. B206, 364 (1989).
22. J. Collins et al., Phys. Rev. D18, 242 (1981); R. Mohapatra et al., ibid. 19, 2165 (1979); S. Oneda et al., Phys. Lett. B88, 343 (1979); L. Wolfenstein, Phys. Rev. D19, 3450 (1979).
23. D. B. Kaplan and A. Manohar, Nucl. Phys. B310, 527 (1988); D.H. Beck, Phys. Rev. D39, 3248 (1989).
24. CERN Expt. NA47 (SMC); SLAC Expts. E142, E143; HERA Expt. (HERMES); Bates Expt 89-06 (SAMPLE).
25. B. A. Campbell, John Ellis, and R. A. Flores, Phys. Lett. B225, 419 (1989).
26. Los Alamos National Laboratory (LAMPF) Proposal LA-11842P, (1990), spokesman: W. C. Louis.
27. Private communication from G.T. Garvey, W. C. Louis, and D. H. White.
28. G. T. Garvey, W. C. Louis, and D. H. White, "Determination of Proton Strange Form Factors from νp Elastic Scattering", LAMPF preprint, December, 1992. Submitted to Phys. Lett.

NEUTRINO-ELECTRON SCATTERING

Jaap Panman

CERN
1211 Geneva 23, Switzerland

Contents

1 Introduction and theoretical framework

Two decades ago the first anti-neutrino-electron scattering event was observed in the Gargamelle bubble chamber at the CERN Proton Synchrotron [1]. Together with the first evidence for neutral-currents in semi-leptonic reactions [2] this event marked the start of the extensive experimental study of neutral-current phenomena, and of the successes of the Standard Model [3]. The cross-sections for neutrino-electron scattering, an interaction of two point-like particles is readily calculable in the framework of the Standard Model [4].

The study of neutrino-electron scattering can provide sensitive tests of the Standard Model; several attempts to combine data from multiple experiments have been made [5, 6, 7, 8, 9]. Improved precision on the measurements of neutral-current processes can reveal small deviations from the predictions of the Standard Model, and can give information on unknown parameters in the model.

Neutrino-electron scattering is a purely leptonic process, and the theoretical calculations are straightforward without being affected by uncertainties such as those that limit the accuracy on electroweak parameters obtained with semileptonic scattering, which are more model-dependent in their interpretation, due to the description of the quarks in the interaction. Neutrino-electron scattering forms an invaluable ingredient in the comparisons made of the different experimental results in our tests of the Standard Model. The study of this process has given a convincing confirmation of the ability of the Standard Model to describe electroweak processes at the tree level and the last generation of experiments allows us to confront the next-order corrections with the measurements.

A study of electron-neutrino and electron-antineutrino-electron scattering can determine the interference of the purely leptonic neutral-current (NC) and charged-current (CC) amplitudes, both of which contribute to this process.

1.1 *Kinematics*

The neutrino-electron reactions are two-body elastic scattering processes. These can be characterized by a single kinematic variable, for which we choose θ_e, the angle of the outgoing electron with respect to the incoming neutrino.

Defining $\mathbf{k}$ and $\mathbf{k}'$ the four-momenta of the incoming and outgoing neutrino and, likewise, $\mathbf{p}$ and $\mathbf{p}'$ for the electron, energy and momentum conservation are expressed as $\mathbf{k} - \mathbf{p}' = \mathbf{k}' - \mathbf{p}$. Remembering for elastic scattering, $\mathbf{k}'^2 = \mathbf{k}^2$ and $\mathbf{p}'^2 = \mathbf{p}^2$, one finds $\mathbf{k}\cdot\mathbf{p}' = \mathbf{k}'\cdot\mathbf{p}$. With the definitions: E_ν, the incoming neutrino energy, E_e and p_e, the outgoing electron energy and momentum and, m_e, the electron mass, one finds:

$$E_\nu E_e - E_\nu p_e \cos\theta_e = m_e(E_\nu - E_e + m_e).$$

With the approximation $m_e \ll E_e$ one can write

$$E_\nu E_e(1 - \cos\theta_e) = m_e(E_\nu - E_e).$$

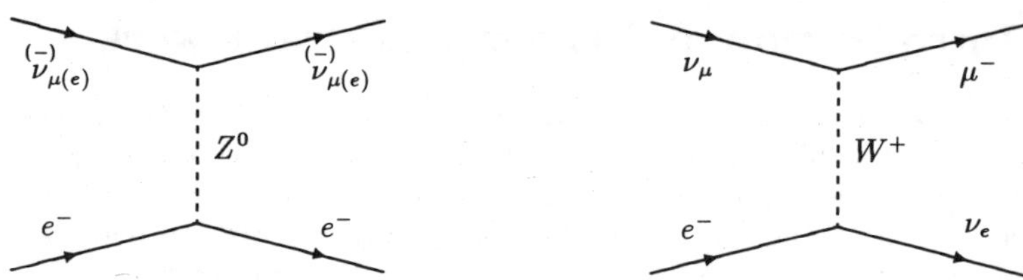

Fig. 1: *Feynman graphs of neutrino-electron scattering. Left, the neutral-current process and right, the charged-current process for muon neutrinos.*

Introducing the frequently used kinematic variably y, the fractional energy loss of the neutrino in the laboratory system,

$$y = \frac{E_e}{E_\nu}, \tag{1}$$

and substituting the small angle approximation for $\cos\theta_e$ one finds,

$$E_e\theta_e{}^2 = 2m_e(1 - y). \tag{2}$$

The bound $0 < y < 1$ translates into a kinematic limit for the recoil angle:

$$E_e\theta_e{}^2 < 2m_e. \tag{3}$$

This kinematical bound provides one of the most significant methods to isolate νe-scattering events from experimental backgrounds, e.g. at high energy accelerators the angle θ_e is limited to $\theta_e < 0.01$ rad for $E_\nu = 10$ GeV. At the same time this fact has prevented a direct measurement of y on an event-by-event basis, since the experimental angular resolutions prove to be of the same order of magnitude as the kinematical bound. When m_e cannot be neglected compared to E_ν and E_e, the kinematical bound takes the form $E_e(1 - \sqrt{1 - m_e{}^2/E_e{}^2}\cos\theta_e) < m_e$.

It is useful to consider the four-momentum transfer $q^2 = (\mathbf{k} - \mathbf{k}')^2 = (\mathbf{p} - \mathbf{p}')^2$. Defining $Q^2 = -q^2$ one obtains the positive quantity $Q^2 = 2\mathbf{p}\cdot\mathbf{p}' \approx 2m_eE_e$. For the present experiments this quantity ranges from 10^{-6} GeV^2 for $\overline{\nu}_e e$-scattering at nuclear reactors to 10^{-2} GeV^2 at high energy accelerators, small compared to $m_Z{}^2$.

1.2 *Cross-sections*

The muon-neutrino induced processes $\nu_\mu e \to \nu_\mu e$ and $\overline{\nu}_\mu e \to \overline{\nu}_\mu e$ can only proceed through the neutral-current interaction; the charged-current produces a muon in the final state (see Fig. 1). To write down the effective Lagrangian we use the fact that neutrinos have only left-handed interactions, while the electron has a left-handed and a right-handed coupling. The expression is further simplified by only considering processes with $Q^2 \ll m_Z{}^2$, such that the propagator reduces to a constant:

$$\frac{G_F}{\sqrt{2}} = \frac{e^2}{8m_W{}^2\sin^2\theta_W}, \tag{4}$$

where the Fermi constant, G_F, is measured in muon decay. With these simplifications, the NC Lagrangian reads:

$$\mathcal{L}^{NC}_{eff} = \frac{G_F}{\sqrt{2}}\rho\left[\overline{\nu}\gamma^\alpha(1-\gamma^5)\nu\right]\left[\overline{e}\gamma^\alpha(g_V-\gamma^5 g_A)e\right] + h.c. \tag{5}$$

Here we use a notation with the vector (V) and axial vector (A) coupling constants of the electron, g_V and g_A, which can be related to the chiral coupling constants:

$$2g_L = g_V + g_A \quad \text{and} \quad 2g_R = g_V - g_A. \tag{6}$$

The ρ-factor describes the relative strength of the neutral-current compared to the charged-current, and is (at the tree level) equal to unity in the Standard Model. We will frequently use a notation where ρ is absorbed into the coupling constants, $\rho g_V \rightarrow g_V$ and $\rho g_A \rightarrow g_A$. Another useful set of parameters to describe the observations is $\sin^2\theta_W$ and ρ, which can be related to the other ones by:

$$g_V = \rho(-\frac{1}{2} + 2\sin^2\theta_W) \quad \text{and} \quad g_A = -\frac{1}{2}\rho. \tag{7}$$

The electron-neutrino induced νe processes can proceed through both the charged-

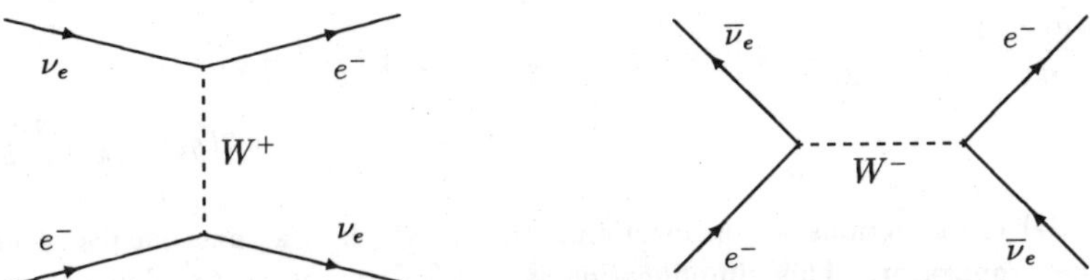

Fig. 2: *Feynman graphs for the charged current electron neutrino-electron scattering process.*

current and neutral-current interaction (Fig. 2). The effective charged-current Lagrangian describes a purely left-handed interaction:

$$\mathcal{L}^{CC}_{eff} = \frac{G_F}{\sqrt{2}}\rho\left[\overline{\nu}\gamma^\alpha(1-\gamma^5)\nu\right]\left[\overline{e}\gamma^\alpha(1-\gamma^5)e\right], \tag{8}$$

which should be added to the neutral-current amplitude to obtain the $\nu_e e$ and $\overline{\nu}_e e$ cross-sections.

The differential cross-section is obtained by squaring the amplitudes (see e.g. [10]). One obtains a left-handed term proportional to $g_L{}^2 m_e E_\nu$, a right-handed term ${g_R}^2 m_e(E_\nu - E_e)^2/E_\nu$, and a left-right interference term $g_L g_R\, {m_e}^2 E_e/E_\nu$, which vanishes as m_e/E_ν for large E_ν as expected for a term connecting left-handed and right-handed states. One observes that the right-handed term is proportional to $(1-y)^2$. In the centre-of-mass system y is nothing else than the scattering angle $\cos\theta^0$. Thus $y=0$ describes forward scattering and $y=1$ backward scattering. For right-handed scattering, the total angular momentum has a value of unity, which suppresses the backward scattering amplitude by angular momentum conservation. In the left-handed process the total angular momentum is zero and one expects an isotropic distribution (reflected by the absence of a $\cos\theta^0$-dependence in the cross-section).

The general form of the differential cross-sections of the neutrino-electron scattering processes can then be written as a superposition of the left-handed, the right-handed and the left-right interference terms, where we use a notation in which ρ is implicit in the couplings:

$$
\begin{aligned}
\frac{d\sigma(\nu_\mu e)}{dy} &= \frac{G_F^2 m_e}{2\pi} E_\nu \left[(g_V+g_A)^2 + (g_V-g_A)^2(1-y)^2 - ({g_V}^2-{g_A}^2)\frac{m_e}{E_\nu}y\right] \\
\frac{d\sigma(\overline{\nu}_\mu e)}{dy} &= \frac{G_F^2 m_e}{2\pi} E_\nu \left[(g_V-g_A)^2 + (g_V+g_A)^2(1-y)^2 - ({g_V}^2-{g_A}^2)\frac{m_e}{E_\nu}y\right] \\
\frac{d\sigma(\nu_e e)}{dy} &= \frac{G_F^2 m_e}{2\pi} E_\nu \left[(g_V+g_A+2)^2 + (g_V-g_A)^2(1-y)^2 - \right. \\
&\qquad \left. ({g_V}^2-{g_A}^2+2(g_V-g_A))\frac{m_e}{E_\nu}y\right] \\
\frac{d\sigma(\overline{\nu}_e e)}{dy} &= \frac{G_F^2 m_e}{2\pi} E_\nu \left[(g_V-g_A)^2 + (g_V+g_A+2)^2(1-y)^2 - \right. \\
&\qquad \left. ({g_V}^2-{g_A}^2+2(g_V-g_A))\frac{m_e}{E_\nu}y\right] \qquad (9)
\end{aligned}
$$

The total cross-sections are proportional to E_ν, which is a consequence of neglecting the propagator. This simplification is justified in the energy-domain in which present experiments are performed. The proportionality factor is $\sigma^0 = G_F^2 m_e E_\nu/2\pi = G_F^2 s/4\pi$, where s is the energy in the centre-of-mass; this behaviour is expected for a point-like cross-section. The value of $\sigma^0 = 4.3\times10^{-42}\, E_\nu\,\mathrm{cm}^2\,\mathrm{GeV}^{-1}$ sets the overall scale for the νe-cross-sections.

In the above discussion the assumption was made that the weak interaction contains only V and A contributions. No scalar (S), pseudo-scalar (P) or tensor (T) contribution has been observed. In the νe cross-section such terms can be described by an admixture of a y^2-dependent term. Experiments have not reached sufficient precision to be sensitive to such terms. In experiments where only the event rates are measured, a suitable combination of S, P and T can mimic the behaviour expected from V and A interactions.

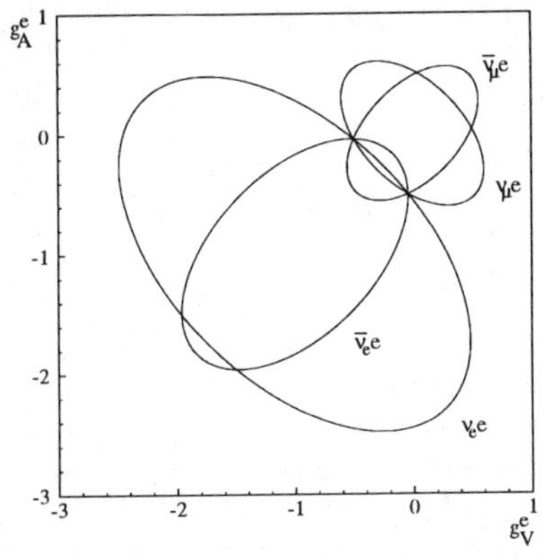

Fig. 3: *Ellipses defined in the g_V-g_A-plane by measurements of the total cross-sections. The curves are drawn for hypothetical experiments, measuring a total cross-section predicted with* $\sin^2\theta_W = 0.23$.

1.3 *Experimentally accessible quantities*

Total cross-sections

Measurements of the total cross-sections are sensitive to the electroweak mixing angle and define a relation between g_V and g_A. Integration over y gives for the $\nu_\mu e$ cross-section, dropping the left-right interference term;

$$\sigma(\nu_\mu e \to \nu_\mu e) = \frac{G_F^2 m_e}{2\pi} E_\nu \left[(g_V + g_A)^2 + \frac{1}{3}(g_V - g_A)^2\right]. \tag{10}$$

This relation defines an ellipse in the g_V-g_A-plane. Similarly, other ellipses are defined by measurements of the other reactions, as shown in Fig. 3. Comparison of different νe-reactions single out a region in the plane, up to a two-fold ambiguity given by interchanging g_V and g_A. This ambiguity can be removed by comparisons to e^+e^--scattering results. Present experimental results have unambiguously selected the region consistent with the Standard Model, namely $g_A = -\frac{1}{2}$ and g_V small, which is expected for a value of $\sin^2\theta_W$ close to 0.25, as indicated by the experimental observations. If only muon-neutrino and anti-neutrino cross-sections are observed, a fourfold ambiguity is present.

Expressing Eq. (10) in terms of $\sin^2\theta_W$ one finds:

$$\sigma(\nu_\mu e \to \nu_\mu e) = \frac{G_F^2 m_e}{2\pi} E_\nu \left[1 - 4\sin^2\theta_W + \frac{16}{3}\sin^4\theta_W\right]. \tag{11}$$

Similar relations can be derived for the other νe processes, and are displayed in Fig. 4. It is shown that the ν and $\bar{\nu}$ cross-sections have a minimum in the physical region for $\sin^2\theta_W$, and thus a measurement of a single cross-section can give an ambiguous result. For this reason, it is often better to measure a ratio or combination of cross-sections to obtain maximum sensitivity to $\sin^2\theta_W$. In practice, experiments measure a *visible* cross-section in a limited kinematical domain, and have to use a model to extrapolate to a total cross-section.

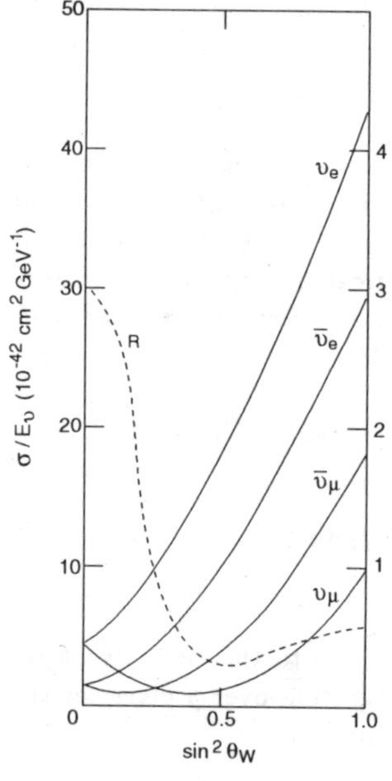

Fig. 4: *Total cross-sections of νe-scattering processes and the cross-section ratio* $R^{\nu\overline{\nu}}$ *as function of* $\sin^2\theta_W$*. The left-hand scale applies to the cross-sections, and the right-hand scale to the ratio.*

Cross-section ratio

The cross-section ratio $R^{\nu\overline{\nu}}$ has been determined by several experiments, and has a large sensitivity to $\sin^2\theta_W$:

$$R^{\nu\overline{\nu}} = \frac{\sigma(\nu_\mu e \to \nu_\mu e)}{\sigma(\overline{\nu}_\mu e \to \overline{\nu}_\mu e)} = 3\frac{\left[1 - 4\sin^2\theta_W + \frac{16}{3}\sin^4\theta_W\right]}{\left[1 - 4\sin^2\theta_W + 16\sin^4\theta_W\right]}. \tag{12}$$

Its maximum sensitivity to $\sin^2\theta_W$ occurs in the region of experimental interest, near $\sin^2\theta_W = 0.25$. In this region one finds $\Delta R^{\nu\overline{\nu}} = 8\Delta\sin^2\theta_W$. There are also experimental advantages of the measurement of the cross-section ratio, since relative normalization of the neutrino flux and relative knowledge of efficiencies are more readily available than absolute determinations. $R^{\nu\overline{\nu}}$ is not directly measured, but a *visible* ratio is determined, including kinematical cuts. In addition, neutrino beams usually contain admixtures of different neutrino types, and $R^{\nu\overline{\nu}}$ becomes an *effective* ratio of normalized event rates in two different beams. For this reason these direct measurements only have a meaning in the context of a specific experiment. We have chosen to give the values of the derived physical parameters rather than the experimental values of $R^{\nu\overline{\nu}}$.

Differential cross-sections

The study of the differential cross-sections $d\sigma/dy$ can provide more information than the total cross-sections alone. To perform a measurements of y for each event, one needs to measure either the ratio of the energy of the incoming neutrino and the energy of the outgoing electron, or the angle of the outgoing particle.

The first method is very difficult in practice, because the energy of the incoming neutrino is not known (except for a monochromatic neutrino source such as pion decay at rest), while to explore the second method the angular resolution of the experiment has to be an order of magnitude better than the kinematical bound. Nevertheless, the shape of the electron energy distribution and the shape of the $E_e\theta_e{}^2$-distribution can be used to *statistically* extract the information from the y-distribution. The shape of the distribution (Eq. 9) defines the coupling constants, without making use of the normalizations of the cross-sections; this provides a more direct test of the chiral structure of the interaction, since the variable y is directly related to the recoil angle in the centre-of-mass system. The distributions in kinematic variables, such as E_e and $E_e\theta_e{}^2$, when used in conjunction with the total cross-sections, provide an additional sensitivity on the couplings and improve the experimental precision.

Charged- and neutral-current interference

The total cross-sections of the electron-(anti)neutrino induced reactions $\nu_e e \to \nu_e e$ and $\overline{\nu}_e e \to \overline{\nu}_e e$ are sensitive to the interference of the CC and NC amplitude. A comparison of the observed total cross-section, σ^{TOT}, with the *incoherent* sum of the expected NC and CC cross-sections defines a measurement of the contribution of the interference, σ^I:

$$\sigma^I = \sigma^{TOT} - \sigma^{NC} - \sigma^{CC}. \tag{13}$$

The CC cross-section follows directly from $\sigma^{CC} = 4\sigma^0$, while the NC cross-section depends on $\sin^2\theta_W$, obtainable from the ν_μ and $\overline{\nu}_\mu$ processes. It is usual to introduce an interference strength, I, with

$$I = \sigma^I/2\sigma^0, \tag{14}$$

which is predicted to be $I = -2 + 4\sin^2\theta_W$, negative and close to -1 for $\sin^2\theta_W$ close to 0.25. Hence, the interference is predicted to be destructive. An unambiguous demonstration of the interference proves that the incoming and outgoing neutrino in NC interactions are identical. This can be used to set upper limits on flavour changing neutral-currents. The exchange of particles with spin different from one would also change the interference and hence a measurement of the interference close to the Standard Model prediction implies that the interaction is mainly V and A rather than S, P and T.

1.4 *Higher order corrections*

Up to now our discussion was limited to the *tree-level* cross-sections. In this approximation no ambiguity exists in the definition of $\sin^2\theta_W$ and all measurements of this quantity should give the same result. Higher order corrections change this picture, and different definitions of the electro-weak parameters are possible. One way of defining $\sin^2\theta_W$ is to use the on-shell (OS) scheme [11], where, to all orders,

$$\sin^2\theta_W = 1 - \frac{m_W{}^2}{m_Z{}^2} \tag{15}$$

Other choices are the $\overline{MS}$ scheme, where the tree level relation for m_W Eq. (4) is used as definition [12]. Also other schemes are a possible choice.

The one-loop corrections modify the cross-sections by a change of the overall scale, a change of the effective coupling and a pure QED term [13].

A correction to the overall scale of the cross-section, multiplying the amplitude, can be expressed as a q^2-dependent departure of ρ from unity. One retains the notation of the Born approximation by the substitution [11, 13, 14]:

$$G_F \to \rho_{\nu e}^{NC}(q^2)G_F. \tag{16}$$

The coupling constants are redefined as effective couplings, for $\sin^2\theta_W$:

$$\sin^2\theta_{\nu e} \to \kappa_{\nu e}(q^2)\sin^2\theta_W \tag{17}$$

Both $\rho_{\nu e}^{NC}$ and $\kappa_{\nu e}$ depend on the mass of the top quark, m_{top}, and the Higgs mass, m_H.

An additional QED term is introduced, which changes the y-distributions depending on experimental details. In this context it is important to know whether additional real photons, emitted at the interaction vertex are included in the measurement of the electron energy or whether the bare electron energy is observed. The former is the case for calorimetric detectors, while the latter prevails at bubble chambers and produces much larger corrections. The formalism can be found in [13]. When radiated photons are included in the electron energy-measurements, the shape of the y-distribution is hardly modified, only shifted, while in the other case a much larger deviation from the tree-level cross-section is observed, especially near the kinematical boundaries.

The recommended procedure is to analyse experimental data correcting for the pure QED-effects, since the shapes of the distributions change in an experiment-dependent manner. The other electroweak effects should be singled out separately. They can be absorbed in changes of the effective parameters. Radiative corrections to all experimental results in this field have been calculated [7, 13, 15].

1.5 *Electromagnetic properties*

The possibility of a neutrino *magnetic moment* has been discussed by various authors [16].

A non-vanishing magnetic moment introduces a helicity change in the interaction and contributes an extra term, σ^μ, to the cross-section which is detectable by a change of total cross-section and a change of the shape of the y-distribution:

$$\frac{d\sigma^\mu}{dy} = \mu_\nu^2 \mu_{Bohr}^2 \frac{1-y}{y} \tag{18}$$

The divergence of this term is only apparent; the lower kinematic bound on the outgoing electron energy acts as a cut-off. The neutrino magnetic moment, μ_ν, is expressed in Bohr-magnetons, $\mu_{Bohr} = \sqrt{\pi}\alpha/m_e$. The total cross-section is changed by integrating Eq. (18) within the kinematic limits, where $E_e{}^{min}$ is the lower experimental bound on the electron energy:

$$\sigma^\mu = \mu_\nu^2 \mu_{Bohr}^2 \left[\log\left(E_\nu/E_e{}^{min}\right) - 1 + E_e{}^{min}/E_\nu\right]. \tag{19}$$

The latter equation can be used by comparing a measured cross-section with a prediction from measurements not involving neutrinos.

Similarly, a non-zero *charge-radius* of the neutrino can induce a modification of the cross-sections. In this case, the interaction does not change the spin, and adds coherently to the Standard Model expression. In the framework of the Standard Model, higher order corrections introduce a small effective charge-radius. It looks, however, somewhat artificial to consider this term independently from other effects of the higher order electro-weak corrections. Several authors have proposed a formulation to describe Standard Model charge-radii, see e.g. [17].

An *anomalous* charge radius, $\langle r^2\rangle_{anom}$, has the effect of modifying the vector part of the interaction, since it is mediated by photon exchange. It can therefore be described as a modification of the vector coupling constant [18]:

$$g_V \rightarrow -\frac{1}{2} + 2(\sin^2\theta_W + \delta), \qquad \text{where} \quad \delta = \frac{\sqrt{2}\pi\alpha}{3G_F}\langle r^2\rangle_{anom} \tag{20}$$

and manifests itself as a change of $\sin^2\theta_W$. Limits for the charge radius have been obtained comparing $\sin^2\theta_W$ obtained with neutrino scattering with $\sin^2\theta_W$ from other processes.

1.6 *Limits on additional Z-bosons*

Neutrino-electron scattering cross-sections are sensitive to the existence of additional Z-bosons. The combination of the results of such experiments with measurements with less sensitivity, such as e^+e^--asymmetries, can give interesting lower limits on the masses of these bosons. It is, however, not possible to define a universal mass-limit valid for all models for additional Z-bosons, the relation of the coupling and the mass being dependent on the details of the models. With the definition, $\Delta g_{\rm A} = g_{\rm A}{}^{\nu e} - g_{\rm A}$ and $\Delta g_{\rm V} = g_{\rm V}{}^{\nu e} - g_{\rm V}$, one can write for a coupling g' (motivated by $SO(10)$) [19]:

$$m_{Z'}{}^2 = \left[\frac{g'}{g^{SM}}\right]^2 m_Z{}^2 \left[\frac{10}{3}\Delta g_A\right]^{-1} \qquad m_{Z'}{}^2 = \left[\frac{g'}{g^{SM}}\right]^2 m_Z{}^2 \left[\frac{5}{3}\Delta g_V\right]^{-1}. \quad (21)$$

We will only discuss a generic boson, Z', with a coupling, $g' = g^{SM}$, equal to the Standard Model Z. For the formalism in other models the reader is referred to [19, 20, 21].

2 Experimental results

The signature of νe-scattering is a single recoiling electron in an angular range allowed by the kinematical limit Eq. (3). The experimental difficulties are greatly enhanced by the low rate of νe-events compared to the total event rate induced by neutrinos on nuclear targets. At high energies, the ratio is between 10^{-4} and 10^{-3}, and effective methods to remove the other events from the samples have to be developed.

The electron recognition employed in the experiments depends on the detectors used and the electron-energy regime. There are two main classes of detectors, bubble chambers and electronic detectors. In bubble chambers, electrons and their shower products can be distinguished as individual tracks, and the magnetic field provides a measurement of the charge of the particle, both of which allow for a strong reduction of contaminating events. Together with a good angular resolution, very clean samples of events were obtained. Problems in bubble chambers are caused by the poor time resolution and the small volume of the chambers, such that background from photons induced by interactions outside the chamber play a role. However, the main disadvantage is the relatively low effective target mass and hence a correspondingly low event rate. In bubble chambers operating in beams of a few GeV neutrino energy, the main backgrounds are caused by: converted γ-rays (Compton electrons or asymmetric e^+e^--pairs) from interactions outside the chamber, or from bremsstrahlung by muons, and the quasi-elastic reaction $\nu_e n \rightarrow e^- p$ and nuclear processes such as resonance excitation and coherent π^0-production.

In the electronic detectors, backgrounds are totally different depending on the neutrino energy range of the beam. No magnetic field is present, and positrons also contribute to the background. At low energies, with neutrino energies of a few MeV at reactors and a few tens of MeV at stopped pion beams, the largest background is formed by cosmic-ray induced activity, and requires sophisticated triggering schemes and massive shielding to be reduced. Beam-induced background is present in the form of neutron-induced γ-rays, neutrino-induced nuclear resonance excitation, and inverse β-decay.

At high energies, the backgrounds are dominated by coherent and diffractive π^0-production [22, 23, 24], electron-(anti)neutrino quasi-elastic scattering and nuclear resonance excitation, and to a lesser extent, inclusive semileptonic interactions with electromagnetic final states. At even higher neutrino energies than used for the νe-experiments up to now, the quasi-elastic ν_e interaction is potentially very dangerous. The approximate p_T^2-invariance of the cross-section causes the angular distribution to shrink faster towards zero than the νe process, spoiling its unique kinematical signature.

To make use of the characteristic signature of the process, detectors have to combine good electron-hadron separation and capabilities for electron-photon discrimination with a good angular resolution. In calorimetric detectors, good angular resolution can only be achieved with low-Z target materials, which result in the development of long electromagnetic showers. In addition, the low cross-section imposes the use of massive targets.

High intensity beams are required to achieve sufficiently large numbers of events. Only when intense muon-neutrino beams could be produced by focusing the secondary pions and kaons in extracted proton beams with the help of magnetic "horns" [25], the study of $\nu_\mu e$-scattering could be undertaken. Horns are made of two cylindrical conductors, in which a pulsed electric field produces a toroidal magnetic field. Electron-neutrino beams are produced by the decay of muons at rest. Sufficient intensity requires the use of the full intensity of the proton beams in meson factories. High fluxes of electron anti-neutrinos at low energies are available at nuclear power reactors.

The measurement of cross-sections also requires the knowledge of the neutrino flux. Specific efforts are to be spent to be able to calculate the beam intensity directly from the beam parameters, or indirectly by recording events from processes with well-known cross-sections.

In the following, experiments which have contributed to our knowledge of neutrino-electron scattering will be described. We first examine experiments in ν_μ and $\overline{\nu}_\mu$ beams, and later we will discuss the experiments using electron-type neutrinos.

2.1 *The Gargamelle experiment*

The Gargamelle bubble chamber was originally designed for operation at the 26 GeV CERN proton synchrotron (PS) neutrino beam.

The secondary pions and kaons were focused by a system of two "horns" [25]. The pions and kaons decay in a 70 m long decay tunnel. A shielding at the end of the tunnel absorbs the remaining pions and kaons. The muons produced together with the neutrinos in the pion and kaon decays are absorbed after 22 m of iron shield. The neutrino energy spectrum was peaked at 2 GeV.

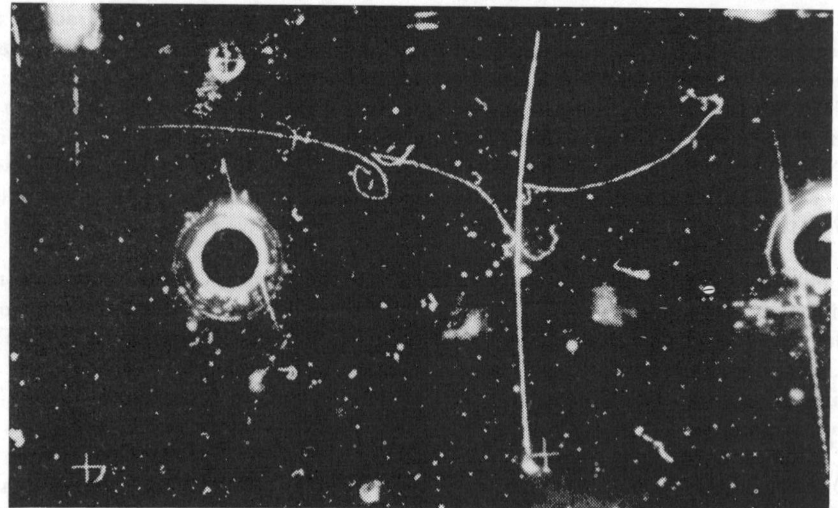

Fig. 5: *The first neutrino-electron scattering candidate found in Gargamelle.*

Gargamelle was a large cylindrical bubble chamber with a diameter of 1.9 m and a length of 4.8 m. The axis of the cylinder coincided with the neutrino beam direction. The 12 m^2 volume could contain 18 tons of freon CF_3Br. A set of copper coils provided a horizontal magnetic field of 2 T perpendicular to the beam. Electrons could be recognized by their characteristic showering in the liquid.

The first evidence for the (anti)neutrino electron scattering process was found by Gargamelle in the PS beam [1]; this result will be discussed separately. In further exposures [26] the total number of candidates in the $\overline{\nu}$-beam of the PS was increased to three with negligible background, while in the ν-beam only one candidate was found compatible with the expected background. Gargamelle has also been exposed to the SPS neutrino beams at CERN where nine $\nu_\mu e \to \nu_\mu e$ candidates were found and no $\overline{\nu}_\mu e \to \overline{\nu}_\mu e$ events.

The PS exposure

In the first exposure to the PS beams a total of 375000 ν pictures and 360000 $\overline{\nu}$ pictures were obtained. One single electron event satisfying all criteria was found in the antineutrino film. This event is displayed in Fig. 5. The required signature was a single electron without nuclear fragments, hadrons or gamma rays correlated to the vertex. The electron energy was required to be larger than 300 MeV and an angle with respect to the beam smaller than 5^0.

The unique candidate had an electron energy of 385 ± 100 MeV and an angle of

$(1.4^{+1.6}_{-1.4})^0$ with respect to the beam.

This event provided the first evidence for the reaction $\overline{\nu}_\mu e \to \overline{\nu}_\mu e$ with a probability that the event was due to non-neutral-current background less than 3%.

The total $\overline{\nu}$ sample was obtained with 1.5 million pictures. In total, three unambiguous events were found (including the candidate described previously). The background was carefully evaluated to be due to isolated γ's converted in the chamber into an asymmetric e^+e^--pair (0.26 ± 0.11 events), the $\nu_e n \to e^- p$ reaction (0.07 ± 0.04 events) and to $\overline{\nu}_e e \to \overline{\nu}_e e$ scattering (estimated to 0.11 events). The total cross-section was calculated after corrections for scanning losses, selection efficiencies and for the kinematical cut efficiency to be

$$\sigma(\overline{\nu}_\mu e \to \overline{\nu}_\mu e) = (1.0^{+2.1}_{-0.9}) \times 10^{-42}\, E_\nu\, \mathrm{cm}^2\, \mathrm{GeV}^{-1},$$

where the error corresponds to a 90% confidence level.

One candidate was found in the ν beam consistent with the estimated background of 0.5 ± 0.3 events An upper limit at 90% confidence level was determined $\sigma(\nu_\mu e \to \nu_\mu e) < 1.4 \times 10^{-42}\, E_\nu\, \mathrm{cm}^2\, \mathrm{GeV}^{-1}$. The combined analysis of ν and $\overline{\nu}$ data gave a 90% confidence interval $0.1 < \sin^2\theta_W < 0.4$ for the electroweak mixing angle.

The SPS exposure

In the exposure to the wide band beam of the CERN SPS 410000 pictures were obtained in the ν beam [27, 28] and 230000 in the $\overline{\nu}$ beam [29]. Cuts were used to prevent radiation from muons to fake single electrons, a single track at the vertex was required and cuts against converted γ's were made. Nine single electron events in the ν beam with $E_e > 2$ GeV satisfied $E_e\theta_e{}^2 < 2m_e$ within the experimental resolution. The backgrounds from converted γ's, the reaction $\nu_e n \to e^- p$ and from nuclear resonance excitation were estimated to be 0.5 ± 0.2 events. The neutrino flux normalization was obtained comparing to the observed ν_μ charged-current events. This yielded a cross-section measurement:

$$\sigma(\nu_\mu e \to \nu_\mu e) = (2.4^{+1.2}_{-0.9}) \times 10^{-42}\, E_\nu\, \mathrm{cm}^2\, \mathrm{GeV}^{-1}.$$

This value corresponds to $\sin^2\theta_W = 0.12^{+0.11}_{-0.07}$ retaining only the solution allowed by other experiments.

In the analysis of $\overline{\nu}$ data no candidate was found, giving an upper limit for the cross-section $\sigma(\overline{\nu}_\mu e \to \overline{\nu}_\mu e) < 2.7 \times 10^{-42}\, E_\nu\, \mathrm{cm}^2\, \mathrm{GeV}^{-1}$, at 90 % confidence level [29], consistent with $\sin^2\theta_W < 0.39$.

An excellent review of the rich harvest of physics results with Gargamelle can be found in [30].

2.2 *The Aachen-Padova experiment*

This detector was exposed to the neutrino beam of the CERN PS.

The detector was an array of 127 optical spark chambers consisting of 1 cm thick aluminium plates, 2×2 m^2 in cross-section and 8 m length with a total effective mass of 19 tons. The chamber was fired each accelerator spill. The sparks were photographed in two orthogonal projections with the help of two large mirrors. The detector had a relative energy resolution of $\Delta E/E = 0.22$ and an angular resolution of $\Delta\theta_{proj} = 12/\sqrt{E/GeV}$ mrad. The radiation length was about 20 cm.

Candidate events were selected on the basis of a showering criterion typical for electromagnetic showers and were rejected if they were associated to other event-like structures in the same picture [31]. No separation between electrons and γ's could be performed.

With selection criteria applied on the scattering angle of the outgoing electron ($\theta_e < 5^0$), the shower energy (0.2 GeV $< E_e <$ 2.0 GeV) and $E_e{\theta_e}^2 < 4m_e$, 32(17) candidates were found in the $\nu_\mu(\overline{\nu}_\mu)$ exposure.

The background was attributed mainly to π^0-production without visible recoil. The background of other sources was negligible. The total number of background events satisfying the criteria for the neutrino-electron scattering candidates were found to be 20.5 ± 2.0 and 7.4 ± 1.0 in the ν_μ and $\overline{\nu}_\mu$ beam, respectively.

Total cross-sections were determined to be:

$$\begin{aligned} \sigma(\nu_\mu e \to \nu_\mu e) &= (1.1 \pm 0.6) \times 10^{-42}\, E_\nu\, \mathrm{cm}^2\, \mathrm{GeV}^{-1} \\ \sigma(\overline{\nu}_\mu e \to \overline{\nu}_\mu e) &= (2.2 \pm 1.0) \times 10^{-42}\, E_\nu\, \mathrm{cm}^2\, \mathrm{GeV}^{-1} \end{aligned}$$

with a cross-section ratio of $R = 2.0^{+3.0}_{-1.0}$. This corresponds to a determination of the electro-weak mixing angle of $\sin^2\theta_W = 0.35 \pm 0.08$.

2.3 *The VMWOF experiment*

The VMWOF experiment was performed in the single horn focused wide-band ν_μ beam at FNAL [32]. The average energy of the neutrino beam was about 20 GeV. The ν_μ beam had a contamination of $\approx 11\%$ $\overline{\nu}_\mu$ and less than 0.5% of both ν_e and $\overline{\nu}_e$. The detector was located at a distance of about 500 m from the decay tunnel.

The apparatus was composed of 49 basic modules, each consisting of a one radiation length thick aluminium plate, a 1×1 m^2 multi wire proportional chamber (MWPC) and a plane of plastic scintillators. Both coordinates transverse to the beam direction were measured with the MWPC plane, by means of a delay line on the cathode planes read out by TDC's. The scintillators measured the charge deposition and were used for triggering. A scintillator counter plane upstream of the apparatus was used as a charged particle veto and a set of two planes separated by steel were used downstream of the apparatus to identify muons.

The trigger was formed by the requirement of a coincidence of an energy deposition of at least 2 GeV and any row of four consecutive scintillation planes fired, with the upstream scintillator veto plane in anticoincidence. This trigger generated a dead time of ≈ 33 %.

The analysis of the triggers was performed by means of a scan using a visual display of the events. Charged-current events were rejected by the requirement that no outgoing muon was observed; semileptonic neutral-current induced events were removed by the criterion that only one electromagnetic shower should be observed in the event. The event sample was further reduced by a cut on the dE/dx and by the requirement that at least one projected angle was smaller than 50 mrad.

To extract the event numbers, an analysis was performed using the distribution of the events in the $(E_e - \theta_e{}^2)$ plane. Of 46 candidate events 34 were attributed to neutrino electron scattering and 12 to background processes. The background was equally distributed between $\nu_e n \to e^- p$ induced by electron neutrinos in the beam and neutral current processes.

The energy spectrum of the beam was calculated from the hadronic energy of charged-current interactions, since no muon momentum measurement was performed. The total neutrino flux was calculated from the measured number of protons on target. The total cross-section was derived to be

$$\sigma(\nu_\mu e \to \nu_\mu e) = (1.4 \pm 0.3_{events} \pm 0.2_{beam}) \times 10^{-42}\, E_\nu\, \mathrm{cm}^2\, \mathrm{GeV}^{-1},$$

which corresponds to $\sin^2\theta_W = 0.25^{+0.07}_{-0.05}{}_{events} \pm 0.04_{beam}$. The uncertainty coming from the beam normalization is given separately, and looks to be underestimated given the coarse method used for its determination. Also the uncertainty attributed to *event*-counting seems to contain only the statistical error, which is an underestimation given the difficulty of this type of experiment. This result will be used with care when averaged with the results of other experiments. It was suggested [6] to increase the systematic error in the cross-section from 0.2 to 0.4 to account for uncertainties in background subtraction, and the determination of the efficiency and normalization. A similar increase of the error should be applied to $\sin^2\theta_W$.

2.4 *The 15 foot bubble chamber*

The Fermilab (FNAL) 15 foot bubble chamber was exposed to the FNAL wide band beam (see section 2.3). The chamber has an almost spherical shape. The fiducial volume was 21 m^2, which gave a mass of 23 tons with the heavy neon-hydrogen mixture used for this experiment. The radiation length was 40 cm and the interaction length 125 cm. The 3.0 T magnetic field was produced by a superconducting magnet.

The exposures to the neutrino and antineutrino beams were analysed separately by two different groups.

In the neutrino exposure (BNL-COL group) [33], approximately 394 000 pictures were taken. Events which had electromagnetic showers within 52 mrad with respect to the beam, and with an energy above 2 GeV were retained in the analysis. Criteria were developed to separate single e^- and e^+ tracks from photon conversion. A total of 22 single e^- candidates were found satisfying all criteria. The backgrounds were mainly converted photons from interactions outside the chamber, and from the electron neutrino induced $\nu_e n \to e^- p$ reaction.

After correction for efficiencies, 13.9 ± 5.4 and 11.4 ± 3.6 events were attributed to νe-scattering in the two exposures, respectively. The normalization of the neutrino flux was obtained through the study of inclusive charged-current ν_μ induced events, using total cross-section values obtained in other beams.

A total cross-section value was measured:

$$\sigma(\nu_\mu e \to \nu_\mu e) = (1.67 \pm 0.44) \times 10^{-42}\, E_\nu\, \mathrm{cm}^2\, \mathrm{GeV}^{-1},$$

corresponding to a value of $\sin^2\theta_W = 0.20^{+0.06}_{-0.05}$, eliminating the solution $0.54^{+0.04}_{-0.06}$ using input from other experiments.

In the antineutrino exposure (FMMS group) [34], 85000 pictures were scanned with an estimated efficiency of 83%. A search for single e^+ was performed with an energy above 0.8 GeV. No candidates were found. The efficiency was checked by a study of ν_e-induced interactions. The antineutrino flux was obtained using the reaction $\overline{\nu}_\mu N \to \mu^+ X$ with a known cross-section:

$$\sigma(\overline{\nu}_\mu e \to \overline{\nu}_\mu e, E_e > 0.8\ \mathrm{GeV}) \leq 2.1 \times 10^{-42}\, E_\nu\, \mathrm{cm}^2\, \mathrm{GeV}^{-1}, \text{ at 90\% C.L.}$$

This corresponds to an upper limit $\sin^2\theta_W < 0.37$.

2.5 *The CHARM experiment*

The CHARM detector [35] was exposed to the CERN wide-band neutrino beam, produced by the 450 GeV proton beam of the SPS. The secondaries were focused by two horns [36]. The experiment was located at about 900 m from the production target. The neutrino energy spectra (Fig. 6) peaked at $E_\nu \approx 20$ GeV, with an admixture of wrong helicity neutrinos of 6%(9%) in the (anti)neutrino beam (in terms of energy-weighted fluxes). The admixture of ν_e and $\overline{\nu}_e$ was of the order of 1%.

The detector combined the functions of a calorimetric detector and a fine-grain tracking device. It consisted of a *target calorimeter* followed by a *muon spectrometer*. The *target calorimeter* was made of 78 units, each made of a target plane, a scintillator plane and a proportional drift tube plane. The *target planes* were made of 8 cm thick marble slabs of 3×3 m^2 cross-section surrounded by a frame of magnetized iron.

Data were taken in two distinct periods; between these two periods, the detector was improved by the installation of aluminium streamer tube planes between each scintillator and proportional drift tube plane. The streamer tubes of 1×1 cm^2 cross-section improved the measurement of the interaction vertex. The angular resolution was $\sigma(\theta_{proj}) = 46\sqrt{E/\mathrm{GeV}}$ mrad [37] without streamer tubes and $\sigma(\theta_{proj}) = 32\sqrt{E/\mathrm{GeV}}$ mrad with streamer tubes [38]. The energy resolution for electrons was $\sigma(E) = (0.18 \pm 0.01)\sqrt{E/\mathrm{GeV}}$.

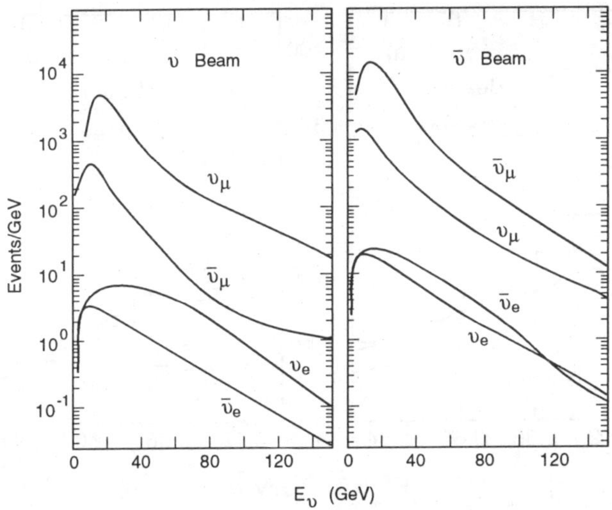

Fig. 6: *Energy spectrum of neutrinos at the CERN 450 GeV SPS. Spectra for all four neutrino species are shown for (left) the neutrino beam and (right) the antineutrino beam, weighted with the neutrino energy to reflect the event rate. The intensity scale is in arbitrary units.*

The muon spectrometer was made of magnetized iron discs, with proportional drift tube planes as detecting elements.

A plane of scintillation counters positioned upstream of the target calorimeter served as a charged particle veto.

Neutrino-electron scattering candidates were selected with criteria exploiting the regular structure of electromagnetic showers in the detector, narrow compared to hadronic showers. Events with a visible muon track and with too much energy deposition near the vertex were removed.

After kinematical cuts: $7.5 < E_e < 30$ GeV in the first exposure [39, 40] and $4.0 < E_e < 30$ GeV in the second exposure [41], and $E_e^2\theta_e^2 < 0.54$ GeV^2, an analysis was performed on the shape of the $E_e^2\theta_e^2$ distribution. The distributions of $E_e^2\theta_e^2$ were used rather than $E_e\theta_e{}^2$ to exploit the knowledge of the $p_T^2 \approx E_e^2\theta_e^2$ distribution of the quasi-elastic electron-neutrino background. The background consisted of coherent π^0 production, electron-neutrino induced quasi-elastic scattering ($\nu_e n \to e^- p$ and $\overline{\nu}_e p \to e^+ n$), and inclusive semileptonic events with predominantly electromagnetic final states. These backgrounds are characterized by a much broader distribution in $E_e^2\theta_e^2$ than the signal of νe events. The $E_e^2\theta_e^2$-distributions are shown in Fig. 7; $83 \pm 16(112 \pm 21)$ $\nu_\mu e(\overline{\nu}_\mu e)$ events were found;

The relative normalization of the ν and $\overline{\nu}$ flux were obtained from a study of

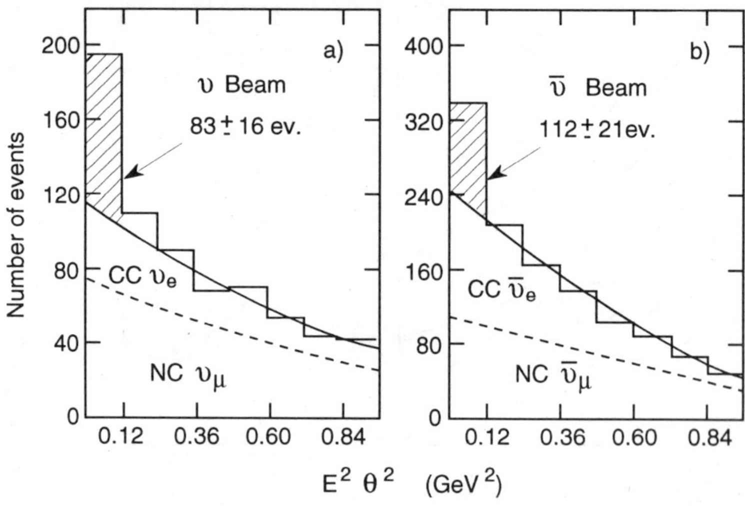

Fig. 7: *Distribution of a) neutrino and b) antineutrino events in the variable* $E_e^2\theta_e^2$ *for the CHARM experiment. The total background is shown as a full line, while the contribution of semi-leptonic neutral-current events are shown as a dashed line.*

Tab. 1: *CHARM results*

quantity	method	value
$\sigma(\nu_\mu e \to \nu_\mu e)$		$(2.2 \pm 0.4_{stat} \pm 0.4_{syst}) \times 10^{-42}\, E_\nu\, \mathrm{cm}^2\, \mathrm{GeV}^{-1}$
$\sigma(\overline{\nu}_\mu e \to \overline{\nu}_\mu e)$		$(1.6 \pm 0.3_{stat} \pm 0.3_{syst}) \times 10^{-42}\, E_\nu\, \mathrm{cm}^2\, \mathrm{GeV}^{-1}$
$\sin^2\theta_\mathrm{W}$	ratio $\nu,\overline{\nu}$	$0.211 \pm 0.035_{stat} \pm 0.011_{syst}$
ρ	sum $\nu,\overline{\nu}$	$1.14 \pm 0.07_{stat} \pm 0.12_{syst}$
g_V	ν and $\overline{\nu}$	$-0.06 \pm 0.07_{stat} \pm 0.02_{syst}$
g_A	ν and $\overline{\nu}$	$-0.57 \pm 0.04_{stat} \pm 0.06_{syst}$
$m_{\mathrm{Z}'}$		> 280 GeV (95% C.L.)
μ_ν	$\nu_\mu,\overline{\nu}_\mu$	$< 1.0 \times 10^{-8} \mu_B$ (95% C.L.)
$\lvert\langle r^2\rangle_{anom}\rvert$		$< 1.0 \times 10^{-32}$ cm^2 (95% C.L.)

inclusive semileptonic charged-current and neutral-current events, and independently from quasi-elastic events, $\nu_\mu n \to \mu^- p$ and $\overline{\nu}_\mu p \to \mu^+ n$. The experimental results are represented in Table 1.

The cross-section ratio of νe events in the ν and the $\overline{\nu}$ beam provided the most precise value of $\sin^2\theta_\mathrm{W}$ [42]. To this end, the observed events were attributed to the four νe reactions ($\nu_\mu e$, $\overline{\nu}_\mu e$, $\nu_e e$ and $\overline{\nu}_e e$) according to the abundances of ν_μ, $\overline{\nu}_\mu$, ν_e and $\overline{\nu}_e$ in the two beams, parametrizing the four cross-sections as function of $\sin^2\theta_W$, avoiding a model dependent subtraction of the background neutrino components in the beam. The total cross-section of $\nu_\mu e$ and $\overline{\nu}_\mu e$ were highly correlated due to the common error on the absolute normalization of the neutrino flux and the determination of the efficiencies of the event selection. From the cross-sections the vector and axial-vector coupling constants of the electron were determined up to a fourfold ambiguity. The ambiguity could be removed using measurements from other leptonic processes, $\nu_e e \to \nu_e e$ and $e^+e^- \to \mu^+\mu^-$, leaving a unique solution. The systematic error on g_A reflects the uncertainty in the absolute normalization. The correlation between these values is small. A value for ρ was obtained from a comparison of the sum of the total cross-sections and other purely leptonic data [43].

An analysis was performed applying an additional condition on the pulse-height deposition in the first scintillator plane, removing virtually all showers initiated by photons. The remaining number of events in the forward scattering peak was consistent with the expected efficiency for electron induced showers, supporting the hypothesis that the events were due to a single recoiling electron.

An analysis of the data along the lines represented by Eq. (21) gave the limit on $m_{Z'}$ shown in the table; also limits on the neutrino magnetic moment and anomalous charge-radius were given.

2.6 *The BBKOPS experiment*

The BBKOPS (E734) experiment was performed at the Brookhaven National Laboratory (BNL). It was exposed to the horn-focused wide band neutrino and antineutrino beams at the 28.3 GeV AGS. The mean neutrino energy was 1.3 GeV. The proton spill of approximately 10^{13} protons was composed of twelve short bunches of 30 ns length, 224 ns apart, extracted in one turn of the accelerator. This structure was repeated each 2 seconds.

The E734 detector was designed to have particle identification, high angular resolution and a large total mass [44]. The main part of the detector was formed by a target calorimeter (*main detector*), followed by a *gamma catcher* and a *muon spectrometer*. The *main detector* consisted of 112 planar *modules* with a 4×4 m^2 cross-section. A module contained a plane of liquid scintillators, providing 80% of the target mass, and two planes of proportional drift tubes mounted in orthogonal orientations. The module thickness was 0.22 radiation lengths, which provided multiple dE/dx measurements per radiation length for particle identification. In particular, this feature made it possible to make a separation between electrons and photons, and to reduce the γ-induced background.

The total mass of the main detector was 170 tons. The target was followed by a 12 radiation lengths gamma catcher for the containment of electromagnetic energy. The end of the detector was formed by a muon spectrometer, made of two sets of proportional drift chamber planes separated by an air gap magnet to measure the momenta of forward going muons. The relevant experimental resolutions for electrons were $\sigma(\theta_{proj}) = (13 \pm 1)\ \sqrt{E_e/GeV}$ mrad and $\sigma(E_e)/E = 0.13\ \sqrt{E_e/GeV}$.

Data were taken in three running periods from 1981 to 1986. About $340(150) \times 10^3$ events were recorded in the $\nu(\overline{\nu})$ beam. During the 3 μs spill, pulse-height signals were recorded in 40 ns time slots, and the time of hits in the PDT system was recorded. Thus no trigger dead-time was incurred. The first step of the analysis was to combine the data for each interaction into one event based on the recorded time in the spill.

The electron showers were selected requiring an energy above 200 MeV, and requiring a pattern characteristic for electromagnetic showers. The analysis of the νe-candidates proceeded with a visual scan, necessary by the complexity of the events and the difficulty of pattern recognition. A fiducial volume cut was applied to eliminate entering tracks, and a cut on the total shower energy $210 < E_e < 2100$ MeV removed the low energy region dominated by π^0 background and the high energy region with ν_e-quasielastic background. The energy near the vertex was limited to enhance events starting with a single track. A dE/dx criterion was used to separate electrons from γ's. Finally, the requirement $\theta_e{}^2 < 0.03$ rad^2 defined a sample of νe candidates.

The background consisted of coherent and incoherent π^0-production ($\nu N \rightarrow \nu N \pi^0$ and $\nu n(p) \rightarrow \nu n(p)\pi^0$) and electron-neutrino quasi-elastics ($\nu_e n \rightarrow e^- p$ and $\overline{\nu}_e p \rightarrow e^+ n$). A description of the background processes was determined in the variables $\theta_e{}^2$ and E_e.

A maximum likelihood fit gave $N(\nu_\mu e) = 160 \pm 17_{stat} \pm 4_{syst}$ and $N(\overline{\nu}_\mu e) = 97 \pm 13_{stat} \pm 5_{syst}$.

The total cross-sections and cross-section ratio

To obtain the total cross-sections the efficiencies for extracting the signal were determined and a normalization of the neutrino flux was measured using the quasi-elastic reactions $\nu_\mu n \rightarrow \mu^- p$ and $\overline{\nu}_\mu p \rightarrow \mu^+ n$. A subtraction of the wrong helicity components of the beam and of the contributions from ν_e and $\overline{\nu}_e$ was performed to obtain the number of events induced by the main components of the beam, in-

Tab. 2: Results from BBKOPS using the total cross-sections and cross-section ratio.

quantity	method	value
$\sigma(\nu_\mu e \rightarrow \nu_\mu e)$		$1.80 \pm 0.20_{stat} \pm 0.25_{syst} \times 10^{-42}\ E_\nu$ cm^2 GeV^{-1}
$\sigma(\overline{\nu}_\mu e \rightarrow \overline{\nu}_\mu e)$		$1.17 \pm 0.16_{stat} \pm 0.13_{syst} \times 10^{-42}\ E_\nu$ cm^2 GeV^{-1}
$\sin^2 \theta_W$	ν and $\overline{\nu}$	$0.199 \pm 0.020_{stat} \pm 0.013_{syst}$
ρ	ν and $\overline{\nu}$	$1.005^{+0.070}_{-0.075}$

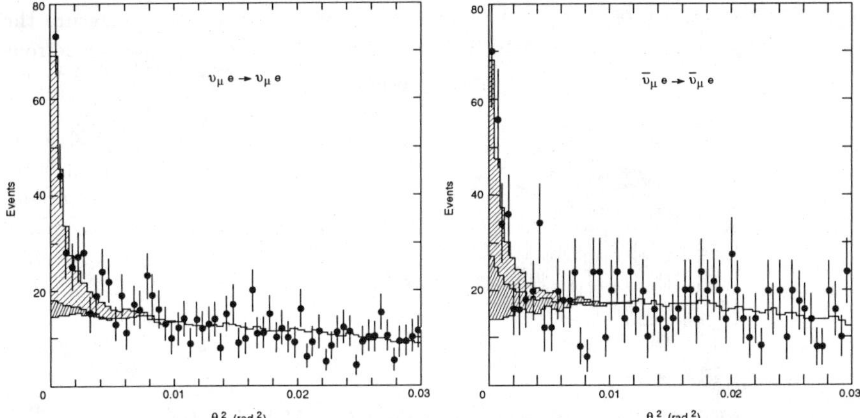

Fig. 8: *Experimental data and the result of the best fit for the BBKOPS experiment. Data are shown plotted against $\theta_e{}^2$ as full circles and the fit results are displayed as a line. The y-independent term is light-shaded; the $(1-y)^2$-term is dark-shaded.*

troducing a small model dependence in the result. In the cross-section ratio many systematic uncertainties cancel, thereby reducing the error in the final result. The results obtained using the total cross-section and the cross-section ratio are shown in Table 2 [45, 46, 47].

The differential cross-sections

An improved result was obtained by making use of all the information contained in the $\theta_e{}^2$-distribution. The signal was decomposed into two y-dependent components, corresponding to the left-handed and right-handed terms in the expression for the differential cross-sections Eq. (9). The contributions from the wrong helicity components and the electron-neutrinos in the beam were neglected. The data are shown in Fig. 8. Values of the electroweak couplings were obtained by directly fitting the

Tab. 3: Results from BBKOPS using the differential cross-sections.

quantity	method	value
g_V	ν and $\overline{\nu}$ diff. cross-section	$-0.107 \pm 0.035_{stat} \pm 0.028_{syst}$
g_A	ν and $\overline{\nu}$ diff. cross-section	$-0.503 \pm 0.023_{stat} \pm 0.028_{syst}$
$\sin^2\theta_W$	ν and $\overline{\nu}$ diff. cross-section	$0.195 \pm 0.018_{stat} \pm 0.013_{syst}$
μ_ν	$\nu_\mu, \overline{\nu}_\mu$	$< 0.85 \times 10^{-9} \mu_B$ (90% C.L.)
$-2.11 \times 10^{-32} < \langle r^2 \rangle_{anom} < 0.24 \times 10^{-32}$ cm^2 (90% C.L.)		

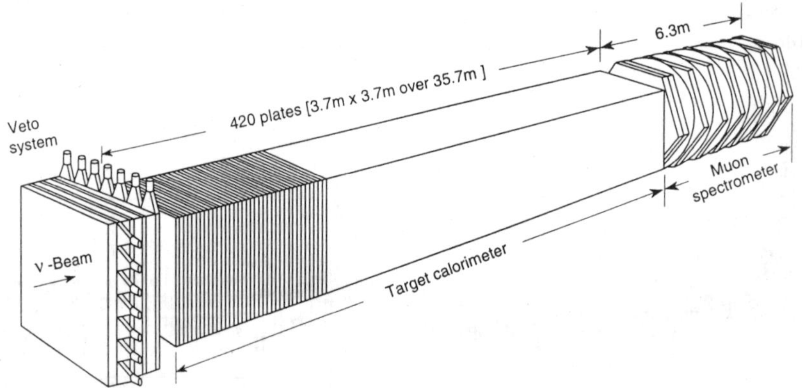

Fig. 9: *Schematic drawing of the CHARM-II detector.*

parameters g_V and g_A. The information contained in the cross-section ratio and the absolute cross-sections is in this way supplemented by the information of the angular distributions. Since the angular resolution is similar to the characteristic kinematical distributions, this provides a small improvement of the sensitivity. This method provided the most accurate measurement of $\sin^2\theta_W$ and was the first to use differential cross-sections of νe-scattering for this purpose; the results are given in Table 3.

2.7 *The CHARM-II experiment*

The CHARM-II detector [50] consisted of a massive target calorimeter followed by a muon spectrometer (Fig. 9). The detector was exposed to the horn focused wide band neutrino beam (WBB) at the CERN 450 GeV Super Proton Synchrotron (SPS). The neutrino energy spectrum had its maximum around 20 GeV. Admixtures of 7%(13%) of the opposite helicity components were observed in the $\nu_\mu(\overline{\nu}_\mu)$-beam, with about 1% abundance of electron-type neutrinos.

The *target calorimeter* was composed of 420 equal units with a total target mass of 600 tons. Each unit contained a 5 cm (0.5 radiation length) glass plate followed by a plane of streamer tubes. The lateral size of a unit was (370×370 cm^2). Both the wire signals of the streamer tubes and the charge on 2 cm cathode strips was recorded. The streamer tubes served as tracking detectors and were used to recognize and measure electromagnetic showers. Every fifth unit a plane of scintillators was inserted to measure the dE/dx of tracks for electron-γ separation. The calorimeter was followed by a toroidal iron muon spectrometer with drift chambers for particle tracking and scintillators for triggering on muons. The target was preceded by a charged particle veto system, made of a sandwich of iron and scintillator planes.

The trigger used the topological patterns of the streamer tube information and selected electron candidates and candidates for monitoring reactions; the dead-time was about 20%. The majority of charged-current induced semileptonic events were rejected at the trigger level.

The energy resolution for electrons was found to be $\Delta E/E = 0.23/\sqrt{E/\mathrm{GeV}} + 0.05$ [51]. The absolute energy scale uncertainty was estimated to be 5%. The angular resolution was equivalent to $\Delta\theta_{proj} \approx 17$ mrad/$\sqrt{E/\mathrm{GeV}}$ in the energy range of the analysis.

The ratio of the neutrino and antineutrino fluxes was obtained with five different methods, which gave consistent results. Four of these were formed by event-rates of processes with a known cross-section ratio for neutrinos and antineutrinos. The ratio of the muon fluxes in the shielding downstream of the decay region gave an independent measurement. The resulting precision in the flux ratio was 2.2%.

The absolute normalization of the neutrino flux was obtained from inclusive neutrino-nucleon scattering used as a monitor reaction. The total uncertainty on the flux measurement was found to be about 5%.

The first selection used only the hit-pattern in the streamer tubes and narrow, dense showers were selected, while broad showers with distinct tracks, characteristic for hadronic interactions, were removed. The contamination of hadronic showers was further reduced by rejecting events with backscattering tracks and by a set of criteria based on the lateral and longitudinal shower profile. Only events with one streamer tube hit in the first plane were considered. The total selection efficiency was 75%, nearly energy independent. All cuts were applied by a fully automatic pattern recognition procedure.

The background consisted of semileptonic neutrino reactions producing predominantly electromagnetic final states. The main contribution was coming from coherent and diffractive neutrino production of single neutral pions in NC interactions [22]. Electromagnetic showers were also produced in quasi-elastic neutrino-nucleon reactions of electron-neutrinos. A small fraction of the background was due to inclusive neutrino reactions with a large electromagnetic component in the final state.

The differential distributions corresponding to the different y-dependent terms in the cross-section were simulated for all four neutrino species in the beam according to Eq. (9). The discrimination was achieved between signal and background in the variable $E_e\theta_e{}^2$, and the background composition was determined owing to their different energy (E_e) distributions. Several analyses were performed, each with different underlying assumptions, reflected in different sets of parameters in the fit. The determination of the y-distribution used a fundamentally different approach, in which no *a priori* shape was assumed.

The theoretical prediction for neutrino-electron scattering was corrected for higher order QED effects [13, 15]. The coupling constants determined from these fits are therefore effective values including, but not corrected for, higher order electroweak effects.

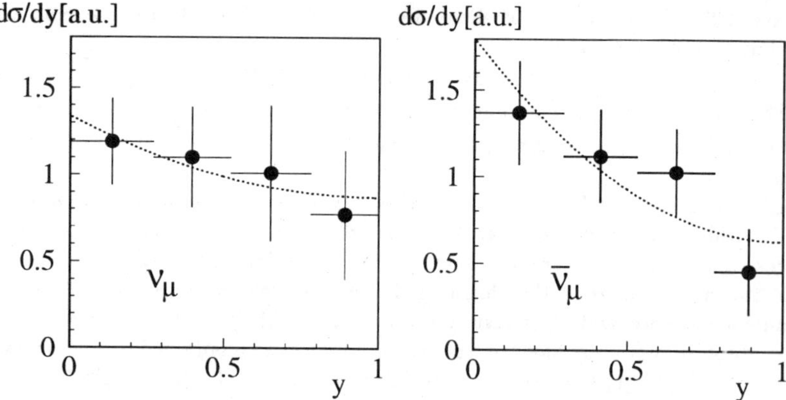

Fig. 10: *Differential cross-sections for* ν_μ*e-scattering (left), and* $\overline{\nu}_\mu$*e-scattering (right) in arbitrary units. The dashed line corresponds to the prediction of the Standard Model for a value of the electroweak mixing angle of* $\sin^2\theta_W = 0.212$.

The y-distribution and right-handed coupling

The shape of the y-distribution of $\nu_\mu e$ and $\overline{\nu}_\mu e$ interactions was determined with a minimum of model dependence. No use was made of the knowledge of the relative normalization of the beams, and no assumption on the shape of the distribution was made other than assuming a smooth behaviour at the scale of the experimental resolution [52, 53]. The data sample used here was collected during the years 1987 – 1990 and represented $\approx 80\%$ of the final statistics.

The differential cross-sections $d\sigma/dy$ were determined from the two-dimensional event distributions in the variables E_e and $E_e\theta_e{}^2$, with a procedure for regularized unfolding [54].

Two different classes of background contributed. One was due to semileptonic processes, as already described above, and was evaluated from a fit to the measured $(E_e, E_e\theta_e{}^2)$ distributions.

The second class of background consisted of events produced by neutrinos of the wrong helicity or by electron-type neutrinos scattering on electrons. For the subtraction $\sin^2\theta_W = 0.2337$ was used [55]. The unfolded differential cross-sections for $\nu_\mu e$ and $\overline{\nu}_\mu e$-scattering are shown in Fig. 10.

The ratio of the right-handed and left-handed coupling constants (g_R/g_L) was determined using the shape of the y-distributions only (without using the knowledge of the normalization) [52, 53]:

$$g_R{}^2/g_L{}^2 = 0.60 \pm 0.19_{stat} \pm 0.09_{syst},$$

which confirmed the existence of the right-handed coupling of the electron to the Z

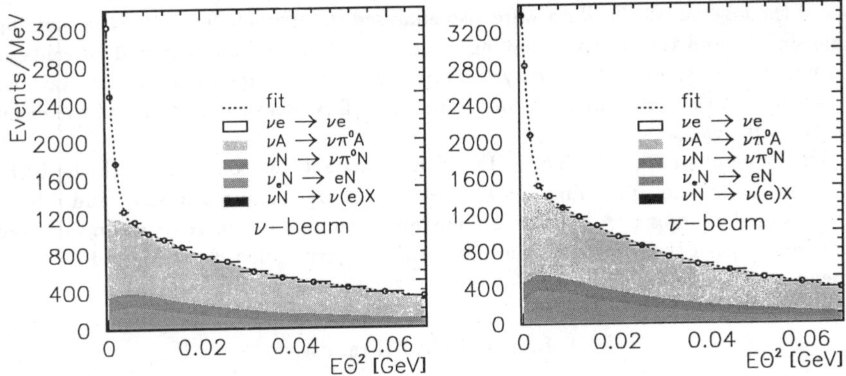

Fig. 11: *Experimental data and the result of the best fit: data are shown as circles and the fit results are displayed as a dashed line. Only the projections in* $E_e\theta_e^2$ *of the 2-dimensional distributions are shown. The different background components are cumulative.*

by three standard deviations. This result is entirely independent of the determination of the absolute cross-section, and of cross-section ratios and is hence independent of any assumed value for ρ.

This result compares well with an electroweak mixing angle $\sin^2\theta_W = 0.212 \pm 0.027_{stat} \pm 0.006_{syst}$, obtained from a direct fit to the shape of the two-dimensional distributions in E_e and $E_e\theta_e^2$. This result was the first measurement of the electroweak mixing angle using only the information from the shape of kinematical distributions from neutrino-electron data.

Coupling constants and $\sin^2\theta_W$

A *preliminary* analysis using the full data sample taken in 1987 – 1991, was based on the measurement of the cross-section ratio Eq. (12) and provided a precise determination of $\sin^2\theta_W$ [53]. The same method was applied previously to partial data samples [56, 57]. The pronounced peaks at low $E_e\theta_e^2$ contain each nearly 3000 νe events as shown in Fig. 11. From these values a *preliminary* determination of $\sin^2\theta_W$ was deduced.

The inclusion of the knowledge on absolute ν-fluxes allowed for the simultaneous determination of two electroweak parameters, e.g. g_V and g_A or $\sin^2\theta_W$ and ρ. The experimental sensitivity was increased by the use of energy distributions of the events [58]. The analysis was based on the data taken from 1987 – 1990.

The fourfold ambiguity in the determination of g_V and g_A was reduced to a two-fold one owing to the presence of ν_e and $\overline{\nu}_e$ components in the beam, inducing about

10% of the νe events. The data were also analysed in terms of the electroweak mixing angle $\sin^2\theta_W$ and the relative coupling strength ρ of neutral and charged currents as free parameters (Table 4). The systematic errors were dominated by uncertainties of the background determination, of the neutrino flux measurement and of the event selection efficiency.

The value of $\sin^2\theta_W$ obtained with the full data set was compared with the LEP-value [55] to give an upper limit on the neutrino charge radius, Eq. (20), and a limit on additional Z-bosons. A limit on the magnetic moment of the neutrino was derived from the *shape* of the y-distribution only, using the expression Eq. (18), and does not depend on an external value for the cross-sections.

Tab. 4: CHARM-II results.

quantity	method	value
$\sin^2\theta_W$	ratio (all data)	$0.232 \pm 0.006_{stat} \pm 0.007_{syst}$
	partial data	
g_V	ν and $\overline{\nu}$	$-0.025 \pm 0.014_{stat} \pm 0.014_{syst}$
g_A	ν and $\overline{\nu}$	$-0.503 \pm 0.007_{stat} \pm 0.016_{syst}$
$\sin^2\theta_W$	ν and $\overline{\nu}$	$0.237 \pm 0.007_{stat} \pm 0.007_{syst}$
ρ	ν and $\overline{\nu}$	$1.006 \pm 0.014_{stat} \pm 0.033_{syst}$
$m_{Z'}$	(all data)	> 375 GeV (90% C.L.)
μ_ν	$d\sigma/dy$	$< 3.3 \times 10^{-9}\mu_B$ (90% C.L.)
$\lvert\langle r^2\rangle_{anom}\rvert$	(all data)	$< 0.45 \times 10^{-32}$ cm^2 (90% C.L.)

These measurements are the most precise values obtained with neutrino-electron scattering.

Comparison with inverse muon decay

The charged-current reaction $\nu_\mu e \to \mu^- \nu_e$ (inverse muon decay) was observed with the same detector, and allowed for a purely leptonic determination of ρ, which has no theoretical uncertainties [59]. The analysis could only be made for a partial sample of the full statistics obtained with the detector. The ratio of this reaction and the $\nu_\mu e \to \nu_\mu e$ and $\overline{\nu}_\mu e \to \overline{\nu}_\mu e$ reaction,

$$R^{NC/CC} = \frac{\sigma(\nu_\mu e \to \nu_\mu e)}{\sigma(\nu_\mu e \to \mu^- \nu_e)} \quad \text{and} \quad R^{\overline{NC}/CC} = \frac{\sigma(\overline{\nu}_\mu e \to \overline{\nu}_\mu e)}{\sigma(\nu_\mu e \to \mu^- \nu_e)}$$

determine two relations between ρ and $\sin^2\theta_W$. From the measurement of these two NC/CC ratios and the $\nu/\overline{\nu}$ flux ratio in the $\overline{\nu}$-beam a simultaneous determination of $\sin^2\theta_W$ and ρ was obtained:

$$\sin^2\theta_W = 0.237 \pm 0.014, \quad \text{and} \quad \rho = 0.987 \pm 0.042.$$

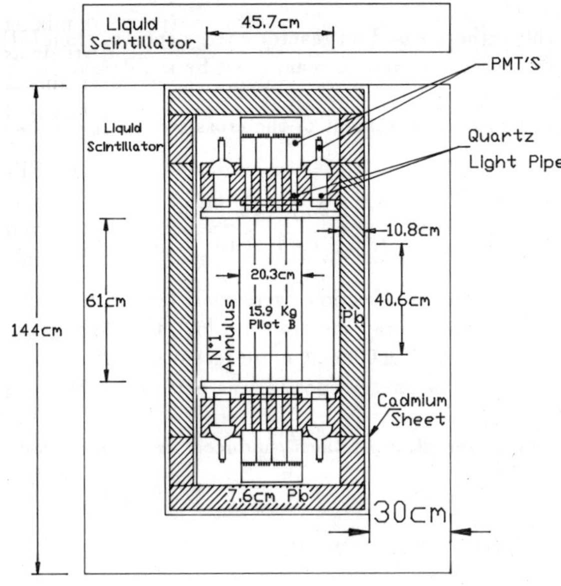

Fig. 12: *Schematic of the Savannah River detector showing the plastic target region, NaI light pipes and annulus, surrounded by the lead and cadmium shields and liquid scintillator anticoincidence detector.*

The error combines statistical and systematic errors. These results are less precise than the measurements using the normalization of the neutrino fluxes, but have the advantage of using exclusively leptonic processes.

2.8 *The Savannah River Reactor experiment*

This experiment was performed using the high $\overline{\nu}_e$ flux (2.2×10^{13}cm^{-2}s^{-1}) of the 1800 MW Savannah River fission reactor [60].

The interactions of e^- from the $\overline{\nu}_e e \rightarrow \overline{\nu}_e e$ reaction were detected with an apparatus consisting of a target of 16 kg plastic scintillator divided into sixteen elements (Fig. 12). The plastic scintillator was surrounded by 300 kg NaI scintillation detectors. A Pb-Cd shield with a 2200 l liquid scintillator enclosed the target. Pulse-height and timing information of the photo-multiplier signals were recorded by photographing an oscilloscope screen. The events were analysed in two energy bins, $1.5 < E < 3$ MeV and $3 < E < 4.5$ MeV, respectively.

The *inverse β-decay* background was identified by one of the two signatures: a prompt e^+ annihilation pulse or the delayed neutron capture. The remaining unidentified events of this type could only account for a few percent compared to the reactor related rate. The contribution of *γ-ray* and *neutron* background was estimated to be well under 10% of the reactor rate, with the help of the observation of signals in the NaI anticoincidence shield and, independently, from the multiplicity distribution of the plastic scintillators in the target.

The data were analysed by the authors based on reactor energy spectra from [61] and later, when new reactor spectra were available, reanalysed by a different group of authors [62].

The original results were given as a measurement of the cross-section in the two energy windows:

$$\sigma(\overline{\nu}_e e \to \overline{\nu}_e e; 1.5 - 3 \text{ MeV}) = (7.6 \pm 2.2) \times 10^{-46} \text{cm}^2$$
$$\sigma(\overline{\nu}_e e \to \overline{\nu}_e e; 3 - 4.5 \text{ MeV}) = (1.86 \pm 0.48) \times 10^{-46} \text{cm}^2$$

(We use the notation given in [62].) These cross-sections were consistent with a value $\sin^2\theta_W = 0.29 \pm 0.05$. With the reactor $\overline{\nu}_e$-energy spectra from [62], the electroweak mixing angle was quoted to be $\sin^2\theta_W = 0.25 \pm 0.05$. A more recent re-evaluation of the experiment with more accurate reactor energy spectra was performed, and can be found in [6].

This experiment made the first observation of the $\overline{\nu}_e$-induced neutrino-electron scattering reaction.

2.9 *The Kurchatov Reactor experiment*

A group from the Kurchatov institute have reported a *preliminary* result of a $\overline{\nu}_e e \to \overline{\nu}_e e$ experiment at a fission reactor with a $\overline{\nu}_e$ flux of 3.4×10^{12} cm^{-2} s^{-1} [63]. The main improvement attempted in this experiment compared to the Savannah River detector [60] was to avoid completely the use of materials containing hydrogen. This was done to suppress the $\nu_e n \to e^- p$ background reaction. As scintillator material organofluoric scintillators were used, segmented into seven hexahedral prisms, each viewed by two photomultipliers. The detector volume contained 103 kg of scintillator. To shield against cosmic-ray muons, the apparatus was placed inside a magnetic shield surrounded by a passive shielding and an active shield of scintillator plates. A small amount of hydrogen-containing material was needed as scintillating dopant.

Events were defined by a coincident signal in the two photomultipliers viewing the same segment, not accompanied by any other signal within 170 μs. A cross-section measurement was performed by subtraction of the signal during periods that the reactor was on and off. The energy window was $3150 < E < 5175$ keV. The *preliminary* result was: $\sigma(\overline{\nu}_e e \to \overline{\nu}_e e) = (6.8 \pm 4.5) \times 10^{-46}$ cm^2/fission, which corresponds to $\sin^2\theta_W = 0.29 \pm 0.10$.

2.10 *The ILM experiment*

The ILM experiment (E225) was performed with neutrinos produced at the beam stop of the Clinton P. Anderson Meson Physics Facility (LAMPF). Pions produced in the beam stop of the 800 MeV proton beam decay at rest and produce an isotropic

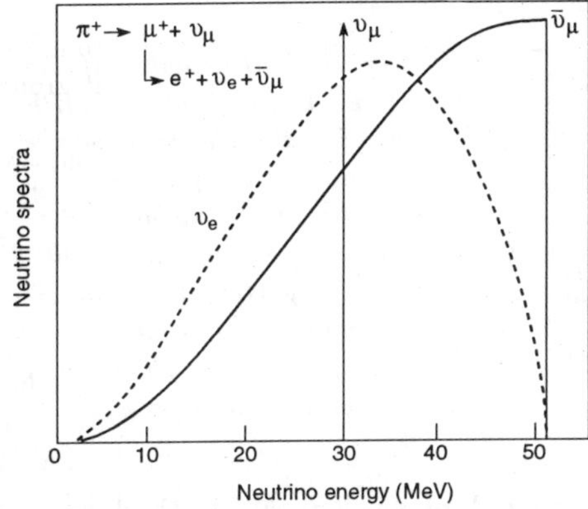

Fig. 13: *Energy spectrum of neutrinos produced by pion and muon decay at rest in the proton beam stop at LAMPF. Spectra for all three neutrino species are shown.*

flux of neutrinos. A monochromatic muon neutrino source is produced by the reaction $\pi^+ \rightarrow \mu^+\nu_\mu$ with an energy of 30 MeV; the subsequent muon decay reaction $\mu^+ \rightarrow e^+\nu_e\overline{\nu}_\mu$ produces a continuous spectrum of ν_e and $\overline{\nu}_\mu$, as shown in Fig. 13. The beam spill length of 750 μs did not allow for a separation of the neutrino species by the different decay times. The rate of $\overline{\nu}_e$ was suppressed owing to nuclear absorption of π^- and μ^-, and was negligibly small. The neutrino spectra follow from straightforward decay kinematics; the absolute intensity required calibration experiments in instrumented beam stops to be determined. An uncertainty of 7.3% on the absolute neutrino flux was achieved.

The neutrino detector viewed the beam stop at 90^0 [64]. It consisted of a 15 ton central active target, segmented in 40 plastic scintillator planes with 305×305 cm^2 cross-section, see Fig. 14. These scintillator planes were interspersed with multiplane flash chamber modules with tubes of approximately 0.5 cm diameter. The scintillator pulse-heights were used for dE/dx measurements, while the flash tubes determined the position and direction of the tracks. An active anticoincidence of multiwire proportional chambers was needed to reduce the cosmic-ray rate of 10^8 muons per day by many orders of magnitude. It was used to set a time window of ≈ 30 μs around each trigger, in which each other activity would veto the event. The trigger required at least three consecutive scintillator planes to be fired, with a veto from the anticoincidence shield, a threshold corresponding to $E_e \geq 14$ MeV in the scintillators and an upper limit of about 18 MeV in any of the scintillation counters. The latter requirement was used to eliminate elastic np scattering. This experiment was the first to demonstrate $\nu_e e \rightarrow \nu_e e$-scattering [65].

To remove cosmic-ray and neutron induced background a set of criteria was applied. A track fit was required in the flash chambers, fiducial volume, total energy, dE/dx and activity cuts in the pre- and post-trigger times were applied. After these

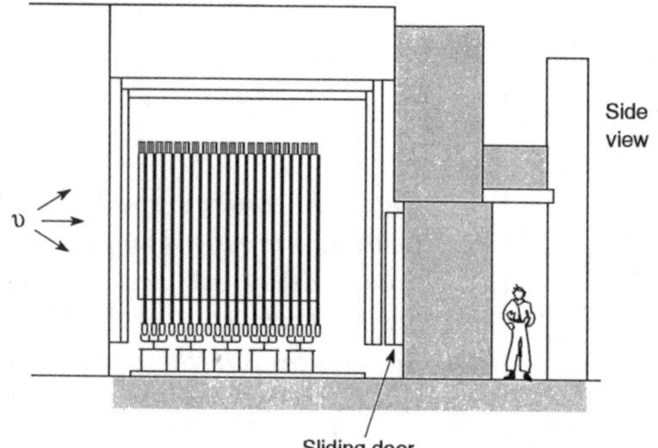

Fig. 14: *Schematic of the ILM detector at LAMPF showing the planar structure of the detector, surrounded by the steel shields (inner open boxes) and the multi-wire proportional anticoincidence detector (outer open boxes). Concrete shielding is shown as shaded boxes.*

cuts a beam-on sample of 7752 events remained, of which about one-third were induced by neutron- rather then ν-interactions. The sample was further purified by cuts on very low pulse-height scintillators, track fit quality and the requirement on the electron candidate to travel in the direction of the forward hemisphere. A statistical subtraction of the beam-off triggers from the final sample of 4880 beam-on candidates defined a beam-related candidate signal of 1492 ± 76 events. The excess of events shown in Fig. 15 in the forward direction is an unambiguous demonstration of the neutrino electron elastic scattering.

The νe signal was extracted by a multiparameter maximum likelihood fit to the kinematic distributions of visible energy, E_{vis}, and the recoil angle, $cos\theta_e$. The beam-related background processes, $\nu_e{}^{12}C$ and $\nu_e{}^{13}C$-scattering and *neutron-induced* γ*-rays* were taken into account. The measured νe rate was 295 ± 35 events, where the error was dominated by the statistical uncertainty due to the large cosmic-ray background. The νe event rate had to be corrected for the contribution from $\nu_\mu e$ and $\overline{\nu}_\mu e$ events in order to obtain the number of $\nu_e e$ events. The $\nu_\mu e$ and $\overline{\nu}_\mu e$ contributions were calculated from measured absolute cross-sections [49, 42], corrected for the efficiencies. The net rate of $\nu_e e$ events was determined to be 236 ± 35.

The total cross-section was found to be [66]

$$\sigma(\nu_e e \to \nu_e e) = (10.0 \pm 1.5_{stat} \pm 0.9_{syst}) \times 10^{-42}\, E_\nu\, \mathrm{cm}^2\, \mathrm{GeV}^{-1},$$

or derived in terms of $\sigma_0 = G_F^2 s/4\pi$, $\sigma(\nu_e e \to \nu_e e) = (2.32 \pm 0.35_{stat} \pm 0.21_{syst})\,\sigma_0$. The systematic error was composed of the uncertainty of the neutrino flux mentioned above, the efficiency (5.2%), and the number of target electrons (1.5%).

To demonstrate the existence of the CC-NC interference term, the difference, N^I, is constructed between the observed rate, $N^{\nu_e e}$, and the predicted CC and NC rate, N^{CC} and N^{NC}. The NC rate was derived from $\nu_\mu e$ measurements [49, 42], while the CC rate was calculated from the theoretical cross-section, using G_F from muon decay. The contribution of the interference was measured to be:

$$N^I = N^{\nu_e e} - N^{CC} - N^{NC} = -221 \pm 35_{stat} \pm 21_{syst}.$$

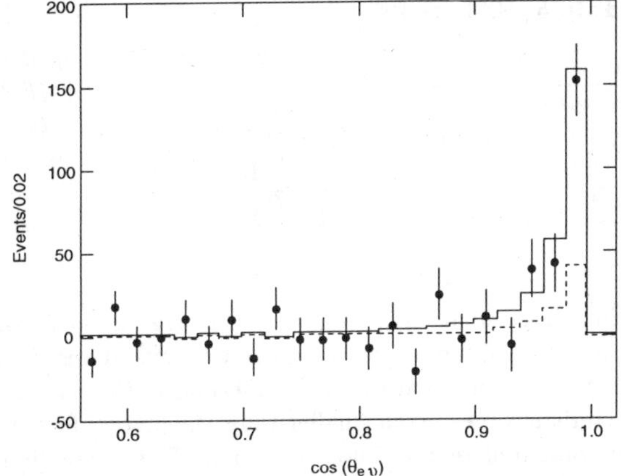

Fig. 15: *Angular distribution of the measured elastic scattering signal in the ILM experiment. The solid line is the result of the best fit,* 295 ± 35 *events. The dashed line is the background contribution from 59.2* $\nu_\mu e$ *and* $\overline{\nu}_\mu e$*-scattering events.*

The existence of the destructive interference was demonstrated with five standard deviation significance [67, 66]. An interference coefficient was defined:

$$I = \sigma^I/2\sigma_0 = -1.07 \pm 0.21,$$

where the error combines both the statistical and systematical uncertainties. This measurement is in excellent agreement with the Standard Model prediction $I = -1.08$ for $\sin^2\theta_W = 0.23$.

Within the framework of the Standard Model a value of the electroweak mixing angle could be obtained from the total cross-section.

A limit on additional Z-bosons was derived from the interference measurement [68], together with various limits in different models [66]. An upper limit for the magnetic moment of the neutrino, μ_ν, was deduced [69] using the equation for the total cross-section Eq. (19). By a comparison of the measurement of $\sin^2\theta_W$ a limit for the electron-neutrino charge-radius was deduced [70, 66]. Similarly, a limit on flavour changing neutral currents of the electron type into μ or τ-type was deduced from the measurement of the interference term [71]. These results are summarized in Table 5.

2.11 *Outlook*

The data of most experiments described above have been published in their final form, with the exception of a few. New publications on data from $\overline{\nu}_e e$-scattering at reactors are expected in the future (see section 2.9). The final analysis of the full data sample of the CHARM-II experiment is not yet available, and an improvement of a number of measured quantities is to be expected (see section 2.7).

A proposal for a new experiment has been submitted at LAMPF [72], with the aim to measure $\sin^2\theta_W$ with a 1% precision. The proposed method is different from

Tab. 5: ILM results.

quantity	method	value
$\sin^2\theta_W$	$\sigma(\nu_e e \to \nu_e e)$	0.249 ± 0.063
μ_ν	ν_e,ν_μ,$\overline{\nu}_\mu$	$< 0.61 \times 10^{-9}\mu_B$ (90% C.L.)
$f_{e\mu}^2 + f_{e\tau}^2$	interference	< 0.35 (90% C.L.)
$-3.56 \times 10^{-32} < \langle r^2 \rangle_{anom} < 5.44 \times 10^{-32}$ cm^2 (90% C.L.)		

the one used in previous precision determinations of $\sin^2\theta_W$ with neutrino-electron scattering. The neutrino beam is obtained from pion and muon decays at rest in a proton beam dump (section 2.10). A special feature of the method is the use of a pulsed proton beam, with characteristic times much smaller than the muon life time, and a time between pulses long compared to the muon life time. The interactions of neutrinos induced by the different neutrino species can then be separated by their different time of arrival with respect to the pulsed proton beam. The prompt ν_μ-component is produced from π-decays, while the $\overline{\nu}_\mu$ and ν_e-components come from μ-decays. The measurement is based on the ratio:

$$R^{\pi\mu} = \frac{(\nu_\mu e \to \nu_\mu e)}{(\overline{\nu}_\mu e \to \overline{\nu}_\mu e) + (\nu_e e \to \nu_e e)},$$

which can be expressed in terms of $\sin^2\theta_W$.

The major experimental challenges are the knowledge of the energy thresholds, of the efficiency of the timing separation and the calculation of the backgrounds. The detection technique is based on the measurement of Cherenkov-radiation in a large cylindrical water reservoir. The beam is dumped in the centre of the cylinder, from the top, and suitable shielding should prevent neutrons and other beam-related particles to enter the neutrino target volume. The experimental techniques have been pioneered by the large water-Cherenkov proton-decay experiments on the one hand and by the ILM $\nu_e e$-experiment (see section 2.10) on the other hand. One of the special features which make the high precision possible is the natural relation between the fluxes of the different neutrino species.

3 Discussion

A compilation of measurements of the cross-sections is given in Table 6, together with the most precise value of $\sin^2\theta_W$ obtained by the experiments. One observes, in general, a good agreement among the measurements. It should be noted that the total cross-sections of *all* experiments are model-dependent in two ways. First of all, events are observed in a limited kinematical domain. The corrections to derive total cross-sections from these *visible* cross-sections depend on the assumed shape of the

Tab. 6: Compilation of total cross-section and $\sin^2\theta_W$ measurements.

Experiment	Beam	$\sigma(\nu_\mu e)$	$\sigma(\overline{\nu}_\mu e)$	$\sin^2\theta_W$
		$(\times 10^{-42}\,E_\nu\,\text{cm}^2\,\text{GeV}^{-1})$		
GGM [26]	CERN PS	< 1.4	$1.0^{+2.1}_{-0.9}$	$0.1 < x < 0.4$
AC-PD [31]	CERN PS	1.1 ± 0.6	2.2 ± 1.0	0.35 ± 0.08
GGM [28, 29]	CERN SPS	$2.4^{+1.2}_{-0.9}$	< 2.7	$0.12^{+0.11}_{-0.07}$
VMWOF [32]	FNAL	$1.4 \pm 0.3 \pm 0.4$		$0.25^{+0.07}_{-0.05} \pm 0.8$
BNL-COL [33]	FNAL	1.67 ± 0.44		$0.20^{+0.06}_{-0.05}$
FMMS [34]	FNAL		< 2.1	< 0.37
BEBC-TST [73]	CERN SPS		< 3.4	< 0.45
CHARM [42]	CERN SPS	$2.2 \pm 0.4 \pm 0.4$	$1.6 \pm 0.3 \pm 0.3$	0.211 $\pm 0.035 \pm 0.011$
BBKOPS [49]	BNL AGS	$1.8 \pm 0.2 \pm 0.25$	$1.17 \pm 0.16 \pm 0.13$	0.195 $\pm 0.018 \pm 0.013$
CHARM-II [58, 53]	CERN SPS	$1.53 \pm 0.04 \pm 0.12$	$1.39 \pm 0.04 \pm 0.10$	0.232 $\pm 0.006 \pm 0.007$
		$\sigma(\nu_e e)$	$\sigma(\overline{\nu}_e e)$	$\sin^2\theta_W$
		$(\times 10^{-42}\,E_\nu\,\text{cm}^2\,\text{GeV}^{-1})$	$(\times 10^{-46}\text{cm}^2)$	
Savannah River [60]	reactor			0.25 ± 0.05
	(1.5 – 3.0) MeV:		7.6 ± 2.2	
	(3.0 – 4.5) MeV:		1.86 ± 0.48	
ILM [66]	LAMPF	$10.0 \pm 1.5 \pm 0.9$		0.249 ± 0.063

distributions, and can only be calculated within a model. In all cases the Standard Model is assumed, but with different values for $\sin^2\theta_W$. An alternative is to derive σ^{tot} from the measurements of g_V and g_A, with reduces the model-dependence to the assumption of a model with any mixture of V and A terms. A second objection is that for most experiments, the beam is not pure, and the neutrino-electron scattering candidates contain admixtures from other neutrino types. When these admixtures are *subtracted* a model dependence is introduced. This problem cannot be neglected for the higher precision experiments, since the subtraction can be as high as 20%. A subtraction using measured cross-sections from other experiments solves this problem up to the next order of approximation. In the measurements of g_V and g_A these problems can be solved by treating all neutrino species in the beam on an equal footing and expressing the observed number of events as a superposition of all components in the beam. This is, however, not always done which makes the comparison of the results difficult. In particular, when average values of many experiments are quoted, the errors are reduced and these considerations become important. This is the reason why the CHARM-II collaboration for its measurements with improved

Tab. 7: Results of two-parameter fits to $\nu_\mu e$ and $\overline{\nu}_\mu e$ data.

	two-parameter fits	
Experiment	g_V	g_A
CHARM	$-0.06 \pm 0.07 \pm 0.02$	$-0.57 \pm 0.04 \pm 0.06$
BBKOPS	$-0.107 \pm 0.35 \pm 0.028$	$-0.503 \pm 0.023 \pm 0.028$
CHARM-II	$-0.025 \pm 0.014 \pm 0.014$	$-0.503 \pm 0.007 \pm 0.016$
average	-0.034 ± 0.016	-0.504 ± 0.014
	$\sin^2\theta_W$	ρ
CHARM	$0.211 \pm 0.035 \pm 0.011$	$1.14 \pm 0.07 \pm 0.12$
BBKOPS	$0.199 \pm 0.020 \pm 0.013$	$1.005^{+0.070}_{-0.075}$
CHARM-II	$0.237 \pm 0.007 \pm 0.007$	$1.006 \pm 0.014 \pm 0.033$
average	0.233 ± 0.008	1.008 ± 0.029

precision preferred to quote g_V and g_A rather than total cross-sections; the total cross-section values appearing in the table for this experiment were derived from the measurements of g_V and g_A.

A summary of cross-section measurements and measurements of $\sin^2\theta_W$ is given in Table 6; limits are given at the 90% C.L., which was also the case for the first cross-section measurement by Gargamelle at the PS.

Only for experiments measuring two cross-sections a determination of g_V and g_A can be obtained. A compilation of these measurements is given in the Table 7. A global fit to all available data has been performed, using radiative corrections given by [7] for the earlier experiments and [15] for the later ones [74, 75]. The constraints of all experiments have been used to obtain the combined result including ν_μ, $\overline{\nu}_\mu$, ν_e and $\overline{\nu}_e$ scattering data. The unavoidable two-fold ambiguity has been resolved with the help of e^+e^--data. The resulting limits in the g_V-g_A-plane are displayed in Fig. 16 and given in Table 7.

In order to be able to compare the νe results with the high precision LEP results, radiative corrections have to be applied. Differences between the two measurements at different Q^2-values are expected to arise from the q^2-dependence of α, the running fine structure constant, and from the q^2-dependent corrections to νe-scattering. These effects turn out to cancel to a large extent:

$$\Delta g_V(\nu e - LEP) = -0.002.$$

This is the result of a cancellation of two contributions, which both depend on the assumed value of m_{top} and m_H. In the calculations [15] the values of these the masses were fixed to $m_{top} = 150$ GeV, $m_H = 100$ GeV. For numerical values we use here the $\overline{MS}$ renormalization scheme [12], which has the advantage to minimize for this process the m_{top} and m_H-dependence; the net correction to $\sin^2\theta_W$ is negligible

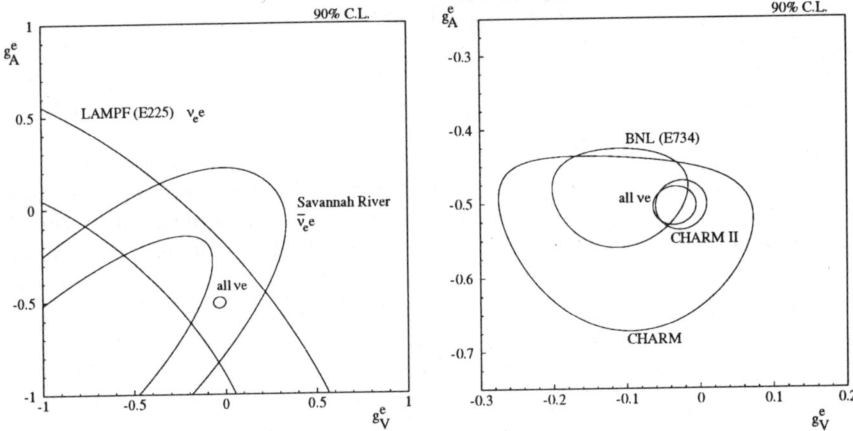

Fig. 16: *a) Comparison of results from $\nu_e e \to \nu_e e$ and $\overline{\nu}_e e \to \overline{\nu}_e e$ scattering with the result of the fit to all νe-data in the g_V-g_A plane. b) Comparison of the most recent muon-neutrino electron-scattering experiments, together with the fit to all data. The ν_e and $\overline{\nu}_e$ data select from the solutions allowed by the ν_μ and $\overline{\nu}_\mu$ data the ones with $g_V \approx -0.5$ or $g_A \approx -0.5$; the solution $g_V \approx -0.5$ is excluded by e^+e^- data.*

and for ρ is -0.005, with a theoretical uncertainty of ± 0.002 and ± 0.004, respectively, induced by assuming a reasonable range of masses ($m_{\text{top}} = [80..180]$ GeV and $m_{\text{H}} = [50..1000]$ GeV). Applying these corrections one finds for the most precise νe experiment (CHARM-II) at $Q^2 = m_{\text{Z}}{}^2$

$$\begin{aligned}\sin^2\bar{\theta}(\nu e) &= 0.237 \pm 0.010_{exp} \pm 0.002_{theor} \\ \bar{\rho}(\nu e) &= 1.001 \pm 0.038_{exp} \pm 0.004_{theor}\end{aligned}$$

The result from νe-scattering compares well with the Standard Model prediction using as input parameters α, G_F and the mass measurements of the Z boson performed at LEP [55]:

$$\begin{aligned}\sin^2\bar{\theta}(\alpha, G_F, m_{\text{Z}}) &= 0.233 \pm 0.002_{theor} \\ \bar{\rho}(\alpha, G_F, m_{\text{Z}}) &= 1.001 \pm 0.004_{theor}\end{aligned}$$

The theoretical errors account for the assumed mass ranges. The experimental error from the uncertainty of the Z mass measurement is negligible at this scale. The data are compared in the g_{V}-g_{A}-plane in Fig. 17.

The good agreement between the e^+e^--data implies that limits can be derived for contributions from terms beyond the Standard Model description. It is not straightforward to derive a combined upper limit for the magnetic moment, since the cross-section has a different dependence on E_ν and E_e compared to the Standard Model

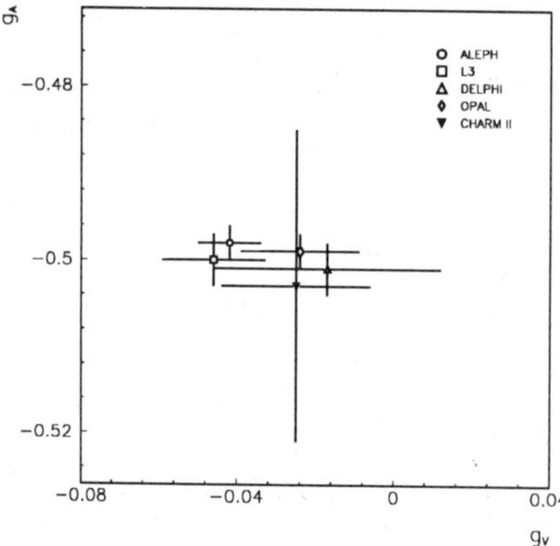

Fig. 17: *Comparison of results from neutrino-electron scattering and from* $e^+e^- \to e^+e^-$ *annihilation at the Z pole in the* g_V-g_A *plane. The crosses show different experimental data points [55, 58].*

process. The experiments have been performed in different beams and the analyses have used different strategies. The best limit obtained from total cross-sections is $\mu_\nu < 0.61 \times 10^{-9}\mu_B$ (90% C.L.). The best limit using exclusively the energy-dependence of the cross-section is $\mu_\nu < 3.3 \times 10^{-9}\mu_B$ (90% C.L.), and is therefore less dependent on an external cross-section prediction. Limits on the anomalous charge radius and on additional Z'-bosons with the assumption of equal coupling compared to the standard Z are given below for the most precise experiment and for the combined data.

experiment	quantity	value
CHARM-II	$m_{Z'}$	> 375 GeV (90% C.L.)
Combined	$m_{Z'}$	> 462 GeV (90% C.L.)
CHARM-II	$\lvert\langle r^2\rangle_{anom}\rvert$	$< 0.45 \times 10^{-32}$ cm^2 (90% C.L.)
Combined	$\lvert\langle r^2\rangle_{anom}\rvert$	$< 0.40 \times 10^{-32}$ cm^2 (90% C.L.)

It was recently pointed out [76] that the coupling constants g_V and g_A, measured in ν_μe-scattering are in fact a product of the muon-neutrino coupling to the Z and the electron coupling to the Z, while the parameters measured in the process $e^+e^- \to e^+e^-$ are sensitive to the e-Z coupling, g_V^e, only. The measurement of the invisible width of the Z at LEP determines the coupling to a mixture of all three neutrino species. Hence if the assumption of lepton universality is not made the ν_μ-data provide a unique measurement of the muon-neutrino coupling. When we use the notation $g_V^{\nu e}$ and $g_A^{\nu e}$ for the coupling measured in ν_μe-scattering, one finds

$$
\begin{aligned}
g_V^{\nu e} &= 2g^\nu g_V^e \\
g_A^{\nu e} &= 2g^\nu g_A^e
\end{aligned} \tag{22}
$$

where g^ν is the ν-Z coupling, predicted to be 0.5 by the weak isospin structure of the Standard Model. If no assumption of lepton universality is made, the combination of

LEP-data and the CHARM-II data give:

$$|2g^{\nu}| = 1.006 \pm 0.036,$$

valid for muon-neutrinos, while the LEP-result from the invisible width gives:

$$|2g^{\nu}| = 1.006 \pm 0.006,$$

for the mixture of the three contributing neutrinos. The fact that both numbers agree among each other confirms lepton universality in the neutrino sector for neutral-current interactions. The values also confirm the prediction based on the weak isospin structure of the Standard model.

4 Conclusion

In conclusion, electroweak parameters determined from the differential cross-sections of neutrino-electron scattering are in very good agreement with those from LEP experiments. The observed agreement of measurements spanning a factor 10^6 in Q^2 is a remarkable confirmation of the Standard Model.

Two decades of neutrino-electron scattering research have given an essential contribution to our understanding of the Standard Model. All four possible reactions have been observed and studied. From the samples of a handful of events in the early experiments, dedicated experiments can now collect many thousands of νe-interactions. The predictions of the Standard Model have been confirmed over a wide energy range as far as the total cross-sections is concerned. The weak-isospin structure of the Standard Model has been verified in the neutral-current sector. In particular, the value and the sign of the axial-vector coupling constants are a direct prediction of the doublet structure of the model, and have been confirmed. Lepton-universality in the neutral-current sector has been demonstrated for neutrinos. A first experimental verification of the relation between total cross-sections and the shape of the angular distribution, a direct prediction of the space-time structure of the neutral current, has been obtained. The ability of the Standard Model to describe higher order corrections to the Born-approximation has been shown by comparisons with other processes. Meaningful limits on departures from the Standard Model have been obtained and no indication for any of such has been found up to now.

References

[1] Gargamelle collab., F.J. Hasert et al., *Phys. Lett.* **B 46** (1973) 121.

[2] Gargamelle collab., F.J. Hasert et al., *Phys. Lett.* **B 46** (1973) 138.

[3] S. Weinberg, *Phys. Rev. Lett.* **19** (1967) 1264; A. Salam in *Elementary Particle Theory*, ed. N. Svartholm (Almquist and Wiksells, Stockholm, 1969) 367; S.L. Glashow, J. Iliopoulos and L. Maiani, *Phys. Rev.* **D 2** (1970) 1285.

[4] G. t'Hooft, *Phys. Lett.* **B 37** (1971) 195.

[5] L.F. Abbott and R.M. Barnett, *Phys Rev.* **D 18** (1978) 3214; **D 19** (1979)3230; I. Liede and M. Roos, *Phys. Lett.* **B 82** (1979); *Nucl. Phys.* **B 167** (1980) 397; J.E. Kim, P. Langacker, M. Levine, and H.H. Williams, *Rev. Mod. Phys.* **53** (1981) 211; P.Q. Hung and J.J. Sakurai, *ARNPS* **31** (1981) 375; L.M. Sehgal, *Prog. Nucl. Part. Phys.* **14** (1985) 1; M. Klein and S. Schlenstedt, *Z. Phys.* **C 29** (1985) 235;

[6] U. Amaldi et al., *Phys. Rev.* **D 36** (1987) 1385.

[7] G.L. Fogli, *Europhys. Lett.* **4** (1987) 527.

[8] G. Costa et al., *Nucl. Phys.* B 297 (1988) 244.

[9] J. Ellis and G.L. Fogli, *Phys. Lett.* **B 249** (1990) 543.

[10] L.B. Okun, *Leptons and Quarks*, North-Holland, Amsterdam (1982).

[11] A. Sirlin, *Phys. Rev.* **D 22** (1980) 971; W.J. Marciano and A. Sirlin, *Phys. Rev.* **D 22** (1980) 2695.

[12] M. Consoli and W. Hollik, in: G. Altarelli et al. (edts), Report CERN 89-08 (1989) 7.

[13] D.Yu. Bardin and V.A. Dokuchaeva, *Nucl. Phys.* **B 246** (1984) 221; Preprint JINR E2-86-260 (1986).

[14] S. Sarantakos, A Sirlin and W.J. Marciano, *Nucl. Phys.* **B 217** (1983) 84.

[15] D.Yu. Bardin, NUFITTER, *a program to calculate electroweak radiative corrections for neutrino-electron scattering.*

[16] G. Domogatskii and D. Nadezhin, *Sov. J. Nucl. Phys.* **12** (1971) 678; A.V. Kyuldjiev, *Nucl. Phys.* **B 243** (1984) 387.

[17] W.A. Bardeen, R. Gastmans and B. Lautrup, *Nucl. Phys.* **B 46** (1972) 319; G. Degrassi, A. Sirlin and W.J. Marciano, *Phys. Rev.* **D 39** (1989) 287; W.J. Marciano and A. Sirlin , *Phys. Rev.* **D 22** (1980) 2695; M. Kuroda and D. Schildknecht, Univ. Bielefeld preprint BI-TP 92/49.

[18] J.E. Kim, V.S. Mathur and S. Okubo, *Phys. Rev.* **D 9** (1974) 3050.

[19] J.A. Grifols and S. Peris, *Phys. Lett.* **B 168** (1986) 264; J.L. Rosner, *Comm. Nucl. Part. Phys.* **14** (1985) 229.

[20] P. Langacker, Mingxing Luo and A.K. Mann, *Rev. Mod. Phys.* **64** (1992) 87.

[21] J.J. Ross, *Proc 1987 Symposium on Lepton and Photon Interactions at High Energies*, DESY, Hamburg, 1987, W. Bartel, R. Rueckl (eds.), 743, Amsterdam, North-Holland (1988).

[22] A.A. Bel'kov and B.Z. Kopeliovich, *Yad. Fiz.* **46** (1987) 874.

[23] K.S. Lackner, *Nucl. Phys.* **B 153** (1979) 526.

[24] D. Rein and L.M. Sehgal, *Nucl. Phys.* **B 223** (1983) 29.

[25] S. van der Meer, CERN 62-16 (1962).

[26] Gargamelle collab., J. Blietschau et al., *Nucl. Phys.* **B 114** (1976) 189.

[27] Gargamelle collab., P. Alibran et al., *Phys. Lett.* **B 74** (1978) 422.

[28] Gargamelle collab., N. Armenise et al., *Phys. Lett.* **B 86** (1979) 225.

[29] Gargamelle collab., D. Bertrand et al., *Phys. Lett.* **B 84** (1979) 354.

[30] P. Musset and J-P. Vialle, *Phys. Reports* **C 39** (1978) 1.

[31] H. Faissner et al., *Phys. Rev. Lett.* **D 41** (1978) 213.

[32] R.H. Heisterberg et al., *Phys. Rev. Lett.* **44** (1980) 635.

[33] BNL-COL collab., A.M. Cnops et al., *Phys. Rev. Lett.* **41** (1978) 367 *and* N.J. Baker et al., *Phys. Rev.* **D 40** (1989) 2753.

[34] FMMS collab., J.P. Berge et al., *Phys. Lett.* **B 84** (1979) 357.

[35] CHARM collab., A.N. Diddens et al., *Nucl. Inst. & Meth.* **176** (1980) 189.

[36] E.H.M Heijne, CERN 83-06 (1986).

[37] CHARM collab., A.N. Diddens et al., *Nucl. Inst. & Meth.* **178** (1980) 27.

[38] CHARM collab., J. Dorenbosch et al., *Nucl. Inst. & Meth.* **253** (1987) 203.

[39] CHARM collab., M. Jonker et al., *Phys. Lett.* **B 105** (1981) 242.

[40] CHARM collab., F. Bergsma et al., *Phys. Lett.* **B 117** (1982) 272.

[41] CHARM collab., F. Bergsma et al., *Phys. Lett.* **B 147** (1984) 481.

[42] CHARM collab., J. Dorenbosch et al., *Z. Phys.* **C 41** (1989) 567.

[43] H.J. Behrends et al., *Phys. Lett.* **B 144** (1982) 187.

[44] BBKOPS collab., L.A. Ahrens et al., *Nucl. Inst. & Meth.* **A 254** (1987) 515.

[45] BBKOPS collab., L.A. Ahrens et al., *Phys. Rev. Lett.* **51** (1983) 1514.

[46] BBKOPS collab., L.A. Ahrens et al., *Phys. Rev. Lett.* **54** (1985) 18.

[47] BBKOPS collab., K. Abe et al., *Phys. Rev. Lett.* **58** (1987) 636.

[48] BBKOPS collab., K. Abe et al., *Phys. Rev. Lett.* **62** (1989) 1709.

[49] BBKOPS collab., L.A. Ahrens et al., *Phys. Rev.* **D 41** (1990) 3297.

[50] CHARM II collab., K. de Winter et al., *Nucl. Inst. & Meth.* **A 278** (1989) 670.

[51] CHARM II collab., D. Geiregat et al., *Nucl. Inst. & Meth.* **A 325** (1993) 92.

[52] G. Raedel, Ph.D. Thesis, Univ. Hamburg (1992), unpublished; CHARM-II collab., B. Akkus et al., CERN-PPE/93-13, submitted to *Phys. Lett.* **B**.

[53] CHARM-II collab., B. Akkus et al., *Neutral Current Coupling Constants from Neutrino-Electron scattering*, proceedings XV Int. Conf. Neutrino Physics and Astrophysics, Granada, Spain (1992), to be published in *Nucl. Phys.* **B**.

[54] V. Blobel, DESY 84/118, Hamburg 1984.

[55] The LEP collaborations: ALEPH, DELPHI, L3 and OPAL, *Phys. Lett.* **B 276** (1992) 247.

[56] CHARM II collab., D. Geiregat et al., *Phys. Lett.* **B 232** (1989) 539.

[57] CHARM II collab., D. Geiregat et al., *Phys. Lett.* **B 259** (1991) 499.

[58] CHARM II collab., P. Vilain et al., *Phys. Lett.* **B 281** (1992) 159.

[59] CHARM-II collab., B. Akkus et al., *New results from Inverse Muon Decay and Low Multiplicity reactions*, proceedings XV Int. Conf. Neutrino Physics and Astrophysics, Granada, Spain (1992), to be published in *Nucl. Phys.* **B**.

[60] F. Reines, H.S. Gurr and H.W. Sobel, *Phys. Rev. Lett.* **37** (1976) 315.

[61] F.T. Avignone, III, *Phys. Rev.* **D 2** (1970) 2609.

[62] F.T. Avignone, III and Z.D. Greenwood, *Phys. Rev.* **D 16** (1977) 2383.

[63] G.S. Vidyakin et al., *JETP Lett.* **49** (1989) 740.

[64] ILM collab., R.C. Allen et al., *Nucl. Inst. & Meth.* **A 269** (1988) 177.

[65] ILM collab., R.C. Allen et al., *Phys. Rev. Lett.* **55** (1985) 2401.

[66] ILM collab., R.C. Allen et al., *Phys. Rev.* **D 47** (1993) 11.

[67] ILM collab., R.C. Allen et al., *Phys. Rev. Lett.* **64** (1990) 1330.

[68] J.A. Grifols, *Mod. Phys. Lett.* **A 31** (1990) 2657.

[69] ILM collab., D.A. Krakauer et al., *Phys. Lett.* **B 252** (1990) 177.

[70] ILM collab., R.C. Allen et al., *Phys. Rev.* **D 43** (1991) R1.

[71] ILM collab., D.A. Krakauer et al., *Phys. Rev. Lett.* **D 45** (1991) 975.

[72] R.C. Allen et al., *A proposal for a Precision Test of the Standard Model by Neutrino-Electron Scattering (Large Cherenkov Detector Project)*, LANL Proposal, LA-11 300-P, 1988.

[73] BEBC-TST collab., N. Armenise et al., *Phys. Lett.* **B 81** (1979) 385.

[74] R. Beyer, Ph.D. Thesis, Univ. Hamburg (1991), unpublished.

[75] The author wishes to thank R. Beyer for the fits performed to the combined experimental data.

[76] V.A. Novikov, L.B. Okun and M.I. Vysotsky, *Phys. Lett.* **B 298** (1993) 453.

ATOMIC PARITY NONCONSERVATION EXPERIMENTS

B. Patrick Masterson and Carl E. Wieman

Joint Institute for Laboratory Astrophysics
University of Colorado and National Institute of Standards and Technology
and Department of Physics, University of Colorado
Boulder, Colorado 80309-0440, USA

Contents

1. Introduction

As discussed throughout this volume, the Weinberg-Salam-Glashow theory of electroweak interactions has been the subject of many experimental tests. One class of these experiments is unique in two respects: the experiments fit on a single table and the energy scale of interest is eV. These experiments are ones which study parity nonconservation (PNC) in atoms arising from the neutral weak currents. A key feature of all such experiments is that they exploit the parity-violating nature of the weak interaction to separate it from the electromagnetic and strong phenomena that are the dominant contributors to atomic structure.

This chapter describes past and ongoing experiments that measure parity-violating effects in atoms, concentrating in later sections on the experiments performed on atomic cesium at JILA. PNC results in a number of atomic systems are presently in good agreement with the predictions of the Standard Model and place interesting limits on proposed extensions of the Standard Model. New experiments are intended to severely test the Standard Model.

We have divided this chapter into five sections. The first presents a general discussion of PNC interactions in atoms and how they are measured. Section 2 concentrates on cesium, an especially attractive atom for both PNC experiments and PNC structure calculations. Section 3 describes the JILA cesium experiment in some detail. Since achieving a high overall signal-to-noise ratio in measuring the PNC modulation is an overriding concern in PNC experiments, we put particular emphasis on this in the discussion. Section 4 describes another primary concern of PNC experiments, the control of systematic errors, along with the results of the 1988 JILA parity experiment. The final section describes the status of an improved cesium parity experiment now under way, and possible future experiments involving different isotopes of cesium confined in an optical trap.

1.1. Parity Nonconservation and Atomic Physics

One result of the neutral weak currents predicted by the Weinberg-Salam-Glashow model is parity nonconservation in atoms. Because electrons and nucleons may exchange virtual Z_0 particles (Fig. 1), just as they interact through virtual photons, and because parity is not conserved in these interactions, stable atoms are not in exact parity eigenstates. This mixing of atomic parity eigenstates is unobservably small in most cases, but Bouchiat and Bouchiat[1] pointed out in 1974 that the mixing is much greater in heavy atoms. To follow their argument we may consider a simple parity-odd Hamiltonian due to Z^0 exchange between electrons and quarks in an atom:

$$\langle n'L'|H_{PNC}|nL\rangle = \langle n'L'|\frac{G_F Q_w}{2\sqrt{2}m_e c}\delta^3(x)\sigma\cdot p|nL\rangle \quad . \tag{1.1}$$

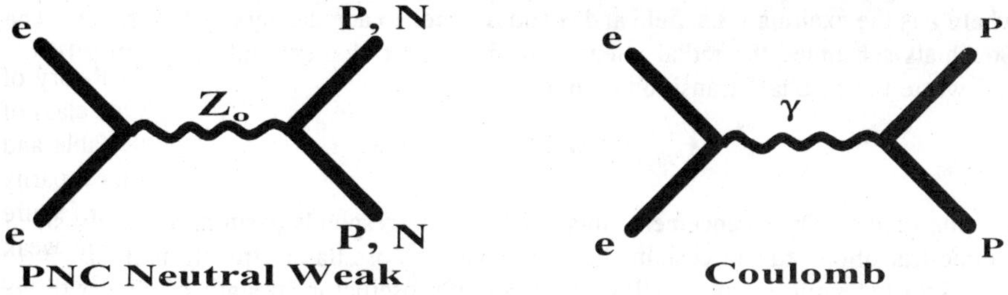

Fig. 1. Neutral electron-nucleon interactions in the Weinberg-Salam-Glashow model.

(This matrix element assumes a point nucleus and nonrelativistic electrons (both untrue for heavy atoms) but retains the characteristics important for the Bouchiats' argument. More valid expressions for the parity violating interaction are given elsewhere in this volume.) The weak charge Q_W is roughly proportional to atomic number Z and is predicted by the Weinberg-Salam model. With $\mathbf{p}_e=(h/i)\nabla$, the matrix element becomes

$$\langle H_{PNC}\rangle \propto \langle n'L'|\frac{\partial}{\partial r}|nL\rangle|_{x=0} \quad . \tag{1.2}$$

With $|n\ L\rangle \propto r^L Z^{L+1/2}$ as $r\rightarrow 0$, this matrix element can be nonvanishing only for L = 1 and L′ = 0, i.e., S and P states are mixed. Moreover, the $|n\ S\rangle$ state contributes $Z^{1/2}$ and the $|n\ P\rangle$ contributes $Z^{3/2}$ to the matrix element, making the matrix element grow as Z^3 overall.

The S-P mixing results in the possibility of electric dipole (E1) transitions between states of the same parity. For example, S states are perturbed:

$$\overline{|nS\rangle} = |nS\rangle + \sum_{n''} \frac{|n''P\rangle\langle n''P|H_{PNC}|nS\rangle}{E_{nS}-E_{n''P}} \quad . \tag{1.3}$$

E1 transitions between these perturbed S states can occur, with the transition amplitude being

$$A_{PNC} = \overline{\langle n'S|}-e\epsilon\cdot r\overline{|nS\rangle} \tag{1.4}$$

$$= \sum_{n''P} \left\{ \frac{\langle n'S\alpha'|-e\epsilon\cdot r|n''P\rangle\langle n''P|H_{PNC}|nS\alpha\rangle}{E_{nS}-E_{n''P}} + \frac{\langle n'S\alpha'|H\dagger_{PNC}|n''P\rangle\langle n''P|-e\epsilon\cdot r|nS\alpha\rangle}{E_{n'S}-E_{n''P}} \right\}$$

where ϵ is the exciting laser field and α and α' refer to angular momentum states. The Bouchiats subsumed the radial integrals in the matrix elements into a quantity $E1_{PNC}$ and wrote the resultant transition amplitude

$$A_{PNC} = iIm(E1_{PNC})\epsilon\langle\alpha'|\sigma|\alpha\rangle \quad . \tag{1.5}$$

In spite of the $(Z)^3$ enhancement, this amplitude corresponds to an exceedingly small atomic transition rate: in cesium, the corresponding oscillator strength is 3×10^{-22} (to be contrasted with a typical allowed transition's oscillator strength of 1), so direct observation is impossible. Following the ideas of Michel and other investigations of neutral current phenomena, the Bouchiats proposed instead to look for the interference between this tiny transition and a larger, parity-conserving electromagnetic transition

$$S \propto |A_{PC} + e^{i\theta}A_{PNC}|^2 \approx |A_{PC}|^2 - 2\sin\theta A_{PC}A_{PNC} \quad , \tag{1.6}$$

where $e^{i\theta}$ is the relative phase between A_{PC} and A_{PNC} and depends on the experimental geometry. $|A_{PC}|^2$ dominates the transition rate (typically by a factor of 700,000) but the second term changes sign under a parity reversal, allowing it to be isolated. The difference in transition rate when the handedness of the atom's coordinate system has been changed, scaled by the overall transition rate $|A_{PC}|^2$, is the ratio between the parity-violating transition amplitude and the parity-conserving amplitude:

$$\frac{\Delta\text{Rate}}{\langle\text{Rate}\rangle} = \frac{4\sin\theta A_{PNC}}{A_{PC}} \quad . \tag{1.7}$$

To test the Weinberg-Salam model, the experimental result must be interpreted free of uncertainties from atomic and nuclear structure calculations. The measured quantity is

$$A_{PNC} \propto G_F Q_W \int d^3x \rho_{nucl}(x)\overline{\psi}_e\gamma_5\psi_e \quad . \tag{1.8}$$

The quantity of interest is the weak charge

$$\begin{aligned} Q_W &= 2[(A+Z)C_{1u} + (2A-Z)C_{1d}] \\ &= \rho[-N + Z(1-4\kappa\sin^2\theta_w(m_W))] \end{aligned} \quad , \tag{1.9}$$

where $\rho = 0.980 + 0.002\,(m_t/m_N)^2$ and $\kappa = 1.003$ with first-order radiative corrections included.[2] To extract Q_W, the nuclear density $\rho_{nucl}(x)$ and the matrix element $\overline{\psi}_e\gamma_5\psi_e$ at the nucleus must be accurately known. This matrix element is found by calculations of the atomic structure and is discussed in detail by Sapirstein et al. elsewhere in this volume.

The Bouchiats' proposal stimulated a number of experiments in lead, bismuth, thallium, and cesium. The results of all of these experiments are in rough agreement with the Weinberg-Salam model, complementing high-energy physics tests. For testing electroweak theory at the 1% level and beyond using atoms, experiments in cesium offer the greatest hope because of cesium's hydrogenlike character: it has one 6S electron outside a tightly bound xenon core, allowing precise atomic structure calculations and thereby direct comparison of experimental results to electroweak theoretical predictions. We shall concentrate on cesium in Section 2 after a brief review of other atomic systems. More thorough discussion of other systems may be found in the review articles by Fortson and Lewis,[3] Commins,[4] and Stacey.[5]

1.2. Atomic PNC Experiments

There are two general choices to be made in the construction of an atomic PNC experiment. The first is the choice of parity-conserving transition amplitude, A_{PC}, with which the parity-violating transition amplitude will interfere, and the second is the choice of atom. The first experiments to produce results looked at interference with an allowed M1 amplitude. These experiments, for technical reasons, were designed to detect the optical rotation of the polarization of light as it passes through a sample of atomic vapor. The second interference amplitude which has been used is a so-called "Stark-induced" electric dipole amplitude produced by applying a dc electric field to the atomic sample. In subsequent discussions these two types of experiments will be labeled as "optical rotation" and "Stark interference" experiments.

The other important experimental issue is the choice of atom. Heavy atoms are desirable because of the Z^3 dependence of the mixing. In addition, appropriate atomic transitions must be available: for allowed magnetic dipole (optical rotation) experiments an NP to NP′ transition is needed, while experiments using Stark-induced E1 amplitudes require either an NS to N′S or NP to N′P transition. (In this case it is desirable to have the initial and final N′s different to keep the magnetic dipole amplitude small so that it does not introduce systematic errors and background noise.) In all cases the transition wavelengths should be accessible to moderately high-power, low-noise tunable lasers. Moreover, if fluorescence detection is used, as in Stark-induced experiments, the fluorescence light wavelength should be far different from the excitation light to allow adequate suppression of scattered light from the exciting laser. From a practical standpoint, it must also be possible to obtain a dense, monatomic vapor of the chosen atom. Finally, the last, and at this fairly well-developed state of the field, probably the most important issue, is the accuracy with which one can determine the γ_5 matrix element for the atom. This has grown in importance as interest has evolved from simply detecting parity nonconservation to making precision tests of the Standard Model. The details of this question of calculating the relevant atomic structure are discussed elsewhere in this volume. Here we will simply summarize by pointing out that, of the atoms been studied, the calculations can be done most accurately for cesium because of its single electron alkali character; calculations for thallium are somewhat less accurate; and the structure of bismuth and lead is least well known.

Optical rotation experiments have been done on two different transitions in bismuth,[6,7,8,9,10] one in lead,[11] and recently one in thallium.[12] In these experiments the interference between the parity-conserving M1 amplitude and the parity-violating amplitude gives rise to a difference in refractive index for left and right circularly polarized light. This approach produces relatively large signal sizes because the M1 amplitude is large, but the modulation fraction is thereby quite small (polarization rotations of less than 10^{-7} radians must be measured) and there are fewer parity reversals to suppress potential systematic effects than in Stark interference experiments.

The basic setup for an optical rotation experiment is shown in Fig. 2. The laser beam (produced by a tuneable dye or diode laser) is prepared in a state of very clean linear polarization, then passes through a vapor cell that contains the atom of interest and finally through a second, nearly crossed polarizer. This polarizer blocks out almost all the laser light unless its polarization has been rotated in the vapor, in which case some light passes through and can be seen at the detector. If the laser frequency is tuned over the appropriate atomic transition, the optical rotation is observed as a tiny change in the crossed polarizer's transmission.

Several steps have been taken to improve the signal-to-noise ratio and to test for potential systematic errors. First, to improve the signal-to-noise ratio, the polarizers have been rotated slightly from perfectly orthogonal, and the incident polarization is modulated using a Faraday rotator. This latter step reduces the noise by shifting the detection bandwidth away from dc. To eliminate sources of potential systematic errors some or all of the following steps have been taken in the various experiments: (1) alternating between an oven containing atomic vapor and an identical oven with no vapor, (2) reversing the direction of the light through the vapor, and (3) careful fitting to the atomic line shape. The parity nonconservation signal is dispersion shaped and thus has quite a different dependence on laser frequency from the absorption. The statistical uncertainty in all experiments of this type is quite small, and the limiting uncertainty has always been systematic errors. Over the years these errors have become increasingly well understood, culminating with the recent experiments at Oxford[10] which have measured PNC optical rotation in bismuth with a fractional uncertainty of 2%.

The first Stark interference experiments were done on the 6P-7P transition in thallium at Berkeley,[13] and shortly thereafter on the 6S-7S transition in cesium at ENS.[14] Our experiments at JILA subsequently used Stark interference on this same transition in cesium. The Stark interference approach allows one to choose the size of the parity-conserving transition to optimize the tradeoff between signal size, which is proportional to the product of PNC and parity conserving amplitudes, and fractional change in the signal when the parity of the experiment is reversed, which is the ratio of the two amplitudes. Relative to the optical rotation experiments, these experiments suffer from complexity, rather low signal-to-noise ratio because of the tiny signals measured, and must also contend with parity-mimicking signals due to unwanted

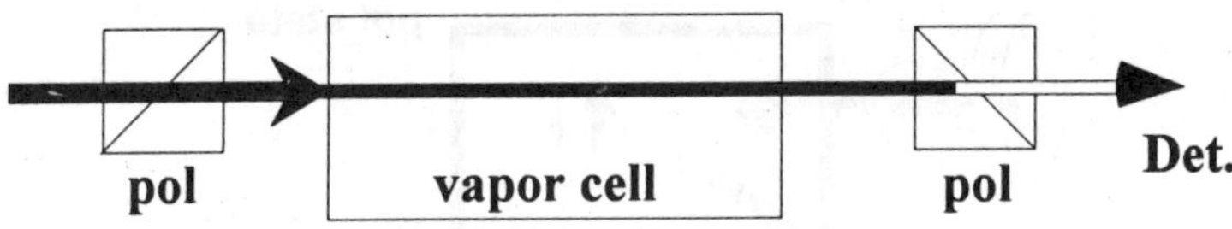

Fig. 2. Optical rotation experimental setup.

interferences between the Stark-induced transition amplitude and magnetic dipole amplitudes. Although forbidden, the magnetic dipole amplitudes are much larger than the PNC amplitudes (by a factor of 20,000 in cesium) and strong suppression of M1 effects is essential. Fortunately, strong suppression is possible since the M1 interference which mimics the PNC signal depends on the direction of laser propagation. Thus by reversing the direction of the light exciting the transition, or reflecting it back on itself, one strongly suppresses this undesirable term. Moreover, there are several other parity reversals (electric field reversal, magnetic field reversal, laser polarization reversal) which greatly suppress systematic errors in general.

We would now like to discuss the first generation of Stark interference experiments which were carried out in Paris and Berkeley. The basic schematic for these experiments is shown in Fig. 3. A circularly polarized laser beam was sent into an atomic vapor cell and excited the transitions of interest in the presence of a dc electric field. In the case of the Paris experiment, the laser was a cw dye laser operating at 540 nm. The laser went into a multipass cell in which the light was reflected back and forth between two mirrors about 200 times. The transition rate was monitored by observing the fluorescence. Since the Zeeman transitions between individual *m* levels were not resolved, the parity nonconserving interference term could not be observed directly in the total atomic transition rate, but it causes a polarization of the excited state. This polarization was detected by looking at the degree of circular polarization of the fluorescence light, as shown in Fig. 3, emitted when the 7S state decayed to the 6P state. These experiments were successful at detecting a small circular polarization and hence a parity violation on two different hyperfine transitions with fractional uncertainties as given in Table 1. One of the major limitations in this experiment was background light from a variety of sources which introduced noise. In order to reduce this background to acceptable levels careful filtering of the florescence was required, which limited the detection efficiency for the desired fluorescence. While the thallium experiment was conceptually quite similar, the experimental details were somewhat different.[13] First, the transition wavelength is in the near UV, which necessitated the use of a pulsed dye laser which was then frequency doubled to drive the 6P-7P transition. In the thallium case the light was not multipassed because of the limitations of the thallium vapor cell which had to be very hot. Two interaction regions were used which had oppositely oriented electric fields, and the light could be reflected back on itself. Rather than measuring the polarization of the florescence light, a second

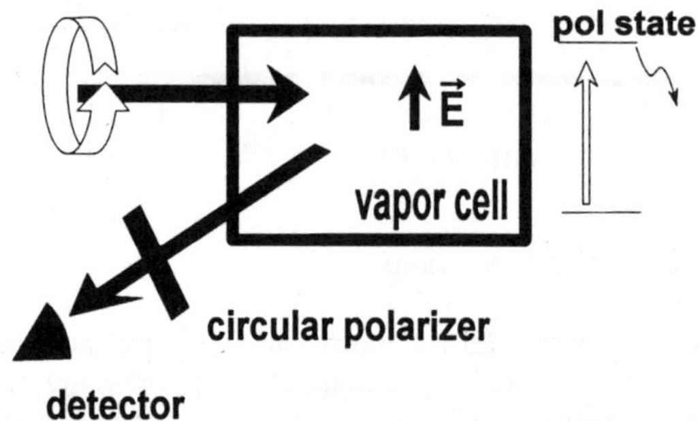

Fig. 3. Vapor cell-based Stark interference PNC experiment.

Table 1. Summary of PNC results

Ancient (controversial) history		Exper.	At. theory
Bismuth	Oxford[6] Univ. of Wash[7] Novosobirsk[8]	*Wide variations*	
Modern era:			
Pb	(Wash. '83)[11]	±28%	±10%[15]
Bi	(Wash. '81)[9]	±18%	±15%[16]
	(Oxford '91)[10]	±2%	"
Tl 1.3μ	(Oxford '91)[12]	±15%	±3%[17]
Stark induced interference:			
Thall	293 nm (Berkeley '85)[18]	±28%	±6%[17]
Cs			
	(Paris '84-'86)[19]	±12%	±1%[20]
	(Col. '85)[21]	±12%	"
⇒	(Col. '88)[22]	±2%	"
		all agree with Standard Model	

laser was used to excite the atoms out of the 7P state to the 8S state, and the subsequent fluorescence was detected as the 8S state decayed. The polarization of the 7P state was detected by using circularly polarized light to excite it. This measurement obtained a fractional uncertainty of about 30%. This was subsequently improved upon in a somewhat different experiment carried out by Drell and Commins,[18] who applied a large magnetic field to the vapor cell, thus exciting the transition with linearly polarized light. Because of the addition of the magnetic field, the PNC could be observed directly in the transition rate and thus in this experiment it was not necessary to detect the polarization of the excited state, only the excitation rate. Using this approach the fractional uncertainty was reduced to 18%. As in the Paris experiment, the Berkeley experiments suffered from low detection efficiency. However, in both experiments the use of multiple reversals (with help from the statistical signal-to-noise ratio) resulted in the systematic uncertainties being smaller than the statistical uncertainties.

Before discussing the PNC work at Colorado we might mention that there have been a number of other proposals and a few experiments attempted involving other atomic species. Probably the most notable was the use of atomic hydrogen in an effort to observe the mixing of the 2S and 2P states of hydrogen. Although Z is 1 and therefore the PNC matrix element is quite small, this is largely offset by the energy denominator which enters into the mixing and is nearly zero for these two states. Hence, the actual S-P mixing for N=2 hydrogen is nearly the same as that for heavy atoms. Because of this, a number of experimental programs were initiated to study PNC in hydrogen. However, the great problem with the hydrogen case is that the systematic errors, such as stray electric fields which can also cause mixing of the 2S and 2P states, are amplified by the same near-zero energy dominator. As a result, the systematic errors relative to the PNC signal are enhanced by a factor of Z^3 compared to those in heavy atoms. Effectively, this means that instead of needing to worry about millivolt/cm stray fields, one must worry about nanovolt/cm fields. This is a nearly impossible problem and therefore, to our knowledge, all the experiments on hydrogen PNC have now been abandoned.

Another set of proposed experiments which is being pursued to varying degrees involves the use of ions or muonic atoms for studying PNC. In these cases, the overlap of the electrons at the nucleus is much larger than for a normal atom and therefore, Δ_{PNC} can be relatively large. However, this is more than offset by the fact that the sample size is very small. At the present time the technology is not available to produce large enough samples to allow meaningful PNC measurements. However, this is very much a function of technology, and is likely to change.

2. 6S→ 7S Transitions in Cesium

The JILA measurement of A_{PNC} in Cs involves exciting 6S→7S transitions in a sample of Cs atoms in a region of crossed electric and magnetic fields as depicted in Fig. 4. **E**, **B**, and the angular momentum of the exciting laser photons, **k**, comprise a handed coordinate system, and reversing the directions of the dc fields or the circular polarization of the exciting laser changes the handedness of the coordinate system. Any corresponding change in the 6S→7S transition rate would indicate parity non-conservation.

When $|m>$ and $|-m>$ states are equally populated (an unpolarized sample) and equally excited there is no net change in the overall transition rate under a parity reversal. This is because the PNC contributions to the rate are equal and opposite for

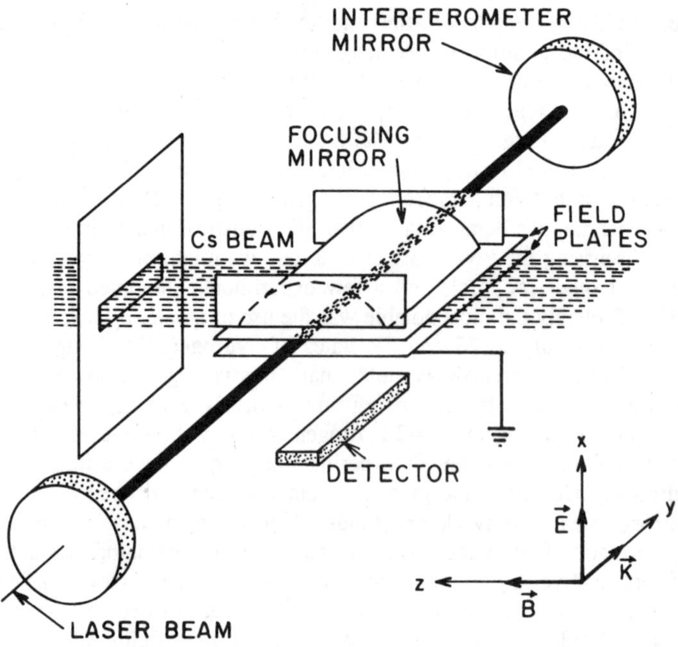

Fig. 4. JILA PNC experiment in cesium.

excitation of |m> and |-m> states. Bouchiat and coworkers at Ecole Normale Superieure monitored the polarization of the fluorescence emitted by excited Cs atoms to recover a change with parity reversals. This approach allows the use of a cesium cell so that very high Cs densities can be reached, but polarization changes in small amounts of light must be measured.

The second approach is to excite the |m⟩ atoms but not the |-m⟩ atoms. This may be accomplished by applying a magnetic field large enough to resolve m→m′ transitions where the ′ refers to the m value of the 7S state. To fully resolve the transitions in a reasonably small (< 100 gauss) magnetic field, a transversely excited beam of atomic cesium must be used. Relative to a cell, this results in a much lower density of resonant atoms but allows a parity-violating modulation to be observed directly in the 6S→7S transition rate. This is the approach taken at JILA, where the most accurate measurement of A_{PNC} has been performed. An increase in signal size is also obtained in this experiment by using a resonant interferometer, the "power buildup cavity," to drastically increase the optical power used to excited the 6S→7S transition.

2.1. *The Cesium Spectrum*

In the presence of a dc electric field there are three 6S→7S transition amplitudes in cesium: the tiny parity-nonconserving amplitude A_{PNC}; the magnetic dipole amplitude A_{M1}; and the "Stark-induced" transition amplitude A_{E1}, which is proportional to the applied electric field. The resultant transition rate and interferences between these amplitudes depend on the relative orientations of dc electric field and laser propagation direction and polarization; this dictates the experimental geometry, which is chosen to give the largest PNC modulation with strong suppression of non-PNC modulations.

The low-lying energy levels in cesium are depicted in Fig. 5. With nuclear spin I=7/2, $^2S_{1/2}$ states are split into hyperfine levels F=3 and F=4. The 6S→7S transition is at 540 nm and is excited with green light from a cw dye laser. The 7S→6P and 6P→6S transitions at 1.4 and 0.8 μm respectively are allowed electric dipole transitions: atoms excited to the 7S state quickly decay back to the ground level with the emission of these two infrared photons.

In addition to its theoretical tractability, cesium presents a most favorable case spectroscopically: the 6S→7S transition may be excited with a single-frequency dye laser using a relatively high-power, long-lived dye; and the 0.8μm 6P→6S fluorescence is ideal for detection since it is near the responsivity peak of silicon photodiodes, is well-separated from the 540 nm excitation wavelength, and is of short enough wavelength to allow detection without detecting a large blackbody radiation background.

From Section 1, the PNC transition amplitude is

$$A_{PNC} = iIm(E1_{PNC})\epsilon\langle F'm_{F'}|\sigma|Fm_F\rangle \quad . \tag{2.1}$$

The Zeeman interaction which splits the m levels is

$$H = -\mu\cdot B = -\frac{\mu_B}{\hbar}(L+g_sS)\cdot B - \frac{\mu_{nm}}{\hbar}g_I\, I\cdot B \tag{2.2}$$

where μ_B is the Bohr magneton. Ignoring the tiny nuclear spin interaction and considering the matrix element between the 6S and 7S states, we have

$$A_{M1}\propto \langle 7SF'm_{F'}|\frac{2\mu_B}{\hbar}S|6SFm_F\rangle\cdot M \tag{2.3}$$

where **M** is the magnetic field of the exciting radiation, with the exciting laser beam propagating along the direction **k**, **M**=k×ϵ. The absence of any spatial dependence means that this matrix usually nearly vanishes between states of different principal quantum number. A nonzero component arises due to spin-orbit effects[23] and hyperfine interactions.[24] Incorporating radial integrals into the constant M1 ≈ $-4\times10^{-5}\mu_B$, we have

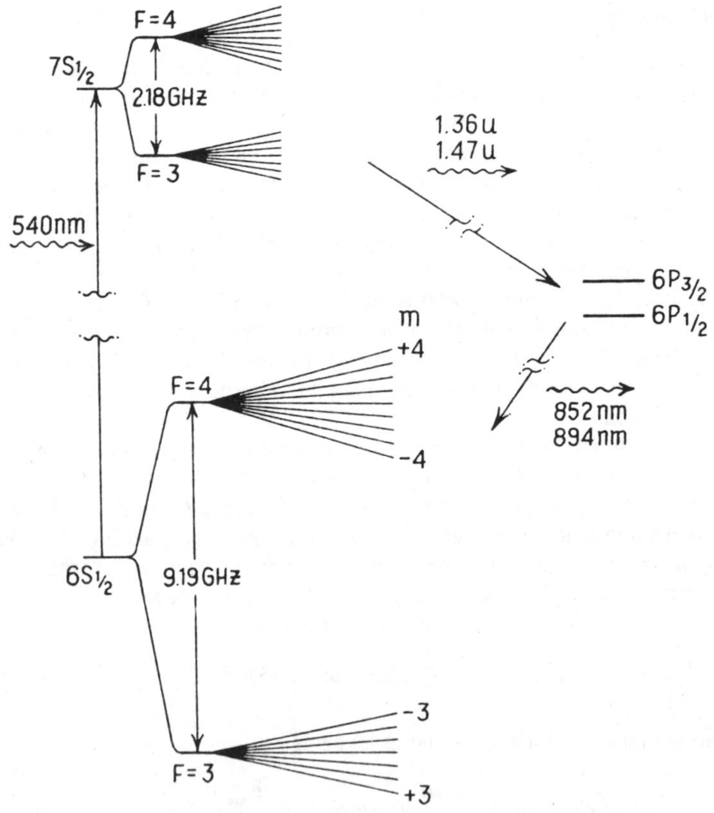

Fig. 5. Low-lying energy levels of cesium (Z=55, I=7/2).

$$A_{M1} = M1(k \times \epsilon) \langle F' m_{F'} | \sigma | F m_F \rangle \quad . \tag{2.4}$$

Apart from the dependence on the direction of the exciting laser beam k, this amplitude behaves very much like A_{PNC} but is approximately 20,000 times larger in spite of its being highly forbidden by electromagnetic standards.

An external electric field mixes S and P states similarly to the PNC interaction:

$$\overline{|nSFm\rangle} = |nSFm\rangle + \sum_{n''} \frac{|n''P\rangle\langle n''P| -eE \cdot r |nSFm\rangle}{E_{nS} - E_{n''P}} . \tag{2.5}$$

This leads to an induced electric dipole transition amplitude

$$A_E = \overline{\langle 7SF'm'|-e\epsilon\cdot r|\overline{6SFm}\rangle}$$

$$= \sum_{n''F''m''} \left[\frac{\langle 7SF'm'|-e\epsilon\cdot r|n''P\rangle\langle n''P|-eE\cdot r|6SFm\rangle}{E_{6S}-E_{n''P}}\right. \tag{2.6}$$

$$\left. + \frac{\langle 7SF'm'|-eE\cdot r|n''P\rangle\langle n''P|-e\epsilon\cdot r|6SFm\rangle}{E_{7S}-E_{n''P}}\right] \quad .$$

This expression may be separated[24] according to its angular momentum dependence into transitions without and with a spin flip:

$$A_E = \alpha E\cdot\epsilon\delta_{FF'}\delta_{mm'} + i\beta\epsilon\langle F'm'|\sigma\times E|Fm\rangle$$

$$= \alpha E\cdot\epsilon\delta_{FF'}\delta_{mm'} + i\beta(E\times\epsilon)\langle F'm'|\sigma|Fm\rangle \quad . \tag{2.7}$$

The quantities α and $\beta\approx\alpha/10$ are scalar and tensor transition polarizabilities, respectively, and incorporate the sums over the radial matrix elements.

Taken altogether, the 6S→7S transition amplitude is

$$A_{6SFm}^{7SF'm'} = \alpha E\cdot\epsilon\delta_{FF'}\delta_{mm'} + [i\beta(E\times\epsilon) + M1(k\times\epsilon) + E1_{PNC}\epsilon]\langle F'm'|\sigma|Fm\rangle \,. \tag{2.8}$$

The square of the absolute value of Eq. (2.8) yields the transition rate that is measured.

Since the sign of the PNC amplitude is opposite for m and -m states, the overall transition rate will not change on a parity reversal if the atomic sample is unpolarized and m and -m atoms are excited equally. To recover a PNC modulation in the transition rate, a magnetic field may be applied to shift the transition frequencies for m and -m atoms. The Zeeman interaction

$$H = -\mu\cdot B_{ext} = \frac{2\mu_B}{\hbar}S\cdot B \tag{2.9}$$

leads to energy shifts of the magnetic sublevels:

$$\Delta E = g_F\mu_B B_z m_F \tag{2.10}$$

where g_F = -1/4 and 1/4 for F=3 and 4, respectively. Since the Zeeman shifts are in opposite directions for the different hyperfine levels, the 6S→7S ΔF=0 spectra and the |ΔF|=1 spectra are very different. In the ΔF=0 case, there are just three lines, spaced 0.35 MHz/G apart, corresponding to Δm=+1, -1, and 0. The Δm = 0 line is much larger than the other two lines since this is the only line on which an α transition is allowed. For |ΔF| = 1, there are seven Δm=0 lines alternating with nine |Δm=1| lines. (See Fig. 6.) Since these are β transitions, they are far smaller than

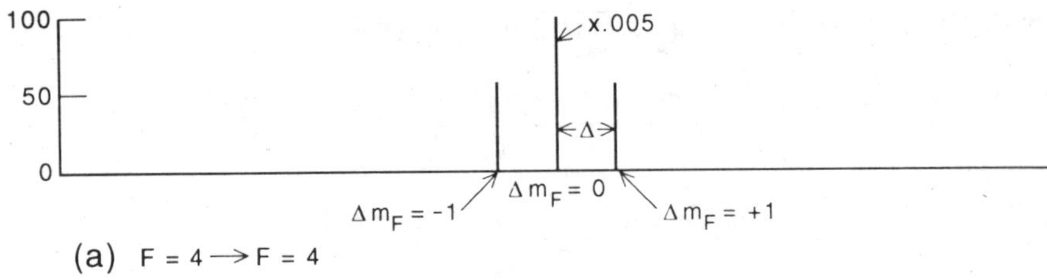

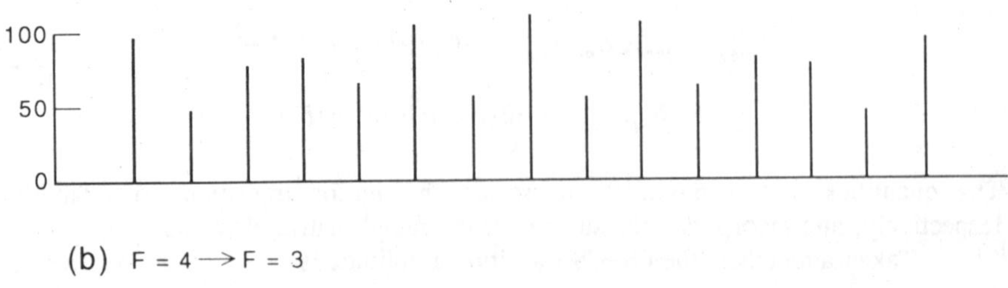

Fig. 6. Theoretical spectra of (a) $\Delta F=0$ and (b) $\Delta F=\pm 1$ 6S→7S transitions in a magnetic field.

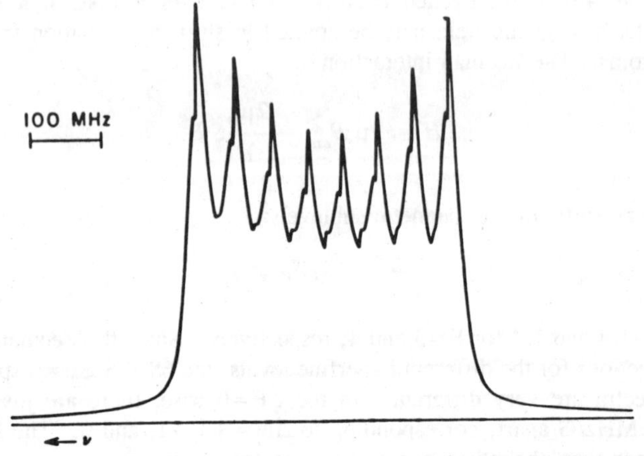

Fig. 7. Cs $6S_{F=3}$ to $7S_{F=4}$ Zeeman spectrum in 74 G magnetic field.

the $\Delta F=0$, $\Delta m=0$ α transitions. The observed $\Delta F=1$ spectrum in a 74 G magnetic field is shown in Fig. 7. This magnetic field strength sufficiently resolves transitions between different magnetic sublevels but is low enough so that F remains a good quantum number to the 5% level. The line shapes are asymmetric because of the off-resonant ac Stark effect in the intense standing wave field of the exciting laser.[25] These asymmetries are of concern because they affect the detuning of the exciting laser from the true transition frequencies but do not otherwise affect the PNC measurements.

2.2. *Experimental Geometry of the Colorado Experiment*

The directions of the applied fields are chosen to allow a PNC modulation to arise in the transition rate while being amenable to experimental control. Since it is not straightforward to change the direction of the 6S→7S exciting laser beam, this direction defines the y axis. The dc electric field is applied nominally along the x axis; although the electric field plates may be tilted relative to the laser beam to give a small y component to the electric field, its z component is defined to be zero, thereby fixing the direction of the z axis. The applied magnetic field is adjusted to lie along this axis. The z axis is also the direction in which the cesium atoms are moving, as shown in Fig. 4.

We now have a handed orthogonal coordinate system whose handedness reverses if one of the applied fields is reversed. To obtain the desired PNC modulation, the Stark-induced β transition amplitude and the PNC amplitude must be made to interfere, i.e., they must be relatively real. This can be arranged by using circular (or, more generally, elliptical) polarized light. The PNC modulation is then maximum on the outermost lines of the Zeeman multiplet (see Fig. 8).

Since the 6S→7S transition is excited in a standing wave optical field, transition rate terms linear in the magnetic dipole amplitude are strongly suppressed, to the extent that the field in the cavity is a perfect standing wave (or, equivalently, to the extent that the forward-going and backward-going travelling waves in the optical cavity are equal in amplitude and polarization). Terms such as the product of the Stark-induced amplitude and M1 cancel out. We may therefore write down the ideal form of the transition rate:

$$\mathrm{Rate}(6S-7S) = \beta^2 E^2 \epsilon_z^2 \left[1 + K \frac{E1_{PNC}}{\beta E} \frac{\epsilon_x}{\epsilon_z}\right] \tag{2.11}$$

where K depends on the specific magnetic quantum levels of the initial 6S and final 7S states and is on the order of unity. The second term is the quantity of interest. It changes sign as the electric field **E** is reversed, as magnetic quantum number m is replaced by -m (or as the applied magnetic field direction **B** is reversed), and as the laser polarization ϵ is reversed. These reversals allow it to be isolated from the much-larger scalar term in the transition rate. The magnitude of the problem of measuring the second term should not be underestimated, however: the fractional modulation is typically a few parts per million at the optimum electric field strengths. Moreover, at

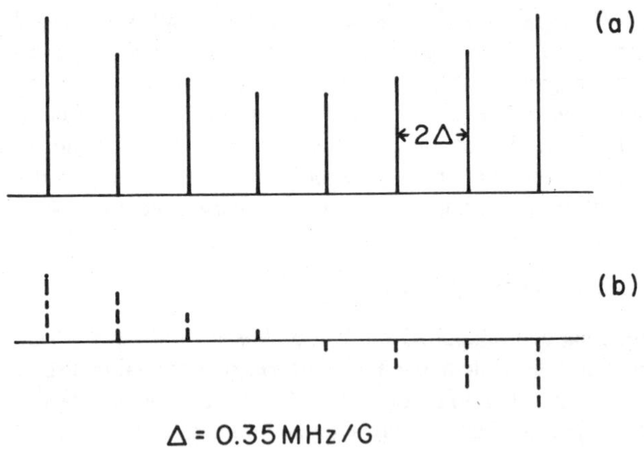

Fig. 8. (a) 6S→7S $|\Delta F| = 1$ Stark-induced transition spectrum. (b) PNC contributions on vastly expanded scale.

these field strengths the overall Stark-induced oscillator strength is on the order of 10^{-11}. The experimental program is therefore attempting to measure an extremely tiny modulation on top of a highly-forbidden transition with high immunity to systematic errors caused by the magnetic dipole transition and by departures of the coordinate system from a perfectly orthogonal, perfectly reversing coordinate system. The program carried out at JILA to do this is described in further detail in the following section. While the details are different, the general issues and problems discussed for this experiment are similar in the other Stark interference experiments. This has made it possible to use much of the analysis and some experimental procedures used in the earlier work.

3. Apparatus

As indicated above, in the JILA experiment the 6S→7S transition is excited in a region of crossed dc electric and magnetic fields in an intense, well-collimated cesium beam. The exciting light is a standing wave inside a high efficiency interferometer dubbed the "power buildup cavity" (PBC). This experimental approach (depicted schematically in Fig. 4) allows for a high signal-to-noise ratio due to the excitation power within the PBC, the high cesium beam density, and the efficient, low-background detection of 6S→7S transitions. At the same time, the experimental geometry is quite clean because the fields that define the atoms' handed coordinate system are well known, and these fields may be measured using the atoms themselves. We now describe the technology of this experiment in some detail.

3.1. *The Atomic Cesium Beam*

The measured parity-violating modulation is proportional to the optical power, the dc electric field, the number of cesium atoms in resonance with the driving laser field, and the transition rate detection efficiency. It is therefore desirable to have a cesium beam with high density, low Doppler width, and a cross section allowing a long interaction path with the exciting laser field and efficient fluorescence detection at the same time. This is produced by an effusive oven with a microchannel plate nozzle (Fig. 9) followed by a movable one-dimensional collimator made up of closely-spaced titanium vanes.

The ability to adjust precisely the direction of the cesium beam is essential to the experiment: those atoms which pass through the PBC at a right angle see the laser field as a standing wave in which the E1-M1 interference is strongly suppressed, while atoms with velocity components along the PBC beam see two travelling waves whose frequencies are Doppler-shifted in opposite directions. In the presence of such a tilt, the amount of E1-M1 suppression will depend on the exciting laser's detuning relative to each of two Doppler-split 6S→7S peaks -- one peak excited by the incident beam and one by the counterpropagating beam. Tilting the atomic beam so that the two Doppler-split peaks are resolved allows the birefringence of the output mirror of the PBC to be measured.[14]

3.2. *$6S_{1/2}$- $7S_{1/2}$ Excitation*

3.2.1. *Interaction Region*

The heart of the experiment is the region where the intense 6S→7S exciting field intersects the atomic beam. Here atoms undergo 6S→7S transitions in an orthogonal coordinate system defined by the laser beam direction, the static electric field, and the static magnetic field. Excited atoms quickly decay via allowed E1 transitions to the $6P_{1/2}$ and $6P_{3/2}$ states and then back to the $6S_{1/2}$ ground state. The 852 and 894 nm photons emitted during the 6P→6S decay are then detected with a liquid nitrogen-cooled silicon photodiode. Since the intersection of the atomic beam and the PBC standing wave field is a line roughly 2 cm long, fluorescence from this region can be efficiently imaged by a cylindrical mirror onto the 0.5 cm by 5 cm photodetector. Because of the kilowatt-level laser beam passing 0.4 cm above the photodiode, careful spatial and spectral filtering of 540 nm light is necessary to avoid excessive scattered light backgrounds.

The electric field is applied in the interaction region by transparent conductively coated quartz plates. The electric field plates, along with the photodetector and collection optics, are mounted on a tilt stage which allows the electric field to be set parallel to the laser beam (or to a specific tilt for certain systematic measurements) while the chamber is evacuated.

Four sets of coils establish a uniform magnetic field precisely along the z direction. Two large Helmholtz coils within the vacuum chamber produce a field of 74 G at the interaction region. Three orthogonal pairs of coils on the outside of the

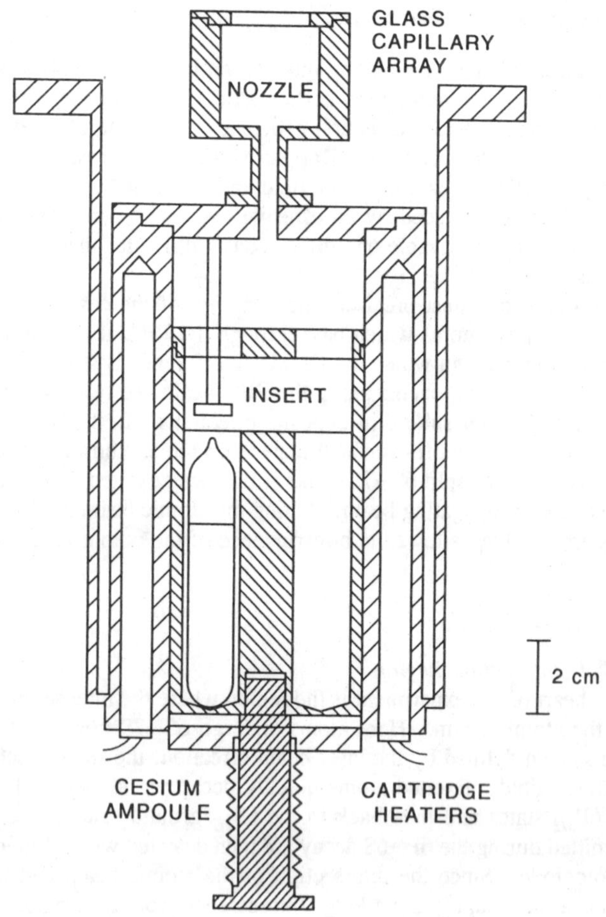

Fig. 9. High intensity cesium oven.

vacuum chamber allow shimming of the magnetic field direction. The current through the Helmholtz pair is reversed by a solid state relay under computer control. The external shimming coils are switched by relays between pairs of independent current supplies, allowing dc fields such as the Earth's and flipping fields due to tilts of the Helmholtz pair to be cancelled.

3.2.2. Power Buildup Cavity

The 6S→7S transitions are so weak (the β transition in a 500 V/cm field, for example, has an oscillator strength of 3×10^{-11}) that the exciting laser field can be made as intense as possible without saturating the transition. (Other effects, such as the ac Stark effect and photoionization, may lead to distortion of the transition line shape, however.) To achieve the highest cw intensity, the standing-wave field inside a resonant Fabry-Perot interferometer excites the atoms. The use of the PBC has several important advantages over other approaches. First, the optical field inside a resonant interferometer may be far more intense than the field of the incident cw laser beam: the one-way power buildup for the 1988 Colorado experiment was 1300, and new mirrors currently provide buildups of 15,000 or more. This is ideally suited to the excitation of the extremely weak 6S→7S transition since atomic absorption is negligibly small. Second, the intracavity power and polarization can be accurately measured and controlled during data-taking by monitoring the light transmitted through the output mirror of the PBC. Third, the high-finesse PBC serves as the reference interferometer to which the dye laser is locked, and the residual laser linewidth relative to the PBC is under 10 kHz rms. This low level of frequency jitter is important for obtaining a stable 6S→7S transition rate. Also, since the laser electric and magnetic fields are 90° out of phase in a standing wave field, the interference between the Stark-induced and magnetic dipole transition amplitudes (which can mimic the PNC signal) is suppressed to the extent that the PBC field is a perfect standing-wave field. Finally, since the intersection between the atomic beam and the laser beam inside the cavity is a thin line, the atomic fluorescence can be imaged with high efficiency onto the photodiode.

Figure 10 sketches the beam path from the dye laser to the PBC. The beam from the dye laser passes through an intensity-controlling Pockels cell, then has its spot size and curvature set to mode-match into the PBC by an adjustable telescope. The light then passes through a Faraday isolator to stop light reflected off the PBC from re-entering the dye laser, then through a polarization-controlling Pockels cell and finally into the PBC. The dye laser itself is a commercial ring dye laser capable of generating 0.3 W single-mode at 540 nm. Frequency tuning and stabilization are accomplished by an intracavity electro-optic modulator, a piezo-mounted mirror, and a pair of galvanometer-mounted Brewster plates.

3.2.3. Frequency Stabilization

The dye laser frequency must be held very stable both to excite the 6S→7S transition with low noise and to remain in resonance with the PBC. Three frequency stabilization systems are employed to set the dye laser frequency precisely on resonance both with the PBC and the 6S→7S transition. The dye laser frequency is locked to the PBC resonance frequency using rf modulation. The PBC resonant frequency is then controlled to match that of a stable reference cavity by feeding an appropriate error signal to the piezo under the input mirror. The reference cavity resonance frequency is then locked to the peak of the 6S→7S transition.

3.2.4. Intensity Control

The optical power inside the PBC must be stabilized to reduce the 6S→7S transition noise. This power is monitored by a photodiode placed after the output

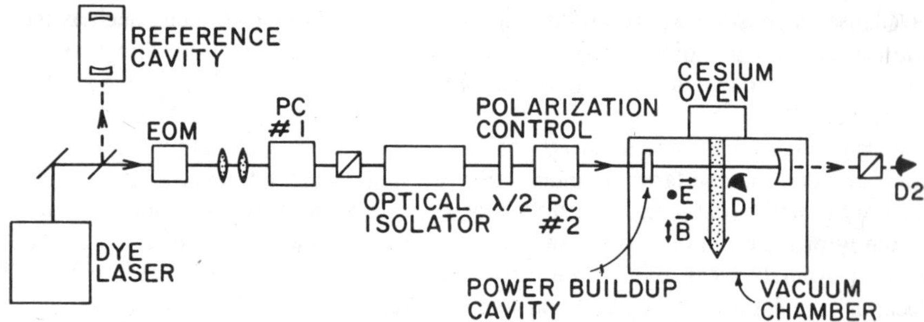

Fig. 10. Layout of laser optics.

mirror of the PBC. This photodiode must be carefully positioned so that spurious noise due to local imperfections on the photodiode, or to interferometric effects between the photodiode surface and its glass cover, is eliminated.

To stabilize the PBC power, the PBC transmission photodiode signal is compared to a stable reference and the difference is amplified, filtered, and sent to a Pockels cell which transfers more or less light into the PBC as the monitored power increases or decreases. The PBC power is thus stabilized to below 1×10^{-5} $Hz^{-1/2}$.

Since it is, in fact, the z component of the polarization that excites the 6S→7S transition, this is the component that is monitored and stabilized by placing a linear polarizer after the PBC and in front of the photodiode. Without this polarizer, the 6S→7S rate may fluctuate because of the birefringence of the PBC. In a birefringent cavity the resonant frequencies of the two orthogonal polarizations are not equal, so the residual frequency jitter of the dye laser relative to the PBC will result in fluctuating amplitudes along the orthogonal polarizations even if the total power is constant.

3.2.5. Polarization Control

The 6S→7S transition is excited by circular polarized light whose helicity may be reversed by reversing the voltage on a Pockels cell located after the Faraday isolator. With temperature control, the Pockels cell can provide circular polarization good to 10^{-4}.

Since suppression of E1-M1 interference relies on cancellation of two terms, one driven by the forward-going travelling wave in the PBC and the other by the counterpropagating wave, the circular polarization must be preserved upon reflection by the output mirror of the PBC. To measure the birefringence of the output mirror, the atomic beam is tilted away from the perpendicular to the laser beam. This splits the 6S→7S resonance into two peaks as a function of laser frequency, one driven by the forward wave and the other by the reflected wave. A difference in transition rate modulation of these two peaks as the polarization is reversed indicates that the output mirror is birefringent. The birefringence of the PBC output mirror used in the 1988

PNC experiment was roughly 10^{-5} radians/reflection, whereas newer mirrors show birefringence below 5×10^{-6} radians/reflection.

3.3. System Performance

In order to measure precisely the parity violating transition amplitude, we must maximize the ratio of the measured parity-violating modulation to the total noise on the signal. To do this, laser power, resonant atom density, and collection efficiency are made as large as possible; the signal-to-noise ratio continues to improve until shot noise becomes negligible. Boosting the electric field, on the other hand, increases the PNC signal linearly but increases the total 6S→7S rate quadratically. One therefore expects an optimum electric field where background noise (which is independent of electric field) and 6S→7S technical noise, which is proportional to the 6S→7S rate, balance, i.e.

$$\text{(background noise)} = \text{(6S}\rightarrow\text{7S rate)} \times \text{(fractional technical noise)} \quad . \tag{3.1}$$

The measured signal is 0.2 E^2 fA where E is the applied electric field in V/cm. The shot noise on this signal is 0.006 E fA $Hz^{-1/2}$ while the PNC modulation, $2\beta E\cdot \mathrm{Im}(E1_{PNC})\epsilon^2$, is 0.0006 E fA $Hz^{-1/2}$. The shot noise-limited SNR is therefore $0.1/Hz^{1/2}$.

To reach the shot noise limit, we must reduce all background noise and non-shot transition noise. There are two sources of background noise: (1) photodetector noise which is below 5 fA $Hz^{-1/2}$ near 5 Hz, and (2) the background from PBC scattered light which is 2 pA with noise below 2 fA $Hz^{-1/2}$.

Two primary classes of technical noise or non-statistical transition noise are resonant atom density fluctuations and 6S→7S excitation fluctuations.

(1) Resonant density fluctuations. Under favorable conditions, the atomic beam density is extremely constant with fluctuations probably well below 3×10^{-6} $Hz^{-1/2}$ fractional. In the presence of pressure fluctuations in the vacuum chamber, however, the beam density may become very noisy. Great care is taken to avoid pressure bursts inside the vacuum chamber, and the base pressure is typically 3×10^{-7} torr.

(2) 6S→7S excitation rate. The 6S→7S excitation rate will be noisy if either the dye laser frequency or the PBC intracavity power fluctuate. With dye laser frequency jitter below 1500 Hz $Hz^{-1/2}$, we do not expect to see fluctuation in the 6S→7S excitation rate above the 5×10^{-6} $Hz^{-1/2}$ level, and, in fact, this appears to be the case.

While the PBC intensity control servo nominally stabilizes the PBC intracavity power to well below 10^{-6} $Hz^{-1/2}$, amplitude fluctuations are possible if the monitored PBC transmission does not correlate with intracavity power. We have observed this on occasion; for example, if some light can leak around the coating of the output mirror, or through it because of a coating imperfection, PBC transmission and intracavity power (measured by monitoring the light scattered off one of the PBC mirrors) may differ by 3×10^{-5} or more. We find that the light scattered off the mirrors can be a more accurate monitor of intracavity power than transmission, although it is not ideal for use in PNC data-taking because intracavity polarization as well as power must be monitored.

The quadrature sum of the various excitation rate fluctuations is approximately 2.2×10^{-5} fractional. We may now use Eq. (3.1) to find the optimum electric field (where background noise and technical noise are equal). The background noise and technical noise balance at a field of 1100 V/cm. At this field, the signal photocurrent is 0.2 nA, the PNC modulation is 0.6 fA, and the total noise is 8.5 fA $Hz^{-1/2}$. We thus have a PNC measurement signal-to-noise ratio of 0.07 in 1 s, allowing a 2% measurement uncertainty in approximately 160 hours of data integration.

4. Systematic Effects and Results

The miniscule size of the PNC effect means that numerous effects associated with an imperfectly aligned or reversing coordinate system, the magnetic dipole amplitude, and unwanted intensity modulations may lead to modulations of the transition rate which mimic the PNC modulation. The measurement and control of potential systematic errors therefore takes the majority of time in a PNC experiment. The goal of the JILA experiments has been to reduce the uncertainty in systematic errors to well below the statistical uncertainty. Since the uncertainties in the systematic corrections to the PNC results are smaller than the statistical uncertainties of the results, and since the systematic uncertainties are themselves limited only by statistical uncertainty in the measurement of the systematic effects, this goal has been attained.

4.1 Controlling Systematic Effects

The general approach to the control of systematic effects is exactly the same as in the earlier (Paris) cesium and (Berkeley) thallium Stark interference experiments. The 6S-7S transition rate is written down in its most general form, with the electric and magnetic fields and the laser polarization allowed to have small components along all unwanted directions; these components may or may not reverse as the corresponding large field is reversed, so that the total number of components considered for a given field is six, or, in the case of laser polarization, eight, since the polarization is complex. (Once the experiment is oriented in a particular direction in space, however, three of these degrees of freedom among all of the components of all of the fields are eliminated.) The resultant transition rate will then contain the large Stark-induced transition, which is constant under all parity reversals, the PNC modulation, which changes sign under all parity reversals, and a large number of terms which change sign under one, two, or three parity reversals and which are proportional to various combinations of the nine field misalignments.

The simplicity of our interaction region geometry allows the transition rate, the interferences between transition amplitudes, and the corresponding modulations with parity reversals to be written down explicitly with a high degree of confidence that the observed modulations should be as expected. The various modulations can then be considered according to which fields cause them to reverse. Modulations under only one or two parity reversals do not constitute PNC systematic errors but can be used to measure the sizes of various misaligned fields. The misaligned field components can then be shimmed out. In this way misaligned fields, both flipping and non-flipping, have been reduced to 10^{-4} or 10^{-5} of the corresponding large field, yielding a precisely orthogonal, precisely reversing coordinate system.

Non-PNC transition rate terms which modulate with all parity reversals ("pseudo" pseudoscalars) must be measured, and, if possible, eliminated. Any such terms that involve three misaligned fields are negligible compared to the size of the PNC term (10^{-5} - 10^{-6}) after the misaligned fields have been minimized. There are three terms, however, that involve only two misaligned fields and can therefore be much larger. (See Table 2.) The first of these terms involves a non-reversing ("stray") electric field in the Y direction times a misaligned magnetic field in the X direction. The second term involves a stray electric field in the Z direction times a misaligned component of electric field in the Y direction. And the third term is a product of E1 and M1 transition amplitudes times a mirror birefrigence factor.

We measure each of the fields and the birefringence involved in these terms while the PNC experiment is actually running so that we can subtract off their contributions. To do this we run a set of auxiliary experiments simultaneously, or interleaved with the PNC data acquisition. These auxiliary experiments involve looking at the 6S-7S atomic transition rate but looking at different hyperfine transitions, using different laser polarizations, and/or applying additional E or B fields and looking at their effects. Two points should be emphasized about dealing with systematic errors in this manner. First, it is important to use the atoms themselves as probes so that exactly the same region of space is sampled, and at nearly the same time as the PNC experiment. Second, the auxiliary experiments must be designed to allow systematic corrections to be measured to an uncertainty that is much less than the statistical uncertainty in the parity nonconservation experiment. This must require a measurement

Table 2. Corrections to PNC data as a percentage of PNC signal.[a)]

Systematic contribution	Range	Average all data	Daily uncertainty
$(\Delta E_y/E)(B_x/B)$	−0.3%→+1.1%	+0.3%	0.4%
$(\Delta E_z/E)(E_y/E)$	−1.3%→+0.4%	−0.1%	0.4%
E1·M1(Δm=±1)	−0.8%→+4.8%	+1.7%	0.6% (ΔF=−1)
	−1.1%→+6.8%	+2.4%	0.9% (ΔF=+1)
E1·M1(Δm=0)	−0.3%→+0.6%	+0.04%	0.04% (ΔF=−1)
	−1.6%→+0.1%	−0.23%	0.06% (ΔF=+1)

[a] ΔE_y and ΔE_z are nonreversing field components, while B_x and E_y are misaligned reversing components. The range column shows the largest and smallest daily corrections.

time that is much shorter than is required to take the parity violation data. If one fails to achieve this, then the uncertainty of an experiment rapidly goes up because too much time is spent taking data on systematic errors rather than on the measurement itself. In the experiments we have designed, the necessary uncertainty requires a tiny fraction of PNC integration time. In Table 2, we show the different sizes of the systematic uncertainties for our 1988 experiment, how much they vary from one run to another, and the average correction, along with the uncertainties. It can be seen that the typical corrections are a few percent or less, and most importantly, the uncertainties in all of these corrections in a given day are less than 1%, and thus much smaller than the statistical uncertainty.

In even the most carefully designed experiment, the possibility remains that something has been overlooked. Statistical checks of the plausibility of the data set can then be useful in searching for unsuspected systematic errors. In the PNC experiment this has been done by performing a number of χ^2 tests on the data set as follows. On short time scales the measured 6S→7S rate modulations show a scatter consistent with the amount of noise in the experiment; this scatter is random and statistical, although the average values may contain unknown systematic contributions. If a systematic contribution changes over longer time scales or in response to some experimental change, then χ^2 tests performed on data sets in which the systematic contribution changed will differ from such tests performed on data sets in which the systematic effect was constant. By subdividing the data according to day, time of day, experimental conditions, etc., the consistency of the data set can be carefully checked and any unknown systematic effect exposed if the effect varies among the data subsets. In our case, a new systematic error (an E1-M1 interference associated with off-resonance excitation of nominally forbidden $\Delta m=0$ transitions) was detected in a preliminary data run in this way.

Having carried out all the detailed studies of systematic errors and χ^2 tests we finally achieved the result

$$\frac{I_m \delta_{PNC}}{\beta} = \begin{cases} -1.639(47)(08)\ \mathrm{mV/cm} & F=4 \to F'=3 \\ -1.513(49)(08)\ \mathrm{mV/cm} & F=3 \to F'=4 \\ -1.576(34)(08)\ \mathrm{mV/cm} & \text{(average)} \end{cases} \tag{4.1}$$

The size of the parity nonconserving mixing is given in terms of the equivalent amount of dc electric field that would be necessary to give the same mixing of s and p states. As shown, we have measured this mixing for two different hyperfine transitions, the 6S $F = 4$ to 7S $F' = 3$ and the 6S $F = 3$ to the 7S $F' = 4$. In both cases, the amount of mixing corresponds to about 1.5 mV/cm. The average of these two is the most important, as we will discuss below, and has a 2% uncertainty dominated by the 0.034 millivolt/cm statistical uncertainty. The systematic uncertainty is about a quarter as large, but it should be noted that this systematic uncertainty is different from many

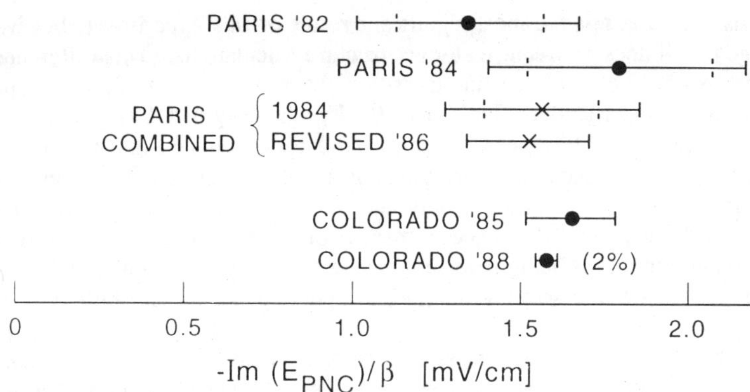

Fig. 11. Cesium PNC measurements.

systematic uncertainties, in that it is actually a true statistical uncertainty in the evaluation of the systematic correction. Therefore if the statistical signal-to-noise in the experiment was improved, this uncertainty would be reduced.

4.2. Atomic PNC Results

In Fig. 11, we show a comparison of the different experimental measurements of parity nonconservation in cesium. It is now the most thoroughly measured atom. On top are the two experimental results of the Paris group in '82 and '84, below them is our 1985 result, and our 1988 result, with its 2% error bar. There is good, but not unreasonably good, agreement among all these numbers. This gives one a certain amount of confidence that no serious systematic errors are being overlooked.

Note in the summary of results from all atomic parity nonconservation experiments in Table 1, the optical rotation experiments are at the top. The first set of these experiments looked at the 648 nm line of bismuth. These results are somewhat controversial in that the findings of the three groups showed substantial deviation. In retrospect, this was probably due to systematic errors that were not sufficiently well controlled. More recent optical rotation experiments have shown better consistency and one can see the uncertainties are mostly in the 15 - 30% error range. The one exception is the recent Oxford measurement in bismuth which has an uncertainty of only 2%. The Stark induced interference experiments are given at the bottom of this table. Most of these are the cesium measurements we have already mentioned, along with the thallium result from Berkeley with a most recent uncertainty of 18%.

The Colorado measurement of PNC in cesium has implications for the Standard Model and for weak interactions in nuclear physics. We will only discuss here the nuclear physics issues probed by the comparison of the two different hyperfine transitions. This is not the most significant result of this experiment, but it has the advantage that it does not require atomic structure calculations. This difference of the measurements on the two hyperfine transitions, $\Delta = 0.126$ (68) mV/cm, is probably not zero. More specifically, it indicates a 97% probability that Δ is greater than zero. When we originally did this measurement, we did not anticipate a non-zero result at this level, and therefore looked very hard at the data and the analysis to try and make it go away, but it stubbornly persisted. Only later did we discover that an effect of nearly this size had been predicted.[26] The primary difference between these two transitions is that the nuclear spin is being reversed relative to the electron spin. Thus, in looking at the difference we are isolating the nuclear spin-dependent contribution to the PNC signal.

There have been two processes discussed which would cause a nuclear spin dependent parity nonconservation. The first is simply the electron-quark portion of the neutral weak current which depends on the spin of the quarks. This interaction is characterized by the C_{2u} and C_{2d} coefficients. Because of the small size of these coefficients, and the fact that they are multiplied by the total nuclear spin, which is much smaller than the number of quarks, their contribution to atomic PNC is very much smaller than the weak charge contribution. However, it has also been pointed out that there is a substantially larger nuclear spin dependent contribution, called the nuclear anapole moment, which arises from weak interactions within the nucleus.[26] The effect of these weak interactions (both charged and neutral) is to mix the parity eigenstates of the nucleus and this leads to a parity nonconserving electromagnetic current in the nucleus. This current takes the form of toroidal helix, and therefore, has no long range electric or magnetic fields. Thus, it has gained the name "anapole moment." This phenomenon was actually first proposed by Zeldovich in 1957 in the general context of parity violation in charged systems.[27] It is not well known because people shortly thereafter decided such an effect could never be measured. However, because the cesium electrons can penetrate the nucleus, they go inside the toroidal helix and thereby detect its existence. The coupling to the electrons is purely electromagnetic (leading to the effect being of order α smaller than the main PNC effect), but because the underlying nuclear currents are parity violating, it leads to parity violation in the electrons.

The nuclear physics community has shown significant interest in this nuclear anapole moment and several authors have calculated the expected size. The first calculation was by Flambaum and Khripovlich[26] and their estimates are consistent with our observations. Haxton et al.[28] have also done similar calculations, in principle, but treated the nuclear physics rather differently. Finally, Bouchiat and Piketty[29] have done a calculation which they say is not consistent with our result. However, the differences in these calculations are apparently not due to any fundamental difference in the theory, but rather to the problem of the basic interpretation of nuclear PNC from other experiments. Depending on how one chooses to interpret the other experiments it is possible to obtain very different constants to characterize PNC interactions in the nucleus, and this is the source of the discrepancy. This emphasizes the need for more good data in this field. There is hope that in the future such nuclear anapole moment measurements can provide these data.

Obviously, this uncertainty from the nuclear physics is a serious issue in the interpretation of atomic parity nonconservation. If we had measured only a single transition it would seriously compromise our ability to test the Standard Model. Fortunately, if we take the sum of the measurements on the two hyperfine transitions, as opposed to the difference, the nuclear spin dependent part is cancelled out. In this way, we also cancel out any questions involving the nuclear structure, which is critical in allowing a precision test of the Standard Model. The implications of the sum, and the value of the weak mixing angle which it implies, are discussed elsewhere in this volume.

5. Future Cesium PNC Experiments

While atomic PNC measurements, particularly the 2% result in cesium, are providing useful information, it is clear that more precise results would be valuable. The mass of the Z is now known to around 1 part in 10^3 and therefore, if atomic PNC results could be improved to that level, they would provide a tenfold improvement in the test of the Standard Model, and correspondingly improved sensitivity to possible new physics. With this in mind, we would like to discuss our efforts to improve the cesium atomic PNC results. We should mention that there is also work under way to improve atomic parity nonconservation measurements by a number of groups. In Paris, the Bouchiat group is building a new experiment which involves stimulated emission probing of the excited state in cesium. At Oxford and Washington, experiments are under way to obtain more precise optical rotation measurements in thallium. At Berkeley, efforts continue to obtain a more precise Stark inference measurement in thallium. All of these experiments have been under development for a number of years, and we can anticipate results, we hope, in the not too distant future.

5.1. PNC Experiment with a Spin-Polarized Cesium Beam

In our efforts to improve the Colorado 1988 experiment, our primary focus has been on getting a better signal-to-noise ratio. This is clearly the major limitation since the statistical uncertainty was much larger than any systematic uncertainties. With this issue in mind we built a new apparatus that uses an optically pumped atomic beam which in principle should provide 16 times more atoms, since there are 16 possible M levels of the 6S state. We also have better mirrors for our power buildup cavity, so that we improve our buildup by about a factor of 10, reaching the number of 15,000 mentioned earlier. A third improvement is our use of downstream detection of the 6S→7S excitation. This concept is illustrated in Fig. 12, which shows the schematic of the improved apparatus. After leaving the oven, the cesium atomic beam is optically pumped into a single F and M_F level by light from two diode lasers which drive two hyperfine transitions of the 6S to $6P_{3/2}$ transition (see Fig. 5). The atoms in the single

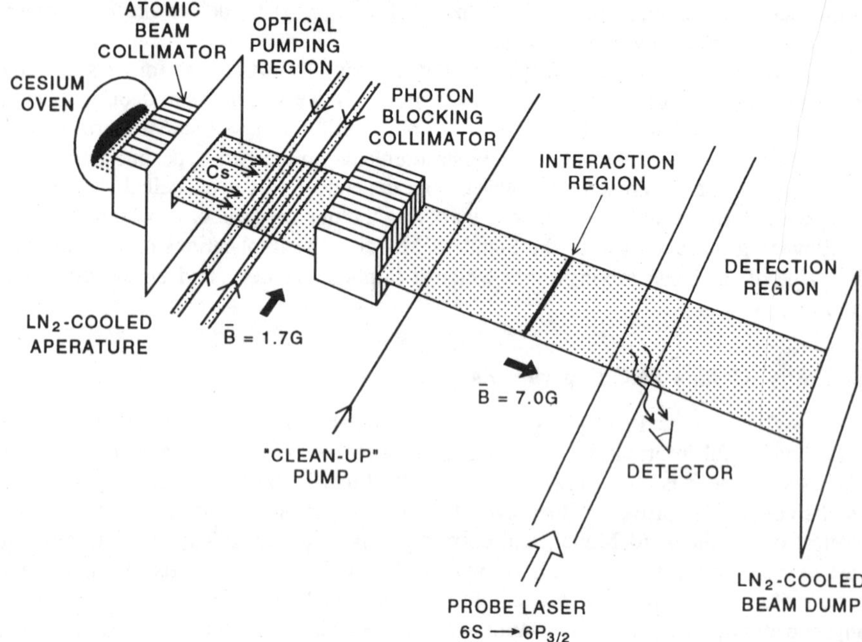

Fig. 12. PNC experiment with spin-polarized cesium beam.

m level then propagate down the atomic beam and intersect the beam from the dye laser where they are excited to the 7S state. They then have a 70% probability of decaying back down into the 6S hyperfine level which was previously depleted. They continue down the optical beam in this state until they reach the probe region. In this region, light from another diode laser again excites the $6S \rightarrow 6P_{3/2}$ transition. However, here we excite them to the particular $6P_{3/2}$ hyperfine state which is a cycling transition ($F'=5$ for the $F=4$ state, and $F'=2$ for the $F=3$ state). On such cycling transitions the atom returns only to the same initial state and hence can be excited many times. Typically 1000 photons are scattered for each $6S \rightarrow 7S$ excitation, of which 200 are actually detected. We detect this fluorescence to determine the 6S to 7S excitation rate. This detection scheme provides a great deal of low noise amplification since we are detecting 200 photons per $6S \rightarrow 7S$ transition instead of the 0.3 detected in the previous apparatus. In addition, since the detection takes place at a different point in space from the actual excitation region, we can now construct our electric field plates out of any material, which greatly simplifies their construction and increases their longevity. Finally, scattered light from the green laser light is now negligible, as is detector noise, because the signal size is so much larger.

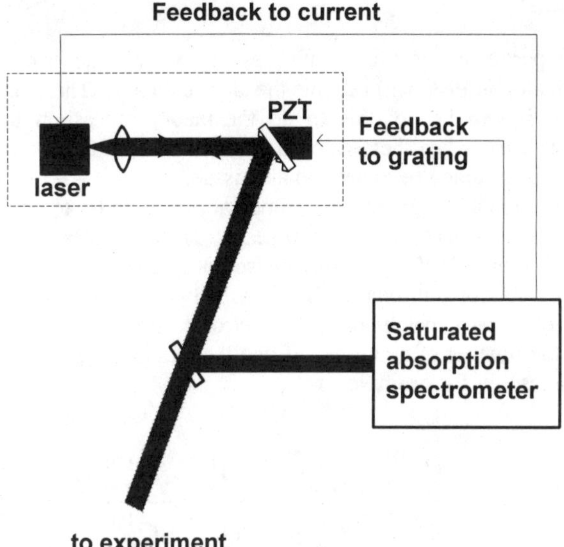

Fig. 13. Probe diode laser schematic.

While offering these significant improvements, the new experiment suffers from new difficulties which have occupied considerable time. First, the excitation of 6S→6P transitions by the probe laser must be quiet enough not to increase the transition noise significantly; this places severe limits on the frequency and intensity fluctuations permitted in the probe diode laser. As shown in Fig. 13, the probe diode laser used has its linewidth narrowed by optical feedback from a diffraction grating and is stabilized, by feedback to grating position and to laser current, to a cesium absorption cell-based saturated absorption spectrometer. The excitation of the 10 MHz-wide 6S→6P transition by this laser has a fractional noise of 3×10^{-6} $Hz^{-1/2}$. Another serious problem is the background of atoms in the nominally depleted hyperfine level at the probe region. These unwanted atoms (typically 2×10^{-4} of the total atomic beam) end up in the wrong hyperfine level after absorbing light scattered by other atoms during the optical pumping process; since the atoms placed into the depleted hyperfine level by 6S→7S transitions number only three times the number of these unwanted atoms, the shot noise on these unwanted atoms now constitutes the dominant background noise in the experiment. The overall signal-to-noise ratio of the new experiment is several times larger than that of the 1988 experiment, however, and preliminary data taking is presently under way.

5.2. The Future

As the experimental accuracy improves beyond 1%, the principal limitation on the usefulness of atomic PNC will become the atomic theory. There have been credible speculations that it will be possible to do the theory in cesium to a part in 10^3. However it is not clear when these calculations will be completed and the question of how to check their accuracy becomes a major issue.

We have begun a longer term experimental project to try to deal with the atomic theory question. This is likely to take 5-10 years. The basic idea is to compare precise measurements of atomic PNC for different isotopes of cesium. The weak charge is sensitive to the number of neutrons and hence will change for different isotopes, but the atomic matrix element depends on the electronic structure and hence is almost independent of the number of neutrons. If one then looks at appropriate combinations of experimental results, for example

$$\frac{\delta_{PNC}^{Cs^{130}} - \delta_{PNC}^{Cs^{150}}}{\delta_{PNC}^{Cs^{130}} + \delta_{PNC}^{cs^{150}}} = \frac{Q_w^{130} - Q_w^{150}}{Q_w^{130} + Q_w^{150}}$$

where δ is the experimentally measured S-P mixing, the atomic matrix element will drop out, leaving a ratio of weak charge which can be directly compared with Standard Model predictions. In this manner we hope to achieve measurements that can be compared with the standard model predictions at the parts in 10^3 level. We should emphasize that this work will not eliminate the value of improved atomic structure calculations. There is always more information to be obtained and used to test the standard model if Q_w is known absolutely for various isotopes, rather than only knowing ratios. It is probably best to consider the comparison of isotopes as a way to test the accuracy of precise calculations, and thus enhance rather than diminish their importance.

There are two major obstacles to carrying out these experiments. First is the need for even better signal-to-noise ratios. Second, and most critical, is the need to be able to carry out PNC measurements with small atomic samples, rather than the many grams used in the atomic beam measurements. This requirement is necessary because all the other isotopes of cesium are radioactive and can only be obtained and used in small quantities. We propose to overcome both of these obstacles with the new technology of laser trapping. This will improve the signal-to-noise ratio because it is possible, even easy, to get optical depths (and hence signals) in trapped atom samples which are 10 or 100 times larger than what we can achieve in our atomic beam.

It is more difficult to show that optical trapping will allow the experiments to be done with very small atomic samples (10^{10} atoms). We are currently working on this problem. The approach we are using is to start with a very small sample of a

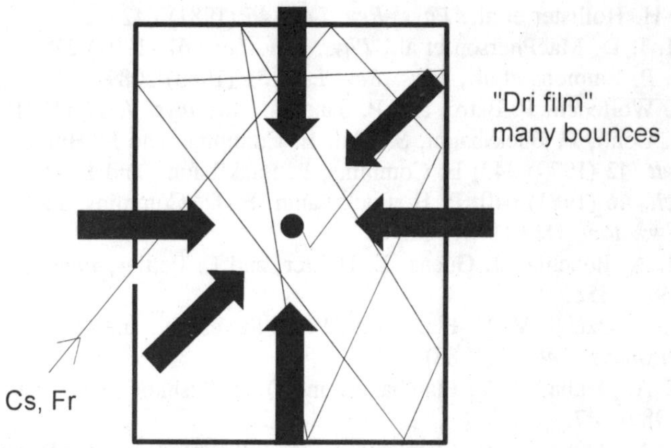

Fig. 14. High-efficiency optical trap.

given isotope (short-lived isotopes will be produced at an accelerator, while longer lived isotopes can be brought to our lab) and inject it into a special cell where the atoms will be efficiently captured by a laser trap (Fig. 14). We have already carried out detailed studies on capturing atoms from a vapor. We are currently developing wall coatings which will allow the cesium atoms to bounce around inside the cell until they are captured, without becoming stuck on the walls. This is an ongoing project, however preliminary work with silicon polymer coatings has been quite encouraging.

Once the atoms are captured, PNC measurements can then be carried out in the cold dense samples. If all goes according to plan, the next decade will see high precision measurements of PNC in a number of cesium isotopes. This will provide detailed information on the nuclear anapole moment and a very precise test of the Standard Model.

This work was supported by the National Science Foundation.

References

1. M.A. Bouchiat and C.C. Bouchiat, *Phys. Lett.* **48B** (1974) 111.
2. W. Marciano, TASI 1990, Boulder, CO.
3. E. N. Fortson and L. L. Lewis, *Phys. Repts.* **113** (1984) 289.
4. E. D. Commins, *Physica Scripta* **36** (1987) 468.
5. D. Stacey, *Physica Scripta* **T40** (1992) 15.
6. P. E. G. Baird et al., *Phy. Rev. Lett.* **39** (1977) 798.
7. L. L. Lewis et al., *Phy. Rev. Lett.* **39** (1977) 795.
8. L. M. Barkov and M. S. Zolorotev, *JETP* **27** (1978) 357.

9. J. H. Hollister et al., *Phys. Rev. Lett.* **46** (1981) 642.
10. M. J. D. MacPherson et al., *Phys. Rev. Lett.* **67** (1991) 2784.
11. T. P. Emmons et al., *Phys. Rev. Lett.* **51** (1983) 2089.
12. T. Wolfeden, P. Baird, and P. Sandars, *Europhys. J.* **15** (1991) 731.
13. R. Conti, P. Bucksbaum, S. Chu, E. Commins, and L. Hunter, *Phys. Rev. Lett.* **42** (1979) 343; E. Commins, P. Bucksbaum, and L. Hunter, *Phys. Rev. Lett.* **46** (1981) 640; P. H. Bucksbaum, E. D. Commins, and L. R. Hunter, *Phys. Rev.* **D24** (1981) 1134.
14. M. A. Bouchiat, J. Guena, L. Hunter, and L. Pottier, *Phys. Lett.* **1178** (1982) 358.
15. V. A. Dzuba, V. V. Flambaum, P. G. Silvestrov, and O. P. Sushkov, *Europhys. Lett.* **7** (1988) 413.
16. V. A. Dzuba, V. V. Flambaum, and O. P. Sushkov, *Phys. Lett. A* **141** (1989) 147.
17. V. A. Dzuba, V. V. Flambaum, P. G. Silvestrov, and O. P. Sushkov, *J. Phys. B* **20** (1987) 3297.
18. P. S. Drell and E. D. Commins, *Phys. Rev. Lett.* **53** (1984) 968.
19. M. A. Bouchiat et al., *J. Phys. (Paris)* **47** (1986) 1709.
20. S. A. Blundell, W. R. Johnson, and J. Sapirstein, *Phys. Rev. Lett.* **65** (1990) 141.
21. S. L. Gilbert, M. C. Noecker, R. N. Watts, and C. E. Wieman, *Phys. Rev. Lett.* **55** (1985) 2680.
22. M. C. Noecker et al., *Phys. Rev. Lett.* **61** (1988) 310.
23. V. V. Flambaum, I. B. Khriplovich, and O.P. Sushkov, *Phys. Lett.* **67A** (1978) 177.
24. M. A. Bouchiat and C. Bouchiat, *J. Phys. (Paris)* **35** (1974) 899.
25. C. Wieman et al., *Phys. Rev. Lett.* **58** (1987) 1738.
26. V. V. Flambaum and I. B. Khriplovich, *JETP* **52** (1980) 835; V. V. Flambaum, I. B. Khriplovich, and O. P. Sushkov, *Phys. Lett.* B **146** (1984) 367.
27. Ya. B. Zel'Dovich, *Zh. Eksp. Teor. Fiz.* **33** (1958) 1531 [Sov. Phys. JETP **7** (1957) 1184].
28. W. C. Haxton, E. M. Henley, and M. J. Musolf, *Phys. Rev. Lett.* **63** (1989) 949.
29. C. Bouchiat and C. Piketty, *Z. Phys. C* **49** (1991) 91.

THE THEORY OF ATOMIC PARITY VIOLATION

S.A. Blundell
Département de Recherche Fondamentale/LI2A, Centre d'Etudes Nucléaires de Grenoble, B.P. 85X, F-38041 Grenoble Cedex, France

and

W. R. Johnson and J. Sapirstein
Department of Physics, University of Notre Dame, Notre Dame, IN 46556 USA

Contents

1 Introduction

The role of atomic physics in precision tests of the QED part of unified theories of weak and electromagnetic interactions is well known. However, atomic physics has not until recently played much of a role in testing the weak interactions because of the extremely small ratio of the energy scale of atoms to the Z mass. Indeed, the first discussion of parity violating effects in atoms arising from neutral weak current interactions between the nucleus and electrons by Zel'dovich [1] concluded that these effects were probably too small to observe. However, after the existence of neutral currents was experimentally established, the Bouchiats [2] showed that parity nonconserving (PNC) transitions in heavy atoms with atomic number Z were enhanced by a factor of Z^3. While still very small, this effect has been observed in a variety of heavy atoms, specifically cesium ($Z = 55$) [3, 4], thallium ($Z = 81$) [5, 6], lead ($Z = 82$) [7], and bismuth ($Z = 83$) [8]. The experiments deal with neutral atoms, and for this reason the electronic structure of the atoms must be understood with some accuracy before the experiments can be interpreted in terms of particle physics. This problem is not present for hydrogen, but PNC experiments in that atom have not proved successful. Unfortunately, while the theory of the atomic structure of many-electron atoms is founded in QED, and is in principle completely understood, practical calculations that are accurate to the 1% level, the level of interest for particle physics, are quite difficult to carry out. It is the purpose of this chapter to describe a calculation of PNC in cesium, the most theoretically tractable of the atoms treated so far experimentally, and one in which an experiment accurate to 2% has been carried out [4]. While the chapter is intended to be self-contained, we note that a more complete account of this calculation has been given in Ref. [9]. The chapter is organized as follows. Following this introduction, a brief review of the experimental and theoretical situation for cesium is given in Section 2. Then the main calculational tool, many-body perturbation theory (MBPT) is introduced in Section 3 and applied to calculations of parity conserving properties of cesium. Section 4 describes more accurate methods that sum infinite classes of MBPT diagrams. This is followed by a detailed description of the PNC calculation in Section 5, and a discussion of a variety of smaller contributions to PNC is given in Section 6. A comparison of theory with experiment and a brief discussion of the particle physics implications is given in Section 7. Finally, Section 8 contains the prospects for more accurate calculations of cesium PNC.

2 General Overview

The PNC transition $6s_{1/2} \rightarrow 7s_{1/2}$ has been observed in atomic cesium by the Bouchiats [3] and by Wieman, *et al.* [4]. Cesium is a 55 electron atom with a nucleus consisting of 78 neutrons and 55 protons with nuclear spin $I = 7/2$. (The experiment is insensitive to other isotopes.) The total angular momentum of atomic s-states is then $F = 3$ or $F = 4$. Both of the transitions, $6s_{1/2}(F = 4) \rightarrow 7s_{1/2}(F = 3)$ and $6s_{1/2}(F = 3) \rightarrow 7s_{1/2}(F = 4)$, have been measured, allowing the isolation of PNC effects that depend on the spin of the nucleus. The structure of this atom is the simplest of those in which PNC has been measured, as it can be described as a single electron outside a closed xenonlike core which is relatively unpolarizable. This should be contrasted with, for example, thallium. While thallium nominally also consists of one $6p_{1/2}$ electron outside a closed core, part of that core is a filled $6s^2_{1/2}$ subshell. It is quite easy to polarize the outer subshell, so that one really has three electrons outside a closed core. This leads to distinctly poorer convergence properties of many-body perturbation theory, the theoretical method used for these calculations, and consequently less accurate atomic theory predictions. Similar considerations apply to lead and bismuth. While it is still possible that the more sophisticated all-orders methods discussed below can allow accurate calculations, we concentrate here on the simplest case of cesium.

The physics that leads to this PNC transition is the exchange of a virtual Z_0 either between a quark in the nucleus and an electron, or between two electrons. The latter effect is extremely small, and is considered in section 6.4. PNC arises when the Z_0 matrix element is vector on the nucleus and axial on the electron ($V_N A_e$), or vice versa ($A_N V_e$). The dominant PNC contribution comes from the former case, because all the quarks contribute coherently. Because of the conserved vector current (CVC), it is possible to define a related conserved charge, the *weak charge*, Q_W,

$$Q_W = 2Z(2C_{1u} + C_{1d}) + 2N(C_{1u} + 2C_{1d}). \tag{1}$$

Here Z is the number of protons, N the number of neutrons, and $C_{1u,d}$ the vector part of the Z_0-quark vertex for the up and down quarks. Putting in the tree level values of C_{1u} and C_{1d} gives

$$Q_W = Z(1 - 4\sin^2\theta_W) - N\ . \tag{2}$$

When radiative corrections are included in an analysis that uses the very precise measurement of the Z_0 mass, a standard model prediction for Q_W results [10]

$$Q_W = -73.2(0.2)\,, \tag{3}$$

that is remarkably independent of the top quark mass, as will be discussed in more detail in Section 7. It is this quantity that one wants to extract from the experiment. The timelike contribution of the ($V_N A_e$) exchange can be described by the effective atomic Hamiltonian

$$H_W = \frac{G_F}{\sqrt{8}} Q_W \rho_{\rm nuc}(\mathbf{r})\gamma_5. \tag{4}$$

Here $\rho_{\rm nuc}(\mathbf{r})$ is a weighted average of the neutron and proton distributions in the nucleus, which leads to nuclear structure uncertainties that will be discussed in section 6.2. The calculation that will be described here leads from this Hamiltonian to the prediction for the nuclear-spin-independent part of the PNC transition

$$E_{\rm PNC} = -0.905(9) \times 10^{-11} i|e|a_0(-Q_W/N)\ . \tag{5}$$

Here the unknown Q_W has been factored out, divided by its approximate value $-N$. This result is in good agreement with a similar calculation by the Novosibirsk group [11]

$$E_{\rm PNC} = -0.91(1) \times 10^{-11} i|e|a_0(-Q_W/N)\ . \tag{6}$$

and other semiempirical calculations discussed further in Ref. [9]. When it is compared with the experimental measurement

$$E_{\rm PNC}^{\rm exp} = -0.8252(184)[61] \times 10^{-11} i|e|a_0\ , \tag{7}$$

there results a determination of Q_W as

$$Q_W = -71.04(1.58)[0.88] \tag{8}$$

where the first error is experimental and the second theoretical. The spacelike part of $(V_N A_e)$ exchange and the timelike part of $(A_N V_e)$ exchange are negligible, but the spacelike part of the latter gives a nuclear spin-dependent effect that will be discussed in section 6.4. Also discussed there is an interesting nuclear physics source of PNC known as the *anapole* moment [12], where arises from photon exchange with weak radiative corrections on the nuclear vertex. This effect enters at the several percent level, but in a way that can be subtracted out as described in Section 6.4. The anapole moment by itself is not gauge invariant, but taken together with a full set of diagrams describing spin-dependent weak radiative corrections, is a real physical effect.

3 Many-Body Perturbation Theory

The technique we will use for calculating various properties of cesium is many-body perturbation theory (MBPT). The perturbation expansion can be expressed in terms of Goldstone diagrams, which are time-ordered Feynman diagrams in the presence of a filled core, where the core states are treated analogously to positron states. MBPT calculations can be restricted to a finite set of diagrams up to a given order of perturbation theory, or can include infinite classes of diagrams, in which case we use the terminology *all-orders* MBPT. The starting point of all our MBPT calculations is the relativistic generalization of the many particle Schrödinger equation, $H\psi = E\psi$, where

$$H = \sum_i (c\vec{\alpha}_i \cdot \vec{p}_i + \beta_i mc^2 - \frac{Ze^2}{r_i}) + \sum_{i<j} \frac{e^2}{|\vec{r} - \vec{r}\,'|}. \tag{9}$$

The finite size of the nucleus modifies the nuclear charge Z for small values of r, and is modeled with a Fermi distribution discussed in Section 6.2. Because the last term is too complex to handle directly, the Hamiltonian is broken up into two parts, $H = H_0 + V_C$, with

$$H_0 = \sum_i (c\vec{\alpha}_i \cdot \vec{p}_i + \beta_i mc^2 - \frac{Ze^2}{r_i} + U(r_i)) \tag{10}$$

and

$$V_C = \sum_{i<j} \frac{e^2}{|\vec{r} - \vec{r}\,'|} - \sum_i U(r_i). \tag{11}$$

Magnetic interactions between electrons (the Breit interaction) can be treated perturbatively; their effect on PNC is discussed in Section 6.1. The potential $U(r)$ is in principle general, but in practice is almost always chosen to be the Hartree-Fock potential, defined as

$$(V_{\mathrm{HF}})_{ij} \equiv \sum_a \tilde{g}_{iaja}, \tag{12}$$

where the Coulomb matrix elements g are defined by

$$g_{ijkl} \equiv e^2 \int \frac{d^3r d^3r'}{|\vec{r} - \vec{r}\,'|} \bar{\psi}_i(\vec{r}) \gamma_0 \psi_k(\vec{r}) \bar{\psi}_j(\vec{r}\,') \gamma_0 \psi_l(\vec{r}\,') \tag{13}$$

and $\tilde{g}_{ijkl} \equiv g_{ijkl} - g_{ijlk}$. The summation over a refers to summing over all electrons in a closed core, for this case the 54 electrons forming the xenonlike core of cesium.

It is now trivial to solve $H_0\psi_0 = E_0\psi_0$ in terms of a Slater determinant of the occupied orbitals. These orbitals satisfy the Dirac equation

$$[c\vec{\alpha}\cdot\vec{p} + \beta mc^2 - \frac{Ze^2}{r} + U(r)]\phi_i = \epsilon_i\phi_i \,. \tag{14}$$

This equation has solutions associated with bound electrons, continuum electrons and positron states. We generally exclude positron states so as to avoid a problem with the Hamiltonian (9) known as *continuum dissolution* [13], in which electrons make transitions to the negative energy sea. A fully field theoretical treatment of course excludes such unphysical transitions, but also leads to small well-defined radiative corrections associated with positron states that can be put in perturbatively if needed. We do, however, consider virtual positron state contributions in some parts of the calculation. Designating electron creation and annihilation operators by $a_i^\dagger$ and a_i, respectively, we may write our Hamiltonian in second-quantized form as $H = H_0 + V_C$, where

$$H_0 = \sum_i \epsilon_i a_i^\dagger a_i, \tag{15}$$

and

$$V_C = \frac{1}{2}\sum_{ijkl} g_{ijk\ell} a_i^\dagger a_j^\dagger a_\ell a_k - \sum_{ij} (U)_{ij} a_i^\dagger a_j. \tag{16}$$

The model of the atom provided by lowest order perturbation theory is rather inaccurate when the HF potential is used: valence removal energies disagree with experiment by on the order of 10%, and matrix elements of the hyperfine operator by about 40%. Thus it is essential for accurate calculations to include the effects of V_C as fully as possible. MBPT proceeds by expanding the many-body wave function $\Psi(v)$ and the energy $E(v)$ in powers of V_C,

$$\Psi(v) = \Psi_0(v) + \Psi_1(v) + \Psi_2(v) + \cdots \,, \tag{17}$$

and

$$E(v) = E_0(v) + E_1(v) + E_2(v) + \cdots \,. \tag{18}$$

The lowest-order wave function, which is an eigenfunction of H_0, is given by

$$\Psi_0(v) = a_v^\dagger |0_c\rangle \,, \tag{19}$$

where $|0_c\rangle$ is the core wave function. The corresponding energy is

$$E_0(v) = \epsilon_v + \sum_a \epsilon_a \,, \tag{20}$$

where ϵ_v is the valence eigenvalue and ϵ_a is the eigenvalue of the ath core electron. The Schrödinger equation leads to an hierarchy of equations for $\Psi_n(v)$ and $E_n(v)$. The solution to the first of these equations is

$$\Psi_1(v) = (H_0 - E_0(v))^{-1}(E_1(v) - V)\Psi_0(v) \,, \tag{21}$$

where

$$E_1(v) = \epsilon_v - \frac{1}{2}\sum_a (V_{HF})_{aa} \,. \tag{22}$$

Some of the main issues of MBPT can be illustrated by considering the second-order many-body self-energy $\Sigma^{(2)}(\epsilon)$, which in the HF potential is defined by its matrix elements

$$\langle i|\Sigma^{(2)}(\epsilon)|j\rangle = -\sum_{abm} \frac{g_{abjm}\tilde{g}_{imab}}{\epsilon_a + \epsilon_b - \epsilon_m - \epsilon} + \sum_{amn} \frac{g_{aimn}\tilde{g}_{mnaj}}{\epsilon_a + \epsilon - \epsilon_m - \epsilon_n}. \tag{23}$$

This object is related to the second-order energy by $E_2(v) = \langle v|\Sigma^{(2)}(\epsilon_v)|v\rangle$. Sums over excited states are first encountered in this order. The sums over m and n indicate summations over all positive energy states excluding the occupied core states a, but including valence states v. The reason for the previously mentioned exclusion of positron states is that were they included in the double summation over m and n with one state of positive energy and the other of negative energy, vanishing energy denominators would result.

A very important practical matter in MBPT calculations is the efficient and accurate evaluation of sums such as encountered in E_2. As a first step the atom is considered as being at the center of a large sphere that confines electrons with MIT bag model boundary conditions; this serves to discretize the continuum states. The radius R is chosen to be large compared to the atom, typically around $R = 50 - 70$ a.u.. For a given value of the angular momentum quantum number κ, the Dirac equation for an electron of energy ϵ with upper and lower components $P_\kappa(r)$ and $Q_\kappa(r)$ respectively, can be obtained by requiring $\delta S = 0$, where

$$S = \frac{1}{2}\int_0^R [cP_\kappa(r)(d/dr - \kappa/r)Q_\kappa(r) - cQ_\kappa(r)(d/dr + \kappa/r)P_\kappa(r) + \tag{24}$$

$$V(r)[P_\kappa^2(r) + Q_\kappa^2(r)) - 2mc^2 Q_\kappa^2(r)]dr - \frac{\epsilon}{2}\int_0^R [P_\kappa^2(r) + Q_\kappa^2(r)]dr. \tag{25}$$

appropriately modified to enforce the boundary conditions. A finite basis set can be introduced by minimizing this action in terms of coefficients p_i and q_i, where

$$P_\kappa(r) = \sum_{i=1}^{n} p_i B_i(r) \tag{26}$$

and

$$Q_\kappa(r) = \sum_{i=1}^{n} q_i B_i(r). \tag{27}$$

Here the $B_i(r)$ are B-splines [14], functions that are piecewise polynomials (typically of sixth to eighth order in our applications), which vanish for most values of r. B-splines provide great flexibility in representing arbitrary functions. They are defined between *knot points*, which can be chosen to fit the physics of the problem. In particular, we use this freedom to put many knot points within the nucleus, where the PNC effect originates, putting the remaining knot points on an exponential grid appropriate for atomic wave functions. The number n above is typically around 50

Table 1: Ionization energies (a.u.) for valence states of cesium calculated in second-order perturbation theory. The quantity $\epsilon_v^{\rm HF}$ is the HF energy and $\epsilon_v^{(2)}$ is the correction from second-order perturbation theory. See Ref. [15]

Orbital	$\epsilon_v^{\rm HF}$	$\epsilon_v^{(2)}$	Theory	Exp.
$6s_{1/2}$	-0.12737	-0.01775	-0.14512	-0.14310
$6p_{1/2}$	-0.08562	-0.00691	-0.09253	-0.09217
$6p_{3/2}$	-0.08378	-0.00618	-0.08997	-0.08964

for the calculations reported here. When these forms are used in the action, the radial integrations can be performed and the action becomes a quadratic form in p_i and q_i. Requiring $\delta S = 0$ then leads to a $2n \times 2n$ eigenvalue equation that generates n positive energy and n negative energy eigenvectors and eigenvalues. These form a relativistic *finite basis set.* For κ values corresponding to occupied core states, the first few eigenvalues accurately reproduce the known solutions to the Dirac equation. Because of the finite radius of the cavity, there are only a finite number of bound states, and the least strongly bound are not realistic because they are influenced by the cavity boundary conditions. The remaining states are unbound, and form a representation of the continuum. Extensive tests have shown that this finite basis set gives answers correct to six digits or more when used to represent summations such as those encountered in E_2. In this way, MBPT expressions can be carried out in an automatic fashion.

We show in Table 1 the results of applying MBPT through 2nd order for valence removal energies of cesium. The lowest order results disagree with experiment at the 10% level, and the first order corrections vanish for the HF potential. The agreement is seen to improve substantially, however, with the inclusion of E_2, to the 1% level. The calculation proceeds by making a partial wave expansion in Eq. 13, which allows the angular integrations to be done at the expense of infinite sums over partial waves. The remaining radial integrals are carried out numerically, and the partial wave summation is extrapolated to infinity using about the first seven partial waves, which is possible because for high l the series generally behaves as $1/l^4$. A major numerical difficulty in higher order MBPT is that at high l more and more angular momentum channels contribute. Nevertheless, second-order energies can be evaluated in a few minutes on modern workstations.

While calculations of energies are a useful monitor of the behavior of MBPT, we wish to accurately predict a parity violating transition amplitude. For this reason it is important to calculate standard parity-conserving amplitudes and compare them to experiment. We use hyperfine splittings and oscillator strengths for this purpose, illustrating the calculations here with dipole transition amplitudes. They are determined by evaluating matrix elements of the dipole operator,

$$eZ = \sum_{ij} \langle i|ez|j\rangle a_i^\dagger a_j \tag{28}$$

in perturbation theory. The first two terms in the expansion are

$$\begin{aligned}\langle\Psi(w)|eZ|\Psi(v)\rangle &= \langle w|ez|v\rangle \\ &+ \langle\Psi_1(w)|eZ|\Psi_0(v)\rangle + \langle\Psi_0(w)|eZ|\Psi_1(v)\rangle \\ &+ \cdots . \end{aligned} \tag{29}$$

If we let d_{ij} designate the first-order dipole matrix element $\langle i|ez|j\rangle$, then the second-order correction from Eq. (29) can be written

$$d^{(2)}_{wv} = \sum_{an} \frac{d_{an}\tilde{g}_{wnav}}{\epsilon_n - \epsilon_a + \omega} + \sum_{an} \frac{\tilde{g}_{wanv} d_{na}}{\epsilon_n - \epsilon_a - \omega} . \tag{30}$$

While there are a large number of corrections from the next order, two particularly important ones are given by

$$\begin{aligned} d^{(3)}_{wv}(\text{RPA}) = &\sum_{abmn} [\frac{\tilde{g}_{wnva} d_{bm} \tilde{g}_{amnb}}{(\epsilon_{mw} - \epsilon_{bv})(\epsilon_{nw} - \epsilon_{av})} + c.c.] + \\ &\sum_{abmn} [\frac{\tilde{g}_{mnab} d_{bm} \tilde{g}_{awvn}}{(\epsilon_{nv} - \epsilon_{aw})(\epsilon_{mw} - \epsilon_{bv})} + c.c.] , \end{aligned} \tag{31}$$

and

$$\begin{aligned} d^{(3)}_{wv}(\text{BO}) = &\sum_{abmi} [\frac{g_{abmv} d_{wi} \tilde{g}_{miba}}{(\epsilon_i - \epsilon_v)(\epsilon_{mv} - \epsilon_{ab})} + c.c.] + \\ &\sum_{amni} [\frac{g_{aimn} d_{wi} \tilde{g}_{mnav}}{(\epsilon_i - \epsilon_v)(\epsilon_{mn} - \epsilon_{av})} + c.c.] . \end{aligned} \tag{32}$$

The terms $d^{(2)}_{wv}$ and $d^{(3)}_{wv}(\text{RPA})$ are parts of the random-phase approximation (RPA), which describes the shielding effect of the core on an externally applied electric field. The RPA is obtained by replacing d_{an} in Eq. (29) by t_{an}, where

$$t_{an} = d_{an} + \sum_{bm} \frac{t_{bm}\tilde{g}_{ambn}}{\epsilon_{mw} - \epsilon_{bv}} + \sum_{bm} \frac{t_{mb}\tilde{g}_{nbma}}{\epsilon_{mv} - \epsilon_{bw}} . \tag{33}$$

The contributions from $d^{(2)}$ and $d^{(3)}(\text{RPA})$ are automatically picked up along with an infinite class of higher order RPA terms by solving these equations iteratively. This is an example of an all-orders method. Note that the sums over excited states for the RPA illustrate a case in which negative energy states should be included, although they contribute a numerically insignificant effect. The second contribution $d^{(3)}_{wv}(\text{BO})$ is relatively large because the energy denominator $\epsilon_i - \epsilon_v$ can be small when i is another valence state of similar energy. It can be rewritten as

$$d^{(3)}_{wv}(\text{BO}) = \langle\phi_{\text{w}}|\text{ez}|\delta\phi_{\text{v}}\rangle + \langle\delta\phi_{\text{w}}|\text{ez}|\phi_{\text{v}}\rangle , \tag{34}$$

where $\delta\phi_{w,v}$ are lowest-order Brueckner orbital (BO) corrections to the HF valence states, defined by

$$\delta\phi_v = \sum_{i\neq v} \phi_i \frac{\langle i|\Sigma^{(2)}(\epsilon_v)|v\rangle}{\epsilon_v - \epsilon_i} . \tag{35}$$

Table 2: Convergence of MBPT for ordinary matrix elements of cesium. A designates the hyperfine splitting in MHz, and D represents the $6p_{3/2} - 6s_{1/2}$ reduced dipole matrix element in units $|e|a_0$.

Order	$A(6s_{1/2})$	$A(7s_{1/2})$	D
(1)	5740	1576	-7.426
(2) RPA	1156	320	0.413
(3) BO	2568	392	0.842
Total	9464	2288	-6.171
Expt	9192	2184	-6.32(8)

The BO corrections arise from the polarization potential of the core, which is represented formally by the many-body self-energy operator. The BO effect can be included to higher order as part of an infinite set of diagrams, as will be described in section 5.1. However, at this point we will examine the behavior of MBPT for matrix elements when $d^{(3)}_{wv}(\mathrm{BO})$ and all orders of RPA are included.

In Table 2, we present the behavior of MBPT for hyperfine splittings and oscillator strengths. In this case, the lowest order results differ from experiment by up to 40%. However, both the RPA and BO corrections are substantial, and bring theory and experiment into agreement at the few-percent level. We can take as an informal error estimate the largest remaining discrepancy, the 7s hyperfine constant, which disagrees with experiment by 5%. It is possible to carry out a precisely analogous calculation for the PNC transition. When this is done [16], we find

$$E_{\mathrm{PNC}} = -0.95(5) \times 10^{-11} i|e|a_0(-Q_W/N)\,. \tag{36}$$

This result will change by 4% when higher-order terms are included, so in this case the error estimate is seen to be reliable.

The results of Table 2 indicate that while the frozen-core Hartree Fock potential gives relatively inaccurate results, MBPT through third order improves agreement with experiment to the few-percent level. Unfortunately, to be useful in precision tests of the standard model, this accuracy must be improved to the order of 1%. One possible approach is to include the next order of MBPT. However, when this is done for energies, it is found that the predictions actually worsen. For example, the $6s_{1/2}$ removal energy starts off 11% too low, becomes 1.4% too high after second order is included, but then becomes 2.5% too low when the third order is calculated. The reason for this is the very large second-order correction: this correction can be considered to be the first term of an infinite set, the next element of which enters in fourth order. In fact, when that fourth order term is added, agreement with experiment improves to the few tenths of a percent level [17]. It is, however, a somewhat dangerous procedure to include selected diagrams from higher order, since it is always possible that neglected diagrams may be larger than expected. It is clearly desirable to find an efficient way to include as many higher order diagrams as possible. Rather than going to the next higher order of MBPT, we have chosen to introduce all-orders methods, similar to the RPA but including a much wider class of higher-order MBPT diagrams. If chosen properly, these approaches can automatically pick up the first few orders of perturbation theory, at the same time accounting for infinite orders of certain types of diagram. We now turn to a discussion of the all-orders method used for our final calculations.

4 All-orders Calculations

To illustrate all-orders methods in a simple case, let us consider helium. We describe the helium ground-state in lowest order by the wave function

$$\Psi_0 = a_a^\dagger a_b^\dagger |0\rangle \, , \tag{37}$$

where the subscripts a and b refer to the $1s_{1/2}$ states with magnetic quantum numbers $m = 1/2$ and $m = -1/2$, respectively. The corresponding lowest-order energy is

$$E_0 = 2\epsilon_{1s} \, . \tag{38}$$

We add a correction to Ψ_0 which is a superposition of states in which each electron is excited to every possible state: $\Psi = \Psi_0 + \Delta\Psi$, with

$$\Delta\Psi = \left(\sum_{mn} \rho_{mnab} a_m^\dagger a_n^\dagger a_a a_b \right) \Psi_0 \, , \tag{39}$$

where the pair m, n is restricted to be different from a, b. Substituting Eq. (39) into the Schrödinger equation we obtain a set of coupled equations for the expansion coefficients

$$\begin{aligned}(\epsilon_m + \epsilon_n - E_0 - \Delta E)\rho_{mnab} = {} & \frac{1}{2} g_{mnab} \\ & - \sum_{k\ell} g_{mnk\ell}\rho_{k\ell ab} \, ,\end{aligned} \tag{40}$$

where

$$\Delta E = \frac{1}{2}\sum_a (V_{\rm HF})_{aa} - \sum_{abmn} \rho_{mnab}\tilde{g}_{abmn} \, . \tag{41}$$

The angular integrals in Eqs. (40) and (41) can performed analytically, and the equations solved iteratively. The value of the energy obtained in the $n^{\rm th}$ iteration is approximately the same as the energy obtained from $n^{\rm th}$-order perturbation theory. In Table 3, we show how the iteration solution to Eqs. (40-41) converges to the exact ground-state energy of helium [18].

All-orders techniques of this type can also be applied to cesium but the analysis becomes considerably more difficult in this case. We assume that the lowest-order wave function Ψ_0 for cesium is a a frozen-core HF wave function. We restrict our consideration to a correction $\Delta\Psi$ that is a linear combination of states

Table 3: Contributions to the ground-state energy of helium in an iterative solution to the all-order equations. Units: a.u. See Ref. [18]

Iteration	Energy	Cumulative
1	-4.00021	-4.00021
2	1.25010	-2.75011
3	-0.15768	-2.90779
4	0.00434	-2.90345
5	-0.00021	-2.90366
...	...	...
All-order		-2.90386

with either one or two orbitals in the HF wave function is excited. With this restriction, $\Delta\Psi$ can be written

$$\begin{aligned}\Delta\Psi &= \Big(\sum_{am} \rho_{ma} a_m^\dagger a_a + \sum_{abmn} \rho_{mnab} a_m^\dagger a_n^\dagger a_a a_b \\ &+ \sum_{m} \rho_{mv} a_m^\dagger a_v + \sum_{amn} \rho_{mnav} a_m^\dagger a_n^\dagger a_a a_v \Big) \Psi_0 \,. \end{aligned} \tag{42}$$

The terms on the first line of Eq. (42) describe single and double excitations of the closed core, while those on the second line describe single and double excitations of the atom where the valence orbital is also excited. Substituting Eq. (42) into the Schrödinger equation one obtains a set of coupled equations for the expansion coefficients that can be found in Ref. [19]. The first and second iterations of the equations for the expansion coefficients leads to results that are identical to first- and second-order perturbation theory. In third-order perturbation theory, terms associated with triple excitations contribute to the energy. These terms have no counterpart in the iterative solution to the equations under consideration.

To account for such terms, we add to $\Delta\Psi$ a triple-excitation correction of the specific form:

$$\begin{aligned}\delta\Psi &= \Big(\sum_{abcmnr} \rho_{mnrabc} a_n^\dagger a_m^\dagger a_r^\dagger a_a a_b a_c \\ &+ \sum_{abmnr} \rho_{mnrabv} a_n^\dagger a_m^\dagger a_r^\dagger a_a a_b a_v \Big) \Psi_0 \,. \end{aligned} \tag{43}$$

Such a term enters in two ways. First, there are a set of equations giving the triple-excitation coefficients in terms of the single-, double-, and triple-excitation coefficients. Second, the triple-excitation coefficients enter on the right-hand side of equations for the single- and double-excitation coefficients. We solve for the triple-excitation coefficients in terms of the singles and doubles, ignoring the triples on the right-hand sides of these equations. We then use these approximate expressions for the triples on the right-hand sides of the equations for the singles. This procedure

Table 4: Valence removal energies for cesium from an all-order calculation. Units: a.u. See Ref. [19]

State	ϵ_v	$\delta\epsilon_v$	Sum	Experiment
$6s_{1/2}$	0.12737	0.01521	0.14257	0.14310
$6p_{1/2}$	0.08562	0.00636	0.09198	0.09217
$6p_{3/2}$	0.08379	0.00572	0.08951	0.08964
$7s_{1/2}$	0.05519	0.00326	0.05845	0.05865
$7p_{1/2}$	0.04202	0.00183	0.04385	0.04393
$7p_{3/2}$	0.04137	0.00166	0.04303	0.04310

leads to equations which, when iterated to third order, include all of the terms from MBPT. Moreover, when the iteration is continued, the single and double excitations are included to all orders. However, the triples also modify the the doubles equation, and this more difficult step has not yet been implemented. As the effects of this modification enter first in fourth-order MBPT, our calculation is complete through third order, but still misses some fourth-order contributions.

In Table 4, we compare the energies of a few states of cesium determined by solving the all-order equations with experiment. It can be seen that the energies of these states all agree with the observed energies to better than 0.5%. The corresponding wave functions can be used to evaluate properties of the various excited states such as hyperfine constants and transition amplitudes between the excited states. However, matrix elements calculated using the all-order wave functions miss some RPA corrections of fourth and higher order. Since some of the matrix elements of interest for the PNC problem are sensitive to the omitted RPA terms, the present formalism is modified slightly to include the RPA exactly. The required modifications are described in detail in Ref. [19] . In Table 5, we show values of hyperfine constants for valence states of cesium from the all-order calculation and we compare the theoretical values with experiment. As with the energies, the hyperfine constants are found to agree well with experiment. Thus we expect that the analogous PNC calculations will be of similar accuracy. The actual calculation was carried out in two different fashions, which we refer to as mixed-parity MBPT and the sum-over-states approach. We now describe these two methods in turn.

5 PNC Calculations

5.1 Mixed-Parity MBPT

In mixed-parity MBPT, we modify H_0 by adding the weak-interaction h_W to the HF potential. This approach leads to a generalization of the single-particle states in which each state acquires an opposite-parity admixture,

$$\phi_k \rightarrow \phi_k + \tilde{\phi}_k \, . \tag{44}$$

Table 5: Hyperfine constants (MHz) for ^{133}Cs with $I = \frac{7}{2}$, $g_I = 0.7377208$ and reduced dipole matrix elements (a.u.) for cesium from an all-order calculation. See Ref. [19]

State	$A^{\rm HF}$	δA	Sum	Experiment
$6s_{1/2}$	1426.81	864.19	2291.00	2298.16
$7s_{1/2}$	392.05	151.99	544.04	545.90(9)
$6p_{1/2}$	161.09	131.58	292.67	291.90(13)
$7p_{1/2}$	57.68	35.53	94.21	94.35(4)
$6p_{3/2}$	23.944	25.841	49.785	50.275(3)
$7p_{3/2}$	8.650	7.605	16.255	16.605(6)

Thus, for example, each $s_{1/2}$ orbital will pick up a small $p_{1/2}$ state admixture. Treating the weak interaction to lowest order (which is certainly justified), the induced correction $\tilde{\phi}_v$ satisfies the equation

$$(h + V_{\rm HF} - \epsilon_v)\tilde{\phi}_v = -h_W\,\phi_v\ , \tag{45}$$

in a first approximation. It is a simple matter to solve this equation for the $6s$ and $7s$ perturbed orbitals and to evaluate the the PNC amplitude to lowest order. We obtain in this first approximation

$$\begin{aligned} E_{\rm PNC} &= \langle\phi_{7s}|ez|\tilde{\phi}_{6s}\rangle + \langle\tilde{\phi}_{7s}|ez|\phi_{6s}\rangle \\ &= -0.740 \times 10^{-11} i|e|a_0(-Q_W/N)\ . \end{aligned} \tag{46}$$

The approximation made in Eq. (45) ignores the dependence of $V_{\rm HF}$ on the core orbitals, which themselves acquire small opposite-parity admixtures. If we take this dependence into account, then the HF potential is also modified

$$V_{\rm HF} \to V_{\rm HF} + \tilde{V}_{\rm HF}\ ,$$

and Eq. (45) is replaced by the system of equations

$$(h + V_{\rm HF} - \epsilon_a)\tilde{\phi}_a = -(h_W + \tilde{V}_{\rm HF})\phi_a\ , \tag{47}$$

$$(h + V_{\rm HF} - \epsilon_v)\tilde{\phi}_v = -(h_W + \tilde{V}_{\rm HF})\phi_v\ . \tag{48}$$

One solves the system Eqs. (47) self-consistently to obtain the perturbation to each core orbital $\tilde{\phi}_a$ and the perturbed HF potential $\tilde{V}_{\rm HF}$. Eq. (48) is then solved to obtain the perturbed valence orbitals $\tilde{\phi}_v$. The corrections to the results of (46) obtained in this way are referred to as weak RPA corrections. We obtain in this approximation

$$E_{\rm PNC}^{\rm HF} = -0.927 \times 10^{-11} i|e|a_0(-Q_W/N)\ . \tag{49}$$

We now proceed to carry out a consistent implementation of MBPT based on parity-mixed single-particle states. The above result (49) is the lowest-order result in such a perturbation theory, corresponding to the "parity-mixed HF". To evaluate the second-order corrections, we linearize the second-order amplitude from Eq. (30) in the weak interaction. As discussed in the previous section, we use RPA amplitudes, $d_{na}^{\rm RPA}$ rather than lowest-order amplitudes on the RHS of Eq. (30). Evaluating the resulting expression, one obtains the correction

$$\delta E_{\rm PNC}^{\rm RPA-ext} = 0.035 \times 10^{-11} i|e|a_0(-Q_W/N)\,.$$

This part of the correction arises from the use of parity-mixed *valence* orbitals in the expression for the RPA amplitude. There is also a much smaller cross term caused by the parity-mixing of the RPA amplitudes themselves, $d_{na}^{\rm RPA} \to d_{na}^{\rm RPA} + \tilde{d}_{na}^{\rm RPA}$. This cross term contributes

$$\delta E_{\rm PNC}^{\rm RPA-int} = 0.002 \times 10^{-11} i|e|a_0(-Q_W/N)\,.$$

The dominant third-order correction to the PNC amplitude arises from the BO terms. If we linearize Eq. (34) in the weak interaction, we obtain

$$\begin{aligned} \delta E_{\rm PNC}^{\rm BO} &= \langle \delta\phi_{7s}|ez|\tilde{\phi}_{6s}\rangle + \langle \delta\tilde{\phi}_{7s}|ez|\phi_{6s}\rangle \\ &+ \langle \tilde{\phi}_{7s}|ez|\delta\phi_{6s}\rangle + \langle \phi_{7s}|ez|\delta\tilde{\phi}_{6s}\rangle\,, \end{aligned} \tag{50}$$

where $\delta\tilde{\phi}_v$ is the weak perturbation to the BO correction for the orbital ϕ_v. Each $\delta\tilde{\phi}_v$ consists of two parts, an "external" part arising from weak corrections to the valence orbitals in Eq. (35), and an "internal" part arising from weak corrections to $\Sigma^{(2)}(\epsilon_v)$,

$$\delta\tilde{\phi}_v = \delta\tilde{\phi}_v^{\rm ext} + \delta\tilde{\phi}_v^{\rm int}\,.$$

The external part leads to the dominant correction

$$\delta E_{\rm PNC}^{\rm BO-ext} = -0.058 \times 10^{-11} i|e|a_0(-Q_W/N)\,, \tag{51}$$

while the internal part gives the much smaller contribution

$$\delta E_{\rm PNC}^{\rm BO-int} = -0.003 \times 10^{-11} i|e|a_0(-Q_W/N)\,.$$

To complete the third-order calculation, we evaluate the tiny corrections from "structural-radiation" [9]

$$\delta E_{\rm PNC}^{\rm StRad} = -0.004 \times 10^{-11} i|e|a_0(-Q_W/N)\,,$$

and normalization effects

$$\delta E_{\rm PNC}^{\rm Norm} = 0.008 \times 10^{-11} i|e|a_0(-Q_W/N)\,.$$

Now we turn to the evaluation of fourth- and higher-order corrections. The largest of these is the correction that arises when the approximate Brueckner orbitals obtained by solving Eq. (35) for $\delta\phi_v$ are replaced by the *chained* Brueckner orbitals determined by solving the second-order quasiparticle equation

$$(h + V_{\rm HF} + \Sigma^{(2)}(\epsilon_v))\phi_v = \epsilon_v\phi_v \tag{52}$$

exactly. Eq. (35) corresponds to a perturbative treatment of the above equation. Solving (52) to all orders in $\Sigma^{(2)}$ reduces the external BO corrections in Eq. (51) by a factor close to 2, leading to the value

$$\delta E_{\rm PNC}^{\rm ChBO-ext} = -0.029(9) \times 10^{-11} i|e|a_0(-Q_W/N)\,.$$

The error in this term was estimated by various methods [9]. In one, the self-energy was scaled, $\Sigma^{(2)}(\epsilon_v) \rightarrow \lambda\Sigma^{(2)}(\epsilon_v)$, with λ chosen to reproduce experimental removal energies. Although the values of $\lambda \sim 0.8$ thus obtained are quite different from unity, $E_{\rm PNC}$ was found to vary by only a few tenths of a percent. This behavior could be traced to accidental cancelations between numerator and denominator contributions in the implicit sum over states (see next section) in the present approach. The cancellations are fortunate, however, because they imply that corrections to the many-body self-energy beyond second order play a reduced role in the PNC effect.

We next replace the valence HF orbitals by chained Brueckner orbitals in an RPA calculation of the transition amplitude, accounting for a set of fourth-order corrections that correspond to core shielding of the Brueckner orbital corrections. This leads to a further modification of the amplitude

$$\delta E_{\rm PNC}^{\rm RPA\times BO} = 0.014 \times 10^{-11} i|e|a_0(-Q_W/N)\,. \tag{53}$$

Putting together these different effects then gives our final prediction for the parity-mixed calculation

$$\delta E_{\rm PNC} = -0.904(9) \times 10^{-11} i|e|a_0(-Q_W/N)\,. \tag{54}$$

This result is in good agreement with other calculations, as discussed in Ref. [9]. We now turn to a second method we used to calculate this amplitude.

5.2 Sum-Over-States for PNC Amplitude

An alternative and very direct way of calculating PNC transition amplitudes is to saturate the sum over exact many-body states X in

$$\begin{aligned} E_{\rm PNC} &= \sum_X \frac{\langle 7S|H_W|X\rangle\langle X|ez|6S\rangle}{E_{7S}-E_X} \\ &+ \sum_X \frac{\langle 7S|ez|X\rangle\langle X|H_W|6S\rangle}{E_{6S}-E_X}\,. \end{aligned} \tag{55}$$

One finds that about 98% of this sum is contributed by the states $X = |6P_{1/2}\rangle \cdots |9P_{1/2}\rangle$, and the remaining 2% by the states with $n \geq 10$, and by states in which X involves excitation of the core, called autoionizing states. If the exact wavefunctions X are replaced by the corresponding lowest-order Slater determinants, this expression can be shown to reproduce the result (46). Here, however, we evaluate the contributions from the states with valence principle quantum numbers $n = 6 \cdots 9$ using our all-order wave functions; the remaining contributions are estimated using perturbation theory. We find [9]:

1. Sum $n = 6 - 9$

$$E_{\rm PNC} = -0.893(7) \times 10^{-11} i|e|a_0(-Q_W/N) \,,$$

2. Tail $n = 10 - \infty$

$$\delta E_{\rm PNC} = -0.018(5) \times 10^{-11} i|e|a_0(-Q_W/N) \,,$$

3. Autoionizing states

$$\delta E_{\rm PNC} = 0.002(2) \times 10^{-11} i|e|a_0(-Q_W/N) \,,$$

Adding these contributions, we obtain

$$E_{\rm PNC} = -0.909(9) \times 10^{-11} i|e|a_0(-Q_W/N) \,. \tag{56}$$

This result is consistent with the mixed-parity determination. Our final result Eq. 5 is an average of the two methods taken together with the effect of the Breit interaction. We now turn to a discussion of that and other small PNC effects.

6 Smaller PNC Contributions

6.1 Breit Interaction

The Breit interaction is taken into account by replacing the Coulomb interaction V in the basic Hamiltonian of Eq. (16) by the sum of the Coulomb and Breit interactions,

$$g_{ijkl} \to g_{ijkl} + b_{ijkl} \,. \tag{57}$$

With this replacement, the HF equations for the single-particle orbitals become

$$(h + V_{\rm HF} + B_{\rm HF})\phi_k = \epsilon_k \phi_k \,. \tag{58}$$

Since the dominant contribution to the PNC amplitude is the PNC-HF contribution, it is sufficient to carry out a PNC-HF calculation including the Breit interaction in addition to the Coulomb interaction in order to evaluate the Breit correction. For this purpose, we solve the equations:

$$(h + V_{\rm HF} + B_{\rm HF} - \epsilon_k)\tilde{\phi}_k = -(h_W + \tilde{V}_{\rm HF})\phi_k \,, \tag{59}$$

and use the resulting perturbed orbitals to evaluate the PNC amplitude as described in the section on mixed-parity calculations. This calculation leads to a 0.2% correction which has been included in Eq. 5.

6.2 Nuclear Density

As mentioned in the introduction, the function $\rho_{\rm nuc}(\mathbf{r})$ in the PNC Hamiltonian is a nuclear density function, close to the neutron density. Since there are no experimental values for the neutron density of ^{133}Cs, we use instead an experimental proton density function. This proton density is taken to be a Fermi distribution

$$\rho_Z(r) = \frac{\rho_0}{1 + e^{-(r-c)/a}}, \tag{60}$$

with parameters $a = 0.523$ fm and $c = 5.674(1)$ fm determined from muonic x-ray measurements [20]. The lowest-order PNC amplitude calculated using this distribution instead of the neutron density is

$$E_{\rm PNC} = -0.7396 \times 10^{-11} i|e|a_0(-Q_W/N). \tag{61}$$

In the absence of an experimental neutron density, we use the theoretical neutron distribution function from a calculation that reproduces the experimental charge radius [21]

$$\rho_N(r) = \frac{\rho_0'}{(1 + e^{-(r-c')/a'})^{b'}}, \tag{62}$$

with $a' = 0.6842$ fm, $b' = 1.589$, and $c' = 6.153$ fm. Calculating the lowest-order PNC amplitude with this distribution gives

$$E_{\rm PNC} = -0.7390 \times 10^{-11} i|e|a_0(-Q_W/N), \tag{63}$$

a difference of only -0.08% from the value determined using the experimental proton distribution. At the 1% level of precision of interest here, we can obviously ignore the uncertainty in $E_{\rm PNC}$ caused by the lack of a precise understanding of the nuclear matter distribution. The uncertainty, however, does play a role when different isotopes are considered. The suggestion has been made to measure PNC in different isotopes, and to obtain information relatively free of electronic structure uncertainties by taking ratios. However, although the electronic structure is certainly almost unchanged, the neutron distribution in different nuclei is more uncertain, and taking the ratio enhances the nuclear physics uncertainty. This issue has been addressed recently by Pollock *et al.* [22], who find significant effects when different nuclear models are used for the case of lead ($Z = 82$).

6.3 Nuclear spin-dependent effects

In addition to the dominant PNC interaction given in Eq. (4), there are other smaller PNC interactions that must be considered. First, there is the interaction between the nuclear axial-vector current and the electron vector current from Z exchange. In the limit of nonrelativistic nucleon motion, this interaction is given by the spin-dependent Hamiltonian

$$h_W^{(2)} = -\frac{G}{\sqrt{2}} K_2 \frac{\kappa - 1/2}{I(I+1)} \alpha \cdot \mathbf{I}\rho(r). \tag{64}$$

Here, $\kappa = 4$, $I = 7/2$ and $K_2 \approx= -0.05$ for the valence proton of ^{133}Cs. Additionally, parity violation in the nucleus leads to to a parity-violating nuclear moment, the anapole moment mentioned in Section 2, that couples electromagnetically to the atomic electrons. The anapole-electron interaction is described by a Hamiltonian similar to (64)

$$h_W^a = \frac{G}{\sqrt{2}} K_a \frac{\kappa}{I(I+1)} \alpha \cdot \mathbf{I} \rho(r) . \tag{65}$$

The parameter $K_a = 0.24 - 0.33$ is determined from nuclear model calculations [23]. These two interactions can be treated together using (65) with $K_a \to K = K_a - K_2(\kappa - 1/2)/\kappa$. The resulting spin-dependent correction was evaluated in the Dirac-Fock approximation including weak core-polarization corrections. Combining that calculation with the previous spin-independent result, we obtain

$$E_{\text{PNC}} = -0.905(9) \times 10^{-11} i|e|a_0 \left[(-Q_W/N) + A(F', F)K\right] , \tag{66}$$

where the matrix $A(F', F)$ is found to be

$$\begin{pmatrix} A(3,3), & A(3,4) \\ A(4,3), & A(4,4) \end{pmatrix} = \begin{pmatrix} 0.029, & 0.048 \\ -0.041, & -0.022 \end{pmatrix} . \tag{67}$$

These values of $A(F', F)$ agree to within 10% with results of semiempirical [24] and MBPT [25] calculations. Linear combinations of amplitudes in (66) can be used to isolate either the spin-dependent or spin-independent parts of the interaction as will be discussed below.

The interference between the hyperfine interaction and the spin-independent PNC interaction leads to a tiny spin-dependent interaction [26, 27] that can also be included in the above analysis by adjusting the value of K in (66) slightly.

6.4 e-e Weak Interaction

The effect of Z exchange between electrons can be taken into account by adding a weak correction $g_{ijk\ell}^w$ to the electron-electron Coulomb interaction. This correction takes the form of a contact interaction

$$g_{ijk\ell}^w = \sqrt{2} G \int \bar{\phi}_i (\gamma_\mu C_{1e} + \gamma_\mu \gamma_5 C_{2e}) \phi_k \bar{\phi}_j (\gamma^\mu C_{1e} + \gamma^\mu \gamma_5 C_{2e}) \phi_\ell \, d^3x , \tag{68}$$

with $C_{1e} = -\frac{1}{2}(1 - 4\sin^2\theta_W)$, and $C_{2e} = \frac{1}{2}$. Only the cross term proportional to $C_{1e}C_{2e}$ contributes to PNC. Treating this interaction in lowest-order perturbation theory leads to the following correction to the PNC amplitude:

$$E_{\text{PNC}}^{e-e} = \sum_{ai} \frac{d_{wi} \tilde{g}_{iava}^w}{\epsilon_v - \epsilon_a} + \sum_{ai} \frac{d_{iv} \tilde{g}_{wiai}^w}{\epsilon_w - \epsilon_a} + \sum_{am} \frac{d_{am} \tilde{g}_{wmva}^w}{\epsilon_{av} - \epsilon_{mw}} + \sum_{am} \frac{d_{am} \tilde{g}_{wavm}^w}{\epsilon_{aw} - \epsilon_{mv}} . \tag{69}$$

Retaining the cross term only, one obtains

$$\begin{aligned} E_{\text{PNC}}^{e-e} &= -0.0172 \, C_{1e} C_{2e} \times 10^{-11} i|e|a_0(-Q_W/N) \\ &= -0.0003 \times 10^{-11} i|e|a_0(-Q_W/N) . \end{aligned} \tag{70}$$

This small nuclear-spin-independent contribution is masked by the much larger uncertainty in the dominant term (4).

7 Comparison with Experiment

We can now make use of the above analysis to extract the value of the weak charge Q_W from experiment. The PNC amplitudes measured by Noecker, Masterson and Wieman [4] in 1988 are

$$\Im(E_{\rm PNC})/\beta = \begin{cases} -1.639(47)(08) & 4 \to 3 \\ -1.513(49)(08) & 3 \to 4 \end{cases} \tag{71}$$

in units of mV/cm. The quantity β is the vector part of the stark induced polarizability for the $6s \to 7s$ transition in cesium. This quantity has also been calculated with an accuracy of better than 1% using the all-order techniques outlined above, giving $\beta = 27.00(20)a_0^3$. Eliminating the spin-dependence from (71) with the aid of Eq. (66) and using the theoretical value for β, one finds

$$\Im(E_{\rm PNC}^{\rm exp}) = -0.8252(184)[61]10^{-11}|e|a_0\,, \tag{72}$$

where the the first error is from experiment and the second from theory. Combining this result with our calculation of the spin-independent amplitude given in Eq. (5), we obtain

$$Q_W = -71.04(1.58)[0.88]\,. \tag{73}$$

Alternatively, if we use (66) to eliminate the spin-independent terms in (71), we obtain the value

$$K = 0.83 \pm 0.46 \tag{74}$$

for the constant governing the spin-dependent interaction.

Radiative corrections to the weak charge Q_W incorporating a parameterization of new physics beyond the standard model have been worked out by Marciano and Rosner [10], who find

$$Q_W(^{133}\mathrm{Cs}_{55}) = -73.20 - 0.8S - 0.005T \pm 0.13\,, \tag{75}$$

assuming the values $m_t = 140$ GeV for the top quark mass and $m_H = 100$ GeV for the Higgs particle mass. The parameters S and T in Eq. (75) are associated partly with deviations of the top quark and Higgs mass from their assumed values and partly with new physics beyond the standard model. The small factor multiplying T makes this prediction very insensitive to the top quark mass in the absence of new physics. Unfortunately, both the experimental and theoretical errors are presently too large to make atomic PNC in cesium a precision test of the standard model. However, there are two features of cesium PNC that even at the present accuracy lead to particle physics implications. The first is the possibility of a large positive values of S in technicolor theories [28]. A value of S=2 moves the theoretical prediction for Q_W more than 2 experimental standard deviations away from the experimental value. The second is the effect of extra Z bosons, which is not accounted for in Eq. 75. Exchange of new Z's can be shown to be strongly constrained by atomic PNC [29]. Of course there is also the possibility of having both a nonvanishing S parameter together with extra Z's, and in addition perhaps entirely new physics that has not been thought of. Since new physics affects different weak interaction tests differently, it is important to have as many such tests as possible. The value of atomic PNC tests will increase when the next stage of accuracy is reached. The prospects for improvement in the experiment are discussed by Wieman elsewhere in this book. We conclude this chapter with a discussion of the prospect for higher theoretical accuracy.

8 Prospects for Higher Theoretical Accuracy

The many-body problem in atomic physics is quite complicated because of the lack of a small expansion parameter. While MBPT provides an unambiguous set of diagrams, unlike Feynman diagrams for $g-2$, which are accompanied by an extra power of α for each additional order, any order in MBPT contributes nominally at the same level. This is because while an additional order is accompanied by powers of α, compensating inverse powers of α arise from the bound electron propagators. However, the numerical factors of higher order diagrams can be small, and the MBPT expansion can be shown empirically to work. This was shown in Section 3, where the first few orders of MBPT were shown to reproduce experiment at the few percent level. However, to get to higher precision, a more complete set of diagrams must be included. Our approach to this problem is to include as many diagrams as possible, using the all-orders techniques described in Section 4. The hope is that while one can never expect to evaluate all of the diagrams in, say, sixth order, the ones picked up automatically by an all-orders method are the largest ones. A sign that this is the case would be finding good agreement with experiment, though there is of course always the danger that that agreement results simply from cancellations between neglected diagrams. It is for this reason that we monitor as many atomic properties as possible, with the reasoning that such a cancellation would not be expected for all of them. Our present method, when applied to energies, picks up every diagram through third order and a wide class of fourth-order diagrams. However, individual fourth-order diagrams that we miss in our method have been directly evaluated for sodium in Ref. [30], and found to enter at the tenth of a percent level. The next stage of PNC calculations, in which we are now engaged, is including completely the triple excitations discussed in Section 4. When these are included, all fourth-order energy diagrams, along with a very extensive class of higher-order diagrams, will have been accounted for. Comparison with experiment will then allow an empirical determination of the size of diagrams of fifth and higher orders left out by the method. If the agreement with experiment gets below the 0.1% level for energies, and to the 0.1% levels for ordinary matrix elements, it seems likely that the PNC calculation could be trusted at roughly the same level. It is probably not useful to aim at much better than this level because of the nuclear physics uncertainties discussed in Section 6.2. At the present 1% level, cesium PNC is still only playing a qualitative role in testing radiative corrections in the standard model, although it does place limits on new physics such as technicolor or extra Z bosons. However, when the next stage of accuracy in the experimental and theoretical determination of PNC in atomic cesium is reached, this atom is likely to play a significant role in precision tests of the standard model.

9 Acknowledgements

This research of W.R.J. and J.S. is being supported by NSF grant PHY92-04089. Many of the calculations described herein were carried out on the Cray 2 computer at the National Center for Supercomputer Applications. We would like to thank P. Langacker, W. Marciano, and J. Rosner for helpful conversations on the particle physics implications of atomic PNC.

10 References

1. Ya. B. Zel'dovich, Zh. Eksperim. i. Theor. Fiz., **36**, 964, (1959) (transl. Soviet Phys. JETP, **9**, 682 (1959)).
2. M.A. Bouchiat and C.C. Bouchiat, J. Phys. (Paris) **35**, 899 (1974).
3. M.A. Bouchiat, *et al.*, J. Phys. (Paris) **47**, 1709 (1986).
4. M.C. Noecker, B.P. Masterson and C.E. Wieman, Phys. Rev. Lett. **61**, 310 (1988).
5. P.S. Drell and E.D. Commins, Phys. Rev. Lett. **53**, 968 (1984).
6. T.M. Wolfenden, P.E.G. Baird, and P.G.H. Sandars, Europhysics Letters **15**, 731 (1991).
7. T.P. Emmons, J.M. Reeves, and E.N. Fortson, Phys. Rev. Lett. **51**, 2089 (1983).
8. M.J.D. Macpherson, K.P. Zetie, R.B. Warrington, D.N. Stacey, and J.P. Hoare, Phys. Rev. Lett. **67**, 2784 (1991).
9. S.A. Blundell, J. Sapirstein and W.R. Johnson, Phys. Rev. D **45**, 1602 (1992).
10. W. Marciano and J. Rosner, Phys. Rev. Lett. **65**, 2963 (1990): see also the contributions of Marciano and Langacker in this book.
11. V.A. Dzuba, V.V. Flambaum, and O.P. Sushkov, Phys. Lett. A **141**, 147 (1989).
12. Ya.Zel'dovich, Zh. Eksp. Teor. Fiz. **33**, 1531 (1957) [Sov. Phys. JETP **6**, 1184 (1958)].
13. J. Sucher, Phys. Rev. A **30**, 703 (1980).
14. C. deBoor, *A Practical Guide to Splines* (Springer, New York, 1978).
15. W.R. Johnson, M. Idrees, and J. Sapirstein, Phys. Rev. A **35**, 3218 (1987).

16. W.R. Johnson, S.A. Blundell, Z.W. Liu, and J. Sapirstein, Phys. Rev. A **37**, 1395 (1988).
17. S.A. Blundell, W.R. Johnson, and J. Sapirstein, Phys. Rev. A **42**, 3751 (1990).
18. S.A. Blundell, W.R. Johnson, Z.W. Liu, and J. Sapirstein, Phys. Rev. A **39**, 3768 (1989).
19. S.A. Blundell, W.R. Johnson and J. Sapirstein, Phys. Rev. A **43**, 3407 (1991).
20. R. Engfer *et al.*, At. Data Nucl. Data Tables **14**, 479 (1974).
21. M. Brack, C. Guet, and H.-B. Hakansson, Phys. Rep. **123**, 275 (1985).
22. S.J. Pollock, E.N. Fortson, and L. Wilets, Phys. Rev. C **46**, 2587 (1992).
23. V.V.Flambaum, I.B.Khriplovich and O.P.Sushkov, Phys. Lett. **146B**, 367 (1984).
24. P.A. Frantsuzov and I. Khriplovich, Z. Phys. D **7**, 297 (1988).
25. A.Ya.Kraftmakher, Phys. Lett. A **132**, 167 (1988).
26. V.V. Flambaum and I.B. Khriplovich, Zh. Eksp. Teor. Fiz. **89**, 1505 (1985) [Sov. Phys. JETP **62**, 872 (1985)].
27. M.G. Kozlov, Phys. Lett. A **130**, 426 (1988).
28. M.E. Peskin and T. Takeuchi, Phys. Rev. Lett. **65**, 964 (1990).
29. P. Langacker and M. Luo, Phys. Rev. D **45**, 278 (1992).
30. S. Salomonsen and A. Ynnerman, Phys. Rev. A **43**, 88 (1991).

Charged Lepton-Hadron Asymmetries in Fixed Target Experiments

P. A. SOUDER
Department of Physics, Syracuse University
Syracuse, NY 13210

Contents

1. Introduction

The measurement of asymmetries in the scattering of polarized charged leptons from fixed targets has provided unique information about weak neutral currents. Indeed, an experiment measuring parity violation in deep-inelastic scattering from deuterium performed at SLAC[1,2] was central in establishing the Standard Model[3] from among a number of models consistent with the neutral current data then available.

To date, a total of four such experiments have been published, including the one cited above, deep inelastic scattering of muons performed at CERN,[4,5] quasi-elastic scattering from nucleons in ^{9}Be performed at Mainz,[6] and elastic scattering from ^{12}C nuclei performed at MIT-Bates.[7] The data from these experiments test an important prediction of the Standard Model, namely parity violation. The vast majority of data about neutral currents is insensitive to this feature. Apart from the experiments described here, only the atomic physics experiments and the study of asymmetries in $Z \to \tau$ probe the parity-violating aspects of the theory.

1.1. General Considerations

In these polarized lepton experiments, the relevant quantity measured is an asymmetry A of one of two types. The first is a pure polarization asymmetry

$$A = (\sigma_R - \sigma_L)/(\sigma_R + \sigma_L), \tag{1}$$

where $\sigma_L(\sigma_R)$ is the differential cross section for the scattering of electrons with left(right) helicity. This asymmetry is strictly zero if parity is conserved. If, in addition to the helicity, the charge of the lepton is reversed, as is typical for muon beams, the relevant asymmetry is

$$B = (\sigma_L^+ - \sigma_R^-)/(\sigma_L^+ + \sigma_R^-), \tag{2}$$

where $\sigma_L^+(\sigma_R^-)$ is the differential cross section for the scattering of positive(negative) muons with left(right) helicity. This asymmetry contains both C-violating and parity-violating terms.

These symmetry-violating asymmetries result from an interference between the exchange of a photon, ordinary electromagnetic scattering, and the exchange of a Z. Consequently, at low Q^2, the asymmetry is of the form

$$A_e(\text{or } B_e) = k_e Q^2/M_Z^2 \sim 10^{-4} Q^2 (\text{GeV/c})^2, \tag{3}$$

where k_e is a quantity on the order of unity that depends on the details of the theory for a particular experiment. Usually k_e can be expressed in terms of model-independent coupling constants from a four fermion interaction as discussed in detail below. The coupling constants can in turn be calculated from a more complete theory such as the Standard Model or, for example, a speculative model involving extra Z's. The goal of precision tests in this sector is then to measure as many of these constants to as high a precision as possible. This information, together with results from other experiments, may be combined[8–12] in a global analysis to provide information about radiative corrections or to provide evidence for physics beyond the Standard Model.

A big experimental advantage to measuring an asymmetry is many systematic errors can be made to cancel when taking the ratio. Consequently, extremely small asymmetries may be measured. For the experiments discussed here, $A \sim 10^{-2} - 10^{-6}$

The following criteria are relevant in designing a precision asymmetry experiment:

1 Maximize the Q^2 to maximize the measured asymmetry.

2 Choose a reaction with large cross section to accumulate statistics rapidly.

3 Choose a reaction with small backgrounds.

4 Develop techniques to minimize the systematic differences between σ_R and σ_L.

5 Find a reaction where the theoretical prediction is independent of hadronic uncertainties.

Unfortunately, these criteria often conflict with one another. Although the asymmetries increase with Q^2, the cross sections at high Q^2 fall at least as fast as Q^4. Therefore the fractional error $\delta A/A$ in the result is not improved by using higher Q^2. One of the cleanest reactions in terms of theory, elastic scattering from nuclei, must be done at low Q^2, both to reduce backgrounds and to keep the theory reliable. Thus a central theme in this field is balancing these various criteria.

1.2. Choosing Suitable Reactions

Most experimental tests of the weak interactions exploit reactions with small cross sections, namely neutrino interactions, or equivalently, the decay of relatively stable particles such as kaons, muons or neutrons. Asymmetry experiments, in contrast, usually measure the reaction with the largest cross section, usually some type of elastic scattering. Elastic scattering of leptons has the feature that the events occupy a region of phase space with little background. The events may be integrated instead of counted. This allows the accumulation of the unprecedentedly large statistics required in a feasible time. Most experiments in particle physics during the past 50 years have taken advantage of the method of coincidence counting, a very powerful way to isolate interesting signals among a huge background. Parity experiments are among the few exceptions to this trend.

Actually, there are many types of elastic scattering, each characterized by a range of momentum transfer. At $\sqrt{Q^2}$ on the order of eV, the scattering is from the atom as a whole. Around 100 Mev/c, elastic scattering is coherent from the nucleus. Around 1 GeV/c, elastic scattering is from the nucleon or quasielastic scattering from individual nucleons in a nucleus. For $\sqrt{Q^2} \gg 1$ GeV/c, the "elastic" scattering is from individual quarks, a process called deep inelastic scattering. At a given angle, these elastic events represent the highest momentum particles that are copiously produced. Inelastic channels yield lower energy particles and *superelastic* channels, such as elastic scattering as a background to deep inelastic scattering, are negligibly rare due to the exponential falloff of form factors with Q^2. Thus the design of the experiment requires resolution sufficient to reject the nearest inelastic channel and high enough Q^2 to sufficiently suppress superelastic channels.

Although elastic scattering may sound quite restrictive, significant variety is possible. By varying the target (nucleus, nucleon, or quarks) and the kinematic variables, one can achieve sensitivity to different aspects of the theory and measure different combinations of the model-independent coupling constants.

There has been some discussion in the literature of using true inelastic channels, such as the Δ resonance or inelastic levels in carbon nuclei, in order to better isolate various coupling constants or obtain larger asymmetries. However, isolating a single inelastic channel requires a more elaborate apparatus. In addition, radiative tails produce too much background, especially for suppressed transitions which otherwise should have large asymmetries. For these reasons, no experiments exploiting inelastic channels have yet been attempted.

1.3. Muons versus Electrons

For unpolarized electromagnetic scattering, electrons and muons each have their advantages and disadvantages. For polarized beams, there is an additional criterion, namely how quickly and how effectively the helicity can be reversed. If the helicity of the beam can be reversed rapidly *while changing no other beam parameters*, such as intensity or energy, one can eliminate most errors, including target thickness, drifts in the response of the apparatus, and beam and spectrometer energies. Consequently, extremely small asymmetries can be measured.

With polarized electrons, it is possible to approach the ideal of perfect helicity reversal by using photoemission as the electron source. The helicity of the electron beam is determined by the helicity of a laser striking a photocathode, and the helicity of the laser is cleanly controlled by a Pockels Cell.

Reversing the polarization of muons is much more difficult. The normal method is to change the sign of the muons by reversing the magnets in the beam line. This cannot be done rapidly and also causes differences in other beam properties. In addition, particles of different charge interact slightly differently with the apparatus. This results in larger systematic errors. Muon beams, however, have the advantage of having high energy and being able to probe high Q^2, where the asymmetries are large and thus the systematic errors are relatively much less important.

1.4. Theoretical Considerations

Fixed targets, which are comprised primarily of light quarks, are not ideal for rigorous and precise theory. However, by using data from electromagnetic (or charged weak) scattering and by exploiting isospin symmetry, one can sometimes make precise predictions. The reliability of these predictons turns out to be limited in part by the fact that isospin symmetry is not exact. In particular, to change a proton into a neuteron, one exchanges the up quarks for down quarks and visa-versa. However, strange quarks are present at some level in the proton, and one cannot exchange them for charmed quarks in the neutron. This breaks the symmetry, and corrections based on the details of strange quark wavefunctions are required at some level. Recently there has been a great deal of interest in the role that strange quarks play in the nucleon,[13] and their effect on measurements of parity violation in the scattering of polarized electrons.[14] Indeed the major motivation for planned future experiments in this field is to provide information about strange quarks.

Radiative corrections[15–18] provide additional complications. Some terms involve box diagrams with light quarks, which are very difficult to evaluate quantitatively. The significance of these corrections[19] is influenced by the fact that the vector current for the lepton is suppressed by a factor of $(\frac{1}{2} - 2\sin^2\theta_W)$ relative to the axial vector current. Thus the largest terms have an axial lepton current and vector hadron currents. The radiative corrections to these terms, however, have a vector lepton current and are small. These terms are best suited for experimental study. The opposite is true for terms with axial hadronic currents. Their contributions are small because they are multiplied by the vector current of the lepton, yet their radiative corrections involve the axial lepton current and are not suppressed. Thus they are less useful for precision tests but rather appear to be a major nuisance for experiments in the field.

2. Phenomenology of Parity Violation

A good reference point for the analysis of low-energy neutral current experiments is a phenomenological Lagrangian.[20] For the experiments described here, the following terms are relevant:

$$\begin{aligned} L^{PV} = (G_F\sqrt{2})\big(&\bar{l}\gamma_\mu\gamma_5 l[\frac{1}{2}\tilde{\alpha}(\bar{u}\gamma_\mu u - \bar{d}\gamma_\mu d) + \frac{1}{2}\tilde{\gamma}(\bar{u}\gamma_\mu u + \bar{d}\gamma_\mu d)] \\ &+\bar{l}\gamma_\mu l[\frac{1}{2}\tilde{\beta}(\bar{u}\gamma_\mu\gamma_5 u - \bar{d}\gamma_\mu\gamma_5 d) + \frac{1}{2}\tilde{\delta}(\bar{u}\gamma_\mu\gamma_5 u + \bar{d}\gamma_\mu\gamma_5 d)] \\ &+\bar{l}\gamma_\mu\gamma_5 l[h^u_{AA}\bar{u}\gamma_\mu\gamma_5 u + h^d_{AA}\bar{d}\gamma_\mu\gamma_5 d]\big), \end{aligned} \tag{4}$$

where $\tilde{\alpha}$, $\tilde{\gamma}$, $\tilde{\beta}$, $\tilde{\delta}$, h^u_{AA}, and h^d_{AA} are six independent couplings. The first four are for parity-violating terms and the last two are for C-violating terms. In this notation l is the spinor for the lepton and $u(d,s)$ represents the spinor for the up (down, strange) quark. This form, which emphasizes isospin, is convenient for fixed target experiments because of the major role that isospin symmetry plays in the theory. All of the predictons for the experiments may be expressed in terms of these coupling constants. In turn, values for these constants may by computed for any theory. Another common notation in the literature uses couplings to the individual quarks, such as C_{1d} for the axial-electron coupling to the d-quarks. Expressions for the coupling constants in the Standard Model, their values in the Standard Model, and their expressions in terms of the C_{iq}'s are given in Table I.

Table I. Model independent coupling constants. Values are given in terms of the C_{iq} coupling constants, for the Standard Model in terms of ρ and $\sin^2\theta_W$, and numerically in the Standard Model for $\sin^2\theta_W = 0.23$.

Constant	C_{iq}	Standard Model	$\sin^2\theta_W = 0.23$
$\tilde{\alpha}$	$C_{1u} - C_{1d}$	$-\rho(1 - 2\sin^2\theta_W)$	-0.54
$\tilde{\gamma}$	$C_{1u} + C_{1d}$	$\frac{2}{3}\rho\sin^2\theta_W$	0.15
$\tilde{\beta}$	$C_{2u} - C_{2d}$	$-\rho(1 - 4\sin^2\theta_W)$	-0.08
$\tilde{\delta}$	$C_{2u} + C_{2d}$	0.	0.00
C_{1s}	C_{1s}	$\rho(\frac{1}{2} - \frac{2}{3}\sin^2\theta_W)$	0.35
h^u_{AA}		$\frac{1}{2}$	0.50
h^d_{AA}		$-\frac{1}{2}$	-0.50

In addition, there are small contributions from heavy quarks. Additional terms are needed:

$$\begin{aligned} L^{PV} = (G_F\sqrt{2})\big(&\bar{l}\gamma_\mu\gamma_5 l[C_{1s}\bar{s}\gamma_\mu s + C_{1c}\bar{c}\gamma_\mu c] \\ &+\bar{l}\gamma_\mu l[C_{2s}\bar{s}\gamma_\mu\gamma_5 s + C_{2c}\bar{c}\gamma_\mu\gamma_5 c]\big). \end{aligned} \tag{5}$$

For simplicity, charmed quarks are neglected, and the Standard Model predictions for the strange quark coefficients ($C_{1s} = C_{1d}, C_{2s} \sim 0$) are used. Due to the small size of the expected corrections, this approximation should be reasonable.

Any asymmetry will be to first order proportional to a sum of these constants. Of course, computing the radiative corrections requires a complete, renormalizable theory. In the following sections, the theoretical expressions for the asymmetries will be given in terms of these constants. In addition, the dominant theoretical uncertainties in these expressions will be discussed. Although the focus will be on the completed experiments, comments will be made on possibilities for future experiments using the same reactions.

2.1. Deep Inelastic Scattering

The expressions for the asymmetries for experimentally accessible processes are now presented. The first is deep inelastic scattering, i.e., elastic scattering incoherently from the individual quarks. The result[21–27] may be expressed, following the work of Ref. 16

$$A^{DIES} = Q^2 \frac{G_F}{2\sqrt{2}\pi\alpha}\left[a(x) + \frac{1-(1-y)^2}{1+(1-y)^2} b(x)\right] \tag{6}$$

where x and y are the usual scaling variables. The coefficients $a(x)$ and $b(x)$ depend upon quark wavefunctions; in this case their x-distributions are:

$$a(x) = \frac{\sum_i f_i(x) C_{1i} Q_i}{\sum_i f_i(x) Q_i^2} \text{ and } b(x) = \frac{\sum_i f_i(x) C_{2i} Q_i}{\sum_i f_i(x) Q_i^2}, \tag{7}$$

where $f_i(x)$ is the momentum density of quarks of type i and the sum runs over all quarks and antiquarks. Listing the relevant flavors, $u(x)$, $d(x)$, $s(x)$, and $c(x)$ are the quark densities in the *proton* and $\overline{u}(x)$... the antiquark densities. For the neutron, the density of up quarks is $d(x)$ and the density of down quarks is $u(x)$ by isospin symmetry. Unfortunately, the strange quark density for the neutron is still $s(x)$ rather than $c(x)$. If it were the latter, the result would be simpler and independent of the $f_i(x)$. Next, isospin is exploited again by considering a target with the same number of protons as neutrons. For such targets, electromagnetic scattering measures the following combination, which appears in the denominator:

$$D(u) = u(x) + d(x) + \overline{u}(x) + \overline{d}(x) + \frac{2}{5}[s(x) + \overline{s}(x)] + \frac{8}{5}[c(x) + \overline{c}(x)]. \tag{8}$$

Also

$$q(x) = u(x) + d(x) + s(x) + c(x). \tag{9}$$

Then

$$a(x) = \frac{9}{5}\left[\frac{\tilde{\alpha} + \frac{1}{3}\tilde{\gamma}}{2} + \frac{2}{15}\frac{s(x) + \overline{s}(x) - c(x) - \overline{c}(x)}{D}\right] \tag{10}$$

and

$$b(x) = \frac{9}{5}\left[\frac{\tilde{\beta} + \frac{1}{3}\tilde{\delta}}{2}\frac{q(x) - \overline{q}(x)}{D} + \text{heavy quark terms}\right]. \tag{11}$$

There are a number of key features that appear here that are typical of all results:

1 $a(x)$, which arises from vector hadronic currents, is independent of hadron physics when heavy quarks are neglected. This was obtained by exploiting isospin, requiring that the number of protons equals the number of neutrons. The corresponding expression for the proton, for example, requires knowledge of $u(x)$ and $d(x)$ separately.

2 $a(x)$ would have no contribution from heavy quark momentum distributions if $s(x) = c(x)$. The large mass difference between these quarks breaks isospin symmetry and complicates the results.

3 The axial vector hadronic current term, $b(x)$, is more complex. It involves a special structure function $xG_3(x)$ proportional to $q(x) - \bar{q}(x)$. In the Standard Model, it is related to xW_3, which may be measured in neutrino scattering. This problem is ameleorated by the fact that the weak coupling is small in this case, $\sim 1 - 4\sin^2\theta_W$, but is typical of problems with axial hadronic currents.

4 The coefficient of the $s(x)$ term in $a(x)$ is actually the complicated expression $(5C_{1s} + \frac{3}{2}\tilde{\alpha} + \frac{1}{2}\tilde{\gamma})$. For simplicity, it is set to the Standard Model value of unity. The presence of light quark coupling constants in this coefficient arises from the fact that the electromagnetic denominator $D(x)$ contains contributions from strange quarks that must be subtracted. The presence of $\tilde{\alpha}$ and $\tilde{\gamma}$ make the $s(x)$ coefficient independent of $\sin^2\theta_W$. A similar cancellation also occurs for elastic scattering. Based on neutrino data involving dimuons, strange quarks are estimated to be present at about the 5% level.

Although the above expression assumes the parton model, it should be quite accurate, especially at low y.[28–31] At low Q^2, there may be additional terms due to correlations between up and down quarks. It is estimated that these correlations change the results of references 1 and 2 by at most 6%. In addition, the $1 + (1-y)^2$ term in the denominator of the $b(x)$ coefficient in the A^{DIES} expression should be $2[1 - y + \frac{1}{2}y^2/(1+R)]$, where R is the ratio of cross sections σ_L/σ_T for longitudinal versus transverse photons. For the $b(x)$ term, $[q(x) - \bar{q}(x)]/D = 0.76$ is appropriate for the SLAC data. An overall 7% error for the uncertainty in the theoretical prediction for the asymmetry, which turns out to be comparable to the experimental errors, will be used.

For muon scattering where the charge as well as helicity is reversed,

$$B = -Q^2 \frac{3G_F}{10\sqrt{2}\pi\alpha} \left[\frac{1-(1-y)^2}{1+(1-y)^2} \left(2h^u_{AA} - h^d_{AA} + \frac{3}{2}|P|(\tilde{\beta} + \frac{1}{3}\tilde{\delta}) \right) \frac{q(x) - \bar{q}(x)}{D} \right] \quad (12)$$

where $|P|$ is the beam polarization. Muon-electron universality is also assumed for simplicity.

This expression is identical to Eq. (6), except that reversing the sign of the lepton cancels the $a(x)$ term and adds the C-violating coupling constants to the $b(x)$ term. For the purpose of the analysis here, $[q(x) - \bar{q}(x)]/D = 0.93$ is used. The uncertainty in the theory should be small compared to the quoted experimental errors for the CERN data.

2.2. Elastic Scattering from the Nucleon

The most general form for a neutral vector and axial vector coupling to the proton is given by

$$\langle p|J^Z_\mu|p\rangle = \overline{U}\left(\gamma_\mu F^Z_{1p} + \frac{i\sigma_{\mu\nu}q^\nu}{2M_N}F^Z_{2p} + \gamma_\mu\gamma_5 G_A\right)U \tag{13}$$

where U is the nucleon spinor. This is analogous to the electromagnetic current, where the form factors are F_{1p} and F_{2p} and the axial vector piece is missing.

By assuming CVC and isospin invariance in the n-p system, the weak neutral form factors can be related to known form factors from electromagnetic scattering.[32–35] A useful form[36] for the asymmetry is given by

$$\begin{aligned} A^{eP} = 3.167\times 10^{-4}\tau\Bigg[&(3\tilde{\gamma}-\tilde{\alpha})\left(\frac{\varepsilon G_{Ep}G_{En}+\tau G_{Mp}G_{Mn}}{\varepsilon G^2_{Ep}+\tau G^2_{Mp}}\right)\\ &+(3\tilde{\gamma}+\tilde{\alpha})\left(\quad 1 \quad\right)\\ &+(\tilde{\beta}+\frac{3}{5}\tilde{\delta})\left(\frac{\sqrt{1-\varepsilon^2}\sqrt{\tau(\tau+1)}G_{Mp}G_A}{\varepsilon G^2_{Ep}+\tau G^2_{Mp}}\right)\\ &+\left(\frac{\varepsilon G_{Ep}G_{Es}+\tau G_{Mp}G_{Ms}}{\varepsilon G^2_{Ep}+\tau G^2_{Mp}}\right)\Bigg] \end{aligned} \tag{14}$$

where $\varepsilon = \left(1+2(1+\tau)\tan^2(\frac{\theta}{2})\right)^{-1}$, $G_E = F_1 - \tau F_2$, $G_M = F_1 + F_2$, and $\tau = Q^2/4M_p$. The coefficient of $\tilde{\delta}$ is estimated by using SU(6).[34] G_{Ms} and G_{Es} are new form factors describing the contributions from strange quarks.[13] Strange quarks in the axial current, as well as the isoscalar part of G_A, whose effects on A^{eP} are suppressed in the Standard Model, are neglected for simplicity. The expression for the neutron is the same except that the subscripts p and n are interchanged and $\tilde{\beta} \to -\tilde{\beta}$.

The asymmetry for the nucleon given above is quite complex, and there are a number of important limitations relevant to using it to test the Standard Model.[37] Some of the issues are:

1 Experimental precision of the electromagnetic nucleon form factors, especially G_{En} and G_{Mn}. Present data are inconsistent and the analyses are model dependent. However, a number of new experiments have been proposed for Mainz and CEBAF which promise to solve this problem in the future.

2 Contributions from strange quarks. Instead of just the momentum distributions $s(x)$ and $\bar{s}(x)$, the matrix elements $\bar{s}\gamma_\mu s$ (for G_{Es}), $\bar{s}\gamma_\mu\gamma_5 s$ (for G_{As}) and $\bar{s}\sigma_{\mu\nu}q^\nu s$

(for G_{Ms}) are required. There are a number of speculative predictions about the sizes of these effects[38–40] that would make significant contributions to the asymmetry. Whether or not these terms are important must be determined experimentally.

3 The uncertainty in the radiative correction to G_A is large, about 20% of itself. At backward angles and moderately small Q^2 values, where this term is most significant, it contributes about 20% to the asymmetry.

One can avoid all of the above problems by working at low Q^2 at forward angles so that the second term in eq (14), where all of the form factors cancel, dominates. Unfortunately, in this region, the experimental figure of merit is well below the maximum, and the kinematics is sufficiently unusual that dedicated spectrometers would have to be built.

Experiments at reasonable Q^2 values, from 0.1-1.5 $(\mathrm{GeV})^2$, at both forward and backward angles, have been proposed. These kinematics have been chosen to maximize the possibility of measuring the strange quark form factors, and thus are unlikely to provide a precision test of the Standard Model.

The experiment performed on ^{9}Be at Mainz measured quasielastic scattering, which, to first approximation, is just the incoherent sum of the elastic asymmetries given above from the proton and neutron. The validity of this approximation has been discussed for the simple case of the deuteron[41] and for other light nuclei.[42] The uncertainties arising from errors in form factors and from lack of knowledge of the effects of strange quarks tend to cancel for this case, and could change the asymmetry by at most a few percent. Uncertainties in the corrections for Fermi motion are also small. Effects which arise from meson exchange currents, off-shell effects, and the possibility of unknown processes hidden under the quasi-elastic peak are more difficult to estimate and are possibly larger. The fact that the experiment at Mainz integrated over a large kinematic range adds to the problem. An overall theoretical uncertainty of 10% on the value for A is probably appropriate for the Mainz experiment.

2.3. Elastic scattering from a Nucleus

The asymmetry for elastic electron scattering from nuclei with spin and isospin equal to zero is given by[43,44]

$$A^{eC} = \frac{3}{2} G_F Q^2 (\sqrt{2}\pi\alpha)^{-1} \left[\tilde{\gamma} + \frac{1}{3} \frac{G_E^s}{G_{Ep} + G_{En}}\right], \tag{15}$$

The result, proportional to $(\sum_i C_{1i})(\sum_i Q_i)$, as is appropriate for coherent scattering from all the quarks, differs from the deep inelastic result, which involves the incoherent sum $\sum_i C_{1i} Q_i$ which appears in Eq. (7). A remarkable feature of this result is that if the strange quarks are neglected, the asymmetry is independent of form factors.

This simple result is expected to be quite reliable at low Q^2. Uncertainties due to parity admixtures in nuclear states and isospin mixing are less than 1% at appropriate kinematics. The strange quark form factor G_E^s,[47] is unknown but should be proportional to Q^2 at low Q^2. Speculative models suggest that it may contribute as much as $\sim$ 20% at Q^2=0.1 (GeV/c)^2. Dispersive radiative corrections,[48] involving the exhange of both a photon and a Z, have not been calculated but may appear at the few percent level. Experiments measuring parity violation in atoms, especially Cs, are also sensitive mainly to $\tilde{\gamma}$ and have been performed with greater precision than electron scattering experiments.[49–51]

3. Description of the Experiments

The experimental methods used for the three electron experiments are similar to each other. I will discuss the common features of these experiments first, and then discuss features unique to each. The muon experiment, which measures a much larger effect, has different features and will be discussed separately.

3.1. Parity Experiments with Electrons

The electron parity experiments have the following elements: a polarized electron source with special attention paid to how the helicity is reversed, a pulsed accelerator, a well instrumented beam line which can detect any helicity correlations in beam parameters, a spectrometer to detect the scattered particles, special electronics to handle the high rates, and a polarimeter to measure the electron polarization.

The polarized electron sources are all based on photoemission from a crystal such as GaAs. These sources, which can provide large beam currents comparable to what can be obtained from a conventional thermionic source, have electron polarizations P on the order of 40%. Thus the experimental asymmetry is reduced by the factor:

$$A_{exp} = PA. \tag{16}$$

The helicity of the beam is controlled by the helicity of a laser beam, which is in turn controlled by an electro-optical device called a Pockels cell. Changing the voltage applied to the Pockels cell reverses the helicity without changing any other beam parameters to first order.

Given the high precision of the parity experiments, however, there are small but important problems with the helicity reversal of the source. A major concern is a helicity correlated intensity difference. This may be caused by the fact that the laser light is not purely in helicity states, but has small linear components. These induce intensity asymmetries due to the linear polarization dependence of reflection and transmission coefficients of optical elements in the laser transport line. This is denoted the Polarization Induced Transport Asymmetry (PITA) effect. The resulting intensity asymmetries may be removed from the data with intensity monitors with sufficient precision. However, due to the beam loading of the accelerator, the intensity asymmetry induces an energy asymmetry. Since electromagnetic cross sections are highly energy dependent, an energy asymmetry is quite serious. In addition, the beam transport lines are not perfectly achromatic, so the energy asymmetries also induce asymmetries in other beam parameters such as the position of the beam.

The tiny differences that do result in the beam parameters can be accurately measured with precision microwave position monitors. Their noise is on the order of 10μ per pulse. When placed at appropriate locations along the beam transport line, including one with reasonable momentum dispersion, they can determine the five first-order beam differences, including position (x and y), angle (θ_x and θ_y), and energy with ample precision. They can either be used to set limits on systematic errors, or, if necessary, be accurately calibrated to make quantitative corrections to the asymmetries A_{raw}.

Given the high rates required for these experiments, all of the data were obtained by integrating the analog signals instead of counting individual events. When integrating, the ratio of average pulse height h of the RMS spread σ to the pulse heights influences the statistical error

$$\delta A/A = (1/\sqrt{N})\sqrt{1 + (\frac{\sigma}{h}^2)} \tag{17}$$

where N is the number of detected events. The $\sqrt{1/N}$ factor is the same as for counting experiments. The second factor represents the cost of the information lost due to integrating. For example, a detector such as a Čerenkov counter which averages n photoelectrons per event with Poisson statistics has $(\sigma/h)^2 = 1/n$. If n is reasonably large, the second term is negligible, and the error is mainly that due to the number of counts. Additional contributions to the error, such as electronic noise or shower fluctuations, can usually be kept small. An additional requirement is that the contribution to the integrated signal from background processes be small. Čerenkov counters, which are sensitive mainly to electrons, are ideal. Depending on the kinematics, gas, plastic or lead glass shower Čerenkov counters may be used. Thin plastic scintillators, which have a large Landau tail and thus a large value for $(\sigma/h)^2$, are unsuitable.

The polarization of the beam is measured by Møller (elastic electron-electron) scattering. A thin ferromagnetic foil provides a target of $\sim 6\%$ polarized electrons. Elastic scattering from these electrons at 90° in the center of mass frame has the characteristic in the lab of half energy electrons scattered at small angles. Their unique kinematics make them easy to detect; a major background comes from radiative elastic scattering from the nuclei in the target. The polarization of the beam can be measured with a precision of about 5% with this method. Limitations include backgrounds, measurement of the magnetization of the foils, and uncertainties in how the magnetization relates to spin polarization (as opposed to orbital angular momentum) of the target electrons. Improvements in measuring the polarization of an electron beam are one prerequisite for major advances in this field.

3.2. The SLAC Experiment

The first and the highest energy electron experiment was performed at SLAC. It measured the asymmetry for deep inelastic scattering from deuterons. The energy of the beam was varied between 16 and 22 GeV. A schematic diagram of the apparatus is given in Fig. 1. The experiment used a single arm magnetic spectrometer comprised of two dipoles and a quadrupole, oriented to have a central angle acceptance of 4°. Sufficient kinematic range was obtained by varying the magnets and also by segmenting the detector into two parts. The Q^2 range was 0.92-1.96 $(\mathrm{GeV/c})^2$. The kinematics were chosen as a compromise among the following requirements:

1 Maximize the figure of merit.

2 Keep the $Q^2 > 1$ $(\mathrm{GeV/c})^2$ so that the theory for deep inelastic scattering applies.

3 Reduce the pion contamination to an acceptable level.

4 Obtain as much range in y as possible to extract both constants from Eq. (6).

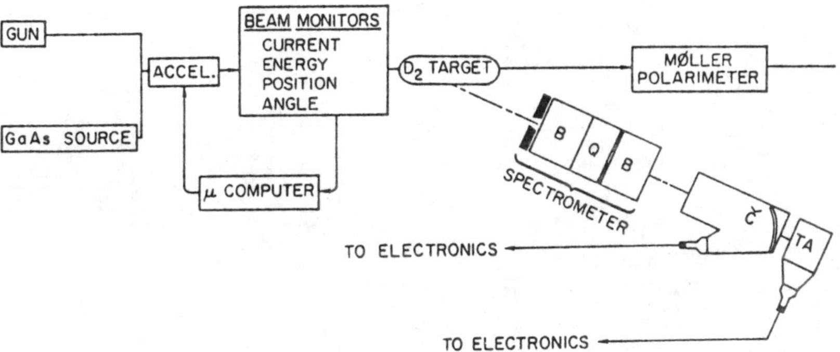

Fig. 1. Schematic layout of the SLAC experiment. Polarized electrons from the GaAs source are accelerated by the linac, analyzed and stablized in a computer-controlled beam transport system, and scattered from a liquid D_2 target. Electrons scattered at 4° are analyzed by a spectrometer and detected by a Čerenkov counter and a total absorption (TA) counter.

The primary detector was a lead glass shower detector. The beam was pulsed, with 120 1.5μs pulses per second. Typically 1000 electrons were detected per pulse. A gas Čerenkov counter provided an independent measure of the asymmetry, but lacked segmentation. Pion contamination was negligible ($< 6\%$) at all but the most extreme kinematic points, where it reached 24%. The target was a 30-cm long liquid deuterium cell, a nontrivial piece of apparatus when operated at high beam current. The experimental asymmetries were on the order of 100 ppm.

3.3. Quasielastic Scattering at Mainz

The parity experiment at Mainz measured quasi-elastic scattering from the nucleons in a Be target in the backward direction. It was designed to provide information about $\tilde{\beta}$, although it is also sensitive to $\tilde{\alpha}$ and $\tilde{\gamma}$. The beam was about 7μA of 300 MeV electrons, and the average scattering angle was 130°. Taking advantage of the fact that kinematics tend to vary slowly at angles much greater than 90°, the experiment featured an impressive solid angle > 2sr as shown in Fig. 2. The accelerator produced 50 3.5μs pulses per second. The target was Beryllium, which is mechanically convenient and also rich in neutrons to enhance the asymmetry. A total of 220 hours of beam on target were obtained.

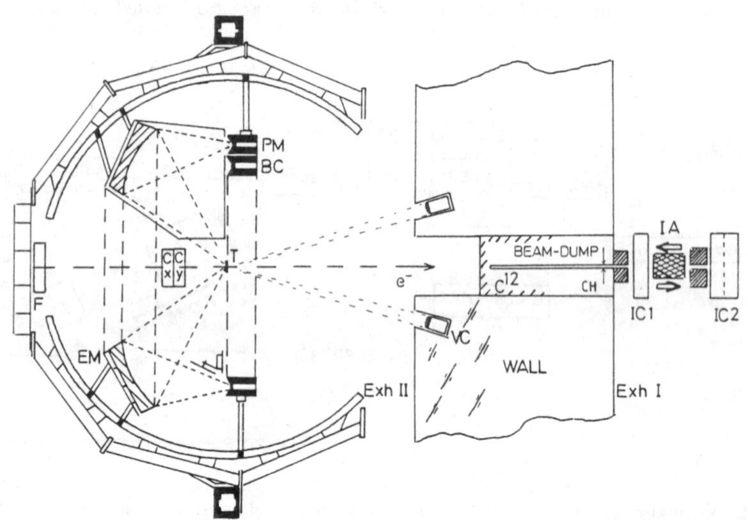

Fig. 2. Side view of the Mainz detector system and the Compton polarimeter. The detector consists of : EM – ellipsoidal mirrors; PM–phototubes; BC–background counters; VC–forward angle lucite detectors; T–2.4 g/cm^2 ^{9}Be target. Beam monitors are: F–ferrite; C_x and C_y–microwave position monitors. The Compton polarimeter consists of: IC1, IC2–ionization chambers; IA–magnetized iron absorber.

The detectors were imaging gas Čerenkov counters; indeed they used the air in the room as the radiating medium. The Čerenkov light was collected by large mirrors and focussed onto phototubes. The energy cutoff of the counters was ~25 MeV, so events from unwanted processes also contributed to the signal. From the radiative tail of elastic scattering from the nucleus as a whole came 22% of the events. Production of the Δ-resonance contributed 6.5%. The "dip region" between the quasielastic peak and the beginning of the Δ contributed 12.5% of the events. The raw asymmetry was on the order of 4 ppm, much smaller than for the SLAC experiment.

The experiment had a number of special features. The photoemitting crystal was GaAsP instead of GaAs. This allowed for high polarization (~44%) with a convenient 643 nm laser. The output of the pulsed laser was controlled by a Pockels cell feedback system that produced a square pulse with uniform intensity and reduced helicity-correlated intensity. The polarization was monitored continuously with a Compton polarimeter that was calibrated periodically with Møller scattering. This is instead of the usual procedure of just taking periodic Møller data.

3.4. Elastic Scattering from ^{12}C at Bates

At the MIT-Bates Linear Accelerator Center, 250 MeV electons were scattered elastically from ^{12}C. The motivation for this experiment was to measure the isoscalar coupling constant $\tilde{\gamma}$ with minimal theoretical ambiguity. The experiment was performed at a Q^2 of 0.0225 (GeV/c)^2, which was a compromise between maximizing the Q^2 to maximize the asymmetry yet keeping Q^2 small enough to keep the elastic form factor large.

Scattered electrons were collected by a pair of single quadrupole spectrometers as shown in Fig. 3. The spectrometers, oriented at $\sim 35°$ to the incident beam and each with a 15 msr solid angle acceptance, focussed elastically scattered electrons onto slabs of lucite which served as Čerenkov counters. About 10^5 electrons were detected each pulse. The resolution was such that events losing more that ~15 MeV were rejected by the spectrometers. Inelastic events within the acceptance of the spectrometer (the first excited state in carbon is at 4 MeV) have a small cross section at the chosen kinematics and also were expected to have an asymmetry similar to that of elastic scattering.

The accelerator at Bates operated with a $\sim$ 1% duty factor; 600 Hz of $\sim$ 16μs pulses. To get a laser to operate with this duty factor, a cw Kr-ion laser was chopped. The average current on target, 30-60μA, was about an order of magnitude larger than for the other experiments. This combination placed a premium in the quantum efficiency of the photocathode, and great effort was taken to maintain the current without unacceptably shortening the lifetime of the photocathode. Although chopping a cw laser resulted in low laser power, it did result in a stable beam. The measured asymmetry was 0.6 ppm, more than two orders of magnitude smaller than those of the SLAC experiment.

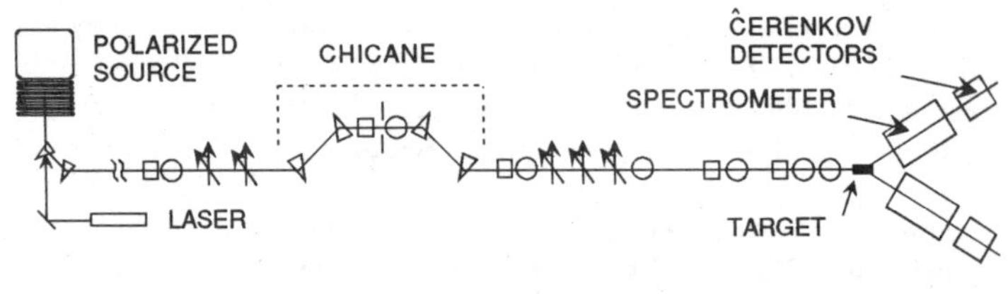

Fig. 3. Schematic diagram of the apparatus at Bates. Emphasized is the beamline instrumentation, which is typical of all of the electron experiments.

3.5. Muon Scattering at CERN

The experiment measuring the electroweak asymmetry in deep inelastic scattering of muons from carbon at CERN was run with two beam energy settings, 120 and 200 GeV. The range in Q^2 was between 15 and 180 $(\mathrm{GeV/c})^2$, with the most sensitive data coming at $Q^2 \sim 50$ $(\mathrm{GeV/c})^2$. The range of y was 0.2-0.85. The beam is naturally polarized in its normal operating mode where forward decay muons are selected. Monte Carlo simulations give $|P| = 0.81 \pm 0.04$ at 200 GeV and 0.66 ± 0.05 at 120 GeV. The 200 GeV calculation was checked by measuring the energy spectrum of the electrons from beam muon decays.

The apparatus, which is shown in Fig. 4, was massive when compared to the electron experiments. The muons were detected with a set of ten saturated iron toroids surrounding a 40-m long carbon target. The toroids were 5 m long and 2.75 m in diameter. The detectors included trigger counters and multiwire proportional counters to determine the trajectories.

Not only were individual scattered events tracked and counted, but even the individual incident beam particles were tracked, quite a contrast from electron experiments. Their momenta were individually measured with a hodoscope positioned after a precisely monitored bending magnet in the beam line. The sign and helicity of the beam were reversed every six days, some eight orders of magnitude more slowly than is done for electron beams. On the other hand, the large Q^2 values, typically 50 $(\mathrm{GeV/c})^2$, result in asymmetries on the order of $\frac{1}{2}\%$, two to four orders of magnitude larger than for electron experiments. The data consisted of three runs, two at 120 GeV and one at 200 GeV. Each run was binned into six Q^2 points which had statistical errors on the order of 0.15-0.30%.

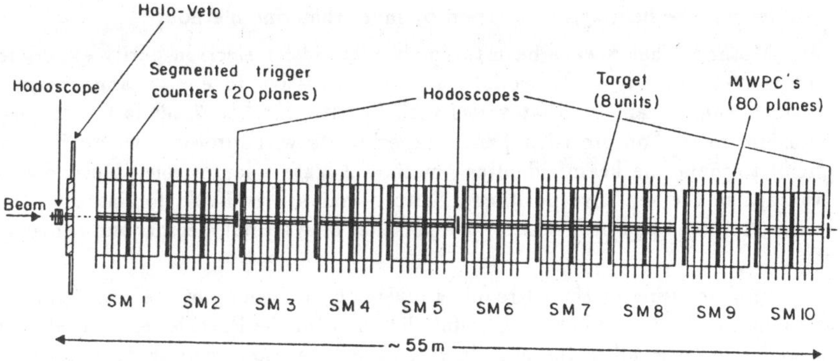

Fig. 4. Schematic layout of the CERN apparatus. Shown are the ten 5-m long saturated iron toroids surrounding the carbon target and instrumented with trigger counters and multiwire proportional chambers.

4. Systematic Errors

The systematic errors in an asymmetry experiment may be divided into two types: *difference* errors that influence data of different helicity differently and *scale* errors that multiply the overall asymmetry. A typical difference error results from a difference in the energy of the + and − beams. Difference errors limit how small an asymmetry can be measured. Scale errors, on the other hand, become important only when the asymmetry is statistically different from zero. A typical scale error is the error in the measurement of the polarization of the beam.

The electron experiments, with their very small asymmetries, have stringent requirements on reducing difference errors. However, the technique of rapidly reversing the beam helicity is effective in dealing with this problem, and it turns out that the largest systematic errors are often the scale errors.

4.1. Electrons

The requirements on systematic errors for the electron experiments are orders of magnitude more stringent than those for standard cross section experiments or even for the muon asymmetry experiment. As a consequence, the methods are unique.

There are four basic methods to reduce systematic errors in an asymmetry experiment:

1 Make the beams of different helicity as identical as possible.

2 Make the apparatus insensitive to the differences in beam parameters.

3 Precisely measure the differences in beam parameters as well as the effects of these differences on the asymmetry and make acurate corrections.

4 Reverse the helicity of the beam by more than one method.

Method 1 has always been the main method for electron parity experiments. Since electromagnetic cross sections are very sensitive to energy and energy is a common beam parameter that varies with helicity, method 2 tends not to play an important role. For precision parity experiments with protons, where it is more difficult to make the beams identical, method 2 plays a larger role. Method 3 takes advantage of the sensitive position monitors that are available for intense electron beams. Method 4 can both test for the presence of systematic errors and cancel them out.

Implementing method 1 requires reversing the helicity of the laser beam without changing other parameters. Careful alignment of the Pockels cell is needed to do this well. However, all of the experiments have observed a helicity-dependent intensity which is difficult to reduce below the 10 ppm level. This is presumably caused by the PITA effect mentioned above, in which the transmission of the laser transport line becomes helicity-dependent due to imperfect circular polarization. Insertion of a half-wave plate, one of the commonest ways to implement method 4, makes the problem worse. In addition, the effect tends to drift slowly with time. Finally, the intensity jitter in the beam makes it difficult to measure the effect rapidly. As a consequence, managing the PITA effect is one of the challenges for these experiments.

An independent problem is that the Pockels cell may deflect the laser beam slightly. This may be reduced by positioning the laser beam at an optimal position on the Pockels cell and also by using a lens to focus the Pockels cell onto the photocathode. Then, even if there is a deflection, it will not effect the position on the cathode.

For the experiment at Bates, which measured the smallest asymmetry, an effective method using a kind of feedback was developed to reduce the PITA effect. It was observed that the PITA effect depended linearly on a voltage difference ΔV_0 with a slope that could be measured. Data were obtained for three minutes, and the size of the PITA effect for that sample was computed. Then ΔV_0 was changed by the appropriate amount to cancel the effect for the next three minutes. By continuing in this way, not only was the PITA effect removed, but also the noise due to fluctuations in the laser intensity were removed. This procedure reduced the PITA effect to less than 1 ppm.

The implementation of method 3 is rather elaborate. The differences in the five first-order beam parameters are detected by a set of five beam monitor differences δM_i, which must used to compute a correction to the cross section $\delta\sigma$. Thus

$$\delta\sigma = \sum_{i=1}^{5} K_i \delta M_i. \qquad (18)$$

The problem is to compute the calibration coefficients K_i. At Bates, data were taken while beam steering coils C_i were being ramped, and the quantities $\partial\sigma/\partial C_i$ and

$\partial M_i/\partial C_j$ were measured. Then

$$K_i = \frac{\partial \sigma}{\partial M_i} = \frac{\partial \sigma}{\partial C_j}\frac{\partial C_j}{\partial M_i} \tag{19}$$

which is obtained by inverting the matrix $\partial M_i/\partial C_j$. Thus the constants K_i are obtained simultaneously with the data and apply for the exact running conditions. The corrections were typically smaller than the statistical errors. The average correction turned out to be 0.040 ppm with an estimated error of 0.006 ppm.

For the Mainz data, a correlation analysis was used. It was found that all of the monitors were correlated, so that all could be expressed in terms of a single monitor δM_1. The correlation $\partial\sigma/\partial M_1$ was measured and the data were corrected

$$\delta\sigma = \frac{\partial \sigma}{\partial M_1}\delta M_1. \tag{20}$$

For the other monitors, it was observed that

$$\delta M_i = \frac{\partial M_i}{\partial M_1}\delta M_1. \tag{21}$$

This is not surprising since there were ~ 10 ppm asymmetries in the beam intensity, and the dominant noise in the accelerator was presumably intensity or energy. This method would not work if there were a second independent and large beam difference, such as a position difference caused by a deflection by the Pockels cell. Typical corrections were < 1 ppm and only slightly larger than the statistical errors. For the data at SLAC, the corresponding corrections were on the order of 2.5% of the asymmetry for each of the 11 data points taken and were thus negligible.

There are a number of other small effects that might create a false asymmetry. Among them are electronic cross-talk between the helicity controlling electronics and the sensitive integrating electronics, asymmetries due to transverse components of the beam polrization, nonlinearities in the electronics coupled to intensity asymmetries, spin-dependent backgrounds from polarized electrons striking magnetized iron, and differences in higher order beam parameters. With care, all of these can be kept to a negligible level.

A powerful method (# 4 above) to detect and eliminate systematic errors is to use an independent method of reversing the helicity of the beam. The most common method, used in all of the electron experiments, is to insert a half-wave plate in the laser beam. This reverses the helicity of the beam, reversing the sign of the parity asymmetry leaving the systematic effects unchanged. Ideally, one has asymmetries

$A_1(A_2)$ for setting 1(2) with:

$$A_1 = A^{PV} + A_{sys}; \quad A_2 = -A^{PV} + A_{sys}. \tag{22}$$

Then

$$A^{PV} = \frac{1}{2}(A_1 - A_2); \quad A_{sys} = \frac{1}{2}(A_1 + A_2). \tag{23}$$

This works well for some systematic effects such as electronic cross talk. The systematic error can be measured to be small, and any possible effect cancells in the average. Unfortunately, inserting a half wave plate changes the PITA effect in a complicated way, and the cancellation of systematic errors induced by the PITA effect is probably imperfect.

At SLAC, an addititional and more nearly ideal method of helicity reversal is possible. The electron spin direction precesses relative to the momentum direction in a bend of $\theta_{bend} = 24.5°$ in the beam switchyard by an angle

$$\theta_{prec} = \frac{E}{m_e c^2}\frac{g-2}{2}\theta_{bend} = \frac{E(GeV)}{3.237}\pi \mathrm{rad}, \tag{24}$$

where E is the beam energy and m_e is the mass and g is the gyromagnetic ratio of the electron. Thus

$$A_{exp} = PA\cos[(E(GeV)/3.237)\pi] \tag{25}$$

may be nulled or reversed by running at the appropriate energies. The SLAC data behaved perfectly under this reversal, lending great credence to their results.

4.2. Muons

The asymmetry B requires that the charge of the beam be reversed as well as the helicity. This cannot be done rapidly, so there is a premium on maintaining the stability of the response of the apparatus with time. The beam reversal was achieved by reversing all of the magnets in the experiment, both for the beam and detector. Thus the two polarities were studied under almost identical conditions. By using the same apparatus for both μ^+ and μ^- beams, many of the possible systematic errors can be made to cancel when computing the asymmetry. Included are errors in absolute factors such as target thickness, solid angle, and magnetic field.

Measuring the beam intensity with the desired precision proved to be difficult. The rates for both beams were kept at 2×10^7 per spill, as nearly identical as possible to cancel errors from deadtime loses, etc. Nonetheless, there was as estimated uncertainty of 0.4% in the relative normalization, which is comparable to the expected asymmetry. To eliminate this problem, the data were fit to the form

$$B = a + bg(y)Q^2 \tag{26}$$

where $g(y) = [1 - (1-y)^2]/[1 + (1-y)^2]$. The quantity a was found to be zero to within the statisitcal error of 0.2%. However, the fact that this parameter had to be included did cause a significant loss of statistical precision in b over what the result would have been if the data were precisely normalized.

Systematic errors due to differences in the magnetic fields, differences in the beam phase space and halo, and differences in the interactions between μ^+ and μ^- with the material in the apparatus caused an error Δb of $0.02(0.03) \times 10^{-3}(\mathrm{GeV/c})^{-2}$ for the 200(100) GeV data. These are about half the size of the statistical errors. An internal test of the validity of the data was studying the x-dependence of the asymmetry. It was observed to be almost flat, as expected in the Standard Model, but would vary strongly if there were errors in the beam energy or spectometer field as large as 0.3%.

5. Results

The results of the four experiments are summarized in Table II. For comparison, they are presented both in terms of A and A/Q^2. For the deep inelastic experiments, where there is a wide kinematic range, A is given as an average $\langle A/Q^2 \rangle \times \overline{Q}^2$. A, which is the quantity relevant to the experimenter, varies by almost four orders of magnitude, whereas $|A/Q^2|$ happens to be almost constant.

Table II. Measured asymmetries. For the CERN data $\langle g(y) \rangle \sim \frac{1}{2}$ and $\overline{Q}^2 \sim 50(\mathrm{GeV/c})^2$. Also given is the value of $\sin^2\theta_W$ obtained from the experiments.

Experiment	Q^2 $(\mathrm{GeV/c})^2$	A (ppm)	A/Q^2 (ppm)/$(\mathrm{GeV/c})^2$	$\sin^2\theta_W$
SLAC	1.4	$119 \pm 9 \pm 6$	$-85.0 \pm 6.2 \pm 4.2$	$0.224 \pm 0.012 \pm 0.08$
CERN	~ 50	$-7600 \pm 1700 \pm 800$	$-76 \pm 16 \pm 8$	$0.23 \pm 0.07 \pm 0.04$
Mainz	0.17	$-9.4 \pm 1.8 \pm 0.5$	$-55 \pm 11 \pm 3$	$0.221 \pm 0.014 \pm 0.004$
Bates	0.023	$1.62 \pm 0.38 \pm 0.08$	$72 \pm 14 \pm \pm 3$	$0.204 \pm 0.048 \pm 0.014$

Errors are quoted as statistical followed by systematic, except for the SLAC data, where some systematic errors are included in the first error as mentioned above, and the second error is the systematic error arising from Møller scattering. Also given is the value for $\sin^2\theta_W$ obtained by the authors by analyzing each experiment in terms of the Standard Model. The theoretical uncertainties in $\sin^2\theta_W$ are smaller than the quoted experimental errors and have not been included.

Perhaps the best use of the results of the four above experiments is to include them in a global fit as mentioned above. However, here the data from these experiments alone will be used to extract values for the relevant model independent coupling constants. The results of the experiments listed in Table II may be expressed in terms of the coupling constants, as given in Table III. Theory uncertainties are neglected for the Bates and CERN data, and are 7% for the SLAC data and 10% for the Mainz data. The statistical, systematic, and theory uncertainties are added in quadrature. To include antiquarks in the analysis for the deep inelastic scattering $[q(x) - \bar{q}(x)]/D = 0.76$ for the SLAC data and 0.93 for the CERN data.

Table III. Model independent parameters measured by the experiments. For the CERN data, the first(second) line is the 120(200) GeV data.

Experiment	Combination of Parameters	Result
SLAC	$\tilde{\alpha} + 0.33\tilde{\gamma} + 0.19\tilde{\beta} + 0.06\tilde{\delta}$	-0.52 ± 0.06
	$\tilde{\beta} + 0.33\tilde{\delta}$	$+0.41 \pm 0.66$
	$\tilde{\alpha} + 0.33\tilde{\gamma}$	-0.60 ± 0.16
CERN	$h^u_{AA} - \frac{1}{2}h^d_{AA} + 0.50(\tilde{\beta} + 0.33\tilde{\delta})$	$+0.87 \pm 0.40$
	$h^u_{AA} - \frac{1}{2}h^d_{AA} + 0.61(\tilde{\beta} + 0.33\tilde{\delta})$	$+0.73 \pm 0.20$
Mainz	$\tilde{\beta} + 0.80\tilde{\alpha} + 0.49\tilde{\gamma} + 0.04\tilde{\delta}$	-0.45 ± 0.10
Bates	$\tilde{\gamma}$	$+0.14 \pm 0.03$

For the SLAC experiment, the first line is the average of all the data. The next two lines are a fit of the data to $a(x)$ and $b(x)$ as done by the authors. This latter method gives two independent, although correlated numbers. When the experiment was performed, a vital issue was if $a(x)$ was zero; this possibility was clearly ruled out by the data. For the present analysis, the first two entries, which are uncorrelated, are more convenient to use.

The results of fits to the constants are presented in Table IV. By using the two parameter fit of the SLAC data, values for each of the constants are obtained. However, the result for $\tilde{\delta}$ has a very large error which makes it of little value. The fact is that none of these experiments are very sensitive to $\tilde{\delta}$. Even if $\tilde{\delta}$ were even as large as $\tilde{\gamma}$, its contribution would be negligible. Therefore it is interesting to fit the results for just three constants with $\tilde{\delta}$ fixed at zero. These results are also given in Table IV, where the uncertainty in $\tilde{\alpha}$ is observed to decrease by a factor of two. Finally, by using the results of the three parameter fit, a value for $h^u_{AA} - \frac{1}{2}h^d_{AA} = 0.79 \pm 0.19$ from the CERN data may be obtained, consistent with the Standard Model prediction of 0.75.

All of the results are consistent with the Standard Model. Most of the information for $\tilde{\alpha}$ comes from the SLAC experiment, for $\tilde{\gamma}$ from the Bates experiment, and for $\tilde{\beta}$ from the Mainz experiment. It is amusing that the smallest percent error is for

$\tilde{\alpha}$, the smallest absolute error is for $\tilde{\gamma}$, and the smallest error on $\sin^2\theta_W$ (if the errors are added linearly) comes from the Mainz experiment. This illustrates the subtleties in evaluating the relative sensitivities of these experiments.

Table IV. Values for the model independent coupling constants obtained by combining the experiments.

Constant	Four Parameter	Three Parameter	Theory
$\tilde{\alpha}$	-0.65 ± 0.14	-0.56 ± 0.07	-0.54
$\tilde{\beta}$	-0.04 ± 0.13	-0.07 ± 0.13	-0.08
$\tilde{\gamma}$	$+0.14 \pm 0.03$	$+0.14 \pm 0.03$	$+0.15$
$\tilde{\delta}$	$+1.38 \pm 2.00$	–	$+0.00$

6. Future Directions

The four experiments described above have made quantitative determinations of the electroweak interference effects due to neutral weak currents, thereby providing a detailed confirmation of the predictions of the Standard Model. In addition, this work has established that it is possible to measure the small asymmetries of interest without large systematic errors. This success is leading towards a new generation of experiments with expanded physics goals. One new direction is to use the Standard Model as a tool to probe hadrons. In particular, parity-violating asymmetries provide a unique probe to study the role of strange quarks in the nucleon.

A number of specific parity experiments are in progress or being planned. The SAMPLE collaboration at Bates is presently setting up to measure the asymmetry for electrons scattered backward from hydrogen.[52] Experiments at CEBAF plan to study the proton (forward and backward) and ^{4}He.[53–55] An experiment to study the proton is also being designed at Mainz. In addition, proposals have been submitted to SLAC[56] to study elastic scattering. The dominant motivation for most of these experiments is measuring the role of strange quarks in the nucleon. These planned experiments have as their goal errors typically at the 5% level, comparable to what was achieved at SLAC. Since the effects of the strange quarks on the measured asymmetries could be as large as 50% or so, this level of precision is appropriate.

A more difficult question is whether scattering polarized leptons from fixed targets can contribute further to our knowlege of the Standard Model in light of precision data being obtained at other laboratories. An analysis of how extensions of the Standard Model would effect $\tilde{\alpha}$ and $\tilde{\gamma}$ as well as other measurable parameters has been published.[12] From that work, it appears that measurements of $\tilde{\alpha}$ and $\tilde{\gamma}$ would have to be at the 1-2% level to be useful. An interesting feature of that analysis is that there are a number of extensions that are best observed by measuring $\tilde{\gamma}$. Extensions that modify $\tilde{\alpha}$ at the 1% level also modify the results of many other experiments at their projected limits. Presently, the most promising experiments for doing this are deep inelastic scattering ($\tilde{\alpha}$) and elastic scattering from C or He ($\tilde{\gamma}$). Precision measurements of $\tilde{\beta}$ and $\tilde{\delta}$ are less attractive because of the uncertainties in the radiative corrections.

The first question is can sufficient statistics be collected. For deep inelastic scattering, about a factor of 6 in statistical error is needed. Improvements in polarized source technology[57] suggest that polarizations of 80% might be achieved in the future, giving a factor of 2. By improving the detector and running longer, the other factor of 3 might be achieved. A letter of intent for such an experiment has recently been submitted to SLAC.[58] For elastic scattering, much larger gains are needed. However, the experiment at Bates had a very small solid angle; an optimal apparatus, such as a large solenoidal spectrometer, would provide ample statistics.[59] An alternative is to study helium at CEBAF with the conventional spectrometers; 1% statistics could be collected in ~2000 hours.

The next issue is systematic errors. The most obvious problem is measuring the beam polarization, which yields 5% errors with current methods. Improving this technique is presently one of the most important experimental challenges in the field of polarized electron physics. There are some new approaches that might yield smaller errors, including as a laser Compton polarimeter or better foil calibration. It might be ultimately possible to keep systematic errors below the 1% level.

Theoretical uncertainties are another important issue. The dominant error for the SLAC work is higher twist terms, contributing $\sim$ 6%. Increasing the Q^2 to 5 or 10 $(\mathrm{GeV/c})^2$ can reduce this problem. The contributions from strange quarks can be estimated from neutrino scattering data. It is possible, although certainly not demonstrated, that the theoretical errors could be controlled if sufficient care were taken. In particular, issues such as the symmetry of the u and d quarks in the sea[60] might become important. For elastic scattering, the main problem is the role of strange quarks, which can be minimized by keeping the Q^2 small. In addition, there are contributions from dispersive corrections that are extremely hard to evaluate that might occur at the few percent level.

In conclusion, measuring $\tilde{\alpha}$ or $\tilde{\gamma}$ at the 1% level will be difficult. Given the projection that $\tilde{\gamma}$ can be determined to 1% or perhaps even 0.3% by measuring parity violation in Cs, as well as the projection that the new physics contained in $\tilde{\alpha}$ can be detected by other planned experiments, it is not clear how much effort will be devoted to these tests. However, the ongoing parity experiments studying hadronic physics are improving the technology, and the task may appear less daunting in the near future. Moreover, if some experiment were to see a deviation from the Standard Model, the motivation would be much greater. Measuring $\tilde{\gamma}$ might be one of the few ways to test the Cs results, and measuring $\tilde{\alpha}$ might determine which extension to the

Standard Model is the correct one.

REFERENCES

1. C. Y. Prescott *et al.*, *Phys. Lett.* **B77** (1978) 347.
2. C. Y. Prescott *et al.*, *Phys. Lett.* **B84** (1979) 524.
3. S. L. Glashow, *Nucl. Phys.* **22** (1961) 579; S. Weinberg, *Phys. Rev. Lett.* **19** (1967) 1264; A. Salam, in *Elementary Particle Theory: Relativistic Groups and Analyticity* (Nobel Symposium No. 8), ed. by N. Svartholm (Almquist and Wicksell, Stockholm, 1978), 367.
4. A. Argento *et al.*, *Phys. Lett.* **120B** (1983) 245.
5. A. Argento *et al.*, *Phys. Lett.* **140B** (1984) 142.
6. W. Heil *et al.*, *Nucl. Phys.* **B327** (1989) 1.
7. P. A. Souder, *et al. Phys. Rev. Lett.* **65** (1990) 694.
8. U. Amaldi *et al.*, *Phys. Rev.* **D36** (1987) 1385.
9. G. Costa *et al.*, *Nucl. Phys.* **B297** (1988) 244.
10. G. L. Folgi and D. Haidt, Z. Phys. C **40** (1988) 379.
11. P. Langacker and M. Luo, *Phys. Rev.* **D44** (1991) 817.
12. P. Langacker, M. Luo, and A. K. Mann, *Rev. Mod. Phys.* **64** (1992) 87.
13. D. B. Kaplan and A. Manohar, *Nucl. Phys.* **B310** (1988) 527.
14. R. D. McKeown, *Phys. Lett.* **219B** (1989) 140.
15. W. J. Marciano, *Phys. Rev.* **D20** (1979) 274; W. J. Marciano and A. Sirlin, *Phys. Rev. Lett.* **46** (1981) 163, A. Sirlin, *Phys. Rev.* **D22** (1980) 971.
16. S. Bellucci, M. Lusignoli, and L. Maiani, *Nucl. Phys.* **B189** (1981) 329.
17. J. F. Wheater, *Phys. Lett.* **105B** (1981) 483.
18. D. Yu. Bardin, P. Ch. Christova, and O. M. Fedorenko, *Nucl. Phys.* **B197** (1982) 1.
19. W. J. Marciano and A. Sirlin, *Phys. Rev.* **D27** (1983) 552; **29** (1984) 75.
20. P. Q. Hung and J. J. Sakurai, *Ann. Rev. of Nucl. and Part. Science* **31** (1981) 375.
21. A. Love *et al.*, *Nucl. Phys.* **B49** (1972) 513.
22. E. Derman, *Phys. Rev.* **D7** (1973) 2755.
23. W. J. Wilson, *Phys. Rev.* **D10** (1974) 218.
24. E. Derman, *Phys. Rev.* **D7** (1973) 2755.

25. S. M. Berman and J. R. Primack, *Phys. Rev.* **D9** (1974) 2171; **D10** (1974) 3895.

26. M. A. B. Bég and G. Feinberg, *Phys. Rev. Lett.* **33** (1974) 606.

27. S. M. Bilenkii *et al.*, Sov. J. Nucl. Phys. **21** (1975) 189.

28. J. D. Bjorken, *Phys. Rev.* **D18** (1978) 3239.

29. L. Wolfenstien, *Nucl. Phys.* **B146** (1978) 477.

30. E. Derman, *Phys. Rev.* **D19** (1979) 133.

31. H. Fritzsch, Z. Phys. C **1** (1979) 321.

32. E. Reya and K. Schilcher, *Phys. Rev.* **D10** (1974) 952.

33. R. N. Cahn and F. J. Gilman, *Phys. Rev.* **D17** (1978) 1313.

34. E. Hoffman and E. Reya, *Phys. Rev.* **D18** (1978) 3230.

35. E. Ch. Christova and S. Petcov, *Phys. Lett.* **B84** (1979) 250.

36. T. W. Donnelly, J. Dubach, and Ingo Sick, *Phys. Rev.* **C37** (1988) 2320.

37. M. J. Musolf and T. W. Donnelly, *Nucl. Phys.* **A546** (1992) 509.

38. R. L. Jaffe, *Phys. Lett.* **229B** (1989) 275.

39. N. W. Park, J Schecter, and H. Weigel, *Phys. Rev.* **D43** (1991) 869.

40. N. W. Park and H. Weigel, *Nucl. Phys.* **541A** (1992) 453.

41. E. Hadjimichael, G. I. Poulis, and T. W. Donnelly, *Phys. Rev.* **C45** (1992) 2666.

42. T. W. Donnelly, *et al. Nucl. Phys.* **A541** (1992) 525.

43. G. Feinberg, *Phys. Rev.* **D12** (1975) 3575.

44. J. D. Walecka, *Nucl. Phys.* **A285** (1977) 349.

45. B. D. Serot *Nucl. Phys.* **A322** (1979) 408.

46. T. W. Donnelly, J. Dubach, and Ingo Sick, *Nucl. Phys.* **A503** (1989) 589.

47. D. H. Beck, *Phys. Rev.* **D39** (1989) 3248.

48. M. J. Musolf and B. R. Holstein, *Phys. Lett.* **B242** (1990) 461; *Phys. Rev.* **D43** (1991) 2956.

49. M. C. Noecker *et al.*, *Phys. Rev. Lett.* **61** (1988) 310.

50. M. A. Bouchiat *et al.*, *Phys. Lett.* **134B** (1984) 463.

51. S. A. Blundel, W. R. Johnson, and J. Sapirstein, *Phys. Rev. Lett.* **65** (1990) 1411.

52. MIT-Bates proposal # 89-06, R. D. McKeown and D. H. Beck, contact people.

53. CEBAF proposal #PR-91-010, J. M. Finn and P. A. Souder, spokespersons.

54. CEBAF proposal #PR-91-004, E. J. Beise, spokesperson.

55. CEBAF proposal #PR-91-017, D. H. Beck, spokesperson.

56. SLAC proposal E-148, R. W. Lourie, spokesperson.

57. T. Maruyama *et al.*, *Phys. Rev. Lett.* **66** (1991) 2376.

58. SLAC proposal E-149, P. Bosted, spokesperson.

59. P. A. Souder and R. Holmes, in *Proceedings of the Workshop on Parity Violation in Electon Scattering*, E. J. Beise and R. D. McKeown, eds., World Scientific, Singapore (1990) p. 137.

60. E. J. Eichten, I. Hinchliffe, and C. Quigg, *Phys. Rev.* **D45** (1992) 2269.

PRECISION ELECTROWEAK TESTS AT HERA

H. SPIESBERGER
Fakultät für Physik, University Bielefeld
4800 Bielefeld, Germany

Contents

1 Introduction

The electron proton storage ring HERA, operating since spring 1992, opens the possibility to study collisions of electrons and protons at very high energies. With electron energies of $E_e = 15 - 30\,GeV$ and proton energies between 300 and $820\,GeV$, the center-of-mass energy ranges from $\sqrt{s} = 134 - 314\,GeV$. A possible HERA upgrade could provide beams with $E_e = 35\,GeV$ and $E_p = 1.2\,TeV$. Both neutral and charged current interactions can be studied in the region of large spacelike momentum transfers Q^2 up to some $10^4\,GeV^2$. With the projected luminosity of $\mathcal{L} = 1.5 \cdot 10^{31} cm^{-2} sec^{-1}$, i.e., about $200\,pb^{-1}$ per year, one expects event numbers in the order of 10^7 for neutral current scattering and 10^4 for charged current scattering in the deep inelastic regime defined by $Q^2 > 4\,GeV^2$ and $x > 10^{-3}$. Both electron and positron beams will be available and experimentation with longitudinal polarization of up to 80 % is being discussed.

One of the main physics subjects at HERA will be the precision measurement of structure functions, in particular at low x, and their understanding in the framework of QCD. However, HERA will also contribute to an understanding of electroweak physics obtained in a regime complementary to what is probed at LEP or at hadron colliders. In particular, the measurement of charged current interactions at large spacelike momentum transfers is unique to HERA.

In many respects, HERA will not be able to compete in precision with other measurements. Therefore, at HERA 'precision' is rather a qualitative than a quantitative property of measurements. Besides the fact that ep cross sections are rapidly decreasing at large Q^2, the low precision as compared to LEP 1 is due to the following two main reasons:

- The incomplete knowledge of structure functions which enter into the cross section formulae turns into systematic uncertainties for the measurement of electroweak physics.
- Due to the dominating $1/Q^2$ behaviour of the cross section, uncertainties in the absolute energy calibration lead to large uncertainties. This, together with the luminosity measurement accounts for a major part of the experimental uncertainties.

Since the two experiments at HERA, H1 [1] and ZEUS [2], just started data taking and no experimental results of electroweak physics measurements are available yet, this contribution has to be restricted to reporting theoretical expectations. This article is based to a large extent on the results of the DESY workshop on HERA physics held in 1991 which updated earlier work [3, 4]. Many more details than can be given here are found in the proceedings of this workshop [5].

A first part of this article provides theoretical predictions for cross sections and derived observables allowing for precision tests. Then, measurements of deep inelastic cross sections to determine standard model parameters are discussed. In particular, ratios of NC / CC cross sections and asymmetries are investigated with respect to the possibility of using them for a determination of the W boson mass as well as of the top quark and the Higgs boson masses.

Then we discuss measurements of the $WW\gamma$ couplings and possible deviations from their standard model values, both in W production and in radiative charged current scattering.

A final part will review some topics of non–standard electroweak physics: Is there a possibility to constrain new physics appearing in virtual corrections? What are the virtues of HERA with respect to the discovery of extensions of the gauge group involving new heavy Z' bosons or right–handed charged currents? Many other topics of more exotic physics will not be covered in this article. For this, the interested reader is referred to the HERA workshop proceedings [3, 5] (see also [6]).

2 Precise Standard Model Predictions for HERA

2.1 *Lowest Order Results*

We study deep inelastic scattering of electrons or positrons off protons

$$e(l) + p(p) \rightarrow e'(l') + X(p_X), \tag{1}$$

with $e' = e$ for neutral current scattering (NC) and $e' = \nu_e$ for charged current scattering (CC). The particle momenta are given in parentheses. A useful set of kinematic variables to describe the process (1) is given by

$$Q^2 = -(l-l')^2, \quad S = (l+p)^2, \quad x = \frac{Q^2}{2p\cdot(l-l')}, \quad y = \frac{Q^2}{xS}. \tag{2}$$

In the region of large x and Q^2 where the contribution of the longitudinal structure function F_L and proton mass effects can be neglected, the differential cross section for the neutral current process for left–handed (L) and right–handed (R) polarized electrons is given by

$$\frac{d^2\sigma^{NC}}{dx\,dQ^2}(e^-_{L,R}) = \frac{2\pi\alpha^2}{xQ^4}\left[\left(1+(1-y)^2\right)F_2^{L,R} + \left(1-(1-y)^2\right)xF_3^{L,R}\right]. \tag{3}$$

For positron scattering one has to replace $F_2^{L,R} \rightarrow F_2^{R,L}$ and $F_3^{L,R} \rightarrow -F_3^{R,L}$. The structure functions are derived in the quark–parton model from basic 4–fermion scattering cross sections:

$$\begin{aligned} F_2^{L,R} &= \sum_f [xq_f(x,Q^2) + x\bar{q}_f(x,Q^2)]\cdot A_f^{L,R}, \\ xF_3^{L,R} &= \sum_f [xq_f(x,Q^2) - x\bar{q}_f(x,Q^2)]\cdot B_f^{L,R}. \end{aligned} \tag{4}$$

They contain the quark (q_f) and anti-quark ($\bar{q}_f$) distribution functions as well as coupling constants and propagators corresponding to photon and Z boson exchange ($L=+, R=-$):

$$\begin{aligned} A_f^{L,R} &= Q_f^2 + 2Q_eQ_f(v_e \pm a_e)v_f\frac{Q^2}{Q^2+M_Z^2} + (v_e \pm a_e)^2(v_f^2+a_f^2)\left(\frac{Q^2}{Q^2+M_Z^2}\right)^2, \\ B_f^{L,R} &= -2Q_eQ_f(v_e \pm a_e)a_f\frac{Q^2}{Q^2+M_Z^2} + 2(v_e \pm a_e)v_fa_f\left(\frac{Q^2}{Q^2+M_Z^2}\right)^2. \end{aligned} \tag{5}$$

In Eq. (3), Higgs boson exchange is neglected because of the small couplings to light fermions. By using Q^2–dependent parton distribution functions, QCD corrections in the leading–logarithmic approximation are included.

The parametrization of the lowest order cross section in terms of vector and axial vector coupling constants $v_{e,f}$, $a_{e,f}$ is quite general and applies to any model where a vector boson with v and a couplings is exchanged together with the photon. In the standard model, the couplings of the fermions to the bosons are given by the electric charge (in units of e) for the photon exchange and by

$$\begin{aligned} a_f &= \frac{1}{2\sin\theta_W\cos\theta_W} I_3^f, \\ v_f &= \frac{1}{2\sin\theta_W\cos\theta_W}\left(I_3^f - 2Q_f\sin^2\theta_W\right) \end{aligned} \tag{6}$$

where Q_f and I_3^f denote the charge and the third isospin component of the fermion f. In the standard model with an arbitrary Higgs system, the mixing angle is related to the vector boson masses by

$$\sin^2\theta_W = 1 - \frac{M_W^2}{\rho_0 M_Z^2}. \tag{7}$$

In the following we restrict ourselves to the minimal model with $\rho_0 = 1$ at the tree level and use the abbreviation

$$s_W^2 = 1 - \frac{M_W^2}{M_Z^2}. \tag{8}$$

More general situations will be discussed in section 5.1.
The muon decay constant G_μ, in lowest order given by

$$G_\mu = \frac{\pi\alpha}{\sqrt{2}s_W^2 M_W^2}, \tag{9}$$

can be used to rewrite the coupling constants

$$\begin{aligned} a_f &= \left(\frac{\sqrt{2}G_\mu M_Z^2}{4\pi\alpha}\right)^{1/2} I_3^f, \\ v_f &= \left(\frac{\sqrt{2}G_\mu M_Z^2}{4\pi\alpha}\right)^{1/2}\left(I_3^f - 2\,Q_f\,s_W^2\right). \end{aligned} \tag{10}$$

The discrepancies between numerical results obtained from the two parametrizations Eqs. (6) and (10) disappear after incorporating consistently electroweak radiative corrections in relation Eq. (9).

In a similar way one can write for the charged current differential cross section

$$\frac{d^2\sigma^{CC}}{dx\,dQ^2}(e^-) = (1-\mathcal{P})\frac{\pi\alpha^2}{4s_W^4}\cdot\frac{1}{(Q^2+M_W^2)^2}\left[u+c+(1-y)^2(\bar{d}+\bar{s}+\bar{b})\right], \tag{11}$$

or

$$\frac{d^2\sigma^{CC}}{dx\,dQ^2}(e^-) = (1-\mathcal{P})\frac{G_\mu^2}{2\pi}\left(\frac{M_W^2}{Q^2+M_W^2}\right)^2\left[u+c+(1-y)^2(\bar{d}+\bar{s}+\bar{b})\right], \tag{12}$$

with $\mathcal{P}$ the degree of longitudinal polarization ($\mathcal{P} = +1$ for right–handed polarization). For positron scattering one has to replace $u+c \to \bar{u}+\bar{c}$ and $\bar{d}+\bar{s}+\bar{b} \to d+s+b$ and change the sign in front of $\mathcal{P}$. The second form has the advantage that it is not changed by large radiative corrections as we will see in the next subsection.

2.2 *Higher Order Corrections*

The calculations of higher order corrections to deep inelastic scattering at HERA [7, 8, 9, 10, 11] have been reviewed in [12, 13] (see also references therein). They are performed in the on–shell scheme, where M_W and M_Z are treated symmetrically as basic parameters together with m_t and M_H (besides the fine structure constant α and the fermion masses). Amplitudes are functions of these basic parameters. The Fermi constant G_μ derived from the μ lifetime is treated as a constraint

$$G_\mu = G_\mu(\alpha, M_Z, M_W, M_H, m_t) \tag{13}$$

on the value of the W boson mass M_W. Here, non–photonic radiative corrections to the μ decay are included in terms of the well–known Δr [14].

We do not discuss QED corrections which are large and indispensable, in particular for NC scattering at small x and large y. They constitute an 'uninteresting', but well–understood part of physics which has to be separated from genuine electroweak effects. Careful comparisons of different calculations during the HERA workshop in 1991 have shown good agreement at the permille level between results for the first order corrections obtained by various authors [13]. We can thus be confident that these corrections do not introduce additional uncertainties. Higher–order corrections, however, are not known with the same level of precision. They have been estimated with the help of the leading logarithmic approximation in [15] and may reach the order of 10 % with respect to the Born cross section at large y and small x; at large Q^2, however, i.e., at large y and large x they are small. In particular, QED corrections will hardly change the sensitivity on electroweak parameters. Therefore we will not go into further details and concentrate on the genuine electroweak corrections.

QCD corrections are also not discussed here and only in the G_μ–M_W relation (in Δr) corrections of order $\mathcal{O}(\alpha\alpha_s m_t^2)$ are included. Since we are dealing with small effects, as we will see below, it is, however, not excluded that QCD corrections to the electroweak form factors to be discussed below could be non–negligible.

The corrections are collected in form factors that change the amplitudes in a simple form:

The dressed *photon exchange amplitude* can be written in the following way:

$$\mathcal{M}_\gamma = \frac{e^2}{1-\hat{\Pi}_f^\gamma(Q^2)} \cdot \frac{1}{Q^2} \cdot Q_e\gamma_\mu \otimes Q_f\gamma^\mu. \tag{14}$$

$\hat{\Pi}_f^\gamma$ is the fermionic part of the photon vacuum polarization. Writing it in the denominator resums the leading logarithmic corrections according to the renormalization group equation and the first factor in Eq. (14) constitutes the running fine structure constant. Bosonic contributions to the vacuum polarization have to be added, combined with vertex and box corrections.

The weak one-loop corrections to the Z *exchange amplitude* $\mathcal{M}_Z$ for $eq \to eq$ can be expressed in terms of four weak form factors (ρ_{eq}, κ_e, κ_q, and κ_{eq}) in the following way, making use of dressed vector couplings [16, 17]:

$$\mathcal{M}_Z(\hat{s},Q^2) \sim \frac{1}{Q^2+M_Z^2}\left[\frac{G_\mu}{\sqrt{2}}M_Z^2\rho_{eq}\right]\left[I_3^e I_3^q \gamma_\mu\gamma_5 \otimes \gamma^\mu\gamma_5 + \bar{v}_e I_3^q \gamma_\mu \otimes \gamma^\mu\gamma_5 \right.$$
$$\left. + I_3^e \bar{v}_q \gamma_\mu\gamma_5 \otimes \gamma^\mu + \bar{v}_{eq}\gamma_\mu \otimes \gamma^\mu\right], \quad (15)$$

$$\bar{v}_f = I_3^f\left[1-4|Q_f|s_W^2\,\kappa_f(\hat{s},Q^2)\right], \quad f=e,q\,, \quad (16)$$
$$\bar{v}_{eq} = I_3^e\bar{v}_q + \bar{v}_e I_3^q - I_3^e I_3^q\left[1-16|Q_eQ_q|s_W^4\,\kappa_{eq}(\hat{s},Q^2)\right], \quad (17)$$

with $\hat{s} = xS$ and s_W^2 from Eq. (8). In the Born approximation, $\rho = \kappa = 1$, and $\bar{v}_{eq} = \bar{v}_e\bar{v}_q \propto v_e v_q$. The above parametrization has the form of a Born–like expression except that the coupling $\bar{v}_{eq}$ does not factorize which it does at the Born level. The axial vector couplings can be kept in their form $a_f = I_3^f$ by absorbing corresponding vertex corrections into the normalization factor $\rho_{eq} = \rho_{eq}(\hat{s},Q^2)$ which receives also contributions from self energies. The form factors κ_f and κ_{eq} combined with s_W^2 give rise to effective mixing angles which depend through the κ's on the fermion species and on the kinematic variables. The form factors can be separated into a universal part independent of the fermion species and a non–universal remainder term. The universal parts contain the dependence on the masses of the top quark and the Higgs boson. The dominating contribution to them is proportional to m_t^2 and with

$$\Delta\rho = \frac{3\alpha}{16\pi s_W^2 c_W^2}\frac{m_t^2}{M_Z^2} \quad (18)$$

one has to one–loop precision

$$\rho_{eq} = 1 + \Delta\rho + \Delta\rho_{eq}^{rem}, \quad (19)$$

$$\kappa_f = 1 + \frac{c_W^2}{s_W^2}\Delta\rho + \Delta\kappa_f^{rem}. \quad (20)$$

Resummation of higher–order terms can be performed by the replacement [16, 18]

$$\rho_{eq} \to \frac{1+\Delta\rho_{eq}^{rem}}{1-\Delta\bar{\rho}}, \quad (21)$$

$$\kappa_f \to (1+\Delta\kappa_f^{rem})\left(1+\frac{c_W^2}{s_W^2}\Delta\bar{\rho}\right), \quad (22)$$

$$\kappa_{eq} \to (1+\Delta\kappa_{eq}^{rem})\left(1+\frac{c_W^2}{s_W^2}\Delta\bar{\rho}\right)^2. \quad (23)$$

Here, $\Delta\bar{\rho}$ contains irreducible two–loop contributions and is defined below in Eq. (28). To $\mathcal{O}(\alpha)$, it is equal to $\Delta\rho$. The remainder terms (index *rem*) are the one–loop expressions with the corresponding leading m_t^2 terms subtracted.

Note that the form factors are functions of the kinematic variables x and Q^2 and depend on the fermion species. The x–dependence is due to box diagram contributions which also lead to $\rho_{eq} \neq \rho_{e\bar{q}}$. The measurement of deep inelastic cross sections covers a large range of x and Q^2; therefore, this dependence cannot be

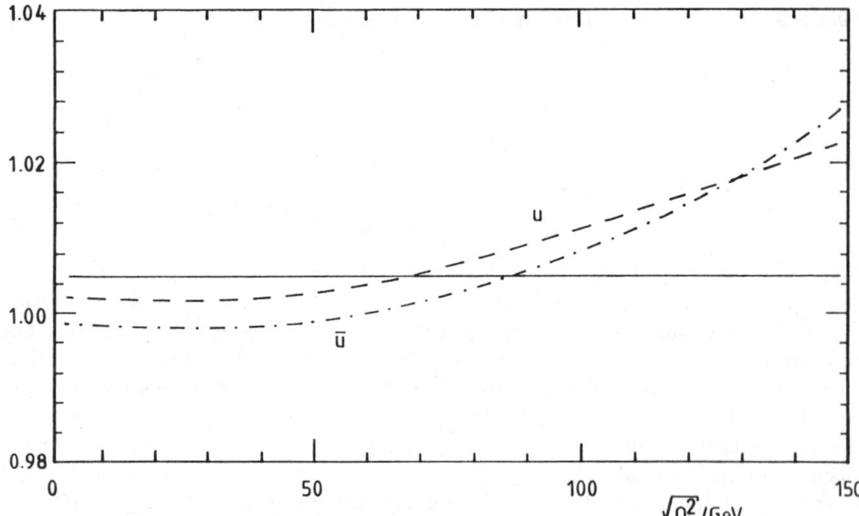

Figure 1: NC form factors $\rho_{eu}(x,Q^2)$ and $\rho_{e\bar{u}}(x,Q^2)$ for $x = 0.3$ as a function of Q^2. $m_t = 130\,GeV$. The horizontal line corresponds to $1+\Delta\rho$.

neglected (LEP measurements are, in contrast to this, performed at a fixed $Q^2 = -M_Z^2$). This dependence is as large as the constant m_t^2 terms. It is shown for ρ_{eu} and $\rho_{e\bar{u}}$ in Fig. 1. The effective mixing angle $s^2_{\text{eff},e} = \kappa_e(x,Q^2)s_W^2$ is displayed in Fig. 2.

The W *exchange amplitude* including higher–order contributions can be written in the following way:

$$\mathcal{M}_W(s,Q^2) = \frac{G_\mu}{4\sqrt{2}}\,\rho^W_{eq}(s,Q^2)\,\frac{M_W^2}{Q^2+M_W^2}\,\gamma_\mu\,[1-\gamma_5]\otimes\gamma^\mu\,[1-\gamma_5]\,. \tag{24}$$

Differently from the neutral current, only a single form factor ρ^W for each parton scattering process is required to accommodate the higher–order contributions. In this representation, the weak radiative corrections are very small with very little dependence on m_t and M_H. The leading t–quark contributions are already absorbed when the matrix element is normalized with G_μ instead of α/s_W^2. Hence, we have

$$\rho^W_{eq} \equiv (\rho^W_{eq})^{rem} \tag{25}$$

which deviates from 1 typically by a few permille.

2.3 *Standard Model Parameters: The G_μ constraint*

The G_μ constraint is an important experimental input which allows one to remove the less well–known M_W. After its consistent inclusion, radiative corrections are

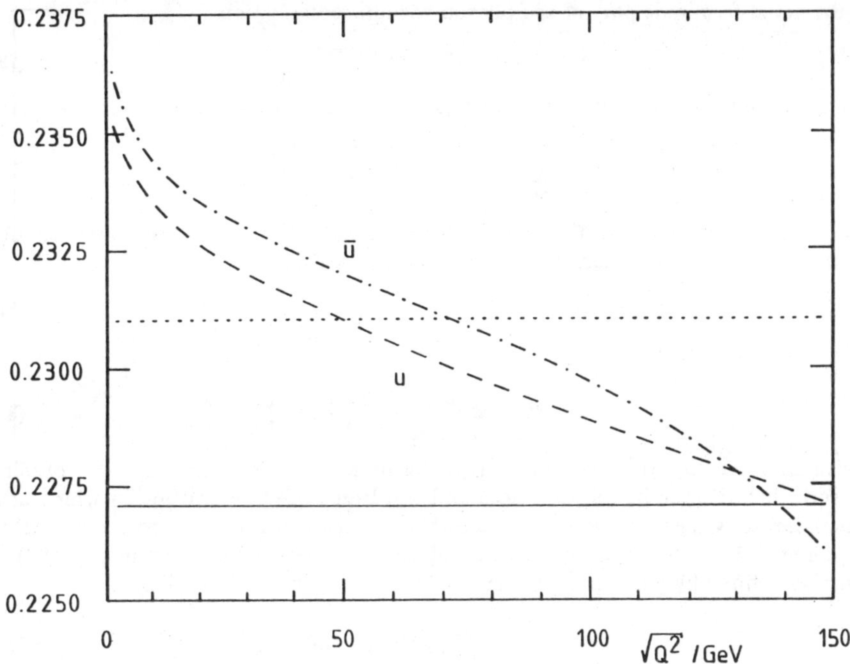

Figure 2: Effective weak mixing angle $s^2_{\text{eff},e}(x,Q^2)$ for the channels $eu \to eu$ and $e\bar{u} \to e\bar{u}$ including heavy box diagrams for $x = 0.3$ as a function of Q^2. $m_t = 130\,GeV$. The horizontal full line corresponds to s^2_W, the dotted line to $s^2_W + c^2_W \Delta\rho$.

smaller. To be consistent, one has to include radiative corrections to the μ decay in relation Eq. (9). In the following we collect the relevant formulae which contain resummed higher-order effects. More details and references to the original literature can be found in the contribution of W. Hollik in this book. After radiative corrections, Eq. (9) is changed to

$$G_\mu = \frac{\pi\alpha}{\sqrt{2}s^2_W M^2_W} \frac{1}{1 - \Delta r(\alpha, M_W, M_Z, M_H, m_t)} \tag{26}$$

where

$$1 - \Delta r = (1 - \Delta\alpha) \cdot \left(1 + \frac{c^2_W}{s^2_W}\Delta\bar{\rho}\right) - \Delta r_{rem} \tag{27}$$

and

$$\Delta\bar{\rho} = 3\,\frac{G_\mu m_t^2}{8\pi^2\sqrt{2}} \cdot \left[1 + \frac{G_\mu m_t^2}{8\pi^2\sqrt{2}} \cdot r_2\right] + \Delta\rho^{\alpha\alpha_s}. \tag{28}$$

For a light Higgs, $r_2 = 19 - 2\pi^2$. In our calculation we use the result of [19] which is correct for an arbitrary value of M_H. Then r_2 is a function of m_t/M_H. $\Delta\alpha$ is the subtracted fermionic part of the photon vacuum polarization

$$\Delta\alpha = \Pi_f^\gamma(0) - \mathrm{Re}\,\Pi_f^\gamma(M_Z^2) \tag{29}$$

and contains large logarithmic corrections from light fermions:

$$\Delta\alpha = \sum_{f=e,\mu,\tau} \frac{\alpha}{3\pi}\left(\log\frac{M_Z^2}{m_f^2} - \frac{5}{3}\right) + \Delta\alpha_{had}\,. \tag{30}$$

$\Delta\alpha_{had}$ is determined form the total hadronic cross section in e^+e^- annihilation and thus an experimental quantity. A recent fit [20] led to:

$$\Delta\alpha = 0.0595 \pm 0.0009 \;\text{ for }\; M_Z = 91.175\,GeV. \tag{31}$$

The leading correction of $\mathcal{O}(\alpha\alpha_s)$ is given by

$$\Delta\rho^{\alpha\alpha_s} = -\Delta\rho\frac{\alpha_s(m_t^2)}{\pi}\cdot\frac{2}{3}\left(\frac{\pi^2}{3}+1\right) \tag{32}$$

with $\Delta\rho$ from Eq. (18). All other $\mathcal{O}(\alpha)$ terms are collected in Δr_{rem}. It contains a term logarithmic in the top mass and the Higgs boson contribution which also increases only logarithmically for large M_H at one–loop according to the screening theorem [21]. The treatment of the higher–order reducible terms in Eq. (26) is further refined by performing in Δr_{rem} the following substitution [22]

$$\frac{\alpha}{s_W^2} \to \frac{\sqrt{2}}{\pi}G_\mu M_W^2(1-\Delta\alpha) \tag{33}$$

in the expansion parameter of the combination

$$\left(\frac{\delta M_Z^2}{M_Z^2} - \frac{\delta M_W^2}{M_W^2}\right) - \Delta\rho$$

after removing the UV singularity according to the $\overline{MS}$ scheme with $\mu = M_Z$. The changes induced by this are, however, at the level of theoretical uncertainties associated with unknown higher–order corrections.

3 Precision Measurements of Deep Inelastic Scattering

3.1 *Observables*

From the given formulae one can calculate differential cross sections for electron or positron scattering, also for polarized beams, or integrated cross sections and ratios of them. After removing M_W by the use of Eq. (26), they will depend on the input parameters α, G_μ, M_Z, and m_t and M_H. In particular we will discuss the ratio of neutral to charged current cross sections

$$R_\pm = \frac{\sigma^{NC}(e^\pm)}{\sigma^{CC}(e^\pm)} \tag{34}$$

and various asymmetries for NC scattering with different polarization and charge:

$$A_{\pm} = \frac{\sigma^{NC}(e_L^{\pm}) - \sigma^{NC}(e_R^{\pm})}{\sigma^{NC}(e_L^{\pm}) + \sigma^{NC}(e_R^{\pm})}, \tag{35}$$

$$B_{\pm} = \frac{\sigma^{NC}(e_L^{\pm}) - \sigma^{NC}(e_R^{\mp})}{\sigma^{NC}(e_L^{\pm}) + \sigma^{NC}(e_R^{\mp})}, \tag{36}$$

$$C_{L,R} = \frac{\sigma^{NC}(e_{L,R}^{-}) - \sigma^{NC}(e_{L,R}^{+})}{\sigma^{NC}(e_{L,R}^{-}) + \sigma^{NC}(e_{L,R}^{+})}. \tag{37}$$

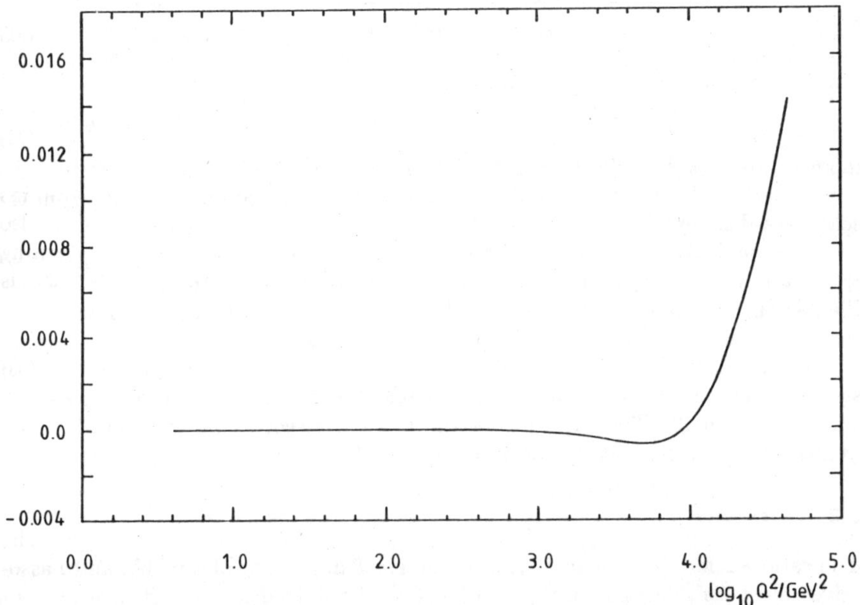

Figure 3: Influence of the form factor ρ_{eq} on the neutral current cross section: $d\sigma^{NC}(\rho)/d\sigma^{NC}(\rho = 1) - 1$ for $x = 0.5$ as a function of Q^2.

In principal, these ratios could be considered for differential as well as for integrated cross sections. The NC cross section is dominated by photon exchange at low Q^2 and therefore determined mainly by the running fine structure constant. Interesting electroweak effects show up at large Q^2. Fig. 3 shows the effect of the electroweak corrections contained in ρ_{eq} on the differential NC cross section. Only above $Q^2 = 10^4\,GeV^2$ the corrections exceed the level of one permille. Since at large Q^2 the event rates are small, a measurement of the differential cross sections will be difficult. In the following, our discussion will be based on integrated cross sections obtained after cuts to remove the low Q^2 contribution. Typically, cuts of the order of $Q^2 > 1000\,GeV^2$ are used. Equivalently, a cut on the hadronic transverse

m_t [GeV]	M_H [GeV]	M_W [GeV]	σ^{NC} [nb]	σ^{CC} [nb]	R_-
90	60	79.932	0.3161	0.0404	7.82
90	1000	79.740	0.3160	0.0403	7.85
120	60	80.091	0.3161	0.0406	7.80
120	1000	79.898	0.3161	0.0404	7.82
150	60	80.258	0.3162	0.0407	7.77
150	1000	80.062	0.3161	0.0405	7.80
200	60	80.584	0.3164	0.0409	7.73
200	1000	80.373	0.3163	0.0408	7.76
250	60	80.978	0.3168	0.0412	7.69
250	1000	80.733	0.3166	0.0410	7.72

Table 1: Unpolarized NC and CC electron scattering cross sections and their ratio R_- for several values of m_t and M_H. $M_Z = 91.175$ GeV. Column 3 displays the corresponding value of M_W. A cut of $p_t^{had} > 25\,GeV$ was used. No QED corrections.

momentum of $p_t^{had} > p_t^{min}$ with $p_t^{min} = \mathcal{O}(25\,GeV)$ can also be used. This excludes automatically also the low x region and reduces systematic uncertainties from the sea quark distribution.

In table 1, cross sections for NC and CC electron scattering are given for $p_t^{had} > 25\,GeV$ for various Higgs and top masses. Here and in the following we use the parton distribution functions from [23], set 1. The results do not contain QED corrections and should not be used for quantitative purposes, but they indicate the correct dependence on electroweak parameters. The charged current cross section shows variations at the percent level, whereas the NC cross section is very stable since ρ_{eq} is small. The variation of the CC cross section is determined almost exclusively by M_W and its variation with m_t and M_H since $\sigma_{CC} \propto G_\mu^2 M_W^4$.

3.2 *The Ratio R_-*

The ratio R_- for electron scattering is the most important observable and has received most consideration in the past [24, 25]. It will lead to the best measurement of standard model parameters. This is first of all due to the fact that it does not require separate machine runs and therefore the largest possible integrated luminosity will contribute to the R_- measurement. A second reason is the following: absolute cross sections are of limited use for the measurement of electroweak parameters since they are affected by additional systematic uncertainties, mainly from the overall normalization (energy calibration and luminosity measurement), which can be reduced by taking ratios like $R_\pm$. Theoretically, the charged current cross section and the ratio $R_\pm$ are equivalent and both determined essentially by M_W. By considering $R_\pm$ one gains through the normalization to the neutral current cross section a suppression of systematic uncertainties. However, σ^{NC} also depends on M_W and s_W^2 and this leads to a reduction of the sensitivity to these parameters. Consequently, very small effects may become important like higher–order corrections,

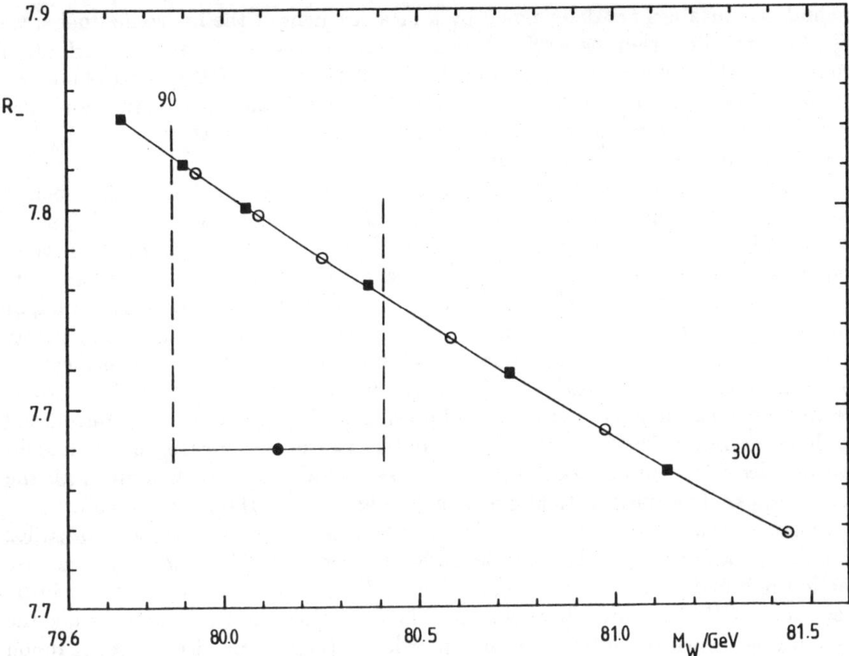

Figure 4: R_- as a function of M_W for $M_H = 60\,GeV$ ($\circ$) and $M_H = 1\,TeV$ ($\blacksquare$). The marked points correspond to the values $m_t = 90$, 120, 150, 200, 250, and $300\,GeV$. $p_t^{had} > 25\,GeV$. Also shown is the value for M_W with its uncertainty from LEP and $p\bar{p}$ data [26].

e.g., of $\mathcal{O}(\alpha\alpha_s)$ which are, however, not included in our calculation.

In Fig. 4, R_- is shown, derived from the total cross sections in the region $p_t^{had} \geq 25\,GeV$, as a function of M_W (see also the last column of table 1). Different values for the Higgs mass are not separated in this representation. Therefore, a measurement of R_- allows for a determination of M_W which is practically unique. With the help of the G_μ–M_W relation, this can be translated into a value for m_t to be then compared with direct measurements. This latter relation is, however, not unique but depends on M_H. A shift in R_- of 1% is obtained from a change in the W boson mass of $\Delta M_W = 630\,MeV$, corresponding to an increase of m_t by about $90\,GeV$. Also shown in the figure is the 1991 result of LEP + $p\bar{p}$ measurements [26].

The measurement of R_- can be interpreted as a direct M_W measurement. Note that the value for M_W obtained via G_μ is indirect and requires model assumptions and, in particular, depends on m_t in the standard model.

A careful analysis of R_- has to take into account also QED and QCD corrections. Purely electromagnetic corrections to both NC and CC cross sections are well–understood [13]. For the ratio R_- they are small of the order of 1 – 2% when kinematical variables are determined by a measurement of the hadronic final state [25]. In particular, they depend only weakly on the choice of parton distribution functions, if the low x region is excluded by suitable cuts. QCD corrections to R_- turn out to be smaller than $0.5\,\%$ [25]. Since both QED and QCD corrections also depend only weakly on the choice of electroweak parameters, they do not lead to additional contributions to systematic uncertainties.

During the HERA workshop in 1991, a careful study of the experimental feasibility of an R_- measurement was performed [27]. There it was argued that R_- determined from integrated cross sections with a cut on the hadronic transverse momentum, $\sigma(p_t^{had} > 25\,GeV)$ leads to an optimal sensitivity on M_W. Based on Monte Carlo simulations of deep inelastic scattering and photoproduction background, as well as a complete H1 detector simulation, a detailed investigation of effects from event selection, NC – CC separation, detector resolution, event misidentification, and background contamination was done. Systematic errors have been shown to be due to two effects mainly: i) imperfect knowledge of the absolute calibration and the detector resolution leading to uncertainties in the p_t^{had} measurement, and ii) inadequacies of the simulation, in particular systematics due to structure function uncertainties. The most critical effect in this analysis is the absolute calibration which leads to a one percent error for R_- which is comparable to the statistical error for $\int \mathcal{L}dt = 200\,pb^{-1}$. Resolution adds another 0.5%. A variety of structure function parametrizations have been applied and found to cause uncertainties in R_- of about $0.5\,\%$. Most critical in this respect is the ratio of down to up–type quarks. It should be known to better than $10\,\%$ in order to keep R_- precise to $1\,\%$. A direct measurement of the absolute size of parton distributions in the kinematical range under investigation [28] should therefore be used to limit the structure function uncertainties.

It was shown that the HERA data with the G_μ constraint superimposed will yield a determination of M_W with an accuracy of $860\,MeV$ or, equivalently, $\Delta m_t = \pm 120\,GeV$. This confrontation of high energy data from HERA with the low energy constraint from the muon lifetime constitutes a non–trivial test of the standard model. When considering R_- as a function of M_W with m_t as a parameter and M_H fixed to $100\,GeV$, the combination of HERA results with collider data for M_W would allow a top mass determination with $\Delta m_t = \pm 40\,GeV$. This precision would be comparable to what can be obtained at present from the LEP data and the G_μ constraint.

A refined analysis assuming several years of running with the high statistics of $\int \mathcal{L}dt = 1000\,pb^{-1}$, a lower p_t–cut of $15\,GeV$, and the use of additional information obtained from a shape analysis of the differential cross section $d\sigma^{CC}/dp_t^{had}$ would allow one to constrain M_W from the HERA data alone and leads to a closed region in the (M_W, m_t) plane as is shown in Fig. 5. Superimposing the G_μ constraint it was found that $\Delta M_W = \pm 200\,MeV$ (or $\Delta m_t = \pm 25\,GeV$) is feasible for a fixed Higgs mass. A variation of M_H from $100\,GeV$ to $800\,GeV$ would shift the central value by less than $50\,MeV$ (or $\Delta m_t = 15\,GeV$). Anticipating a measurement of the top quark mass with $\Delta m_t = \pm 10\,GeV$ from the Tevatron would yield $\Delta M_W = \pm 60\,MeV$ for a fixed

Higgs mass and a variation of M_H from $100\,GeV$ to $800\,GeV$ would give an additional uncertainty of $\Delta M_W = 135\,MeV$. The bounds for different M_H are separated in Fig. 5, so the level of being sensitive to M_H will be reached.

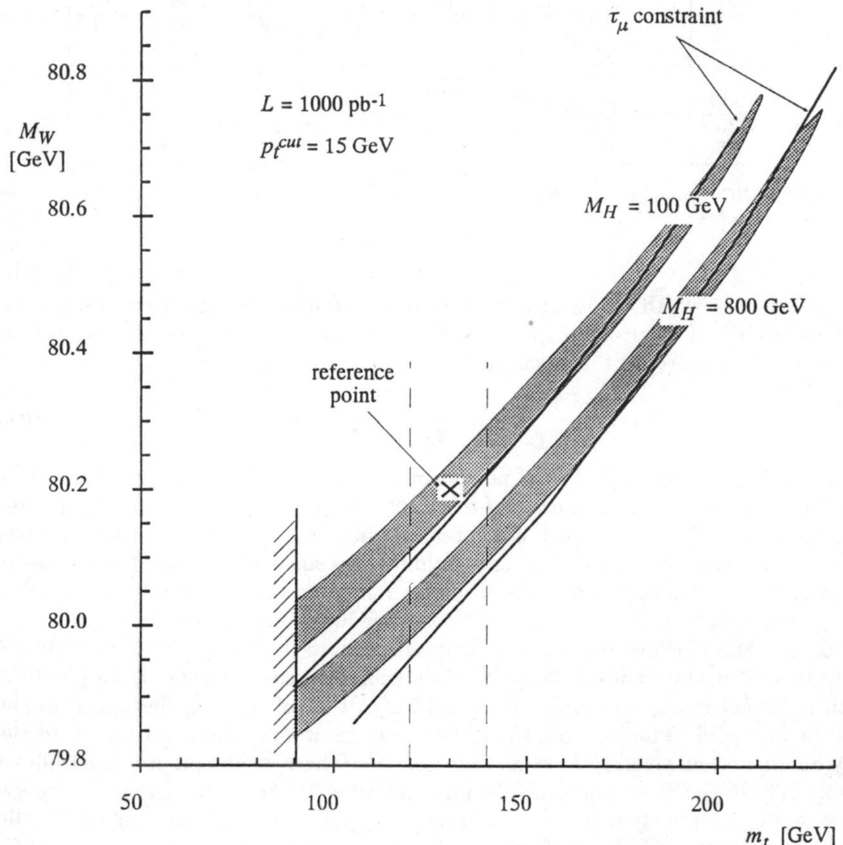

Figure 5: The $1\,\sigma$–contour for a fit to R_- and $d\sigma^{CC}/dp_t^{had}$ for two values of M_H. The reference point $M_W = 80.2\,GeV$ and $m_t = 130\,GeV$ for $M_H = 100\,GeV$ has been chosen arbitrarily but consistent with present experiments. The constraint from the muon lifetime is also indicated (assuming $M_Z = 91.174\,GeV$). Taken from [27].

3.3 *NC Asymmetries*

A lot of success in preparing beams with transverse polarization at HERA was achieved already and polarized lepton beams will be made available if it will have been demonstrated that a degree of polarization of more than 60 % can be reached.

m_t [GeV]	M_H [GeV]	A_- [%]	A_+ [%]	B_- [%]	B_+ [%]	C_L [%]	C_R [%]
90	60	7.67	-7.29	3.05	-2.67	10.32	-4.62
90	1000	7.65	-7.29	3.06	-2.70	10.33	-4.60
120	60	7.68	-7.29	3.05	-2.67	10.32	-4.64
120	1000	7.66	-7.30	3.06	-2.70	10.34	-4.61
150	60	7.71	-7.31	3.06	-2.66	10.35	-4.66
150	1000	7.69	-7.32	3.07	-2.69	10.36	-4.63
200	60	7.80	-7.37	3.08	-2.65	10.43	-4.73
200	1000	7.76	-7.37	3.09	-2.69	10.43	-4.68
250	60	7.93	-7.47	3.11	-2.65	10.55	-4.83
250	1000	7.84	-7.43	3.11	-2.69	10.51	-4.75

Table 2: The NC asymmetries $A_\pm$, $B_\pm$ and $C_{L,R}$ in % for several values of m_t and M_H. $M_Z = 91.175$ GeV. A cut of $p_t^{had} > 25\,GeV$ was used. No QED corrections.

Therefore it is worth studying the potential of asymmetry measurements for the determination of electroweak parameters. The purely statistical precision of an asymmetry measurement is given by

$$\Delta A = \frac{1 - A^2}{\sqrt{1 - A}} \frac{1}{\sqrt{2N}} \tag{38}$$

where N is the number of events per beam setting. For $\int \mathcal{L}dt = 100\,pb^{-1}$ and $p_t^{had} > 25\,GeV$ this gives a statistical error of a few 10^{-3}. No investigation of the experimentally reachable accuracy of such measurements has been done yet. Large uncertainties have to be expected for the charge asymmetries since in this case measurements of runs with different beams have to be combined and no automatic cancellation of systematic effects from energy calibration or luminosity measurement will take place. For this one has to rely on the normalization to the low Q^2 data which should help to reduce systematic effects. The precision of measurements of the polarization asymmetries $A_\pm$ will probably be similar to that of $R_\pm$. Smaller uncertainties can be expected to follow from the p_t^{had} measurement since the contribution to the asymmetries from small p_t^{had} events vanishes. Also there will be no problem from an imperfect NC – CC separation. The uncertainty of the polarization measurement, however, is completely unknown. In principal, the asymmetries are also affected by QED corrections, but these effects are largely cancelled. The asymmetries are also rather insensitive to uncertainties in the structure functions input.

Table 2 shows predicted values for the various asymmetries derived from integrated cross sections with a cut on p_t^{had} of $25\,GeV$ for various values of the top and Higgs mass. The electron polarization asymmetry A_-, displayed in Fig. 6, is more sensitive to changes in m_t than R_-. For example, a variation of 1% in A_- corresponds to a variation of $65 - 80\,GeV$ in m_t, or $380 - 460\,MeV$ in M_W, depending on the value of M_H. However, in all cases the changes with m_t and M_H are small when compared with the statistical accuracy. Only in the case of the electron polarization asymmetry, the variation with m_t between 90 and $250\,GeV$ comes close to the statistical accuracy for $\int \mathcal{L}dt = 200\,pb^{-1}$. Of special interest is the mixed

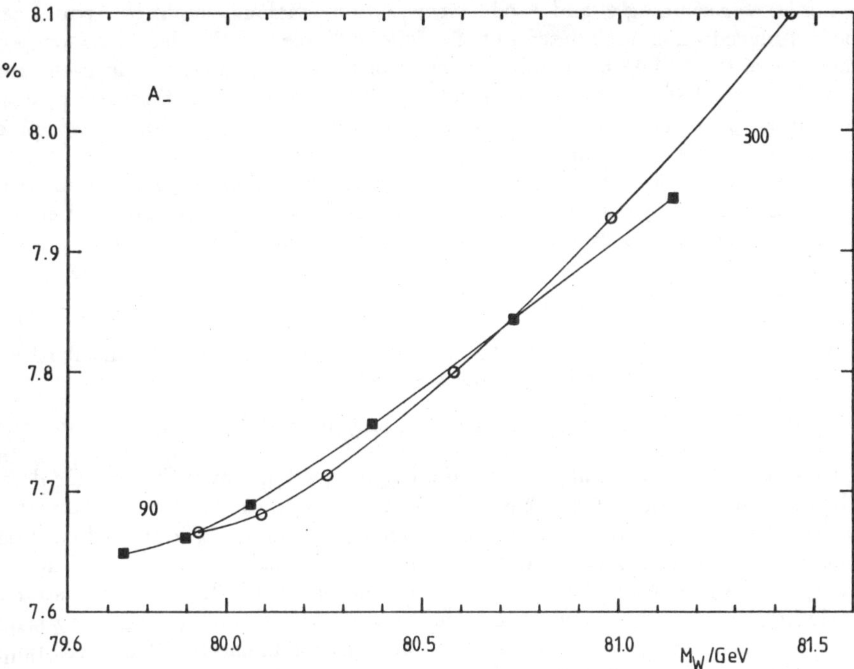

Figure 6: A_- as a function of M_W for $M_H = 60\,GeV$ (○) and $M_H = 1\,TeV$ (■). The marked points correspond to the values $m_t = 90$, 120, 150, 200, 250, and $300\,GeV$. $p_t^{had} > 25\,GeV$.

asymmetry B_+: its independence of the standard model parameters is completely negligible also for larger m_t than shown in the table but it is sensitive to extensions of the gauge group with an additional Z' boson.

4 Measurements of the $WW\gamma$ Coupling

Besides the conventional measurements of deep inelastic cross sections which can be used to determine the free parameters M_W, m_t, or M_H of the standard model, HERA has also some potential to test the three–boson couplings.

In the standard model, these couplings are unambiguously fixed by the non–abelian nature of the $SU(2) \times U(1)$ symmetry. In order to interpret a measurement of the three–boson couplings and to quantify possible deviations from the standard model, one has to generalize the standard model Lagrangian and allow for some ad hoc introduced non–standard interaction. The widely used procedure is to release the restrictions imposed by $SU(2)$ gauge invariance and to consider the most general Lorentz–invariant three–boson interactions preserving electromagnetic $U(1)$ gauge symmetry. Excluding C and P odd terms, two anomalous couplings κ and λ can be introduced which are related to the magnetic dipole and the electric quadrupole moment of the W boson. The introduction of these anomalous couplings violates unitarity and destroys the renormalizability of the theory. It is therefore justified for purely phenomenological reasons only and without changing other aspects of the theory at the same time does not correspond to a viable field theoretic model. We therefore do not consider the measurement of $WW\gamma$ couplings as a measurement of non–standard physics, but rather as a quantified statement of how precisely the three–boson couplings are tested inside the standard model framework.

4.1 *W Production at HERA*

The effect of anomalous $WW\gamma$ couplings on the production of $W^{\pm}$ bosons at HERA has been studied in [29, 30]. The processes

$$ep \to e^{-}W^{\pm}X, \quad ep \to \nu_e W^{-}X \tag{39}$$

have small cross sections at the level of a few $10^{-2}\,pb$. Since only the leptonic decay channels into $e^{\pm}$ and $\mu^{\pm}$ can easily be identified, one cannot expect more than about 50 events [31]. However, already moderate deviations of κ and λ from their standard model values $\kappa_{SM} = 1$, $\lambda_{SM} = 0$ destroy the gauge cancellations and can therefore lead to observable effects. In [30], the processes (39) have been studied including realistic experimental cuts and taking into account the branching ratios of $W^{\pm} \to l^{\pm}\nu_l$ with $l = e, \mu$. In order to account for unknown QCD corrections, normalization errors of 10 % (30 %) have been allowed for and with an assumed integrated luminosity of $1000\,pb^{-1}$ (at least five years of running) the following 90 % confidence level sensitivities have been found:

$$\begin{aligned} \Delta\kappa = \kappa - 1 &\simeq \pm 0.6 \;\; (\pm 0.9), \\ \lambda &\simeq \pm 1.2 \;\; (\pm 1.3). \end{aligned} \tag{40}$$

4.2 *Radiative Charged Current Scattering*

The $WW\gamma$ coupling enters also in a diagram contributing to the cross section for charged current radiative scattering

$$ep \to \nu\gamma X. \tag{41}$$

In contrast to the process (39), this is a charged current process which is suppressed as compared to neutral current processes like $ep \to e^{-}W^{\pm}X$. However, photons are observed directly and no reduction by an additional branching ratio takes place. Moreover, due to the vanishing photon mass, the phase space for the process (41) is larger than for (39).

Process (41) is dominated by photons which are either soft or collinear to the incoming lepton and the in– and outgoing quarks. These events are insensitive to the $WW\gamma$ coupling. Sensitivity to the three–boson coupling can be expected only for events where the photon has large transverse momentum and is well–separated from the jet produced by the scattered quark. Therefore the signal is found in the large p_t^γ end of the spectrum of events with an isolated hard photon, missing transverse momentum and a jet. The requirement of isolation can be implemented as a cut on the distance in the pseudorapidity–azimuthal angle plane.

The potential of (41) for tests of the $WW\gamma$ couplings was investigated in [32, 33, 34]. It was shown that after realistic experimental cuts, a study of this process can be based on a sample of about 300 events so that an investigation of differential cross sections is not unrealistic. Most sensitive is the distribution with respect to the transverse momentum of the bremsstrahlung photon. A likelihood fit to this distribution measured at HERA with $\int \mathcal{L}dt = 1000pb^{-1}$ should be able to establish $95\,\%CL$ limits of

$$\begin{aligned} \Delta\kappa &\simeq \pm 1.8\,, \\ \lambda &\simeq \pm 2.0\,. \end{aligned} \tag{42}$$

A HERA upgrade with $E_e = 35\,GeV$, $E_p = 1.2\,TeV$ and the same luminosity could improve the limits for $\Delta\kappa$ somewhat to $\Delta\kappa \simeq \pm 1.4$ and those for λ by almost a factor of 2 to $\lambda \simeq \pm 1.1$. Since the dominating terms in the cross section originate from the interference of anomalous couplings with standard model ones, which are linear in $\Delta\kappa$ and λ, the bounds scale roughly with the square–root of the luminosity. Due to the smaller cross section, positron scattering is less sensitive than electron scattering.

Somewhat weaker bounds (by a factor of 1.2 to 1.7) have been found in [33, 34] due to a different method to determine the sensitivity (in [33]) or due to taking into account a normalization error of 30 % (in [34])[1].

At higher center-of-mass energies as could be realized at LEP $\times$ LHC ($\sqrt{s} = 1.4\,TeV$), the process (41) would lead to bounds on $\Delta\kappa$ which are comparable to those obtained from $e^+e^- \to W^+W^-$ at LEP II. Bounds on λ would be even superior. But they are still less sensitive than those from SSC or LHC.

5 Electroweak Physics Beyond the Standard Model

5.1 *Virtual New Physics*

The self energy corrections receive contributions from the complete electroweak standard model. Also beyond the standard model, any new particle will contribute a loop diagram to the self energies. Consequently, measurements with a precision high enough for being sensitive to radiative corrections will offer the possibility to access new physics degrees of freedom without the need to provide enough energy to produce them directly in an experiment.

[1] Since the lower end of the p_t^γ spectrum is insensitive to the $WW\gamma$ couplings, a measurement of low p_t^γ photons should allow one to reduce this error.

We consider the expansions

$$\Sigma^{ij}(q^2) = \Sigma^{ij}(0) + q^2 F^{ij}(q^2) \tag{43}$$

of the unrenormalized self energies with $(ij) = (\gamma\gamma)$, (γZ), (ZZ), (WW). Alternatively to (γZ) and (ZZ), also the isospin components (3γ) and (33) can be used with[2] $\Sigma^{3\gamma} = s_W \Sigma^{\gamma\gamma} - c_W \Sigma^{\gamma Z}$ and $\Sigma^{33} = c_W^2 \Sigma^{ZZ} - 2c_W s_W \Sigma^{\gamma Z} + s_W^2 \Sigma^{\gamma\gamma}$. Neglecting the q^2 dependence of F^{ij}, there are 8 constant quantities. Due to the QED Ward identity, one has $\Sigma^{\gamma\gamma}(0) = 0$ and in the on–mass–shell renormalization scheme 4 other combinations of Σ^{ij}'s and F^{ij}'s are absorbed into renormalization constants. Thus, three remaining quantities can be found which are finite and consequently observable. Using the notation of [35] one can choose them as

$$\begin{aligned}
\Delta\rho &= \frac{\Sigma^{ZZ}(0)}{M_Z^2} - \frac{\Sigma^{WW}(0)}{M_W^2} - \frac{s_W}{c_W}\frac{2\Sigma^{\gamma Z}(0)}{M_Z^2}, \\
\Delta_1 &= \frac{1}{s_W}F^{3\gamma}(M_Z^2) - F^{33}(M_Z^2), \\
\Delta_2 &= F^{33}(M_Z^2) - F^{WW}(M_W^2). \qquad (44)
\end{aligned}$$

(45)

Several other definitions are being used in the literature. The S, T, U parameters of [36] are related to Eq. (44) by

$$S = \frac{4s_W^2}{\alpha}\Delta_1, \quad T = \frac{1}{\alpha}\Delta\rho, \quad U = \frac{4s_W^2}{\alpha}\Delta_2, \tag{46}$$

and the ϵ_i's of [37] by

$$\epsilon_1 = \Delta\rho, \quad \epsilon_2 = -\Delta_2, \quad \epsilon_3 = \Delta_1. \tag{47}$$

The classification of self energy corrections into the above given combinations is motivated by their properties with respect to isospin symmetry. $\Delta\rho$ is a measure of the violation of the custodial isospin symmetry and enters directly into the relation between the weak mixing angle and the gauge boson masses (see Eq. (7) with $\rho_0 = 1 + \Delta\rho$). The Δ_i's contribute universally, i.e., in a process–independent way, to any observable. To one–loop precision one can write

$$\mathcal{O}^{1-loop} = (1 + c_\rho\Delta\rho + c_1\Delta_1 + c_2\Delta_2 + \delta_{rem})\mathcal{O}^0. \tag{48}$$

The coefficients c_i are specific to the process under consideration and δ_{rem} contains other contributions from vertex corrections and box diagrams. δ_{rem} has also to account for q^2 dependent terms in the self energies that have been neglected in the definition of the Δ_i.

An important observable which has been used for a determination of the Δ_i's is Δr in the G_μ–M_W relation. Again neglecting higher than $\mathcal{O}(\alpha)$ terms, one has

$$\Delta r = \Delta\alpha - \frac{c_W^2}{s_W^2}\Delta\rho - \frac{c_W^2 - s_W^2}{s_W^2}\Delta_2 + 2\Delta_1 + \delta r_{rem}. \tag{49}$$

---

[2]Note that we use here and in the following the standard model value of the weak mixing angle although we are going to discuss effects of non–standard model physics which will affect also the value of $\sin\theta_W$. This is justified since the error is formally of higher order and numerically small as long as we consider small non–standard effects.

Another observable is the effective weak mixing angle determined from the NC amplitude on the Z resonance. The universal contribution appearing in its definition has the following form:

$$s^2_{W,\mathrm{eff}} = (1+\Delta\kappa)s^2_W + ... \tag{50}$$

with

$$\Delta\kappa = \frac{c^2_W}{s^2_W}(\Delta\rho + \Delta_2) - \Delta_1. \tag{51}$$

Since s^2_W is defined via the M_W/M_Z mass ratio, its value itself depends on Δ_i when M_W, G_μ and m_t is used as input. It is therefore useful to define the effective mixing angle in relation to s^2_0 determined from

$$G_\mu = \frac{\pi\alpha(M^2_Z)}{\sqrt{2}s^2_0 c^2_0 M^2_Z}. \tag{52}$$

s^2_0 includes the running of α $\left(\alpha(M^2_Z) = \alpha(1+\Delta\alpha)\text{, see Eq. (29)}\right)$ but no purely weak corrections (compare Eqs. (26, 27)). To one–loop order one has then

$$s^2_{W,\mathrm{eff}} = (1+\Delta k)s^2_0 + ... \tag{53}$$

with

$$\Delta k = -\frac{c^2_W}{c^2_W - s^2_W}\Delta\rho + \frac{\Delta_1}{c^2_W - s^2_W}. \tag{54}$$

This shows that $s^2_{W,\mathrm{eff}}$ does not receive a contribution from Δ_2 and consequently the measurement of NC amplitudes does not allow one to determine Δ_2 (or U). NC amplitudes are sensitive only to Δ_1 and $\Delta\rho$, the latter one entering also directly into their normalization. In the CC amplitude, the sensitivity to virtual new physics enters only through M_W which is determined by Δr.

Assuming the standard model to be valid at present energies and keeping fixed the yet unknown masses of the top quark and the Higgs boson at reference values m^0_t, M^0_H, one can ask whether additional contributions from beyond the standard model to the Δ_i's have an observable effect on present day's experiments. Following the procedure of [38] we write

$$\Delta_i = \Delta^{SM}_i(m_t = m^0_t, M_H = M^0_H) + \Delta^{nonSM}_i \tag{55}$$

and take as reference values

$$m^0_t = 140\,GeV, \quad M^0_H = 300\,GeV. \tag{56}$$

Considering small non–standard contributions Δ^{nonSM}_i, one can write for an observable

$$\mathcal{O} = \mathcal{O}^{SM}(1 + c_\rho\Delta\rho^{nonSM} + c_1\Delta^{nonSM}_1 + c_2\Delta^{nonSM}_2), \tag{57}$$

where $\mathcal{O}^{SM}$ is the prediction for the observable including standard model radiative corrections for the reference values m^0_t and M^0_H.

In table 3 we have listed the sensitivities c_i for the NC and CC total cross sections, the ratios $R_\pm$ and the NC asymmetries. For example, a change of $\Delta\rho$ by

	c_ρ	c_1	c_2
$\sigma_{NC}(e^-)$	0.24	-0.21	–
$\sigma_{NC}(e^+)$	0.17	-0.19	–
$\sigma_{CC}(e^-)$	1.1	-0.67	0.80
$\sigma_{CC}(e^+)$	0.93	-0.55	0.65
R_-	-0.89	0.46	-0.79
R_+	-0.76	0.36	-0.65
A_-	3.4	-3.1	–
A_+	2.6	-2.2	–
B_-	2.2	-1.5	–
B_+	-0.26	1.4	–
C_L	2.5	-1.9	–
C_R	4.2	-4.2	–

Table 3: Sensitivities to non–standard contributions Δ_i^{nonSM} according to Eq. (57). Cross sections were obtained with $p_t^{had} > 25\,GeV$.

0.01 would reduce R_- by 0.89 % and A_- would increase by 3.4 %. As already explained, the NC observables have no sensitivity on Δ_2 and for them there is no entry in the last column of the table.

The unpolarized NC cross sections, being dominated by the one–photon exchange at low Q^2, have only small sensitivities to the Δ_i's. They have the same sign as those of the CC cross sections, thus reducing slightly the sensitivity of the ratios $R_\pm$. The value of $c_\rho = -0.89$ for R_- is, of course, in agreement with what was discussed in section 3.2: a change of m_t from $140\,GeV$ to $230\,GeV$, corresponding to an increase of the standard model contribution to $\Delta\rho$ by 0.01, would reduce R_- by 0.89 %. These two top mass values correspond to $M_W = 80.690\,GeV$ and $80.101\,GeV$, resp. (for $M_H = 300\,GeV$).

The sensitivity of R_- on Δ_2 is not much worse than that on $\Delta\rho$ and a 1 % measurement of R_- would determine Δ_2 with an uncertainty of $\pm 1.25 \cdot 10^{-2}$. This is somewhat worse than what has been achieved recently from a combined fit on LEP and collider data [38].

The NC asymmetries are much more sensitive than $R_\pm$ to variations of both $\Delta\rho$ or Δ_1. Ignoring systematic uncertainties, the large sensitivity 3.4 for A_- on $\Delta\rho$ would allow a measurement of $\Delta\rho$ with a precision of ± 0.01 for an integrated luminosity of $\int \mathcal{L}dt = 230\,pb^{-1}$ and about $1000\,pb^{-1}$ per beam would be needed to achieve a precision of $\Delta\rho = \pm 0.005$. This high sensitivity is due to the fact that the pure photon exchange contribution to the NC cross sections is suppressed in the asymmetries. The sensitivity to Δ_1 is via the weak mixing angle only, that to $\Delta\rho$ is due partly to its effect on $s^2_{W,\text{eff}}$ and partly to the normalization of the Z exchange part of the NC amplitude. Again it is seen that the mixed charge–polarization asymmetry B_+ is exceptional: it is almost insensitive to $\Delta\rho$ and would be an interesting observable for disentangling virtual non–standard physics contributions and the effect of a heavy standard model top from other sources as for example new gauge bosons. The table also shows that electron scattering gives slightly better results than positron scattering. Note that also due to its larger cross section, electron scattering is superior to positron scattering. One has to wait for the future to see whether the higher sensitivities of the NC asymmetries $A_\pm$ and C_R can be utilized for precision measurements of the Δ_i's.

The above discussion assumed only constant non–standard contributions to F^{ij}. This simplifying assumption was motivated from similar studies at LEP where $\Delta\rho$ and Δ_1 have already been very much constrained. The next simplest case would be to assume

$$F^{ij} \propto \frac{\alpha}{\pi} c^{ij} \frac{q^2}{\Lambda^2}. \tag{58}$$

A possible extra contribution of this form to R_-

$$\Delta\rho_W = -\frac{\alpha}{\pi} c^{WW} \frac{q^2}{\Lambda^2} \tag{59}$$

would, however, require a rather low lying new–physics scale of $\Lambda \simeq 300\,GeV$ together with an unnatural large coupling strength of $c^{WW} = 40$ in order to lead to a visible effect of $1\,\%$ in R_-.

One should keep in mind that the measurement of non–zero values for the Δ_i's does not automatically imply the presence of new physics. A comparison of different measurements is needed in order to separate non–standard effects from a standard model heavy top mass. In addition, unknown higher–order effects may blur the interpretation of the measurements.

5.2 *Extra Heavy Z' Bosons*

In grand–unified theories the standard model is embedded in a larger gauge group which predicts the existence of additional gauge bosons. Most important and theoretically interesting possibilities are left–right symmetric models and simple $U(1)$ extensions of the gauge group. The good prospects of HERA in the search for new gauge bosons has been recognized since long [6, 39]. It is beyond the scope of this article to discuss specific models which all lead to largely differing discovery limits. Rather we focus on general features which are common to most of the standard model extensions.

The direct production of new gauge bosons is excluded at HERA. Also the effect of Z–Z' mixing which modifies the standard Z couplings and thus changes the event rates is already very much constrained from LEP data[3], whereas the present bounds on Z' masses are rather weak, typically in the range of $160 - 400\,GeV$, depending on the model. Thus one has to focus on the modification of the NC amplitude due to virtual Z' exchange. Since the additional terms are dominated by the γ–Z' interference contribution which is proportional to $Q^2/(Q^2+M_{Z'}^2)$, observable effects can be expected only at the highest possible Q^2. Thus, precision measurements and high luminosities are required. It may turn out that for smaller center-of-mass energies, the loss in the reach of Q^2 can be overcompensated by a gain in the luminosity.

[3] HERA could have reached limits on Z–Z' mixing of not better than 2°.

An additional Z' boson leads to additional terms in the NC effective Lagrangian of the following form

$$-\mathcal{L}_{NC}^{eff} = eJ_{em}^{\mu}A_{\mu} + \frac{e}{\sin\theta_W \cos\theta_W} J_Z^{\mu} Z_{\mu} + g' J_{Z'}^{\mu} Z'_{\mu} \tag{60}$$

where J_{em}, J_Z are determined by the couplings in Eq. (6) and

$$J_{Z'}^{\mu} = \sum_f \bar{f}\gamma^{\mu}(v'_f - a'_f\gamma_5)f. \tag{61}$$

$J_{Z'}$ contributes additional terms to the $A_f^{L,R}$, $B_f^{L,R}$ in Eq. (5). As can be seen from these equations, the couplings v'_f and a'_f appear in the following combinations

$$\begin{aligned} \eta_{VV}^q &= g'^2 v'_e v'_q, & \eta_{AA}^q &= g'^2 a'_e a'_q, \\ \eta_{VA}^q &= g'^2 v'_e a'_q, & \eta_{AV}^q &= g'^2 a'_e v'_q. \end{aligned} \tag{62}$$

For unpolarized cross sections, the dominating γ–Z' interference contains only η_{VV}^q with the factor $[1+(1-y)^2]$ (in F_2) and η_{AA}^q with $[1-(1-y)^2]$ (in F_3). Since F_2 dominates F_3 and the up–type quark distribution is considerably bigger than the down–type ones, the most important term is thus determined by η_{VV}^u. As an immediate consequence, the string–motivated models which always have $v'_u = 0$, lead to small changes of unpolarized cross sections. Indeed, the discovery limits for Z' masses in this important class of models are relatively small, usually below $200\,GeV$. Asymmetry measurements are more powerful in these cases and would lead to improved discovery limits.

A model–independent analysis has to deal in general with a 6–dimensional space of coupling constants. Models with $SU(2)_L$ invariant couplings have only 5 parameters due to the relation $v'_u + a'_u = v'_d + a'_d$. In [40], 2σ exclusion regions and directions of largest sensitivity in the parameter space have been determined. The result can be expressed most generally with the help of the quantity

$$m_{Z'} = M_{Z'} \cdot \frac{e}{g' \cos\theta_W \sin\theta_W} \cdot \frac{1/4}{c} \tag{63}$$

where c is a measure of the overall normalization of the Z' couplings. Assuming $c = 1/4$, i.e., coupling strengths of the Z' of equal size as the standard model axial couplings, it was shown that HERA can reach masses of up to $850\,GeV$ in the most favorable cases, and in large parameter space regions $m_{Z'} = 400 - 800\,GeV$ can be reached for measurements with polarized electron and positron beams (each contributing $\int \mathcal{L}dt = 100\,pb^{-1}$ and $\mathcal{P} = 0.8$). This study motivated the construction of so–called 'HERA–tailored' sets of Z' couplings. They have been studied in more detail in [41] where additional effects have been considered as for example QED corrections in the leading logarithmic approximation, weak standard model loop corrections and realistic experimental conditions. The results have shown that when compared with Tevatron, there are models where HERA is superior, others where both are comparable. If deviations from the standard model predictions are found, HERA will be able to disentangle different models since ep cross sections are sensitive to relative signs of the couplings whereas Tevatron cross sections are determined by squares of them. To achieve this, the measurement of the various NC asymmetries is indispensable. The most sensitive one was found to be B_+ [42, 43] where even in specific string–inspired models Z' masses of up to about $300\,GeV$ can lead to an observable effect. For very heavy Z' the strongest constraints come from $A-$.

In extended models which contain both a Z' and an additional charged gauge boson W', for example in left–right symmetric models, the ratio R_- can be utilized for a determination of mass limits [44]. The amplitude for charged current scattering is changed from

$$\mathcal{M}_W \propto \frac{M_W^2 G_\mu/\sqrt{2}}{Q^2 + M_W^2} \tag{64}$$

in the standard model to

$$\mathcal{M}_{W_1W_2} \propto \frac{g_1^2}{8}\frac{1}{Q^2+M_1^2} + \frac{g_2^2}{8}\frac{1}{Q^2+M_2^2} \tag{65}$$

in a model with two charged gauge bosons W_1 and W_2 with masses M_1 and M_2. At low energies, the G_μ constraint has to be obeyed which gives the condition

$$\frac{G_\mu}{\sqrt{2}} = \frac{g_1^2}{8M_1^2} + \frac{g_2^2}{8M_2^2}. \tag{66}$$

Identifying M_1 with the standard W boson mass M_W, one obtains for the enhancement of the differential charged current cross section

$$r = \frac{d^2\sigma^{CC}/dx\,dQ^2(W_1+W_2)}{d^2\sigma^{CC}/dx\,dQ^2(SM)} = \left[1 + x\frac{M_W^2}{Q^2+M_2^2}\left(\frac{Q^2}{M_W^2} - \frac{Q^2}{M_2^2}\right)\right]^2 \tag{67}$$

where $x = g_2^2/(4\sqrt{2}M_W^2/G_\mu)$ is the ratio of the gauge couplings of W_2 and W_1.

The direct comparison of the measured differential charged current cross section to the standard model prediction as in Eq. (67) is spoiled by large normalization uncertainties. In order to take advantage of the high precision of the R_- measurement, one has to take into account simultaneous changes in the NC amplitude induced by the presence of the Z'. An important precondition for turning this measurement successfully into mass limits for new gauge bosons is the existence of a relation between the W' and Z' masses. For some models, such a measurement substantially increases the HERA discovery limits beyond those from the NC asymmetries and beyond those from direct searches at Tevatron [44].

5.3 *Right–Handed Charged Currents*

Extended models with gauge groups

$$SU(2)_L \times SU(2)_R \times U(1) \tag{68}$$

predict the existence of right–handed charged currents. Such models require additional neutrinos. A minimal possibility is embedding (68) as a subgroup in a $SO(10)$

grand–unified theory. There, only one additional right–handed neutrino is required for each generation to obtain anomaly–free fermion representations.

Right–handed W_R exchange contributes to the charged current cross section at HERA and is accompanied by the production of right–handed neutrinos

$$e + p \rightarrow \nu_R + X. \tag{69}$$

If ν_R is a light Dirac neutrino, only indirect searches are possible: one has to look for small deviations of the cross section for charged current events characterized by missing transverse momentum. For experiments with right–handed polarized electrons, limits for the mass of the right–handed charged boson M_{W_R}, even with optimistic assumptions concerning experimental errors, are not better than $400\,GeV$ [45]. This is below the present limit of $450\,GeV$ obtained from low energy data [46] and from LEP data [47]. For models with spontaneous parity breaking, the present limit is even bigger, $1-3\,TeV$, [48].

However, if ν_R is sufficiently heavy and a Majorana neutrino, decays into both electron and positron are possible. This would lead to spectacular signatures at HERA consisting of a positron accompanied by missing transverse momentum. A detailed Monte Carlo study taking into account the full hadronic final state and a detector simulation with realistic experimental conditions [49] has shown that the electron decay channel cannot be used since no way was found to reduce the background from standard deep inelastic scattering. However, the positron decay channel is background–free after cuts on missing transverse momentum and isolation. Also charge identification was shown to be feasible. Asking for more than 5 events, the discovery limits at HERA design values were derived to be

$$m_{\nu_R} > 120\,GeV, \tag{70}$$

and

$$M_{W_R} > 700\,GeV. \tag{71}$$

An HERA upgrade with $\sqrt{s} = 450\,GeV$ and $\int dt\mathcal{L} = 4000\,pb^{-1}$ could reach masses up to $1\,TeV$.

6 Conclusions

Precision measurements at HERA, making use of the high luminosity of several years of experimentation with electron proton collisions, will contribute to an improved understanding of electroweak physics, provided a careful study of experimental uncertainties succeeds in keeping systematic effects at the presently expected low level.

The most interesting observable is the ratio of neutral to charged current cross sections R_- which is favorable since it will profit of a high statistic and of the cancellation of systematic uncertainties. With an integrated luminosity of one year of running, a combination of HERA data on R_- with the G_μ constraint will yield a determination of M_W with an error of $860\,MeV$. This measurement can be interpreted as a direct M_W measurement. In particular, it does not require an assumption about the value of the top mass. The latter can be determined from R_- using G_μ with an accuracy of $120\,GeV$. Combining R_- with present day's collider data on M_W would yield the top mass with an error of $\Delta m_t = \pm 40\,GeV$. It seems feasible in the long run, that an ultimate precision of $\Delta M_W = \pm 200\,MeV$, or equivalently $\Delta m_t = \pm 25\,GeV$, will be reached from an analysis of HERA data with an integrated luminosity of $\int \mathcal{L}dt = 1000\,pb^{-1}$.

Polarization and charge asymmetries of the neutral current scattering process are in principal more sensitive to electroweak parameters than R_-. However, they require special runs for each charge and polarization and they will therefore not be measured with a luminosity as high as can be used for R_-. Moreover, it has not yet been studied to what extent systematic effects can be kept under control.

We have also discussed the prospects of ep scattering for a measurement of $WW\gamma$ couplings. At HERA, the measurement of W boson production turned out to be somewhat superior to charged current radiative scattering. However, the limits for anomalous couplings $\Delta\kappa$ and λ are rather large and these measurements will at most serve as a consistency check of measurements at LEP 2 and hadron colliders.

In addition to these purely standard model topics, we also asked to what extent precision measurements of R_- or NC asymmetries might be able to detect new physics effects that come through virtual corrections. Assuming that physics beyond the standard model will not change too much the q^2 dependence of electroweak form factors, $\Delta\rho$, and with a slightly smaller accuracy also Δ_2, can be determined from R_- with a precision at the level of 10^{-2}.

In the search for new heavy gauge bosons, HERA will improve the mass limits of Z' bosons considerably. It was shown that in the most favorable cases Z' masses of up to $850\,GeV$, and in large regions of the space of coupling constants in the range of $400 - 800\,GeV$, will either be seen or excluded. If a Z' should be found at not too high a mass, ep scattering with polarized electron and positron beams will be an important tool for the study of its nature. Finally, right–handed currents, if they are coupled to a heavy Majorana neutrino, are observable at HERA for W_R masses up to $700\,GeV$ and neutrino masses up to $120\,GeV$.

HERA for itself will most likely not compete in precision of electroweak measurements with other experiments. This does not mean that electroweak studies at HERA are not worth being done. It is indispensable to know whether our understanding of the electroweak interaction is correct in all high energy reactions. In many cases, the interpretation of a measurement is ambiguous. Important examples mentioned in this article are the measurement of virtual new physics with the help of the Δ_i's or the search for Z' bosons. In these cases, HERA will certainly help to clarify the interpretation of other experiments, although it may probably not contribute to an increase of the precision with which electroweak parameters will have been measured elsewhere by the end of this century.

7 Acknowledgements

I am indebted to my colleagues W. Hollik, D. Bardin, J. Blümlein, B. Kniehl, T. Riemann and D. Haidt for the fruitful collaboration with them during the HERA workshop 1991 which made this article possible.

8 References

1. H1 collaboration, *Technical Proposal for the H1 Detector*, DESY 1986; H1 collaboration, *Technical Progress Report, H1 Detector*, DESY 1987.
2. ZEUS collaboration, *The ZEUS Detector. Technical Proposal*, DESY, March 1986; ZEUS collaboration, *The ZEUS Detector. Status Report 1989*, DESY, March 1989.
3. R. D. Peccei, Ed., *Proceedings of the HERA Workshop, Hamburg, Oct. 12 - 14, 1987*, Vol. 2 (Hamburg 1987).
4. D. Atwood et al. in *Research Directions for the Decade*, Proceedings of the Snowmass Summer Study on High Energy Phsysics, Snowmass, Colorado, 1990, Eds. E. L. Berger and I. Butler (World Scientific, Singapore, 1992).
5. W. Buchmüller and G. Ingelman, Eds., *Physics at HERA, Proceedings of the Workshop, Hamburg, Oct. 29 - 30, 1991*, Vol. 2 (Hamburg 1991).
6. J. Cashmore et al., *Phys. Rep.* **122 C** (1985) 275.
7. D. Bardin, C. Burdik, P. Christova and T. Riemann, *Z. Phys.* **C42** (1989) 679.
8. D. Bardin, C. Burdik, P. Christova and T. Riemann, *Z. Phys.* **C44** (1989) 149.
9. M. Böhm and H. Spiesberger, *Nucl. Phys.* **B294** (1987) 1081.
10. M. Böhm and H. Spiesberger, *Nucl. Phys.* **B304** (1988) 749.
11. H. Spiesberger, *Nucl. Phys.* **B349** (1991) 109.
12. G. Kramer and H. Spiesberger, in *HERA proceedings 1991* [5], Vol. 2, p. 789.
13. H. Spiesberger et al., in *HERA proceedings 1991* [5], Vol. 2, p. 798.
14. A. Sirlin, *Phys. Rev.* **D22** (1980) 971.
15. J. Kripfganz, H.-J. Möhring and H. Spiesberger, *Z. Phys.* **C49** (1991) 501.
16. D. Yu. Bardin, M. S. Bilenky, G. V. Mitselmakher, T. Riemann and M. Sachwitz, *Z. Phys.* **C44** (1989) 493.
17. D. Yu. Bardin, W. Hollik and T. Riemann, *Z. Phys.* **C49** (1991) 485.
18. A. Akhundov et al., *package DIZET, version 4.04 (21 Aug 1991)*; version 2 was described in: D. Bardin et al., *Comput. Phys. Commun.* **59** (1990) 303.
19. R. Barbieri et al., *Phys. Lett.* **B288** (1992) 95.

20. F. Jegerlehner, in *Progress in Particle and Nuclear Physics*, ed. A. Fässler, (Pergamon Press, Oxford, U. K. 1991); updated from: H. Burkhardt, F. Jegerlehner, G. Penso and C. Verzegnassi, *Z. Phys.* **C43** (1989) 497.
21. M. Veltman, *Acta Phys. Polon.* 8 (1977) 475; *Phys. Lett.* **B70** (1977) 253.
22. M. Consoli, W. Hollik and F. Jegerlehner, *Phys. Lett.* **B227** (1989) 167;
W. Hollik, *Fortschr. Phys.* **38** (1990) 165;
S. Franchiotti and A. Sirlin in *M.A.B. Bég Memorial Volume*, Eds. A. Ali and P. Hoodbhoy (World Scientific, Singapore, 1991) p. 58.
23. D. W. Duke and J. F. Owens, *Phys. Rev.* **D30** (1984) 49.
24. J. F. Wheater, *Nucl. Phys.* **B233** (1984) 365;
J. Blümlein, M. Klein and T. Riemann, in *HERA proceedings 1987* [3], Vol. 2, p. 687;
G. Cozzika, D. Haidt and G. Ingelman, in *HERA proceedings 1987* [3], Vol. 2, p. 713;
O. Gry, G. D. Heath and E. Paul, in *HERA proceedings 1987* [3], Vol. 2, p. 719.
25. W. Hollik et al., in *HERA proceedings 1991* [5], Vol. 2, p. 923.
26. H. Plothow-Besch, in *Proceedings of the Joint International Lepton–Photon Symposium and Europhysics Conference on High Energy Physics, Geneva 1991*, Eds. S. Hegarty, K. Potter and E. Quercigh, (World Scientific, Singapore, 1991) Vol. 1, p. 36.
27. V. Brisson et al., in *HERA proceedings 1991* [5], Vol. 2, p. 947.
28. G. Ingelman and R. Rückl, *Z. Phys.* **C44** (1989) 291;
J. Blümlein, M. Klein, T. Naumann and T. Riemann, in *HERA proceedings 1987* [3], Vol. 1, p. 67;
J. Blümlein and M. Klein, in *HERA proceedings 1991* [5], Vol. 1, p. 101.
29. M. Böhm and A. Rosado, *Z. Phys.* **C42** (1989) 479.
30. U. Baur and D. Zeppenfeld, *Nucl. Phys.* **B325** (1989) 253.
31. U. Baur, J. A. M. Vermaseren and D. Zeppenfeld, *Nucl. Phys.* **B375** (1992) 3.
32. T. Helbig and H. Spiesberger, *Nucl. Phys.* **B373** (1992) 73.
33. S. Godfrey, *Z. Phys.* **C55** (1992) 619.
34. U. Baur and M. Doncheski, preprint, Florida State University, Tallahassee, FSU-HEP-920225 (1992).
35. G. Burgers and F. Jegerlehner, in *Z Physics at LEP1*, CERN 89-08, Eds. G. Altarelli, R. Kleiss and C. Verzegnassi.
36. M. E. Peskin and T. Takeuchi, *Phys. Rev. Lett.* **65** (1990) 964; *Phys. Rev.* **D46** (1992) 381.
37. G. Altarelli and R. Barbieri, *Phys. Lett.* **B253** (1991) 161.

38. J. Ellis, G. L. Fogli, and E. Lisi, *Phys. Lett.* **B285** (1992) 238; *Phys. Lett.* **B292** (1992) 427;

39. G. Altarelli, B. Mele and R. Rückl, in CERN report 84–10, Ed. M. Jacob (1984) p. 549.

40. P. Haberl, F. Schrempp, H.-U. Martyn and B. Schrempp, in *HERA proceedings 1991* [5], Vol. 2, p. 980.

41. H.-U. Martyn et al., in *HERA proceedings 1991* [5], Vol. 2, p. 987.

42. F. Cornet and R. Rückl, *Phys. Lett.* **B184** (1987) 263.

43. S. Capstick and S. Godfrey, *Phys. Rev.* **D35** (1987) 3351.

44. T. G. Rizzo, *Phys. Rev.* **D46** (1992) 3751.

45. F. Cornet and R. Rückl, in *HERA Proceedings 1987* [3], Vol. 2, p. 771.

46. P. Langacker and S. Uma Sankar, *Phys. Rev.* **D40** (1989) 1569.

47. W. Buchmüller and C. Greub, DESY 92-023 (1992).

48. G. Beall, M. Bender and A. Soni, *Phys. Rev. Lett.* **48** (1982) 848; G. Ecker and W. Grimus, *Nucl. Phys.* **B258** (1985) 328.

49. W. Buchmüller et al., in *HERA proceedings 1991* [5], Vol. 2, p. 1003.

V. THE WEAK CHARGED CURRENT

PRECISION MEASUREMENTS IN MUON AND TAU DECAYS

W. Fetscher and H.-J. Gerber
Institut für Mittelenergiephysik (IMP), ETH Zürich,
CH-5232 Villigen PSI, Switzerland

Contents

1. Introduction

The standard model [1, 2, 3] introduces the $V-A$ form of the charged leptonic weak interaction *by construction*. It has been shown recently that $V-A$ *follows* from a small set of experiments [4]. We will present the corresponding methods and discuss the room left open by the experimental errors for interactions other than $V-A$. For muons, all of these experiments have been performed.

The most general, derivative-free, lepton-number conserving four-fermion interaction [5] contains ten complex coupling constants which represent nineteen free parameters to be determined by experiment. Based on fields with definite chiralities and using the freedom of the Fierz transformations, a set of coupling constants may be chosen such that the $V-A$ interaction corresponds to *one single constant* only [6, 7]. The experimental proof of $V-A$ then consists in the construction of measurable quantities which evaluate the remaining eighteen constants to be zero [4]. To achieve this, use is made of the general fact (pointed out in Ref. [8]) that a null-result for a sum of positive semi-definite terms requires each term to be zero.

For the muon decay interaction, the ten complex planes of Fig. 2.2 represent a lower limit for $V-A$ and upper limits for the nine remaining couplings.

If lepton-number conservation is not assumed, other decays like $\mu \to e\gamma$ become possible, which open windows to new physics. Experiments looking for such forbidden decays have been driven to a level of 10^{-13} for the muon and 10^{-5} for the tau (see reviews [9] and [10], respectively). The lepton-number non-conserving four-fermion interaction has been studied in detail [11, 12]. The authors of Ref. [12] arrive at the interesting result that it is not possible, even in principle, to test lepton-number conservation in muon decay if the final neutrinos are massless and are not observed.

In this review we wish to present *examples* of typical thoughts which inspired experiments and their analysis and which led to the present knowledge of the Lorentz structure of the charged leptonic weak interaction.

Since there exist complete reviews on experimental results [13], on the future of muon physics in general [14], with emphasis on rare and forbidden processes [9], on theoretical [15, 16, 17] and historical [18] aspects, we allow ourselves to be complementary rather than to aim at completeness.

2. Muon Decay

2.1. Hamiltonian

The three leptonic decays $\mu^+ \to \bar{\nu}_\mu e^+ \nu_e$, $\tau^+ \to \bar{\nu}_\tau \mu^+ \nu_\mu$ and $\tau^+ \to \bar{\nu}_\tau e^+ \nu_e$, as well as their charge conjugate decays, can be described by the most general, local, derivative-free and lepton-number conserving four-fermion point interaction Hamiltonian. The point interaction permits one to use equivalent Hamiltonians which differ in the way the fermions are grouped together [5, 19]. The older literature preferred a "charge retention" form with parity-odd and parity-even terms in which e^+ and μ^+ as the usually detected particles were grouped together [20, 15]. This had the advantage that limits to some single coupling constants could be obtained from then existing experimental results. The disadvantage was that this (charge retention) Hamiltonian represented interactions proceeding via the exchange of a neutral boson X which would carry the lepton numbers both of muon and electron and would thus not be universal. The use of a "charge changing" form, where the charged leptons are grouped with their neutrino and which is adapted to a charged boson exchange, resulted in absolute values of differences of coupling constants. Both forms mentioned above are in addition complicated by the fact that a fully parity-violating interaction like e.g. the $V - A$ interaction is represented by *four* coupling constants C_V, C'_V, C_A and C'_A.

In the following we will use a charge-changing Hamiltonian characterized by fields of definite handedness [6, 7]. The matrix element for μ decay may be denoted as [4, 21]

$$M = 4\,\frac{G_F}{\sqrt{2}} \sum_{\substack{\gamma=S,V,T\\ \varepsilon,\mu=R,L}} g^{\gamma}_{\varepsilon\mu} < \bar{e}_{\varepsilon}\, |\Gamma^{\gamma}|(\nu_e)_n >< (\bar{\nu_{\mu}})_m |\Gamma_{\gamma}| \mu_{\mu} > \quad . \tag{2.1}$$

Here G_F is the Fermi coupling constant, while γ labels the type of interaction: Γ^S, Γ^V, Γ^T (4-scalar, 4-vector, 4-tensor). The indices ε and μ indicate the chirality (left- or right-handed) of the spinors of the charged leptons, $\varepsilon \mathrel{\hat{=}}$ electron, $\mu \mathrel{\hat{=}}$ muon. The chiralities n and m of the ν_e and the $\bar{\nu}_{\mu}$ spinors, respectively, are uniquely determined for given γ, ε and μ. In this picture, the coupling constants $g^{\gamma}_{\varepsilon\mu}$ have a simple physical interpretation: $n_{\gamma}|g^{\gamma}_{\varepsilon\mu}|^2$ is equal to the (relative) probability for a μ-handed muon to decay into an ε-handed electron by the interaction Γ^{γ}; the factors $n_S = 1/4$, $n_V = 1$ and $n_T = 3$ take care of the proper normalization. The standard model thus corresponds to $g^V_{LL} = 1$, all other couplings being zero.
For leptonic τ decays, μ should be substituted by τ and e by μ or e.

2.2. Observables

2.2.1. Electron Decay Distribution

In the following we give the distribution of electrons from polarized muons including the effects of the electron mass; the generalization to leptonic τ decays is obvious. We consider the decay $\mu^+ \to \bar{\nu}_{\mu} e^+ \nu_e$ and its charge conjugate $\mu^- \to \nu_{\mu} e^- \bar{\nu}_e$. In the case of double signs the upper sign refers to μ^+, the lower to μ^- decay. The kinematical range of the electron energy E_e is given by

$$m_e \leqq E_e \leqq W_{\mu e} \equiv \frac{m_{\mu}^2 + m_e^2}{2m_{\mu}} \quad . \tag{2.2}$$

With the standard reduced energy variable $x = E_e/W_{\mu e}$ this leads to

$$x_0 \leqq x \leqq 1 \tag{2.3}$$

with $x_0 = m_e/W_{\mu e}$. We note that x_0 is small for muon decay ($x_0 = 9.67 \times 10^{-3}$) and for the electronic τ decay ($x_0 = 0.29 \times 10^{-3}$), but it is not small for the muonic τ decay ($x_0 = 59.2 \times 10^{-3}$).

The differential decay probability for an $e^{\pm}$ with reduced energy between x and $x + dx$, emitted at an angle between ϑ and $\vartheta + d\vartheta$ with respect to the muon's polarization $\vec{\wp}_\mu$, and having its spin pointing into the direction of the arbitrary unit vector $\hat{\zeta}$ (see Fig. 2.1) is given by

$$\frac{d^2\Gamma}{dx\,dcos\vartheta} = \frac{1}{4} m_\mu W_{\mu e}^4 G_F^2 \sqrt{x^2 - x_0^2}\left(F_{IS}(x) \pm \wp_\mu cos\vartheta \cdot F_{AS}(x)\right)\left(1 + \vec{\wp}_e \cdot \hat{\zeta}\right) \tag{2.4}$$

where we have used $\wp_\mu = |\vec{\wp}_\mu|$, and where $\vec{\wp}_e$ is the polarization vector of the $e^{\pm}$:

$$\vec{\wp}_e = P_{T_1} \cdot \hat{x} + P_{T_2} \cdot \hat{y} + P_L \cdot \hat{z} \, . \tag{2.5}$$

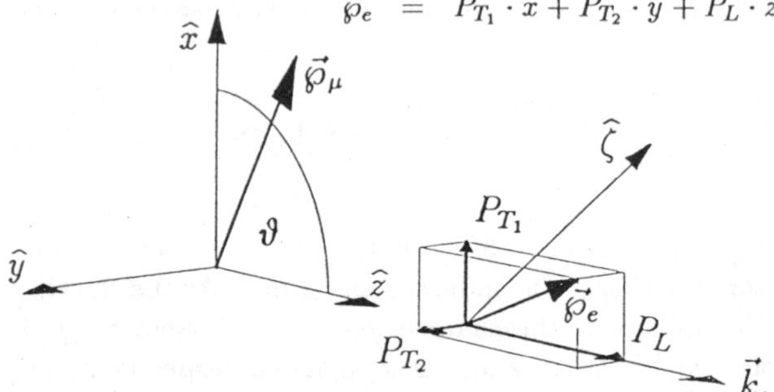

Fig. 2.1 Kinematic variables for the decay $\mu \to e\nu\bar{\nu}$ in the muon rest system.

We have defined a right-handed coordinate system with

$$\hat{z} = \frac{\vec{k}_e}{|\vec{k}_e|}, \quad \hat{y} = \frac{\vec{k}_e \times \vec{\wp}_\mu}{|\vec{k}_e \times \vec{\wp}_\mu|}, \quad \hat{x} = \hat{y} \times \hat{z} \quad . \tag{2.6}$$

Here $\vec{k}_e$ is the momentum vector of the electron, while P_L designates the longitudinal polarization, P_{T_1} the transverse component of $\vec{\wp}_e$ lying in the plane defined by $\vec{k}_e$ and $\vec{\wp}_\mu$, and P_{T_2} the component perpendicular to that plane. $P_{T_2} \neq 0$ signals violation of time reversal symmetry. These polarization components are

$$P_{T_1}(x,\vartheta) = \frac{\wp_\mu sin\vartheta \cdot F_{T_1}(x)}{F_{IS}(x) \pm \wp_\mu cos\vartheta \cdot F_{AS}(x)} \tag{2.7}$$

$$P_{T_2}(x,\vartheta) = \frac{\wp_\mu sin\vartheta \cdot F_{T_2}(x)}{F_{IS}(x) \pm \wp_\mu cos\vartheta \cdot F_{AS}(x)} \tag{2.8}$$

$$P_L(x,\vartheta) = \frac{\pm F_{IP}(x) + \wp_\mu cos\vartheta \cdot F_{AP}(x)}{F_{IS}(x) \pm \wp_\mu cos\vartheta \cdot F_{AS}(x)} \tag{2.9}$$

The functions $F_\nu(x)$ shall be decomposed as

$$F_\nu(x) = F_\nu^{V-A}(x) + G_\nu(x) , \tag{2.10}$$

where $G_\nu(x) \equiv 0$ for $g_{LL}^V = 1$ ("$V - A$"). Physics beyond the standard model would thus be contained *exclusively* in the $G_\nu(x)$. The index ν stands for IS (isotropic part of energy spectrum), AS (anisotropic part of energy spectrum), T_1 (transverse polarization P_{T_1}), T_2 (transverse polarization P_{T_2}), IP (isotropic part of longitudinal polarization) and AP (anisotropic part of longitudinal polarization). The F_ν^{V-A} do not depend on specific decay parameters:

$$F_{IS}^{V-A}(x) = \tfrac{1}{6}\{-2x^2 + 3x - x_0^2\} \tag{2.11}$$

$$F_{AS}^{V-A}(x) = \tfrac{1}{6}(x^2 - x_0^2)^{1/2}\{2x - 2 + (1 - x_0^2)^{1/2}\} \tag{2.12}$$

$$F_{T_1}^{V-A}(x) = -\tfrac{1}{6}(1 - x)x_0 \tag{2.13}$$

$$F_{T_2}^{V-A}(x) = 0 \tag{2.14}$$

$$F_{IP}^{V-A}(x) = \tfrac{1}{6}(x^2 - x_0^2)^{1/2}\{-2x + 2 + (1 - x_0^2)^{1/2}\} \tag{2.15}$$

$$F_{AP}^{V-A}(x) = \tfrac{1}{6}\{2x^2 - x - x_0^2\} \tag{2.16}$$

The functions $G_\nu(x)$ depend on the decay parameters $(\varrho, \xi'', \xi', \xi, \delta, \eta, \eta'', \alpha'/A, \beta'/A)$, where $\eta = (\alpha - 2\beta)/A$ and $\eta'' = (3\alpha + 2\beta)/A$:

$$G_{IS}(x) = \tfrac{1}{9}\left\{2(\varrho - \tfrac{3}{4})(4x^2 - 3x - x_0^2) + 9\eta(1 - x)x_0\right\} \tag{2.17}$$

$$\begin{aligned} G_{AS}(x) = {} & \tfrac{1}{9}(x^2 - x_0^2)^{1/2}\{3(\xi - 1)(1 - x) \\ & +2(\xi\delta - \tfrac{3}{4})(4x - 4 + (1 - x_0^2)^{1/2})\} \end{aligned} \tag{2.18}$$

$$\begin{aligned} G_{T_1}(x) = {} & \tfrac{1}{12}\left\{-2[(\xi'' - 1) + 12(\varrho - \tfrac{3}{4})](1 - x)x_0 - 3\eta(x^2 - x_0^2)\right. \\ & \left.+\eta''(-3x^2 + 4x - x_0^2)\right\} \end{aligned} \tag{2.19}$$

$$G_{T_2}(x) = \tfrac{1}{3}(x^2 - x_0^2)^{1/2}\{3\frac{\alpha'}{A}(1 - x) + 2\frac{\beta'}{A}(1 - x_0^2)^{1/2}\} \tag{2.20}$$

$$\begin{aligned} G_{IP}(x) = {} & \tfrac{1}{54}(x^2 - x_0^2)^{1/2}\{9(\xi' - 1)(-2x + 2 + (1 - x_0^2)^{1/2}) \\ & +4\xi(\delta - \tfrac{3}{4})(4x - 4 + (1 - x_0^2)^{1/2})\} \end{aligned} \tag{2.21}$$

$$G_{AP}(x) = \tfrac{1}{6}\{(\xi'' - 1)(2x^2 - x - x_0^2) + 4(\varrho - \tfrac{3}{4})(4x^2 - 3x - x_0^2) + 2\eta''(1 - x)x_0\} \tag{2.22}$$

Table 2.1 gives an overview on the influence of the decay parameters to the various observables.

Table 2.1 Dependence of the functions G_ν on the various decay parameters. The G_ν describe the effects of non-$V-A$ contributions to the observables of the electron. Full dependence of a G_ν to a decay parameter is expressed by an "x", restricted dependence by an "(x)". Restricted dependence means proportionality to $x_0 = m_e/W_{\mu e} \approx 10^{-2}$.

Observable	ν	ϱ	ξ''	ξ'	ξ	δ	η	η''	α'/A	β'/A
Isotropic part spectrum	IS	x					(x)			
Anisotropic part of spectrum	AS				x	x				
Transverse electron polarization P_{T_1}	T_1	(x)	(x)				x	x		
Transverse electron polarization P_{T_2} (T-violating)	T_2								x	x
Isotropic part of longitudinal polarization P_L	IP			x	x	x				
Anisotropic part of longitudinal polarization P_L	AP	x	x					(x)		

2.2.2. Electron Neutrino Energy Distribution

It has recently been realized that present experiments which detect the ν_e from the decay of unpolarized μ^+ by the reaction $^{12}C(\nu_e, e^-)^{12}N(g.s.)$ not only determine the neutrino absorption cross section but also measure the ν_e energy spectrum [22]. The energy spectrum can be described by spectrum shape parameters ω_L and η_L for left-handed and ω_R and η_R for right-handed ν_e. In contrast to the energy spectrum of the electrons it allows a new null-test of the standard model [22] (see Sect. 2.4.6). The right-handed ν_e cannot be detected as they are sterile in matter. For the energy spectrum of the left-handed ν_e one obtains [23]

$$\frac{d\Gamma_L}{dy} = \frac{m_\mu^5 G_F^2}{16\pi^3} \cdot Q_L^{\nu_e} \cdot \{F_1(y) + \omega_L \cdot F_2(y) + \eta_L x_0 F_3(y)\} \tag{2.23}$$

Here $d\Gamma_L/dy$ is the probability of a left-handed ν_e to be emitted with the reduced energy $y = 2E_\nu/m_\mu$. The probability $Q_L^{\nu_e}$ of the ν_e to be left-handed, the spectral shape parameter ω_L and the low energy parameter η_L are [23]

$$
\begin{aligned}
Q_L^{\nu_e} &= \tfrac{1}{4}|g_{RL}^S|^2 + \tfrac{1}{4}|g_{RR}^S|^2 + |g_{LL}^V|^2 + |g_{LR}^V|^2 + 3|g_{RL}^T|^2 \\
&= \tfrac{1}{2}(1 - \wp_{\nu_e})
\end{aligned}
\tag{2.24}
$$

$$
\omega_L = \frac{3}{4}\frac{\left\{|g_{RR}^S|^2 + 4|g_{LR}^V|^2 + |g_{RL}^S + 2g_{RL}^T|^2\right\}}{\{|g_{RL}^S|^2 + |g_{RR}^S|^2 + 4|g_{LL}^V|^2 + 4|g_{LR}^V|^2 + 12|g_{RL}^T|^2\}} \tag{2.25}
$$

$$
\eta_L = 2\frac{Re\left\{g_{LL}^V g_{RR}^{S*} + g_{LR}^V(g_{RL}^{S*} + 6g_{RL}^{T*})\right\}}{\{|g_{RL}^S|^2 + |g_{RR}^S|^2 + 4|g_{LL}^V|^2 + 4|g_{LR}^V|^2 + 12|g_{RL}^T|^2\}}, \tag{2.26}
$$

where $\wp_{\nu_e}$ denotes the longitudinal polarization of the ν_e. The functions $F_1(y)$, $F_2(y)$ and $F_3(y)$ are given by

$$
F_1(y) = \frac{(1 - x_0^2 - y)^2 y^2}{1 - y} \tag{2.27}
$$

$$
F_2(y) = \frac{2}{9}\frac{(1 - x_0^2 - y)^2 y^2}{(1 - y)^3}(-4y^2 + y(7 - x_0^2) - 3 + 3x_0^2) \tag{2.28}
$$

$$
F_3(y) = \frac{(1 - x_0^2 - y)^2 y^2}{(1 - y)^2} \tag{2.29}
$$

The corresponding quantities $d\Gamma_R/dy$, $Q_R^{\nu_e}$, ω_R and η_R for the energy spectrum of the right-handed ν_e may be obtained from Eqs. 2.23-2.26 simply by the substitutions $R \Leftrightarrow L$ for every chirality index and by $\wp_{\nu_e} \to -\wp_{\nu_e}$. The size of ω_k and η_k $(k = R, L)$ is constrained by the fact that $g_{LL}^V \approx 1$, and $g_{\epsilon\mu}^\gamma \approx 0$ for all other interactions. One therefore finds $\omega_L \gtrapprox 0$ and $\eta_L \approx 0$, but $0 \leqq \omega_R \leqq 1$ and $-1 \leqq \eta_R \leqq 1$. The term with $\eta_L x_0 F_3(y)$ can be neglected because both η_L and the factor $x_0 \approx 10^{-2}$ are small. Thus the energy spectrum of left-handed ν_e which *can be* detected by absorption on ^{12}C is described effectively by *two* observables, namely $Q_L^{\nu_e} = 1 - Q_R^{\nu_e}$ which is a measure of the total rate and therefore of the ν_e polarization, and ω_L which describes the shape of the spectrum. The probability $Q_L^{\nu_e}$ can be determined from the total absorption rate, if the absolute absorption cross section σ_A is known with sufficient precision.

2.2.3. Inverse Muon Decay

The reaction $\nu_\mu e^- \to \mu^- \nu_e$, usually called *inverse muon decay*, shall be governed by the same Hamiltonian as (normal) muon decay. Its measurement will, in contrast to normal muon decay, allow to separate g_{LL}^S from g_{LL}^V and thus to complete the experimental proof of $V-A$.

The normalized total cross section S is obtained by integrating over the electron energy and the energy spectrum of the incoming ν_μ and by dividing by the theoretical value for a pure $V-A$ interaction [24]. S has been calculated for the general decay interaction [25, 7]. In terms of our coupling constants it is given by [26]

$$S = \tfrac{1}{2}(1+\wp_{\nu_\mu})\cdot S_R + \tfrac{1}{2}(1-\wp_{\nu_\mu})\cdot S_L, \tag{2.30}$$

where $\wp_{\nu_\mu}$ is the longitudinal polarization of the incoming ν_μ. The quantities S_R and S_L are the normalized cross sections for right- respectively left-handed ν_μ:

$$S_R = \tfrac{3}{32}|g^S_{LL}|^2 + |g^V_{RR}|^2 + \tfrac{3}{8}|g^V_{LR}|^2 + \tfrac{3}{32}|g^S_{RL} - \tfrac{10}{3}g^T_{RL}|^2 + \tfrac{4}{3}|g^T_{RL}|^2 \tag{2.31}$$

$$S_L = \tfrac{3}{32}|g^S_{RR}|^2 + |g^V_{LL}|^2 + \tfrac{3}{8}|g^V_{RL}|^2 + \tfrac{3}{32}|g^S_{LR} - \tfrac{10}{3}g^T_{LR}|^2 + \tfrac{4}{3}|g^T_{LR}|^2 \tag{2.32}$$

We will see that the longitudinal polarization $\wp_{\nu_\mu}$ of the ν_μ from π^+ decay is close to -1 within 3.2×10^{-3} [27, 28] so that the couplings in S_R cannot contribute to S. We will further see that normal muon decay places stringent limits on all the couplings in S_L except on g^V_{LL}. Therefore S just measures $|g^V_{LL}|^2$. This has made it possible to derive a lower limit for $|g^V_{LL}|$ for the first time and thus to establish $V-A$ as the dominant interaction in muon decay [4].

2.2.4. Radiative Muon Decays

Radiative muon decay, $\mu^+ \to \bar{\nu}_\mu e^+ \nu_e \gamma$, is treated in detail in Ref. [29]. The electron and gamma spectra depend on the decay parameters ϱ and δ as well as on the two new combinations $\bar{\eta}$ and $\xi\cdot\kappa$ [30]:

$$\bar{\eta} = \tfrac{1}{A}(a+2c) = \tfrac{1}{4}(7-12\varrho-\xi'') \tag{2.33}$$

$$\xi\cdot\kappa = \tfrac{1}{A}(a'+2c') = -\tfrac{1}{12}(3\xi+3\xi'-8\xi\delta) \tag{2.34}$$

The parameter $\bar{\eta}$, in particular, is positive semidefinite and zero in the standard model. It can be expressed by a sum of absolute squares of combinations of coupling constants:

$$\begin{aligned}\bar{\eta} = {} & (|g^V_{RL}|^2 + |g^V_{LR}|^2) + \tfrac{1}{8}(|g^S_{LR} + 2g^T_{LR}|^2 + |g^S_{RL} + 2g^T_{RL}|^2) \\ & +2(|g^T_{LR}|^2 + |g^T_{RL}|^2)\,.\end{aligned} \tag{2.35}$$

This allows one to get upper limits for each term separately.

It is also worthwhile to mention the decay $\mu^+ \to \bar{\nu}_\mu e^+ \nu_e e^+ e^-$ which has been calculated recently in terms of the most general decay interaction (Eq. 2.1) [31]. The decay distribution depends on the probabilities $Q_{\varepsilon\mu}$ and on the bilinear quantities $B_{RL} = (a+a')/(2A)$, $B_{LR} = (a-a')/(2A)$ defined in Ref. [4].

2.3. Lorentz Structure

2.3.1. Decay Parameters

The nine parameters (ϱ, ξ'', ξ', ξ, δ, η, η'', α'/A, β'/A) describing the *electron's spectrum, decay asymmetry* and *polarization vector* can be represented [20] by the intermediate quantities (a, a', α, α', b, b', β, β', c, c'), whose values are known from experiment [32]. They are all real, bilinear combinations of the coupling constants:

$$a = 16(|g^V_{RL}|^2 + |g^V_{LR}|^2) + |g^S_{RL} + 6g^T_{RL}|^2 + |g^S_{LR} + 6g^T_{LR}|^2 \quad (2.36)$$

$$a' = 16(|g^V_{RL}|^2 - |g^V_{LR}|^2) + |g^S_{RL} + 6g^T_{RL}|^2 - |g^S_{LR} + 6g^T_{LR}|^2 \quad (2.37)$$

$$\alpha = 8Re\left\{g^V_{RL}(g^{S*}_{LR} + 6g^{T*}_{LR}) + g^V_{LR}(g^{S*}_{RL} + 6g^{T*}_{RL})\right\} \quad (2.38)$$

$$\alpha' = 8Im\left\{g^V_{LR}(g^{S*}_{RL} + 6g^{T*}_{RL}) - g^V_{RL}(g^{S*}_{LR} + 6g^{T*}_{LR})\right\} \quad (2.39)$$

$$b = 4(|g^V_{RR}|^2 + |g^V_{LL}|^2) + |g^S_{RR}|^2 + |g^S_{LL}|^2 \quad (2.40)$$

$$b' = 4(|g^V_{RR}|^2 - |g^V_{LL}|^2) + |g^S_{RR}|^2 - |g^S_{LL}|^2 \quad (2.41)$$

$$\beta = -4Re\left\{g^V_{RR}g^{S*}_{LL} + g^V_{LL}g^{S*}_{RR}\right\} \quad (2.42)$$

$$\beta' = 4Im\left\{g^V_{RR}g^{S*}_{LL} - g^V_{LL}g^{S*}_{RR}\right\} \quad (2.43)$$

$$c = \tfrac{1}{2}\left\{|g^S_{RL} - 2g^T_{RL}|^2 + |g^S_{LR} - 2g^T_{LR}|^2\right\} \quad (2.44)$$

$$c' = \tfrac{1}{2}\left\{|g^S_{RL} - 2g^T_{RL}|^2 - |g^S_{LR} - 2g^T_{LR}|^2\right\} \quad (2.45)$$

From Eqs. 2.36-2.45 it can be seen that these quantities are not completely independent. The transformation from the 20-dimensional space of the $g^\gamma_{\varepsilon\mu}$ to the 10-dimensional space of the $\{a, ..., c'\}$ leads to the following six constraints [32]:

$$a \geqq 0 \quad (2.46) \qquad a^2 \geqq a'^2 + \alpha^2 + \alpha'^2 \quad (2.49)$$

$$b \geqq 0 \quad (2.47) \qquad b^2 \geqq b'^2 + \beta^2 + \beta'^2 \quad (2.50)$$

$$c \geqq 0 \quad (2.48) \qquad c^2 \geqq c'^2 \quad (2.51)$$

These constraints are very important for any general analysis of muon decay, since they strongly influence the final errors of the quantities they relate.

Table 2.2 Allowed ranges, V-A values and experimental results (in units of 10^{-3}) for the muon decay parameters. Upper and lower limits are given with 90% $c.\ell.$

Decay parameter	Minimum	Maximum	$V-A$ value	Experimental result $[10^{-3}]$	Ref.	Comments
ϱ	0	1	3/4	751.8 ± 2.6	[33]	
ξ''	-7/3	3	1	650 ± 360	[34]	
ξ	-3	3	1	1004.5 ± 8.6	[35, 36]	
$\wp_\mu^\pi \xi$	-3	3	1	1002.7 ± 8.4	[35]	$\pi^+ \to \mu^+ \nu_\mu$
$\wp_\mu^K \xi$	-3	3	1	1001.3 ± 6.1	[37]	$K^+ \to \mu^+ \nu_\mu$
$\xi\delta$	-1	1	3/4			
δ	-∞	∞	3/4	748.6 ± 3.8	[28]	
ξ'	-1	1	1	998 ± 45	[34]	
$\xi\delta/\varrho$	-1	1	1	> 996.8	[36]	$\wp_\mu \xi\delta/\varrho$
η	-1	1	0	-7 ± 13	[32]	
η''	-3	3	0	12 ± 16	[32]	
α'/A	-1	1	0	-0.2 ± 4.3	[32]	
β'/A	-1/4	1/4	0	1.5 ± 6.3	[32]	
ω_L	0	1	0		[23, 38]	$^{12}C(\nu_e, e^-)^{12}N$
S_L	0	1	1	1006 ± 47	[39, 40]	$\nu_\mu e^- \to \mu^- \nu_e$
$\bar{\eta}$	0	1	0	-30 ± 100	[29]	$\mu^+ \to \bar{\nu}_\mu e^+ \nu_e \gamma$
Q_{RR}	0	1	0	< 2.0	[4]	
Q_{LR}	0	1	0	< 3.9	[4]	
Q_{RL}	0	1	0	< 45.0	[4]	
Q_{LL}	0	1	1	> 949.0	[4]	
Q_R^μ	0	1	0	< 7.7		
Q_R^e	0	1	0	< 38.0		
$Q_R^{\nu_\mu}$	0	1	0	< 80.0		
$Q_R^{\nu_e}$	0	1	0	< 80.0	[23]	

The decay parameters are given by

$$\varrho = \frac{1}{A}(3b + 6c) \tag{2.52}$$

$$\xi'' = \frac{1}{A}(3a + 4b - 14c) \tag{2.53}$$

$$\xi = -\frac{1}{A}(3a' + 4b' - 14c') \tag{2.54}$$

$$\xi\delta = \frac{1}{A}(-3b' + 6c') \tag{2.55}$$

$$\xi' = -\frac{1}{A}(a' + 4b' + 6c') \tag{2.56}$$

$$\eta = \frac{1}{A}(\alpha - 2\beta) \tag{2.57}$$

$$\eta'' = \frac{1}{A}(3\alpha + 2\beta) \tag{2.58}$$

The parameters α'/A and β'/A are determined directly. The allowed range, the

$V-A$ predictions and the experimental results of the muon decay parameters, which are not all independent, are given in Table 2.2.

2.3.2. Complete Determination of the Lorentz Structure

The precise measurement of individual decay parameters alone generally does not give conclusive information about the decay interaction due to the many different couplings and the interference terms between them. An example is the famous Michel parameter ϱ. A precise measurement yielding the V-A value of 3/4 by no means establishes the V-A interaction. In fact any interaction consisting of an arbitrary combination of g^S_{LL}, g^S_{LR}, g^S_{RL}, g^S_{RR}, g^V_{RR} and g^V_{LL} will yield exactly $\varrho = \frac{3}{4}$ [41]. This can be seen if we write ϱ in the form [42]

$$\varrho - \tfrac{3}{4} = -\tfrac{3}{4}\left\{|g^V_{LR}|^2 + |g^V_{RL}|^2 + 2(|g^T_{LR}|^2 + |g^T_{RL}|^2) + Re(g^S_{LR}g^{T*}_{LR} + g^S_{RL}g^{T*}_{RL})\right\} . \tag{2.59}$$

For $\varrho = \frac{3}{4}$ and $g^T_{LR} = g^T_{RL} = 0$ (no tensor interaction) we find $g^V_{LR} = g^V_{RL} = 0$, with all the remaining six couplings being arbitrary !

The magnitude of the decay interaction is contained in the Fermi coupling constant G_F. Thus the $g^\gamma_{\varepsilon\mu}$ may be normalized, dimensionless coupling constants, resulting in

$$A \equiv a + 4b + 6c = 16 \tag{2.60}$$

or equivalently,

$$\begin{array}{ccccccccc} & \frac{1}{4}|g^S_{RR}|^2 & + & \frac{1}{4}|g^S_{RL}|^2 & + & \frac{1}{4}|g^S_{LR}|^2 & + & \frac{1}{4}|g^S_{LL}|^2 & \\ + & |g^V_{RR}|^2 & + & |g^V_{RL}|^2 & + & |g^V_{LR}|^2 & + & |g^V_{LL}|^2 & \\ & & + & 3|g^T_{RL}|^2 & + & 3|g^T_{LR}|^2 & & & = 1 \end{array} \tag{2.61}$$

By rearranging the terms in Eq. 2.61 according to the chiralities ε and μ of the electron and the muon (ε, $\mu = R, L$) it is possible to define the four quantities $Q_{\varepsilon\mu}$:

$$Q_{RR} = \tfrac{1}{4}|g^S_{RR}|^2 + |g^V_{RR}|^2 \tag{2.62}$$

$$Q_{LR} = \tfrac{1}{4}|g^S_{LR}|^2 + |g^V_{LR}|^2 + 3|g^T_{LR}|^2 \tag{2.63}$$

$$Q_{RL} = \tfrac{1}{4}|g^S_{RL}|^2 + |g^V_{RL}|^2 + 3|g^T_{RL}|^2 \tag{2.64}$$

$$Q_{LL} = \tfrac{1}{4}|g^S_{LL}|^2 + |g^V_{LL}|^2 \tag{2.65}$$

We note that $0 \leqq Q_{\varepsilon\mu} \leqq 1$ and $\sum_{\varepsilon,\mu} Q_{\varepsilon\mu} = 1$. $Q_{\varepsilon\mu}$ is then the probability for the decay of a muon of handedness μ into an electron of handedness ε. The main point is now that the $Q_{\varepsilon\mu}$ can be expressed by the known quantities $\{a, ..., c'\}$ [4]:

$$Q_{RR} = 2(b+b')/A \tag{2.66}$$

$$Q_{LR} = [(a-a')+6(c-c')]/(2A) \tag{2.67}$$

$$Q_{RL} = [(a+a')+6(c+c')]/(2A) \tag{2.68}$$

$$Q_{LL} = 2(b-b')/A \tag{2.69}$$

The existing measurements show that the three quantities Q_{RR}, Q_{LR} and Q_{RL} are zero, within errors. This gives upper limits to the absolute values of eight of the ten complex coupling constants $g^{\gamma}_{\epsilon\mu}$. Furthermore we find that Q_{LL} is bounded by a lower limit which confirms that both muon and electron are left-handed. It can be seen from Eq. 2.65, however, that the data from the measurements of the μ and the e do not allow one to distinguish vector (g^V_{LL}) from scalar (g^S_{LL}) interaction. This type of ambiguity has been noted before in the context of a different Hamiltonian [8], and electron - neutrino correlation measurements (not performed up to date) have been proposed. It has been shown that one can resolve this ambiguity also with the information from inverse muon decay [4]. The total rate S, normalized to the rate predicted by $V-A$, for the reaction $\nu_\mu + e^- \rightarrow \mu^- + \nu_e$ with ν_μ of negative helicity has been found to be close to 1 [39, 40]. S depends effectively only on those five coupling constants g^V_{LL}, g^V_{RL}, g^S_{LR}, g^T_{LR} and g^S_{RR} which describe interactions with left-handed ν_μ. The four latter ones are found to be small. One thus obtains [4]

$$S = |g^V_{LL}|^2 \tag{2.70}$$

which yields a *lower* limit for $|g^V_{LL}|$, and through the normalization requirement Eq. 2.61 an *upper* limit for the remaining $|g^S_{LL}|$:

$$|g^S_{LL}| < 2\sqrt{1-S} \tag{2.71}$$

Thus the weak interaction has been completely determined for muon decay using only data from this purely leptonic interaction. The results are shown in Fig. 2.2 where each of the ten complex normalized coupling constants $g'^{\gamma}_{\epsilon\mu} = g^{\gamma}_{\epsilon\mu}/\max(|g^{\gamma}_{\epsilon\mu}|)$ is given within one of the squares defined uniquely by the handednesses of the electron and the muon and by the type of interaction. The outer circles display the mathematical limits for the $g'^{\gamma}_{\epsilon\mu}$ in the complex plane, while the inner circles for nine of the $g'^{\gamma}_{\epsilon\mu}$ show the areas still allowed by experiments (90% *c.ℓ.*). For g'^V_{LL}, which has been chosen to be real, one gets the small line close to $g'^V_{LL} = 1$ in agreement with the standard model.

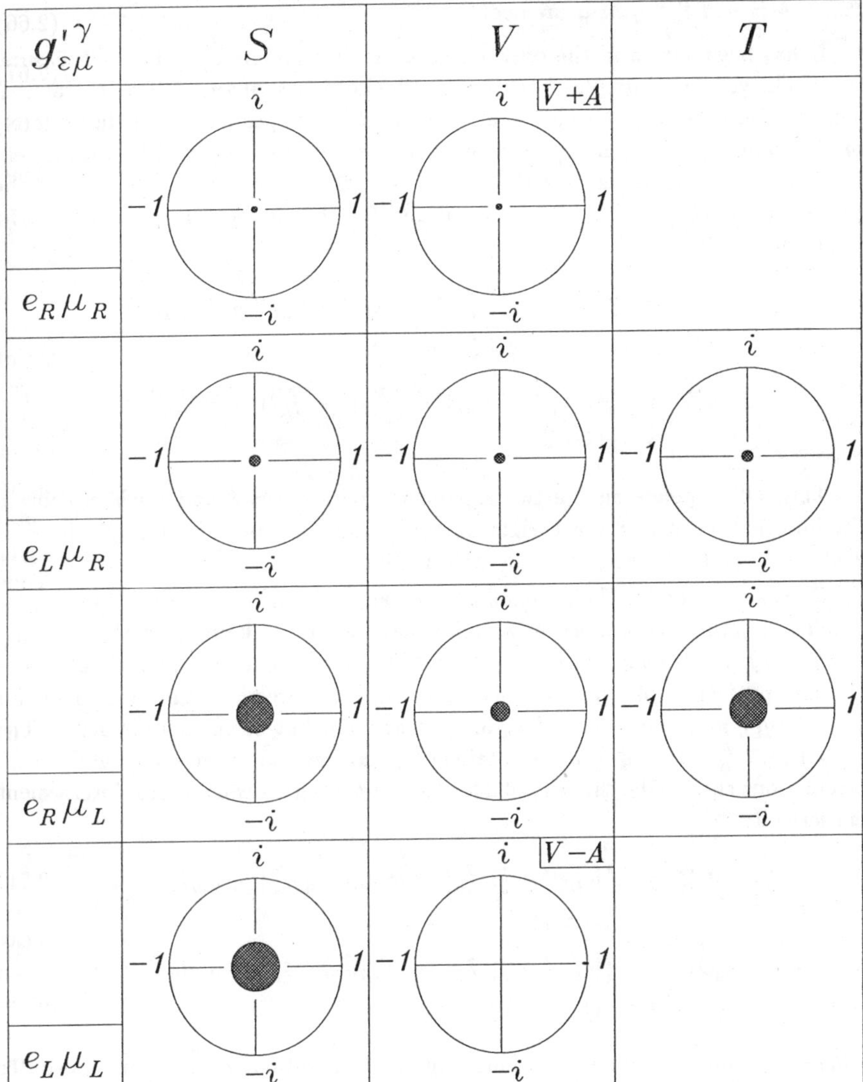

Fig. 2.2 90% *c.ℓ.* limits for the reduced coupling constants $g'^{\gamma}_{\varepsilon\mu} = g^{\gamma}_{\varepsilon\mu}/\max(|g^{\gamma}_{\varepsilon\mu}|)$ describing the decay $\mu^+ \to \bar{\nu}_\mu e^+ \nu_e$ (From Ref. [26]. Updated limits from [28, 40]). Each coupling is uniquely determined by the handednesses ε and μ of the electron and the muon, respectively, and by the type of interaction $\gamma = S$, V or T. The maximal possible values of the coupling constants are 2, 1 and $1/\sqrt{3}$ for $\gamma = S$, V resp. T.

2.3.3. Minimal Set of Measurements

It has been shown in the previous section that a limited amount of measurements allows one to completely determine the Lorentz structure of the interaction. This has been possible because the quantities $Q_{\epsilon\mu}$ are positive semidefinite forms of the coupling constants $g^{\gamma}_{\epsilon\mu}$, and because they can be expressed by experimental results. By summing over the first or over the second index one obtains the probabilities Q^{μ}_{R} and Q^{e}_{R} for the muon and for the electron, respectively, to be right-handed [26]:

$$\begin{aligned} Q^{\mu}_{R} &= \tfrac{1}{4}|g^{S}_{RR}|^2 + \tfrac{1}{4}|g^{S}_{LR}|^2 + |g^{V}_{RR}|^2 + |g^{V}_{LR}|^2 + 3|g^{T}_{LR}|^2 \\ &= \tfrac{1}{2}\left\{1 + \tfrac{1}{9}(3\xi - 16\xi\delta)\right\} \end{aligned} \tag{2.72}$$

$$\begin{aligned} Q^{e}_{R} &= \tfrac{1}{4}|g^{S}_{RR}|^2 + \tfrac{1}{4}|g^{S}_{RL}|^2 + |g^{V}_{RR}|^2 + |g^{V}_{RL}|^2 + 3|g^{T}_{RL}|^2 \\ &= \tfrac{1}{2}(1 - \xi') \end{aligned} \tag{2.73}$$

Thus Q^{μ}_{R} depends on the two asymmetry parameters ξ and δ which contain the information of how much right-handed the muon can be, while Q^{e}_{R} depends exclusively on the average electron polarization ξ'.

One can also derive the probabilities $Q^{\bar{\nu}_e}_{R}$ and $Q^{\nu_\mu}_{R}$ for the $\bar{\nu}_e$ and ν_μ to be right-handed [23]. In this connection we note that we strictly keep apart the concepts of *chirality* as a transformation property of the fermion spinors and of *helicity* as the spin projection of a pure particle state onto the direction of motion. Thus for $V - A$ ($g^{V}_{LL} = 1$) both $\bar{\nu}_\mu$ and ν_e are left-handed, but of opposite helicity. The quantities $Q^{\bar{\nu}_e}_{R}$ and $Q^{\nu_\mu}_{R}$ can be obtained by remembering that scalar and tensor interactions change the chirality of a fermion at the vertex while vector interactions conserve it:

$$\begin{aligned} Q^{\bar{\nu}_e}_{R} &= \tfrac{1}{4}|g^{S}_{LL}|^2 + \tfrac{1}{4}|g^{S}_{LR}|^2 + |g^{V}_{RR}|^2 + |g^{V}_{RL}|^2 + 3|g^{T}_{LR}|^2 \\ &= \tfrac{1}{2}(1 - \wp_{\bar{\nu}_e}) \end{aligned} \tag{2.74}$$

$$\begin{aligned} Q^{\nu_\mu}_{R} &= \tfrac{1}{4}|g^{S}_{LL}|^2 + \tfrac{1}{4}|g^{S}_{RL}|^2 + |g^{V}_{RR}|^2 + |g^{V}_{LR}|^2 + 3|g^{T}_{RL}|^2 \\ &= \tfrac{1}{2}(1 + \wp_{\nu_\mu}) \ , \end{aligned} \tag{2.75}$$

where $\wp_{\bar{\nu}_e}$ and $\wp_{\nu_\mu}$ denote the longitudinal polarizations of the neutrinos. To determine the interaction it is sufficient that three of the Q^{α}_{R} are measured and found to be zero, within errors. This amounts to measuring the polarization of at least one of the neutrinos precisely, as has been proposed for $\wp_{\nu_e}$ which can be deduced from a measurement of the total rate of absorption of the ν_e on ^{12}C (see Sect. 2.2.2). An alternative to *measuring* the polarization of the neutrinos from μ decay is to induce inverse muon decay with ν_μ from π^+ decay [39, 40] whose polarization is *precisely known* [27] (see Sect. 2.3.2). We conclude that only five of nineteen possible measurements are necessary to completely determine the $(V-A)$ interaction:

(1) Muon lifetime which yields the magnitude G_F

(2,3) Asymmetry parameters ξ and δ which yield the muon chirality

(4) Electron longitudinal polarization which yields the electron chirality, and

(5) Inverse muon decay with ν_μ of known helicity or rate of absorption of ν_e from muon decay.

2.4. Measurements

2.4.1. Lifetime

The total rate Γ of muon decay exhibits the strength G_F which is generally assumed to be universal for the charged weak interaction:

$$\Gamma = \frac{G_F^2 m_\mu^5}{192\pi^3} \cdot \{1 + 4\eta \frac{m_e}{m_\mu} - 8(\frac{m_e}{m_\mu})^2\} \cdot f_W \cdot f_r, \qquad (2.76)$$

where $f_W = 1 + \frac{3}{5}(m_\mu/m_W)^2$, $f_r = 1 - \frac{\alpha}{2\pi}(\pi^2 - \frac{25}{4})$ and $m_e/m_\mu = 4.84 \times 10^{-3}$. The factor f_W represents the influence of the finite mass m_W of the intermediate boson. The smallness of this term for muon (and for leptonic tau) decay justifies the analysis based on the four-fermion point interaction of Eq. 2.1. The radiative corrections are finite for the vector type interaction $\gamma = V$ (See Ref. [15],[43]-[51]). To first order, $f_r = 0.996$, independent of the lepton masses. The parameter η influences the rate [52], the spectrum and the polarization of the electron. The *uncertainty in G_F* derived from lepton decay is *dominated by the uncertainty in η*. This fact *is most often ignored*, for muon as well as for tau decay. (See also Sects. 2.4.2. and 3.2.)

Experimental difficulties in the measurement of the muon lifetime arise from the different fluctuations of the physical processes which detect the moment of the birth of the muon and of the occurrence of the decay electron. Special care is needed because muon spin rotation and the range of electron energies (see e.g. [53]) tend to distort the time spectrum. The average muon mean life is

$$\tau_\mu = (2\,197.03 \pm 0.04)\ ns$$

which corresponds to

$$G_F = (11\,664.1 \pm 0.2) \times 10^{-9} (\hbar c)^3 GeV^{-2},$$

where $\eta = 0$ has been assumed. The present uncertainty due to the term $4\eta m_e/m_\mu$ is 20 *times larger*. As seen from

$$\eta = \tfrac{1}{2} Re \left\{ g^V_{LL} g^{S*}_{RR} + g^V_{RR} g^{S*}_{LL} + g^V_{LR}(g^{S*}_{RL} + 6g^{T*}_{RL}) + g^V_{RL}(g^{S*}_{LR} + 6g^{T*}_{LR}) \right\} \qquad (2.77)$$

a right-handed scalar interaction ($Re\ g^S_{RR} \neq 0$) would cause $\eta \neq 0$ in first order.

2.4.2. Electron Energy Spectrum

The continuous electron energy spectrum shows that the muon decays into three light particles (at least). The electrons with high energies are most probable. This reflects the fact that the two unobserved neutral particles are emitted preferentially in the same direction, and suggests that they are not identical fermions [52]. The existence of the reaction ("inverse muon decay" [24, 39, 40])

$$\nu_\mu e^- \rightarrow \mu^- + neutral$$

and of the chain [54]

$$\begin{aligned} \mu^+ &\rightarrow e^+ + neutral(1) + neutral(2) \\ neutral(1) + n &\rightarrow e^- + X \end{aligned}$$

confirms that $neutral(1)$ is ν_e and $neutral(2)$ is $\bar{\nu}_\mu$. The identity of the neutrinos in muon decay with those in pion or in nuclear beta decay has been studied in Refs. [12, 55].

The electron energy spectrum derived from Eq. 2.1 for the decay of unpolarized muons ($\wp_\mu = 0$) recorded by a spectrometer insensitive to electron polarization ($<\hat{\zeta}>= 0$) is given by Eq. 2.4:

$$\frac{d\Gamma}{dx} \sim \sqrt{x^2 - x_0^2} \left\{ (-x^2 + x) + \tfrac{2}{9}\varrho(4x^2 - 3x - x_0^2) + \eta(1 - x)x_0 \right\} \qquad (2.78)$$

Radiative corrections are sizeable and have in addition to be included (See Ref. [15],[43]-[51]). ϱ is of influence at the upper end, η at the lower end of the spectrum. Precise experimental determinations of η from the spectrum shape have been difficult because the sensitivity to η is diminished by a factor of $x_0 \approx 10^{-2}$ and because it is not entirely possible to suppress all unwanted bremsstrahlung processes in the detector material which tend to populate the low energy region in a way that cannot be precisely predicted.

η has also been determined from the energy dependence of the transverse polarization P_{T_1} (see Eq. 2.19 and Sect. 2.4.5), where it occurs *without* the suppression factor x_0. The results are

$$\begin{aligned} \eta &= (-120 \pm 210) \times 10^{-3} && \text{[33] from spectrum measurement} \\ &= (\ -11 \pm\ 85) \times 10^{-3} && \text{[32] from } P_{T_1} \\ &= (\ \ -7 \pm\ 13) \times 10^{-3} && \text{[32] from global fit to all measurements} \end{aligned}$$

in agreement with $\eta = 0$.

The knowledge of η is important for a precise determination not only of the Fermi coupling constant but also of ϱ. Given a measured spectrum with data in a limited energy range (such as Fig. 2.3 [56]), the statistical accuracy depends strongly on any prior knowledge of η, as pointed out by Ref. [57] and as further

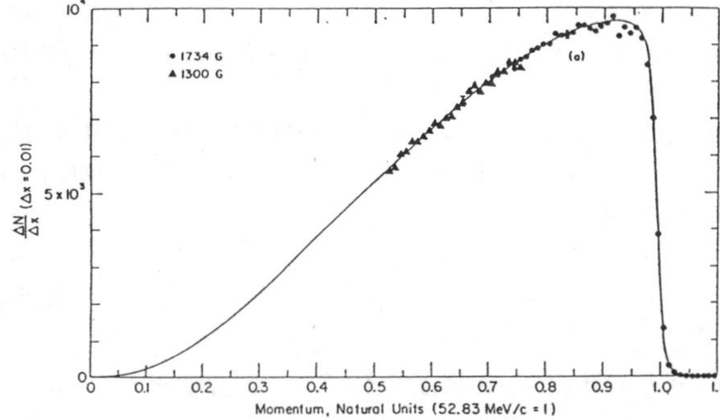

Fig. 2.3 Positron energy spectrum of μ^+ decay [56]. Solid line: Eq. 2.76 with $\varrho = 0.760$, $\eta = 0$, and radiative corrections included.

discussed in Ref. [56]. The experimental difficulties include the precise knowledge of the resolution function of the spectrometer, i.e. the calibration over a wide range of energies. Thin-walled acoustical [58, 59] or wire spark chambers [56] have been used to ensure that the positrons follow well defined trajectories through a homogeneous magnetic field. The present average value

$$\varrho = (751.8 \pm 2.6) \times 10^{-3}$$

agrees with $\varrho = 3/4$. It is mainly given by an experiment of 1966 [59]. For the determination of the muon decay interaction, ϱ is of secondary importance, as discussed in Sect. 2.3.2. It is, however, sensitive to $W_R - W_L$ mixing in left-right symmetric models and to supersymmetric muon decays (see Fig. 2.10 and Table 2.3)

2.4.3. Electron Decay Asymmetry

The measurement of the electron decay asymmetry $\mathcal{A}(x)$ from polarized muons determines how strong the chiral components (L, R) of the muon take part in the interaction. It has been used to search for right-handed currents and other muon decay modes outside the standard model.

If the combination

$$\begin{aligned} \tfrac{1}{2}(1 + \tfrac{1}{3}\xi - \tfrac{16}{9} \cdot \xi \cdot \delta) &= \tfrac{1}{4}|g^S_{RR}|^2 + \tfrac{1}{4}|g^S_{LR}|^2 + |g^V_{RR}|^2 + |g^V_{LR}|^2 + 3|g^T_{LR}|^2 \\ &\equiv Q_{RR} + Q_{LR} \equiv Q^\mu_R \end{aligned} \tag{2.79}$$

would take a value different from zero, then a coupling to the right-handed component of the muon would have to exist, i.e. at least one $g^\gamma_{\epsilon R} \neq 0$. Conversely, the finding that $Q^{\mu}_{R} = 0$ proves that the coupling acts exclusively on the left-handed component of the muon.

The distribution of the directions of flight of the positrons (electrons) is given by Eq. 2.4

$$\frac{d^2\Gamma}{dx d\cos\vartheta}(x,\vartheta) \equiv \mathrm{w}(x,\vartheta) \sim \left(F_{IS}(x) \;\pm\; \wp_\mu \cdot cos\vartheta \cdot F_{AS}(x)\right) \quad . \tag{2.80}$$

It depends on the reduced energy x, the angle ϑ between the muon polarization and the positron momentum, as chosen by the detector, and on the amount of polarization $\wp_\mu \geq 0$. The asymmetry

$$\mathcal{A}(x) \equiv \frac{\mathrm{w}(x,o) - \mathrm{w}(x,\pi)}{\mathrm{w}(x,o) + \mathrm{w}(x,\pi)} = \wp_\mu \cdot \frac{F_{AS}(x)}{F_{IS}(x)} \tag{2.81}$$

depends on the parameters ϱ, η, ξ and $\xi\delta$ (see Eqs. 2.10-2.12, 2.17 and 2.18).

We discuss now the experimental situations in which the parameters δ and ξ have their special influence. The energy x_δ where the asymmetry vanishes, $\mathcal{A}(x_\delta) = 0$, depends on the parameter δ only. Solving x_δ for δ, we obtain

$$(\delta - \frac{3}{4}) \;\approx\; \frac{3}{4} \cdot \frac{1 - 2x_\delta - (\sqrt{1 - x_0^2} - 1)}{4x_\delta - 3 + \sqrt{1 - x_0^2} - 1} \tag{2.82}$$

This allows one to determine δ from an asymmetry measurement as a function of the energy using polarized muons ($\wp_\mu \neq 0$). The knowledge of the magnitude $\wp_\mu$ of their polarization is not required.

The distributions of the directions of flight of the positrons (electrons) as seen by an apparatus which is equally sensitive to positrons of all energies is given by

$$\begin{aligned} \frac{d\Gamma}{d\cos\vartheta}(\vartheta) \;&\sim\; \int_{x_0}^{1} dx \cdot \sqrt{x^2 - x_0^2} \cdot F_{IS}(x) \pm \wp_\mu \cdot cos\vartheta \cdot \int_{x_0}^{1} dx \cdot \sqrt{x^2 - x_0^2} \cdot F_{AS}(x) \\ &\sim\; (1 \pm \mathcal{A}' \cdot cos\vartheta) \end{aligned} \tag{2.83}$$

The integral asymmetry $\mathcal{A}'$ is proportional to $\wp_\mu \cdot \xi$ and depends on η in first and on δ in second order of x_0. Neglecting x_0 $(x_0 = 0)$ one obtains

$$\mathcal{A}' \;=\; \tfrac{1}{3} \cdot \wp_\mu \cdot \xi \quad . \tag{2.84}$$

This allows one to determine ξ from an experiment using muons of known polarization. In the analysis the knowledge of the values of other muon decay parameters is unimportant.

The special case $\mathcal{A}(1) = \wp_\mu \cdot \xi\delta/\varrho$ has been the basis of a null-experiment for a model independent precise absolute determination of the muon polarization $\wp_\mu$ and in turn of the ν_μ polarization, $\wp_{\nu_\mu}$, as well as for a determination of $\xi\delta/\varrho$. Suppose positive muons from the decay of π^+ at rest, $\pi^+ \to \mu^+\nu_\mu$ are investigated. The μ^+ helicity then is equal to the ν_μ helicity h_{ν_μ}, which is known to be negative. These muons then are polarized with spins opposite to their line of flight with $\wp_\mu = |\wp_{\nu_\mu}|$. From Eq. 2.80 we find

$$\mathrm{w}(1,\vartheta) \quad \sim \quad (1 + \wp_\mu \cdot (\xi\delta/\varrho) \cdot cos\vartheta) \quad , \tag{2.85}$$

where ϑ is again the angle between the polarization of the muon and its decay positron. With the standard model values $\wp_{\nu_\mu} = -1$ and $\xi\delta/\varrho = +1$, we conclude that a positive muon from a pion decaying at rest is forbidden to emit a positron of maximum energy exactly along its line of flight ($\vartheta = 180^0$). This has been the basis of an ingenious precision experiment by the LBL-Berkeley-Northwestern-Triumf collaboration [60, 36] (see Fig 2.4).

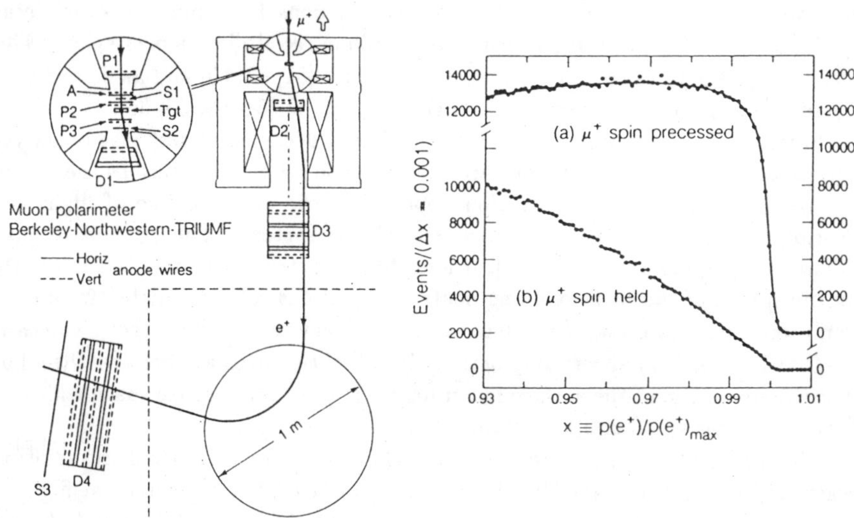

Fig. 2.4 Left: Positron momentum spectrometer. Positive muons fully backward polarized stop at "Tgt", where their spins either are kept fixed in a longitudinal holding field or where they precess in a transverse field. **Right:** Positive muons *do not* emit positrons of maximum energy along the direction opposite to their spins. This is demonstrated by curve (b) at $x = 1$. Curve (a) confirms that muons with isotropically distributed spins *do* emit positrons at $x = 1$. From Ref. [28].

The experiment may be interpreted in an entirely model independent way as follows [27]:

Since w(1,0) is a non-negative number and since $\wp_\mu$ may take its maximum value $\wp_\mu = 1$ independently of $(\xi\delta/\varrho)$, we must have $|\xi\delta/\varrho| \leqq 1$. If an experiment using Eq. 2.85 yields the value $\mathcal{A}_{exp}$ for

$$\wp_\mu \cdot |\xi\delta/\varrho| \;=\; \mathcal{A}_{exp} \leqq 1 \quad , \tag{2.86}$$

then we can draw *two independent* conclusions:

$$|\xi\delta/\varrho| \;\geqq\; \mathcal{A}_{exp} \tag{2.87}$$

$$\wp_\mu \;\geqq\; \mathcal{A}_{exp} \qquad \Rightarrow \qquad |\wp_{\nu_\mu}| \geqq \mathcal{A}_{exp} \tag{2.88}$$

The art of the experiment is the design of an apparatus which selects undisturbed muons from pion or kaon decay and which then would yield a result of $\mathcal{A}_{exp}$ as close as possible to one. Then $\wp_\mu$, $|\wp_{\nu_\mu}|$, and $|\xi\delta/\varrho|$ would be constrained between $\mathcal{A}_{exp}$ and one.

Beams of muons with a precisely known relation between the polarization $\wp_\mu$ and $\wp_{\nu_\mu}$ are produced by collecting those muons from pion or kaon decay $\pi^+ \to \mu^+\nu_\mu$ or $K^+ \to \mu^+\nu_\mu$, which have a unique recoil direction in space. This has been realized in two ways: as "surface muons" [61, 62, 63] and as muons from a parallel beam of pions decaying in flight in vacuum [64, 65, 66, 35].

Surface muons originate from pions (or kaons) decaying at rest just below the surface of the hadron production target. The recoil momentum pushes the muons out of this target into the direction opposite to the neutrino line of flight and possibly towards the ion optical beam transport system. The amount of their polarization equals $\wp_\mu = G \cdot |\wp_{\nu_\mu}|$, where the geometrical factor G is determined by the spread of the corresponding neutrino directions. G is estimated from Coulomb multiple scattering of the muons on their way out of the target. Values of $1-G \approx 10^{-3}$ with an uncertainty of roughly half of this size have been reached [60]. A great advantage of the surface muon beams is their small phase space which allows one to use small and thin stopping targets.

Muon beams from pions decaying in flight in vacuum avoid Coulomb multiple scattering. The muon spin lies in the plane of the laboratory lines of flight of the original pion, $\hat{k}_\pi$, and its decay muon, $\hat{k}_\mu$ and points inwards (towards $\hat{k}_\pi$) for μ^+ and outwards for μ^- (See Fig. 2.5). The transverse and longitudinal muon spin components ζ_T and ζ_L with respect to the muon's laboratory line-of-flight are simply given by

$$\zeta_T \;=\; sin\vartheta_\mu / sin\Theta_\mu \tag{2.89}$$

$$\zeta_L \;=\; \mp\sqrt{1-\zeta_T^2}\;, \tag{2.90}$$

where the upper (lower) sign applies for the muon emitted with smaller (larger) momentum for the given angle of emission ϑ_μ, and where

ϑ_μ = laboratory angle between $\widehat{k}_\pi$ and $\widehat{k}_\mu$
Θ_μ = maximum laboratory angle by kinematics (Jacobian peak angle)
$sin\Theta_\mu = (m_\pi^2 + m_\mu^2)/(2m_\pi k_\pi)$
k_π = pion beam momentum.

The selection of a small slice of muon energy in the laboratory in the vicinity of the Jacobian peak corresponds to a choice of a small range of neutrino directions and thus of a degree of polarization $\wp_\mu = G \cdot \wp_{\nu_\mu}$. Again, the geometrical factor G, which also has been studied experimentally [66], is close to one (> 0.99), and it is known with an uncertainty of $< 10^{-3}$ [35].

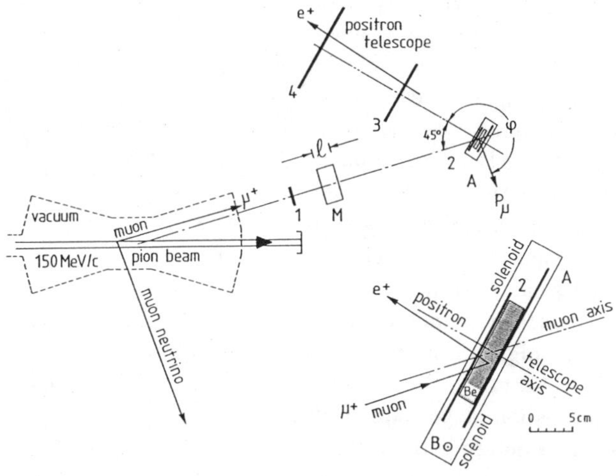

Fig. 2.5 Muon Spin Rotation apparatus to measure the integral asymmetry of the e^+ directional distribution following the decay of highly polarized muons. A parallel beam of monoenergetic (150 MeV/c) pions decay in flight in vacuum. Muons with energies within a well determined interval are selected to stop in a beryllium plate, Be, employing the moderator of length ℓ. The original orientation of the muons' polarization vector P_μ is thus defined. A rectangular solenoid produces a vertical magnetic field $B = 3\ mT$ causing the polarization of the stopped muons to precess in the horizontal plane. This gives rise to a sinusoidal modulation of the exponential decrease of the positron rate. The amplitude of the modulation ($\approx 1/3$) is proportional to the quantity desired, $P_\mu\xi$ [35].

In order to measure the decay asymmetry, the muons are stopped in a metal (Be, Al) immersed in a transverse magnetic field where the spins precess. Detectors track the muon and the decay positron momenta. The positron intensity shows a time modulation corresponding to the decay asymmetry. It is a favourable circumstance that there exist substances (Al, Cu, Ag, Au, bromoform) that barely influence the spin direction of muons which reside inside of them or which they do slow down. The disappearance of muon polarization during slowing down [67, 68] and thermalization [69], i.e. at early times compared to the muon precession time, fakes a smaller $\mathcal{A}_{exp}$. Depolarization at later times is seen in the data [70, 36]. This "muon spin relaxation" is a central subject of "μSR spectroscopy" [71]. It is a complicated solid state phenomenon whose time dependence cannot be reliably predicted with high precision. It can be accounted for essentially by extrapolation of the precession signal amplitude to zero time. The determination of the parameters of the extrapolating function in the same experiment generally reduces the statistical significance of the data considerably due to their strong correlation with the signal. The relaxation time in pure metals at room temperature is often conveniently big compared to the muon lifetime (See Fig. 2.6). For the measurement of $\wp_\mu \xi \delta / \varrho$, and of δ a precise positron energy spectrometer with well known resolution (0.15% rms) and acceptance has been used [36]. The results are

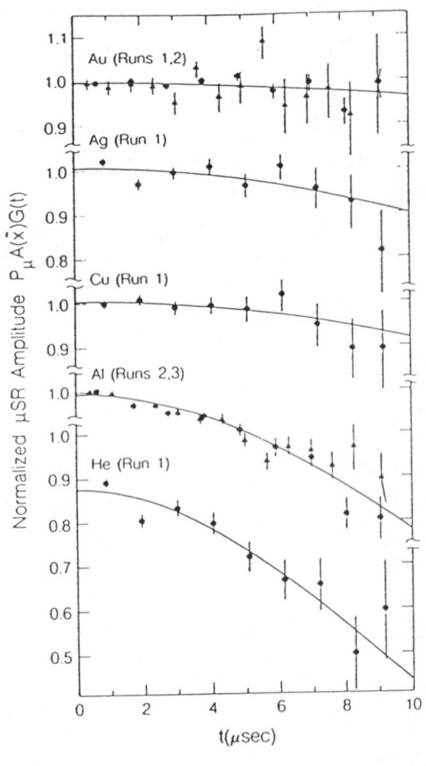

Fig. 2.6 Decay of the precession signal amplitude due to loss of phase coherence between the rotating muon spins. The curves assume a Gaussian time dependence of the relaxation [70]. Data obtained with the apparatus of Fig. 2.4.

$$\delta = (748.6 \pm 2.6_{\text{stat.}} \pm 2.8_{\text{syst.}}) \times 10^{-3}$$

$$\wp_\mu \xi \delta / \varrho = (997.90 \pm 0.46_{\text{stat.}} \pm 0.75_{\text{syst.}}) \times 10^{-3}$$

$$\text{and } \wp_\mu \xi \delta / \varrho > 996.82 \times 10^{-3} \ (90\% \ c.\ell.)$$

With Exprs. 2.86-2.88:

$$\begin{aligned} \wp_\mu &> 996.82 \times 10^{-3} \text{ (90\% } c.\ell.) \\ \wp_{\nu_\mu} &> 996.82 \times 10^{-3} \text{ (90\% } c.\ell.) \\ |\xi\delta/\varrho| &> 996.82 \times 10^{-3} \text{ (90\% } c.\ell.) \end{aligned}$$

For the measurement of $\wp_\mu \cdot \xi$, positron detectors with low energy thresholds are used. The results obtained from the decays $\pi^+ \to \mu^+\nu_\mu$ ($\wp_\mu^\pi$) and $K^+ \to \mu^+\nu_\mu$ ($\wp_\mu^K$) are:

$$\begin{aligned} \wp_\mu^\pi \cdot \xi &= (1002.7 \pm 7.9_{\text{stat.}} \pm 3.0_{\text{syst.}}) \times 10^{-3} \text{ [35]} \\ \wp_\mu^K \cdot \xi &= (1001.3 \pm 3.0_{\text{stat.}} \pm 5.3_{\text{syst.}}) \times 10^{-3} \text{ [37]} \\ \wp_\mu^K \cdot \xi &> 990 \times 10^{-3} \text{ (90\% } c.\ell.) \text{ [37]} \end{aligned}$$

Since ξ is not limited close to the measured value of $\wp_\mu \cdot \xi$, we cannot draw any specific conclusion on $\wp_\mu$ and ξ separately, contrary to the case of $\wp_\mu\xi\delta/\varrho$. In fact, $-3 \leqq \xi \leqq +3$, and thus the authors of ref. [35] do not quote an upper limit on $\wp_\mu\xi$. In order to isolate ξ from $\wp_\mu\xi$, one has to deduce $\wp_\mu$ from the measurement of $\wp_\mu\xi\delta/\varrho$ of ref. [36]. Examples of experimental data from the most simple apparatus (Fig. 2.5) as well as from the most sophisticated one (Fig. 2.4) are shown in Figs. 2.7 and 2.8.

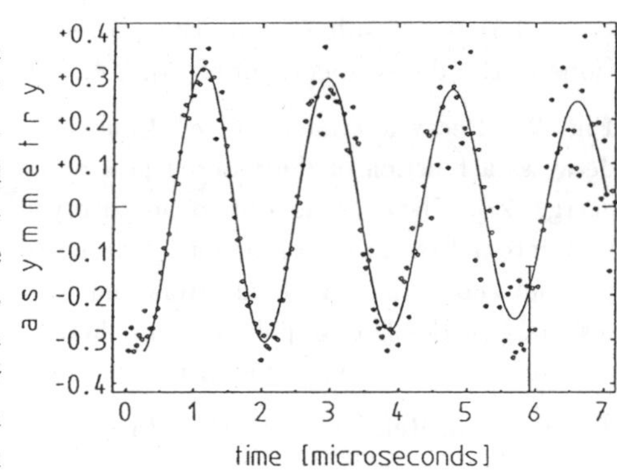

Fig. 2.7 Measured time distribution between the stop of highly polarized μ^+ in Be and the observation of the decay positron [35, 66]. Data obtained with the apparatus of Fig. 2.5. Positrons from a low threshold up to the maximum energy have been accepted. The exponential μ decay time has been factored out. The spatial asymmetry of the emitted e^+ relative to the muon spin rotates due to the spin's precession in a weak magnetic field. This generates the oscillation of the time spectrum. The solid line is the result of a best fit to the data. The slow decrease in amplitude (relaxation) is due to internal fields in Be.

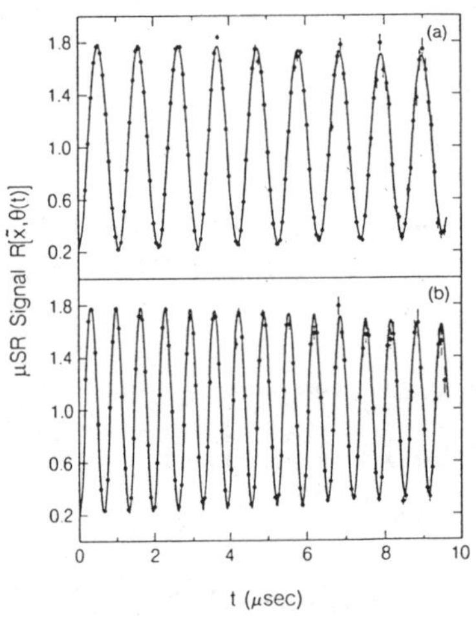

For the $\wp_\mu \xi$ experiments (e.g. Fig. 2.7 or Ref. [37]) it is typical that corrections are small. The raw data show an amplitude, which is close to the final experimental result. These experiments also constitute a simple means to measure $\wp_\mu$.

Fig. 2.8 Measured distributions of the time intervals between the stop of highly polarized and precessing muons in very pure metals and the acceptance of the decay positron [70]. In this experiment only the positrons of highest momentum are selected which results in a large asymmetry. Curve (a) has been taken at a field of 7 mT, curve (b) at 11 mT. The exponential decay time has been factored out. Data obtained with the apparatus of Fig. 2.4.

Fig. 2.9 shows that the measured decay asymmetry, as derived from data similar to those of Fig. 2.8 but with positrons selected at various momenta, vanishes near $x = 0.5$. The small but finite shift $(x_\delta - 0.5)$ is a result of radiative corrections to the decay spectrum of Eq. 2.4.

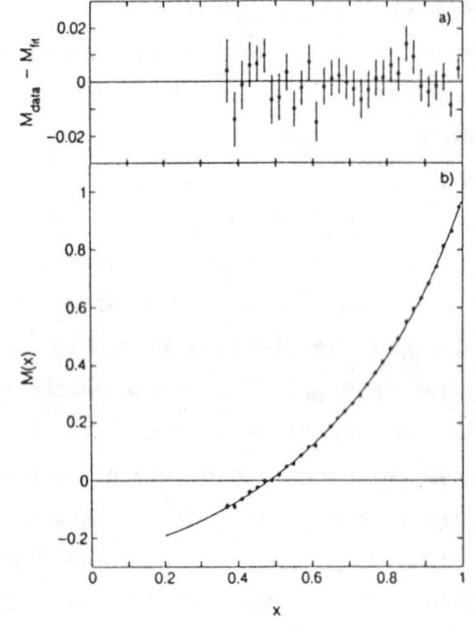

Fig. 2.9 Decay asymmetry of e^+ from μ^+ decay as a function of the reduced positron energy [28]. The curve is a fit to the theory with internal radiative corrections. The upper plot shows the statistical errors and fit residuals for the 32 data points. Data obtained with the apparatus of Fig. 2.4.

The result of Ref. [36] is of importance in several respects:

(1) It constitutes by far the best measurement of the magnitude of a neutrino polarization, in particular of $\wp_{\nu_\mu}$ [27]. The sign of $\wp_{\nu_\mu}$ has been measured in separate experiments [72]- [77].

(2) Surface muon beams from pion decay can be prepared with a high and well known polarization $\wp_\mu > 0.99682$ (90% *c.ℓ.*)

(3) The polarization $\wp_\mu$ of muons stopped in various metals is preserved to a high accuracy during several lifetimes under the influence of an external magnetic holding field of 1.1 T [60] or 0.3 T [36].

(4) The precise value of $\xi\delta/\varrho$ gives the strongest constraint for the absence of couplings with both right-handed electrons and muons, g^V_{RR} and g^S_{RR}. Together with δ and ϱ, it constrains g^γ_{LR} , $\gamma = S, V, T$.

(5) The product $\wp_\mu\xi\delta/\varrho$ is of interest in the search for right-handed currents introduced to models which include also the pion (or kaon) decay.

For the theoretical significance of the results we refer to Refs. [17, 78, 79, 80, 81]. We restrict ourselves to mentioning two special cases of left-right symmetric models which have been used to interpret experimental results.

Right-handed currents admixed to the interaction change $\mathcal{A}_{exp}$, because the decay muon is no more fully polarized ($\wp_\mu < 1$) and because high energy positron emission opposite to the muon spin is no more strictly forbidden ($\xi\delta/\varrho < 1$, $\xi < 1$). Such a deviation from the standard model may e.g. be due to a right-handed intermediate vector boson W_R of mass $\approx m_R$. With the assumption of left-right symmetry, the parity violation of weak interactions at low energies would still be preserved, if $m_R \gg m_L$. We now expect $g^V_{RR} \neq 0$ dependent on m_R, and $g^V_{RL} \neq 0$, $g^V_{LR} \neq 0$ dependent on a possible mixing of the states W_R and W_L. In such a model in which $g^V_{RL} = g^V_{LR}$ and in which all of the scalar and tensor couplings are zero and therefore also $c = c' = a' = 0$ we find

$$(1 - \xi\delta/\varrho) \approx 2|g^V_{RR}|^2 \tag{2.91}$$

and

$$(1 - \xi) = 2|g^V_{RR}|^2 + 2|g^V_{RL}|^2. \tag{2.92}$$

The relations of the coupling constants $g^V_{\varepsilon\mu}$ to the model parameters $\varepsilon = (m_L/m_R)^2$ and the $W_R - W_L$ mixing angle ζ are:

$$g^V_{LL} \approx 1 \tag{2.93}$$

$$g^V_{RR} \approx \varepsilon \ll 1 \tag{2.94}$$

$$g^V_{RL} = g^V_{LR} \approx \zeta \ll 1 \quad . \tag{2.95}$$

Right-handed currents decrease the muon polarization in kaon decay more strongly than in pion decay if the right-handed quarks mix more than the usual left-handed ones, i.e. if $\vartheta_R > \vartheta_L$ [82]. The measurement of $\wp_\mu\xi$ with muons from $K^+ \rightarrow \mu^+\nu_\mu$ may thus be somewhat more sensitive to ε in this case. (We use

$\vartheta_R \equiv \theta_1^R, \vartheta_L \equiv \theta_1^L$ and $\theta_3^R = \theta_3^L = 0$, where $\theta_i^\varepsilon, i = 1, 3, \varepsilon = R, L$ is the notation of ref. [82]).

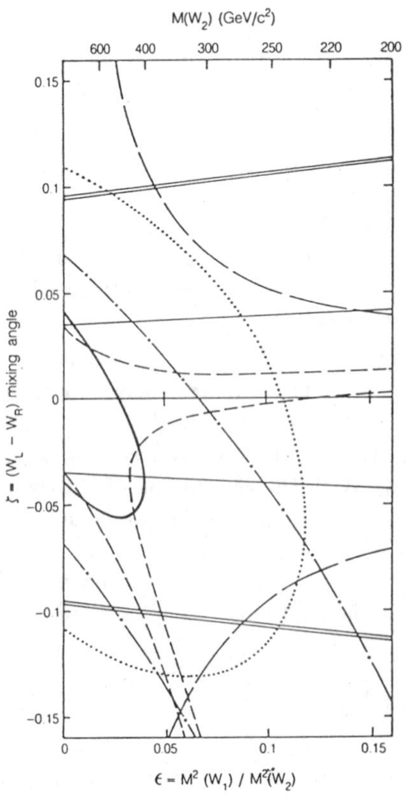

Fig. 2.10 Experimental 90% confidence limits on the mass-squared ratio ε and mixing angle ζ for the gauge bosons W_R and W_L [36]. The allowed regions include $\varepsilon = \zeta = 0$. Bold curve from $\wp_\mu \xi \delta / \varrho$ [36], dotted curve from $\wp_\mu \xi = (972 \pm 14) \times 10^{-3}$ as of 1968, solid lines from ϱ [59], double lines from νN and $\bar{\nu} N$ scattering [83]. Other curves from nuclear β decay. For more details, see Ref. [36].

Table 2.3 summarizes the basic relations. LRS1 is the special case $\vartheta_L = \vartheta_R$ of LRS2, as described in Ref. [17] but with equal strengths $g_R = g_L$ and CP conservation $\omega = \alpha = 0$. A measured value of $\varrho, \xi, \wp_\mu$ or some combination defines a contour plot in the plane of ε vs. ζ. If it is compatible with the standard model value, its error places limits on m_R and on the mixing angle (See Fig. 2.10). In the strangeness changing decay $K^+ \rightarrow \mu^+ \nu_\mu$ the quark mixing angle enters as $sin\vartheta_R$, in $\pi^+ \rightarrow \mu^+ \nu_\mu$ as $cos\vartheta_R$. With the usual assumption that the muons from kaon and from pion decay are identical, the difference $\vartheta_R \neq \vartheta_L$ just leads to a different muon polarization $\wp_\mu$.

Since most of the muon decay experiments do not observe the identity of the neutrinos, the decay $\mu^+ \rightarrow e^+ \tilde{\bar{\nu}}_\mu \tilde{\nu}_e$ to the supersymmetric scalar partners of the neutrinos, mediated by a wino, may admix positrons with deviating properties [84, 85]. These influence the measured values of ξ, δ and ϱ but leave $\xi\delta/\varrho$ unchanged. Under the assumption of light scalar neutrinos ($m_{\tilde{\nu}} < 10\ MeV$) which would not energetically inhibit the decay, the sensitivities to a hypothetical wino of mass $\tilde{m}$ are also given in Table 2.3. We note that a supersymmetric contribution is not the result of a four-fermion interaction. In general, the expressions derived from Eq. 2.1 then do not provide the correct degrees of freedom for the corresponding observables [4]. Some results are (90% $c.\ell.$): $m_R > 482\ GeV/c^2$ ($\zeta = 0$), $|\zeta| < 0.040$ ($m_R = \infty$) [36]; $m_R > 635\ GeV/c^2$ ($\zeta = 0, \sin\vartheta_R = 1$) [37]; $\tilde{m} > 270\ GeV/c^2$ (light sneutrinos) [35]. For limits on composite leptons and on familons, see Ref. [36].

Table 2.3 Muon decay parameters beyond the standard model. $\varepsilon = (m_L/m_R)^2$, $\zeta = W_L - W_R$ mixing angle. The muon polarization also depends on the quark mixing angles ϑ_R *and* ϑ_L.
LRS1, LRS2: Left-right symmetric models [17, 78, 79, 80, 81, 82]
SUSY [84, 85]: $\lambda = (m_L/\,\widetilde{m})^2$. $\widetilde{m}$ = mass of the wino. Neutrino masses are assumed to be sufficiently small and of no influence on the energy spectrum.
The parametrization of other measurements may be read off this table immediately, e.g. $1 - \wp_\mu\xi \approx (1 - \wp_\mu) + (1 - \xi)$. Thus $1 - \xi\delta/\varrho$ becomes independent of λ.

Measurement	Standard Model	LRS1 $\vartheta_R = \vartheta_L$	LRS2 ϑ_L, ϑ_R	SUSY
$\frac{4}{3}(\varrho - \frac{3}{4})$	0	$-2\zeta^2$	$-2\zeta^2$	$\frac{1}{2}\lambda$
$\frac{4}{3}(\delta - \frac{3}{4})$	0	0	0	$-\frac{3}{2}\lambda$
$1 - \xi$	0	$2(\varepsilon^2 + \zeta^2)$	$2(\varepsilon^2 + \zeta^2)$	-2λ
$1 - \wp_\mu$	0	$2(\varepsilon + \zeta)^2$		
$1 - \wp_\mu^\pi$	0		$2(\varepsilon \cdot \frac{\cos\vartheta_R}{\cos\vartheta_L} + \zeta)^2$	
$1 - \wp_\mu^K$	0		$2(\varepsilon \cdot \frac{\sin\vartheta_R}{\sin\vartheta_L} + \zeta)^2$	

2.4.4. Longitudinal Electron Polarization

The measurement of the longitudinal polarization P_L of the electrons from the decay of polarized or unpolarized muons allows one to determine the parameters ξ'' and ξ', as can be seen from Eqs. 2.9, 2.10, 2.21 and 2.22. The parameter ξ' is of special interest. In terms of the coupling constants $g^\gamma_{\varepsilon\mu}$ we have

$$\begin{aligned} 1 - \xi' &= \tfrac{1}{2}(4 \cdot (|g^V_{RR}|^2 + |g^V_{RL}|^2) + |g^S_{RR}|^2 + |g^S_{RL}|^2 + 12 \cdot |g^T_{RL}|^2) \\ &= 2(Q_{RR} + Q_{RL}) \equiv 2Q^e_R \quad , \end{aligned} \tag{2.96}$$

where $Q_{\varepsilon\mu}$ is the probability of the decay of a muon with chirality μ into an electron with chirality ε. Note that Eq. 2.96 is a sum of absolute squares where only coupling constants with $\varepsilon = R$ appear. A deviation of ξ' from 1 would require the existence of a coupling with the right-handed components of the electron, i.e. at least one $g^\gamma_{R\mu} \neq 0$. Conversely, a measurement with the result $\xi' = 1$ proves that the coupling acts exclusively on the left-handed component of the electron.

To determine ξ', the longitudinal polarization P_L of the electrons from unpolarized muons has been measured. For the purpose of illustration, we neglect the electron mass m_e and use the experimentally well confirmed values $\varrho = \delta = \frac{3}{4}$ and obtain from Eq. 2.9

$$\xi' = P_L \quad .$$

The measurement of the electron's longitudinal polarization P_L consists of its comparison with the spin polarization of the electrons contained in a piece of saturated ferromagnetic material [86]-[95], [67]. The comparison is done by scattering the decay electrons from the electrons of a ferromagnet, using the fact that relativistic electron-electron scattering most often occurs when the two spins have opposite directions. See Fig. 2.11.

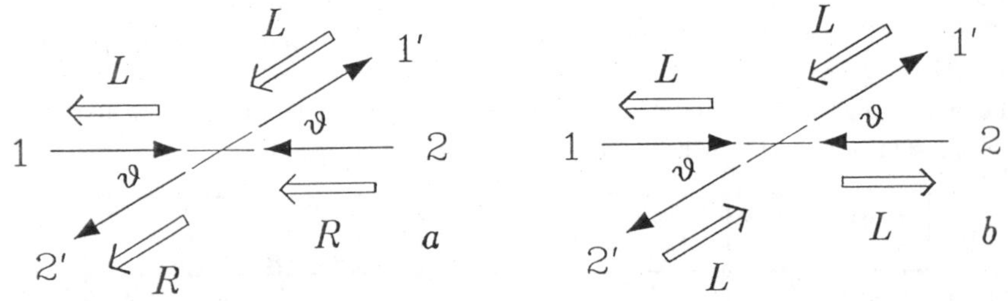

Fig 2.11 Comparison of the longitudinal polarizations of electrons 1 and 2. Scattering of two fast electrons $1+2 \to 1'+2'$ occurs most often when the spins look opposite, as in *case b*. In scattering due to a vector interaction (such as electromagnetic), *chirality* R, L is conserved. In case of *relativistic* particles, this means, that 1 and $1'$, and also 2 and $2'$ have the same *helicity*. Angular momentum conservation partially inhibits the polarization configuration $LR \to LR$ (*case a*) mainly for big scattering angles ϑ, whereas it is of no influence for the configuration $LL \to LL$, (*case b*). The $LL \to LL$ configuration is furthermore favoured by the fact, that the *indistinguished* reaction $1+2 \to 2'+1'$ leads to *distinguishable* and thus incoherent final states in *case a*, but to *indistinguishable* and thus coherent final states in *case b*. For $\vartheta = 90^0$ e.g. LL is *eight* times more probable. The same conclusion applies, when particle 1 is a relativistic positron, which scatters elastically from an electron. The annihilation in flight $e^+e^- \to \gamma\gamma$ of relativistic positrons with electrons, however, occurs mainly when their longitudinal spins are parallel (RL).

The spin polarization of the ferromagnet's electrons is deduced from their total spin angular momentum, which becomes macroscopically measurable, when all electron spins are flipped simultaneously, as done in Einstein-de Haas experiments. For the comparison of the longitudinal polarization of *positrons* from $\mu^+ \to \bar{\nu}_\mu e^+ \nu_e$ with the polarization of a ferromagnet's electrons, elastic scattering as well as annihilation in flight into two gamma rays, $e^+e^- \to \gamma\gamma$, are most suitable. The results of measurements of P_L are displayed in Fig. 2.12. They yield an average of $< |P_L| > = \ 0.998 \pm 0.042$. Radiative corrections to the electron polarization $\vec{\wp}_e$ are negligibly small [96, 97].

From the resulting error of ξ', which is dominated by the error of this $< |P_L| >$, upper limits for all couplings of right-handed *electrons* to muons (of any handedness) $|g^\gamma_{R\mu}|$, $\mu = R, L$, follow, in principle, from Eq. 2.96. Improved

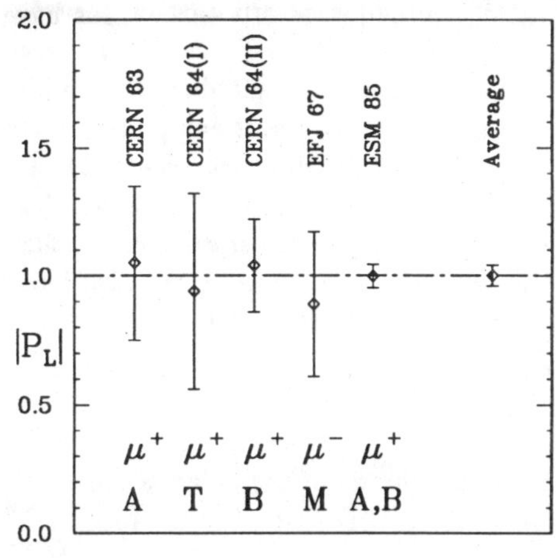

Fig. 2.12 Longitudinal polarization of electrons and positrons in the decay of unpolarized muons. The experiments (CERN 63 [98], CERN 64(I)[99], CERN 64(II)[100], EFJ 67[101], ESM 85[34]) confirm full forward polarization for positrons and full backward polarization for electrons, i.e. weak coupling to *left-handed* electrons only.
A (Annihilation), T (Transmission of polarization in bremsstrahlung), B (Bhabha scattering), and M (Møller scattering) indicate the methods used to compare the decay particle's polarization with the electron polarization in a ferromagnet.

values of these limits are obtained for $|g^V_{RL}|$ and $|g^S_{RL} + 6g^T_{RL}|$ by considering also

$$B_{RL} = \tfrac{1}{16}|g^S_{RL} + 6g^T_{RL}|^2 + |g^V_{RL}|^2 = \tfrac{1}{2A}(a + a') \quad . \tag{2.97}$$

For all couplings of right-handed *muons* to electrons (of any handedness) $|g^\gamma_{\varepsilon R}|$, $\varepsilon = R, L$, better limits have been reached from the experiments with polarized muons.

The parameter ξ'' in positive muon decay has been determined from a measurement of $P_L(x, \vartheta)$ as a function of the positron reduced energy x and the angle ϑ between the muon spin and the positron momentum [34]. The present precision of the measured combination $(\xi'' - \xi \cdot \xi')/\xi = -0.35 \pm 0.33$ does, however, not lead to better constraints of the couplings.

Experiments measuring P_L and $P_L(x, \vartheta)$ are described in Fig. 2.13. Positive pions decay inside the stopping target, where they create unpolarized muons, which also decay. The positrons encounter the electrons of the magnetized iron foil. Both effects, elastic scattering ($e^+e^- \to e^+e^-$) or annihilation in flight ($e^+e^- \to \gamma\gamma$) are observed in order to quantitatively compare the positron polarization with the spin polarization of the electrons in the ferromagnet. The four NaI counters identify and measure the energies of the final state particles e^+e^- or $\gamma\gamma$. The apparatus thus performs simultaneously two independent experiments, one is most sensitive to antiparallel, the other to parallel spins. The backgrounds and the sources of the systematic errors are widely different in the two cases. Since the multiple scattering in the iron foil alters the directions of flight of the charged particles, but leaves their energies mostly unchanged, the measured energies allow the determi-

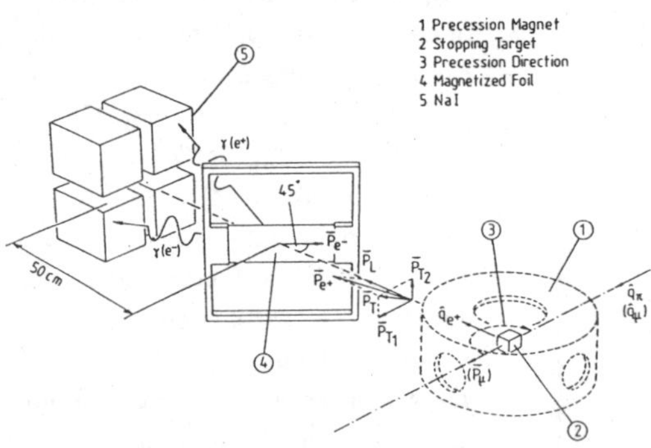

Fig. 2.13 Measurement of the longitudinal polarization of positrons from polarized and unpolarized muons. Longitudinally polarized muons stop at 2 and precess. (For unpolarized muons, pions stop at 2 and the magnet is removed). The decay positron scatters or annihilates at 4. e^+e^- or $\gamma\gamma$ pairs are detected at 5. From Ref. [34]

nation of the energy of the incoming decay positron and of the polarization sensitivity of the cross section. For both reactions the cross section has the form

$$\sigma_i(E_1, E_2) = \sigma_{oi}(E_1, E_2) \cdot [1 + A_i(E_1, E_2) \cdot \vec{\wp}_{e^+} \cdot \vec{\wp}_{e^-}] \quad , \tag{2.98}$$

where i stands for $(e^+e^- \to e^+e^-)$ or $(e^+e^- \to \gamma\gamma)$. E_1 and E_2 are the laboratory energies of the final state particles. $\vec{\wp}_{e^+}$ and $\vec{\wp}_{e^-}$ are the polarizations to be compared. Since $\vec{\wp}_{e^-}$ lies in the plane of the foil, as does the magnetization, the foil is inclined (under 45^0) to the positron's line of flight.

The analyzing power $\mathcal{A}_i(E_1, E_2)$ reaches high values (-0.78 for scattering and +0.89 for annihilation), however the fraction of electrons which take part in ferromagnetism is small. $\wp_{e^-}$ has typical values of 55×10^{-3}. Therefore, upon changing the foil inclination from 45^0 to 135^0 or reversing $\vec{\wp}_{e^-}$, correspondingly small signals in $\sigma_i(E_1, E_2)$ arise.

$\wp_{e^-}$ is determined from the macroscopic gyromagnetic ratio g' of iron as deduced from the angular momentum an iron piece displays when its magnetization is reversed. The electrons in iron may contribute to the magnetization and to the angular momentum with a gyromagnetic ratio $g = 2$, when they are polarized, and with $g = 1$, when they perform an orbital motion. Unpolarized electrons at rest contribute to neither of them. For pure iron the Einstein-De Haas effect gives $g' = 1.919 \pm 0.002$ [102]. This result, being not far from the free electron's g value shows that most of the magnetization comes from the polarized electrons, namely $f \equiv$ (magnetization due to spin orientation) / (total magnetization) = $(1-1/g')/(1-1/g) = 0.958 \pm 0.001$. From the measured magnetization $M \approx B/\mu_0$ the electron polarization becomes

$$|\vec{\wp}_e| = \frac{f \cdot B}{\mu_0 \cdot N \cdot \mu_B} = (54.44 \pm 0.56) \times 10^{-3}, \tag{2.99}$$

where N is the number density of electrons in iron and μ_B the Bohr magneton.

In order to measure the longitudinal polarization $P_L(x,\vartheta)$ also as a function of the angle ϑ between the muon spin and the electron momentum, *polarized muons* are stopped in the stopping target of Fig. 2.13. Due to a vertical magnetic field in the target region, they precess in a horizontal plane. A counter telescope, not shown in Fig. 2.13, detects their precession phase. From the individual decay times the angle ϑ is known for each recorded event.

With the experimental techniques available today and with considerable effort in instrumentation and beamtime, only moderate improvements (factor 3, say) in the polarization results seem possible. The small effective analyzing powers and their absolute calibration are a difficulty.

2.4.5. Transverse Electron Polarization

Transverse electron polarization P_T (P_{T_1}, P_{T_2}) is defined in Fig. 2.1. Independent of any assumption about the mechanism of muon decay or even the nature of the two unobserved neutral particles, time reversal invariance (disregarding the negligible final state interactions) requires$P_{T_2} = 0$.

In a decay into light photinos, $\mu \to e\tilde{\gamma}\tilde{\gamma}$, P_T of observable size may arise, if the two effective masses $\mathcal{M}_k$, $k = R, L$, of the intermediate scalar leptons are not too heavy: $|\mathcal{M}_L| \cdot |\mathcal{M}_R| \gtrapprox 4m_W^2$ (90% *c.ℓ.*) [103, 32].

Within the four fermion interaction Eq. 2.1 the measurement of $\vec{P}_T$ as a function of energy allows one to determine the parameters $\eta \equiv (\alpha - 2\beta)/A$, $\eta'' \equiv (3\alpha + 2\beta)/A$, α' and β' (see Eqs. 2.13, 2.14, 2.19 and 2.20). η is of special interest. Although η describes, together with the Michel parameter ϱ, the shape of the (isotropic) positron energy spectrum, it is practically difficult to deduce its value from a spectrum measurement, since its influence there is suppressed by a factor $x_0 \approx 10^{-2}$. On the other hand, its value has to be known for a precise determination of ϱ, since η and ϱ are statistically highly correlated. In Eq. 2.19 for P_{T_1}, η arises without suppression factor. It is interesting to note that P_{T_1} does not vanish in the standard model interaction, as may be seen from Eq. 2.13, and it may take sizeable values ($|P_{T_1}| \leqq 1/3$) for positron energies $E_e <$ few times m_e. The parameters α' and β' determine P_{T_2}.

Since, by definition, probabilities and polarizations cannot exceed unity, there exist bounds such as Exprs. 2.46-2.51. From 2.49 and 2.50 we deduce

$$\alpha^2 + \alpha'^2 \leqq (a - a') \cdot (a + a') \tag{2.100}$$

$$\beta^2 + \beta'^2 \leqq (b - b') \cdot (b + b') \quad . \tag{2.101}$$

The values of the factors on the r.h.s., as obtained from the precision measurements of ξ', $\xi\delta/\varrho$, ξ, δ (or ϱ), are small, except for $(b - b')$, which is of order $A/2 = 8$. At present, both products on the r.h.s. are below the experimental uncertainties of α, α', β and β' whose values are compatible with zero. They vanish, if ξ' takes the standard model value, $\xi' = 1$.

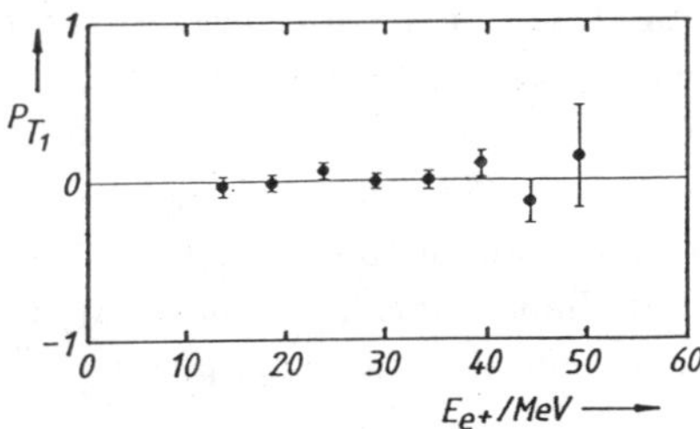

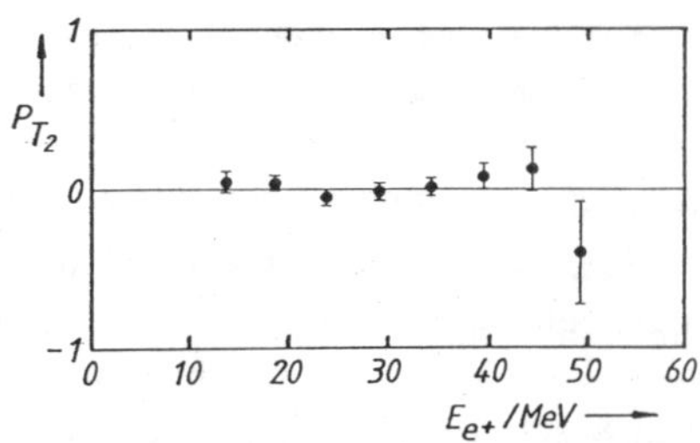

Fig. 2.14 The transverse polarization components P_{T_1} and P_{T_2} as obtained from an analysis of the annihilation in flight events as a function of the positron total energy E_{e^+}. The errors include both statistical and systematic contributions. Data obtained from the apparatus of Fig. 2.15. From Ref. [32].

The experimental results are displayed in Fig. 2.14. They are compatible with being independent of the energy between 9.5 MeV and maximum positron energy. The averages over the energy are

$$\begin{aligned} P_{T_1} &= (16 \pm 23) \times 10^{-3} \\ P_{T_2} &= (\ 7 \pm 23) \times 10^{-3} \end{aligned}$$

The values of $\alpha, \alpha', \beta, \beta'$ are collected in Table 2.4.

Table 2.4 Results for α/A, α'/A, β/A, β'/A, η and η'' from the experiment of Fig. 2.15. The errors are dominated by statistics. A value of zero for a parameter indicates that it was set identically equal to zero in the least squares analysis. Also included are the low energy shape parameter $\eta = (\alpha - 2\beta)/A$ and the parameter $\eta'' = (3\alpha + 2\beta)/A$. Their errors include the correlation between α/A and β/A. The correlation coefficient between α/A and β/A (or α'/A and β'/A) is $\rho = -0.894$. All other correlation coefficients are ≈ 0. In all cases, the χ^2 per degree of freedom was 0.97. (All values are given in units of 10^{-3}). From Ref. [32].

α/A	β/A	α'/A	β'/A	η	η''
15 ± 52	2 ± 18	−47 ± 52	17 ± 18	11 ± 85	48 ± 125
0	6 ± 8	0	3 ± 8	−12 ± 16	12 ± 16
19 ± 23	0	−1 ± 23	0	19 ± 23	58 ± 70
16 ± 52	2 ± 18	0	0	13 ± 85	51 ± 125
0	0	−46 ± 52	16 ± 18	0	0

From Eqns. 2.38, 2.39, 2.42 and 2.43 we see that α and α' are of second order in deviations from the standard model, whereas

$$\begin{aligned} \beta &= -4Re\left\{g^V_{RR}g^{S*}_{LL} + g^V_{LL}g^{S*}_{RR}\right\} \\ \beta' &= 4Im\left\{g^V_{RR}g^{S*}_{LL} - g^V_{LL}g^{S*}_{RR}\right\} \end{aligned} \quad (2.102)$$

are sensitive to first order in (magnitude and phase of) g^S_{RR}.

The experiment is described in Fig. 2.15. It also consists of a comparison with the spin polarized electrons in a ferromagnetic foil, using annihilation in flight $e^+e^- \rightarrow \gamma\gamma$. It is based on the fact that the photons from the annihilation of a relativistic and transversely polarized positron electron pair are preferentially emitted in the plane spanned by the particle line of flight $\hat{q}_{e^+}$ and the bisector $\vec{b}$ between the (transverse) polarization directions $\vec{p}_T$ and $\vec{\wp}_{e^-}$.

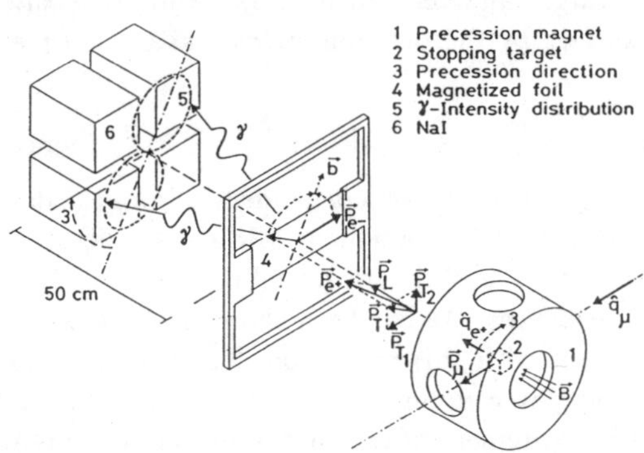

Fig. 2.15 Measurement of the transverse components P_{T_1} and P_{T_2} of the positron in muon decay. Longitudinally polarized muons at 2 precess in a vertical plane. The decay positrons annihilate with transversely polarized electrons at 4. The plane of the two annihilation gamma rays tends to contain the bisector $\vec{b}$ of the spin directions of the e^+e^- pair, as indicated by 5. The data are shown in Fig. 2.14. From Ref. [32].

The distribution as a function of the azimuthal angle φ between this plane and the actual plane of the two photons is given by

$$(d\sigma/d\Omega) \sim (1 - A_T(E_1, E_2) \cdot P_T \cdot \wp_{e^-} \cdot sin^2\varphi) \quad . \quad (2.103)$$

The polar diagram of this function is indicated in Fig. 2.15 as a "figure of eight". The analysing power $A_T(E_1, E_2)$ as a function of the photon energies E_1 and E_2 takes conveniently high values (up to 0.92), but $\wp_{e^-}$ is again small. P_T is measured by probing the above intensity distribution with the four NaI crystals detecting pairs of gamma rays. Instrumental asymmetries are averaged out by precessing the muon spins in a plane perpendicular to the symmetry axis of the apparatus. This causes the bisector $\vec{b}$ to precess with half of the muon spin rotation frequency. After a rotation of 180^0, however, the situation is identical for the detectors because of the two photons in the final state. Therefore the *effect* shows again the μSR frequency. In order to make use of the high intensity of the polarized muon beam at SIN (now PSI), the muon precession is synchronized to the arrivals of the muon beam packets. This allows several muons to rotate with parallel spins in the stopping target at the same time.

The average positron direction of flight of the events detected by the non-diagonal NaI pairs is slightly off axis. Therefore the pair coincidence rate shows a small μSR modulation which marks the precession phase. For two NaI pairs of the same type (left and right or upper and lower pairs) this precession phase is shifted by 180^0, while for a signal due to a transverse polarization the phases are equal. From this one can deduce both magnitude and phase of $\vec{P}_T$ with respect to the muon spin. The data recorded for each event include information on the time of decay, on the positron's direction of flight and on the energies of the photons E_1 and E_2.

A finite positron transverse polarization would make the positron spins precess rapidly in the magnetic field in the target and tend to destroy itself. It is the Lorentz contraction of the magnet as seen by the positron which makes the time spent in the field sufficiently short.

2.4.6. Electron Neutrino Energy Spectrum

The shape parameter ω_L of the energy spectrum of left-handed ν_e from the decay of unpolarized μ^+ is completely analogous to the famous Michel parameter ϱ. There is an important difference, however: The standard model predicts $\varrho = 3/4$, but $\omega_L = 0$. This has intriguing consequences: As has been shown in Sect. 2.3.2 a precise measurement of ϱ in agreement with this prediction does not allow to derive any limits on six of the ten complex coupling constants describing muon decay. On the other hand a precise measurement of ω_L puts upper limits to the coupling constants $|g^S_{RR}|$, $|g^V_{LR}|$ and $|g^S_{RL} + 2g^T_{RL}|$ [22, 23].

From an experimental point of view it is exciting to realize that the measurement of ω_L is a null test of the standard model, comparable to the measurement of $P_\mu \xi \delta / \varrho$ [36] where one looks for events where none are predicted. For a purely left-handed (or a purely right-handed) interaction angular momentum conservation prohibits events at the maximum ν_e energy, while for all of the remaining interactions the rate is even enhanced at the spectrum endpoint. This leads to $\omega_L > 0$ and makes the method very sensitive (see Fig. 2.16).

The ν_e spectrum and therefore the parameter ω_L can be measured with a neutrino spectrometer of sufficient energy resolution and known energy dependence of the detection efficiency. One possible process is the reaction ${}^{12}C(\nu_e, e^-){}^{12}N(g.s.)$ The KARMEN collaboration has built such a detector with 56 t liquid scintillator as target [38]. There, π^+ stop in a beam dump and decay into an unpolarized sample of μ^+. The μ^+ decay with a mean lifetime of 2.2 μs, and ν_e are converted promptly in the scintillator to e^-. The observation of the subsequent β^+ decay of ${}^{12}N(g.s.)$ with a mean lifetime of 15.9 ms, together with the required spatial co-

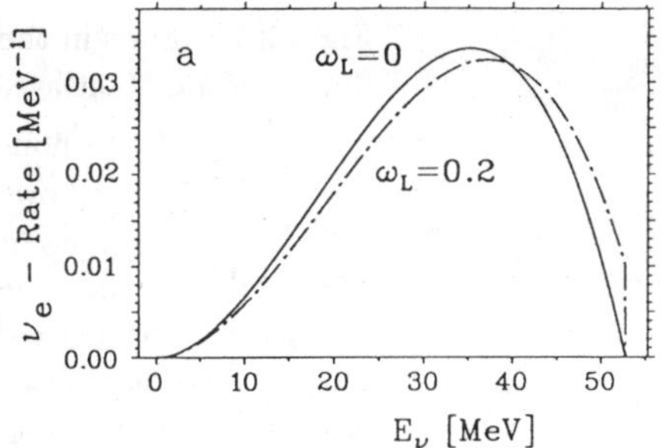

Fig. 2.16 Normalized energy distributions of left-handed ν_e from the decay of unpolarized μ^+. The spectrum shape parameter ω_L is the analog of the Michel parameter ϱ of the e^+. For a pure $V-A$ interaction ω_L is equal to zero. From Ref. [22].

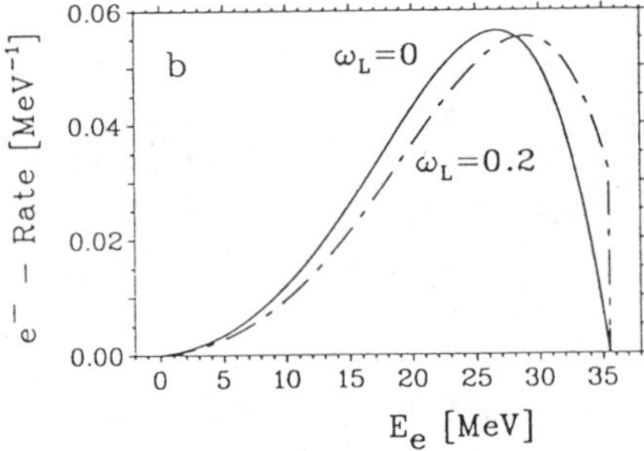

Fig. 2.17. Sensitivity of the normalized energy distribution of e^- from the reaction $^{12}C(\nu_e, e^-)^{12}N(g.s.)$ to ω_L. The curves represent the product of the left-handed ν_e distribution of Fig. 2.16. with an absorption cross section σ_A parametrized by a polynomial of second order in the electron energy E_e. From Ref. [22].

incidence, greatly reduces background (see Fig. 2.18). The authors assumed the pure $V-A$ law for the decay and obtained the energy dependence of the absorption cross section. They were able to measure the energy E_e of the e^- with the high precision $\sigma(E_e)/E_e = 11.5\%/\sqrt{E_e}$ (E_e in MeV). It has been pointed out recently that one can use the same measurement to derive independently the shape $d\Gamma_L/dy$ of the neutrino spectrum [22]. There it was shown that the pronounced structure at the spectrum endpoint for $\omega \neq 0$ can easily be separated from the smoothly varying absorption cross section σ_A. The sensitivity is further enhanced by the fact that σ_A rises strongly with energy, so that the region of interest is magnified as with a looking class (see Fig.2.17). This boosts the sensitivity to $\Delta\omega_L = 0.26/\sqrt{N}$, where N is the number of events.

In the following we assume that the KARMEN experiment will reach a final sensitivity of $\Delta\omega_L = 1\%$. This will improve existing upper limits to $|g^S_{RL} + 2g^T_{RL}|$ according to Eq. 2.25 [23]:

$$|g^S_{RL} + 2g^T_{RL}| \leqq \sqrt{\tfrac{16}{3}\Delta\omega_L} \tag{2.104}$$

The effects of contributions from right-handed ν_e lead to the effective spectrum shape parameter

$$\omega_{eff} = \frac{\omega_L Q^{\nu_e}_L + \varepsilon(y)\omega_R Q^{\nu_e}_R}{Q^{\nu_e}_L + \varepsilon(y)Q^{\nu_e}_R} \approx \omega_L + \varepsilon(y)\frac{Q^{\nu_e}_R}{Q^{\nu_e}_L}(\omega_R - \omega_L), \tag{2.105}$$

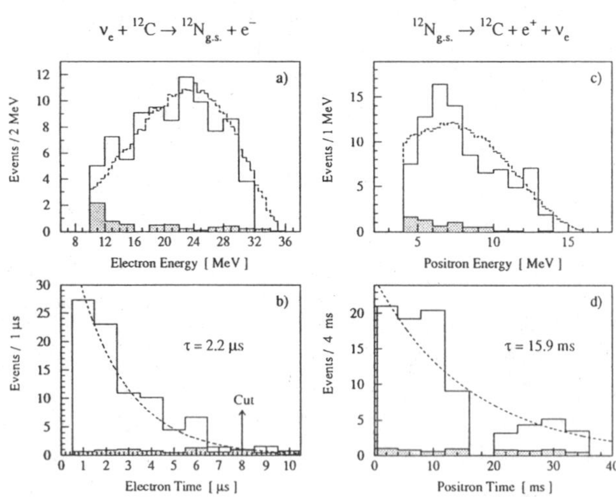

Fig. 2.18 Energy and time spectra of delayed coincidences from the reaction $^{12}C(\nu_e, e^-)^{12}N_{g.s.}$ and $^{12}N \rightarrow {}^{12}Ce^+\nu_e$. The visible energies of prompt electrons (a) and delayed positrons (c) are compared to simulations (broken lines). The corresponding time distributions are shown in (b) and (d) with the decay curves of μ^+ and ^{12}N superimposed. The normalized "beam off" background is shown as a shaded area. From Ref. [38].

where $\varepsilon(y)$ is the relative amount of right-handed ν_e contributing to the absorption cross section σ_A. From muon decay one finds that *emission* is suppressed by the factor $Q_R^{\nu_e}/Q_L^{\nu_e} < 0.087$ (90% *c.ℓ.*), and from nuclear beta decay that *absorption* is suppressed by $\varepsilon(y) < 7.4 \times 10^{-3}$ (90% *c.ℓ.*) [23]. The total contribution to the value of ω_L is therefore $< 0.6 \times 10^{-3}$ and can safely be neglected for the presently achievable precision.

The second observable of the neutrino distribution is the probability $Q_L^{\nu_e} = 1 - Q_R^{\nu_e}$ of the ν_e to be left-handed. A measurement of the total rate with an error of $< 5\%$ would improve present experimental limits which would lead to improved limits for the coupling $|g_{LL}^S|$, which can hardly be detected directly because both neutrinos are right-handed. The measurement of $Q_L^{\nu_e}$, however, does not seem to be feasible at present, mainly because the absorption cross section has to be determined absolutely either by theoretical calculation or by suitable calibration. The experimental error on the rate is presently in the order of 10% [38] and might be improved in a future, dedicated experiment.

2.4.7. Inverse Muon Decay

Measurements of inverse muon decay (IMD) are difficult to perform because a large ν_μ laboratory energy $E_{\nu_\mu} > 10.93\ GeV$ is transformed to the small c.m. energy $\sqrt{s} \approx \sqrt{2m_e E_{\nu_\mu}}$ resulting in extremely small cross sections. Thus in the CHARM II experiment [39] with its average neutrino energies of $\approx 20\ GeV$ the IMD total cross section is more than 1000 times smaller than the total neutrino cross section on nucleons. The CCFR experiment [40] at the Fermilab Tevatron has used neutrinos with energies up to 600 GeV corresponding to a total energy of 780 MeV in the c.m. system. The three main problems in determining the absolute cross section precisely are, as usual, to identify the good events with well-known efficiency, to subtract correctly any background and to calibrate the neutrino flux with sufficient precision. For the identification kinematics is favourable because the small c.m. energy leads to events with correspondingly small transverse momen-

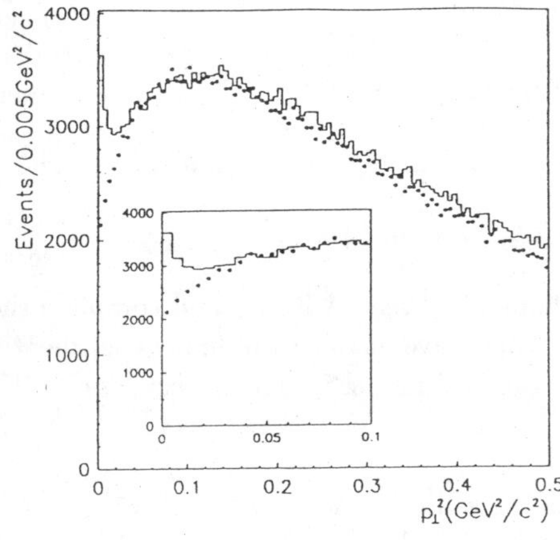

Fig. 2.19 $p_\perp^2$ distribution for events with $E_\mu > 10.9$ GeV and $E_{had} <$ 1.5 GeV for incident ν_e (solid line) and $\bar{\nu}_\mu$ (marked with dots) [39]. The $\bar{\nu}_\mu$ distribution was normalized to the ν_e distribution in the range $0.05 < p_\perp^2 <$ 0.1 GeV^2/c^2. The insertion shows the small $p_\perp^2$ region with the peak due to the inverse muon decay reaction. From Ref. [39].

ta and muon emission angles. Thus the trigger condition is given by single muons emitted nearly in forward direction and with negligible hadron energy. Although the background shapes cannot be reliably predicted, they should be equal for ν_μ and $\bar{\nu}_\mu$ reactions at $p_\perp^2 = 0$ [104]. Background comes from a broad continuum of quasi-elastic processes (QEP, charged current events on nuclei). The signal can be seen clearly as a narrow peak on the $p_\perp^2$ distribution over a huge broad background of QEP events (see Fig. 2.19). Since there is no IMD for $\bar{\nu}_\mu$ on e^- one uses the background for antineutrinos, properly

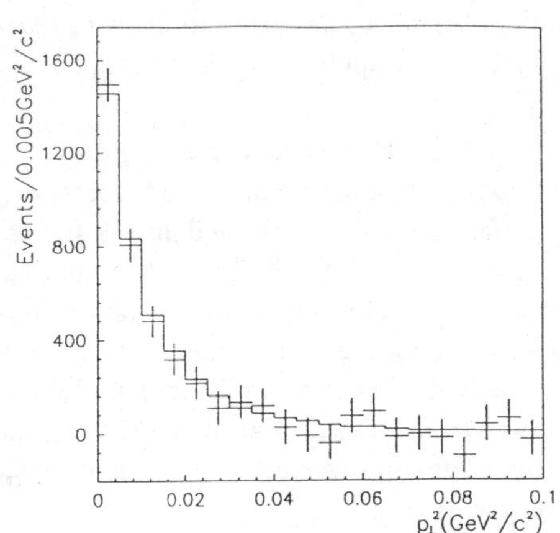

Fig. 2.20 Distribution of inverse muon decay events as a function of $p_\perp^2$ after subtraction of background. The solid line represents the expected distribution. From Ref. [39].

normalized, for background subtraction at small momentum transfer. The resulting IMD $p_\perp^2$ distribution agrees well with the expected distribution (see Fig. 2.20). For the CHARM II experiment the systematic error on the detection of the events is 4% and equal to the statistical error. The error on the total neutrino flux and therefore on the overall normalization is 5%. After subtracting radiative corrections, adding the errors in quadrature and dividing by the Born term of the standard model prediction they obtain

$$S = (1054 \pm 79) \times 10^{-3} \; [39]$$

The result of the CCFR measurement, obtained at higher neutrino energies, is

$$S = (981 \pm 49_{\text{stat.}} \pm 30_{\text{syst.}}) \times 10^{-3} \; [40]$$

These results tell how strongly left-handed ν_μ react with e^- and complete the informations from normal muon decay. They have enabled the first experimental confirmation of the assumptions of the standard model for muon decay [4].

2.4.8. Radiative Muon Decays

The measurement of the parameter $\bar{\eta}$ in the radiative decay, $\mu^+ \to \bar{\nu}_\mu e^+ \nu_e \gamma$, is of interest because it gives upper limits on $|g^V_{RL}|$, $|g^V_{LR}|$, $|g^S_{LR}+2g^T_{LR}|$ and $|g^S_{RL}+2g^T_{RL}|$. The branching ratio for this process, however,is only 1.4%, and a measurement has been performed as a by-product of searching for the process $\mu \to e\gamma$ [29, 105]. There e^+ and γ with energies $> 25\ MeV$ and at an angle of 180^0 were detected in coincidence. The decays of 6×10^{11} μ^+ resulted in the detection of only about 7500 events which yielded the value of $\bar{\eta} = -0.014 \pm 0.090$. We conclude that although an improved measurement of $\bar{\eta}$ is desirable, it seems to be rather difficult to reach a precision of, say, 0.01 which is necessary to be competitive with standard muon decay experiments.

The measurement of the decay $\mu^+ \to \bar{\nu}_\mu e^+ \nu_e e^+ e^-$ similarly is a by-product of the search for the decay $\mu \to 3e$ [31]. This experiment has verified the theoretically predicted branching ratio of $B = 36 \times 10^3$ and has determined limits for the probabilities $Q_{\epsilon\mu}$ and for B_{LR} and B_{RL} (see Sect. 2.2.4). With 2723 events the limits for Q_{RL} and B_{RL} are only a factor 3.3 less accurate than the limits from normal muon decay [31, 21], whereas Q_{RR} and Q_{LR} are by an order of magnitude less accurate. We conclude that a measurement of that decay is very useful due to its high sensitivity to the important decay parameters, but that the very low branching ratio makes it unrealistic to hope for improvements on the knowledge about the decay interaction from this particular decay.

3. Leptonic Tau Decays

3.1. General Remarks

The standard model considers the tau as a heavy muon or as a heavy electron, having its own associated neutrino. We discuss the experiments needed in order to prove that the tau (and its neutrino) shares the same weak interaction as the muon and the electron (and their neutrinos) do. We indicate to what extent the present experiments contribute to answer that question. For general reviews, see Refs. [10, 106, 107].

Leptonic τ decays shall be described by the same Hamiltonian and decay parameters as μ decay. They are insensitive to effects of the finite mass of the intermediate vector boson, since the lifetime and the decay parameters are modified by terms of the order of $(m_\tau/m_W)^2 \approx 0.5 \times 10^{-3}$ [108]. Effects of a finite τ-neutrino mass m_{ν_τ} are in the order of $m_{\nu_\tau}^2/m_\tau^2$ [109]. With the present experimental limit of $m_{\nu_\tau} < 31\ MeV/c^2$ [110, 111, 112] any possible effect is $\lessapprox 0.4 \times 10^{-3}$. Radiative corrections, however, do have to be taken into account in analyzing experimental results (See Ref. [15], [43]-[51]).

3.2. Universality

The universality of the charged weak interaction for the three leptonic decays $\mu \to \nu_\mu e \bar{\nu}_e$, $\tau \to \nu_\tau \mu \bar{\nu}_\mu$ and $\tau \to \nu_\tau e \bar{\nu}_e$ can be checked by testing whether

a) the couplings $g^\gamma_{\varepsilon\mu}$ are equal and

b) the strength G_F is the same.

The Fermi constant G_F as the measure of the strength is given by [52]

$$G_F^2 = \frac{1}{\tau_\ell} \cdot \frac{192\pi^3}{m_\ell^5} \cdot \frac{1}{1 + 4\eta \cdot m_{\ell'}/m_\ell} \quad , \tag{3.1}$$

where τ_ℓ is the lifetime of the mother lepton ℓ, m_ℓ its mass and $m_{\ell'}$ the mass of the charged daughter lepton. Radiative and higher order corrections have been neglected here. (See also Eq. 2.76). Instead we point out the importance of the so-called low energy parameter η for the muonic decay of τ [113, 41], since $m_\mu/m_\tau \approx 1/17$. With the allowed range of $|\eta_\mu| \leqq 1$ one obtains

$$0.9 < \frac{G_F(\eta_\mu \neq 0)}{G_F(\eta_\mu = 0)} < 1.15 \tag{3.2}$$

if η_μ is not known ! The parameter η_μ can be determined either by analyzing the muon momentum distribution or by deriving upper limits for the coupling constants by measuring the decay parameters ξ'_μ, ξ_μ and δ_μ which will constrain the value of η_μ.

We note that the parameter η is due to the interference of interactions leading to the same helicities of the leptons in the final state, but to opposite chiralities for the charged leptons μ and e (see Eq. 2.77). Therefore the terms with η in the spectrum or the total decay rate Γ are proportional to the corresponding masses m_μ or m_e. With a dominant left-handed vector interaction a scalar interaction could contribute in *first* order according to

$$\eta \approx \tfrac{1}{2} Re\, g^S_{RR} \quad . \tag{3.3}$$

If we further assume the coupling to be caused by a charged Higgs particle with a strength proportional to the mass of the charged daughter lepton, then we obtain $\eta_\mu/\eta_e = m_\mu/m_e \approx 207$. The sensitivity for this kind of scalar coupling would therefore be $\approx 43\,000$ times larger for the decay of a τ to a μ than for the decay into an electron ! We conclude that the measurement of η_μ is of special interest and note that it should be possible to extract a value from existing data (see Sect. 3.3.2).

3.3. Measurements

The experimental methods used to measure the observables of leptonic τ decays differ from those of μ decay due to several reasons:

a) Muons from $\pi^\pm$ or $K^\pm$ decays are strongly polarized due to the weak interaction causing the meson decay. Decay asymmetry experiments then yield the product of the polarization with the asymmetry parameters. τ leptons, in contrast, used to be produced in pairs of τ^+ and τ^- from colliding e^+ and e^- beams at energies below the Z^0 resonance. The τ's are then unpolarized, being produced via electromagnetic interaction. Their spins, however, are correlated [114] so that the asymmetry parameters can be obtained by measuring correlations between the decay products of the two τ's [41, 113]. Recently τ leptons are also produced from Z^0 decay at the Z^0 resonance. They are polarized due to the Z^0 decay, so that again the product of the asymmetry parameters of both decays can be measured [115].

b) The lifetime of the τ is much shorter than the lifetime of the μ, because the leptonic decay rate is proportional to m_ℓ^5 and because the τ lepton can also decay into hadrons. Except when they are produced at threshold the τ leptons decay therefore in flight due to their short lifetime, and their decay spectra generally are not measured in the rest system. It is therefore not possible, for example, to stop τ's in matter and let their spins precess in a weak magnetic field to measure the decay asymmetry, because they are not polarized and because they decay before stopping.

3.3.1. Spectrum Shape

The Michel parameter ϱ is the only decay parameter measured up to date. Earlier measurements left sufficient room for speculation because ϱ_e looked systematically smaller than ϱ_μ (see Table 3.1).

Table 3.1 Measurements of the two Michel parameters ϱ_e and ϱ_μ from leptonic tau decays. All values are in units of 10^{-3}. When two errors are given, the first is statistical and the second systematic.

Experiment	Ref.	ϱ_e	ϱ_μ
DELCO 79	[116]	720 ± 150	
CLEO 85	[117]	600 ± 130	810 ± 130
MAC 87	[118]	$620 \pm 170 \pm 140$	$890 \pm 140 \pm 80$
CB 89	[119]	$640 \pm 60 \pm 70$	
average		640 ± 60	840 ± 110
ARGUS 90	[120]	$747 \pm 45 \pm 28$	$734 \pm 55 \pm 27$

The latest results [120], however, agree well with $\varrho_e = \varrho_\mu$ and with the standard model value. One of the difficulties in these experiments in contrast to the measurement of ϱ in muon decay is the correct identification of the tau. In the ARGUS experiment this was done by identifying the accompanying tau through the decays $\tau^+ \to \bar{\nu}_\tau \pi^+ \pi^+ \pi^-$ and $\tau^+ \to \bar{\nu}_\tau \pi^+ \pi^+ \pi^- \pi^0$. For the energy spectra in the c.m. system the maximum rate is at the highest energies if radiative corrections

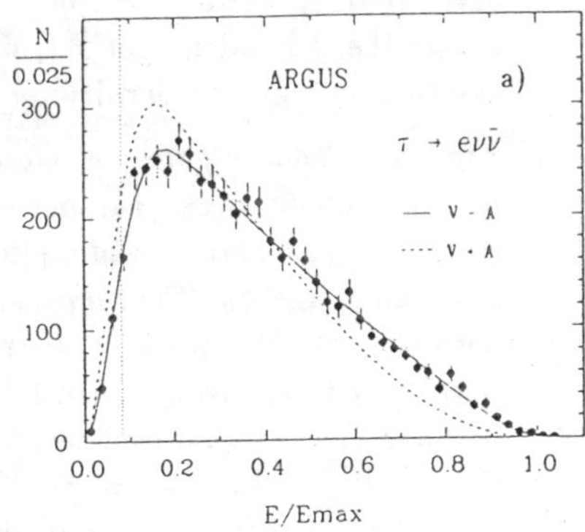

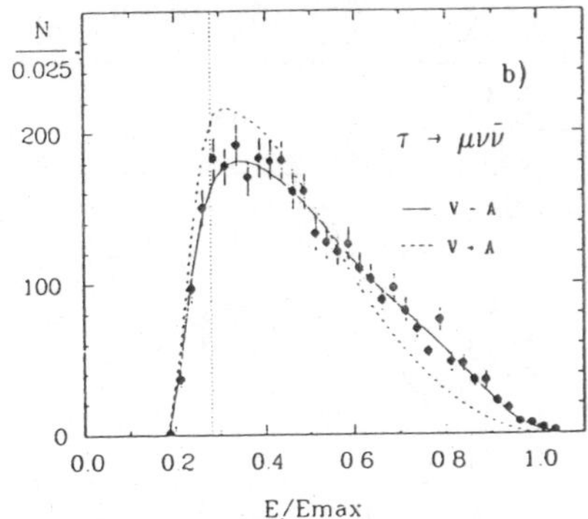

Fig. 3.1 Normalized electron (a) and muon (b) energy distributions for leptonic tau decays show agreement with the standard model and also with a purely right-handed interaction. The dashed curves, labelled "$V+A$", correspond to a coupling $g^V_{LR} = 1$. This has implications for left-right symmetric and for supersymmetric models. See Section 2.3.2 and Table 2.3. From Ref. [120].

are neglected. Due to the Lorentz transformation the maximum in the laboratory system is shifted to lower energies (see Fig. 3.1.).

In the analysis, $\eta = 0$ has been *assumed.* As in μ decay the spectrum depends both on ϱ and η, which are correlated ($\rho_{\varrho\eta} = 0.46$) [41]. Taking this into account would lead to slightly larger errors. On the other hand it has been shown that η_μ can be determined together with the Michel parameter ϱ_μ with a comparable precision because of the favourable mass ratio of m_μ/m_τ. From Eq. 3.1 we see that the measurement of η_μ is essential for a precise test of universality for the muonic τ decay. Implications of ϱ measurements for the Lorentz structure of the interaction are discussed in Sect. 2.3.2 and in Ref. [42]. There exist many four-fermion interactions, including the purely right-handed one with $g^V_{RR} \neq 0$, which lead to $\varrho = 3/4$. See also Table 2.3 and Refs. [17], [78]-[82].

3.3.2. Decay Asymmetry

The four asymmetry parameters ξ_μ, δ_μ, ξ_e and δ_e can be determined by analyzing the decay distributions of spin-correlated τ pairs. It has been shown that the combined measurement of, for example, the e^+ momentum k_e, the μ^- momentum k_μ and the opening angle $\vartheta_{e\mu}$ between the two momenta contains sufficient information not only about η and ϱ but also on the asymmetry parameters ξ and δ [41, 113]. There the figures of merit $\mathcal{M}_i$ were calculated for the decay parameters $x_i = \varrho, \eta, \delta, \xi$ by integration of the distributions $R(k_e, k_\mu, \cos\vartheta_{e\mu})$ and $S(k_e, k_\mu)$ over phase space for 10^7 events. For uncorrelated decay parameters x_i the error Δx_i on the parameter x_i is simply given by $\Delta x_i = \mathcal{M}_i^{-1/2}$. Fig. 3.2 shows the dependence of $\mathcal{M}_i$ on the total energy E_0. Sensitivity is highest for the scalar parameters ϱ_μ and η_μ at threshold, while it increases with energy for the pseudo-scalar parameters δ_μ and ξ_μ. However, measurements are possible for all four quantities both at threshold and at higher energy, e.g. at the Υ resonance.

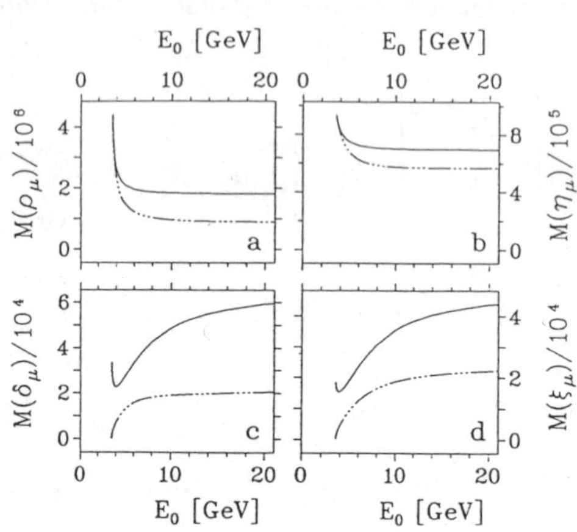

Fig. 3.2 Total Figures of Merit for the leptonic decay parameters $\varrho_\mu, \eta_\mu, \delta_\mu, \xi_\mu$ as a function of the total beam energy E_0. The curves are derived from the correlated distributions $R(k_e, k_\mu, \cos\vartheta_{e\mu})$ (solid lines) and $S(k_e, k_\mu)$ (dashed lines) for the reactions $e^+e^- \to \tau^+\tau^- \to (\mu^+\nu_\mu\bar{\nu}_\tau)(e^-\bar{\nu}_e\nu_\tau)$ by numerical integration. From Ref. [41, 113].

In a joint analysis of spin-correlated $\tau^+\tau^-$ pairs decaying into $\mu^\pm, e^\pm, K^\pm$ or $\pi^\pm$ the statistical errors were calculated for an ideal detector [41] taking into account the measured branching ratios. For 10^7 τ pairs produced at a c.m. energy of 10.55 GeV the statistical errors are $\Delta\varrho = 2\times10^{-3}$, $\Delta\eta_\mu = 3\times10^{-3}$, $\Delta\xi = 13\times10^{-3}$ and $\Delta\delta = 11\times10^{-3}$. Instead of evaluating with respect to ξ and δ it is highly recommended to use instead the linear combination Q_R^τ which is the probability of the τ to be right-handed in the decay (see Sect. 2.3.3). Q_R^τ can be determined much more precisely than either ξ or δ alone, because these two parameters are strongly negatively correlated which leads to $\Delta Q_R^\tau = 6.7\times10^{-3}$. Q_R^τ is equal to the sum of the squares of five complex coupling constants (see Eq. 2.72) and supposed to be zero in the standard model so that a precise measurement puts upper limits on each of the coupling constants individually.

3.3.3. Muon Polarization

In muon decay the polarization of positrons from stopped muons is measured by annihilation in flight or Bhabha scattering on longitudinally polarized electrons. The analyzing power is rather large for both processes and can be close to one, but the final measured asymmetry is of the order of a few percent because only every fourteenth electron in the magnetized iron foil is polarized. For the decay $\tau^+ \to \bar\nu_\tau\mu^+\nu_\mu$ one can use the decay asymmetry of the stopped μ^+ with the ten times higher *effective* analyzing power of 1/3 ! To perform this kind of experiment one needs outside of the main detector a dedicated polarimeter consisting of several alternating planes of nondepolarizing material like aluminium or marble plates, scintillators and an array of drift tubes or proportional chambers to measure the direction of the muons and their decay positrons. The stopped muons would precess in a small magnetic field $B \approx 6\ mT$ which should be tangential to the (assumed) cylindrical surface of the main detector. By measuring the time distribution of the decay one can derive the decay asymmetry and from this the longitudinal polarization of the muon.

Such a measurement has been performed at muon momenta of 16 GeV/c for the reaction $\bar\nu_\mu Fe \to X\mu^+$ [121] where the chirality transfer from the $\bar\nu_\mu$ onto the μ^+ was measured. There by detecting 3400 events a statistical error on the μ^+ polarization of 20% was reached. Using this result the presumable statistical error which could be reached in a B meson factory for 10^7 produced τ pairs was estimated to 15% [122]. According to Eq. 2.73 this corresponds to an error of 7.5 % for Q_R^μ which is the probability for the μ to be right-handed in τ decay and allows one to put upper limits to $|g_{RR}^S|$, $|g_{RL}^S|$, $|g_{RR}^V|$, $|g_{RL}^V|$ and $|g_{RL}^T|$ simultaneously. With this very important measurement one can get information about three types of interactions not accessible by the asymmetry measurement and put upper limits on any *type* of interaction where the right-handed component of the τ takes part in leptonic τ decays.

References

[1] S.L. Glashow, Nucl. Phys. **22** (1961) 579.

[2] S. Weinberg, Phys. Rev. Lett. **19** (1967) 1264

[3] A. Salam, in *Elementary Particle Theory*, ed. N. Svartholm (Almquist and Wiksells, Stockholm, 1969), p. 367.

[4] W. Fetscher, H.-J. Gerber and K.F. Johnson, Phys. Lett. **173B** (1986) 102.

[5] L. Michel, Proc. Phys. Soc. **A63** (1950) 514.

[6] F. Scheck, *Lepton, Hadrons and Nuclei* (North-Holland, Amsterdam, 1983).

[7] K. Mursula and F. Scheck, Nucl. Phys. **B253** (1985) 189.

[8] C. Jarlskog, Nucl. Phys. **75** (1966) 659.

[9] R. Engfer and H.K. Walter, Ann. Rev. Nucl. Part. Sci. **36** (1986) 327.

[10] Martin L. Perl, Rep. Prog. Phys. **55** (1992) 653.

[11] W. Pauli, Nuov. Cim. **6** (1957) 204.

[12] P. Langacker and D. London, Phys. Rev. **D 39** (1989) 266.

[13] Particle Data Group, K. Hikasa *et al.*, Phys. Rev. **D 45** (1 June 1992).

[14] K. Jungmann, V.W. Hughes and G. zu Putlitz (Eds.), *The Future of Muon Physics* (Springer-Verlag, Berlin Heidelberg New York,1992);
Z. Phys. **C56** (1992) 1.

[15] F. Scheck, *Muon Physics*, Phys. Rep. **44** (1978).

[16] P. Langacker, Comments Nucl. Part. Phys. **19** (1989) 1.

[17] P. Herczeg, Phys. Rev. **D 34** (1986) 3449.

[18] *International Colloquium on the History of Particle Physics* , Journal de Physique **43** (1982).

[19] M. Fierz, Z. Physik **101** (1937) 553.

[20] T. Kinoshita and A. Sirlin, Phys. Rev. **108** (1957) 844.

[21] W. Fetscher and H.-J. Gerber, in Particle Data Group, K. Hikasa *et al.*, Phys. Rev. **D 45** (1 June 1992) VI.16.

[22] Wulf Fetscher, Phys. Rev. Lett. **69** (1992) 2758.

[23] Wulf Fetscher, submitted for publication (January, 1993)

[24] M. Jonker *et al.* (CHARM I collaboration), Phys. Lett. **93 B** (1980) 203; F. Bergsma *et al.* (CHARM I collaboration), Phys. Lett. **122 B** (1983) 465.

[25] K. Mursula, M. Roos and F. Scheck, Nucl. Phys. **B219** (1983) 321.

[26] W. Fetscher, in *Neutrino 86: Neutrino Physics and Astrophysics*, proceedings of the 12th International Conference, Sendai, Japan, 1986, edited by T. Kitagaki and H. Juta (World Scientific, Singapore, 1986).

[27] W. Fetscher, Phys. Lett. **140B** (1984) 117.

[28] B. Balke *et al.*, Phys. Rev. **D 37** (1988) 587.

[29] W. Eichenberger, R. Engfer and A. van der Schaaf, Nucl. Phys. **A412** (1984) 523.

[30] R.H. Pratt, Phys. Rev. **111** (1958) 649.

[31] A. Kersch, N. Kraus and R. Engfer , Nucl. Phys. **A485** (1988) 523.

[32] H. Burkard *et al.*, Phys. Lett. **160 B** (1985) 343.

[33] S.E. Derenzo, Phys. Rev. **181** (1969) 1854.

[34] H. Burkard *et al.*, Phys. Lett. **150 B** (1985) 242.

[35] I. Beltrami *et al.*, Phys. Lett. **194 B** (1987) 326.

[36] A. Jodidio *et al.*, Phys. Rev. **D 34** (1986) 1967; A. Jodidio *et al.*, Phys. Rev. **D 37** (1988) 237.

[37] J. Imazato *et al.*, Phys. Rev. Lett. **69** (1992) 877.

[38] B. Bodmann *et al.* (KARMEN collaboration), Phys. Lett. **280 B** (1992) 198.

[39] D. Geiregat *et al.* (CHARM II collaboration), Phys. Lett. **247B** (1990) 131.

[40] S. R. Mishra *et al.* (CCFR collaboration), Phys. Lett. **252B** (1990) 170.

[41] Wulf Fetscher, Phys. Rev. **D 42** (1990) 1544.

[42] H.-J. Gerber, "Lepton Properties", Proceedings International Europhysics Conference on High Energy Physics, Uppsala 1987. Olga Botner (Ed.), Uppsala University.

[43] A.M. Sachs and A. Sirlin, in *Muon Physics*, V. Hughes and C.S. Wu (ed.), (Academic Press, N.Y. 1975, Vol. II).

[44] R.E. Behrends, R.J. Finkelstein and A. Sirlin, Phys. Rev. **101** (1956) 866.

[45] T. Kinoshita and A. Sirlin, Phys. Rev. **107** (1957) 593.

[46] S.M. Berman, Phys. Rev. **112** (1958) 267.

[47] T. Kinoshita and A. Sirlin, Phys. Rev. **113** (1959) 1652.

[48] S.M. Berman and A. Sirlin,Ann. Phys. (N.Y.) **20** (1962) 20.

[49] H. Grotch, Phys. Rev. **168** (1968) 1872.

[50] G. Källén, Springer Tracts in Mod. Phys. **46** (1968) 67

[51] V. Florescu and O. Kamei, Nuov. Cim. **61A** (1968) 967.

[52] C. Bouchiat and L. Michel, Phys. Rev. **106** (1957) 170.

[53] K.L. Giovanetti *et al.*, Phys. Rev. **D 29** (1984) 343.

[54] S.E. Willis *et al.*, Phys. Rev. Lett. **44** (1980) 522.

[55] P. Herczeg, in *The Future of Muon Physics*, eds. K. Jungmann, V.W. Hughes and G. zu Putlitz (Springer-Verlag, Berlin Heidelberg New York,1992); Z.Phys. **C56** (1992) 129.

[56] Bruce Arne Sherwood, Phys. Rev. **156** (1967) 1475.

[57] W.F. Dudziak, R. Sagane and J. Vedder, Phys. Rev. **114** (1959) 336.

[58] M. Bardon, P. Norton, J. Peoples, A.M. Sachs and J.L. Franzini, Phys. Rev. Lett. **14** (1965) 449.

[59] J. Peoples, Nevis Report **147** (1966) (unpublished).

[60] J. Carr *et al.*, Phys. Rev. Lett. **51** (1983) 627.

[61] A.E. Pifer, T. Bowen and K.R. Kendall, Nucl. Instr. Meth. **135** (1976) 39.

[62] C.J. Oram, J.B. Warren, G.M. Marshall and J. Dornboos, Nucl. Instr. Meth. **179** (1981) 95.

[63] K.H. Tanaka *et al.*, Nucl. Instr. Meth. **A316** (1992) 134.

[64] G. Ascoli, Z. Physik **150** (1958) 407.

[65] M. Bardon, D. Berley and L.M. Lederman, Phys. Rev. Lett. **2** (1959) 56.

[66] I. Beltrami *et al.*, Helv. Phys. Act. **60** (1987) 611.

[67] J. Ford and M. Mullin, Phys. Rev. **108** (1957) 477.
Their formula (21) is incorrect. See footnote no. 2 of Ref. [35] and Ref. [36].

[68] J. Heintze, Z. Physik **148** (1957) 560.

[69] J.H. Brewer, K.M. Crowe, F.N. Gygax and A. Schenck, in *Muon Physics*, ed. V. Hughes and C.S. Wu (Academic, New York, 1975), Vol. III.

[70] D. Stoker *et al.*, Phys. Rev. Lett. **54** (1985) 1887.

[71] A. Schenck, *Muon Spin Spectroscopy* (Adam Hilger Ltd., Bristol and Boston,1985)

[72] A.I. Alikhanov, Yu.V. Galaktionov, Yu.V. Gorodkov, G.P. Eliseev and V.A. Lyubimov, JETP **11** (1960) 1380.

[73] G. Backenstoss, B.D. Hyams, G. Knop, P.C. Marin and U. Stierlin, Phys. Rev. Lett. **6** (1961) 415.

[74] M. Bardon, P. Franzini and S. Lett, Phys. Rev. Lett. **7** (1961) 23.

[75] A. Possoz *et al.*, Phys. Lett. **B 70** (1977) 265.

[76] R. Abela, G. Backenstoss, W. Kunold, L.M. Simons and R. Metzner, Nucl. Phys. **A395** (1983) 413.

[77] L.Ph. Roesch, V.L. Telegdi, P. Truttmann, A. Zehnder, L. Grenacs and L. Palffy, Helv. Phys. Act. **55** (1982) 844

[78] P. Langacker and S.U. Sankar, Phys. Rev. **D 40** (1989) 1569.

[79] J.C. Pati and A. Salam, Phys. Rev. **D 10** (1974) 275.

[80] R.N. Mohapatra and J.C. Pati, Phys. Rev. **D 10** (1974) 566.

[81] M.A.B. Beg, R.V. Budny, R. Mohapatra and A. Sirlin, Phys. Rev. Lett. **38** (1977) 1252.

[82] T. Oka, Phys. Rev. Lett. **50** (1983) 1423.

[83] H. Abramowicz *et al.*, Z. Physik **C12** (1982) 225.

[84] W. Buchmüller and F. Scheck, , Phys. Lett. **145B** (1984) 421.

[85] H. Haber and G. Kane, Phys. Rep. **117** (1985) 75.

[86] H.A. Tolhoek, Rev. Mod. Phys. **28** (1956) 277.

[87] L.A. Page, Rev. Mod. Phys. **31** (1959) 759.

[88] H. Frauenfelder and A. Rossi, *Determination of the Polarization of Electrons and Photons*, in "Nuclear Physics", L.C.L. Yuan and C.S. Wu (ed.) (Academic Press, N.Y. 1963).

[89] A. Ashkin, L.A. Page and W.M. Woodward, Phys. Rev. **94** (1954) 357.

[90] L.A. Page, Phys. Rev. **106** (1957) 394.

[91] W.H. McMaster, Nuov. Cim. **17** (1960) 395.

[92] A.M. Bincer, Phys. Rev. **107** (1957) 1434.

[93] K. Bockmann, G. Kramer and W.R. Theis, Z. Phys. **150** (1958) 201.

[94] J. Ullman, H. Frauenfelder, H.J. Lipkin and A. Rossi, Phys. Rev. **122** (1961) 963.

[95] Lester L. DeRaad, Jr. and Yee Jack Ng, Phys. Rev. **D 11** (1975) 1586.

[96] W.E. Fischer and F. Scheck, Nucl. Phys. **B83** (1974) 25.

[97] M.T. Mehr and F. Scheck, Nucl. Phys. **B149** (1979) 123.

[98] A. Bühler, N. Cabibbo, M. Fidecaro, T. Massam, Th. Müller, M. Schneegans and A. Zichichi, Phys. Lett. **7** (1963) 368.

[99] S. Bloom, L.A. Dick, L. Feuvrais, G.R. Henry, P.C. Macq and M. Spighel, Phys. Lett. **8** (1964) 87.

[100] J. Duclos, J. Heintze, A. de Rújula and V. Soergel, Phys. Lett. **9** (1964) 62.

[101] Daniel M. Schwartz, Phys. Rev. **162** (1967) 1306.

[102] G.G. Scott, Rev. Mod. Phys. **34** (1962) 102.

[103] James S. Barber and Robert S. Shrock, Phys. Lett. **139B** (1984) 427.

[104] S.L. Adler, Phys. Rev. **B 135** (1964) 963.

[105] A. van der Schaaf, R. Engfer, H.P. Povel, W. Dey, H.K. Walter and C. Petitjean, Nucl. Phys. **A340** (1980) 249.

[106] B.C. Barish and R. Stroynowski, Phys. Rep. **157** (1988) 1.

[107] K.K. Gan and M.L. Perl, Int. J. Mod. Phys. **A3** (1988) 531.

[108] T.D. Lee and C.N. Yang, Phys. Rev. **108** (1957) 1611.

[109] J. Missimer, F. Scheck and R. Tegen, Nucl. Phys. **B188** (1981) 29

[110] H. Albrecht *et al.* (ARGUS collaboration), Phys. Lett. **202B** (1988) 149.

[111] H. Albrecht *et al.* (ARGUS collaboration), Phys. Lett. **292B** (1992) 221.

[112] J.Z. Bai *et al.* (BES collaboration), Phys. Rev. Lett. **69** (1992) 3021.

[113] W. Fetscher, *Leptonic τ Lepton Decays*, in *Proposal for a B-Meson-Factory*, PSI Report **PR-88-09** (1988).

[114] Y.S. Tsai, Phys. Rev. **D 4** (1971) 2821.

[115] C.A. Nelson, Phys. Rev. **D 40** (1989) 123.

[116] W. Bacino *et al.* (DELCO collaboration), Phys. Rev. Lett. **42** (1979) 749.

[117] S. Behrends *et al.* (CLEO collaboration), Phys. Rev. **D 32** (1985) 2468.

[118] W.T. Ford *et al.* (MAC collaboration), Phys. Rev. **D 36** (1987) 1971.

[119] H. Janssen *et al.* (Crystal Ball collaboration), Phys. Lett. **228B** (1989) 273.

[120] H. Albrecht *et al.* (ARGUS collaboration), Phys. Lett. **246B** (1990) 278.

[121] M. Jonker *et al.* (CHARM I collaboration), Phys. Lett. **86B** (1979) 229.

[122] K. Wacker *et al.*, *Proposal for a B-Meson-Factory*, PSI Report **PR-88-09** (1988).

SYMMETRY-TESTS IN SEMILEPTONIC WEAK INTERACTIONS: A SEARCH FOR NEW PHYSICS[†]

Jules Deutsch
Institut de Physique Nucléaire, Université Catholique de Louvain
B-1348, Louvain-la-Neuve, Belgium
and
Paul Quin
Department of Physics, University of Wisconsin, Madison, 53706, USA

Contents

[†] "The Art and Science of Finding Nothing" [E. Vogt]

1 Introduction

In this chapter we wish to present to a readership composed mostly of particle physicists a critical overview of symmetry-tests, past, present and future, performed mostly in experiments on complex nuclei.

It is a popular belief that the glories of semileptonic experiments on complex nuclei, such as the discoveries of the neutrino and parity-violation, belong to history and that the relative "softness" of the nucleus, the complexity of its structure, prevent us from drawing unquestionable information from nuclear β-decay or from muon-capture on modern issues of concern to particle physicists. Would one not be tempted, for instance, to discard with suspicion any sign of new physics beyond the Standard Model emerging solely from nuclear-physics experiments?

In what follows we hope to illustrate the complementarity of the low-energy nuclear physics experiments to those performed at higher energies in testing the symmetry-properties of the Standard Model and - hopefully - in finding indications for new physics beyond it. This complementarity is due mainly to the relative ease of performing experiments with high statistical accuracy in a low-energy physics environment and to the multiplicity of various quantum-states provided by complex nuclei. We shall stress also that for experiments with well-chosen nuclei the uncertainties introduced by their structure, the so-called recoil-order corrections, are well beyond the precision of the actual experiments and are far from hindering us in extracting reliable information.

As many experiments provide information on several issues, and a given issue requires simultaneous consideration of different experiments, our topic can not be presented in a completely "linear" fashion. We will follow, instead, the following segmentation.

In Sect. 2 we shall introduce the Standard Model's semileptonic weak interaction and its appropriate generalization for nuclear β-decay and muon-capture. We then proceed to examine the various possible experiments, commenting on the nuclear-physics limitations induced by recoil-order and other corrections, presenting the performance achieved and the challenges and vistas for improvements both for nucleons and nuclei. In this section the results of the various experiments will be discussed in terms of the precision by which they measure the correlation coefficients and their sensitivity to interactions beyond the Standard Model.

In Sect. 3 we shall consider the information the experiments we discussed in Sect. 2 can provide, individually or in combination, on the various phenomenological four-fermion coupling constants. We shall assume first a time-reversal conserving vector–axial-vector interaction and discuss the constraints on the couplings we have actually and those we can hope for. After that we shall relax the pure V-A assumption and discuss the constraints one obtains on the various four-fermion couplings assumed to be non-zero one-by-one. We shall terminate the section by considering the constraints on T-invariance both for the V-A description and more general formulations.

In Sect. 4 we consider particular scenarios beyond the Standard Model and discuss the constraints readily obtained or expected in the future. Our discussion shall start with that of the manifest left-right symmetric models, followed by relaxing some of its constraints. We shall comment also on the constraints on more exotic scenarios, such as models with leptoquarks.

In Sect. 5, finally, we shall attempt to draw some conclusions on the perspectives of the field.

The reader may note that we do not touch in this chapter on several important issues, such as searches for finite neutrino-mass [1-4], solar neutrinos [2,5-9], double β-decay [10,11], and forbidden lepton-number violating muon-capture modes [12], such as $\mu^- + \mathrm{X} \rightarrow \mathrm{X} + e^-$. Not only are we not experts in these fields but also the community has the excellent and recent reviews [cited above] on these most important research fields.

2 The Nuclear Weak Interaction and Experiments

2.1 The Nuclear β-Decay Interaction

A fundamental premise of the standard electroweak gauge theory is the universality of lepton and quark couplings to the charged gauge bosons $\mathrm{W}_{\mathrm{L}}^{\pm}$, where the subscript L refers to the assumed left-handed nature of the coupling. For nuclear β-decay the Standard Model proposes a maximally parity violating, left-handed $V-A$ current-current interaction to describe the $\{d\text{-}u\}$$-$W and W$-$$\{\nu_e\text{-}e\}$ vertices; *e.g.*, for the transition taking place in neutron decay [13]

$$\mathcal{H}_{\mathrm{SM}}^{\ell q} \quad = \quad \frac{\mathrm{G_F}}{\sqrt{2}}\, \mathrm{V}_{ud}\, \bar{e}\gamma_\lambda(1-\gamma_5)\nu_e\, \bar{u}\gamma^\lambda(1-\gamma_5)d\,, \tag{1}$$

where $\mathrm{G_F}$ is the Fermi coupling constant, linked to the gauge coupling constant g and the charged gauge boson mass $\mathrm{M_W}$ by $\mathrm{G_F}/\sqrt{2} = g^2/8\mathrm{M_W^2}$, and V_{ud} is the $\{u\text{-}d\}$ element of the CKM [Cabibbo-Kobayashi-Maskawa] quark mixing-matrix. Whereas $\mathrm{G_F}$ is obtained from the muon lifetime, nuclear β-decay experiments determine V_{ud} and explore the helicity structure of the weak Hamiltonian through measurements of spacial and spin dependent correlations.[†]

The quark level interaction given in Eq. (1) does not describe directly β-decay transitions between free nucleons or complex nuclei, and the minimal replacement of the quark spinors by those appropriate for the neutron and proton, *i.e.*, $\bar{u} \rightarrow \bar{\mathrm{p}}$ and $d \rightarrow \mathrm{n}$, is not sufficient because the hadronic axial-vector current is renormalized by the strong interaction [14]. The renormalization of the hadronic axial-vector current

[†] Recoil-order, radiative, and nuclear-mismatch corrections are discussed later in this subsection and in Subsect. 2.3.

is determined to good precision by neutron β-decay. However, for complex nuclei we must introduce effective matrix elements, either measured by experiment or calculated in the impulse approximation. Alternatively one can use phenomenological form factors extracted from electromagnetic observables connecting the initial and final states or their isobar analogues [15]. Fortunately, this procedure presents no fundamental difficulties for the nuclear β-decay experiments discussed below.

The modification of the hadronic current $\mathcal{J}^{\lambda}_{\rm quark} = \bar{u}\gamma^{\lambda}(1-\gamma_5)d$ required to treat quarks bound in nucleons, viz. $\mathcal{J}^{\lambda} = \langle {\rm p}|\bar{u}\,\gamma^{\lambda}(1-\gamma_5)\,d|{\rm n}\rangle$, is straightforward for the so-called superallowed transitions [14]; *i.e.* for Fermi [vector current] transitions between mirror nuclei [*e.g.*, neutron decay] and between $\rm J^{\pi} = 0^+$ members of isobaric triplets. In these transitions the conserved-vector-current [CVC] hypothesis gives the vector nuclear matrix element as the weak transition within the first quark generation unrenormalized by the strong interaction. Explicitly, we write this matrix element for the $d \to u$ transition for quarks bound in nucleons [or complex nuclei] as

$$\mathcal{J}^{\lambda}_{\rm V} = \langle {\rm p}|\bar{u}\gamma^{\lambda}d|{\rm n}\rangle = g_{\rm V}(q^2)\langle {\rm p}|\gamma^{\lambda}|{\rm n}\rangle \to g_{\rm V}(q^2){\rm M_F} = g_{\rm V}{\rm M_F}\,, \tag{2}$$

where $g_{\rm V}(q^2)$ is the vector form factor at momentum transfer q. In the $q \to 0$ limit appropriate for nuclear β-decay the CVC approximation gives the form factor as $g_{\rm V} = g_{\rm V}(0) \equiv 1$, with the Fermi matrix element calculated from isospin selection rules; viz., ${\rm M_F} = \sqrt{\rm T(T+1) - T_{3\mathit{i}}T_{3\mathit{f}}}\,\delta^{\rm J}_{f,i}\delta^{\rm T}_{f,i}$, where $\rm T_{\mathit{i}}$ [$\rm T_{\mathit{f}}$] denote the isospin of the initial [final] state [14]. Eqs. (1) and (2) provide the complete Standard Model prescription for the $\rm J^{\pi} = 0^+$ pure Fermi transitions,† while transitions between mirror nuclei receive an additional contribution from the axial-vector current.

The hadronic contribution to Eq. (1) for Gamow-Teller [axial-vector] transitions is more complicated than for the vector current transitions discussed above, although the modification of $\mathcal{J}^{\lambda}_{\rm quark}$ is treated in the same fashion; viz.,

$$\mathcal{J}^{\lambda}_{\rm A} = \langle {\rm p}|\bar{u}\gamma^{\lambda}\gamma_5 d|{\rm n}\rangle = g_{\rm A}(q^2)\langle {\rm p}|\gamma^{\lambda}\gamma_5|{\rm n}\rangle \to g_{\rm A}(q^2){\rm M_{GT}} = g_{\rm A}{\rm M_{GT}}\,. \tag{3}$$

The complications are as follows. First, the axial-vector current is renormalized by the strong interaction, with the result that $g_{\rm A} = g_{\rm A}(0)$ is not unity. However, the value of $g_{\rm A}$ deduced from neutron β-decay is expected to be in reasonable agreement with the value $g_{\rm A}^{\rm PCAC} = 1.31(1)$ [16-18] calculated from the Goldberger-Treiman [partially-conserved-axial-vector-current hypothesis, PCAC] relation [14,19]. Second, the Gamow-Teller matrix element ${\rm M_{GT}} = \langle {\rm p}|\gamma^{\lambda}\gamma_5|{\rm n}\rangle$ cannot be calculated reliably except for the neutron, where ${\rm M^n_{GT}} = \sqrt{3}$. Despite this handicap of an arbitrary parameter, the product $g_{\rm A}{\rm M_{GT}}$ which must be determined from experiments, mirror-nucleus transitions play an important role because of the presence of vector–axial-vector interference terms in the spin dependent correlations and

† Recoil-order, radiative, and nuclear-mismatch corrections are discussed in Subsect. 2.3.

the availability of precise measurements for the neutron and ^{19}Ne decays. The axial-vector current is entirely responsible for pure Gamow-Teller transitions, the product $g_A M_{GT}$ however cancels in the correlation coefficients and introduces no impediment in the analysis of the corresponding experiments.

The Standard Model interaction given in Eq. (1) is time reversal invariant and excludes contributions from scalar, pseudoscalar, and tensor couplings. A general formulation of the nuclear β-decay transition probability with no assumptions regarding invariance with respect to parity, charge conjugation, and time reversal was published by Jackson, Treiman, and Wyld [20], referred to as JTW, shortly after Lee and Yang's conjecture that parity is violated in nuclear β-decay. The interaction $\mathcal{H}_\beta$ considered by JTW is the most general local, nonderivative, four-fermion interaction with left- $[V - A]$ and right-handed $[V + A]$ vector and axial-vector currents as well as scalar and tensor couplings. Here we assume, in addition, that all neutrinos are sufficiently light so that we can ignore phase-space distortions caused by neutrino mass. The transition probability $\mathcal{W}$ derived from the JTW interaction contains many correlation terms constructed from the β and ν spins, momenta, and energies as well as the initial and final nuclear polarizations. Precision measurements have only been made for a small subset, and for the present purposes $d\mathcal{W}$ can be written in terms of the β and ν momenta and total energies [p and E], the nuclear vector polarization of the initial state [$\langle \mathbf{J} \rangle$], the β spin [$\vec{\sigma}$], a Coulomb factor $\Gamma = \sqrt{1-(\alpha Z)^2}$, and the electron mass m_e as a sum of scalar and pseudoscalar terms [20], viz.,

$$d\mathcal{W} = d\mathcal{W}_0\, \xi \left(1 + \frac{\vec{p}\cdot\vec{p}_\nu}{E\,E_\nu} a + \frac{\Gamma m_e}{E} b + \langle \mathbf{J} \rangle \cdot \left[\frac{\vec{p}}{E} A + \frac{\vec{p}_\nu}{E_\nu} B + \frac{\vec{p}\times\vec{p}_\nu}{E\,E_\nu} D \right] \right.$$
$$\left. + \ \vec{\sigma}\cdot \left[\frac{\vec{p}}{E} G + \hat{p}\hat{p}\cdot\langle \mathbf{J} \rangle Q' + \langle \mathbf{J} \rangle \times \frac{\vec{p}}{E} R \right] \right). \tag{4}$$

In the Standard Model, the Fierz interference b and the time reversal violating D- and R-parameter terms are zero, while the β-ν correlation a, β- and ν-asymmetry parameters A and B, respectively, β-polarization G, and polarization-asymmetry correlation Q' parameters are at most simple functions of angular momentum and a single parameter $\lambda = g_A M_{GT}/g_V M_F$. Expressions for the correlation coefficients are given in the appendix, and the generalization of the Standard Model interaction given in Eq. (1) is presented in Subsects. 3.1 and 3.2. Although the selection rules for the hadronic matrix elements are the same for the vector and scalar couplings, as well as for the axial-vector and tensor couplings, we must introduce scalar and tensor form factors similar to those defined in Eqs. (2) and (3); viz.,

$$\langle p|\bar{u}d|n\rangle = g_S(q^2)\langle p||n\rangle \rightarrow \ g_S M_F \tag{5}$$

and

$$\langle \mathrm{p}|\bar{u}\,\frac{\sigma^{\lambda\mu}}{\sqrt{2}}\,d|\mathrm{n}\rangle = g_{\mathrm{T}}(q^2)\langle \mathrm{p}|\,\frac{\sigma^{\lambda\mu}}{\sqrt{2}}\,|\mathrm{n}\rangle \rightarrow\; g_{\mathrm{T}}\mathrm{M}_{\mathrm{GT}}\,. \tag{6}$$

A summary of the β-decay selection rules, properties of the correlation terms under parity and time reversal, and the classification of allowed transitions based on their $\log_{10} ft$-values is given in Table 1.

Table 1: A summary of the selection rules for allowed nuclear β-decay, an approximate classification ranging from superallowed to hindered based on $\log_{10} ft$, and the properties of the correlation coefficients under parity [P, $\vec{\mathrm{r}} \rightarrow -\vec{\mathrm{r}}$] and time reversal [T, t $\rightarrow$ −t] is given below. Hindered transitions are prone to sizeable and unknown recoil corrections.

	selection rules			classification	correlation coefficients		
Interaction	ΔJ	ΔT	$\Delta\pi$	$\log_{10} ft$		P-even	P-odd
Fermi/mirror	0	0	0		T-even	a, b, Q'	A, B, G
Gamow-Teller	0 , ±1	0 , ±1	0		T-odd	D	R
	no 0→0						
superallowed				3-4			
allowed				4-6			
hindered				>6			

The $d\mathcal{W}_0\,\xi$ term in Eq. (4) contains the coupling strength, Coulomb, and phase space factors; viz.,

$$d\mathcal{W}_0\,\xi = \frac{\mathrm{G}_{\mathrm{F}}^2\mathrm{V}_{ud}^2}{(2\pi)^5}\,\xi\,\mathrm{F}(\pm\mathrm{Z},E)\,(E-E_0)^2 Ep\,dEd\Omega_e d\Omega_\nu\,, \tag{7}$$

where $\mathrm{F}(\pm\mathrm{Z},E)$ is the Fermi function. The integral of $d\mathcal{W}$, the inverse lifetime τ^{-1}, is used to obtain the comparative half life or ft-value, where f is the integral of the Coulomb and phase space factors and t is the half life. At the present 0.1% precision it is essential to apply radiative [δ^{R}] and nuclear-mismatch [δ^{C}] corrections to ft which yields [21]

$$t^{-1} \;=\; \frac{\mathrm{G}_{\mathrm{F}}^2\mathrm{V}_{ud}^2}{K(\hbar c)^6}\,f\left(1+\delta^{\mathrm{R}}\right)\left(1-\delta^{\mathrm{C}}\right)\frac{\xi}{4}\left(1+\langle\frac{\Gamma}{\mathrm{W}}\rangle\,b\right) \tag{8}$$

or

$$\mathcal{F}t \;=\; \frac{K(\hbar c)^6}{\mathrm{G}_{\mathrm{F}}^2\mathrm{V}_{ud}^2}\,\frac{4}{\xi\left(1+\langle\frac{\Gamma}{\mathrm{W}}\rangle\,b\right)} \;=\; \mathcal{F}t_0\,\frac{4}{\xi\left(1+\langle\frac{\Gamma}{\mathrm{W}}\rangle\,b\right)}\,, \tag{9}$$

where $\mathrm{W} = E/\mathrm{m_e}$. Here, $K = \pi^3\hbar\ln 2/(\mathrm{m_e c^2})^5 = 4.06014\times 10^{-7}\,\mathrm{GeV^{-4}s}$, $\mathrm{G_F}/(\hbar c)^3 = 1.16639(2)\times 10^{-5}\,\mathrm{GeV^{-2}}$, and $K(\hbar c)^6/\mathrm{G_F^2} = 2984.37(11)\,\mathrm{s}$ [22]. Assuming that the unitarity of the CKM [Cabibbo-Kobayashi-Maskawa] quark mixing-matrix is realized by three generations, the Particle Data Group [22] evaluation of V_{us} and V_{ub} gives $V_{us}^2 + V_{ub}^2 = 0.04864(79) = 1 - V_{ud}^2$, or $V_{ud}^2 = 0.95138(79)$. Taking into account the uncertainty in the radiative corrections, estimated to be 0.08% [23,24], this value of V_{ud}^2 gives the Standard Model prediction†

$$\mathcal{F}t_0^{\rm SM} = 3136.9(36)\,{\rm s}\,. \quad \text{[Standard Model]} \tag{10}$$

The Standard Model implies $b = 0$, and for the $J^\pi = 0^+$ pure Fermi transitions $4/\xi \equiv 1$, so the numerical value given in Eq. (10) is the predicted $\mathcal{F}t$ value for these transitions. The main interest in determining precisely the $J^\pi = 0^+$ $\mathcal{F}t$-values is to check the unitarity of the first row of the CKM matrix assuming three generations, *i.e.* using Eq. (9) [with $b = 0$ and $\xi = 4$] to obtain V_{ud}^2. Corroborating information can be obtained from the $\mathcal{F}t$-values for superallowed transitions between mirror nuclei provided that the axial-vector contribution can be determined with sufficient precision from other measurements, such as the β-asymmetry parameter. A second aspect of Eq. (9) is that CVC alone predicts the constancy of $\mathcal{F}t^{\,(J^\pi=0^+)}$ for all the corresponding transitions. This feature can be used to limit various deviations from the Standard Model, such as the presence of scalar couplings [21,26] or heavy neutrino admixture to the electron flavor [21,27].

The precision of superallowed nuclear β-decay experiments has improved dramatically during the past five years. We discuss below refined analyses of the ft-values for $J^\pi = 0^+$ transitions, complementary information obtained from new measurements of the neutron half life and β-asymmetry parameter, as well as advances in positron polarization experiments. The present situation is in stark contrast to the latest [28] comprehensive analysis of nuclear β-decay constraints on the "standard $V - A$ model" performed in 1984, which was dominated by now quite old measurements for long-lived, highly-hindered Gamow-Teller transitions.

Recoil-order corrections have received extensive exposition in the literature [see Ref. [15] and references therein], and because they always enter relative to the main Fermi and Gamow-Teller strengths $g_V^2 M_F^2$ and $g_A^2 M_{GT}^2$, respectively, we expect the corrections to be smallest for "fast" transitions with low values of $\log_{10} ft$. These are the superallowed and allowed transitions, and the dominant recoil-order contribution to the correlation coefficients is expected to arise from the weak-magnetism form factor [the contribution to $\mathcal{F}t$ is small and well determined]. Fortunately, the weak-magnetism contribution can be calculated from electromagnetic observables using CVC, with a typical reliability of 10-20% [15,29]. One also expects smaller, higher order contributions for those transitions where the nuclear spin J is ≥ 1, but again these corrections can be estimated with reasonable precision be-

† For a vector–axial-vector interaction more general than that of the Standard Model [see Subsect. 3.2] V_{us} must be obtained from the K_{e3}–decays alone [25]. In this case $V_{us}^2 + V_{ub}^2 = 0.0482(10)$ [22] and $V_{ud}^2 = 0.9518(10)$. The result given in Eq. (10) becomes $\mathcal{F}t_0^{\rm SM} = 3135.5(42)\,{\rm s}$.

cause the nuclear structure is relatively simple [29-32]. The final result is that for the superallowed and allowed transitions used here, the recoil-order corrections are smaller than 2%, typically known to better than 0.3%, and always contribute less than 20% to the uncertainties of the results shown in Table 2 for transitions with $\log_{10} ft \lesssim 6$. The situation is much more uncertain for hindered transitions, such as the pure Gamow-Teller electron polarization measurements, where the recoil-order contributions cannot be reliably estimated and may be as large as the uncertainty given in Table 2. However, new measurements of the neutron ν-asymmetry [Subsect. 2.4] and the polarization-asymmetry correlation [Subsect. 2.6] will eventually render these electron polarization measurements obsolete.

One class of recoil-order terms which have received considerable attention are "second-class" current [33] contributions which have G-parity opposite to the Standard Model's vector and axial-vector currents. Second class currents have been sought in a number of experiments [29], and the most recent results [34,35] limit the second class current contribution in nuclear β-decay to be about an order-of-magnitude smaller than weak-magnetism. We refer the interested reader to the review article by Grenacs [29] and the more recent analysis of Morita [36].

The β-decay measurements to be discussed below are tabulated in Table 2. Radiative, nuclear-mismatch and recoil-order corrections [including the weak-magnetism contribution calculated using CVC] have been evaluated or applied to all entries except for the hindered Gamow-Teller electron polarization [G] measurements. For several entries in Table 2 we use weighted means from several experiments. A fit to the 54 nuclear β-decay measurements used to construct Table 2, taking the Standard Model interaction for the coefficients in Eq. (4), $g_{\rm V} = 1$, the numerical value $\mathcal{F}t_0 = 2984.37(11)/\mathrm{V}_{ud}^2\,\mathrm{s}$, and adjusting V_{ud}, $g_{\rm A}$, and the ^{19}Ne Gamow-Teller matrix element to minimize χ^2, gave $\chi^2 = 48.8$ for 51 degrees of freedom [ν], which corresponds [22] to a confidence level [C.L.] of 55%. However, the fit is not in particularly good agreement [$2\,\sigma$] with CKM unitarity; an additional nuclear-mismatch correction to the $\mathcal{F}t$-values [see Subsect. 2.3] and several extensions of the Standard Model interaction provide a significantly lower χ^2. The implications of these "improved" fits will be discussed below and in Sect. 3.

2.2 The Nuclear Muon-Capture Interaction

In the Standard Model nuclear muon-capture is described by the same $V - A$ current-current interaction as nuclear β-decay, except that here one explores simultaneously the $\{d\text{-}u\}$–$\mathrm{W_L}$ vertex of β-decay and the $\mathrm{W_L}$–$\{\nu_\mu\text{-}\mu\}$ vertex discussed in the chapter of this volume devoted to muon-decay. So in the Standard Model no new information is expected from muon-capture, and it is precisely for this reason that it is interesting to pin down the specific features of new physics beyond the Standard Model to which muon-capture could be sensitive.

Table 2: Results from the various experiments discussed in Sect. 2.

Quantity	Transition	$\log_{10}ft$	Value (Error)	References
$\mathcal{F}t$	$0^+ \xrightarrow[\mathrm{F}]{\beta^+} 0^+$	3.5	3138.4(32)	Subsect. 2.3
	$\mathrm{n} \xrightarrow[\mathrm{mixed}]{\beta^-} \mathrm{p}$	3.0	1082.4(22)	Subsect. 2.4
	${}^{19}_{10}\mathrm{Ne} \xrightarrow[\mathrm{mixed}]{\beta^+} {}^{19}_{9}\mathrm{F}$	3.2	1756.3(27)	Subsect. 2.5
a	$\mathrm{n} \xrightarrow[\mathrm{mixed}]{\beta^-} \mathrm{p}$	3.0	−0.1012(46)	[37]
	${}^{6}_{2}\mathrm{He} \xrightarrow[\mathrm{GT}]{\beta^-} {}^{6}_{3}\mathrm{Li}$	2.9	−0.3343(30)	[38]
	${}^{32}_{18}\mathrm{Ar} \xrightarrow[\mathrm{F}]{\beta^+} {}^{32}_{17}\mathrm{Cl}$	3.2	1.00(4)	[39]
	${}^{33}_{18}\mathrm{Ar} \xrightarrow[\mathrm{F}]{\beta^+} {}^{33}_{17}\mathrm{Cl}$	3.3	1.02(2)	[39]
A	$\mathrm{n} \xrightarrow[\mathrm{mixed}]{\beta^-} \mathrm{p}$	3.0	−0.1127(11)	Subsect. 2.4
	${}^{19}_{10}\mathrm{Ne} \xrightarrow[\mathrm{mixed}]{\beta^+} {}^{19}_{9}\mathrm{F}$	3.2	−0.03808(46)	Subsect. 2.5
B	$\mathrm{n} \xrightarrow[\mathrm{mixed}]{\beta^-} \mathrm{p}$	3.0	0.997(28)	[40]
D	$\mathrm{n} \xrightarrow[\mathrm{mixed}]{\beta^-} \mathrm{p}$	3.0	−0.0005(14)	[22]
	${}^{19}_{10}\mathrm{Ne} \xrightarrow[\mathrm{mixed}]{\beta^+} {}^{19}_{9}\mathrm{F}$	3.2	0.00010(62)	[41]
G	${}^{3}_{1}\mathrm{H} \xrightarrow[\mathrm{mixed}]{\beta^-} {}^{3}_{2}\mathrm{He}$	3.1	−1.005(26)	[42]
	Gamow-Teller	≈ 7.4	−0.998(14)	[43-45]
$G_{\mathrm{F}}/G_{\mathrm{GT}}$	${}^{14}_{8}\mathrm{O} \xrightarrow[\mathrm{F}]{\beta^+} {}^{14}_{7}\mathrm{N}$ / ${}^{10}_{6}\mathrm{C} \xrightarrow[\mathrm{GT}]{\beta^+} {}^{10}_{5}\mathrm{B}$	3.5/3.0	0.9996(37)	[46]
	${}^{26}_{13}\mathrm{Al}^m \xrightarrow[\mathrm{F}]{\beta^+} {}^{26}_{12}\mathrm{Mg}$ / ${}^{30}_{15}\mathrm{P} \xrightarrow[\mathrm{GT}]{\beta^+} {}^{30}_{14}\mathrm{Si}$	3.5/4.8	1.003(4)	[47]
$\mathcal{B}(\mathcal{PA})$	${}^{107}_{49}\mathrm{In} \xrightarrow[\mathrm{GT}]{\beta^+} {}^{107}_{48}\mathrm{Cd}$	5.7	0.0080(52)	[48]
R	${}^{8}_{3}\mathrm{Li} \xrightarrow[\mathrm{GT}]{\beta^-} {}^{8}_{4}\mathrm{Be}$	5.6	0.004(7)	[49]
	${}^{19}_{10}\mathrm{Ne} \xrightarrow[\mathrm{mixed}]{\beta^+} {}^{19}_{9}\mathrm{F}$	3.2	−0.079(53)	[50]
ρ	$\mu^+ \to e^+$		0.7518(26)	[22]
δ	$\mu^+ \to e^+$		0.7486(38)	[22]
$P_\mu\xi$	$[\pi \to \mu]\ \mu^+ \to e^+$		1.0027(85)	[22]
	$[K \to \mu]\ \mu^+ \to e^+$		1.0013(61)	[51]
$P_\mu\xi\delta/\rho$	$[\pi \to \mu]\ \mu^+ \to e^+$		0.99790(88)	[52]

The quark-level $V - A$ muon-capture interaction of the Standard Model has a structure similar to that for β-decay, as given in Eq. (1); viz.,

$$\mathcal{H}_{\rm SM}^{\ell q} \quad = \quad \frac{\rm G_F}{\sqrt{2}}\, {\rm V}_{ud}\, \bar{\mu}\gamma_\lambda(1-\gamma_5)\nu_\mu\, \bar{u}\gamma^\lambda(1-\gamma_5)d\,. \tag{11}$$

However on the nucleon level the structure of the muon-capture hadronic current is more complex than that for β-decay due to the high momentum transfer provided by the muon mass. The result is that the hadronic vector and axial-vector Lorentz covariants are constructed not only with the Dirac-matrices but also the four-momentum transfer q_λ; viz.,

$$\mathcal{J}^\lambda = \langle {\rm p}|{\rm V}^\lambda - {\rm A}^\lambda|{\rm n}\rangle\,, \tag{12}$$

with

$${\rm V}^\lambda = g_{\rm V}(q^2)\gamma^\lambda + g_{\rm M}(q^2)\sigma^{\lambda\nu}q_\nu/2{\rm m_N} \tag{13}$$

and

$${\rm A}^\lambda = g_{\rm A}(q^2)\gamma^\lambda\gamma_5 - g_{\rm P}(q^2)q^\lambda\gamma_5/{\rm m_N}\,, \tag{14}$$

where $\rm m_N$ can be chosen to be the nucleon mass.

The values of the coupling constants $g_i(q^2 \simeq {\rm m}_\mu^2)$ can be predicted, using the CVC and PCAC prescriptions, from the electromagnetic properties of the nucleon, the pion-nucleon coupling constant, and the pion β-decay constant [53]. In the Standard Model no novel features emerge and the tests using muon-capture of the nucleon provide - in principle - only a check of the $V - A$ structure described above. Problems associated with muonic molecular hydrogen at all but the lowest pressures render these tests very delicate [29,54]. However precision experiments have been proposed using high-intensity beams and storage-devices to provide high stopped-muon densities.

For complex nuclei, which are much easier to access experimentally than the proton, the hadronic covariants can be evaluated using the nucleon covariants and appropriate nuclear wave functions calculated in the impulse approximation [53] or, in the "elementary particle approach", using phenomenological form factors extracted from the electromagnetic and β-decay observables of the corresponding states and their isobar analogues [53,55]. This latter approach assumes universality in the lepton-couplings [e-μ universality], an ingredient of the Standard Model which can be tested in muon-capture as discussed below. We now consider extensions of the Standard Model interaction and examine to what extent we could observe an indication for the "new physics" they imply in muon capture. We assume the presence of a single deviation from the Standard Model at one time, and disallow potential "conspiracies" in which several deviations would mask mutually their presence.

The first anomalous interaction we consider is the presence of a $V+A$ component in Eq. (11); *i.e.*, right-handed currents which introduce incomplete parity violation in the muon-capture process [56]. The constraints obtained in Ref. [56] rely on correlation measurements in muon-capture of the type we discuss below, and they are less stringent than the limits deduced from the π-μ-e decay chain and from nuclear β-decay. It is unlikely that these constraints on $V+A$ components will be markedly improved in the near future by muon-capture experiments.

A second type of deviation would include an "induced scalar" term of the form $ig_S(q^2)q^\lambda/\mathrm{m_N}$ in the vector covariant V^λ. This term is a "second-class" current [33], having G-parity opposite to the main terms, and it is also forbidden by CVC. Similarly, an "induced tensor" term of the form $g_T(q^2)\sigma^{\lambda\nu}q_\nu\gamma_5/\mathrm{m_N}$ can be introduced into the axial-vector covariant A^λ. This term is also second class, and as a result would be difficult to accommodate in classical gauge theories [57]. Limits on second class induced tensor terms are discussed in Ref. [29]. Because of the comparatively higher momentum transfer these induced terms should be, in principle, easier to observe in muon-capture than in β-decay. However, their extraction, or even obtaining limits, requires independent information on the induced pseudoscalar contribution which has a similar q-dependence [g_P can be calculated [14] assuming the validity of PCAC]. In addition to these induced terms one could also introduce genuine scalar, tensor and pseudoscalar interactions [58], but these contributions cannot be disentangled from the induced scalar, induced tensor, and momentum-factorized axial-vector interactions. Although information on these deviations from the Standard Model can in principle be extracted also from neutrino-nucleus scattering, no major development is expected in this domain in the coming years.

Many extensions of the Standard Model anticipate non-zero neutrino masses and, consequently, generation-mixing similar to that observed in the quark-sector. In these scenarios the muon, for example, couples to neutrinos of various masses with a coupling strength characteristic of the amplitude of each corresponding neutrino mass-eigenstate. The coupling strengths of the charged leptons to massive neutrinos is expressed by the elements of a neutrino mixing-matrix similar to the CKM matrix in the quark-sector [1]. Excepting the possible deficit in solar neutrinos, tentatively attributed to neutrino mixing in the Sun, searches for neutrino oscillations and neutrino decay have produced only upper limits to the elements of the neutrino mixing-matrix [1-9]. Following a suggestion to consider charged particle spectroscopy in neutrino-emitting systems [59], attention was called to the possibility of a muon-capture experiment to search for a heavy-neutrino admixture in the muon-flavor [60]. This experiment is complementary to the searches discussed above, featuring a generally higher sensitivity albeit for more massive neutrinos.

The last extension of the Standard Model interaction we shall discuss in more detail below [Subsect. 2.10] is the assumption of relative imaginary phases between the hadronic pieces; *i.e.*, time reversal violation [61]. We shall not discuss here the $\pi \to \mu\nu_\mu$ decay-channel or reactions induced by ν_μ; these reaction imply the same vertexes as nuclear muon-capture but are discussed in other chapters of this volume.

2.3 The $\mathcal{F}t$-values for Pure Fermi Transitions

Measurements of endpoint energies, lifetimes and branching ratios are used to obtain the comparative half life ft for the superallowed $J^\pi = 0^+$ transitions, and here the Chalk River group [21] has played a major role in the experiments and their evaluation, with the result that the ft-values for eight transitions have been measured with a typical precision of 0.1% [21,62]. The very difficult measurement for ^{10}C decay, the transition of lowest Z, has recently been improved to 0.5% precision [63], and the f and t values for these nine transitions, taken from Ref. [24], are given in Table 3.

Table 3: The ft input data for superallowed $J^\pi = 0^+$ transitions and ^{19}Ne.

Parent	f [a]	t [a] (s)	$\frac{\alpha}{\pi}C_{NS}^{THH}$ [b] (%)	$\frac{\alpha}{\pi}C_{NS}^{OB}$ [c] (%)	δ_C^{THH} [d] (%)	δ_C^{OB} [d] (%)
$^{10}_{6}$C	2.2972(10)	1313.6(63)		−0.402(46)	0.17(2)	
$^{14}_{8}$O	42.675(20)	71.137(19)	−0.267(70)	−0.330(46)	0.28(3)	0.19(5)
$^{26}_{13}$Al^m	477.89(18)	6.3502(19)	0.058(12)	0.098(23)	0.33(4)	0.24(6)
$^{34}_{17}$Cl	1996.2(5)	1.52704(89)	−0.040(14)	−0.030(23)	0.64(7)	0.48(7)
$^{38}_{19}$K^m	3294.6(19)	0.92471(65)	−0.023(23)	−0.023(23)	0.70(7)	0.49(12)
$^{42}_{21}$Sc	4466.0(12)	0.68059(42)	0.116(23)	0.144(12)	0.39(6)	0.39(5)
$^{46}_{23}$V	7195.8(41)	0.42280(20)	0.037(07)	0.072(23)	0.45(6)	0.21(6)
$^{50}_{25}$Mn	10723.4(31)	0.28336(36)	0.037(07)	0.072(23)	0.50(9)	0.28(6)
$^{54}_{27}$Co	15747.0(38)	0.19344(14)	0.046(07)	0.083(23)	0.59(6)	0.35(6)
$^{19}_{10}$Ne [e]	98.49(12)	17.250(11)		−0.090(20)	0.38(4)	0.25(6)

[a] From Table 2 in Ref. [24]. [b] From Table 6 in Ref. [24]. [c] From Table 4 in Ref. [23]. [d] From Table 2 in Ref. [64]. See also Refs. [21,65]. [e] See Subsect. 2.5.

As we discussed in Subsect. 2.1, CVC alone predicts a constant $\mathcal{F}t$-value for the superallowed $J^\pi = 0^+$ transitions [see Eq. (10)]. However in 1984 Towner and Hardy noticed a systematic difference in the corrected ft-values, with $\mathcal{F}t$ increasing with the charge Z of the daughter nucleus. They attributed [66] this systematic trend to an error in the radiative corrections, which typically amount to a 4% increase in f. These radiative corrections have undergone a complete reevaluation in the past several years [23,24,67-70], and there is presently good agreement on their evaluation. Referring to Eq. (8), we use the standard decomposition $\delta^{\rm R} = \delta_{\rm r}+\Delta_{\beta\mu}+(\alpha/\pi)C_{\rm NS}$, where $\delta_{\rm r}$ is the usual "outer" correction to f, $\Delta_{\beta\mu}$ is the "inner" radiative correction used to compare nuclear β-decay to muon decay, and $(\alpha/\pi)C_{\rm NS}$ is a recently discovered [69] nuclear structure dependent axial-vector two-nucleon correction term. The results of two recent calculations [23,24] of $(\alpha/\pi)C_{\rm NS}$ are given in Table 3, with δ_r and $\Delta_{\beta\mu}$ from Ref. [69] listed in Table 4. The calculated values of $(\alpha/\pi)C_{\rm NS}$ may be systematically too large because unquenched magnetic moments were used [23,24] in their evaluation, but the size of this systematic uncertainty has not been investigated.

Table 4: Calculation of the $\mathcal{F}t$-values for $J^\pi = 0^+$ transitions and ^{19}Ne.

Parent	ft [a] (s)	$\delta_{\rm r}$ [b] (%)	$\Delta_{\beta\mu}$ [c] (%)	$\frac{\alpha}{\pi}\overline{C_{\rm NS}}$ [d] (%)	$\overline{\delta_{\rm C}}$ [e] (%)	$\mathcal{F}t$ [f] (s)
$^{10}_{6}$C	3018(15)	1.66	2.50	−0.402(46)	0.17(4)	3125 (15)
$^{14}_{8}$O	3035.8(16)	1.52	2.50	−0.311(45)	0.256(49)	3140.4(27)
$^{26}_{13}$Al^m	3034.7(15)	1.45	2.49	0.066(12)	0.302(63)	3146.8(25)
$^{34}_{17}$Cl	3048.3(19)	1.43	2.49	−0.037(14)	0.560(94)	3148.9(36)
$^{38}_{19}$K^m	3046.5(28)	1.42	2.49	−0.023(19)	0.65(11)	3144.7(47)
$^{42}_{21}$Sc	3039.5(20)	1.43	2.48	0.138(12)	0.390(73)	3150.2(31)
$^{46}_{23}$V	3042.4(23)	1.43	2.48	0.040(08)	0.330(81)	3152.2(35)
$^{50}_{25}$Mn	3038.6(40)	1.42	2.48	0.040(08)	0.348(95)	3147.3(51)
$^{54}_{27}$Co	3046.1(23)	1.42	2.48	0.050(08)	0.470(81)	3151.5(35)
$^{19}_{10}$Ne [g]	1699.0(23)	1.49	2.49	−0.090(20)	0.340(60)	1759.0(26)

[a] From Table 3. [b] $\delta_{\rm r} = (\alpha/2\pi)\overline{\rm g} + \delta_2 + \delta_3$ from Table I in Ref. [69].
[c] $\Delta_{\beta\mu} = \Delta{\rm S} - (\alpha/2\pi)\overline{\rm g}$ from Table I in Ref. [69].
[d] Weighted mean from Table 3 with errors inflated by a factor of 1.15.
[e] Weighted mean from Table 3 with errors inflated by a factor of 1.9.
[f] $\mathcal{F}{\rm t} = {\rm ft}(1 + \delta_{\rm r} + \Delta_{\beta\mu} + (\alpha/\pi)\overline{C_{\rm NS}})(1 - \overline{\delta_{\rm C}})$. [g] See Subsect. 2.5.

As can be seen in Eq. (8), a further ingredient in the $\mathcal{F}t$ calculation is the evaluation of the nuclear-mismatch corrections $\delta^{\rm C}$, which include configuration mixing, isospin non-conservation, and radial overlap effects. The two sets of comprehensive calculations of the valence-nucleon contribution to $\delta^{\rm C}$, referred to as $\delta_{\rm C}^{\rm THH}$ [Refs. [21,71], Towner, Hardy, and Harvey] and $\delta_{\rm C}^{\rm OB}$ [Ref. [65], Ormand and Brown], are typically $\delta_{\rm C} \approx 0.4\%$. These calculations however differ systematically by 0.16%, which is larger than the precision of the ft measurements. Three approaches have been used to reconcile these $\delta_{\rm C}$ calculations. Towner and Hardy *et al.* [21,23,64] use unweighted averages of $\delta_{\rm C}^{\rm THH}$ and $\delta_{\rm C}^{\rm OB}$ with "statistical" errors and than add an overall systematic error, while Barker *et al.* and Ormand and Brown [24,65] include systematic and *ad hoc* errors in the individual δ_C^{THH} and δ_C^{OB} values. An alternative approach, suggested by Wilkinson [72,73], is to treat δ_C^{THH} and δ_C^{OB} as estimates of the valence-nucleon contribution to δ^C and include a separate phenomenological core-nucleon correction term Δ_C^Z; *i.e.*, $\delta^C = \delta_C + \Delta_C^Z Z^\alpha$. This approach will be discussed later. The results of Towner [23], who uses the average values of δ_C^{THH} and δ_C^{OB}, and the calculation of Barker *et al.* [24] using δ_C^{OB} are plotted as triangular points in panels A and B, respectively, of Fig. 1. Their weighted means $\mathcal{F}t_0^{mean}$, shown by the dashed lines and circular points, are $\mathcal{F}t_0^{mean} = 3149.6(10)$ s [23] and $\mathcal{F}t_0^{mean} = 3150.8(17)$ s [24]. Towner's result [23] has an additional systematic error of 0.08%. Neither of these results is in good agreement with the Standard Model prediction $\mathcal{F}t_0^{SM} = 3136.9(36)$ s given in Eq. (10), which is shown by the star symbol in both panels A and B.

The results shown in Fig. 1 are plotted *vs* $\langle\Gamma/W\rangle$ because the endpoint energy dependence of the $\mathcal{F}t$-values can be used to determine or to constrain [21,26] the Fierz interference contribution in Eq. (9). A fit to the results represented in Fig. 1 obtained using $\mathcal{F}t(\langle\Gamma/W\rangle) = \mathcal{F}t_0^{b_F}/(1 + \langle\Gamma/W\rangle\, b_F)$ is shown by the bold lines and the square intercept at $\langle\Gamma/W\rangle = 0$. Here b_F is the Fermi Fierz-interference term between the vector and scalar contributions to $\mathcal{H}_\beta$. The evidence for a b_F contribution is marginal, especially for the results of Towner [23] shown in panel A of Fig. 1, where only the ^{14}O measurement with $\langle\Gamma/W\rangle = 0.44$ hints to a non-zero slope. However, Barker *et al.* [24] include a calculation [74] for ^{10}C decay with $\langle\Gamma/W\rangle = 0.62$, and as shown in panel B the indication for non-zero b_F may possibly be considered as significant. A strong argument against the interpretation of the results shown in Fig. 1 in terms of a non-zero b_F comes however from an analysis of the neutron, ^{19}Ne, and Fermi/Gamow-Teller positron polarization ratio measurements [see Subsects. 2.4 to 2.6 and 3.3]. The results of this analysis disagree with the value of b_F obtained from the $\mathcal{F}t(\langle\Gamma/W\rangle)$ fits shown in Fig. 1. As a consequence, it seems unlikely that the non-zero value of the Fierz interference term b_F is at the origin of the discrepancy between the $\mathcal{F}t_0^{mean}$-values obtained by Towner [23] and Barker *et al.* [24] and the Standard Model prediction $\mathcal{F}t_0^{SM}$ given in Eq. (10).

The resolution of this dilemma, proposed some time ago on rather fundamental grounds by Wilkinson [72], is that the Z-dependence of the nuclear-mismatch corrections evaluated using the methods of Refs. [65,71] is too small because the contribution of the core-nucleons is ignored. Wilkinson [73] and others [75,76] have used instead a nuclear-mismatch correction with the form $(1 - \delta_C - \Delta_C^Z Z^\alpha)$, where Δ_C^Z is a phenomenological core-nucleon correction term whose value is empirically determined by a fit to the $\mathcal{F}t$ measurements. The value chosen for α depends on how the valence-nucleon contribution to δ^C is evaluated [73,76], and here we will employ a linear fit [76] with $\alpha = 1$. Following Wilkinson's lead, we suggest that

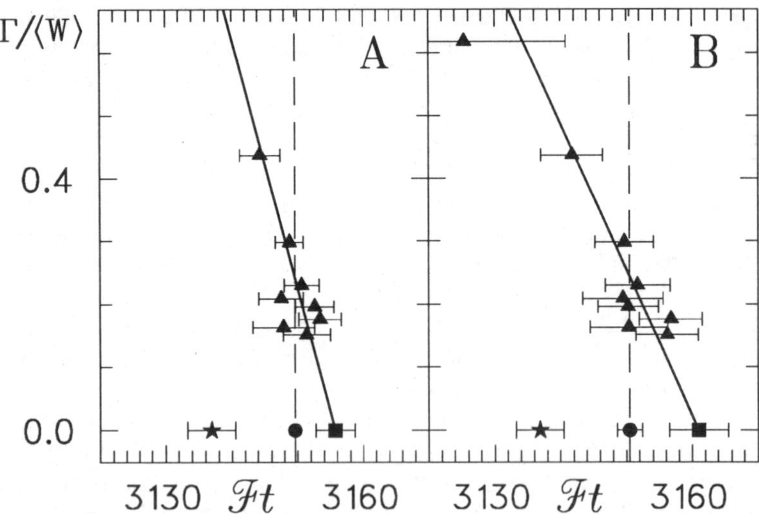

Figure 1: The $\mathcal{F}t$-values for $J^{\pi} = 0^{+}$ pure Fermi transitions from Refs. [23] and [24], plotted *vs* $\langle\Gamma/W\rangle$ where $\Gamma = \sqrt{1-\alpha^2 Z^2}$ and $W = E/m_e$, are shown in panels A and B, respectively. The lines and intercepts at $\langle\Gamma/W\rangle = 0$ are discussed in the text. The Standard Model prediction $\mathcal{F}t_0^{\rm SM}$ is shown by the star symbol.

$\delta_C^{\rm THH}$ and $\delta_C^{\rm OB}$ represent a reasonable statistical interval for the valence-nucleon contribution to δ^{C}. Consequently, we evaluated the $\overline{\delta_C}$ values shown in Table 4 as the weighted means of $\delta_C^{\rm THH}$ and $\delta_C^{\rm OB}$ calculated using the "statistical" errors of Hardy *et al.* [64] uniformly inflated by a factor of 1.9 to give a combined $\chi^2 = 8$ for the eight pairs of calculated values about their respective weighted means. We use the same procedure for $(\alpha/\pi)C_{\rm NS}$, obtaining $(\alpha/\pi)\overline{C_{\rm NS}}$ from the two recent calculations [23,24] with the calculated errors uniformly inflated by a factor of 1.15 to obtain $\chi^2 = 8$. The resulting $\mathcal{F}t$-values are given in Table 4 and plotted as triangular points *vs* Z in panel A of Fig. 2.

A fit to the eight most precise measurements [^{10}C excluded] with the expression $\mathcal{F}t(Z) = \mathcal{F}t_0^{(Z=0)}/\left(1-\Delta_C^Z Z\right)$ gives $\mathcal{F}t_0^{(J^\pi=0^+)}\Delta_C^Z = 0.53(18)$s and a Z = 0 intercept

$$\mathcal{F}t_0^{(Z=0)} = \mathcal{F}t_0^{(J^\pi=0^+)} = 3138.4(32)\,\text{s}, \quad [Z = 0 \text{ intercept, best estimate}] \qquad (15)$$

which we adopt as our most reliable estimate of $\mathcal{F}t_0^{(J^\pi=0^+)}$ in the absence of a scalar contribution in $\mathcal{H}_\beta$. The fit and Z = 0 intercept are shown by the bold line and square point, respectively, in panel A of Fig. 2, and the weighted mean of the eight most precise measurements, $\mathcal{F}t_0^{\rm mean} = 3147.3(15)\,$s, is shown for comparison by the circular point.

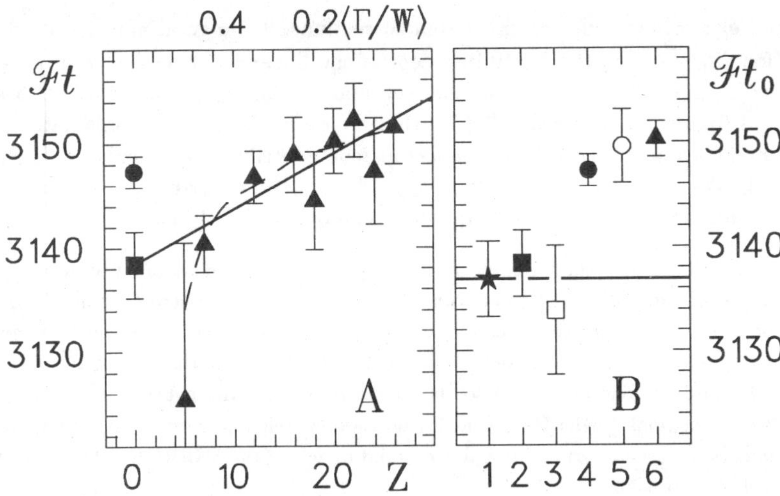

Figure 2: In panel A the $\mathcal{F}t$-values for $J^\pi = 0^+$ pure Fermi transitions from Table 4 are plotted as triangles *vs* the charge Z of the daughter nucleus, and a corresponding approximate ⟨Γ/W⟩ scale is also shown. The weighted mean is shown by the filled circle at Z = 0. The bold, straight line shows the fit obtained using a phenomenological nuclear-mismatch correction term Δ_C^Z; the corresponding value of the intercept, shown as a filled square, is given in Eq. (15). The dashed curve is a fit obtained using a scalar Fierz interference term. Panel B shows as a star the Standard Model prediction $\mathcal{F}t_0^{SM}$ given in Eq. (10), with the Z = 0 intercept from panel A shown by the filled square. The result deduced from experiments performed on mirror-nuclei, $\mathcal{F}t_0^{mirror}$ given in Eq. (24), is shown by the open square. The last three points are mean values $\mathcal{F}t_0^{mean}$, *i.e.*, obtained using $\Delta_C^Z = 0$, taken from the present evaluation [solid circle], from Towner's evaluation [23] with the systematic error included [open circle], and from the Barker *et al.*'s evaluation [24] using δ_C^{OB} [solid triangle]. These results show the necessity of using the Δ_C^Z correction term to reconcile the available data with the three generation Standard Model.

The alternate fit with non-zero b_F is shown by the dashed line, with an approximate ⟨Γ/W⟩ scale given at the top of panel A. Constraints on the scalar interaction are discussed in Subsect. 3.3.

Some illustrative results are gathered in Table 5, which shows the values of $\mathcal{F}t_0^{mean}$, $\mathcal{F}t_0^{b_F}$, and $\mathcal{F}t_0^{(Z=0)}$ for various treatments of the nuclear-mismatch corrections. We have used for input data the $J^\pi = 0^+$ $\mathcal{F}t$-values from: (*i*) Towner [23], labeled THH+OB; (*ii*) Barker *et al.* [24] with $\delta^C = \delta_C^{THH}$ or δ_C^{OB}, labeled THH and OB, respectively; and (*iii*) the present results using $\delta^C = \overline{\delta_C}$, which is labeled $\overline{\delta_C}$. The calculations of Barker *et al.* show the systematic difference in $\mathcal{F}t_0^{mean}$ for the

δ_C^{THH} and δ_C^{OB} valence-nucleon mismatch corrections, while good agreement is obtained for $\mathcal{F}t_0^{(Z=0)}$; *i.e.*, the Z = 0 intercept of the linear fit using the core-nucleon Δ_C^Z contribution. This fact, which has been noted in previous analyses [73,75,76], supports Wilkinson's contention [72,73] that a core-nucleon nuclear-mismatch correction needs to be added to the valence-nucleon contributions calculated in Refs. [21,65,71]. We also show the results obtained from the neutron and ^{19}Ne mirror-nucleus transitions [see Subsect. 2.4], which is labeled MIRROR.

Table 5: We show below the weighted mean [$\mathcal{F}t_0^{mean}$] and fits to the $\mathcal{F}t$-values using either a non-zero Fermi Fierz interference contribution [$\mathcal{F}t_0^{b_F}$] or a non-zero phenomenological core-nucleon nuclear-mismatch correction term [$\mathcal{F}t_0^{(Z=0)}$] obtained using various treatments of the valence-nucleon mismatch corrections. The χ^2/ν values for the fit results shown in columns 4 and 6 are one or less. Also shown are results for the neutron and ^{19}Ne mirror-nucleus transitions [MIRROR]. The Fermi/Gamow-Teller positron polarization ratio measurements are used as an additional constraint on b_F for the MIRROR entries. See the text for a full discussion.

	$b_F = \Delta_C^Z = 0$		$\Delta_C^Z = 0$		$b_F = 0$	
δ_C	$\mathcal{F}t_0^{mean}$ (s)	χ^2/ν	$\mathcal{F}t_0^{b_F}$ (s)	$\Re e(a_{R+}^S)$ [a]	$\mathcal{F}t_0^{(Z=0)}$ (s)	$\mathcal{F}t_0^{(Z=0)}\Delta_C^Z$ (s)
THH+OB$^{Ref.\ [23]}$	3149.6(10)	0.94	3155.7(30)	−0.004(2)	3143.8(32)	0.34(19)
THH$^{Ref.\ [24]}$	3145.7(15)	0.94	3153.0(40)	−0.005(3)	3138.8(44)	0.40(25)
OB$^{Ref.\ [24]}$	3150.3(17)	0.81	3161.0(45)	−0.007(3)	3139.7(51)	0.62(29)
$\overline{\delta_C}$	3147.3(15)	1.62	3156.5(29)	−0.006(2)	3138.4(32)	0.53(18)
MIRROR	3138.0(82)	2.08	3133.9(66)	0.007(5)	3133.9(63)	0.53 [b]
$\overline{\delta_C}$+MIRROR	3147.1(16)	1.78	3151.1(35)	−0.003(2)	3137.2(29)	0.61(16)

[a] $b_F = \pm 2\,\Re e(a_{R+}^S)$ for $e^{\mp}$ transitions [see Subsect. 3.1 and Eq. (38)].
[b] $\mathcal{F}t_0^{(Z=0)}\Delta_C^Z = 0.53(18)$ is taken from the fit to the $J^\pi = 0^+$ $\mathcal{F}t$-values.

The evaluation of $\mathcal{F}t_0^{(J^\pi=0^+)}$ presented above is in good agreement with the Standard Model prediction [Eq. (10)] and with the value obtained from the neutron and ^{19}Ne mirror-nucleus transitions [as discussed in Subsect. 2.4], viz. $\mathcal{F}t_0^{mirror} =$ 3133.9(63) s. We believe that the agreement between $\mathcal{F}t_0^{(J^\pi=0^+)}$ and $\mathcal{F}t_0^{mirror}$ provides at the present time convincing evidence for the phenomenological core-nucleon correction term proposed by Wilkinson. We further illustrate in panel B of Fig. 2 the necessity of this correction term to reconcile the available data with the three generation Standard Model. The first point, shown as a star, is the Standard Model prediction $\mathcal{F}t_0^{SM}$ given in Eq. (10). The second point, show as a solid square, is $\mathcal{F}t_0^{(J^\pi=0^+)}$ from Eq. (15); *i.e.*, obtained including the Δ_C^Z correction. It is in excellent agreement with the Standard Model prediction. Although less precise, $\mathcal{F}t_0^{\rm mirror}$, given in Eq. (24) and shown by the open square, is in good agreement with these first two results. The last three points are $\mathcal{F}t_0^{\rm mean}$-values, *i.e.*, evaluated without including the core-nucleon mismatch correction $\Delta_{\rm C}^{\rm Z}$, taken from the present evaluation [solid circle], from Towner's evaluation [23] with the systematic error included [open circle], and from the Barker *et al.*'s evaluation [24] using $\delta_{\rm C}^{\rm OB}$ [solid triangle]. It is obvious that all three $\mathcal{F}t_0^{\rm mean}$ results, obtained disregarding the $\Delta_{\rm C}^{\rm Z}$-correction, disagree with the Standard Model prediction [star]. The deviations are 2.7, 2.5, and 3.4 σ, respectively.

We evaluated $\mathcal{F}t_0^{\rm expt}$ from a fit both to the $\mathcal{F}t^{\,({\rm J}^\pi=0^+)}$ measurements, including the $\Delta_{\rm C}^{\rm Z}$ correction, and mirror-nucleus β-decay experiments. The result is

$$\mathcal{F}t_0^{\rm expt} = 3137.2(2.9)\,{\rm s}\,. \quad \text{[best estimate]} \tag{16}$$

Then, using Eq. (9) with $\xi = 4$ and $b = 0$ we obtain

$$\mathrm{V}_{ud}^2 = 0.95128 \pm 0.00088 \pm 0.00080\,, \quad \text{[best estimate]} \tag{17}$$

where the first error is statistical and the second error is the uncertainty [23,24] in the "inner" radiative correction $\Delta_{\beta\mu}$. Taking $\mathrm{V}_{us}^2 + \mathrm{V}_{ub}^2 = 0.04862(79)$ from the evaluation of the Particle Data Group [22], we obtain

$$\mathrm{V}_{ud}^2 + \mathrm{V}_{us}^2 + \mathrm{V}_{ub}^2 = 0.9999(14)\,. \quad \text{[experiment]} \tag{18}$$

This is in excellent agreement with the Standard Model expectation of unity.[†]

As far as the future is concerned, a more precise measurement for ^{10}C will reduce the uncertainty in $\Delta_{\rm C}^{\rm Z}$ and increase the reliability and accuracy of $\mathcal{F}t_0^{({\rm J}^\pi=0^+)}$. It will be however difficult to achieve a final uncertainty near the approximately 1 s precision one should be able to obtain from the $\mathcal{F}t$-values given in Table 4 without a reliable shell-model calculation of the core-nucleon contribution to the nuclear-mismatch correction. Several new ideas for experimental measurements [64] and theoretical calculations [77,78] which may improve our understanding of the nuclear-mismatch corrections are being actively pursued. In addition, two new, precise experiments for the mirror-nucleus decays of ^{21}Na [79] and ^{37}K [80] will extend these measurements over the same range of Z as the $\mathrm{J}^\pi = 0^+$ experiments and considerably reduce the error in $\mathcal{F}t_0^{\rm mirror}$. Finally, a more precise measurement of the $\pi^+ \to e^+\nu_e\pi^0$ branching ratio would provide an independent measurement of V_{ud} without the uncertainties of nuclear-mismatch corrections; such a measurement has been initiated at PSI [81]. It should be noted, however, that a significant improvement in the CKM unitarity test, Eq. (18), will require more precise values not

[†] For a vector–axial-vector interaction more general than that of the Standard Model [see Subsects. 3.1 and 3.2] V_{us} must be obtained from the K_{e3}–decays alone [25], which gives $\mathrm{V}_{us}^2 + \mathrm{V}_{ub}^2 = 0.0482(10)$ [22] and $\mathrm{V}_{ud}^2 + \mathrm{V}_{us}^2 + \mathrm{V}_{ub}^2 = 0.9995(16)$.

only of V_{ud}^2 but also of V_{us}^2 and $\Delta_{\beta\mu}$, as these three contributions have comparable uncertainties.

2.4 Neutron-Decay Experiments

The most complete set of precision experiments is available for the decay of the free neutron. This is a relatively recent and fortunate occurrence, because the neutron measurements are sensitive to all extensions of the Standard Model which contribute to $\mathcal{H}_\beta$, including $V + A$ currents and scalar and tensor couplings. The neutron $\mathcal{F}t$, β-asymmetry parameter A, and time reversal violating D-parameter correlation in particular provide significant constraints on deviations from the Standard Model. The D-parameter measurements are discussed in Subsect. 2.10.

The recent review by Freedman [82] shows the remarkable impact of cold and ultracold neutron experiments on measurements of the neutron lifetime which have occurred since 1988. An update of the results for now eight measurements of τ_n [83-90] is shown in panel A of Fig. 3. The measurements are plotted chronologically, with the oldest measurement at Y = 1 and the weighted mean at Y = 9. The ideogram [21,82] shown is a sum of normalized gaussians with centroids at τ_i, widths reflecting the uncertainties $[\sigma_i]$ of the measurements, and areas weighted by $1/\sigma_i$. The curve gives a visual impression of the consistency of the measurements and the weighting, which differs from the usual $1/\sigma_i^2$ used for the probability distribution, accentuates [21] systematic errors. For these τ_n measurements the scatter is statistical, and the weighted mean is $\tau_n = 888.7(18)\,\mathrm{s}$ with $\chi^2/\nu = 0.55$. Using the very precisely known phase space factor $f_n = 1.71465(15)$ [91] and inner radiative correction $\Delta_{\beta\mu} = 2.49\%$ we obtain the value cited in Table 2,

$$\mathcal{F}t_n = 1082.5(22)\,\mathrm{s}\,. \tag{19}$$

The two most precise [87,90] τ_n measurements shown in Fig. 3 result from experiments which used stored ultracold neutrons [92]. We have not included a preliminary result [93] also obtained using stored ultracold neutrons which would lower the weighted mean to $\tau_n = 887.0(15)\,\mathrm{s}$ with $\chi^2/\nu = 0.80$; two new experiments of this type are in preparation at ILL [94] and Gatchina [95]. New, so-called direct measurements which aim for a precision comparable to the stored ultracold neutron experiments are also planned [96,97]. Because $\mathcal{F}t_n$ has considerable impact on a wide range of Standard Model constraints, the additional measurements which reduce the uncertainty will certainly be of high significance.

Panel B in Fig. 3 shows measurements and ideogram for the neutron β-asymmetry parameter A in the same format as the τ_n measurements in panel A, and here again the agreement of the measurements is good. The weak-magnetism contribution to A is about 1.5%, and the two most precise experiments [102,103] correct for this contribution. The older measurements have been reanalyzed and

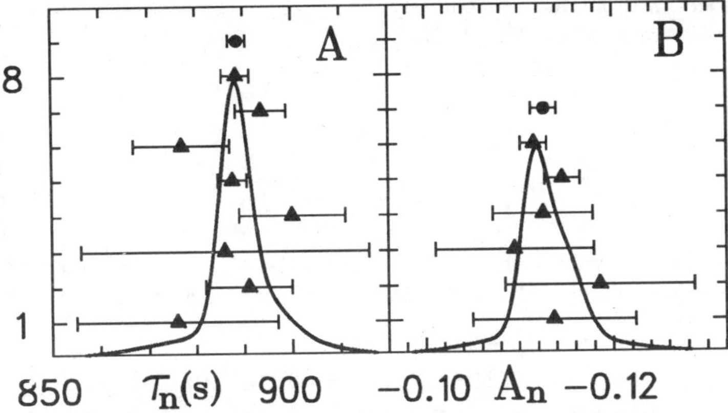

Figure 3: Eight recent measurements of the neutron lifetime τ are shown by the triangular data points in panel A, with the weighted mean shown by the ninth, circular data point. The curve is an ideogram which gives a visual impression of the consistency of the measurements. Panel B shows two recent [small error bars] and four older measurements of the neutron β-asymmetry parameter using the same notation as in panel A. These measurements are discussed in Subsect. 2.4.

corrected by Klemt *et al.* [98], and we use their evaluation to calculate the four less precise data points represented in Fig. 3. The weighted mean is given in Table 2. The LNPI-IAE β-asymmetry measurement [103], the data point at Y = 6 in panel B of Fig. 3, has caused considerable discussion [99-101] because it is $2.5\,\sigma$ smaller than the Standard Model prediction obtained from the combination of the $\mathcal{F}t_0^{(J^\pi=0^+)}$ and $\mathcal{F}t_{\rm n}$ measurements as discussed below. Although this measurement required significant corrections for the relatively low neutron polarization and spatially non-uniform polarization profile, the experiment was carefully done and the result is in statistical agreement with the other correlation measurements.

Because the analysis of the $\mathcal{F}t$ and asymmetry parameter measurements for mirror-nucleus transitions play a particularly important role in nuclear β-decay tests for new physics beyond the Standard Model, we describe here how this analysis proceeds using the neutron data as an example. The basic idea is to use ratios of $\mathcal{F}t$ measurements in combination with asymmetry parameter or other correlation measurements to obtain the best value for $g_{\rm A}{\rm M}_{\rm GT}/g_{\rm V}{\rm M}_{\rm F}$. In the present case the matrix elements for the $J^\pi = 0^+$ pure Fermi transitions [14] and the neutron [91] are known [viz., ${\rm M}_{\rm F}^{0^+} = \sqrt{2}$, ${\rm M}_{\rm F}^{\rm n} = 1$, and ${\rm M}_{\rm GT}^{\rm n} = \sqrt{3}$], and using the expressions for ξ and $A\,\xi$ in the appendix evaluated in the Standard Model's $V - A$ interaction [see Subsect. 3.1] one has immediately

$$\mathcal{R}_{\mathrm{n}}^{\mathcal{F}t} = \frac{2\mathcal{F}t_0^{(J^\pi=0^+)}}{\mathcal{F}t_{\mathrm{n}}} = 1 + 3\left(\frac{g_{\mathrm{A}}}{g_{\mathrm{V}}}\right)^2 \tag{20}$$

and

$$A_{\mathrm{n}} = -2\,\frac{(g_{\mathrm{A}}/g_{\mathrm{V}})^2 - g_{\mathrm{A}}/g_{\mathrm{V}}}{1 + 3\,(g_{\mathrm{A}}/g_{\mathrm{V}})^2}\,. \tag{21}$$

Below we list the values of $g_{\mathrm{A}}/g_{\mathrm{V}}$ obtained from $\mathcal{R}_{\mathrm{n}}^{\mathcal{F}t}$, the two most recent and precise measurements of A_{n}, and the weighted mean of A_{n} given in Table 2;

$$\begin{aligned} g_{\mathrm{A}}/g_{\mathrm{V}} &= 1.2648(17)\,, \quad [\mathcal{R}_{\mathrm{n}}^{\mathcal{F}t}] \\ &= 1.2624(50)\,, \quad [A_{\mathrm{n}} = -0.1146(19),\ \text{Ref. [102]}] \\ &= 1.2545(37)\,, \quad [A_{\mathrm{n}} = -0.1116(14),\ \text{Ref. [103]}] \\ &= 1.2574(29)\,. \quad [A_{\mathrm{n}} = -0.1127(11),\ \text{Table 2}] \end{aligned}$$

The values of $g_{\mathrm{A}}/g_{\mathrm{V}}$ obtained from $\mathcal{R}_{\mathrm{n}}^{\mathcal{F}t}$ and the A_{n} from Ref. [103] are not in good agreement [2.5 σ], and choosing the value of $g_{\mathrm{A}}/g_{\mathrm{V}}$ which minimizes to both $\mathcal{R}_{\mathrm{n}}^{\mathcal{F}t}$ and A_{n} [from Table 2] χ^2 gives the rather large value of $\chi^2 = 4.8$ [$\nu = 1$, 2.8% C.L.].

Although the measurements of ν-asymmetry parameter B_{n} and β-ν correlation parameter a_{n} for neutron decay are not nearly as precise as the β-asymmetry parameter measurement, the ν-asymmetry parameter in particular provides significant constraints on several interactions beyond the Standard Model. The β-ν correlation is measured observing the energy spectrum of the recoil protons [37], and although the most precise measurement of a_{n} has a statistical accuracy of 2%, the final result suffers from a systematic uncertainty of 4.2% due mostly to the low recoil-energy of the protons. We have reevaluated the weak-magnetism contribution to a_{n} and find that the published value should be reduced by less than 1.5%, which is significantly smaller than the correction given in Ref. [46]. Because the measured values are not tabulated, the uncertainty on the slope of a_{n} obtained from the graphical data is comparable to our weak-magnetism correction. Consequently, in reporting this value in Table 2 we have not included this correction on a_{n}.

Our best estimate of $g_{\mathrm{A}}/g_{\mathrm{V}}$ is provided by a simultaneous fit to $\mathcal{R}_{\mathrm{n}}^{\mathcal{F}t}$ and the a_{n}, A_{n}, and B_{n} correlation measurements listed in Table 2. The result is

$$g_{\mathrm{A}}/g_{\mathrm{V}} = 1.2628(19)\,, \quad \text{[best estimate]} \tag{22}$$

where the error has been increased [22] by $\sqrt{\chi^2/\nu}$, with $\chi^2 = 5.0$ for $\nu = 3$ [17% C.L.].

The procedure outlined above is used when the neutron β-decay data are evaluated in isolation, and a similar set of relations can be written for the ^{19}Ne experiments discussed in the next subsection [for ^{19}Ne the product $g_A M_{GT}^{Ne}/g_V$ is used to minimize χ^2].

As mentioned in Subsect. 2.3, mirror-nucleus transitions can also be used to provide an independent test of CKM unitarity. We illustrate this by rewriting Eq. (20) as

$$\mathcal{R}_n^{\mathcal{F}t_0} = \frac{2\mathcal{F}t_0^n}{\mathcal{F}t_n} = 1 + 3\left(\frac{g_A}{g_V}\right)^2, \tag{23}$$

where $\mathcal{F}t_0^n$ is a value extracted from neutron-decay alone for comparison with the Standard Model prediction $\mathcal{F}t_0^{SM}$ [Eq. (10)]. Eqs. (21) and (23) can be solved to obtain $\mathcal{F}t_0^n$ and g_A/g_V, but this result is not useful unless a resulting precision can be assigned to $\mathcal{F}t_0^n$. The straightforward approach is to include the ν-asymmetry parameter and β-ν correlation measurements in the analysis, in which case $\mathcal{F}t_0^n$ and g_A/g_V are overdetermined. Using this procedure we obtain the result $\mathcal{F}t_0^n = 3108(13)$ s. As expected, and mainly due to the new world average of A_n [see above], this result is not in good agreement [$2.1\,\sigma$] with $\mathcal{F}t_0^{SM} = 3136.9(36)$ s. We extend this analysis to include the ^{19}Ne measurements discussed in the next subsection, using a common value $\mathcal{F}t_0^{mirror}$ in the respective neutron and ^{19}Ne $\mathcal{R}_{mirror}^{\mathcal{F}t_0}$ equations and adjusting g_A/g_V and M_{GT}^{Ne} to obtain the lowest χ^2; we obtain the result

$$\mathcal{F}t_0^{mirror} = 3133.9(63)\,\text{s} \quad \text{[best estimate]} \tag{24}$$

with $\chi^2/\nu = 1.17$, 32% C.L. This value was cited in Subsect. 2.3, and this calculation also provides the value $g_A/g_V = 1.2622(18)$.

The procedure for calculating $\mathcal{R}_n^{\mathcal{F}t}$ and A_n when $\mathcal{H}_\beta$ includes interactions beyond the Standard Model is straightforward, and one finds for example that introducing $V + A$ currents into $\mathcal{H}_\beta$ reduces the discrepancy between the values of g_A/g_V obtained from $\mathcal{R}_n^{\mathcal{F}t}$ and A_n, while including scalar couplings generally increases the discrepancy. In Subsect. 3.1 we introduce a parameterization of $\mathcal{H}_\beta$ in terms of helicity coefficients $a_{\ell q}$, where the helicity indicies refer to the leptons and quarks, respectively. Here we use this parameterization in lowest order to display the sensitivity of $\mathcal{R}_n^{\mathcal{F}t}$ and A_n to $V + A$ currents. Eqs. (20) and (21) describe the relations implied by the Standard Model with the coefficient for a purely left-handed lepton-quark interaction $a_{LL}^V = 1$ and all other $a_{\ell q}$'s zero. However, for a generalized vector–axial-vector interaction, using the notation defined in Eq. (35) of Subsect. 3.1, we have the expressions

$$\mathcal{R}_n^{\mathcal{F}t} = 1 + 3\left(\frac{g'_A}{g_V}\right)^2 \left(\frac{1 + |a_{RL}^V - a_{RR}^V|^2}{1 + |a_{RL}^V + a_{RR}^V|^2}\right) \tag{25}$$

and

$$A_n = -2\,\frac{(g'_A/g_V)^2\,(1-|a^V_{RL}-a^V_{RR}|^2) - (g'_A/g_V)\,(1+|a^V_{RL}|^2-|a^V_{RR}|^2)}{1+|a^V_{RL}+a^V_{RR}|^2+3\,(g'_A/g_V)^2\,(1+|a^V_{RL}-a^V_{RR}|^2)}\,, \tag{26}$$

where

$$\left(\frac{g'_A}{g_V}\right)^2 = \left(\frac{g_A}{g_V}\right)^2 \frac{|a^V_{LL}-a^V_{LR}|^2}{|a^V_{LL}+a^V_{LR}|^2}\,. \tag{27}$$

We see immediately that the neutron measurements are sensitive to the presence of vector or axial-vector interactions which couple right-handed electrons to both left- and right-handed quarks, while the coupling of left-handed electrons to right-handed quarks $[a^V_{LR}]$ appears only as a renormalization of the axial-vector form factor g_A. These are general features of the β-decay correlations; we will discuss the constraints on a^V_{LR} obtained from the unitarity of the CKM matrix for three generations and from the time reversal violating D-parameter correlation in Subsect. 3.2.

Another important feature of mirror-nucleus transitions are the interference terms between the vector and axial-vector currents which significantly constrain the right-handed-lepton–quark helicity coefficients $|a^V_{RL} \pm a^V_{RR}|$. For example, this term enters with only a negative sign in the purely axial-vector pseudoscalar meson decay, and more importantly, the constraint on $|a^V_{RL} - a^V_{RR}|$ from the $(\pi^+ \to e^+\nu_e)/(\pi^+ \to \mu^+\nu_\mu)$ branching ratio measurement vanishes if the contribution is the same for both branches [104]. We consider the constraints on these helicity coefficients and on several scenarios for extensions to the Standard Model in Sects. 3 and 4.

The perspectives for new neutron decay experiments are excellent. There are several proposals for correlation measurements [105,106], with the resolution of the $\mathcal{R}^{\mathcal{F}t}$–$A_n$ discrepancy topping the list. In this regard, the proposed measurements of a_n [97] and of the ratio A_n/B_n [105] will provide determinations of g_A/g_V which are independent of $\mathcal{F}t_0^{(J^\pi=0^+)}$ and the precise knowledge of the neutron polarization.

2.5 Other Mirror-Nucleus Transitions

Several $\mathcal{F}t$ and asymmetry parameter measurements are available for mirror-nucleus transitions between complex nuclei [see Ref. [107] for a survey], however only the experiments for ^{19}Ne are as yet sufficiently precise to provide a significant test of the Standard Model. We have updated the ^{19}Ne $\mathcal{F}t$ calculation of Ref. [100] using a shell model calculation of $(\alpha/\pi)C_{NS}$ [108] and only the two most precise lifetime measurements [109]. The input data are given in Tables 3 and 4, and the final value given in Table 2 includes the $\mathcal{F}t_0^{(J^\pi=0^+)}\Delta_C^Z = 0.53(18)\,$s core-nucleon nuclear-mismatch correction term determined from the fit to the $\mathcal{F}t^{(J^\pi=0^+)}$ measurements.

The unpublished 1983 Princeton [110] measurement of the ^{19}Ne asymmetry parameter is often used in the literature [14,47,52,100]. The experimental measurements when extrapolated to $E = \mathrm{m_e}$, as quoted in Ref. [110], and corrected for recoil-order contributions, give $A(^{19}\mathrm{Ne}) = -0.03669(83)$ with $\chi^2/\nu = 1.72$ [100]. However, the slope dA/dE is larger than the recoil-order corrections [15] predict. Recent reevaluations which (*i*) integrate the recoil-order corrections for each point of the experimental spectrum and take the weighted mean of the recoil-corrected measurements [76] or (*ii*) fit the measurements to the energy dependence predicted by the recoil-order calculations [111] give results which are in mutual agreement. As the evidence for an additional slope of unknown origin [$dA/dE = -0.0011(8)$] as assumed in the original work [110] is not convincing, we prefer to use the values obtained in Refs. [76] and [111]. Following the prescription in Ref. [76], $A(^{19}\mathrm{Ne}) = -0.03808(26)$, where the uncertainty has been inflated to take into account the large $\chi^2/\nu = 1.90$. An additional systematic error of $\pm 1\%$ in the nuclear polarization is included in the value given in Table 2. This transition is especially sensitive to the right-handed lepton, left-handed quark coefficient $a^{\mathrm{V}}_{\mathrm{RL}}$ in the generalized vector–axial-vector interaction [see Subsect. 3.1].

Time reversal violating ^{19}Ne correlation measurements are discussed in Subsect. 2.10.

Precision measurements for mirror-nucleus transitions are few in number because the asymmetry parameter experiments must achieve a large and well measured [to better than 1%] nuclear polarization. Two experiments are being developed [79,80] which use laser optical traps and optical pumping to provide localized, highly polarized samples of ^{21}Na and ^{37}K. These transitions have large asymmetry parameters and are mainly sensitive to the $a^{\mathrm{V}}_{\mathrm{RR}}$ coefficient in the generalized vector–axial-vector interaction. When completed these experiments will complement the constraints provided by the neutron and ^{19}Ne experiments, and these four transitions will permit an independent test of $\mathcal{F}t_0$ and CKM unitarity.

2.6 Longitudinal Polarization Experiments

Measurements of absolute electron polarizations and β–γ-ray circular polarization experiments formed the major weak interaction industry in the 1960's and played an important role in establishing the $V - A$ interaction. However, the often quoted result that $P(e^{\mp}) = \mp v/c$ with 0.5% accuracy is unrealistic by today's standards, and also untrue if systematic errors are included [113]. In addition, most of these experiments studied long-lived, hindered Gamow-Teller transitions for which there is no reliable estimate of the nuclear physics [recoil-order] corrections [see Subsect. 2.1]. We report in Table 2 the weighted mean of the ten most reliable measurements of $G_{\mathrm{GT}} = \mathrm{P}/(v/c)$ with carefully evaluated systematic errors [43,44]; the measurements used are listed in Ref. [45]. Nine of these measurements, from the classic experiments of van Klinken [44], have a combined statistical precision of 0.9% and absolute accuracy of 1.6%. However, because of the uncertainties in the nuclear physics corrections, we do not use these data to constrain subtle contributions, such as the Fierz term or the T-odd contribution of an imaginary tensor current [76,112]. The one measurement free of these problems, also listed in Table 2, is the electron polarization for the tritium mirror-nucleus transition, although here some additional concern regarding the accuracy of Mott scattering polarimetery for very low electron energies [113] may make the error estimate of this measurement optimistic.

A entirely different class of experiments are two recent measurements of the ratio of positron polarizations $G_{\rm F}/G_{\rm GT}$ for pure Fermi and pure Gamow-Teller transitions. These are believed to be exceedingly reliable measurements, with accuracies of less than 0.5% for the polarization ratio achieved using both Bhabba scattering [47] and time resolved spectroscopy of positronium formed in a strong magnetic field [46]. In addition to being free of absolute polarimeter analyzing power uncertainties, the techniques employed in these measurements reduce many other systematic errors. In terms of the helicity coefficients given in Eq. (35) of Subsect. 3.1, the positron polarization ratio is

$$\frac{G_{\rm F}}{G_{\rm GT}} = \frac{1-2\,|a^{\rm V}_{\rm RL}+a^{\rm V}_{\rm RR}|^2}{1-2\,|a^{\rm V}_{\rm RL}-a^{\rm V}_{\rm RR}|^2} \simeq 1-8\,\Re e(a^{\rm V}_{\rm RL}a^{\rm V\,*}_{\rm RR}) \tag{28}$$

in the generalized vector–axial-vector interaction. The $G_{\rm F}/G_{\rm GT}$ measurements provide a very important constraint on $a^{\rm V}_{\rm RL}$ and $a^{\rm V}_{\rm RR}$, especially in view of the high precision achieved: viz., $\Re e(a^{\rm V}_{\rm RL}a^{\rm V\,*}_{\rm RR}) = -0.00015(34)$.

The $G_{\rm F}/G_{\rm GT}$ measurements also provide significant constraints on scalar and tensor interactions, and here we find in lowest order using the notation introduced in Subsect. 3.1

$$\frac{G_{\rm F}}{G_{\rm GT}} = \left(\frac{1-4\,(g_{\rm S}/g_{\rm V})^2\,|a^{\rm S}_{\rm R+}|^2}{1-2\,(g_{\rm S}/g_{\rm V})\,\Re e(a^{\rm S}_{\rm R+})}\right)\left(\frac{1+4\,(g_{\rm T}/g_{\rm A})\,\Re e(a^{\rm T}_{\rm RL})}{1-16\,(g_{\rm T}/g_{\rm A})^2\,|a^{\rm T}_{\rm RL}|^2}\right). \tag{29}$$

The relative positronium polarization measurement is clearly capable of high precision, and this technique is also being used for the measurements discussed next.

Polarization-asymmetry $[\mathcal{PA}]$ correlation experiments are a variant of the $G_{\rm F}/G_{\rm GT}$ experiments in which the ratio of the positron polarization $[\mathcal{P}^-/\mathcal{P}^+]$ for two orientations of the emitter's nuclear spin is measured. These experiments require only relative measurements of the polarimeter analyzing power and nuclear polarization, and they offer significantly enhanced sensitivity to several extensions of the Standard Model if the asymmetry is large [114]: viz.,

$$\frac{\mathcal{P}^-}{\mathcal{P}^+} = \mathcal{R}_0\left[1-\frac{8\beta^2\mathcal{A}}{\beta^4-\mathcal{A}^2}\,\mathcal{B}(\mathcal{PA})\right] \tag{30}$$

where

$$\mathcal{R}_0 = \frac{\beta^2 - \mathcal{A}}{1 - \mathcal{A}} \frac{1 + \mathcal{A}}{\beta^2 + \mathcal{A}}, \tag{31}$$

β is the measured v/c, $\mathcal{A}$ is the measured asymmetry $\vec{\beta} \cdot \vec{\mathrm{J}}A$, and $\mathcal{B}(\mathcal{PA})$ is the contribution beyond the Standard Model. The one experiment completed sofar, a 5% measurement of $\mathcal{P}^-/\mathcal{P}^+$ for ^{107}In [48], provides a 0.5% constraint on $\mathcal{B}$. In some models this precision is comparable to a 1% absolute measurement of the electron polarization [G] or an asymmetry parameter [A]. Depending on the transition, this correlation is mainly sensitive to either $|a^{\mathrm{V}}_{\mathrm{RL}}|^2$ or $|a^{\mathrm{V}}_{\mathrm{RR}}|^2$ in the generalized vector–axial-vector interaction, and so provides constraints which complement the $G_{\mathrm{F}}/G_{\mathrm{GT}}$ experiments. Two new $\mathcal{P}^-/\mathcal{P}^+$ experiments, for ^{12}N [115] and ^{21}Na [116] will considerably improve on the precision of this type of experiment.

2.7 The β-ν Correlation for Pure Fermi and Gamow-Teller Transitions

The β-ν correlation for pure transitions is potentially useful in searches for scalar or tensor couplings because the deviation from $V - A$ is independent of their helicity structure and time reversal invariance properties. Specifically, in terms of the helicity coefficients introduced in Subsect. 3.1, the β-ν correlation parameters a_{F} and a_{GT} are, respectively

$$a_{\mathrm{F}} = 1 - 4\,\frac{g_{\mathrm{S}}^2}{g_{\mathrm{V}}^2}\left(|a^{\mathrm{S}}_{\mathrm{L+}}|^2 + |a^{\mathrm{S}}_{\mathrm{R+}}|^2\right) \tag{32}$$

$$a_{\mathrm{GT}} = -\frac{1}{3}\Big[1 - 16\,\frac{g_{\mathrm{T}}^2}{g_{\mathrm{A}}^2}\left(|a^{\mathrm{T}}_{\mathrm{LR}}|^2 + |a^{\mathrm{T}}_{\mathrm{RL}}|^2\right)\Big]. \tag{33}$$

The β-ν correlation can be measured from the energy spectrum of the recoil-ions, and the 1% measurement [38] of a_{GT} for ^{6}He decay sets a standard which has not been equaled. This measurement benefits from the relatively low-mass and large energy release, and the only other direct measurement of comparable precision is the 5% measurement for neutron decay.

However, in addition to direct recoil-energy measurements one may resort to indirect determinations by observing the Doppler displacement or broadening of a γ-ray or charged particle emitted by the recoil nucleus. The recent measurements of a_{F} for ^{32}Ar and ^{33}Ar have achieved [39] accuracies of 4% and 2% respectively [the ^{33}Ar decay is essentially [39,117] a pure Fermi transition, and the recoil order corrections for the Ar experiments are negligible]. These transitions lead to proton-unstable final states, and the isospin-forbidden feature of the proton-decay makes the states sufficiently narrow that a_F can be obtained from the Doppler width. An analysis by Adelberger [117] shows that these experiments give considerably improved limits on scalar couplings in nuclear β-decay. Measurements for α- [118]

and γ-ray [119] decay of recoil nuclei are promising but still insufficiently accurate to provide constraints comparable with these measurements.

In addition to the prospect of improved precision from the Doppler width measurements [117], it is particularly interesting to improve the precision of the β-ν correlation for neutron decay. The neutron experiment provides independent information on g_A/g_V [see Subsect. 2.4], and it therefore can be used to check the reliability of the polarization sensitive electron- and neutrino-asymmetry parameter experiments.

2.8 *Measurements of the Fierz-Interference Terms*

Measurements of the spectrum shape alone – in principle – provide a direct measure of the Fierz term b which arises from the interference of the vector and scalar or axial-vector and tensor currents. However, these measurements are notoriously difficult, as witness the recent controversy regarding the existence of a 17 keV neutrino [22,120,121], and poorly understood forbidden-corrections to highly-hindered decays [122] render the most precise measurements unuseable. As a consequence, the constraint on b arises mostly from $\mathcal{F}t$-values constrained by correlation measurements and from the Fermi/Gamow-Teller polarization-ratio experiments discussed in Subsect. 2.6.

The strength of the constraints obtained in Sect. 3 is reduced by the fact that the contribution of the energy dependent Fierz term cannot be included in many correlation measurements [*e.g.,* the neutron asymmetry parameter] because experimenters no longer publish their measurements in tabular form. Archival journals appear most happy to do this, and referees should insist on it!

2.9 *Experiments on Nuclear Muon-Capture*

We shall briefly describe the various muon-capture experiments, proceeding in order of increasing complexity and commenting on the physics they address, emphasizing experiments relevant to the subject of this chapter. For a detailed discussion of this topic we refer the reader to the review article by Mukhopadhyay [53] and the updated comments of Ref. [123].

The "elementary particle" approach mentioned in Subsect. 2.2 predicts the partial muon-capture rate between well-defined nuclear states using, among others inputs, the transition probability of the inverse β-decay transition. Such a muon-capture rate measurement with 1% precision will provide a useful test of μ-e universality if the nuclear physics uncertainties and the induced pseudoscalar contribution are understood at a comparable level of precision. The ^{3}He-^{3}H transition is a favorable candidate because the nuclear physics uncertainties are dominated by meson-exchange effects which have been obtained from electromagnetic observables at the 1% level [124], and correlation measurements presently underway at

LAMPF [125] and TRIUMF [126] will provide an accurate determination of the induced pseudoscalar contribution. The precision measurement of the ^{3}He-^{3}H muon capture rate is being performed at PSI [127].

In this context one may note that the classical test of μ-e universality is the e/μ branching ratio in π-decay, and two recent experiments [128,129] have tested the equality of the electron and muon couplings with a precision of 0.3%. One has to emphasize, however, that the well-known helicity-suppression of the electron-decay channel renders this test very sensitive to the presence of any hypothetical pseudoscalar coupling beyond the Standard Model. The "second-best" test is the e/μ branching ratio in τ-decay, which presently has a precision of 1.1% [130]. However, it has been pointed out [131] that although the branching ratio is in agreement with the Standard Model the actual rates are not, and the τ-decay constraint should be viewed with caution.

The proposal [60] to use muon-capture in the search for a heavy-neutrino admixture in the muon flavor requires a high-resolution measurement of the energy of the charged particle emitted together with the neutrino in a muon-capture reaction with two bodies in the final state. This experiment has been realized for the $^3\text{He} + \mu^- \rightarrow ^3\text{H} + \nu_\mu$ capture, in which a 1.9 MeV triton is emitted in conjunction with the ν_μ. The admixture of a massive neutrino into the dominant channel would produce a second peak of lower energy in the recoil-spectrum of the triton. This search is complementary to the spectroscopy of the emitted muon in π- and K-decays, which left uncovered the neutrino mass region between 30 and 80 MeV/c^2. The preliminary results from a PSI experiment have been published [132], and a refined analysis of additional measurements is in progress [133].

Considering the recent developments in the field, such as the measurements of the Z^0 width at LEP which restricts the number of neutrino generations significantly lighter than the Z^0 to three [22] and the analysis of rare τ-decay modes which restrict the mass of the predominant admixed mass-eigenstate to below 32 MeV [134], the interest of these searches drifted into the neutrino mass region below this value. A mixing limit of $< 7 \times 10^{-5}$ obtained from π-decay [135] applies to the mass region $m_\nu \lesssim 30\,\text{MeV/c}^2$, and the possibility of improving on this limit in muon-capture is under investigation at PSI [127].

The most intricate muon-capture experiments involve correlations between the neutrino-[recoil-ion]-momentum and the spins of the initial and final nuclear states [polarization and/or alignment]. The first experiments of this type [136,137], as discussed in Ref. [29], were mainly used to extract information on the induced pseudoscalar coupling, although the sign of the muon neutrino-helicity can also be obtained from these correlations [29,56]. Some of these experiments are now under refinement, but their main purpose remains the investigation of the induced pseudoscalar coupling, and they are not expected to provide competitive tests of the Standard Model. Measurements of the muon capture-rate ratio for different hyperfine levels are also mainly sensitive to the induced pseudoscalar coupling, and they are not directly related to tests of the Standard Model.

We shall discuss in the following subsection the use of muon-capture in the search for time reversal violation [53].

2.10 Time Reversal Violating Correlations

If the only source of CP-violation would be the one offered by the Standard Model [22], effects in β-decay and muon-capture would be second order in the weak interaction and so vanishingly small. So any observation of time reversal violation [TRV] in these processes indicates the presence of new physics beyond the Standard Model. The importance of elucidating the origin of CP-violation is stressed elsewhere in this volume, and there are several excellent review articles [138,139] which discuss the prospects for observable T-violation in nuclear systems.

CP-violation is observed only in the K_{L}-system, although there have been extensive searches in the strong-, weak-, and electromagnetic-sectors [22,138,139]. We shall discuss searches for TRV in the semileptonic transitions, restricting first to a complex phase [ϕ_{VA}] in the vector–axial-vector interaction and then enlarging the discussion to include potential scalar and tensor couplings which have an imaginary phase difference with the main $V-A$ interaction.

Mirror-nucleus transitions, which have vector and axial-vector contributions of comparable size, are the best place to search for ϕ_{VA}. The P-even, T-odd D-parameter correlation [see Eq. (4)] provides the most sensitive test, especially for the coupling of left-handed leptons to right-handed quarks [see Subsect. 3.5], and null results with precisions of 0.15% and 0.06% have been obtained for the neutron [22] and ^{19}Ne [41], respectively. Proposed new experiments [140,141] may achieve significant improvements in these results [*e.g.,* 0.01% for the neutron]. At these precision levels electromagnetic final-state interactions, which though T-conserving can mimic TRV [142], are important for ^{19}Ne where $D^{\mathrm{em}} \simeq 0.01\%$. For the neutron, $D^{\mathrm{em}} \simeq 10^{-6}$.

Even much less precise results could be of interest in the muon-sector. Phenomenologically there is no need to have similar effects in the electron- and muon-sectors, as exemplified by the Higgs-coupling which is trivially stronger in the muon-sector than in the electron-sector. Other scenarios are discussed in the contribution by Herczeg [13], and the observables can be calculated using the nuclear form factors given in Ref. [143].

The P-even, T-odd correlation is exemplified by the $\left(\hat{\sigma}_\mu \cdot \vec{\mathrm{k}}_{\mathrm{R}} \times \vec{\mathrm{k}}_\gamma\right)\left(\vec{\mathrm{k}}_{\mathrm{R}} \cdot \vec{\mathrm{k}}_\gamma\right)$ correlation, where $\hat{\sigma}_\mu$ is the muon spin direction, $\vec{k}_{\mathrm{R}}$ is the nuclear-recoil direction, and $\vec{k}_\gamma$ is the direction of the γ-ray which de-excites the nucleus formed in muon-capture. The γ-ray anisotropy is used to measure the alignment of the nuclear state formed in the capture process [144]. The feasibility of an experiment to measure this TRV correlation in $^{16}\mathrm{O}+\mu^- \rightarrow ^{16}\mathrm{N}^{(1^-,397\,\mathrm{keV})}+\nu_\mu$ by direct observation of the recoil in a drift-chamber is being investigated at PSI [145]. The five-fold correlation given above is sensitive to ϕ_{VA} for this transition.

For transitions of sufficiently high momentum-transfer the search for ϕ_{VA} can also be performed using interference terms between the axial-vector matrix element and the recoil-order weak-magnetism vector matrix element, or the respective corresponding form factors. A β-decay experiment using ^{8}Li is underway in Seattle [146,147]. A corresponding P-odd, T-odd muon-capture experiment could use the $\left(\hat{\sigma}_\mu \cdot \vec{\mathrm{J}}_{\mathrm{R}} \times \vec{\mathrm{k}}_{\mathrm{R}}\right)$ correlation, where $\vec{\mathrm{J}}_{\mathrm{R}}$ is the polarization nucleus formed in the muon-capture which can be measured from the asymmetry in its subsequent β-decay [148]. Directional sensitivity to $\vec{k}_R$ is achieved with the selective depolarization method pioneered by Grenacs [29] for ^{12}B, and a feasibility test for $^{16}\mathrm{O}+\mu^- \rightarrow ^{16}\mathrm{N}+\nu_\mu$ is underway at PSI [148].

It is not yet clear to what precision the muon-capture tests could be pushed. Further theoretical work to relate the eventual results to those of the semileptonic K-decay TRV triple correlations, which have been measured with a precision of about 2% [22,149], would also be welcome. The K-decay results will be considerably improved by new experiments at KEK [150].

Turning now to TRV imaginary scalar and tensor couplings, it was pointed out in Subsect. 2.7 that the β-ν correlations receive contributions only from the magnitudes of these couplings, and therefore can set limits on both real and imaginary scalar and tensor interaction. Adelberger [117] has shown that improved limits on an imaginary scalar coupling in particular can be obtained from the new a_{F} measurements [39] for ^{32}Ar and ^{33}Ar. Slightly weaker constraints [76] can be obtained from the $\mathcal{F}t$, and asymmetry parameter, and $G_{\mathrm{F}}/G_{\mathrm{GT}}$ measurements discussed in Subsects. 2.4 to 2.6. The β-asymmetry parameter and electron polarization coefficients A and G, respectively, also receive TRV contributions from first-order final-state interactions [20], and polarization measurements have been used [112] to set limits on the imaginary tensor contribution. However, these first-order estimates are deemed unreliable, especially for hindered transitions [15], and we do not consider them here.

A perhaps more promising experiment is the P-odd, T-odd R-parameter correlation, which receives a first-order contribution from an interference between the main $V-A$ interaction and imaginary scalar and tensor couplings of right-handed leptons [see Subsect. 3.1]. A recent measurement [49] for the Gamow-Teller decay of ^{8}Li, employing a novel Mott polarimeter, has achieved a null result with a precision of 0.7%. A new preliminary result [151], $R=0.001(4)$, will set unprecedented limits on the TRV tensor interaction, although it should be pointed out that the final-state interaction contributions which mimic TRV may be as large as $R^{\mathrm{em}} \simeq 0.001$ for this transition.

A potentially more interesting measurement is the R-parameter correlation for mirror-nucleus transitions, which can be used to search for an interference between the axial-vector current and a TRV scalar interaction. Unfortunately, these experiments suffer from the small Mott analyzing power for positrons, and the only such experiment to date is a 5% measurement for ^{19}Ne [50]. At the present time, a mirror-nucleus R-parameter measurement with a precision on the order of 0.5% is required to provide limits which are competitive with the constraints set by the other measurements discussed above, and the prospects for such an experiment are not good. We do not use the ^{19}Ne R-parameter measurement to deduce the results presented in Sect. 3.

2.11 Neutrino-Induced Reactions

Because of experimental reasons, namely the increase of cross-sections with energy, most neutrino-reaction Standard Model tests are performed with high-energy neutrinos [152] and are discussed elsewhere in this volume. At low- and intermediate-energies the prime source of information is provided by muon-decay neutrinos, which can be either scattered purely leptonically on the target electrons [153] or give rise to charge-changing semileptonic reactions, *i.e.,* inverse β-decay [154,155]. The results can be then used to constrain anomalous interactions at play in muon-decay [153]. In this regard Fetscher [156] has called attention to the information one can extract from the KARMEN-collaboration experiment [155], and the interested reader will find more details on this issue in the chapter on muon-decay.

Important constraints on left-right symmetric extensions of the Standard Model were deduced from the energetics of supernova SN1987a [157], following the line of reasoning pioneered by Raffelt and Seckel [158]. Because both the production of the neutronization neutrinos and their eventual reabsorption in the supernova are governed by the semileptonic weak-coupling, the observed energetics of the supernova will signal any deviation from the standard left-handed coupling of known strength. The absence of such a deviation allows one to conclude, assuming that potential right-handed neutrinos are light enough, that either they are scarcely emitted or strongly reabsorbed. These two alternatives place upper and lower bounds on the coupling of a right-handed gauge boson. The corresponding limits obtained in Ref. [157] will be used in Sect. 4. An evaluation of the supernova constraints on left-right symmetric models based on a recent analysis of the SN1987a energetics [160] is underway [159].

2.12 Summary

The measurements discussed above and several muon-decay experiments which in some models provide related information are presented in Table 2. The muon-decay parameters ρ, δ, and ξ refer to the Michel spectrum for polarized-muon decay [22], and P_μ is the polarization of the muon in π- or K-meson decay. The muon-decay measurement of the asymmetry at the endpoint of the $\mu^+ \to e^+\nu_e\bar{\nu}_\mu$-decay spectrum, $P_\mu\delta\xi/\rho$, requires additional discussion because the authors [52] regard this measurement as a lower bound, with $P_\mu\delta\xi/\rho > 0.99677$ [90% C.L.]. Although this is reasonable when the measurement is treated in isolation, it is not clear what one should do in a global analysis because the experimental result, $P_\mu\delta\xi/\rho = 0.99790 \pm 0.00046(\text{stat}) \pm 0.00075(\text{syst})$, is $2.4\,\sigma$ below the Standard Model prediction of unity and the error is claimed to include an estimate of the various systematic contributions. The possible upward corrections to the result, as discussed in Ref. [52], were estimated to amount to about $+0.00030 \pm 0.00045$ [76] but were specifically not applied by the authors [52] of the original research paper. Therefore we have decided to use the experimental result without consideration of these upward corrections, but with the caveat that any deviation from the Standard Model which can be directly attributed to this measurement should be considered cautiously, realizing that possible corrections to this result may reduce the discrepancy.

A fit to the 24 combined nuclear β-decay and muon-decay measurements listed in Table 2, using the Standard Model interaction for the correlation coefficients and $\mathcal{F}t^{\text{SM}}$ from Eq. (10), adjusting g_{A} and the ^{19}Ne Gamow-Teller matrix element to minimize χ^2, gives excellent agreement with the Standard Model; viz., $\chi^2 = 18$ for $\nu = 22$ [C.L. = 71%]. The fit is considerably worse [$\chi^2 = 25$, C.L. = 30%] if the phenomenological core-nucleon nuclear-mismatch correction Δ_C^Z discussed in Subsect. 2.3 is not included in the $\mathcal{F}t_0^{(\text{J}^\pi=0^+)}$ calculation. We believe that this difference is significant and entirely explained by Δ_C^Z, although alternative explanations in terms of interactions beyond the Standard Model have been discussed elsewhere [23,24]. Right-handed $[V + A]$ currents give an additional large decrease in the χ^2 contribution from the correlation measurements: this scenario will be discussed in Subsect. 3.2 and Sect. 4.

3 β-Decay Constraints on the Weak Interaction

3.1 The Generalized β-Decay Hamiltonian

The transition probability $d\mathcal{W}$ given in Eq. (4) assumes a Hamiltonian considerably more general than the Standard Model interaction given in Eq. (1). In the discussion which follows we will use the interaction considered by JTW [20], which for neutron decay gives

$$\begin{aligned}\mathcal{H}_\beta &= \frac{\text{G}_\text{F}}{\sqrt{2}}\,\tilde{\text{V}}_{ud}\Big[\,\bar{e}\,\gamma_\lambda\,(\text{G}_\text{V} + \text{G}'_\text{V}\gamma_5)\,\nu_e\,\langle \text{p}|\gamma^\lambda|\text{n}\rangle \\ &\qquad\qquad + \;\bar{e}\,\gamma_\lambda\gamma_5\,(\text{G}_\text{A} + \text{G}'_\text{A}\gamma_5)\,\nu_e\,\langle \text{p}|\gamma^\lambda\gamma_5|\text{n}\rangle\end{aligned}$$

$$+\ \bar{e}\,(G_S + G'_S\gamma_5)\,\nu_e\,\langle \mathrm{p}||\mathrm{n}\rangle$$
$$+\ \bar{e}\,\frac{\sigma_{\lambda\mu}}{\sqrt{2}}\,(\mathrm{G_T} + \mathrm{G'_T}\gamma_5)\,\nu_e\,\langle \mathrm{p}|\,\frac{\sigma^{\lambda\mu}}{\sqrt{2}}\,|\mathrm{n}\rangle\,\Big]\,, \tag{34}$$

where $\widetilde{\mathrm{V}}_{ud}$ denotes a convenient but unknown normalization of the generalized β-decay Hamiltonian which can be related to the V_{ud} element of the CKM matrix in restricted circumstances [see Subsect. 3.2]. Although this Hamiltonian leads to the most concise expressions for the correlation coefficients, it gives little insight into the structure of the weak interaction beyond the trivial Standard Model expectation $\mathrm{G_V}/g_\mathrm{V} = -\mathrm{G'_V}/g_\mathrm{V} = \mathrm{G_A}/g_\mathrm{A} = -\mathrm{G'_A}/g_\mathrm{A} = 1$. An alternative formulation, the four fermion lepton-quark interaction expressed in the helicity projection form [HPF], explicitly displays the handedness of the couplings in the lepton- and quark-sectors [161], and it is picturesque, informative, and often directly connected to model calculations.[†] This form is especially useful for the lepton-lepton interaction [$\mu^- \to e^-\nu_\mu\bar{\nu}_e$], and here we follow the notation of Fetscher *et al.* [162] which is used in the discussion of the muon-decay parameters in the periodic Review of Particle Properties [22]. Using the HPF, the Hamiltonian given above for neutron decay becomes [161-163]

$$\begin{aligned}
\mathcal{H}_\beta^{\ell q} \;=\; \frac{\mathrm{G_F}}{\sqrt{2}}\,\widetilde{\mathrm{V}}_{ud}\,\Big[\, & a^{\mathrm{V}}_{\mathrm{LL}}\,\bar{e}\,\gamma_\lambda(1-\gamma_5)\,\nu\,\bar{u}\,\gamma^\lambda(1-\gamma_5)\,d \\
&+ a^{\mathrm{V}}_{\mathrm{LR}}\,\bar{e}\,\gamma_\lambda(1-\gamma_5)\,\nu\,\bar{u}\,\gamma^\lambda(1+\gamma_5)\,d \\
&+ a^{\mathrm{V}}_{\mathrm{RL}}\,\bar{e}\,\gamma_\lambda(1+\gamma_5)\,\nu\,\bar{u}\,\gamma^\lambda(1-\gamma_5)\,d \\
&+ a^{\mathrm{V}}_{\mathrm{RR}}\,\bar{e}\,\gamma_\lambda(1+\gamma_5)\,\nu\,\bar{u}\,\gamma^\lambda(1+\gamma_5)\,d \\
&+ a^{\mathrm{S}}_{\mathrm{RL}}\,\bar{e}\,(1-\gamma_5)\,\nu\,\bar{u}\,(1-\gamma_5)\,d \\
&+ a^{\mathrm{S}}_{\mathrm{RR}}\,\bar{e}\,(1-\gamma_5)\,\nu\,\bar{u}\,(1+\gamma_5)\,d \\
&+ a^{\mathrm{S}}_{\mathrm{LL}}\,\bar{e}\,(1+\gamma_5)\,\nu\,\bar{u}\,(1-\gamma_5)\,d \\
&+ a^{\mathrm{S}}_{\mathrm{LR}}\,\bar{e}\,(1+\gamma_5)\,\nu\,\bar{u}\,(1+\gamma_5)\,d \\
&+ a^{\mathrm{T}}_{\mathrm{RL}}\,\bar{e}\,\frac{\sigma_{\lambda\mu}}{\sqrt{2}}\,(1-\gamma_5)\,\nu\,\bar{u}\,\frac{\sigma^{\lambda\mu}}{\sqrt{2}}\,(1-\gamma_5)\,d \\
&+ a^{\mathrm{T}}_{\mathrm{LR}}\,\bar{e}\,\frac{\sigma_{\lambda\mu}}{\sqrt{2}}\,(1+\gamma_5)\,\nu\,\bar{u}\,\frac{\sigma^{\lambda\mu}}{\sqrt{2}}\,(1+\gamma_5)\,d\,\Big]\,.
\end{aligned} \tag{35}$$

The subscript indices on the $a_{\ell q}$'s specify the helicity of the emitted charged lepton and the initial quark [161,162], and there is a one-to-one relationship between the leptonic coefficients $g_{e\mu}$ introduced by Fetscher *et al.* [162] and the semileptonic coefficients $a_{\ell q}$ given above. These $a_{\ell q}$ coefficients are in general complex numbers, with time reversal invariance requiring $a_{\ell q}$ real. The $V-A$ Standard Model predicts $a^{\mathrm{V}}_{\mathrm{LL}} \equiv 1$ and all other $a_{\ell q}$'s zero.

[†] We wish to thank Dr. Peter Herczeg for illuminating discussions regarding the HPF Hamiltonian for semileptonic transitions.

The simplicity and quadratic relations [22,162] that the HPF brings to the description of μ-decay does not carry over to semileptonic transitions because of the more complex hadronic current and the presence of pure Fermi and pure Gamow-Teller transitions. In addition, the pseudoscalar interaction is practically absent in nuclear β-decay because $\langle \mathrm{p}|\gamma_5|\mathrm{n}\rangle \to 0$ in the nonrelativistic limit. Consequently, we introduce the symbols

$$a^{\mathrm{S}}_{\mathrm{R+}} = a^{\mathrm{S}}_{\mathrm{RL}} + a^{\mathrm{S}}_{\mathrm{RR}} \tag{36}$$

and

$$a^{\mathrm{S}}_{\mathrm{L+}} = a^{\mathrm{S}}_{\mathrm{LL}} + a^{\mathrm{S}}_{\mathrm{LR}} \tag{37}$$

to denote the fact that only these sums of the fundamental lepton-quark scalar coefficients are constrained by nuclear β-decay. The relations between the $a_{\ell q}$'s and the G coefficients in Eq. (34) can be directly read off from the two expressions, *e.g.* $\mathrm{G_V} = g_{\mathrm{V}}(a^{\mathrm{V}}_{\mathrm{LL}} + a^{\mathrm{V}}_{\mathrm{LR}} + a^{\mathrm{V}}_{\mathrm{RL}} + a^{\mathrm{V}}_{\mathrm{RR}})$, and are gathered in the appendix, while the form factors are defined in Eqs. (2), (3), (5), and (6). For simplicity in presenting numerical results we have normalized the form factors so that $g_{\mathrm{S}}/g_{\mathrm{V}} = g_{\mathrm{T}}/g_{\mathrm{A}} = g_{\mathrm{V}} = 1$. In general we adopt the Standard Model normalization $a^{\mathrm{V}}_{\mathrm{LL}} = a^{\mathrm{SM}}_{\mathrm{LL}} \equiv 1$, however a modification of Eq. (35) which is appropriate for constraining a generalized vector–axial-vector interaction by CKM unitarity is discussed in Subsect. 3.2.

We now proceed to show how the β-decay measurements discussed in Sect. 2 place constraints on the various coefficients in Eq. (35) which would indicate physics beyond the Standard Model. Especially strong limits can be obtained from interference terms involving $a^{\mathrm{V}}_{\mathrm{LL}}$ and one of the other coefficients; *e.g.*, in lowest order the expressions for the Fierz interference terms and the TRV D- and R-parameter correlations are

$$b_{\mathrm{F}} = \pm 2 \frac{g_{\mathrm{S}}}{g_{\mathrm{V}}} \frac{\Re e\,(a^{\mathrm{V}}_{\mathrm{LL}}\, a^{\mathrm{S}*}_{\mathrm{R+}})}{|a^{\mathrm{V}}_{\mathrm{LL}}|^2}, \tag{38}$$

$$b_{\mathrm{GT}} = \mp 4 \frac{g_{\mathrm{T}}}{g_{\mathrm{A}}} \frac{\Re e\,(a^{\mathrm{V}}_{\mathrm{LL}}\, a^{\mathrm{T}*}_{\mathrm{RL}})}{|a^{\mathrm{V}}_{\mathrm{LL}}|^2}, \tag{39}$$

$$D = -\frac{4 g_{\mathrm{V}} g_{\mathrm{A}} \mathrm{M_F M_{GT}} \sqrt{\frac{J}{J+1}}\; \Im m(a^{\mathrm{V}}_{\mathrm{LL}} a^{\mathrm{V}*}_{\mathrm{LR}} + a^{\mathrm{V}*}_{\mathrm{RL}} a^{\mathrm{V}}_{\mathrm{RR}})}{(g^2_{\mathrm{V}} \mathrm{M^2_F} + g^2_{\mathrm{A}} \mathrm{M^2_{GT}})\,|a^{\mathrm{V}}_{\mathrm{LL}}|^2}, \tag{40}$$

$$R_{scalar} = \frac{2 g_{\mathrm{A}} g_{\mathrm{S}} \mathrm{M_F M_{GT}} \sqrt{\frac{J}{J+1}}\; \Im m(a^{\mathrm{V}}_{\mathrm{LL}} a^{\mathrm{S}*}_{\mathrm{R+}})}{(g^2_{\mathrm{V}} \mathrm{M^2_F} + g^2_{\mathrm{A}} \mathrm{M^2_{GT}})\,|a^{\mathrm{V}}_{\mathrm{LL}}|^2}, \tag{41}$$

and

$$R_{\mathrm{GT}} = \pm 4 \frac{g_{\mathrm{T}}}{g_{\mathrm{A}}} \lambda_{\mathrm{JJ'}} \frac{\Im m\,(a^{\mathrm{V}}_{\mathrm{LL}}\, a^{\mathrm{T}*}_{\mathrm{RL}})}{|a^{\mathrm{V}}_{\mathrm{LL}}|^2}, \tag{42}$$

where upper (lower) sign is for $e^{\mp}$ and $\lambda_{\mathrm{JJ'}}$ is defined in the appendix. We do not include tensor contributions in Eq. (41) because any imaginary tensor contribution is already tightly constrained by the R-parameter measurement for the ^{8}Li pure Gamow-Teller transition [see Table 2]. While $a^{\rm T}_{\rm LR}$ in particular is well constrained by these interference terms, the limits on $|a^{\rm T}_{\rm LR}|$ and $|a^{\rm S}_{\rm L+}|$ come from quadratic contributions to the $\mathcal{F}t$-values and β-ν correlations [see Subsect. 2.7].

The empirical limits on these coefficients in Eq. (35) are correlated because the "best fit" description of the measurements occurs for non-zero values of several $a_{\ell q}$'s other than $a^{\rm V}_{\rm LL}$. In fact, the best fit excludes the $V-A$ solution at more than 95% C.L. Whether this discrepancy is significant will be discussed in Sect. 4, where we confront the nuclear β-decay results with constraints from the leptonic and nonleptonic sectors. In this section we ignore this discrepancy and place limits on the extreme deviation of the $a_{\ell q}$ coefficients from their $V-A$ values allowed at 90% C.L. and $3\,\sigma$ from the best fit obtained using a subset of the coefficients in Eq. (35). The procedure we use to obtain the limits on the $a_{\ell q}$'s is illustrated for $a^{\rm V}_{\rm RL}$ and $a^{\rm V}_{\rm RL}$ in the next subsection.

In the discussion which follows we first consider, in turn, time reversal invariant contributions from (*i*) a generalized vector–axial-vector interaction, (*ii*) the scalar interaction, and (*iii*) the tensor interaction. Following this we allow each of these interactions to be complex, and seek nuclear β-decay constraints on time reversal violation. Finally, a model independent analysis of the Hamiltonian is compared to similar results obtained from an analysis [22] of the muon-decay parameters. The results of this analysis are given in Table 6, and for numerical tabulations we assume everywhere that $g_{\rm S}=g_{\rm V}=1$ and $g_{\rm T}=g_{\rm A}$.

3.2 The Real Vector–Axial-Vector Interaction

In the Standard Model the generalized β-decay Hamiltonians of Eqs. (34) and (35) are considerably simplified: viz., ${\rm G_V}/g_{\rm V}=-{\rm G'_V}/g_{\rm V}={\rm G_A}/g_{\rm A}=-{\rm G'_A}/g_{\rm A}=a^{\rm V}_{\rm LL}=g_{\rm V}=1$ all other $a_{\ell q}=0$. If, moreover, the unitarity of the CKM matrix is realized by three generations, we have ${\rm V}^2_{ud}+{\rm V}^2_{us}+{\rm V}^2_{ub}=1$.

Following our strategy outlined in the introduction we derive constraints on the coefficients in Eq. (35) assuming a minimal one-by-one deviation from the Standard Model. We shall assume in this subsection that the scalar and tensor contributions are zero and separate out moreover from the sum $(a^{\rm V}_{\rm LL}+a^{\rm V}_{\rm LR})$ a term $a^{\rm V}_{\rm L+}$ which arises from physics beyond the Standard Model; viz., with $a^{\rm SM}_{\rm LL}\equiv 1$

$$a^{\rm V}_{\rm LL}+a^{\rm V}_{\rm LR}=a^{\rm SM}_{\rm LL}+a^{\rm V'}_{\rm LL}+a^{\rm V}_{\rm LR}=1+a^{\rm V}_{\rm L+}\,. \tag{43}$$

We first consider the constraints on the β-decay coefficients $a^{\rm V}_{\rm L+}$, $a^{\rm V}_{\rm RL}$, and $a^{\rm V}_{\rm RR}$ which arise in part from the unitarity of the CKM matrix. Recalling the definitions of the hadronic currents given in Eqs. (2) and (3), the β-decay Hamiltonian Eq. (35) becomes

$$\mathcal{H}_\beta \;=\; \frac{\rm G_F}{\sqrt{2}}\,\tilde{\rm V}_{ud}\left[\left(1+a^{\rm V}_{\rm L+}\right)\bar{e}\,\gamma_\lambda(1-\gamma_5)\,\nu\,\langle{\rm p}|\gamma^\lambda|{\rm n}\rangle\right.$$

$$+ \left(a^{\rm V}_{\rm RL} + a^{\rm V}_{\rm RR}\right) \bar{e}\, \gamma_\lambda (1+\gamma_5)\, \nu \, \langle {\rm p}|\gamma^\lambda|{\rm n}\rangle$$
$$+ g_{\rm A}\, \bar{e}\, \gamma_\lambda \gamma_5 (1-\gamma_5)\, \nu \, \langle {\rm p}|\gamma^\lambda\gamma_5|{\rm n}\rangle$$
$$- g_{\rm A}\left(a^{\rm V}_{\rm RL} - a^{\rm V}_{\rm RR}\right) \bar{e}\, \gamma_\lambda \gamma_5 (1+\gamma_5)\, \nu \, \langle {\rm p}|\gamma^\lambda\gamma_5|{\rm n}\rangle \Big] . \qquad (44)$$

Here we have assumed from CVC that $g_{\rm V} = 1$ and utilized the result, shown in Eq. (27), that a deviation of $(a^{\rm V}_{\rm LL} - a^{\rm V}_{\rm LR})/(a^{\rm V}_{\rm LL} + a^{\rm V}_{\rm LR})$ from unity cannot be distinguished from a renormalization of the axial-vector current form factor $g_{\rm A}$. Using Eqs. (9), (16), and (43) we find the experimental constraint deduced from β-decay [see Subsect. 2.3] gives

$$\mathrm{V}^2_{ud} = \frac{K(\hbar c)^6}{\mathrm{G}^2_{\rm F}\mathcal{F}t_0^{\rm expt}} = \widetilde{\mathrm{V}}^2_{ud}\left[|1 + a^{\rm V}_{\rm L+}|^2 + |a^{\rm V}_{\rm RL} + a^{\rm V}_{\rm RR}|^2\right]_{ud} = 0.9513(12). \qquad (45)$$

Similar relations hold for V^2_{us} and V^2_{ub}.

In attempting to constrain the coefficients $a^{\rm V}_{\rm L+}$, $a^{\rm V}_{\rm RL}$, and $a^{\rm V}_{\rm RR}$ by the unitarity requirement of the CKM matrix, one is forced [164] to adopt model dependent simplifying assumptions to circumvent the following difficulties: (*i*) there is no compelling way to predict the effect of new physics, considered phenomenologically in this subsection, on the extraction of the $\{u\text{-}s\}$ and $\{u\text{-}b\}$ coupling strengths from experiment; and (*ii*) there is no compelling prescription on how to apply the unitarity constraint on the quark coupling-coefficients $a^{\rm V}_{\ell q}$ of Eq. (45). As a possible approach we assume that: (*i*) there is no family-mixing in the right-handed quark sector; (*ii*) only the quarks of the first family have right-handed couplings; and (*iii*) any eventual new left-handed coupling $a^{\rm V}_{\rm L+}$ is the same for all families. Using then the three generation constraint

$$\widetilde{\mathrm{V}}^2_{ud} + \widetilde{\mathrm{V}}^2_{us} + \widetilde{\mathrm{V}}^2_{ub} = 1 \qquad (46)$$

and Eqs. (45) and (18) yields

$$2\Re e(a^{\rm V}_{\rm L+}) + |a^{\rm V}_{\rm L+}|^2 + \widetilde{\mathrm{V}}^2_{ud}\,|a^{\rm V}_{\rm RL} + a^{\rm V}_{\rm RR}|^2_{ud} = -0.0001(14)\,. \qquad (47)$$

The limits on $\Re e(a^{\rm V}_{\rm L+})$, obtained using Eq. (47) combined with the analysis of the constraints on $a^{\rm V}_{\rm RL}$ and $a^{\rm V}_{\rm RR}$ discussed below, are given in Table 6. The constraint given in Eq. (47) remains approximately valid in less restrictive scenarios provided that $a^{\rm V}_{\rm L+} \ll 1$ and that new contributions do not modify the value of V^2_{us} by more than about 10^{-3}.

In specific scenarios where strangeness-changing processes receive contributions from new physics, such as the manifestly left-right symmetric models we discuss in Subsect. 4.2, one should extract V_{us} from the pure Fermi K_{e3}–decays alone, as discussed in Ref. [25]. We then obtain from Ref. [22] $\mathrm{V}^2_{us} + \mathrm{V}^2_{ub} = 0.0482(10)$ and $\mathrm{V}^2_{ud} + \mathrm{V}^2_{us} + \mathrm{V}^2_{ub} = 0.9995(16)$. Additional constraints on $a^{\rm V}_{\rm LL}$ and $a^{\rm V}_{\rm LR}$, given in

Table 6: Summary of the constraints on the various coefficients $a_{\ell q}$ in the generalized β-decay interaction, Eq. (35). The "Restricted Fits" include only a single additional interaction [*i.e.*, right-handed currents or scalar or tensor contributions], while the "General Fit" uses a real generalized vector–axial-vector interaction [with no CKM unitarity constraint and $a^{\rm V}_{\rm L+} = 0$] and complex scalar and tensor couplings. We have assumed the Standard Model value $a^{\rm SM}_{\rm LL} \equiv 1$, with the definition $a^{\rm V}_{\rm LL} + a^{\rm V}_{\rm LR} = a^{\rm SM}_{\rm LL} + a^{\rm V}_{\rm L+}$, and we have normalized the form factors so that $g_{\rm S}/g_{\rm V} = g_{\rm T}/g_{\rm A} = 1$. The entry for $g_{\rm A}$ shows the allowed range of variation for this parameter. The last column shows the results obtained by Fetscher and Gerber from the analysis of muon-decay [22]. In general the constraints are not linear; *i.e.*, the $3\,\sigma$ constraint is not equal to three times the $1\,\sigma$ constraint.

		Restricted Fit		General Fit	μ-Decay
Interaction	Coefficient	90% C.L.	$3\,\sigma$	90% C.L.	90% C.L.
Vector and	$a^{\rm SM}_{\rm LL} \equiv 1$				$\lvert g^{\rm V}_{\rm LL}\rvert > 0.96$
Axial-Vector	$\lvert a^{\rm V}_{\rm L+}\rvert$	< 0.0069	0.0097		$\lvert g^{\rm V}_{\rm LR}\rvert < 0.060$
	$\lvert a^{\rm V}_{\rm RL}\rvert$	< 0.044	0.056	0.087	$\lvert g^{\rm V}_{\rm RL}\rvert < 0.110$
	$\lvert a^{\rm V}_{\rm RR}\rvert$	< 0.104	0.122	0.104	$\lvert g^{\rm V}_{\rm RR}\rvert < 0.033$
	$\lvert \Re e\,(a^{\rm V}_{\rm L+})\rvert$	< 0.0058	0.0082		
	$\lvert \Re e\,(a^{\rm V}_{\rm RL})\rvert$	< 0.010	0.047		
	$\lvert \Re e\,(a^{\rm V}_{\rm RR})\rvert$	< 0.104	0.122		
	$\lvert \Im m\,(a^{\rm V}_{\rm L+})\rvert$	< 0.0038	0.0052		
	$\lvert \Im m\,(a^{\rm V}_{\rm RL})\rvert$	< 0.043	0.056		
	$\lvert \Im m\,(a^{\rm V}_{\rm RR})\rvert$	< 0.104	0.122		
Scalar	$\lvert a^{\rm S}_{\rm L+}\rvert$	< 0.072	0.101	0.072	$\lvert g^{\rm S}_{\rm LL}\rvert < 0.55$
					$\lvert g^{\rm S}_{\rm LR}\rvert < 0.125$
	$\lvert a^{\rm S}_{\rm R+}\rvert$	< 0.035	0.052	0.061	$\lvert g^{\rm S}_{\rm RL}\rvert < 0.424$
					$\lvert g^{\rm S}_{\rm RR}\rvert < 0.066$
	$\lvert \Re e\,(a^{\rm S}_{\rm R+})\rvert$	< 0.007	0.012		
Tensor	$\lvert a^{\rm T}_{\rm LR}\rvert$	< 0.025	0.037	0.030	$\lvert g^{\rm T}_{\rm LR}\rvert < 0.036$
	$\lvert a^{\rm T}_{\rm RL}\rvert$	< 0.013	0.020	0.013	$\lvert g^{\rm T}_{\rm RL}\rvert < 0.122$
	$\lvert \Re e\,(a^{\rm T}_{\rm RL})\rvert$	< 0.0035	0.0052		
	$\lvert \Im m\,(a^{\rm T}_{\rm RL})\rvert$	< 0.013	0.020		
Axial-Vector	$g_{\rm A}$	1.261–1.266	1.259–1.268	1.26–1.32	

terms of the various parameters of several models, are discussed in Refs. [165-171].

Turning now to the correlation coefficient measurements, several recent articles [99-101,172] have discussed the evidence for non-maximal parity violation in the vector–axial-vector interaction. The question was raised by three asymmetry measurements: that for the neutron [Ref. [103], see Subsect. 2.4], for ^{19}Ne [Ref. [110], see Subsect. 2.5], and for positrons emitted near the endpoint of the $\mu^+ \rightarrow e^+\nu_e\bar{\nu}_\mu$-decay spectrum [Ref. [52], see Subsect. 2.12]. In the analyses cited above these measurements are respectively $2.6\,\sigma$, $1.9\,\sigma$ and $2.4\,\sigma$ smaller than the values of $A_{\rm n}$ and $A(^{19}{\rm Ne})$ calculated in the Standard Model from the respective $\mathcal{R}^{\mathcal{F}t}$ values [see Subsect. 2.4], and from the Standard Model prediction $P_\mu\delta\xi/\rho = 1$ [22,163,165]. The size of the neutron and ^{19}Ne discrepancies depends on the values of A and $\mathcal{R}^{\mathcal{F}t}$ considered in the various analyses; in this review we use the recent reevaluation of the ^{19}Ne asymmetry parameter, discussed in Subsect. 2.5, which is in excellent agreement with the $\mathcal{R}^{\mathcal{F}t}$ prediction of the Standard Model, leaving us with only two anomalous results. Although it is not surprising, and in fact expected, that two of the 24 measurements in the combined β- and μ-decay data sets compiled in Table 2 lie about $2\,\sigma$ away from the $V-A$ prediction, it is also not surprising that models for new physics which reconcile several discrepant measurements can produce a significant decrease in the χ^2-value of a global fit to the data and generate considerable speculation.

We illustrate the procedure used to constrain the helicity coefficients in Eq. (35) using the generalized vector–axial-vector interaction.† We described in Subsect. 2.4 how $\mathcal{F}t$ and correlation measurements are combined to provide the best estimate of $g_{\rm A}/g_{\rm V}$ in the standard $V-A$ description of the weak interaction. Generalizing the procedure to include $a^{\rm V}_{\rm RL}$ and $a^{\rm V}_{\rm RR}$ is straightforward, and the contributions of these terms to, for example, $\mathcal{R}^{\mathcal{F}t}$, $A_{\rm n}$, and $G_{\rm F}/G_{\rm GT}$ are given in Subsects. 2.4 and

† The justification of our approach is discussed in the "PROBABILITY, STATISTICS, AND MONTE-CARLO" section of the Review of Particle Properties [22], to which we shall refer in the following. For a fit to N data points using n' free parameters the number of degrees of freedom ν is $\nu = N - n'$. We adjust the values of the free parameters to obtain the best fit $\chi^2 = \chi^2_{\rm min}$, and give in the text the "reduced chisquared" defined as $\chi^2_{\rm min}/\nu$. The reader can verify in Fig. 3 [page III.34] of Ref. [22] whether this best fit can be considered as satisfactory. We generally comment on this in the text. The errors on the parameters are determined using the prescription of Eq. 2.20 [page III.37] of Ref. [22]: the one standard deviation [σ] contour is given by the curve with $\chi^2 = \chi^2_{\rm min} + 1$, independent of the value of $\chi^2_{\rm min}/\nu$. The 2 and $3\,\sigma$ contours are at $\chi^2 = \chi^2_{\rm min} + 4$ and $\chi^2 = \chi^2_{\rm min} + 9$, respectively, and the 90% C.L. contour corresponds to $1.64\,\sigma$ [$\chi^2 = \chi^2_{\rm min} + 2.7$]. It should be noted that whenever the number of free parameters is greater than two, the constraint represented in the bi-parametric plot corresponding to the two (relevant) parameters is obtained assigning to the other (irrelevant) parameters their value which gives $\chi^2_{\rm min}$. The 90% C.L. limits of the individual parameters as shown at various places in the text and in Table 6 are obtained from the projection on the axes of the 90% C.L. contour, as illustrated in Fig. 5 [page III.39] of Ref. [22]. The procedure is similar in the hyper-space of n' dimensions if the number of free parameters is greater than two.

2.6. The basic idea is to adjust the value of the unknown g_A/g_V ratio to minimize the χ^2 at each point on a grid of a^V_{RL} *vs.* a^V_{RR} values. Smooth curves are then drawn through points which produce a fixed value of χ^2 above the minimum χ^2_{min}. The curves [or contours] for $\chi^2 = \chi^2_{min} +$ 1, 2.7, 4 and 9 correspond to deviations from the "best fit" value of $1\,\sigma$, 90% C.L., $2\,\sigma$, and $3\,\sigma$, respectively. The results of this procedure for the β-asymmetry parameter and $\mathcal{R}^{\mathcal{F}t}$ measurements both for the neutron and ^{19}Ne [with the product $g_A M^{Ne}_{GT}/g_V$ adjusted to give the best χ^2 at each point] are shown in panels A and B of Fig. 4. The bold contour lines correspond to deviations from $\chi^2_{min} = 0$ of $3\,\sigma$, with the thick dotted curves showing the parameter combinations which give $\chi^2 = \chi^2_{min} = 0$. The allowed region is between the Standard Model $a^V_{RL} = a^V_{RR} = 0$ origins and the contour lines. The bold and dotted curves, shown only for positive values of a^V_{RR}, are symmetric through the origin; *i.e.*, for the simultaneous inversion $a^V_{RL} \to -a^V_{RL}$ and $a^V_{RR} \to -a^V_{RR}$. The Standard Model origins are 2.2 and $0.2\,\sigma$ away from the $\chi^2_{min} = 0$ solutions for the neutron and ^{19}Ne measurements, respectively.

It should be stressed that the constraints shown in Fig. 4 do not use the model dependent treatment of $\widetilde{V}_{ud}$ discussed at the beginning of this subsection.

The remainder of the data in Table 2 are treated in a similar fashion except that there are no adjustable parameters, such as $g_A M_{GT}/g_V$, for the measurements on pure Fermi and Gamow-Teller transitions. The results for various measurements are shown at the $3\,\sigma$ exclusion level in panel C of Fig. 4. They are (*i*) bold curves for the combined G_F/G_{GT} polarization ratio measurements, which are sensitive to the product $a^V_{RL} a^V_{RR}$, (*ii*) the dot-dash and dashed lines which show the limits deduced from the Gamow-Teller electron polarization [G_{GT}] and polarization-asymmetry correlation [$\mathcal{P}^-/\mathcal{P}^+$] measurement, respectively, and (*iii*) the bold straight line which is the limit obtained from a combined analysis of G_{GT}, $G(^3\mathrm{He})$, $\mathcal{P}^-/\mathcal{P}^+$, B_n and $\mathcal{R}^{\mathcal{F}t}_n$ [these measurements constrain a^V_{RR} at small values of a^V_{RL}].

The constraint provided by the full nuclear β-decay data set, shown in panel D of Fig. 4 as 90% C.L. [inner, dashed] and $3\,\sigma$ [outer, bold] contours, is similar to the result of Carnoy *et al.* [100]. The present analysis of the semileptonic data gives the χ^2 minimum, shown by the solid point in Fig. 4 panel D, at $a^V_{RL} = -0.003$ and $a^V_{RR} = 0.077$, with the Standard Model origin excluded at $2.3\,\sigma$. The dashed contour encloses the region which is within $1.64\,\sigma$ of χ^2_{min} [*i.e.*, the boundary is $\chi^2 = \chi^2_{min} + 2.7$], while the bold curve shows the limits for $\chi^2 = \chi^2_{min} + 9$.

We note that for left-right symmetric models with neutrinos of negligible mass the large value of a^V_{RR} at the χ^2 minimum is not compatible with the muon-decay correlation parameters or with limits obtained from nonleptonic K-decays [see Subsect. 4.2]. The extreme values of a^V_{RL} and a^V_{RL} on the 90% C.L. and $3\,\sigma$ contours are reported in Table 6. The limits on a^V_{L+} presented in Table 6 are obtained using CKM unitarity and the TRV D-parameter, with a^V_{RL} and a^V_{RR}, in addition to g_A/g_V and $g_A M^{Ne}_{GT}/g_V$, adjusted to minimize the χ^2 at each step in the search.

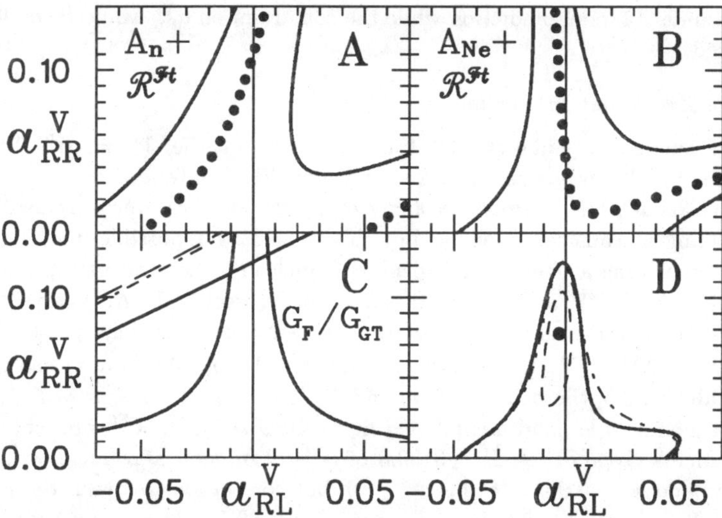

Figure 4: We show here the constraints on the generalized vector–axial-vector interaction coefficients $a_{\rm RL}^{\rm V}$ and $a_{\rm RR}^{\rm V}$ for selected nuclear β-decay measurements. Except for the enclosed 90% C.L. region shown by the dashed line in panel D, the allowed regions are between the Standard Model $a_{\rm RL}^{\rm V} = a_{\rm RR}^{\rm V} = 0$ origins and the bold $3\,\sigma$ contour lines shown in panels A to D. In panels A and B we show the constraints provided by the $\mathcal{R}^{\mathcal{F}t}$ and β-asymmetry parameter measurements for the neutron and ^{19}Ne, respectively. The thick dotted curves show the parameter values which give $\chi^2 = \chi^2_{\rm min} = 0$. Panel C shows the constraint from the G_F/G_{GT} polarization ratio measurements by the bold curve, while the dot-dash and dashed lines give the limits for the Gamow-Teller electron polarization [G_{GT}] and polarization-asymmetry correlation [$\mathcal{P}^-/\mathcal{P}^+$] measurements, respectively. The bold straight line is the $3\,\sigma$ limit obtained from a combined analysis of G_{GT}, $G(^3{\rm He})$, $\mathcal{P}^-/\mathcal{P}^+$, B_n and $\mathcal{R}_{\rm n}^{\mathcal{F}t}$. Finally, in panel D we show the constraint provided by the full β-decay data set as 90% C.L. [dashed] and $3\,\sigma$ [bold] contours. The $3\,\sigma$ constraint with the neutron β-asymmetry measurement excluded from the fit is shown by the dot-dashed line.

It is important to note that while the neutron β-asymmetry parameter [$A_{\rm n}$] and $\mathcal{R}_{\rm n}^{\mathcal{F}t}$ measurements give χ^2 minima $2.2\,\sigma$ from the Standard Model origin, as shown by the thick dotted line in panel A of Fig. 4, the $A_{\rm n}$ measurement plays only a minor role in establishing the $3\,\sigma$ constraints for the generalized vector–axial-vector interaction. This can be seen by comparing the bold [full data set] and dot-dashed [$A_{\rm n}$ excluded] $3\,\sigma$ contours shown in panel D of Fig. 4. In addition, a calculation with $A_{\rm n}$ set equal to the Standard Model value calculated from $\mathcal{R}_{\rm n}^{\mathcal{F}t}$ [see Subsect. 2.4] but with the error taken from Table 2 shows that the constraint on $a_{\rm RR}^{\rm V}$ will be tightened by only 10% if new measurements of $A_{\rm n}$ bring the value closer to the Standard Model prediction while the constraint on $a^{\mathrm{V}}_{\mathrm{RL}}$ would be essentially unchanged.

3.3 The Real Scalar Interaction

We discussed in Subsect. 2.3 the sensitivity of the $\mathrm{J}^\pi = 0^+$ $\mathcal{F}t$ measurements to the presence of a vector-scalar Fierz interference term, $b_{\mathrm{F}} \simeq -2(g_{\mathrm{S}}/g_{\mathrm{V}})\Re e(a^{\mathrm{S}}_{\mathrm{R+}})/a^{\mathrm{V}}_{\mathrm{LL}}$. Previous analyses [21,26] of these measurements have placed stringent constraints on b_{F}, but this is no longer possible due to the necessity of including a phenomenological core-nucleon nuclear-mismatch correction to reconcile $\mathcal{F}t_0^{(\mathrm{J}^\pi=0+)}$ and $\mathcal{F}t_0^{\mathrm{n}}$ as discussed in Subsect. 2.3. An analysis of the $\mathcal{F}t^{(\mathrm{J}^\pi=0+)}$ measurements with non-zero b_{F} and Δ_C^Z contributions yields the ungainly result $\mathcal{F}t_0^{(\mathrm{J}^\pi=0+)} = 3161(22)\,\mathrm{s}$ and $b_{\mathrm{F}} = -0.030(26)$. However, as pointed out in Subsect. 2.3, the neutron, ^{19}Ne and $G_{\mathrm{F}}/G_{\mathrm{GT}}$ experiments provide evidence against a non-zero b_{F}, and a combined fit to the $\mathcal{F}t^{(\mathrm{J}^\pi=0+)}$, $\mathcal{R}^{\mathcal{F}t}$, and correlation measurements gives $\mathcal{F}t_0^{(\mathrm{J}^\pi=0+)} = 3130.8(69)\,\mathrm{s}$ [still in good agreement with CKM unitarity] and $b_{\mathrm{F}} = 0.0047(46)$ for $a^{\mathrm{S}}_{\mathrm{L+}} = 0$, but correlations between the possibly non-zero $\Re e(a^{\mathrm{S}}_{\mathrm{R+}})$ and $|a^{\mathrm{S}}_{\mathrm{L+}}|$ give the more general 90% C.L. and $3\,\sigma$ limits shown in Table 6.

The nuclear β-decay constraints on a real scalar coupling give the limits on $\Re e(a^{\mathrm{S}}_{\mathrm{R+}})$ and $|a^{\mathrm{S}}_{\mathrm{L+}}|$ shown in Table 6, while the bounds on $|a^{\mathrm{S}}_{\mathrm{R+}}|$ come from considering a possible TRV imaginary scalar coupling [see Subsect. 3.5]. Although there is no indication for a scalar interaction, the present constraint on b_{F} from the $\mathcal{F}t^{(\mathrm{J}^\pi=0+)}$ measurements alone is considerably weaker than the limits set a few years ago [21,26]. It is doubtful that this limit can be significantly improved without a refined understanding of the nuclear-mismatch corrections or new, high precision measurements for mirror-nucleus transitions. On the other hand, the constraints on $|a^{\mathrm{S}}_{\mathrm{L+}}|$ and $|a^{\mathrm{S}}_{\mathrm{R+}}|$, as previously discussed in Refs. [76,117], are a significant improvement on the bounds obtained elsewhere [28,171].

The bounds on possible scalar and tensor couplings we obtain from the nuclear β-decay data are generally more stringent than those deduced from an equivalent analysis [22] of muon-decay [see Table 6]. The reasons for this are twofold. First, the pseudoscalar interaction is practically absent in β-decay because $\langle \mathrm{p}|\gamma_5|\mathrm{n}\rangle \to 0$ in the nonrelativistic limit, which means that there are fewer couplings to constrain. Second, the β-decay constraints are insensitive to scalar-tensor interference terms in the correlation parameters, which means that one avoids possible cancellations which allow large values for the individual parameters. Although it is not possible to simply relate in a model independent way the limits on the scalar and tensor interactions from β- and muon-decay, the β-decay constraints are significant for models which propose universality of the lepton- and quark-couplings to new bosons.

3.4 The Real Tensor Interaction

The observed branching ratios for the pion decays $\pi \to \mu\nu_\mu$ and $\pi \to e\nu_e$ provide strong evidence for the V-A interaction, μ-e universality, and absence of a sizeable pseudoscalar interaction in semileptonic weak decays [128,129,173]. While only the axial-vector current contributes to these π_{ℓ_2} decays, the radiative $\pi \to e\nu_e\gamma$ decay receives contributions from both the vector and axial-vector currents and admits the possibility of a tensor current. A recent measurement of the $\pi^- \to e^-\bar{\nu}_e\gamma$ branching ratio [174], over a wider kinematic range than heretofore observed, gave a value more than three standard deviations below the $V-A$ expectation. Poblaguev [175] has suggested that a tensor current which interferes with the dominant QED inner bremsstrahlung correction to $\pi \to e\nu_e$ decay can effectively account for this discrepancy. However, the strength obtained by Poblaguev is several orders of magnitude larger than radiative corrections or SUSY and leptoquark extensions of the Standard Model can accommodate [13,171,176], and if genuine would require a fundamental restructuring of the electroweak interaction.

Two evaluations of the tensor interaction introduced by Poblaguev, using the quark model [175] and PCAC [176] to evaluate the tensor form factor $\langle\pi|\bar{u}\sigma_{\lambda\mu}d|\gamma\rangle = g_{\rm T}^{\pi}\,\langle\pi|\sigma_{\lambda\mu}|\gamma\rangle$, give $a_{\rm RL}^{\rm T} = 0.042(13)$ and $0.014(4)$, respectively. The nuclear β-decay data used here provide quite tight limits on the Gamow-Teller Fierz interference term $b_{\rm GT} = 4\Re e(g_{\rm T}/g_{\rm A})\,a_{\rm RL}^{\rm T}/a_{\rm LL}^{\rm V}$, viz., $b_{\rm GT} = -0.0056(51)$ or $a_{\rm RL}^{\rm T} = -0.0014(13)$, if we assume a typical value [172,177,178] $g_{\rm T}/g_{\rm A} = 1$. Although the calculation of $g_{\rm T}$ for β-decay faces the same factor of three uncertainty as $g_{\rm T}^{\pi}$ [177], the difference between the $a_{\rm RL}^{\rm T}$ limits obtained from $\pi^- \to e^-\bar{\nu}_e\gamma$ and β-decay remains at the $3\,\sigma$ level [179], which makes Poblaguev's tensor interaction scenario unlikely. On the other hand, Poblaguev [175] obtains a much smaller value for $g_{\rm T}$, and he has recently suggested [180] that chiral models give $g_{\rm T} \approx 0$ for semileptonic transitions between nucleons. In this case nuclear β-decay provides no constraints on the quark-level tensor interaction, and at the present time no clear cut resolution of this controversy.

The results from nuclear β-decay find no evidence for a real tensor contribution in semileptonic decays. The limits obtained for $|a_{\rm RL}^{\rm T}|$ and $|a_{\rm RL}^{\rm T}|$ given in Table 6, which assume $g_{\rm T} = g_{\rm A}$, are discussed in the next subsection.

3.5 Time Reversal Violating Interactions

As we discussed in Subsect. 2.10, the Standard Model's [22] CP-violation scenario is second order in the weak interaction, and so vanishingly small in β-decay and muon-capture experiments. Herczeg [13,138,171] and others [14,139,143] have recently reviewed the prospects for observing TRV in nuclear systems and the constraints various models impose on TRV observables in nuclear β-decay. We discuss here the experimental β-decay constraints on the reality of the helicity coefficients, beginning first with the generalized vector–axial-vector interaction. We mentioned in Subsect. 2.7 the TRV limits provided by the modulus of the coefficients, and have also discussed in Subsect. 3.1 the direct constraints on the phase between $a^{\rm V}_{\rm LL}$ and $a^{\rm V}_{\rm LR}$ and $a^{\rm T}_{\rm RL}$, respectively, deduced from the TRV D- and R-parameter correlations.

We shall show first the rather remarkable result that the coupling of right-handed leptons to quarks can maximally violate time reversal with little change in the $3\,\sigma$ bounds found for a real vector–axial-vector interaction; i.e., $a^{\rm V}_{\rm RL}$ and $a^{\rm V}_{\rm RR}$ can independently range between being purely real and purely imaginary.

The limits on time reversal violation in the vector–axial-vector helicity coefficients come from two sources. First, from the expression [Eq. (40)] for the D-parameter correlation we find [with $a^{\rm V}_{\rm LL} a^{\rm V*}_{\rm LR} = a^{\rm SM}_{\rm LL} a^{\rm V*}_{\rm L+}$]

$$\Im m(a^{\rm SM}_{\rm LL} a^{\rm V*}_{\rm L+} + a^{\rm V*}_{\rm RL} a^{\rm V}_{\rm RR}) = 0.00016(56)\,, \tag{48}$$

where the numerical value is obtained from neutron and ^{19}Ne D-parameter measurements [see Table 2]. The second constraint, on the phase between $a^{\rm V}_{\rm RL}$ and $a^{\rm V}_{\rm RR}$, comes from interference terms in the correlation parameters, especially for the ratio of the positron polarization for pure Fermi and Gamow-Teller transitions [Eq. (28)] and the β-asymmetry parameter and $\mathcal{F}t$ ratio [shown for neutron decay in Eqs. (26) and (25)]. Following the procedure outlined in Subsects. 3.1 and 3.2, but now searching on complex coefficients, we obtain the limits

$$|\Im m(a^{\rm SM}_{\rm LL} a^{\rm V*}_{\rm L+})| \quad < 0.0038\,, \quad [90\%\ \text{C.L.}] \tag{49}$$

and

$$|\Im m(a^{\rm V*}_{\rm RL} a^{\rm V}_{\rm RR})| \quad < 0.0038\,. \quad [90\%\ \text{C.L.}] \tag{50}$$

The results given above are quite reliable because they are derived mostly from the D-parameter and the $G_{\rm F}/G_{\rm GT}$ polarization-ratio measurements. Additional results are given in Table 6.

Turning now to the scalar and tensor interactions, it is straightforward to extend the analysis discussed in Subsects. 3.3 and 3.4 to complex couplings. The most significant effect of complex couplings is to reduce the impact of the rather tight constraints which the Fierz interference terms [Eqs. (38) and (39)] impose on $\Re e(a^{\rm S}_{\rm R+})$ and $\Re e(a^{\rm T}_{\rm RL})$, and one sees this effect dramatically by comparing the limits on $\Re e(a^{\rm S}_{\rm R+})$ and $|a^{\rm S}_{\rm R+}|$ as shown in Table 6. In addition, the $\mathcal{F}t$ and correlation measurements are insensitive to the phases of $a^{\rm S}_{\rm L+}$ and $a^{\rm T}_{\rm LR}$, which enter in lowest order only as the modulus. The results for complex couplings give the limits for $|a^{\rm S}_{\rm R+}|$, $|a^{\rm T}_{\rm LR}|$, and $|a^{\rm T}_{\rm RL}|$ shown in Table 6. These results are much more stringent than those obtained a few years ago [28,171]. However, Herczeg [138,171] has investigated scalar and tensor couplings in models with additional charged Higgs doublets and leptoquarks, and the model-dependent constraints he obtains are more stringent than those found here.

3.6 Comparison with Muon-Decay

Fetscher and Gerber provide periodically in the Review of Particle Properties [22] the constraints on the HPF helicity coefficients for muon-decay. It should be noted that their analysis ignores semileptonic contributions which can only be included in a model dependent way: *e.g.*, the muon polarization in pion decay is taken to be $P_\mu = 1$.

We performed a similar general fit of this type to the nuclear β-decay observables, with the restrictions that the $a^{\mathrm{V}}_{\mathrm{RL}}$ and $a^{\mathrm{V}}_{\mathrm{RR}}$ vector-axial-vector coefficients are taken to be real and that $a^{\mathrm{V}}_{\mathrm{L+}} = 0$. We introduced this last restriction because $\Re e(a^{\mathrm{V}}_{\mathrm{L+}})$ is constrained by CKM unitarity alone [see Subsect. 3.2], and so is not constrained in the general fit in a model independent way. Although computer limitations prevented a complete search, the limits shown in Table 6 would not change dramatically using complex $a^{\mathrm{V}}_{\mathrm{RL}}$ and $a^{\mathrm{V}}_{\mathrm{RR}}$. Our results for nuclear β-decay, which are given in the "General Fit" column in Table 6, compare favorably with the muon-decay limits, which are also shown. This was not the case several years ago [28], and as discussed in Sect. 2 the prospects of further improvements in the β-decay limits is quite favorable.

4 Physics Beyond The Standard Model

4.1 Introduction

The minimal Standard Model provides a consistent description of all phenomena encompassed by three generations of quarks and leptons. It is often stressed however, as discussed elsewhere in this volume, that because of numerous free parameters and *ad hoc* assumptions the Standard Model should not be considered the ultimate description of particles and their interactions. At some energy scale, and perhaps at some precision level even at the energies available to us, we expect to find deviations from the Standard Model's predictions, deviations which will provide us with hints for "new physics" and eventually for new and encompassing models hopefully more satisfying than their predecessor.

The best examples of precision tests of the Standard Model in the domain of low- and medium-energy nuclear physics are often considered to be experiments which search for deviations from maximal parity violation or time reversal invariance. However, perhaps the most significant result identified up to now is the stringent constraint of CKM unitarity, Eq. (18), obtained from $\mathcal{F}t$-values for nuclear β-decay and from the corresponding information from K_{e3}– and hyperon-decays [22]. All models for "new physics" must respect this unquestionably fundamental constraint, and we note its importance for several models envisaged in the literature, such as models with exotic fermions [167] and supersymmetric models without automatic baryon- and lepton-number conservation [168,181].

In the following two subsections we shall briefly discuss how the limits on the various HPF four-fermion interaction coefficients obtained in Sect. 3 are related to several models, and also point out how the constraints on the coefficients obtained in different sectors, such as β- and muon-decay, can be related in various scenarios. It is not our intention to be complete, and in particular we do not include or survey the multitude of often more severe but model dependent constraints which arise, for example, from the K_L–K_S mass difference and nonleptonic K-decays. For a more complete discussion of these models and constraints we refer the reader to the companion chapter by Herczeg [13], and to recent articles [70,72,138,139,165-172,182] and monographs [14,181].

4.2 Left-Right Symmetric Models

Maximal parity-violation is one of the unsatisfactory, built in by hand features of the Standard Model. Left-right model extensions of the Standard Model restore the parity-symmetry of nature by extending the electroweak gauge group to $SU(2)_L \otimes SU(2)_R \otimes U(1)$ [see Ref. [165] for a recent, comprehensive discussion]. In the charged-current sector these models introduce additional, predominantly right-handed charged gauge bosons $W_2^\pm$ to complement the predominantly left-handed gauge bosons $W_1^\pm$. The right-handed bosons acquire, via spontaneous symmetry breaking at moderate energies, a mass m_2 which is larger than the $m_1 \simeq 81\,\mathrm{GeV}$ mass of the observed $W^\pm$. By construction the left-handed couplings dominate at low-energies, with parity-symmetry restored at a higher energy scale. The weak-interaction eigenstates W_L, which corresponds to the Standard Model's W gauge boson, and W_R are linear combinations of the physical bosons W_1 and W_2, with

$$W_L = W_1 \cos\zeta - W_2 \sin\zeta \tag{51}$$

and

$$W_R = e^{i\omega} \left(W_1 \sin\zeta + W_2 \cos\zeta \right), \tag{52}$$

where ω is a CP-violating phase. The restoration of parity-symmetry by the $SU(2)_R$ group proposed in left-right models can also be obtained in various symmetry-breaking chains built on larger gauge groups [181].

There is a hierarchy of left-right symmetric models [165,181], and we discuss first the manifest [183] left-right symmetric one [MLRS] in which the constraints deduced from β- and muon-decay are often represented [46,47,52,99,100]. To start, we shall assume that both the left- and right-handed neutrinos are light enough to be emitted without any phase-space restriction, and we shall neglect any CP-violation [165,181] beyond that proposed by the Standard Model. In these models both the gauge-couplings and quark mixing-matrices $V^{L(R)}$ are assumed to be identical in the left- and right-handed sectors; *i.e.*, $g_L = g_R$ and $V^L = V^R$. The parameters of the MLRS model are the mixing angle ζ and a mass-ratio parameter $\delta = (m_1/m_2)^2$.

In terms of these parameters and in lowest order the helicity coefficients of Sect. 3 are $a^{\rm V}_{\rm LL} = 1$ and

$$a^{\rm V}_{\rm LR} = a^{\rm V}_{\rm RL} = g^{\rm V}_{\rm LR} = g^{\rm V}_{\rm RL} = -\zeta \tag{53}$$

and

$$a^{\rm V}_{\rm RR} = g^{\rm V}_{\rm RR} = \delta\,. \tag{54}$$

These expressions equate the nuclear β-decay and muon-decay helicity coefficients, but this equality is broken in generalized left-right models which introduce unequal gauge coupling constants and separate quark mixing-matrices in the left- and right-handed sectors, although $\rm V^L$ remains the usual CKM matrix [165].

We show in panel A of Fig. 5 the $1.65\,\sigma$ [90% C.L., dashed curve] and $3\,\sigma$ [bold curve] constraints obtained on the parameters ζ and δ both from muon-decay [52] and nuclear β-decay. The procedure used to construct the contours is discussed in Subsect. 3.2, and the β-decay constraints are equivalent to those shown in panel D of Fig. 4 because the CKM unitarity constraint [$\zeta = -a^{\rm V}_{\rm LR}$] is not applied in this separate analysis of the muon- and β-decay experiments. However, the muon-decay analysis differs from the approach used by Fetscher and Gerber [22], cited in Subsect. 3.6, because the constraint on the right-handed current contribution to the muon polarization in π- and K-decay [163] is included here. The "V"-shaped diagonal lines show in both panels A and B are the limits of the constraint $\zeta \lesssim \delta$ which derives from rather general properties of the Higgs structure of the interaction [165,184]. Although this constraint has little impact on the β-decay solution, a large fraction of the parameter-space allowed by muon-decay is excluded.

It should be noted that both the β- and muon-decay data sets individually exclude the $V - A$ Standard Model solution $\zeta = \delta = 0$ at greater than $2\,\sigma$, although the parameter-regions corresponding to the small χ^2-values are mutually exclusive at 90% C.L. This discrepancy disappears at the $3\,\sigma$ level, and we show the corresponding combined constraint as the bold curve in panel B of Fig. 5 [note the difference in scales for panels A and B]. The shaded area outside of the filled dot shown near the origin in panel B of Fig. 5 is the region excluded by the neutrino luminosity of supernova SN1987a [see Subsect. 2.11], and the filled areas near the origin and for $\delta \simeq 0.025$ and $\zeta \simeq 0$ show the allowed regions for MLRS models with the CKM unitarity constraint included. We note that the unitarity constraint can be applied in a rigorous way in MLRS models [165].

If the difference between the muon- and β-decay constraints is confirmed by further experimental scrutiny, the most "economic" [165] explanation considered is the existence of a right-handed muon-neutrino too heavy to influence significantly the asymmetry result [52] which dominates the muon-decay constraint shown in panel A of Fig. 5. As noted in Ref. [52], a right-handed muon neutrino with mass $m_{\bar{\nu}_\mu} \gtrsim 7.5\,{\rm MeV/c^2}$ is sufficient to remove the apparent discrepancy. A potential

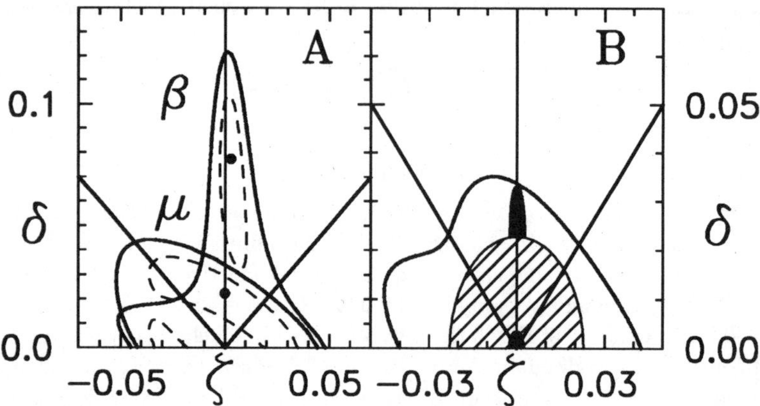

Figure 5: Panel A shows the 90% C.L. [dashed] and $3\,\sigma$ [bold] limits on the MLRS model parameters δ and ζ obtained from the nuclear β-decay and muon-decay measurements, respectively, ignoring the CKM unitarity constraint on ζ. The solid dots indicate the best fit values for the β- and muon-decay results, respectively. The disagreement between the constraints deduced from these data sets is discussed in Subsect. 4.2. The bold curve in panel B is the combined β- and muon-decay $3\,\sigma$ constraint on δ and ζ [note the difference in scales for panels A and B]. The shaded area outside of the filled dot shown near the origin in panel B is the region excluded by the neutrino luminosity of supernova SN1987a [see Subsect. 2.11], and the filled areas near the origin and for $\delta \simeq 0.025$ and $\zeta \simeq 0$ show the allowed regions for MLRS models with the CKM unitarity constraint included. The "V"-shaped diagonal lines shown in both panels A and B are the limits of the constraint $\zeta \lesssim \delta$ which derives from the Higgs structure of the interaction. See Subsect. 4.2 for a detailed discussion.

difficulty is that this "sterilization" of the right-handed neutrinos in the muon sector may destroy the agreement with CKM unitarity given in Eq. (18). However, a straightforward calculation with $\zeta = 0$, $\delta = 0.08$, and $m_{\bar{\nu}_\mu} = 7.5\,\mathrm{MeV/c^2}$ gives $V_{ud}^2 + V_{us}^2 + V_{ub}^2 = 0.9980(14)$, which still provides good agreement. The constraint [104] provided by the $(\pi^+ \to e^+\nu_e)/(\pi^+ \to \mu^+\nu_\mu)$ branching ratio measurement [128,129] is weaker that the unitarity constraint discussed above.

If the right-handed muon-neutrino is heavy, and there is considerable theoretical support [181] for this scenario, then the β-decay constraints shown in panel A and the supernova constraint shown in panel B of Fig. 5 constrain δ to the range $0.02 \lesssim \delta \lesssim 0.1$ with $\zeta \approx 0$ or δ and $\zeta \lesssim 0.003$ at 90% C.L. Nuclear β-decay experiments presently underway [115,116] hope to eventually explore the entire $0.02 \lesssim \delta \lesssim 0.1$ range. If these experiments fail to confirm a right-handed gauge boson in this mass region, then the combination of the β-decay and supernova results will exclude a possible right-handed gauge boson with mass less than several TeV.

In the MLRS model, in addition to the results shown in panel B of Fig. 5, one can obtain more severe limits on the parameters from other sources. For example, inclusion K_L–K_S mass difference constraint [165,169,170] eliminates the filled region of possible solutions near $\delta \simeq 0.025$. We refer the interested reader to the paper by Langacker and Sankar [165], which provides a full discussion in the context of general class of $SU(2)_L \otimes SU(2)_R \otimes U(1)$ models. These generalized left-right models typically relax the constraints on the MLRS model parameters from, for example, the K_L–K_S mass difference.

In these generalized left-right symmetric models one allows for unequal gauge-coupling constants $[g_R/g_L = r]$ and quark mixing-matrices $[V^R_{ud}/V^L_{ud} = R]$. One obtains the following results for the semileptonic sector [see Ref. [170]]; $a^V_{RR} = r^2 R\delta$, $a^V_{LR} = rR\zeta$, and $a^V_{RL} = R\zeta$. Similar results are obtained in the muon-sector [163]. The number of free parameters clearly precludes any precise constraint or characterization of the right-handed gauge boson itself. As noted in Refs. [163,165,169,170],there is considerable freedom in these models and the existence of even rather light right-handed gauge bosons cannot be excluded. As far as the future is concerned, a new measurement of the Michel ρ-parameter underway at LAMPF is expected to significantly improve the constraint on ζ extracted from muon-decay [185].

4.3 Models with Leptoquarks

Leptoquarks are fractionally charged bosons which mediate lepton–quark transitions, and they appear in several extensions of the Standard Model [166,181]. As they do not contribute in lowest-order to purely leptonic and purely hadronic transitions, leptoquarks avoid the rather stringent constraints that these processes impose on left-right symmetric models. However, strong limits prevail for leptoquark interactions which cause proton decay or mediate lepton family-number violating transitions, and we refer the reader to Refs. [13,104,138,166,171,182] for a more complete discussion of leptoquark interactions relevant to the present chapter.

Leptoquarks can contribute of course to semileptonic transitions, but at the present time there are no compelling models for leptoquark interactions which would influence the low-energy helicity structure of these processes. For the most general leptoquark interaction [13,138,171], where there are different leptoquarks for each generation, nuclear β-decay provides only the "General Fit" limits on the first generation four-fermion interaction helicity coefficients $a_{\ell q}$ given in Table 6 [see Refs. [104,171] and the chapter in this volume by Herczeg]. In particular, we note that the CKM unitarity constraint [Eq. (47)] on a^V_{L+} does not hold in this scenario. Direct searches for scalar leptoquarks have produced mass limits of

$M_{\mathcal{L}} \gtrsim 45\,\mathrm{GeV/c^2}$ for these particles [22].

Herczeg [13,104,138,171] has investigated the constraints on a general SU(3)$\otimes$ SU(2)$\otimes$U(1) invariant leptoquark interaction for spin-one [vector] and spin-zero [scalar] leptoquark exchange which may have an impact on low-energy phenomenology, with a leptoquark mass $M_{\mathcal{L}}$ perhaps as low as $100\,\mathrm{GeV/c^2}$ [182]. This globally baryon and lepton-number conserving interaction extends the Lagrangian introduced by Buchmüller *et al.* [182] to include right-handed neutrinos, and the constraints on several leptoquark scenarios provided by pseudoscalar-meson decays and nuclear β-decay are reviewed in Refs. [13,104,138,166,171].

For simplicity we consider only vector leptoquarks, termed $X^{(Q)}$-type in Ref. [171], where Q = 2/3, -1/3 is the electric charge of the leptoquark boson, and we refer the reader to Refs. [13,138,166,171] for a discussion of the results for scalar leptoquarks. The X-type leptoquark quark-level interaction for transitions such as neutron decay is [13,171]

$$\mathcal{H}_{\mathrm{X}} = \sum_{i,j=\mathrm{V,A}}^{\mathrm{Q}=2/3} f_{ij}\, \bar{e}\Gamma_i d\, \bar{u}\Gamma_j \nu_e + \sum_{i,j=\mathrm{V,A}}^{\mathrm{Q}=-1/3} h_{ij}\, \bar{e}\Gamma_i u^c\, \bar{d^c}\Gamma_j \nu_e\,, \tag{55}$$

where $\Gamma_{\mathrm{V}} = \gamma_\lambda$, $\Gamma_{\mathrm{A}} = \gamma_\lambda\gamma_5$, and $\{u^c, d^c\}$ are the charge-conjugates of $\{u, d\}$. A Fierz transformation relates the vector and scalar helicity coefficients $a_{\ell q}$ given in Eq. (35) to the f_{ij} and h_{ij} leptoquark coefficients, and explicit formulas are given in Refs. [13,171]. The results we seek are more transparent if we write the leptoquark interaction in HPF notation [161], and using the helicity subscripts on $f_{\ell\nu}^{\mathrm{V}}$ to denote the charged-lepton and neutrino helicities one has

$$\begin{aligned}\mathcal{H}_{\mathrm{X}(2/3)} \;=\;\; & f_{\mathrm{LL}}^{\mathrm{V}} \bar{e}\,\gamma_\lambda(1-\gamma_5)\,d\,\bar{u}\,\gamma^\lambda(1-\gamma_5)\,\nu_e \\ & + f_{\mathrm{LR}}^{\mathrm{V}} \bar{e}\,\gamma_\lambda(1-\gamma_5)\,d\,\bar{u}\,\gamma^\lambda(1+\gamma_5)\,\nu_e \\ & + f_{\mathrm{RL}}^{\mathrm{V}} \bar{e}\,\gamma_\lambda(1+\gamma_5)\,d\,\bar{u}\,\gamma^\lambda(1-\gamma_5)\,\nu_e \\ & + f_{\mathrm{RR}}^{\mathrm{V}} \bar{e}\,\gamma_\lambda(1+\gamma_5)\,d\,\bar{u}\,\gamma^\lambda(1+\gamma_5)\,\nu_e\,. \end{aligned} \tag{56}$$

A similar expression can be written for the $\mathcal{H}_{\mathrm{X}(-1/3)}$ leptoquark coefficients $h_{\ell\nu}^{\mathrm{V}}$.

If we use the expressions given in Ref. [171], then it is straightforward to show that $f_{\mathrm{LR}}^{\mathrm{V}}$ and $f_{\mathrm{RL}}^{\mathrm{V}}$ are related to the scalar $a_{\ell q}^{\mathrm{S}}$ coefficients given in Eq. (35); *e.g.*, $f_{\mathrm{RL}}^{\mathrm{V}} = -2\,(\mathrm{G_F V}_{ud}/\sqrt{2})\,g_{\mathrm{S}} a_{\mathrm{R}+}^{\mathrm{S}}$. Pseudoscalar-meson decays provide stringent constraints on $f_{\mathrm{LR}}^{\mathrm{V}}$ and $f_{\mathrm{RL}}^{\mathrm{V}}$ [13,104,166,171], and if we ignore possible cancellations, such as the scenario where the leptoquark couplings are inversely proportional to the lepton masses [13,104,171], then the $(\pi^+ \to e^+\nu_e)/(\pi^+ \to \mu^+\nu_\mu)$ branching ratio measurement [128,129] gives the 90% C.L. constraints [13]

$$|f_{\mathrm{LR}}^{\mathrm{V}}|,\ |f_{\mathrm{RL}}^{\mathrm{V}}|,\ |h_{\mathrm{LR}}^{\mathrm{V}}|,\ \text{and}\ |h_{\mathrm{RL}}^{\mathrm{V}}| \;\lesssim\; 1.2\times 10^{-3}\, g_{\mathrm{S}}\, \frac{\mathrm{G_F V}_{ud}}{\sqrt{2}}\,. \tag{57}$$

In general, the pseudoscalar-meson constraints provide limits on the scalar helicity coefficients $a_{\ell q}^{\mathrm{S}}$ which are more than an order of magnitude smaller than the results given in Table 6 [similar limits apply to the tensor coefficients $a^{\rm T}_{\ell q}$ generated by the scalar leptoquark interaction]. As a consequence, the contribution to nuclear β-decay observables from scalar and tensor couplings which arise from leptoquark interactions can be ignored at the present level of precision.

We consider next the contributions to the vector helicity coefficients $a^{\rm V}_{\ell q}$ in Eq. (35). Following the development in Ref. [171] we obtain [with $g_{\rm V} = 1$]

$$\frac{{\rm G_F V}_{ud}}{\sqrt{2}}\, a^{\rm V}_{\rm LL} = \frac{g^2{\rm V}_{ud}}{8{\rm M}^2_{\rm W}}\, a^{\rm V}_{\rm LL} \;=\; f^{\rm V}_{\rm LL} + \frac{g^2{\rm V}_{ud}}{8{\rm M}^2_{\rm W}} = \frac{g^2_1}{8{\rm M}^2_1} + \frac{g^2{\rm V}_{ud}}{8{\rm M}^2_{\rm W}}\,, \tag{58}$$

$$\frac{{\rm G_F V}_{ud}}{\sqrt{2}}\, a^{\rm V}_{\rm LR} = \frac{g^2{\rm V}_{ud}}{8{\rm M}^2_{\rm W}}\, a^{\rm V}_{\rm LR} \;=\; -h^{\rm V}_{\rm LL} = -\frac{g^2_2}{8{\rm M}^2_2}\,, \tag{59}$$

$$\frac{{\rm G_F V}_{ud}}{\sqrt{2}}\, a^{\rm V}_{\rm RL} = \frac{g^2{\rm V}_{ud}}{8{\rm M}^2_{\rm W}}\, a^{\rm V}_{\rm RL} \;=\; -h^{\rm V}_{\rm RR} = -\frac{g^2_3}{8{\rm M}^2_3}\,, \tag{60}$$

$$\frac{{\rm G_F V}_{ud}}{\sqrt{2}}\, a^{\rm V}_{\rm RR} = \frac{g^2{\rm V}_{ud}}{8{\rm M}^2_{\rm W}}\, a^{\rm V}_{\rm RR} \;=\; f^{\rm V}_{\rm RR} = \frac{g^2_4}{8{\rm M}^2_4}\,, \tag{61}$$

where g and ${\rm M_W}$ are the gauge-coupling and mass of the Standard Model ${\rm W_L}$-boson, and where g_i and ${\rm M}_i$ represent effective leptoquark coupling constants and masses [see Refs. [138,166,182] and especially the chapter in this volume by Herczeg].

Herczeg has noted [13,104] that the $(\pi^+ \to e^+\nu_e)/(\pi^+ \to \mu^+\nu_\mu)$ branching ratio constraint on the $a^{\rm V}_{\ell q}$ helicity coefficients disappears if the coefficients are the same for the first two lepton generations; *i.e.*, $(a^{\rm V}_{\ell q})^{[e]} = (a^{\rm V}_{\ell q})^{[\mu]}$. This condition is realized in the MLRS models discussed in the previous subsection; what differentiates the leptoquark interaction from left-right-symmetric models, as far as our analysis is concerned, is that the leptoquark contribution to the "muon-decay" data in Table 2 is restricted to the muon polarization P_μ in $\pi \to \mu\nu_\mu$ decay. The constraints on $a^{\rm V}_{\rm RL}$ and $a^{\rm V}_{\rm RR}$, which we assume here to be real, from a fit to the β-decay and muon-polarization data are shown by the 90% C.L.[dashed] and $3\,\sigma$ [bold] contours in the $a^{\rm V}_{\rm RL}$ *vs* $a^{\rm V}_{\rm RR}$ plane in panel A of Fig. 6 and in the $(a^{\rm V}_{\rm RL} + a^{\rm V}_{\rm RR})$ *vs* $(a^{\rm V}_{\rm RL} - a^{\rm V}_{\rm RR})$ plane in panel B of Fig. 6. The curves are symmetric through the origin; *i.e.*, for the simultaneous inversion $a^{\rm V}_{\rm RL} \to -a^{\rm V}_{\rm RL}$ and $a^{\rm V}_{\rm RR} \to -a^{\rm V}_{\rm RR}$. The χ^2 minimum is $2.8\,\sigma$ away from the Standard Model origin, and the maximum values of the helicity coefficients at 90% C.L. are

$$|\,a^{\rm V}_{\rm RL}\,| \;<\; 0.043\,, \tag{62}$$

$$|\,a^{\rm V}_{\rm RR}\,| \;<\; 0.042\,, \tag{63}$$

$$|\,a^{\rm V}_{\rm RL} + a^{\rm V}_{\rm RR}\,| \;<\; 0.051\,, \tag{64}$$

$$|\,a^{\rm V}_{\rm RL} - a^{\rm V}_{\rm RR}\,| \;<\; 0.042\,, \tag{65}$$

$$|\,a^{\rm V}_{\rm L+}| \;<\; 0.0019\,. \tag{66}$$

The stringent constraint given in Eq. (66) results from CKM unitarity [see Ref. [166] and Subsect. 3.2], with both $f^{\rm V}_{\rm LL}$ and $h^{\rm V}_{\rm LL}$ contributing to $a^{\rm V}_{\rm L+}$ [see Eqs. (58),

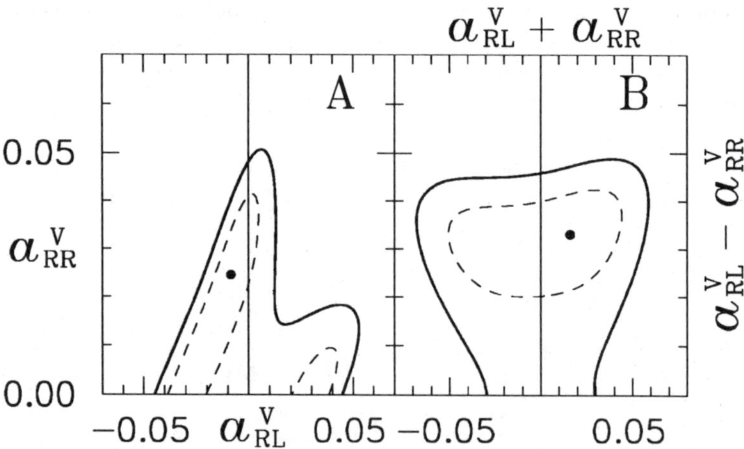

Figure 6: We show in panel A the 90% C.L. [dashed] and 3σ [bold] constraints on the generalized vector–axial-vector interaction helicity coefficients a^V_{RL} and a^V_{RR} for a leptoquark scenario where the leptoquark interaction is the same for the first two generations. These constraints are shown in the $(a^V_{RL} + a^V_{RR})$ *vs* $(a^V_{RL} - a^V_{RR})$ plane in panel B, where the result suggests a non-zero value for the helicity amplitude combination $|a^V_{RL} - a^V_{RR}|$. See the text for a full discussion.

(59), and (43)]. If we assume that only one of f^V_{LL} and h^V_{LL} is non-zero, then from Eqs. (58), (59), and (66) we find at 90% C.L.

$$|f^V_{LL}| \text{ and } |h^V_{LL}| \lesssim 1.9 \times 10^{-3} \frac{G_F V_{ud}}{\sqrt{2}} \tag{67}$$

and

$$M_i \gtrsim 1.0\,\mathrm{TeV}/c^2 \, \frac{g_i}{g} \quad [i = 1, 2]\,. \tag{68}$$

We quote the constraints deduced from the unitarity of the CKM matrix, as given in Eqs. (66), (67), and (68), for completeness only, and stress that the simplifying assumptions made in the derivation of Eq. (47) are rather artificial in a leptoquark scenario. Note however that the constraints given in Eqs. (62) to (65), and also shown in Fig. 6, are independent of the unitarity constraint and so stand on much firmer ground.

The results shown in Fig. 6 suggest the existence of at least one leptoquark interaction with purely right-handed coupling [see Eqs. (60) and (61)]; i.e., at least one of f^V_{RR} and h^V_{RR} is non-zero at 90% C.L. Considering only the upper bounds shown in panel A of Fig. 6, and using the 90% C.L. limits from Eqs. (62) and (63) in Eqs. (60) and (61), one finds

$$\mathrm{M}_i \gtrsim 390\,\mathrm{GeV/c^2}\,\frac{g_i}{g} \quad [i = 3,4]\,. \tag{69}$$

We have not found constraints on $f^{\mathrm{V}}_{\mathrm{RR}}$- and $h^{\mathrm{V}}_{\mathrm{RR}}$-type leptoquark interactions elsewhere in the literature.

As mentioned already, the result illustrated in panel B of Fig. 6 suggests a non-zero value only for the helicity coefficient combination $a^{\mathrm{V}}_{\mathrm{RL}} - a^{\mathrm{V}}_{\mathrm{RR}}$ [the χ^2 minimum is $0.9\,\sigma$ from $a^{\mathrm{V}}_{\mathrm{RL}} + a^{\mathrm{V}}_{\mathrm{RR}} = 0$]. This solution proposes the existence of leptoquarks with both Q = 2/3 and -1/3, and moreover requires a cancellation of leptoquark coefficients, with $f^{\mathrm{V}}_{\mathrm{RR}} \approx h^{\mathrm{V}}_{\mathrm{RR}}$.

Although the χ^2 minimum near $|\,a^{\mathrm{V}}_{\mathrm{RL}} - a^{\mathrm{V}}_{\mathrm{RR}}\,| \simeq 0.032$ in panel B of Fig. 6 is dominated by the muon polarization P_μ measured in the $[\pi \to \mu\nu_\mu]$ $P_\mu\xi\delta/\rho$ experiment [52] [and is therefore subject to the caveat discussed in Subsect. 2.12], the nuclear β-decay measurements alone give also the χ^2 minimum at $|\,a^{\mathrm{V}}_{\mathrm{RL}} - a^{\mathrm{V}}_{\mathrm{RR}}\,| \simeq 0.032$, and so preserve the indication of an anomalous coupling. The β-decay data alone provide the 90% C.L. limit $|\,a^{\mathrm{V}}_{\mathrm{RL}} - a^{\mathrm{V}}_{\mathrm{RR}}\,| < 0.047$ for $|\,a^{\mathrm{V}}_{\mathrm{RL}} + a^{\mathrm{V}}_{\mathrm{RR}}\,| \simeq 0$. In the scenario where $|a^{\mathrm{V}}_{\mathrm{RL}} + a^{\mathrm{V}}_{\mathrm{RR}}| = 0$, and here considering only the upper bounds shown in panel B of Fig. 6, the 90% C.L. limit of Eq. (69) gives $\mathrm{M}_i \gtrsim 560\,\mathrm{GeV/c^2}\, g_i/g$.

Finally, we note that in this scenario the constraint on $|a^{\mathrm{V}}_{\mathrm{RL}} + a^{\mathrm{V}}_{\mathrm{RR}}|$ comes entirely from the nuclear β-decay measurements.

5 Conclusions

This chapter on search for new physics beyond the Standard Model in low- and medium-energy semileptonic weak interactions concludes unfortunately as do the other chapters of this volume: the Standard Model is discouragingly successful, and wherever we look, high in energy or low in error-limits, we do not perceive any signposts of new physics. Wishful thinking tempts us of course, and from time-to-time we read reports of encouraging experimental anomalies. These reports raise always a lot of excitement, and the community reacts by improving the corresponding experimental techniques and precision; up to now, however, these strongly improved experiments resulted in the conclusion that the reported anomalies were experimental artifacts and that the Standard Model stands even stronger than before.

One example we discussed is the reported deviation from unitarity of the three-generation Cabibbo-Kobayashi-Maskawa quark mixing-matrix as deduced from the study of the superallowed pure Fermi transitions. Our conclusion was that this apparent deviation may be ascribed to a "theoretical" artifact in the evaluation of nuclear corrections. The anomaly in the $V - A$ helicity structure suggested by asymmetry parameter measurements may also result from "experimental" artifacts. As a matter of fact, we are in the middle of a feverish activity to improve the experiments on the β-decays of both the neutron and of selected mirror-nucleus transitions. When these experiments bear fruit, they will teach us whether our hopes to see the first indications of some new physics will again be deceived, as they have been so often in the past.

Beyond these exciting issues of high topical interest, there is a steady and important though less glamorous activity going on, as we discussed in detail. On the side of the experimentalists, we implement new ideas and technical developments which may lead to significant improvements in the precision of various symmetry-tests. This effort to put more and more digits behind a decimal point or to measure new and more sensitive correlations is the very essence of experimental physics, and we discussed many promises to do so in low- and intermediate-energy physics of weak-interactions. We hinted also to the important activity going on in the community of theoretical physicists working out various scenarios of new physics beyond the standard model and how these scenarios can be confirmed, constrained or excluded by existing or possible experiments at high-, intermediate-, or low-energy. As we discussed, these important theoretical analysis stress the complementarity of approaches offered by the various energy-domains.

We hope to have shared with the reader the enthusiasm of the community of physicists working in the field of low- and intermediate energy weak interactions. We hope to have shared also our conviction of the importance of this activity and its complementarity to high-energy physics, and our conviction, finally, that the numerous and often ingenious efforts going on in this healthy and active field of physics are fully justified.

Many colleagues did not spare of their time and efforts to update us on their research. We wish to thank, in particular, R.C. Allen, M. Allet, V.D. Barger, F.C. Barker, K. Bodek, F. Boehm, L. De Braeckeleer, B.A. Brown, D.A. Bryman, J. Byrne, S. Ciechanowicz, M.D. Cooper, D. Dubbers, J. Egger, W. Fetscher, S.J. Freedman, R. Gandhi, Yu.V. Gaponov, G.L. Greene, J.C. Hardy, O. Hausser, W.C. Haxton, E.M. Henley, B.R. Holstein, W. Jaus, M. Kadono, I.B. Khriplovich, A. Morales, M. Morita, V.I. Morozov, O. Naviliat-Cuncic, J.M. Pendlebury, C.L. Petitjean, D. Pocanic, G. Rasche, K. Riisager, R.G.H. Robertson, D. Schardt, A.P. Serebrov, N. Severijns, P.A. Souder, J. Sromicki, I.S. Towner, R.E. Tribble, P. Vogel, T.G. Walker, and D.H. Wilkinson who contributed important material we included in our review. Particular thanks are due to E.G. Adelberger, A.S. Carnoy, L. Grenacs, P. Herczeg, R. Prieels, J.E. Schewe, and P.A. Voytas: we profited very much from enlightening discussions with them.

6 Appendix

We present below the relations between the G-coefficients in Eq. (34) and the helicity coefficients $a_{\ell q}$ of Eq. (35). The form factors [g] are defined in Eqs. (2), (3), (5), and (6).

$$\begin{aligned}
\mathrm{G_V} &= g_\mathrm{V}\left(a^\mathrm{V}_\mathrm{LL} + a^\mathrm{V}_\mathrm{LR} + a^\mathrm{V}_\mathrm{RL} + a^\mathrm{V}_\mathrm{RR}\right)\\
-\mathrm{G'_V} &= g_\mathrm{V}\left(a^\mathrm{V}_\mathrm{LL} + a^\mathrm{V}_\mathrm{LR} - a^\mathrm{V}_\mathrm{RL} - a^\mathrm{V}_\mathrm{RR}\right)\\
\mathrm{G_A} &= g_\mathrm{A}\left(a^\mathrm{V}_\mathrm{LL} - a^\mathrm{V}_\mathrm{LR} - a^\mathrm{V}_\mathrm{RL} + a^\mathrm{V}_\mathrm{RR}\right)\\
-\mathrm{G'_A} &= g_\mathrm{A}\left(a^\mathrm{V}_\mathrm{LL} - a^\mathrm{V}_\mathrm{LR} + a^\mathrm{V}_\mathrm{RL} - a^\mathrm{V}_\mathrm{RR}\right)\\
\mathrm{G_S} &= g_\mathrm{S}\left(a^\mathrm{S}_\mathrm{L+} + a^\mathrm{S}_\mathrm{R+}\right)\\
\mathrm{G'_S} &= g_\mathrm{S}\left(a^\mathrm{S}_\mathrm{L+} - a^\mathrm{S}_\mathrm{R+}\right)\\
\mathrm{G_T} &= 2\,g_\mathrm{T}\left(a^\mathrm{T}_\mathrm{LR} + a^\mathrm{T}_\mathrm{RL}\right)\\
\mathrm{G'_T} &= 2\,g_\mathrm{T}\left(a^\mathrm{T}_\mathrm{LR} - a^\mathrm{T}_\mathrm{RL}\right)
\end{aligned}$$

The phase convention we use in Eq. (34) is the one suggested by Herczeg [171] which uses the metric and γ-matrices of Bjorken and Drell [187]. This leads to $\mathrm{G'_V}$, $\mathrm{G_A}$, $\mathrm{G'_S}$, $\mathrm{G'_P}$, and $\mathrm{G'_T}$ having the opposite sign from the corresponding C-coefficients used by JTW [20]. The correlation coefficients in Eq. (4) are given below, with the upper [lower] sign corresponding to a β^-- [β^+]-transition. For a transition from an initial state with nuclear spin J to a state with nuclear spin J′, the symbol $\lambda_{\mathrm{J'J}}$ is

$$\begin{aligned}
\lambda_{\mathrm{J'J}} &= 1 \quad [J' = J-1]\,,\\
\lambda_{\mathrm{J'J}} &= \frac{1}{J+1} \quad [J' = J]\,,
\end{aligned}$$

and

$$\lambda_{\mathrm{J'J}} = -\frac{J}{J+1} \quad [J' = J+1]\,.$$

We do not include the first order Coulomb final-state interaction corrections or the scalar-tensor interference terms in these expressions.

$$\begin{aligned}
\xi &= \mathrm{M_F^2}\Big(|\mathrm{G_V}|^2 + |\mathrm{G'_V}|^2 + |\mathrm{G_S}|^2 + |\mathrm{G'_S}|^2\Big)\\
&\quad + \mathrm{M_{GT}^2}\Big(|\mathrm{G_A}|^2 + |\mathrm{G'_A}|^2 + |\mathrm{G_T}|^2 + |\mathrm{G'_T}|^2\Big)\\
a\,\xi &= \mathrm{M_F^2}\Big(|\mathrm{G_V}|^2 + |\mathrm{G'_V}|^2 - |\mathrm{G_S}|^2 - |\mathrm{G'_S}|^2\Big)\\
&\quad - \frac{\mathrm{M_{GT}^2}}{3}\Big(|\mathrm{G_A}|^2 + |\mathrm{G'_A}|^2 - |\mathrm{G_T}|^2 - |\mathrm{G'_T}|^2\Big)\\
b\,\xi &= \pm 2\,\Re e\Big[\mathrm{M_F^2}\Big(\mathrm{G_V G_S^*} + \mathrm{G'_V G'^*_S}\Big) - \mathrm{M_{GT}^2}\Big(\mathrm{G_A G_T^*} + \mathrm{G'_A G'^*_T}\Big)\Big]
\end{aligned}$$

$$
\begin{aligned}
A\xi &= 2\,\Re e\Big[\pm\lambda_{\mathrm{J'J}}\mathrm{M}_{\mathrm{GT}}^2\big(\mathrm{G_A G_A'^*} - \mathrm{G_T G_T'^*}\big) \\
&\quad -\delta_{\mathrm{J'J}}\mathrm{M_F M_{GT}}\sqrt{\frac{\mathrm{J}}{\mathrm{J}+1}}\,\big(\mathrm{G_V G_A'^*} + \mathrm{G_V' G_A^*}\big)\Big] \\
B\xi &= 2\,\Re e\Big\{\mp\lambda_{\mathrm{J'J}}\mathrm{M}_{\mathrm{GT}}^2\Big[\big(\mathrm{G_A G_A'^*} + \mathrm{G_T G_T'^*}\big) \mp \frac{\Gamma m}{E}\big(\mathrm{G_A G_T'^*} + \mathrm{G_A' G_T^*}\big)\Big] \\
&\quad -\delta_{\mathrm{J'J}}\mathrm{M_F M_{GT}}\sqrt{\frac{\mathrm{J}}{\mathrm{J}+1}}\,\Big[\big(\mathrm{G_V G_A'^*} + \mathrm{G_V' G_A^*}\big) \\
&\quad \mp\frac{\Gamma m}{E}\big(\mathrm{G_V G_T'^*} + \mathrm{G_V' G_T^*} - \mathrm{G_A G_S'^*} - \mathrm{G_A' G_S^*}\big)\Big]\Big\} \\
D\xi &= 2\,\Im m\Big[\delta_{\mathrm{J'J}}\mathrm{M_F M_{GT}}\sqrt{\frac{\mathrm{J}}{\mathrm{J}+1}}\,\big(\mathrm{G_V G_A^*} + \mathrm{G_V' G_A'^*}\big)\Big] \\
G\xi &= \pm 2\,\Re e\Big[\mathrm{M_F^2}\big(\mathrm{G_V G_V'^*} - \mathrm{G_S G_S'^*}\big) + \mathrm{M_{GT}^2}\big(\mathrm{G_A G_A'^*} - \mathrm{G_T G_T'^*}\big)\Big] \\
Q'\xi &= \Re e\Big\{\lambda_{\mathrm{J'J}}\mathrm{M}_{\mathrm{GT}}^2\Big[\big(|\mathrm{G_A}|^2 + |\mathrm{G_A'}|^2 + |\mathrm{G_T}|^2 + |\mathrm{G_T'}|^2\big) \mp \frac{\Gamma m}{E}\big(\mathrm{G_A G_T^*} + \mathrm{G_A' G_T'^*}\big)\Big] \\
&\quad \mp 2\delta_{\mathrm{J'J}}\mathrm{M_F M_{GT}}\sqrt{\frac{\mathrm{J}}{\mathrm{J}+1}}\,\Big[\big(\mathrm{G_V G_A^*} + \mathrm{G_V' G_A'^*}\big) \\
&\quad \mp\frac{\Gamma m}{E}\big(\mathrm{G_V G_T^*} + \mathrm{G_V' G_T'^*} - \mathrm{G_A G_S^*} - \mathrm{G_A' G_S'^*}\big)\Big]\Big\} \\
R\xi &= 2\,\Im m\Big[\mp\lambda_{\mathrm{J'J}}\mathrm{M}_{\mathrm{GT}}^2\big(\mathrm{G_A G_T'^*} + \mathrm{G_A' G_T^*}\big) \\
&\quad +\delta_{\mathrm{J'J}}\mathrm{M_F M_{GT}}\sqrt{\frac{\mathrm{J}}{\mathrm{J}+1}}\big(\mathrm{G_V G_T'^*} + \mathrm{G_V' G_T^*} - \mathrm{G_A G_S'^*} - \mathrm{G_A' G_S^*}\big)\Big]
\end{aligned}
$$

References

[1] F. Boehm and P. Vogel, *Physics of Massive Neutrinos* (Cambridge University Press, 1987).

[2] *Neutrinos* (ed. H.V. Klapdor, Springer-Verlag, 1988).

[3] B. Kayser, F. Gibrat-Debru, and F. Perrier, *The Physics of Massive Neutrinos* (World Scientific, 1989).

[4] J. Wilkerson, in *Proc.* 15th *Int. Conf. on Neutrino Physics and Astrophysics,* (ed. A. Morales), to be published in Nucl. Phys. B.

[5] J.N. Bahcall, *Neutrino Astrophysics* (Cambridge University Press, 1989).

[6] J.N. Bahcall and M. Pinsonneault, Rev. Mod. Phys. **64**, 885 (1992).

[7] W. Haxton, Nucl. Phys. **B28A**, 88 (1992).

[8] S. Turck-Chieze, in *Proc.* 15th *Int. Conf. on Neutrino Physics and Astrophysics* (ed. A. Morales), to be published in Nucl. Phys. B.

[9] S. Petcov, in *Proc.* 15th *Int. Conf. on Neutrino Physics and Astrophysics* (ed. A. Morales), to be published in Nucl. Phys. B.

[10] A. Morales, Nucl. Phys. **B28A**, 181 (1992).

[11] T. Tomoda, Rep. Prog. Phys. **54**, 53 (1991).

[12] A. van de Schaaf, *Progress in Particle and Nuclear Physics* (ed. A. Faessler, Pergamon Press, 1993) to be published; and J.D. Vergados, Phys. Rep. **133**, 1 (1986).
[13] P. Herczeg, this volume.
[14] B.R. Holstein, *Weak Interactions in Nuclei* (Princeton University Press, 1989).
[15] B.R. Holstein, Rev. Mod. Phys. **46**, 789 (1974); **48**, 673(E) (1976).
[16] B.R. Holstein, Phys. Lett. B **244**, 83 (1990).
[17] R.G.E. Timmermans *et al.*, Phys. Rev. Lett. **67**, 1074 (1991).
[18] The calculation used $F_\pi = 92.4(2)\,\mathrm{MeV}$ [16], $g_{NN\pi} = 13.34(11)$ [17], and $g_A = F_\pi g_{NN\pi}/m_\mathrm{N}$.
[19] M.L. Goldberger and S.B. Treiman, Phys. Rev. **111**, 354 (1958).
[20] J.D. Jackson, S.B. Treiman and H.W. Wyld, Phys. Rev. **106**, 517 (1957) and Nucl. Phys. **4**, 206 (1957).
[21] J.C. Hardy *et al.*, Nucl. Phys. **A509** 429 (1990).
[22] Particle Data Group, Phys. Rev. D **45**, S1 (1992).
[23] I.S. Towner, Nucl. Phys. **A540**, 478 (1992).
[24] F.C. Barker *et al.*, Nucl. Phys. **A540**, 501 (1992).
[25] J. Deutsch, Act. Phys. Hung. **68**, 129 (1990).
[26] W.E. Ormand *et al.*, Phys. Rev. C **40**, 2914 (1989).
[27] J. Deutsch *et al.*, Nucl. Phys. **A518**, 149 (1990) and L. Van Elmbt *et al.*, Nucl. Phys. **A469**, 531 (1987).
[28] A.I. Boothroyd *et al.*, Phys. Rev. C **29**, 603 (1984).
[29] L. Grenacs, Ann. Rev. Nucl. Part. Sci. **35**, 455 (1985).
[30] B.A. Brown, private communication.
[31] F.P. Calaprice and B.R. Holstein, Nucl. Phys. **A273**, 301 (1976).
[32] F.P. Calaprice *et al.*, Phys. Rev. C **15**, 2178 (1977).
[33] S. Weinberg, Phys. Rev. **112**, 1375 (1958).
[34] T. Minamisono *et al.*, J. Phys. Soc. Japan **55**, Suppl. 1012 (1986).
[35] K. Koshigiri *et al.*, J. Phys. Soc. Japan **55**, Suppl. 1014 (1986).
[36] M. Morita, Few Body Systems, Suppl. **5**, 165 (1992),
[37] Chr. Stratowa *et al.*, Phys. Rev. D **18**, 3970 (1978) and R. Dobrozemsky *et al.*, Phys. Rev. D **11**, 510 (1975).
[38] C.H. Johnson *et al.*, Phys. Rev. **132**, 1149 (1963).
[39] D. Schardt and K. Riisager, submitted to Z. Phys. C.
[40] B.G. Erozolimski *et al.*, Phys. Lett. B **33**, 351 (1970) and C.J. Christensen *et al.*, Phys. Rev. C **1**, 1693 (1970).
[41] A.L. Hallin *et al.*, Phys. Rev. Lett. **52**, 337 (1984); **52**, 1054(E) (1984).
[42] F.W.J. Koks and J. van Klinken, Nucl. Phys. **A272**, 61 (1976).
[43] A.R. Brosi *et al.*, Nucl. Phys. **33**, 353 (1962).
[44] J. van Klinken, Nucl. Phys. **75**, 145 (1966).
[45] The ^{32}P measurement from Ref. [43] gave $G = -0.990(9)$. The reported systematic error of ± 0.02 was added in quadrature. The measurements from Ref. [44] give: [$G = -1.01(4)$ and $-0.97(3)$ ^{32}P], [$G = -1.05(4)$, $-0.96(3)$, $-0.99(3)$, and $-1.01(3)$ ^{60}Co], [$G = -096(4)$ ^{114}In], and [$G = -1.03(2)$, and $-1.016(16)$ ^{153}Sm]. These measurements give a weighted mean $G = -1.0032(88)$. The reported systematic error of ± 0.016 was added in quadrature. The weighted mean of the two data sets is given in Table 2.
[46] A.S. Carnoy *et al.*, Phys. Rev. Lett. **65**, 3249 (1991).

[47] V.A. Wichers *et al.*, Phys. Rev. Lett. **58**, 1821 (1987).
[48] N. Severijns *et al.*, in *Proc. Int. Nucl. Phys. Conf.* (ed. R. Bock *et al.*) Nucl. Phys. **A553**, 827c (1993); and N. Severijns *et al.*, to be published in Phys. Rev. Lett.
[49] M. Allet *et al.*, Phys. Rev. Lett. **68**, 572 (1991) and M. Allet *et al.*, in Proc. of the Conf. on Weak and Electromagnetic Interactions in Nuclei (Dubna, 1992), to be published.
[50] M.B. Schneider *et al.*, Phys. Rev. Lett. **51**, 1239 (1983); **52**, 695(E) (1984).
[51] J. Imazato *et al.*, Phys. Rev. Lett. **69**, 877 (1992).
[52] A. Jodidio *et al.*, Phys. Rev. D **34**, 1967 (1986); **37**, 237(E) (1988).
[53] N.C. Mukhopadhyay, Phys. Reports **30C**, 1 (1977).
[54] E. Zavattini, in *Muon Physics* (ed. V.W. Hughes and C.S. Wu, Academic Press, 1975).
[55] W-Y Hwang and H. Primakoff, Phys. Rev. C **16**, 397 (1977).
[56] F.J. Botella and J. Bernabeu, Nucl. Phys. **A414**, 456 (1984) and Phys. Rev. D **32**, 1755 (1985); and F.J. Botella, Phys. Rev. D **32**, 1760 (1985).
[57] P. Langacker, Phys. Rev. D **15**, 2386 (1977).
[58] T.D. Lee and C.N. Yang, Phys. Rev. **104**, 254 (1956).
[59] R. Schrock, Phys. Lett. **96B**, 159, 1980; Phys. Rev. D **24**, 1233 (1981); and Phys. Rev. D **24**, 1273 (1981).
[60] J. Deutsch *et al.*, Phys. Rev. D **27**, 1644 (1983).
[61] S. Chiechanowicz, Z. Phys A **337**, 97 (1990) and references therein.
[62] S.W. Kikstra *et al.*, Nucl. Phys. **A529**, 39 (1991).
[63] Y. Nagai *et al.*, Phys. Rev. C **43**, R9 (1991).
[64] I.S. Towner *et al.*, in Proc. of the Conf. on Weak and Electromagnetic Interactions in Nuclei (Dubna, 1992), to be published.
[65] W.E. Ormand and B.A. Brown, Phys. Rev. Lett. **62**, 866 (1989); **63**, 103(E) (1989).
[66] I.S. Towner and J.C. Hardy, in *Proc.* 7th *Int. Conf. on Atomic Masses and Fundamental Constants* (ed. O. Klepper GSI, Darmstadt, 1984), p. 564.
[67] W.J. Marciano and A. Sirlin, Phys. Rev. Lett. **56**, 22 (1986).
[68] A. Sirlin and R. Zucchini, Phys. Rev. Lett. **57**, 1994 (1986); A. Sirlin, Phys. Rev. D **35**, 3423 (1987); and W. Jaus and G. Rasche, Phys. Rev. D **35**, 3420 (1987).
[69] W. Jaus and G. Rasche, Phys. Rev. D **41**, 166 (1990).
[70] W.J. Marciano, Ann. Rev. Nuc. Part. Sci. **41**, 469 (1991).
[71] I.S. Towner *et al.*, Nucl. Phys. **A284**, 269 (1977) and I.S. Towner in *Symmetry violations in subatomic physics* (ed. B. Castel and P.J. O'Donnel, World Scientific, 1989) p. 211.
[72] D.H. Wilkinson in *Nuclear Weak Processes and Nuclear Structure* (ed. M. Morita *et al.*, World Scientific, 1989) p. 1.
[73] D.H. Wilkinson, Nucl. Phys. **A511**, 301 (1990).
[74] We associate the ^{10}C result with the Ormand-Brown data set because all inputs except δ_C are based on the calculations in Ref. [24].
[75] G. Rasche and W.S. Wollcock, Mod. Phys. Lett. **A5**, 1273 (1990).
[76] P.A. Quin, in *Proc. Int. Nucl. Phys. Conf.* (ed. R. Bock *et al.*) Nucl. Phys. **A553**, 319c (1993).
[77] F.C. Barker, Nucl. Phys. **A537**, 134 (1992) and private communication.
[78] W. Haxton, private communication.
[79] S.J. Freedman, private communication.

[80] T.G. Walker, private communication.
[81] D. Pocanic, private communication.
[82] S.J. Freedman, Comments Nucl. Part. Phys. **19**, 209 (1990).
[83] Yu.Yu. Kostvintsev *et al.*, Piz'ma Zh. Eksp. Teor. Fiz. **44**, 444 (1986) [JETP Lett. **44**, 571 (1986)] and V.I. Morozov, Nucl. Instrum. Methods **A284**, 108 (1989).
[84] P.E.Spivak, Zh. Eksp. Teor. Fiz. **94**, 1 (1988) [JETP**67**, 1735 (1988)].
[85] J. Last *et al.*, Phys. Rev. Lett. **60**, 995 (1988).
[86] R. Kossakowski *et al.*, Nucl. Phys. **A503**, 473 (1989).
[87] W. Mampe *et al.*, Phys. Rev. Lett. **63**, 593 (1989).
[88] W. Paul *et al.*, Z. Phys. C **45**, 31 (1989).
[89] J. Byrne *et al.*, Phys. Rev. Lett. **65**, 289 (1990).
[90] V.P. Alfimenkov *et al.*, Piz'ma Zh. Eksp. Teor. Fiz. **52**, 984 (1990) [JETP Lett. **52**, 373 (1990)].
[91] D.H. Wilkinson, Nucl. Phys. **A377**, 474 (1982).
[92] D. Dubbers, in *Progress in Particle and Nuclear Physics* (ed. A. Faessler, Pergamon Press, 1991) p. 173.
[93] V.I. Morozov *et al.*, in *Proc. of the Int. School "Low-Energy Weak Interactions"* (LEWI'90, Dubna, 1990) p. 346.
[94] J.M. Pendlebury, private communication.
[95] A.P. Serebrov, private communication.
[96] J. Byrne, private communication.
[97] G.L. Greene, private communication.
[98] E. Klemt *et al.*, Z. Phys. C **37** (1988) 179.
[99] Yu.V. Gaponov *et al.*, Phys. Lett. B **253**, 283 (1991).
[100] A.S. Carnoy *et al.*, J. Phys. G **18**, 823 (1992).
[101] B.G. Erozolimski and Yu.A. Mostovi, Sov. J. Nucl. Phys. **53**, 260 (1991).
[102] P. Bopp *et al.* Phys. Rev. Lett. **56**, 919 (1986).
[103] B.G. Erozolimskiĭ*et al.*, Yad. Fiz. **52**, 1583 (1990) [Sov. J. Nucl. Phys. **52**, 999 (1990)] and B.G. Erozolimskiĭ *et al.*, Phys. Lett. B **263**, 33 (1991).
[104] P. Herczeg in *Rare Decay Symposium* (ed. D. Bryman *et al.*, World Scientific, 1989) p. 24.
[105] A.P. Serebrov and N.V. Romanenko, in Proc. of the Conf. on Weak and Electromagnetic Interactions in Nuclei (Dubna, 1992), to be published.
[106] J. Byrne, D. Dubbers, G.L. Greene, J.M. Pendlebury, and R.G.H. Robertson, private communications.
[107] G.S. Masson and P.A. Quin, Phys. Rev. C**41**, 1110 (1990).
[108] F.C. Barker, private communication, and I.S. Towner, private communication.
[109] G. Azuelos and J.E. Kitching, Phys. Rev. C**12**, 563 (1975) and L.E. Piilonen, Ph.D. dissertation, Princeton University (1985), unpublished.
[110] D.F. Schreiber, Ph.D. dissertation, Princeton University (1983), unpublished.
[111] R. Prieels, in Proc. of the Conf. on Weak and Electromagnetic Interactions in Nuclei (Dubna, 1992), to be published.
[112] M. Skalsey and M.S. Hatamian, Phys. Rev. C **31**, 2218 (1985).
[113] G.D. Fletcher, Phys. Rev. A **34**, 911 (1986).
[114] P.A. Quin and T.A. Girard, Phys. Lett. B **229** (1989) 29.
[115] O. Naviliat-Cuncic, private communication.
[116] J.E. Schewe, private communication.

[117] E.G. Adelberger, private communication and to be published.
[118] E.T.H. Clifford *et al.,* Nucl. Phys. **A493**, 293 (1989).
[119] Y.G. Egorov *et al.,* Nucl. Phys. **A524**, 425 (1991).
[120] M. Chen *et al.,* Phys. Rev. Lett. **69**, 3151 (1992).
[121] J.L. Mortara *et al.,* Phys. Rev. Lett. **70**, 394 (1993).
[122] B.R. Holstein, Phys. Rev. C **16** (1977) 753; **20** (1979) 387.
[123] J. Deutsch, Z. Phys. C **56**, S143 (1992).
[124] J. Congleton *et al.,* to be published in Nucl. Phys. B, and references cited therein.
[125] P.A. Souder, private communication.
[126] O. Hausser, private communication.
[127] C.L. Petitjean, private communication.
[128] D.I. Britton *et al.,* Phys. Rev. Lett. **68**, 3000 (1992).
[129] G. Czapek *et al.,* Phys. Rev. Lett. **70**, 17 (1993).
[130] D.A. Bryman, TRIUMF preprint TRI-PP-92-3.
[131] W.J. Marciano, Phys. Rev. D **45**, R721, 1992.
[132] B. Tasiaux *et al.,* Particle World **2**, 81 1991.
[133] S. Lontie, private communication.
[134] ARGUS collaboration, in *Proc.* 15th *Int. Conf. on Neutrino Physics and Astrophysics* (ed. A. Morales), to be published in Nucl. Phys. B.
[135] R. Abela *et al.*, Phys. Lett. B **105**, 263 (1981).
[136] G.H. Miller *et al.,* Phys. Lett. B **41**, 50 (1972).
[137] P. Truttman *et al.,* Phys. Lett. **B83**, 48 (1979).
[138] P. Herczeg, in *Progress in Nuclear Physics* (ed. W.-Y. Pauchy Hwang *et al,* Elsevier, 1991) p. 171.
[139] P. Herczeg, Hyp. Int. **43** (1988) 77; F. Boehm, Hyp. Int. **43** (1988) 95; E.M. Henley, in *Weak and Electromagnetic Interactions in Nuclei* (ed. P. Depommier, Edition Frontieres, 1989) p. 181; B.R. Holstein, Art Rich Memorial Symposium, 1991, to be published; and E.M. Henley, Chinese J. Phys. **30**, 1 (1992).
[140] T.J. Bowles *et al.*, J. de Phys. **45**, C3-27 (1984).
[141] F.P. Calaprice, private communication.
[142] C.G. Callen and S.B. Treiman, Phys. Rev. **162**, 1494 (1967) and B.R. Holstein, Phys. Rev. C **5**, 1529 (1972).
[143] S. Ciechanowicz, Z. Phys. C **337**, 97 (1990) and references cited therein; S. Ciechanowicz and N.P.Popov, submitted to Nucl. Phys.
[144] A.S. Carnoy, thèse annexe, Université Catholique de Louvain, 1991, unpublished and private communication.
[145] J. Egger, private communication.
[146] L. De Braeckeleer, Phys. Rev. C **45**, 1935 (1992) and private communication.
[147] E.M. Henley and I.B. Khriplovich, private communication and to be published.
[148] K. Bodek, private communication.
[149] M.K. Cambell *et al.,* Phys. Rev. Lett. **47**, 1032 (1981).
[150] M. Kadono, private communication.
[151] J. Sromicki, private communication.
[152] D. Geiregat *et al.,* Phys. Lett. B **259**, 499 (1991) and references therein.
[153] R.C. Allen *et al.,* LAMPF preprint LA-UR-92-1843.
[154] R.C. Allen *et al.,* Phys. Rev. Lett. **64**, 1871 (1990).
[155] B. Bodmann *et al.,* KARMEN collaboration, Phys. Lett. B **280**, 198 (1992).

[156] W. Fetscher, Phys. Rev. Lett. **69**, 2758 (1992); for a criticism on some of the reported constraints see the contribution of W. Fetscher and H.-J. Gerber to this volume.
[157] R. Barbieri and R.N. Mohapatra, Phys. Rev. D **39**, 1229 (1989).
[158] G. Raffelt and D. Seckel, Phys. Rev. Lett. **60**, 1793 (1988).
[159] R. Gandhi, private communication.
[160] A. Burrows *et al.,* Phys. Rev. Lett. **68**, 3834 (1992).
[161] K. Mursula and F. Scheck, Nucl. Phys. **B253**, 189 (1985).
[162] W. Fetscher *et al.,* Phys. Lett. B **173**, 102 (1986).
[163] P. Herczeg, Phys. Rev. D **34**, 3449 (1986).
[164] P. Herczeg, private communication.
[165] P. Langacker and S.U. Sankar, Phys. Rev. D **40**, 1569 (1989).
[166] W. Buchmüller and D. Wyler, Phys. Lett. B **177**, 377 (1986).
[167] P. Langacker and D. London, Phys. Rev. D **38**, 886 (1988).
[168] V. Barger *et al.,* Phys. Rev. D **40**, 2987 (1989).
[169] P. Herczeg in *Neutrino Mass and Low Energy Weak Interactions* (ed. V. Barger and D. Cline, World Scientific, 1985) p. 288.
[170] P. Herczeg, in Weak and Electromagnetic Interactions in Nuclei (ed. H.V. Klapdor, Springer-Verlag, 1986), p. 528.
[171] P. Herczeg, in *Fundamental Symmetries in Nuclei and Particles* (ed. H. Henrikson and P. Vogel, World Scientific, 1990) p. 46.
[172] P. Herczeg, in Proc. of the Conf. on Weak and Electromagnetic Interactions in Nuclei (Dubna, 1992), to be published.
[173] R.E. Marshak, Riazzudin and C.P. Ryan, *Theory of Weak Interactions in Particle Physics* (Wiley, 1969).
[174] V.N. Bolotov *et al.,* Phys. Lett. B **243**, 308 (1991).
[175] A.A. Poblaguev, Phys. Lett. B **238**, 108 (1990).
[176] V.M. Belyaev and I.I. Kogan Phys. Lett. B **280**, 238 (1992).
[177] S.L. Adler *et al.,* Phys. Rev. D **11**, 3309 (1975).
[178] B. McWilliams and L.-F. Li, Nucl. Phys. B **179**, 62 (1981).
[179] P.A. Quin *et al.,* Phys. Rev. D **47**, 1247 (1993).
[180] A.A. Poblaguev, private communication.
[181] R.N. Mohapatra, *Unification and Supersymmetry* (Springer-Verlag, 1992).
[182] W. Buchmüller *et al.,* Phys. Lett. B **191**, 442 (1987).
[183] M.A.B. Bég *et al.,* Phys. Rev. Lett. 38, 1252 (1977).
[184] E. Massó, Phys. Rev. Lett. 52, 1956 (1984).
[185] M.D. Cooper and R.E. Tribble, MEGA collaboration, private communication.
[186] See for example B. Schrempp in *Physics at HERA* (ed. W. Buchmüller and G. Ingelman, Hamburg, 1991) v. 2, p. 1034.
[187] J.D. Bjorken and S.D. Drell, *Relativistic Quantum Mechanics* (McGraw-Hill, 1964).

UNIVERSALITY OF THE WEAK INTERACTIONS

A. Sirlin
Department of Physics, New York University
New York, NY 10003, USA

Abstract

The idea of universality of the weak interactions and its impact on the development of particle physics are discussed. Recent results, constraints on new physics, and the current situation concerning the comparison of theory and experiment are analyzed.

1. Introduction

The principle of universality of the weak interactions is a concept of deep and enduring significance. Indeed, several important milestones in the development of particle physics have been partially motivated by the need to rest this principle on firmer theoretical grounds.

The idea arose originally in the period 1947-49, when several authors [1] proposed a universal weak interaction among the doublets (p,n), (ν_e, e) and (ν_μ, μ). In fact, it was noted that the three basic processes $n \rightarrow p + e^- + \overline{\nu}_e$, $\mu^- \rightarrow e^- + \nu_\mu + \overline{\nu}_e$, $\mu^- + p \rightarrow n + \nu_\mu$ are characterized roughly by a single coupling constant, of magnitude $\approx 10^{-49}$erg cm^3. In 1951, Enrico Fermi, the founder of weak interaction theory, stated that this similarity is probably not accidental and has a deep meaning, not understood at the time [2]. He further suggested a possible analogy with the universality of electric charge [2].

The proposal by Lee and Yang [3] that parity is not conserved in the weak interactions and its subsequent experimental verification [4], led to two further developments of great importance: it brought the study of weak interactions to the forefront of physics and, by greatly enlarging the number of observables available for experimental and theoretical scrutiny, paved the way to the determination of the basic phenomenological interaction. In conjunction with elegant theoretical arguments, this led Marshak and Sudarshan and Feynman and Gell-Mann to propose a universal V-A Fermi Interaction for charged current processes [5,6].

In the framework of this theory, it was noted [6] that $G_\mu = G_V$ within roughly 1%, where G_μ is the μ decay coupling constant and G_V is the vector coupling in ß decay, as determined from the superallowed Fermi transitions. Even if one assumed that $G_\mu = G_V$ at the Lagrangian level as a manifestation of some principle of universality, a precise equality of the two couplings was not expected. In fact, the particles involved in ß decay are obviously affected by strong interactions, while this is not the case in muon decay. This prompted Feynman and Gell-Mann to invoke the conserved vector current [CVC] hypothesis [6-8], previously discussed by Gershtein and Zeldovich [9]. Namely, the vector current in ß decay is assumed to be conserved in the presence of the strong interactions. As such conservation laws are usually associated with symmetries of the theory, they further identified the relevant operator with the $\Delta I_3 = 1$ isospin current. The near equality $G_\mu \approx G_V$ could then be understood as a consequence of two ideas: a principle of universality that states that $G_\mu = G_V$ at the Lagrangian level, and CVC that guarantees that the strong interactions do not renormalize G_V at $q^2 = 0$ in the limit of isospin invariance. The analogy with the universality of electric charge, anticipated by Fermi, is clear.

The CVC hypothesis, in turn, had another far reaching consequence. If the $\Delta S = 0$ vector current is conserved, it would be only natural to expect that the $\Delta S = 1$ vector current in semileptonic decays is also conserved, at least in some suitable limit. This was one of the basic arguments that led to the search for higher partial symmetries of the Strong Interactions. A number of possibilities were explored [10], culminating in the phenomenologically successful $SU(3)_{flavor}$ theory [11]. Gell-Mann also pointed out that, in order to implement universality, a normalization of the hadronic currents is required. This observation was one of the important motivations for Current Algebra [12,13]. Indeed, the non-linearity of the basic relation

$$[J_o^a(t,\vec{x}), J_o^b(t,\vec{y})] = if^{abc} J_o^c(t,\vec{x})\, \delta^3(\vec{x}-\vec{y}), \qquad [1]$$

where f^{abc} (a,b,c = 1...8) are the SU(3) structure constants, fixes the normalization of these currents.

Another theoretical development, that was to have an intimate connection with the analysis of universality, began a few years earlier. In 1956, Behrends, Finkelstein, and Sirlin initiated a program to study the radiative corrections to weak interactions [14]. It continues to the present day, in the analysis of precision electroweak physics. The first application was the accurate study of the energy spectrum and other observables in muon decay and the precise determination of the parameters ρ_{Michel}, ξ and δ, as those are crucial quantities to verify the V-A theory. When the CVC hypothesis was formulated, it was natural to suspect that the small difference between G_μ and G_V was due to electromagnetic corrections, since these effects break the conservation of isospin symmetry. We have in mind here electromagnetic contributions not contained in the Fermi Coulomb-function. However, when the $O(\alpha)$ corrections to the decay probabilities of ß and muon decays were analyzed in the framework of the local V-A theory by Berman [15] and by Kinoshita and Sirlin [16], they were found to be convergent in μ-decay but logarithmically divergent in ß decay! A simple argument to understand this result is given in Ref. [17]. In fact, it can be shown that the result for muon decay can be generalized: to all orders in e but first order in G_μ, the corrections to μ decay in the local V-A theory are convergent after mass and charge renormalizations [17-19]. A further complication at the time was that, for any reasonable value of the cutoff $\Lambda \geq 1\mathrm{GeV}$, the radiative corrections increased the difference between G_μ and G_V! On the other hand, the study of these higher order effects led to an interesting result in field theory: the discovery that mass singularities cancel in total decay rates [16]. This, in turn, provided the motivation for the derivation of the Kinoshita-Lee-Nauenberg (KLN) theorem [20]. The situation, as it existed in 1960, was summarized by Feynman [21].

The framework for the discussion of universality was suddenly changed in 1963 when Cabibbo proposed his theory of semileptonic decays, constructed on the basis of SU(3) currents [22]. One of the crucial elements in this theory was a departure in the statement of universality. Rather than demanding $G_\mu = G_V(\Delta S=0)$ at the Lagrangian level, the principle of universality was expressed as

$$G_\mu^2 = G_V^2(\Delta S=0) + G_V^2(\Delta S=1), \qquad (2)$$

or, alternatively, $G_V(\Delta S=0) = G_\mu \cos\theta_C$, $G_V(\Delta S=1) = G_\mu \sin\theta_C$, where θ_C is the Cabibbo angle. Eq. (2) had two important consequences: it successfully incorporated the fact that $\Delta S=1$ semileptonic processes are suppressed relative to $\Delta S=0$ reactions and, furthermore, the radiative corrections had an effect that was at least in the right direction to ensure the new concept of universality.

During the sixties there were a number of developments related to the problem of universality. In 1960 Behrends and Sirlin found that if the conservation of the SU(2) vector currents is broken by mass splittings, their matrix elements at zero momentum transfer are not renormalized to first order in the symmetry breaking [23]. They also suggested the generalization of this theorem to higher partial symmetries, which at the time had not been established. The results were confirmed by Terent'ev, who employed a different argument [24]. In 1964, Ademollo and Gatto independently derived the analogous theorem in the SU(3) context [25]. This non-renormalization theorem plays an important role in the analysis of universality, both in the SU(2) framework, where it applies to ß decay, and in the SU(3) case, where it is relevant to the $\Delta S=1$ semileptonic processes. In 1966, Bjorken, using current algebra methods, gave convincing arguments indicating that the strong interactions do not soften the short distance behaviour of the radiative corrections to the Fermi transitions in ß decay [26]. Thus, in all probability the cutoff was not due to the strong interactions. This analysis was extended by Abers, Norton and Dicus, who considered also the divergent part of the corrections to the Fermi amplitude arising from the axial vector current [27]. The coefficient of this last contribution was calculated by applying the Bjorken-Johnson-Low limit [26,28] with naive, i.e. canonical, evaluation of the relevant commutators. As we will see, the situation regarding the axial vector contribution is more complicated, but this only became clear later. Meanwhile, using a completely different approach, the present author showed that the radiative corrections to some important observables, such as the energy spectrum in allowed ß decay, can be accurately evaluated in spite of the complications arising from the strong interactions [29]. The two approaches reached common ground when, in a subsequent paper [30], Abers, Dicus, Norton and Quinn were able to obtain not only the divergent parts, but also the corrections to the energy spectrum deduced in Ref. [29]. Although other formulations of the radiative corrections were proposed, notably by Källén [30], current algebra became the prevalent approach.

The situation in 1967 was both interesting and perplexing. The current algebra method had been the basis of great technical progress; in fact, it had led to the conclusion that the corrections to the Fermi amplitude involving the vector current are independent of strong interaction dynamics and it appeared that the divergent contributions arising from the axial vector current were also known. The great difficulty, of course, was that in the local Fermi theory this answer was divergent! Two different solutions to this dilemma were suggested at the time: i) Cabibbo, Maiani, and Preparata [32] and Johnson, Low, and Suura [33] proposed a change in the space-space part of the current algebra of hadronic currents, in such a manner that the radiative corrections to ß decay become convergent ii) the present author proposed that the solution to the problem lies in the intermediate boson theory [34]. The arguments in Refs. [30,34] indicated that in this case the natural cutoff is $\Lambda \approx m_W$, but they were not complete: the theory employed was not renormalizable and logarithmic divergences, albeit with very small coefficients, remained in the analysis; moreover, there was no indication of the magnitude of m_W. Analogous results had been previously found by Lee, Shaffer, Dorman, and Bailin, who had studied the corrections in the intermediate boson theory in the case of a "bare" nucleon, i.e. a nucleon devoid of strong interactions [35]. The situation, in 1968, was summarized in Ref. [36]. Shortly afterwards, Adler and W-K. Tung showed that the naive evaluation of commutators involved in the divergent contribution of the axial vector current is modified perturbatively by the strong interactions in a class of Abelian models [19,37]. Thus, the theoretical status of the first solution (i) was uncertain and the second proposal (ii) became increasingly attractive. Another problem that was discussed at the time was the Z-dependence of the radiative corrections to the Fermi amplitude. Characterizing the order of magnitude of the current interactions by the value of the corresponding matrix elements at zero-momentum transfer, Bég, Bernstein, and Sirlin concluded that the expansion parameter is $Z\alpha$, which is not obvious by naive power counting [38]. Furthermore, they showed that the only $Z\alpha$ contribution is that contained in the Fermi Coulomb-function. No general theorems are known regarding the corrections of $O((Z\alpha)^2)$ and higher.

Several years before, another line of development was initiated that was to provide a more fundamental basis for the discussion of universality. In 1954, Yang and Mills proposed the idea of non-abelian gauge invariance [39,40]. In the late sixties this principle, applied to the $SU(2)_L xU(1)$ theory, was combined with the concept of spontaneous symmetry breaking. A realistic gauge theory, the Standard Model (SM) of Glashow, Salam and Weinberg [41], was born! The theory incorporates universality in a fundamental way. To illustrate this point, we may consider a simplified world with two generations. The part of the Lagrangian density that describes the kinematics of fermions and their interactions with the gauge fields is

$$\mathcal{L}^{(f)} = (\overline{u}\,\overline{d}')_L \gamma^\mu i D_\mu^{(f)} \begin{pmatrix} u \\ d' \end{pmatrix}_L + (\overline{c}\,\overline{s}')_L \gamma^\mu i D_\mu^{(f)} \begin{pmatrix} c \\ s' \end{pmatrix}_L$$

$$+ (\overline{\nu_e}\,\overline{e})_L \gamma^\mu i D_\mu^{(f)} \begin{pmatrix} \nu_e \\ e \end{pmatrix}_L + (\overline{\nu_\mu}\,\overline{\mu})_L \gamma^\mu i D_\mu^{(f)} \begin{pmatrix} \nu_\mu \\ \mu \end{pmatrix}_L , \quad (3)$$

where

$$i D_\mu^{(f)} = i\partial_\mu - \frac{g_o \vec{\tau}.\vec{W}_\mu}{2} - \frac{g_o' \, Y^{(f)} B_\mu}{2} \quad (4)$$

is the covariant derivative, $d' = d \cos\theta_C + s \sin\theta_C$, $s' = s \cos\theta_C - d \sin\theta_C$ the weak eigenstates, $\vec{\tau}$ the Pauli matrices, $Y^{(f)}$ the weak hypercharge of the f doublet, g_o and g_o' the $SU(2)_L$ and $U(1)$ coupling constants, respectively, and a summation over the color degree of freedom is understood. The principle of non-abelian gauge invariance tells us that g_o is a universal parameter, independent of the nature of the fermion doublet f! Thus, it seems likely that a deep reason for universality has been unveiled and that Fermi's question [2] has perhaps been answered. It is one of the ironies in the history of these developments that in the SM there is no such compelling argument for the quantization of electric charge. In fact, except for the constraint $\Sigma_f Y^{(f)} = 0$ necessary for the cancellation of anomalies, the $Y^{(f)}$ are, in principle, arbitrary. In order to have a fundamental reason for the quantization of electric charge, it appears that embedding in a larger, simple group, is necessary. Indeed, this provides an important motivation for GUT's.

The discovery of a third fermion generation led, in the context of the SM, to a generalization of Eq. (2). In fact, the unitarity of the Cabibbo–Kobayashi–Maskawa (CKM) matrix [42] may be regarded as the present statement of universality.

As the SM is renormalizable, it was natural to expect that it could provide the basis for the solution of the long-standing problem of radiative corrections to ß-decay. This was finally achieved in the period 1974-1978 [43,44]. In order to carry out a realistic calculation, it was necessary once more to control the effects of the strong interactions. This required the development of a current algebra formulation [44], the generalization to gauge theories of the methods used in the sixties to treat the electromagnetic corrections in the local Fermi theory. Fortunately, in the seventies the theoretical framework was based on $SU(3)_C \times SU(2)_L \times U(1)$, with an asymptotically free theory of strong interactions. In his treatment of electroweak corrections to strong interaction symmetries, such as parity and strangeness conservation, Weinberg emphasized the great advantages of this formulation [45]. Similarly, in Ref. [44], which deals with electroweak corrections to leptonic and semileptonic amplitudes, the new framework was extensively invoked.

In the most popular versions of the SM, involving Higgs doublets and singlets, the O(α) corrections to the Fermi amplitude turn out to be surprisingly simple: aside from very small asymptotic effects induced by the strong interactions, the answer is the same as in the local Fermi theory, with the cutoff Λ set equal to m_Z. For other versions (with Higgs triplets, for example) the theoretical expression is slightly more complicated. However, invoking the phenomenological fact that $\cos^2\theta_W \approx m_W^2/m_Z^2$, both results are essentially identical. The calculation assumes that the couplings of the Higgs scalars to ordinary fermions are of O(gm/m_W), where m represents a generic fermion mass. This is of course the case in the standard version with one Higgs-doublet. The simplicity of the relation between the calculations carried out in the intermediate boson and local Fermi theories is not completely understood at this time.

The dominant contribution in the corrections to the superallowed Fermi transitions turns out to be

$$P = P^{o}\left\{1 + \frac{3\alpha}{2\pi}\left[\ln\left(\frac{m_Z}{2E_m}\right) + 2\overline{Q}\,\ln\left(\frac{m_Z}{M}\right)\right] + \ldots\right\}, \qquad (5)$$

where P^o is the zeroth order probability expressed in terms of G_μ, E_m is the end-point energy of the electron or positron, $\ln(m_Z/M)$ is a short distance contribution arising from the axial vector current, M is a hadronic mass of O(1GeV) and $\overline{Q}$ is the average charge of the underlying fields in the appropriate current. For ß decay the underlying fields are the u and d quarks and $\overline{Q}=1/6$. Eq. (5) is a particular example of an asymptotic theorem that governs the short distance behaviour of the radiative corrections to arbitrary semileptonic decays mediated by $W^{\pm}$[46]. It is interesting to note that if Eq. (5) is applied to μ-decay, we have $\overline{Q}=-1/2$ (the underlying fields are μ and ν_μ) and the m_Z dependence disappears, in conformity with the convergence of the radiative corrections in the local V-A theory. The first term in Eq. (5) is quite large, ≈3.45% in ^{14}O, and is crucial for the phenomenological viability of the SM. The fact that the contemporary data required a large cutoff, $m_W \geq 20 m_p$, was already emphasized by Blin-Stoyle and Freeman in 1970 [47]. It is indeed an important success of the SM that it automatically leads to a large radiative correction, a result which, as we will see later in greater detail, is necessary for consistency with the unitarity of the CKM matrix.

In summary, in this brief survey we have seen how the idea of universality of the weak interactions has been directly or indirectly related to a number of fundamental developments such as CVC, the search for higher partial symmetries of the Strong Interactions, Current Algebra, Cabibbo universality, study of the radiative corrections in weak interactions, KLN and non-renormalization theorems, non-abelian gauge symmetry and the development and consistency of the SM.

2. Recent Developments

It is convenient to start our discussion by writing down the radiatively corrected expression for the transition probability in superallowed Fermi transitions [44]:

$$\Delta P\, d^3p_e = \Delta P^o\, d^3p_e \left\{ 1 + \frac{\alpha}{2\pi}\left[3\ln\left(\frac{m_Z}{m_p}\right) + g(E,E_m) + 6\bar{Q}\ln\left(\frac{m_Z}{M}\right) + 2C + A_g\right]\right\}. \quad (6)$$

The uncorrected decay probability ΔP^o is expressed in terms of G_μ, defined from the muon lifetime [16,48] by

$$\frac{1}{\tau_\mu} = \frac{G_\mu^2 m_\mu^5}{192\pi^3} f\left(\frac{m_e^2}{m_\mu^2}\right)\left[1 + \frac{3}{5}\frac{m_\mu^2}{m_W^2}\right] \times\left[1 + \frac{\alpha}{2\pi}\left(\frac{25}{4} - \pi^2\right)\left(1 + \frac{2\alpha}{3\pi}\ln\left(\frac{m_\mu}{m_e}\right)\right)\right], \quad (7)$$

where

$$f(x) = 1 - 8x - 12x^2 \ln x + 8x^3 - x^4. \quad (8)$$

Numerically, $G_\mu = 1.16639(2)\times10^{-5}\ \mathrm{GeV}^{-2}$. The probability ΔP^o incorporates the Fermi function F(Z,E) that describes the Coulomb interaction between the nuclei and the emerging positron. It is also proportional to $|V_{ud}|^2$ where V_{ij} ($i=u,c,t$; $j=d,s,b$)

denotes the elements of the CKM matrix. The function $g(E, E_m)$ represents the radiative correction to the electron or positron spectrum in allowed ß decay [29]. In the total decay probability, it is replaced by the average value $\overline{g}(E_m) = \int d^3p_e \Delta P^\circ g(E, E_m) / \int d^3p_e \Delta P^\circ$. The first two terms between square brackets in Eq. (6) arise from the vector current and are independent of hadron dynamics. In particular, the proton mass m_p cancels in their sum. The third term represents short distance effects from the axial vector current, while 2C stands for the corresponding non-asymptotic part. The choice of M and 2C are model dependent. In Ref. [49] they were estimated by varying M in the range 400MeV≤M≤1600MeV, which spans mass scales from $3m_\pi$ throughout the A_1 resonance region, and evaluating C in the Born approximation using two different models of the nuclear transition. Specifically, the $0^+ \rightarrow 0^+$ decay of the nucleus as a whole and the independent-nucleon model were considered, leading to C=0 [30] and C=0.885 [49], respectively. Combining these calculations and recalling $\overline{Q}=1/6$, the estimate $(\alpha/2\pi)[\ln(m_P/M)+2C]=0.0012\pm0.0018$ was obtained. Recently, Jaus and Rasche have proposed a more detailed evaluation of C [50]. Writing $C=C_{Born}+C_{NS}$, they identify C_{Born} with the Born approximation calculation of the diagram in which the axial vector insertion and the emission of the photon involves the same nucleon, leading again to $C_{Born}\approx0.885$, and C_{NS} with the amplitudes in which the two processes originate in different nucleons. The calculation of C_{Born} was confirmed by Towner, who recommends $C_{Born}=0.881\pm0.030$ [51]. The analysis of C_{NS} depends on nuclear structure shell-model calculations. Recent evaluations of C_{NS} have been reported by Towner and by Barker, Brown, Jaus and Rasche [51-53]. Towner, for example, finds $\alpha C_{NS}/\pi \leq 0.06\%$ in six of the eight accurately measured transitions, $\approx -0.3\%$ for ^{14}O and $\approx +0.1\%$ for ^{42}Sc. On the other hand, $\alpha C_{Born}/\pi \approx 0.2\%$ so that, except for ^{14}O, this is the dominant part of $\alpha C/\pi$. Barker et al. obtain results of the same sign and order of magnitude, but the details differ somewhat. The quantity $A_{\overline{g}} \approx -0.34$ is a very small contribution induced asymptotically by QCD.

Further corrections of $O(Z\alpha^2)$ and $O(Z^2\alpha^3)$ arising from the nucleus-positron interaction and not contained in the Fermi function, have been included since the pioneering work of Jaus and Rasche [54]. They are called δ_2 and δ_3, respectively. These studies were motivated by an ambiguity noted in early discussions [55]: it was not clear whether the $O(\alpha)$ corrections should be added to the Fermi function or the two effects factorized. The difference between the two approaches is of $O(Z\alpha^2)$ and the explicit evaluation of the terms of this order, after the factorization displayed in Eq. (6), clarifies in principle this problem. Around 1985 the detailed analysis of the Ft values was not in satisfactory agreement with CVC. The situation was greatly improved when Sirlin and Zucchini reexamined δ_2 with analytic methods [56] and found significantly different results from those obtained previously. Their work was subsequently confirmed by new numerical calculations of Jaus and Rasche [57]. Quantitatively, δ_2 varies from 0.22% for ^{14}O to 0.50% for ^{54}Co, while δ_3 is much smaller.

There is also a correction, δ_C , that reflects the lack of perfect overlap between the wave functions of the parent and daughter nuclei due to Coulomb forces. The largest effect is related to the fact that protons are typically less bound than neutrons. As a consequence, the radial wave function for protons extends further than that for neutrons and the overlap of the parent and daughter nuclei is reduced from unity. A second effect, considerably smaller, is that the degree of configuration mixing in the shell-model wave functions varies among the members of an isospin multiplet. There are two modern calculations of δ_C , by Towner, Hardy, and Harvey [58], and by Ormand and Brown [59]. Both exhibit similar nucleus to nucleus variations, but there is a difference in the actual numbers, with the values in Ref. [58] being systematically larger. Averaging over the eight accurately measured nuclear transitions, the difference between the two calculations of δ_C is 0.16%. In a recent analysis, Hardy, Towner, Koslowsky, Hagberg, and Schmeing employ the average of the two δ_C calculations for each nuclear decay [60]. This average value is 0.24% for ^{14}O, reaches 0.60% for ^{38m}K and is 0.47% for ^{54}Co.

In Ref. [49], Marciano and Sirlin incorporated leading logarithms of $O((\alpha \ln(m_Z/m_p))^n)$ ($n \geq 2$) by means of a renormalization group analysis. Including the δ_2, δ_3 and δ_C corrections, this leads to

$$P = P^o \left\{ 1 + \frac{\alpha}{2\pi} \left(\ln\left(\frac{m_p}{M}\right) + 2C \right) + \frac{\alpha(m_p)}{2\pi} \left[\bar{g}(E_m) + A_{\bar{g}} \right] + \delta_2 + \delta_3 \right\} \times S(m_p, m_Z)\,(1 - \delta_C)\,, \qquad (9)$$

where P and P^o refer to the total decay rates and $S(m_p, \mu)$ satisfies an appropriate renormalization group equation. The short distance contribution $S(m_p, m_Z)$ is numerically equal to 1.0225. For ^{14}O, the incorporation of the leading higher order logarithms increases P by $\approx$0.18%. In order to verify CVC, it is advantageous to factor out the nuclear-dependent part of the radiative corrections. A simple way of doing this is to replace

$$\{\quad\} S(m_p, m_Z) \to (1+\delta_R)(1+\Delta_R), \qquad (10a)$$

where { } in the l.h.s. represents the expression between curly brackets in Eq. (9),

$$1+\delta_R = 1+\frac{\alpha(m_p)}{2\pi}\bar{g}(E_m)+\frac{\alpha}{\pi}C_{NS}+\delta_2+\delta_3 \qquad (10b)$$

and

$$1+\Delta_R = \left\{1+\frac{\alpha}{2\pi}\left[\ln\left(\frac{m_p}{M}\right)+2C_{BORN}+A_{\bar{g}}\right]\right\} S(m_p, m_Z). \qquad (10c)$$

The difference between the two sides of Eq. (10a) is negligible. One can then introduce a radiatively corrected Ft value:

$$Ft = ft(1+\delta_R)(1-\delta_C) = K/(2G_V'^2), \qquad (11a)$$

where

$$K = 2\pi^3 \ln 2\hbar^7/m_e^5c^4 = 8.12027\times10^{-7}(\hbar c)^6 GeV^{-4}s, \qquad (11b)$$

$$G_V'^2 = G_V^2(1+\Delta_R), \qquad (11c)$$

and $G_V=G_\mu|V_{ud}|$. In Eq. (11a), ft is the usual ft value, without the corrections of $O(\alpha)$ and higher we have discussed, while Ft includes the nuclear-dependent correction $(1+\delta_R)(1-\delta_C)$. The short distance nuclear-independent correction $1+\Delta_R$ has been absorbed in the effective coupling constant G_V'. The CVC test consists of checking the constancy of the Ft values in the eight accurately measured transitions. The analysis of Ref. [53] finds good agreement with CVC with a weighted average of 3074.0±1.0s and a χ^2 per degree of freedom of 0.95. The largest variation occurs for ^{14}O for which Ft=3068.8±3.0s, which is $\approx 1.7\sigma$ lower. Actually, the authors of Ref. [53] employ a more complicated and slightly different separation than that shown in Eqs. (10b,c). Adjusting for the difference, the values corresponding to Eqs. (10b,c) are $(Ft)_{ave}$=3074.7±1.0s and $(Ft)_{^{14}O}$=3069.5±3.0s. For the unitarity test one inserts $(Ft)_{ave}$ in Eqs. (11a,b), determines G_V', and, via Eq. (11c), G_V. In conjunction with G_μ, this leads to $|V_{ud}|$. Employing the above $(Ft)_{ave}$ and the range $400MeV \leq M \leq 1600MeV$ with central value M=800MeV, one finds $\Delta_R = 0.0244 \pm 0.0009$ and [53]

$$|V_{ud}| = 0.9734 \pm 0.0006 \pm 0.0004. \qquad (12a)$$

The first error in Eq. (12a) arises from the statistical one in Ft plus a systematic uncertainty of 0.08% that reflects the difference between the δ_C calculations of Refs. [58] and [59]. The second error essentially represents the uncertainty in M.

The analysis of universality requires a precise determination of V_{us}. A detailed study of the theoretical framework, based on chiral perturbation theory, is given by Leutwyler and Roos in Ref. [61], which contains also references to important earlier work. Combining the errors in quadrature so that Eq. (12a) becomes $V_{ud} = 0.9734 \pm 0.0007$, and employing $V_{us} = 0.2205 \pm 0.0018$, $V_{ub} = 0.004 \pm 0.002$ [62], the authors of Ref. [53] find

$$V_{ud}^2 + V_{us}^2 + V_{ub}^2 = 0.9962 \pm 0.0016, \qquad (12b)$$

which falls short of unity by 2.4 times the estimated error. The magnitude of the difference depends somewhat on the range chosen for M. For example, if $m_{A1}/2 \leq M \leq 2m_{A1}$ with the central value $M = m_{A1} = 1.26\text{GeV}$ is employed, a reasonable procedure since the A_1 resonance is now well established [62], one finds $\Delta_R = 0.0238 \pm 0.0009$,

$$V_{ud} = 0.9737 \pm 0.0007, \qquad (13a)$$

$$V_{ud}^2 + V_{us}^2 + V_{ub}^2 = 0.9967 \pm 0.0016, \qquad (13b)$$

which differs from unity by 2.1 times the estimated uncertainty.

The evaluation of δ_C has been, for a long time, the subject of considerable discussions. Recently, Wilkinson [63] has pointed out that the calculations of Refs. [58] and [59] take into account the charge-dependent effects on the valence nucleons and that there should be additional contributions to δ_C from the core nucleons. He has also developed a heuristic procedure to extract this core contribution from the data, as a smooth function of the nuclear charge Z. The objective is to extrapolate to $Z = 0$ and thus obtain the Ft value corresponding to an idealized world without Coulombic distortions due to the core nucleons. The analysis leads to [63]

$$V_{ud} = 0.9750 \pm 0.0004 \pm 0.0004, \qquad (14a)$$

where the first error is associated with Wilkinson's approach and we have added the second to reflect the uncertainty of M in the short distance contribution. From Eq. (14a) one finds:

$$V_{ud}^2 + V_{us}^2 + V_{ub}^2 = 0.9992 \pm 0.0014. \qquad (14b)$$

Eq. (14b) is in very good agreement with universality. It should be remembered, however, that this is an heuristic procedure, not based on an ab initio calculation. It would be highly desirable to have a theoretical formulation that explains the origin and magnitude of the core contributions.

It is interesting to note that Wilkinson's results are not far from those obtained by using only the data from ^{14}O, the lowest-Z accurately measured transition. Using again the range $m_{A1}/2 \leq M \leq 2m_{A1}$ and the $(Ft)_{^{14}O}$ value reported before, we find

$$V_{ud} = 0.9745 \pm 0.0005 \pm 0.0004 \quad (^{14}O), \qquad (15a)$$

$$V_{ud}^2 + V_{us}^2 + V_{ub}^2 = 0.9983 \pm 0.0015 \quad (^{14}O). \qquad (15b)$$

Thus, when based on ^{14}O alone, the analysis is in resonable agreement with universality. The rationale for focussing attention on ^{14}O is that one intuitively expects this to be the case less affected by Coulombic distortions, since it has the lowest Z value among the accurately measured decays. We also note that the error in V_{ud} has not increased, relative to Eq. (12a), by considering only one nuclear transition rather than the average of the eight. The reason is that the uncertainty assigned to δ_C in ^{14}O by comparing the calculations of Refs. [58] and [59] is 0.04% and this is already included in the $(Ft)_{^{14}O}$ value.

In summary, using the analysis of Refs. [53] and [63], allowing two different ranges of values of M and the possibility of extracting V_{ud} from ^{14}O alone, we have obtained values of V_{ud} in the range $0.9734 \leq V_{ud} \leq 0.9750$. The lowest value 0.9734, reported in Ref. [53], leads to a difference with the SM of 2.4 times the estimated error. However, if the range $m_{A1}/2 \leq M \leq 2m_{A1}$ is employed, the deviation is reduced to 2.1 times the estimated error. Finally, 0.9745, extracted from ^{14}O alone, and 0.9750, obtained in Wilkinson's analysis, are in reasonable to very good agreement with universality. Since the evaluation of δ_C is a significant source of uncertainty, these observations suggest, once more, the importance of accurate measurements of the rates of pion [64] and ^{10}C [52,65] ß decays. In particular, it is worthwhile to note that the current central value of V_{ud} extracted from ^{10}C is even larger than that derived from ^{14}O.

It is important to note the very significant role played by the radiative corrections in the analysis of the phenomenological consistency of the SM. For example, on the basis of Eqs. [7,9] one finds that the radiative corrections decrease the value of $|V_{ud}|^2$ extracted from ^{14}O and μ decays by about 4.1%. If these corrections were not applied, the r.h.s. of Eq. (15b) would be ≈1.039 and the SM would not be tenable!

3. Neutron Decay

With the advent of cold neutron techniques, there have been important advances in the measurement of neutron decay observables. From the neutron lifetime one can determine $G_V^2+3G_A^2$. If this is combined with a measurement of G_A/G_V, one can find G_V and, therefore, V_{ud}. A recent precision experiment by Erozolimski et al. [66] leads to an average value $|g_A/g_V|=1.257\pm0.003$. Combining this result with $\tau_n=887.4\pm1.7s$ [67] and using $\Delta_R=0.0238\pm0.0009$, one finds

$$V_{ud} = 0.9813 \pm 0.0021 \text{ (neutron decay)}, \qquad (16a)$$

and

$$V_{ud}^2 + V_{us}^2 + V_{ub}^2 = 1.0116 \pm 0.0041 \quad \text{(neutron decay)}, \qquad (16b)$$

which is 2.8σ above unity. It is also clear that Eq. (16a) has a considerably larger error than values derived from the super-allowed Fermi transitions and differ from them by more than 3σ. These are serious discrepancies and it is to be hoped that they will be clarified in the future, perhaps by new determinations of g_A/g_V.

4. Constraints on New Physics

Over the years, the test of universality has been used to derive constraints on new physics. A problem in carrying out such analysis is that it requires a precise value of $\Sigma_j|V_{uj}|^2$, as derived in the SM, and its uncertainty. Comparison of Eqs. (12b,13b,14b,15b) indicates that, at present, various determinations of this quantity from the superallowed Fermi transitions differ by as much as 0.3%. Nonetheless, we will illustrate this important subject with a few examples by choosing the determination from ^{14}O (Eq. (15b)).

a) *4th Generation.* If there exists a fourth generation, Eq. (15b) tells us that $|V_{ub'}|<0.06$ (90% C.L.). This is certainly not very restrictive, since V_{ub} is much smaller than the bound for $V_{ub'}$.

b) *Additional Z's.* In some models with additional U(1) factors, the extra Z's have different couplings to quarks and leptons, and therefore give rise to new loop corrections (box diagrams) that distinguish μ and semileptonic decays [68]. In frequently discussed theories, these contributions have the effect of decreasing the value of $\Sigma_j|V_{uj}|^2$ extracted from experiments relative to the corresponding quantity in the SM. For example, in the case of Z_χ (the additional Z occurring in SO(10)→SU(5)xU(1)$_\chi$), $\Sigma_j|V_{uj}|^2$ is no longer equal to Eq.(15b) but, instead, we have

$$|\Delta|+\Sigma_j|V_{uj}|^2=0.9983\pm0.0015, \qquad (17a)$$

$$|\Delta|=\frac{3\alpha}{2\pi\cos^2\theta_W}\frac{\ln x}{x-1}, \qquad (17b)$$

where $x=m_{Z_\chi}{}^2/m_W{}^2$. Using unitarity, i.e. $\Sigma_j|V_{uj}|^2=1$, and the method recommended in Ref. [62] for the statistical treatment of upper limits when considering bounded physical regions, one finds $|\Delta|\leq0.0016$ at 90% CL, leading to $m_{Z_\chi}\geq195\,GeV$. This is not competitive with the bound $m_{Z_\chi}>$ 340 GeV (95%CL) from $p\bar{p}$ direct searches [62], but it is conceptually interesting because it arises from loop effects. For m_{Z_χ} =340GeV, $|\Delta|=8\text{x}10^{-4}$, which worsens somewhat the test of universality based on the superallowed Fermi transitions. On the other hand, Z_ψ, the additional Z occurring in E_6→SO(10)xU(1)$_\psi$, has a unique coupling to the light quarks and leptons and, therefore, does not affect the SM conclusions.

c) *Compositeness.* Manifestations of compositeness are frequently discussed in terms of residual 4-Fermi interactions characterized by a coupling $1/\Lambda^2$, where Λ represents the composite mass scale. This is to be compared with the coupling $G_\mu/\sqrt{2}$ of the usual 4-Fermi amplitudes. If, for some reason, the new interaction involves only particles of the same generation, it would affect ß transitions while leaving μ decay unaltered. Therefore, the test of universality would be also modified.

Eq. (15b) leads then to $2\sqrt{2}/(G_\mu\Lambda^2)<0.0036$ or $\Lambda>8$TeV (90% C.L.).

Veltman has also argued that the hypothetical compositeness of W bosons would affect the test of universality, because one should expect additional diagrams in ß decay, not present in μ-decay [69]. Presumably, the magnitude of the effect depends on the composite mass scale.

d) *Left-Right Symmetry.* In the "manifest" left-right symmetric models [70], there are two small parameters: the mixing angle ζ that relates the W_1 and W_2 mass eigenstates to the left and right handed fields W_L and W_R, and $\delta=(m_1/m_2)^2$, where $m_i(i=1,2)$ are the corresponding masses. To second order in the small parameters ζ and δ, $G_V/G_\mu=(1-\zeta)V_{ud}$, with analogous shifts for the other semileptonic decays [70,71]. As a consequence, Eq.(15b) becomes

$$\Sigma_j|V_{uj}|^2=0.9983\pm0.0015+2\zeta \qquad (18)$$

Thus, universality can be satisfied by choosing $\zeta=0.0009\pm0.0008$. An analysis of these models that combines the available data from neutron and nuclear ß decays is given in Ref. [72]. These studies favor the existence of W_R with a mass between 207 and 369 GeV and a very small value of ζ. However, this solution is not consistent with the constraints on δ derived from μ decay in the framework of manifest left-right models. Additional experimental information on neutron and nuclear ß decays is clearly highly desirable.

5. Electron-Muon Universality

In the previous Sections we have focussed our attention on CVC and the unitarity of the CKM matrix. It is clear that the theory leads to other, more specific, manifestations of universality. A particularly beautiful example is the verification of e-μ universality from $R_{e/\mu}=\Gamma(\pi\to e\bar{\nu}_e+e\bar{\nu}_e\gamma)/\Gamma(\pi\to\mu\bar{\nu}_\mu+\mu\bar{\nu}_\mu\gamma)$. Since the classical paper of Ruderman and Finkelstein [73], it has been known that, to $O(\alpha^0)$, $R_{e/\mu}$ is not affected by the dynamics of the strong interactions. $O(\alpha)$ corrections were computed in the pioneering papers of Refs. [74,75] but, for a long time, it was not clear to what extent these calculations are realistic. More recently, Terent'ev [76] and Marciano and Sirlin [77] gave complementary arguments that indicate that the $O(\alpha)$ corrections to $R_{e/\mu}$ can be evaluated with high precision, in spite of the complications caused by the strong interactions. When compared with the theoretical expectations, the combination of two recent precise experiments [78,79] leads to $g_e/g_\mu=0.9985\pm0.0015$ [80], a verification of e-μ universality at the 0.2% level!

References

1. B. Pontecorvo, *Phys. Rev.* **72** (1947) 246; J. Tiomno and J.A. Wheeler, *Rev. Mod. Phys.* **21** (1949) 144; O. Klein, *Nature* **161** (1948) 897; G. Puppi, *Nuovo Cim.* **5** (1948) 587 and **6** (1949) 194; T.D. Lee, M. Rosenbluth and C.N. Yang, *Phys. Rev.* **75** (1949) 905.
2. E. Fermi, *Elementary Particles* (Yale University Press, New Haven, 1951).
3. T.D. Lee and C.N. Yang, *Phys. Rev.* **104** (1956) 254.
4. C.S. Wu, E. Ambler, R.W. Hayward, D.D. Hoppes and R.P. Hudson, *Phys. Rev.* **105** (1957) 1413; R. Garwin, L. Lederman and M. Weinrich, *Phys. Rev.* **105** (1957) 1415; J.I. Friedman and V.L. Telegdi, *Phys. Rev.* **105** (1957) 1681; H. Frauenfelder et al., *Phys. Rev.* **106** (1957) 386.
5. E.C.G. Sudarshan and R.E. Marshak, in *Proc. Padua-Venice Conf. on Mesons and Recently Discovered Particles (1957)*, reprinted in *Development of the Weak Interaction Theory*, ed. P.K. Kabir (Gordon and Breach, New York, 1963) p.118; E.C.G. Sudarshan and R.E. Marshak, *Phys. Rev.* **109** (1958) 1860.
6. R.P. Feynman and M. Gell-Mann, *Phys. Rev.* **109** (1958) 193.
7. M. Gell-Mann, *Phys. Rev.* **111** (1958) 362.
8. For a recent discussion of CVC and its role in the history of weak interactions, see E.C. Sudarshan and R.E. Marshak, *U. of Texas (Austin) Report* DOE-ER-40200-154, *Virginia Tech Report* VPI-IHEP-88/7.
9. S.S. Gershtein and Ya.B. Zeldovich, *Zhur. Eksptl. i Teort. Fiz.* **29** (1955) 698 [translation: *Soviet Phys.* JETP **2** (1957) 576].
10. R.E. Behrends, J. Dreitlein, C. Fronsdal, and B.W. Lee, *Rev. Mod. Phys.* **34** (1962) 1 and references cited therein.
11. M. Gell-Mann, *Phys. Rev.* **125** (1962) 1067 and Caltech Report CTSL-20 (1961) (reprinted in M. Gell-Mann and Y. Ne'eman, *The Eightfold Way* (W.A. Benjamin, New York, 1964)); Y. Ne'eman, *Nucl. Phys.* **26** (1961) 222.
12. M. Gell-Mann, *Physics*, **1** (1964) 63.
13. S.L. Adler and R.F. Dashen, *Current Algebras and Applications to Particle Physics* (W.A. Benjamin, New York, 1968).
14. R.E. Behrends, R.J. Finkelstein, and A. Sirlin, *Phys. Rev.* **101** (1956) 866.
15. S.M. Berman, *Phys. Rev.* **112** (1958) 267.
16. T. Kinoshita and A. Sirlin, *Phys. Rev.* **113** (1959) 1652.
17. S.M. Berman and A. Sirlin, *Ann. Phys.* **20** (1962) 20.
18. G. Preparata and W.I. Weisberger, *Phys. Rev.* **175** (1968) 1965.

19. S.L. Adler, in *Lectures on Elementary Particles and Quantum Field Theory, 1970 Brandeis Univ. Summer Institute in Theoretical Physics, Vol. 1*, ed. S. Deser, M. Grisaru and H. Pendleton (M.I.T. Press, Cambridge, Massachusetts, 1970) p. 1.
20. T. Kinoshita, *J. Math. Phys.* **3**, (1962) 650; T.D. Lee and M. Nauenberg, *Phys. Rev.* **133B** (1964) 1549.
21. R.P. Feynman, in *Proceedings of the 1960 Annual International Conference on High Energy Physics at Rochester*, ed. E.C.G. Sudarshan, J.H. Tinlot, and A.C. Melissinos (Interscience Publishers, New York, 1960) p. 501.
22. N. Cabibbo, *Phys. Rev. Lett.* **10** (1963) 10.
23. R.E. Behrends and A. Sirlin, *Phys. Rev. Lett.* **4** (1960) 186.
24. M.V. Terent'ev, *Zhur. Eksptl. i Teort. Fiz.* **44** (1963) 1320 [translation: *Sov. Phys. JETP* **17** (1963) 890].
25. M. Ademollo and R. Gatto, *Phys. Rev. Lett.* **13** (1964) 264.
26. J.D. Bjorken, *Phys. Rev.* **148** (1966) 1467.
27. E.S. Abers, R.E. Norton, and D.A. Dicus, *Phys. Rev. Lett.* **18** (1967) 676.
28. K. Johnson and F.E. Low, *Progr. Theoret. Phys. (Kyoto) Suppl.* **37-38** (1966) 74.
29. A. Sirlin, *Phys. Rev.* *164* (1967) 1767.
30. E.S. Abers, D.A. Dicus, R.E. Norton and H.R. Quinn, *Phys. Rev.* **167** (1968) 1461.
31. G. Källén, *Nucl. Phys.* **B1** (1967) 225.
32. N. Cabibbo, L. Maiani, and G. Preparata, *Phys. Lett.* **25B** (1967) 31 and **25B** (1967) 132.
33. K. Johnson, F.E. Low, and H. Suura, *Phys. Rev. Lett.* **18** (1967) 1224.
34. A. Sirlin, *Phys. Rev. Lett.* **19** (1967) 877.
35. T.D. Lee, *Phys. Rev.* **128** (1962) 899; R.A. Shaffer, *Phys. Rev.* **128** (1962) 1452 and **131** (1963) 2203; G. Dorman, *Nuovo Cim.* **32** (1964) 1226; D. Bailin, *Phys. Rev.* **135** (1964)B166 and *Nuovo Cim.* **40A** (1965) 822.
36. A. Sirlin, in *Proceedings of the 14th International Conference on High-Energy Physics, Vienna 1968*, ed. J. Prentki and J. Steinberger (CERN, Scientific Information Service, Geneva, 1968) p. 321.
37. S.L. Adler and W.-K. Tung, *Phys. Rev. Lett.* **22** (1969) 75.
38. M.A.B. Bég, J. Bernstein and A. Sirlin, *Phys. Rev. Lett.* **23** (1969) 270 and *Phys. Rev.* **D6** (1972) 2597.
39. C.N. Yang and R.L. Mills, *Phys. Rev.* **96** (1954) 191.

40. See also the prescient work of O. Klein, in *New Theories in Physics* (International Institute of Intellectual Cooperation, League of Nations, 1938) p. 77.
41. S.L. Glashow, *Nucl. Phys.* **22** (1961) 579; A. Salam, in *Elementary Particle Theory*, ed. N. Svartholm (Almqvist and Wiksells, Stockholm, 1969) p. 367; S. Weinberg, *Phys. Rev. Lett.* **19** (1967) 1264.
42. M. Kobayashi and M. Maskawa, *Progr. Theor. Phys.* **49** (1973) 652.
43. A. Sirlin, *Nucl. Phys.* **B71** (1974) 29, *Nucl. Phys* **B100** (1975) 291, and in *American Institute of Physics Conference Proceedings No.23, Particles and Fields Subseries No. 10*, ed. E. Carlson (AIP, New York, 1974) p. 114.
44. A. Sirlin, *Rev. Mod. Phys.* **50** (1978) 573.
45. S. Weinberg, *Phys. Rev.* **D8** (1973) 4482.
46. A. Sirlin, *Nucl. Phys.* **B196** (1982) 83.
47. R.J. Blin-Stoyle and J.M. Freeman, *Nucl. Phys.* **A150** (1970) 369.
48. M. Roos and A. Sirlin, *Nucl. Phys.* **B29** (1971) 296.
49. W.J. Marciano and A. Sirlin, *Phys. Rev. Lett.* **56** (1988) 22.
50. W. Jaus and G. Rasche, *Phys. Rev.* **D41** (1990) 166.
51. I.S. Towner, *Nucl. Phys.*, **A540** (1992) 478.
52. F.C. Barker, B.A. Brown, W. Jaus and G. Rasche, *Nucl. Phys.* **A540** (1992) 501.
53. I.S. Towner, J.C. Hardy, E. Haberg and V.T. Koslowsky, *Superallowed Fermi β-Decay - a Status Report*, invited talk at the Int. Symp. on Weak and Electromagnetic Interactions in Nuclei (WEIN-92), Dubna, Russia (1992) and Chalk River Laboratories Report (1992).
54. W. Jaus and G. Rasche, *Nucl. Phys.* **A143** (1970) 202; W. Jaus, *Phys. Lett.* **40** (1972) 616.
55. N. Brene, M. Roos, and A. Sirlin, *Nucl. Phys.* **B6** (1968) 255.
56. A. Sirlin and R. Zucchini, *Phys. Rev. Lett.* **57** (1986) 1994; A. Sirlin, *Phys. Rev.* **D35** (1987) 3423.
57. W. Jaus and G. Rasche, *Phys. Rev.* **D35** (1987) 3420.
58. I.S. Towner, J.C. Hardy and M. Harvey, *Nucl. Phys.* **A284** (1977) 269.
59. W.E. Ormand and B.A. Brown, *Nucl. Phys.* **A440** (1985) 274; *Phys. Rev. Lett.* **62** (1989) 866.
60. J.C. Hardy, I.S. Towner, V.T. Koslowsky, E. Hagberg and H. Schmeing, *Nucl. Phys.* **A509** (1990) 429. See also Ref. [53].

61. H. Leutwyler and M. Roos, *Z. Phys.* **C25** (1984) 91.

62. K. Hikasa et al., Review of Particle Properties, *Phys. Rev.* **D45**, 1 June 1992, Part II.

63. D.H. Wilkinson, *Nucl. Phys.* **A511** (1990) 301; *Superallowed Fermi Beta-Decay: Applications*, TRIUMF Report [1990], contribution to the PANIC Conference, Boston, Massachusetts (1990). See also G. Rasche and W.S. Woolcock, *Mod. Phys. Lett.* **A5** (1990) 1273.

64. D. Počanić, private communication.

65. G. Rasche, D. Robustelli and F.C. Barker, *Inst. für Theor. Phys. Universität Zurich Report* [1991].

66. B.G. Erozolimski et al., *Phys. Lett.* **B263** (1991) 33.

67. K. Schreckenbach and W. Mampe, *J. Phys.* **G18** (1992) 1.

68. W.J. Marciano and A. Sirlin, *Phys. Rev.* **D35** (1987) 1672.

69. M. Veltman, in *Proceedings of the NATO Advanced Research Workshop on Radiative Corrections: Results and Perspectives, Brighton, United Kingdom, July 1989, NATO ASI Series B, Vol. 233*, ed. N. Dombey and F. Boudjema (Plenum Press, New York, 1990) p. 1.

70. M.A.B. Bég, R.V. Budny, R. Mohapatra, and A. Sirlin, *Phys. Rev. Lett.* **38** (1977) 1252 and references cited therein.

71. B.R. Holstein and S.B. Treiman, *Phys. Rev.* **D16** (1977) 2369.

72. A.S. Carnoy, J. Deutsch, R. Prieels, N. Severijns and P.A. Quin, *J. Phys.* **G18** (1992) 823; A.S. Carnoy, J. Deutsch, T.A. Girard and R. Prieels, *Phys. Rev. Lett.* **65** (1990) 3249 and *Phys. Rev.* **C43** (1991) 2825; J. Deutsch and P.A. Quin, this volume.

73. M. Ruderman and R. Finkelstein, *Phys. Rev.* **76** (1949) 1458.

74. S.M. Berman, *Phys. Rev. Lett.* **1** (1958) 468.

75. T. Kinoshita, *Phys. Rev. Lett.* **2** (1959) 477.

76. M.V. Terent'ev, *Yad. Fiz.* **18** (1973) 870 [translation: *Sov. J. Nucl. Phys.* **18** (1974) 449].

77. W.J. Marciano and A. Sirlin, *Phys. Rev. Lett.* **36** (1976) 1425.

78. D.I. Britton et al., *Phys. Rev. Lett.* **68** (1992) 3000.

79. G. Czapek et al., *Phys. Rev. Lett.* **70** (1993) 17.

80. D.A. Bryman, TRIUMF Report TRI-PP-93-4 (1993).

BETA DECAY AND MUON DECAY BEYOND THE STANDARD MODEL

Peter Herczeg
Theoretical Division, Los Alamos National Laboratory
Los Alamos, NM 87545, USA

Contents

I. INTRODUCTION

Nuclear beta decay and the usual decay of the muon are among the oldest tools of particle physics. Their study helped to develop the electroweak component of the Standard Model (SM) [1,2]. Today their main role is to probe for possible deviations from the predictions of this model. Although the SM is consistent with all observations, for many theoretical reasons, and especially because of the large number of undetermined parameters in the model, the existence of new physics is expected.

In beta decay and muon decay new physics can manifest itself through the effects of neutrino mass and mixing, and through new contributions to the decay interaction. Comparison of the superallowed $0^+ \to 0^+$ beta decay rates and the muon decay rate provides the ud-element of the Kobayashi-Maskawa (KM) matrix. Limits on charged-current universality, expressed through the unitarity relation for the matrix elements of the KM matrix, set constraints on some classes of new decay interactions, and also on some additional types of new physics.

In this article we shall review and discuss new beta decay and muon decay interactions, and the role of the beta decay and muon decay experiments in obtaining information on them. Experimental aspects of searches for new interactions in beta decay and muon decay are discussed in Ref. [3] and Ref. [4], respectively. The subject of charged-current universality is reviewed in Ref. [5]. For the status of searches for neutrino mass and heavy neutrinos in beta decay and for analyses of limits on neutrino mixing from muon decay we refer the reader to the articles in Ref. [6].

Section II and Section III of this review deals with beta decay and muon decay, respectively. In Section IV we summarize our conclusions.

II. BETA DECAY

II.1. Introduction

In the SM the $d \to u\, e^- \overline{\nu}_e$ (and $u \to d\, e^+ \nu_e$) transition underlying beta decay arises from W-exchange, and has the V-A form [7]

$$H \;=\; (G\, U_{ud}/\sqrt{2})\; \overline{e}\gamma_\lambda(1-\gamma_5)\nu_e\; \overline{u}\gamma^\lambda(1-\gamma_5)d \;+\; \text{H.c.}\;, \tag{1}$$

where $G/\sqrt{2} = g^2/8M_W^2$, and U_{ud} is the ud-element of the Kobayashi-Maskawa matrix. The field $\frac{1}{2}(1-\gamma_5)\,\nu_e$ in the interaction (1) represents a massless two-component neutrino, which is the $T_z = +1/2$ state of the $\mathrm{SU}(2)_L$ doublet involving the electron.

Although the SM interaction dominates beta decay, the data still allow sizable contributions from possible new interactions [3]. We shall consider here only such contributions from new physics which appear already at the tree-level [8]. Tree-level contributions can be described at the quark level by nonderivative local four-fermion couplings [9]. The most general form of such interactions for the $d \to u\, e^- \nu_e$ transition can be written as

$$H_\beta \;=\; H_{V,A} \;+\; H_{S,P} \;+\; H_T\;, \tag{2}$$

where

$$\begin{aligned} H_{V,A} &= \bar{e}\gamma^{\lambda}(1-\gamma_5)\,\nu_e^{(L)}\ [a_{LL}\,\bar{u}\,\gamma_\lambda(1-\gamma_5)d\ +\ a_{LR}\,\bar{u}\,\gamma_\lambda(1+\gamma_5)d] \\ &+\ \bar{e}\,\gamma^{\lambda}(1+\gamma_5)\,\nu_e^{(R)}\ [a_{RR}\,\bar{u}\,\gamma_\lambda(1+\gamma_5)d\ +\ a_{RL}\,\bar{u}\,\gamma_\lambda(1-\gamma_5)d] \\ &\qquad +\ \text{H.c.}\ , \end{aligned} \tag{3}$$

$$\begin{aligned} H_{S,P} &= \bar{e}(1-\gamma_5)\,\nu_e^{(L)}\ [A_{LL}\,\bar{u}(1-\gamma_5)d\ +\ A_{LR}\,\bar{u}\ (1+\gamma_5)d] \\ &+\ \bar{e}(1+\gamma_5)\,\nu_e^{(R)}\ [A_{RR}\,\bar{u}(1+\gamma_5)d\ +\ A_{RL}\,\bar{u}(1-\gamma_5)d] \\ &\qquad +\ \text{H.c.}\ , \end{aligned} \tag{4}$$

$$H_T = \alpha_{LL}\,\bar{e}\,\frac{\sigma_{\lambda\mu}}{\sqrt{2}}\,(1-\gamma_5)\,\nu_e^{(L)}\,\bar{u}\,\frac{\sigma_{\lambda\mu}}{\sqrt{2}}\,(1-\gamma_5)d \tag{5}$$

$$+\ \alpha_{RR}\,\bar{e}\,\frac{\sigma_{\lambda\mu}}{\sqrt{2}}\,(1+\gamma_5)\,\nu_e^{(R)}\,\bar{u}\,\frac{\sigma_{\lambda\mu}}{\sqrt{2}}\,(1+\gamma_5)d\ +\ \text{H.c.}$$

Our notation in Eqs. (3) - (5) is such that the first and the second subscript on the coupling constants gives, respectively, the chirality of the neutrino and of the d-quark. The fields e, u and d are mass-eigenstates. Note that there are no tensor couplings of the α_{LR}- and α_{RL}-type, due to the relation $\sigma_{\lambda\mu}\gamma_5 = \frac{1}{2}\,i\epsilon_{\lambda\mu\alpha\beta}\,\sigma^{\alpha\beta}$. The interactions (3) - (5) are time reversal invariant only if all the coupling constants can be made real.

In Eqs. (3) - (5) we have assumed that in all interaction terms the electron couples only to the neutrino states $\nu_e^{(L)}$ and $\nu_e^{(R)}$, where $\nu_e^{(L)}$ is the neutrino state in the $W^+ \to e^+\,\nu_e^{(L)}$ amplitude and $\nu_e^{(R)}$ is a right-handed singlet state. Couplings involving other neutrino states are possible, but for these in most cases additional constraints apply [10].

In general $\nu_e^{(L)}$ and $\nu_e^{(R)}$ are, respectively, linear combinations of the left-handed and the right-handed components of the neutrino mass-eigenstates ν_i

$$\nu_e^{(L)} = \sum_i U_{ei}\,\nu_{iL}\ , \tag{6}$$

$$\nu_e^{(R)} = \sum_i V_{ei}\,\nu_{iR}\ , \tag{7}$$

where $\nu_{iL} = \frac{1}{2}(1-\gamma_5)\,\nu_i,\ \nu_{iR} = \frac{1}{2}(1+\gamma_5)\,\nu_i$; U_{ei} and V_{ei} are (in a basis where the charged leptons are diagonal) elements of the neutrino mixing matrix [11].

Let us consider the decay of the nucleon due to the interaction (2) in the case when only a single neutrino mass-eigenstate is involved, i.e. when $\nu_e^{(L)}$ and $\nu_e^{(R)}$ are, respectively, the left-handed and right-handed components of a (Dirac) mass-eigenstate ν_e. Neglecting the induced form factors (see Ref. [12]), the effective interaction describing $n \to p\,e^-\bar{\nu}_e$ is given by

$$H_\beta^{(N)} \simeq H_{V,A}^{(N)} + H_S^{(N)} + H_T^{(N)}\ , \tag{8}$$

where

$$H_{V,A}^{(N)} = \bar{e}\gamma_\lambda(C_V + C'_V\gamma_5)\,\nu_e\,\bar{p}\gamma^\lambda n$$
$$+\ \bar{e}\gamma_\lambda\gamma_5(C_A + C'_A\gamma_5)\,\nu_e\,\bar{p}\gamma^\lambda\gamma_5 n\ +\ \text{H.c.}\ , \tag{9}$$

$$H_S^{(N)} = \bar{e}(C_S\ +\ C'_S\gamma_5)\,\nu_e\,\bar{p}n\ +\ \text{H.c.} \tag{10}$$

$$H_T^{(N)} = \bar{e}\,\frac{\sigma_{\lambda\mu}}{\sqrt{2}}\,(C_T + C'_T\gamma_5)\,\nu_e\,\bar{p}\,\frac{\sigma_{\lambda\mu}}{\sqrt{2}}\,n\ +\ \text{H.c.} \tag{11}$$

In Eqs. (9) - (11)

$$C_V = g_V\,(a_{LL}\ +\ a_{LR}\ + a_{RR}\ +\ a_{RL})\ , \tag{12}$$

$$C'_V = g_V(-a_{LL}\ - a_{LR}\ +\ a_{RR}\ +\ a_{RL})\ , \tag{13}$$

$$C_A = g_A(a_{LL}\ -\ a_{LR}\ +\ a_{RR}\ - a_{RL})\ , \tag{14}$$

$$C'_A = g_A(-a_{LL}\ +\ a_{LR}\ +\ a_{RR}\ -\ a_{RL})\ , \tag{15}$$

$$C_S = g_S(A_{LL}\ +\ A_{LR}\ +\ A_{RR}\ +\ A_{RL})\ . \tag{16}$$

$$C'_S = g_S(-A_{LL}\ -\ A_{LR}\ +\ A_{RR}\ +\ A_{RL})\ , \tag{17}$$

$$C_T = 2g_T(\alpha_{LL}\ +\ \alpha_{RR})\ , \tag{18}$$

$$C'_T = 2g_T(-\alpha_{LL}\ +\ \alpha_{RR})\ , \tag{19}$$

where the constants $g_V \equiv g_V(0)$, $g_S \equiv g_S(0)$ and $g_T \equiv g_T(0)$ are defined by

$$\langle p|\bar{u}\,\gamma_\lambda d|n\rangle = g_V(q^2)\bar{u}_p\,\gamma_\lambda\,u_n \tag{20}$$

$$\langle p|\bar{u}\,\gamma_\lambda\gamma_5 d|n\rangle = g_A(q^2)\bar{u}_p\,\gamma_\lambda\,\gamma_5\,u_n \tag{21}$$

$$\langle p|\bar{u}d|u\rangle = g_S(q^2)\bar{u}_p\,u_n \tag{22}$$

and

$$\langle p|\bar{u}\sigma_{\lambda\mu}d|n\rangle = g_T(q^2)\bar{u}_p\,\sigma_{\lambda\mu}\,u_n\ . \tag{23}$$

CVC predicts $g_V = 1$, and in the absence of new interactions the experimental value of g_A is $g_A = -1.2573 \pm 0.0028$ [13]. In the Hamiltonian (8) we did not include the combination from the part of $H_{S,P}$ (Eq. (4)) involving the pseudoscalar quark current, since these terms give no contribution to the beta decay observables in the nonrelativistic approximation for the nucleons. The interaction (8) is identical with the general beta decay interaction considered in Ref. [14].

In the general case when $\nu_e^{(L)}$ and $\nu_e^{(R)}$ are linear combinations of the mass-eigenstates, the observed beta decay probability is the sum of the probabilities of decays into the energetically allowed neutrino mass-eigenstates. In the following we shall assume that the neutrinos that can be produced in beta-decay are light enough that the effect of their masses on the decay probability can be neglected. In particular, we shall neglect the terms arising from the interference between amplitudes involving neutrinos of different chirality. As it is easily seen, under the above assumption the effect of neutrino mixing can be taken into account by multiplying in observables the coupling constants $\frac{1}{2}(C_K - C'_K)$ $(K = V, A, S, T)$ (which describe the interactions of $\nu_e^{(L)}$) by $\sqrt{u_e}$, and the coupling constants $\frac{1}{2}$ $(C_K+C'_K)$ $(K = V, A, S, T)$ (describing the interactions of $\nu_e^{(R)}$) by $\sqrt{v_e}$, where

$$u_e = \sum_i{}' |U_{ei}|^2 \tag{24}$$

$$v_e = \sum_i{}' |V_{ei}|^2 \tag{25}$$

The prime on the summation in Eqs. (24) and (25) indicates that the sum extends only over the neutrinos that are light enough to be produced in beta decay. Without changing the notation, we shall understand in the following that in the expressions for the observables given in Ref. [14] the substitutions $\frac{1}{2}(C_K - C'_K) \to \frac{1}{2}(C_K - C'_K)\sqrt{u_e}$, $\frac{1}{2}(C_K + C'_K) \to \frac{1}{2}(C_K + C'_K)\sqrt{v_e}$ have always been made. Note that if all the neutrinos are light, we have $u_e = v_e = 1$, as a consequence of the unitarity of the neutrino mixing matrix.

The terms in the Hamiltonians (9) - (11) involving the right-handed neutrino state $\nu_e^{(R)}$ can manifest themselves in beta decay only if either the right-handed neutrinos are sufficiently light, or (for Majorana neutrinos) if there is mixing between the heavy right-handed neutrinos and the light ones. In the latter scenario the effects of the $\nu_e^{(R)}$-terms are expected to be suppressed by the light-heavy neutrino mixing angles, which should be small. For the interactions of light right-handed neutrinos there are stringent constraints from nucleosynthesis in the standard big bang model, and from the energetics of supernova 1987A.

In the standard big bang cosmology an experimental upper limit on the primordial ^{4}He abundance gives a constraint on the number of light ($\lesssim 1$ MeV) neutrino species [15]. The most recent analysis [16] of nucleosynthesis set the limit $N_\nu \leq 3.3$ on the effective number of light neutrinos. An implication of this result is that if light right-handed neutrinos exist, their interactions must be weaker than the strength of the weak interactions [17]. For one light right-handed neutrino (in addition to three known left-handed neutrinos) the limit $N_\nu \leq 3.3$ yields a decoupling temperature of $T_d \gtrsim 200$ MeV [17]. Since one has roughly $T_d \simeq (G'/G)^{2/3}$ [17], where G' is the strength of the interactions involving the right-handed neutrino, this implies $G' \lesssim 4 \times 10^{-3}G$ for the $\nu_e^{(R)}$-terms in the Hamiltonian (8). A more quantitative estimate of the upper limit on the strength of the beta decay interactions involving $\nu_e^{(R)}$ would require an analysis of the interaction of $\nu_e^{(R)}$ with the pions which, to our knowledge, has not been yet done.

A stringent constraint on the interactions of light ($\lesssim$ 10 MeV) right-handed neutrinos comes from the observed neutrino pulse from the supernova 1987A [18,19]. The observed $\overline{\nu}_e$-luminosity is consistent with the standard supernova model, and this implies severe constraints on possible new cooling mechanisms of the supernova core. The requirement that the process $e^- \, p \to \nu_e^{(R)} \, n$ does not carry away most of the energy that can be radiated by the supernova leads for V,A interactions to the conclusion [19] that the coupling constants have to satisfy either the upper bound

$$\left(\tfrac{1}{6}\, g_V^2 |\eta_{RR}^{(e)} + \eta_{RL}^{(e)}|^2 \; + \; \tfrac{1}{2}\, g_A^2 |\eta_{RR}^{(e)} - \eta_{RL}^{(e)}|^2\right)^{1/2} \lesssim 1.2 \times 10^{-5} \,, \tag{26}$$

or the lower bound

$$\left(\tfrac{1}{6}\, g_V^2 |\eta_{RR}^{(e)} + \eta_{RL}^{(e)}|^2 \; + \; \tfrac{1}{2}\, g_A^2 |\eta_{RR}^{(e)} - \eta_{RL}^{(e)}|^2\right)^{1/2} \gtrsim \; 2 \times 10^{-2} \,, \tag{27}$$

where $\eta_{Rk} = a_{Rk}\sqrt{\overline{v}_e}/a_{LL}\sqrt{\overline{u}_e}$ $(k = L, R)$. For other types of beta decay interactions involving $\nu_e^{(R)}$ the constraints are probably similar. The present experimental limits are not far from ruling out the range (27) (see Section II.2.1). The bounds from the supernova on the beta decay interactions could be evaded if the right-handed neutrino has some additional interaction with electrons or nucleons which can trap them. A special interaction of this kind, which moreover does not cause conflict with the limit from nucleosynthesis on the effective number of light neutrinos, has been suggested in Ref. [20].

In the following while we shall bear in mind the constraints on right-handed neutrinos from nucleosynthesis and the supernova, we shall not invoke them in our discussions, since they do not diminish the importance of terrestrial experiments.

To conclude this section, we shall list the expressions from Ref. [14] for the few observables which we shall need to refer to in the subsequent discussions.

For allowed decays the $e^{\mp}$ longitudinal polarization P_L in the direction parallel to $\vec{p}_e$ in allowed transitions is

$$P_L \; = \; G\, \frac{p_e}{E_e}/(1 \; + \; b\, \frac{m_e}{E_e}) \,, \tag{28}$$

where

$$\begin{aligned} G\xi \; &= \; 2|M_F|^2[\mp Re(C_S C_S'^* \; - \; C_V C_V'^*) \; - \; \frac{\alpha Z m_e}{p_e}\; Im(C_S C_V'^* \; + \; C_S' C_V^*)] \\ &+ \; 2|M_{GT}|^2[\mp Re(C_T C_T'^* \; - \; C_A C_A'^*) \\ &+ \; \frac{\alpha Z m_e}{p_e}\; Im(C_T C_A'^* \; + \; C_T' C_A^*) \,, \end{aligned} \tag{29}$$

$$\begin{aligned} \xi \; &= \; |M_F|^2(|C_S|^2 \; + \; |C_S'|^2 \; + \; |C_V|^2 \; + \; |C_V'|^2) \\ &+ \; |M_{GT}|^2(|C_A|^2 \; + \; |C_A'|^2 \; + |C_T|^2 \; + |C_T'|^2) \end{aligned} \tag{30}$$

and b is the Fierz interference term, given by

$$b\xi = \pm 2(1-\alpha^2 Z^2)^{1/2}\, Re[|M_F|^2(C_S C_V^* + C_S' C_V'^*) - |M_{GT}|^2 (C_T C_A^* + C_T' C_A'^*)] \,. \tag{31}$$

The coefficients D and R of the correlations $\langle\vec{J}\rangle \cdot \vec{p}_e \times \vec{p}_\nu / J E_e E_\nu$ and $\vec{\sigma} \cdot \langle\vec{J}\rangle \times \vec{p}_e / J E_e$ ($\vec{\sigma}$ = electron spin, $\vec{J}$ = nuclear spin) can be written as $D = D_t + D_f$, $R = R_t + R_f$, where D_t, R_t represent the T-violating contributions and D_f, R_f are the T-invariant contributions due to electromagnetic final state interactions. D_t and R_t are given by

$$D_t\xi = 2\delta_{J'J} M_F M_{GT} \left(\frac{J}{J+1}\right)^{1/2} Im(C_S C_T^* + C_S' C_T'^* + C_V C_A^* + C_V' C_A'^*) \,, \tag{32}$$

$$\begin{aligned} R_t\xi = & \pm 2\lambda_{J'J} |M_{GT}|^2\, Im(C_T C_A'^* + C_T' C_A^*) \\ & + 2\delta_{JJ'} M_F M_{GT} \left(\frac{J}{J+1}\right)^{1/2} Im(C_S C_A'^* \\ & + C_S' C_A^* + C_V C_T'^* + C_V' C_T^*) \,, \end{aligned} \tag{33}$$

where $\lambda_{J'J}$ is an angular momentum factor, defined in Ref. [14].

In the next section we shall consider new beta decay interactions of V,A structure, first the model independent aspects (Section II.2.1) and then V,A interactions in left-right symmetric models (Section II.2.2), in models with exotic quarks and leptons (Section II.2.3), and the V,A interactions arising from the exchange of leptoquarks (Section II.2.4). In Sections II.3 and II.4 we discuss, respectively, scalar and tensor interactions added to the SM interaction. The special case of scalar and tensor interactions from leptoquark exchange is considered in Section II.5.

II.2 New V,A Interactions

II.2.1 Model Independent Considerations

In this section we shall discuss the parameters which describe the general beta decay interaction involving vector and axial-vector currents, and consider the constraints on them which come from beta decay itself, and from processes or observables to which the new V,A beta decay interactions contribute in first order.

The most general form of the Hamiltonian for $d \to u e^- \nu_e^{(L,R)}$ constructed from vector and axial-vector currents is given in Eq. (3). For given neutrino states $\nu_e^{(L)}$ and $\nu_e^{(R)}$ the Hamiltonian (3) contains 8 real parameters (four complex coupling constants). One of these is an overall phase, which does not enter the observables. We can choose therefore a_{LL} to be real and positive. Defining $\eta_{ik} = a_{ik}/a_{LL}$ ($ik = LR, RR, RL$), a set of the remaining six parameters is, for example, $|\eta_{LR}|, |\eta_{RR}|, |\eta_{RL}|$, the phases $e^{i\varphi_L} = \eta_{LR}/|\eta_{LR}|$, $e^{i\varphi_R} = \eta_{RL}\eta_{RR}^*/|\eta_{RR}||\eta_{RL}|$ and $e^{i\varphi_{RL}} = \eta_{RL}^*/\eta_{RL}$.

We shall not consider further the phase φ_{RL} since in observables terms proportional to sin φ_{RL} are proportional to neutrino mass. The reason is that such terms arise from the interference of amplitudes involving neutrinos of different chiralities.

Let us consider the beta decay of the nucleon in the framework of the Hamiltonian (9). We can write (9) as (cf. Eqs. (12) - (15))

$$H_\beta^{(N)} = a_{LL} g_V (1+\eta_{LR})[\bar{e}\gamma_\mu(1-\gamma_5)\,\nu_e^{(L)}\,\bar{p}\gamma^\mu(1-\lambda\gamma_5)n + \bar{e}\gamma_\mu(1+\gamma_5)\,\nu_e^{(R)}\,\bar{p}\gamma^\mu(x+\gamma_5\lambda y)n] + \text{H.c.}\,, \tag{34}$$

where [21]

$$\lambda = \left(\frac{g_A}{g_V}\right)\frac{1-\eta_{LR}}{1+\eta_{LR}}\,, \tag{35}$$

$$x = \frac{\eta_{RR}+\eta_{RL}}{1+\eta_{LR}}\,, \tag{36}$$

and

$$\lambda y = \left(\frac{g_A}{g_V}\right)\frac{\eta_{RR}-\eta_{RL}}{1+\eta_{LR}}\,. \tag{37}$$

As follows from Eq. (34), normalized observables (such as asymmetries or polarizations) can involve 5 parameters: $|\lambda|, |x|, |y|$, the phase of λ, and the relative phase of x and λy. The rate depends also on a_{LL}. As seen from Eq. (34), as long as the induced form factors are neglected (as we do here), the number of parameters at the nucleon level remains the same as at the quark level, since the only change is that $1+\eta_{LR}$ is replaced at the nucleon level by $g_V(1+\eta_{LR})$, and $(1-\eta_{LR})$ and $(\eta_{RR}-\eta_{RL})$ get multiplied by g_A.

In the following we shall keep in λ, x, y and λy only the lowest order terms in the η_{ik}'s. In this approximation we have

$$Re\,\lambda \simeq (g_A/g_V)(1-2\,Re\,\eta_{LR})\,, \tag{38}$$

$$Im\,\lambda \simeq -2(g_A/g_V)\,Im\,\eta_{LR} \simeq -2(Re\,\lambda)Im\,\eta_{LR}\,, \tag{39}$$

$$x \simeq \eta_{RR} + \eta_{RL}\,, \tag{40}$$

$$y \simeq \eta_{RR} - \eta_{RL}\,, \tag{41}$$

$$\lambda y \simeq (Re\,\lambda)(\eta_{RR}-\eta_{RL})\,, \tag{42}$$

$$Re\,x^*\lambda y \simeq (Re\,\lambda)(|\eta_{RR}|^2-|\eta_{RL}|^2)\,, \tag{43}$$

$$Im\,x^*\lambda y \simeq -(Re\,\lambda)Im\,\eta_{RR}^*\eta_{RL}\,. \tag{44}$$

For a_{LL} and the η_{ik}'s the substitutions in the observables required to take into account the presence of more than one neutrino mass-eigenstate and neutrino mixing are

$$\begin{aligned} a_{LL} &\to a_{LL}^{(e)} \equiv a_{LL}\sqrt{u_e} \\ \eta_{RR} &\to \eta_{RR}^{(e)} \equiv \eta_{RR}\sqrt{\tilde{v}_e} \\ \eta_{RL} &\to \eta_{RL}^{(e)} \equiv \eta_{RL}\sqrt{\tilde{v}_e} \end{aligned} \tag{45}$$

In Eq. (45) $\tilde{v}_e = v_e/u_e$, where u_e and v_e have already been defined in Eqs. (24) and (25).

Observables in pure Fermi transitions can depend only on $Re\,C_V C_V'^*$, $Im\,C_V C_V'^*$ and $|C_V|^2 \pm |C_V'|^2$, and observables in pure Gamow-Teller transitions on $Re\,C_A C_A'^*$, $Im\,C_A C_A'^*$, and $|C_A|^2 \pm |C_A'|^2$. From these the only ones which are not associated with interference between amplitudes involving neutrinos of different chirality are $Re\,C_K C_K'^*$ and $|C_K|^2+|C_K'|^2$ $(K = V, A)$. Thus normalized observables in pure beta decays can involve only

$$\frac{Re(C_V C_V'^*)}{|C_V|^2+|C_V'|^2} = -\frac{1}{2}\frac{1-|x^{(e)}|^2}{1+|x^{(e)}|^2} \simeq -\frac{1}{2}(1-2|\eta_{RR}^{(e)}+\eta_{RL}^{(e)}|^2)\,, \tag{46}$$

or

$$\frac{Re(C_A C_A'^*)}{|C_A|^2+|C_A'|^2} = -\frac{1}{2}\frac{1-|y^{(e)}|^2}{1+|y^{(e)}|^2} \simeq -\frac{1}{2}(1-2|\eta_{RR}^{(e)}-\eta_{RL}^{(e)}|^2)\,, \tag{47}$$

where $x^{(e)}$ and $y^{(e)}$ are the x and y in which the substitutions (45) have been made. The rates involve

$$|C_V|^2 + |C_V'|^2 \simeq 2|a_{LL}^{(e)}|^2 g_V^2\,|1+\eta_{LR}|^2\left(1+|\eta_{RR}^{(e)}+\eta_{RL}^{(e)}|^2\right), \tag{48}$$

or

$$|C_A|^2 + |C_A'|^2 \simeq 2|a_{LL}^{(e)}|^2 g_A^2\,|1-\eta_{LR}|^2\left(1+|\eta_{RR}^{(e)}-\eta_{RL}^{(e)}|^2\right). \tag{49}$$

In mixed transitions normalized observables can also depend on

$$\frac{|C_A|^2+|C_A'|^2}{|C_V|^2+|C_V'|^2} = |\lambda|^2\frac{1+|y^{(e)}|^2}{1+|x^{(e)}|^2} \simeq |\lambda|^2(1-4Re\,\eta_{RR}^{(e)*}\,\eta_{RL}^{(e)})\,, \tag{50}$$

$$\begin{aligned} Re(C_V C_A^* + C_V' C_A'^*)/(|C_V|^2+|C_V'|^2) &= (Re\,\lambda + Re\,x^{(e)*}\lambda y^{(e)})/(1+|x^{(e)}|^2) \\ &\simeq (Re\,\lambda)(1-2|\eta_{RL}^{(e)}|^2 - 2Re\,\eta_{RR}^{(e)*}\,\eta_{RL}^{(e)})\,, \end{aligned} \tag{51}$$

$$\begin{aligned} Re(C_V C_A'^* + C_V' C_A^*)/(|C_V|^2+|C_V'|^2) &= (-Re\,\lambda + Re\,x^{(e)*}\lambda y^{(e)})/(1+|x^{(e)}|^2) \\ &\simeq -(Re\,\lambda)(1-2|\eta_{RR}^{(e)}|^2 - 2Re\,\eta_{RR}^{(e)*}\eta_{RL}^{(e)})\,, \end{aligned}$$

(52)

$$Im(C_V C_A^* + C_V' C_A'^*)/(|C_V|^2 + |C_V'|^2) = -(Im\,\lambda + Im\,x^{(e)*}\lambda y^{(e)})/(1+|x^{(e)}|^2)$$
$$\simeq 2(Re\,\lambda)(Im\,\eta_{LR} + Im\,\eta_{RR}^{(e)*}\,\eta_{RL}^{(e)}) \,,$$

(53)

and

$$Im(C_V C_A'^* + C_A' C_A^*)/(|C_V|^2 + |C_V'|^2) = (Im\,\lambda - Im\,x^{(e)*}\lambda y^{(e)})/(1+|x^{(e)}|^2)$$
$$\simeq -2(Re\,\lambda)(Im\,\eta_{LR} - Im\,\eta_{RR}^{(e)*}\,\eta_{RL}^{(e)}) \,.$$

(54)

In nuclear beta decay observables g_V and g_A appear multiplied by the Fermi matrix element M_F and the Gamow-Teller matrix element M_{GT}, respectively.

$Im\,\eta_{LR}$, $Im\,\eta_{RR}^{(e)*}\,\eta_{RL}^{(e)}$. A general V, A interaction contributes to the time reversal violating component D_t of the T-odd D-correlation (cf. Eq. (32)). D_t is given by

$$D_t \simeq a_D\, Im(\eta_{LR} + \eta_{RR}^{(e)*}\,\eta_{RL}^{(e)}) \,. \tag{55}$$

The quantity a_D is proportional to $(r\,Re\,\lambda)/(1+r^2|\lambda|^2)$, where $r = M_{GT}/M_F$. For ^{19}Ne and for n-decay $a_D \simeq -1.03$ and $a_D \simeq 0.87$, respectively.

The best limit on D_t/a_D comes at present from ^{19}Ne-decay. The experimental value $D = (0.1 \pm 0.6) \times 10^{-3}$ [22] yields

$$|Im(\eta_{LR} + \eta_{RR}^{(e)*}\,\eta_{RL}^{(e)})| < 1.05 \times 10^{-3} \qquad (90\%\ c.l.) \,. \tag{56}$$

The contribution D_f of the electromagnetic final-state interactions to D has been estimated for this case to be of the order of $2 \times 10^{-4} p_e/(p_e)_{max}$ [23]. The present experimental result for the D-coefficient in neutron decay is $D = -0.0005 \pm 0.0014$ [13]. The final-state interaction contribution D_f is smaller than for ^{19}Ne decay by an order of magnitude [23].

The couplings that contribute to D will also give contributions to the electric dipole moment of the neutron (D_n) and of the electron (D_e) (through two-loop diagrams involving the a_{LR} and $(a_{LL})_{SM}$, and/or the a_{RR} and a_{RL} couplings). The upper limits on $|Im\,\eta_{LR}|$ and $|Im\,\eta_{RR}^{(e)*}\,\eta_{RL}^{(e)}|$ from the experimental limits [13] on these observables are not likely to be stronger than $\sim 10^{-3}$. The $Im\,\eta_{LR}$-term combined with the weak interaction contributes also to ϵ'/ϵ. Again, the corresponding limit on $|Im\,\eta_{LR}|$ from the experimental value of ϵ'/ϵ is not likely to be more stringent than $\sim 10^{-3}$.

$|\eta_{RR}^{(e)}|$, $|\eta_{RL}^{(e)}|$. A constraint on $|\eta_{RR}^{(e)} - \eta_{RL}^{(e)}|$ comes from measurements of the longitudinal polarization (P_L) of the charged lepton in Gamow-Teller beta decays. The experimental result (see Ref. [3]) $P_L^{GT}(E_e/p_e) = -0.998 \pm 0.014$ yields

$$|\eta_{RR}^{(e)} - \eta_{RL}^{(e)}| < 0.110 \qquad (90\%\ c.l.) \tag{57}$$

A novel type of experiment which can be very sensitive to the presence of $\eta_{RR}^{(e)}$ and/or $\eta_{RL}^{(e)}$ is the measurement of the ratio of the longitudinal polarizations of positrons emitted parallel and antiparallel to the direction of the spin of a polarized nucleus [24]. The first experiment of this kind, which measured the polarization of the positron from polarized ^{107}In decay, yielded $|\eta_{RR}^{(e)} - \eta_{RL}^{(e)}|^2 = 0.0080 \pm 0.0052$ [25] , implying

$$|\eta_{RR}^{(e)} - \eta_{RL}^{(e)}| < 0.122 \qquad (90\%\ c.l.)\ . \tag{58}$$

Comparisons of e^+-longitudinal polarizations in Fermi and Gamow-Teller decays [26] yielded $P_L^F/P_L^{GT} = 1.0010 \pm 0.0027$, implying (note that $P_L^F/P_L^{GT} - 1 \simeq -8Re\,\eta_{RR}^{(e)*}\,\eta_{RL}^{(e)}$)

$$-6.8 \times 10^{-4} \ < \ Re\,\eta_{RR}^{(e)*}\,\eta_{RL}^{(e)} \ < \ 4.3 \times 10^{-4} \qquad (90\%\ c.l.)\ . \tag{59}$$

$Re\,\eta_{LR}$. As $Re\,\eta_{LR}$ appears in lowest order only in the parameter λ, the uncertainties in the theoretical value of g_A (and in nuclear beta decay also the theoretical uncertainties in the ratio $|M_{GT}|^2/|M_F|^2$) prevent the possibility of setting stringent limits on $Re\,\eta_{LR}$ from beta decay.

In models where $a_{LL} \simeq (a_{LL})_{SM}$ a bound on $Re\,\eta_{LR}$ can be derived from the experimental value of the ratio

$$R_\pi \ = \ \frac{\Gamma(\pi \to e\nu_e) + \Gamma(\pi \to e\nu_e\gamma)}{\Gamma(\pi \to \mu\nu_\mu) \ + \ \Gamma(\pi \to \mu\nu_\mu\gamma)}\ . \tag{60}$$

R_π for the interaction (3) is then given by [27]

$$R_\pi \ = \ (R_\pi)_{SM}\ (u_e/u_\mu)\ [(1 - Re\,\eta_{LR})^2 \ + \ (Im\,\eta_{LR})^2 \ + \ |\eta_{RR}^{(e)} - \eta_{RL}^{(e)}|^2]\ , \tag{61}$$

where $(R_\pi)_{SM} = 1.234 \pm 0.001$ [28] is the value of R_π in the SM. A recent experiment [29] measured R_π with an improved precision, obtaining

$$(R_\pi)_{expt} = (1.2265 \ \pm \ 0.0034(\mathrm{stat}) \pm 0.004(\mathrm{sys})) \times 10^{-4}\ . \tag{62}$$

Neglecting the last two terms in Eq. (61), the result (62) implies

$$0.9932\ (u_\mu/u_e)^{1/2} \ < |1 - Re\,\eta_{LR}| < 1.0007\ (u_\mu/u_e)^{1/2} \qquad (90\%\ c.l.)\ , \tag{63}$$

where u_μ is defined as u_e (Eq. (24)) except for the replacement $U_{ei} \to U_{\mu i}$, and we have assumed for simplicity that the u_e's for $\pi \to e\,\nu_e$ and beta decay are equal.

For $u_e = u_\mu$ (63) allows $Re\,\eta_{LR}$ to be [30] either in the range

$$-0.0007 < Re\,\eta_{LR} < 0.0068\ , \tag{64}$$

or in the range

$$1.9932 < Re\,\eta_{LR} < 2.0007\ . \tag{65}$$

We note that the range (65) is ruled out by the experimental value of $|C_A|^2 + |C'_A|^2/(|C_V|^2 + C'_V|^2)$ ($\simeq (1.27)^2$ (see Ref. [3])). The latter, deduced from the neutron lifetime, would require $g_A/g_V \simeq 4$ (see Eqs. (50) and (59)) which is unreasonably large with respect to the theoretical prediction of g_A based on the Adler-Weisberger relation.

If we allow $|\eta^{(e)}_{RR} - \eta^{(e)}_{RL}|$ and $|Im\,\eta_{LR}|$ to be as large as the upper limits in Eqs. (75) and (73) below, or in Eqs. (75) and (57), the only appreciable effect is the increase of the upper bound in (64) to 1.3 $\times 10^{-2}$.

Comparison of the predicted value of M_W (using the values of G_F, $\sin^2\theta_W$, and Δr given in Ref. [13]) with the experimental one shows that the factor $(u_e\,u_\mu)^{1/2}$ cannot be smaller than unity by more than $\sim 1.5\%$. To account for possible new V,A contributions to muon decay we shall allow $(u_e\,u_\mu)^{1/2}$ to deviate from unity by 2%, which implies $0.98 \leq (u_e/u_\mu)^{1/2} \leq 1.02$. Allowing $(u_e/u_\mu)^{1/2}$ to take any value in this range has no appreciable effect on the range (65), but (64) becomes $(-2.1 \times 10^{-2}) < Re\,\eta_{LR} < 2.7 \times 10^{-2}$. If $|\eta^{(e)}_{RR} - \eta^{(e)}_{RL}|^2$ is also present (the effects of $Im\,\eta_{LR}$ are negligible), we obtain finally for $Re\,\eta_{LR}$ the bound

$$-2.1 \times 10^{-4} \ < \ Re\,\eta_{LR} \ < \ 3.3 \times 10^{-2} \ . \tag{66}$$

The limit on $Re\,\eta_{LR}$ from R_π may be weaker if a new V,A interaction contributing to $\pi \to \mu\nu_\mu$ is also present. Note that one would obtain $R_\pi = (R_\pi)_{SM}$ if the coupling constants in the muonic interaction are equal to the coupling constants in the interaction involving the electron.

The Results of a Comprehensive Analysis of Beta Decay Data. A comprehensive analysis of beta decay data [3], which included experimental results on neutron-decay and ^{19}Ne-decay, yielded the following 90% c.l. upper limits:

$$|\eta^{(e)}_{RR}| \ < \ 0.104 \ , \tag{67}$$

$$|\eta^{(e)}_{RL}| \ < \ 0.044 \ , \tag{68}$$

$$|Re\,\eta^{(e)}_{RR}| \ < \ 0.104 \ , \tag{69}$$

$$|Re\,\eta^{(e)}_{RL}| \ < \ 0.010 \ , \tag{70}$$

$$|Im\,\eta^{(e)}_{RR}| \ < \ 0.104 \ , \tag{71}$$

$$|Im\,\eta^{(e)}_{RL}| \ < \ 0.043 \ , \tag{72}$$

$$|Im\,\eta_{LR}| \ < \ 0.0038 \ , \tag{73}$$

and also [31]

$$|\eta^{(e)}_{RR} + \eta^{(e)}_{RL}| \ < \ 0.106 \ , \tag{74}$$

$$|\eta_{RR}^{(e)} - \eta_{RL}^{(e)}| < 0.106 \ . \quad (75)$$

This analysis was done in the framework of the Hamiltonian (8), with $H_{S,P}^{(N)}$ and $H_T^{(N)}$ absent.

It should be noted that the central value for $\eta_{RR}^{(e)}$ ($(\eta_{RR}^{(e)})_{central} = 0.077$) turns out to be 2.3 σ away from zero [3]. It should also be noted that if the experimental result of Ref. [32] on the asymmetry parameter A_n in neutron decay is removed from the data set, $|\eta_{RR}^{(e)}|$ becomes consistent with zero at the 1σ level [33].

The constraints we have considered in this section are independent of the source of the interaction (3) (except for the constraint from R_π, as explained in the text). For specific mechanisms that give rise to an interaction of the form (3) and for particular models additional constraints will apply in general.

Right-handed V,A beta decay interactions can arise at the tree-level if there are new charged gauge bosons which have right-handed couplings to the ordinary leptons and/or quarks, or if there are new quarks and leptons which have right-handed couplings to the W and which mix with the ordinary quarks and leptons, or in models involving leptoquarks. We shall discuss these mechanisms in the subsequent sections.

II.2.2. Left-Right Symmetric Models

Left-right symmetric models [34] are attractive extensions of the standard electroweak model, which provide a framework for the understanding of the origin of parity violation in the weak interaction. The simplest models are based on the gauge group $SU(2)_L \times SU(2)_R \times U(1)_{B-L}$ [34,35]. $SU(2)_L \times SU(2)_R \times U(1)_{B-L}$ models involve an additional neutral gauge boson, and also a new charged gauge boson. The coupling of the charged gauge bosons W_L and W_R to the quarks and the leptons is given by

$$\begin{aligned} \mathcal{L} &= (g_L/\sqrt{2})(\overline{Q}_L^{(u)}\gamma_\lambda U_L Q_L^{(d)} + \overline{n}_L \gamma_\lambda U^\dagger E_L)W_L \\ &+ (g_R/\sqrt{2})(\overline{Q}_R^{(u)}\gamma_\lambda U_R Q_R^{(d)} + \overline{n}_R \gamma_\lambda V^\dagger E_R)W_R + \text{H.c.} \ , \end{aligned} \quad (76)$$

where g_L and g_R are the $SU(2)_L$ and the $SU(2)_R$ gauge coupling constants, respectively, $\overline{Q}^{(u)} \equiv (\overline{u}, \overline{c}, \overline{t})$, $\overline{Q}^{(d)} \equiv (\overline{d}, \overline{s}, \overline{b})$, $\overline{E} = (\overline{e}, \overline{\mu}, \overline{\tau})$; $\overline{n}_L \equiv (\overline{\nu}_{1L}, \overline{\nu}_{2L}, \ldots)$ and $\overline{n}_R \equiv (\overline{\nu}_{1R}, \overline{\nu}_{2R}, \ldots)$ contain all the neutrino mass-eigenstates (three in the case of Dirac neutrinos, and six if the neutrinos are Majorana fermions); $\psi_{L,R} = \frac{1}{2}(1 \mp \gamma_5)\psi$ ($\psi = Q^{(u)}, Q^{(d)}, \ldots$). The matrices U_L, U_R and U, V are the quark and leptonic mixing matrices, respectively. The fields W_L and W_R are linear combinations of the mass-eigenstates W_1 and W_2:

$$\begin{aligned} W_L &= \cos\zeta\, W_1 + \sin\zeta\, W_2 \\ W_R &= e^{i\omega}(-\sin\zeta\, W_1 + \cos\zeta\, W_2) \ , \end{aligned} \quad (77)$$

where ζ is a mixing angle and ω is a CP-violating phase.

In $SU(2)_L \times SU(2)_R \times U(1)_{B-L}$ models CP-violation is present already for two-quark generations [36], due to CP-violating phases in U_R. For n generations U_L and U_R contain together $n(n-1)$ mixing angles and n^2-n+1 CP-violating phases [37].

The Hamiltonian responsible for nuclear beta decay [38] resulting from (76) is of the form (3) with

$$a_{LL}^{(e)} \simeq (g_L^2 \cos\theta_1^L / 8m_1^2)\sqrt{u_e}\ , \tag{78}$$

$$\eta_{RR}^{(e)} \simeq e^{i\alpha}(\cos\theta_1^R/\cos\theta_1^L)(g_R^2 m_1^2/g_L^2\, m_2^2)\sqrt{\tilde{v}_e}\ , \tag{79}$$

$$\eta_{LR} \simeq -e^{i(\alpha+\omega)}(\cos\theta_1^R/\cos\theta_1^L)(g_R\zeta/g_L) \tag{80}$$

$$\eta_{RL}^{(e)} \simeq -e^{-i\omega}(g_R\zeta/g_L)\sqrt{\tilde{v}_e}\ , \tag{81}$$

where m_1, m_2 are the masses of W_1, W_2; $\cos\theta_1^L = (U_L)_{ud}$ and $e^{i\alpha}\ \cos\theta_1^R = (U_R)_{ud}$. Note that for the phases φ_L and φ_R (see Section II.2.1) one has the relation $\varphi_R = -\varphi_L (= -\alpha - \omega)$.

A comprehensive analysis of [39] of the constraints on general $SU(2)_L \times SU(2)_R \times U(1)$ models led to the bound

$$(g_R^2 m_1^2/g_L^2 m_2^2) \lesssim 7.5 \times 10^{-2}\ , \tag{82}$$

valid for any type of right-handed neutrinos. The limit (82) has been obtained from the $K_L - K_S$ mass difference Δm_K requiring that each individual contribution to Δm_K, corresponding to box diagrams with a given pair of internal quarks, is smaller than the experimental value of Δm_K, and assuming some reasonable restrictions on fine-tuned cancellations. In the presence of CP-violation there is also a constraint from ϵ [40]. The limit on $g_R^2\, m_1^2/g_L^2\, m_2^2$ would become then more stringent than (82) if the CP-violating phases are not small.

The limit (82) implies

$$|\eta_{RR}^{(e)}| \lesssim 7.5 \times 10^{-2}\ , \tag{83}$$

where we have used $u_e \simeq 1$, and $\cos\theta_1^L \simeq 1$.

A search [41] for the processes $W' \to e\nu$ and $W' \to \mu\nu$ in $\overline{p}p$ collisions, where W' is a heavy vector boson, led to the lower bound $m_{W'} > 520$ GeV/c^2 (95% c.l.) (and therefore $m_2 > 520$ GeV/c^2) for $m_{\nu_R} \lesssim 15$ GeV/c^2, assuming Standard Model-strength couplings to the three fermion families. To deduce from this result a limit for η_{RR} does not appear to be straightforward and has not yet been to our knowledge considered.

A limit on $\eta_{RL}^{(e)}$ is provided by muon decay. The constants g_{LL}^V and κ_{ij}^V ($ij = RR, LR, RL$) (see Eqs. (223)-(225) in Section III.2) involved in the muon decay Hamiltonian generated by the leptonic couplings in (76) are related to the beta decay constants (78)-(81) as

$$(G_F/\sqrt{2})g^V_{LL} = a_{LL}/\cos\theta^L_1 \ , \tag{84}$$

$$\kappa^V_{RR} = \eta_{RR}[(\cos\theta^R_1/\cos\theta^L_1)e^{i\alpha}]^{-1} \ , \tag{85}$$

$$\kappa^V_{LR} = \eta_{LR}[(\cos\theta^R_1/\cos\theta^L_1)e^{i\alpha}]^{-1} \ , \tag{86}$$

$$\kappa^V_{RL} = \eta_{RL} \ . \tag{87}$$

The best limit on $|\kappa^V_{RL}\sqrt{\tilde{v}_e}|$ comes from the experimental value of the ρ-parameter (see Section III.2), implying $|\kappa^V_{RL}\sqrt{\tilde{v}_e}| < 0.067$ (90% c.l.), and therefore

$$|\eta^{(e)}_{RL}| < 0.067 \qquad\qquad (90\%\ c.l.). \tag{88}$$

Note that $\eta^{(e)}_{RR}$ is not constrained by muon decay data [42], since muon decay gives a bound on $\kappa^V_{RR}\sqrt{\tilde{v}_e\,\tilde{v}_\mu}$, and $\tilde{v}_\mu$ could be much smaller than $\tilde{v}_e$ [43].

Turning to the D-coefficient (Eq. (55)), the contribution from $Im\,\eta^{(e)*}_{RR}\,\eta^{(e)}_{RL}$ is relatively small [42] due to the relation $Im\,\eta^{(e)*}_{RR}\,\eta^{(e)}_{RL} = (g^2_R m^2_1/g^2_L m^2_2)Im\,\eta_{LR}$, and the limit (82). One has therefore [38]

$$D/a_D \simeq Im\,\eta_{LR} \simeq -(g_R\zeta/g_L)(\cos\theta^R_1/\cos\theta^L_1)\sin(\alpha+\omega) \ , \tag{89}$$

and thus from (56)

$$|Im\,\eta_{LR}| < 1.1\times 10^{-3} \ . \tag{90}$$

The phase $\alpha+\omega$ is constrained also by the experimental limit on the neutron electric dipole moment D_n and on ϵ'/ϵ, which require $|Im\,\eta_{LR}| \lesssim 3\times 10^{-5}$ [44]. These limits are however not as reliable as the limit (90), since the corresponding calculations may involve unknown uncertainties. Combining the bound (59) from P^F_L/P^{GT}_L with the limit on $Im\,\eta^{(e)*}_{RR}\,\eta^{(e)}_{RL}$ yields

$$|\eta^{(e)}_{RR}||\eta^{(e)}_{RL}| < 6.9\times 10^{-4} \ . \tag{91}$$

Information on η_{LR} in $SU(2)_L\times SU(2)_R\times U(1)$ models can be obtained from data on inclusive $\overline{\nu}_\mu$- and ν_μ-scattering on nuclei [45], since the constant η_{LR} for the $\nu_\mu\, d \to \mu\, u$ interaction is the same as for $d \to u\, e^-\overline{\nu}_e$ [46]. The ratio of the antineutrino to neutrino cross-sections is sensitive to $|\eta_{LR}|$ for x and y both large. A recent experiment by the CCFR collaboration yielded [47]

$$|\eta_{LR}| < 0.039 \qquad\qquad (90\%\ c.l.) \ . \tag{92}$$

For $Re\,\eta_{LR}$ one has also the limit [48]

$$|Re\,\eta_{LR}| < 6\times 10^{-3} \tag{93}$$

from the requirement that the PCAC predictions for the $K \to 3\pi$ amplitudes in terms of the $K \to 2\pi$ amplitudes, which hold to an accuracy of $\sim 10\%$, would be retained even in the presence of right-handed currents.

Combining (90) and (93) gives

$$|\eta_{LR}| < 6.1 \times 10^{-3} . \tag{94}$$

We comment yet on a relation involving $Re\,\eta_{LR}$ which follows from charged-current universality. The unitarity of the Kobayashi-Maskawa matrix for three families leads to the relation [49]

$$(g_R\zeta/g_L)[(U_L)_{ud}\,Re\,e^{i\omega}(U_R)_{ud} + (U_L)_{us}\,Re\,e^{i\omega}(U_R)_{us}] \simeq \tfrac{1}{2}(S_u - 1) , \tag{95}$$

where $S_u \equiv \sum_{i=d,s,b} |(\tilde{U}_L)_{ui}|^2$, and $|(\tilde{U}_L)_{ui}|$ is the apparent value of $|(U_L)_{ui}|$.

For U_R which yields the limit (82) one has $(U_R)_{ud} \neq 0$, $(U_R)_{us} = 0$, and therefore the relation (95) sets a bound on $Re\,\eta_{LR}$. Using for S_u the values obtained in various treatments of the superallowed $0^+ \to 0^+$ beta decays (see Refs. [3,5]), one obtains $|Re\,\eta_{LR}| <$ (1.6 to 3.4) $\times 10^{-3}$. Combining this with the limit (90) yields $|\eta_{LR}| < 3.6 \times 10^{-3}$. For the cases when $(U_R)_{us} \neq 0, (U_R)_{ud} = 0$ there is no significant constraint from (95) on $(g_R\xi/g_L)Re\,e^{i\omega}(U_R)_{us}$, since the present experimental result [50] on the D-coefficient in hyperon decays sets only a weak limit on the imaginary part of this term [51].

Nonmanifest models with $\tilde{v}_e = \tilde{v}_\mu = 1$. If in the models discussed above (referred to as models with nonmanifest left-right symmetry [52], which allow $g_R \neq g_L$ and $\theta_i^R \neq \theta_i^L$) $\tilde{v}_e = \tilde{v}_\mu = 1$, there is a limit on $|\eta_{RR}^{(e)}|$ which is somewhat stronger than (83), from the experimental lower bound on the quantity $R = 1 - \delta\,\xi\,P_\mu/\rho$ in muon decay (see Section III.2). The latter implies $|\kappa_{RR}^V| < 0.040$, and therefore (since $|\eta_{RR}^{(e)}| = |\eta_{RR}| \lesssim |\kappa_{RR}^V|$)

$$|\eta_{RR}^{(e)}| \lesssim 0.040 . \tag{96}$$

From the muon decay ρ-parameter one obtains in this case

$$|\eta_{RL}^{(e)}| \lesssim 0.047 . \tag{97}$$

Also, as $|\eta_{LR}| \lesssim |\eta_{RL}|$, and since $\eta_{RL} = \eta_{RL}^{(e)}$ for $\tilde{v}_e = 1$, Eq. (97) implies the additional limit $|\eta_{LR}| \lesssim 0.047$ on $|\eta_{LR}|$. An example of models with $\tilde{v}_e = \tilde{v}_\mu = 1$ is the class where U = V (such as $SU(2)_L \times SU(2)_R \times U(1)$ models with Dirac neutrinos and a discrete symmetry). One has $\tilde{v}_e = \tilde{v}_\mu = 1$ also in models where all the neutrinos are sufficiently light to be produced in beta decay.

Models with pseudomanifest or manifest left-right symmetry. In such models $g_R = g_L$ and $\theta_i^R = \theta_i^L$ [52]. In manifestly left-right symmetric models [53] in addition the relative phases between the right-handed and the left-handed couplings vanish, and $\omega = 0$ (so that $\sin(\alpha + \omega) = 0$) [54].

In manifestly left-right symmetric models the $K_L - K_S$ mass difference implies the bound $|\eta_{RR}| = m_1^2/m_2^2 \lesssim 6 \times 10^{-3}$ [55], and therefore

$$|\eta_{RR}^{(e)}| \lesssim 6 \times 10^{-3} . \tag{98}$$

In pseudomanifest models the constraint from ϵ has to be also considered [40]. Assuming that there are no fine-tuned cancellations among the contributions of the various CP-violating phases, the limit from ϵ is more stringent than (98) if the CP-violating phases are not small.

For $\eta_{RL}^{(e)}(= -\zeta\sqrt{\tilde{v}_e})$ the limit (97) holds. Applying the relation $|\zeta| \lesssim m_1^2/m_2^2$ [56], which fails to hold only if the Higgs sector of the model contains Higgs bosons in representations with $T_R \gg 1$ [39], one has also

$$|\eta_{RL}^{(e)}| \lesssim 6\times 10^{-3}\,. \tag{99}$$

For $|\eta_{LR}|(= |\zeta|)$ the limits (92) and (94) hold. In manifestly left-right symmetric models one has in addition the bound $|\zeta| < 3.4\times 10^{-3}$ from charged-current universality (Eq. (95)).

II.2.3. Exotic Fermions

The interaction which gives rise to the $d \to u\, e^-\overline{\nu}_e$ transition can contain terms involving right-handed currents even if the electroweak gauge group is just the standard model SU(2)×U(1) group. This happens if new quarks and leptons exist whose right-handed components are in non-singlet representations of SU(2), and which mix with the usual quarks and leptons [57]. Fermions with noncanonical SU(2)×U(1) assignments are referred to as "exotic." Such fermions occur in many extensions of the SM. If the electric charge and color assignments of the new fermions are the standard ones, the only possibility for the non-singlet right-handed fermions is to be in doublet representations (i.e. they can only be mirror fermions or vector doublets) [58]. The new charged fermions have to be heavy (heavier than $\sim$ 45 GeV) in view of limits on direct production at LEP [13].

The coupling of the W to charged currents involving the usual quarks and charged leptons and the light neutrinos is given by

$$\begin{aligned}\mathcal{L} \;=\; (g/\sqrt{2})[\overline{Q}_L^{(u)}\,\gamma_\lambda(A_L^{u\dagger}\,A_L^d)\,Q_L^d \;+\; \overline{n}_{\ell L}\,\gamma_\lambda(A_L^{\nu\dagger}\,A_L^e)E \\ +\; \overline{Q}_R^u\,\gamma_\lambda(F_R^{u\dagger}\,F_R^d)\,Q_R^{(d)} \\ +\; \overline{n}_{\ell R}^c\,\gamma_\lambda(F_R^{\nu\dagger}\,F_R^e)E]W^\lambda \;+\; \text{H.c.}\,,\end{aligned} \tag{100}$$

where all the fields correspond to mass-eigenstates, $\overline{Q}^{(u)} \equiv (\overline{u},\overline{c},\overline{t})$, $\overline{Q}^{(d)} \equiv (\overline{d},\overline{s},\overline{b})$, $\overline{E} \equiv (e,\mu,\tau)$, $\overline{n}_{\ell L} \equiv (\nu_{1L},\nu_{2L},\ldots)$ and $\overline{n}_{\ell R}^c \equiv (\overline{\nu}_{1R}^c,\overline{\nu}_{2R}^c,\ldots) = C(\overline{n}_{\ell L})^T$. In Eq. (100) the matrices A_L^k and $F_R^k (k = u,d,e,\nu)$ relate, respectively, the ordinary and the exotic fermion weak eigenstates to the light fermion mass-eigenstates.

The mixing of the exotic fermions with the ordinary ones leads in general to flavor-changing neutral currents (FCNC) between ordinary fermions. For FCNC transitions among charged fermions there are stringent constraints on the strength of the corresponding interactions from experimental limits on processes such as for example $\mu \to 3e$ and $K_L \to \mu\mu$. Exotic-ordinary fermion mixing gives rise also to deviations from the SM predictions in flavor-conserving neutral current processes and in charged-current reactions. To analyze the constraints on the pertinent parameters one can work in the limit where the FCNC transitions are absent [58]. The matrices A_L^k and F_R^k $(k = u,d,e)$ have then the greatly simplified forms [58]

$$A^k_L = \hat{A}^k_L\, c^k_L \qquad (k = u, d, e)\ ,$$
$$F^k_R = \hat{F}^k_R\, s^k_R \qquad (k = u, d, e)\ , \tag{101}$$

where the $\hat{A}^k_L$ are unitary matrices which describe usual intergenerational mixing, and the $\hat{F}^k_R$ are unitary matrices in the special case when the number of the exotic states is equal to the number of ordinary states. The c^k_L and s^k_R are diagonal matrices of $\cos\theta^i_L$ and $\sin\theta^i_R$, respectively (the matrix elements of c^u_L, for example, are $(c^u_L)_{ij} = \delta_{ij}\cos\theta^i_L$ $(i = u, c, t)$). The angles $\theta^i_{L,R}$ are the light-heavy mixing angles. The weak eigenstates of the charged leptons can be chosen so that each corresponds to a unique light mass-eigenstate [58]. Allowing for CP-violation in the mixing described by $F^{(e)}_R$, one has then $A^e_L = c^e_L$ and $F^e_R = s^e_R\, e^{i\varphi_{(e)}}$, where $e^{i\varphi_{(e)}}$ is a diagonal matrix of $e^{i\varphi_i}$ $(i = e, \mu, \tau)$.

It follows that the beta decay interaction arising from the couplings (100) is of the form (3) with

$$a^{(e)}_{LL} \simeq (g^2 U_{ud}/8m^2_W)\sqrt{u_e}\ , \tag{102}$$
$$\eta_{LR} \simeq s^u_R\, s^d_R\, (\hat{V}_R)_{ud}\ , \tag{103}$$
$$\eta^{(e)}_{RL} \simeq e^{i\varphi_e}\, s^e_R\, \sqrt{\tilde{v}_e}\ , \tag{104}$$
$$\eta^{(e)}_{RR} \simeq \eta_{LR}\,\eta^{(e)}_{RL}\ , \tag{105}$$

where in a given coupling constant we have kept only the terms lowest order in the light-heavy mixing. The quantities u_e and v_e are given here by $u_e = \sum_i |(A^\nu_L)_{ei}|^2$ and $v_e = \sum_i |(F^\nu_R)_{ei}|^2$ (denoted in Ref. [58] by $(c^{\nu_e}_L)^2$ and $(s^{\nu_e}_R)^2$, respectively). In Eq. (103) $\hat{V}_R$ is a matrix which is unitary for $n^u_R = n^d_R = m^u_R$, where n^u_R and n^d_R are, respectively, the number of the $Q(\equiv$ electric charge$) = 2/3$ and $Q = -1/3$ right-handed singlet states, and m^u_R is the number of the $Q = 2/3$ right-handed doublet states. The phase φ_e (which is the phase φ_{RL} here) has no detectable effect, as discussed in Section II.2.1.

A global analysis of the constraints on ordinary-exotic fermion mixings was made in Ref. [58]. The results for the parameters $\eta^{(e)}_{RL}$ and $Re\,\eta_{LR}$ are [59]

$$|\eta^{(e)}_{RL}| < 4.2\times 10^{-2} \qquad (90\%\ c.l.)\ , \tag{106}$$
$$|Re\,\eta_{LR}| < 6\times 10^{-3} \qquad (90\%\ c.l.)\ . \tag{107}$$

The limit (106) originates from muon decay data. $Re\,\eta_{LR}$ is constrained by the PCAC prediction for $K \to 3\pi$ decays (Eq. (93)), by $\overline{\nu}_\mu$ and ν_μ scattering on nuclei (Eq. (92)), and by charged current universality [58].

A new global analysis of the constraints on ordinary-exotic fermion mixings, which updates the analysis of Ref. [58], was carried out in Ref. [60]. The limits on $|\eta_{RL}^{(e)}|$ and $|Re\,\eta_{LR}|$ remain the same as in the previous analysis (Eqs. (106) and (107)).

The elements of $\hat{V}_R$ are complex. The phase in $(\hat{V}_R)_{ud}$ contributes to D_t (Ref. [61]; see also Ref. [62]). Note that the phases φ_R and φ_L (see Section II.2.1) are here related ($\varphi_R = -\varphi_L$). Writing $(\hat{V}_R)_{ud} = e^{i\phi}(\hat{V}_R)'_{ud}$, where $(\hat{V}_R)'_{ud}$ is real, we have (see Eqs. (55)).

$$\begin{aligned} D_t/a_D \;&\simeq\; [1-(s^e_R\,\sqrt{\tilde{v}_e})^2]\;Im\,\eta_{LR} \;\simeq\; Im\,\eta_{LR} \\ &\simeq\; s^u_R\, s^d_R (\hat{V}_R)'_{ud}\sin\phi\;. \end{aligned} \tag{108}$$

From the experimental limit on D we have

$$|Im\,\eta_{LR}| \;<\; 1.1\times 10^{-3}\;. \tag{109}$$

The phase ϕ contributes also to the electric dipole moment of the neutron D_n and to ϵ'/ϵ [61]. The upper limits on $|D_t/a_D|$ from these observables are $\sim 3\times 10^{-5}$. However, as mentioned in Section II.2.2, the limits from D_n and ϵ'/ϵ are not as reliable as the limit (109).

Combining (107) and (109) one obtains

$$|\eta_{LR}| \;<\; 6.1\times 10^{-3}\;. \tag{110}$$

From Eqs. (105), (106) and (110) one has

$$|\eta^{(e)}_{RR}| \;<\; 2.5\times 10^{-4}\;. \tag{111}$$

Finally, Eq. (105) and the bounds (106), (107) and (109) yield

$$|Re\,\eta^{(e)*}_{RR}\,\eta^{(e)}_{RL}| \;<\; 1.1\times 10^{-5}\;, \tag{112}$$

$$|Im\,\eta^{(e)*}_{RR}\,\eta^{(e)}_{RL}| \;<\; 1.9\times 10^{-6}\;. \tag{113}$$

From (112) and (113) one obtains

$$|\eta^{(e)}_{RR}|\;|\eta^{(e)}_{RL}| \;<\; 1.1\times 10^{-5}\;. \tag{114}$$

II.2.4. V,A Interactions from Leptoquark Exchange

Leptoquarks are bosons which couple to lepton-quark pairs. They appear in many extensions of the SM [63]. In some models they can be light enough to cause observable effects in low-energy processes. Both spin-one and spin-zero leptoquarks occur.

The $d\to u\,e^-\overline{\nu}_e^{(L,R)}$ transitions can be mediated by leptoquarks of electric charge $|Q|=2/3$ and $|Q|=1/3$ (see Refs. [64,62]). We shall denote the spin-one and spin-zero leptoquarks generically as $X_{|Q|}$ and $Y_{|Q|}$, respectively.

The exchange of $X_{|Q|}$ and $Y_{|Q|}$ leptoquark states which contribute to the $d \to u\, e^- \overline{\nu}_e^{(L,R)}$ transitions generate four fermion interactions of the general forms

$$\begin{aligned} H_{X(2/3)} &= \overline{u}\gamma_\lambda(1-\gamma_5)\nu_e^{(L)}[f_{LL}\,\overline{e}\gamma^\lambda(1-\gamma_5)d \;+\; f_{LR}\,\overline{e}\gamma^\lambda(1+\gamma_5)d] \\ &+ \overline{u}\gamma_\lambda(1+\gamma_5)\nu_e^{(R)}[f_{RL}\,\overline{e}\gamma^\lambda(1-\gamma_5)d \;+\; f_{RR}\,\overline{e}\gamma^\lambda(1+\gamma_5)d] \;+\; \text{H.c.} \end{aligned} \tag{115}$$

$$\begin{aligned} H_{X(1/3)} &= \overline{d}^c\gamma_\lambda(1-\gamma_5)\nu_e^{(L)}[h_{LL}\,\overline{e}\gamma^\lambda(1-\gamma_5)u^c \;+\; h_{LR}\,\overline{e}\gamma^\lambda(1+\gamma_5)u^c] \\ &+ \overline{d}^c\gamma_\lambda(1+\gamma_5)\nu_e^{(R)}[h_{RL}\,\overline{e}\gamma^\lambda(1-\gamma_5)u^c \;+\; h_{RR}\,\overline{e}\gamma^\lambda(1+\gamma_5)u^c] \;+\; \text{H.c.} \end{aligned} \tag{116}$$

$$\begin{aligned} H_{Y(2/3)} &= \overline{u}(1-\gamma_5)\nu_e^{(L)}[F_{LL}\,\overline{e}(1-\gamma_5)d \;+\; F_{LR}\,\overline{e}(1+\gamma_5)d] \\ &+ \overline{u}(1+\gamma_5)\nu_e^{(R)}[F_{RL}\,\overline{e}(1-\gamma_5)d \;+\; F_{RR}\,\overline{e}(1+\gamma_5)d] \;+\; \text{H.c.} \end{aligned} \tag{117}$$

$$\begin{aligned} H_{Y(1/3)} &= \overline{d}^c(1-\gamma_5)\nu_e^{(L)}[H_{LL}\,\overline{e}(1-\gamma_5)u^c \;+\; H_{LR}\,\overline{e}(1+\gamma_5)u^c] \\ &+ \overline{d}^c(1+\gamma_5)\nu_e^{(R)}[H_{RL}\,\overline{e}(1-\gamma_5)u^c \;+\; H_{RR}\,\overline{e}(1+\gamma_5)u^c] \;+\; \text{H.c.} \end{aligned} \tag{118}$$

In Eqs. (115) - (118) the first superscript on the coupling constants denotes always the neutrino chirality, while the second subscript indicates the chirality of the fourth fermion in the coupling. The fields u, d and e in Eqs. (115) - (118) are mass-eigenstates. The constants f_{ij}, h_{ij}, F_{ij} and $H_{ij}(i = L, R; j = L, R)$ are sums of ratios of the form $x_{ik}x'_{jk}/m^2_{Y_k}$, where x_{ik}, x'_{jk} are the effective fermion-leptoquark coupling constants and m_{Y_k} are the leptoquark masses.

A Fierz transformation takes the Hamiltonians (115) and (116) into quark-lepton interactions of the form (2) with the tensor interaction absent, and the Hamiltonians (117) and (118) into interactions of the form (2) involving all the possible Lorentz covariants (i.e. V,A,S,P and T) [65]. The V,A and S,P (or S,P,T) parts are in general governed by different combinations of the coupling constants. In this section we shall consider the V,A part of the leptoquark-exchange interaction. The S,P component, and for spin-zero leptoquarks the S,P,T component, will be discussed in Section II.5.

The V,A components of the four-fermion interaction resulting from the exchange of $X_{|Q|}$ and $Y_{|Q|}$ ($|Q| = 2/3, 1/3$) are of the form (3) with (Ref. [64]; see also Ref. [62]).

$$\begin{aligned} a_{LL} &= f_{LL} \\ a_{RR} &= f_{RR} \qquad\qquad (X_{(2/3)} - \text{exchange}) \\ a_{LR} &= a_{RL} = 0 \end{aligned} \tag{119}$$

$$\begin{aligned} a_{LL} &= a_{RR} = 0 \\ a_{LR} &= -h_{LL} \\ a_{RL} &= -h_{RR} \end{aligned} \qquad (X_{(1/3)} - \text{exchange}) \qquad (120)$$

$$\begin{aligned} a_{LL} &= a_{RR} = 0 \\ a_{LR} &= -\tfrac{1}{2}\, F_{LR} \\ a_{RL} &= -\tfrac{1}{2}\, F_{RL} \end{aligned} \qquad (Y_{(2/3)} - \text{exchange}) \qquad (121)$$

and

$$\begin{aligned} a_{LL} &= -\tfrac{1}{2}\, H_{LR} \\ a_{RR} &= -\tfrac{1}{2}\, H_{RL} \\ a_{LR} &= a_{RL} = 0 \end{aligned} \qquad (Y_{(1/3)} - \text{exchange}) \qquad (122)$$

To obtain the full V,A interaction we have to add to a_{LL} in each case the SM contribution $(a_{LL})_{SM} = g^2 U_{ud}/8m_W^2$.

A consequence of Eqs. (119) - (122) is that for all cases the product $a^*_{RR}a_{RL}$ vanishes. It follows that there is no contribution from the V,A part of the $X_{|Q|}$-exchange or the $Y_{|Q|}$-exchange interaction to P_L^F/P_L^{GT} (Ref. [64]; see also Ref. [62]), and also that the second term in Eq. (55) for the D-coefficient is absent. A nonzero value of $a^*_{RR}a_{RL}$ could arise however if leptoquarks of the same spin but different charges, or leptoquarks of the same charge but different spins simultaneously contribute. We note that in such cases the phases φ_R and φ_L (see Section II.2.1) are different in general (Ref. [61]; see also Ref. [66]).

In the presence of only one of the four leptoquark types $X_{|Q|}$ and $Y_{|Q|}$, D_t can be nonzero for $X_{(1/3)}$- and $Y_{(2/3)}$-exchange (Ref. [61]; see also Ref. [62]). We have

$$D_t = -\, a_D\, Im\, h_{LL}/(a_{LL})_{SM} \qquad (X_{(1/3)} - \text{exchange})\ , \qquad (123)$$

and

$$D_t = -\,\frac{1}{2}\, a_D\, Im\, F_{LR}/(a_{LL})_{SM} \qquad (Y_{(2/3)} - \text{exchange})\ . \qquad (124)$$

The Lagrangian describing the most general $SU(2)_L \times U(1) \times SU(3)_c$ invariant couplings of spin-zero leptoquarks to a fermion family of the SM contains 9 different leptoquark states (10 states if a right-handed neutrino is included in the family) characterized by a definite fermion number and definite quantum numbers with respect to the SM gauge group [67,68]. Similarly, the most general such Lagrangian for the spin-one leptoquarks has 9 distinct leptoquark states [68] (10 if a right-handed neutrino is included). Inspection shows that from the spin-zero leptoquarks the ones that can contribute to the $d \to u e^- \overline{\nu}_e^{(L,R)}$ transitions are (using the notation of Ref. [68]) the $|Q| = 1/3$ states S_1 and $(S_3)_{T_z=0}$, and the $|Q| = 2/3$ states $(R_2)_{T_z=-1/2}$ and $(\tilde{R}_2)_{T_z=+1/2}$. The spin-one leptoquarks contributing to $d \to u e^- \overline{\nu}_e^{(L,R)}$ are the $|Q| = 1/3$ states $(V_{2\mu})_{T_z=-1/2}$ and $(\tilde{V}_{2\mu})_{T_z=+1/2}$ and the $|Q| = 2/3$ states $U_{1\mu}$ and $(U_{3\mu})_{T_z=0}$. Terms of the a_{LL}-type can arise from $S_1, (S_3)_{T_z=0}, U_{1\mu}$ or $(U_{3\mu})_{T_z=0}$ exchange, while a_{RR}-type couplings can be generated only by S_1 and $U_{1\mu}$ exchange [64]. None of the leptoquark states of definite $SU(2)_L \times U(1)$ quantum numbers, contributing to $d \to u e^- \overline{\nu}_e^{(L,R)}$ gives rise to an a_{LR}- or a_{RL}-type interaction. Such couplings can however be generated if (as generally expected) the states $(R_2)_{T_z=-1/2}$ and $(\tilde{R}_2)_{T_z=+1/2}$ or the states $(V_{2\mu})_{T_z=-1/2}$ and $(\tilde{V}_{2\mu})_{T_z=+1/2}$ mix (Ref. [61]; see also Ref. [69]).

We shall consider as an example the interaction generated by the S_1 leptoquark [70]. The coupling of the S_1 to the first fermion family is given by

$$\begin{aligned}\mathcal{L} = & [\tfrac{1}{2} g_{1L} (\overline{u}^{c\prime}(1-\gamma_5)\, e - \overline{d}^{c\prime}(1-\gamma_5)\, \nu_e^{(L)}) \\ & + \tfrac{1}{2}\, g_{1R}\, \overline{u}^{c\prime}(1+\gamma_5)\, e + \tfrac{1}{2} g_{1R}^{(\nu)}\, \overline{d}^{c\prime}(1+\gamma_5)\, \nu_e^{(R)}]\, S_1 + \text{H.c.}\end{aligned} \tag{125}$$

In Eq. (125) the primed fields and $\nu_e^{(L)}$, $\nu_e^{(R)}$ are weak eigenstates. Neglecting for simplicity generation mixing and neutrino mixing, the coupling constants for the beta decay interaction resulting from (125) are [71]

$$a_{LL} = |g_{1L}|^2/8m_1^2\,, \tag{126}$$
$$a_{RR} = (-g_{1R}^{*}\, g_{1R}^{(\nu)}/8m_1^2)\,, \tag{127}$$
$$a_{LR} = a_{RL} = 0\,, \tag{128}$$
$$(C_S - C'_S)/g_S = g_{1L}\, g_{1R}^{*}/4m_1^2\,, \tag{129}$$
$$(C_S + C'_S)/g_S = g_{1L}^{*}\, g_{1R}^{(\nu)}/4m_1^2 \tag{130}$$
$$(C_T - C'_T)/g_T = -(C_S - C'_S)/g_S = -(C_P - C'_P)/g_P\,, \tag{131}$$
$$(C_T + C'_T)/g_T = -(C_S + C'_S)/g_S = -(C_P - C'_P)/g_P\,, \tag{132}$$

where m_1 is the mass of the S_1.

From the constraints we have considered for the V,A beta decay interactions only the model independent ones (Section II.2.1) apply for the general interaction from leptoquark exchange. The limit (66) on $Re\,\eta_{LR}$ holds only if, in addition to the conditions already given for (66), there are no accidental cancellations in R_π between the $Re\,\eta_{LR}$ contribution and the contributions from the a_{LL} and the $Re\,(C_P - C'_P)/g_P$-part of the interaction.

The existing limits on the masses and the couplings of spin-zero leptoquarks from constraints on direct production at accelerators still allow the strength of the leptoquark-exchange interaction to be comparable with the strength of the weak interaction, and even somewhat stronger than G [13]. The limits on the masses and the couplings of spin-one leptoquarks, which are not quoted in Ref. [13], are probably similar. The exchange of leptoquarks which couple to $Q = -1/3$ quarks can contribute to the decay $K_L \to \mu e$ (at the tree-level) and to the $K^o \to \overline{K}^o$ amplitude (through box diagrams involving two leptoquark propagators) [72]. However, the constraints from these observables (which in principal could be stringent) may not apply since the leptoquark-fermion coupling constants involved, which are not identical to those appearing in beta decay, could be suppressed. For muon decay there is no contribution from leptoquarks at the tree level. The size of the loop-level contributions is most likely considerably below the present experimental limits; moreover, the contributing diagrams contain coupling constants different from those relevant for beta decay.

II.3. Scalar Interactions

The general form of the effective $d \to u\, e^- \overline{\nu}_e$ Hamiltonian involving scalar and pseudoscalar currents is given in Eq. (4). For nuclear beta decay we can restrict our attention to the terms in (4) involving the scalar current $\overline{u}d$ (see Eq. (8)). This part of (4) can be written as

$$H_S \;=\; [a_{LS}\, \overline{e}(1-\gamma_5)\nu_e^{(L)} \;+\; a_{RS}\, \overline{e}(1+\gamma_5)\nu_e^{(R)}]\overline{u}d \;+\; \text{H.c.}\,, \tag{133}$$

where $a_{LS} = A_{LL} + A_{LR}$, $a_{RS} = A_{RR} + A_{RL}$. The constants C_S and C'_S are given by $C_S - C'_S = 2g_S\, a_{LS}$, $C_S + C'_S = 2g_S\, a_{RS}$, where g_S has been defined in Eq. (22). We shall define also $\eta_{kS} \equiv a_{kS}/a_{LL} \simeq a_{kS}/(G\, U_{ud}/\sqrt{2})$ $(k = L, R)$ and $\eta_{RS}^{(e)} \equiv \eta_{RS}\, \sqrt{\tilde{v}_e}$.

A possible source of scalar-type beta decay couplings is the exchange of charged Higgs bosons [73]. The bounds on charged Higgs masses and couplings from limits on direct production still allow the strength of the scalar contributions to be comparable and even somewhat stronger than the strength of the weak interactions [13]. Charged Higgs bosons are present in many extensions of the SM, and also in the standard $\mathrm{SU}(2)_L \times \mathrm{U}(1)$ model if the Higgs sector contains (for example) two Higgs doublets. In supersymmetric models the presence of two Higgs doublets is required. In multi-Higgs models the couplings of the Higgs bosons to the fermions is in general undetermined. It should be noted however that in models which use discrete symmetries to eliminate flavor changing neutral currents and thus allow light neutral Higgs bosons, the couplings of the charged Higgs bosons to the first family are usually small, suppressed by the small fermion masses, although some enhancement could come from ratios of Higgs vacuum expectation values. But in general scalar contributions from Higgs-exchange to the beta decay interaction of strength near the present experimental limits on scalar couplings are not ruled out.

In the supersymmetric SM with R-parity violation [74] scalar contributions to the $d \to u\, e^- \overline{\nu}_e$ transition can arise at the tree-level from the exchange of sleptons. Scalar beta decay interactions can also come from the exchange of leptoquarks. This class of scalar interactions will be discussed in Section II.5. The scalar-type interactions would show up in the allowed approximation only in Fermi (or mixed) transitions. Restricting attention, as before, only to coupling constant combinations which do not result from the interference of amplitudes involving neutrinos of different chirality, the observables in Fermi transitions can depend on

$$K_{V,S}^2 \equiv |C_V|^2+|C_V'|^2+|C_S|^2+|C_S'|^2 = 2[1+g_S^2(|\eta_{LS}|^2+|\eta_{RS}^{(e)}|^2](GU_{ud}/\sqrt{2})^2 \quad (134)$$

$$Re\, C_S C_S'^* / K_{V,S}^2 \simeq \frac{1}{2} g_S^2(-|\eta_{LS}|^2 + |\eta_{RS}^{(e)}|^2) \,, \quad (135)$$

$$Re(C_S C_V^* + C_S' C_V'^*)/K_{V,S}^2 \simeq 2g_S\, Re\, \eta_{LS} \,, \quad (136)$$

$$Re(C_S C_V'^* + C_S' C_V^*)/K_{V,S}^2 \simeq -2g_S\, Re\, \eta_{LS} \,, \quad (137)$$

$$Im(C_S C_V^* + C_S' C_V'^*)/K_{V,S}^2 \simeq 2g_S\, Im\, \eta_{LS} \,, \quad (138)$$

$$Im(C_S C_V'^* + C_S' C_V^*)/K_{V,S}^2 \simeq -2g_S\, Im\, \eta_{LS} \,. \quad (139)$$

In observables in mixed transitions one can have in addition combinations proportional to

$$Re(C_S C_A^* + C_S' C_A'^*)/K_{V,S}^2 \simeq 2g_A g_S\, Re\, \eta_{LS} \,, \quad (140)$$

$$Re(C_S C_A'^* + C_S' C_A^*)/K_{V,S}^2 \simeq -2g_A g_S\, Re\, \eta_{LS} \,, \quad (141)$$

$$Im(C_S C_A^* + C_S' C_A'^*)/K_{V,S}^2 \simeq 2g_A g_S\, Im\, \eta_{LS} \,, \quad (142)$$

$$Im(C_S C_A'^* + C_S' C_A^*)/K_{V,S}^2 \simeq -2g_A g_S\, Im\, \eta_{LS} \,. \quad (143)$$

We shall consider now the available constraints on $Re\,\eta_{kS}$ and $Im\,\eta_{kS}$ $(k = L, R)$, assuming that the Hamiltonian (133) is the only new beta decay interaction.

$Im\,\eta_{LS}$. A constraint on the CP-violating coupling constant $Im\,\eta_{LS}$ comes from the experimental limit on the parity and time reversal violating tensor-type electron-nucleon interaction $(G/\sqrt{2})C_T^{eN}\, \bar{e}\sigma_{\lambda\mu} e\, \overline{N}\gamma_5\, \sigma^{\lambda\mu} N$. The constant $Im\,\eta_{LS}$ contributes to C_T^{eN} through diagrams involving W-exchange in addition to the scalar interaction [75]. The present limit on C_T^{eN} is $|C_T^{eN}| < 4 \times 10^{-8}$ [76], which follows from the experimental bound on the electric dipole moment of the ^{199}Hg atom. This limit on C_T^{eN} implies [75]

$$|Im\,\eta_{LS}| \lesssim 8 \times 10^{-5} \,. \quad (144)$$

Direct limits on $Im\,\eta_{LS}$ can be obtained from searches for the T-odd R-correlation (Eq. (33)). For a scalar-type interaction R_t is given by

$$R_t \simeq 2a_R g_S\, Im\,\eta_{LS} \,, \quad (145)$$

where the quantity a_R contains the ratio $g_A M_{GT}/g_V M_F$. A measurement of R in ^{19}Ne-decay yielded $R = 0.079 \pm 0.053$ [77]. For ^{19}Ne-decay $a_R \simeq 0.26$ [77], so that one obtains

$$-0.015 < g_S\, Im\, \eta_{LS} < 0.32 \quad (90\%\ c.l.) \tag{146}$$

The contribution R_f of electromagnetic final-state interactions is small, about 10^{-3} [78].

Limits on $Im\, \eta_{LS}$ follow also from experimental results on time-reversal-even observables sensitive to the combination $|C_S|^2 + |C'_S|^2$. The most stringent limit on $g_S^2(|\eta_{LS}|^2 + |\eta_{RS}^{(e)}|^2)$ $(= \frac{1}{2}(|C_S|^2 + |C'_S|^2))$ among limits provided by individual observables is

$$g_S^2(|\eta_{LS}|^2 + |\eta_{RS}^{(e)}|^2) < 1.1 \times 10^{-2} \quad (90\%\ c.l.)\ , \tag{147}$$

deduced from data on the broadening of the delayed proton peak following the superallowed $1/2^+ \to 1/2^+$ decay of ^{32}Ar and ^{33}Ar [79]. The limit (147) yields

$$|g_S\, Im\, \eta_{LS}| < 0.105 \qquad (90\%\ c.l.)\ . \tag{148}$$

Bounds on $Im\, \eta_{LS}$ can be obtained in principle also from experiments measuring the $e^{\pm}$-longitudinal polarization in Fermi transitions. The constant $Im\, \eta_{LS}$ enters P_L through the Coulomb corrections (see Eq. (29)). Limits on $Im\, \eta_{LS}$ obtained this way have not been so far reported.

$Re\, \eta_{LS}$. A source of constraints on $g_S\, Re\, \eta_{LS}$ are observables which are sensitive to the Fierz interference term in Fermi transitions or mixed transitions. One of these is the charged lepton longitudinal polarization P_L^F. The measured ratios P_L^F/P_L^{GT} imply [26] $Re(C_S - C'_S) = (0.0027 \pm 0.0109)(G U_{ud}/\sqrt{2})$, yielding

$$-7.7 \times 10^{-3} < g_S\, Re\, \eta_{LS} < 1.0 \times 10^{-2} \quad (90\%\ c.l.) \tag{149}$$

The Fierz interference term affects also the ft-values. For $g_S\, Re\, \eta_{LS}$ the bound

$$-3.5 \times 10^{-3} < g_S\, Re\, \eta_{LS} < 4.7 \times 10^{-3} \quad (90\%\ c.l.) \tag{150}$$

was deduced from experimental ft-values of superallowed beta decays [80]. It should be noted however, that in recent studies of superallowed beta transitions it has been observed that the uncertainties in the effects of charge-dependent nuclear forces on the ft-values are larger than previously considered [81]. Consequently, the constraint on $g_S\, Re\, \eta_{LS}$ may be weaker than that in Eq. (150) [82,3].

The limit (147) on $|C_S|^2 + |C'_S|^2$ yields

$$|g_S\, Re\, \eta_{LS}| < 0.105 \qquad (90\%\ c.l.)\ . \tag{151}$$

$\eta_{RS}^{(e)}$. Neglecting the interference term between the SM and the η_{RS} amplitudes (which is proportional to neutrino mass), the parameter $\eta_{RS}^{(e)}$ enters the beta decay observables always quadratically. Since the contribution of $Im\, \eta_{RS}^{(e)}$ to the parity and time-reversal violating C_T^{eN}-interaction (through diagrams involving $Im\, \eta_{RS}^{(e)}$ and the weak interaction) is also proportional to the neutrino mass, the best limit on both $Re\, \eta_{RS}^{(e)}$ and $Im\, \eta_{RS}^{(e)}$ comes from bounds on $|C_S|^2 + |C'_S|^2$. Eq. (147) implies

$$|g_S \, Re \, \eta_{RS}^{(e)}| < 0.105 \qquad (90\% \; c.l.) \; , \tag{152}$$

$$|g_S \, Im \, \eta_{RS}^{(e)}| < 0.105 \qquad (90\% \; c.l.) \; . \tag{153}$$

The Results of a Comprehensive Analysis of Beta Decay Data. A global analysis of beta decay data [3], which included experimental data on neutron decay, ^{19}Ne-decay, and the *ft*-values of superallowed $0^+ \to 0^+$ beta decays yielded the following 90% c.l. upper limits:

$$|g_S \, \eta_{LS}| \; < \; 0.035 \; , \tag{154}$$

$$|g_S \, Re \, \eta_{LS}| \; < \; 0.007 \; , \tag{155}$$

$$|g_S \, \eta_{RS}^{(e)}| \; < \; 0.072 \; , \tag{156}$$

and [31]

$$g_S^2 (|\eta_{LS}|^2 + |\eta_{RS}^{(e)}|^2) \; < \; 0.0055 \; . \tag{157}$$

All the limits on η_{LS} and $\eta_{RS}^{(e)}$ from beta decay depend on the value of g_S (Eq. (22)). The constant g_S was calculated in Ref. [83] in connection with a study of neutral current interactions of a general Lorentz structure. Employing a quark model with spherically symmetric wave function, g_S can be expressed as $g_S = -\frac{1}{2} + \frac{9}{10} \, g_A \simeq 0.6$. The uncertainty in this prediction has been estimated to be about 30% to 60% [83]. Including an uncertainty of this size, one has

$$1/4 \lesssim g_S \lesssim 1 \; . \tag{158}$$

II.4. Tensor Interactions

The general tensor interaction for $d \to u \, e^- \nu_e^{(L,R)}$ is given in Eq. (5). We shall write it in the form

$$H_T = \left(a_{LT} \, \overline{e} \, \frac{\sigma_{\lambda\mu}}{\sqrt{2}} \, (1 - \gamma_5) \, \nu_e^{(L)} \; + \; a_{RT} \, \overline{e} \, \frac{\sigma_{\lambda\mu}}{\sqrt{2}} \, (1 + \gamma_5) \, \nu_e^{(R)} \right) \overline{u} \, \frac{\sigma^{\lambda\mu}}{\sqrt{2}} \, d \\ + \; \text{H.c.} \tag{159}$$

where $a_{LT} = 2\alpha_{LR}$, $a_{RT} = 2\alpha_{RL}$. Note that $C_T - C_T' = 2 g_T \, a_{LT}$, $C_T + C_T' = 2 g_T \, a_{RT}$, where the constant g_T has been defined in Eq. (23). We shall define also $\eta_{kT} \equiv \eta_{kT}/a_{LL} \simeq a_{kT}/(G \, U_{ud}/\sqrt{2})$ $(k = L, R)$ and $\eta_{RT}^{(e)} = \eta_{RT} \sqrt{\tilde{v}_e}$.

One of the ways in which tensor interactions can arise in gauge theories is through loop corrections to the tree-level amplitudes. In the SM this mechanism is the only possibility, and yields an interaction of the form (159) with $a_{RT} = 0$ and $\eta_{LT} \simeq 10^{-8} - 10^{-9}$ [84]. In renormalizable gauge-theories the only mechanism that can generate tensor quark-lepton interactions at the tree-level is the exchange of spin-zero leptoquarks [65]. The constraints discussed in this section apply to tensor interactions of any origin. For the tensor interactions arising from leptoquarks exchange additional constraints apply. These will be discussed in the next Section.

In the allowed approximation tensor interactions can manifest themselves only in Gamow-Teller (or mixed) beta decays. The observables in pure Gamow-Teller transitions can depend on combinations of coupling constants, which can be obtained from Eqs. (134)-(139) by replacing $C_S, C'_S, C_V, C'_V, \eta_{LS}, \eta_{RS}, g_V$ and g_S with $C_T, C'_T, C_A, C'_A, \eta_{LT}, \eta_{RT}, g_A$ and g_T, respectively. In mixed transitions the observables can depend in addition on the coupling constant combinations (140)-(143) with $C_S, C'_S, C_A, C'_A, \eta_{LS}, \eta_{RS}$ and $g_A g_S$ replaced by $C_T, C'_T, C_V, C'_V, \eta_{LT}, \eta_{RT}$ and $g_V g_T$, respectively.

$Im\,\eta_{LT}$. Like $Im\,\eta_{LS}$, the parameter $Im\,\eta_{LT}$ is constrained by the experimental limit on the parity and time reversal violating tensor-type electron-nucleon interaction [75]. The limit $|C_T^{eN}| < 4 \times 10^{-8}$ (see Section II.3) implies [75]

$$|Im\,\eta_{LT}| \lesssim 2 \times 10^{-5} \,. \tag{160}$$

$Im\,\eta_{LT}$ contributes to the T-odd R-correlation. R_t is given by

$$R_t \;=\; 2a'_R\, g_T\, Im\,\eta_{LT} \;, \tag{161}$$

where the constant a'_R contains for mixed transitions the ratio of the nuclear matrix elements. For ^{19}Ne-decay $a'_R \simeq 0.18$, so that the experiment of Ref. [77] (see Section II.3.) sets the limit $|g_T\, Im\,\eta_{LT}| < 0.5$. A more stringent bound on $g_T\, Im\,\eta_{LT}$ comes from a recent measurement of R in the decay $^8Li \;\to\; ^8Be$ (2.9 MeV) $+\, e^- \,+\, \overline{\nu}_e$ [85]. R_t for this decay is given by Eq. (161) with $a'_R = (3g_A)^{-1} \simeq 0.26$. The experiment yielded $R = (0 \pm 4) \times 10^{-3}$, which implies [85]

$$|g_T\, Im\,\eta_{LT}| < 1.3 \times 10^{-2} \quad (90\%\ c.l.) \,. \tag{162}$$

The contribution R_f of the electromagnetic final-state interactions is expected to show up at the level of $\sim 10^{-3}$ [86].

Limits on $g_T\, Im\,\eta_{LT}$ follow also from bounds on $|C_T|^2 + |C'_T|^2$. From the results of a comprehensive analysis of beta decay data [3], listed further on in this section, the limit on $g_T^2(|\eta_{LT}|^2 + |\eta_{RT}^{(e)}|^2)$ $(= \frac{1}{2}(|C_T|^2 + |C'_T|^2))$ is (Eq. (181))

$$g_T^2(|\eta_{LT}|^2 + |\eta_{RT}^{(e)}|^2) < 4.3 \times 10^{-3} \quad (90\%\ c.l.) \,. \tag{163}$$

This implies

$$|g_T\, Im\,\eta_{LT}| < 6.6 \times 10^{-2} \quad (90\%\ c.l.) \,. \tag{164}$$

Data on $e^{\pm}$-longitudinal polarizations yielded [87]

$$-4.5 \times 10^{-2} < g_T\, Im\,\eta_{LT} < 2.9 \times 10^{-2} \quad (90\%\ c.l.) \,. \tag{165}$$

$Re\,\eta_{LT}$. It was pointed out in Ref. [88] that the tensor interaction (159) is constrained by the experimental value of the ratio R_π (Eq. (60)) of the $\pi \to e\nu_e$ to $\pi \to \mu\nu_\mu$ rates, since electromagnetic corrections to the tensor interaction (159) induce an effective interaction involving pseudoscalar quark currents [89].

The induced S, P interaction for the operator (159) is of the form [88]

$$\begin{aligned} H &= \tfrac{1}{4}\, a_{LT}\, k_o\, \overline{e}(1-\gamma_5)\nu_e^{(L)}\, \overline{u}(1-\gamma_5)d \\ &+ \tfrac{1}{4}\, a_{RT}\, k_o\, \overline{e}(1+\gamma_5)\nu_e^{(R)}\, \overline{u}(1+\gamma_5)d \;+\; \text{H.c.}\,, \end{aligned} \tag{166}$$

where $k_o \simeq -2.8 \times 10^{-2}$. The contribution of the Hamiltonian (159) to R_π is [27]

$$R_\pi = (R_\pi)_{SM}(u_e/u_\mu)[(1-\tfrac{1}{4}\,k_o\,\omega\,Re\,\eta_{LT})^2+(\tfrac{1}{4}\,k_o\,\omega\,Im\,\eta_{LT})^2+|\tfrac{1}{4}\,k_o\,\omega\,\overset{(e)}{\eta_{RT}}|^2]\,, \tag{167}$$

where $\omega \equiv (m_\pi/m_e)(m_\pi/(m_u+m_d)) \simeq 3.3 \times 10^3$ [90],

Let us consider the case when $\overset{(e)}{\eta_{RT}} = 0$. Neglecting $Im\,\eta_{LT}$, we have from the experimental result (62)

$$0.9932\ (u_\mu/u_e)^{1/2} \;<\; \left|1 \;-\; \tfrac{1}{4}\,k_o\omega\,Re\,\eta_{LT}\right| \;<\; 1.0007\ (u_\mu/u_e)^{1/2} \tag{168}$$

Allowing for $(u_e/u_\mu)^{1/2}$ the range $0.98 < (u_e/u_\mu)^{1/2} < 1.02$ (see Section II.2.1) $(R_\pi)_{expt}$ (Eq. (30)) requires [30] $Re\,\eta_{LT}$ to be either in the range

$$-1.2 \times 10^{-3} < Re\,\eta_{LT} < 9.1 \times 10^{-4}\,, \tag{169}$$

or in the range

$$-8.86 \times 10^{-2} < Re\,\eta_{LT} < -8.75 \times 10^{-2}\,. \tag{170}$$

Since it seems highly unlikely that $Re\,\eta_{LT}$ would have a nonzero value in a fine-tuned interval like (170), we have to conclude that most probably $|Re\,\eta_{LT}|$ obeys

$$|Re\,\eta_{LT}| < 1.2 \times 10^{-3}\,. \tag{171}$$

The bounds on $Re\,\eta_{LT}$ from $(R_\pi)_{expt}$ could be weaker if there is also a tensor contribution to $d\overline{u} \to \mu\nu_\mu$. Note that $R_\pi = (R_\pi)_{SM}$ if $u_e = u_\mu$ and the ratio $(Re\,\eta_{LT})_e/(Re\,\eta_{LT})_\mu$ of the tensor coupling constants for the electron and the muon family is m_e/m_μ. There is however no reason for such a scenario.

The range of values of $Re\,\eta_{LT}$ allowed by $(R_\pi)_{expt}$ could be wider if both the a_{LT} and the a_{RT} terms in (159) are present [91]. R_π is then given by Eq. (167). Since the present upper limit (Eqs. (180), (181)) on $|\overset{(e)}{\eta_{RT}}|^2$ allows $|\tfrac{1}{4}\,k_o\omega\,\overset{(e)}{\eta_{RT}}|^2$ to be as large as ~ 5, the lower bound on $[R_\pi/(R_\pi)_{SM}]$ from $(R_\pi)_{expt}$ can be satisfied even with $|1-\tfrac{1}{4}\,k_o\,\omega\,Re\,\eta_{LT}| = 0$. It follows that the values of $Re\,\eta_{LT}$ allowed by $(R_\pi)_{expt}$ span now the whole interval

$$-8.9 \times 10^{-2} < Re\, \eta_{LT} < 9.1 \times 10^{-4} \,. \tag{172}$$

A limit $Re(C_T - C_T') = g_A(0.0027 \pm 0.0109)(G\, U_{ud}/\sqrt{2})$ is provided by the experimental results on the ratio P_L^F / P_L^{GT} of longitudinal polarization in beta decays [26] yielding

$$-9.6 \times 10^{-3} < g_T\, Re\, \eta_{LT} < 1.3 \times 10^{-2} \quad (90\%\ c.l.) \,. \tag{173}$$

$\eta_{RT}^{(e)}$. The best limits on $Re\, \eta_{RT}^{(e)}$ and $Im\, \eta_{RT}^{(e)}$ come from bounds on $|C_T|^2 + |C_T'|^2$. The upper bound (181) implies

$$|g_T\, Re\, \eta_{RT}^{(e)}| \ < \ 6.6 \times 10^{-2} \qquad (90\%\ c.l.) \,, \tag{174}$$

$$|g_T\, Im\, \eta_{RT}^{(e)}| \ < \ 6.6 \times 10^{-2} \qquad (90\%\ c.l.) \,. \tag{175}$$

The constant $|\eta_{RT}^{(e)}|$ is constrained also by $(R_\pi)_{expt}$. From Eq. (167) and (62) we obtain

$$|\eta_{RT}^{(e)}| \ < \ 4.5 \times 10^{-2} \tag{176}$$

($|\eta_{RT}^{(e)}| \ < \ 1.7 \times 10^{-3}$ if $Re\, \eta_{LT}$ is absent). The same limit follows from (167) and (62) for $|Im\, \eta_{LT}|$.

The Results of a Comprehensive Analysis of Beta Decay Data. For the Hamiltonian consisting of the SM Hamiltonian and the Hamiltonian (159), the global analysis of beta decay data in Ref. [3] yielded the following 90% c.l. limits:

$$|g_T\, \eta_{LT}| \ < \ 0.026 \tag{177}$$

$$|g_T\, Re\, \eta_{LT}| \ < \ 0.007 \tag{178}$$

$$|g_T\, Im\, \eta_{LT}| \ < \ 0.026 \tag{179}$$

$$|g_T\, \eta_{RT}^{(e)}| \ < \ 0.050 \tag{180}$$

and [31]

$$g_T^2(|\eta_{LT}|^2 + |\eta_{RT}^{(e)}|^2) \ < \ 0.0043 \,. \tag{181}$$

The limits on η_{LT} and η_{RT} from beta decay depend on the constant g_T. This was estimated in Ref. [83] in the same framework as g_S, with the result $g_T = \frac{5}{3}(\frac{1}{2} + \frac{3}{10}\, g_A) \simeq 1.46$. Allowing for the same uncertainty in this prediction as in the prediction for g_S, we have

$$0.6 \lesssim g_T \lesssim 2.3 \,. \tag{182}$$

Data from a recent experiment on $\pi \to e\nu_e\gamma$ decay [92] appear to be in disagreement with the SM description of this decay. It has been suggested [93] that the discrepancy may be due to the presence of an a_{LT}-type tensor interaction. This interpretation has been investigated further in Refs. [84], [88], [94], [95], and [91]. To account for the discrepancy, the value of $|Re\,\eta_{LT}|$ is required to be in the range [96]

$$1.5 \times 10^{-2} \;<\; |Re\,\eta_{LT}| \;<\; 4.2 \times 10^{-2} \,. \tag{183}$$

For this range to have an overlap with the values of $|Re\,\eta_{LT}|$ allowed by the limit (178) from beta decay, the tensor constant $|g_T|$ cannot be larger than ~ 0.5 [97]. Not considering the solution (170), $(R_\pi)_{expt}$ allows values of $|Re\,\eta_{LT}|$ in the range (183) only if there is a sufficiently large additional contribution to R_π from new interactions. This could be an a_{RT}-tensor interaction as described around Eq. (172).

II.5. S,T Interactions from Leptoquark-Exchange

In Section II.2.4. we have considered the leptoquark states that can mediate the $d \to u\,e^-\overline{\nu}_e^{L,R}$ transition, and discussed the V,A part of the resulting interactions. Here we shall discuss the S,P (or S,P,T) part of these interactions. As mentioned in Section II.2.4, the coupling constants of the S,P (or S,P,T) part of the leptoquark-exchange interactions are in general independent of the coupling constants in the V,A parts.

Spin-one leptoquarks. The S,P part of the $d \to u\,e^-\overline{\nu}_e^{(L,R)}$ interaction from $X_{(2/3)}$ and $X_{(1/3)}$ exchange is of the form (4) with (Ref. [64]; see also Ref. [62])

$$a_{LS} \;=\; a_{LP} \;=\; -2f_{LR} \tag{184}$$

$$(X_{(2/3)} - exchange)$$

$$a_{RS} \;=\; -\,a_{RP} \;=\; -2f_{RL} \tag{185}$$

and

$$a_{LS} \;=\; a_{LP} \;=\; -2h_{LR} \tag{186}$$

$$(X_{(1/3)} - exchange)$$

$$a_{RS} \;=\; -\,a_{RP} \;=\; 2h_{RL} \tag{187}$$

where a_{LS} and a_{RS} are defined in Eq. (133), and the coupling constants $a_{LP} = A_{LR} - A_{LL}$ and $a_{RP} = A_{RR} - A_{RL}$ describe the part of the interaction (4) involving the pseudoscalar current $\overline{u}\gamma_5 d$; the constants f_{LR}, f_{RL} and h_{LR}, h_{RL} have been defined in Eqs. (115)-(116).

Since the a_{LP} and a_{RP} terms contribute to R_π, the scalar interaction from $X_{(2/3)}$ and $X_{(1/3)}$ exchange is constrained by $(R_\pi)_{expt}$. Using the relations (184), (185) and (186), (187) R_π can be expressed as

$$R_\pi \;=\; (R_\pi)_{SM}(u_e/u_\mu)[(1+\omega\,Re\,\eta_{LS})^2 + \;(\omega\,Im\,\eta_{LS})^2 + |\omega\,\eta_{RS}^{(e)}|^2] \,. \tag{188}$$

Eq. (188) with $0.98 \leq (u_e/u_\mu)^{1/2} \leq 1.02$, and the experimental result (62) imply, assuming that both the a_{LP} and the a_{RP} terms are present (see the discussion around Eq. (172)),

$$-6.2 \times 10^{-4} < Re\,\eta_{LS} < 6.4 \times 10^{-6} \qquad (90\%.c.l.)\ . \tag{189}$$

In Eq. (188) we have left out the contribution of the A-component of the interaction, assuming that if it is not small (note that this contribution is not enhanced by ω), there is no accidental cancellation in R_π between the $\omega\, Re\,\eta_{LS}$-term and the terms from the A-component involving the left-handed neutrino.

For an S,P interaction the tensor electron-nucleon coupling constant C_T^{eN} constrains the combination $Im(\eta_{LS} - \eta_{LP})$ [75], which vanishes due to the relations (184) and (186). A limit on $Im\,\eta_{LS}$ comes in this case from the experimental limit on the scalar-type parity and time-reversal violating electron-nucleon interaction $(G/\sqrt{2})\, C_S^{eN}\, \overline{e}\, i\gamma_5 e\, \overline{N}\, N$. The present limit on C_S^{eN} is $|C_S^{eN}| < 1.4 \times 10^{-6}$ (90% c.l.) [99], obtained from the experimental bound on the electric dipole moment of the Tl atom. This implies $|Im\,\eta_{LP}| \lesssim 10^{-4}$ [75], and therefore

$$|Im\,\eta_{LS}| \lesssim 10^{-4}\ . \tag{190}$$

The limit on $|Im\,\eta_{LS}|$ from $(R_\pi)_{expt}$ is $|Im\,\eta_{LS}| < 3.1 \times 10^{-4}$ (6.3×10^{-5} if $Re\,\eta_{LS}$ is absent). For $|\eta_{RS}^{(e)}|$ the same limits are obtained. Thus

$$|\eta_{RS}^{(e)}| < 3.1 \times 10^{-4} \tag{191}$$

($|\eta_{RS}^{(e)}| < 6.3 \times 10^{-5}$ if $Re\,\eta_{LS}$ is absent). For $u_e = u_\mu$ one would have $|\eta_{RS}^{(e)}| < 3.1 \times 10^{-4}$ ($|\eta_{RS}^{(e)}| < 1.2 \times 10^{-5}$ if $Re\,\eta_{LS}$ is absent).

The further constraints on η_{LS} and $\eta_{RS}^{(e)}$ are those considered in Section II.3.

Spin-zero leptoquarks. The general form of S,P,T, part of the $d \to u\, e^- \overline{\nu}_e^{(L,R)}$ interaction mediated by $Y_{(2/3)}$ and $Y_{(1/3)}$ leptoquarks is the sum of the Hamiltonians (4) and (5), with (Ref. [64]; see also Ref. [62])

$$a_{LS} = -a_{LP} = -\tfrac{1}{2}\, F_{LL} \tag{192}$$
$$a_{RS} = a_{RP} = -\tfrac{1}{2}\, F_{RR} \tag{193}$$

$$(Y_{(2/3)} - exchange)$$

$$a_{LT} = a_{LS} \tag{194}$$
$$a_{RT} = a_{RS}\ , \tag{195}$$

and

$$a_{LS} = -a_{LP} = -\tfrac{1}{2}\, H_{LL} \tag{196}$$
$$a_{RS} = a_{RP} = -\tfrac{1}{2}\, H_{RR} \tag{197}$$

$$(Y_{(1/3)} - exchange)$$

$$a_{LT} = -a_{LS} \tag{198}$$
$$a_{RT} = -a_{RS}\ , \tag{199}$$

respectively. The constants a_{LT}, a_{RT} have been defined in Eq. (159), and the constants $F_{LL}, F_{RR}, H_{LL}, H_{RR}$ in Eqs. (117) and (118).

The $\overline{u}\gamma_5 d$-component of the S,P,T part of the interaction contributes to R_π. Leaving out (as in Eq. (188)) the contribution of the axial-vector interaction, R_π is given by

$$R_\pi \; = \; (R_\pi)_{SM} \; (u_e/u_\mu)[(1+\omega \, Re \, \eta_{LP}^{(|Q|)})^2 \; + \; (\omega \, Im \, \eta_{LP}^{(|Q|)})^2 \; + \; |\omega \, \eta_{RP}^{(e)(|Q|)}|^2] \quad (200)$$

$(|Q|) = 2/3,\ 1/3)$, where $\eta_{LP}^{|Q|} = a_{LP}^{(|Q|)}/a_{LL}$, $\eta_{RP}^{(e)(|Q|)} = a_{RP}^{(e)(|Q|)}/a_{LL}$. In Eq. (200) we have neglected the small contribution of the P-interaction generated from the tensor part of the interaction by electromagnetic corrections. Using the relations (192) - (195) and (196) - (199) and allowing $0.98 < (u_e/u_\mu)^{1/2} < 1.02$, one has from $(R_\pi)_{expt}$

$$-6.4 \times 10^{-6} \; < \; Re \, \eta_{LT}^{(2/3)} \; < \; 6.2 \times 10^{-4} \; , \quad (201)$$

$$-6.2 \times 10^{-4} \; < \; Re \, \eta_{LT}^{(1/3)} \; < \; 6.4 \times 10^{-6} \quad (202)$$

The same limits hold also for $Re \, \eta_{LS}^{|Q|}$ $(|Q| = 2/3,\ 1/3)$.

For S,P and T interactions one has from C_T^{eN} (see Section II.3) the limit $|Im(6\eta_{LT} + \eta_{LS} - \eta_{LP})| \lesssim 8 \times 10^{-5}$ [75]. Using the relations (192), (194) and (196), (198) this yields

$$|Im \, \eta_{LT}^{(2/3)}| \; \lesssim \; 1 \times 10^{-5} \; , \quad (203)$$

$$|Im \, \eta_{LT}^{(1/3)}| \; \lesssim \; 2 \times 10^{-5} \; . \quad (204)$$

The limit on $|Im \, \eta_{LS}|$ $(= |Im \, \eta_{LT}|)$ from $(R_\pi)_{expt}$ is the same as in the case of spin-one leptoquark exchange. The same limit holds also for $|\eta_{RS}^{(e)(|Q|)}|$ $(= |\eta_{RT}^{(e)(|Q|)}|)$ $(Q = 2/3, 1/3)$. Thus

$$|\eta_{RT}^{(e)(|Q|)}| \; = \; |\eta_{RS}^{(e)(|Q|)}| \; < \; 3.1 \times 10^{-4} \; . \quad (205)$$

In the presence of both scalar and tensor couplings the constraint from the ratio P_L^F/P_L^{GT} of positron polarizations is given by [98] (recall that since $\eta_{RR}^{(e)*} \, \eta_{RL}^{(e)} = 0$, the V,A part of the interaction does not contribute to P_L^F/P_L^{GT})

$$Re(C_S - C_S') + g_A^{-1} \, Re(C_T - C_T') = (0.0027 \pm 0.0109)(G \, U_{ud}/\sqrt{2}) \; . \quad (206)$$

Let us denote the η_{LT} from the $Y_{(|Q|)}$-exchange interaction by $\eta_{LT}^{(|Q|)}$. Using the relations (194) and (198), Eq. (206) gives

$$-9.6\times10^{-3} \; < \; (g_T + g_S\, g_A)\; Re\,\eta_{LT}^{(2/3)} \; < \; 1.3\times10^{-2} \qquad (90\%\ c.l.)\ , \quad (207)$$

and

$$-9.6\times10^{-3} \; < \; (g_T - g_S\, g_A)\; Re\,\eta_{LT}^{(1/3)} \; < \; 1.3\times10^{-2} \qquad (90\%\ c.l.)\ . \quad (208)$$

The constraints (207) and (208) are the same as (173), except for the factors $(g_T \pm g_S\, g_A)$, which replace g_T in Eq. (173).

The bounds from the ft-values of superallowed beta decays considered in Section II.3 hold here not only for $g_S\, Re\,\eta_{LS}$, but also for $g_S\, \eta_{LT}^{(2/3)}$ and $g_S\,(-\eta_{LT}^{(1/3)})$.

In the presence of both S- and T-couplings the D-coefficient contains a term proportional to $Im(C_S C_T^* + C_S' C_T'^*) = 2 g_S g_T Im(a_{LS}\, a_{LT}^* + a_{RS}^{(e)}\, a_{RT}^{(e)*})$ [14]. This term vanishes however for leptoquark interactions due to the relations (194), (195) and (198), (199) [100].

Interaction terms of the a_{LS}-type can arise from S_1, $(R_2)_{T_z=-1/2}$, $U_{1\mu}$ and $(V_{2\mu})_{T_z=-1/2}$ exchange, while a_{RS}-type couplings can be generated by S_1, $(\tilde{R}_2)_{T_z=+1/2}$, $U_{1\mu}$ and $(\tilde{V}_{2\mu})_{T_z=1/2}$ exchange [64]. As discussed earlier, the exchange of spin-one leptoquark states will generate simultaneously P-type couplings, and the exchange of spin-zero states simultaneously P,T couplings.

The limits (201) and (202) make $Re\,\eta_{LT}$ from $Y_{(2/3)}$ or $Y_{(1/3)}$ exchange too small to be able to account for the discrepancy between theory and experiment in $\pi \to e\nu_e\gamma$ decay [91], mentioned at the end of Section II.4. The constraints on S,T couplings from $(R_\pi)_{expt}$ could be weaker if there is a cancellation in R_π between leptoquark contributions to $\pi \to e\nu_e$ and $\pi \to \mu\nu_\mu$, but there is no known model in which this could happen naturally. The constraints from $(R_\pi)_{expt}$ on the tensor couplings could be weaker also if $Y_{(2/3)}$ and $Y_{(1/3)}$ contribute simultaneously [66]. Then $(R_\pi)_{expt}$ constrains only $Re(\eta_{LT}^{(2/3)} - \eta_{LT}^{(1/3)})$, while $Re\,\eta_{LT} = Re(\eta_{LT}^{(2/3)} + \eta_{LT}^{(1/3)})$ could in principal be sufficiently large. Similarly, the limits on the scalar couplings could be weaker than (189) and (191) if spin-one and spin-zero leptoquarks contribute simultaneously [66]. But again, there is no known model in which such cancellations could take place in a way other than accidentally.

III. MUON DECAY

III.1. Introduction

The main decay mode of the muon is the decay into two neutrinos: $\mu^+ \to e^+ + n + n'$ [101]. In the SM $n = \nu_{eL}$ and $n' = \overline{\nu}_{\mu L}$, where ν_{eL} and $\nu_{\mu L}$ are massless two-component neutrinos which are, respectively, the $T_z = 1/2$ states of the $SU(2)_L$ doublets involving the electron and the muon. The interaction responsible for this decay is due to W-exchange and has the $V - A$ form

$$H_{SM}^{(\mu)} = (G/\sqrt{2})\, \overline{e}\, \gamma_\lambda (1-\gamma_5)\, \nu_e\, \overline{\nu}_\mu\, \gamma^\lambda (1-\gamma_5) \mu \; + \; \text{H.c.} \qquad (209)$$

where $G = (g^2/8M_W^2)(1+\Delta r)$; Δr represents radiative corrections [102].

In extensions of the SM there may be new interactions contributing to the decay mode $\mu^+ \to e^+ + \nu_e + \overline{\nu}_\mu$, and also new interactions giving rise to other decays of the type $\mu^+ \to e^+ + n + n'$. In the presence of the new interactions the neutrinos are expected to be massive, and the neutrino gauge group eigenstates are not expected to coincide with the mass eigenstates.

In models where lepton family numbers are conserved, or conserved to a good approximation, it is sufficient to consider for muon decay the most general lepton family number conserving (LC) four-fermion interaction. This can be written in the helicity projection form [103] as

$$
\begin{aligned}
H_{LC}^{(\mu)} &= 4(G_F/\sqrt{2})(g_{LL}^V\,\overline{e}_L\,\gamma_\lambda\,\nu_{eL}\,\overline{\nu}_{\mu L}\,\gamma^\lambda\,\mu_L \\
&+ g_{RR}^V\,\overline{e}_R\,\gamma_\lambda\,\nu_{eR}\,\overline{\nu}_{\mu R}\,\gamma^\lambda\,\mu_R \;+\; g_{LR}^V\,\overline{e}_L\,\gamma_\lambda\,\nu_{eL}\,\overline{\nu}_{\mu R}\,\gamma^\lambda\,\mu_R \\
&+ g_{RL}^V\,\overline{e}_R\,\gamma_\lambda\,\nu_{eR}\,\overline{\nu}_{\mu L}\,\gamma^\lambda\,\mu_L \;+\; g_{LL}^S\,\overline{e}_L\,\nu_{eR}\,\overline{\nu}_{\mu R}\,\mu_L \;+\; g_{RR}^S\,\overline{e}_R\,\nu_{eL}\,\overline{\nu}_{\mu L}\,\mu_R \\
&+ g_{LR}^S\,\overline{e}_L\,\nu_{eR}\,\overline{\nu}_{\mu L}\,\mu_R \;+\; g_{RL}^S\,\overline{e}_R\,\nu_{eL}\,\overline{\nu}_{\mu R}\,\mu_L \\
&+ g_{LR}^T\,\overline{e}_L\,t_{\alpha\beta}\,\nu_{eR}\,\overline{\nu}_{\mu L}\,t^{\alpha\beta}\,\mu_R \;+\; g_{RL}^T\,\overline{e}_R\,t_{\alpha\beta}\,\nu_{eL}\,\overline{\nu}_{\mu R}\,t^{\alpha\beta}\,\mu_L) \;+\; \text{H.c.} \qquad (210)
\end{aligned}
$$

In Eq. (210) we have followed the notation and the normalization of the coupling constants of Ref. [104] (used also in Refs. [105] and [4]). Thus the first and the second subscript on the coupling constants indicates the handedness of the electron and of the muon, respectively. G_F is the Fermi constant, and $t_{\alpha\beta} = (1/\sqrt{2})\,\sigma_{\alpha\beta}$. The neutrino states $\nu_{eL}(=\frac{1}{2}(1-\gamma_5)\nu_e)$ and $\nu_{eR}(=\frac{1}{2}(1+\gamma_5)\nu_e)$ are the left-handed and the right-handed components of the mass-eigenstate ν_e.

The Hamiltonian (210) contains 19 real parameters (10 complex coupling constants minus an overall phase). Neglecting neutrino masses, the $e^\pm$-observables depend on 10 real constants [106], denoted in the literature by $a, a', b, b', c, c', \alpha, \alpha', \beta$ and β'. The constants $a, a', \ldots$ are bilinear combinations of the coupling constants. The transverse polarization parameters α, α', β and β' are measured directly, while the remaining six constants are determined through measurements of the muon lifetime, the spectrum parameters ρ, ξ, δ and the longitudinal polarization parameters ξ' and ξ''.

In Ref. [104] (see also Ref. [4]) limits have been set on all the coupling constants of the Hamiltonian (210) using the experimental results on the muon decay parameters and on the inverse muon decay cross-section. One of the results is the lower bound

$$
Q_{LL} \equiv (\tfrac{1}{4}|g_{LL}^S|^2 + |g_{LL}^V|^2) \;>\; 0.949 \qquad (90\%\ c.l.)\,, \qquad (211)
$$

obtained from muon decay data alone, on the quantity Q_{LL} which contains the SM contribution. To obtain limits on the individual constants g_{LL}^V and g_{LL}^S in (211) one needs additional information. This is provided by the inverse muon decay process $\nu_\mu e^- \to \mu^- \nu_e$ [104]. Taking into account the constraints on the coupling constants from muon decay data, and using the result $1 + h < 0.00318$ for the

ν_μ-helicity h (deduced [107] from the experimental lower bound on $P_\mu \xi \delta / \rho$ [108], where ($-P_\mu$) is the degree of longitudinal polarization of the μ^+ at the instant of μ^+-decay), one obtains [104] for the ratio of the $\nu_\mu e^- \to \mu^- \nu_e$ cross-section and the $\nu_\mu e^- \to \mu^- \nu_e$ cross section predicted by the SM

$$S \simeq |g^V_{LL}|^2 . \tag{212}$$

The cross-section for inverse muon decay has been measured recently by the CHARM II collaboration [109] and by the CCFR collaboration [110], obtaining

$$S = 1.054 \pm 0.079 \qquad \text{(CHARM II)} , \tag{213}$$

$$S = 0.981 \pm 0.057 \qquad \text{(CCFR)} . \tag{214}$$

The result (214) yields (Ref. [110]; see also Ref. [105])

$$|g^V_{LL}| > 0.96 \qquad (90\% \ c.l.) , \tag{215}$$

and since $Q_{LL} \leq 1$, one has also (Ref. [110]; see also Ref. [105])

$$|g^S_{LL}| < 0.55 \qquad (90\% \ c.l.) . \tag{216}$$

The limits from (213) are only slightly weaker.

Thus the conclusion for the Hamiltonian (210) is that the only term in (210) which we know to be nonzero is the SM term. Moreover, this term is responsible for at least 92.5% of the observed muon decay rate. For the absolute values of the non-standard coupling constants the upper limits obtained are in the range 0.033 to 0.55 (Refs. [110, 111]; see also Ref. [105]).

The most general Hamiltonian [112], which allows for lepton family number violation and total lepton number violation can be obtained from the Hamiltonian (210) by the replacements

$$\begin{aligned} & g^V_{LL} \, \overline{e}_L \gamma_\lambda \nu_{eL} \overline{\nu}_{\mu L} \gamma^\lambda \mu_L \\ & \qquad \to \sum_{i,j} (g^V_{LL})_{ij} \, \overline{e}_L \gamma_\lambda n_{iL} \overline{n}_{jL} \gamma^\lambda \mu_L , \\ & g^V_{RR} \, \overline{e}_R \gamma_\lambda \nu_{eR} \overline{\nu}_{\mu R} \gamma^\lambda \mu_R \\ & \qquad \to \sum_{i,j} (g^V_{RR})_{ij} \, \overline{e}_R \gamma_\lambda n^c_{iR} n^c_{jR} \gamma_\lambda \mu_R , \end{aligned} \tag{217}$$

and analogous replacements for all the other terms in (210). As in (210) the fermion fields in (217) are mass eigenstates. The indices i, j run over all the neutrino states that can be emitted in the decay. The set n_{iL} includes all the left-handed neutrinos ($n_{1L} \equiv \nu_{eL}$, $n_{2L} \equiv \nu^c_{eL}$, $n_{3L} \equiv \nu_{\mu L}$, etc.), and the set n^c_{iR} all the right-handed ones ($n^c_{1R} \equiv \nu^c_{eR}$, $n^c_{2R} \equiv \nu_{eR}$, $n^c_{3R} \equiv \nu^c_{\mu R}$, etc.).

For the general case the $e^{\pm}$-spectrum and polarizations can again be described (assuming that the masses of the neutrinos that can be emitted in the decay can be neglected) by the constants $a, a', b, b', c, c', \alpha, \alpha', \beta$ and β', which have the same physical meaning as in the LC case [112]. The constants $a, a', \ldots$ are obtained by replacing the set of coupling constant combinations appearing in the LC case by a new set involving the coupling constants of the general Hamiltonian. There is a one-to-one correspondence between the coupling constant combinations pertinent to the LC case and the coupling constant combinations for the general case [112]. Consequently, the limits set on the coupling constant combinations of the LC case apply for the corresponding combinations of the general case. Thus, since [112]

$$\begin{aligned} \tfrac{1}{4}\,|g^S_{LL}|^2 &\leftrightarrow \sum_{i>j} |(g^V_{LL})_{ij} + \tfrac{1}{2}(g^S_{LL})_{ji}|^2 \,, \\ |g^V_{LL}|^2 &\leftrightarrow \sum_{i\leq j} |(g^V_{LL})_{ij} + \tfrac{1}{2}\,(g^S_{LL})_{ji}|^2 \,, \end{aligned} \tag{218}$$

the constraint (211) becomes

$$Q_{LL} \equiv \sum_{i,j} |(g^V_{LL})_{ij} + \tfrac{1}{2}(g^S_{LL})_{ji}|^2 > 0.949 \qquad (90\%\ c.l.) \tag{219}$$

The inverse muon decay process is now $\nu_\pi e^- \to \mu^- n_i$, where $\nu_\pi = \sum_j c_j n_{jL}$ is the neutrino state emitted in $\pi^+ \to \mu^+ \nu_\pi$ decay, and n_i are some neutrino states. The ratio S is given by [112]

$$S \simeq \sum_i |(g^V_{LL})_{i3} + \tfrac{1}{2}(g^S_{LL})_{3i}|^2 \,, \tag{220}$$

where the neutrino state ν_π has been denoted as n_{3L}.

The limits corresponding to (215) and (216) are

$$\sum_i |(g^V_{LL})_{i3} + \tfrac{1}{2}(g^S_{LL})_{3i}|^2 > 0.925 \qquad (90\%\ c.l.)\,, \tag{221}$$

$$\sum_{\substack{i,j \\ j\neq 3}} |(g^V_{LL})_{ij} + \tfrac{1}{2}(g^S_{LL})_{ji}|^2 < 0.075 \qquad (90\%\ c.l.)\,. \tag{222}$$

The sum in Eq. (221) contains the SM contribution, but the contributions of the scalar and the vector coupling constants in (221) cannot be separated. What one learns in this case from the bound (221) is that at least one of the μ^+ decay modes which involves the neutrino $\overline{\nu}_\pi$ produced in $\pi^- \to \mu^- \overline{\nu}_\pi$ decay dominates the μ^+ decay rate [112].

We note yet that there is some information also on the second neutrino in muon decay [113]. This follows from the experiment of Ref. [114], where neutrinos

(n_e) from μ^+ decay have been observed through the reaction $n_e\, D \to ppe^-$. The good agreement of the measured $n_e\, D \to ppe^-$ cross section with the calculated one in the SM indicates that the total muon decay rate contains a substantial contribution from μ^+ decay into a final state in which one of the neutrinos is the one accompanying the positron in nuclear beta decay. From a search [115] for $e^{\pm}$ production by ν_π on nucleons one has in addition some evidence that $\nu_\pi \neq n_e$ and $\nu_\pi \neq \overline{n}_e$.

In the subsequent two sections we shall discuss new contributions to muon decay in left-right symmetric models and in models with exotic fermions. Tree-level contributions to $\mu^+ \to e^+ + n + n'$ decays from new interactions arise also in many other theoretical schemes. These include models involving charged Higgs bosons [116], models with neutral flavor-changing gauge bosons or Higgs bosons [117], R-parity violating supersymmetric models [74], models with dileptonic gauge bosons [118], and models with dileptonic Higgs bosons [119]. New contributions to $\mu^+ \to e^+ + n + n'$ are present also in models with composite leptons, generated by constituent exchange [120]. The strength of these interactions is of the order of g^2/Λ_c^2 where g is an effective strong coupling constant and Λ_c is the compositeness scale. Assuming $g^2/4\pi \simeq 1$, muon decay provides a lower bound of a few TeV on Λ_c for some types of couplings.

We note yet that the undetected weakly interacting particles in the process $\mu^+ \to e^+ + \textit{missing neutrals}$ may include also light particles other than neutrinos [121].

III.2. Left-Right Symmetric Models

Neglecting mixing in the leptonic sector, the leptonic couplings in the Lagrangian (76) give rise to a V,A muon decay interaction of the form included in Eq. (210) with [122]

$$(G_F/\sqrt{2})\, g^V_{LL} \simeq g_L^2/8m_1^2 \,, \tag{223}$$

$$\kappa^V_{RR} \simeq g_R^2\, m_1^2/g_L^2\, m_2^2 \,, \tag{224}$$

$$\kappa^V_{LR} = \kappa^{V*}_{RL} \simeq -e^{i\omega}\, (g_R\zeta/g_L) \,, \tag{225}$$

where $\kappa^V_{ik} = g^V_{ik}/g^V_{LL}$ ($ik = RR, LR, RL$). Note that

$$|g^V_{LL}|^2 = (1 + |\kappa^V_{RR}|^2 + |\kappa^V_{LR}|^2 + |\kappa^V_{RL}|^2)^{-1} \,, \tag{226}$$

$$\begin{aligned} |g^V_{ij}|^2 &= |\kappa^V_{ij}|^2\, (1 + |\kappa^V_{RR}|^2 + |\kappa^V_{LR}|^2 + |\kappa^V_{RL}|^2)^{-1} \\ &\simeq |\kappa^V_{ij}|^2 \qquad (ij = RR, LR, RL) \,. \end{aligned} \tag{227}$$

In $SU(2)_L \times SU(2)_R \times U(1)_{B-L}$ models the neutrinos are massive in general, and the neutrinos are therefore expected to mix.

Let us consider the case when the neutrinos are Dirac fermions. Neglecting the kinematic effects of the neutrino masses (including interference terms proportional to neutrino mass) the mixing can be taken into account by making in observables the substitutions

$$g^V_{LL} \rightarrow g^V_{LL}\sqrt{u_e u_\mu}\ , \tag{228}$$

$$\kappa^V_{RR} \rightarrow \kappa^V_{RR}\sqrt{\tilde{v}_e \tilde{v}_\mu}\ , \tag{229}$$

$$\kappa^V_{LR} \rightarrow \kappa^V_{LR}\sqrt{\tilde{v}_\mu}\ , \tag{230}$$

$$\kappa^V_{RL} \rightarrow \kappa^V_{RL}\sqrt{\tilde{v}_e}\ , \tag{231}$$

where

$$u_l = \sum_i{}' |U_{li}|^2 \qquad (l = e, \mu)\ , \tag{232}$$

$$v_l = \sum_i{}' |V_{li}|^2 \qquad (l = e, \mu)\ , \tag{233}$$

and $\tilde{v}_l = v_l/u_l$ $(l = e, \mu)$; for n generations U and V are $n \times n$ unitary matrices. The prime on the summation signs in Eqs. (232) and (233) indicates that the sum is over the neutrinos that can be emitted in muon decay. Note that the u_e and v_e in Eqs. (232) and (233) are not equal in general to the analogous quantities for beta decay, but we shall assume here for simplicity that they are the same, and use the same symbol for them. Some of the muon decay observables involve the muon polarization. The polarization of the μ^+ from $\pi^+ \rightarrow \mu^+ \nu_\mu$ decay is $(-P_\mu)$, where

$$P_\mu \simeq 1 - 2|\eta_{RR} - \eta_{RL}|^2\,\tilde{v}_\mu\ . \tag{234}$$

The coupling constants η_{RR} and η_{RL} in Eq. (234) are $\eta_{RR} = \eta^{(e)}_{RR}/\sqrt{\tilde{v}_e}$ and $\eta_{RL} = \eta^{(e)}_{RL}/\sqrt{\tilde{v}_e}$, where $\eta^{(e)}_{RR}$ and $\eta^{(e)}_{RL}$ are given in Eqs. (79) and (81).

It follows that the $e^\pm$ spectrum and polarizations are described by four parameters: $\kappa^V_{RR}\sqrt{\tilde{v}_e \tilde{v}_\mu}$, $|\kappa^V_{LR}|\sqrt{\tilde{v}_\mu}$, $|\kappa^V_{LR}|\sqrt{\tilde{v}_e}$ (where we have used $|\kappa^V_{RL}| = |\kappa^V_{LR}|$) and $|\eta_{RR} - \eta_{RL}|\sqrt{\tilde{v}_\mu}$.

Among limits provided by individual observables the best one on $|\kappa^V_{LR}|\sqrt{\tilde{v}_\mu}$ and $|\kappa^V_{LR}|\sqrt{\tilde{v}_e}$ comes from the parameter ρ, and the best limit on $\kappa_{RR}\sqrt{\tilde{v}_e \tilde{v}_\mu}$ from measurements of the quantity $R_{(\mu)} \equiv 1 - (\delta\xi/\rho)P_\mu$ [108,4].

The parameter ρ is given by

$$\rho \simeq \tfrac{3}{4}(1 - |\kappa^V_{LR}|^2\,\tilde{v}_\mu - |\kappa^V_{LR}|^2\,\tilde{v}_e)\ . \tag{235}$$

The experimental value $\rho = 0.7518 \pm 0.0026$ [13] implies [123]

$$|\kappa^V_{LR}|\,\sqrt{\tilde{v}_\mu} \;<\; 6.7\times 10^{-2} \qquad (90\%\ c.l.)\ , \tag{236}$$

$$|\kappa^V_{LR}|\,\sqrt{\tilde{v}_e} \;<\; 6.7\times 10^{-2} \qquad (90\%\ c.l.)\ . \tag{237}$$

$R_{(\mu)}$ is given by

$$R_{(\mu)} \;\simeq\; 2|\kappa^V_{RR}|^2\,\tilde{v}_e\,\tilde{v}_\mu \;+\; 2|\eta_{RR}-\eta_{RL}|^2\,\tilde{v}_\mu\ . \tag{238}$$

The experimental result $R_{(\mu)} < 0.00323$ [108] yields

$$|\kappa^V_{RR}|\,\sqrt{\tilde{v}_e\,\tilde{v}_\mu} \;<\; 0.040 \qquad (90\%\ c.l.)\ , \tag{239}$$

and also

$$|\eta_{RR}-\eta_{RL}|\,\sqrt{\tilde{v}_\mu} \;<\; 0.040 \qquad (90\%\ c.l.)\ . \tag{240}$$

Combining (240) and (236) (noting that $|\eta_{RL}| = |\kappa^V_{LR}|$), we obtain

$$|\eta_{RR}|\,\sqrt{\tilde{v}_\mu} \;<\; 0.11\ . \tag{241}$$

A slightly better bound than (241) follows from the limit (82).

The remaining spectrum parameters and the polarization parameters are given by

$$\delta \;\simeq\; \tfrac{3}{4}(1-3|\kappa^V_{LR}|^2\,\tilde{v}_\mu \;+\; 3|\kappa^V_{LR}|^2\,\tilde{v}_e)\ , \tag{242}$$

$$\xi \;\simeq\; 1-2|\kappa^V_{RR}|^2\,\tilde{v}_e\,\tilde{v}_\mu \;+\; 2|\kappa_{LR}|^2\,\tilde{v}_\mu - 4|\kappa^V_{LR}|^2\,\tilde{v}_e\ , \tag{243}$$

$$\xi' \;\simeq\; 1-2|\kappa^V_{RR}|^2\,\tilde{v}_e\,\tilde{v}_\mu \;-\; 2|\kappa^V_{LR}|^2\,\tilde{v}_e\ , \tag{244}$$

$$\xi'' \;\simeq\; 1+2|\kappa_{LR}|^2\,\tilde{v}_e \;+\; 2|\kappa^V_{RL}|^2\,\tilde{v}_\mu\ , \tag{245}$$

$$\alpha = \alpha' = \beta = \beta' = 0\ . \tag{246}$$

In models where $\tilde{v}_e = \tilde{v}_\mu = 1$ one obtains from the experimental limit on $R_{(\mu)}$

$$|\kappa^V_{RR}| \;<\; 0.040 \qquad (90\%\ c.l.) \tag{247}$$

(which is a slightly better limit than the bound (82)), and (using $|\eta_{RR}| \lesssim \kappa^V_{RR}$)

$$4|\eta_{RR}|^2 \;+\; 2|\eta_{RL}|^2 \;+\; 4Re\,\eta_{RR}\,\eta^*_{RL} \;\lesssim\; 0.00323\ . \tag{248}$$

Eq. (248) yields $|\eta_{RR}| \lesssim 0.040$ and $|\eta_{LR}| = |\kappa^V_{LR}| \lesssim 0.057$. From the ρ-parameter and from the relation $|\kappa^V_{LR}| \lesssim |\kappa^V_{RR}|$ [56] (which is valid unless Higgs boson with $T_R \gg 1$ are present) one has the better limits

$$|\kappa^V_{LR}| \;<\; 0.047 \qquad (90\%\ c.l.) \tag{249}$$

and

$$|\kappa^V_{LR}| \lesssim 0.040, \tag{250}$$

respectively.

In manifestly left-right symmetric models (where $\kappa^V_{RR} \simeq m_1^2/m_2^2$ and $\kappa^V_{LR} = \kappa^V_{RL} = -\zeta$) the best limits are $|\kappa^V_{RR}| \lesssim 6\times10^{-3}$ (from the $K_L - K_S$ mass difference) and $|\kappa^V_{LR}| \lesssim 3.4\times10^{-3}$ (from charged-current universality) (see Section II.2.2). For pseudomanifest models one has again $|\kappa^V_{RR}| \lesssim 6\times10^{-3}$ (see the text after Eq. (98)) and, barring a cancellation in Eq. (95), $|\kappa^V_{LR}| < 3.6\times10^{-3}$ (obtained by combining the limits from (95) and (90)).

So far we have been dealing with the case when the neutrinos are Dirac fermions. For Majorana neutrinos $\nu_\ell^{(L)} - (\nu_{\ell'}^{(R)})^c$ ($\ell' = \ell$ or $\ell' \neq \ell$) mixing can also take place. (U and V are then $n\times 2n$ matrices; $(U, V^*)^T$ is unitary [11]), and as a consequence new terms appear in the muon decay spectrum [124]. The new terms affect the parameters α, α', β and β', which no longer vanish, and the coupling constants $|\kappa^V_{LR}|^2\ \tilde{v}_\mu$ and $|\kappa^V_{LR}|^2\ \tilde{v}_e$, which have to be replaced by $|\kappa^V_{LR}|^2\ (\tilde{v}_\mu + |\omega_{\mu e}|^2/u_e\, u_\mu)$ and $|\kappa^V_{LR}|^2\ (\tilde{v}_e + |\omega_{e\mu}|^2/u_e\, u_\mu)$, where $\omega_{\ell\ell'} = \sum_k' V_{\ell k}\, U_{\ell' k}$ [124]. The upper limits in Eqs. (236) and (237) remain valid, and the same limits apply for $|\omega_{\mu e}|^2/u_e\, u_\mu$ and $|\omega_{e\mu}|^2/u_e\, u_\mu$. For α, α', β and β' one has, considering for simplicity only terms first order in the new interactions, $\alpha \simeq 0$, $\alpha' \simeq 0$, and [124]

$$\beta \simeq -8\, \kappa^V_{RR}\ Re\, \omega_{e\mu}\, \omega^*_{\mu e}/u_e\, u_\mu\ , \tag{251}$$

$$\beta' \simeq -8\, \kappa^V_{RR}\ Im\, \omega_{e\mu}\, \omega^*_{\mu e}/u_e\, u_\mu\ . \tag{252}$$

The experimental values $\beta/A = (3.9 \pm 6.2)\times10^{-3}$ and $\beta'/A = (1.5\pm6.3)\times10^{-3}$ yield

$$-2.8\times10^{-2} < \kappa^V_{RR}\ Re\, \omega_{e\mu}\, \omega^*_{\mu e}/u_e\, u_\mu < 1.3\times10^{-2} \qquad (90\%\ c.l.)\ , \tag{253}$$

$$-2.4\times10^{-2} < \kappa^V_{RR}\ Im\, \omega_{e\mu}\, \omega^*_{\mu e}/u_e\, u_\mu < 1.8\times10^{-2} \qquad (90\%\ c.l.)\ . \tag{254}$$

In addition to new contributions from the gauge bosons, muon decay receives in $SU(2)_L\times SU(2)_R\times U(1)_{B-L}$ models also contributions from the exchange of charged Higgs bosons. One such contribution is from the physical charged Higgs bosons of the bidoublet field $\phi(2T_L+1=2, 2T_R+1=2, Y=2)$, which generates the Dirac masses of fermions [35]. The size of this contribution depends among others on the allowed range of the right-handed scale in the model and on unknown Yukawa couplings.

A further Higgs contribution is present in an attractive class of $SU(2)_L\times SU(2)_R\times U(1)_{B-L}$ models [125], which employs the triplet Higgs boson Δ_R (1,3,2) to induce part of the symmetry breaking. These models provide a framework for the understanding of the smallness of the masses of the usual neutrinos. The Δ_R couples to the right-handed leptons, and generates a large Majorana mass term for the right-handed neutrinos. The models contain also the triplet Higgs field Δ_L (3,1,2), required by the discrete left-right symmetry included in the model. The singly charged Higgs boson Δ_L^+ mediates the exotic muon decay $\mu^+ \to e^+ \overline{\nu}_e \nu_\mu$ [126]. The corresponding Hamiltonian is of the form

$$\begin{aligned} H &= 2(G_\mu^{(e)}/\sqrt{2})\, \overline{\nu}_e^c\, (1-\gamma_5)\, e\, \overline{\mu}(1+\gamma_5)\, \nu_\mu^c \;+\; \text{H.c.} \\ &= (G_\mu^{(e)}/\sqrt{2})\, \overline{\mu}\, \gamma_\lambda(1-\gamma_5)\, e\, \overline{\nu}_\mu \gamma^\lambda\, (1-\gamma_5)\, \nu_e \;+\; \text{H.c.}\,, \end{aligned} \tag{255}$$

where $G_\mu^{(e)} = \sqrt{2}\, f_{ee}\, f_{\mu\mu}^*/4m_+^2$; the $f_{\ell\ell}$ ($\ell = e, \mu$) are lepton-Δ_L Yukawa couplings and m_+ is the mass of the Δ_L^+. It can be shown [127] that $|G_\mu^{(e)}| \gtrsim 2 \times 10^{-4}\, G_F$ for muon neutrino masses in the range 35 keV $\lesssim m_{\nu_\mu} \lesssim$ 270 keV (= the present experimental upper limit for m_{ν_μ}), which in the model is the allowed range of m_{ν_μ} for which ν_μ is required by cosmological considerations to be unstable. The lower bound on $|G_\mu^{(e)}|$ in the range 35 keV $\lesssim m_{\nu_\mu} \lesssim A$ ($<$ 270 keV) increases with decreasing A. For $A = 35$ keV one has $|G_\mu^{(e)}| \gtrsim 2 \times 10^{-2}\, G_F$.

The best limit on $|G_\mu^{(e)}|$ is [128]

$$|G_\mu^{(e)}| \;<\; 0.16\, G_F \qquad\qquad (90\%\ c.l.)\,, \tag{256}$$

obtained in an experiment searching for the decay $\mu^+ \to e^+ \overline{\nu}_e \nu_\mu$. $|G_\mu^{(e)}|$ can be as large as the upper limit (256) [129]. The Δ_L^+ contributes also to other two-neutrino muon decay modes, but the amplitudes for these are proportional to leptonic mixing angles. The non-standard contributions to muon decay from the gauge bosons are expected to be small, since in this class of models the limit on $g_R^2\, m_1^2/g_L^2\, m_2^2$ is $\sim$ 0.015 [39] and the contribution of the light mass eigenstates in $\nu_e^{(R)}$ is suppressed by the small light-heavy mixing angles.

III.3. Models with Exotic Fermions

The Langrangian (100) (Section II.2.3) generates a muon decay interaction of V,A structure. In the case of Dirac neutrinos muon decay is described by the effective coupling constants $g_{LL}^V \sqrt{u_e\, u_\mu}$, $\kappa_{RR}^V \sqrt{\tilde{v}_e\, \tilde{v}_\mu}$, $\kappa_{LR}^V \sqrt{\tilde{v}_\mu}$, and $\kappa_{RL}^V \sqrt{\tilde{v}_e}$, which are given by

$$g_{LL}^V \sqrt{u_e\, u_\mu} \;=\; (g^2/8M_W^2)\, \sqrt{u_e\, u_\mu}\,, \tag{257}$$

$$\kappa_{LR}^V\, \sqrt{\tilde{v}_\mu} \;\simeq\; s_R^\mu\, \sqrt{\tilde{v}_\mu}\,, \tag{258}$$

$$\kappa_{RL}^V\, \sqrt{\tilde{v}_e} \;\simeq\; s_R^e\, \sqrt{\tilde{v}_e}\,, \tag{259}$$

$$\kappa^V_{RR}\sqrt{\tilde{v}_e\,\tilde{v}_\mu} \simeq s^e_R\, s^\mu_R\sqrt{\tilde{v}_e\,\tilde{v}_\mu} \simeq (\kappa^V_{LR}\sqrt{\tilde{v}_\mu})\,(\kappa^V_{RL}\sqrt{\tilde{v}_e})\,. \tag{260}$$

The quantities u_ℓ and $v_\ell = \tilde{v}_\ell u_\ell$ ($\ell = e, \mu$) are here given by $u_\ell = \sum_i |(A^\nu_L)_{\ell i}|^2$ and $v_\ell = \sum_i |(F^\nu_R)_{\ell i}|^2$ (see Section II.2.3). The muon decay parameters are given by expressions identical to those in Eqs. (235), (238), and (242)-(246), except that $|\kappa^V_{LR}|\sqrt{\tilde{v}_e}$ has to be replaced everywhere by $|\kappa^V_{RL}|\sqrt{\tilde{v}_e}$, since here $|\kappa^V_{RL}| \neq |\kappa^V_{LR}|$.

The global analysis in Ref. [58] of the contraints on ordinary-exotic fermion mixings yielded

$$|\kappa^V_{LR}|\sqrt{\tilde{v}_\mu} < 0.039 \qquad (90\%\ c.l.)\,, \tag{261}$$

$$|\kappa^V_{RL}|\sqrt{\tilde{v}_e} < 0.042 \qquad (90\%\ c.l.)\,. \tag{262}$$

The sources of these limits are data on muon decay [58]. For $\kappa^V_{RR}\sqrt{\tilde{v}_\mu}$ the relation (260) and the limits (261) and (262) imply

$$|\kappa^V_{RR}|\sqrt{\tilde{v}_e\tilde{v}_\mu} < 1.7\times 10^{-3}\,. \tag{263}$$

As $|s^{(e)}_R\sqrt{\tilde{v}_e}| \lesssim |s^{(e)}_R|$ and $|s^{(\mu)}_R\sqrt{\tilde{v}_\mu}| \lesssim |s^{(\mu)}_R|$, the coupling constants $|\kappa^V_{LR}|\sqrt{\tilde{v}_\mu}$ and $|\kappa^V_{RL}\sqrt{\tilde{v}_e}|$ are constrained also by limits on $|s^e_R|$ and $|s^\mu_R|$, which come predominantly from neutral current data (both low energy and at the Z-peak). The present limits on $|s^e_R|$ and $|s^\mu_R|$ are $|s^e_R| < 0.10$ and $|s^\mu_R| < 0.12$ [60], which give weaker limits on $|\kappa^V_{LR}|\sqrt{\tilde{v}_\mu}$ and $|\kappa^V_{RL}|\sqrt{\tilde{v}_e}$ than (261) and (262).

For Majorana neutrinos $\nu^{(L)}_e - (\nu^{(R)}_e)^c$ and $\nu^{(L)}_\mu - (\nu^{(R)}_e)^c$ mixing can be present, induced by mixing between the ordinary and the exotic doublet neutrinos. The muon decay parameters are obtained in this case formally in the same way as in left-right symmetry models. The quantity $\omega_{\ell\ell'}$ (see Section III.2) is given here by $\omega_{\ell\ell'} = (A^\nu_L\, F^{\nu\dagger}_L)_{\ell'\ell}$ [58], where $F^{\nu *}_L = F^\nu_R$.

IV. CONCLUSIONS

In this article we have reviewed and discussed possible new interactions in beta decay and muon decay, and the available constraints on the associated coupling constants. We can summarize the main points as follows:

Beta Decay

For new V,A interactions beta decay provides information on the coupling constants $\eta^{(e)}_{RR}$ and $\eta^{(e)}_{RL}$ (including the T-violating combination $Im\,\eta^{(e)*}_{RR}\,\eta^{(e)}_{RL}$ sensitive to their relative phase) describing the interactions involving right-handed neutrinos, and on the T-violating constant $Im\,\eta_{LR}$ which is sensitive to the relative phase of the a_{LR} and the SM interaction.

New V,A interactions involving right-handed currents can arise at the tree-level through the exchange of new gauge bosons with right-handed couplings to the fermions (such as the W_R in left-right symmetric models), as a result of mixing of the usual fermions with exotic ones with right-handed couplings to the W, and from the exchange of leptoquarks.

All the above mechanisms can give a T-violating contribution to the D-coefficient. In models involving leptoquarks a nonzero $Im\,\eta_{LR}$ can be generated by the exchange of $X_{(1/3)}$ or $Y_{(2/3)}$ leptoquarks. A nonzero $Im\,\eta_{RR}^{(e)*}\,\eta_{RL}^{(e)}$ term can arise from leptoquark exchange only if leptoquarks of the same spin but different charges or leptoquarks of the same charge but different spins contribute simultaneously. Both $|Im\,\eta_{LR}|$ and $|Im\,\eta_{RR}^{(e)*}\,\eta_{RL}^{(e)}|$ can be as large as the present experimental limit (10^{-3}) on $|Im\,(\eta_{LR}+\eta_{RR}^{(e)*}\,\eta_{RL}^{(e)})|$. In left-right symmetric models and in models with exotic fermions the D-coefficient is dominated by the $Im\,\eta_{LR}$ term. In these models more stringent limits ($\sim 10^{-4}$) on $|Im\,\eta_{LR}|$ than from the D-coefficient follow from the experimental bounds on the electric dipole moment of the neutron D_n and on ϵ'/ϵ. However in view of the uncertainties in the calculations of D_n and ϵ'/ϵ these limits are not as reliable as the direct limit.

For leptoquark models the best limits on $|\eta_{RR}^{(e)}|$ and $|\eta_{RL}^{(e)}|$ are those from beta decay ($|\eta_{RR}^{(e)}| < 0.10$, $|\eta_{RL}^{(e)}| < 0.04$). Already in the most general version of $SU(2)_L\times SU(2)_R\times U(1)_{B-L}$ models $|\eta_{RL}^{(e)}|$ is constrained also by muon decay data. In $SU(2)_L\times SU(2)_R\times U(1)_{B-L}$ models with manifest or pseudomanifest left-right symmetry there is a stringent limit on $|\eta_{RR}^{(e)}|$ from the $K^\circ - \overline{K}^\circ$ amplitude, and on $|\eta_{RR}^{(e)}|$ from $K \to 3\pi$ decays, from charged-current universality, and also from the relation $|\zeta| \lesssim m_1^2/m_2^2$. In models with exotic fermions limits for $\eta_{RL}^{(e)}$ and $\eta_{RR}^{(e)}$ follow from muon decay data and also from neutral current experiments. At present the best limit on $|\eta_{RL}^{(e)}|$ comes from muon decay, and the best limit on $|\eta_{RR}^{(e)}|$ from combining the constraints from $K \to 3\pi$ and muon decay.

Scalar beta decay interactions can arise at the tree level through the exchange of charged Higgs bosons, the exchange of leptoquarks, and in R-parity violating supersymmetric models also by the exchange of sleptons. Beta decay is sensitive to the coupling constants $g_S\,\eta_{LS}$ and $g_S\,\eta_{RS}$, and with the exception of $g_S\,Im\,\eta_{LS}$ provides the most stringent limits on them ($|g_S\,Re\,\eta_{LS}| < 7\times10^{-3}$, $|g_S\,\eta_{RS}^{(e)}| < 7\times 10^{-2}$). $Im\,\eta_{LS}$ can be probed in beta decay through the T-odd R-coefficient and also through some T-even observables. The most stringent limit on $Im\,\eta_{LS}$ ($|Im\,\eta_{LS}| \lesssim 10^{-4}$, which is three orders of magnitude smaller than the upper limit from beta decay) has been deduced from the experimental bound on the P,T-violating tensor electron-nucleon interaction. However this limit may involve unknown theoretical uncertainties and therefore a limit from beta decay, even an order of magnitude weaker, would be valuable.

Tensor beta decay interactions are described by the coupling constants $g_T\,\eta_{LT}$ and $g_T\,\eta_{RT}$. Except for $Im\,\eta_{LT}$, for a tensor interaction of unspecified origin the best limits on these constants are from beta decay ($|g_T\,Re\,\eta_{LT}| < 7\times10^{-3}$, $|g_T\,\eta_{RT}| < 5\times10^{-2}$). The most stringent limit on $Im\,\eta_{LT}$ ($|Im\,\eta_{LT}| \lesssim 2\times10^{-5}$) comes from the experimental bound on the tensor electron-nucleon interaction. The next best limit ($|g_T\,Im\,\eta_{LT}| \lesssim 10^{-2}$) is from a search for the R-correlation. Again, improved limits from beta decay would be valuable.

The exchange of spin-one and spin-zero leptoquarks generates S,P and S,P,T beta decay interactions (in addition to V,A couplings), respectively. In renormalizable gauge theories the exchange of spin-zero leptoquarks is the only possible source of tensor interactions at the tree level. For the S and T interactions from leptoquark exchange stringent limits on η_{LS}, $\eta_{RS}^{(e)}$, η_{LT} and $\eta_{RT}^{(e)}$ follow from the $\pi \to e\nu_e/\pi \to \mu\nu_\mu$ ratio ($|Re\,\eta_{Lk}| < 6 \times 10^{-4}$, $|Im\,\eta_{Lk}| < 3 \times 10^{-4}$, $|\eta_{Rk}^{(e)}| < 3 \times 10^{-4}$ $(k = S, T)$. These limits would be weaker if there is some cancellation between the electronic and the muonic pseudoscalar contribution to R_π. But there is no known model in which this could happen naturally. The limits from $(R_\pi)_{expt}$ for the tensor couplings could also be weaker if there is a pseudoscalar contribution from both $Y_{(2/3)}$ and $Y_{(1/3)}$ leptoquarks, and for the scalar couplings if there is such a contribution from both spin-one and spin-zero leptoquarks. But again, there is no known model in which the required cancellations could take place in a way other than accidentally.

New beta decay interactions could arise also in composite models (which we mentioned here only in connection with muon decay), as a result of constituent exchange. The strength of such interactions is of the order of $4\pi/\Lambda_c^2$, where Λ_c is the compositeness scale. With $\Lambda_c \simeq$ a few TeV (the current lower bound from other processes), the corresponding beta decay coupling constants would have values near their present upper bounds.

Muon Decay

Experiment indicates that among the decays $\mu^+ \to e^+ + n + n'$ the decay mode which dominates the total $\mu^+ \to e^+ + n + n'$ rate involves the neutrino species of the Standard Model scenario. The combination of the coupling constants which dominates the rate includes the coupling constant of the Standard Model interaction.

Muon decay gives information on left-right symmetric models and on models with exotic fermions which is complementary to the information provided by beta decay.

In left-right symmetric models the gauge bosons generate a V,A interaction described for Dirac neutrinos by the four effective coupling constants $g_{LL}^V\,\sqrt{u_e\,u_\mu}$, $\kappa_{RR}^V\,\sqrt{\tilde{v}_e\,\tilde{v}_\mu}$, $\kappa_{LR}^V\,\sqrt{\tilde{v}_\mu}$ and $\kappa_{LR}^V\,\sqrt{\tilde{v}_e}$. For the most general version of SU(2)$_L\times$ SU(2)$_R\times$U(1)$_{B-L}$ models the best limits on $\kappa_{RR}^V\,\sqrt{\tilde{v}_e\,\tilde{v}_\mu}$, $\kappa_{LR}^V\,\sqrt{\tilde{v}_\mu}$ and $\kappa_{LR}^V\,\sqrt{\tilde{v}_e}$ come from muon decay itself ($\kappa_{RR}^V\,\sqrt{\tilde{v}_e\,\tilde{v}_\mu} < 4 \times 10^{-2}$, $|\kappa_{LR}^V|\,\sqrt{\tilde{v}_\mu} < 7 \times 10^{-2}$, $|\kappa_{LR}|\,\sqrt{\tilde{v}_e} < 7 \times 10^{-2}$). For Majorana neutrinos some additional parameters are needed to account for the effects of neutrino mixing, but these do not affect the above limits. In SU(2)$_L$ $\times$ SU(2)$_R$ $\times$ U(1)$_{B-L}$ models with triplet Higgs bosons the new contributions from the gauge bosons are expected to be small. A new effect for muon decay is the presence of the exotic decay mode $\mu^+ \to e^+\,\overline{\nu}_e\nu_\mu$, mediated by the Δ_L^+ Higgs boson. This decay can probe the mass of the muon neutrino in the range where cosmological considerations require ν_μ to be unstable.

In models with exotic fermions the V,A interaction is described formally by the same four effective coupling constants and the same additional parameters for the Majorana neutrino case as in left-right symmetric models, except that now $|\kappa_{RL}^V| \neq |\kappa_{LR}^V|$. At present the best limits on $|\kappa_{LR}^V|\,\sqrt{\tilde{v}_\mu}$, $|\kappa_{RL}^V|\,\sqrt{\tilde{v}_e}$ and $\kappa_{RR}^V\,\sqrt{\tilde{v}_e\tilde{v}_\mu}$ ($|\kappa_{LR}^V|\,\sqrt{\tilde{v}_\mu} < 4 \times 10^{-2}$, $|\kappa_{RL}^V|\,\sqrt{\tilde{v}_e} < 4 \times 10^{-2}$, $\kappa_{RR}^V\,\sqrt{\tilde{v}_e\tilde{v}_\mu} < 2 \times 10^{-3}$) come from muon decay. These coupling constants are constrained also by neutral current experiments.

Muon decay probes new leptonic interactions also in many other extensions of the Standard Model.

Our overall conclusion is that beta decay and muon decay gives unique information on some classes of possible interactions beyond the Standard Model. It is important to improve the sensitivity of the pertinent experiments by as much as possible.

Note Added. After most of this article was completed I learned that for the $\pi \to e\nu_e/\pi \to \mu\nu_\mu$ ratio R_π a new experimental result [130] has appeared. The constraint on $R_\pi/(R_\pi)_{SM}$ from this experiment is $0.9967 \ (u_\mu/u_e)^{1/2} < R_\pi/(R_\pi)_{SM} < 1.0033 \ (u_\mu/u_e)^{1/2}$, to be compared with the bound in Eq. (63). Inspection shows that the bound corresponding to Eq. (64) is $(-3.3 \times 10^{-2}) < Re\eta_{LR} < 3.3 \times 10^{-2}$ and that the difference between the bounds (65), (66), (169), (176), (189), and (191) and the corresponding bounds implied by the result of Ref. [130] is negligible.

ACKNOWLEDGEMENTS

I would like to thank T. J. Bowles, R. L. Burman, P. Langacker, D. London, S. R. Mishra, I. S. Towner, J. F. Wilkerson, and especially J. Deutsch, R. N. Mohapatra, and P. A. Quin for enlightening conversations. To P. Langacker I am indebted also for useful comments on the manuscript. This work was supported by the United States Department of Energy.

References

[1] S. Weinberg, Phys. Rev. Lett. **19**, 1264 (1967); A. Salam, in *Elementary Particle Theory: Relativistic Groups and Analycity (Nobel Symposium No. 8)*, edited by N. Svartholm (Almqvist and Wiksell, Stockholm, 1969), p. 367; S. L. Glashow, Nucl. Phys. **22**, 579 (1961); S. L. Glashow, J. Iliopoulos, and L. Maiani, Phys. Rev. D **2**, 1285 (1970).

[2] The electroweak component of the Standard Model will be understood here to be the minimal version of the $SU(2)_L \times U(1)$ gauge theory (Ref. [1]), containing one Higgs doublet and only left-handed neutrinos.

[3] J. Deutsch and P. A. Quin, this volume.

[4] W. Fetscher and H.-J. Gerber, this volume.

[5] A. Sirlin, this volume.

[6] For a review of the current status of searches for neutrino mass in beta decay see J. F. Wilkerson, to appear in the Proceedings of the XV International Conference on Neutrino Physics and Astrophysics - NEUTRINO '92, Grenada, Andalucia, Spain, June 7-12, 1992 (to be published in Nucl. Phys. B, Proceedings supplements); Reviews of the searches for heavy neutrinos in beta decay include J. F. Wilkerson, to appear in the Proceedings of the 7th meeting of the Division of Particles and Fields of the APS (DPF 92), Fermilab, Batavia,

Illinois, Nov. 10-14, 1992; J. Deutsch, M. Lebrun, and R. Prieels, Nucl. Phys. **A518**, 149 (1990); see also R. E. Shrock, Phys. Lett. B **96**, 159 (1980). For both the above subjects see also F. Boehm and P. Vogel, *Physics of Massive Neutrinos* (Cambridge University Press, Cambridge, England, 1992). Analyses of limits on neutrino mixing from muon decay data have been made in R. E. Shrock, Phys. Rev. D **24**, 1275 (1981) and in M. S. Dixit, P. Kalyniak and J. N. Ng, Phys. Rev. D **27**, 2216 (1983).

[7] Our metric, γ-matrices and $\sigma_{\lambda\mu}$ are the same as in J. D. Bjorken and S. D. Drell, *Relativistic Quantum Mechanics* (McGraw-Hill, New York, 1963).

[8] Loop-contributions from new physics in beta decay have been, to our knowledge, considered so far only in connection with tests of the unitarity relation for the measured elements of the Kobayashi-Maskawa matrix. The unitarity relation can set a limit on the mass of possible new neutral gauge bosons, which contribute through radiative corrections [W. J. Marciano and A. Sirlin, Phys. Rev. D **35**, 1972 (1987)]. Generally loop-effects from new interactions are not likely to be large enough compared with the theoretical uncertainties in the description of beta decay in the SM.

[9] Exceptions are beta decay interactions involving second-class currents [S. Weinberg, Phys. Rev. **112**, 1375 (1958)]. Interactions involving second-class currents cannot be introduced without spoiling the renormalizability of the theory or having to face severe theoretical and phenomenological difficulties [B. R. Holstein and S. B. Treiman, Phys. Rev. D **13**, 3059 (1976); P. Langacker, Phys. Rev. D **14**, 2340 (1976) and Phys. Rev. D **15**, 2386 (1977)]. For a review of the present experimental limits on second-class currents see L. Grenacs, Ann. Rev. Nucl. Part. Sci. **35**, 455 (1985).

[10] Examples are the couplings involving $\overline{e}\,\Gamma(1+\gamma_5)(\nu_e^{(L)})^c$ and $\overline{e}\,\Gamma(1-\gamma_5)\,\nu_\mu^{(L)}$, where $(\nu_e^{(L)})^c = C\,\overline{\nu}_e^{(L)T}$ and $\nu_\mu^{(L)}$ is the neutrino state in the $W^+ \to \mu^+\,\nu_\mu^{(L)}$ amplitude. The former coupling is constrained by double beta decay, and the latter by muon-number violating processes. On the other hand, there are as yet no significant constraints on couplings involving $\overline{e}\,\Gamma(1-\gamma_5)\,\nu_\tau^{(L)}$.

[11] S. M. Bilenky and B. Pontecorvo, Lett. Nuovo Cimento **17**, 569 (1976); S. M. Bilenky, J. Hosek, and S. T. Petcov, Phys. Lett. B **94**, 495 (1980); T. Yanagida and M. Yoshimura, Prog. Theor. Phys. **64**, 1870 (1980); J. Schechter and J. W. F. Valle, Phys. Rev. D **22**, 2227 (1980); For reviews see P. Langacker, in *Neutrino Physics*, edited by H. V. Klapdor (Springer Verlag, Berlin, 1990); B. Kayser, F. Gibrat-Debu, and F. Perrier, *The Physics of Massive Neutrinos* (World Scientific, Singapore, 1989); R. N. Mohapatra and P. B. Pal, *Massive Neutrinos in Physics and Astrophysics* (World Scientific, Singapore, 1991).

[12] S. Weinberg, Ref. [9].

[13] Particle Data Group, "Review of Particle Properties," Phys. Rev. D **45**, Number 11, June 1, 1992, Part II.

[14] J. D. Jackson, S. B. Treiman, and H. W. Wyld, Jr., Phys. Rev. **106**, 517 (1957). Our constants C_K, C'_K are the same as in this reference, except for the opposite sign of C'_V, C_A, C'_S, C'_P and C'_T.

[15] G. Steigman, D. N. Schramm, and J. E. Gunn, Phys. Lett. B **66**, 202 (1977).

[16] T. P. Walker, G. Steigman, D. N. Schramm, K. A. Olive and H.-S. Kang, Astrophys. J. **376**, 51 (1991).

[17] G. Steigman, K. A. Olive and D. N. Schramm, Phys. Rev. Lett. **43**, 239

(1979); K. A. Olive, D. N. Schramm and G. Steigman, Nucl. Phys. **B180**, 497 (1981). For a review see R. N. Mohapatra and P. Pal, Ref. [11].

[18] G. Raffelt and D. Seckel, Phys. Rev. Lett. **60**, 1793 (1988).

[19] R. Barbieri and R. N. Mohapatra, Phys. Rev. D **39**, 1229 (1989).

[20] K. S. Babu, R. N. Mohapatra, and I. Z. Rotstein, Phys. Rev. D **45**, R3312 (1992).

[21] Note that the parameters x and y (Eqs. (36) and (37)) are identical to the parameters x and y introduced by B. R. Holstein and S. B. Treiman, [Phys. Rev. D **16**, 2369 (1977)] only for $\eta_{RL} = \eta_{LR}$. Note also that the coupling constants (and therefore x and y) are assumed to be real in the above paper.

[22] A. L. Hallin, F. P. Calaprice, D. MacArthur, L. E. Piilonen, M. B. Schneider, and D. F. Schreiber, Phys. Rev. Lett. **52**, 337 (1984).

[23] C. G. Callan, Jr., and S. B. Treiman, Phys. Rev. **162**, 1494 (1967).

[24] P. A. Quin and T. A. Girard, Phys. Lett. B **229**, 29 (1989).

[25] N. Severijns *et al.*, to be published in Phys. Rev. Lett.

[26] A. S. Carnoy, J. Deutsch, T. A. Girard, and R. Prieels, Phys. Rev. C **43**, 2825 (1991); V. A. Wichers, T. R. Hageman, J. van Klinken, and H. W. Wilschut, Phys. Rev. Lett. **58**, 1821 (1987).

[27] For a review of $\pi \to \ell\nu_\ell$ decays see P. Herczeg, in *Weak and Electromagnetic Interactions in Nuclei*, Proceedings of the Third International Symposium on Weak and Electromagnetic Interactions in Nuclei (WEIN-92), edited by Ts. Vylov (World Scientific, 1993).

[28] W. J. Marciano, private communication to D. I. Britton, et al., Ref. [29].

[29] D. I. Britton *et al.*, Phys. Rev. Lett. **68**, 3000 (1992).

[30] We follow here the treatment in Ref. [27].

[31] P. Quin, private communication.

[32] B. G. Erozolimskii *et al.*, Yad. Fiz. **52**, 1583 (1990) [Sov. J. Nucl. Phys. **52**, 999 (1990)]; B. G. Erozolimskii *et al.*, Phys. Lett. B **263**, 33 (1991).

[33] P. Quin, private communication. For a discussion of neutron decay and for the pertinent references, see Ref. [3].

[34] J. C. Pati and A. Salam, Phys. Rev. D **10**, 275 (1974); R. N. Mohapatra and J. C. Pati, Phys. Rev. D **11**, 566, 2558 (1975); G. Senjanović and R. N. Mohapatra, Phys. Rev. D **12**, 1502 (1975).

[35] For a review of $SU(2)_L \times SU(2)_R \times U(1)_{B-L}$ models see R. N. Mohapatra, *Unification and Supersymmetry* (Springer-Verlag, New York, 1986).

[36] R. N. Mohapatra and J. C. Pati, Phys. Rev. D **11**, 566 (1975).

[37] R. N. Mohapatra and D. P. Sidhu, Phys. Rev. D **17**, 1876 (1978); D. Chang, Nucl. Phys. B **214**, 435 (1983); P. Herczeg, Phys. Rev. D **28**, 200 (1983).

[38] P. Herczeg, Ref. [37]. See also P. Herczeg, in *Neutrino Mass and Low-Energy Weak Interactions*, edited by V. Barger and D. Cline (World Scientific, Singapore, 1985), p. 288.

[39] P. Langacker and S. Uma Sankar, Phys. Rev. D **40**, 1569 (1989).

[40] P. Herczeg, Ref. [37] and in *Rare Meson Decays*, collected transparencies for the 7th Workshop on Particles and Nuclei, Heidelberg, December 2-3, 1986, Max Planck Institute für Kernphysik report, 1986; H. Harari and M. Leurer, Nucl. Phys. B **233**, 221 (1984); D. London and D. Wyler, Phys. Lett. B **232**, 503 (1989).

[41] F. Abe *et al.*, Phys. Rev. Lett. **67**, 2609 (1991).

[42] P. Herczeg, in *Weak and Electromagnetic Interactions in Nuclei*, edited by H. V. Klapdor (Springer-Verlag, 1986), p. 528.

[43] The special case of a heavy right-handed muon neutrino, corresponding to $v_\mu = 0$, was considered by A. P. Serebrov and N. V. Romanenko [St. Petersburg Nuclear Physics Institute preprint No. 1775, March 1992] to reconcile muon-decay and neutron decay data. J. Deutsch and P. A. Quin (this volume) considered for the same reason the case of an intermediate-mass right-handed neutrino, noting on the basis of the calculation by A. Jodidio *et al.*, [Phys. Rev. D **34**, 1967 (1986); Phys. Rev. D **37**, 237 (1988) (E)] that for $m_{\nu_R} \gtrsim 7.5$ MeV the limit on m_1^2/m_2^2 from muon decay becomes sufficiently weak that the conflict is avoided.

[44] For a recent discussion of ϵ'/ϵ in $SU(2)_L \times SU(2)_R \times U(1)_{B-L}$ models see X.-G. He, B. H. J. McKellar, and S. Pakvasa, Phys. Rev. Lett. **61**, 1267 (1988). The limit $|D_t/a_D| \lesssim 3 \times 10^{-5}$ from D_n is implied by the calculation of G. Beall and A. Soni, Phys. Rev. Lett. **47**, 552 (1981), and of G. Ecker, W. Grimus, and H. Neufeld, Nucl. Phys. **B299**, 421 (1983).

[45] H. Abramowicz et al., Z. Phys. C **12**, 225 (1982).

[46] Due to an oversight, the limit on $|\eta_{LR}|$ from inclusive neutrino and antineutrino scattering on nuclei was included in Refs. [62,66] (see below) among the model independjent constraints for beta decay. Since in the experiments of Refs. [45,47] the neutrino beams consist predominantly of ν_μ and $\overline{\nu}_\mu$, the constraint holds for the beta decay interaction only if, like in left-right symmetric models, the η_{LR}'s in the muonic and the electronic interactions are the same. Note that in such models R_π is not sensitive to η_{LR}.

[47] S. R. Mishra *et al.*, Phys. Rev. Lett. **68**, 3499 (1992).

[48] J. F. Donoghue and B. R. Holstein, Phys. Rev. Lett. B **113**, 382 (1982). Following Ref. [39] we interpret their limit as a 1σ error.

[49] P. Langacker and S. Uma Sankar, Ref. [39]. The relation (95) for the case of pseudomanifest $SU(2)_L \times SU(2)_R \times U(1)_{B-L}$ models with heavy right-handed neutrinos was derived, and used to obtain a limit on $|\zeta|$, in L. Wolfenstein, Phys. Rev. D **29**, 2130 (1984).

[50] J. Lindquist *et al.*, Phys. Rev. D **16**, 2104 (1977); V. G. Lind *et al.*, Phys. Rev. **135**, B1483 (1964).

[51] See the second paper of Ref. [38].

[52] R. N. Mohapatra, F. E. Paige, and D. P. Sidhu, Phys. Rev. D **17**, 2642 (1987). See also R. N. Mohapatra in *New Frontiers in High Energy Physics*, edited by A. Perlmutter and L. F. Scott (Plenum, New York, 1978), p. 337.

[53] M. A. B. Bég, R. V. Budny, R. N. Mohapatra, and A. Sirlin, Phys. Rev. Lett. **38**, 1252 (1977).

[54] Investigations of beta decay in manifestly left-right symmetric models include M. A. B. Bég *et al.*, Ref. [53]; B. R. Holstein and S. B. Treiman, Ref. [21]; A. S. Carnoy, J. Deutsch, and B. R. Holstein, Phys. Rev. D **38**, 1636 (1988).

[55] G. Beall, M. Bander, and A. Soni, Phys. Rev. Lett. **48**, 848 (1982). See also Ref. [39].

[56] E. Massó, Phys. Rev. Lett. **52**, 1956 (1984). The generalization of this relation for non-manifest models was given in Ref. [39].

[57] P. Langacker and D. London, Phys. Rev. D **38**, 886 (1988); This paper contains also references to earlier work on mixings between the usual fermions and various types of exotic ones. For a review see D. London, this volume.

[58] P. Langacker and D. London, Ref. [57].

[59] The limits (106) and (107) result when all the mixing angles are allowed to vary

simultaneously. When only one angle at a time is allowed to vary, the upper limit on $|\eta_{RL}^{(e)}|$ remains the same, and the limit on $|Re\,\eta_{LR}|$ becomes better by a factor of ~ 3 (see Ref. [58]).
[60] E. Nardi, E. Roulet, and D. Tommasini, Nucl. Phys. **B386**, 239 (1992). For a review see D. London, this volume.
[61] P. Herczeg, "The T-Odd D-Correlation in Beta Decay," in preparation.
[62] P. Herczeg, in *Fundamental Symmetries in Nuclei and Particles*, edited by H. Henrikson and P. Vogel (World Scientific, Singapore, 1990), p. 46.
[63] A few of the papers dealing with leptoquarks and their phenomenology are O. Shanker. Nucl. Phys. **B206**, 253 (1982); O. Shanker, Nucl. Phys. **B204**, 375 (1982); W. Buchmüller and D. Wyler, Phys. Lett. B **177**, 377 (1986); Y. Kizikuri, Phys. Lett. B **185**, 183 (1987); W. Buckmüller, R. Rückl, and D. Wyler, Phys. Lett. B **191**, 442 (1987); R. N. Mohapatra, G. Segré, and L. Wolfenstein, Phys. Lett. B **145**, 433 (1984); R. N. Mohapatra, Phys. Rev. D **34**, 3457 (1986); L. J. Hall and L. J. Randall, Nucl. Phys. **B274**, 157 (1986); S. M. Barr, Phys. Rev. D **34**, 1567 (1986); S. M. Barr and E. M. Freire, Phys. Rev. D **41**, 2129 (1990); M. Claudson, E. Farhi, and R. L. Jaffe, Phys. Rev. D **34**, 873 (1986); P. Langacker and R. Rückl, 1991, unpublished (see P. Langacker, M. Luo and A. K. Mann, Rev. Mod. Phys. **64**, 87 (1992).
[64] P. Herczeg, "Effects of Leptoquarks in Beta Decay," in preparation.
[65] The fact that a Fierz transform of a V,A four-fermion interaction does not contain a tensor-type coupling, but a Fierz transform of an S,P interaction does, was observed in F. Scheck, *Leptons, Hadrons and Nuclei* (North-Holland, Amsterdam, 1983). The same features have been observed in O. Shanker, Nucl. Phys. **B204**, 375 (1982) in connection with the four-fermion interactions generated by the exchange of spin-one and spin-zero leptoquarks.
[66] P. Herczeg, Hyp. Int. **75**, 127 (1992).
[67] S. M. Barr, Ref. [63].
[68] W. Buchmüller, R. Rückl, and D. Wyler, Ref. [63].
[69] P. Herczeg, in *Progress in Nuclear Physics*, edited by W-Y. Pauchy Hwang, S.-C. Lee, C.-E. Lee and D. J. Ernst (North-Holland, New York, 1991), p. 171.
[70] The beta decay interactions generated by the other leptoquark states will be included in Ref. [64].
[71] The complete quark-lepton four-fermion interaction arising from S_1-exchange in the absence of right-handed neutrinos has been written down in P. Langacker, M. Luo, and A. K. Mann, Ref. [63].
[72] O. Shanker, Nucl. Phys. **B204**, 375 (1982); S. M. Barr and E. M. Freire, Phys. Rev. D **41**, 2129 (1990).
[73] Effects of charged Higgs bosons in beta decay have been studied in B. McWilliams and L.-F. Li, Nucl. Phys. **B179**, 62 (1981); H. E. Haber, G. L. Kane, and T. Sterling, Nucl. Phys. **B161**, 493 (1979).
[74] Papers dealing with the phenomenology of R-parity violating supersymmetric models include V. Barger, G. F. Giudice, and T. Han, Phys. Rev. D **40**, 2987 (1989); R. N. Mohapatra, invited talk at the International Workshop on Low Energy Muon Science, Santa Fe, NM, March 1993, University of Maryland preprint UMD-PP-93-138, to be published in the Proceedings. These papers include also references to earlier work on this class of models.
[75] I. B. Khriplovich, Nucl. Phys. **B352**, 385 (1991).
[76] S. K. Lamoreaux, in *Weak and Electromagnetic Interactions in Nuclei*, Pro-

ceedings of the Third International Conference on Weak and Electromagnetic Interactions (WEIN-92), edited by Ts. Vylov (World Scientific, Singapore, 1993). This paper reports a new limit $|d(^{199}Hg)| < 2.5 \times 10^{-27}$ ecm (95% c.l.) on the electric dipole moment of the ^{199}Hg atom, which implies the quoted limit on C_T^{eN}.

[77] M. B. Schneider, F. P. Calaprice, A. T. Hallin, D. W. MacArthur, and D. F. Schreiber, Phys. Rev. Lett. **51**, 1239 (1983).

[78] P. Vogel and B. Werner, Nucl. Phys. **A404**, 345 (1983); B. R. Holstein, Phys. Rev. C **28**, 342 (1983).

[79] E. G. Adelberger, Phys. Rev. Lett. **70**, 2856 (1993), and private communication.

[80] W. E. Ormand, B. A. Brown, and B. R. Holstein, Phys. Rev. C **40**, 2914 (1989).

[81] See I. S. Towner, J. C. Hardy, E. Hagberg, and V. T. Koslowsky, in *Weak and Electromagnetic Interactions in Nuclei*, Proceedings of the Third International Symposium on Weak and Electromagnetic Interactions in Nuclei (WEIN-92), edited by Ts. Vylov (World Scientific, Singapore, 1993).

[82] I. S. Towner, private communication.

[83] S. L. Adler *et al.*, Phys. Rev. D **11**, 3309 (1975).

[84] V. M. Belyaev and I. I. Kogan, Phys. Lett. B **280**, 238 (1992).

[85] M. Allet *et al.*, to appear in the Proceedings of the High Energy Spin 92 Symposium, Nagoya, Japan, Nov. 9-14, 1992. The earlier result of this experiment appeared in M. Allet *et al.*, Phys. Rev. Lett. **68**, 572 (1992).

[86] P. Vogel and B. Werner, Ref. [78]. See also M. Allet et al., Proceedings of the XIIth Moriond Workshop in Tests of Fundamental Symmetries, to be published by Editions Frontières.

[87] M. Skalsey and S. Hatamian, Bull. Am. Phys. Soc. **30**, 1277 (1985); Phys. Rev. C **31**, 2218 (1985).

[88] M. B. Voloshin, Phys. Lett. B **283**, 120 (1992).

[89] In Ref. [88] only a tensor interaction involving a left-handed neutrino was considered, but a tensor interaction with a right-handed neutrino is affected in the same way.

[90] For the quark, masses we have taken $m_u = 4.2$ MeV, $m_d = 7.5$ MeV [S. Weinberg, Trans. N. Y. Acad. Sci. **38**, 185 (1977)]. These values are consistent with the values given in J. Gasser and H. Leutwyler, Phys. Rev. **87**, 77 (1982).

[91] P. Herczeg, "On the Question of a Tensor Interaction in $\pi \to e\,\nu_e\,\gamma$ Decay," to be published. See also Ref. [27].

[92] V. N. Bolotov *et al.*, Phys. Lett B **243**, 308 (1990).

[93] A. A. Poblaguev, Phys. Lett. B **238**, 108 (1990).

[94] A. A. Poblaguev, Phys. Lett. B **286**, 169 (1992).

[95] P. A. Quin, J. Deutsch, T. E. Pickering, J. E. Schewe, and P. A. Voytas, Phys. Rev. D **47**, 1247 (1993).

[96] The range (183) corresponds to a PCAC calculation of the tensor form factor in $\pi \to e\nu_e\gamma$. A quark model calculation yielded a range for $|f_T|$ which has bounds larger than those in (183) by a factor of ~ 3.

[97] The limit on $Re\,\eta_{LT}$ from beta decay quoted in Ref. [95], is somewhat larger than (178), requiring $|g_T| \lesssim 0.6$. It was obtained from a data set, from which some of the older experimental results on the neutron asymmetry parameter A_n were excluded [31].

[98] A. S. Carnoy *et al.*, Ref. [26].
[99] K. Abdullah *et al.*, Phys. Rev. Lett. **65**, 2347 (1990).
[100] I am indebted to John Wilkerson for asking me about the S-T term in the *D*-coefficient.
[101] For reviews of aspects of muon decay see F. Scheck, Phys. Rev. **44**, 187 (1978); F. Scheck, Ref. [65]; R. Engfer and H. K. Walter, Ann. Rev. Nucl. Part. Sci. **36**, 327 (1986); W. Fetscher and H.-J. Gerber, this volume.
[102] A. Sirlin, Phys. Rev. D **22**, 970 (1980).
[103] F. Scheck, Ref. [65]; K. Mursula and F. Scheck, Nucl. Phys. **B253**, 189 (1985).
[104] W. Fetscher and H.-J. Gerber, and K. F. Johnson, Phys. Lett. B **173**, 102 (1986).
[105] W. Fetscher and H.-J. Gerber, in Particle Data Group, "Review of Particle Properties," Phys. Rev. D **45**, Number 11, June 1, 1992, p. VI.16.
[106] T. Kinoshita and A. Sirlin, Phys. Rev. **108**, 884 (1957).
[107] W. Fetscher, Phys. Lett. B **140**, 117 (1984).
[108] A. Jodidio *et al.*, Phys. Rev. D **34**, 1967 (1986); Phys. Rev. D **37**, 237 (1988) (E); J. Carr *et al.*, Phys. Rev. Lett. **51**, 627 (1983); D. P. Stoker *et al.*, Phys. Rev. Lett. **54**, 1887 (1985).
[109] D. Geiregat *et al.*, Phys. Lett. B **247**, 131 (1990).
[110] S. R. Mishra *et al.*, Phys. Lett. B **252**, 170 (1990).
[111] B. Balke *et al.*, Phys. Rev. D **37**, 587 (1988).
[112] P. Langacker and D. London, Phys. Rev. D **39**, 266 (1989).
[113] For reviews of the available information on the identity of the neutrinos emitted in muon decay see S. P. Rosen, in Proceedings of the Workshop on New Directions in Neutrino Physics at Fermilab, Fermilab, Sept. 14-16, 1988, edited by R. H. Bernstein (Fermilab, 1988), unpublished; P. Herczeg, Z. Phys. C **56**, S129 (1992).
[114] S. E. Willis *et al.*, Phys. Rev. Lett. **44**, 522 (1980).
[115] A. M. Cooper *et al.*, Phys. Lett. B **112**, 97 (1982).
[116] B. McWilliams and L.-F. Li, Ref. [73]; H. E. Haber, G. L. Kane and T. Sterling, Ref. [73].
[117] K. Mursula and F. Scheck, Ref. [103].
[118] E. D. Carlson and P. H. Frampton, Phys. Lett. B **283**, 123 (1992).
[119] An example is the Δ_L^+ Higgs boson in a class of left-right symmetric models (see Section III.2).
[120] E. J. Eichten, K. D. Lane, and M. E. Peskin, Phys. Rev. Lett. **50**, 811 (1983); A. Jodidio *et al.*, Ref. [108] and references quoted therein.
[121] An example is decays into superpartners, such as $\mu^+ \to e^+ \tilde{\gamma}\tilde{\gamma}$ [R. E. Schrock, Phys. Lett. B **139**, 427 (1984)] and $\mu^+ \to e^+ \tilde{\nu}_e \tilde{\overline{\nu}}_\mu$ [W. Buchmüller and F. Scheck, Phys. Lett. B **145**, 421 (1984)]. It should be noted that the possibility that the superpartners of the usual neutrinos are light enough to be emitted in muon decay is now ruled out by the experimental result on the invisible width of the Z.
[122] Muon decay in manifestly left-right symmetric models has been discussed in M. A. Bég *et al.*, Ref. [53]; M. Doi, T. Kotani, and E. Takasugi, Progr. Theor. Phys. **71**, 1440 (1984); M. Doi, T. Kotani, H. Nishiura, K. Okuda, and E. Takasugi, Progr. Theor. Phys. **67**, 281 (1982); Sci. Rep. Col. Gen. Educ. Osaka Univ. **30**, 119 (1981), and in nonmanifest left-right symmetric models in P. Herczeg, Phys. Rev. D **34**, 3449 (1986).

[123] Unlike in P. Herczeg, Ref. [122], we used here the method suggested in Ref. [13] for setting limits on parameters which are constrained to lie within a bounded physical region.

[124] M. Doi *et al.* (1981, 1982, 1984), Ref. [116]; R. E. Shrock, Phys. Lett. B **112**, 382 (1982); P. Langacker and D. London, Ref. [112].

[125] R. N. Mohapatra and G. Senjanović, Phys. Rev. Lett. **44**, 912 (1980); Phys. Rev. D **23**, 165 (1981).

[126] P. Herczeg and R. N. Mohapatra (unpublished), reported in P. Herczeg, in *Rare Decay Symposium*, edited by D. Bryman *et al.* (World Scientific, Singapore, 1989), p. 24.

[127] P. Herczeg and R. N. Mohapatra, Phys. Rev. Lett. **69**, 2475 (1992).

[128] D. A. Krakauer *et al.*, Phys. Lett. B **263**, 534 (1991).

[129] Indirect empirical limits on $G_\mu^{(e)}$ come from the bound (222), from the W-mass, and from charged-current universality (Ref. [127]); for a review see P. Herczeg, Ref. [113]. We would like to warn the reader that in the latter reference we called G_F the constant $(g^2/8M_W^2)(1+\Delta r)$ and G_μ the constant whose value is deduced from the muon decay rate ($G_\mu = 1.6639(2) \times 10^{-5} GeV^{-2}$). This can be confusing, since Ref. [13] defines G_F as $G_F = 1.16639(2) \times 10^{-5}\ GeV^{-2}$. In the present article we define G_F as in Ref. [13] and use the symbol G for $(g^2/8M_W^2)(1+\Delta r)$.

[130] G. Czapek *et al.*, Phys. Rev. Lett. **70**, 17 (1993).

VI. PRECISION TESTS AT HADRON COLLIDERS

Precision Electroweak Tests at Hadron Colliders

K. Einsweiler
Lawrence Berkeley Laboratory
Berkeley, California 94720, USA

ABSTRACT

An overview of tests of the Standard Model of electroweak interactions in hadron colliders is given. The subjects include the properties of the W and Z bosons, with particular emphasis placed on the measurement of the W mass and width, searches for additional heavy gauge bosons, and the search for the t quark.

Contents

1 Introduction

The use of hadron colliders to investigate the weak interaction directly was first proposed [1] in 1976, shortly after the first neutral current results provided a useful measurement of $\sin^2\theta_W$, and hence a strong indication that the W and Z masses were in the range of 50 to 100 GeV. The proposal to create a proton anti-proton collider was based on the necessity of using a single ring to contain both beams, as well as the desire to have valence anti-quarks in order to produce the gauge bosons with the largest possible cross sections. The proposal to convert the CERN SPS into a proton-antiproton collider was approved in 1978, and first collisions, at $\sqrt{s} = 546$ GeV, were produced in 1981. The production and decay of the charged boson $W^\pm \to \ell^\pm\nu$ was first observed [2] by the two general purpose experiments UA1 and UA2 in data taken during 1982, with the observation of the neutral boson $Z \to \ell^+\ell^-$ following in early 1983 [3]. The energy of the machine grew to $\sqrt{s} = 630$ GeV, and the data samples grew to several hundred W's and several dozen Z's by the end of the initial runs of the SPS Collider in 1985. During that period, a total of 0.9 pb^{-1} (0.7 pb^{-1}) of integrated luminosity was accumulated by the UA2 (UA1) experiments. These data samples allowed the first precision studies of Electroweak interactions in hadron colliders [4].

It has now been ten years since those first direct observations of W's and Z's, and the precision study of Electroweak interactions in hadron colliders has been steadily progressing. The production of significant luminosity by the Fermilab Tevatron collider and the commissioning of the CDF detector, combined with the high luminosity CERN SPS running with the improved ACOL anti-proton source, initiated a new era. The UA2 experiment collected a total of 13.0 pb^{-1} of data during the period 1988 to 1990, while CDF accumulated 4.1 pb^{-1} during the 1988-1989 period. These two data samples for these experiments form the basis for the results presented in this chapter.

Before proceeding to the detailed discussion of results, we remind the reader of the features of the two detectors. In its upgraded form (post–1987), the UA2 detector [5], shown in Fig. 1, is a non-magnetic detector with a small tracking volume and high quality hermetic calorimeter coverage for the region $|\eta| < 3$, where the pseudo-rapidity η is defined to be:

$$\eta = -\ln\tan(\theta/2)$$

and θ is the usual polar angle relative to the beam direction. The UA2 detector has excellent electron identification over the range $|\eta| < 1.6$, good missing-E_T resolution, but no muon detection capability. The CDF detector [6], shown in Fig. 2, is a magnetic detector with a large, high-quality tracking system providing excellent momentum resolution for $|\eta| < 1$. High quality calorimetry covers the region $|\eta| < 1.1$, with the full coverage extending out to $|\eta| < 4.2$. The electron identification is excellent for $|\eta| < 1$, and remains good out to $|\eta| = 1.8$. The missing-E_T resolution is more modest due to magnetic sweeping effects and calorimeter non-uniformities.

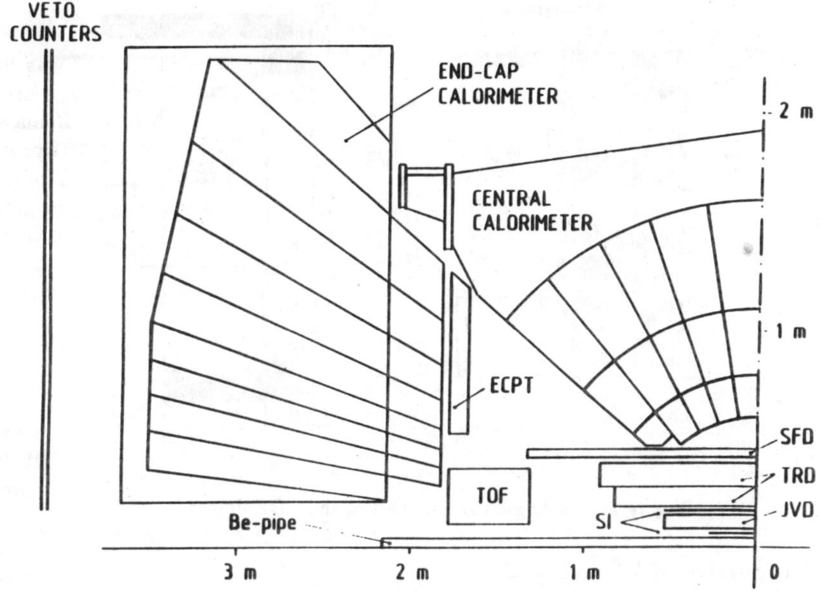

Figure 1: A side view of one half of the UA2 detector.

The muon system has limited coverage ($|\eta| < 0.6$), but uses the high momentum resolution of the central tracking system to achieve its excellent performance.

The present Tevatron collider run at Fermilab includes the new D0 detector, and is expected to provide data samples of roughly 100 pb^{-1} by the end of 1994. This order of magnitude increase will lead to still more precise tests of the Standard Model of the Electroweak interactions. With the new Main Injector Upgrade at Fermilab, it should be possible to acquire a data sample of 1000 pb^{-1}. Such a sample would allow thorough testing of the Electroweak theory through precision measurements. The natural extension of this line of research should then unfold during the following decade with the unraveling of the Electroweak symmetry breaking mechanism at the SSC and LHC accelerators.

In this chapter, an overview of measurements which test Electroweak theory in hadron colliders is given. This overview is divided into three major sections. The first reviews our knowledge of the properties of the W and Z bosons. The second describes searches for additional, heavy W and Z bosons. The final subject is the search for the t quark, which due to its very large mass, plays a significant role in precision Electroweak tests.

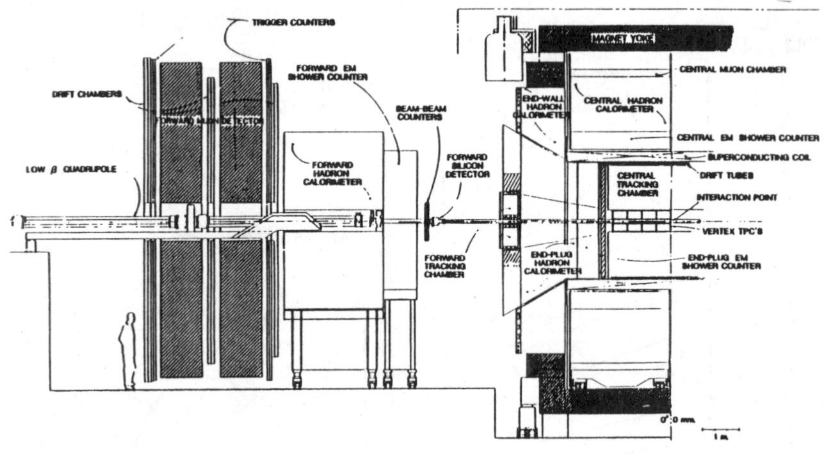

Figure 2: A side view of one half of the CDF detector.

2 Properties of W and Z Bosons

2.1 *Overview*

2.1.1 *Production*

The production of a W or Z boson in a hadron collider occurs predominantly via the Drell–Yan mechanism [7], whereby a quark from one incoming hadron annihilates with an anti-quark from the other hadron to form the gauge boson (e.g., $u\bar{d} \rightarrow W^+$ or $u\bar{u} \rightarrow Z^0$). The large masses of the W and Z make it useful (but not mandatory) to produce them using valence anti-quarks. The expected production cross sections, evaluated to $\mathcal{O}(\alpha_s^2)$, are given in Table 1.

The longitudinal momentum distribution for the bosons produced by this lowest order process is simply :

$$P_L(W) = \frac{\sqrt{s}}{2}(x_q - x_{\bar{q}})$$

where x_q and $x_{\bar{q}}$ are the Feynman x values for the quark and anti-quark. The transverse momentum is approximately zero. Higher order QCD corrections modify this simple picture somewhat, leading to observed $P_T(W)$ distributions which have a mean of 5–10 GeV. Note that the mean $P_T(W)$ increases slowly with the center of mass energy (approximately logarithmically). There is a large peak in the low $P_T(W)$ region which arises from the lowest order Drell-Yan process with the additional emission of many soft gluons from the incoming quark and anti-quark. At larger

Table 1: A summary of the predicted cross sections for production of W and Z bosons in hadron colliders. The results are derived from a complete calculation [8] to $\mathcal{O}(\alpha_s^2)$, and use the HMRSB parton distribution functions with $\Lambda_{QCD} = 190$ MeV. The cross sections are in nb, and do not include the leptonic branching ratios of the W or Z.

	$\sqrt{s} = 630$ GeV		$\sqrt{s} = 1800$ GeV	
	pp	$p\bar{p}$	pp	$p\bar{p}$
W	3.50	7.03	17.15	21.11
Z	0.82	2.14	4.89	6.37

$P_T(W)$, there is a long tail of events arising from the higher order QCD process of $W + jets$ production. A consistent calculation of the $P_T(W)$ distribution to $\mathcal{O}(\alpha_s)$ as a function of rapidity was performed in 1984 [9], and found to be in good agreement with the experimental data. The complete $\mathcal{O}(\alpha_s^2)$ calculation has been completed recently [10], allowing a considerable reduction in the Q^2 scale dependence of the theoretical prediction. Again, the agreement with the data is good.

2.1.2 *Decay*

The W and Z both decay approximately 70% of the time to $q\bar{q}$ final states for which the separation of the signal from the large 2-jet background is extremely difficult. The UA2 experiment has succeeded in detecting a significant signal in the 2-jet final state, with a signal-to-noise ratio of 1 to 100, but great perseverance was required [11]. For this reason, the study of the W and Z bosons in hadron colliders is primarily a study of their leptonic decay modes: $W \to \ell\nu$ and $Z \to \ell\ell$, where $\ell = e$ or μ. The decays $W \to \tau\nu$; $\tau \to \ell\nu\bar{\nu}$ produce indistinguishable final states, and hence also add a small contribution. The Standard Model branching ratios (for a heavy t quark) are : $BR(W \to \ell\nu) = 0.108$ and $BR(Z \to \ell\ell) = 0.033$.

The cross section times branching ratio for the W and Z have been carefully measured to be :

$$\begin{aligned} \sigma \cdot BR(W \to \ell\nu) &= 0.682 \pm 0.012 \pm 0.040 \text{ nb} \quad UA2\ [12] \quad \sqrt{s} = 630 \text{ GeV} \\ \sigma \cdot BR(Z \to \ell\ell) &= 0.066 \pm 0.004 \pm 0.004 \text{ nb} \end{aligned}$$

$$\begin{aligned} \sigma \cdot BR(W \to \ell\nu) &= 2.19 \pm 0.04 \pm 0.21 \text{ nb} \quad CDF\ [13] \quad \sqrt{s} = 1800 \text{ GeV} \\ \sigma \cdot BR(Z \to \ell\ell) &= 0.209 \pm 0.013 \pm 0.017 \text{ nb} \end{aligned}$$

The dominant systematic error in these measurements arises from the measurement of the absolute luminosity. Since there is no exclusive process with a precisely determined cross section, it is necessary to reference all measured cross sections to the total inelastic cross section, which has relatively large uncertainties (roughly 5% at this time). Note that the cross section and branching ratio for the $Z \to \ell\ell$ process are each a factor of three smaller than those for the $W \to \ell\nu$ process, leading to useful Z data samples which are typically an order of magnitude smaller than those for the W. At the maximum initial luminosity achieved so far at the SPS (and very recently at the Tevatron) of approximately 5×10^{30} cm^{-2}sec^{-1}, the observed cross sections lead to a production rate of one W every 90 seconds at the Tevatron collider.

2.1.3 *Detection*

The identification of electrons and muons in a hadron collider environment has now become a well-established art. Typically, rejections of $\sim 10^5$ against jet backgrounds are achievable with efficiencies of about 85% for leptons with $P_T > 15$ GeV. This allows very unambiguous identification of W and Z events in the hadron collider environment (typical background levels are at or below a few percent).

For the $W \to \ell\nu$ decay, it is also vital to reconstruct the missing neutrino via energy conservation. It is essentially impossible to reconstruct the total energy of an event in a hadron collider, as typically only about 5% of the total center of mass energy for a typical "minimum bias" event[1] appears in the central rapidity plateau covered by active calorimetry in the collider detectors ($|\eta| \leq 3$ at the SPS, $|\eta| \leq 4$ at the Tevatron). Thus, we are forced to rely on transverse energy balance to reconstruct $P_T(\nu)$ in W events :

$$\vec{P}_T(W) = \vec{P}_T(\ell) + \vec{P}_T(\nu) \sim -\vec{P}_T(\text{hadrons}) \qquad (1)$$

where

$$\vec{P}_T(\text{hadrons}) = \left\{\sum E_{\text{tower}}\hat{v}_{\text{tower}}\right\}_T \qquad (2)$$

with E_{tower} being the energy in a given calorimeter tower, and $\hat{v}_{\text{tower}}$ being a unit vector from the event vertex to the center of the given tower. Note that it is important to exclude any energy deposited in the calorimeter by the lepton in order to avoid double-counting. Thus, the measured value for the neutrino momentum is given by:

$$\vec{P}_T(\nu) = -\vec{P}_T(\ell) - \vec{P}_T(\text{hadrons}) \qquad (3)$$

This procedure for reconstructing the neutrino is marred by two major difficulties. First, it is very difficult to estimate the fraction of the total transverse energy recoiling against the W which is actually observed by the detector, due to the presence of the multiple soft gluon emission which produces particles over a very large region of rapidity. Second, the recoiling energy is carried by a large number of soft particles ($dn/d\eta \sim 6$ over 6–8 units of rapidity with a mean P_T of 400 MeV). The accurate measurement of the direction and energy of these soft particles is a very challenging experimental problem. Fully reconstructed Z events can be used to study these issues (the neutrino in Eq. 1 is replaced by a lepton, and hence $P_T(Z)$ is accurately measured using the two leptons and can be compared directly with

[1] So-called "minimum bias" events are selected by making very minimal requirements, such as the presence of at least two charged particles in a large region of η. Such triggers generally see $\geq 95\%$ of the total inelastic cross section in a hadron collider, and thus provide an unbiased sample of "typical" events.

the measurement from P_T(hadrons)). Unfortunately, present Z statistics allow only a limited understanding of this complex measurement.

It is possible to determine $P_L(\nu)$ by imposing the W mass constraint (usually ignoring the finite width of the W) on the sum of the lepton and neutrino 4-vectors, and then solving the resulting quadratic equation. Various techniques have been used to eliminate the wrong solution (which may sometimes be unphysical, since $P_L(W) > \sqrt{s}/2$ is kinematically forbidden). One such technique is to choose the solution which has the smallest $P_L(W)$. This leads to a correct choice about 70% of the time at the SPS and 60% of the time at the Tevatron (a random choice would give 50%...). Such techniques are useful when it is necessary to derive values for quantities in the W center-of-mass frame, but the relatively large number of wrong choices and the assumed W mass do limit the utility of the resulting "reconstruction".

2.2 *Measurement of $M(W)$ and $M(Z)$*

The single most important property of the W and Z bosons from the perspective of Electroweak theory is their mass. The $Z \rightarrow \ell\ell$ decay, with the observation and accurate reconstruction of both leptons, provides a direct measurement of the Z mass. The $W \rightarrow \ell\nu$ decay, with its unobserved neutrino, requires an indirect measurement of the mass using variables which are strongly correlated with $M(W)$.

The basic variables are the 3-vectors of the lepton and the neutrino. Since the longitudinal momentum of the neutrino is not measurable, that of the electron is also of little use, and we are reduced to relying on $\vec{P}_T(\ell)$ and $\vec{P}_T(\nu)$. Since the W decay is a 2-body decay, the distributions of these momenta carry substantial information about the W mass. Unfortunately, they are also very sensitive to assumptions about the transverse momentum of the W. A very useful combination, with greatly reduced sensitivity to the transverse momentum of the W, is the transverse mass:

$$M_T^2 = 2P_T(\ell)P_T(\nu)\left[1 - \cos\phi_{\ell\nu}\right] \tag{4}$$

where $\phi_{\ell\nu}$ is the angle between the two vectors in the transverse plane. To better understand this variable, it is useful to introduce $P_{||}$ and $P_{\perp}$, the projections of the observed value of $P_T(W)$ (referred to as P_T(hadrons), see Eq. 2) onto the lepton direction. In that case, one can write the following expansion in powers of the W transverse momentum, valid for small $P_T(W)$:

$$M_T^2 \sim 4P_T(\ell)^2 + 2P_T(\ell)P_{||} + P_{||}^2 + P_{\perp}^2 + \mathcal{O}\left(\frac{(P_{||}^2 + P_{\perp}^2)^2}{P_T(\ell)^2}\right) + \cdots \tag{5}$$

It is apparent that the measurement of the recoiling hadrons opposite the lepton is being used to improve the information in the lepton P_T by partially accounting for the transverse motion of the W. The resulting variable is not a mass (it is not Lorentz invariant), but it is a useful phenomenological variable.

The three variables mentioned above ($P_T(\ell)$, $P_T(\nu)$, and M_T) essentially exhaust the list of useful measurements, and the subsequent discussion will concentrate on them. Note that for those W's whose decay products have large longitudinal momentum (and are therefore detected at large η), there is little mass information available in transverse variables. This makes the measurement intrinsically central, with most useful information contained in the region $|\eta| < 1$ at the SPS, and $|\eta| < 1.5$ at the Tevatron. In addition, the highest quality energy and momentum information is only available over a similar, central region of the detectors.

The strategy for the measurement of $M(W)$ is significantly different for the non-magnetic UA2 detector than for the magnetic CDF detector. In UA2, religious attention to calorimeter calibration permitted a 1% determination of the energy scale [14]. This is still inadequate for a precision mass measurement, and hence UA2 really measures the ratio $M(W)/M(Z)$ by simultaneously measuring the individual masses, and then cancelling the scale uncertainty after a careful evaluation of calorimeter non-linearities. The W mass is then deduced by using the precise LEP determination of $M(Z)$ to define the scale. In CDF, the presence of a precise magnetic field (known to about 0.05%) and an intricate yet elegant calibration procedure allow a determination of the momentum scale to roughly 0.1%. For the $W \to \mu\nu$ analysis, the absolute W mass is derived directly from this momentum scale. For the $W \to e\nu$ analysis, the W sample itself can be used to transfer the momentum scale to the calorimetry (which has better resolution and less sensitivity to bremsstrahlung). Hence, CDF can measure the absolute W mass. This is in principle a large advantage, because the limited Z statistics no longer play a role.

Finally, it is worth mentioning that there have been numerous suggestions for making a direct measurement of quantities such as $M(Z) - M(W)$, in which the unpleasant systematic errors associated with the direct $M(W)$ measurement would largely cancel. There are two problems with this approach. First, due to the limited Z statistics (an order of magnitude fewer events than for the W), any attempt to measure $M(Z)$ which doesn't use the full information available in the invariant mass will have a substantially larger statistical error. For example, the statistical error on $M(Z)$ derived from fitting M_T is typically about three times larger for a given event sample than the statistical error derived from fitting the invariant mass directly. This limits the use of this technique to a distant future era when only systematic errors play an important role in this analysis. Moreover, when the possible systematic errors in a mass difference measurement are studied, they do not turn out to be dramatically smaller than those in the direct $M(W)$ measurement. Hence, while such measurements remain useful as cross-checks, they are unlikely to provide higher precision in the near future.

2.2.1 *Event Selection*

The event selection must satisfy several complex requirements. It must provide a largely background-free sample of well-measured W decays. Furthermore, this sample should not contain any selection biases which would make it difficult to estimate the systematic errors associated with the physics and detector models used in fitting for $M(W)$. In order to do this, both experiments require high P_T leptons — $P_T(\ell)$ and $P_T(\nu)$ above 20 (25) GeV for UA2 (CDF). The leptons must also be inside fiducial regions in the central detector where they will be well-measured. In UA2, the events were required to satisfy $P_T(W) < 20$ GeV to insure that they did not contain large jets which could degrade the missing-E_T resolution. In CDF, the events were required to contain no reconstructed jet with $P_T > 7$ GeV. The selection criteria must reduce the background level in the signal to below the 1% level to avoid any significant systematic error, as a 1% flat background can shift the measured mass by about 50 MeV. The final event samples consist of 2065 $W \to e\nu$ in UA2, and 1130 $W \to e\nu$ and 592 $W \to \mu\nu$ in CDF. Note that the corresponding cross section measurements, where one strives for the largest possible samples, contain roughly twice as many events in both experiments.

The Z sample is also important for the UA2 analysis where $M(W)/M(Z)$ is measured. In this case, there are two samples defined. The first consists of events in which both electrons are contained in the fiducial regions of the central calorimeter. The second sample has one electron in the same region, and the other anywhere else within the acceptance (it is an independent sample). The direction of the second electron is used, but not its energy. The energy is instead computed by imposing the constraint of transverse energy balance using the rest of the energy in the event. This gives worse mass resolution and some small additional systematics, but an energy scale which is still determined by the electron in the central fiducial region. The first sample contains 95 events, the second contains 156, all with a mass between 70 and 110 GeV.

2.2.2 *Reconstruction and Systematics*

<u>Electron:</u>

The primary problem is extracting a stable, well-defined energy scale for the calorimeter measurement. The electron is measured in a minimal number of calorimeter towers in order to reduce the effect of the underlying event, and then various corrections are applied to reconstruct the best estimate of the true electron energy. Several effects are important.

1. There is energy leakage outside the towers used to measure the electron energy. Typically this effect is about 1% of the electron energy, and must be accurately estimated. There is also underlying event energy sitting in the towers used to reconstruct the electron, which should be corrected out. This is typically a 50–100 MeV effect.

2. It is necessary to correct for the effects of material traversed by the electron before it reaches the active calorimeter medium. There are two aspects to this problem. There are significant layers of material immediately in front of the calorimeter (the 1.5 X_0 pre-radiator in UA2 and the solenoidal coil

in CDF) which absorb roughly 1% of the incident electron energy on average. In addition, the smaller amount of material distributed throughout the tracking volume produces bremsstrahlung photons which can affect the energy deposited in the immediate region of the electron in a magnetic detector such as CDF.

3. The calorimeter's response is slightly non-uniform. Both UA2 and CDF use scintillating plate calorimeters, which typically have variations of several percent in light output (and hence measured energy) over the face of a given tower. This transverse non-uniformity must be corrected to give a uniform energy scale. There are also potential longitudinal non-uniformities arising from the different construction of the electromagnetic and hadronic calorimeters, which can induce small non-linearities in the energy measurement.

A magnetic detector has a significant advantage in that, with adequate statistics, all of these effects can be studied *in situ*. In a non-magnetic detector such as UA2 or D0, one must rely on careful studies of detector behavior in test-beams and Monte Carlo simulations. However, even in the CDF case, it is necessary to understand the amount of material in the tracking volume in detail in order to transfer the absolute momentum scale supplied by the magnetic field to the calorimeter energy scale. The difficulty arises because of the presence of bremsstrahlung which distorts the E/P comparison. This is illustrated in Fig. 3, where the tail for $E/P > 1$ arises from bremsstrahlung in the material in the tracking volume. After applying the requisite amount of effort, the systematic errors associated with the electron energy measurement do not play a dominant role in the ultimate precision. In UA2, these effects (including non-cancellation of non-linearities in the energy scale between $M(W)$ and $M(Z)$) were about 100 MeV. In CDF, the E scale uncertainty was ultimately taken to be 190 MeV, with the limited statistics for the E/P transfer dominating the underlying 80 MeV momentum scale error.

Muon:

The CDF momentum scale determination relies on the elegant tilted-cell geometry of the central tracking drift chamber [16]. This design allows each track to give a direct measure of the time-distance relationship for the chamber, allowing excellent control of these otherwise very bothersome systematics. The alignment of the chamber to achieve the ultimate momentum resolution was performed using the high P_T electron tracks from W decays. The $\psi(3097)$ and $\Upsilon(9460)$ decays into muon pairs serve as convenient cross checks, limiting the potential scale errors to less than 0.1%, or 80 MeV.

Neutrino:

The transverse momentum of the neutrino is determined using transverse energy balance in the W event. This calculation has been outlined previously, (see Eq. 3), and suffers from a number of technical difficulties. It is the major contributor to the transverse mass resolution, and lies at the heart of precision measurements of $M(W)$. This measurement is heavily based on the calorimetry, and typically involves dividing the observed energy depositions into several classes (e.g., electron

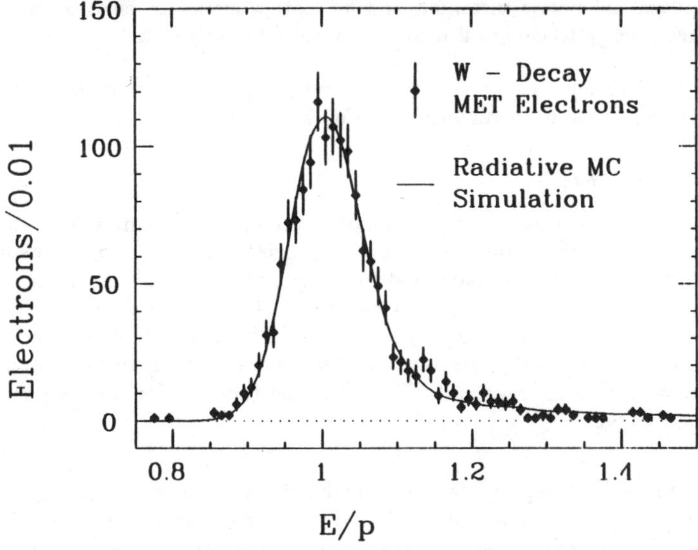

Figure 3: The distribution of E/P for W electrons compared with a simulation which includes radiative effects. The matching of E and P is used to transfer the precise momentum scale to the calorimeter energy scale.

or muon, underlying event, jet), each receiving appropriate corrections. At the end, all of the corrected values are summed, being careful to avoid double-counting. Many of the issues which play a role are intricately related to the models used for the production and measurement of the W, and their discussion is deferred to that section. Most effects are uncorrelated with the W decay, and hence degrade the neutrino resolution without affecting the neutrino energy scale. There are some subtle effects which require special attention:

1. The excess energy deposited by the lepton in the calorimetry must be accounted for. For the muon case, this is fairly straightforward as the energy deposition is normally all contained in a single tower. For the electron case, somewhat less than 1% of the electron energy leaks out into the adjacent calorimeter towers. This contribution has already been included in the corrected electron energy, and must be removed from the P_T(hadrons) sum (see Eq. 2).

2. The underlying event energy sitting under the lepton must be recovered. This process is rendered slightly more subtle by the use of finite readout thresholds (of order 50 MeV) for calorimeter towers. Towers with underlying event energy

will often fall below threshold and be lost, whereas an electron will deposit enough energy to cause all nearby towers to be read out.

Note that both of these effects are parallel to the lepton direction, and hence affect $P_{\|}$ and the transverse mass directly (see Eq. 4 and Eq. 5).

2.2.3 *Fitting Procedures*

In its simplest form, the problem consists of taking a set of measurements of interest (for example, the transverse mass) and their associated measurement errors, and describing them with a likelihood function which contains all of the available information about the relationship between the observable quantity and the true physics quantities ($M(W)$ and $\Gamma(W)$). These likelihood functions are too complex to be evaluated analytically, and are usually generated by a simple simulation of the physics and detector response. The likelihood function for the transverse mass fit is defined to be:

$$\mathcal{L} = f(M_T; M(W), \Gamma(W)) \tag{6}$$

where the function f represents the probability that the particular value of M_T would be observed for a given value of the W mass and width. This likelihood function must include all of the necessary information about the physics of W production and decay, as well as the expected detector resolution effects. The function is usually created by using a highly efficient and tunable parametrization of the physics and detector (via a Monte Carlo) to generate a multi-dimensional histogram, which may be numerically smoothed to reduce the effects of finite Monte Carlo statistics, and which approximates the actual likelihood (see Eq. 6). A general purpose minimization package is then used to maximize the likelihood function for the observed data sample by varying $M(W)$ and $\Gamma(W)$, thereby determining the best values for these parameters and their corresponding statistical errors.

Due to the finite Monte Carlo statistics and numerical imperfections, this technique itself contributes slightly to the systematic error. Such effects may be studied by fitting to large simulated data samples with known values of the W mass and width. Typically these studies indicate systematic errors of less than or the order of 50 MeV arising from the fitting procedure.

2.2.4 *Physics and Detector Models*

The key elements in the fitting procedure described previously are the models used to create distributions of the W and Z kinematic variables, and the models used to define the effects of the detector on the observed variables. These models must be flexible enough to accurately reproduce the distributions observed in the experiment, while at the same time allowing systematic studies of the effects of all of the relevant uncertainties on the mass measurements.

The starting point is the production of W or Z bosons with the proper P_T and P_L distributions. Typically, the W or Z is produced using a standard set of parton distribution functions and the lowest order (Born) production process. This boson is then given a P_T distribution based either on empirical parametrizations of the data or on QCD higher order predictions (which depend on the P_L of the produced boson). Finally, a small correction function is implemented to allow further detailed tuning of the predicted $P_T(W)$ distribution to agree with the data. As the mass measurement takes place in the low $P_T(W)$ region (typically $P_T(W) < 10$ GeV), the theoretical predictions suffer from many uncertainties. The predictions involve integrating over the infinite number of very soft gluons which may be produced. These integrals extend to very small Q^2, and rely on the parton distribution functions and the strong coupling constant in this almost non-perturbative region of phase space. As a result, there is no truly correct physics model to use. One uses as much "QCD-inspired" calculation as possible, and then tries to parametrize the remaining uncertainties and discrepancies. The idea is to capture the key physics ingredients correctly, and then develop flexible models to quantify the residual uncertainties in the physics.

The next step in the event generation process, once the W or Z has been created with the correct kinematics, is to place the event in the detector and simulate the measurement process. The observables whose measurement must be simulated are $P_T(\ell)$, $P_T(\nu)$, and M_T. For the lepton, the measurement can be modelled using the detector acceptance and an accurate description of the resolution function. For the neutrino (and the transverse mass), it is necessary to simulate $P_T(\text{hadrons})$, as Eq. 3 and Eq. 4 make clear. Modelling this variable requires a model for the rest of the event (i.e., all of the soft particles which accompany the W or Z). Unfortunately, this involves a great deal of poorly understood physics (production of the underlying event plus fragmentation of all of the soft gluons accompanying the boson). This physics is sufficiently tenuous that experimentalists prefer to take a more empirical approach. The general method used by both UA2 and CDF is to construct measurement models which depend on global quantities that affect the detector response, for example, the total scalar E_T in the event:

$$\Sigma E_T = \sum |\{E_{\text{tower}} \hat{v}_{\text{tower}}\}_T| \tag{7}$$

where the T refers to the transverse component relative to the primary vertex in the event, and the calorimeter energy deposits associated with the W decay lepton have been excluded from the sum. Although the response of the detector to such global quantities cannot be deduced from first principles, one can usually adjust the details of the model to reproduce the observed data. The alternative would be to generate local quantities (e.g., the multiplicity and P_T distributions for all of the accompanying tracks) for which the detector response can be deduced from test-beams and Monte Carlo simulation, but where no theoretical model exists to guide the generation process.

In order to simulate $P_T(\text{hadrons})$, one must start from the underlying kinematic variable which is responsible, namely $P_T(W)$. A phenomenological approach is taken to creating the observed distributions from the generated one. Numerous studies have shown that the behavior of many observables in W events (total scalar E_T, track multiplicity, track momentum, etc.) when examined as a function of $P_T(W)$ approaches that observed in minimum bias events as $P_T(W)$ approaches zero. This suggests a model which incorporates a standard minimum bias event plus additional energy flow which is associated with the transverse momentum of the W. For example, one can study the excess scalar E_T, defined to be:

$$\Sigma\tilde{E}_T = \Sigma E_T - P_T(W) \tag{8}$$

The mean value of this quantity is observed to increase linearly with the W boson transverse momentum with a typical slope of about 0.4, indicating that there is additional recoil energy emitted beyond that required to simply balance the P_T of the W boson. This additional energy will look much like a minimum bias event when $P_T(W)$ is low (i.e., it will be carried by many low momentum hadrons), but as $P_T(W)$ increases, the recoiling energy will take on the behavior of a jet, with a correspondingly harder momentum spectrum for the hadrons. Both UA2 and CDF use this conceptual framework in building more detailed models.

There are two basic aspects to understanding the measurement of $P_T(\text{hadrons})$. The first aspect involves the resolution for reconstructing the individual components of $P_T(\text{hadrons})$. This is generally studied in minimum bias events, where one finds a resolution function proportional to the total scalar E_T, typically with a form (all units are GeV):

$$\sigma\left(P_{x,y}(\text{hadrons})\right) = 0.5 \times \sqrt{\Sigma E_T}$$

for each component. The resolution for the combined magnitude is then 1.4 times worse. Typical jet resolutions are also proportional to the total E_T of the jet, with slightly poorer resolution due to the highly collimated nature of the energy deposits. The expected resolution for low P_T W production should be somewhere in between, as the total scalar E_T has contributions from the underlying minimum bias event as well as the hadron energy recoiling against the W.

The second aspect involves the scale for $P_T(\text{hadrons})$. Since minimum bias events have no net P_T imbalance, scale errors are not visible. Instead, one must study the balance between the P_T of a Z boson and the total P_T carried by the recoiling hadrons. The single most powerful distribution for this study is the sum of the P_T of the two leptons from the Z decay plus the P_T of the recoiling hadrons. This vector sum is projected onto a special direction in the transverse plane which minimizes the contributions of lepton measurement errors to the result. This direction, referred to as the η axis, is defined to be the bisector of the two lepton directions (and hence depends only on their angles, not on their energies, which are more susceptible to measurement errors). The results of such studies of P_T balance in Z events can be summarized in terms of the observed degradation in response. In UA2, the typical response of the detector to the recoiling hadrons is low by about 30% for $P_T(Z) < 5$ GeV, with an asymptotic value of roughly 0.95 for $P_T(Z) > 10$ GeV. In CDF, one finds that the response is about 50% low in the $P_T < 5$ GeV region, rising to an asymptotic value of about 0.65 for $P_T(Z) > 10$ GeV. Note that this low response is due to a combination of acceptance (not all recoiling hadrons lie within the detector) and calorimeter response (resolution, non-linearity, magnetic sweeping, etc.). The larger scale error in CDF has several sources, including the magnetic field, calorimeter calibrations and non-linearities. This scale error appears as an additional smearing in variables like M_T, and must be included as part of the detector measurement model.

In order to guide the simulation process, it is vital to compare the model predictions with the data wherever possible. The Z events are very important in this respect because they allow a detailed study of the production properties of the Z as a function of its transverse momentum. Since the $P_T(Z)$ can be accurately reconstructed using the two lepton momenta, these measurements are un-ambiguous and can be directly compared to a simulation. In particular, the observed mean of the P_T balance projected onto the η axis is related to the scale factor aspect of the measurement model described above, whereas the observed RMS is related to the resolution aspect of the measurement model.

In the UA2 analysis, a theoretical calculation is used to derive the initial $P_T(Z)$ distribution. The Z data is used to evaluate a small correction function which allows the simulation to reproduce the observed $P_T(Z)$ distribution. The resolution in minimum bias events, and the energy balance in Z events is then used to constrain the resolution smearing and the scale aspects of the measurement model. It is found that the resolution observed in minimum bias events gives an excellent description of the RMS of the P_T balance in Z events. Finally, the resulting tuned simulation is compared with the observed $P_T(W)$ distribution as a cross check. The results are shown in Fig. 4 and Fig. 5.

In the CDF analysis, a parametrization of the underlying $P_T(W)$ is created, and then passed through the smearing model described above in order to compare it against the observed $P_T(W)$ distribution. The input parametrization is iterated until good agreement with the observed W data is obtained. The smearing model contains an underlying event whose resolution is determined from minimum bias data, plus the recoiling hadrons, which are treated as though they were a jet with correspondingly poorer resolution. In this case, the agreement between the input parametrization and the observed $P_T(Z)$ distribution serves as a cross check on the model. The comparison with the Z data is shown in Fig. 6.

2.2.5 *Results and Error Analysis*

Now that all of the ingredients of the analysis have been dissected in detail, the final step is to put everything together again and describe the results. In both the UA2 and the CDF analyses, fits (and the accompanying systematic error analysis) are performed for the three principle kinematic variables (M_T, $P_T(\ell)$, and $P_T(\nu)$). In the end, the fit to the transverse mass gives both smaller statistical and smaller systematic errors. The analysis of the other two variables serves to support the detailed analysis of the systematic errors, as they are each sensitive to different aspects of the analysis uncertainties. The $P_T(\ell)$ fit is most sensitive to the $P_T(W)$ distribution and effects that influence the lepton energy scale. The $P_T(\nu)$ fit is most

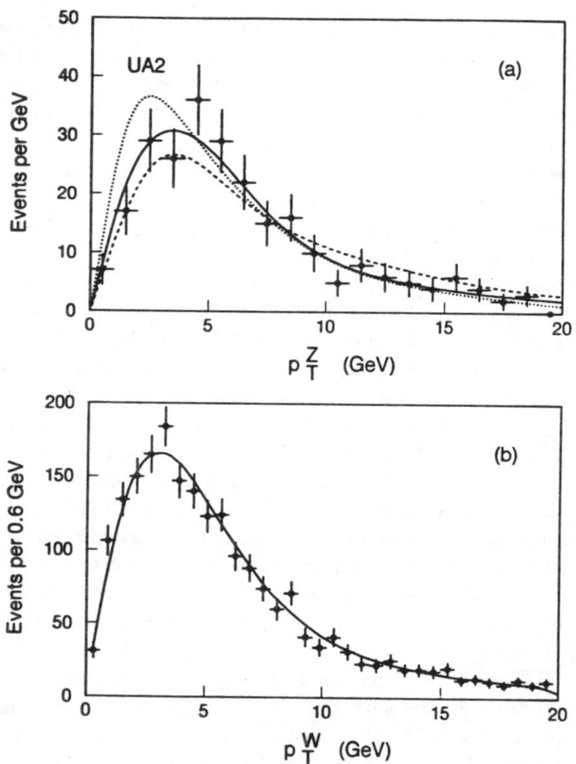

Figure 4: (a) The UA2 data for $P_T(ee)$, including the predictions of the central model (solid), and the soft (dotted) and hard (dashed) variations used for systematic error studies. (b) The UA2 data for $P_T(W)$, including the prediction of the central model.

sensitive to the effects which influence the neutrino scale and resolution. In both CDF and UA2, these additional fits do give results which are in agreement with the transverse mass fits.

Rather than present a detailed synopsis of each experiment's error analysis, the results of the analyses are summarized in Table 2 and Table 3. The basic classification of the entries in the tables is defined below.

UA2 Analysis:

The UA2 analysis of the transverse mass spectrum, taking into account some correlations among the relevant uncertainties, results in a measurement of the ratio

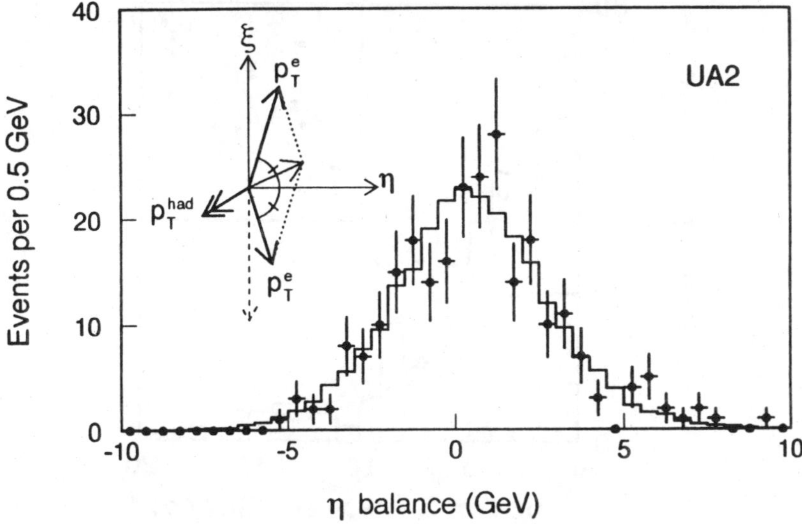

Figure 5: The momentum balance along the η axis in UA2 $Z \to ee$ events. The points show the data, while the histogram shows the model prediction.

of the masses:

$$M(W)/M(Z) = 0.8813 \pm 0.0036\ (\mathit{stat}) \pm 0.0019\ (\mathit{syst})$$

This measurement can be combined with the LEP value for the Z mass to give the final value:

$$M(W) = 80.35 \pm 0.33\ (\mathit{stat}) \pm 0.17\ (\mathit{syst})$$

The results of the fits to the three kinematic variables are displayed in Fig. 7. The uncertainties in Table 2 are briefly summarized below.

1. The parton distribution function contribution was evaluated by studying the variations in the fitted mass when using any of the available next-to-leading order parton distribution fits. The HMRSB set was used as the central value. The effect occurs because of the induced variations in the $P_L(W)$ distribution which then distort the predicted lineshape. The sensitivity was reduced by fitting in the region $60 < M_T < 120$ GeV rather than in the region $40 < M_T < 120$ GeV.

2. The electron energy resolution was studied in test-beams, and the effects of tower-to-tower gain variations were also included. The result is that the expected resolution for a 40 GeV electron is $3.3 \pm 0.5\%$.

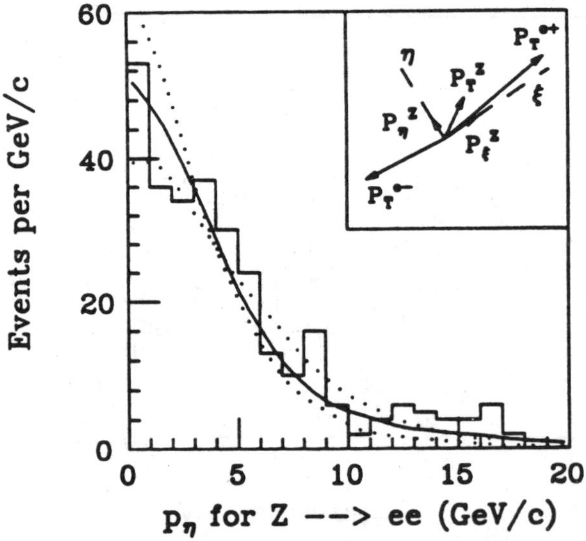

Figure 6: The momentum balance along the η axis in CDF $Z \to ee$ events. The histogram shows the data, while the solid curve shows the model prediction, and the dotted curves show the allowed variations used in the systematic error studies.

3. The neutrino scale generally follows the electron energy scale, but receives additional contributions from electron energy leakage, which are corrected out in the electron energy measurement.

4. The effects of the $P_T(W)$ distribution and the resolution for P_T(hadrons) were discussed in the previous section. The allowed variations in the model parameters were defined using minimum bias events, Z events, and the observed mean value of $P_T(W)$.

5. The underlying event contributes energy underneath the electron which modifies the original test-beam calibration, and hence adds further uncertainty.

6. The potential errors induced by the fitting procedure are studied using large simulated data samples with known W mass.

7. The effects of the radiative decays $W \to e\nu\gamma$ and $Z \to ee\gamma$ were included using a GEANT simulation of the detector response to low energy photons, which includes the effects of energy clustering in the calorimeter.

8. Any P_T dependence to the electron identification efficiency can distort the

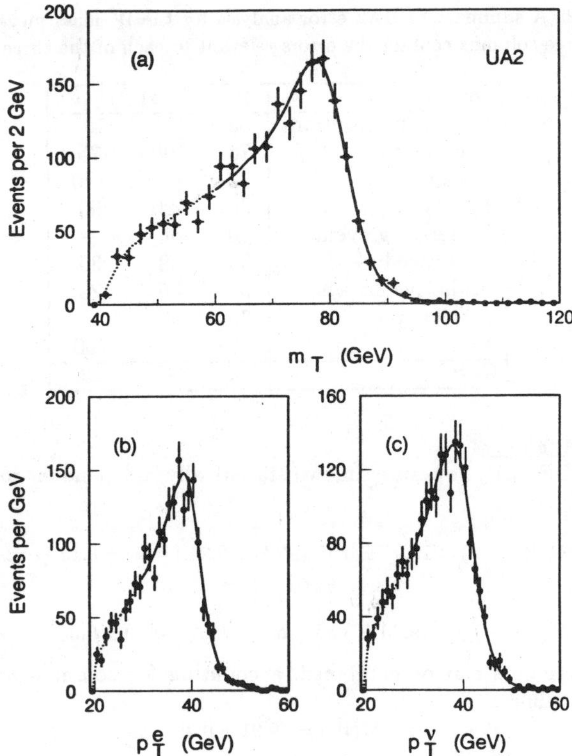

Figure 7: The UA2 fits for $M(W)$ to (a) the M_T spectrum, (b) the $P_T(e)$ spectrum, and (c) the $P_T(\nu)$ spectrum. The points show the data while the curves represent the fits. The solid portion of the curve indicates the region over which the fits are performed.

observed lineshape. Studies of this efficiency over the relevant P_T range give an estimated $-5 \pm 5\%$ for the allowed variation.

9. The variation in electron detection efficiency as a function of $P_{||}$ has been studied. Clearly, when the hadronic energy recoiling against the W lies along the electron direction, the electron will become less isolated and their will be a corresponding reduction in detection efficiency. This results in a significant change in the $P_T(\nu)$ lineshape since events with large positive $P_{||}$ have large $P_T(\nu)$.

Table 2: A summary of UA2 error analysis for the W mass measurement. The three columns contain the errors relevant to each of the three fits.

Origin	M_T	$P_T(e)$	$P_T(\nu)$
Parton distributions	85	135	105
e resolution	75	100	75
ν scale	70	—	140
$P_T(W)$	60	120	90
Underlying event	30	50	—
Fit procedure	30	40	30
Radiative decays	30	50	20
Efficiency	30	40	30
$P_{\|\|}$	25	95	350
Total	160	240	420

CDF Analysis:

The CDF analysis results in two statistically independent measurements of the W mass:

$$M(W \to e) = 79.91 \pm 0.35\ (stat) \pm 0.24\ (syst) \pm 0.19\ (scale)$$

and

$$M(W \to \mu) = 79.90 \pm 0.53\ (stat) \pm 0.32\ (syst) \pm 0.08\ (scale)$$

These measurements can be combined, accounting for common uncertainties, to give the final value:

$$M(W) = 79.91 \pm 0.39$$

The results of the fits to the transverse mass in the electron and muon channel are shown in Fig. 8.
The uncertainties in Table 3 are briefly summarized below.

1. The tracking scale error was determined to be 0.1% based on an analysis of the maximum possible residual curvature errors.

2. The calorimeter scale error is the result of transferring the tracking scale to the calorimeter using the E/P distribution. It receives contributions of 130 MeV from the finite statistics of the W sample used for calibration, 80 MeV from the uncertainty on the amount of material in the tracking volume which in turn affects the shape of the E/P distribution, and 90 MeV from uncertainties associated with the data selection, fitting procedures and resolutions in the E/P analysis.

3. The parton distribution function effects were evaluated by using many recent fits, and taking the observed spread in fit mass values for large simulated data samples.

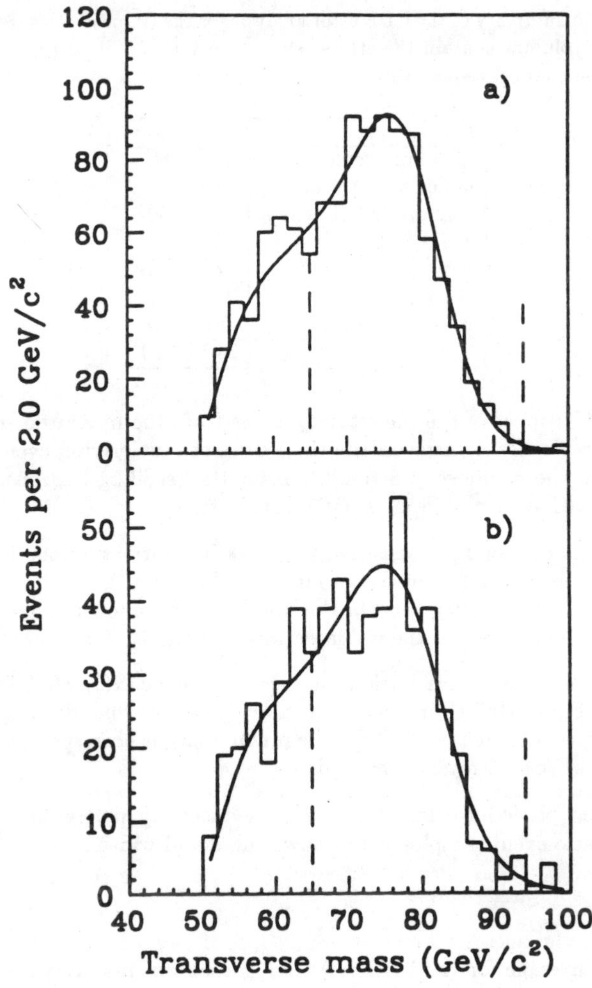

Figure 8: The CDF fits for $M(W)$ to the M_T spectrum. (a) the electron channel, and (b) the muon channel. The histogram shows the data while the curves represent the fits. The dashed lines indicate the region over which the fits are performed.

Table 3: A summary of the CDF error analysis for the W mass measurement. The two columns contain the errors relevant to the M_T fit to the electron and muon channel respectively.

Origin	e	μ
Tracking scale	80	80
Calorimeter scale	175	—
Parton distributions	60	60
$P_T(W)$	145	150
$P_{\|\|}$	170	240
Background	50	110
Fit procedure	50	50
Total	310	320

4. The $P_T(W)$ aspect of the uncertainty arises from the measurement model described previously. It has contributions from the underlying event energy and resolution, the response and resolution for the recoiling hadrons, and the allowed variations in the input $P_T(W)$ distribution.

5. The distribution of $P_{||}$ is sensitive to errors in subtracting the lepton energy from P_T(hadrons) and from biases introduced by the "no-jet" event selection. The quoted uncertainty reflects the statistical error on the mean of the $P_{||}$ component compared to the unbiased mean of the $P_{\perp}$ component.

6. The presence of possible backgrounds can systematically shift the measured mass. The uncertainty in the electron analysis corresponds roughly to a 1% flat background, whereas that for the muon analysis corresponds to a possible 1% residual cosmic ray background.

7. The fitting procedure uncertainty was deduced by performing fits to large Monte Carlo event samples with known mass and width.

2.2.6 *Future Prospects*

The combined average for the UA2 and CDF measurements gives a best estimate for the W mass of 80.14 ± 0.27 GeV. The present CDF analysis achieved a combined systematic error on the W mass of 390 MeV. For the Run 1a data sample, expected to be about 25 pb^{-1}, one would expect this to scale to about 200 MeV. For the total Run 1 data sample of about 100 pb^{-1}, one could hope to achieve a 100 MeV error. The new D0 experiment has some advantages (better missing-E_T resolution and a non-magnetic configuration) and some dis-advantages (no momentum measurement which implies that calibration and detailed studies must rely on test-beam and Monte Carlo simulation). The result is that one would expect similar measurement capabilities with somewhat complementary experimental systematic errors. For the ultimate Tevatron data sample of 500-1000 pb^{-1}, it may be possible to reach the level of 50 MeV combined error (each of the major experiments would acquire close to 10^6 W events and 10^5 Z events).

To justify this assessment in somewhat more detail, it is useful to go through the detailed results of the present uncertainty analyses shown in Table 2 and Table 3, and attempt to scale the values to larger data samples. We will carry out this assessment for a non-magnetic experiment (UA2 representing D0) and a magnetic experiment (CDF).

For the UA2 analysis, which should be representative of future D0 analyses, most uncertainties can be expected to scale with the acquired data sample. The major exception is the parton distribution function uncertainty. This uncertainty arises largely because of the empirical fact that the u quarks in the proton have, on average, slightly higher momentum than do the d quarks. This leads to the W^+ being preferentially boosted along the proton direction and the W^- along the anti-proton direction, or an overall broadening of the W rapidity distribution. The magnitude of the difference $\langle x \rangle_u - \langle x \rangle_d$ for a given set of parton distribution functions is directly correlated with the observed shift in the fitted W mass. Improved data, including measurements of the W asymmetry in hadron colliders (described later in this chapter) should improve the present situation, pushing the systematic error below the 50 MeV level. For each of the other contributions, a sufficient number of Z events, coupled with very zealous attention to understanding calorimeter behavior, could allow a non-magnetic experiment to keep improving its W mass measurement. Of particular concern is the detailed response of the calorimeter to electrons, including effects such as resolution, leakage, and tower-to-tower gain variations. A very large effort in test-beam and Monte Carlo simulation will be required to keep them under control.

For the CDF analysis, the errors which don't necessarily improve with time again include the parton distribution functions, plus the tracking scale. The tracking scale can ultimately be related to the LEP Z mass scale whose present uncertainty is below the 10 MeV level. That is, in the limit of infinite Z statistics it will be better to measure $M(W)/M(Z)$ and use the LEP $M(Z)$ rather than measure the absolute value for $M(W)$ directly. The cross-over occurs when the statistical error on the Z mass is smaller than the scale error. In the present CDF electron analysis the total scale error was 190 MeV and the statistical error on the Z mass was 430 MeV. With a data sample of 25 pb^{-1}, the absolute measurement is probably still better, whereas with a data sample of 100 pb^{-1}, the ratio is probably better. Again, for each of the other contributions, a large Z event sample should allow continued improvement. In general, CDF could have a significant advantage because although its raw performance in certain key areas may be worse (for example, the P_T balance in Z events is significantly better in UA2 than in CDF), the enormous advantage of an *in situ* calibration in controlling systematic errors is likely to ultimately outweigh the slightly poorer raw performance.

In summary, it appears likely (though by no means to be taken for granted) that by early 1995, there will be two W mass measurements from the Tevatron with errors on the 100 MeV level. Although it would appear that there are no insurmountable obstacles, only further experience will indicate whether the Tevatron can continue to improve its measurements to reach the 50 MeV level by the end of the decade in order to remain competitive with the high quality measurements which should then become available from the LEP-200 program. Furthermore, it is only at this 50 MeV level that the W mass achieves its full potential as a probe of the Standard Model.

2.3 *Measurement of* $\Gamma(W)$

In the W mass measurement analysis described previously, the W width is also a parameter of the fit. Unfortunately, it is very strongly correlated with the detector resolution, and the large systematic uncertainty associated with this resolution translates directly into a large uncertainty on the W width itself. It is possible to make a significantly more precise measurement of the W width by an indirect procedure. The quantity measured is :

$$R = \frac{\sigma \cdot BR(W \to \ell\nu)}{\sigma \cdot BR(Z \to \ell\ell)} \tag{9}$$

This ratio can be re-expressed in the following manner :

$$R = \frac{\sigma(W)}{\sigma(Z)} \frac{\Gamma(W \to \ell\nu)}{\Gamma(Z \to \ell\ell)} \frac{\Gamma(Z)}{\Gamma(W)} \tag{10}$$

The first two ratios can be calculated under the general assumption that the gauge couplings for quarks and leptons are those of the Standard Model. The Z width has been accurately measured at LEP, allowing one to deduce $\Gamma(W)$ once R has been measured. Higher order QCD corrections largely cancel in the ratio [17], making the result very insensitive to variations in α_s. The major uncertainty arises from the parton distribution function dependence of the ratio of the total W and Z cross sections. This is normally referred to as a theoretical error, despite its origin in the experimentally determined parton distribution functions.

There are three measurements of R available, and each takes a slightly different approach towards evaluating the theoretical systematic errors. Nevertheless, the overall errors on the measurements are dominated by the statistical error arising from the small number of Z decays observed. For this measurement, one tries to select the largest possible W and Z samples for which the detection efficiency and acceptances can be reliably determined. Furthermore, one attempts to minimize the remaining systematics after taking the ratio of cross sections by using very similar event selection criteria for the W and Z events. The major uncertainty in the absolute cross section measurements, which is the luminosity, cancels completely in the ratio.

The results are :

$$\begin{array}{rcll} \Gamma(W) & = & 2.18^{+0.26}_{-0.24}\,(expt) \pm 0.04\,(theory) & \text{UA1 [18]} \\ \Gamma(W) & = & 2.10^{+0.14}_{-0.13}\,(stat) \pm 0.06\,(syst) \pm 0.06\,(theory) & \text{UA2 [19]} \\ \Gamma(W) & = & 2.16 \pm 0.17\,(stat + syst) & \text{CDF [20]} \end{array}$$

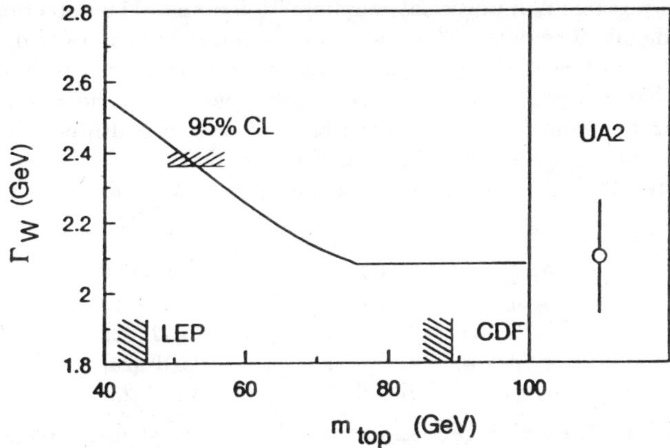

Figure 9: A comparison of the Standard Model prediction for $\Gamma(W)$ as a function of the t quark mass with the UA2 measurement of $\Gamma(W)$.

The expected value for $\Gamma(W)$ in the Standard Model with a heavy t quark (in which case the decay $W \to t\bar{b}$ is kinematically forbidden) is 2.08 GeV. The measurement of the total W width is sensitive to all decay modes, whether or not they are detectable. For example, it is possible that the t quark is light but that its semi-leptonic decays are suppressed, rendering its observation difficult. The measurements above can be combined to give an upper limit on $\Gamma(W)$ which implies $M_{\rm top} > 55$ GeV at 95% C.L., independent of the manner in which the t quark decays. This result is indicated in Fig. 9, where the Standard Model prediction for $\Gamma(W)$ versus $M_{\rm top}$ is compared with the UA2 R measurement.

The prospects for future improvement are good. The major limitation in the present measurements arises from the limited Z statistics. The other sources of uncertainty are the efficiencies, acceptances, and backgrounds. All of these can be quantified more carefully with an increased data sample (especially of Z events). It should be possible to reduce the present experimental error to 2-3% with 100 pb^{-1}, and perhaps reach an ultimate limit of 1% with 1000 pb^{-1} of data. It remains to be seen whether a commensurate improvement in our knowledge of the parton distribution functions, which dominate the theoretical error, is possible. Reaching the 1% level remains a very challenging goal.

2.4 *Universality of $e/\mu/\tau$ Couplings*

The universality of lepton couplings to the vector bosons is a consequence of the $SU(2)$ gauge invariance of the Standard Model. By comparing the relative branching ratios for the W into different leptonic final states ($W \to e\nu$, $\mu\nu$, $\tau\nu$), it is possible to constrain possible non-universal couplings to leptons. The detection of $W \to \tau\nu$ is rather difficult. The decay $W \to \tau\nu \to \ell\nu\bar{\nu}\nu$ is not experimentally distinguishable from $W \to \ell\nu$. Instead, it is necessary to rely on the decay $\tau \to$ hadrons, where the τ is required to decay into one or three charged particles. The $\tau \to$ hadrons decay can be separated from QCD jets on the basis of its low multiplicity and low mass (causing the corresponding jet to appear very narrow).

The results, presented in terms of the ratio of W lepton couplings to the W boson are :

$$
\begin{array}{lcll}
g_\mu/g_e & = & 1.00 \pm 0.07 \pm 0.04 & \text{UA1 [21]} \\
g_\tau/g_e & = & 1.01 \pm 0.10 \pm 0.06 & \text{UA1 [21]} \\
g_\tau/g_e & = & 0.997 \pm 0.056 \pm 0.042 & \text{UA2 [22]} \\
g_\mu/g_e & = & 1.01 \pm 0.04 & \text{CDF [20]} \\
g_\tau/g_e & = & 0.97 \pm 0.07 & \text{CDF [23]}
\end{array}
$$

These results have been largely superseded by improved measurements made at LEP on the Z peak. Nevertheless, all measurements support lepton universality.

2.5 *Forward/Backward Asymmetry in W and Z Decays*

The parity violation inherent in the Standard Model of the Electroweak interactions manifests itself in hadron colliders in the form of forward/backward charge asymmetries in W and Z leptonic decays. The charged current associated with W production and decay is pure $V - A$, leading to maximal parity violation in this case. At low $P_T(W)$, W's are produced almost fully polarized in a hadron collider. Furthermore, as most W production arises from valence quarks and anti-quarks, the decay electron from a W^- will tend to appear in the proton direction, and vice-versa for the positron from the decay of a W^+. This feature was verified at the CERN SPS collider shortly after the discovery of the W boson. The result is most easily seen by examining the distribution of $\cos\theta^*$, the lepton angle in the W center of mass frame (note that the longitudinal momentum of the ν must be determined using the W mass constraint). The expected distribution is of the form :

$$dN/d\cos\theta^* \sim (1 + \cos\theta^*)^2$$

A further test of the W properties may be carried out by measuring the mean value of the charge-weighted polar angle :

$$\langle Q\cos\theta^* \rangle \sim \frac{1}{J(J+1)}$$

where Q is the W charge. Note that the mean value will be zero for the case of a W boson having spin 0. This formula is correct in the approximation that the contribution to the W production cross section from cases where both quark and anti-quark arise from the sea is negligible. The expected value at the CERN SPS Collider, after correction for the sea-sea contribution, and including the uncertainties due to higher order QCD corrections (multiple soft gluon emission and $W + jets$

production), is claimed by UA1 to lie in the range 0.42 to 0.48. The measured value reported by UA1 [18] is :

$$\langle Q \cos\theta^* \rangle = 0.43 \pm 0.07$$

which is consistent with a W boson having spin 1.

A more subtle asymmetry, which relies on the underlying Electroweak interaction, but is itself a QCD effect, is the rapidity asymmetry. This asymmetry arises because the u valence quarks in the proton have, on average, slightly higher momentum than do the d quarks. This leads to the W^+ being preferentially boosted along the proton direction and the W^- along the anti-proton direction. The resulting asymmetry is defined as :

$$A(y) = \frac{d\sigma(\ell^+)/dy - d\sigma(\ell^-)/dy}{d\sigma(\ell^+)/dy + d\sigma(\ell^-)/dy}$$

and is sensitive to the ratio of the u and d components of the parton distribution function. The CDF experiment has performed a first measurement of this quantity [24], but greater statistics are required in order to place meaningful constraints on the parton distribution functions.

In the Z case, the neutral current is a mixture of the weak and electromagnetic currents, and hence depends on $\sin^2\theta_W$. For values of $\sin^2\theta_W$ close to 0.25, the expected asymmetry is very small. It has been measured by CDF [25] in the $Z \to e^+e^-$ final state to be :

$$A_{FB} = 5.2 \pm 5.9\ (stat) \pm 0.4\ (syst)\ \%$$

corresponding to a value of the mixing angle :

$$\sin^2\theta_W = 0.228^{+0.017}_{-0.015}\ (stat) \pm 0.002\ (syst)$$

The distribution of the angle of the lepton in the Z center of mass is shown in Fig. 10.

The uncertainty on the measurement of the Z asymmetry is almost totally statistical, so it will clearly improve with more data. The small systematic error is possible because CDF is very charge symmetric. An error of ± 0.005 on $\sin^2\theta_W$ should be achievable with 100 pb^{-1}, and perhaps an ultimate error of ± 0.002 could be reached with a 1000 pb^{-1} data sample. The LEP experiments have already achieved comparable precision (± 0.002) in this measurement, thereby substantially reducing the interest in a precise measurement in a hadron collider.

2.6 *Study of $W + \gamma$ and $Z + \gamma$ Events*

The interactions of W and Z bosons with quarks and leptons have been carefully studied in recent years. The self-interactions of the gauge bosons (3-boson and 4-boson vertices exist in the Standard Model due to the non-Abelian nature of the underlying theory) are much more difficult to test directly in experiments. The

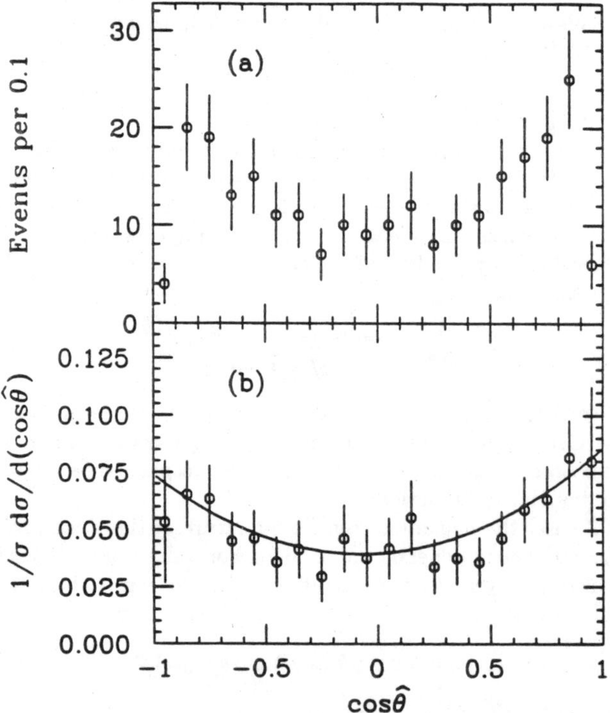

Figure 10: The distribution of the polar angle of the lepton in the Z center of mass frame for the Z forward/backward asymmetry measurement. The lower curve includes acceptance corrections and indicates the result of the likelihood fit for the asymmetry.

production of $W\gamma$ and WZ boson pairs is an excellent tool to study the $WW\gamma$ and WWZ 3-boson couplings which exist in the Standard Model. The study of $Z\gamma$ events is of less interest, since there is no corresponding 3-boson coupling. However, this process is sensitive to possible internal structure (compositeness) of the gauge bosons. The picture becomes substantially more complex if CP violating 3-boson couplings are allowed, but for this discussion, only CP conserving couplings are considered.

The process $p\bar{p} \rightarrow \ell\nu\gamma$ receives contributions both from QED processes and from the 3-boson $WW\gamma$ vertex [26]. In the most general case, assuming CP invariance, there are two free parameters in the Lagrangian, denoted κ and λ. They are related to the magnetic dipole and electric quadrupole moments of the W boson. In the Standard Model, $\kappa = 1$ and $\lambda = 0$. These values correspond to a minimum in the predicted cross section. Any deviation from these Standard Model values will cause the total cross section to increase. It is convenient to divide the contributions to the $\ell\nu\gamma$ final state into two classes: those arising from $W + \gamma$ production, which truly probe the $WW\gamma$ vertex, and those arising from radiative W decay where $W \to \ell\nu\gamma$. This classification is not strictly gauge invariant, but in the limit that the W is narrow (i.e., that its width is much smaller than its mass), the two contributions approximately factorize, and can be experimentally separated by looking at the 3-body transverse mass $M_T(\ell\nu\gamma)$. The radiative decays tend to lie below the W mass and have the photon emitted close to the lepton, whereas the $W + \gamma$ events lie above the W mass, and have the photon well-separated from the lepton. The $W + \gamma$ events show much greater sensitivity to anomalous W couplings than do the radiative decays.

In addition to the total cross section, there are several kinematic variables which are sensitive to deviations from the Standard Model values of the parameters κ and λ. To lowest order, for the process $u\bar{d} \to W^+\gamma$, the cross section goes to zero at $\cos\theta^* = -1/3$. This so-called "radiation-zero" is somewhat diluted by kinematic cuts and by the smearing that takes place when one reconstructs the neutrino longitudinal momentum in order to boost to the W center of mass frame. It appears as a dip near $\cos\theta^* = 0$ when the final state is summed over charges. This dip disappears quickly when the κ and λ parameters deviate from their Standard Model values, and hence provides great sensitivity. Furthermore, higher-order QCD corrections do not alter this picture [27]. Experimentally, the best variable to study is the P_T of the photon, as it is directly measurable and doesn't require neutrino reconstruction. The basic technique is then to examine the distribution of $P_T(\gamma)$ at large P_T, looking for an excess of events relative to the Standard Model prediction.

Experimentally, the event selection is similar to that for inclusive W and Z events, with the addition of an isolated, soft photon [28]. The UA2 analysis required both the electron and the neutrino from the W decay to satisfy $P_T > 20$ GeV. The photon candidate was required to have $P_T > 4.5$ GeV. To reduce backgrounds, the photon was required to be separated by at least 15° from the electron, and to have no charged track detected with 10°. The dominant background comes from $W + jet$ production, where the jet passes the photon selection criteria. The UA2 analysis found 16 candidate events with an expected background of 6.8 ± 1 event. The distributions of kinematic variables are shown in Fig. 11. The Standard Model prediction, including the expected background, has been superimposed.

This result can be converted into an upper limit on the cross section for $W+\gamma$ production, and hence into a limit on the values of κ and λ. However, a more sensitive test involves using the $P_T(\gamma)$ distribution. Given the large backgrounds present in the analysis, it is necessary to measure the P_T distribution of the background, and add it to the Monte Carlo prediction for $W+\gamma$ production as a function of κ and λ. After doing this, a likelihood fit is performed and leads to the 95% confidence

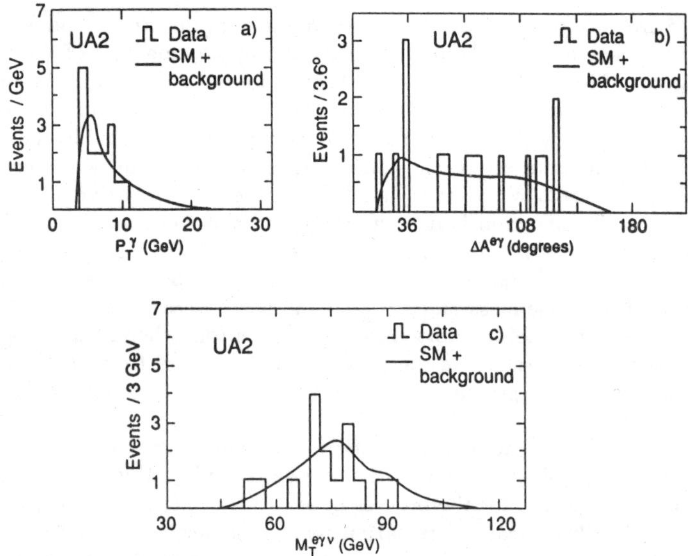

Figure 11: Distributions of: (a) $P_T(\gamma)$, (b) the opening angle between the e and the γ, $A_{e\gamma}$, and (c) the transverse mass $M_T(e\nu\gamma)$ for the 16 UA2 $W+\gamma$ candidate events. The curves represent the Standard Model prediction including the expected background contributions.

limits:

$$-3.5 < \kappa < 5.9 \text{ assuming } \lambda = 0$$
$$-3.6 < \lambda < 3.5 \text{ assuming } \kappa = 0$$

The two parameters are correlated, and a large negative value for one parameter can partially compensate a large positive value for the other, weakening the limits in the case where both parameters deviate from their Standard Model values. Ref. [28] contains contour plots which describe the correlated limits in the (κ, λ) plane. There is no published result for the $W+\gamma$ process at the Tevatron, but a phenomenological analysis [26] suggests that the study of $W+\gamma$ events starts to become significant for a 100 pb^{-1} data sample, where precisions of $\Delta\kappa = \pm 1$ and $\Delta\lambda = \pm 0.3$ should be possible. The cross section at large $P_T(\gamma)$ is sensitive to the square of the anomalous couplings, so increased statistics improve the measurement only like the fourth root of the integrated luminosity, leading to small additional gains with 1000 pb^{-1} of data. The higher energy of the LHC and SSC machines is required to probe the couplings to the several percent level, or equivalently to be sensitive to anomalous gauge boson couplings at mass scales of order 10 TeV.

3 Searches for Additional Heavy Bosons

Given the tremendous success of the $SU(2) \times U(1)$ non-Abelian gauge theory, it is natural to wonder whether Nature has found need for additional bosons beyond the W and Z contained in this model. For example, left-right symmetric models contain an additional $SU(2)$ group with a corresponding heavy right-handed W'. Many Grand Unified theories predict additional $U(1)$ groups with corresponding heavy Z' bosons. Hadron colliders, because of their large center of mass energy, allow a particularly direct search for additional W' and Z' bosons. Rather than searching for small deviations from the expected Standard Model behavior of the charged and neutral currents at low energy, it is possible to look directly for evidence of the processes $W' \to \ell\nu$ and $Z' \to \ell^+\ell^-$.

In producing heavy objects, the higher energy of the Tevatron gives CDF an overwhelming advantage, ruling out masses which are more than twice as large as those excluded by searches at the CERN SPS Collider.

To carry out the search, it is necessary to select high P_T leptons with the best possible efficiency. Since the expected backgrounds are lower at high P_T, the requirements can be slightly looser than those used at lower P_T. Furthermore, it is necessary to carefully understand the expected contribution from the Standard Model W and Z resonances as well as the Drell-Yan continuum. With these goals in mind, CDF has analyzed the transverse mass distribution [29] in the search for $W' \to \ell\nu$, and the di-lepton invariant mass distribution [30] in the search for $Z' \to \ell\ell$.

3.1 *Searches for W'*

For the W' search, a central lepton with $P_T > 30$ GeV and missing transverse energy greater than 30 GeV were both required. The electron selection efficiency was estimated to be 75% for 250 GeV electrons (and somewhat higher for electrons with lower P_T). For muons, the efficiency was estimated to be 75%, independent of the muon P_T. As the mass of the W' increases, the geometric acceptance for the central selection criteria also increases, reaching 55% for $W' \to e\nu$ and 34% for $W' \to \mu\nu$ at a mass greater than 200 GeV.

The W' analysis uses a fit to the observed transverse mass distribution of the form :

$$dN/dM_T = \alpha W + \beta W' \tag{11}$$

where W and W' represent the expected contributions from the Standard Model W boson and from an additional W' boson with mass $M_{W'}$ and width :

$$\Gamma(W') = \Gamma(W)\frac{M_{W'}}{M_W}$$

The data sample used for this fit is shown in Fig. 12. In this type of fit, where one is looking for small deviations from the Standard Model, it is critical to use the

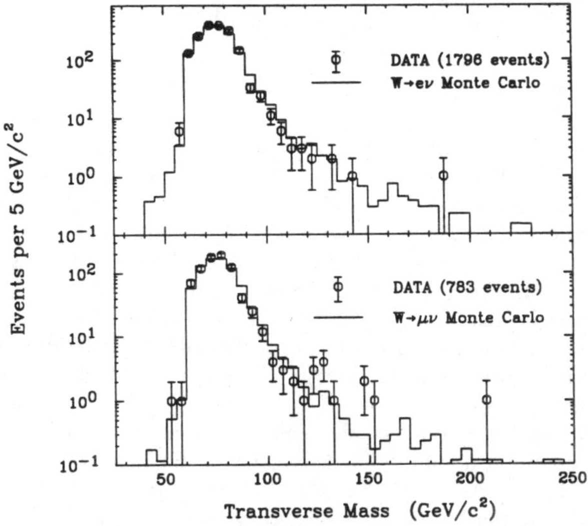

Figure 12: The observed transverse mass distribution for the $W' \to \ell\nu$ search for the (a) electron and (b) muon samples. The histogram indicates the Monte Carlo prediction for the Standard Model W, normalized to the fraction α from the fit defined in Eq. 11.

correct lineshape for the W. Any errors in the assumed shape can create a false signal (or perhaps obscure a real signal). The lineshape at large transverse mass is influenced by the uncertainties on the W transverse momentum distribution and the resulting resolution smearing of the W Jacobean peak. A complete $\mathcal{O}(\alpha_s^2)$ calculation with the corresponding systematic error analysis, was used [10] to quantify this effect. Additional systematic uncertainties arise from possible mass-dependence of the K factor, the detection efficiencies, and the luminosity. All of the systematic uncertainties have been combined and incorporated into the likelihood function for the fit.

The resulting cross section limit is shown in Fig. 13. A curve indicates the expectation for a W' with standard strength couplings and a branching ratio of 1/12 for each lepton family (i.e., the decay $W' \to t\bar{b}$ is assumed to occur). This leads to a limit of $M_{W'} > 520$ GeV at 95% C.L. for such an additional heavy boson.

3.2 *Searches for Z'*

For the Z' analysis, a similar procedure was used. In this case, the events are easily identified, so adequate background rejection may be obtained by imposing

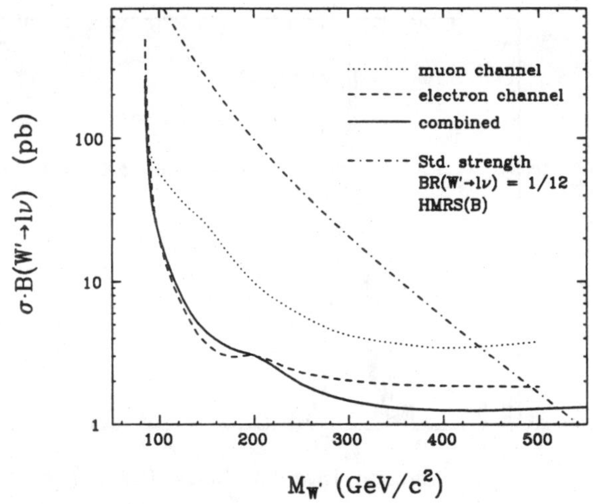

Figure 13: The 95% C.L. cross section limit for the $W' \to \ell\nu$ search. The dot-dashed line indicates the prediction for a W' with a standard branching ratio of 1/12 to each lepton family.

reasonable criteria on one lepton and very loose criteria on the second lepton. For the electron case, one good central electron with $P_T > 15$ GeV was required. The second electron could either be an identified energy cluster in the calorimeter with $E_T > 7$ GeV, or a stiff track with $P_T > 20$ GeV. For the muon case, one identified muon was required, and the second muon had to have a stiff track with $P_T > 20$ GeV and energy deposition in the calorimeter consistent with a single minimum-ionizing particle. In both cases, isolation (limited additional energy in a cone around the lepton or a limited number of tracks in a cone around the lepton) was used to control backgrounds while retaining high efficiency.

The muon data sample is displayed in Fig. 14. A likelihood fit, including the expected contribution from the Standard Model Z, the Drell-Yan continuum, plus a heavy Z' has been performed to both the muon and electron data samples. The systematic uncertainties are included in the likelihood function. The resulting cross section limit as a function of mass is shown in Fig. 15. Assuming a Z' with Standard Model couplings, this corresponds to a limit of $M_{Z'} > 412$ GeV at 95% C.L.

It is also instructive to consider a family of possible Z' which frequently appears in Grand Unified theories, namely those arising from models based on the E_6 gauge group [31]. The results depend strongly on whether the Z' couples only to

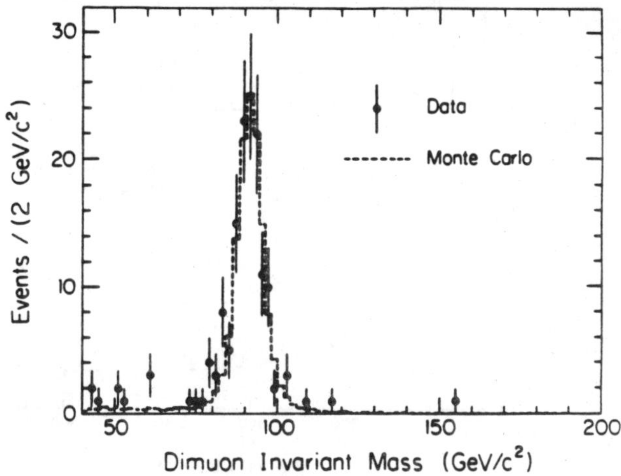

Figure 14: The observed invariant mass distribution for di-muon events in the $Z' \rightarrow \ell\ell$ search. The histogram indicates the Monte Carlo prediction for the Standard Model Z plus Drell-Yan contributions.

the standard quarks and leptons, or whether it also couples to the new supersymmetric fermions in the model. Four cases were considered by CDF, including the Z_ψ, Z_χ, Z_η, and Z_{LR}. Qualitatively, the limits in the scenario that the Z' decays only to standard quarks and leptons are reduced to 300–350 GeV. If additional supersymmetric decays are allowed, the limits further decrease to 200–250 GeV.

As one extends the searches to higher masses, the muon resolution starts to deteriorate, but not enough to significantly impair the ability to see a signal. The expected sensitivity to a Z' is about 650 GeV with a 100 pb^{-1} data sample, reaching an ultimate limit of about 750 GeV for a 1000 pb^{-1} data sample.

4 Searches for the t Quark

The Standard Model predicts the existence of a sixth quark, which is the weak-isospin partner of the bottom quark. Despite searches spanning more than one decade, no direct evidence for this quark has been uncovered. Due to its large expected mass (and hence large Yukawa couplings in the Standard Model Lagrangian), the t quark plays a very important role in precision tests of Electroweak theory.

In a hadron collider, this quark is produced in pairs via gluon fusion. If its mass is small enough ($M_t < M_W - M_b$), it may also appear in the decays of W bosons. A previous generation of searches by UA1, UA2, and CDF excluded this

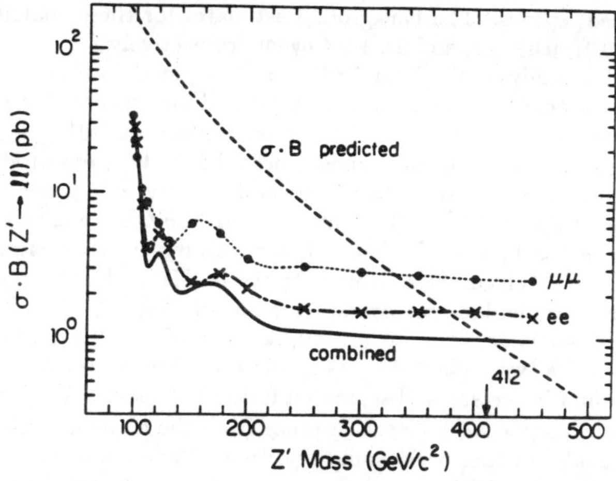

Figure 15: The 95% C.L. cross section limit for the $Z' \to \ell\ell$ search. The dashed line indicates the prediction for a Z' with Standard Model couplings to quarks and leptons.

light t quark region [32]. A heavier t quark is expected to decay predominantly via the charged current decay $t \to W + b$. In the present discussion, the most recent CDF results are summarized [33], followed by a brief discussion of future prospects.

4.1 *Present Limits*

In order to improve the signal to noise ratio in the search, at least one high P_T lepton is required. Two separate analyses have been performed by CDF. The first uses the high-P_T di-lepton final states $e\mu$, ee, and $\mu\mu$ which arise from the process $t\bar{t} \to WbWb$, with both $W \to \ell\nu$, for which the expected branching ratio from a $t\bar{t}$ pair is approximately 4/81. The second analysis starts from the one lepton final state where $t\bar{t} \to WbWb$ with one $W \to \ell\nu$ and the other $W \to q\bar{q}$, for which the expected branching ratio is approximately 24/81, ignoring τ contributions. To improve the signal to noise ratio in this lepton plus jets final state, CDF has searched for an additional low-P_T muon as a tag for one of the b quarks.

In the high-P_T di-lepton analysis, both leptons were required to have $P_T > 15$ GeV. By requiring that the azimuthal angle between the two leptons, $\Delta\phi_{\ell\ell}$, be less than 160° and that the missing transverse energy is consistent with a W decay (missing-$E_T > 20$ GeV), one is left with a very small sample. There is one $e\mu$ event observed in the signal region, which is described in detail in Ref. [33]. The major sources of background are $Z \to \tau\tau$, WW pair production, QCD production of $b\bar{b}$, and mis-identified leptons. The background estimates for these contributions are 0.2 ± 0.1, 0.12 ± 0.01, 0.3 ± 0.2, and 0.6 ± 0.4 events respectively.

In the b-tag analysis, the high-P_T lepton is required to have $P_T > 20$ GeV. Missing transverse energy greater than 20 GeV and two jets with $P_T > 10$ GeV are also required. One is left with 104 $e + jets$ events and 91 $\mu + jets$ events, which are consistent with the backgrounds expected from $W + jets$ production. For $M_t \sim 90 - 100$ GeV, the b quark from t decay will be very soft, producing a muon with a mean P_T of about 3 GeV. The soft muon in the CDF analysis is required to lie in the range $2 < P_T < 15$ GeV. The lower requirement is determined by the minimum muon P_T which can be reliably detected in CDF. The upper requirement ensures that this sample does not overlap that of the high-P_T di-lepton analysis. The present analysis is sensitive to a t quark mass of roughly 90 GeV. In this kinematic region, the two highest P_T jets in the event are almost always from the decay of the W, and hence are well-separated from the b muon. An examination of the variable $\Delta R = (\Delta\eta^2 + \Delta\phi^2)^{1/2}$, the distance from the b muon to the nearest of the two jets, provides further background rejection. There are no candidate events with $\Delta R > 0.5$, whereas 75% of the $t\bar{t}$ events would be expected to satisfy this requirement.

In simple terms, the analysis can be summarized by stating that for $M_t =$ 80, (90, 100) GeV, one would have expected to see 7.9, (4.9, 3.0) events from $t\bar{t}$ production in the high-P_T di-lepton channel. One $e\mu$ candidate was observed, consistent with the expected background in this channel of 1.2 ± 0.5 events. Similarly, for the b-tag search, 2.4, (1.6, 1.1) events from $t\bar{t}$ production should have been seen.

Combining the results from these separate searches in the di-lepton channels gives the result shown in Fig. 16. The 95% C.L. on the cross section for $t\bar{t}$ production is plotted as a function of the t quark mass. The shaded band represents the theoretical prediction. The width of the band is determined by varying the Q^2 scale and parton distribution functions in the next-to-leading order calculation [34]. The experimental uncertainties have been evaluated using a detailed simulation of $t\bar{t}$ production and decay in the CDF detector. The total uncertainty arising from lepton identification, geometric acceptance, isolation requirements, and missing transverse energy requirements was estimated to be 11%. An additional 6.8% uncertainty from the luminosity measurement must also be included. The final result is that $M_t < 91$ GeV is excluded at the 95% C.L. for a t quark decaying with the expected Standard Model semi-leptonic branching fractions.

4.2 *Future Prospects*

Searches in the $e\mu$ channel will remain relatively background-free until the expected signal from a t quark is roughly the same magnitude as that from direct WW pair production which provides the background. This happens for a t quark mass of about 160 GeV. At that point, studies indicate that by simply requiring additional jets in the event, as expected for the $t\bar{t}$ signal, the sensitivity can be extended to about 200 GeV. The complementary lepton plus jets plus b-tag analysis will play

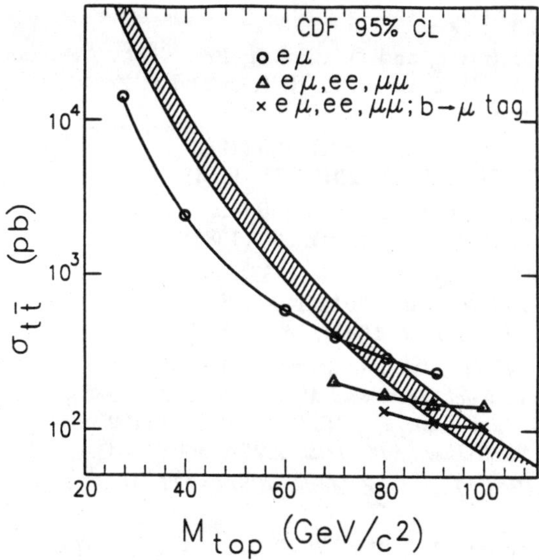

Figure 16: The 95% C.L. cross section limit for the $t\bar{t}$ search compared with a band of theoretical predictions. The intersection of the experimental limit with the lowest allowed theoretical cross section gives the t quark mass limit.

a very important role in confirming any signal seen in the di-lepton channel. The present CDF analysis used a b-tag based on the observation of a soft muon, but in the future, through the use of silicon vertex detectors, a b-tag based on the non-zero b lifetime should be more efficient and more robust against backgrounds. If one assumes that approximately 10 events are needed for reliable discovery, then the mass reach for a 100 pb^{-1} data sample is approximately 150 GeV, and that of a 1000 pb^{-1} data sample is about 200 GeV. Thus, with the Main Injector Upgrade at Fermilab, it should be possible to acquire the data necessary to observe a Standard Model t quark within the range allowed by current global fits to precision Electroweak measurements.

References

1. C. Rubbia, P. McIntyre, and D. Cline in *Proc. Int. Neutrino Conference*, Aachen, 1976, eds. H. Faissner, H. Reithler and P. Zerwas (Vieweg, Braunschweig, 1977), p. 683.
2. G. Arnison et al., *Phys. Lett.* **122B**, 103 (1983); M. Banner et al., *Phys. Lett.* **122B**, 476 (1983).
3. G. Arnison et al., *Phys. Lett.* **126B**, 398 (1983); P. Bagnaia et al., *Phys. Lett.* **129B**, 130 (1983).
4. C. Albajar et al., *Z. Phys.* **C44**, 15 (1989); R. Ansari et al., *Phys. Lett.* **186B**, 440 (1987), R. Ansari et al., *Phys. Lett.* **194B**, 158 (1987).
5. R. Ansari et al., *Nucl. Inst. and Meth.* **A279**, 388 (1989); R. Ansorge et al., *Nucl. Inst. and Meth.* **A265**, 33 (1988); K. Borer et al., *Nucl. Inst. and Meth.* **A286**, 128 (1990); A. Beer et al., *Nucl. Inst. and Meth.* **A224**, 360 (1984).
6. F. Abe et al., *Nucl. Inst. and Meth.* **A271**, 387 (1988) and references contained therein.
7. S. D. Drell and T. M. Yan, *Phys. Rev. Lett.* **25**, 316 (1970).
8. R. Hamberg, W. L. van Neerven, and T. Matsuura, *Nucl. Phys.* **B359**, 343 (1991).
9. G. Altarelli, R. K. Ellis, M. Greco, and G. Martinelli, *Nucl. Phys.* **B246**, 12 (1984); G. Altarelli, R. K. Ellis, and G. Martinelli, *Z. Phys.* **C27**, 617 (1985).
10. P. B. Arnold and M. H. Reno, *Nucl. Phys.* **B319**, 37 (1989); P. B. Arnold and R. P. Kauffman, *Nucl. Phys.* **B349**, 381 (1991).
11. J. Alitti et al., *Z. Phys.* **C49**, 17 (1991).
12. J. Alitti et al., *Phys. Lett.* **B276**, 365, (1992).
13. F. Abe et al., *Phys. Rev.* **D44**, 29, (1991).
14. J. Alitti et al., *Phys. Lett.* **B241**, 150, (1990).
15. J. Alitti et al., *Phys. Lett.* **B276**, 354, (1992).
16. F. Abe et al., *Phys. Rev. Lett.* **65**, 2243, (1990); F. Abe et al., *Phys. Rev.* **D43**, 2070, (1991).
17. K. Hikasa, *Phys. Rev.* **D29**, 1939, (1984); D. A. Dicus and S. S. D. Willenbrock, *Phys. Rev.* **D34**, 148, (1986).
18. C. Albajar et al., *Phys. Lett.* **B253**, 503, (1991).
19. J. Alitti et al., *Phys. Lett.* **B276**, 365, (1992).
20. F. Abe et al., *Phys. Rev. Lett.* **69**, 28, (1992).
21. C. Albajar et al., *Z. Phys.* **C44**, 15, (1989).
22. J. Alitti et al., *Z. Phys.* **C52**, 209, (1991).

23. F. Abe et al., *Phys. Rev. Lett.* **68**, 3398, (1992).

24. F. Abe et al., *Phys. Rev. Lett.* **68**, 1458, (1992).

25. F. Abe et al., *Phys. Rev. Lett.* **67**, 1502, (1991).

26. K. O. Mikelian *Phys. Rev.* **D17**, 750, (1978);
K. O. Mikelian, M. A. Samuel, and D. Sahdev, *Phys. Rev. Lett.* **43**, 746, (1979);
J. Cortes, K. Hagiwara, and F. Herzog *Nucl. Phys.* **B278**, 26, (1986);
U. Baur and D. Zeppenfeld *Nucl. Phys.* **B308**, 127, (1988);
U. Baur and E. L. Berger *Phys. Rev.* **D41**, 1476, (1990).

27. J. Smith, D. Thomas, and W. L. van Neerven, *Z. Phys.* **C44**, 267, (1989).

28. J. Alitti et al., *Phys. Lett.* **B277**, 194, (1992).

29. F. Abe et al., *Phys. Rev. Lett.* **67**, 2418, (1991);
F. Abe et al., *Phys. Rev. Lett.* **67**, 2609, (1991).

30. F. Abe et al., *Phys. Rev. Lett.* **68**, 1463, (1992).

31. F. del Aguila, M. Quiros, and F. Zwirner, *Nucl. Phys.* **B287**, 457 (1987);
D. London and J. L. Rosner, *Phys. Rev.* **D34**, 1530 (1986);
F. del Aguila, J. M. Moreno, and M. Quiros, *Phys. Rev.* **D41**, 134 (1990).

32. C. Albajar et al., *Z. Phys.* **C48**, 1, (1990);
T. Åkesson et al., *Z. Phys.* **C46**, 179, (1990);
F. Abe et al., *Phys. Rev. Lett.* **64**, 142 and 147, (1990);
F. Abe et al., *Phys. Rev.* **D43**, 664, (1991).

33. F. Abe et al., *Phys. Rev. Lett.* **68**, 447, (1992);
F. Abe et al., *Phys. Rev.* **D45**, 3921, (1992).

34. P. Nason, S. Dawson, and R. K. Ellis *Nucl. Phys.* **B303**, 607 (1988);
G. Altarelli, M. Diemoz, G. Martinelli, and P. Nason, *Nucl. Phys.* **B308**, 724 (1988);
R. K. Ellis, *Phys. Lett.* **B259**, 492, (1991).

VII. IMPLICATIONS OF PRECISION EXPERIMENTS

TESTS OF THE STANDARD MODEL AND SEARCHES FOR NEW PHYSICS

PAUL LANGACKER
Department of Physics, University of Pennsylvania,
Philadelphia, Pennsylvania, USA 19104-6396

Contents

1 Introduction

Despite its many successes, the standard model cannot be taken seriously as a candidate for the ultimate theory of matter. As described in the article *Structure of the Standard Electroweak Model* earlier in this volume, it is a complicated theory with many free parameters, several fine-tuning problems, and many arbitrary and unexplained features. Historically, precision electroweak experiments were crucial for establishing the standard electroweak model (especially the unification aspects) as correct to first approximation and excluding alternatives. At present and in the future they will continue to be important for establishing the domain of validity of the standard model, determining its parameters, and searching for and excluding various possibilities for underlying new physics.

Earlier chapters of this volume have described the standard model and its renormalization, the various types of precision experiments, and their implications in detail. This chapter is devoted to global analyses of the Z-pole, M_W, and neutral current data[1], which contains more information than any one class of experiment. The subsequent sections will summarize some of the relevant data and theoretical formulas, the status of the standard model tests and parameter determinations, the possible classes of new physics, and the implications of the precision experiments. In particular, the model independent analysis of neutral current couplings (which establishes the standard model to first approximation); the implications of supersymmetry; (supersymmetric) grand unification; and a number of specific types of new physics, including heavy Z' bosons[2] , new souces of SU_2 breaking, new contributions to the gauge boson self-energies, $Zb\bar{b}$ vertex corrections, certain types of new 4-Fermi operators and leptoquarks, and exotic fermions are described. Leptoquarks are covered in much more detail in the chapters by Deutsch and Quin and by Herczeg, and exotic fermions in the chapter by London. Future prospects are described in the chapter by Luo and in [2].

2 The Standard Model and its Parameters

2.1 *Recent Data*

Recent results from Z-pole experiments are shown in Table 1. These include the results of the four LEP experiments ALEPH, DELPHI, L3, and OPAL (including preliminary results from the 1993 LEP energy scan), averaged including a proper

[1]In most cases, the types of new physics constrained by charged current data (Chapter V) are separate and do not require a simultaneous analysis with the gauge boson and neutral current data. One major exception is mixing between ordinary and exotic fermions. See the article by D. London in this volume.

[2]$SU_{2L} \times SU_{2R} \times U_1$ models (which involve an additional heavy W' boson coupling to right-handed currents) are described in the chapters by Deutsch and Quin and by Herczeg, and in [1].

Quantity	Value	Standard Model
M_Z (GeV)	91.1888 ± 0.0044	input
Γ_Z (GeV)	2.4974 ± 0.0038	$2.497 \pm 0.001 \pm 0.003 \pm [0.002]$
$R = \Gamma(\text{had})/\Gamma(\ell\bar{\ell})$	20.795 ± 0.040	$20.784 \pm 0.006 \pm 0.003 \pm [0.03]$
$\sigma_{\text{had}} = \frac{12\pi}{M_Z^2} \frac{\Gamma(e\bar{e})\Gamma(\text{had})}{\Gamma_Z^2}$ (nb)	41.49 ± 0.12	$41.44 \pm 0.004 \pm 0.01 \pm [0.02]$
$R_b = \Gamma(b\bar{b})/\Gamma(\text{had})$	0.2202 ± 0.0020	$0.2156 \pm 0 \pm 0.0004$
$R_c = \Gamma(c\bar{c})/\Gamma(\text{had})$	0.1583 ± 0.0098	$0.171 \pm 0 \pm 0$
$A_{FB}^{0\ell} = \frac{3}{4}\left(A_\ell^0\right)^2$	0.0170 ± 0.0016	$0.0151 \pm 0.0005 \pm 0.0006$
$A_\tau^0\,(P_\tau)$	0.143 ± 0.010	$0.142 \pm 0.003 \pm 0.003$
$A_e^0\,(P_\tau)$	0.135 ± 0.011	$0.142 \pm 0.003 \pm 0.003$
$A_{FB}^{0b} = \frac{3}{4}A_e^0 A_b^0$	0.0967 ± 0.0038	$0.0994 \pm 0.002 \pm 0.002$
$A_{FB}^{0c} = \frac{3}{4}A_e^0 A_c^0$	0.0760 ± 0.0091	$0.071 \pm 0.001 \pm 0.001$
$\bar{s}_\ell^2\left(A_{FB}^Q\right)$	0.2320 ± 0.0016	$0.2322 \pm 0.0003 \pm 0.0004$
$A_e^0\,(A_{LR}^0)$ (SLD)	0.1637 ± 0.0075 $(92+93)$ $(0.1656 \pm 0.0076$ $(93))$	$0.142 \pm 0.003 \pm 0.003$
N_ν	2.988 ± 0.023	3

Table 1: Z-pole observables from LEP and SLD compared to their standard model expectations. The standard model prediction is based on M_Z and uses the global best fit values for m_t and α_s, with M_H in the range $60 - 1000$ GeV. The $R_b - R_c$ correlation is -0.4. The lineshape correlations are given in [4].

treatment of common systematic uncertainties [3, 4]. In addition, the result from the SLD experiment at SLAC [5] on the left-right asymmetry A_{LR} is shown. The first row in Table 1 gives the value of the Z mass, which is now known to remarkable precision. Also shown are the lineshape variables Γ_Z, R, and σ_{had}, which are respectively the total Z width, the ratio of the Z width into hadrons to the width into a single charged lepton, and the peak hadronic cross section after removing QED effects; the heavy quark production rates; various forward-backward asymmetries, A_{FB}; quantities derived from the τ polarization P_τ and its angular distribution; and the effective weak angle $\bar{s}_\ell^2$ obtained from the jet charge asymmetry. N_ν is the number of effective active neutrino flavors with masses light enough to be produced in Z decays. It is obtained by subtracting the widths for decays into hadrons and charged leptons from the total width Γ_Z obtained from the lineshape. The asymmetries are expressed in terms of the quantity

$$A_f^o = \frac{2\bar{g}_{Vf}\,\bar{g}_{Af}}{\bar{g}_{Vf}^2 + \bar{g}_{Af}^2}, \tag{1}$$

where $\bar{g}_{V,Af}$ are the vector and axial vector couplings to fermion f.

From the Z mass one can predict the other observables, including electroweak loop effects. The predictions also depend on the top quark and Higgs mass, and α_s is needed for the QCD corrections to the hadronic widths and the relation between M_Z and $\sin^2\theta_W$. The predictions are shown in the third column of Table 1, using

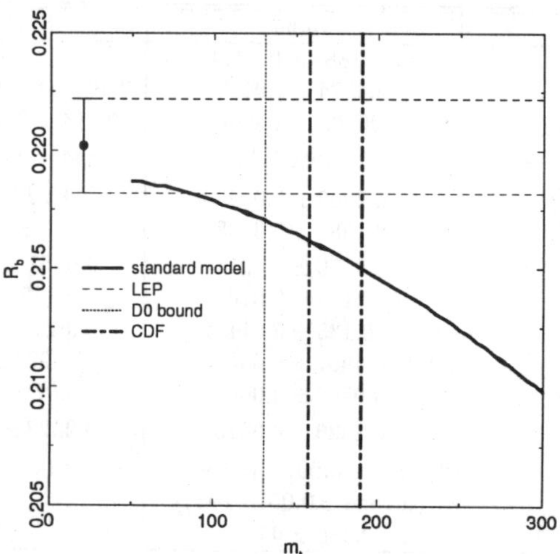

Figure 1: Standard model prediction for $R_b \equiv \Gamma(b\bar{b})/\Gamma(\text{had})$ as a function of m_t, compared with the LEP experimental value. Also shown are the D0 lower bound of 131 GeV and the CDF range 174 ± 16 GeV.

the value $m_t = 175 \pm 11$ GeV obtained for $M_H = 300$ GeV in a global best fit to all data. The first uncertainty is from M_Z and Δr (related to the running of α up to M_Z), while the second is from m_t and M_H, allowing the Higgs mass to vary in the range $60 - 1000$ GeV. The last uncertainty is the QCD uncertainty from the value of α_s. Here the value and uncertainty are given by $\alpha_s = 0.127 \pm 0.005$, obtained from the global fit to the lineshape.

The data is in excellent agreement with the standard model predictions except for two observables. The first is

$$R_b = \frac{\Gamma(b\bar{b})}{\Gamma(\text{had})} = 0.2202 \pm 0.0020. \tag{2}$$

This is some 2.3σ higher than the standard model expectation 0.2156 ± 0.0004. Because of special vertex corrections, the $b\bar{b}$ width actually decreases with m_t, as opposed to the other widths which all increase. It is apparent from Figure 1 that R_b favors a small value of m_t. By itself R_b is insensitive to M_H. However, when combined with other observables, for which m_t and M_H are strongly correlated, the effect is to favor a smaller Higgs mass. Another possibility, if the effect is more than a statistical fluctuation, is that it may be due to some sort of new physics. Many types of new physics will couple preferentially to the third generation, so this is a serious possibility.

Quantity	Experiment	SM	Topless	Mirror	Vector
R_b	0.2202 ±0.0020	0.2156	0.017	0.2156	0.35
A_{FB}^{0b}	0.0967 ±0.0038	0.0994	0	−0.0994	0

Table 2: Predictions of the standard model (SM), topless models, a mirror model with $(t\ b)_R$ in a doublet, and a vector model with left and right-handed doublets, for R_b and A_{FB}^{0b}, compared with the experimental values.

Despite the small discrepancy, R_b and the forward-backward asymmetry A_{FB}^{0b} (corrected for $B\bar{B}$ oscillations) are sufficient to establish that the left-handed b belongs to a weak doublet. From the fit to these and other data one obtains uniquely

$$t_{3L}(b) = -0.500 \pm 0.005 \qquad t_{3R}(b) = 0.026 \pm 0.018 \tag{3}$$

for the third component of the weak isospin of the $b_{L,R}$, respectively, updating an analysis by Schaile and Zerwas [6]. This is in agreement with the standard model expectations of $-1/2$ and 0 and excludes topless models[3]. The values of R_b and A_{FB}^{0b} are also compared with the predictions of various alternative $SU_2 \times U_1$ models in Table 2. Only the standard model is in agreement with the data.

The other discrepancy is the value of the left-right asymmetry

$$A_{LR}^0 = A_e^0 = \frac{2\bar{g}_{Ve}\bar{g}_{Ae}}{\bar{g}_{Ve}^2 + \bar{g}_{Ae}^2} = 0.164 \pm 0.008 \tag{4}$$

obtained by the SLD collaboration. A_{LR}^0 is very clean both experimentally [8] (most systematic effects other than the absolute beam polarization cancel in the ratio) and theoretically (most radiative corrections cancel), and is very sensitive to both the weak angle and new physics. The SLD value is some 2.5σ higher than the standard model expectation of 0.142 ± 0.004. Assuming the standard model, A_{LR}^0 (combined with M_Z) predicts a top quark mass around 250 GeV, much higher than other determinations. Unless this is a statistical fluctuation, the obvious possibility is that the high value of A_{LR}^0 is due to new physics, such as $S < 0$, where S is a parameter describing certain types of heavy new physics (see Section 4.8). In addition, there are possible tree-level physics such as heavy Z' bosons or mixing with heavy exotic doublet leptons, E'_R, which could significantly affect the asymmetry. However, new physics probably cannot for the entire discrepancy, because the LEP observables A_{FB}^{0e} and $A_e^0(P_\tau)$ (obtained from the angular distribution of the τ polarization) measure the same quantity[4] A_e^0 as A_{LR}^0. Together, the LEP values imply

$$A_e^0|_{\rm LEP} = 0.138 \pm 0.009 \, , \tag{5}$$

[3]Earlier indirect arguments for the existence of the t quark are summarized in [7].

[4]The relation makes use only of the assumption that the LEP and SLD observables are dominated by the Z-pole. The one (unlikely) loophole is the possibility of an important contribution from other sources, such as new 4-fermi operators. These are mainly significant slightly away from the pole (at the pole they are out of phase with the Z amplitude and do not interfere).

in agreement with the standard model expectation. Thus, there is a direct experimental conflict between the LEP and SLD values of A_e^0 at the 2.2σ level.

It will take more time and more statistics to see whether the values of R_b and A_{LR}^0 constitute true discrepancies with the standard model (and between experiments in the latter case) or are due to statistical fluctuations at the 2-2.5 σ level or other experimental problems. I will generally take the view that the effects are consistent with (large) fluctuations.

There are many other precision observables. Some recent ones are shown in Table 3. These include the D0 limit [9] $m_t > 131$ GeV and the value $m_t = 174 \pm 16$ GeV suggested by the CDF candidate events [10]. There are new observations of the W mass [11, 12] from both D0, which has presented a preliminary new value 79.86 ± 0.40 GeV, and from CDF, which finds 80.38 ± 0.23 GeV. Combining these and earlier data one obtains the results shown. Other observables include M_W/M_Z from UA2[5] [13], atomic parity violation from Boulder [14], recent results on neutrino electron scattering from CHARM II [15], and new measurements of $s_W^2 \equiv 1 - M_W^2/M_Z^2$ from the CCFR collaboration at Fermilab [16]. This on-shell definition of the weak angle is determined from deep inelastic neutrino scattering with small sensitivity[6] to the top quark mass. The result combined with earlier experiments [17]-[19] is also shown. All of these quantities are in excellent agreement with the standard model predictions.

In the global fits to be described [20], all of the earlier low energy observables [18]-[21] not listed in the table are fully incorporated, as are full treatments of statistical, systematic, and theoretical uncertainties, and correlations between the experiments.

2.2 *Theoretical Expressions and Radiative Corrections*

2.2.1 *The Z and W Masses*

In the electroweak theory one defines the weak angle by

$$\sin^2\theta_W \equiv \frac{g'^2}{g^2+g'^2} \longrightarrow \sin^2\hat{\theta}_W(M_Z) \quad (\overline{MS}) \tag{6}$$

where g' and g are respectively the gauge couplings of the U_1 and SU_2 gauge groups. Although initially defined in terms of the gauge couplings, after spontaneous sym-

[5]One could, of course, multiply M_W/M_Z by the LEP M_Z and include the result in the M_W average. (In fact, such a procedure was carried out in the D0 analysis.) I do not do so because, in principle, it would introduce a correlation between M_Z and M_W. In practice, the effect is negligible because of the tiny uncertainty in M_Z.

[6]The sensitivity is small but not zero. The quoted CCFR value is for $(m_t, M_H) = (150,100)$ GeV. The combined result is for (m_t, M_H) in the allowed range.

Quantity	Value	Standard Model
M_W (GeV)	80.17 ± 0.18	$80.31 \pm 0.02 \pm 0.07$
$M_W/M_Z(UA2)$	0.8813 ± 0.0041	$0.8807 \pm 0.0002 \pm 0.0007$
$Q_W(C_S)$	$-71.04 \pm 1.58 \pm [0.88]$	$-72.93 \pm 0.07 \pm 0.04$
$g_A^{\nu e}$ (CHARM II)	-0.503 ± 0.017	$-0.506 \pm 0 \pm 0.001$
$g_V^{\nu e}$ (CHARM II)	-0.035 ± 0.017	$-0.037 \pm 0.001 \pm 0$
$s_W^2 \equiv 1 - \frac{M_W^2}{M_Z^2}$	0.2218 ± 0.0059 [CCFR] 0.2260 ± 0.0048 [All]	$0.2245 \pm 0.0003 \pm 0.0013$
M_H (GeV)	≥ 60 LEP	$< \begin{cases} 0(600), \text{ theory} \\ 0(800), \text{ indirect} \end{cases}$
m_t	> 131 D0 174 ± 16 CDF	$175 \pm 11^{+17}_{-19}$ [indirect]
$\alpha_s(M_Z)$	0.123 ± 0.006 LEP jets 0.116 ± 0.005 jets + low energy	$0.127 \pm 0.005 \pm 0.002$ [Z lineshape]

Table 3: Recent observables from the W mass and other non-Z-pole observations compared with the standard model expectations. Direct values and limits on M_H, m_t, and α_s are also shown.

metry breaking one can relate the weak angle to the W and Z masses by

$$M_W^2 = \frac{A^2}{\sin^2\theta_W} \longrightarrow \frac{A^2}{\sin^2\hat{\theta}_W(1-\Delta\hat{r}_W)} \tag{7}$$

and

$$M_Z^2 = \frac{M_W^2}{\cos^2\theta_W} \longrightarrow \frac{M_W^2}{\hat{\rho}\cos^2\hat{\theta}_W} \tag{8}$$

where

$$A^2 \equiv \frac{\pi\alpha}{\sqrt{2}G_F} = (37.2802 \text{ GeV})^2 \,. \tag{9}$$

The first form of equations (6)–(8) are valid at tree level. However, the data is sufficiently precise that one must include full one loop radiative corrections, which means that one must replace the quantities by the expressions shown in the last part of equations. There are a number of possible ways of defining the renormalized weak angle. Here I am using the quantity $\sin^2\hat{\theta}_W(M_Z) \equiv \hat{s}_Z^2$, which is renormalized according to modified minimal subtraction, $\overline{MS}$ [22]. This basically means that one removes the $\frac{1}{n-4}$ poles and some associated constants (artifacts of dimensional regularization) from the gauge couplings.

In equation (7) the quantity $\Delta\hat{r}_W$ contains the finite radiative corrections which relate the W and Z masses, muon decay, and QED. The dominant contribution is given by the running of the fine structure constant α from low energies, where it is defined in QED, up to the Z-pole, which is the scale relevant for electroweak interactions,

$$\frac{1}{1-\Delta\hat{r}_W} \simeq \frac{\alpha(M_Z)}{\alpha} \sim \frac{1/128}{1/137}. \tag{10}$$

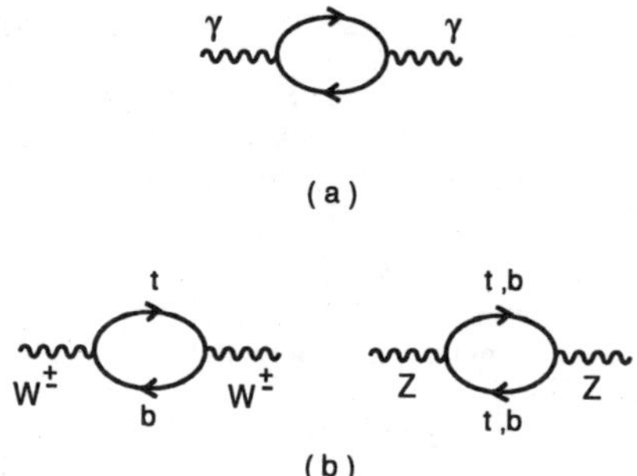

Figure 2: (a) Photon self-energy diagram leading to the running of α. (b) Contributions of the top and bottom quarks to the W and Z self-energies.

There is only a weak dependence on the top quark mass in this scheme, leading to a value

$$\Delta\hat{r}_W \sim 0.07 \tag{11}$$

dominated by the running of α. More precisely, in the $\overline{\text{MS}}$ scheme [23]

$$\alpha^{-1}(M_Z) = 127.9 \pm 0.1, \tag{12}$$

where the uncertainty is from the contribution of light hadrons to the photon self-energy diagrams. This leads to a theoretical uncertainty of ± 0.0009 [24] in $\Delta\hat{r}_W$, which turns out to be the dominant theoretical uncertainty in the precision electroweak tests and, in particular, in the expressions relating the Z mass to other observables. A similar effect leads to a significant theoretical uncertainty in the anomalous magnetic moment of the muon, $g_\mu - 2$, which will dominate the experimental uncertainties in the new Brookhaven experiment unless associated measurements are made of the cross-section for $e^+e^- \rightarrow$ hadrons at low energies. Including other small contributions, one predicts

$$\Delta\hat{r}_W = 0.0706 \pm 0.0002 \pm 0.0009, \tag{13}$$

where the central value is for $(m_t, M_H) = (175,\ 300)$ GeV, the first uncertainty is from m_t and M_H varying in the global best fit range, and the second is from $\alpha(M_Z)$.

Because m_t is so much heavier than the bottom quark mass there is large SU_2 breaking generated by loop diagrams involving the top and bottom quarks, in particular from the W and Z self-energy diagrams shown in Figure 2. There is little shift in the W mass, because that effect is already absorbed into the observed value of the Fermi constant, so $\Delta\hat{r}_W$ has no large m_t dependence. However, the Z mass prediction is shifted down. In particular, the quantity $\hat{\rho}$ in equation (8) depends quadratically on m_t. It is given by [25]

$$\hat{\rho} \sim 1 + \rho_t, \tag{14}$$

where

$$\rho_t = \frac{3G_F m_t^2}{8\sqrt{2}\pi^2} \sim 0.0031(m_t/100 \text{ GeV})^2. \tag{15}$$

Here and throughout, m_t refers to the pole mass, which corresponds approximately to the kinematic mass relevant to direct searches at colliders. (The running mass is discussed by Marciano in this volume.) For m_t in the range 100 – 200 GeV the effect on $\hat{\rho}$ can be quite significant. ρ_t propagates to other observables and generates most of the m_t dependence. (The one important exception is the vertex correction to $Z \to b\bar{b}$ decay.) $\hat{\rho}$ contains additional contributions from bosonic loops, including $\ln(M_H/M_Z)$ terms that are strongly correlated with ρ_t.

From the precise value $M_Z = 91.1888 \pm 0.0044$ GeV from LEP one has (using the expressions in [23])

$$\sin^2 \hat{\theta}_W(M_Z) = 0.2319 \pm 0.0005. \tag{16}$$

The uncertainty is an order of magnitude smaller than one had prior to the Z-pole experiments at LEP. The uncertainty from the experimental error in the Z mass is negligible, of order 0.00003. The theoretical uncertainty 0.0003 coming from $\Delta\hat{r}_W$ is much larger. The largest uncertainty, however, is from m_t and M_H, ~ 0.0004. Here I have used the range of m_t from the global best fit, and 60 GeV $< M_H <$ 1000 GeV. If one knew m_t one would have a more precise value of the weak angle. The sensitivity is displayed in Figure 3. Clearly, one cannot determine the weak angle from M_Z alone because of the m_t dependence. One must have either other indirect observables with a different dependence on m_t or a direct measurement. Before discussing other possibilities, I will digress somewhat on the radiative corrections [22].

The radiative corrections fall into three categories. First, there are the reduced QED corrections, which involve the emission of real photons and the exchange of virtual photons but do not include vacuum polarization diagrams. These constitute a gauge invariant set but depend on the details of the experimental acceptances and cuts. They generally are removed from the data by the experimenters. The second class has already been described. It is the electromagnetic vacuum polarization diagrams, which lead to the running from $\alpha^{-1} \sim 137$ at low energies to $\alpha(M_Z)^{-1} \sim 128$ at the Z-pole. As we have seen this leads to a significant uncertainty $\Delta\Delta\hat{r}_W \sim \Delta\alpha(M_Z)/\alpha \sim 0.0009$ [24], which can lead to a shift of approximately 3 GeV in the predicted value of m_t.

The electroweak corrections are now quite important. One must include full 1-loop corrections as well as dominant 2-loop effects. The electroweak corrections include and are dominated by the gauge self-energy diagrams for the W, Z, and γZ mixing. In addition, there are box diagrams and vertex corrections, which are smaller but which have to be included. Recently there has been some progress on

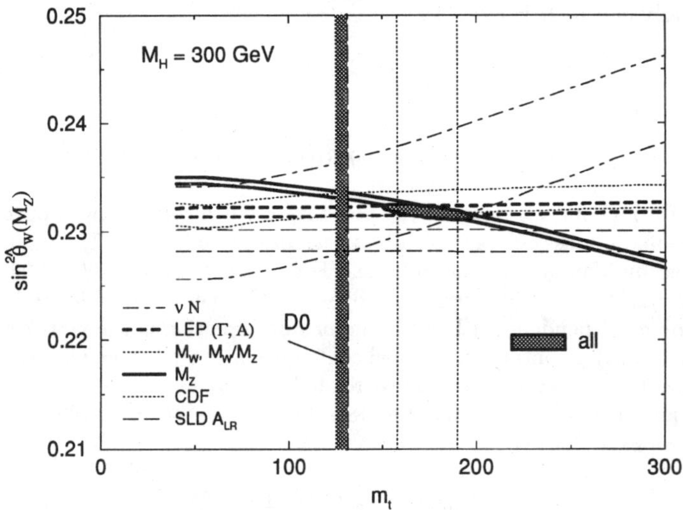

Figure 3: Values of $\sin^2 \hat{\theta}_W(M_Z)$ as a function of m_t from various observables.

the dominant 2-loop effects. In particular, the dominant terms of order $\alpha^2 m_t^4$ are included. The net effect is to replace (14) by [26]

$$\hat{\rho} \rightarrow 1 + \rho_t \left[1 + \rho_t R\left(\frac{M_H}{m_t}\right)\right], \tag{17}$$

where R, which comes from 2-loop diagrams, is strongly dependent on M_H, with $R(0) = 19 - 2\pi^2$. There are additional smaller contributions which must be included in the numerical analysis.

There are also significant mixed QCD-electroweak diagrams, such as those obtained by the exchange of the gluon across the quarks in a self-energy diagram. The dominant contribution involves top quark loops and is of order $\alpha\alpha_s m_t^2$. This leads to the replacement [27].

$$\hat{\rho} \rightarrow 1 + \rho_t \left[1 - 2\alpha_s(m_t)\frac{\pi^2 + 3}{9\pi}\right] \sim 1 + 0.9\rho_t, \tag{18}$$

which raises the predicted value of m_t by approximately 5%. Recently there have been discussions and estimates of $t\bar{t}$ threshold corrections, which are $O(\alpha\alpha_s^2 m_t^2)$. These have been estimated using both perturbative [28] methods and by dispersion relations [23]. One estimate [28] is that the effect is mainly to shift the scale at which α_s should be evaluated for the t quark loop, namely $\alpha_s(m_t) \rightarrow \alpha_s(0.15m_t)$. This is in good numerical agreement with the dispersion relation estimate, which is used here. The threshold corrections raise the predicted values of m_t by +3 GeV.

2.2.2 *Renormalization of* $\sin^2\theta_W$

There are a number of definitions of the renormalized weak angle used in the literature, each with its advantages and disadvantages. At tree-level there are several equivalent expressions, namely

$$\sin^2\theta_W = \frac{g'^2}{g^2+g'^2} = 1 - \frac{M_W^2}{M_Z^2} = \frac{\pi\alpha}{\sqrt{2}G_F M_W^2}. \tag{19}$$

The first definition is based on the coupling constants; the last two take meaning only after spontaneous symmetry breaking has occurred, and therefore mix in parts of the theory other than the gauge vertices. At higher order one must define a renormalized angle. One can use the various expressions in equation (19) as starting points, and the resulting definitions differ by finite terms of order α, which also depend on m_t and M_H. This has led to considerable confusion (and heat).

Two common definitions are based on the spontaneous symmetry breaking (SSB) of the theory, namely on the gauge boson masses. The most famous is the on-shell definition [29, 22]

$$s_W^2 = 1 - \frac{M_W^2}{M_Z^2} = 0.2243 \pm 0.0012. \tag{20}$$

This is very simple conceptually. However, the W mass is not determined as precisely as M_Z, so s_W^2 must actually be extracted from other data and not from the defining relation (20). This leads to a strong dependence on m_t, which accounts for almost all of the uncertainty in s_W^2. (The value given for s_W^2 and the other definitions is from a global fit to all data.)

In the on-shell scheme one defines the radiative correction parameter Δr by

$$M_W^2 = M_Z^2 c_W^2 = \frac{A^2}{s_W^2(1-\Delta r)}, \tag{21}$$

where $c_W^2 = 1 - s_W^2$, A is defined in (9), and

$$\Delta r \sim 1 - \frac{\alpha}{\alpha(M_Z)} - \rho_t/\tan^2\theta_W + \text{small terms}, \tag{22}$$

which depends sensitively on m_t. For the allowed range, one expects

$$\Delta r = 0.040 \pm 0.004 \pm 0.0009, \tag{23}$$

where the second uncertainty is from $\alpha(M_Z)$.

The Z-mass definition [30],

$$s_{M_Z}^2\left(1 - s_{M_Z}^2\right) = \frac{\pi\bar{\alpha}(M_Z)}{\sqrt{2}G_F M_Z^2} = 0.2312 \pm 0.0003, \tag{24}$$

is obtained by simply removing the m_t dependence from the expression for the Z mass. $\bar{\alpha}^{-1}(M_Z) = 128.87 \pm 0.12$ differs by finite constants from the $\overline{\rm MS}$ quantity $\alpha^{-1}(M_Z)$. This is the most precise – the uncertainty is mainly from $\alpha(M_Z)$. The use of $s^2_{M_Z}$ is essentially equivalent to using the Z mass as a renormalized parameter, introducing the weak angle as a useful derived quantity. This scheme is simple and precise, and by definition there is no m_t dependence in the relation betwen M_Z and $s^2_{M_Z}$. However, the m_t dependence and uncertainties enter as soon as one predicts other quantities in terms of it.

Both of the definitions based on spontaneous symmetry breaking are awkward in the presence of any type of new physics that shifts the values of the gauge boson masses. There are other definitions based on the gauge coupling constants. These are especially useful for applications to grand unification, and they tend to be less sensitive to the presence of new physics. One is the modified minimal subtraction or $(\overline{MS})$ definition [31, 22]

$$\hat{s}^2_Z = \frac{\hat{g}'^2(M_Z)}{\hat{g}'^2(M_Z) + \hat{g}^2(M_Z)} = 0.2317 \pm 0.0004, \tag{25}$$

defined by removing the poles and associated constants from the gauge couplings. As we have seen, the uncertainty is mainly from $\alpha(M_Z)$ and m_t. It is useful to also define $\hat{c}^2_Z = 1 - \hat{s}^2_Z$. There are variant definitions[7] of $\hat{s}^2_Z$, depending on the treatment of $\alpha \ln(m_t/M_Z)$ terms. One cannot decouple all such terms because $m_t \gg m_b$ breaks SU_2. The version used here [32, 23] decouples them from $\gamma - Z$ mixing, essentially eliminating any m_t dependence from the Z-pole asymmetry formulas. The on-shell and $\overline{\rm MS}$ definitions are related by

$$\hat{s}^2_Z = \kappa_W s^2_W, \tag{26}$$

where κ_W depends on m_t and M_H. For example, $\kappa_W = 1.033$ for $(m_t, M_H) =$ (175,300) GeV, and the dominant m_t dependence given by

$$\kappa_W \sim 1 + \rho_t / \tan^2\theta_W. \tag{27}$$

The detailed relation is given in [23, 33].

Finally, the experimental groups at LEP and SLC have made extensive use of

$$\begin{aligned} \bar{g}_{Af} &= \sqrt{\rho_f}\, t_{3f} \\ \bar{g}_{Vf} &= \sqrt{\rho_f}\left[t_{3f} - 2\bar{s}^2_f q_f\right]. \end{aligned} \tag{28}$$

These are the effective axial and vector couplings of the Z to fermion f. In equation (28) $t_{3f} = \pm\frac{1}{2}$ is the weak isospin of the left-handed component of fermion f and q_f is

[7] An alternate form, $\hat{s}^2_{\rm ND}$ [33], which was used frequently in earlier literature, does not decouple the $\ln m_t$ terms. Its numerical value is close to the effective angle $\bar{s}^2_\ell$ for the favored m_t range. The precise translations are given in [23, 19]. Another variant, used in the program ZFITTER [34], is described by Hollik in this volume.

its electric charge. The electroweak self-energy and vertex corrections are absorbed into the coefficient ρ_f and the effective weak angle $\bar{s}_f^2$. The $\bar{g}_{V,Af}$ are obtained from the data after removing all photonic contributions. In principle there are also electroweak box contributions. However, these are very small at the Z pole, and are typically ignored or removed from the data.

There is a different effective weak angle for each type of fermion. $\bar{s}_f^2$ is related to the $\overline{MS}$ angle by

$$\bar{s}_f^2 = \kappa_f \hat{s}_Z^2, \tag{29}$$

where κ_f is a form factor. The best measured is for the charged leptons. For the relevant m_t, $\kappa_\ell \sim 1.0013$ [35], so that

$$\bar{s}_\ell^2 \sim \hat{s}_Z^2 + 0.00028 = 0.2320 \pm 0.0004. \tag{30}$$

There is an additional theoretical uncertainty of ± 0.0001 from the precise definition of the angles and higher order effects. These effective angles are very simple for the discussion of the Z-pole data, but are difficult to relate to other types of observables. All of these definitions have advantages and disadvantages, some of which are listed in Table 4. Other definitions and schemes, such as the *-scheme, are described by Hollik in this volume.

2.2.3 *Other Z-Pole Observables*

The other Z-pole observables can also be computed. For example, the partial width for Z to decay into fermions $f\bar{f}$ is given approximately by [36]

$$\Gamma(f\bar{f}) \simeq C_f \frac{G_F M_Z^3}{6\sqrt{2}\pi} \left[|\bar{g}_{Af}|^2 + |\bar{g}_{Vf}|^2 \right]. \tag{31}$$

For the heavier quarks and leptons kinematic mass corrections must be applied, although they are only important for $\Gamma(\bar{b}b)$. Effective couplings are proportional to $\sqrt{\rho}$ so that each partial width increases quadratically with m_t. This comes from the replacement

$$\frac{M_Z g^2}{8\cos^2\hat{\theta}_W} \to \hat{\rho}\frac{G_F}{\sqrt{2}} M_Z^3, \tag{32}$$

which incorporates many of the low energy corrections. In equation (31) there is an additional coefficient

$$C_f = \begin{cases} 1 + \frac{3\alpha}{4\pi} q_f^2 & \text{(leptons)} \\ 3\left(1 + \frac{3\alpha}{4\pi} q_f^2\right) \left(1 + \frac{\alpha_s}{\pi} + 1.409\left(\frac{\alpha_s}{\pi}\right)^2 - 12.77\left(\frac{\alpha_s}{\pi}\right)^3\right) & \text{(quarks)}, \end{cases} \tag{33}$$

which includes QED and QCD corrections. (α_s is the strong coupling in the $\overline{\text{MS}}$ scheme, evaluated at M_Z.) For $\Gamma(\bar{b}b)$ there is additional m_b and m_t dependence in the QCD corrections [37]. The α_s dependence of the hadronic widths leads

On-shell : $s_W^2 = 1 - \frac{M_W^2}{M_Z^2} = 0.2243\,(12)$
+ most familiar + simple conceptually – large m_t dependence from Z-pole observables – depends on SSB mechanism – awkward for new physics
Z-mass : $s_{M_Z}^2 = 0.2312(3)$
+ most precise (no m_t dependence) + simple conceptually – m_t reenters when predicting other observables – depends on SSB mechanism – awkward for new physics
$\overline{MS}$: $\hat{s}_Z^2 = 0.2317\,(4)$
+ based on coupling constants + convenient for GUTs + usually insensitive to new physics + Z asymmetries $\sim$ independent of m_t – theorists definition; not simple conceptually – usually determined by global fit – some sensitivity to m_t – variant forms (m_t cannot be decoupled in all processes; $\hat{s}_{ND}^2$ larger by 0.0001 – 0.0002)
effective : $\bar{s}_\ell^2 = 0.2320\,(4)$
+ simple + Z asymmetry independent of m_t + Z widths: m_t in ρ_f only – phenomenological; exact definition in computer code – different for each f – hard to relate to non Z-pole observables

Table 4: Advantages and disadvantages of several definitions of the weak angle.

to a determination of $\alpha_s = 0.127 \pm 0.005$. (There is also some sensitivity to the mixed QCD-electroweak contributions to $\hat{\rho}$, which relates M_Z to the other observables.) For fixed M_Z most of the m_t dependence is in the $\hat{\rho}$ factor. One major exception[8] is that $\Gamma(b\bar{b})$ decreases with m_t due to special m_t-dependent vertex corrections [38], [39]. These are included in the ρ_b and κ_b factors, but to an excellent numerical approximation $\Gamma(b\bar{b})$ can be written as [39],

$$\Gamma(b\bar{b}) \to \Gamma^0(b\bar{b})\left(1+\delta_{bb}^{SM}\right) \sim \Gamma^0(b\bar{b})\left[1-10^{-2}\left(\frac{m_t^2}{2M_Z^2}-\frac{1}{5}\right)\right], \tag{34}$$

where $\Gamma^0(b\bar{b})$ is the standard model expression without the corrections. This special dependence is useful for separating m_t from the Higgs mass, and (especially) from such new physics as higher-dimensional Higgs representations.

For describing the lineshape, it is convenient to use the total width Γ_Z; the hadronic peak cross section $\sigma_{\mathbf{had}} = \frac{12\pi}{M_Z^2}\frac{\Gamma(e\bar{e})\Gamma(\mathbf{had})}{\Gamma_Z^2}(\mathrm{nb})$ (after removing QED effects); and the ratio $R \equiv \Gamma(\mathrm{had})/\Gamma(\ell\bar{\ell})$ (where $\Gamma(\ell\bar{\ell})$ is the average of the e, μ, and τ widths after verifying lepton-family universality and removing small m_τ effects). These are weakly correlated, although in practice one generally includes the full M_Z, Γ_Z, $\sigma_{\mathbf{had}}$, R, $A_{FB}^{0\ell}$ correlation matrix in fits[9]. Γ_Z is sensitive to both m_t and α_s. Both R and $\sigma_{\mathbf{had}}$ are insensitive to m_t because the $\hat{\rho}$ factor cancels, while both (especially R) are sensitive to α_s.

The standard model predictions for Γ_Z, $\sigma_{\mathbf{had}}$, and R as a function of m_t are compared with the experimental results in Figures 4. ($\hat{s}_Z^2$ in $\bar{g}_{Vf}$ is obtained from M_Z). One sees that the agreement is excellent for m_t in the 100 – 200 GeV range. The results of fits to the Z widths are listed in Table 5. The prediction for R_b is shown in Figure 1. As already discussed, it is higher that the standard model prediction for all allowed m_t, but favors smaller m_t.

The invisible width,

$$\begin{aligned}\Gamma(\mathrm{inv}) &= \Gamma_Z - \Gamma(\mathrm{had}) - \sum_i \Gamma(\ell_i\bar{\ell}_i) \\ &\equiv N_\nu \Gamma(\nu\bar{\nu})\end{aligned} \tag{35}$$

in Figure 5 is clearly in agreement with $N_\nu = 3$ but not $N_\nu = 4$. In fact, the result [3] $N_\nu = 2.988 \pm 0.023$ not only eliminates extra fermion families with $m_\nu \ll M_Z/2$, but also supersymmetric models with light sneutrinos ($\Delta N_\nu = 0.5$) and models with triplet ($\Delta N_\nu = 2$) or doublet ($\Delta N_\nu = 0.5$) Majorons [40]. N_ν does not include sterile (SU_2-singlet) neutrinos. However, the complementary bound $N'_\nu < 3.3$ (95% CL) from nucleosynthesis [41] *does* include sterile neutrinos for a wide range of masses and mixings [40], provided their mass is less than ~ 30 MeV.

In addition there are various asymmetries observed at LEP and SLD. The forward-backward asymmetry for $e^+e^- \to Z \to f\bar{f}$ is given, after removing photonic

---

[8] There is also an indirect m_t dependence in $\bar{s}_f^2$ if one regards M_Z as fixed.

[9] M_W is also correlated to M_Z due to the theoretical uncertainty in $\alpha(M_Z)$.

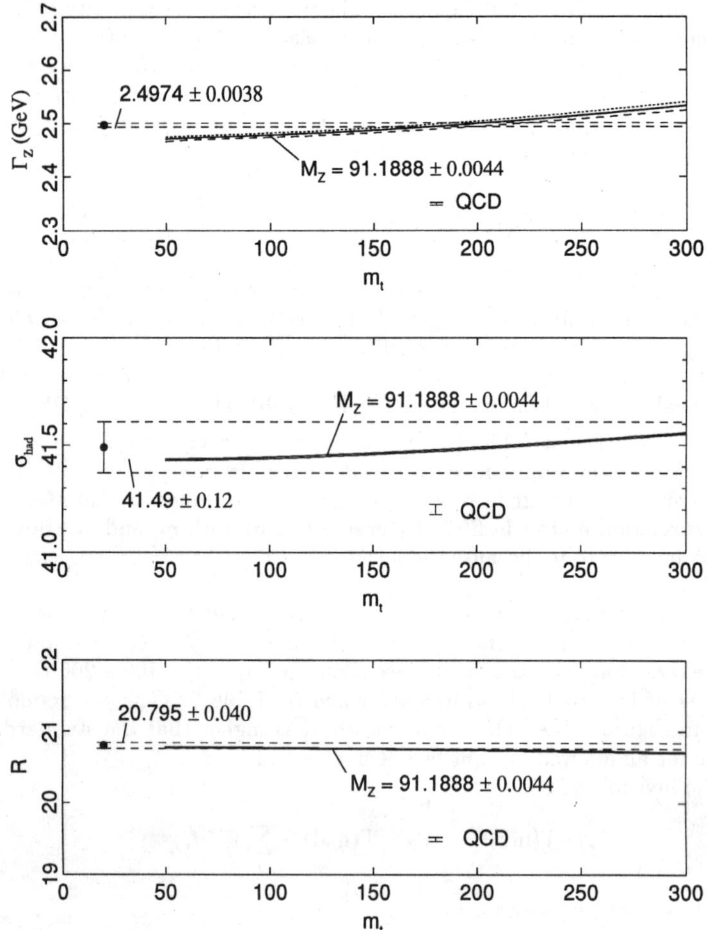

Figure 4: The standard model predictions for Γ_Z, σ_{had}, and R as a function of m_t, compared with the experimental results. For Γ_Z the dotted, solid, and dashed lines are for $M_H = 60$, 300, and 1000 GeV, respectively. The M_H dependence is too small to see for the σ_{had} and R graphs. The QCD uncertainties are indicated.

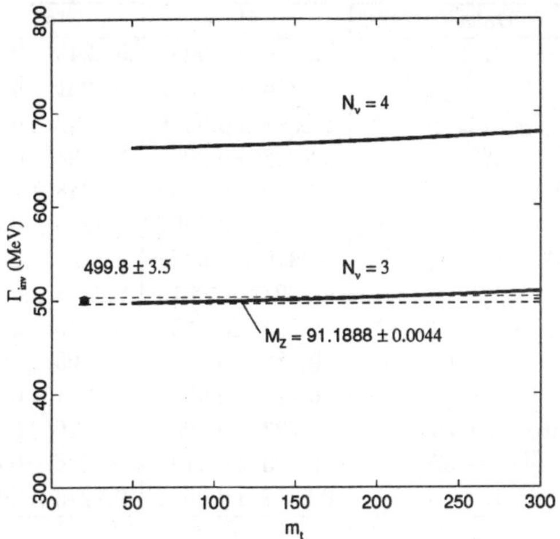

Figure 5: Theoretical prediction for $\Gamma(\text{inv})$ in the standard model with $N_\nu = 3$ and 4, compared with the experimental value 499.8 ± 3.5 MeV.

effects and boxes, by

$$A_{FB}^{0f} \simeq \frac{3}{4} A_e^0 A_f^0, \tag{36}$$

where A_f^0 is defined in (1). $A_{FB}^{0\ell}$ is the average of the e, μ, and τ asymmetries after verifying lepton-family universality. A_{FB}^{0b}, which is determined after correcting for $B\bar{B}$ oscillations, is mainly sensitive to $\hat{s}_Z^2$ (or small new physics effects) in A_e^0. The LEP experiments also extract the weak angle from the jet charge asymmetry [3]. Other asymmetries include the polarization of produced τ's. The polarization is given as a function of the scattering angle $z \equiv \cos\theta$ by

$$P_\tau^0 = -\frac{A_\tau^0 + A_e^0 \frac{2z}{1+z^2}}{1 + A_\tau^0 A_e^0 \frac{2z}{1+z^2}}. \tag{37}$$

From the angular distribution one can obtain A_τ^0 and A_e^0 with little correlation, with A_τ^0 coming mainly from the average polarization and A_e^0 mainly from its forward-backward asymmetry. The SLD collaboration has polarized electrons; from the left-right asymmetry as the polarization is reversed one can also determine A_e^0, namely $A_{LR}^0 = A_e^0$.

All of these asymmetries are independent of m_t when expressed in terms of the effective angles $\bar{s}_f^2$ and almost independent of m_t when expressed in terms of the $\overline{MS}$ angle $\hat{s}_Z^2$. One can therefore determine $\bar{s}_\ell^2$ or $\hat{s}_Z^2$ from the data without

Data	$\hat{s}^2_Z$	s^2_W
M_Z	0.2319 ±0.0005	0.2245 ±0.0013
$M_W, \frac{M_W}{M_Z}$	0.2326 ±0.0011	0.2251 ±0.0014
$\Gamma_Z, \sigma_{\rm had}, R$	0.2317 ±0.0018	0.2242 ±0.0017
$A^{0\ell}_{FB}$	0.2308 ±0.0009	0.2234 ±0.0013
$A^0_\tau(P_\tau), A^0_e(P_\tau)$	0.2322 ±0.0009	0.2248 ±0.0013
A^{0b}_{FB}	0.2324 ±0.0007	0.2249 ±0.0012
all LEP asymmetries	0.2319 ±0.0004	0.2242 ±0.0011
A^0_{LR}	0.2291 ±0.0010	0.2218 ±0.0013
$\nu_\mu(\bar{\nu}_\mu)N \to \nu_\mu(\bar{\nu}_\mu)X$	0.234 ±0.005	0.226 ±0.005
$\nu_\mu(\bar{\nu}_\mu)p \to \nu_\mu(\bar{\nu}_\mu)p$	0.212 ±0.032	0.205 ±0.030
$\nu_\mu(\bar{\nu}_\mu)e \to \nu_\mu(\bar{\nu}_\mu)e$	0.228 ±0.008	0.221 ±0.007
atomic parity	0.223 ±0.008	0.216 ±0.008
$e^{\uparrow\downarrow}D \to eX$	0.223 ±0.018	0.216 ±0.017
All	0.2317 ±0.0004	0.2243 ±0.0012

Table 5: Values of $\hat{s}^2_Z$ ($\overline{\rm MS}$) and s^2_W (on-shell) obtained from various inputs, assuming the global best fit values $m_t = 175 \pm 11$ GeV (for $M_H = 300$ GeV) and $\alpha_s = 0.127 \pm 0.005$, correlated with 60 GeV $< M_H <$ 1000 GeV. For Γ_Z and $\sigma_{\rm had}$ the experimental M_Z is used, so that $\sin^2\theta_W$ is determined from the electroweak vertices. (The uncertainty in $\hat{s}^2_Z$ would be reduced to 0.0004 if one used the theoretical formulas, but the additional sensitivity is mainly due to the M^3_Z factor in Γ_Z.) For deep inelastic scattering ($\nu_\mu(\bar{\nu}_\mu)N \to \nu_\mu(\bar{\nu}_\mu)X$) from (approximately) isoscalar targets (Perrier, this volume) the uncertainty includes 0.003 (experiment) and 0.005 (theory). $\nu_\mu(\bar{\nu}_\mu)p$ and $\nu_\mu(\bar{\nu}_\mu)e$ refer respectively to elastic scattering from nucleons (Mann, this volume) and electrons (Panman, this volume). $e^{\uparrow\downarrow}D$ refers to the SLAC polarized eD asymmetry (Souder, this volume). For atomic parity violation (Masterson and Wieman, this volume), the experimental and theoretical (Blundell, Johnson, and Sapirstein, this volume) components of the error are 0.007 and 0.004 respectively.

theoretical uncertainties from m_t. On the other hand, in the on-shell or Z-mass schemes the formulas involve quadratic m_t dependence.

The predictions for $A^{0\ell}_{FB}$, A^0_ℓ, and A^{0b}_{FB} are compared with the experimental data in Figure 6. Again, the agreement is excellent.

The values obtained for $\sin^2\theta_W$ in the $\overline{\rm MS}$ and on-shell schemes from various Z-pole observables, M_W, and low energy neutral current processes are listed in Table 5 and displayed in Figure 7. The low energy values are not as precise as those from the Z-pole and M_W. However, they are still important as they probe different couplings and kinematic ranges, and are sensitive to certain types of new physics to which the W and Z are blind.

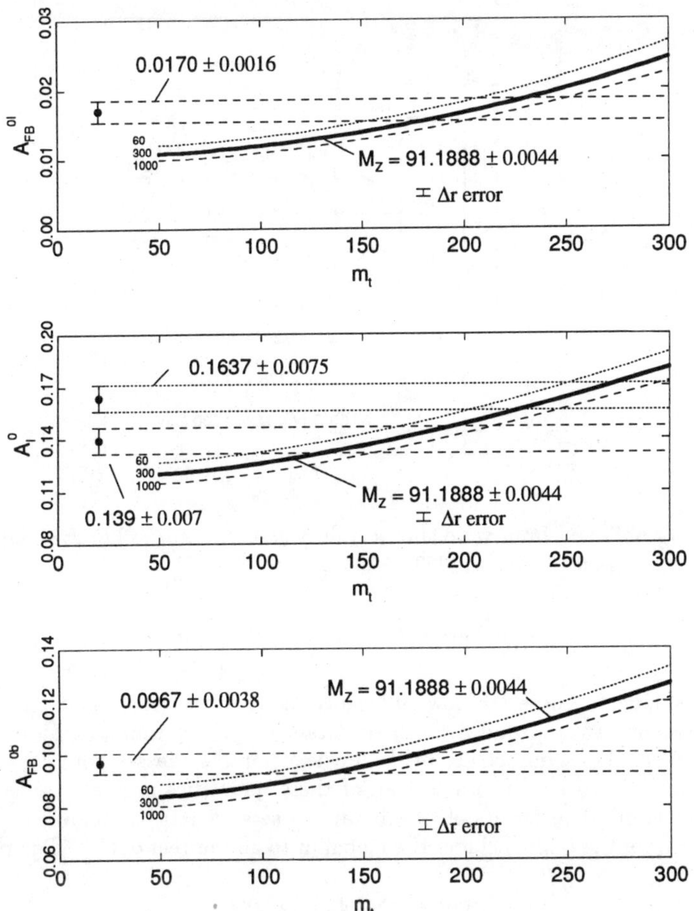

Figure 6: Theoretical prediction for $A_{FB}^{0\ell}$, A_ℓ^0, and A_{FB}^{0b} in the standard model as a function of m_t for $M_H = 60$ (dotted line), 300 (solid), and 1000 (dashed) GeV, compared with the experimental values. The theoretical uncertainties from $\Delta\Delta\hat{r}_W = \pm 0.0009$ are also indicated. For A_ℓ^0 the value 0.139 ± 0.007 is the average of the LEP values from $A_e^0(P_\tau)$ and $A_\tau^0(P_\tau)$, while 0.1637 ± 0.0075 is the SLD value from A_{LR}^0.

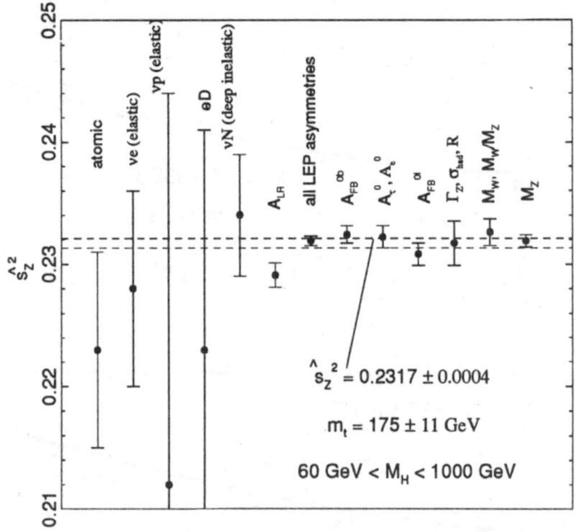

Figure 7: $\hat{s}_Z^2$ obtained from various observables assuming $m_t = 175 \pm 11$ GeV, $\alpha_s = 0.127 \pm 0.005$, and $60 < M_H < 1000$ GeV.

2.3 *The Standard Model Parameters:* m_t, α_s, $\sin^2\theta_W$

(m_t, and $\sin^2\theta_W$): There are now sufficiently many observables that one can precisely determine $\hat{s}_Z^2$, m_t, and $\alpha_s(M_Z)$ simultaneously. For example, $\hat{s}_Z^2$ can be determined from the asymmetries, m_t from the W and Z masses, and $\alpha_s(M_Z)$ from the hadronic Z-widths. In practice all of these quantities are determined from a simultaneous fit. The results of fits to various sets of data are shown in Table 6. The first row of the table includes the global fit to all indirect data[10]. The predicted value,

$$m_t = 175 \pm 11^{+17}_{-19} \text{ GeV}, \tag{38}$$

is in remarkable agreement with the value 174 ± 16 GeV suggested by the CDF candidate events [10]. The second row includes the direct (CDF) value for m_t as a separate constraint. Since the indirect data is consistent with the CDF value, adding it has little effect on the standard model fits. (It is very important for the beyond the standard model fits, however.) The third row includes the value $\alpha_s^{\rm other} = 0.116 \pm 0.005$ of $\alpha_s(M_Z)$ obtained from jet and low energy data [42] as a separate constraint. The resulting $\alpha_s = 0.122(3)(1)$ may be viewed as a weighted average of these other determinations with the lineshape value of 0.127(5)(2). The

[10]The correlation coefficients are $\rho_{\hat{s}_Z^2 m_t} = -0.67$, $\rho_{\hat{s}_Z^2 \alpha_s} = 0.30$, and $\rho_{\alpha_s m_t} = -0.20$. The overall χ^2 of the fit is 181 for 206 d.f., which is low (mainly due to the older neutral current data) but acceptable: the probability of $\chi^2 \leq 181$ is 10%. The correlations for the other data sets are similar.

Set	$\hat{s}^2_Z$	$\alpha_s(M_Z)$	m_t (GeV)	$\Delta\chi^2_H$
All indirect	0.2317(3)(2)	0.127(5)(2)	$175 \pm 11^{+17}_{-19}$	4.4
Indirect + CDF (174 ± 16)	0.2317(3)(3)	0.127(5)(2)	$175 \pm 9^{+12}_{-13}$	4.4
Indirect $+\alpha_s^{\text{other}}(0.116 \pm 0.005)$	0.2316(3)(2)	0.122(3)(1)	$178^{+10}_{-11}{}^{+17}_{-19}$	6.0
LEP + low energy	0.2320(3)(2)	0.128(5)(2)	$168^{+11}_{-12}{}^{+17}_{-19}$	2.7
All indirect ($S = 2.2$)	0.2319(3)(2)	0.128(5)(2)	$170^{+11}_{-12}{}^{+17}_{-19}$	3.3
Z-pole	0.2316(3)(1)	0.126(5)(2)	$179^{+11}_{-12}{}^{+17}_{-19}$	4.2
LEP	0.2320(4)(2)	0.128(5)(2)	$170^{+12}_{-13}{}^{+18}_{-20}$	2.6
SLD + M_Z	0.2291(10)(0)	—	$251^{+24}_{-26}{}^{+21}_{-23}$	

Table 6: Results for the electroweak parameters in the standard model from various sets of data. The central values assume $M_H = 300$ GeV, while the second errors are for $M_H \to 1000$ (+) and 60 (−). The last column is the increase in the overall χ^2 of the fit as M_H increases from 60 to 1000. From [20].

other fits show the sensitivity to the various data sets. The fourth row includes the LEP results and the low energy data but not SLD. Comparing with the first row one sees that the predicted m_t is pulled up significantly (by ~ 7 GeV) by the SLD result. The next row combines the LEP and SLD measurements of A_e, increasing the error in the weighted average by the scale factor[11] $S = 2.2$. The last rows are the result of the Z-pole, LEP, and SLD observables by themselves.

The central value in (38) is for $M_H = 300$ GeV, while the second uncertainty is from M_H varying from 60 to 1000 GeV. This reflects the strong correlation between the m_t^2 and $\ln(M_H)$ terms in the $\hat{\rho}$ parameter (but not in the $Zb\bar{b}$ vertex corrections). For other values of M_H one finds the approximate prediction

$$m_t \sim 175 \pm 11 + 13 \ln\left(\frac{M_H}{300\text{GeV}}\right), \tag{39}$$

where the coefficient 13 is obtained from the fits (see Figure 8) and should be viewed as an approximate interpolation of more complicated formulae. In particular, supersymmetric extensions of the standard model, which involve a light standard model-like Higgs, favor the lower part of the range in (38).

The χ^2 distributions for the fit of the indirect data as a function of m_t are shown for various values of M_H in Figure 9. One again observes the strong correlation between m_t and M_H. From the indirect data one can obtain upper and lower limits on m_t. The weakest upper limit is for $M_H = 1000$ GeV, from which one finds $m_t < 205$ (209) GeV at 90 (95)% CL. The corresponding limits for other M_H are 170 (174) GeV for $M_H = 60$ GeV and 188 (192) GeV for $M_H = 300$. Similarly, the indirect data alone set significant lower limits on m_t, which would continue to hold even in the presence of nonstandard t decays which could invalidate the direct

[11] S is the square root of the χ^2/df. This is the procedure recommended by the Particle Data Group [43] when there is a discrepancy between experiments.

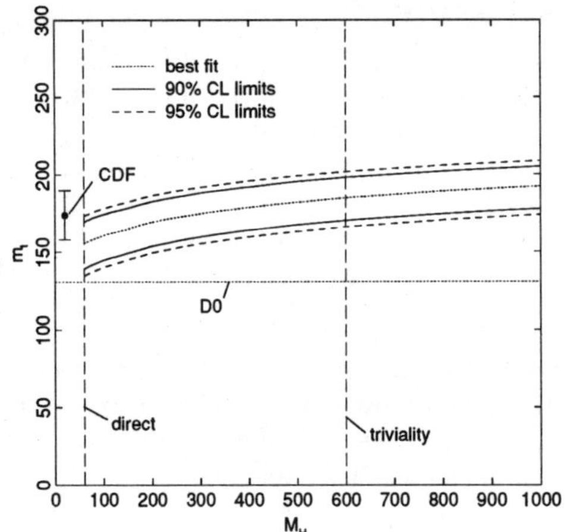

Figure 8: Best fit value for m_t and upper and lower limits as a function of M_H. The direct lower limit $M_H > 60$ GeV [44] and the approximate triviality limit [45] $M_H <$ 600 GeV are also indicated. The latter becomes $M_H < 200$ GeV if one requires that the standard model holds up to the Planck scale. The CDF value $m_t = 174 \pm 16$ GeV and the D0 bound $m_t > 131$ GeV are also indicated.

collider searches. The weakest limit is for $M_H = 60$ GeV, for which one obtains $m_t > 140$ (135) GeV at 90 (95)% CL.

From the indirect data one obtains

$$\begin{aligned} \hat{s}_Z^2 &= 0.2317 \pm 0.0003 \pm 0.0002 \\ s_W^2 &= 0.2243 \pm 0.0012 \\ \bar{s}_\ell^2 &= 0.2320 \pm 0.0003 \pm 0.0002 \end{aligned} \tag{40}$$

for the the $\overline{\text{MS}}$, on-shell, and effective weak angles, respectively. The first uncertainties are mainly from m_t and $\alpha(M_Z)$, while the second is from M_H in the range 60 - 1000 GeV. $\hat{s}_Z^2$ and $\bar{s}_\ell^2$ are much less sensitive to m_t and M_H than s_W^2. With the exception of A_{LR}^0 the values obtained from individual observables are in excellent agreement with (40). In particular, the $\hat{s}_Z^2$ values obtained assuming $m_t = 175 \pm 11$ GeV, $\alpha_s = 0.127 \pm 0.005$, and 60 GeV $< M_H <$ 1000 GeV are shown in Table 5 and in Figure 7. The agreement is remarkable.

One can also extract the radiative correction parameter $\Delta\hat{r}_W$ defined in (7). Fitting to all indirect data and keeping the full m_t dependence in $\hat{\rho}$ but leaving $\Delta\hat{r}_W$ free, one finds

$$\Delta\hat{r}_W = 0.067 \pm 0.002, \tag{41}$$

compared with the expectation 0.0706(2)(9) in (13). Including the CDF m_t value, $\Delta\hat{r}_W = 0.068 \pm 0.002$. For the analogous parameter Δr in the on-shell scheme, eqn.

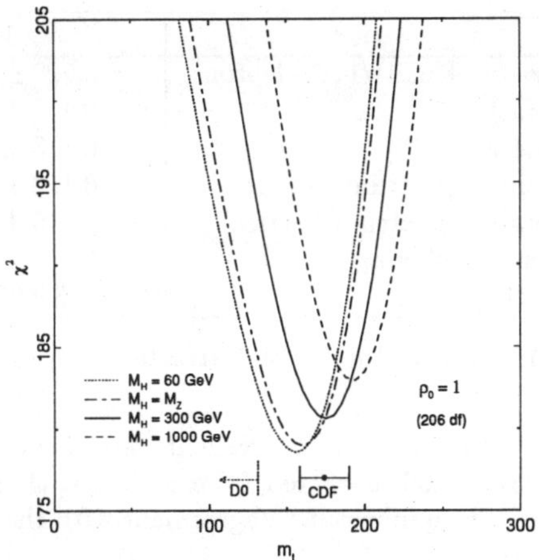

Figure 9: χ^2 distribution for all indirect data (206 df) in the standard model as a function of m_t, for $M_H = 60$, M_Z, 300, and 1000 GeV. The direct constraints on m_t from CDF and D0 are displayed but are not included in the χ^2.

(21), one obtains

$$\Delta r = 0.044 \pm 0.005 \tag{42}$$

from the indirect data, or 0.041 ± 0.003 including CDF, compared to the expectation 0.040(4)(1) in (23).

(α_s): Using the results of the 1993 LEP energy scan one can now extract the strong coupling constant α_s at the Z-pole with a small experimental and theoretical error,

$$\alpha_s(M_Z) = 0.127 \pm 0.005 \pm 0.002 \quad \text{(lineshape)}, \tag{43}$$

where the second uncertainty is from M_H. α_s is only weakly correlated with the other parameters. It is determined mainly from the ratio $R \equiv \Gamma(\text{had})/\Gamma(\ell\bar{\ell})$, which is insensitive to m_t (except in the $b\bar{b}$ vertex). There is also sensitivity from Γ_Z, σ_{had}, and in the mixed QCD-electroweak corrections which relate M_Z, $\hat{s}^2_Z$, and M_W. This determination is very clean theoretically, at least within the standard model. It is the Z-pole version of the long held view that the ratio of hadronic to leptonic rates in e^+e^- would be a "gold-plated" extraction of α_s and test of QCD. Using a recent estimate [46] of the $(\alpha_s/\pi)^4$ corrections to C_f in (33), *i.e.* $-90(\alpha_s/\pi)^4$, one can estimate that higher-order terms lead to an additional uncertainty $\sim \pm 0.001$ in the $\alpha_s(M_Z)$ value in (43). It should be cautioned, however, that the lineshape value is sensitive to the presence of new physics which affects the hadronic Z decays. In particular, it will be discussed in Section 4.6 that if one allows for new physics in the $Zb\bar{b}$ vertex to account for R_b, the extracted value of α_s decreases to 0.111(9)(1).

Source	$\alpha_s(M_Z)$
$R_\tau = \Gamma(\tau\to\nu_\tau + \text{had})/\Gamma(\tau\to\text{leptons})$	0.122 ± 0.005
Deep inelastic	0.112 ± 0.005
Υ, J/Ψ decays	0.113 ± 0.006
Charmonium spectrum (lattice)	0.110 ± 0.006
Bottomonium spectrum (lattice)	0.115 ± 0.002
LEP, event topologies	0.123 ± 0.006
LEP, lineshape	$0.127 \pm 0.005 \pm 0.002$

Table 7: Values of α_s at the Z-pole extracted from various methods.

The lineshape value of α_s is an excellent agreement with the independent value $\alpha_s(M_Z) = 0.123 \pm 0.006$ extracted from jet event shapes at LEP using resummed QCD [42]. It is also in excellent agreement with the prediction [47]

$$\alpha_s(M_Z) \sim 0.129 \pm 0.008, \qquad \text{SUSY} - \text{GUT} \tag{44}$$

of supersymmetric grand unification. As can be seen in Table 7, however, it is somewhat larger than some of the low energy determinations of α_s (which are then extrapolated theoretically to the Z-pole), in particular those from deep inelastic scattering, the lattice calculation of the charmonium spectrum[12], and a recent lattice calculation of the bottomonium spectrum which claims a very small uncertainty [50]. This has led some authors to speculate that there might be a light gluino which would modify the running of α_s [51]. It should be noted, however, that there is an independent low energy LEP determination from the ratio R_τ of hadronic to leptonic τ decays, which gives a larger value.

The third row of Table 6 includes the value $\alpha_s^{\text{other}} = 0.116 \pm 0.005$ obtained from low energy and jet data [42]. The resulting value $\alpha_s = 0.122(3)(1)$ can be regarded as a weighted fit to all data, including the Z lineshape. However, given the discrepancies between the individual determinations, caution is advised in accepting either this value or the small error. For this reason, when discussing grand unification in Section 4.4, α_s will generally be taken as a prediction rather than an input, or else the more conservative range 0.12 ± 0.01 will be used.

The value of $\alpha_s(M_Z)$ from the precision experiments is anticorrelated with m_t, as can be seen in Figure 10. In particular, larger m_t corresponds to slightly smaller $\alpha_s(M_Z)$, in better agreement with the low energy data. As mentioned above, one would also obtain a smaller value if there is new physics which enhances the $Zb\bar{b}$ vertex.

[12]The value 0.110 ± 0.006 [48] has increased somewhat from the published value of 0.105 ± 0.004 [49], reducing the discrepancy.

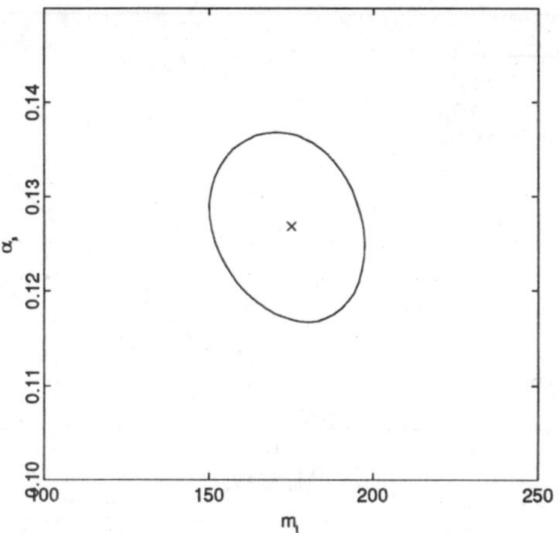

Figure 10: 90% CL allowed region in $\alpha_s(M_Z)$ and m_t from a combined fit to precision Z-pole and other data (but not including event topology and low energy determinations of $\alpha_s(M_Z)$).

2.4 *The Higgs Mass*

One can attempt to use the precision data to constrain the Higgs boson mass. This enters $\hat{\rho}$ logarithmically and is strongly correlated with the quadratic m_t dependence in everything but the $Z \to b\bar{b}$ vertex correction. The χ^2 distributions as a function of the Higgs mass are shown with and without the additional CDF constraint $m_t = 174 \pm 16$ GeV in Figure 11. In both cases, the minimum occurs at or near the lower limit, 60 GeV, allowed by direct searches at LEP. (The increase in χ^2 for $M_H = 1000$ GeV is shown in Table 6.) A low value for M_H is consistent with the minimal supersymmetric extension of the standard model, which generally predicts a relatively light standard model-like Higgs scalar. However, the constraint is weak statistically. From the χ^2 distribution one obtains the weak upper limits

$$\text{indirect}: \quad M_H < 570 \ (880) \text{ GeV} \tag{45}$$

at 90 (95)% CL from the indirect precision data, and

$$\text{indirect} + \text{CDF} \ : \ M_H < 510 \ (730) \text{ GeV} \tag{46}$$

including the CDF direct constraint from m_t. (These results include the direct LEP limit $M_H > 60$ GeV [45].) Clearly, no definitive conclusion can be drawn. Furthermore, the sensitivity to M_H is driven almost entirely by the experimental values of R_b and A^0_{LR}, both of which are well above the standard model expectations. Omitting these leads to an almost flat χ^2 distribution, as can be seen in Figure 11.

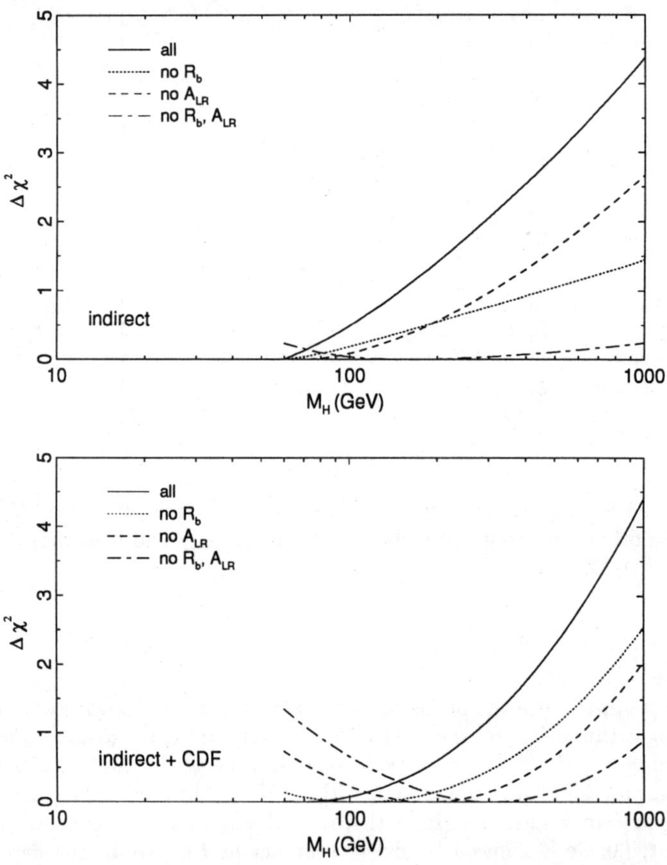

Figure 11: Increase in χ^2 from the best fits as a function of M_H, with and without the CDF constraint $m_t = 174 \pm 16$ GeV, and distributions omitting R_b and/or A_{LR}^0.

If these are due to large statistical fluctuations or new physics the constraint on M_H essentially disappears.

The weak M_H dependence does not imply that the data is insensitive to the spontaneous symmetry breaking mechanisms. Alternative schemes generally yield large effects on the precision observables, as will be described below.

2.5 *Have Electroweak Corrections Been Seen?*

The data can also be interpreted in terms of whether one has actually observed the electroweak (as opposed to the simple running α) corrections. Novikov *et al.* [30] have noted that there is a large cancellation between the fermionic and bosonic contributions to the W and Z self-energies, and that until recently the data could actually be fit by a properly interpreted Born theory. However, the data is now sufficiently good that even given the cancellations these electroweak loops are needed at the 2σ level. Gambino and Sirlin [52] and Dittmaier *et al.* [53] have interpreted the data in somewhat different way. They have argued that the fermionic loops, both in the running of α and the t, b loops, are unambiguous theoretically, and certainly should be there if the theory is to make any sense. However, the bosonic loops, which involve triple-gauge vertices, gauge-Higgs vertices, etc., have never been independently tested in other processes. They have shown that the data are inconsistent if one simply ignores bosonic loops (which are a gauge-invariant subset of diagrams), thus providing convincing though indirect evidence for their existence.

3 Model Independent Analyses

Long before the LEP era, the standard model was strongly favored compared to competing gauge theories by model independent analyses of neutrino scattering, atomic parity violation, and e^+e^- below the Z pole [54]. In a model independent analysis one writes an effective Lagrangian involving all of the four-fermi operators that can be obtained at tree-level in a gauge theory[13], and then tries to determine their coefficients directly from the data. Each electroweak gauge theory makes a prediction for the coefficients. One can therefore see whether the coefficients are uniquely determined and whether they are in agreement with the standard model predictions.

The model independent analyses were important historically in establishing the uniqueness of the standard model. They are still important today because the four-fermi parameters that they determine are not quite the same as those measured in the (more accurate) Z-pole experiments, at least in the presence of new physics. The former are sensitive to all standard model and new physics contributions to the

[13] In practice one usually assumes family universality.

Quantity	Standard Model Expression
$\epsilon_L(u)$	$\rho_{\nu N}^{NC}\left(\frac{1}{2}-\frac{2}{3}\kappa_{\nu N}\sin^2\theta_W+\lambda_{uL}\right)$
$\epsilon_L(d)$	$\rho_{\nu N}^{NC}\left(-\frac{1}{2}+\frac{1}{3}\kappa_{\nu N}\sin^2\theta_W+\lambda_{dL}\right)$
$\epsilon_R(u)$	$\rho_{\nu N}^{NC}\left(-\frac{2}{3}\kappa_{\nu N}\sin^2\theta_W+\lambda_{uR}\right)$
$\epsilon_R(d)$	$\rho_{\nu N}^{NC}\left(\frac{1}{3}\kappa_{\nu N}\sin^2\theta_W+\lambda_{dR}\right)$
$g_V^{\nu e}$	$\rho_{\nu e}\left(-\frac{1}{2}+2\kappa_{\nu e}\sin^2\theta_W\right)$
$g_A^{\nu e}$	$\rho_{\nu e}\left(-\frac{1}{2}\right)$
C_{1u}	$\rho'_{eq}\left(-\frac{1}{2}+\frac{4}{3}\kappa'_{eq}\sin^2\theta_W\right)$
C_{1d}	$\rho'_{eq}\left(\frac{1}{2}-\frac{2}{3}\kappa'_{eq}\sin^2\theta_W\right)$
C_{2u}	$\rho_{eq}\left(-\frac{1}{2}+2\kappa_{eq}\sin^2\theta_W\right)+\lambda_{2u}$
C_{2d}	$\rho_{eq}\left(\frac{1}{2}-2\kappa_{eq}\sin^2\theta_W\right)+\lambda_{2d}$

Table 8: Standard model expressions for the neutral-current parameters for ν-hadron, νe, and e-hadron processes. If radiative corrections are ignored, $\rho=\kappa=1$, $\lambda=0$. At $O(\alpha)$ in the on-shell scheme, $\rho_{\nu N}^{NC}=1.0089$, $\kappa_{\nu N}=1.0356$, $\lambda_{uL}=-0.0032$, $\lambda_{dL}=-0.0026$, and $\lambda_{uR}=1/2\lambda_{dR}=3.6\times10^{-5}$ for $m_t=175$ GeV, $M_H=300$ GeV, $M_Z=91.1888$ GeV, and $\langle Q^2\rangle=20$ GeV^2. For νe scattering, $\rho_{\nu e}=1.0137$ and $\kappa_{\nu e}=1.0358$ (at $\langle Q^2\rangle=0$). For atomic parity violation, $\rho'_{eq}=0.9880$ and $\kappa'_{eq}=1.034$. For the SLAC polarized electron experiment, $\rho'_{eq}=0.979$, $\kappa'_{eq}=1.032$, $\rho_{eq}=1.001$, and $\kappa_{eq}=1.06$ after incorporating additional QED corrections, while $\lambda_{2u}=-0.013$, $\lambda_{2d}=0.003$. The dominant m_t dependence is given by $\rho\sim1+\Delta\rho_t$, while $\kappa\sim 1+\Delta\rho_t/\tan^2\theta_W$ (on-shell) or $\kappa\sim1$ ($\overline{\text{MS}}$).

process. The latter are the actual couplings of the Z to the corresponding fermions, and are sensitive only to the types of new physics which directly affect the Z.

Previous to LEP the most precise neutral current tests involved the neutrino-quark interactions, as described in the article by Perrier and in [54, 55]. There have been deep inelastic experiments on (approximately) isoscalar and p and n targets, $\nu N\to\nu X$, $\nu p\to\nu X$, $\nu n\to\nu X$, elastic scattering such as $\nu p\to\nu p$, and various inelastic reactions, such as coherent $\nu N\to\nu\pi^0 N$. In particular, the deep inelastic cross sections for neutral current scattering divided by the corresponding charged current cross sections for targets such as iron and carbon, for which most of the strong interaction uncertainties[14] [59] and those involving the neutrino flux cancel in the ratio, have been measured at the 1% level. The less precise measurements on proton and neutron targets constrain the isospin structure of the current. Assuming family universality and left-handed neutrinos, the most general effective four-fermi interaction for νq scattering that can be generated from gauge interactions, *i.e.*,

[14]The largest residual uncertainty is in the charm quark threshold in the charged current denominator [16, 54, 58].

Quantity	Experimental Value	SM	Correlation
$\epsilon_L(u)$	0.332 ± 0.016	0.345 ± 0.001	
$\epsilon_L(d)$	-0.438 ± 0.012	-0.429 ± 0.001	non -
$\epsilon_R(u)$	-0.178 ± 0.013	-0.156	Gaussian
$\epsilon_R(d)$	$-0.026^{+0.075}_{-0.048}$	0.078	
g_L^2	0.3017 ± 0.0033	0.303 ± 0.001	
g_R^2	0.0326 ± 0.0033	0.030	small
θ_L	2.50 ± 0.035	2.46	
θ_R	$4.58^{+0.46}_{-0.28}$	5.18	
$g_A^{\nu e}$	-0.507 ± 0.014	-0.506 ± 0.001	-0.04
$g_V^{\nu e}$	-0.041 ± 0.015	-0.037 ± 0.001	
C_{1u}	-0.214 ± 0.046	-0.189 ± 0.001	-0.995 -0.79
C_{1d}	0.359 ± 0.041	0.341 ± 0.001	0.79
$C_{2u} - \frac{1}{2}C_{2d}$	-0.04 ± 0.13	-0.051 ± 0.002	

Table 9: Values of the model-independent neutral current parameters, compared with the standard model predictions (SM) using $M_Z = 91.1888 \pm 0.0044$ GeV and $m_t = 175 \pm 11$ GeV for $M_H = 300$ GeV. $g_{L,R}^2$ are defined by $g_{L,R}^2 = \epsilon_{L,R}(u)^2 + \epsilon_{L,R}(d)^2$, while θ_i, $i = L$ or R, is defined as $\tan^{-1}[\epsilon_i(u)/\epsilon_i(d)]$.

allowing only vector and axial interactions, is[15]

$$-L^{\nu N} = \frac{G_F}{\sqrt{2}}\bar{\nu}\gamma^\mu(1-\gamma^5)\nu \sum_{i=u,d}\left[\epsilon_L(i)\bar{q}_i\gamma_\mu(1-\gamma^5)q_i + \epsilon_R(i)\bar{q}_i\gamma_\mu(1+\gamma^5)q_i\right], \quad (47)$$

where the parameters $\epsilon_L(i)$ and $\epsilon_R(i)$ refer to the interactions of left- and right-handed quark i with neutrinos. The standard model expressions for the ϵ's are listed, including the radiative corrections, in Table 8. (Specific values of the radiative corrections are given in the on-shell scheme, which was used in most of the analyses [54]. They can be translated to the $\overline{\text{MS}}$ scheme using (26).) Other gauge theories would give other values. The data is sufficient to determine the four ϵ's uniquely[16]. The results are shown in Figure 12. The left-hand couplings give the most precise low energy determination of $\sin^2\theta_W$. The values of the parameters are listed along with the standard model predictions in Table 9. Also given are the values and predictions for the quantities g_L^2, g_R^2, θ_L, θ_R, which are the squares of the radii and the angles in the ϵ_L and ϵ_R planes. These quantities are much more weakly correlated than the ϵ's themselves.

There have been a number of neutrino-electron experiments [54, 15], as described in the article by Panman. The best measured are $\nu_\mu e \to \nu_\mu e$ and $\bar{\nu}_\mu e \to \bar{\nu}_\mu e$. The most general Lagrangian allowed by a gauge theory with family universality

[15] Alternate parametrizations are described in [60].

[16] The deep inelastic experiments usually presented their results in terms of $\sin^2\theta_W$ only and required considerable reanalysis for the model independent studies [54].

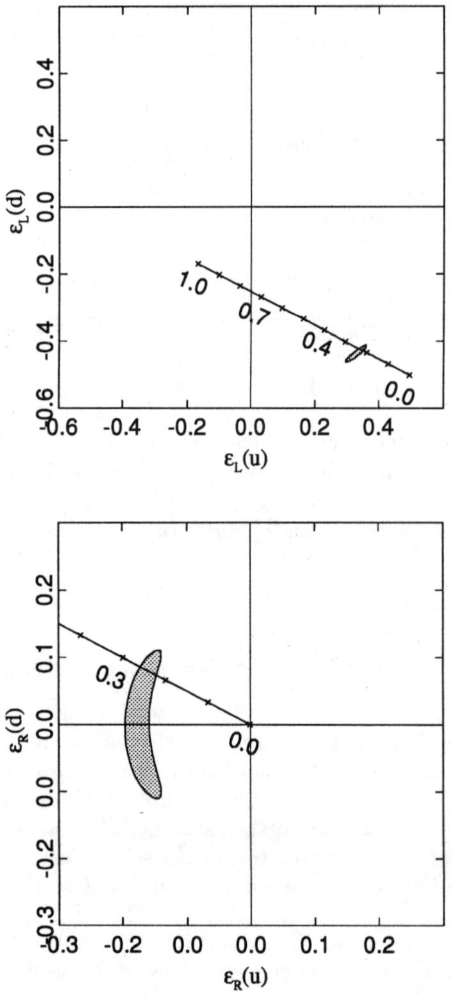

Figure 12: The allowed regions in $\epsilon_L(u)$, $\epsilon_L(d)$, $\epsilon_R(u)$ and $\epsilon_R(d)$, compared with the standard model predictions as a function of s_W^2. The agreement is excellent.

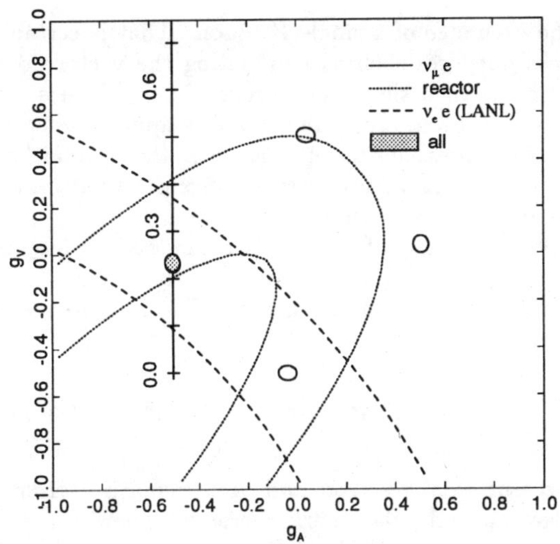

Figure 13: Allowed regions in the $g_V^{\nu e}$, $g_A^{\nu e}$ plane from neutrino-electron data, compared with the standard model prediction as a function of s_W^2.

and left-handed neutrinos is

$$- L^{\nu e} = \frac{G_F}{\sqrt{2}} \bar{\nu} \gamma^\mu (1 - \gamma^5) \nu J_\mu^e, \tag{48}$$

where

$$J_\mu^e = \bar{e} \gamma_\mu \left[g_V^{\nu e} - g_A^{\nu e} \gamma^5 \right] e, \tag{49}$$

and $g_V^{\nu e}$ and $g_A^{\nu e}$ refer to the vector an axial vector couplings of the electron in the four-fermi interaction. The ν_μ reactions determine $g_V^{\nu e}$ and $g_A^{\nu e}$ up to a four-fold ambiguity, as is shown in Figure 13. One of the solutions corresponds to the standard model and the others to the interchange of $g_V^{\nu e}$ with $g_A^{\nu e}$ and an overall sign change.

Some of the ambiguity can be eliminated by considering the additional reactions $\nu_e e \rightarrow \nu_e e$ and $\bar{\nu}_e e \rightarrow \bar{\nu}_e e$, for which both charged and neutral currents contribute and interfere. The $\nu_e e$ reaction has been measured at LANL [61]; it yields the additional constraint shown in Figure 13. Finally, there is the Savannah River reactor $\bar{\nu}_e e$ experiment [62], which yields a different allowed contour. From these one sees that the ν_e data determine the couplings up to a two-fold ambiguity, one of which (the axial-vector dominant) corresponds to the standard model. It is possible to eliminate the second (vector dominant) solution by simultaneously considering $e^+e^- \rightarrow \mu^+\mu^-$ data if one assumes that the amplitude factorizes into neutrino and charged-lepton factors. This is true if the neutral current amplitude is dominated by the exchange of a single Z boson. That is certainly a reasonable assumption for gross purposes such as eliminating the vector-dominant solution. However, it should be warned that many types of new physics, such as extra Z' bosons, break the factorization, and care should be applied when relating the two types of reactions. The expressions for $g_V^{\nu e}$ and $g_A^{\nu e}$ in the standard model are shown in Table 8, and the numerical values extracted from the data are compared with the standard model predictions in Table 9.

There have been a number of measurements of the interference between weak and electromagnetic amplitudes in the electron-quark system.

The parity-violating eq interaction generated by Z exchange can be expressed in an arbitrary gauge theory by

$$+\,L^{eH} = \frac{G_F}{\sqrt{2}} \sum_{i=u,d} \left[C_{1i}\bar{e}\gamma^\mu\gamma^5 e\bar{q}_i\gamma_\mu q_i + C_{2i}\bar{e}\gamma^\mu e\bar{q}_i\gamma_\mu\gamma^5 q_i\right], \tag{50}$$

where the C_{1i} represent axial electron and vector quark currents while the C_{2i} represent vector electron and axial quark currents. There are additional parity-conserving pieces, which are negligibly small compared to electromagnetism. There have been several experiments on the electron-quark coupling. Most notably, the polarized electron-deuteron asymmetries $e^{(\uparrow\downarrow)}D \to eX$ from SLAC [63], and, more recently, other experiments, such as an $e^{(\uparrow\downarrow)}Be \to eBe^*$ asymmetry experiment from Mainz [64], $e^{(\uparrow\downarrow)}C$ from Bates [65], and the BCDMS μC asymmetry experiment [66] done at CERN. These are described in detail in the article by Souder.

In addition, parity violation manifests itself in atomic physics by leading to parity-violating mixtures between S and P wave states. The subject has had a long and difficult history, but in recent years the experiments have advanced greatly; in particular, there are now high precision measurements of parity violation in the cesium atom, first in Paris [67] and more recently in Boulder [14], as described in the article by Masterson and Wieman. Cesium is a very clean atom to study theoretically [68]: it is a single electron outside of a tightly-bound core. Recent calculations of the matrix elements needed to interpret the experimental results are accurate at the 1% level[17], as described in detail by Blundell, Johnson, and Sapirstein. The experimental precision should equal this soon. The C_{1i} couplings are much better determined than the C_{2i} because the C_{1i} operators are coherent with respect to the nucleons in a heavy nucleus, while the C_{2i} operators couple to nucleon spin. Also, the C_{2i} operators involve theoretical ambiguities from the s quark content of the nucleon [69] and nuclear anapole moment effects [69]. The experimental constraints are compared with the standard model predictions in Figure 14 and Table 9. The agreement is excellent. It is apparent that the cesium results are especially useful for constraining types of new physics which shift the predicted parameters in a direction perpendicular to the narrow band in Figure 14 [70, 71].

[17] In the future it may be possible to eliminate most theoretical uncertainties by comparing the parity-violating effects in different isotopes of the same atom.

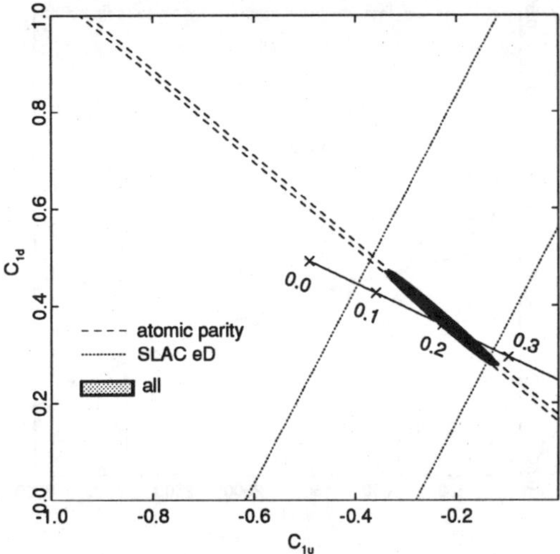

Figure 14: Constraints from eD scattering and atomic parity violation, compared with the standard model prediction as a function of s_W^2.

Weak-electromagnetic interference [72] can also be observed in e^+e^- annihilation into $\mu^+\mu^-$, $\tau^+\tau^-$, $b\bar{b}$, $c\bar{c}, \cdots$. Experiments were done at SLAC, DESY and TRISTAN well below the Z-pole, where the dominant contribution is photon exchange, with the Z a perturbation. The interference can lead to a forward-backward asymmetry, A_{FB}, in the direction of the μ^- with respect to the e^-. Well below the Z-pole the asymmetry is predicted in the standard model to be

$$A_{FB} = \frac{-3G_F}{16\sqrt{2}\pi\alpha} \frac{sM_Z^2}{M_Z^2 - s}, \tag{51}$$

for $\sqrt{s} \ll M_Z$, where $\sqrt{s}$ is the total center of mass energy. This depends only on the axial couplings of the electron and muon, and is therefore independent of $\sin^2\theta_W$ except for a (small) indirect dependence via M_Z. Therefore, it is essentially an absolute prediction of the model[18]. The magnitude of the asymmetry is predicted to increase approximately linearly with s. The data is compared with the prediction in Figure 15; the agreement is excellent. The annihilation into hadrons is discussed in detail in the article by Haidt.

The model independent analyses established that most of the four-fermi interactions are uniquely determined, consistent with the standard model and in dis-

[18]One can extract $\sin^2\theta_W$ by using the tree-level relation $G_F/\sqrt{2}\pi\alpha = (2\cos^2\theta_W \sin^2\theta_W M_Z^2)^{-1}$, and combining A_{FB} with the value of M_Z measured in other experiments. In fact, the asymmetry limits the deviation from the standard model, while M_Z yields $\sin^2\theta_W$.

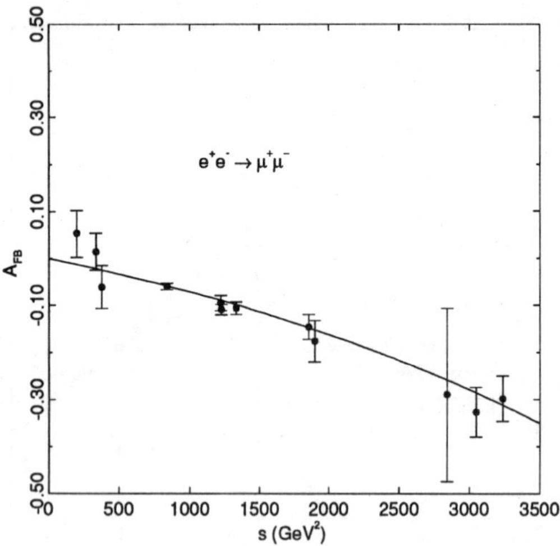

Figure 15: Forward-backward asymmetry for $e^+e^- \rightarrow \mu^+\mu^-$ as a function of the square of the center of mass energy, compared with the prediction of the standard model (which is almost independent of $\sin^2\theta_W$).

agreement with many competing gauge theories. The more precise Z-pole measurements from LEP and SLC subsequently excluded large classes of small deviations from the standard model predictions. However, there are many types of new physics which do not directly affect the Z or its couplings and to which the Z-pole experiments are blind. The low energy experiments therefore still play a significant role in probing for such types of new physics.

4 Beyond the Standard Model

4.1 *Unification or Compositeness*

Most work in particle physics today is directed towards searching for the new physics beyond the standard model. Although there are many theoretical ideas for the nature of such new physics most possibilities fall into one of two general categories.

The first, which I describe as the Bang scenario, involves the unification of the interactions. In such schemes there is generally a grand desert up to a grand unification (GUT) or Planck scale (M_P). This is the natural domain of elementary Higgs fields, supersymmetry, GUTs, and superstring theories. If nature should choose this route there is a possibility of probing to M_P and to the very early universe. There are hints from coupling constant unification that this may be the correct path. Some of the implications are that there should be supersymmetry, which can ultimately be probed by finding the new superpartners at the LHC. Secondly, one expects to have a light Higgs boson, which acts much like the standard model Higgs except that it must be lighter than 110 – 150 GeV, which should be detectable at the LHC or possibly at LEP 2. (The standard model Higgs could be as heavy as 600 – 1000 GeV.) Finally, a very important prediction of at least the simplest cases is that one expects an *absence* of deviations from the standard model predictions for precision electroweak tests, CP violation, or rare K decays, because of the decoupling of the heavy superpartners. Of course, it is hard to take the observed absence of such deviations as compelling evidence for supersymmetric unification, but they are nevertheless suggestive. Some such schemes also lead to predictions for m_b, proton decay, neutrino masses, and rare decays.

If the coupling constant unification is not just an accident there are very few types of new physics other than supersymmetry that could be present without spoiling it (unless two new effects cancel). These include additional heavy Z' bosons, gauge singlets, and a small number of new sequential, mirror, or exotic fermion families.

The other general possibility is the Whimper scenario, in which nature consists of onion-like layers of matter at shorter and shorter distance scales. This is the domain of composite fermions and scalars and of dynamical symmetry breaking. Experimental limits imply that any new layer of compositeness would have to be strong binding, and is therefore not analogous to previously observed levels. If nature should choose this route, then at most one more layer would be accessible to us at the LHC and future colliders. Such schemes generally predict significant rates for rare decays such as $K \to \mu e$. This is a generic feature of almost all such models, and the fact that they have not been observed is a severe problem for the general approach and has made it difficult to construct realistic models. If one somehow evades the problem of rare decays one still generally expects to see significant effects in precision observables, including new 4-fermi operators, decrease of the $Z \to b\bar{b}$ partial width, and modifications to ρ_0 and to the parameters S, T, and U. The fact that these have not been seen constitutes an additional serious difficulty for most such models. In the future one would also expect to see new particles and anomalous interactions among gauge bosons.

4.2 *Searches for New Physics*

Since there is no evidence for deviations from the standard model, the boson and neutral current data (and also precision charged current data) can be used to set limits on many kinds of possible new physics. These include (a) heavy Z' bosons [73]-[76]; (b) the ρ_0 parameter, associated with higher-dimensional Higgs representations or other new sources of SU_2-breaking [77]; and (c) classes of new physics (such as technicolor or new multiplets of fermions or scalars) which only affect the Z, W, and neutral current observables via gauge self-energy diagrams [78]-[85]. Other applications include (a) verifying the canonical (*i.e.*, left-handed doublets, right-handed singlets) weak isospin assignments of the known quarks and leptons [7, 6]; (b) searches for mixing between ordinary and exotic fermions (*e.g.*, left-handed singlets or right-handed doublets, which are predicted in many extensions of the standard model) [86, 2]; (c) searches for leptoquarks or new four-fermi operators associated with compositeness [2, 70, 71]; (d) searching for anomalous contributions to the $Zb\bar{b}$ vertex [87], such as may be generated by light superpartners [88] or extended technicolor interactions [89]-[91]. The sensitivity of existing and projected experiments for various classes of new physics are described in detail in [2].

4.3 *Supersymmetry and Precision Experiments*

Let us now consider how the predictions for the precision observables are modified in the presence of supersymmetry. There are basically three implications for the precision results. The first, and most important, is in the Higgs sector. In the standard model the Higgs mass is arbitrary. It is controlled by an arbitrary quartic Higgs coupling, so that M_H could be as small as 60 GeV (the experimental limit) or as heavy as a TeV. The upper bound is not rigorous: larger values of M_H would correspond to such large quartic couplings that perturbation theory would break down. This cannot be excluded, but would lead to a theory that is qualitatively different from the (perturbative) standard model. In particular, there are fairly convincing triviality arguments, related to the running of the quartic coupling, which exclude a Higgs which acts like a distinct elementary particle for M_H above $O(600\text{ GeV})$ [45].

However, in supersymmetric extensions of the standard model the quartic coupling is no longer a free parameter. It is given by the squares of gauge couplings, with the result that all supersymmetric models have at least one Higgs scalar that is relatively light, typically with a mass similar to the Z mass. In the minimal supersymmetric standard model (MSSM) one has $M_H < 150$ GeV[19], which generally acts just like the standard model Higgs[20] except that it is necessarily light.

In the standard model there is a strong $m_t - M_H$ correlation, and one has the prediction for m_t is (39). For the standard model range $60 < M_H < 1000$ GeV this corresponds to $m_t = 175 \pm 11^{+17}_{-19}$ GeV (SM). However, in MSSM one has the smaller range $60 < M_H < 150$ GeV, leading to the lower prediction

$$m_t = 160^{+11+6}_{-12-5} \text{ (MSSM)}. \tag{52}$$

This is on the low side of the CDF range, (174±16 GeV), but is certainly consistent. Similarly, the indirect data predict

$$\hat{s}^2_Z = 0.2316(3)(1) \text{ (MSSM)} \tag{53}$$

[19] At tree-level, $M_H < M_Z$.

[20] This is true if the second Higgs doublet is much heavier than M_Z.

and

$$\alpha_s = 0.126(5)(1)\ \text{(MSSM)}, \tag{54}$$

which differ slightly from the standard model values in Table 6 because of the lower M_H range.

There can be additional effects on the radiative corrections due to sparticles and the second Higgs doublet that must be present in the MSSM [92]. However, for most of the allowed parameter space one has $M_{\text{new}} \gg M_Z$, and the effects are negligible by the decoupling theorem. For example, a large $\tilde{t} - \tilde{b}$ splitting would contribute to the ρ_0 (SU_2-breaking) parameter to be discussed below, leading to a smaller prediction for m_t, but these effects are negligible for $m_{\tilde{q}} \gg M_Z$. Similarly, there would be new contributions to the $Z \to b\bar{b}$ vertex for $m_{\chi^\pm}$, $m_{\tilde{t}}$, or $M_H^\pm \sim M_Z$.

There are only small windows of allowed parameter space for which the new particles contribute significantly to the radiative corrections. Except for these, the only implications of supersymmetry from the precision observables are: (a) there is a light standard model-like Higgs, which in turn favors a smaller value of m_t. Of course, if a light Higgs were observed it would be consistent with supersymmetry but would not by itself establish it. That would require the direct discovery of the superpartners, probably at the LHC. (b) Another important implication of supersymmetry, at least in the minimal model, is the *absence* of other deviations from the standard model predictions. (c) In supersymmetric grand unification one expects the gauge coupling constants to unify when extrapolated from their low energy values [93]. This is consistent with the data in the MSSM but not in the ordinary standard model (unless other new particles are added). This is not actually a modification of the precision experiments, but a prediction for the observed gauge couplings. Of course, one could have supersymmetry without grand unification.

4.4 *(Supersymmetric) Grand Unification*

It is interesting to compare the value of $\hat{s}^2_Z$ in Table 6 or eqn. (53) with the only models available which predict it, namely grand unified theories [93]-[97].

In a grand unified theory there is only one underlying gauge coupling constant, and when the low energy couplings are extrapolated to high energy they are expected to (approximately) meet at the unification scale M_X above which symmetry breaking can be neglected. We define the couplings $g_s = g_3$, $g = g_2$, and $g' = \sqrt{3/5}g_1$ of the standard model $SU_3 \times SU_2 \times U_1$ group, and the fine-structure constants $\alpha_i = g_i^2/4\pi$. The extra factor in the definition of g_1 is a normalization condition [97]. The couplings are expected to meet only if the corresponding group generators are normalized in the same way. However, the standard model generators are conventionally normalized as $\text{Tr}(Q_s^2) = \text{Tr}(Q_2^2) = 5/3\text{Tr}(Y/2)^2$, so the factor

$\sqrt{3/5}$ is needed to compensate. Thus,

$$\sin^2\theta_W = \frac{g'^2}{g^2+g'^2} = \frac{g_1^2}{\frac{5}{3}g_2^2+g_1^2} \underset{g_1=g_2}{\rightarrow} \frac{3}{8}. \tag{55}$$

One expects $\sin^2\theta_W = 3/8$ at the unification scale [97] for which $g_1 = g_2$.

To test the unification, one starts with the couplings at M_Z, which are now very well known from the LEP and low energy data. Using as inputs $\alpha^{-1}(M_Z) = 127.9 \pm 0.1$ [23], $\hat{s}_Z^2 = 0.2316 \pm 0.0003$, and[21] $\alpha_s(M_Z) = 0.12 \pm 0.01$, one obtains

$$\begin{aligned} \alpha_1^{-1}(M_Z) &\equiv \frac{3}{5}\alpha^{-1}(M_Z)\hat{c}_Z^2 = 58.97 \pm 0.05 \\ \alpha_2^{-1}(M_Z) &\equiv \alpha^{-1}(M_Z)\,\hat{s}_Z^2 = 29.62 \pm 0.04 \\ \alpha_3^{-1}(M_Z) &\equiv \alpha_s^{-1}(M_Z) = 8.3 \pm 0.7. \end{aligned} \tag{56}$$

These may be extrapolated to high energy using the two-loop renormalization group equations

$$\frac{d\alpha_i^{-1}}{d\ln\mu} = -\frac{b_i}{2\pi} - \sum_{j=1}^{3}\frac{b_{ij}\alpha_j}{8\pi^2}. \tag{57}$$

The 1-loop coefficients are

$$b_i = \begin{pmatrix} 0 \\ -\frac{22}{3} \\ -11 \end{pmatrix} + F\begin{pmatrix} \frac{4}{3} \\ \frac{4}{3} \\ \frac{4}{3} \end{pmatrix} + N_H\begin{pmatrix} \frac{1}{10} \\ \frac{1}{6} \\ 0 \end{pmatrix}, \tag{58}$$

assuming the standard model. F is the number of fermion families and N_H is the number of Higgs doublets. In the MSSM,

$$b_i = \begin{pmatrix} 0 \\ -6 \\ -9 \end{pmatrix} + F\begin{pmatrix} 2 \\ 2 \\ 2 \end{pmatrix} + N_H\begin{pmatrix} \frac{3}{10} \\ \frac{1}{2} \\ 0 \end{pmatrix}, \tag{59}$$

where the difference is due to the additional particles in the loops. The 2-loop coefficients can be found in [96]. Equation 57 can be integrated to yield

$$\alpha_i^{-1}(\mu) = \alpha_i^{-1}(M_Z) - \frac{b_i}{2\pi}\ln\left(\frac{\mu}{M_Z}\right) + \sum_{j=1}^{3}\frac{b_{ij}}{4\pi b_j}\ln\left[\frac{\alpha_j^{-1}(\mu)}{\alpha_j^{-1}(M_Z)}\right], \tag{60}$$

for an arbitrary scale μ. To first approximation one can neglect the last (2-loop) term, in which case the inverse coupling constant varies linearly with $\ln\mu$. However,

[21] α_s is considerably less well-determined than $\alpha(M_Z)$ and $\hat{s}_Z^2$. I will therefore usually take α_s as a prediction rather than an input. When it is used as an input, however, I will use the conservative range 0.12 ± 0.01.

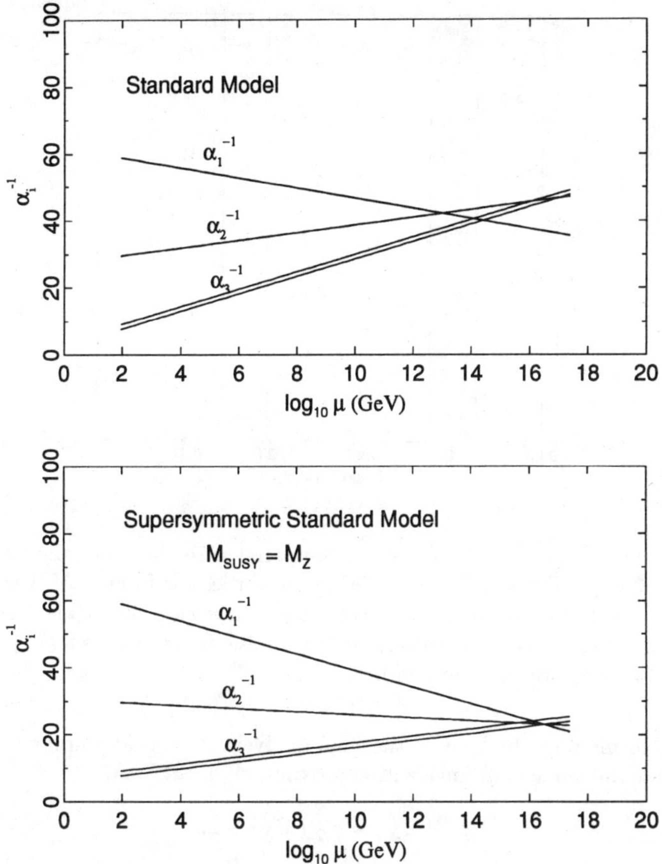

Figure 16: Running couplings in (a) the standard model (SM) and (b) in the minimal supersymmetric extension of the standard model (MSSM) with two Higgs doublets for $M_{SUSY} = M_Z$. The corresponding figure for $M_{SUSY} = 1$ TeV is almost identical. It is seen that the couplings unify at $\simeq 10^{16}$ GeV in the MSSM. The effects of threshold uncertainties are seen in Figures 17 and 18.

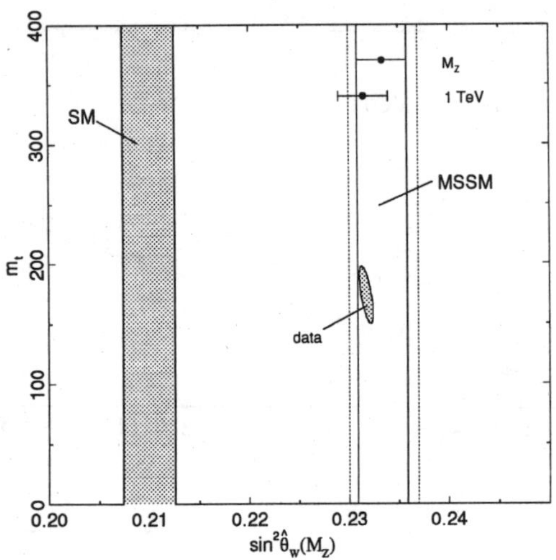

Figure 17: 90% CL regions in $\hat{s}^2_Z$ vs m_t, compared with the predictions of ordinary and SUSY–GUTs. The smaller ranges of uncertainties are from $\alpha(M_Z)$ and $\alpha_s(M_Z)$ only, while the larger range includes the various low and high scale uncertainties, added in quadrature. The predictions for degenerate SUSY masses at M_Z and 1 TeV are shown for comparison. Updated from [47].

the 2-loop terms must be kept in the final analysis. In a grand unified theory one expects that the three couplings will approximately meet at M_X,

$$\alpha_i^{-1}(M_X) = \alpha_G^{-1}(M_X) - \Delta_i. \tag{61}$$

The Δ_i are small corrections [98] associated with the low energy threshold (*i.e.*, m_t and the new sparticles and Higgs not degenerate with M_Z), the high scale thresholds ($m_{\text{heavy}} \neq M_X$), or with non-renormalizable operators.

The running couplings in the standard model are shown in Figure 16a, ignoring threshold corrections. They clearly do not meet at a point, thus ruling out simple grand unified theories such as SU_5, SO_{10}, or E_6 which break in a single step to the standard model [95]. Of course, such models are also excluded by the non-observation of proton decay, but this independent evidence is welcome.

On the other hand, in the minimal supersymmetric extension of the standard model the couplings do meet within the experimental uncertainties [96, 99, 93]. This is illustrated in Figure 16b for the case in which all of the new particles have a common mass $M_{SUSY} = M_Z$. Almost identical curves are obtained for larger M_{SUSY}, such as 1 TeV. (In practice, the splittings between the sparticle masses are more important than the average value [100, 98].) The unification scale M_X is sufficiently large ($> 10^{16}$ GeV) that proton decay by dimension-6 operators is adequately suppressed, although there may still be a problem with dimension-5 operators [101]. This success is encouraging for supersymmetric grand unified theories such as $SUSY\text{–}SU_5$ or $SUSY\text{–}SO_{10}$.

To display the theoretical uncertainties, it is conventional to use $\alpha(M_Z)$ and $\alpha_s(M_Z)$ to predict $\hat{s}^2_Z$. Using $\alpha^{-1}(M_Z) = 127.9 \pm 0.1$ and $\alpha_s(M_Z)= 0.12 \pm 0.01$ one predicts

$$\begin{aligned} \hat{s}^2_Z &= 0.2334 \pm 0.0025 \pm 0.0025 \text{ (MSSM)}, \\ \hat{s}^2_Z &= 0.2100 \pm 0.0025 \pm 0.0007 \text{ (SM)}, \end{aligned} \tag{62}$$

where the first uncertaintly is from α_s and α^{-1}, and the second is an estimate of theoretical uncertainties from m_t, the superspectrum, high-scale thresholds, and possible non-renormalizable operators [47]. The MSSM prediction is in agreement with the experimental value 0.2316(3)(1), while the SM prediction is in conflict with the data. These results are displayed in Figure 17.

Because of the large uncertainty in $\alpha_s(M_Z)$, it is convenient to invert the logic and use the precisely known α^{-1} and $\hat{s}^2_Z$ to predict $\alpha_s(M_Z)$:

$$\begin{aligned} \alpha_s(M_Z) &= 0.129 \pm 0.002 \pm 0.008 \text{ (MSSM)}, \\ \alpha_s(M_Z) &= 0.073 \pm 0.001 \pm 0.001 \text{ (SM)}, \end{aligned} \tag{63}$$

where again the second error is theoretical. It is seen that the SUSY prediction is in agreement with the experimental $\alpha_s(M_Z)= 0.12 \pm 0.01$, while the simplest ordinary GUTs are excluded. The central value prefers the larger values of $\alpha_s(M_Z)$ suggested by the Z-pole data over some of the low energy determinations in Table 7, but the theoretical uncertainties are comparable to the error on the observed $\alpha_s(M_Z)$ (which is also dominated by theory). If the low energy values turn out to be true, supersymmetric unification would require large but not unreasonable threshold corrections. The $\alpha_s(M_Z)$ predictions are shown in Figure 18.

The success of the coupling constant unification, which is insensitive to the gauge group and the number of complete families, provides a hint that supersymmetric grand unification (or some superstring imitator) may be on the right track. Of course it is possible that the success may be an accident. Similarly, there are many more complicated schemes which could yield coupling constant unification, such as those involving a large group breaking in two or more stages to the standard model, or those with ad hoc new representations split into light and heavy components. However, the MSSM is the only scheme in which the unification is a prediction rather than being achieved by adjusting new parameters or representations. Perhaps the coupling constants may indeed prove to be the "first harbinger of supersymmetry" [99].

Unless the apparent coupling constant unification is an accident, there are stringent restrictions on the types of new "SUSY-safe" physics which do not drastically disturb the predictions (unless, of course, one allows two large effects to

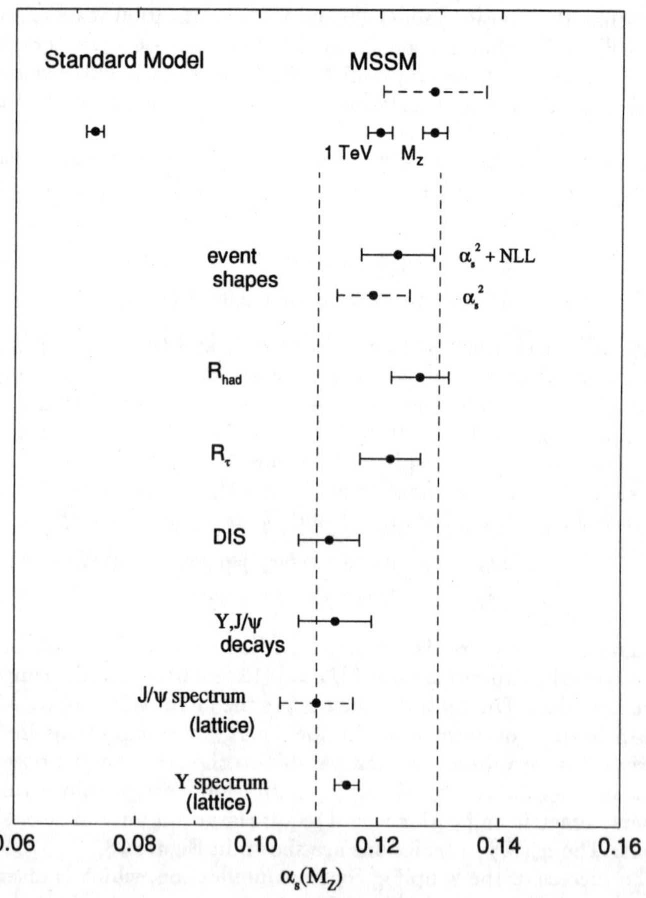

Figure 18: Predictions for $\alpha_s(M_Z)$ from α^{-1} and $\hat{s}_Z^2$ in ordinary and SUSY GUTs. The dashed lines represent the experimental range 0.12 ± 0.01. In the SUSY (MSSM) case the error bar includes the theoretical uncertainties, added in quadrature. The smaller error bars are for degenerate sparticle masses at M_Z and 1 TeV. Updated from [47].

cancel). These are: supersymmetry (required), additional heavy Z' gauge bosons or additional gauge groups which commute with the standard model, additional complete ordinary or mirror fermion families and their superpartners (the neutrinos would have to be very heavy because of the LEP result $N_\nu = 2.988 \pm 0.023$), complete exotic fermion supermultiplets such as occur in some E_6 models, or gauge singlets.

There are many other implications of supersymmetric grand unification. These include proton decay [101]; Yukawa unification (the prediction of m_b/m_τ), which leads to stringent constraints on the ratio $\tan\beta$ of the vacuum expectation values of the two Higgs doublets which give mass to the t and b, respectively [102]; the upper limit on the standard model-like Higgs [103, 104]; cold dark matter [105]; neutrino mass [106]; and a possible connection with superstring theories [107].

4.5 *Extended Technicolor/Compositeness*

In contrast to unification/supersymmetry, the other major class of extensions, which includes compositeness and dynamical symmetry breaking, leads to many implications at low energies. The most important are large flavor changing neutral currents (FCNC). Even if these are somehow evaded one generally expects anomalous contributions to the $Z \to b\bar{b}$ vertex, typically $\Gamma(b\bar{b}) < \Gamma^{SM}(b\bar{b})$ in the simplest extended technicolor (ETC) models [89]. Similarly, one expects $\rho_0 \neq 1$, and $S_{\rm new} \neq 0, T_{\rm new} \neq 0$, where ρ_0, $S_{\rm new}$, and $T_{\rm new}$ parameterize certain types of new physics, as will be described below. Finally, in theories with composite fermions one generally expects new 4-fermi operators generated by constituent interchange, leading to effective interactions of the form

$$L = \pm\frac{4\pi}{\Lambda^2}\bar{f}_1\Gamma f_2\bar{f}_3\hat{\Gamma} f_4. \tag{64}$$

Generally, the Z-pole observables are not sensitive to such operators, since they only measure the properties of the Z and its couplings. However, low energy experiments are sensitive. In particular, FCNC constraints typically set limits of order $\Lambda \geq O(100 \text{ TeV})$ on the scale of the operators unless the flavor-changing effects are fine-tuned away. Even then there are significant limits from other flavor conserving observables. For example, atomic parity violation [14] is sensitive to operators such as [70, 2]

$$L = \pm\frac{4\pi}{\Lambda^2}\bar{e}_L\gamma_\mu e_L\bar{q}_L\gamma^\mu q_L. \tag{65}$$

The existing data already sets limits $\Lambda > O(10 \text{ TeV})$, as described in section 4.11. Future experiments should be sensitive to ~ 40 TeV.

4.6 *The $Zb\bar{b}$ Vertex*

The $Zb\bar{b}$ vertex is especially interesting, both in the standard model and in the presence of new physics. In the standard model there are special vertex contributions, shown in Figure 19, which depend quadratically on the top quark mass. Their value is shown approximately in (34). $\Gamma(b\bar{b})$ actually decreases with m_t, as opposed to other widths which all increase due to the $\hat{\rho}$ parameter. The m_t and M_H dependences in $\hat{\rho}$ are strongly correlated, but the special vertex corrections to $\Gamma(b\bar{b})$ are independent of M_H, allowing a separation of m_t and M_H effects.

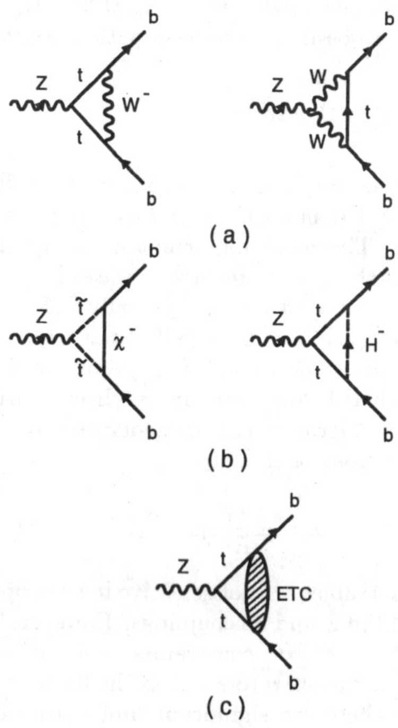

Figure 19: (a) Standard model vertex corrections to $Z \to b\bar{b}$. (b) New contributions in supersymmetry. (c) New contributions in extended technicolor.

The vertex is also sensitive to a number of types of new physics. One can parameterize such effects by [87]

$$\Gamma(b\bar{b}) \to \Gamma^{SM}(b\bar{b})\,(1+\delta_{bb}^{\text{new}}) \sim \Gamma^{0}(b\bar{b})\left(1+\delta_{bb}^{\text{SM}}+\delta_{bb}^{\text{new}}\right). \tag{66}$$

If the new physics gives similar contributions to vector and axial vector vertices then the effects on $A^b_{\rm FB}$ are negligible. In supersymmetry one can have both positive and negative contributions [88]. In particular, light $\tilde{t} - \chi^{\pm}$ can give $\delta^{\rm SUSY}_{bb} > 0$, as is suggested by the data, while light charged Higgs particles yield $\delta^{\rm Higgs}_{bb} < 0$. In practice, both effects are too small to be important in most allowed regions of parameter space [108]. In extended technicolor (ETC) models there are typically new vertex contributions generated by the same ETC interactions that are needed to generate the large top quark mass. It has been argued that these are typically large and negative [89],

$$\delta^{\rm ETC}_{bb} \sim -0.056\xi^2 \left(\frac{m_t}{150{\rm GeV}}\right), \tag{67}$$

where ξ is a model dependent parameter of order unity. They may be smaller in models with walking technicolor, but nevertheless are expected to be negative and significant [90]. This is in contrast to the data, which suggests a positive contribution if any, implying a serious problem for many ETC models. One possible way out are models in which the ETC and electroweak groups do not commute, for which either sign is possible [91].

Another possibility is mixing between the b and exotic heavy fermions with non-canonical weak interaction quantum numbers. Many extensions of the standard model predict, for example, the existence of a heavy D_L, D_R, which are both SU_2 singlet quarks with charge $-1/3$. These can mix with the d, s, or b quarks, but one typically expects such mixing to be largest for the third generation. However, this mechanism gives a negative contribution

$$\delta^{D_L}_{bb} \sim -2.3 s_L^2 \tag{68}$$

to $\delta^{\rm new}_{bb}$, where s_L is the sine of the $b_l - D_L$ mixing angle. R_b can be increased if there is an additional heavy Z' boson which only couples to the third family [109].

One can extract $\delta^{\rm new}_{bb}$ from the data, in a global fit to the standard model parameters as well as $\delta^{\rm new}_{bb}$. This yields [20]

$$\delta^{\rm new}_{bb} = 0.023 \pm 0.011 \pm 0.003, \tag{69}$$

which is $\sim 2\sigma$ above zero. This value is hardly changed when one allows additional new physics, such as described by the S, T, and U parameters. $\delta^{\rm new}_{bb}$ is correlated with $\alpha_s(M_Z)$, because one can describe $R = \Gamma({\rm had})/\Gamma(\ell\bar{\ell})$ with a smaller QCD correction to Γ(had) [20, 109]. Allowing for $\delta^{\rm new}_{bb}$, one obtains $\alpha_s(M_Z) = 0.111 \pm 0.009 \pm 0$, considerably smaller than the standard model value 0.127(5)(2). Allowing $\delta^{\rm new}_{bb} \neq 0$ has negligible effect on $\hat{s}^2_Z$ or m_t. One can also perform more detailed fits allowing separate corrections to the left and right-handed b couplings [20, 110]. Using both R_b and A^{0b}_{FB} (as well as all of the other data) as constraints, one finds that the anomaly should be in the b_R coupling, as can be seen in equation (3).

4.7 ρ_0: *Nonstandard Higgs or Non-degenerate Heavy Multiplets*

One parameterization of certain new types of physics is the parameter

$$\rho_0 \equiv \frac{M_W^2}{M_Z^2 \hat{c}_Z^2 \hat{\rho}}. \tag{70}$$

ρ_0 is exactly unity in the standard model, and any deviation would indicate new sources of SU_2 breaking other than the ordinary Higgs doublets or the top/bottom splitting. New physics can affect ρ_0 at either the tree or loop-level

$$\rho_0 = \rho_0^{\text{tree}} + \rho_0^{\text{loop}}. \tag{71}$$

The tree-level contribution is given by Higgs representations larger than doublets, namely,

$$\rho_0^{\text{tree}} = \frac{\sum_i (t_i^2 - t_{3i}^2 + t_i)\,|\langle\phi_i\rangle|^2}{\sum_i 2t_{3i}^2 |\langle\phi_i\rangle|^2}, \tag{72}$$

where t_i (t_{3i}) is the weak isospin (third component) of the neutral Higgs field ϕ_i. If one has only Higgs singlets and doublets ($t_i = 0, \frac{1}{2}$), then $\rho_0^{\text{tree}} = 1$. However, in the presence of larger representations with non-zero vacuum expectation values

$$\rho_0^{\text{tree}} \simeq 1 + 2\sum_i \left(t_i^2 - 3t_{3i}^2 + t_i\right) \frac{|\langle\phi_i\rangle|^2}{|\langle\phi_{\frac{1}{2}}\rangle|^2}. \tag{73}$$

One can also have loop-induced contributions similar to that from t and b, due to non-degenerate multiplets of fermions or bosons. For new doublets

$$\rho_0^{\text{loop}} = \frac{3G_f}{8\sqrt{2}\pi^2} \sum_i \frac{C_i}{3} F(m_{1i}, m_{2i}), \tag{74}$$

where $C_i = 3(1)$ for color triplets (singlets) and

$$F(m_1, m_2) = m_1^2 + m_2^2 - \frac{4m_1^2\, m_2^2}{m_1^2 - m_2^2} \ln\frac{m_1}{m_2} \geq (m_1 - m_2)^2. \tag{75}$$

Loop contributions to ρ_0 are generally positive[22], and if present would lead to lower values for the predicted m_t. ρ_0^{tree} can be either positive or negative depending on the quantum numbers of the Higgs field. The ρ_0 parameter is extremely important because one expects $\rho_0 \sim 1$ in most superstring theories [113], which generally do not have higher-dimensional Higgs representations[23], while typically $\rho_0 \neq 1$ from many sources in models involving compositeness.

---

[22]One can have $\rho^{\text{loop}} < 0$ for Majorana fermions [111] or boson multiplets with vacuum expectation values [112].

[23]The only known exceptions are string models in which the observed particles are composite, or models with $k > 1$ worldsheet currents [114].

In the presence of ρ_0 the standard model formulas for the observables are modified. As long as ρ_0-1 is sufficiently small, one can simply incorporate the effects of $\rho_0 - 1$ in the tree-level formulas and take $\rho_0 = 1$ in the radiative corrections [115]. At tree level,

$$M_Z \to \frac{1}{\sqrt{\rho_0}} M_Z^{SM}, \Gamma_Z \to \rho_0 \Gamma_Z^{SM}, A_{NC} \to \rho_0 A_{NC}^{SM}, \tag{76}$$

where A_{NC} is a neutral current amplitude. It has long been known that ρ_0 is close to 1. However, until recently it has been difficult to separate ρ_0 from m_t, because most observables only involve the combination $\rho_0\hat{\rho}$. The one exception has been the $Z \to b\bar{b}$ vertex. However, assuming that CDF has really observed the top quark directly one can use the known m_t to calculate $\hat{\rho}$ and therefore separate ρ_0. In practice one fits to m_t, ρ_0 and the other parameters simultaneously, using the CDF value $m_t = 174 \pm 16$ GeV as an additional constraint. One obtains

$$\begin{aligned} \hat{s}_Z^2 &= 0.2316(3)(1) & \rho_0 &= 1.0012 \pm 0.0017 \pm 0.0017 \\ \alpha_s &= 0.125(6)(1) & m_t &= 166 \pm 15 \pm 0 \text{ GeV}, \end{aligned} \tag{77}$$

where the second uncertanty is from M_H. Even in the presence of the classes of new physics parameterized by ρ_0 one still has robust predictions for the weak angle and a good determination of α_s. Most remarkably, given the CDF constraint, ρ_0 is constrained to be extremely close to unity, causing serious problems for compositeness models. The allowed region in ρ_0 vs $\hat{s}_Z^2$ are shown in Figure 20. This places limits $|\langle\phi_i\rangle|/|\langle\phi_{1/2}\rangle| <$ few% on non-doublet vacuum expectation values, and places constraints $\frac{C}{3}F(m_1, m_2) < O((100 \text{ GeV})^2)$ on the splittings of additional fermion or boson multiplets.

One can also consider the possibility that there are both new sources of SU_2 breaking and new contributions to the $Zb\bar{b}$ vertex. A simultaneous fit to all data yields [20]

$$\begin{aligned} \hat{s}_Z^2 &= 0.2316(3)(2) & \rho_0 &= 1.0004 \pm 0.0018 \pm 0.0018 \\ \alpha_s &= 0.111(9)(0) & m_t &= 174 \pm 16^{+1}_{-0} \text{ GeV} \\ \delta_{bb}^{\text{new}} &= 0.022 \pm 0.011 \pm 0. \end{aligned} \tag{78}$$

Just as in the standard model (with $\rho_0 = 1$) the value of α_s decreases significantly if one allow for a nonzero $Zb\bar{b}$ vertex correction. The other parameters are changed moderately or not at all.

Even without the CDF constraint the special m_t dependence of $\Gamma(b\bar{b})$ can be used to obtain an upper limit on m_t for arbitrary Higgs representations, i.e., arbitrary ρ_0 [77, 87]. From the precision data and D0 lower bound ($m_t > 131$ GeV), one finds $m_t < 183(195)$ GeV at 90(95)% C.L., essentially independent of M_H.

4.8 *Heavy Physics by Gauge Self Energies*

A larger class of extensions of the standard model can be parameterized by the S, T and U parameters [78]-[85], which describe that subset of new physics which affect

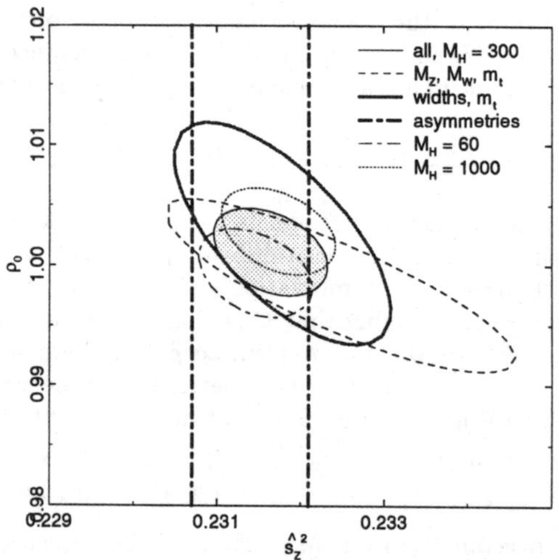

Figure 20: 90% C.L. allowed regions in ρ_0 vs $\hat{s}^2_Z$ for $M_H = 60$, 300, and 1000 GeV.

only the gauge boson self-energies[24] but do not directly affect tree-level amplitudes, vertices, etc. One introduces three parameters

$$\begin{aligned} S &= S_{\rm new} + S_{m_t} + S_{M_H} \\ T &= T_{\rm new} + T_{m_t} + T_{M_H} \\ U &= U_{\rm new} + U_{m_t}. \end{aligned} \tag{79}$$

The new physics contributions are defined by

$$\alpha T^{\rm loop}_{\rm new} \equiv \frac{\Pi^{\rm new}_{WW}(0)}{M_W^2} - \frac{\Pi^{\rm new}_{ZZ}(0)}{M_Z^2}$$

$$\frac{\alpha}{4\hat{s}^2_Z\hat{c}^2_Z} S_{\rm new} \equiv \frac{\Pi^{\rm new}_{ZZ}(M_Z^2) - \Pi^{\rm new}_{ZZ}(0)}{M_Z^2} \tag{80}$$

$$\frac{\alpha}{4\hat{s}^2_Z}(S+U)_{\rm new} \equiv \frac{\Pi^{\rm new}_{WW}(M_W^2) - \Pi^{\rm new}_{WW}(0)}{M_W^2} \tag{81}$$

where $\Pi^{\rm new}_{WW}$ and $\Pi^{\rm new}_{ZZ}$ are the contributions of new physics to the W and Z self-energies. T is associated with the difference between the W and Z self-energies at $Q^2 = 0$ and describes the breaking of the SU_{2V} vector generators. T is equivalent

[24]This formalism assumes that the new physics is much heavier than M_Z. The results can be generalized to new physics scales comparable to M_Z by introducing additional parameters [84].

to the ρ_0 parameter and is induced by mass splitting in multiplets of fermions or bosons. S ($S+U$) are associated with the differences between the Z (W) propagators at $Q^2 = 0$ and M_Z^2 (M_W^2), and describe the breaking of the SU_{2A} axial generators. S is generated, for example, by degenerate heavy chiral families of fermions. U is zero in most extensions of the standard model. S, T, and U are induced by loop corrections and have a factor of α extracted, so they are expected to be $O(1)$ if there is new physics. They are related to equivalent parameters defined in [79] by

$$\begin{aligned} S &= h_{AZ} = S_Z = 4\hat{s}_Z^2\epsilon_3/\alpha \\ T &= h_V = \epsilon_1/\alpha \\ U &= h_{AW} - h_{AZ} = S_W - S_Z = -4\hat{s}_Z^2\epsilon_2/\alpha. \end{aligned} \tag{82}$$

S, T and U were introduced to describe the contributions of new physics. However, they can also parametrize the effects of very heavy m_t and M_H (compared to M_Z). Expressions for S_{m_t} and S_{M_H}, which are respectively the m_t and M_H contributions to S, and similarly for T and U, may be found in [79, 80]. Until recently it was difficult to separate the m_t and new physics contributions. Therefore, most analyses fixed m_t and M_H at arbitrary reference values (e.g., M_Z, or the result of the best fit in the standard model), and fit to the total S, T, and U. The results could then be compared to the standard model expectations for other values of m_t and M_H. Now, however, with the CDF value of m_t it is possible to directly extract the new physics contributions. That is, one can determine $S_{\rm new}$, $T_{\rm new}$, and $U_{\rm new}$ in a simultaneous fit with $\hat{s}_Z^2$, m_t, α_s, and (optionally) $\delta_{bb}^{\rm new}$, with the M_H dependence included in the uncertainties. In practice, one can use the full m_t and M_H dependence of all observables, and not just their contributions to S, T, and U, which are approximations valid for masses much larger than M_Z.

A new multiplet of degenerate chiral fermions will contribute to $S_{\rm new}$ by

$$S_{\rm new}|_{\rm degenerate} = \sum_i C_i |t_{3L}(i) - t_{3R}(i)|^2/3\pi \geq 0, \tag{83}$$

where C_i is the number of colors and t_{3LR} are the t_3 quantum numbers. A fourth family of degenerate fermions would yield $\frac{2}{3\pi} \sim 0.21$, while QCD-like technicolor models, which typically have many particles, can give larger contributions. For example, $S_{\rm new} \sim 0.45$ from an isodoublet of fermions with four technicolors, and an entire technigeneration would yield 1.62 [78]. Non-QCD-like theories such as those involving walking could yield smaller or even negative contributions [81]. Nondegenerate scalars or fermions can contribute to $S_{\rm new}$ with either sign [82].

The T parameter is analogous to $\rho_0^{\rm loop}$. For a non-degenerate family

$$T_{\rm new}^{\rm loop} \sim \frac{\rho_0^{\rm loop}}{\alpha} \sim 0.42 \frac{\Delta m^2}{(100\ GeV)^2}, \tag{84}$$

where

$$\Delta m^2 = \sum_i \frac{C_i}{3} F(m_{1i}, m_{2i}) \geq \sum_i \frac{C_i}{3} (m_{1i} - m_{2i})^2 \tag{85}$$

and $F(m_1, m_2)$ is defined in (75). Usually $T_{\rm new}^{\rm loop} > 0$, although there may be exceptions for theories with Majorana fermions [111] or additional Higgs doublets [112]. In practice, higher-dimensional Higgs multiplets could mimic $T_{\rm new}$ with either sign (see equation (72)), and cannot be separated from loop effects unless they are seen directly or have other effects. That is, $T_{\rm new}^{\rm loop}$ and the contribution $\rho_0^{\rm tree}$ enter observables in the universal combination $\rho_0^{\rm tree}/(1 - \alpha T_{\rm new}^{\rm loop})$. Therefore, although T was originally defined to include loop contributions only, I will extend the definition to include tree-level effects,

$$T_{\rm new} \equiv T_{\rm new}^{\rm loop} + T_{\rm new}^{\rm tree}, \tag{86}$$

where $\rho_0^{\rm tree} = 1 + \alpha T_{\rm new}^{\rm tree}$. Then,

$$\rho_0 = 1 + \alpha T_{\rm new} \tag{87}$$

are equivalent parameters describing new sources of both tree and loop level SU_{2V} breaking.

Usually $U_{\rm new}$ is small, although there are counterexamples, such as anomalous triple-gauge vertices [85]. Supersymmetric extensions of the standard model usually give negligible contributions to S, T, and U [92].

The standard model expressions for observables are replaced by

$$\begin{aligned} M_Z^2 &= \frac{1}{\rho_0}\left(M_Z^{SM}\right)^2 \frac{1}{1 - G_F\left(M_Z^{SM}\right)^2 S_{\rm new}/2\sqrt{2}\pi} \\ M_W^2 &= \left(M_W^{SM}\right)^2 \frac{1}{1 - G_F\left(M_W^{SM}\right)^2 (S+U)_{\rm new}/2\sqrt{2}\pi}, \end{aligned} \tag{88}$$

where $M_{W,Z}^{SM}$ are the standard model expressions (7,8) in terms of $\hat{s}_Z^2$, m_t, M_H, and α_s, and ρ_0 is related to $T_{\rm new}$ by (87). Furthermore,

$$\begin{aligned} \Gamma_Z &= \rho_0 M_Z^3 \beta_Z \\ \Gamma_W &= M_W^3 \beta_W \\ A = &= \rho_0 A^{SM}, \end{aligned} \tag{89}$$

where $\beta_{Z,W}$ is the standard model expression for the reduced width $\Gamma_{Z,W}^{SM}/\left(M_{Z,W}^{SM}\right)^3$, $M_{Z,W}$ is the physical mass, and A (A^{SM}) is a neutral current amplitude (in the standard model).

There is enough data to simultaneously determine the new physics contributions to S, T, and U, the standard model parameters, and also $\delta_{bb}^{\rm new} = \frac{\Gamma(b\bar{b})}{\Gamma^{\rm SM}(b\bar{b})} - 1$. For example, $S_{\rm new}$, $T_{\rm new}$, $U_{\rm new}$, $\delta_{bb}^{\rm new}$, $\hat{s}_Z^2$, $\alpha_s(M_Z)$ and m_t are constrained by M_Z, Γ, M_W, R_b, asymmetries, R, and m_t (CDF), respectively. One obtains [20]

$$\begin{aligned} S_{\rm new} &= -0.21 \pm 0.24^{-0.08}_{+0.17} & \hat{s}_Z^2 &= 0.2313(4)(1) \\ T_{\rm new} &= -0.09 \pm 0.32^{+0.16}_{-0.11} & \alpha_s(M_Z) &= 0.112(9)(0) \\ U_{\rm new} &= -0.53 \pm 0.61 & m_t &= 175 \pm 16 \pm 0 \text{ GeV} \\ \delta_{bb}^{\rm new} &= 0.022 \pm 0.011 \pm 0, \end{aligned} \tag{90}$$

where the second error is from M_H. The $T_{\rm new}$ value corresponds to $\rho_0 = 0.9993 \pm 0.0023^{+0.0012}_{-0.0008}$, which differs from the value in (77) because of the presence of $S_{\rm new}$, $U_{\rm new}$, and $\delta_{bb}^{\rm new}$. The data is consistent with the standard model: $S_{\rm new}$ and $T_{\rm new}$ are close to zero with small errors, and the tendency to find $S < 0$ that existed in earlier data is no longer present (although A_{LR}^0 by itself favors $S < 0$). The constraints on $S_{\rm new}$ are a problem for those classes of new physics such as technicolor which tend to predict $S_{\rm new}$ large and positive, and $S_{\rm new}$ allows, at most, one additional family of ordinary fermions at 90% CL. (Of course the invisible Z width precludes any new families unless the additional neutrinos are heavier than $M_Z/2$.) The allowed regions in $S_{\rm new}$ vs $T_{\rm new}$ are shown in Figure 21. The seven parameter fit still favors a non-zero $Z \to b\bar{b}$ vertex correction $\delta_{bb}^{\rm new}$, almost identical to the value obtained without S, T, and U. The low value of the extracted α_s compared to the standard model value (0.127(5)(2)) is entirely due to $\delta_{bb}^{\rm new}$. The value of $\hat{s}_Z^2$ is slightly lower than the standard model value (0.2317(3)(3)). One can repeat the fits without the $Zb\bar{b}$ correction $\delta_{bb}^{\rm new}$. One finds almost identical values for $S_{\rm new}$, $T_{\rm new}$, and $U_{\rm new}$. Now, however, α_s= 0.125 (6)(0), close to the standard model value.

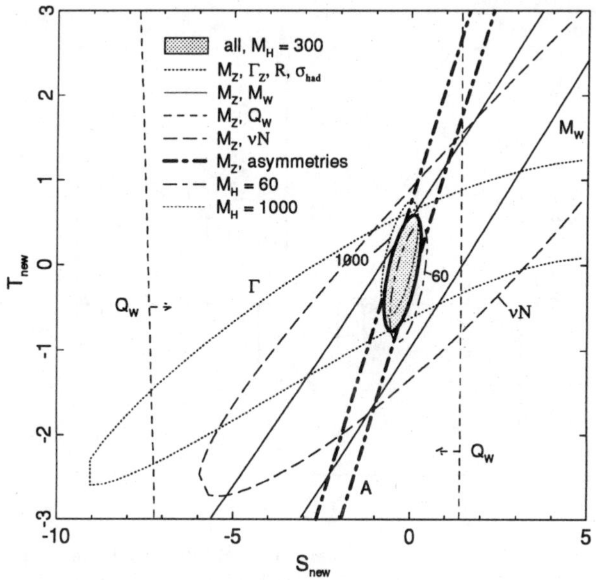

Figure 21: Constraints on $S_{\rm new}$ and $T_{\rm new}$ from various observables and from the global fit to all data. The fit to M_W and M_Z assumes $U_{\rm new} = 0$, while $U_{\rm new}$ is free in the other fits.

There is no simple parametrization which utilizes all experimental information and which describes every type of new physics [80]. The S, T, and U formalism parametrizes important classes of new physics associated with gauge boson self-energies and non-standard Higgs representations. It utilizes all experimental information, but does not apply to such effects as non-universal vertex corrections. An alternative formalism [83] is based on the shifts induced in M_W/M_Z, $\Gamma(\ell\bar{\ell})$, $A_{FB}^{0\ell}$, and R_b. It applies to all types of new physics, but cannot make use of other observables unless extra assumptions are made. A more general possibility involves deviation vectors (M. Luo, this volume, and [2]), as shown in Figure 22. Each type of new physics defines a deviation vector, the components of which are the deviations of each observable from its Standard Model prediction, normalized to the uncertainty. The length (direction) of the vector represents the strength (type) of new physics. The latter would be expecially convenient for diagnosing the origin of new physics if significant deviations were observed. One can also describe new physics by effective Lagrangian techniques [116], which are especially useful for enforcing the correlated effects of different operators that are related by gauge invariance or other symmetries. Many types of new physics, such as heavy Z' bosons or mixing with exotic fermions, are most conveniently described by special parametrizations.

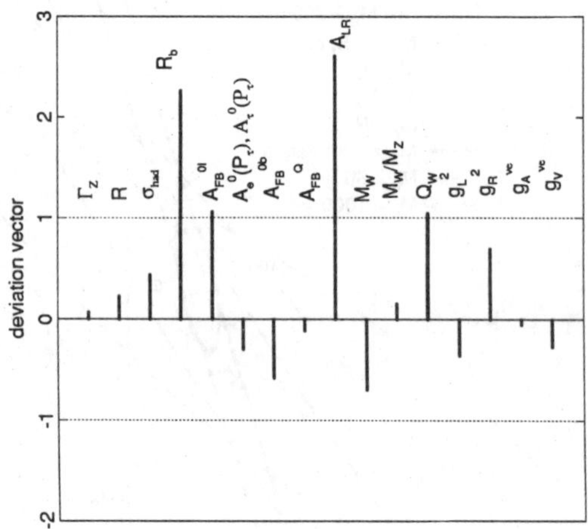

Figure 22: Deviation vectors for various observables. Each bar represents $(O_i - O_i^{SM})/\Delta O_i$, where O_i is the experimental value of the ith observable, O_i^{SM} is the standard model prediction for O_i in terms of M_Z and the global allowed ranges of m_t, α_s, and M_H, and ΔO_i is the total uncertainty, obtained by adding the experimental and theoretical uncertainties in quadrature.

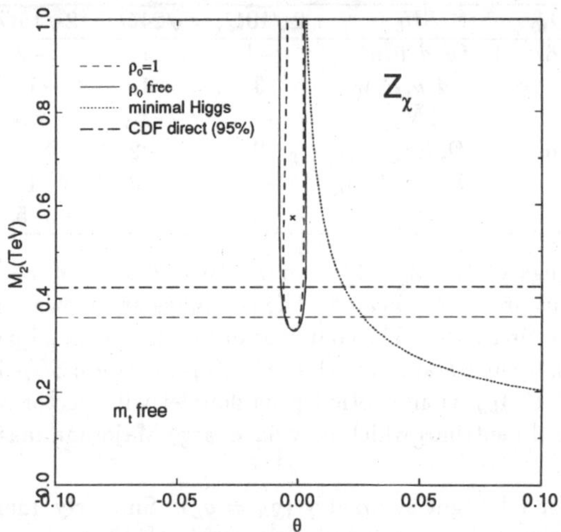

Figure 23: 90% CL ($\Delta\chi^2 = 4.6$) allowed region in M_2 and θ for the SO_{10} boson Z_χ using indirect (WNC, $M_{W,Z}$, Z-pole) data for the cases ρ_0-free and $\rho_0 = 1$. α_s, m_t, and $\hat{s}^2_Z$ are free parameters, $\lambda_g = 1$ is assumed, and the CDF constraint $m_t = 174 \pm 16$ GeV is included. Also shown are the additional constraint in the minimal Higgs case, best fit point, and the band of 95% CL lower limits on M_2 from the direct CDF search. Updated from [124].

4.9 *Additional Z' Bosons*

Many extensions of the standard model predict the existence of additional gauge bosons, which could be light enough to be observable [73]-[76]. For example, there is an extra Z' in $SU_{2L} \times SU_{2R} \times U_1$ models, which couples to the current [73]

$$J_{LR} = \sqrt{\frac{3}{5}} \left(\alpha J_{3R} - \frac{1}{2\alpha} J_{B-L} \right), \tag{91}$$

where T_{3R} is the third component of SU_{2R}, $B - L$ is the U_1 current

$$B - L = 2(Q - T_{3L} - T_{3R}), \tag{92}$$

and

$$\alpha = \left[\left(\frac{g_R}{g_L} \right)^2 \left(\frac{1 - \sin^2\theta_W}{\sin^2\theta_W} \right) - 1 \right]^{1/2}, \tag{93}$$

SO_{10}	SU_5	$2\sqrt{10}Q_\chi$	$\sqrt{24}Q_\psi$	$2\sqrt{15}Q_\eta$
16	$10\ (u,d,\bar{u},e^+)_L$	-1	1	-2
	$5^*(\bar{d},\nu,e^-)_L$	3	1	1
	$1\bar{N}_L$	-5	1	-5
10	$5\ (D,\bar{E}^0,E^+)_L$	2	-2	4
	$5^*(\bar{D},E^0,E^-)_L$	-2	-2	1
1	$1S_L^0$	0	4	-5

Table 10: Couplings of the Z_χ^0, Z_ψ^0, and Z_η^0 to a 27-plet of E_6. The SO_{10} and SU_5 representations are also indicated. The couplings are shown for the left-handed (L) particles and antiparticles. The couplings of the right-handed particles are minus those of the corresponding L antiparticles. The D is an exotic SU_2-singlet quark with charge $-1/3$. $(E^0, E^-)_{L,R}$ is an exotic lepton doublet with vector SU_2 couplings. N and S are new Weyl neutrinos which may have large Majorana masses.

which is $\simeq 2.53$ for left-right symmetry ($g_R = g_L$). Similarly, many grand unified theories lead to extra U_1's when they break, such as [117, 118]

$$\begin{aligned} SO_{10} &\rightarrow SU_5 \times U_{1\chi} \\ E_6 &\rightarrow SO_{10} \times U_{1\psi} \\ E_6 &\rightarrow SU_3 \times SU_2 \times U_1 \times U_{1\eta} \end{aligned} \tag{94}$$

The corresponding charges are shown in Table 10. The Z_χ couplings are a special case of the Z_{LR}, corresponding to $g_R = g_L$ and $\sin^2\theta_W = 3/8$ (*i.e.*, $\alpha = \sqrt{2/3} = 0.82$, while

$$J_\eta = \sqrt{\frac{3}{8}} J_\chi - \sqrt{\frac{5}{8}} J_\psi \tag{95}$$

emerges in a particular E_6-breaking that is motivated by some superstring-inspired models. The gauge coupling g_2 of the extra Z is given by

$$\left(\frac{g_2}{g_1}\right)^2 = \frac{5}{3}\sin^2\theta_W \lambda_g, \tag{96}$$

where $g_1 = g/\cos\theta_W$ and $\lambda_g \leq 1$ depends on the symmetry-breaking pattern [117] ($\lambda_g = 1$ by construction for the Z_{LR}).

The above models all involve family-universal fermion couplings, which ensures the absence of flavor changing neutral currents. The possibility of a Z' which couples only to the third family is discussed in [109].

In the presence of one extra Z, there are two observed mass eigenstate bosons Z_1 and Z_2, with masses M_1, M_2, which are related to the weak eigenstates $Z_{1,2}^0$ by a mixing angle θ:

$$\begin{aligned} Z_1 &= Z_1^0\cos\theta + Z_2^0\sin\theta \\ Z_2 &= -Z_1^0\sin\theta + Z_2^0\cos\theta. \end{aligned} \tag{97}$$

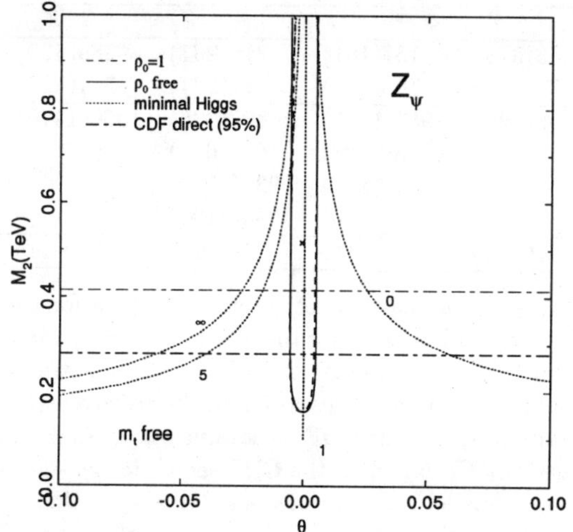

Figure 24: Same as 23, only for Z_ψ. In the minimal-Higgs case, $\sigma = 0$, 1, 5, and ∞.

I assume that Z_1 is the observed Z. The Z_1^0 is the $SU_2 \times U_1$ model boson, and the Z_2^0 has the couplings of the new U_1' group. The presence of the extra Z modifies the neutral current and Z-pole results in three ways: (a) the Z_1 mass M_1 is reduced compared to the standard model prediction due to mixing, (b) the Z_1 couplings are changed by mixing (this can be especially important for g_{Ve}, which is small in the standard model), and (c) Z_2 exchange contributes to neutral current processes. All of these must be included in global fits to the data[25].

Indirect limits on the mass M_2 and mixing angle θ can be obtained from the Z, W, and neutral current data. Because of large number of possible couplings to fermions it is easiest to consider specific models, such as the Z_χ, Z_ψ, Z_η, and Z_{LR}. One can consider three versions of each model, depending on how much is assumed concerning the Higgs sector of the theory.

In each case, one has [120]

$$\tan^2\theta = \frac{M_0^2 - M_1^2}{M_2^2 - M_0^2}, \tag{98}$$

where $M_{1,2}$ are the physical $Z_{1,2}$ masses, and M_0 is the Z_1^0 mass before mixing, *i.e.*,

$$M_0 = \frac{M_W}{\sqrt{\rho_0 \hat{\rho}} \hat{c}_Z} = \frac{M_W}{\sqrt{\rho_0} c_W}, \tag{99}$$

[25]A further extension, in which one allows for the simultaneous effects of an extra Z and mixing between ordinary and exotic fermions, is discussed in [119].

	Z_χ	Z_ψ	Z_η	Z_{LR}	Z'
$\rho_0 =$ free	353(378)	167(181)	216(237)	389(420)	951(1050)
$\rho_0 = 1$	334(352)	168(182)	217(237)	391(422)	958(1050)
$\sigma = 0$	919(1020)	954(1040)	407(436)	1360(1470)	–
$\sigma = 1$	–	167(181)	614(673)	–	–
$\sigma = 5$	–	791(851)	930(1010)	–	–
$\sigma = \infty$	–	1020(1100)	1090(1190)	–	–
direct	360-425	280-415	290-440	350-445	505

Table 11: Lower limits on the mass M_2 of an extra Z boson for the $Z_\chi, Z_\psi, Z_\eta, Z_{LR}$, and Z' models. The indirect limits are for the $\rho_0 =$free, $\rho_0 = 1$, and for minimal Higgs models with $\sigma = 0, 1, 5, \infty$. (For the Z_χ and Z_{LR} the limits are independent of σ.) Both the 95% CL and 90% (in parentheses) CL limits are given. In all cases, $\hat{s}^2_Z$, m_t, and α_s are free parameters, and the CDF constraint $m_t = 175 \pm 16$ GeV is included. The direct limits are 95% CL based on the CDF search for $\bar{p}p \to Z' \to e^+e^-$ [123]. The range of values for each model is for various possibilities for $B(Z_2 \to e^+e^-)$, ranging from the assumption that the only open decay channels are the known fermions (strongest limits), to the case that decays can also occur into exotic fermion and superpartner channels. Updated from [124].

where ρ_0 is given by (72), just as in the standard model. Eqn. (99) is written at tree-level. The normal $SU_2 \times U_1$ radiative corrections must of course be applied to all formulae. Since, we are searching for small tree-level perturbations, it is a good first approximation[26] to ignore contributions of the extra Z to the radiative corrections.

In the unconstrained-Higgs (ρ_0 free) version, no assumption is made concerning the quantum numbers of the Higgs fields which break the $SU_2 \times U_1 \times U_1'$ symmetry. This is analogous to the $\rho_0 \neq 1$ extension of the standard model (due to Higgs triplets, etc.). In this case, ρ_0 and hence M_0 are unconstrained, and $\hat{s}^2_Z$, M_1, M_2, and θ are all free parameters (as are m_t and α_s). There are so many experimental constraints, however, that these can all be simultaneously determined or constrained. The 90 and 95% CL lower limits on M_2 for the χ, ψ, η, and LR models and for a heavy Z' with the same couplings as the ordinary Z (such couplings are not expected in gauge theories, but are useful for comparison) are shown in Table 11, and the allowed regions in M_2 and θ in Figures 23–26. The limits on the mixing angle θ are shown for each model in Table 12. In all fits, the CDF constraint $m_t = 174 \pm 16$ GeV is included. The values of $\hat{s}^2_Z$, α_s, m_t, and $\rho_0 - 1$ are almost identical to the $SU_2 \times U_1$ case.

In the constrained-Higgs ($\rho_0 = 1$) case one assumes that SU_{2L} breaking is due to Higgs doublets, but no assumption is made concerning their U_1' charges. In this case, $\rho_0 = 1$, so M_0 is known. The free parameters are $\hat{s}^2_Z$, M_2, θ (and m_t and

[26]Some care must be applied when using the on-shell definition [121], and there is a small correction due to the Z_2 contribution to muon decay [122].

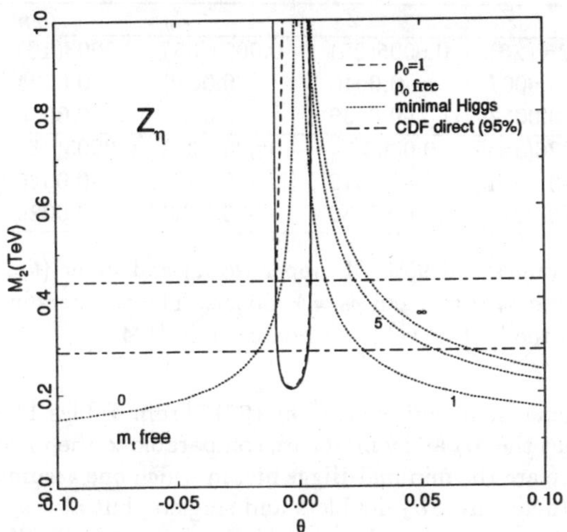

Figure 25: Same as 24, only for Z_η.

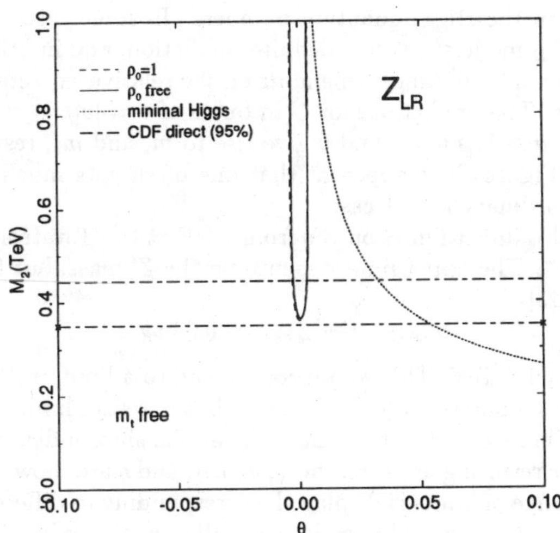

Figure 26: Same as 23, only for Z_{LR}, assuming $g_R = g_L$.

	Z_χ	Z_ψ	Z_η	Z_{LR}	Z'
$\rho_0 =$ free	-0.0025 (26)	$-0.0005(25)$	$-0.0024(35)$	$0.0003(16)$	$-0.0007(12)$
$\theta_{\rm min}$	-0.0067	-0.0046	-0.0080	-0.0024	-0.0027
$\theta_{\rm max}$	$+0.0017$	$+0.0036$	$+0.0034$	$+0.0029$	$+0.0013$
$\rho_0 = 1$	$-0.0026(19)$	0.0000^{+20}_{-32}	$-0.0023(34)$	$0.0003(18)$	$-0.0006(12)$
$\theta_{\rm min}$	-0.0054	-0.0046	-0.0079	-0.0025	-0.0027
$\theta_{\rm max}$	$+0.0013$	$+0.0036$	$+0.0033$	$+0.0029$	$+0.0010$

Table 12: Best fit values and 95% CL upper ($\theta_{\rm max}$) and lower ($\theta_{\rm min}$) limits on the mixing angle θ for the $\rho_0 =$ free and $\rho_0 = 1$ models. The numbers in parentheses are the uncertainties in the best fit values. Updated from [124].

α_s). M_1 is not independent, but is given by (98). From Tables 11 and 12 and the Figures one sees that the M_2 and θ limits are comparable to the unconstrained case.

Finally, there are the minimal-Higgs fits, in which one assumes not only that the relevant Higgs fields are SU_2 doublets and singlets, but also specific values for their U_1' charges. Since the same Higgs fields implement both SU_2 breaking and $Z - Z'$ mixing, such models lead to a relation

$$\theta = C\frac{M_1^2}{M_2^2}, \tag{100}$$

where C depends on the Higgs quantum numbers. In some cases, such as the Z_χ model and some Z_{LR} models[27], C is a definite prediction, and in others, such as Z_ψ and Z_η, it varies over a finite range depending on the relative expectation values σ of two Higgs doublets. The predictions for C in terms of $\sigma = |\bar{v}|/|v|$ are given in [124]. It is expected that $\sigma > 1$, since $\bar{v}$ and v give rise to m_t and m_b, respectively. From Table 11 and the Figures it is apparent that one often gets much more stringent limits than in the ordinary $\rho_0 = 1$ case.

There are also direct limits on Z's from CDF at the Tevatron, from searches for $\bar{p}p \to Z' \to e^+e^-$. The upper limit depends on the Z' mass, but for the 500 GeV range is around [123]

$$\sigma_{\bar{p}p\to Z'}\ B_{Z'\to e^+e^-} < 0.33\, pb \tag{101}$$

at 95% CL at $\sqrt{s} = 1.8$ TeV. This would correspond to a limit of $M > 505$ GeV for a Z' with the same couplings as the ordinary Z, but a weaker limit on more realistic Z's. It is safe to ignore $Z - Z'$ mixing in the analysis, since indirect limits indicate that θ is small. The resulting limits on the χ, ψ, LR, and η are shown in Table 11 and in the figures. A range of limits is displayed, corresponding to different assumptions concerning the $Z' \to e^+e^-$ branching ratio, *i.e.*, whether the only open channels are the known fermions (strongest limits), or whether decays into exotic fermions and superpartners are allowed (weakest limits). The direct limits are currently slightly

[27]Those for which the SU_{2R} breaking is due to Higgs triplets [1].

stronger than the indirect limits for the unconstrained and constrained Higgs cases, but weaker than most of the limits for the minimal-Higgs models.

Altogether, the current limits on the masses of extra Z's are rather weak (typically ~ 400 GeV) except when specific Higgs representations are assumed, and depend strongly on the U_1' charges. However, there are rather strong limits on the mixing angles, mainly from the Z-pole data. Further implications are discussed in [124, 125], future prospects for indirect searches are considered in [2], and the direct discovery and diagnostic potential of future hadron and e^+e^- colliders in [126].

4.10 *Exotic Fermions*

The known fermions are all ordinary [86], *i.e.*, the left-handed particles transform as weak doublets, and the right-handed particles as singlets:

$$\begin{pmatrix} \nu_e \\ e^- \end{pmatrix}_L, \begin{pmatrix} u \\ d \end{pmatrix}_L, \quad e_R^-, \; u_R, \; d_R \tag{102}$$

It is of course possible that there are additional (sequential) families of ordinary fermions. However, the LEP constraint $N_\nu = 2.988 \pm 0.023$ [3] from the invisible Z width excludes sequential families unless the additional neutrinos are far heavier ($m_\nu \gtrsim 40$ GeV) than those of the known families. Another constraint comes from the $S_{\rm new}$ parameter, *i.e.*, the shift in M_Z induced by vacuum polarization diagrams involving the new families (see section 4.8). From the current value $S_{\rm new} = -0.21 \pm 0.24^{-0.08}_{+0.17}$ and the contribution

$$\Delta S = \frac{2}{3\pi}[(N_L - 3) + N_R] \tag{103}$$

of N_L ordinary families and N_R mirror (right-handed doublet, left-handed singlet) families, we obtain the 90% CL limit

$$N_L - 3 + N_R < 1.6, \tag{104}$$

i.e., it is unlikely that there are more than 1 or 2 additional families, even if the associated neutrinos are very heavy. The constraints on exotic fermions from both charged and neutral current processes are described in detail by London and by Herczeg in this volume, and in [86].

4.11 *Four-Fermi Operators and Leptoquarks*

Many types of new physics, such as compositeness, dynamical symmetry breaking, leptoquarks, and supersymmetry lead to new effective four-fermi operators (in supersymmetry they are expected to be small for the interesting models). These generally lead to flavor changing neutral current effects. Even if these are somehow evaded, there are often flavor-conserving effects which contribute to charged and neutral current processes, universality violation, etc. [2, 70, 71].

The Z-pole observables, although most precise, are only sensitive to the properties of the Z and are essentially blind to such operators[28]. However, neutral current processes are often quite sensitive. In particular, atomic parity violation (APV) is sensitive to those new operators which shift the values of C_{1u} and C_{1d} in (50) in a direction orthogonal to the standard model band in Figure 14.

As an example, consider the effective operator

$$L_{\text{new}} = \pm \frac{4\pi}{\Lambda_{\pm}^2} \bar{e}_L \gamma_\mu e_L \bar{q}_L \gamma^\mu q_L, \tag{105}$$

which shifts C_{ij} by

$$\begin{aligned} \Delta C_{1u} &= \Delta C_{1d} = \Delta C_{2u} = \Delta C_{2d} = \frac{\mp\sqrt{2}\pi}{G_F \Lambda_{\pm}^2} \\ &= -0.0029 \pm 0.0023, \end{aligned} \tag{106}$$

where the second line is the constraint from current data [2, 70]. This implies

$$\begin{aligned} \Lambda_+ &< 7.6 \text{ TeV} \\ \Lambda_- &< 21 \text{ TeV}, \end{aligned} \tag{107}$$

at 95% C.L. This is already a stronger limit than will be be attainable at HERA, although HERA would also be sensitive to parity-conserving operators. Future APV experiments (see Masterson and Wieman, this volume) will be even more sensitive.

As another example, consider a charge $-\frac{1}{3}$ scalar leptoquark S, such as in predicted by some grand unified theories [2], with an interaction

$$-L = \left[\eta_L(\bar{u}_R^c e_L - \bar{d}_R^c \nu_L) + \eta_R \bar{u}_L^c e_R\right] S + H.C. \tag{108}$$

This implies

$$\begin{aligned} \Delta C_{1u} &= \Delta C_{2u} = \frac{\mp\sqrt{2}\eta_{L,R}^2}{8 G_F M_S^2} \\ &= -0.0064 \pm 0.0047, \end{aligned} \tag{109}$$

where the current experimental constraint in the second line implies

$$M_S > \begin{cases} 315 \text{ GeV} |\eta_L| / \sqrt{4\pi\alpha} \\ 1040 \text{ GeV} |\eta_R| / \sqrt{4\pi\alpha}. \end{cases} \tag{110}$$

[28]On the Z-pole, new operators are out of phase and do not interfere with the Z amplitude. Off peak there may be interference, but the Z amplitude is suppressed.

(Universality constraints imply that either η_L or η_R should be negligibly small for relevant M_S.) These limits equal or exceed what will be attainable at HERA for electromagnetic-strength leptoquark couplings, although HERA will be sensitive to light ($M_S < O(300$ GeV)) leptoquarks even for very small couplings. The APV limits on M_S should be improved by a factor of 4 or so. There are also significant constraints on leptoquarks from $Z\ell\bar{\ell}$ vertex corrections [127]. Leptoquarks are discussed in more detail in this volume by Herczeg and by Deutsch and Quin.

5 Conclusions

- The precision data have confirmed the standard electroweak model. However, there are possible hints of discrepancies at the 2 – 3 σ level in $\Gamma(b\bar{b})/\Gamma(\text{had})$ and A^0_{LR}.

- The data not only probes the tree-level structure, but the electroweak loops have been observed at the 2σ level. These consist of much larger fermionic pieces involving the top quark and QED, which only partially cancel the bosonic loops. The bosonic loops, which probe non-abelian vertices and gauge-Higgs vertices, are definitely needed to describe the data.

- The global fit to the data within the standard model yields

$$\begin{aligned} \overline{MS}: \hat{s}^2_Z &= 0.2317(3)(2) & m_t &= 175 \pm 11\,^{+17}_{-19} \\ \text{on}-\text{shell}: s^2_W \equiv 1 - \frac{M^2_W}{M^2_Z} &= 0.2243(12) & \alpha_s(M_Z) &= 0.127(5)(2) \\ \text{effective}: \bar{s}^2_\ell &= 0.2320(3)(2) \end{aligned} \tag{111}$$

  where the second uncertainty is from M_H. The prediction for m_t is in remarkable agreement with the value $m_t = 174 \pm 16$ suggested by the CDF events. The data has also allowed, for the first time, a clean and precise extraction of α_s from the lineshape. This is in excellent agreement with the value $\alpha_s(M_Z) = 0.123 \pm 0.005$ from event shapes. Both are larger than many of the low energy determinations when extrapolated to the Z-pole. The lineshape determination, however, is sensitive to the presence of new physics which affects the $Zb\bar{b}$ vertex or the hadronic widths.

- The agreement between the indirect prediction for m_t with the tentative direct CDF observation and of α_s with the various other determinations is an impressive success for the entire program of precision observables.

- Combining the direct CDF value of m_t with the indirect constraints does not make a large difference within the context of the standard model. However, when one goes beyond the standard model, the direct m_t allows a clean extraction of the new physics contributions to ρ_0, which is now shown to be very close to unity, $\rho_0 = 1.0012(17)(17)$. This strongly limits Higgs triplet vacuum expectation values and non-degenerate heavy multiplets. Similarly, it allows an extraction of the new physics contributions to $S_{\rm new}$, $T_{\rm new}$, $U_{\rm new}$, which are consistent with zero. Finally, one can determine the new physics contributions to the $b\bar{b}$ vertex: $\delta_{bb}^{\rm new}$ is approximately 2.3σ away from zero, reflecting the large value of the $b\bar{b}$ width.

- The data exhibit a slight preference for a light Higgs, but this is not very compelling statistically. One finds only $M_H \leq 570(880)$ GeV at 90(95%) CL. Furthermore, the preference depends crucially on the large observed value of $\Gamma(b\bar{b})$, and to a lesser extent on the large SLD value for A_{LR}^0. Omitting these values the M_H dependence of the observables is weak.

- The major prediction of supersymmetry is that one does not expect large deviations in the precision observables. The new particles tend to be heavy and decouple. One implication that is relevant, however, is that supersymmetric theories have a light standard model-like Higgs. They therefore favor the lighter Higgs mass and the lower end of the predicted m_t range. Also, the observed gauge couplings are consistent with the coupling constant unification expected in supersymmetric grand unification, but not with the simplest version of non-supersymmetric unification.

- In compositeness and dynamical symmetry breaking theories one typically expects not only large flavor changing neutral currents but significant deviations of ρ_0 from unity and of $S_{\rm new}$ and $T_{\rm new}$ from zero. One further expects that $\delta_{bb}^{\rm new} < 0$, at least in the simplest models. Therefore, the precision experiments are a major difficulty for this class of models.

1. For reviews, see R.N. Mohapatra, *Unification and Supersymmetry* (Springer, New York, 1986); P. Langacker and S. Uma Sankar, *Phys. Rev.* **D40**, 1569 (1989).
2. P. Langacker, M. Luo, and A. K. Mann, *Rev. Mod. Phys.* **64**, 87 (1992).
3. D. Schaile, this volume, and plenary talk presented at the *27th International Conference on High Energy Physics*, Glasgow, July 1994.
4. The LEP Collaborations, CERN/PPE/93-157.
5. SLD: K. Abe *et al., Phys. Rev. Lett.* **73**, 25 (1994); **70**, 2515 (1993).
6. D. Schaile and P. M. Zerwas, *Phys. Rev.* **D45**, 3262 (1992).
7. P. Langacker, *Comm. Nucl. Part. Sci.* **19**, 1 (1989).
8. A. Blondel, this volume.
9. DØ: S. Abachi *et al., Phys. Rev. Lett.* **72**, 2138 (1994).
10. CDF: F. Abe *et al., Phys. Rev. Lett.* **73**, 225 (1994), *Phys. Rev.* **D50**, 2966 (1994).
11. K. Einsweiler, this volume.

12. K. Hara, *Topical Workshop on $p\bar{p}$ Physics,* Tsukuba, Oct. 1993.
13. UA2: S. Alitti *et al., Phys. Lett.* **B276**, 354 (1992).
14. Boulder: M. C. Noecker *et al., Phys. Rev. Lett.* **61**, 310 (1988).
15. CHARM II: P. Vilain *et al., Phys. Lett.* **B281**, 159 (1992).
16. CCFR: C. G. Arroyo *et al., Phys. Rev. Lett.* **72**, 3452 (1994).
17. F. Perrier, this volume.
18. U. Amaldi *et al., Phys. Rev.* **D36**, 1385 (1987); P. Langacker and M. Luo, *Phys. Rev.* **D44**, 817 (1991); G. Costa *et al., Nucl. Phys.* **B297**, 244 (1988).
19. P. Langacker and J. Erler, in *Reviews of Particle Properties*, *Phys. Rev.* **D50**, 1304, 1312 (1994).
20. J. Erler and P. Langacker, Pennsylvania preprint UPR-0632T.
21. See the articles in Chapters III and IV of this volume.
22. For recent reviews, see the articles by W. Hollik and W. Marciano, this volume. Early references are given in [2, 19].
23. S. Fanchiotti, B. Kniehl, and A. Sirlin, *Phys. Rev.* **D48**, 307 (1993).
24. F. Jegerlehner, in *Testing the Standard Model*, ed. M. Cvetic and P. Langacker (World Scientific, Singapore, 1991), p 476.
25. M. Veltman, *Nucl. Phys.* **B123**, 89 (1977); M. Chanowitz, M. A. Furman, and I. Hinchliffe, *Phys. Lett.* **B87**, 285 (1978).
26. R. Barbieri *et al., Phys. Lett.* **B288**, 95 (1992); *Nucl. Phys.* **B409**, 105 (1993).
27. A. Djouadi and C. Verzegnassi, *Phys. Lett.* **B195**, 265 (1987); *Nuovo Cimento* **100A**, 357 (1988).
28. B. H. Smith and M. B. Voloshin, Minnesota preprint UMN-TH-1241/94.
29. A. Sirlin, *Phys. Rev.* **D22**, 971 (1980); W. Marciano and A. Sirlin, *Phys. Rev.* **D22**, 2697 (1980).
30. V. A. Novikov, L. B. Okun, and M. I. Vysotsky, *Nucl. Phys.* **B397**, 35 (1993).
31. W. Marciano and A. Sirlin, *Phys. Rev.* **D27**, 552 (1983); S. Sarantakos, A. Sirlin, and A. Marciano, *Nucl. Phys.* **B217**, 84 (1983).
32. W. J. Marciano and J. L. Rosner, *Phys. Rev. Lett.* **65**, 2963 (1990).
33. G. Degrassi, S. Fanchiotti, and A. Sirlin, *Nucl. Phys.* **B351**, 49 (1991).
34. ZFITTER: D. Bardin *et al.,* CERN-TH.6443/92.
35. P. Gambino and A. Sirlin, *Phys. Rev.* **D49**, 1160 (1994).
36. References may be found in [19].
37. See J. Fleischer *et al., Phys. Lett.* **B293**, 437 (1992); K. G. Chetyrkin *et al., Mod. Phys. Lett.* **A8**, 2785 (1993), and [19, 20].
38. W. Beenakker and W. Hollik, *Z. Phys.* **C40**, 141 (1988); A. A. Akhundov *et al., Nucl. Phys.* **B276**, 1 (1986).

39. J. Bernabeu, A. Pich, and A. Santamaria, *Nucl. Phys.* **B363**, 326 (1991).
40. For a review, see P. Langacker, in *Testing the Standard Model*, ed. M. Cvetic and P. Langacker (World, Singapore, 1991), p. 863.
41. K. A. Olive *et al., Phys. Lett.* **B236**, 454 (1990); T. P. Walker *et al., Astrophys. J.* **376**, 51 (1991).
42. For reviews, see I. Hinchliffe in *Reviews of Particle Properties, Phys. Rev.* **D50**, 1297 (1994); S. Bethke, in *Radiative Corrections: Status and Outlook*, Gatlinburg, TN, June, 1994; B. R. Webber, in *27th International Conference on High Energy Physics*, Glasgow, July 1994, Cavendish-HEP-94/15.
43. L. Montanet *et al., Phys. Rev.* **D50**, 1180 (1994).
44. P. Janot, invited talk at *Neutrino 94*, Eilat, Israel, May 1994, Orsay LAL-94-59.
45. For reviews, see J. Gunion *et al., The Higgs Hunter's Guide,* (Addison-Wesley, Redwood City, 1990); M. Sher, *Phys. Reports* **179**, 273 (1989).
46. A. L. Kataev and V. V. Starshenko, CERN-TH-7198/94.
47. P. Langacker and N. Polonsky, *Phys. Rev.* **D47**, 4028 (1993); N. Polonsky, *Unification and Low-Energy Supersymmetry at One and Two-Loop Orders*, Univ. of Pennsylvania Ph.D. thesis, 1994.
48. A. X. El-Khadra, Ohio State OHSTPY-HEP-T-93-020.
49. A. X. El-Khadra *et al., Phys. Rev. Lett.* **69**, 729 (1992).
50. C. T. H. Davies *et al.,* OHSTPY-HEP-T-94-013.
51. L. Clavelli, *Phys. Rev.* **D46**, 2112 (1992); L. Clavelli, P. Coulter, and K. Yuan, *Phys. Rev.* **D47**, 1973 (1993); J. Blümlein and J. Botts, *Phys. Lett.* **B325**, 190 (1994). See also I Antoniadis, J. Ellis, and D. V. Nanopoulos, *Phys. Lett.* **B262**, 109 (1991).
52. P. Gambino and A. Sirlin, *Phys. Rev. Lett.* **73**, 621 (1994).
53. S. Dittmaier *et al., Nucl. Phys.* **B426**, 249 (1994); D. Schildknecht, BI-TP 94/18.
54. For reviews and analyses, see J. E. Kim *et al., Rev. Mod. Phys.* **53**, 211 (1981); U. Amaldi *et al.,* [18]; G. Costa *et al.,* [18]; G.L. Fogli and D. Haidt, *Zeit. Phys.* **C40**, 379 (1988); P. Langacker and M. Luo, [18]; P. Langacker and J. Erler, [19, 20].
55. For recent results, see [16, 56, 57].
56. CDHS: H. Abramowicz *et al., Phys. Rev. Lett.* **57**, 298 (1986), A. Blondel *et al., Zeit. Phys.* **C45**, 361 (1990).
57. CHARM: J. V. Allaby *et al., Zeit. Phys.* **C36**, 611 (1987).
58. For a recent discussion, see [19].
59. C. H. Llewellyn-Smith, *Nucl. Phys.* **B228**, 205 (1983).
60. J. E. Kim *et al., Rev. Mod. Phys.* **53**, 211 (1981).

61. LANL: R.C. Allen *et al., Phys. Rev. Lett.* **64**, 1330 (1990); *Phys. Rev.* **D45**, 975 (1992), **D47**, 11 (1993).
62. Savannah River: F. Reines *et al., Phys. Rev. Lett.* **37**, 315 (1976).
63. SLAC *eD*: C. Y. Prescott *et al., Phys. Lett.* **84B**, 524 (1979).
64. Mainz: W. Heil *et al., Nucl. Phys.* **B327**, 1 (1989).
65. Bates: P.A. Souder *et al., Phys. Rev. Lett.* **65**, 694 (1990).
66. BCDMS: A. Argento *et al., Phys. Lett.* **120B**, 245 (1983); **140B**, 142 (1984).
67. M. A. Bouchiat *et al., Phys. Lett.* **134B**, 463 (1984).
68. Atomic theory: S.A. Blundell, W.R. Johnson, and J. Sapirstein, *Phys. Rev. Lett.* **65**, 1411 (1990); V. A. Dzuba *et al., Phys. Lett.* **141A**, 147 (1989).
69. For references, see [2].
70. P. Langacker, *Phys. Lett.* **B256**, 277 (1991).
71. M. Leurer, *Phys. Rev.* **D49**, 333 (1994).
72. C. Kiesling, *Standard Theory of Electroweak Interactions,* (Springer, Berlin, 1988); R. Marshall, *Z. Phys.* **C43**, 607 (1989); Y. Mori *et al., Phys. Lett.* **B218**, 499 (1989).
73. Some recent analyses of extra Z bosons include P. Langacker and M. Luo, *Phys. Rev.* **D45**, 278 (1992); L. S. Durkin and P. Langacker, *Phys. Lett.* **166B**, 436 (1986); U. Amaldi *et al., Phys. Rev.* **D36**, 1385 (1987); the sensitivity of future precision experiments is discussed in [2].
74. G. Altarelli, *et al., Phys. Lett.* **B263**, 459 (1991), *ibid.* **B261**, 146 (1991); J. Layssac, F. M. Renard, and C. Verzegnassi, *Z. Phys.* **C53**, 97 (1992); F. M. Renard and C. Verzegnassi, *Phys. Lett.* **B260**, 225 (1991).
75. K. T. Mahanthappa and P. K. Mohapatra, *Phys. Rev.* **D43**, 3093 (1991); P. Langacker, *Phys. Lett.* **B256**, 277 (1991).
76. F. del Aguila, W. Hollik J. M. Moreno, and M. Quiros, *Nucl. Phys.* **B372**, 3 (1992); F. del Aguila, J. M. Moreno, and M. Quiros, *Nucl. Phys.* **B361**, 45 (1991); *Phys. Lett.* **B254**, 479 (1991); M. C. Gonzalez-Garcia and J. W. F. Valle, *Phys. Lett.* **B259**, 365 (1991).
77. P. Langacker and M. Luo, *Phys. Rev.* **D44**, 817 (1991); J. Erler and P. Langacker, [20].
78. M. Peskin and T. Takeuchi, *Phys. Rev. Lett.* **65**, 964 (1990), *Phys. Rev.* **D46**, 381 (1992); M. Golden and L. Randall, *Nucl. Phys.* **B361**, 3 (1991); B. W. Lynn, M. E. Peskin, and R. G. Stuart, in *Physics at LEP*, CERN 86-02, Vol. I, p. 90.
79. W. Marciano and J. Rosner, *Phys. Rev. Lett.* **65**, 2963 (1990); D. Kennedy and P. Langacker, *Phys. Rev. Lett.* **65**, 2967 (1990), *Phys. Rev.* **D44**, 1591 (1991); G. Altarelli and R. Barbieri, *Phys. Lett.* **B253**, 161 (1990); B. Holdom and J. Terning, *Phys. Lett.* **B247**, 88 (1990).

80. For a more detailed discussion, see P. Langacker and J. Erler, in *Reviews of Particle Properties*, *Phys. Rev.* **D50**, 1312 (1994).
81. R. Sundrum and S. D. H. Hsu, *Nucl. Phys.* **B391**, 127 (1993); R. Sundrum, *Nucl. Phys.* **B395**, 60 (1993); M. Luty and R. Sundrum, *Phys. Rev. Lett.* **70**, 529 (1993); T. Applequist and J. Terning, *Phys. Lett.* **B315**, 139 (1993).
82. H. Georgi, *Nucl. Phys.* **B363**, 301 (1991); M. J. Dugan and L. Randall, *Phys. Lett.* **B264**, 154 (1991); E. Gates and J. Terning, *Phys. Rev. Lett.* **67**, 1840 (1991).
83. For an alternative parameterization, see G. Altarelli, R. Barbieri, and S. Jadach, *Nucl. Phys.* **B369**, 3 (1992), **B376**, 444(E) (1992). See also G. Altarelli *et al.*, [87].
84. C. P. Burgess *et al.*, *Phys. Lett.* **B326**, 276 (1994), *Phys. Rev.* **D50**, 529 (1994).
85. G. Altarelli and R. Barbieri, [79].
86. P. Langacker and D. London, *Phys. Rev.* **D38**, 886 (1988); J. Maalampi and M. Roos, *Phys. Rep.* **186**, 53 (1990); E. Nardi and E. Roulet, *Phys. Lett.* **B248**, 139 (1990); E. Nardi, E. Roulet, and D. Tommasini, *Nucl. Phys.* **B386**, 239 (1992), *Phys. Rev.* **D46**, 3040 (1992), CERN-TH.7443/94.. For recent reviews, see E. Nardi, *Waikoloa 1993, Proceedings, Physics and Experiments with Linear* e^+e^- *Colliders*, vol 2, p 496; D. London, this volume; P. Herczeg, this volume.
87. A. Blondel, A. Djouadi, and C. Verzegnassi, *Phys. Lett.* **B293**, 253 (1992); G. Altarelli, R. Barbieri, and F. Caravaglios, *Nucl. Phys.* **B405**, 3 (1993).
88. A. Djouadi *et al.*, *Nucl. Phys.* **B349**, 48 (1991); M. Boulware and D. Finnell, *Phys. Rev.* **D44**, 2054 (1991); G. Altarelli *et al.*, *Phys. Lett.* **B314**, 357 (1993).
89. R. S. Chivukula, B. Selipsky, and E. H. Simmons, *Phys. Rev. Lett.* **69**, 575 (1992).
90. R. S. Chivukula *et al.*, *Phys. Lett.* **B311**, 157 (1993).
91. R. S. Chivukula, E. H. Simmons, and J. Terning, *Phys. Lett.* **B331**, 383 (1994).
92. R. Barbieri *et al.*, *Nucl. Phys.* **B341**, 309 (1990); J. Ellis, G. L. Fogli, and E. Lisi, *Nucl. Phys.* **B393**, 3 (1993); H. E. Haber, Santa Cruz SCIPP 93/06.
93. Recent studies of the implications of the couplings for grand unification include: P. Langacker and M. Luo, *Phys. Rev.* **D44**, 817 (1991); U. Amaldi, W. de Boer and H. Fürstenau, *Phys. Lett.* **B260**, 447 (1991). J. Ellis, S. Kelley and D. V. Nanopoulos, *Phys. Lett.* **B249**, 441 (1990). F. Anselmo, L. Cifarelli, A. Peterman, and A. Zichichi, *Nuo. Cim.* **104A**, 1817 (1991).
94. H. Georgi and S. L. Glashow, *Phys. Rev. Lett.* **32**, 438 (1974).
95. For reviews, see P. Langacker, *Phys. Rep.* **C72**, 185 (1981); *Ninth Work-*

shop on Grand Unification, ed. R. Barloutaud (World, Singapore, 1988) p3; *Radiative Corrections: Status and Outlook*, Gatlinburg, TN, June, 1994, UPR-0639T; G. G. Ross, *Grand Unified Theories* (Benjamin, 1985).

96. L. E. Ibáñez and G. G. Ross, *Phys. Lett.* **105B**, 439 (1981); S. Dimopoulos, S. Raby, and F. Wilczek, *Phys. Rev.* **D24**, 1681 (1981); D.R.T. Jones, *Phys. Rev.* **D25**, 581 (1982); M.B. Einhorn and D.R.T. Jones, *Nucl. Phys.* **B196**, 475 (1982); W. J. Marciano and G. Senjanovic, *Phys. Rev.* **D25**, 3092 (1982).
97. H. Georgi, H. R. Quinn and S. Weinberg, *Phys. Rev. Lett.* **33**, 451 (1974).
98. P. Langacker and N. Polonsky, [47]; R. Barbieri and L. J. Hall, *Phys. Rev. Lett.* **68**, 752 (1992); L. J. Hall and U. Sarid, *Phys. Rev. Lett.* **70**, 2673 (1993); M. Carena, S. Pokorski, and C. Wagner, *Nucl. Phys.* **B406**, 59 (1993); and references theirin.
99. U. Amaldi *et al.*, [18].
100. G. G. Ross and R. G. Roberts, *Nucl. Phys.* **B377**, 571 (1992).
101. R. Arnowitt and P. Nath, *Phys. Rev. Lett.* **69**, 725 (1992); *Phys. Lett.* **B287**, 89 (1992); *Phys. Rev.* **D46**, 3981 (1992); H. Murayama and T. Yanagida, *Phys. Rev. Lett.* **69**, 1014 (1992); *Nucl. Phys.* **B402**, 46 (1993) and references therein.
102. V. Barger, M. S. Berger and P. Ohmann, *Phys. Rev.* **D47**, 1093 (1993); P. Langacker and N. Polonsky, *Phys. Rev.* **D49**, 1454 (1994); M. Carena *et al.*, [98]; G. L. Kane, C. Kolda, L. Roszkowski, and J. D. Wells, *Phys. Rev.* **D49**, 6173 (1994), **D50**, 3498 (1994).
103. J. Ellis, G. Ridolfi, and F. Zwirner, *Phys. Lett.* **B257**, 83 (1991); **262**, 477 (1991); A. Brignole, *Phys. Lett.* **B281**, 284 (1992); H. E. Haber and R. Hempfling, *Phys. Rev.* **D48**, 4280 (1993).
104. V. Barger, M. S. Berger, P. Ohmann, and R. J. Phillips, *Phys. Lett.* **B314**, 351 (1993); P. Langacker and N. Polonsky, *Phys. Rev.* **D50**, 2199 (1994).
105. For a review, see L. Roszkowski, *32nd Cracow School of Theoretical Physics*, Zakopane, June 1992.
106. M. Cvetic and P. Langacker, *Phys. Rev.* **D46**, 2759 (1992); E. K. Akhmedov, Z. Berezhiani, and G. Senjanovic, *Phys. Rev.* **D47**, 3245 (1993).
107. L. Ibanez, D. Lust, and G. Ross, *Phys. Lett.* **B272**, 251 (1991); L. Ibanez and D. Lust, *Nucl. Phys.* **B382**, 305 (1992).
108. J. D. Wells, C. Kolda, and G. Kane, *Phys. Lett.* **B338**, 219 (1994).
109. B. Holdom, Toronto UTPT-94-20.
110. T. Takeuchi, A. Grant, and J. Rosner, Fermilab-Conf-94/279-T.
111. S. Bertolini and A. Sirlin, *Phys. Lett.* **B257**, 179 (1991).
112. A. Denner *et al.*, *Phys. Lett.* **B240**, 438 (1990).
113. M. Cvetic and P. Langacker, *Phys. Rev.* **D42**, 1797 (1990).

114. D. Lewellen, *Nucl. Phys.* **B337**, 61 (1990).

115. Possible subleties in the renormalization program are discussed by A. Pomarol, Pennsylvania UPR-0603T.

116. A. De Rújula *et al., Nucl. Phys.* **B384**, 3 (1992); C. P. Burgess and D. London, *Phys. Rev.* **D48**, 4337 (1993); C. P. Burgess *et al., Phys. Rev.* **D49**, 6115 (1994).

117. R. Robinett and J. Rosner, *Phys. Rev.* **D25**, 3036 (1982), **D26**, 2396 (1982); R. Robinett, *Phys. Rev.* **D26**, 2388 (1982).

118. P. Langacker, R. Robinett, and J. Rosner, *Phys. Rev.* **D30**, 1470 (1984).

119. E. Nardi, E. Roulet, and D. Tommasini [86].

120. P. Langacker, *Phys. Rev.* **D30**, 2008 (1984).

121. G. Degrassi and A. Sirlin, *Phys. Rev.* **D40**, 3066 (1989).

122. W. J. Marciano and A. Sirlin, *Phys. Rev.* **D35**, 1672 (1987).

123. CDF: F. Abe *et al.,* Fermilab-Pub 94-198-E.

124. P. Langacker and M. Luo, [73].

125. M. Cvetic and P. Langacker, *Phys. Rev.* **D46**, 4943 (1992).

126. F. del Aguila, M. Cvetic, and P. Langacker, *Phys. Rev.* **D48**, 969 (1993); F. del Aguila and M. Cvetic, *Phys. Rev.* **D50**, 3158 (1994).

127. G. Bhattacharyya, J. Ellis, and K. Sridhar, CERN-TH.7280/94.

Exotic Fermions

David London
Laboratoire de Physique Nucléaire
Université de Montréal
C.P. 6128, Montréal, P.Q., Canada H3C 3J7

Contents

1 Introduction

As has often been said, the standard model, although extremely successful in explaining virtually all known experimental data, cannot be the whole story – there are too many questions still left unanswered. By themselves, exotic fermions do not solve any of these problems, but they often appear in models which do address some of these remaining questions. In this chapter I will discuss the constraints which can be put upon exotic fermions, particularly as regards their mixings with the ordinary fermions.

In the standard model, all left-handed (L) fermions transform as doublets under weak $SU(2)_W$, while all right-handed (R) fermions are singlets:

$$\begin{pmatrix}\nu_e\\ e^-\end{pmatrix}_L \quad \begin{pmatrix}\nu_\mu\\ \mu^-\end{pmatrix}_L \quad \begin{pmatrix}\nu_\tau\\ \tau^-\end{pmatrix}_L \quad \begin{pmatrix}u\\ d\end{pmatrix}_L \quad \begin{pmatrix}c\\ s\end{pmatrix}_L \quad \begin{pmatrix}t\\ b\end{pmatrix}_L, \tag{1}$$

$$e^-_R \quad \mu^-_R \quad \tau^-_R \quad \begin{matrix}u_R & c_R & t_R\\ d_R & s_R & b_R\end{matrix}. \tag{2}$$

Many models which go beyond the standard model predict the existence of new fermions which transform in a non-standard way under $SU(2)_W$. In E_6 models, for example, in the 27-plet one finds, in addition to the ordinary particles, vector singlet quarks and vector doublet leptons. Vector singlet (doublet) fermions refer to particles whose L and R components both transform as singlets (doublets) under $SU(2)_W$. One also finds new $SU(2)_W$-singlet Weyl neutrinos in the 27-plet. Mirror fermions are another type of exotic fermion, whose transformation properties under $SU(2)_W$ are opposite those of ordinary fermions, i.e., left-handed singlets and right-handed doublets. These appear, for instance, in grand unified theories which include family unification [1].

The possibilities for new fermions are listed in Table 1. In the following analysis, all particles whose L and R components obey the same transformation properties as those in Eqs. 1 and 2 (i.e., L-handed doublets, R-handed singlets) will be called *ordinary*. These include the standard fermions, as well as any new sequential (e.g., fourth family) fermions. Those particles whose L and/or R components transform differently than those of ordinary fermions are called *exotic*. Note that particles with noncanonical electric or colour charges are not considered here. This restricts the 'exotic' label to mirror fermions, vector doublets and singlets, and Weyl neutrinos.

There are two ways to look for signals of exotic fermions – directly and indirectly. The best limits on direct production of such particles come from LEP [2],

$$M_N, M_{E^-}, M_U, M_D > 45 \text{ GeV}, \tag{3}$$

although the bound on M_N depends on the type of exotic neutrino. For example, the mass limit on exotic singlets can be considerably weaker. As to indirect signals, one possibility is to look for loop-induced effects in rare processes. This is a model-dependent enterprise, depending on the mass of the exotic fermions, their couplings

Sequential Fermions
$\binom{N}{E^-}_L \quad E_R\,, \quad \binom{U}{D}_L \quad \begin{matrix} U_R \\ D_R \end{matrix}$
Non-Canonical $SU(2)_W \times U(1)$ Assignments
a) Mirror Fermions
$E_L^- \quad \binom{N}{E^-}_R\,, \quad \begin{matrix} U_L \\ D_L \end{matrix} \quad \binom{U}{D}_R$
b) Vector Doublets
$\binom{N}{E^-}_L \quad \binom{N}{E^-}_R\,, \quad \binom{U}{D}_L \quad \binom{U}{D}_R$
c) Vector Singlets
$E_L^- \quad E_R^-\,, \quad U_L \quad U_R \quad D_L \quad D_R$
d) Weyl Neutrinos
$N_L \quad N_R$

Table 1: Possible $SU(2)_W \times U(1)$ assignments for new fermions. Pairs of particles enclosed in parentheses indicate $SU(2)_W$-doublets; otherwise they are $SU(2)_W$-singlets. N and E refer to leptons of charge 0 and -1, respectively; U and D are quarks of charge 2/3 and $-1/3$.

to ordinary gauge bosons, the possible existence of other gauge bosons, etc., and will not be discussed here.

The other indirect signal, which is the focus of this chapter, is to look for signs of exotic fermions through their mixings with ordinary fermions. These mixings can be analysed in a model independent way. In general, mixing between ordinary and exotic fermions will induce flavour-changing neutral currents (FCNC). The experimental absence of FCNC places extremely stringent limits on fermion mixing. However, there are directions in parameter space where it is possible to fine-tune away FCNC. Nevertheless, even these regions can be constrained by looking at data involving charged currents and flavour-conserving neutral currents. I will review here the constraints which current experimental data place upon the mixings of ordinary and exotic fermions. The material contained in this chapter comes principally from work by Langacker and London (Ref. [3]), and Nardi, Roulet and Tommasini (Ref. [4]). Where there are holes in the exposition, I refer the reader to these two articles for details.

This chapter is organized as follows. In section 2, I introduce the formalism needed to describe the mixing between ordinary and exotic fermions. For charged fermions, in order to avoid FCNC, it is necessary to consider fine-tuned directions in parameter space in which each ordinary charged fermion mixes with its own exotic fermion. In this way mixing is parametrized by one angle per ordinary (L- or R-handed) charged fermion. Neutrinos are more complicated, both due to the possibility of Dirac and Majorana masses, and because there is no empirical evidence requiring the absence of FCNC between neutrino species. However, due to the fact that neutrinos are unobserved in experiment, it is possible to parametrize mixing by one effective angle, plus one auxiliary parameter, per neutrino species. In section 3, I review the experimental data which is used to constrain mixing. Here I will include the theoretical expressions, including mixing, which are to be fitted to the experimental results. The fact that certain results are normalized to other data must be carefully taken into account to ensure that *all* mixing effects are included. The fits are given in section 4. There are enough constraints to limit all mixings of L- and R-handed ordinary fermions. Two types of fits are presented. In the first, only one particle at a time is allowed to mix. This yields the most stringent limits on fermion mixing. In the second fit, all particles can mix simultaneously, which weakens the constraints due to the possibility of fine-tuned cancellations of the effects of different particle mixings. I conclude in section 5.

2 Mixing Formalism

In this section, I present the formalism for describing the mixing of ordinary and exotic fermions. As mentioned in the introduction, charged fermions and neutrinos must be treated separately. The material in this section is taken almost completely from Ref. [3].

2.1 Charged Fermions

Since electromagnetic gauge invariance is unbroken, fermions with different charges cannot mix. Charged fermions can therefore be divided into three categories – those with $Q_{em} = 2/3$ (u-type), $Q_{em} = -1/3$ (d-type) and $Q_{em} = -1$ (e-type). For each of these types it is convenient to put the L and R gauge eigenstates of both ordinary and exotic fermions into a single vector,

$$\psi^0_{L(R)} = \begin{pmatrix} \psi^0_O \\ \psi^0_E \end{pmatrix}_{L(R)}, \tag{4}$$

in which the subscripts O and E stand for ordinary and exotic, respectively. Here and below, the superscript 0 indicates the weak-interaction basis; the mass basis is denoted by the absence of superscripts. In the above equation there are n_L (n_R) ordinary L-handed (R-handed) fields and m_L (m_R) exotic L-handed (R-handed) fields.

The light (l) and heavy (h) mass eigenstates can be written similarly,

$$\psi_{L(R)} = \begin{pmatrix} \psi_l \\ \psi_h \end{pmatrix}_{L(R)}. \tag{5}$$

The dimensionality of these vectors is as above, that is, there are n_L (n_R) light L (R) states and m_L (m_R) heavy L (R) states. Of course, the labels 'light' and 'heavy' should not be taken literally – for example, fourth generation particles (if they exist) are known to be heavy, yet they are included among the 'light' particles. This decomposition is useful as a reminder that in general we expect the light states to consist mainly of ordinary particles and the heavy states to be principally exotic.

The weak and mass eigenstates are related by a unitary transformation

$$\psi_a^0 = U_a \psi_a, \tag{6}$$

in which $a = L, R$. It is useful to write the matrix U in block form as

$$U_a = \begin{pmatrix} A_a & E_a \\ F_a & G_a \end{pmatrix}. \tag{7}$$

Since all our experimental data concerns only the light eigenstates, the important elements of U_a are the $n_a \times n_a$ submatrix A_a, which relates the light mass states and the ordinary weak states, and F_a, which is $m_a \times n_a$ and describes the overlap of the light eigenstates with the exotic fermions. The unitarity of U_a requires

$$A_a^\dagger A_a + F_a^\dagger F_a = A_a A_a^\dagger + E_a E_a^\dagger = I, \tag{8}$$

which shows that A_a is not by itself unitary. However, since we expect the light (heavy) particles to be mainly ordinary (exotic), we see that the deviation of A_a from unitarity is of second order in the mixing.

Let us now examine the effects of mixing on the neutral currents of the light fermions. In the weak basis, the coupling of the Z^0 to charged fermions can be written

$$\frac{1}{2} J_Z^\mu = \overline{\psi}_{OL}^0 I_{3W} \gamma^\mu \psi_{OL}^0 + \overline{\psi}_{ER}^0 I_{3W} \gamma^\mu \psi_{ER}^0 - \sin^2\theta_W J_{em}^\mu, \tag{9}$$

in which $I_{3W} = +1/2$ for u-type fermions, and $I_{3W} = -1/2$ for d- and e-type fermions. Using Eqs. 6 and 7, the weak neutral current can now be expressed in terms of mass eigenstates. Keeping only those terms which involve just the light states, this gives

$$\frac{1}{2} J_Z^\mu = \overline{\psi}_{lL} \gamma^\mu I_{3W} A_L^\dagger A_L \psi_{lL} + \overline{\psi}_{lR} \gamma^\mu I_{3W} F_R^\dagger F_R \psi_{lR} - \overline{\psi}_l \gamma^\mu Q_{em} \sin^2\theta_W \psi_l. \tag{10}$$

The important point to recognize here is that, since neither A_L nor F_R is unitary (Eq. 8), $A_L^\dagger A_L$ and $F_R^\dagger F_R$ are not necessarily diagonal. In other words, FCNC will in general be induced among the light particles.

It is useful to parametrize the FCNC between the light particles i and j as

$$\lambda_{ij}^L = \left(A_L^\dagger A_L\right)_{ij} = -\left(F_L^\dagger F_L\right)_{ij}, \quad \lambda_{ij}^R = \left(F_R^\dagger F_R\right)_{ij}, \quad i \neq j. \tag{11}$$

Note that these are of second order in light-heavy mixing. As can be seen from Table 2, the constraints on the $\lambda_{ij}^{L,R}$, $i \neq j$ are quite stringent, which strongly limits the mixing of ordinary and exotic fermions. However, it is possible to evade these bounds by considering the fine-tuned cases in which both $A_L^\dagger A_L$ and $F_R^\dagger F_R$ are diagonal. These

Quantity	Upper Limit	Source
$\lvert\lambda_{\mu e}\rvert$	1×10^{-6}	$\mu \not\to 3e$ [2]
$\lvert\lambda_{\mu\tau}\rvert$, $\lvert\lambda_{e\tau}\rvert$	7×10^{-3}	$\tau \not\to 3\ell$ [2]
$\lvert\lambda_{ds}\rvert$	6×10^{-4}	$\Delta m_{K_L K_S}$ [2]
	1×10^{-5}	$K_L \to \mu^+\mu^-$ [2]
$\lvert\lambda_{cu}\rvert$	1×10^{-3}	D^0-$\overline{D^0}$ mixing [2]
$\lvert\lambda_{bd}\rvert$, $\lvert\lambda_{bs}\rvert$	2×10^{-3}	$B \not\to \ell^+\ell^- X$ [5]

Table 2: Limits on the flavour changing neutral current parameters λ_{ij} (Eq. 11). The bounds on leptonic FCNC are taken or adapted from Ref. [6], while the limits on hadronic FCNC are taken, updated or adapted from Ref. [7]. There is no bound on $|\lambda_{bd}|$ from B_d^0-$\overline{B_d^0}$ mixing because this mixing can in principle be explained by a nonzero λ_{bd} [7].

correspond to those directions in mixing parameter space in which each ordinary fermion mixes with its own exotic fermion. For the rest of this chapter, I will assume $\lambda_{ij}^{L,R} = 0$ for $i \neq j$.

With this (strong) assumption, using Eq. 8 one can write

$$\left(A_a^\dagger A_a\right)_{ij} = \left(c_a^i\right)^2 \delta_{ij} , \qquad \left(F_a^\dagger F_a\right)_{ij} = \left(s_a^i\right)^2 \delta_{ij} , \qquad a = L, R, \tag{12}$$

in which $\left(s_a^i\right)^2 \equiv 1 - \left(c_a^i\right)^2 \equiv \sin^2\theta_a^i$, where $\theta_{L(R)}^i$ is the mixing angle of the ith L-handed (R-handed) ordinary fermion and its exotic partner. With this notation, the neutral current in Eq. 10 becomes

$$\frac{1}{2} J_Z^\mu = \sum_i \left[\overline{\psi}_{iL}\gamma^\mu \tilde{\epsilon}_L(i)\psi_{iL} + \overline{\psi}_{iR}\gamma^\mu \tilde{\epsilon}_R(i)\psi_{iR}\right] , \tag{13}$$

where the sum is over the light particles and

$$\begin{aligned} \tilde{\epsilon}_L(i) &= I_{3W}^i \left(c_L^i\right)^2 - Q_{em}^i \sin^2\theta_W , \\ \tilde{\epsilon}_R(i) &= I_{3W}^i \left(s_R^i\right)^2 - Q_{em}^i \sin^2\theta_W . \end{aligned} \tag{14}$$

From Eqs. 13 and 14, the effects of mixing are clear. First, the mixing of ordinary L doublets with exotic L singlets results in a nonuniversal reduction ($(c_L^i)^2$) of the isospin current. Second, mixing in the R-handed sector induces a R-handed current ($(s_R^i)^2$). The electromagnetic current is unchanged, reflecting the simple fact that only particles of the same charge can mix. In the presence of mixing, the vector and axial couplings for fermion i are

$$\begin{aligned} v_i &\equiv \tilde{\epsilon}_L(i) + \tilde{\epsilon}_R(i) = I_{3W}^i \left[\left(c_L^i\right)^2 + \left(s_R^i\right)^2\right] - 2Q_{em}^i \sin^2\theta_W , \\ a_i &\equiv \tilde{\epsilon}_L(i) - \tilde{\epsilon}_R(i) = I_{3W}^i \left[\left(c_L^i\right)^2 - \left(s_R^i\right)^2\right] . \end{aligned} \tag{15}$$

The hadronic charged current involving the light quarks is

$$\frac{1}{2}J_W^{\mu\dagger} = \overline{\psi}_{uL}\gamma^\mu V_L \psi_{dL} + \overline{\psi}_{uR}\gamma^\mu V_R \psi_{dR}, \tag{16}$$

in which ψ_{uL} and ψ_{dL} are column vectors of the light L u-type and d-type quarks, respectively. Recall that 'light' L-handed particles include possible extra sequential or vector doublet quarks. Thus, the first 3 components of ψ_{uL} and ψ_{dL} are the standard quarks, while the remaining $n_l - 3$ quarks are nonstandard. The column vectors ψ_{uR} and ψ_{dR} are defined completely analogously. In Eq. 16, $V_L = A_L^{u\dagger} A_L^d$ is the generalized Cabibbo-Kobayashi-Maskawa (CKM) matrix. The point to observe here, however, is that V_L is non-unitary in the presence of mixing between the ordinary and exotic fermions. It can be decomposed as

$$V_{Lij} = c_L^{u_i} c_L^{d_j} \widehat{V}_{Lij}, \tag{17}$$

where $\widehat{V}_L$ is the true (unitary) CKM matrix. Here and below I use the term 'true' to refer to a quantity in the absence of mixing, and I denote this by a symbol with a caret. 'Apparent' quantities, which are represented by symbols with no caret, are those which are actually measured. In Eq. 17, we see that apparent CKM matrix elements are reduced from their true values by the nonuniversal factor $c_L^{u_i} c_L^{d_j}$. If ψ_{uL} and/or ψ_{dL} contain nonstandard 'light' quarks, this will manifest itself through the apparent nonunitarity of $\widehat{V}_L$. The second term in Eq. 16 is a R-handed charged current, induced when both R-handed u_i and d_j quarks mix with exotic $SU(2)_W$-doublets. Like V_L, the apparent R-handed CKM matrix V_R is non-unitary, but can be written

$$V_{Rij} = s_R^{u_i} s_R^{d_j} \widehat{V}_{Rij}, \tag{18}$$

where $\widehat{V}_R$ is unitary.

2.2 Neutrinos

As mentioned in the introduction, neutrinos must be treated separately for several reasons. First of all, there are three types of L-handed neutrino weak eigenstates:

$$\begin{pmatrix} n_{OL}^0 \\ e_L^{0-} \end{pmatrix}, \quad \begin{pmatrix} e_L^{0+} \\ n_{EL}^0 \end{pmatrix}, \quad n_{SL}^0 . \tag{19}$$

Here, the n_{OL}^0 are ordinary $SU(2)_W$-doublets with $I_{3W} = 1/2$, the n_{EL}^0 are exotic $SU(2)_W$-doublets with $I_{3W} = -1/2$, and the n_{SL}^0 are exotic $SU(2)_W$-singlets. Note that the n_{EL}^0 are usually referred to as antineutrinos. However, Majorana masses are possible for neutrinos, in which case there is no real distinction between particle and antiparticle. This is the second difference between neutrinos and charged particles. In the general Majorana case all three types of ν can mix. Finally, there are no experimental constraints on FCNC involving neutrinos. Despite these differences, mixing between ordinary and exotic neutrinos can be analyzed using a formalism similar to that introduced in Sec. 2.1.

Since in the presence of Majorana masses one does not distinguish between particle and antiparticle, in dealing with neutrinos it is convenient to denote all L states as n_L and all R states as n_R^c. These are related by $n_R^c = C(\overline{n}_L^T)$, where C is the charge conjugation matrix. Thus, in analogy to the charged fermion case, all L-handed weak eigenstate neutrinos are put together into a vector

$$n_L^0 = \begin{pmatrix} n_{OL}^0 \\ n_{EL}^0 \\ n_{SL}^0 \end{pmatrix} . \tag{20}$$

As above, the neutrino mass eigenstates are divided into two classes, 'light' (i.e., essentially massless) and 'heavy':

$$n_L = \begin{pmatrix} n_{lL} \\ n_{hL} \end{pmatrix} . \tag{21}$$

The weak and mass bases are related by a unitary transformation $n_L^0 = U_L n_L$, in which U can be decomposed as

$$U_L = \begin{pmatrix} A & E \\ F & G \\ H & J \end{pmatrix}_L . \tag{22}$$

Similarly, $n_R^{0c} = U_R n_R^c$, with $U_R = U_L^*$. In Eq. 22, the matrices A_L, F_L and H_L describe the overlap of the massless neutrinos with ordinary doublets (n_{OL}^0), exotic doublets (n_{EL}^0), and exotic singlets (n_{SL}^0), respectively. The LEP data has constrained the number of light $SU(2)_W$-doublets to be 3. Thus, exotic doublet ν's must have a mass greater than $M_Z/2$. This implies that the components of F_L are small. As to H_L, I will assume that the light neutrinos are mainly n_{OL}^0. If they are massless or have Majorana masses, then there are no light singlets, and all components of H_L are small. If the n_{lL} have small Dirac masses, then it is necessary to include 3 light singlets in the spectrum. In this case, the components of H_L corresponding to these singlets may be large, but the remaining components must be small. As far as the formalism is concerned, there is little difference between these two possibilities.

Dropping the subscript l, the weak neutral current for the light neutrino states can now be written

$$\frac{1}{2} J_Z^\mu = \frac{1}{2} \overline{n}_L \gamma^\mu \left(A_L^{\nu\dagger} A_L^\nu - F_L^{\nu\dagger} F_L^\nu \right) n_L . \tag{23}$$

The $A_L^{\nu\dagger} A_L^\nu$ and $F_L^{\nu\dagger} F_L^\nu$ terms come from the neutral currents of the n_{OL}^0 and n_{EL}^0, respectively. As in Sec. 2.2, neither A_L nor F_L is unitary. On the other hand, unlike the charged fermion case, there is no experimental evidence to suggest that $A_L^{\nu\dagger} A_L^\nu$ and

$F_L^{\nu\dagger} F_L^{\nu}$ are diagonal. However, as we will see, essentially the same effect is produced when one sums over the unobserved final state ν's in weak processes.

The leptonic charged current is

$$\begin{aligned}\frac{1}{2}J_W^{\mu\dagger} &= \overline{n}_L\gamma^\mu A_L^{\nu\dagger} c_L^e e_L + \overline{n}_R^c \gamma^\mu F_R^{\nu\dagger} s_R^e e_R \\ &= \sum_{ia}\left[\overline{n}_{iL}\gamma^\mu \left(A_L^{\nu\dagger}\right)_{ia} c_L^{e_a} e_{aL} + \overline{n}_{iR}^c \gamma^\mu \left(F_R^{\nu\dagger}\right)_{ia} s_R^{e_a} e_{aR}\right]. \end{aligned} \tag{24}$$

Note that since $F_R^\nu = F_L^{\nu^*}$, the second term in Eq. 24, which is the induced right-handed current, is of second order in light-heavy mixing. This term is produced when both the light neutrino and charged lepton mix with a member of an exotic doublet. The left-handed charged current is reduced in strength by the factor $\left(A_L^{\nu\dagger}\right)_{ia} c_L^{e_a}$ due to ordinary-exotic mixing.

We can now see the effect of summing over the final state ν's in a weak process. In the presence of mixing, the rate for the charged current transition $e_a \to n_i$ relative to its value (Γ_0) in the absence of mixing is

$$\frac{1}{\Gamma_0}\Gamma(e_a \to n_i) = (c_L^{e_a})^2 (A_L^\nu)_{ai}\left(A_L^{\nu\dagger}\right)_{ia} + (s_R^{e_a})^2 (F_R^\nu)_{ai}\left(F_R^{\nu\dagger}\right)_{ia}. \tag{25}$$

However, since the final ν's are unobserved, we must sum over them. The effect of this is to reduce the many parameters describing neutrino mixing to a single mixing angle per neutrino flavour:

$$\frac{1}{\Gamma_0}\sum_i \Gamma(e_a \to n_i) = (c_L^{e_a})^2 (c_L^{\nu_a})^2 + (s_R^{e_a})^2 (s_R^{\nu_a})^2 , \tag{26}$$

where the effective neutrino mixing angles $(c_L^{\nu_a})^2 = \left(A_L^\nu A_L^{\nu\dagger}\right)_{aa}$ and $(s_R^{\nu_a})^2 = \left(F_R^\nu F_R^{\nu\dagger}\right)_{aa}$ have been introduced. The second term in Eq. 26, which comes from the induced right-handed charged current, is of $O(s^4)$. From now on we will be working to second order in light-heavy mixing, so that this term can be dropped.

The final state neutrino produced in Eq. 26 is

$$|n_{aL}\rangle \equiv \frac{\sum_i \left(A_L^{\nu\dagger}\right)_{ia} |n_{iL}\rangle}{c_L^{\nu_a}}, \tag{27}$$

so that the cross section for scattering into the "right" charged lepton (e_{aL}) is

$$\frac{1}{\sigma_0}\sigma(n_{aL} \to e_{aL}) = (c_L^{e_a})^2 (c_L^{\nu_a})^2 . \tag{28}$$

(There is also the possibility of scattering into the "wrong" lepton (Ref. [8]), but this will not be discussed here.) One can also calculate the neutral current cross section for the rescattering of the neutrino in Eq. 27. Summing again over the final unobserved neutrinos, this can be found from Eqs. 23 and 27 to give

$$\begin{aligned}\frac{1}{\sigma_0}\sum_i \sigma(n_{aL}\to n_{iL}) &= \frac{1}{(c_L^{\nu_a})^2}\left[A_L^\nu\left(A_L^{\nu\dagger}A_L^\nu - F_L^{\nu\dagger}F_L^\nu\right)^2 A_L^{\nu\dagger}\right]_{aa} \\ &= 1-\frac{2}{(c_L^{\nu_a})^2}\left[A_L^\nu\left(2F_L^{\nu\dagger}F_L^\nu + H_L^{\nu\dagger}H_L^\nu\right)A_L^{\nu\dagger}\right]_{aa} + O(s^4), \quad (29)\end{aligned}$$

where, in the second line, I have used the unitarity of U_L (Eq. 22) and the fact that the components of F_L and H_L are all of $O(s)$[1]. Thus it is evident that this cross section depends not only on the mixing angle, but also on the type of neutrino(s) with which the ordinary neutrino mixes. Eq. 29 simplifies even further when one realizes that, for $a\epsilon 1,2,...,p$ (p is the number of light ν's), A_L^ν differs from the identity by terms of $O(s)$. In this case we obtain

$$\frac{1}{\sigma_0}\sum_i \sigma(n_{aL}\to n_{iL}) = 1-\Lambda_a\left(s_L^{\nu_a}\right)^2, \quad (30)$$

where the parameter Λ_a is defined to be $\Lambda_a = 4\lambda_F^a + 2\lambda_H^a$, with $(F_L^{\nu\dagger}F_L^\nu)_{aa} \equiv \lambda_F^a\,(s_L^{\nu_a})^2$ and $(H_L^{\nu\dagger}H_L^\nu)_{aa} \equiv \lambda_H^a\,(s_L^{\nu_a})^2$. The λ's are constrained to lie between 0 and 1, so that Λ_a takes values between 0 and 4, depending on the mixing involved.

Finally, using Eq. 23 and the same approximations as above, it is possible to calculate the rate for the decay of the Z^0 into undetected neutrinos. Assuming the existence of 3 light neutrinos in the absence of mixing, and normalizing to the decay rate of the Z^0 into one neutrino, this gives

$$\frac{1}{\Gamma_0^{1\nu}}\Gamma(Z^0\to invisible) = 3-\sum_a \Lambda_a\left(s_L^{\nu_a}\right)^2. \quad (31)$$

Having presented the formalism for the mixing of ordinary and exotic fermions, I will now turn to the experimental data which is used to constrain such mixings.

3 Experimental Data

In this section, I will present the experimental data which are used to constrain the mixing of ordinary and exotic fermions. I must emphasize at the outset that these results, taken from Ref. [4], are somewhat outdated, since the analysis was done in the summer of 1991. However, except for the ν_L^τ, the constraints obtained here would not be much improved if present data were used. I will comment further on this in Section 4.

[1] I have assumed that the light ν's are either massless or Majorana; the case of light Dirac ν's does not change the formalism significantly.

In using the data to constrain fermion mixing, it must be remembered that mixing can cause a discrepancy between the experimental result and the theoretical expression in two ways. Not only can mixing directly affect the process being examined, but it can also appear indirectly. This can happen, for example, when the extraction of a particular result requires normalization to another piece of experimental data. Thus, in putting constraints on mixing, one must be very careful to include *all* mixing effects.

Most of the experimental results are precise enough that it is necessary to include radiative corrections in order that there be agreement with the standard model. In the present analysis, radiative corrections will be included, but only those due to ordinary particles without mixing. (The inclusion of mixing is a second order effect.) Radiative corrections involving exotic fermions are typically much smaller, although it must be acknowledged that in the case of exotic nondegenerate $SU(2)_W$ doublets, the corrections could be large [9].

In order to calculate radiative corrections, it is necessary to choose a set of input parameters. These are typically taken to be the electromagnetic coupling α, measured at $q^2 = 0$, the Fermi constant G_μ, and the Z-mass M_Z, fixed to be $M_Z = 91.175$ GeV [10]. The values of α and M_Z as extracted from experiment are not affected by mixing. On the other hand, since G_μ is obtained directly from μ-decay, there is an effect due to mixing. The measured value of $G_\mu = 1.16637(2) \times 10^{-5}$ GeV^{-2} is related to its true value $\widehat{G}_\mu$ by

$$G_\mu = \widehat{G}_\mu c_L^e c_L^{\nu_e} c_L^\mu c_L^{\nu_\mu} \tag{32}$$

due to the possible mixing of the leptons with exotic fermions. Since many experimental results are normalized to μ-decay, indirect effects of mixing can appear in this way. Finally, it is necessary to include t-quark mass and the Higgs mass in the radiative corrections. These are fixed to be $m_t = 120$ GeV and $m_H = 100$ GeV, respectively.

3.1 M_W

Including radiative corrections, the theoretical expression for M_W as a function of α, M_Z and G_μ is given by [11],[12]

$$M_W^2 = \frac{\rho M_Z^2}{2}\left[1 + \sqrt{1 - \frac{G_\mu}{\widehat{G}_\mu}\frac{4\mathcal{A}}{\rho M_Z^2}\left(\frac{1}{1-\Delta\alpha} + \Delta r^{rem}\right)}\right], \tag{33}$$

where $\mathcal{A} = \pi\alpha/\sqrt{2}G_\mu$. Here, $\rho \simeq 1 + 3G_\mu m_t^2/8\sqrt{2}\pi^2$ contains the leading t-quark effects [9], $1/(1-\Delta\alpha)$ renormalizes the QED coupling to the M_Z scale, including the large logs, and Δr^{rem} includes all remaining small corrections. The sole (indirect) dependence of M_W on fermion mixings is found in the ratio $G_\mu/\widehat{G}_\mu$ (Eq. 32).

The average value of M_W as measured by CDF and UA2 is [13]

$$M_W = 80.13 \pm 0.31 \text{ GeV}, \tag{34}$$

where the LEP result for M_Z has been used to convert the UA2 measurement of M_W/M_Z into a value for M_W.

3.2 Charged Currents

There are a number of experiments involving charged currents which can be used to constrain fermion mixing. In the interest of brevity, I will present only those experimental results which are most important for bounding the mixing. Further information may be found in the articles in Chapter V of this volume.

3.2.1 Lepton Universality

In the standard model, the coupling of the W to each of the lepton doublets $(\nu_e\, e^-)_L$, $(\nu_\mu\, \mu^-)_L$ and $(\nu_\tau\, \tau^-)_L$ is universal, that is, $g_e = g_\mu = g_\tau$. In the presence of mixing, this equality can be altered:

$$\begin{aligned} \left(\frac{g_i}{g_e}\right)^2 &= \frac{\left(c_L^{\ell_i}\right)^2 (c_L^{\nu_i})^2 + \left(s_R^{\ell_i}\right)^2 (s_R^{\nu_i})^2}{(c_L^e)^2 (c_L^{\nu_e})^2 + (s_R^e)^2 (s_R^{\nu_e})^2} \\ &\simeq 1 + (s_L^e)^2 + (s_L^{\nu_e})^2 - \left(s_L^{\ell_i}\right)^2 - (s_L^{\nu_i})^2 , \qquad i = \mu, \tau, \end{aligned} \tag{35}$$

where only terms of $O(s^2)$ have been kept in the second line.

These ratios have been measured in several experiments. The most precise are

- Pion and Kaon decay:

$$\frac{\Gamma(\pi \to \mu\nu)}{\Gamma(\pi \to e\nu)} , \qquad \frac{\Gamma(K \to \mu\nu)}{\Gamma(K \to e\nu)} , \tag{36}$$

- Tau and Muon decay:

$$\frac{\Gamma(\tau \to \mu\bar{\nu}\nu)}{\Gamma(\tau \to e\bar{\nu}\nu)} , \qquad \frac{\Gamma(\tau \to \mu\bar{\nu}\nu)}{\Gamma(\mu \to e\bar{\nu}\nu)} . \tag{37}$$

The experimental data are shown in Table 3. These experiments constrain the left-handed mixing angles of the leptons. However, muon and tau decay have been measured accurately enough to put limits on the right-handed mixing angles of these leptons. The observables relevant to right-handed leptonic currents are all of $O(s^4)$. I will not discuss these here, but rather refer the reader to Refs. [3] and [14] for details.

3.2.2 Quark-Lepton Universality

In order to test quark-lepton universality, one typically compares the rate for the decay of a hadron with that of muon decay. In the standard model, in the absence

Quantity	Measured Value	Source
$(g_\mu/g_e)^2$	1.014 ± 0.011	$\pi \to \ell\nu$ [15]
$(g_\mu/g_e)^2$	1.013 ± 0.046	$K \to \ell\nu$ [15]
$(g_\mu/g_e)^2$	$1.016 \pm 0.026^\dagger$	$\Gamma(\tau \to \mu\bar{\nu}\nu)/\Gamma(\tau \to e\bar{\nu}\nu)$ [15],[16]
$(g_\tau/g_e)^2$	$0.952 \pm 0.031^\dagger$	$\Gamma(\tau \to \mu\bar{\nu}\nu)/\Gamma(\mu \to e\bar{\nu}\nu)$ [15],[16]

Table 3: Experimental constraints on lepton universality (Eq. 35). There is a correlation between the data marked with a †, which has been taken into account in the fits [4].

of mixing, these should be equal, up to factors of CKM matrix elements. However, these matrix elements obey another constraint, namely that of the unitarity of the CKM matrix. In this sense, a test of quark-lepton universality is equivalent to a test of CKM matrix unitarity.

V_{ud} is measured by comparing the rates for β-decay (vector current only) and μ-decay. V_{us} is obtained similarly, except that K_{e3} and hyperon decay are used. In the presence of mixing, the true values of the CKM matrix elements differ from the measured values by [3]

$$V_{ui} = \frac{c_L^u c_L^i \widehat{V}_{Lui} + s_R^u s_R^i \widehat{V}_{Rui}}{c_L^\mu c_L^{\nu_\mu}} \qquad i = d, s. \tag{38}$$

V_{ub} is also related to its true value in this way, but in any case its size is too small to be of interest for this analysis.

Using the fact that $\sum_{i=1}^n |\widehat{V}_{Lui}|^2 = 1$, expanding Eq. 38 to $O(s^2)$, and defining

$$\kappa_{ij} = s_R^{u_i} s_R^{d_j} \frac{\widehat{V}_{Rij}}{\widehat{V}_{Lij}}, \tag{39}$$

one obtains

$$\begin{aligned} \sum_{i=1}^{3} |V_{ui}|^2 &= 1 + (s_L^\mu)^2 + \left(s_L^{\nu_\mu}\right)^2 - (s_L^u)^2 - \sum_{i=4}^{n} |\widehat{V}_{Lui}|^2 \\ &\quad + |V_{ud}|^2 \left(2\mathrm{Re}(\kappa_{ud}) - \left(s_L^d\right)^2\right) + |V_{us}|^2 \left(2\mathrm{Re}(\kappa_{us}) - (s_L^s)^2\right). \end{aligned} \tag{40}$$

The experimental value for this quantity is [15]

$$\sum_{i=1}^{3} |V_{ui}|^2 = 0.9981 \pm 0.0021. \tag{41}$$

For those CKM matrix elements involving the c-quark, the analysis is similar to the above, except that the mixing of the first-generation particles can be neglected since such mixings are constrained considerably better from other processes than from the relatively imprecise measurements of V_{cd} and V_{cs}. Thus we have

$$\begin{aligned} V_{cd} &= c_L^c \widehat{V}_{Lcd} , \\ V_{cs} &= c_L^c c_L^s \widehat{V}_{Lcs} + s_R^c s_R^s \widehat{V}_{Rcs} , \end{aligned} \tag{42}$$

and

$$\sum_{i=1}^{3} |V_{ci}|^2 = 1 - (s_L^c)^2 - \sum_{i=4}^{n} |\widehat{V}_{Lci}|^2 + |V_{cs}|^2 \left(2\mathrm{Re}(\kappa_{cs}) - (s_L^s)^2\right), \tag{43}$$

where $|V_{cb}|^2$ has been neglected. The experimental value is [15]

$$\sum_{i=1}^{3} |V_{ci}|^2 = 1.08 \pm 0.37. \tag{44}$$

The hadronic right-handed currents κ_{ud} and κ_{us} are constrained through the unitarity of the CKM matrix. There are additional, very stringent constraints coming from the predictions of PCAC for nonleptonic $K_{\pi 3}$ amplitudes relative to $K_{\pi 2}$ amplitudes [17]. Interpreting these limits as 1σ errors [3], one has

$$\kappa_{ud}, \kappa_{us} = 0 \pm 0.0037. \tag{45}$$

There are also additional (weak) constraints on κ_{cd} and κ_{cs}, but they will not be discussed here (see Refs. [3] and [4]).

3.3 Neutral Currents (Low Energy)

At low energy, neutral current interactions can be parametrized through effective lagrangians in which the Z^0 has been integrated out. In this subsection, I will discuss three types of scattering processes – νq, νe and eq. In all three cases, radiative corrections are important [18]. For simplicity, these corrections are not shown explicitly, but are included in the fits. The experiments are further discussed in the articles in Chapter IV of this volume.

3.3.1 Deep-Inelastic Neutrino Scattering

The effective lagrangian describing the scattering of neutrinos from quarks can be written as

$$-\mathcal{L}^{\nu q} = \frac{4G_F}{\sqrt{2}} \bar{\nu}_L \gamma^\mu \nu_L \sum_{i=u,d,\dots} \left[\epsilon_L(i) \bar{q}_L^i \gamma_\mu q_L^i + \epsilon_R(i) \bar{q}_R^i \gamma_\mu q_R^i\right]. \tag{46}$$

In order to extract the values of $\epsilon_L(i)$ and $\epsilon_R(i)$, the neutral current processes are normalized to the corresponding charged current processes, that is, the ratios

$$R_\nu = \frac{\sigma(\nu N \to \nu X)}{\sigma(\nu N \to \mu^- X)}, \quad R_{\bar{\nu}} = \frac{\sigma(\bar{\nu} N \to \bar{\nu} X)}{\sigma(\bar{\nu} N \to \mu^+ X)} \tag{47}$$

Quantity	Experimental Value	Source
g_L^2 g_R^2 θ_L θ_R	0.2977 ± 0.0042 0.0317 ± 0.0034 2.50 ± 0.03 $4.59^{+0.44}_{-0.27}$	Deep inelastic [15]
g_V^e g_A^e	-0.10 ± 0.05 -0.50 ± 0.04	Low-energy $\nu_\mu e$: BNL [19]
g_V^e g_A^e	-0.06 ± 0.07 -0.57 ± 0.07	High-energy $\nu_\mu e$: CHARM I [20]
g_V^e/g_A^e	0.047 ± 0.046	CHARM II [21]
C_{1u} C_{1d}	$-0.249 \pm 0.066^\dagger$ $0.391 \pm 0.059^\dagger$	Atomic parity [22]
$C_{2u} - \frac{1}{2}C_{2d}$	0.21 ± 0.37	SLAC e-D [23]

Table 4: Low-energy neutral current data. There are non-negligible correlations between the measurements marked with a †. These have been taken into account in the fits [4].

are used. Thus, mixing effects enter both in the numerator and in the denominator. Taking all effects into account, the values of $\epsilon_L(i)$ and $\epsilon_R(i)$ obtained from deep-inelastic neutrino scattering are

$$\epsilon_{L,R}(i) = F_1(s^2, \kappa)\tilde{\epsilon}_{L,R}(i), \tag{48}$$

where the $\tilde{\epsilon}_{L,R}(i)$ are defined in Eq. 14, and [3].

$$F_1(s^2, \kappa) = \frac{1 - \frac{1}{2}\Lambda_\mu \left(s_L^{\nu_\mu}\right)^2}{1 - (s_L^\mu)^2 - \left(s_L^{\nu_\mu}\right)^2 - \mathrm{Re}(\kappa_{ud})} \tag{49}$$

incorporates the mixing effects in the neutrinos as well as in the normalization. The experimental values of $g_a^2 \equiv \epsilon_a(u)^2 + \epsilon_a(d)^2$ and $\theta_a \equiv \tan^{-1}[\epsilon_a(u)/\epsilon_a(d)]$, $a = L, R$ are given in Table 4.

3.3.2 Neutrino-Electron Scattering

The neutral current interaction of ν_μ and e can be described by

$$-\mathcal{L}^{\nu_\mu e} = \frac{2G_F}{\sqrt{2}}\bar{\nu}_L\gamma^\mu\nu_L\bar{e}\gamma_\mu(g_V^e - g_A^e\gamma_5)\,e. \tag{50}$$

As in deep-inelastic neutrino scattering, the vector- and axial-couplings of the electron are obtained by normalizing the neutral current process (in this case ν_μ-e scattering) to a charged current process (ν_μ-*hadron* scattering). Again, mixing effects appear in both places. The low-energy experiments from BNL normalize to the quasielastic process $\nu_\mu n \to \mu^- p$, leading to

$$\begin{aligned} g_V^e &= F_2(s^2) v_e = F_2(s^2) \left[-\frac{1}{2} (c_L^e)^2 - \frac{1}{2} (s_R^e)^2 + 2 \sin^2 \theta_W \right], \\ g_A^e &= F_2(s^2) a_e = F_2(s^2) \left[-\frac{1}{2} (c_L^e)^2 + \frac{1}{2} (s_R^e)^2 \right], \end{aligned} \tag{51}$$

where [3]

$$F_2(s^2) = \frac{1 - \frac{1}{2}\Lambda_\mu \left(s_L^{\nu_\mu}\right)^2}{1 - (s_L^\mu)^2 - \left(s_L^{\nu_\mu}\right)^2}. \tag{52}$$

The high-energy experiments at CERN and Fermilab normalize to $\nu N \to \mu^- X$ as in deep-inelastic scattering, so that in this case g_V^e and g_A^e are as in Eq. 51, but with $F_2(s^2)$ replaced by $F_1(s^2, \kappa)$ of Eq. 49. The experimental values of g_V^e and g_A^e are shown in Table 4. Note that the CHARM II collaboration has recently measured the ratio g_V^e / g_A^e, in which the dependence on $F_1(s^2, \kappa)$ cancels.

3.3.3 Atomic Parity Violation

Atomic parity violation arises through the interference of the electromagnetic and weak interactions. The parity violating couplings C_{1i} and C_{2i} are defined by

$$-\mathcal{L}^{eq} = \frac{G_F}{\sqrt{2}} \sum_i \left[C_{1i}\, \bar{e}_L \gamma_\mu \gamma_5 e\, \bar{q}^i \gamma^\mu q^i + C_{2i}\, \bar{e}_L \gamma_\mu e\, \bar{q}^i \gamma^\mu \gamma_5 q^i \right]. \tag{53}$$

Including mixing, these couplings are given by

$$C_{1i} = 2 \left(\frac{\widehat{G}_\mu}{G_\mu} \right) a_e v_i \,, \qquad C_{2i} = 2 \left(\frac{\widehat{G}_\mu}{G_\mu} \right) v_e a_i \,, \tag{54}$$

where the vector and axial couplings have been defined in Eq. 15 and $\widehat{G}_\mu / G_\mu$ in Eq. 32. C_{1u} and C_{1d} are measured in parity violating transitions in cesium; the combination $C_{2u} - \frac{1}{2} C_{2d}$ has been determined in polarized e-D scattering at SLAC. All the experimental values are given in Table 4.

3.4 Neutral Currents (Z Peak)

The very accurate measurements at LEP put strong constraints on the mixing of ordinary and exotic fermions, particularly as regards the τ-lepton and heavy quarks. Here I will present the experimental data on the decay widths of the Z^0 as well as the forward-backward asymmetries for leptons and heavy flavours. The material in this subsection comes entirely from Ref. [4]. For more information on the experiments, see the article by Schaile in this volume.

3.4.1 Z^0 Decay Widths

Taking into account all radiative corrections, at one loop the partial width for the decay $Z^0 \to f\bar{f}$ is [12]

$$\gamma_{Z\to f\bar{f}} = N_c^f \frac{M_Z}{12\pi}\sqrt{2}\widehat{G}_\mu M_Z^2 \rho_f \left(v_f^2 + a_f^2\right)\left(1+\delta_{QED}^f\right)\left(1+\delta_{QCD}^f\right), \tag{55}$$

where $N_c^f = 3(1)$ for quarks (leptons), δ_{QCD}^f is the QCD correction for hadronic final states, and δ_{QED}^f is an additional photonic correction. Fermion mixing effects appear in two places – first, in the vector and axial couplings v_f and a_f (see Eq. 15), and also in the effective weak mixing angle which appears in the vector coupling. This weak mixing angle is renormalized by electroweak effects [4]:

$$s_{eff}^2(f) = \frac{1}{2}\left[1 - \sqrt{1 - \frac{G_\mu}{\widehat{G}_\mu}\frac{4\mathcal{A}}{\rho M_Z^2}\left(\frac{1}{1-\Delta\alpha} + \Delta\bar{r}^{rem}\right)}\right]. \tag{56}$$

As in the renormalized expression for M_W (Eq. 33), mixing effects appear indirectly in $G_\mu/\widehat{G}_\mu$ (Eq. 32). There are also electroweak corrections in the ρ_f term: $\rho_f = \rho + \Delta\rho_f^{rem}$, where ρ contains all large t-quark effects and is universal, and $\Delta\rho_f^{rem}$ (and $\Delta\bar{r}^{rem}$ above) include all the nonuniversal flavour-dependent corrections. In doing the fits, all corrections have been taken into account, including the finite mass effects for heavy fermions.

The experimental values of the five partial widths Γ_Z, Γ_h, Γ_e, Γ_μ and Γ_τ [10] have large correlations among themselves. The widths are all shown in Table 5.

3.4.2 Leptonic Asymmetries

On resonance, the forward-backward asymmetry in the process $e^+e^- \to Z^0 \to f\bar{f}$ takes the form

$$A_f^{FB} = 3\frac{v_e a_e}{v_e^2 + a_e^2}\frac{v_f a_f}{v_f^2 + a_f^2}. \tag{57}$$

As in the partial widths, mixing effects enter both in the vector and axial couplings (Eq. 15), and in the renormalized effective weak mixing angle (Eq. 56). In the fits, all QED and QCD (for hadronic final states) corrections have been included [4]. The τ polarization asymmetry has also been measured at LEP. This asymmetry is written

$$A_\tau^{pol} = -2\frac{v_\tau a_\tau}{v_\tau^2 + a_\tau^2}. \tag{58}$$

The experimental values [24] for all leptonic asymmetries are given in Table 5.

Quantity	Experimental Value
Γ_Z	$2487 \pm 10^\dagger$
Γ_h	$1739 \pm 13^\dagger$
Γ_e	$83.2 \pm 0.6^\dagger$
Γ_μ	$83.4 \pm 0.9^\dagger$
Γ_τ	$82.8 \pm 1.1^\dagger$
A_e^{FB}	-0.019 ± 0.014
A_μ^{FB}	0.0070 ± 0.0079
A_τ^{FB}	0.099 ± 0.096
A_τ^{pol}	-0.121 ± 0.040
Γ_b	367 ± 19
Γ_c	299 ± 45
A_b^{FB}	0.123 ± 0.024
A_c^{FB}	0.064 ± 0.049
$a_b^{\gamma Z}$	-0.405 ± 0.095
$a_c^{\gamma Z}$	0.515 ± 0.085
$A_{c,D^*}^{\gamma Z}$ (29 GeV)	-0.101 ± 0.027
$A_{c,D^*}^{\gamma Z}$ (35 GeV)	-0.161 ± 0.034

Table 5: Partial widths (given in MeV) and asymmetries measured at the Z peak. The correlations among the measurements marked with a † have been taken into account in the fits [4]. Also displayed are the axial couplings $a_{b,c}^{\gamma Z}$ and the charm asymmetries with D^* tagging $A_{c,D^*}^{\gamma Z}$, all measured off resonance.

3.4.3 Heavy Flavours

The partial widths for $Z^0 \to b\bar{b}$ [25] and $c\bar{c}$ [26] have also been measured. These are listed in Table 5. The forward-backward asymmetries for these final states have also been measured [27]. For $b\bar{b}$, there is a peculiarity which must be taken into account. Due to the fact that neutral B-mesons can oscillate into $\overline{B}$-mesons, the observed asymmetry is not the true asymmetry but must be corrected:

$$A_b^{FB} = \frac{A_{obs}^{FB}}{1 - 2\chi_B}\,, \tag{59}$$

where χ_B is a measure of the probability for B-$\overline{B}$ oscillations. Experimentally, this parameter has been found to be $\chi_B = 0.146 \pm 0.016$ [28]. The forward-backward asymmetries for both $b\bar{b}$ (corrected) and $c\bar{c}$ final states is given in Table 5.

Finally, the forward-backward asymmetries for $b\bar{b}$ and $c\bar{c}$ final states have also been measured at lower energies at PEP and PETRA. In this region, the asymmetries $A_{b,c}^{\gamma Z}$ include interference between the γ and the Z^0, and essentially measure the product of axial couplings $a_e a_{b(c)}$. Both final states are tagged using high p and

p_T leptons, leading to large correlations between the two measurements [29]. Due to the correlations, these data are only used for those fits where only one mixing angle at a time is allowed to vary. For $c\bar{c}$, there is an additional tagging method not applicable for $b\bar{b}$, namely using D^*'s [30]. The results using this method are used in all the fits. The experimental data are shown in Table 5. Note that the axial couplings a_b and a_c [31] are given in the case of lepton tagging, while for D^* tagging the forward-backward asymmetry is shown [30].

4 Constraints

In the section, I present the constraints which the experimental data shown in the previous section place on the mixing between ordinary and exotic fermions. I will show the results of two fits. In the first (the 'individual fit'), only one mixing angle at a time is allowed to be nonzero, and in the second (the 'joint fit') all mixing angles vary simultaneously.

In both fits, the constraints are obtained by using a least-squares method. One complication is that the mixing angles are bounded, that is, $0 \leq s_{L,R}^2 \leq 1$. In order to deal with this, the following procedure is used. For each parameter s_i^2, the χ^2 distribution is calculated. Then, assuming a probability distribution

$$P(s_i^2)ds_i^2 = N_i e^{-\chi^2(s_i^2)/2} ds_i^2, \tag{60}$$

in which $N_i^{-1} = \int_0^1 exp(-\chi^2(s_i^2)/2)ds_i^2$ (i.e. N_i is chosen such that $P(s_i^2)$ is properly normalized in the domain [0,1]), the 90% C.L. upper bounds on the s_i^2 are calculated from $P(s_i^2)$.

Despite the large number of parameters, the experimental data is comprehensive enough to constrain all mixing angles. The results of the individual and joint fits are shown in Table 6, which is taken from Ref. [4]. In the 'Source' column of this Table are listed those observables which are most important for constraining the mixing angles in the individual fits. However, in the joint fit it is possible to evade the bounds from these observables through fine-tuned cancellations between different mixings. In this case, other observables, which depend on different combinations of the mixings, become important. These new observables, which are denoted by a $*$ in Table 6, are typically less precise, so that the constraints in the joint fit are somewhat weaker than those in the individual fit.

From this Table it is evident that the neutral current data at the Z peak is especially important for bounding all mixings. For the first generation fermions and the μ and ν_μ, the low-energy charged and neutral current results (particularly νq and eq scattering) are also useful. In addition, the asymmetries off the Z peak are helpful in constraining the mixing angles of the c- and b-quarks.

I must again stress that the data used to obtain these constraints are already a bit out of date. For example, only the 1990 LEP data was used; the inclusion of the

	Individual	Joint			Source
		$\Lambda = 2$	$\Lambda = 0$	$\Lambda = 4$	
$(s_L^e)^2$	0.0047	0.015	0.0090	0.015	$\Gamma_e, M_W^*, A_\mu^{FB*}, eq^*, g_e^*$
$(s_R^e)^2$	0.0062	0.010	0.0082	0.010	$\Gamma_e, A_e^{FB}, A_\mu^{FB*}, \nu e^*$
$(s_L^\mu)^2$	0.0017	0.0094	0.0090	0.011	$V_{ui}^2, \nu q, g_\mu, \Gamma_\mu, s_{eff}^{LEP*}, M_W^*$
$(s_R^\mu)^2$	0.0086	0.014	0.014	0.013	Γ_μ, A_μ^{FB}
$(s_L^\tau)^2$	0.011	0.017	0.015	0.017	$\Gamma_\tau, A_\tau^{FB}, g_\tau, A_\tau^{pol*}$
$(s_R^\tau)^2$	0.011	0.012	0.014	0.012	$\Gamma_\tau, A_\tau^{pol}, A_\tau^{FB}, g_\tau^*$
$(s_L^u)^2$	0.0045	0.019	0.015	0.019	$V_{ui}^2, \Gamma_h, \Gamma_Z, eq, \nu q$
$(s_R^u)^2$	0.018	0.024	0.025	0.024	$\nu q, \Gamma_h, \Gamma_Z, eq$
$\left(s_L^d\right)^2$	0.0046	0.019	0.016	0.019	$V_{ui}^2, \Gamma_h, \Gamma_Z, \nu q$
$\left(s_R^d\right)^2$	0.020	0.030	0.028	0.029	$eq, \Gamma_h, \Gamma_Z, \nu q$
$(s_L^s)^2$	0.011	0.038	0.039	0.041	$\Gamma_h, \Gamma_Z, V_{ui}^2$
$(s_R^s)^{2\dagger}$	0.36	0.67	0.63	0.74	Γ_h, Γ_Z
$(s_L^c)^2$	0.013	0.040	0.042	0.042	$\Gamma_h, \Gamma_Z, \Gamma_c^*, A_c^{\gamma Z*}$
$(s_R^c)^2$	0.029	0.097	0.10	0.099	$\Gamma_h, \Gamma_Z, A_c^{\gamma Z*}, \Gamma_c^*, A_c^{FB*}$
$\left(s_L^b\right)^2$	0.011	0.070	0.072	0.069	$\Gamma_h, \Gamma_Z, \Gamma_b, A_b^{FB*}$
$\left(s_R^b\right)^{2\dagger}$	0.33	0.39	0.40	0.39	$\Gamma_b, \Gamma_Z, \Gamma_h, A_b^{\gamma Z}, A_b^{FB*}$
$(s_L^{\nu_e})^2$	0.0097	0.015	0.016	0.014	$s_{eff}^{LEP}, g_e, s_{eff}^{NC}, M_W^*$
$\left(s_L^{\nu_\mu}\right)^2$	0.0019	0.015	0.0087	0.011	$V_{ui}^2, g_\mu, \nu q, s_{eff}^{LEP}, M_W^*$
$(s_L^{\nu_\tau})^{2\dagger}$	0.032	0.064	0.097	0.035	Γ_Z, g_τ
$\sum_{i=4}^n \hat{V}_{ui}^2$	0.0048	0.014	0.010	0.018	V_{ui}^2
$\sum_{i=4}^n \hat{V}_{ci}^2$	0.53	0.76	0.76	0.76	V_{ci}^2
$\lvert\kappa_{ud}\rvert$	0.0011	0.0059	0.0060	0.0058	$V_{ui}^2, \nu q$,RHC's,νe^*
$\lvert\kappa_{us}\rvert$	0.0054	0.0061	0.0061	0.0061	V_{ui}^2,RHC's

Table 6: 90% C.L. upper limits on mixing angles for individual fits (one angle at a time is allowed to vary) and joint fits (all angles allowed to vary simultaneously) [4]. Observables which are most important for the constraints are shown in the 'Source' column (those quantities which contribute only in the joint fits are tagged with an asterisk). s_{eff}^{LEP} and s_{eff}^{NC} refer to the weak mixing angle as extracted in neutral current measurements at the Z peak and at low energy, respectively. See the text for a discussion of the bounds on $(s_R^s)^2$, $\left(s_R^b\right)^2$ and $(s_L^{\nu_\tau})^2$ (marked with a †).

1991 LEP data would surely strengthen most of the bounds somewhat. The most important new development is in τ-decays. The value of $(g_\tau/g_e)^2$ shown in Table 3 differs from its standard model value of 1 by about 1.5 standard deviations. However, the latest measurements of the τ mass and lifetime have removed this discrepancy [32]. Thus, the limits on $(s_L^{\nu_\tau})^2$ shown in Table 6, which depend on the old value of $(g_\tau/g_e)^2$, should be taken with a grain of salt – the new bounds are probably quite a bit better.

In all fits, $\Lambda_e = \Lambda_\mu = \Lambda_\tau$ has been assumed. Furthermore, in the individual fit, $\Lambda = 2$ was taken. Note that, in this fit, only the neutrino mixings can depend on Λ. Since $(s_L^{\nu_e})^2$ and $\left(s_L^{\nu_\mu}\right)^2$ are bounded mainly by charged current data, the dependence on Λ is minimal. On the other hand, the constraint on $(s_L^{\nu_\tau})^2$ does depend on Λ: $(s_L^{\nu_\tau})^2 < 0.098, 0.032, 0.015$ for $\Lambda = 0, 2, 4$. (As I said in the previous paragraph, these numbers should not be taken too seriously. However, even with the new data, the strong dependence of $(s_L^{\nu_\tau})^2$ on Λ will persist.)

The constraints on $(s_R^s)^2$ and $\left(s_R^b\right)^2$ are considerably weaker than those of other angles due to a peculiarity of the observables which bound them. These mixing angles are constrained mainly by the LEP observables, which depend on the couplings $v_q a_q$ and $v_q^2 + a_q^2$ $(q = s, b)$. However, for $(s_R^q)^2 \simeq 0.3$, the s^4 terms cancel against the s^2 terms. Thus there are two minima in the χ^2 distribution, centered around 0 and 0.3. The 90% C.L. bounds of Table 6 are obtained by integrating over both regions. The restriction to the region centered at no mixing gives stronger bounds, $(s_R^s)^2 \lesssim 0.09$ and $\left(s_R^b\right)^2 \lesssim 0.10$.

The bounds on most mixing angles in Table 6 are quite stringent. However, one might argue that the exotic fermions which give rise to these mixings necessarily appear in models with other forms of new physics, extra Z's for instance, and that these new effects might weaken significantly the mixing limits. This seems quite unlikely, given the number and variety of constraints. In fact, such a study has been done [33], in the context of E_6 and $SO(10)$ models. In this paper, the effects of Z-Z' mixing and fermion mixing were analyzed simultaneously. In general, the presence of an extra Z did not much alter the mixing limits. Although not a proof, this analysis lends support to the idea that, regardless of the model, it is rather difficult to evade the constraints on the mixing of ordinary and exotic fermions found in Table 6.

5 Conclusions

In this chapter, I have discussed the constraints which precision measurements put on the mixing of ordinary and exotic fermions. Exotic fermions are defined as new fermions whose left- or right-handed components transform in a non-standard way under $SU(2)_W$, that is, L singlets and/or R doublets. Excluding noncanonical colour and electric charge assignments, there are 4 types of exotic fermions – mirror fermions, vector singlets and doublets, and new Weyl neutrinos.

In general, mixing between ordinary and exotic fermions will lead to flavour-changing neutral currents among the light particles, which are extremely well constrained experimentally. However, if one chooses fine-tuned directions in mixing parameter space such that each ordinary fermion mixes with its own exotic fermion, then the bounds from FCNC can be evaded. I have developed the formalism which describes this mixing – there is one mixing angle per L and R charged fermion. For neutrinos, the situation is more complicated due to the possibility of Majorana masses and the fact that there is no experimental evidence against FCNC involving neutrinos. Nevertheless, because the final neutrinos in any process are unobserved, it is possible to describe mixing in the neutrino sector by one angle, plus one auxiliary parameter, per ordinary neutrino species.

There are enough constraints from low-energy charged and neutral current data, as well as the experimental results from LEP, to constrain all mixing angles. I have described two types of fits. In the first, all mixing angles but one are set to zero, and the non-zero angle is constrained. In the second all angles are allowed to be non-zero simultaneously. The results are shown in Table 6. In the individual fit, most of the mixing parameters $\left(s^f\right)^2$ are constrained to be of order 1%, with some of the angles (such as those for $e_{L,R}$, u_L, d_L, μ_L and $\nu_{\mu L}$) quite a bit smaller. The two exceptions are s_R and b_R, whose mixings are bounded to be only about 0.3. In the joint fit, due to the possibility of accidental cancellations among the mixings, the limits are weakened. Typically, the constraints are relaxed by a factor of 2-3, but this factor can be as much as 6-8 in a few cases.

Future prospects are discussed in the article by Luo in this volume.

Acknowledgements:

I would like to thank E. Nardi for helpful discussions. This work was supported in part by the Natural Sciences and Engineering Research Council of Canada, and by FCAR, Québec.

References

[1] See, for example, F. Wilczek and A. Zee, *Phys. Rev.* **D25** (1982) 553; J. Bagger *et al.*, *Nucl. Phys.* **B244** (1984) 247.

[2] M. Aguilar-Benitez *et al.*, (Particle Data Group), Phys. Rev. **D45**, Part II, (1992) 1.

[3] P. Langacker and D. London, *Phys. Rev.* **D38** (1988) 886.

[4] E. Nardi, E. Roulet and D. Tommasini, *Nucl. Phys.* **B386** (1992) 239.

[5] UA1 Collaboration, C. Albajar *et al.*, *Phys. Lett.* **262B** (1991) 163.

[6] E. Nardi, *Phys. Rev.* **D48** (1993) 1240.

[7] D. Silverman, *Phys. Rev.* **D45** (1992) 1800.

[8] P. Langacker and D. London, *Phys. Rev.* **D38** (1988) 907.

[9] M. Veltman, *Nucl. Phys.* **B123** (1977) 89; M.B. Einhorn, D.R.T. Jones and M. Veltman, *Nucl. Phys.* **B191** (1981) 146.

[10] ALEPH Collaboration, D. Decamp *et al.*, *Zeit. Phys.* **C53** (1992) 1; DELPHI Collaboration, P. Abreu *et al.*, *Nucl. Phys.* **B367** (1991) 511; L3 Collaboration, B. Adeva *et al.*, *Zeit. Phys.* **C51** (1991) 179; OPAL Collaboration, G. Alexander *et al.*, *Phys. Lett.* **264B** (1991) 219.

[11] A. Sirlin, *Phys. Rev.* **D22** (1980) 971, *ibid.* **D29** (1984) 89; W.J. Marciano and A. Sirlin, *Phys. Rev.* **D29** (1984) 945.

[12] M. Consoli and W. Hollik, in *Z physics at LEP, vol. 1*, eds. G. Altarelli *et al.*, CERN 89-08 (1989); G. Burgers and F. Jegerlehner, *ibid.*; W. Hollik, *CERN-JINR School Phys.1989*, p 50, (QCD161:C15:1989); G. Burgers and W. Hollik, in *Polarization at LEP, vol. 1*, eds. J. Ellis and R.D. Peccei, CERN 86-02 (1986); D.C. Kennedy and B.W. Lynn, *Nucl. Phys.* **B322** (1989) 1; J.G. Im, D.C. Kennedy, B.W. Lynn and R.G. Stuart, *Nucl. Phys.* **B321** (1989) 83; J.L. Rosner, EFI 90-18 (1990).

[13] UA2 Collaboration, J. Alitti *et al.*, *Phys. Lett.* **241B** (1990) 150; CDF Collaboration, F. Abe *et al.*, *Phys. Rev. Lett.* **65** (1990) 2243.

[14] P. Langacker and D. London, *Phys. Rev.* **D39** (1989) 266.

[15] M. Aguilar-Benitez *et al.* (Particle Data Group), *Phys. Lett.* **239B** (1990) 1.

[16] L3 Collaboration, B. Adeva *et al.*, *Phys. Lett.* **265B** (1991) 451; OPAL Collaboration, G. Alexander *et al.*, *Phys. Lett.* **266B** (1991) 201.

[17] J.F. Donoghue and B.R. Holstein, *Phys. Lett.* **113B** (1982) 382.

[18] For reviews of low-energy neutral current experiments, see U. Amaldi *et al.*, *Phys. Rev.* **D36** (1987) 2191; G. Costa *et al.*, *Nucl. Phys.* **B297** (1988) 244.

[19] BNL Collaboration, K. Abe *et al.*, *Phys. Rev. Lett.* **62** (1989) 1709.

[20] CHARM Collaboration, J. Dorenbosch *et al.*, *Zeit. Phys.* **C41** (1989) 567.

[21] CHARM-II Collaboration, D. Geiregat *et al.*, *Phys. Lett.* **259B** (1991) 499.

[22] M.C. Noecker, B.P. Masterson and C.E. Wieman, *Phys. Rev. Lett.* **61** (1988) 310.

[23] C.Y. Prescott *et al.*, *Phys. Lett.* **77B** (1978) 347; *ibid.* **84B** (1979) 524.

[24] FB asymmetries: Ref. [10]; τ polarization asymmetry: ALEPH Collaboration, D. Decamp *et al.*, *Phys. Lett.* **265B** (1991) 430; OPAL Collaboration, G. Alexander *et al.*, Ref. [16].

[25] MARK II Collaboration, J.F. Kral *et al.*, *Phys. Rev. Lett.* **64** (1990) 1211; ALEPH Collaboration, D. Decamp *et al.*, *Phys. Lett.* **244B** (1990) 551; L3 Collaboration, B. Adeva *et al.*, *Phys. Lett.* **261B** (1991) 177; OPAL Collaboration, M. Akrawy *et al.*, *Phys. Lett.* **263B** (1991) 311; DELPHI Collaboration, P. Abreu *et al.*, CERN-PPE/90-118.

[26] ALEPH Collaboration, D. Decamp *et al.*, Ref. [25]; OPAL Collaboration, M. Akrawy *et al.*, Ref. [25]; DELPHI Collaboration, P. Abreu *et al.*, *Phys. Lett.* **252B** (1990) 140; OPAL Collaboration, M. Akrawy *et al.*, *Phys. Lett.* **262B** (1991) 341.

[27] A_b^{FB} and A_c^{FB}: ALEPH Collaboration, D. Decamp *et al.*, *Phys. Lett.* **263B** (1991) 325; A_b^{FB} only: L3 Collaboration, B. Adeva *et al.*, *Phys. Lett.* **252B** (1990) 713; OPAL Collaboration, M. Akrawy *et al.*, Ref. [25].

[28] L3 Collaboration, B. Adeva *et al.*, *Phys. Lett.* **252B** (1990) 703; ALEPH Collaboration, D. Decamp *et al.*, *Phys. Lett.* **258B** (1991) 236; UA1 Collaboration, C. Albajar *et al.*, *Phys. Lett.* **262B** (1991) 171.

[29] JADE Collaboration, E. Elsen *et al.*, *Zeit. Phys.* **C46** (1990) 349; CELLO Collaboration, J.J. Behrend *et al.*, *Zeit. Phys.* **C47** (1990) 333.

[30] JADE Collaboration, F. Ould-Saada *et al.*, *Zeit. Phys.* **C44** (1989) 567; JADE Collaboration, E. Elsen *et al.*, Ref. [29].

[31] CELLO Collaboration, J.J. Behrend *et al.*, Ref. [29].

[32] P. Drell, Proceedings of the *XXVI International Conference on High Energy Physics*, 1992.

[33] E. Nardi, E. Roulet and D. Tommasini, *Phys. Rev.* **D46** (1992) 3040.

VIII. THE FUTURE

Future High Precision Experiments and New Physics Beyond Standard Model

Mingxing Luo
Institute for Nuclear Theory, University of Washington
Seattle, WA 98195, USA

ABSTRACT

This paper analyzes the high precision ($\leq 1\%$) electroweak experiments that have been done or are likely to be done in this decade. Starting with the Standard Model (SM) predictions of fourteen weak neutral current observables and fifteen W and Z properties to the one-loop level, we investigated the implications of the corresponding experimental measurements to various types of possible new physics that enter at the tree or loop level. Certain experiments appear to have special promise as probes of the new physics considered here.

Contents

1 Introduction

A few years ago, a theoretical framework was developed for the analysis of the high precision electroweak experiments that were likely to be done in the next ten years [1]. Since then, significant experimental progresses have been made, especially at the e^+e^- collider LEP at CERN [2]. The mass of the Z boson M_Z has been shifted from 91.177 to 91.187 GeV, the error on M_Z, ΔM_Z, has been reduced from 0.031 to 0.007 GeV, which is much better than the error anticipated. The errors on the various Z decay widths have also been improved to be better than anticipated. Furthermore, there are preliminary results of the various asymmetries at the Z pole in the e^+e^- collision. The error on forward-backward asymmetry of μ is again better than anticipated. On the other hand, the global analysis of existing W and Z particle properties and weak-neutral current data [3] yields a new constraint on the mass of the top quark, $m_t = 150^{+17+15}_{-23-25}$. We update the analysis by incorporating these new experimental results. We will choose the value of $M_Z = 91.187 \pm 0.007\ GeV$, $m_t = 150\ GeV$, and $M_H = 250\ GeV$ as inputs for the theoretical prediction, in contrast with the inputs $M_Z = 91.177 \pm 0.021\ GeV$, $m_t = 100\ GeV$, and $M_H = 100\ GeV$ used in [1].

Historically, the electroweak experiments played an important role in the establishment of the the Standard Model (SM). The new experiments considered in this paper, each designed to have a precision of the order of one percent or better, will test the electroweak interaction component of the SM, the Glashow-Weinberg-Salam theory [4], to high precision, and might possibly detect new physics beyond the SM. They provide a complementary alternative to the direct method of studying higher energy phenomena to find departures from the SM.

The establishment of the SM is one of the major accomplishments in particle physics during the past twenty years. The SM is mathematically self-consistent and is compatible with all known experimental data [5]. But there are many theoretical questions that cannot be answered satisfactorily within the framework of the SM. For example, the SM has a complicated gauge structure and many free parameters.

If we believe the unification of all the fundamental forces and the self-determination of physical parameters within the theory, we have to seek a more fundamental theory, which is not subject to these shortcomings but reduces to the SM at present energies. To do so, we construct new theories and models, such as grand unified theories, supersymmetry, and superstrings. Experimentally, we try to detect new physical phenomena which might be induced by such underlying new physics. One way is to build higher energy colliders. There is, however, a complementary approach. That is, to improve the precision of present electroweak experiments. If there exists physics beyond the SM, there might be remnants of it at present energies which would cause small deviations of physical quantities from the SM values. These deviations would be too small to be observed given the present experimental errors, but might become amenable to more precise experiments.

In this decade, there have been and will be performed a series of high precision experiments relating to the properties of the Z- and W-bosons and to weak neutral current observables. As mentioned at the beginning, the Z-boson mass has already been measured very precisely [2]. The decaying widths of Z, the forward-backward and possibly polarization asymmetries in e^+e^- collisions, and the W boson mass have been or will be measured with high but somewhat less precision (see [6]–[9] and references in [1]). High precision measurements of neutrino-electron scattering, atomic parity violation, deep inelastic neutrino scattering, and other neutral current processes are all likely to be performed (see [10]–[13] and references in [1]). In Table 1.1, we have collected the relevant observables, their expectations in the SM, and their present and projected experimental uncertainties.

For the analysis of high precision experiments, we have to take another impor-

Table 1.1: The observables considered in this article, their SM predictions, and their present and future experimental uncertainties (including theoretical uncertainties where they are important). The SM predictions use the observed value of M_Z and assume $m_t = 150\ GeV$, and $M_H = 250\ GeV$. g_L^2, g_R^2, R_ν, θ_L, and θ_R are quantities measured in νN scattering; g_V^e, g_A^e, $\sigma_\nu/\sigma_{\bar{\nu}}$, and $\sigma_\nu/(\sigma_{\bar{\nu}}+\sigma_\nu)$ are relevant to νe scattering; $C_{1\pm}$, C_{2pm} are measured in atomic parity violation, muonic atoms, and lN scattering; A_{LR}, A_{FB}, and A_{pol} are asymmetries at the Z-pole, and the Γ's are the partial and total Z widths.

Quantities		(present)			(future)	
O_a	O_a^{SM}	O_a^{exp}	ΔO_a^{exp}	$\Delta \sin^2\theta_W^{exp}$	ΔO_a^{exp}	$\Delta \sin^2\theta_W^{exp}$
$M_Z(GeV)$	—	91.187	0.007	0.0003	—	—
$M_W(GeV)$	80.267	80.1	0.3	0.0018	0.105	0.0006
g_L^2	0.3025	0.3003	0.0039	0.0052	—	—
g_R^2	0.030	0.0323	0.0033	0.013	—	—
R_ν	0.315	—	—	—	0.001	0.002
θ_L	2.46	2.49	0.035	—	—	—
θ_R	5.18	4.58	0.41	—	—	—
g_V^e	−0.039	−0.035	0.017	0.009	—	—
g_A^e	−0.505	−0.508	0.015	—	—	—
$\sigma_\nu/\sigma_{\bar{\nu}}$	1.164	1.083	0.10	0.012	0.046	0.005
$\sigma_\nu/(\sigma_{\bar{\nu}}+\sigma_\nu)$	0.148	—	—	—	0.0027	0.0025
C_{1+}	0.129	0.126	0.003	0.01	0.0013	0.003
$C_{1+}(iso)$	0.129	—	—	—	0.0003	0.0009
C_{1-}	−0.367	−0.45	0.1	0.07	—	—
C_{2p}	−0.015	—	—	—	0.046	—
$C_{2p}(1)$	−0.015	—	—	—	0.0046	—
C_{2m}	−0.059	—	—	—	0.11	0.03
$2C_{1u}+C_{1d}$	−0.036	—	—	—	0.004	0.002
$A_{LR}(SLC)$	0.140	—	—	—	0.0066	0.0008
$A_{LR}(LEP)$	0.140	—	—	—	0.0041	0.0005
$A_{FB}^{pol}(c)$	0.476	—	—	—	0.025	0.01
$A_{FB}(c)$	0.067	0.072	0.027	0.0065	0.007	0.0017
$A_{FB}^{pol}(b)$	0.697	—	—	—	0.02	0.04
$A_{FB}(b)$	0.098	0.098	0.012	0.002	0.0054	0.001
$A_{FB}^{pol}(\mu)$	0.105	—	—	—	0.009	0.0015
$A_{FB}(\mu)$	0.015	0.0174	0.0027	0.0017	—	—
$A_{pol}(\tau)$	0.140	0.140	0.018	0.0025	0.01	0.0014
$\Gamma_{inv}(GeV)$	0.501	0.505	0.006	0.0025	—	—
$\Gamma_{l\bar{l}}(GeV)$	0.0838	0.0833	0.0003	0.0006	—	—
$\Gamma_{c\bar{c}}(GeV)$	0.298	0.235	0.038	0.02	0.03	0.016
$\Gamma_{b\bar{b}}(GeV)$	0.376	0.373	0.009	0.004	—	—
$\Gamma_Z(GeV)$	2.493	2.492	0.007	0.0005	—	—

Table 1.2: The types of new physics considered in this paper

Tree level physics:	extra Z: χ, ψ, η, Z_{LR}
	non-standard Higgs representations
	leptoquarks: type I, type II
	extra fermions: $u'_{L,R}$, $d'_{L,R}$, $e'_{L,R}$, $\nu'_{L,R}$
	compositeness: four-fermi contact operators
Loop level physics:	m_t, M_H
	extra fermions
	S-T parameters: gauge boson self-energies
	two Higgs doublets
	supersymmetry

tant physical phenomenon into serious consideration, namely radiative corrections within the SM (see [14], [15], and references in [1]). The radiative corrections are generally small with a few exceptions; they are usually in the order of one percent of the Born approximation values. (The radiative corrections to the masses of W and Z bosons are about 4 percent.) With one percent experimental errors, one has to include the radiative corrections for physical observables in the analysis. They are essential to the verification of the SM. Furthermore, they must be included if one is to search for small deviations due to new physics. Radiative corrections at the one-loop level to all of the SM predictions have been included in our analysis. Note that the SM is now established as correct to an excellent first approximation. Hence, we can apply the SM radiative corrections, treating any new physics (whether it enters at tree or loop level) as a perturbation.

In [1], we investigated ten general types of possible new physics that enter at the tree or loop level, as listed in Table 1.2. For new physics entering at the tree level, we considered extra Z bosons, extra scalar bosons, extra fermions, and compositeness. For new physics entering at the loop level, we considered heavy extra fermions which break the custodial symmetry, extra Higgs doublets, S-T parameters, and supersymmetry. The unknown top quark and Higgs boson masses, m_t and M_H, enter predictions of the observables through radiative corrections. We include these in Table 1.2 even though they are not technically "new physics", because the techniques for constraining them are similar to those in the search for new physics and because they lead to complications in the analysis. We also took 27 specific examples to illustrate our formalism [1]. Five of these examples will be updated in this paper, other cases can be updated in a similar manner. A brief summary of the implications to all the 27 specific examples will also be included.

In our formalism we emphasize the interplay of experimental possibilities and theoretical models. We explore how experimental data will constrain theories and how theories may guide future experiments. The testing of the SM and searching for new physics require the comparison of a number of precise experiments. Most observables are not absolute predictions of the SM, but depend on $\sin^2\theta_W$. A single experimental result can usually be accommodated by choosing $\sin^2\theta_W$ appropriately. However, comparison of the values of $\sin^2\theta_W$ obtained by a number of experiments probes for physics beyond the SM. If a series of high precision experiments yield equal values for $\sin^2\theta_W$ (within experimental and theoretical uncertainties) then the SM is successfully tested at that level and limits can be set on the possible contributions of new physics.

On the other hand, observed deviations would be evidence for new physics. In this case one wants as much information as possible to distinguish with clarity the different kinds of new physics. We describe two schemes to distinguish one type of new physics from another which serve as a prescription for the analysis of future high precision experiments. One possibility is to use the $\sin^2\theta_W$ values extracted from the different experiments. The pattern of deviations of the various $\sin^2\theta_W$ from each other may be a distinguishing feature. Another and more powerful possibility is to compare the observables directly with the SM predictions. Again, the pattern of deviations is a diagnostic of the type of new physics. In either case, the most sensitive probe for new physics involves the totality of experiments. The high precision electroweak experiments and our theoretical framework comprise one method to probe incisively and exhaustively for any weakness in the SM, and to recognize and identify it if found.

Following this introductory section, we provide a general formalism for the analysis in Section 2; Section 3 is a survey of high precision experiments; Section 4 is a general discussion of various types of possible new physics; in Section 5 are comparison of experiments and possible new physics; in Section 6 are the conclusions.

2 Formalism

Consider a physical observable O_a and denote its experimental value as O_a^{exp} and the theoretical prediction as O_a^T. Then, within experimental and theoretical uncertainties,

$$O_a^T = O_a^{exp} \tag{1}$$

if both theory and experiment are correct (the superscript ^{T} will be omitted in the following for convenience). Starting from the SM as our best guess, we have the SM prediction O_a^{SM} of O_a. Were the SM the true theory, then

$$O_a \equiv O_a^{SM} = O_a^{exp}. \tag{2}$$

Were there new physics beyond the SM, then

$$O_a = O_a^{SM} + \sum_i \Delta O_a^i. \tag{3}$$

where ΔO_a^i is the contribution to O_a of the new physics i. We have assumed that contributions from new physics are sufficiently small compared with that from the

SM that they can be calculated perturbatively and contributions from different new physics can be simply summed up.

In general, O_a^{SM} depends on the $SU(2) \times U(1)$ gauge structure and the free parameters of the theory. In the GWS theory, besides the fermion masses, there are three free parameters, namely:

$$\alpha, G_F, \sin^2\theta_W, \tag{4}$$

the QED coupling constant, the Fermi weak interaction coupling constant, and the weak mixing angle. α and G_F have already been determined precisely by other experiments. $\sin^2\theta_W$ is the single unspecified parameter in the GWS theory, representing the mixing between the $SU(2)$ and $U(1)$ sectors. At tree level, $\sin^2\theta_W$ is given by $g'^2/(g'^2+g^2)$, where g and g' are respectively the $SU(2)$ and $U(1)$ gauge coupling constants. To go beyond the tree level, one must define a renormalized $\sin^2\theta_W$. One possibility is the on-shell definition $\sin^2\theta_W^M = 1 - M_W^2/M_Z^2$, where M_W and M_Z are the measured masses. Another possibility is the $\overline{MS}$ definition $\sin^2\hat{\theta}_W(M_Z)$. The two definitions are related by $\sin^2\hat{\theta}_W(M_Z) = C(m_t, M_H)\sin^2\theta_W^M$, where $C = 1.029$ for $m_t = 150\ GeV$, $M_H = 250\ GeV$. These definitions should yield the same predictions if some appropriate summations of high order terms are taken[1]. However, since we are dealing with deviations from the SM to the lowest order of perturbation, our analysis is essentially independent of which definition of $\sin^2\theta_W$ is used.

Express both O_a and O_a^{SM} as functions of $\sin^2\theta_W$,

$$O_a = O_a(\sin^2\theta_W), \quad \text{and} \quad O_a^{SM} = O_a^{SM}(\sin^2\theta_W). \tag{5}$$

Since we are ignorant of the exact value of $\sin^2\theta_W$, we define an effective $\sin^2\theta_W^a$ for each O_a by assuming the validity of the SM:

$$O_a(\sin^2\theta_W) = O_a^{SM}(\sin^2\theta_W^a) \tag{6}$$

with

$$\sin^2\theta_W^a = \sin^2\theta_W + \Delta\sin^2\theta_W^a. \tag{7}$$

where $\Delta\sin^2\theta_W^a$ is the shift away from the true $\sin^2\theta_W$ due to new physics. From (3), (5), and (6) we have

$$\Delta\sin^2\theta_W^a = \frac{\sum_i \Delta O_a^i}{\left.\frac{dO_a^{SM}}{dx}\right|_{x^a}}, \tag{8}$$

where $x^a = \sin^2\theta_W^a$. The validity of the SM requires that all $\Delta\sin^2\theta_W^a$ should vanish, and all of the $\sin^2\theta_W^a$ should be equal to $\sin^2\theta_W$ (within uncertainties). The $\sin^2\theta_W^a$ would differ from one another if there is new physics beyond the SM. Thus, one way to test the SM is simply to compare the values of $\sin^2\theta_W^a$ extracted from measurements of different physical quantities O_a.

However, not all physical quantities are sensitive to $\sin^2\theta_W$. For example, the axial part of ν-e couplings is essentially independent of $\sin^2\theta_W$. So when using the $\sin^2\theta_W$ analysis, it is possible to trade a relatively precise experiment for a poor

[1] For a thorough discussion of $\sin^2\theta_W$, see [1] and references within.

result. It is important to have predictions of the physical quantities themselves. We therefore consider two analyses: one in which the O_a are compared directly and one in which the extracted $\sin^2\theta_W^a$ are compared. In the former no information is lost, while the latter is easier to use. To study the O_a, we need a numerical estimate of $\sin^2\theta_W$ to compute the SM prediction. There are two reasonable choices for the estimate of $\sin^2\theta_W$: one is the world average value [3] $\sin^2\theta_W^{avg} = 0.2260 \pm 0.0024$ (on-shell) or $\sin^2\theta_W^{avg} = 0.2325 \pm 0.0007$ ($\overline{MS}$); another possibility is $\sin^2\theta_W^Z$ extracted from M_Z. Both $\sin^2\theta_W^{avg}$ and $\sin^2\theta_W^Z$ may be shifted from the true $\sin^2\theta_W$ by new physics, and that shift must be taken into account in the analysis. This is easier to do for $\sin^2\theta_W^Z$, since it comes from a single measured quantity, and $\sin^2\theta_W^Z$ is used in this paper.

Expressing the equations above in terms of $\sin^2\theta_W^Z$, we have

$$\sin^2\theta_W^a = \sin^2\theta_W^Z + \Delta\sin^2\theta_W^{(a;Z)} \tag{9}$$

where

$$\Delta\sin^2\theta_W^{(a;Z)} = \Delta\sin^2\theta_W^a - \Delta\sin^2\theta_W^Z \tag{10}$$

i.e., $\Delta\sin^2\theta_W^{(a;Z)}$ represents both the shift in $\sin^2\theta_W^a$ and in $\sin^2\theta_W^Z$ due to new physics. A nonzero $\Delta\sin^2\theta_W^{(a;Z)}$ (beyond uncertainties) would indicate new physics. Eqs. (8), (9), and (10) are the basic results for the $\sin^2\theta_W$ analysis. The value of $\Delta\sin^2\theta_W^{(a;Z)}$ obtained from experiments by the differences between the measured and the SM values shown in (9) can be compared with the predictions from (8) and (10) for each type of new physics. For the analysis based directly on the observables $O_a(\sin^2\theta_W)$ one uses:

$$\begin{aligned} O_a(\sin^2\theta_W) &= O_a^{SM}(\sin^2\theta_W) + \sum_i \Delta O_a^i. \\ &= O_a^{SM}(\sin^2\theta_W^Z) - \frac{dO_a^{SM}}{dx}\Big|_{x^Z} \Delta\sin^2\theta_W^Z + \sum_i \Delta O_a^i \\ &= O_a^{SM}(\sin^2\theta_W^Z) + \Delta O_a \end{aligned} \tag{11}$$

We will refer to the SM prediction as:

$$O_a^{SM}(\sin^2\theta_W^Z) \tag{12}$$

while the contribution from the new physics is

$$\Delta O_a = -\frac{dO_a^{SM}}{dx}\Big|_{x^Z} \Delta\sin^2\theta_W^Z + \sum_i \Delta O_a^i. \tag{13}$$

At the one-loop level, ΔO_a is independent of the renormalization scheme. Again, the first term in ΔO_a represents the difference between the (assumed) value of $\sin^2\theta_W^Z$ and the true $\sin^2\theta_W$, while the second term is the direct effect of the new physics on O_a. The calculated $O_a(\sin^2\theta_W)$ in (11) are the quantities to be directly compared with experiment. The experimental values of ΔO_a obtained from (11) can therefore be compared with the predictions for each type of new physics calculated by (7) to (13).

3 Physical Observables and Their Measurements

In this section we discuss the physical observables. It is summarized in Table 1.1 which gives the SM predictions $O_a^{SM}(\sin^2\theta_W^Z)$ for twenty-nine observables and the present and anticipated experimental and theoretical errors (1σ). For details and references, please see [6]–[13], [1] and references within.

M_Z, M_W: In the SM, the masses of Z and W bosons are predicted to be:

$$M_Z^2 = \frac{\pi\alpha/\sqrt{2}G_F}{\rho\cos^2\theta_W\sin^2\theta_W(1-\Delta r)}, \quad M_W^2 = \frac{\pi\alpha/\sqrt{2}G_F}{\sin^2\theta_W(1-\Delta r)}, \tag{14}$$

to the one-loop level; where Δr and $\rho - 1$ are radiative correction parameters which have been calculated and which depend on and the renormalization scheme and m_t, M_H. The LEP experiments yield the extremely precise value $M_Z = 91.187 \pm 0.007\ GeV$. This value of M_Z together with the assumed values $m_t = 150\ GeV$ and $M_H = 250\ GeV$ are used as the input parameters in the numerical calculations that follow, and are referred to as the *standard input parameters* (SIP) in the remainder of the paper. With these inputs $\sin^2\theta_W^Z = 0.2259 \pm 0.0003$ (on-shell) or 0.2325 ± 0.0003 ($\overline{MS}$). The present world-averaged value of M_W is $80.1 \pm 0.3\ GeV$, which leads to an error of 0.0018 in $\sin^2\theta_W$ for the SIP, independent of renormalization scheme. One expects improved measurements by three independent techniques: W production at the Tevatron, W exchange at HERA, and the reaction $e^+e^- \to W^+W^-$ in the second phase of LEP (LEP 200). A reasonable projection of the experimental error on M_W is $100\ MeV$, which would lead to an error of 0.0006 in $\sin^2\theta_W$.

Low energy effective neutrino-quark coupling coefficients: The low energy effective Lagrangian for neutrino-quark interaction can be written as:

$$-L_{eff} = \frac{G_F}{\sqrt{2}}\sum_{\alpha=1}^{n}\bar{\nu}\gamma^\mu(1-\gamma_5)\nu\sum_i\left[\epsilon_L(i)\bar{q}_i\gamma_\mu(1-\gamma_5)q_i + \epsilon_R(i)\bar{q}_i\gamma_\mu(1+\gamma_5)q_i\right]. \tag{15}$$

The explicit expressions of the ϵ's in SM are given in [1]. The quantities extracted from experiments are the following combinations of the $\epsilon_{L,R}$: $g_L^2 = \epsilon_L^2(u) + \epsilon_L^2(d)$, $g_R^2 = \epsilon_R^2(u) + \epsilon_R^2(d)$, $\theta_L = \tan^{-1}\epsilon_L(u)/\epsilon_L(d)$, $\theta_R = \tan^{-1}\epsilon_R(u)/\epsilon_R(d)$. Their SM predictions and experimental status are listed in Table 1.1. Measurements of $\epsilon_{L,R}$, especially the measurements of the deep inelastic scattering cross-section ratios $R_\nu = \sigma(\nu_\mu N \to \nu_\mu X)/\sigma(\nu_\mu N \to \mu^- X)$ and $R_{\bar{\nu}} = \sigma(\bar{\nu}_\mu N \to \bar{\nu}_\mu X)/\sigma(\bar{\nu}_\mu N \to \mu^+ X)$ on approximately isoscalar targets are significant accomplishments of the second and the third generation weak neutral current experiments. In the future, there may be deep inelastic experiments using a high energy ν_μ beam, *e.g.*, at Fermilab. At high energies the theoretical uncertainties will be reduced and one may obtain a measurement of $R_\nu \simeq g_L^2 + 0.4g_R^2$ corresponding to $\Delta\sin^2\theta_W = 0.002$ or better.

Low energy effective ν_μ-electron coupling coefficients: Similar to the $\nu - q$ interaction, one has the effective Lagrangian of the ν_μ-electron interaction:

$$-L_{eff}^{\nu_\mu e} = \frac{G_F}{\sqrt{2}}\bar{\nu_\mu}\gamma^\mu(1-\gamma_5)\nu_\mu\bar{e}\gamma_\mu(g_V^e - g_A^e\gamma_5)e \tag{16}$$

(see [1] for the SM predictions of $g^e_{V,A}$.) Their SM predictions and present experimental values are listed in Table 1.1. The measurement of the ratio g_V/g_A is expected to be improved, but not the separate values. Here we emphasis two types of experiments which minimize possible systematic errors and which, with sufficient statistics, would yield precise values of the measured quantities. One is of the ratio: $R_{\nu/\bar{\nu}} = \sigma_{\nu_\mu e \to \nu_\mu e}/\sigma_{\bar{\nu}_\mu e \to \bar{\nu}_\mu e}$ to be performed at CERN. Another is of the ratio: $R_{\nu/\bar{\nu}\nu_e} = \sigma_{\nu_\mu e \to \nu_\mu e}/(\sigma_{\bar{\nu}_\mu e \to \bar{\nu}_\mu e} + \sigma_{\nu_e e \to \nu_e e})$ proposed at Los Alamos.

Atomic parity violation coupling constants: In general, the parity violating coupling between electron and hadron at low energy can be parameterized as follows:

$$-L^{eH} = -\frac{G_F}{\sqrt{2}}\sum_i (C_{1i}\bar{e}\gamma_\mu\gamma_5 e\bar{q}_i\gamma^\mu q_i + C_{2i}\bar{e}\gamma_\mu e\bar{q}_i\gamma^\mu\gamma_5 q_i). \tag{17}$$

(see [1] for the SM predictions of C's.) There are various ways to measure C's. We consider the following combinations which are the quantities extracted from experiments: $C_{1+} = 0.666C_{1u} + 0.747C_{1d}$, $C_{1-} = 0.747C_{1u} - 0.666C_{1d}$, $C_{2p} = C_{2u} + C_{2d}$ and $C_{2m} = C_{2u} - C_{2d}$, where the combination C_{1+} corresponds to parity-violation in the cesium atom. Their SM predictions and experimental statues are listed in Table 1.1. The present experimental values of the coefficients are: $C_{1+} = 0.126 \pm 0.003$ and $C_{1-} = -0.45 \pm 0.10$; C_{2p} and C_{2m} are poorly determined. The experimental error of C_{1+} may soon be reduced to $< 0.5\%$. The total error would then be dominated by the theoretical uncertainty of $\sim 1\%$ [16]. Future measurements involving different isotopes of cesium or other elements will determine the C_1's more precisely [17]. We project two uncertainties for C_{1+}: a 1% uncertainty expected soon (C_{1+}) and a later possible 0.2% uncertainty from the use of different isotopes ($C_{1+}(iso)$).

Asymmetries of e^+e^- scattering at the Z-pole:

a) The left-right asymmetry with longitudinally-polarized initial electrons is given by:

$$A_{LR} = \frac{\sigma(e^-_L e^+ \to \sum_f f\bar{f}) - \sigma(e^-_R e^+ \to \sum_f f\bar{f})}{\sigma(e^-_L e^+ \to \sum_f f\bar{f}) + \sigma(e^-_R e^+ \to \sum_f f\bar{f})}, \tag{18}$$

where the summation runs through all fermion pairs $f\bar{f}$ comprising the final states. A_{LR} is small but sensitive to $\sin^2\theta_W$, so its precise measurement will provide an excellent determination of $\sin^2\theta_W$. In particular, the extracted $\sin^2\hat{\theta}_W(M_Z)$ is insensitive to m_t and M_H. Within the SIP, $A_{LR} = 0.140$. The projected experimental error of A_{LR} at LEP is 0.003 if LEP goes ahead with polarization, corresponding to an error of 0.0003 in $\sin^2\theta_W$. At SLC, an error in A_{LR} of 0.006 is expected. However, we must fold in the theoretical error $\Delta A^{pred}_{LR} \simeq 0.003$ in the SM prediction of A_{LR}, from ΔM_Z and the uncertainty in Δr. The effective error is therefore $\Delta A_{LR} \simeq 0.004$ (LEP) or 0.0066 (SLC), corresponding to $\Delta\sin^2\theta_W \simeq 0.0005$ (0.0008).

b) The forward-backward asymmetries of final state fermions are:

$$A_{FB}(f) = \frac{(\int_{\cos\theta>0} - \int_{\cos\theta<0})d\Omega\frac{d\sigma}{d\Omega}(e^+e^- \to \bar{f}f)}{(\int_{\cos\theta>0} + \int_{\cos\theta<0})d\Omega\frac{d\sigma}{d\Omega}(e^+e^- \to \bar{f}f)}. \tag{19}$$

with $f = u, d, \mu,$ *etc.* Here θ is the angle between the directions of the incoming electron and outgoing fermion f. The tree level predictions for various A_{FB}'s are given in [1]. Their SM predictions and experimental errors are listed in Table 1.1. There are already some preliminary results of them from LEP [2]. We note that the errors on the forward-backward asymmetry of the μ is 0.0027, which is lower than the anticipated error 0.0035.

c) The polarized-forward-backward asymmetry is:

$$A_{FB}^{pol}(f) = \frac{\sigma_{FB}^L(f) - \sigma_{FB}^R(f)}{2\sigma}, \tag{20}$$

where

$$\sigma_{FB}^{L,R}(f) = \left[\int_{\cos\theta>0} - \int_{\cos\theta<0}\right] d\Omega \frac{d\sigma}{d\Omega}(e^-_{L,R}e^+ \to f\bar{f}), \tag{21}$$

$$\sigma = \int d\Omega \frac{d\sigma}{d\Omega}(e^-e^+ \to f\bar{f}). \tag{22}$$

The tree level predictions for $f = u,\ d,\ \mu$ are included in [1], and the SM predictions and experimental errors in Table 1.1. There are still no experimental results yet.

d) Final state polarization asymmetries of the τ lepton: Define:

$$A_{pol}(\tau) = \frac{\sigma(e^-e^+ \to \tau_L^-\tau^+) - \sigma(e^-e^+ \to \tau_R^-\tau^+)}{\sigma(e^-e^+ \to \tau_L^-\tau^+) + \sigma(e^-e^+ \to \tau_R^-\tau^+)} \tag{23}$$

In the SM, it measures the V and A interference. The preliminary result of $A_{pol}(\tau)$ is 0.140 ± 0.018. The error $\Delta A_{pol}(\tau)$ will be improved to be 0.01.

<u>Z decay widths:</u>

a) Partial Z-decay widths: the Z decays into lepton and quark pairs with different widths:

$$\Gamma_{f\bar{f}}; \quad f = \nu, e, u, d, ... \tag{24}$$

Their tree level SM expressions are listed in [1]. And their SM predictions and experimental status are listed in Table 1.1. The experimental leptonic width is the averaged width of e, μ and τ; we denote it as $\Gamma_{l\bar{l}}$. The width into unobserved (invisible) particles, Γ_{inv} is determined by the difference of the total Z-width (see below) and the Z-width into observable particles (hadrons and charged particles). The errors on most of the partial widths are more precisely than anticipated. Currently, $\Delta\Gamma_{l\bar{l}} = 0.0003\ GeV$, $\Delta\Gamma_{inv} = 0.006\ GeV$, $\Delta\Gamma_{b\bar{b}} = 0.009\ GeV$.

b) Inclusive Z-decay width: The total or inclusive Z-decay width is the summation of all the partial widths of fermions. That is:

$$\Gamma_Z = 3\Gamma_{Z\to\nu\bar{\nu}} + 3\Gamma_{Z\to e^+e^-} + 2\Gamma_{Z\to u\bar{u}} + 3\Gamma_{Z\to d\bar{d}}, \tag{25}$$

where the coefficients are due to the 3 families (the t is too massive to contribute). This quantity is measured by the shape of the energy dependence of the total cross-section of the e^+e^- collision around the Z-pole. With the SIP, $\Gamma_Z = 2.493\ GeV$. Experimentally, $\Gamma_Z = 2.492 \pm 0.007\ GeV$. The errors is significantly lower than the anticipated $0.015\ GeV$.

4 New Physics

In this section we briefly review the new physics considered in our analysis. For the details, please see [1] and references within.

4.1 *Extra Z bosons*

Extra intermediate vector bosons (IVBs) beside the SM prediction of the W and Z arise naturally in Grand Unification Theories (GUTs) and other extensions of the SM, due to their larger gauge groups. These IVBs are usually too heavy to be relevant to physics in the laboratory since their mass are related to the large mass scale of GUTs and other theories. However, there can be extra IVBs whose masses are independent of the large mass scale. And they may gain masses by other Higgs fields at low energy, *e.g.*, in the TeV range. In that case, we may either see them directly in colliders, or find their effects at present energies in high precision experiments. Here we are interested the extra neutral Z bosons. For a theory with the SM as its first order approximation plus $n-1$ extra Z bosons, the gauge group is $SU(2) \times \prod_{\alpha=1}^{n} U(1)_\alpha$. This symmetry is broken, in the process the IVBs acquire masses and there results the mixing among them. The mixing is caused by the diagonalization of the gauge boson mass matrix. These extra IVBs in general open a new channel for the weak neutral current processes, and their mixing with the ordinary Z changes the coupling of fermions with the ordinary Z. The mixing also changes the mass of the ordinary Z. Since $\sin^2\theta_W^Z$ derived from M_Z is used as input in our analysis, the change of M_Z shifts the SM predictions for physical quantities.

To be concrete and to follow the spirit of grand unification, we take the $U(1)$'s as relics of an underlying non-abelian gauge group, in which there is only one overall coupling constant and the fermion charges in one irreducible representation are related. The strength depends on the GUT breaking pattern. In this paper we take the $E(6)$ model as an important example. We also consider the extra Z boson in $SU(2)_L \times SU(2)_R \times U(1)$ models. For these models, there are effectively two extra free parameters in the SM with an extra Z boson,

$$\hat{\rho}_2 = (\frac{g_2}{g_1})^2(\frac{M_Z}{M_{Z_2}})^2 \tag{26}$$

$$\hat{\Theta} = \frac{g_2}{g_1}\Theta, \tag{27}$$

where M_{Z,Z_2} and $g_{1,2}$ are the masses and coupling constants of the ordinary Z and the extra Z_2, respectively; Θ is the Z_1^0-Z_2^0 mixing angle. Typically, $(g_2/g_1)^2 \simeq 5\sin^2\theta_W/3 \simeq 0.38$. Both $\hat{\rho}_2$ and $\hat{\Theta}$ are small numbers, but their ratio $C = \hat{\Theta}/\hat{\rho}_2$ is of the order of unity. C depends on the $U(1)_2$ quantum numbers of the Higgs fields which cause the mixing. In our analysis, $\hat{\rho}_2$ is taken to be the free parameter λ and some typical values of C (usually motivated by simple $E(6)$ models) are selected for each type of Z_2.

4.2 *Extra scalar bosons*

It is always possible to incorporate scalar bosons into gauge theories without spoiling the symmetry. In the minimal SM, one $SU(2)$ scalar doublet employed to realize the Higgs mechanism. After spontaneous symmetry breaking, one scalar particle remains. $SU(3)$ is not broken in QCD, so no colored scalar bosons are introduced. In theories beyond the SM, extra scalar bosons usually cannot be avoided. In Grand Unification Theories, more scalar bosons must be introduced to break the symmetries via the Higgs mechanism. In technicolor theories there are no elementary scalars but scalars are generated as bound states. In supersymmetry, two Higgs doublets are required, and for each ordinary fermion there is a scalar partner. Scalars may have either color or weak charges. In this section, we take the non-standard Higgs fields as an example for color-singlet scalars and leptoquarks for color-non-singlet scalars.

A. Non-standard Higgs boson

In the SM, $\rho_0 = M_W^2/M_Z^2 \cos^2\theta_W$ is unity. In theories beyond the SM, if (color-singlet) scalars which carry non-standard $SU(2)$ charges (*i.e.*, $SU(2)$ triplets) develop vacuum expectation values, ρ_0 will be different from unity. The non-standard Higgs fields thus manifest themselves at tree level. Additional Higgs doublets do not change the SM prediction $\rho_0 = 1$. The shift of ρ_0 changes the prediction of M_Z and quantities related to it are directly affected (except for negligible effects associated with scalar exchange). However, $\sin^2\theta_W^Z$ extracted from M_Z assuming the validity of the SM differs from the true $\sin^2\theta_W$. This gives artificial changes in quantities which are sensitive to $\sin^2\theta_W$ when they are predicted from $\sin^2\theta_W^Z$.

B. Leptoquarks

For colored bosons we consider the leptoquarks. They are color triplets which change quarks to leptons (or antileptons) and vice versa when they are emitted or absorbed. They have many experimental consequences, but we only consider effects on weak neutral current phenomena. The contributions of leptoquarks are suppressed at the Z pole, as are those of photons; since their contributions are $\pi/2$ out of the phase at the Z-pole. Accordingly, e^+e^- collider experiments at the Z-pole will give no information about them. Neither will leptoquarks contribute to pure leptonic processes; since leptoquarks change quarks to leptons and leptons to quarks at tree level. Only the νq coupling constants and atomic parity-violation coefficients can be affected by leptoquarks.

In a general leptoquark theory, there may be many extra parameters. To be concrete, we take the following simple example for illustration: those which are predicted in the $SU(5)$ unification theory. In this example, there are only two relevant parameters: $\sqrt{2}|\eta_L|^2/8M_S^2G_F$ and $\sqrt{2}|\eta_R|^2/8M_S^2G_F$, where $\eta_{L,R}$ are coupling strengths. To be more specific, we take two definite examples: *type-1 leptoquark*: with $\eta_L \neq 0$ and $\eta_R = 0$; *type-2 leptoquark*: with $\eta_L = 0$ and $\eta_R \neq 0$.

4.3 *Extra fermions*

Most extensions of the SM predict the existence of extra fermions [18]. For instance, in the $E(6)$ model fermions in one family are assigned to a 27-plet, which includes the 15 ordinary fermions and 12 extra ones. Among the 12 extra fermions, we have one vector-singlet D quark, D_L, D_R; one vector-doublet lepton: $\begin{pmatrix} E^0 \\ E^- \end{pmatrix}_L$, $\begin{pmatrix} E^0 \\ E^- \end{pmatrix}_R$; and two Weyl neutrinos $\bar{N}_L$, S_L^0. Here we will not consider the actual production of new heavy fermions. Rather, we will focus on their mixing with the known light particles. This can affect weak universality, lead to induced right-handed currents, and affects the relations between the Fermi-constant and $M_{W,Z}$. Specifically, we consider the mixing between ordinary fermions and the exotic right-handed $SU(2)$ doublets and left-handed $SU(2)$ singlets, which modifies the neutral current couplings of the light particles.

The theories of extra fermions are complicated by many free parameters. To simplify, we separate a general theory with many extra fermions into specific ones with only one extra fermion. So we do not deal with a theory with n extra fermions, but n theories with one extra fermion. In each theory, we have a single free parameter, $\sin^2\theta$, where θ is the mixing angle between the ordinary and exotic fermions.

4.4 *Contact operators*

No hint has been seen of lepton or quark form factors or other signs of compositeness. Nevertheless, it is possible that the leptons and quarks have a substructure if the compositeness scale Λ is very large ($\Lambda > O(1\ TeV)$). One consequence should be the generation by constituent interchange of four-fermi or other effective operators at energies small compared to Λ. These must be $SU(3) \times SU(2) \times U(1)$ invariant since new physics at the scale $\Lambda \gg M_Z$ must preserve the low-energy symmetries. One can write a great many invariant operators including those which generate FCNC; S, P, and T operators; and those which contribute to both charged and neutral current phenomena. Even restricting to V, A operators which affect WNC processes only, there are many possibilities. We consider three representative cases of four-fermi contact operators: $\bar{l}_{\mu L}\gamma^\mu l_{\mu L}\bar{q}_L\gamma_\mu q_L$, which will shift the values of $\epsilon_L(u)$, $\epsilon_L(d)$, C_{2u}, and C_{2d}; $\bar{\nu}_{\mu L}\gamma^\mu \nu_{\mu L}\bar{e}_L\gamma_\mu e_L$ which will shift the values of g_V^e and g_A^e; $\bar{e}_L\gamma^\mu e_L\bar{q}_L\gamma_\mu q_L$ which will shift the values of C_{1u} and C_{1d}.

4.5 *Heavy particle loop contributions*

In this section we will discuss those which only affect the neutral current and W and Z observables through radiative corrections. As usual, these types of new physics are related to a high mass scale since they are not observed at the present energy level. Most types of heavy physics affect low energy observables only through inverse powers of the the heavy scale and therefore have little effect except possibly for mediating rare processes. However, that is not true if the heavy physics breaks the symmetries of the low energy theory. These heavy physics will affect the neutral current and W, Z observables by shifting the self-energy diagrams. These shifts can be parametrized in terms of three parameters, $h_V = T$, h_{AW} and h_{AZ}, where h_V (h_A) refers to the breaking of the vector (axial) part of $SU(2)$. In most cases, the axial parameters are approximately equal, $h_{AW} \sim h_{AZ} = S$. h_V is sensitive to non-degenerate $SU(2)$ multiplets which break the vector $SU(2)$ symmetries, including the top quark mass, non-degenerate fourth family fermions, non-degenerate Higgs multiplets, and $\tilde{b}$-$\tilde{t}$ splitting in supersymmetry. The effects of h_V are equivalent to $\rho_0 \neq 1$. It is a very powerful parameterization, many types of new physics which enter at the one-loop level affect only the gauge boson self-energies and can be described by the parameters h_V, h_{AW}, and h_{AZ}. Among them, the important examples are: theories with extra fermion multiplets, the theories with two Higgs doublets, and the minimal supersymmetric standard model (MSSM).

In this paper, $M_H = 250\ GeV$ and $m_t = 150\ GeV$ are used as inputs for the calculation of radiative corrections. The effects of other values of m_t and M_H are treated as if they are new physics. Most of the effects of a Higgs mass M_H or top quark mass m_t differing from their reference values can be described by h_V and h_A. However, they are treated separately in order to take the correlations between h_V, h_{AW}, and h_{AZ} properly into account and to include additional m_t dependence from $Z \to \bar{b}b$ vertex diagrams. The radiative corrections to physical observables depend logarithmatically on M_H and are generally small. The effects of different values of m_t are considerably larger due to the dominant $\alpha m_t^2/M_Z^2$ dependence of the ratio M_W^2/M_Z^2 and of quantities that depend on the ratio.

5 Comparison of experiments vs new physics

In Table 1.1, we have collected the measurements and the projected experimental errors of the observables O_a, ΔO_a^{exp} and $\Delta \sin^2\theta_W^{exp}$, and the SM predictions of the observables O_a up to one-loop radiative corrections. The effects of a variety of possible types of new physics involving extra Z bosons, extra scalar bosons, extra fermions, compositeness, and types of new physics which enter at the loop level were discussed in last section. For the formulas of ΔO_a^i and $\Delta \sin^2\theta_W^a$, please see [1].

This section compares the experiments with new physics. We will take five of the 27 examples considered in [1] for illustration. The results are collected in Tables 5.1 to 5.5 and the accompanying figures. The SM predictions for the quantities in column 1 are given in column 2. Column 3 gives the projected one standard deviation experimental errors, ΔO_a^{exp}, which include the theoretical uncertainties when they are significant. The latter include both the theoretical uncertainties in the extraction of O_a^{exp} and the uncertainties in the SM prediction of O_a due to the uncertainty ~ 0.0003 in $\sin^2\theta_W^Z$. The contributions of new physics of type i to each of the observables, ΔO_a^i, is given in column 4. To ensure accuracy, in each case the ΔO_a^i were determined by two independent calculations. The ΔO_a^i depend on a coupling constant[2], λ, which determines the strength of the new physics. We assume $\lambda = 0.01$ for definiteness in column 4, but λ can be scaled linearly to any value preferred by the reader. In column 5 is the minimum value of λ for a given observable; λ_a^{min} is the value necessary to make the new physics contribution to that observable equal to the projected experimental error. Of course, a one standard deviation effect is not sufficient to either establish or exclude a given type of new physics, but it is a reasonable measure of sensitivity. Columns 6 and 7 play the same role for $\sin^2\theta_W^{(a;Z)}$ as columns 3 and 4 do for O_a, and in most cases the values of λ^{min} in column 5 are also applicable to the values in columns[3] 6 and 7. We reemphasize that although the central values of $\sin^2\theta_W^a$ extracted from each experiment depend on the renormalization scheme, both the experimental uncertainty $\Delta\sin^2\theta_W^{exp}$ in column 6 and the relative shift $\Delta\sin^2\theta_W^{(a;Z)}$ in column 7 are essentially scheme independent.

To see what experiments are sensitive to a given type of new physics i, for each O_a we define the ratio of the calculated deviation from the SM due to the new physics i to the projected experimental error:

$$r_a^i(\lambda) = \frac{|\Delta O_a^i|}{\Delta O_a^{exp}}. \tag{28}$$

Although r_a^i itself is dependent on the value of λ, the relative sensitivities, or the relative values of r_a^i for different O_a, are independent of λ. The larger r_a^i is, the more significantly the i-th new physics manifests itself, and the more likely for the new physics i to be seen in experiments measuring O_a. $r_a^i(\lambda)$ is proportional to λ, with $r_a^i(\lambda_a^{min}) = 1$. To see the relative sensitivity of each O_a graphically, we plot $1/\lambda_a^{min}$ from the numerical results for each type of new physics. As an example, consider Fig. 5.1, in which the new physics is an extra Z_χ with $C = \sqrt{2/5}$. The x-axis is labeled by the physical observables with their explicit definitions to be found in Section 3. The vertical bars represent the relative values of $1/\lambda_a^{min}$ for each O_a. The higher the bar, the more sensitive is O_a to the Z_χ. The values of M_{Z_2} corresponding to $1/\lambda^{min}$ are indicated on the right-hand scale. The height of each bar can be scaled up or down if the precision of the corresponding experiment varies. For example, if the error of M_W is reduced to 50 MeV rather than 100 MeV, then the bar of M_W should be scaled up by a factor of two. On the other hand, if the error of A_{LR} is 0.006 rather than the projected 0.004, then the bar of A_{LR} should be scaled down accordingly. At the moment, let us take the projected precision seriously. Then we see that R_ν is especially sensitive to this type of Z_χ. Looking back to Table 5.1, at $\lambda = 0.01$ the deviation in R_ν is some five times its projected experimental error, and $\lambda_a^{min} = 0.0019$, corresponding to $M_{Z_2} \simeq 1295\ GeV$. M_W, g_L^2, $\sigma_{\nu e}/(\sigma_{\nu_e e}+\sigma_{\bar\nu e})$, $C_{1+}(iso)$, $A_{LR}(LEP)$, Γ_{inv}, $\Gamma_{l\bar l}$, and Γ_Z all have $\lambda^{min} < 0.01$, corresponding to $M_{Z_2} = 565\ GeV$.

[2]In some cases, such as extra Z bosons, there is more than one extra parameter. However, only one is related to the coupling strength, while the others are of order unity. We pick typical values for the latter parameters for definiteness.

[3]That means that the analyses based on $\sin^2\theta_W$ and on the observables themselves are equivalent. That is true provided that the expected variation in the observable can be described by a reasonable variation in $\sin^2\theta_W$ and that the relation is approximately linear within the experimentally prescribed region. This condition is satisfied by all of the observables in this paper except θ_L, θ_R, g_A^e, and C_{2p}, which are insensitive to $\sin^2\theta_W$.

Table 5.1: Deviations from the SM by an Extra Z_χ Boson ($C = (2/5)^{1/2}$). Type: Z_χ Free parameter: $\lambda = (\frac{g_2^2}{M_{Z_2}^2})/(\frac{g_1^2}{M_Z^2})$ with $C = \Theta\frac{g_2}{g_1}/\lambda = (2/5)^{1/2}$; (Section 4.1). g_1, g_2 and M_Z, M_{Z_2} are the coupling constants and masses of the ordinary Z and the extra Z_χ bosons, respectively; Θ is their mixing angle. Inputs: $M_Z = 91.187\ GeV$, $m_t = 150\ GeV$, $M_H = 250\ GeV$, ($\sin^2\theta_W^M = 0.2259$, $\sin^2\hat{\theta}_W = 0.2325$). **Comments:** For the explicit definitions of physical observables, see *Section 3.* λ^{min} is the value of λ for which the change ΔO_a is equal to the projected one standard deviation experimental error. The inverses of the sixth column $1/\lambda^{min}$ are plotted in Fig. 5.1. Following the argument given in the text, the bigger $1/\lambda_a^{min}$ is, the more likely for Z_χ to be detected by measuring O_a. R_ν is sensitive to this particular Z_χ; it has $\lambda_a^{min} = 0.0019$ which corresponds to $M_{Z_2} = 1295\ GeV$ for $g_2^2/g_1^2 = 5/3\sin^2\theta_W$. M_W, g_L^2, $\sigma_{\nu e}/(\sigma_{\nu_e e} + \sigma_{\bar{\nu}e})$, $C_{1+}(iso)$, $A_{LR}(LEP)$, Γ_{inv}, $\Gamma_{l\bar{l}}$, and Γ_Z all have $\lambda^{min} < 0.01$, corresponding to $M_{Z_2} = 565\ GeV$.

Quantities	O_a^{SM}	ΔO_a^{exp}	ΔO_a^i $\lambda = 0.01$	λ_a^{min}	$\Delta\sin^2\theta_W^{exp}$	$\Delta\sin^2\theta_W^{(a;Z)}$ $\lambda = 0.01$
$M_Z(GeV)$	91.1870	0.0070	—	—	0.0003	—
$M_W(GeV)$	80.2672	0.1050	0.2263	0.0046	0.0006	−0.0013
g_L^2	0.3025	0.0039	0.0074	0.0053	0.0052	−0.0099
g_R^2	0.0300	0.0033	−0.0017	0.0198	0.0130	−0.0064
R_ν	0.3145	0.0013	0.0067	0.0019	0.0020	−0.0105
θ_L	2.4634	0.0350	0.0057	0.0617	—	—
θ_R	5.1765	0.4100	−0.0323	0.1271	—	—
g_V^e	−0.0388	0.0170	0.0071	0.0240	0.0085	0.0035
g_A^e	−0.5054	0.0150	−0.0001	—	—	—
$\sigma_\nu/\sigma_{\bar{\nu}}$	1.1636	0.0456	−0.0323	0.0141	0.0050	0.0035
$\sigma_\nu/(\sigma_{\bar{\nu}} + \sigma_\nu)$	0.1480	0.0027	−0.0037	0.0071	0.0025	0.0035
C_{1+}	0.1285	0.0013	−0.0005	0.0262	0.0033	−0.0013
$C_{1+}(iso)$	0.1285	0.0003	−0.0005	0.0068	0.0009	−0.0013
C_{1-}	−0.3665	0.1000	−0.0018	0.5437	0.0687	−0.0013
C_{2p}	−0.0145	0.0460	−0.0001	—	—	—
$C_{2p}(1)$	−0.0145	0.0046	−0.0001	0.4374	—	—
C_{2m}	−0.0588	0.1100	−0.0050	0.2206	0.0273	−0.0012
$2C_{1u} + C_{1d}$	−0.0362	0.0040	−0.0025	0.0157	0.0020	−0.0013
$A_{LR}(SLC)$	0.1399	0.0066	−0.0051	0.0130	0.0008	0.0007
$A_{LR}(LEP)$	0.1399	0.0041	−0.0051	0.0081	0.0005	0.0007
$A_{FB}^{pol}(c)$	0.4759	0.0250	0.0048	0.0520	0.0096	−0.0018
$A_{FB}(c)$	0.0666	0.0070	−0.0017	0.0401	0.0017	0.0004
$A_{FB}^{pol}(b)$	0.6971	0.0200	0.0044	0.0457	0.0428	−0.0094
$A_{FB}(b)$	0.0976	0.0054	−0.0029	0.0184	0.0010	0.0005
$A_{FB}^{pol}(\mu)$	0.1050	0.0090	−0.0038	0.0236	0.0015	0.0007
$A_{FB}(\mu)$	0.0147	0.0027	−0.0011	0.0253	0.0017	0.0007
$A_{pol}(\tau)$	0.1399	0.0110	−0.0051	0.0216	0.0014	0.0007
$\Gamma_{inv}(GeV)$	0.5008	0.0060	0.0080	0.0075	0.0025	−0.0034
$\Gamma_{l\bar{l}}(GeV)$	0.0838	0.0003	−0.0004	0.0085	0.0006	0.0008
$\Gamma_{c\bar{c}}(GeV)$	0.2976	0.0300	−0.0001	—	0.0155	0.0001
$\Gamma_{b\bar{b}}(GeV)$	0.3763	0.0090	0.0029	0.0313	0.0040	−0.0013
$\Gamma_Z(GeV)$	2.4929	0.0070	0.0153	0.0046	0.0005	−0.0011

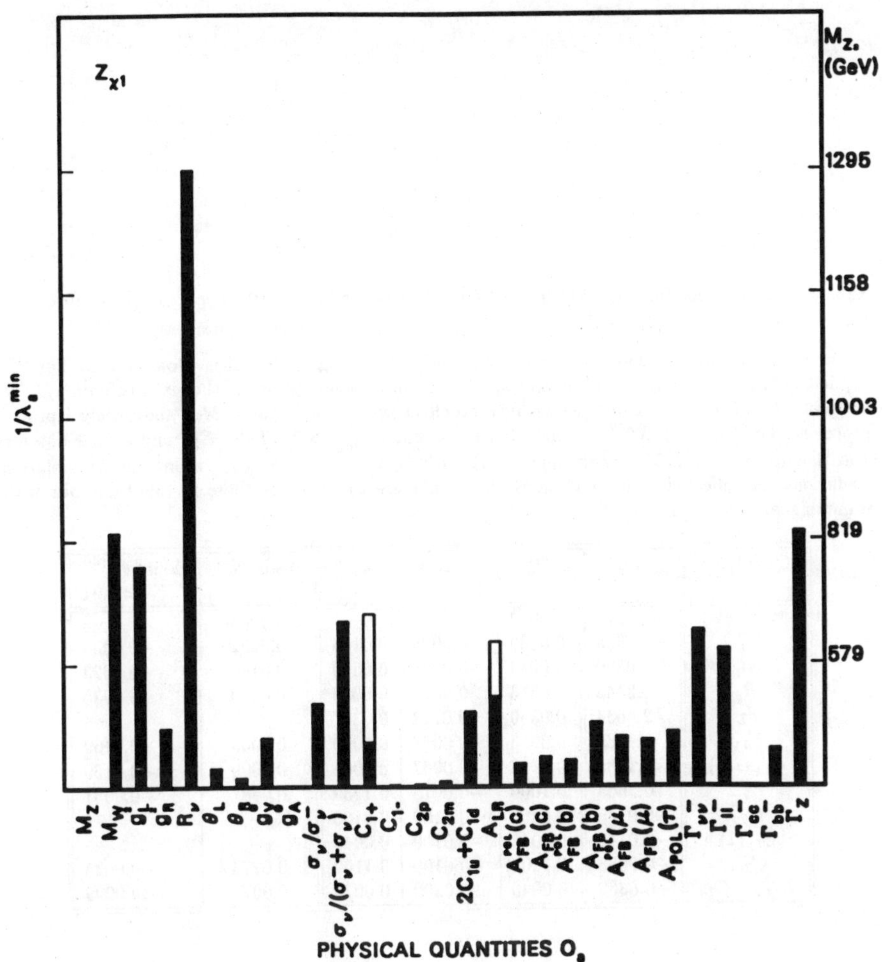

Figure 5.1: The new physics of Z_χ with $C = (2/5)^(1/2)$: solid bar for C_{1+}, C_{2p}, and $A_{LR}(SLC)$; open bar for $C_{1+}(iso)$, $C_{2p}(1)$, and $A_{LR}(LEP)$. The right-hand scale corresponds to M_{Z_2} in GeV.

Table 5.2: Deviations from the SM by an $SU(5)$ Leptoquark ($\eta_L \neq 0$). Type: $SU(5)$ Leptoquark. Free parameter: $\lambda = \frac{2(1/2)|\eta_L|^2}{8M_S^2 G_F}$. ($\eta_L \neq 0$, $\eta_R = 0$); (Section 4.2). Inputs: $M_Z = 91.187\ GeV$, $m_t = 150\ GeV$, $M_H = 250\ GeV$, ($\sin^2\theta_W^M = 0.2259$, $\sin^2\hat{\theta}_W = 0.2325$). **Comments:** For the explicit definitions of physical observables, see *Section 3*. The inverses of the sixth column $1/\lambda^{min}$ are plotted in Fig. 5.2. Leptoquarks have no effect on Z-pole physics, M_W and purely leptonic processes. For $C_{1+}(iso)$, $\lambda^{min} = 0.0005$, corresponding to $\frac{M_S}{|\eta_L|} > 5.5\ TeV$. C_{1+} and $2C_{1u} + C_{1d}$ are sensitive up to $\sim 2.8\ TeV$. For the leptoquark with $\eta_R \neq 0$, $\eta_L = 0$ only the atomic parity-violation coefficients are affected. The deviations on the C's are the same as those in this table but with minus signs.

Quantities	O_a^{SM}	ΔO_a^{exp}	ΔO_a^i $\lambda = 0.01$	λ_a^{min}	$\Delta\sin^2\theta_W^{exp}$	$\Delta\sin^2\theta_W^{(a;Z)}$ $\lambda = 0.01$
$M_Z(GeV)$	91.1870	0.0070	—	—	0.0003	—
g_L^2	0.3025	0.0039	0.0025	0.0155	0.0052	−0.0034
g_R^2	0.0300	0.0033	−0.0006	0.0550	0.0130	−0.0023
R_ν	0.3145	0.0013	0.0023	0.0056	0.0020	−0.0035
θ_L	2.4634	0.0350	0.0114	0.0307	—	—
C_{1+}	0.1285	0.0013	−0.0067	0.0020	0.0033	−0.0169
$C_{1+}(iso)$	0.1285	0.0003	−0.0067	0.0005	0.0009	−0.0169
C_{1-}	−0.3665	0.1000	−0.0075	0.1339	0.0687	−0.0051
C_{2p}	−0.0145	0.0460	−0.0100	0.0460	—	—
$C_{2p}(1)$	−0.0145	0.0046	−0.0100	0.0046	—	—
C_{2m}	−0.0588	0.1100	−0.0100	0.1100	0.0273	−0.0025
$2C_{1u} + C_{1d}$	−0.0362	0.0040	−0.0200	0.0020	0.0020	−0.0099

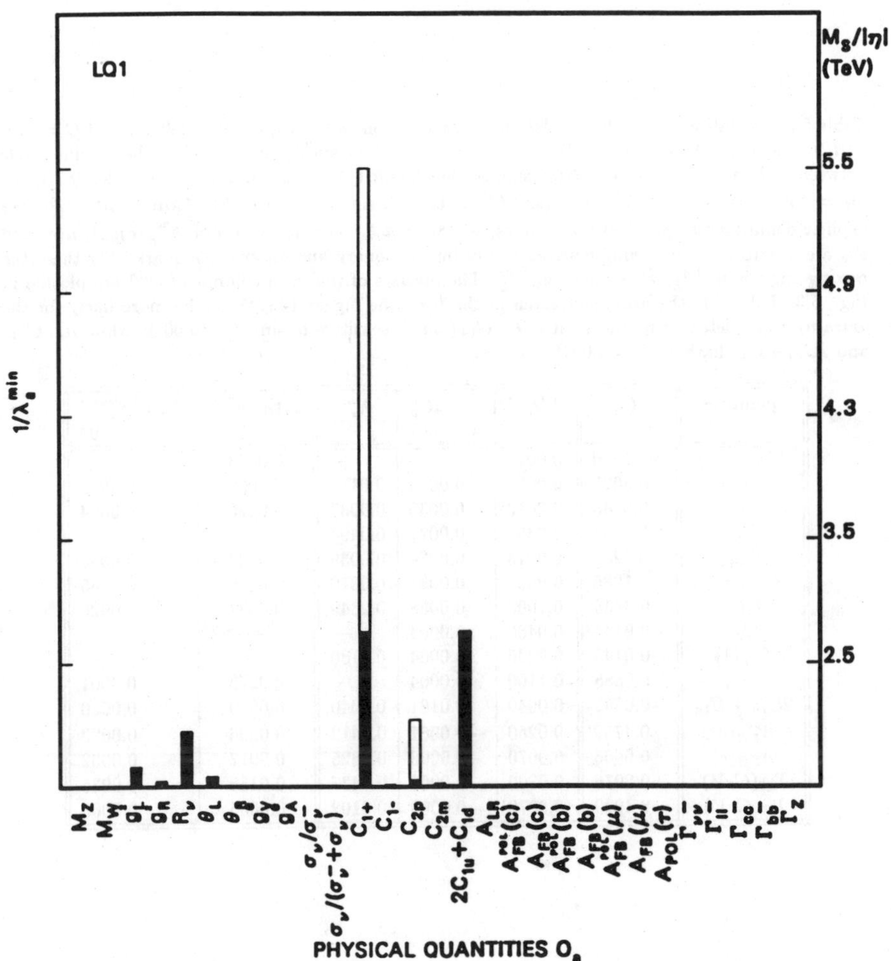

Figure 5.2: The new physics of $SU(5)$ leptoquark, type 1: solid bar for C_{1+} and C_{2p}; open bar for $C_{1+}(iso)$ and $C_{2p}(1)$. The right-hand scale corresponds to $M_S/|\eta|$ in TeV.

Table 5.3: Deviations from the SM by an Extra Fermion (u_L). Type: extra left-handed $Q = 2/3$, $SU(2)$-singlet quark; (Section 4.3). Free parameter: $\lambda = \sin^2\theta_L^u$, where θ_L^u is the mixing angle between ordinary and exotic ($SU(2)$-singlet) left-handed u-quark. Inputs: $M_Z = 91.187\ GeV$, $m_t = 150\ GeV$, $M_H = 250\ GeV$, ($\sin^2\theta_W^M = 0.2259$, $\sin^2\hat{\theta}_W = 0.2325$). **Comments:** For the explicit definitions of physical observables, see *Section 3*. The deviations of $A_{FB}^{pol}(c)$, $A_{FB}(c)$ and $\Gamma_{c\bar{c}}$ are due to the the mixing between the ordinary c-quark and an exotic c-quark. For these the mixing angle is $\sin^2\theta_L^c$ rather than $\sin^2\theta_L^u$. The inverses of the sixth column $1/\lambda^{min}$ are plotted in Fig. 5.3. Following the argument given in the text, the bigger $1/\lambda_a^{min}$ is, the more likely for the extra u_L to be detected by measuring O_a. $C_{1+}(iso)$ is sensitive to $\sin^2\theta_L^u \sim 0.0010$, while R_ν, C_{1+}, and $2C_{1u} + C_{1d}$ have $\lambda^{min} \sim 0.004$.

Quantities	O_a^{SM}	ΔO_a^{exp}	ΔO_a^i $\lambda = 0.01$	λ_a^{min}	$\Delta\sin^2\theta_W^{exp}$	$\Delta\sin^2\theta_W^{(a;Z)}$ $\lambda = 0.01$
$M_Z(GeV)$	91.1870	0.0070	—	—	0.0003	—
g_L^2	0.3025	0.0039	−0.0035	0.0113	0.0052	0.0046
R_ν	0.3145	0.0013	−0.0035	0.0037	0.0020	0.0054
θ_L	2.4634	0.0350	0.0071	0.0494	—	—
C_{1+}	0.1285	0.0013	0.0034	0.0039	0.0033	0.0085
$C_{1+}(iso)$	0.1285	0.0003	0.0034	0.0010	0.0009	0.0085
C_{1-}	−0.3665	0.1000	0.0038	0.2649	0.0687	0.0026
C_{2p}	−0.0145	0.0460	0.0004	—	—	—
$C_{2p}(1)$	−0.0145	0.0046	0.0004	0.1186	—	—
C_{2m}	−0.0588	0.1100	0.0004	—	0.0273	0.0001
$2C_{1u} + C_{1d}$	−0.0362	0.0040	0.0101	0.0040	0.0020	0.0050
$A_{FB}^{pol}(c)$	0.4759	0.0250	−0.0061	0.0412	0.0096	0.0023
$A_{FB}(c)$	0.0666	0.0070	−0.0008	0.0825	0.0017	0.0002
$\Gamma_{c\bar{c}}(GeV)$	0.2976	0.0300	−0.0069	0.0437	0.0155	0.0035
$\Gamma_Z(GeV)$	2.4929	0.0070	−0.0069	0.0102	0.0005	0.0005

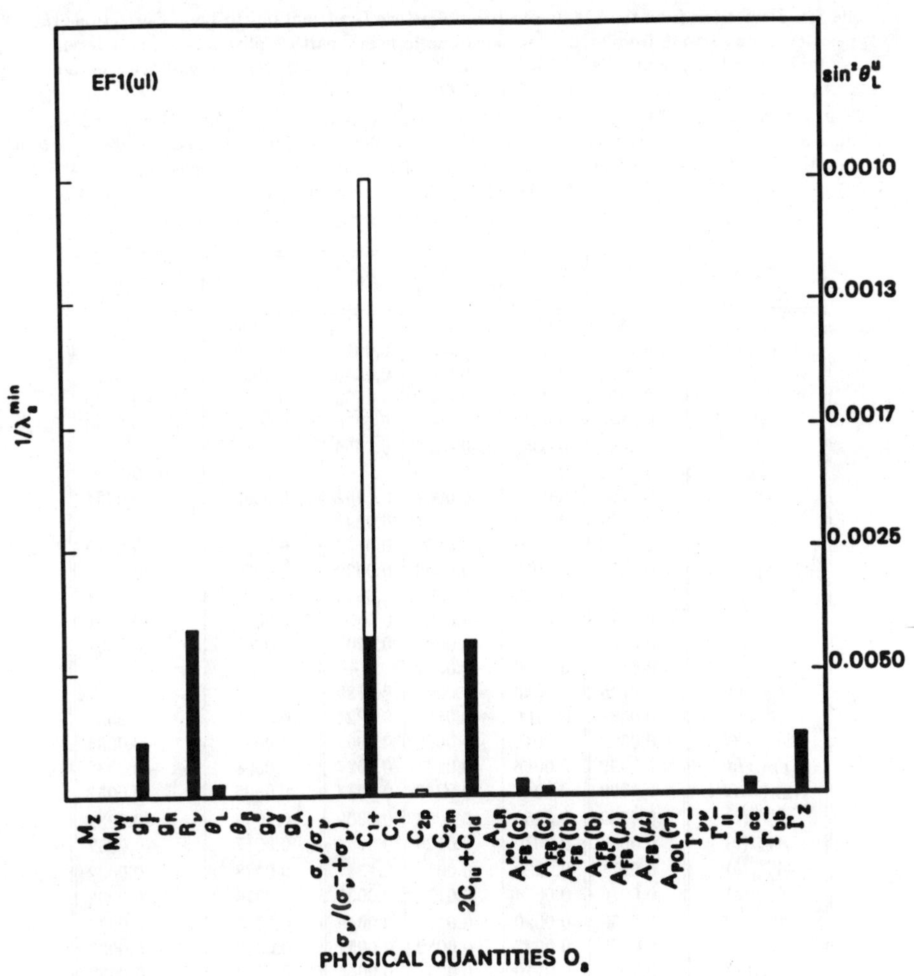

Figure 5.3: The new physics of one extra u_L: solid bar for C_{1+} and C_{2p}; open bar for $C_{1+}(iso)$ and $C_{2p}(1)$. The right-hand scale corresponds to $\sin^2\theta_L^u$.

Table 5.4: Deviations from the SM due to Heavy Particle Contributions to Loop Corrections (h_V). Type: Deviations arising from loop corrections due to heavy particle physics, *e.g.*, non-degenerate heavy fermion or boson multiplets, with $h_V \neq 0$, $h_{AZ} = 0$, $h_{AW} = 0$; (Section 4.5). Free parameter: $\lambda = \alpha h_V$ ($\alpha = 1/137.036$) Inputs: $M_Z = 91.187\ GeV$, $m_t = 150\ GeV$, $M_H = 250\ GeV$, ($\sin^2\theta_W^M = 0.2259$, $\sin^2\hat{\theta}_W = 0.2325$). **Comments:** For the explicit definitions of physical observables, see *Section 3*. The inverses of the sixth column $1/\lambda^{min}$ are plotted in Fig. 5.4. The deviation pattern among the observables is exactly the same as that of non-standard Higgs fields. R_ν has $\lambda^{min} = 0.0015$, corresponding to $h_V = 0.21$; $A_{LR}(LEP)$ has $\lambda^{min} = 0.0017$ ($h_V = 0.23$); and M_W has $\lambda^{min} = 0.0019$ ($h_V = 0.26$). $A_{LR}(SLC)$, $A_{FB}(b)$, $\sigma_\nu/(\sigma_{\bar{\nu}} + \sigma_\nu)$, $\Gamma_{l\bar{l}}$, and Γ_Z all have $\lambda^{min} < 0.004$.

Quantities	O_a^{SM}	ΔO_a^{exp}	ΔO_a^i $\lambda = 0.01$	λ_a^{min}	$\Delta\sin^2\theta_W^{exp}$	$\Delta\sin^2\theta_W^{(a;Z)}$ $\lambda = 0.01$
$M_Z(GeV)$	91.1870	0.0070	—	—	0.0003	—
$M_W(GeV)$	80.2672	0.1050	0.5657	0.0019	0.0006	−0.0032
g_L^2	0.3025	0.0039	0.0084	0.0046	0.0052	−0.0113
g_R^2	0.0300	0.0033	−0.0002	0.1503	0.0130	−0.0009
R_ν	0.3145	0.0013	0.0083	0.0015	0.0020	−0.0130
θ_L	2.4634	0.0350	−0.0018	0.1956	—	—
θ_R	5.1765	0.4100	0.0000	—	—	1000.0000
g_V^e	−0.0388	0.0170	−0.0067	0.0252	0.0085	−0.0034
g_A^e	−0.5054	0.0150	−0.0051	0.0297	—	—
$\sigma_\nu/\sigma_{\bar{\nu}}$	1.1636	0.0456	0.0289	0.0158	0.0050	−0.0032
$\sigma_\nu/(\sigma_{\bar{\nu}} + \sigma_\nu)$	0.1480	0.0027	0.0068	0.0039	0.0025	−0.0064
C_{1+}	0.1285	0.0013	0.0000	0.3817	0.0033	0.0001
$C_{1+}(iso)$	0.1285	0.0003	0.0000	0.0992	0.0009	0.0001
C_{1-}	−0.3665	0.1000	−0.0083	0.1207	0.0687	−0.0057
C_{2p}	−0.0145	0.0460	−0.0001	—	—	—
$C_{2p}(1)$	−0.0145	0.0046	−0.0001	0.3180	—	—
C_{2m}	−0.0588	0.1100	−0.0134	0.0824	0.0273	−0.0033
$2C_{1u} + C_{1d}$	−0.0362	0.0040	−0.0068	0.0059	0.0020	−0.0034
$A_{LR}(SLC)$	0.1399	0.0066	0.0247	0.0027	0.0008	−0.0032
$A_{LR}(LEP)$	0.1399	0.0041	0.0247	0.0017	0.0005	−0.0032
$A_{FB}^{pol}(c)$	0.4759	0.0250	0.0083	0.0302	0.0096	−0.0032
$A_{FB}(c)$	0.0666	0.0070	0.0129	0.0054	0.0017	−0.0032
$A_{FB}^{pol}(b)$	0.6971	0.0200	0.0015	0.1347	0.0428	−0.0032
$A_{FB}(b)$	0.0976	0.0054	0.0174	0.0031	0.0010	−0.0032
$A_{FB}^{pol}(\mu)$	0.1050	0.0090	0.0185	0.0049	0.0015	−0.0032
$A_{FB}(\mu)$	0.0147	0.0027	0.0052	0.0052	0.0017	−0.0032
$A_{pol}(\tau)$	0.1399	0.0110	0.0247	0.0045	0.0014	−0.0032
$\Gamma_{inv}(GeV)$	0.5008	0.0060	0.0050	0.0120	0.0025	−0.0021
$\Gamma_{l\bar{l}}(GeV)$	0.0838	0.0003	0.0010	0.0029	0.0006	−0.0023
$\Gamma_{c\bar{c}}(GeV)$	0.2976	0.0300	0.0047	0.0644	0.0155	−0.0024
$\Gamma_{b\bar{b}}(GeV)$	0.3763	0.0090	0.0052	0.0172	0.0040	−0.0023
$\Gamma_Z(GeV)$	2.4929	0.0070	0.0333	0.0021	0.0005	−0.0023

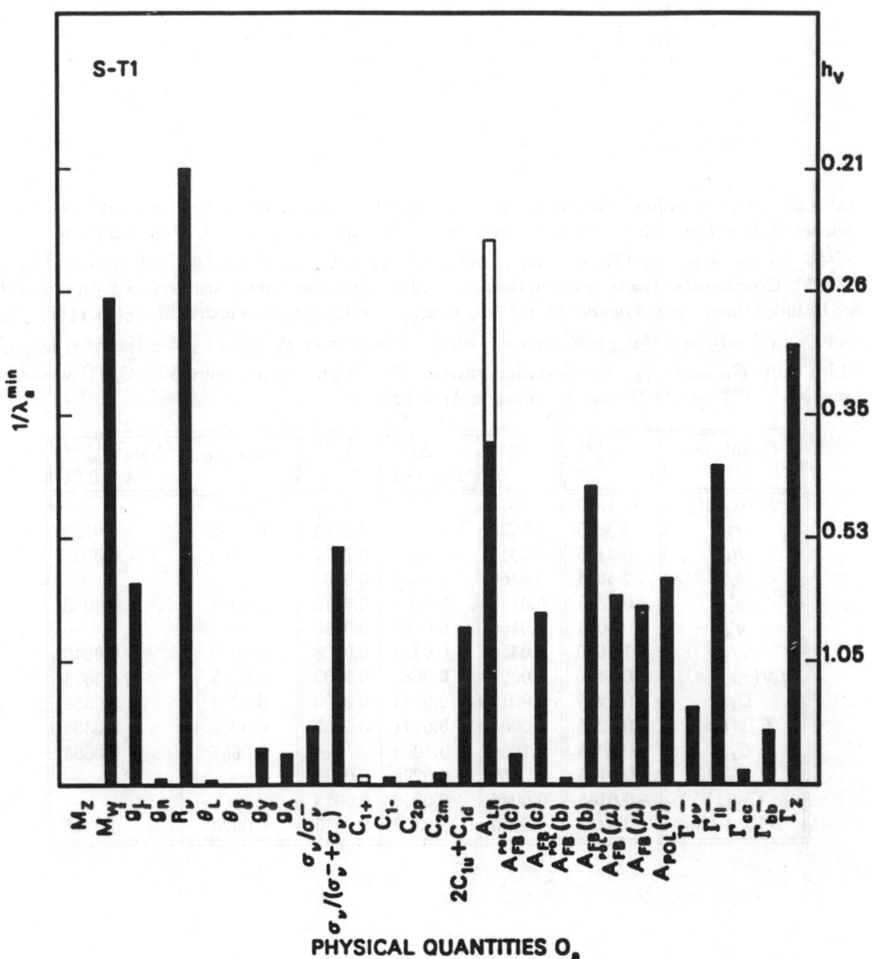

Figure 5.4: Heavy physics through loop corrections (h_v): solid bar for C_{1+}, C_{2p}, and $A_{LR}(SLC)$; open bar for $C_{1+}(iso)$, $C_{2p}(1)$, and $A_{LR}(LEP)$. The right-hand scale is h_v.

Table 5.5: Deviations from the SM due to Four-Fermi Operators induced by Compositeness. Type: four-fermi operators induced by compositeness of fermions; (Section 4.4). Free parameter: $\lambda = \frac{2^{(1/2)}\pi}{G_F\Lambda^2}$ Inputs: $M_Z = 91.187\ GeV$, $m_t = 150\ GeV$, $M_H = 250\ GeV$, ($\sin^2\theta_W^M = 0.2259$, $\sin^2\hat{\theta}_W = 0.2325$). **Comments:** For the explicit definitions of physical observables, see *Section 3*. In this table are included three types of operators: (1) $-L_1 = \pm\frac{4\pi}{\Lambda_1^2}\bar{l}_{\mu L}\gamma^\mu l_{\mu L}\bar{q}_L\gamma_\mu q_L$, which shifts $\epsilon_L(u)$, $\epsilon_L(d)$, C_{2u}, and C_{2d}; (2) $-L_2 = \pm\frac{4\pi}{\Lambda_2^2}\bar{\nu}_{\mu L}\gamma^\mu \nu_{\mu L}\bar{e}_L\gamma_\mu e_L$, which shifts g_V^e and g_A^e; (3) $-L_3 = \pm\frac{4\pi}{\Lambda_3^2}\bar{e}_L\gamma^\mu e_L\bar{q}_L\gamma_\mu q_L$, which shifts C_{1u} and C_{1d}. The deviation pattern of L_1 is plotted in Figure 5.5. $C_{2p}(1)$ would be sensitive to $\lambda^{min} = 0.0023$, corresponding to $\Lambda_1 = 12.9\ TeV$.

Quantities	O_a^{SM}	ΔO_a^{exp}	ΔO_a^i $\lambda = 0.01$	λ_a^{min}	$\Delta\sin^2\theta_W^{exp}$	$\Delta\sin^2\theta_W^{(a;Z)}$ $\lambda = 0.01$
$M_Z(GeV)$	91.1870	0.0070	—	—	0.0003	—
g_L^2	0.3025	0.0039	−0.0017	0.0234	0.0052	0.0022
R_ν	0.3145	0.0013	−0.0017	0.0077	0.0020	0.0026
θ_L	2.4634	0.0350	−0.0256	0.0137	—	—
g_V^e	−0.0388	0.0170	0.0100	0.0170	0.0085	0.0050
g_A^e	−0.5054	0.0150	0.0100	0.0150	—	—
$\sigma_\nu/\sigma_{\bar{\nu}}$	1.1636	0.0456	−0.0421	0.0108	0.0050	0.0046
$\sigma_\nu/(\sigma_{\bar{\nu}}+\sigma_\nu)$	0.1480	0.0027	−0.0083	0.0032	0.0025	0.0078
C_{1+}	0.1285	0.0013	0.0141	0.0009	0.0033	0.0358
$C_{1+}(iso)$	0.1285	0.0003	0.0141	0.0002	0.0009	0.0358
C_{1-}	−0.3665	0.1000	0.0008	—	0.0687	0.0006
C_{2p}	−0.0145	0.0460	0.0200	0.0230	—	—
$C_{2p}(1)$	−0.0145	0.0046	0.0200	0.0023	—	—
$2C_{1u}+C_{1d}$	−0.0362	0.0040	0.0300	0.0013	0.0020	0.0148

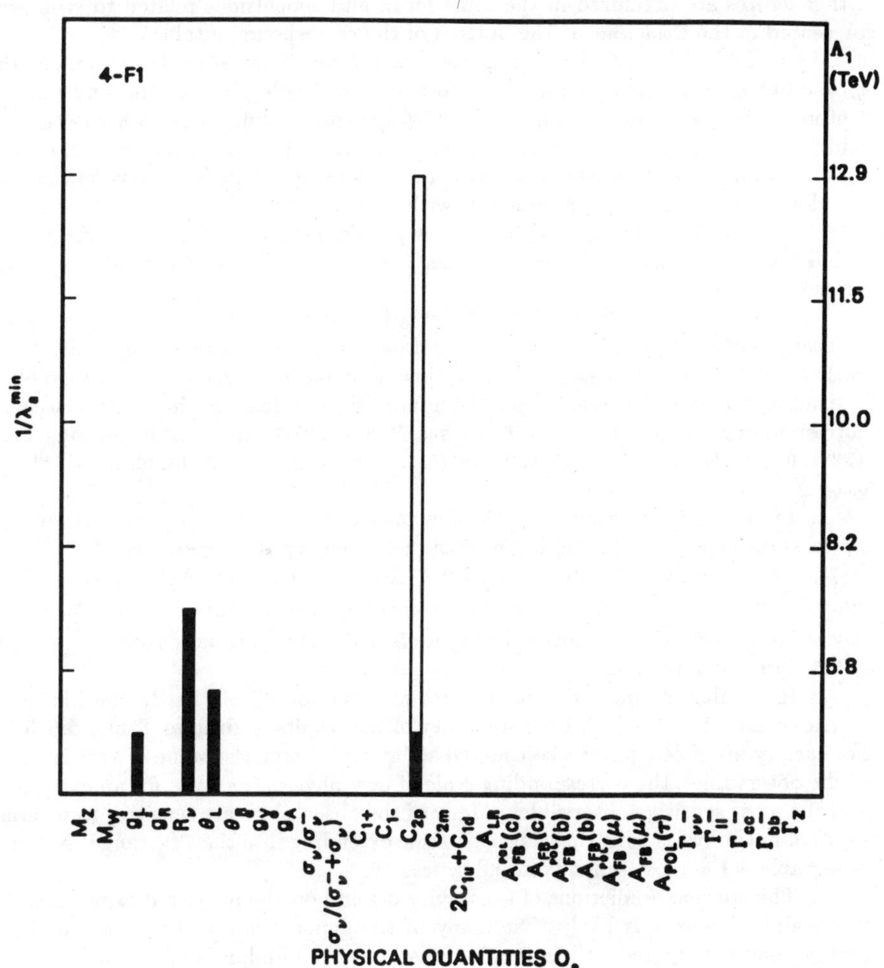

Figure 5.5: The new physics of four-fermi operator, type I: solid bar for C_{2p}, open bar for $C_{2p}(1)$. The right-hand scale corresponds to the compositeness scale $\Lambda\ TeV$.

Other figures are organized in the same form, and conclusions related to each are presented in the *Comments* at the bottom of the corresponding table.

In Table 5.2 and Figure 5.2, the new physics is an $SU(5)$ leptoquark with $\eta_L \neq 0$ but $\eta_R = 0$. Leptoquarks have no effect on Z-pole physics, M_W and purely leptonic processes. For $C_{1+}(iso)$, $\lambda^{min} = 0.0005$, corresponding to $\frac{M_S}{|\eta_L|} > 5.5\ TeV$. C_{1+} and $2C_{1u} + C_{1d}$ are sensitive up to $\sim 2.8\ TeV$. For the leptoquark with $\eta_R \neq 0$, $\eta_L = 0$ only the atomic parity-violation coefficients are affected. The deviations on the C's are the same as those in this table but with minus signs.

In Table 5.3 and Figure 5.3, the new physics is an exotic ($SU(2)$-singlet) left-handed u quark. $C_{1+}(iso)$ is sensitive to $\sin^2\theta_L^u \sim 0.0010$, while R_ν, C_{1+}, and $2C_{1u} + C_{1d}$ have $\lambda^{min} \sim 0.004$.

In Table 5.4 and Figure 5.4, the deviations arising from loop corrections due to heavy particle physics, *e.g.*, non-degenerate heavy fermion or boson multiplets, with $h_V \neq 0$, $h_{AZ} = 0$, $h_{AW} = 0$. The deviation pattern among the observables is exactly the same as that of non-standard Higgs fields. R_ν has $\lambda^{min} = 0.0015$, corresponding to $h_V = 0.21$; $A_{LR}(LEP)$ has $\lambda^{min} = 0.0017$ ($h_V = 0.23$); and M_W has $\lambda^{min} = 0.0018$ ($h_V = 0.25$). $A_{LR}(SLC)$, $A_{FB}(b)$, $\sigma_\nu/(\sigma_{\bar\nu} + \sigma_\nu)$, $\Gamma_{l\bar l}$, and Γ_Z all have $\lambda^{min} < 0.004$.

In Table 5.5 and Figure 5.5, the new physics are four-fermi operators induced by compositeness. In this table are included three types of operators: (1) $-L_1 = \pm\frac{4\pi}{\Lambda_1^2}\bar l_{\mu L}\gamma^\mu l_{\mu L}\bar q_L\gamma_\mu q_L$, which shifts $\epsilon_L(u)$, $\epsilon_L(d)$, C_{2u}, and C_{2d}; (2) $-L_2 = \pm\frac{4\pi}{\Lambda_2^2}\bar\nu_{\mu L}\gamma^\mu\nu_{\mu L}\bar e_L\gamma_\mu e_L$, which shifts g_V^e and g_A^e; (3) $-L_3 = \pm\frac{4\pi}{\Lambda_3^2}\bar e_L\gamma^\mu e_L\bar q_L\gamma_\mu q_L$, which shifts C_{1u} and C_{1d}. The deviation pattern of L_1 is plotted in Figure 5.5. $C_{2p}(1)$ would be sensitive to $\lambda^{min} = 0.0023$, corresponding to $\Lambda_1 = 12.9\ TeV$.

In similar manner, we can update analysis for all of the 27 specific new physics considered in [1]. A brief summary of the results is given in Tables 5.6–5.9. For each types of new physics considered are given the smallest value of λ^{min} for any of the observables, the corresponding scale of new physics (*e.g.*, M_{Z_2} for an extra Z), and the most sensitive observables. Many of the observables are sensitive to several types of new physics with characteristic scale extending into the TeV range. No one observable is the most sensitive to all types.

The specific predictions of sensitivity depend on the projected experimental uncertainties, and it is likely that many of the experiments will be more or less precise than anticipated. While keeping this caveat in mind, it is apparent from the Tables 5.1–5.9 that the observables M_W, R_ν, $C_{1+}(iso)$, $A_{LR}(LEP)$, and Γ_Z are especially promising, each showing considerable sensitivity to a variety of new physics. The observables $\sigma_\nu/(\sigma_{\bar\nu} + \sigma_{\nu_e})$, $A_{FB}(b)$, $A_{pol}(\tau)$, $\Gamma_{l\bar l}$, and the existing g_L^2 are also sensitive; most of the remaining observables are strong probes of one or more types of new physics. New physics at the TeV scale or higher may be observed or excluded if a significant number of these observables are measured precisely. If all the measured values are consistent with the SM predictions then, except for the unlikely possibility of fine-tuned cancellations, various types of new physics will be excluded at scales of the order of those shown in Tables 5.6–5.9.

Table 5.6: Summary of the discussion in section 5: implications for extra Z bosons. The smallest λ^{min} is the value of the overall coupling constant $\lambda^{min} = \frac{g_2^2}{M_{Z_2}^2} / \frac{g_1^2}{M_Z^2}$ at which the extra Z is manifested in the most sensitive observable. Here g_1, g_2, M_Z, and M_{Z_2} are the coupling constants and masses of the ordinary Z and extra Z_2, and $\frac{g_2^2}{g_1^2} = \frac{5}{3}\sin^2\theta_W$ is assumed.

New Physics	Smallest λ^{min}	Implications $M_{Z_2}(GeV)$	Most Sensitive Observables
$Z_\chi(C=\sqrt{2/5})$	0.0019	~ 1295	R_ν; $\sigma_{\nu e}/(\sigma_{\nu_e e}+\sigma_{\bar{\nu}e})$, M_W, g_L^2, Γ_Z, $C_{1+}(iso)$, $A_{LR}^{(LEP)}$, Γ_{inv}, $\Gamma_{l\bar{l}}$
$Z_\chi(C=0)$	0.0011	~ 1700	$C_{1+}(iso)$; C_{1+}, $\sigma_\nu/(\sigma_{\bar{\nu}}+\sigma_\nu)$, $2C_{1u}+C_{1d}$
$Z_\psi(C=\sqrt{2/3})$	0.0017	~ 1370	R_ν; M_W, $C_{1+}(iso)$, A_{LR}, $A_{FB}(b)$, Γ_Z
$Z_\psi(C=-\sqrt{2/3})$	0.0017	~ 1370	$\Gamma_{l\bar{l}}$; M_W, R_ν, $C_{1+}(iso)$, A_{LR}, $A_{FB}(b)$
$Z_\psi(C=0)$	0.025	~ 357	$\sigma_\nu/(\sigma_{\bar{\nu}}+\sigma_\nu)$
$Z_\eta(C=-\frac{1}{\sqrt{15}})$	0.0075	~ 650	$A_{LR}^{(LEP)}$; $C_{2p}(1)$, $A_{LR}^{(SLC)}$, $A_{FB}(b)$
$Z_\eta(C=\frac{4}{\sqrt{15}})$	0.0008	~ 2000	$C_{1+}(iso)$; M_W, R_ν, $\Gamma_{l\bar{l}}$, Γ_Z
$Z_\eta(C=0)$	0.0045	~ 840	$C_{1+}(iso)$; $C_{2p}(1)$
$Z_{3R}(C=\sqrt{3/5}\alpha)$	0.0009	~ 1880	R_ν, $A_{LR}^{(LEP)}$; M_W, $A_{LR}^{(SLC)}$, $C_{1+}(iso)$, $A_{FB}(b)$, Γ_Z
$Z_{3R}(C=-\sqrt{3/5}/\alpha)$	0.0008	~ 2000	$C_{1+}(iso)$; C_{1+}, $\Gamma_{l\bar{l}}$
Z_{3R} $(C=0)$	0.0010	~ 1785	$C_{1+}(iso)$; C_{1+}

If, however, deviations are observed, then as much information as possible will be needed to confirm the discrepancy and to determine the nature of the new physics and its strength. From the pattern of the deviations of the various observables it will be possible in most cases to distinguish which type of new physics is responsible, or at least which is most likely out of the set of possibilities considered here. The deviation pattern can also distinguish true new physics from values of m_t and M_H different from the reference values used in the SM predictions. Identification of the origin of deviations is another reason for carrying out a large variety of precision experiments.

To quantify this, it is useful to consider an n-dimensional vector space, each point of which represents the possible outcome of n precise experiments. Let us choose the origin in this space as the SM prediction for the n observables. Then one can define an n-dimensional vector with components:

$$V_a^{exp} = \frac{O_a^{exp} - O_a^{SM}}{\Delta O_a^{exp}}, \tag{29}$$

where V_a^{exp} is the deviation of the ath observable from the the SM prediction,

Table 5.7: Summary of the discussion in section 5: implications for exotic fermions. The smallest value λ^{min} is the value of $(\sin^2\theta^i_{L,R})^2$ at which the exotic fermion is manifested in the most sensitive observable. Here $\theta^i_{L,R}$ is the mixing angle between the ordinary and exotic fermions.

New Physics	Smallest λ^{min}	Implications	Most Sensitive Observables
u_L	0.0010	$(\theta^u_L)^2 \sim 0.0010$	$C_{1+}(iso)$; R_ν, C_{1+}, $2C_{1u}+C_{1d}$
u_R	0.0010	$(\theta^u_R)^2 \sim 0.0010$	$C_{1+}(iso)$; C_{1+}, $2C_{1u}+C_{1d}$
d_L	0.0009	$(\theta^d_L)^2 \sim 0.0009$	$C_{1+}(iso)$; R_ν, C_{1+}
d_R	0.0009	$(\theta^d_R)^2 \sim 0.0009$	$C_{1+}(iso)$; C_{1+}
e_L	0.0025	$(\theta^e_L)^2 \sim 0.0025$	$C_{1+}(iso)$; $A_{pol}(\tau)$, Γ_Z
e_R	0.0021	$(\theta^e_R)^2 \sim 0.0021$	$A_{LR}^{(LEP)}$, $C_{1+}(iso)$; $A_{LR}^{(SLC)}$, $A_{FB}(b)$, $A_{pol}(\tau)$
ν_{eL}	0.0033	$(\theta^{\nu_e}_L)^2 \sim 0.0033$	$A_{LR}^{(LEP)}$; $\sigma_\nu/(\sigma_{\bar\nu}+\sigma_\nu)$, $A_{LR}^{(SLC)}$, $A_{FB}(b)$, $\Gamma_{l\bar{l}}$, Γ_Z
μ_L	0.0018	$(\theta^\mu_L)^2 \sim 0.0018$	R_ν; g_L^2, $A_{FB}(b)$, A_{LR}, Γ_Z
$\nu_{\mu L}$	0.0033	$(\theta^{\nu_\mu}_L)^2 \sim 0.0033$	$A_{LR}^{(LEP)}$; $A_{LR}^{(SLC)}$, $A_{FB}(b)$, $\Gamma_{l\bar{l}}$, Γ_Z

weighted by the inverse of the experimental uncertainty. Similarly, each type i of new physics predicts a deviation vector with components:

$$V^i_a(\lambda) = \frac{\Delta O^i_a(\lambda)}{\Delta O^{exp}_a}, \quad a = 1, ..., n \tag{30}$$

which has a magnitude proportional to the coupling strength λ and a direction depending on a. If V^{exp}_a is compatible with zero (*i.e.*, the total length is of order n and no single component is much bigger than unity), there is no evidence for new physics. One can then set upper limits on the magnitude of each $V^i_a(\lambda)$ and therefore on each λ. On the other hand, if $V^{exp}_a \gg n$ then there is evidence for new physics. The most likely type or types can be determined, at least among the possibilities considered, by choosing i such that $V^i_a(\lambda)$ and V^{exp}_a are approximately parallel. The strength λ could be estimated from the requirement that $V^i_a(\lambda) = V^{exp}_a$.

The observables considered in this paper would usually be sufficient to distinguish the nature of new physics from the directions of $V^i_a(\lambda)$. Rather than grapple further with an n-dimensional vector space, however, it is productive as illustration to display the projections onto 2-dimensional subspaces. In [1], we have chosen 36 subspaces involving pairs of M_W, R_ν, $C_{1+}(iso)$, $A_{LR}(LEP)$, $\sigma_\nu/(\sigma_{\bar\nu}+\sigma_{\nu_e})$, $A_{FB}(b)$, $A_{pol}(\tau)$, $\Gamma_{l\bar{l}}$, and Γ_Z, which are representative of the most sensitive observables. The result is that it will be possible to distinguish among most of the examples of new physics considered here with reasonable confidence. The types of new physics which manifest themselves at the tree level all have distinctive patterns. These are different from the patterns induced by loop effects with the exception of non-standard Higgs fields, which cannot be distinguished from h_V by the measurements treated here.

Table 5.8: Summary of the discussion in section 5: implications for non-standard Higgs fields ($\lambda = \rho_0 - 1$), leptoquarks ($\lambda = \frac{2^{1/2}|\eta|^2}{8M_S^2 G_F}$), and four fermi operators ($\lambda = \frac{2^{1/2}\pi}{G_F \Lambda^2}$). The smallest value λ^{min} is the value of λ at which these new physics are manifested in the corresponding most sensitive observable.

New Physics	Smallest λ^{min}	Implications	Most Sensitive Observables
NS Higgs	0.0015	$VEV_{NH} \sim 7\ GeV$	R_ν, $A_{LR}^{(LEP)}$, M_W; $A_{LR}^{(SLC)}$, $A_{FB}(b)$, $\Gamma_{l\bar{l}}$, Γ_Z
Leptoquark	0.0005	$\frac{M_S}{\eta_L} \sim 5.5\ TeV$	$C_{1+}(iso)$; C_+, $2C_{1u} + C_{1d}$
4-Fermi op. (I)	0.0023	$\Lambda_1 \sim 12.9\ TeV$	$C_{2p}(1)$; R_ν
4-Fermi op. (II)	0.0032	$\Lambda_2 \sim 10.9\ TeV$	$\sigma_\nu/(\sigma_{\bar{\nu}} + \sigma_\nu)$
4-Fermi op. (III)	0.0002	$\Lambda_3 \sim 44\ TeV$	$C_{1+}(iso)$

Distinguishing new loop effects (h_A and h_V) from M_H and m_t, or even M_H and m_t from each other is more difficult, though not impossible. It is likely, however, that m_t will be accurately known from direct observation at the Tevatron within a few years. It is also possible that the Higgs boson will be directly observed as well. With that information, one could certainly elicit h_A and h_V. If M_H is not known from direct observation its effects on the observables considered here are generally small, though non-negligible for $M_H \leq 1\ TeV$. Assuming the exact validity of the SM (*i.e.*, $h_A = h_V = 0$), precision measurements could yield an estimate of M_H with a sensitivity of a few hundred GeV.

6 Conclusions

Rased upon experimental improvements and the theoretical framework developed in [1], the implications of high precision electroweak experiments to new physics have been studied. The framework was set up by taking several steps: (i) Using M_Z as input, the predictions of the SM for 28 additional observables encompassing weak neutral current phenomena and intermediate vector boson masses and decay widths were calculated numerically to the one-loop level. (ii) The modifications of each of these SM predictions which might be caused by 10 general types of possible new physics were calculated numerically for 27 specific examples. (iii) The latter values, each subject to the specification of a single coupling strength parameter, were compared with the projected numerical experimental errors of future high precision experiments with a three-fold purpose: testing the SM (including radiative corrections) at the highest level of precision, recognizing the relative sensitivities of the several observables to the different types of new physics treated here, and delineating the nature of the new physics if deviations from the SM are found empirically. The detailed numerical results of the analysis were presented in 27 tables and accompanying figures.

Table 5.9: Summary of the discussion in section 5: implications for heavy physics through loops. For h_V, $\lambda = \alpha h_V$; for h_A, $\lambda = 2^{1/2} G_F M_Z^2 h_A / 4\pi$; for the variations of m_t and M_H, $r_a = |\Delta O_a|/\Delta O_a^{exp}$. The smallest value λ^{min} is the value of λ at which these new physics are manifested in the corresponding most sensitive observable.

New Physics	Smallest λ^{min}	Implications	Most Sensitive Observabless
h_V	0.0015	$h_V \sim 0.21$	R_ν, $A_{LR}^{(LEP)}$, M_W; $A_{LR}^{(SLC)}$, $A_{FB}(b)$, $\Gamma_{l\bar{l}}$, Γ_Z
h_A	0.0017	$h_A \sim 0.16$	$A_{LR}^{(LEP)}$; $A_{LR}^{(SLC)}$, $C_{1+}(iso)$, $A_{FB}(b)$
$M_H = 50\ GeV$	—	$r_a \sim 1.75$	$A_{LR}^{(LEP)}$; M_W, R_ν, $A_{LR}^{(SLC)}$, $A_{FB}(b)$
$M_H = 1000\ GeV$	—	$r_a \sim 1.74$	$A_{LR}^{(LEP)}$; M_W, R_ν, $A_{LR}^{(SLC)}$, $A_{FB}(b)$
$m_t = 100\ GeV$	—	$r_a \sim 3.2$	R_ν, $A_{LR}^{(LEP)}$, M_W; $A_{LR}^{(SLC)}$, g_L^2, $\sigma_{\nu e}/(\sigma_{\nu_e e} + \sigma_{\bar{\nu} e})$, asymmetries $2C_{1u} + C_{1d}$, $\Gamma_{l\bar{l}}$, Γ_Z
$m_t = 200\ GeV$	—	$r_a \sim 4.0$	R_ν, $A_{LR}^{(LEP)}$, M_W; asymmetries $A_{LR}^{(SLC)}$, g_L^2, $\sigma_{\nu e}/(\sigma_{\nu_e e} + \sigma_{\bar{\nu} e})$, $2C_{1u} + C_{1d}$, $\Gamma_{l\bar{l}}$, Γ_Z

In this paper, we have changed the inputs for theoretical predictions $M_Z = 91.177 \rightarrow 91.187\ GeV$, $m_t = 100 \rightarrow 150\ GeV$, and $M_H = 100 \rightarrow 250\ GeV$. We have also incorporated other new experimental results in our analysis. The three general conclusions emerged from the original analysis are still intact. They are: First, an analytic framework is necessary for the understanding of future high precision experiments and their interpretation within the SM. As noted earlier, the results of most electroweak experiments depend on the value of the weak mixing angle, $\sin^2\theta_W$. A single precision measurement usually just specifies that angle. To constrain the theory, the results of equivalent precision measurements of other observables are necessary. Furthermore, the sensitivity of different observables to $\sin^2\theta_W$ and to the radiative corrections varies significantly so that comparison of the values of $\sin^2\theta_W$ extracted from several experiments may not by itself provide the most stringent test of their internal consistency. In addition, precise numerical comparison of one experiment with another, or experiment with theory, requires the top quark mass and Higgs mass to be known or at least constrained within reasonably tight limits. All of these difficulties may be overcome with a sufficient number of high precision experiments analyzed within a global or general framework such as proposed in [1]. The absence of the framework will lead to a waste of the high precision of these experiments and limit the conclusions to be drawn from them.

Second, the outcome of the global analysis will include accurate determinations of the radiative corrections within the SM, which will test the gauge nature of the theory at its foundation. The high precision experiments uniquely verify renormalizibility of the theory and the consistency of the calculations of the radiative corrections.

Finally, the variety of observables open to precise experimental study and the precise nature of the electroweak theoretical predictions jointly constitute a powerful means of searching for new physics. The wide range of momentum transfers and the varied sensitivities of the electroweak phenomena to new physics form a network through which new physics in the TeV region is unlikely to pass unnoticed. We had tried to furnish examples of this by considering specific instances of possible new physics. In general, all of the measurements considered show good sensitivity to one or another type of the new physics treated. Possibly, the new physics nature holds in store for us is beyond our imagination right now, which offers still another reason to encourage the performance of a multiplicity of measurements. Precision measurements of the masses of Z and W, precise determinations of neutrino-quark and neutrino-electron scattering cross-sections, precision measurements of the parameters of atomic parity-violation, and precision measurements of the asymmetries in e^+e^- scattering at the Z-pole, combined with precise knowledge of the Z-decay widths are all possible, and to a greater or lesser extent are possible with existing accelerators and, in many cases, with existing detectors.

This circumstance, global in character experimentally and theoretically, presents a window of opportunity through which a view of the future realm of elementary particle physics may be obtained before we are transported there by the forthcoming high energy colliders. We must use our resources wisely to take advantage of the opportunity.

One conspicuous change of the picture is the emerging importance of the Z decay widths in our analysis, which following naturally from their better than anticipated precision. $\Gamma_{l\bar{l}}$ and Γ_Z are especially important. This verifies the wisdom of the precaution stated in section 7 of [1] that "... we do not wish to underestimate the inventiveness and ingenuity of the experimentalists who will carry out the future electroweak experiments." That proves once again the validity of using high precision experiments to test the SM and to find new physics beyond. That also points out the importance of global analysis similar to this one.

Acknowledgements

It is appropriate to thank all my colleagues and friends in physics. Especially I am grateful to Paul G. Langacker, Alfred K. Mann, and William J. Marciano; it has been a great pleasure to know and to work with them.

References

1. P. Langacker, M. Luo, and A. Mann, *Rev. of Mod. Phys.* **Vol. 64** (1992) 87.

2. L. Rolandi, XXVI ICHEP, Aug. 1992.

3. P. Langacker, private communication.

4. Weinberg, S., *Phys. Rev. Lett.* **19** (1967) 1264; Salam, A., 1969, in *Elementary Particle Theory*, ed. N. Svartholm (Almquist and Wiksells, Stockholm, 1969) p. 367; Glashow, S.L., Iliopoulos, J., and Maiani, L., *Phys. Rev.* **D2** (1970) 1285.

5. See, for example, P. Langacker, *Structure of the Standard Model*, this book.

6. D. Schaile, *Z-Pole Experiments*, this book.

7. A. Blondel, *Polarization in e^+e^- annihilation*, this book.

8. D. Treille, *The LEP II Program*, this book.

9. D. Burke, *Future e^+e^- Colliders at High Energy*, this book.

10. F. Perrier, *The Measurement of Electroweak Parameters from Deep Inelastic Neutrino Scattering*, this book.

11. A. K. Mann, *Elastic Neutrino-Nucleon Scattering*, this book.

12. J. Panman, *Neutrino-Electron Scattering*, this book.

13. C. Wieman, *Atomic Parity Violation Experiments*, this book.

14. W. Hollik, *Renormalization of the Standard Model* and *Predictions for e^+e^- Processes*, this book.

15. W. Marciano, *Radiative Corrections to Neutral Current Processes*, this book.

16. J. Sapirstein, *The Theory of Atomic Parity Violation*, this book.

17. See [13]. The uncertainties in the *Cs* nuclear structure may cause extra theoretical errors. A rough estimate shows the effect is negligible, but further investigation is required to get a definite conclusion. See, for example, S. Pollock, E. Fortson, and L. Willets, preprint 40561-050-INT92-00-14, to be published in *Phys. Rev.* **C**.

18. D. London, *Exotic Fermions*, this book. See also [1] and references within.